S0-CAA-887

# Index of Applications

# Finite Mathematics +
# Applied Calculus

Stefan Waner | Steven R. Constenoble

Australia • Brazil • Japan • Korea • Mexico • Singapore • Spain • United Kingdom • United States

**Finite Mathematics + Applied Calculus**

Finite Mathematics and Applied Calculus, Sixth Edition
Stefan Waner | Steven R. Constenoble

© 2014, 2011, 2007 Cengage Learning. All rights reserved.

Library of Congress Control Number: 2012947816

Senior Project Development Manager:
 Linda deStefano

Market Development Manager:
 Heather Kramer

Senior Production/Manufacturing Manager:
 Donna M. Brown

Production Editorial Manager:
 Kim Fry

Sr. Rights Acquisition Account Manager:
 Todd Osborne

For product information and technology assistance, contact us at
**Cengage Learning Customer & Sales Support, 1-800-354-9706**

For permission to use material from this text or product,
submit all requests online at **cengage.com/permissions**
Further permissions questions can be emailed to
**permissionrequest@cengage.com**

This book contains select works from existing Cengage Learning resources and was produced by Cengage Learning Custom Solutions for collegiate use. As such, those adopting and/or contributing to this work are responsible for editorial content accuracy, continuity and completeness.

**Compilation © 2013 Cengage Learning**
ISBN-13: 978-1-305-01016-1
Printed in Mexico

ISBN-10: 1-305-01016-7
**Cengage Learning**
5191 Natorp Boulevard
Mason, Ohio 45040
USA

Cengage Learning is a leading provider of customized learning solutions with office locations around the globe, including Singapore, the United Kingdom, Australia, Mexico, Brazil, and Japan. Locate your local office at:
**international.cengage.com/region.**
Cengage Learning products are represented in Canada by Nelson Education, Ltd.
For your lifelong learning solutions, visit **www.cengage.com/custom.**
Visit our corporate website at **www.cengage.com.**

# Brief Contents

# Preface

*Finite Mathematics and Applied Calculus*, sixth edition, is intended for a one- or two-term course for students majoring in business, the social sciences, or the liberal arts. Like the earlier editions, the sixth edition of *Finite Mathematics and Applied Calculus* is designed to address the challenge of generating enthusiasm and mathematical sophistication in an audience that is often underprepared and lacks motivation for traditional mathematics courses. We meet this challenge by focusing on real-life applications and topics of current interest that students can relate to, by presenting mathematical concepts intuitively and thoroughly, and by employing a writing style that is informal, engaging, and occasionally even humorous.

The sixth edition goes further than earlier editions in implementing support for a wide range of instructional paradigms: from traditional face-to-face courses to on-line distance learning courses, from settings incorporating little or no technology to courses taught in computerized classrooms, and from classes in which a single form of technology is used exclusively to those incorporating several technologies. We fully support three forms of technology in this text: TI-83/84 Plus graphing calculators, spreadsheets, and powerful online utilities we have created for the book. In particular, our comprehensive support for spreadsheet technology, both in the text and online, is highly relevant for students who are studying business and economics, where skill with spreadsheets may be vital to their future careers.

## Our Approach to Pedagogy

**Real-World Orientation** We are confident that you will appreciate the diversity, breadth, and abundance of examples and exercises included in this edition. A large number of these are based on real, referenced data from business, economics, the life sciences, and the social sciences. Examples and exercises based on dated information have generally been replaced by more current versions; applications based on unique or historically interesting data have been kept.

Adapting real data for pedagogical use can be tricky; available data can be numerically complex, intimidating for students, or incomplete. We have modified and streamlined many of the real-world applications, rendering them as tractable as any "made-up" application. At the same time, we have been careful to strike a pedagogically sound balance between applications based on real data and more traditional "generic" applications. Thus, the density and selection of real data–based applications has been tailored to the pedagogical goals and appropriate difficulty level for each section.

**Readability** We would like students to read this book. We would like students to *enjoy* reading this book. Thus, we have written the book in a conversational and student-oriented style, and have made frequent use of question-and-answer dialogues to encourage the development of the student's mathematical curiosity and intuition. We hope that this text will give the student insight into how a mathematician develops and thinks about mathematical ideas and their applications.

**Rigor**  We feel that mathematical rigor need not be antithetical to the kind of applied focus and conceptual approach that are earmarks of this book. We have worked hard to ensure that we are always mathematically honest without being unnecessarily formal. Sometimes we do this through the question-and-answer dialogues and sometimes through the "Before we go on . . ." discussions that follow examples, but always in manner designed to provoke the interest of the student.

**Five Elements of Mathematical Pedagogy to Address Different Learning Styles**  The "Rule of Four" is a common theme in many texts. Implementing this approach, we discuss many of the central concepts **numerically**, **graphically,** and **algebraically** and clearly delineate these distinctions. The fourth element, **verbal communication** of mathematical concepts, is emphasized through our discussions on translating English sentences into mathematical statements and in our extensive Communication and Reasoning exercises at the end of each section. A fifth element, **interactivity**, is implemented through expanded use of question-and-answer dialogues but is seen most dramatically in the student Website. Using this resource, students can interact with the material in several ways: through interactive tutorials in the form of games, chapter summaries, and chapter review exercises, all in reference to concepts and examples covered in sections and with online utilities that automate a variety of tasks, from graphing to regression and matrix algebra.

**Exercise Sets**  Our comprehensive collection of exercises provides a wealth of material that can be used to challenge students at almost every level of preparation and includes everything from straightforward drill exercises to interesting and rather challenging applications. The exercise sets have been carefully graded to move from straightforward basic exercises and exercises that are similar to examples in the text to more interesting and advanced ones, marked as "more advanced" for easy reference. There are also several much more difficult exercises, designated as "challenging." We have also included, in virtually every section of every chapter, interesting applications based on real data, Communication and Reasoning exercises that help students articulate mathematical concepts and recognize common errors, and exercises ideal for the use of technology.

Many of the scenarios used in application examples and exercises are revisited several times throughout the book. Thus, for instance, students will find themselves using a variety of techniques, from solving systems of equations to linear programming, or graphing through the use of derivatives and elasticity, to analyze the same application. Reusing scenarios and important functions provides unifying threads and shows students the complex texture of real-life problems.

## New to This Edition
## Content

- Chapter 1 (page 39): We now include, in Section 1.1, careful discussion of the common practice of representing functions as equations and vice versa; for instance, a cost equation like $C = 10x + 50$ can be thought of as defining a cost *function* $C(x) = 10x + 50$. Instead of rejecting this practice, we encourage the student to see this connection between functions and equations and to be able to switch from one interpretation to the other.

  Our discussion of functions and models in Section 1.2 now includes a careful discussion of the algebra of functions presented through the context of important applications rather than as an abstract concept. Thus, the student will see from the outset *why* we want to talk about sums, products, etc. of functions rather than simply *how* to manipulate them.

- Chapter 2 (page 125): The Mathematics of Finance is now Chapter 2 of the text because the discussion of many important topics in finance relate directly to the first discussions of compound interest and other mathematical models in Chapter 1. Note that our discussion of the Mathematics of Finance does not require the use of logarithmic functions to solve for exponents analytically but instead focuses on numerical solution using the technologies we discuss. However, the use of logarithms is presented as an option for students and instructors who prefer to use them.

- Chapter 10 (page 687): Our discussion of limits now discusses extensively when, and why, substitution can be used to obtain a limit. We now also follow the usual convention of allowing only one-sided limits at endpoints of domains. This approach also applies to derivatives, where we now disallow derivatives at endpoints of domains, as is the normal convention.

- Chapter 11 (page 783): The closed-form formula for the derivative of $|x|$, introduced in Section 11.1, is now more fully integrated into the text, as is that for its antiderivative (in Chapter 13). (It is puzzling that it is not standard fare in other calculus books.)

- Chapter 13 (page 945): The sections on antiderivatives and substitution have been reorganized and streamlined and now include discussion of the closed-form antiderivative of $|x|$ and well as new exercises featuring absolute values.

  The definite integral is now introduced in the realistic context of the volume of oil released in an oil spill comparable in size to the *BP* 2011 Gulf oil spill.

- Chapter 15 (page 1081): The discussion of level curves in Section 15.1 is now more extensive and includes added examples and exercises.

- **Case Studies:** A number of the Case Studies at the ends of the chapters have been extensively revised, using updated real data, and continue to reflect topics of current interest, such as subprime mortgages, hybrid car production, the diet problem (in linear programming), spending on housing construction, modeling tax revenues, and pollution control.

## Current Topics in the Applications

- We have added and updated numerous real data exercises and examples based on topics that are either of intense current interest or of general interest to contemporary students, including Facebook, XBoxes, iPhones, Androids, iPads, foreclosure rates, the housing crisis, subprime mortgages, travel to Cancun, the *BP* 2011 Gulf oil spill, and the U.S. stock market "flash crash" of May 6, 2010. (Also see the list, in the inside back cover, of the corporations we reference in the applications.)

## Exercises

- We have expanded the chapter review exercise sets to be more representative of the material within the chapter. Note that all the applications in the chapter review exercises revolve around the fictitious online bookseller, *OHaganBooks.com,* and the various—often amusing—travails of *OHaganBooks.com* CEO John O'Hagan and his business associate Marjory Duffin.

- We have added many new conceptual Communication and Reasoning exercises, including many dealing with common student errors and misconceptions.

## End-of-Chapter Technology Guides

- Our end-of-chapter detailed Technology Guides now discuss the use of spreadsheets in general rather than focusing exclusively on Microsoft® Excel, thus enabling readers to use any of the several alternatives now available, such as Google's online Google Sheets®, Open Office®, and Apple's Numbers®.

## Continuing Features

**Case Study**  Projecting Market Share

You are the sales director at *Selular*, a cellphone provider, and things are not looking good for your company: The recently launched *iClone* competitor is beginning to chip away at Selular's market share. Particularly disturbing are rumors of fierce brand loyalty by iClone customers, with several bloggers suggesting that iClone retains close to 100% of their customers. Worse, you will shortly be presenting a sales report to the board of directors, and the CEO has "suggested" that your report include 2-, 5-, and 10-year projections of Selular's market share given the recent impact on the market by iClone, and also a "worst-case" scenario projecting what would happen if iClone customers are so loyal that none of them ever switch services.

- **Case Studies**  Each chapter ends with a section entitled "Case Study," an extended application that uses and illustrates the central ideas of the chapter, focusing on the development of mathematical models appropriate to the topics. These applications are ideal for assignment as projects, and to this end we have included groups of exercises at the end of each.

- **Before We Go On**  Most examples are followed by supplementary discussions, which may include a check on the answer, a discussion of the feasibility and significance of a solution, or an in-depth look at what the solution means.

- **Quick Examples**  Most definition boxes include quick, straightforward examples that a student can use to solidify each new concept.

- **Question-and-Answer Dialogue**  We frequently use informal question-and-answer dialogues that anticipate the kinds of questions that may occur to the student and also guide the student through the development of new concepts.

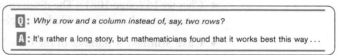

> **Q:** *Why a row and a column instead of, say, two rows?*
>
> **A:** It's rather a long story, but mathematicians found that it works best this way . . .

- **Marginal Technology Notes**  We give brief marginal technology notes to outline the use of graphing calculator, spreadsheet, and Website technology in appropriate examples. When necessary, the reader is referred to more detailed discussion in the end-of-chapter Technology Guides.

- **End-of-Chapter Technology Guides**  We continue to include detailed TI-83/84 Plus and Spreadsheet Guides at the end of each chapter. These Guides are referenced liberally in marginal technology notes at appropriate points in the chapter, so instructors and students can easily use this material or not, as they prefer. Groups of exercises for which the use of technology is suggested or required appear throughout the exercise sets.

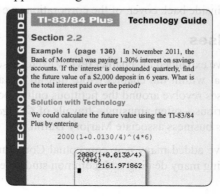

**TI-83/84 Plus**  Technology Guide

**Section 2.2**

**Example 1 (page 136)**  In November 2011, the Bank of Montreal was paying 1.30% interest on savings accounts. If the interest is compounded quarterly, find the future value of a $2,000 deposit in 6 years. What is the total interest paid over the period?

**Solution with Technology**

We could calculate the future value using the TI-83/84 Plus by entering

    2000(1+0.0130/4)^(4*6)

    2000(1+0.0130/4)
    ^(4*6)
             2161.971862

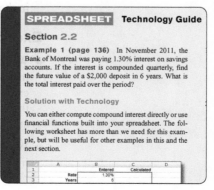

**SPREADSHEET**  Technology Guide

**Section 2.2**

**Example 1 (page 136)**  In November 2011, the Bank of Montreal was paying 1.30% interest on savings accounts. If the interest is compounded quarterly, find the future value of a $2,000 deposit in 6 years. What is the total interest paid over the period?

**Solution with Technology**

You can either compute compound interest directly or use financial functions built into your spreadsheet. The following worksheet has more than we need for this example, but will be useful for other examples in this and the next section.

- **Communication and Reasoning Exercises for Writing and Discussion**
  These are exercises designed to broaden the student's grasp of the mathematical concepts and develop modeling skills. They include exercises in which the student is asked to provide his or her own examples to illustrate a point or design an application with a given solution. They also include "fill in the blank" type exercises, exercises that invite discussion and debate, and exercises in which the student must identify common errors. These exercises often have no single correct answer.

## Supplemental Material

### For Instructors and Students

#### Enhanced WebAssign®

**Content**

Exclusively from Cengage Learning, Enhanced WebAssign® combines the exceptional mathematics content in Waner and Costenoble's text with the most powerful online homework solution, WebAssign. Enhanced WebAssign engages students with immediate feedback, rich tutorial content, videos, animations, and an interactive eBook, helping students to develop a deeper conceptual understanding of the subject matter. The interactive eBook contains helpful search, highlighting, and note-taking features.

Instructors can build online assignments by selecting from thousands of text-specific problems, supplemented if desired with problems from any Cengage Learning textbook. Flexible assignment options give instructors the ability to choose how feedback and tutorial content is released to students as well as the ability to release assignments conditionally based on students' prerequisite assignment scores. Increase student engagement, improve course outcomes, and experience the superior service offered through CourseCare. Visit us at http://webassign.net/cengage or www.cengage.com/ewa to learn more.

**Service**

Your adoption of Enhanced WebAssign® includes CourseCare, Cengage Learning's industry leading service and training program designed to ensure that you have everything that you need to make the most of your use of Enhanced WebAssign. CourseCare provides one-on-one service, from finding the right solutions for your course to training and support. A team of Cengage representatives, including Digital Solutions Managers and Coordinators as well as Service and Training Consultants, assists you every step of the way. For additional information about CourseCare, please visit www.cengage.com/coursecare.

Our Enhanced WebAssign training program provides a comprehensive curriculum of beginner, intermediate, and advanced sessions, designed to get you started and effectively integrate Enhanced WebAssign into your course. We offer a flexible online and recorded training program designed to accommodate your busy schedule. Whether you are using Enhanced WebAssign for the first time or an experienced user, there is a training option to meet your needs.

#### www.WanerMath.com

The authors' Website, accessible through www.WanerMath.com and linked within Enhanced WebAssign, has been evolving for more than a decade and has been receiving increasingly more recognition. Students, raised in an environment in which computers permeate both work and play, now use the Internet to engage with the material in an active way. The following features of the authors' Website are fully integrated with the text and can be used as a personalized study resource as well as a valuable teaching aid for instructors:

- *Interactive tutorials* on almost all topics, with guided exercises, which also can be used in classroom instruction or in distance learning courses

- *More challenging game versions of tutorials* with randomized questions that complement the traditional interactive tutorials and can be used as in-class quizzes

- *Detailed interactive chapter summaries* that review basic definitions and problem-solving techniques and can act as pre-test study tools

- *Downloadable Excel tutorials* keyed to examples given in the text

- *Online utilities for use in solving many of the technology-based application exercises.* The utilities, for instructor use in class and student use out of class, include a function grapher and evaluator that now also graphs derivatives and offers curve-fitting, regression tools, an interactive Riemann sum grapher with an improved numerical integrator, a matrix algebra tool, linear programming tools, and a line entry calculator that calculates permutations and combinations, expands multinomial expressions, does derivatives and numerical integration, and gives Taylor series.

- *Chapter true-false quizzes* with feedback for many incorrect answers

- *Supplemental topics* including interactive text and exercise sets for selected topics not found in the printed texts

- *Spanish versions* of chapter summaries, tutorials, game tutorials, and utilities

## For Students

**Student Solutions Manual** *by Waner and Costenoble*
ISBN: 9781285085661
The student solutions manual provides worked-out solutions to the odd-numbered exercises in the text, plus problem-solving strategies and additional algebra steps and review for selected problems.

To access this and other course materials and companion resources, please visit **www.cengagebrain.com**. At the CengageBrain.com home page, search for the ISBN of your title (from the back cover of your book) using the search box at the top of the page. This will take you to the product page where free companion resources can be found.

## For Instructors

**Complete Solution Manual** *by Waner and Costenoble*
ISBN: 9781285085678
The instructor's solutions manual provides worked-out solutions to all of the exercises in the text.

**Solution Builder** *by Waner and Costenoble*
ISBN: 9781285085685
This time-saving resource offers fully worked instructor solutions to all exercises in the text in customizable online format. Adopting instructors can sign up for access at www.cengage.com/solutionbuilder.

**PowerLecture™ with ExamView® computerized testing** *by Waner and Costenoble*
ISBN: 9781285085708

This CD-ROM provides the instructor with dynamic media tools for teaching, including Microsoft® PowerPoint® lecture slides, figures from the book, and the Test Bank. You can create, deliver, and customize tests (both print and online) in minutes with ExamView® computerized testing, which includes Test Bank items in electronic format. In addition, you can easily build solution sets for homework or exams by linking to Solution Builder's online solutions manual.

**Instructor's Edition**
ISBN: 9781133610595

**www.WanerMath.com**
The Instructor's Resource Page at www.WanerMath.com features an expanded collection of instructor resources, including an updated corrections page, an expanding set of author-created teaching videos for use in distance learning courses, and a utility that automatically updates homework exercise sets from the fifth edition to the sixth.

## Acknowledgments

This project would not have been possible without the contributions and suggestions of numerous colleagues, students, and friends. We are particularly grateful to our colleagues at Hofstra and elsewhere who used and gave us useful feedback on previous editions. We are also grateful to everyone at Cengage for their encouragement and guidance throughout the project. Specifically, we would like to thank Richard Stratton and Jay Campbell for their unflagging enthusiasm and Alison Eigel Zade for whipping the book into shape.

We would also like to thank our accuracy checker, Jerrold Grossman, and the numerous reviewers who provided many helpful suggestions that have shaped the development of this book.

Priscilla Chaffe-Stengel, *California State University, Fresno*

Celeste Hernandez, *Richland College*

Erick Hofacker, *University of Wisconsin–River Falls*

Jan Rychtar, *University of North Carolina at Greensboro*

Jack Narayan, *State University of New York at Oswego*

Mary Wagner-Krankel, *St. Mary's University*

Stefan Waner
Steven R. Costenoble

# O

# Precalculus Review

**Website**

www.WanerMath.com

- At the Website you will find section-by-section interactive tutorials for further study and practice.

DreamPictures/Taxi/Getty Images

1

# Introduction

In this chapter we review some topics from algebra that you need to know to get the most out of this book. This chapter can be used either as a refresher course or as a reference.

There is one crucial fact you must always keep in mind: The letters used in algebraic expressions stand for numbers. All the rules of algebra are just facts about the arithmetic of numbers. If you are not sure whether some algebraic manipulation you are about to do is legitimate, try it first with numbers. If it doesn't work with numbers, it doesn't work.

## **O.1**    Real Numbers

The **real numbers** are the numbers that can be written in decimal notation, including those that require an infinite decimal expansion. The set of real numbers includes all integers, positive and negative; all fractions; and the irrational numbers, those with decimal expansions that never repeat. Examples of irrational numbers are

$$\sqrt{2} = 1.414213562373\ldots$$

and

$$\pi = 3.141592653589\ldots$$

**Figure 1**

It is very useful to picture the real numbers as points on a line. As shown in Figure 1, larger numbers appear to the right, in the sense that if $a < b$ then the point corresponding to $b$ is to the right of the one corresponding to $a$.

### Intervals

Some subsets of the set of real numbers, called **intervals**, show up quite often and so we have a compact notation for them.

### Interval Notation

Here is a list of types of intervals along with examples.

|  | *Interval* | *Description* | *Picture* | *Example* |
|---|---|---|---|---|
| **Closed** | $[a, b]$ | Set of numbers $x$ with $a \le x \le b$ | $a$ $\quad$ $b$ (includes end points) | $[0, 10]$ |
| **Open** | $(a, b)$ | Set of numbers $x$ with $a < x < b$ | $a$ $\quad$ $b$ (excludes end points) | $(-1, 5)$ |
| **Half-Open** | $(a, b]$ | Set of numbers $x$ with $a < x \le b$ | $a$ $\quad$ $b$ | $(-3, 1]$ |
|  | $[a, b)$ | Set of numbers $x$ with $a \le x < b$ | $a$ $\quad$ $b$ | $[0, 5)$ |

| | | | | |
|---|---|---|---|---|
| **Infinite** | $[a, +\infty)$ | Set of numbers $x$ with $a \le x$ | | $[10, +\infty)$ |
| | $(a, +\infty)$ | Set of numbers $x$ with $a < x$ | | $(-3, +\infty)$ |
| | $(-\infty, b]$ | Set of numbers $x$ with $x \le b$ | | $(-\infty, -3]$ |
| | $(-\infty, b)$ | Set of numbers $x$ with $x < b$ | | $(-\infty, 10)$ |
| | $(-\infty, +\infty)$ | Set of all real numbers | | $(-\infty, +\infty)$ |

## Operations

There are five important operations on real numbers: addition, subtraction, multiplication, division, and exponentiation. "Exponentiation" means raising a real number to a power; for instance, $3^2 = 3 \cdot 3 = 9$; $2^3 = 2 \cdot 2 \cdot 2 = 8$.

A note on technology: Most graphing calculators and spreadsheets use an asterisk * for multiplication and a caret sign ^ for exponentiation. Thus, for instance, $3 \times 5$ is entered as $3*5$, $3x$ as $3*x$, and $3^2$ as $3\char94 2$.

When we write an expression involving two or more operations, like

$$2 \cdot 3 + 4$$

or

$$\frac{2 \cdot 3^2 - 5}{4 - (-1)}$$

we need to agree on the order in which to do the operations. Does $2 \cdot 3 + 4$ mean $(2 \cdot 3) + 4 = 10$ or $2 \cdot (3 + 4) = 14$? We all agree to use the following rules for the order in which we do the operations.

### Standard Order of Operations

**Parentheses and Fraction Bars**    First, calculate the values of all expressions inside parentheses or brackets, working from the innermost parentheses out, before using them in other operations. In a fraction, calculate the numerator and denominator separately before doing the division.

#### Quick Examples

1. $6(2 + [3 - 5] - 4) = 6(2 + (-2) - 4) = 6(-4) = -24$

2. $\dfrac{(4 - 2)}{3(-2 + 1)} = \dfrac{2}{3(-1)} = \dfrac{2}{-3} = -\dfrac{2}{3}$

3. $3/(2 + 4) = \dfrac{3}{2 + 4} = \dfrac{3}{6} = \dfrac{1}{2}$

4. $(x + 4x)/(y + 3y) = 5x/(4y)$

**Exponents**   Next, perform exponentiation.

**Quick Examples**

$\left.\begin{array}{l}\textbf{1. } 2 + 4^2 = 2 + 16 = 18 \\ \textbf{2. } (2+4)^2 = 6^2 = 36\end{array}\right\}$   Note the difference.

**3.** $2\left(\dfrac{3}{4-5}\right)^2 = 2\left(\dfrac{3}{-1}\right)^2 = 2(-3)^2 = 2 \times 9 = 18$

**4.** $2(1 + 1/10)^2 = 2(1.1)^2 = 2 \times 1.21 = 2.42$

**Multiplication and Division**   Next, do all multiplications and divisions, from left to right.

**Quick Examples**

**1.** $2(3-5)/4 \cdot 2 = 2(-2)/4 \cdot 2$   Parentheses first

$\qquad = -4/4 \cdot 2$   Left-most product

$\qquad = -1 \cdot 2 = -2$   Multiplications and divisions, left to right

**2.** $2(1 + 1/10)^2 \times 2/10 = 2(1.1)^2 \times 2/10$   Parentheses first

$\qquad = 2 \times 1.21 \times 2/10$   Exponent

$\qquad = 4.84/10 = 0.484$   Multiplications and divisions, left to right

**3.** $4\dfrac{2(4-2)}{3(-2\cdot 5)} = 4\dfrac{2(2)}{3(-10)} = 4\dfrac{4}{-30} = \dfrac{16}{-30} = -\dfrac{8}{15}$

**Addition and Subtraction**   Last, do all additions and subtractions, from left to right.

**Quick Examples**

**1.** $2(3-5)^2 + 6 - 1 = 2(-2)^2 + 6 - 1 = 2(4) + 6 - 1$

$\qquad = 8 + 6 - 1 = 13$

**2.** $\left(\dfrac{1}{2}\right)^2 - (-1)^2 + 4 = \dfrac{1}{4} - 1 + 4 = -\dfrac{3}{4} + 4 = \dfrac{13}{4}$

$\left.\begin{array}{l}\textbf{3. } 3/2 + 4 = 1.5 + 4 = 5.5 \\ \textbf{4. } 3/(2+4) = 3/6 = 1/2 = 0.5\end{array}\right\}$   Note the difference.

**5.** $4/2^2 + (4/2)^2 = 4/2^2 + 2^2 = 4/4 + 4 = 1 + 4 = 5$

## 🄣 Entering Formulas

Any good calculator or spreadsheet will respect the standard order of operations. However, we must be careful with division and exponentiation and use parentheses as necessary. The following table gives some examples of simple mathematical expressions and their equivalents in the functional format used in most graphing calculators, spreadsheets, and computer programs.

| *Mathematical Expression* | *Formula* | *Comments* |
|---|---|---|
| $\dfrac{2}{3-x}$ | `2/(3-x)` | Note the use of parentheses instead of the fraction bar. If we omit the parentheses, we get the expression shown next. |
| $\dfrac{2}{3}-x$ | `2/3-x` | The calculator follows the usual order of operations. |
| $\dfrac{2}{3\times5}$ | `2/(3*5)` | Putting the denominator in parentheses ensures that the multiplication is carried out first. The asterisk is usually used for multiplication in graphing calculators and computers. |
| $\dfrac{2}{x}\times5$ | `(2/x)*5` | Putting the fraction in parentheses ensures that it is calculated first. Some calculators will interpret $2/3*5$ as $\dfrac{2}{3\times5}$, but $2/3\,(5)$ as $\dfrac{2}{3}\times5$. |
| $\dfrac{2-3}{4+5}$ | `(2-3)/(4+5)` | Note once again the use of parentheses in place of the fraction bar. |
| $2^3$ | `2^3` | The caret ^ is commonly used to denote exponentiation. |
| $2^{3-x}$ | `2^(3-x)` | Be careful to use parentheses to tell the calculator where the exponent ends. Enclose the *entire exponent* in parentheses. |
| $2^3-x$ | `2^3-x` | Without parentheses, the calculator will follow the usual order of operations: exponentiation and then subtraction. |
| $3\times2^{-4}$ | `3*2^(-4)` | On some calculators, the negation key is separate from the minus key. |
| $2^{-4\times3}\times5$ | `2^(-4*3)*5` | Note once again how parentheses enclose the entire exponent. |
| $100\left(1+\dfrac{0.05}{12}\right)^{60}$ | `100*(1+0.05/12)^60` | This is a typical calculation for compound interest. |
| $PV\left(1+\dfrac{r}{m}\right)^{mt}$ | `PV*(1+r/m)^(m*t)` | This is the compound interest formula. *PV* is understood to be a single number (present value) and not the product of *P* and *V* (or else we would have used `P*V`). |
| $\dfrac{2^{3-2}\times5}{y-x}$ | `2^(3-2)*5/(y-x)`<br>or<br>`(2^(3-2)*5)/(y-x)` | Notice again the use of parentheses to hold the denominator together. We could also have enclosed the numerator in parentheses, although this is optional. (Why?) |
| $\dfrac{2^y+1}{2-4^{3x}}$ | `(2^y+1)/(2-4^(3*x))` | Here, it is necessary to enclose both the numerator and the denominator in parentheses. |
| $2^y+\dfrac{1}{2}-4^{3x}$ | `2^y+1/2-4^(3*x)` | This is the effect of leaving out the parentheses around the numerator and denominator in the previous expression. |

## Accuracy and Rounding

When we use a calculator or computer, the results of our calculations are often given to far more decimal places than are useful. For example, suppose we are told that a square has an area of 2.0 square feet and we are asked how long its sides are. Each side is the square root of the area, which the calculator tells us is

$$\sqrt{2} \approx 1.414213562$$

However, the measurement of 2.0 square feet is probably accurate to only two digits, so our estimate of the lengths of the sides can be no more accurate than that. Therefore, we round the answer to two digits:

$$\text{Length of one side} \approx 1.4 \text{ feet}$$

The digits that follow 1.4 are meaningless. The following guide makes these ideas more precise.

### Significant Digits, Decimal Places, and Rounding

The number of **significant digits** in a decimal representation of a number is the number of digits that are not leading zeros after the decimal point (as in .0005) or trailing zeros before the decimal point (as in 5,400,000). We say that a value is **accurate to $n$ significant digits** if only the first $n$ significant digits are meaningful.

### When to Round

After doing a computation in which all the quantities are accurate to no more than $n$ significant digits, round the final result to $n$ significant digits.

#### Quick Examples

1. 0.00067 has two significant digits.                The 000 before 67 are leading zeros.

2. 0.000670 has three significant digits.             The 0 after 67 is significant.

3. 5,400,000 has two or more significant digits.      We can't say how many of the zeros are trailing.[1]

4. 5,400,001 has 7 significant digits.                The string of zeros is not trailing.

5. Rounding 63,918 to three significant digits gives 63,900.

6. Rounding 63,958 to three significant digits gives 64,000.

7. $\pi = 3.141592653...$   $\frac{22}{7} = 3.142857142...$ Therefore, $\frac{22}{7}$ is an approximation of $\pi$ that is accurate to only three significant digits (3.14).

8. $4.02(1 + 0.02)^{1.4} \approx 4.13$                We rounded to three significant digits.

---

[1]If we obtained 5,400,000 by rounding 5,401,011, then it has three significant digits because the zero after the 4 is significant. On the other hand, if we obtained it by rounding 5,411,234, then it has only two significant digits. The use of scientific notation avoids this ambiguity: $5.40 \times 10^6$ (or 5.40 E6 on a calculator or computer) is accurate to three digits and $5.4 \times 10^6$ is accurate to two.

One more point, though: If, in a long calculation, you round the intermediate results, your final answer may be even less accurate than you think. As a general rule,

> *When calculating, don't round intermediate results. Rather, use the most accurate results obtainable or have your calculator or computer store them for you.*

When you are done with the calculation, *then* round your answer to the appropriate number of digits of accuracy.

## 0.1 EXERCISES

*Calculate each expression in Exercises 1–24, giving the answer as a whole number or a fraction in lowest terms.*

**1.** $2(4 + (-1))(2 \cdot -4)$

**2.** $3 + ([4 - 2] \cdot 9)$

**3.** `20/(3*4)-1`

**4.** `2-(3*4)/10`

**5.** $\dfrac{3 + ([3 + (-5)])}{3 - 2 \times 2}$

**6.** $\dfrac{12 - (1 - 4)}{2(5 - 1) \cdot 2 - 1}$

**7.** `(2-5*(-1))/1-2*(-1)`

**8.** `2-5*(-1)/(1-2*(-1))`

**9.** $2 \cdot (-1)^2/2$

**10.** $2 + 4 \cdot 3^2$

**11.** $2 \cdot 4^2 + 1$

**12.** $1 - 3 \cdot (-2)^2 \times 2$

**13.** `3^2+2^2+1`

**14.** `2^(2^2-2)`

**15.** $\dfrac{3 - 2(-3)^2}{-6(4 - 1)^2}$

**16.** $\dfrac{1 - 2(1 - 4)^2}{2(5 - 1)^2 \cdot 2}$

**17.** `10*(1+1/10)^3`

**18.** `121/(1+1/10)^2`

**19.** $3\left(\dfrac{-2 \cdot 3^2}{-(4 - 1)^2}\right)$

**20.** $-\left(\dfrac{8(1 - 4)^2}{-9(5 - 1)^2}\right)$

**21.** $3\left(1 - \left(-\dfrac{1}{2}\right)^2\right)^2 + 1$

**22.** $3\left(\dfrac{1}{9} - \left(\dfrac{2}{3}\right)^2\right)^2 + 1$

**23.** `(1/2)^2-1/2^2`

**24.** `2/(1^2)-(2/1)^2`

*Convert each expression in Exercises 25–50 into its technology formula equivalent as in the table in the text.*

**25.** $3 \times (2 - 5)$

**26.** $4 + \dfrac{5}{9}$

**27.** $\dfrac{3}{2 - 5}$

**28.** $\dfrac{4 - 1}{3}$

**29.** $\dfrac{3 - 1}{8 + 6}$

**30.** $3 + \dfrac{3}{2 - 9}$

**31.** $3 - \dfrac{4 + 7}{8}$

**32.** $\dfrac{4 \times 2}{\left(\frac{2}{3}\right)}$

**33.** $\dfrac{2}{3 + x} - xy^2$

**34.** $3 + \dfrac{3 + x}{xy}$

**35.** $3.1x^3 - 4x^{-2} - \dfrac{60}{x^2 - 1}$

**36.** $2.1x^{-3} - x^{-1} + \dfrac{x^2 - 3}{2}$

**37.** $\dfrac{\left(\frac{2}{3}\right)}{5}$

**38.** $\dfrac{2}{\left(\frac{3}{5}\right)}$

**39.** $3^{4-5} \times 6$

**40.** $\dfrac{2}{3 + 5^{7-9}}$

**41.** $3\left(1 + \dfrac{4}{100}\right)^{-3}$

**42.** $3\left(\dfrac{1 + 4}{100}\right)^{-3}$

**43.** $3^{2x-1} + 4^x - 1$

**44.** $2^{x^2} - (2^{2x})^2$

**45.** $2^{2x^2 - x + 1}$

**46.** $2^{2x^2 - x} + 1$

**47.** $\dfrac{4e^{-2x}}{2 - 3e^{-2x}}$

**48.** $\dfrac{e^{2x} + e^{-2x}}{e^{2x} - e^{-2x}}$

**49.** $3\left(1 - \left(-\dfrac{1}{2}\right)^2\right)^2 + 1$

**50.** $3\left(\dfrac{1}{9} - \left(\dfrac{2}{3}\right)^2\right)^2 + 1$

## 0.2 Exponents and Radicals

In Section 1 we discussed exponentiation, or "raising to a power"; for example, $2^3 = 2 \cdot 2 \cdot 2$. In this section we discuss the algebra of exponentials more fully. First, we look at *integer* exponents: cases in which the powers are positive or negative whole numbers.

## Integer Exponents

### Positive Integer Exponents

If $a$ is any real number and $n$ is any positive integer, then by $a^n$ we mean the quantity $a \cdot a \cdot \cdots \cdot a$ ($n$ times); thus, $a^1 = a$, $a^2 = a \cdot a$, $a^5 = a \cdot a \cdot a \cdot a \cdot a$. In the expression $a^n$ the number $n$ is called the **exponent**, and the number $a$ is called the **base**.

#### Quick Examples

$$3^2 = 9 \qquad\qquad 2^3 = 8$$

$$0^{34} = 0 \qquad\qquad (-1)^5 = -1$$

$$10^3 = 1{,}000 \qquad 10^5 = 100{,}000$$

### Negative Integer Exponents

If $a$ is any real number *other than zero* and $n$ is any positive integer, then we define

$$a^{-n} = \frac{1}{a^n} = \frac{1}{\underbrace{a \cdot a \cdot \cdots \cdot a}} \quad (n \text{ times})$$

#### Quick Examples

$$2^{-3} = \frac{1}{2^3} = \frac{1}{8} \qquad\qquad 1^{-27} = \frac{1}{1^{27}} = 1$$

$$x^{-1} = \frac{1}{x^1} = \frac{1}{x} \qquad\qquad (-3)^{-2} = \frac{1}{(-3)^2} = \frac{1}{9}$$

$$y^7 y^{-2} = y^7 \frac{1}{y^2} = y^5 \qquad 0^{-2} \text{ is not defined}$$

### Zero Exponent

If $a$ is any real number other than zero, then we define

$$a^0 = 1$$

#### Quick Examples

$$3^0 = 1 \qquad\qquad\qquad 1{,}000{,}000^0 = 1$$

$$0^0 \text{ is not defined}$$

When combining exponential expressions, we use the following identities.

| **Exponent Identity** | **Quick Examples** |
|---|---|
| **1.** $a^m a^n = a^{m+n}$ | $2^3 2^2 = 2^{3+2} = 2^5 = 32$ |
| | $x^3 x^{-4} = x^{3-4} = x^{-1} = \dfrac{1}{x}$ |
| | $\dfrac{x^3}{x^{-2}} = x^3 \dfrac{1}{x^{-2}} = x^3 x^2 = x^5$ |
| **2.** $\dfrac{a^m}{a^n} = a^{m-n}$ if $a \neq 0$ | $\dfrac{4^3}{4^2} = 4^{3-2} = 4^1 = 4$ |
| | $\dfrac{x^3}{x^{-2}} = x^{3-(-2)} = x^5$ |
| | $\dfrac{3^2}{3^4} = 3^{2-4} = 3^{-2} = \dfrac{1}{9}$ |
| **3.** $(a^n)^m = a^{nm}$ | $(3^2)^2 = 3^4 = 81$ |
| | $(2^x)^2 = 2^{2x}$ |
| **4.** $(ab)^n = a^n b^n$ | $(4 \cdot 2)^2 = 4^2 2^2 = 64$ |
| | $(-2y)^4 = (-2)^4 y^4 = 16y^4$ |
| **5.** $\left(\dfrac{a}{b}\right)^n = \dfrac{a^n}{b^n}$ if $b \neq 0$ | $\left(\dfrac{4}{3}\right)^2 = \dfrac{4^2}{3^2} = \dfrac{16}{9}$ |
| | $\left(\dfrac{x}{-y}\right)^3 = \dfrac{x^3}{(-y)^3} = -\dfrac{x^3}{y^3}$ |

**Caution**

- In the first two identities, the bases of the expressions must be the same. For example, the first gives $3^2 3^4 = 3^6$, but does *not* apply to $3^2 4^2$.

- People sometimes invent their own identities, such as $a^m + a^n = a^{m+n}$, which is wrong! (Try it with $a = m = n = 1$.) If you wind up with something like $2^3 + 2^4$, you are stuck with it; there are no identities around to simplify it further. (You can factor out $2^3$, but whether or not that is a simplification depends on what you are going to do with the expression next.)

---

**EXAMPLE 1  Combining the Identities**

$$\dfrac{(x^2)^3}{x^3} = \dfrac{x^6}{x^3} \qquad \text{By (3)}$$

$$= x^{6-3} \qquad \text{By (2)}$$

$$= x^3$$

$$\dfrac{(x^4 y)^3}{y} = \dfrac{(x^4)^3 y^3}{y} \qquad \text{By (4)}$$

$$= \dfrac{x^{12} y^3}{y} \qquad \text{By (3)}$$

$$= x^{12} y^{3-1} \qquad \text{By (2)}$$

$$= x^{12} y^2$$

### EXAMPLE 2  Eliminating Negative Exponents

Simplify the following and express the answer using no negative exponents.

**a.** $\dfrac{x^4 y^{-3}}{x^5 y^2}$    **b.** $\left(\dfrac{x^{-1}}{x^2 y}\right)^5$

**Solution**

**a.** $\dfrac{x^4 y^{-3}}{x^5 y^2} = x^{4-5} y^{-3-2} = x^{-1} y^{-5} = \dfrac{1}{xy^5}$

**b.** $\left(\dfrac{x^{-1}}{x^2 y}\right)^5 = \dfrac{(x^{-1})^5}{(x^2 y)^5} = \dfrac{x^{-5}}{x^{10} y^5} = \dfrac{1}{x^{15} y^5}$

## Radicals

If $a$ is any non-negative real number, then its **square root** is the non-negative number whose square is $a$. For example, the square root of 16 is 4, because $4^2 = 16$. We write the square root of $n$ as $\sqrt{n}$. (Roots are also referred to as **radicals**.) It is important to remember that $\sqrt{n}$ is never negative. Thus, for instance, $\sqrt{9}$ is 3, and not $-3$, even though $(-3)^2 = 9$. If we want to speak of the "negative square root" of 9, we write it as $-\sqrt{9} = -3$. If we want to write both square roots at once, we write $\pm\sqrt{9} = \pm 3$.

The **cube root** of a real number $a$ is the number whose cube is $a$. The cube root of $a$ is written as $\sqrt[3]{a}$ so that, for example, $\sqrt[3]{8} = 2$ (because $2^3 = 8$). Note that we can take the cube root of any number, positive, negative, or zero. For instance, the cube root of $-8$ is $\sqrt[3]{-8} = -2$ because $(-2)^3 = -8$. Unlike square roots, the cube root of a number may be negative. In fact, the cube root of $a$ always has the same sign as $a$.

Higher roots are defined similarly. The **fourth root** of the *non-negative* number $a$ is defined as the non-negative number whose fourth power is $a$, and written $\sqrt[4]{a}$. The **fifth root** of any number $a$ is the number whose fifth power is $a$, and so on.

**Note**  We cannot take an even-numbered root of a negative number, but we can take an odd-numbered root of any number. Even roots are always positive, whereas odd roots have the same sign as the number we start with.  ∎

### EXAMPLE 3  *n*th Roots

$$\sqrt{4} = 2 \qquad \text{Because } 2^2 = 4$$

$$\sqrt{16} = 4 \qquad \text{Because } 4^2 = 16$$

$$\sqrt{1} = 1 \qquad \text{Because } 1^2 = 1$$

If $x \ge 0$, then $\sqrt{x^2} = x$     Because $x^2 = x^2$

$$\sqrt{2} \approx 1.414213562 \qquad \sqrt{2} \text{ is not a whole number.}$$

$$\sqrt{1+1} = \sqrt{2} \approx 1.414213562 \qquad \text{First add, then take the square root.}^2$$

$$\sqrt{9+16} = \sqrt{25} = 5 \qquad \text{Contrast with } \sqrt{9} + \sqrt{16} = 3 + 4 = 7.$$

---

[2]In general, $\sqrt{a+b}$ means the square root of the *quantity* $(a+b)$. The radical sign acts as a pair of parentheses or a fraction bar, telling us to evaluate what is inside before taking the root. (See the Caution on the next page.)

$$\frac{1}{\sqrt{2}} = \frac{\sqrt{2}}{2}$$ 　　Multiply top and bottom by $\sqrt{2}$.

$$\sqrt[3]{27} = 3$$ 　　Because $3^3 = 27$

$$\sqrt[3]{-64} = -4$$ 　　Because $(-4)^3 = -64$

$$\sqrt[4]{16} = 2$$ 　　Because $2^4 = 16$

$$\sqrt[4]{-16} \text{ is not defined}$$ 　　Even-numbered root of a negative number

$$\sqrt[5]{-1} = -1, \text{ since } (-1)^5 = -1$$ 　　Odd-numbered root of a negative number

$$\sqrt[n]{-1} = -1 \text{ if } n \text{ is any odd number}$$

---

**Q:** *In the example we saw that $\sqrt{x^2} = x$ if $x$ is non-negative. What happens if $x$ is negative?*

**A:** If $x$ is negative, then $x^2$ is positive, and so $\sqrt{x^2}$ is still defined as the non-negative number whose square is $x^2$. This number must be $|x|$, the **absolute value of $x$**, which is the non-negative number with the same size as $x$. For instance, $|-3| = 3$, while $|3| = 3$, and $|0| = 0$. It follows that

$$\sqrt{x^2} = |x|$$

for every real number $x$, positive or negative. For instance,

$$\sqrt{(-3)^2} = \sqrt{9} = 3 = |-3|$$

and 　　$$\sqrt{3^2} = \sqrt{9} = 3 = |3|.$$

In general, we find that

$$\sqrt[n]{x^n} = x \text{ if } n \text{ is odd, and } \sqrt[n]{x^n} = |x| \text{ if } n \text{ is even.}$$

---

We use the following identities to evaluate radicals of products and quotients.

## Radicals of Products and Quotients

If $a$ and $b$ are any real numbers (non-negative in the case of even-numbered roots), then

$$\sqrt[n]{ab} = \sqrt[n]{a}\,\sqrt[n]{b}$$ 　　Radical of a product = Product of radicals

$$\sqrt[n]{\frac{a}{b}} = \frac{\sqrt[n]{a}}{\sqrt[n]{b}} \quad \text{if } b \neq 0$$ 　　Radical of a quotient = Quotient of radicals

### Notes

- The first rule is similar to the rule $(a \cdot b)^2 = a^2 b^2$ for the square of a product, and the second rule is similar to the rule $\left(\dfrac{a}{b}\right)^2 = \dfrac{a^2}{b^2}$ for the square of a quotient.

- *Caution* There is no corresponding identity for addition:

$$\sqrt{a + b} \text{ is } not \text{ equal to } \sqrt{a} + \sqrt{b}$$

(Consider $a = b = 1$, for example.) Equating these expressions is a common error, so be careful! ∎

**Quick Examples**

1. $\sqrt{9 \cdot 4} = \sqrt{9}\sqrt{4} = 3 \times 2 = 6$        Alternatively, $\sqrt{9 \cdot 4} = \sqrt{36} = 6$

2. $\sqrt{\dfrac{9}{4}} = \dfrac{\sqrt{9}}{\sqrt{4}} = \dfrac{3}{2}$

3. $\dfrac{\sqrt{2}}{\sqrt{5}} = \dfrac{\sqrt{2}\sqrt{5}}{\sqrt{5}\sqrt{5}} = \dfrac{\sqrt{10}}{5}$

4. $\sqrt{4(3+13)} = \sqrt{4(16)} = \sqrt{4}\sqrt{16} = 2 \times 4 = 8$

5. $\sqrt[3]{-216} = \sqrt[3]{(-27)8} = \sqrt[3]{-27}\sqrt[3]{8} = (-3)2 = -6$

6. $\sqrt{x^3} = \sqrt{x^2 \cdot x} = \sqrt{x^2}\sqrt{x} = x\sqrt{x}$    if $x \geq 0$

7. $\sqrt{\dfrac{x^2 + y^2}{z^2}} = \dfrac{\sqrt{x^2 + y^2}}{\sqrt{z^2}} = \dfrac{\sqrt{x^2 + y^2}}{|z|}$        We can't simplify the numerator any further.

## Rational Exponents

We already know what we mean by expressions such as $x^4$ and $a^{-6}$. The next step is to make sense of *rational* exponents: exponents of the form $p/q$ with $p$ and $q$ integers as in $a^{1/2}$ and $3^{-2/3}$.

---

**Q** : *What should we mean by $a^{1/2}$ ?*

**A** : The overriding concern here is that all the exponent identities should remain true. In this case the identity to look at is the one that says that $(a^m)^n = a^{mn}$. This identity tells us that

$$(a^{1/2})^2 = a^1 = a.$$

That is, $a^{1/2}$, when squared, gives us $a$. But that must mean that $a^{1/2}$ is the *square root* of $a$, or

$$a^{1/2} = \sqrt{a}.$$

A similar argument tells us that, if $q$ is any positive whole number, then

$$a^{1/q} = \sqrt[q]{a}, \text{ the } q\text{th root of } a.$$

Notice that if $a$ is negative, this makes sense only for $q$ odd. To avoid this problem, we usually stick to positive $a$.

---

**Q** : *If $p$ and $q$ are integers ($q$ positive), what should we mean by $a^{p/q}$ ?*

**A** : By the exponent identities, $a^{p/q}$ should equal both $(a^p)^{1/q}$ and $(a^{1/q})^p$. The first is the $q$th root of $a^p$, and the second is the $p$th power of $a^{1/q}$, which gives us the following.

## Conversion Between Rational Exponents and Radicals

If $a$ is any non-negative number, then

$$a^{p/q} = \sqrt[q]{a^p} = \left(\sqrt[q]{a}\right)^p.$$

    ↑       ↑       ↑

Using exponents    Using radicals

In particular,

$$a^{1/q} = \sqrt[q]{a}, \text{ the } q\text{th root of } a.$$

### Notes

- If $a$ is negative, all of this makes sense only if $q$ is odd.
- All of the exponent identities continue to work when we allow rational exponents $p/q$. In other words, we are free to use all the exponent identities even though the exponents are not integers. ■

### Quick Examples

1. $4^{3/2} = (\sqrt{4})^3 = 2^3 = 8$
2. $8^{2/3} = (\sqrt[3]{8})^2 = 2^2 = 4$
3. $9^{-3/2} = \dfrac{1}{9^{3/2}} = \dfrac{1}{(\sqrt{9})^3} = \dfrac{1}{3^3} = \dfrac{1}{27}$
4. $\dfrac{\sqrt{3}}{\sqrt[3]{3}} = \dfrac{3^{1/2}}{3^{1/3}} = 3^{1/2-1/3} = 3^{1/6} = \sqrt[6]{3}$
5. $2^2 2^{7/2} = 2^2 2^{3+1/2} = 2^2 2^3 2^{1/2} = 2^5 2^{1/2} = 2^5 \sqrt{2}$

---

### EXAMPLE 4 Simplifying Algebraic Expressions

Simplify the following.

**a.** $\dfrac{(x^3)^{5/3}}{x^3}$          **b.** $\sqrt[4]{a^6}$          **c.** $\dfrac{(xy)^{-3}y^{-3/2}}{x^{-2}\sqrt{y}}$

**Solution**

**a.** $\dfrac{(x^3)^{5/3}}{x^3} = \dfrac{x^5}{x^3} = x^2$

**b.** $\sqrt[4]{a^6} = a^{6/4} = a^{3/2} = a \cdot a^{1/2} = a\sqrt{a}$

**c.** $\dfrac{(xy)^{-3}y^{-3/2}}{x^{-2}\sqrt{y}} = \dfrac{x^{-3}y^{-3}y^{-3/2}}{x^{-2}y^{1/2}} = \dfrac{1}{x^{-2+3}y^{1/2+3+3/2}} = \dfrac{1}{xy^5}$

---

## Converting Between Rational, Radical, and Exponent Form

In calculus we must often convert algebraic expressions involving powers of $x$, such as $\dfrac{3}{2x^2}$, into expressions in which $x$ does not appear in the denominator, such as $\dfrac{3}{2}x^{-2}$. Also, we must often convert expressions with radicals, such as $\dfrac{1}{\sqrt{1+x^2}}$, into expressions with no radicals and all powers in the numerator, such as $(1 + x^2)^{-1/2}$. In these cases, we are converting from **rational form** or **radical form** to **exponent form**.

## Rational Form

An expression is in **rational form** if it is written with positive exponents only.

### Quick Examples

1. $\dfrac{2}{3x^2}$ is in rational form.

2. $\dfrac{2x^{-1}}{3}$ is not in rational form because the exponent of $x$ is negative.

3. $\dfrac{x}{6} + \dfrac{6}{x}$ is in rational form.

## Radical Form

An expression is in **radical form** if it is written with integer powers and roots only.

### Quick Examples

1. $\dfrac{2}{5\sqrt[3]{x}} + \dfrac{2}{x}$ is in radical form.

2. $\dfrac{2x^{-1/3}}{5} + 2x^{-1}$ is not in radical form because $x^{-1/3}$ appears.

3. $\dfrac{1}{\sqrt{1+x^2}}$ is in radical form, but $(1 + x^2)^{-1/2}$ is not.

## Exponent Form

An expression is in **exponent form** if there are no radicals and all powers of unknowns occur in the numerator. We write such expressions as sums or differences of terms of the form

$$\text{Constant} \times (\text{Expression with } x)^p \qquad \text{As in } \tfrac{1}{3}x^{-3/2}$$

### Quick Examples

1. $\dfrac{2}{3}x^4 - 3x^{-1/3}$ is in exponent form.

2. $\dfrac{x}{6} + \dfrac{6}{x}$ is not in exponent form because the second expression has $x$ in the denominator.

3. $\sqrt[3]{x}$ is not in exponent form because it has a radical.

4. $(1 + x^2)^{-1/2}$ is in exponent form, but $\dfrac{1}{\sqrt{1+x^2}}$ is not.

---

### EXAMPLE 5  Converting from One Form to Another

Convert the following to rational form:

**a.** $\dfrac{1}{2}x^{-2} + \dfrac{4}{3}x^{-5}$ 　　　　　　　　**b.** $\dfrac{2}{\sqrt{x}} - \dfrac{2}{x^{-4}}$

Convert the following to radical form:

**c.** $\dfrac{1}{2}x^{-1/2} + \dfrac{4}{3}x^{-5/4}$ 　　　　　　**d.** $\dfrac{(3+x)^{-1/3}}{5}$

Convert the following to exponent form:

**e.** $\dfrac{3}{4x^2} - \dfrac{x}{6} + \dfrac{6}{x} + \dfrac{4}{3\sqrt{x}}$ 　　　**f.** $\dfrac{2}{(x+1)^2} - \dfrac{3}{4\sqrt[5]{2x-1}}$

**Solution**  For (a) and (b), we eliminate negative exponents as we did in Example 2:

**a.** $\dfrac{1}{2}x^{-2} + \dfrac{4}{3}x^{-5} = \dfrac{1}{2}\cdot\dfrac{1}{x^2} + \dfrac{4}{3}\cdot\dfrac{1}{x^5} = \dfrac{1}{2x^2} + \dfrac{4}{3x^5}$

**b.** $\dfrac{2}{\sqrt{x}} - \dfrac{2}{x^{-4}} = \dfrac{2}{\sqrt{x}} - 2x^4$

For (c) and (d), we rewrite all terms with fractional exponents as radicals:

**c.** $\dfrac{1}{2}x^{-1/2} + \dfrac{4}{3}x^{-5/4} = \dfrac{1}{2}\cdot\dfrac{1}{x^{1/2}} + \dfrac{4}{3}\cdot\dfrac{1}{x^{5/4}}$

$= \dfrac{1}{2}\cdot\dfrac{1}{\sqrt{x}} + \dfrac{4}{3}\cdot\dfrac{1}{\sqrt[4]{x^5}} = \dfrac{1}{2\sqrt{x}} + \dfrac{4}{3\sqrt[4]{x^5}}$

**d.** $\dfrac{(3+x)^{-1/3}}{5} = \dfrac{1}{5(3+x)^{1/3}} = \dfrac{1}{5\sqrt[3]{3+x}}$

For (e) and (f), we eliminate any radicals and move all expressions involving $x$ to the numerator:

**e.** $\dfrac{3}{4x^2} - \dfrac{x}{6} + \dfrac{6}{x} + \dfrac{4}{3\sqrt{x}} = \dfrac{3}{4}x^{-2} - \dfrac{1}{6}x + 6x^{-1} + \dfrac{4}{3x^{1/2}}$

$= \dfrac{3}{4}x^{-2} - \dfrac{1}{6}x + 6x^{-1} + \dfrac{4}{3}x^{-1/2}$

**f.** $\dfrac{2}{(x+1)^2} - \dfrac{3}{4\sqrt[5]{2x-1}} = 2(x+1)^{-2} - \dfrac{3}{4(2x-1)^{1/5}}$

$= 2(x+1)^{-2} - \dfrac{3}{4}(2x-1)^{-1/5}$

---

## Solving Equations with Exponents

---

### EXAMPLE 6  Solving Equations

Solve the following equations:

**a.** $x^3 + 8 = 0$ 　　　**b.** $x^2 - \dfrac{1}{2} = 0$ 　　　**c.** $x^{3/2} - 64 = 0$

**Solution**

**a.** Subtracting 8 from both sides gives $x^3 = -8$. Taking the cube root of both sides gives $x = -2$.

**b.** Adding $\frac{1}{2}$ to both sides gives $x^2 = \frac{1}{2}$. Thus, $x = \pm\sqrt{\frac{1}{2}} = \pm\frac{1}{\sqrt{2}}$.

**c.** Adding 64 to both sides gives $x^{3/2} = 64$. Taking the reciprocal (2/3) power of both sides gives

$$(x^{3/2})^{2/3} = 64^{2/3}$$
$$x^1 = \left(\sqrt[3]{64}\right)^2 = 4^2 = 16$$

so          $x = 16$.

---

## 0.2    EXERCISES

*Evaluate the expressions in Exercises 1–16.*

**1.** $3^3$          **2.** $(-2)^3$          **3.** $-(2\cdot 3)^2$          **4.** $(4\cdot 2)^2$

**5.** $\left(\dfrac{-2}{3}\right)^2$     **6.** $\left(\dfrac{3}{2}\right)^3$     **7.** $(-2)^{-3}$     **8.** $-2^{-3}$

**9.** $\left(\dfrac{1}{4}\right)^{-2}$     **10.** $\left(\dfrac{-2}{3}\right)^{-2}$     **11.** $2\cdot 3^0$     **12.** $3\cdot(-2)^0$

**13.** $2^3\,2^2$     **14.** $3^2 3$     **15.** $2^2 2^{-1} 2^4 2^{-4}$     **16.** $5^2 5^{-3} 5^2 5^{-2}$

*Simplify each expression in Exercises 17–30, expressing your answer in rational form.*

**17.** $x^3 x^2$          **18.** $x^4 x^{-1}$          **19.** $-x^2 x^{-3} y$          **20.** $-xy^{-1}x^{-1}$

**21.** $\dfrac{x^3}{x^4}$          **22.** $\dfrac{y^5}{y^3}$          **23.** $\dfrac{x^2 y^2}{x^{-1}y}$          **24.** $\dfrac{x^{-1}y}{x^2 y^2}$

**25.** $\dfrac{(xy^{-1}z^3)^2}{x^2 yz^2}$          **26.** $\dfrac{x^2 yz^2}{(xyz^{-1})^{-1}}$          **27.** $\left(\dfrac{xy^{-2}z}{x^{-1}z}\right)^3$

**28.** $\left(\dfrac{x^2 y^{-1}z^0}{xyz}\right)^2$     **29.** $\left(\dfrac{x^{-1}y^{-2}z^2}{xy}\right)^{-2}$     **30.** $\left(\dfrac{xy^{-2}}{x^2 y^{-1}z}\right)^{-3}$

*Convert the expressions in Exercises 31–36 to rational form.*

**31.** $3x^{-4}$          **32.** $\dfrac{1}{2}x^{-4}$          **33.** $\dfrac{3}{4}x^{-2/3}$

**34.** $\dfrac{4}{5}y^{-3/4}$     **35.** $1 - \dfrac{0.3}{x^{-2}} - \dfrac{6}{5}x^{-1}$     **36.** $\dfrac{1}{3x^{-4}} + \dfrac{0.1x^{-2}}{3}$

*Evaluate the expressions in Exercises 37–56, rounding your answer to four significant digits where necessary.*

**37.** $\sqrt{4}$          **38.** $\sqrt{5}$          **39.** $\sqrt{\dfrac{1}{4}}$

**40.** $\sqrt{\dfrac{1}{9}}$          **41.** $\sqrt{\dfrac{16}{9}}$          **42.** $\sqrt{\dfrac{9}{4}}$

**43.** $\dfrac{\sqrt{4}}{5}$          **44.** $\dfrac{6}{\sqrt{25}}$          **45.** $\sqrt{9} + \sqrt{16}$

**46.** $\sqrt{25} - \sqrt{16}$     **47.** $\sqrt{9+16}$     **48.** $\sqrt{25-16}$

**49.** $\sqrt[3]{8-27}$     **50.** $\sqrt[4]{81-16}$     **51.** $\sqrt[3]{27/8}$

**52.** $\sqrt[3]{8 \times 64}$          **53.** $\sqrt{(-2)^2}$          **54.** $\sqrt{(-1)^2}$

**55.** $\sqrt{\dfrac{1}{4}(1+15)}$          **56.** $\sqrt{\dfrac{1}{9}(3+33)}$

*Simplify the expressions in Exercises 57–64, given that x, y, z, a, b, and c are positive real numbers.*

**57.** $\sqrt{a^2 b^2}$          **58.** $\sqrt{\dfrac{a^2}{b^2}}$          **59.** $\sqrt{(x+9)^2}$

**60.** $(\sqrt{x+9})^2$     **61.** $\sqrt[3]{x^3(a^3+b^3)}$     **62.** $\sqrt[4]{\dfrac{x^4}{a^4 b^4}}$

**63.** $\sqrt{\dfrac{4xy^3}{x^2 y}}$          **64.** $\sqrt{\dfrac{4(x^2+y^2)}{c^2}}$

*Convert the expressions in Exercises 65–84 to exponent form.*

**65.** $\sqrt{3}$          **66.** $\sqrt{8}$          **67.** $\sqrt{x^3}$

**68.** $\sqrt[3]{x^2}$          **69.** $\sqrt[3]{xy^2}$          **70.** $\sqrt{x^2 y}$

**71.** $\dfrac{x^2}{\sqrt{x}}$          **72.** $\dfrac{x}{\sqrt{x}}$          **73.** $\dfrac{3}{5x^2}$

**74.** $\dfrac{2}{5x^{-3}}$     **75.** $\dfrac{3x^{-1.2}}{2} - \dfrac{1}{3x^{2.1}}$     **76.** $\dfrac{2}{3x^{-1.2}} - \dfrac{x^{2.1}}{3}$

**77.** $\dfrac{2x}{3} - \dfrac{x^{0.1}}{2} + \dfrac{4}{3x^{1.1}}$          **78.** $\dfrac{4x^2}{3} + \dfrac{x^{3/2}}{6} - \dfrac{2}{3x^2}$

**79.** $\dfrac{3\sqrt{x}}{4} - \dfrac{5}{3\sqrt{x}} + \dfrac{4}{3x\sqrt{x}}$          **80.** $\dfrac{3}{5\sqrt{x}} - \dfrac{5\sqrt{x}}{8} + \dfrac{7}{2\sqrt[3]{x}}$

**81.** $\dfrac{3\sqrt[5]{x^2}}{4} - \dfrac{7}{2\sqrt{x^3}}$          **82.** $\dfrac{1}{8x\sqrt{x}} - \dfrac{2}{3\sqrt[5]{x^3}}$

**83.** $\dfrac{1}{(x^2+1)^3} - \dfrac{3}{4\sqrt[3]{(x^2+1)}}$     **84.** $\dfrac{2}{3(x^2+1)^{-3}} - \dfrac{3\sqrt[3]{(x^2+1)^7}}{4}$

*Convert the expressions in Exercises 85–96 to radical form.*

**85.** $2^{2/3}$          **86.** $3^{4/5}$          **87.** $x^{4/3}$          **88.** $y^{7/4}$

**89.** $(x^{1/2}y^{1/3})^{1/5}$     **90.** $x^{-1/3}y^{3/2}$     **91.** $-\dfrac{3}{2}x^{-1/4}$     **92.** $\dfrac{4}{5}x^{3/2}$

**93.** $0.2x^{-2/3} + \dfrac{3}{7x^{-1/2}}$    **94.** $\dfrac{3.1}{x^{-4/3}} - \dfrac{11}{7}x^{-1/7}$

**95.** $\dfrac{3}{4(1-x)^{5/2}}$    **96.** $\dfrac{9}{4(1-x)^{-7/3}}$

*Simplify the expressions in Exercises 97–106.*

**97.** $4^{-1/2}4^{7/2}$    **98.** $2^{1/a}/2^{2/a}$    **99.** $3^{2/3}3^{-1/6}$

**100.** $2^{1/3}2^{-1}2^{2/3}2^{-1/3}$    **101.** $\dfrac{x^{3/2}}{x^{5/2}}$    **102.** $\dfrac{y^{5/4}}{y^{3/4}}$

**103.** $\dfrac{x^{1/2}y^2}{x^{-1/2}y}$    **104.** $\dfrac{x^{-1/2}y}{x^2y^{3/2}}$

**105.** $\left(\dfrac{x}{y}\right)^{1/3}\left(\dfrac{y}{x}\right)^{2/3}$    **106.** $\left(\dfrac{x}{y}\right)^{-1/3}\left(\dfrac{y}{x}\right)^{1/3}$

*Solve each equation in Exercises 107–120 for x, rounding your answer to four significant digits where necessary.*

**107.** $x^2 - 16 = 0$    **108.** $x^2 - 1 = 0$

**109.** $x^2 - \dfrac{4}{9} = 0$    **110.** $x^2 - \dfrac{1}{10} = 0$

**111.** $x^2 - (1+2x)^2 = 0$    **112.** $x^2 - (2-3x)^2 = 0$

**113.** $x^5 + 32 = 0$    **114.** $x^4 - 81 = 0$

**115.** $x^{1/2} - 4 = 0$    **116.** $x^{1/3} - 2 = 0$

**117.** $1 - \dfrac{1}{x^2} = 0$    **118.** $\dfrac{2}{x^3} - \dfrac{6}{x^4} = 0$

**119.** $(x-4)^{-1/3} = 2$    **120.** $(x-4)^{2/3} + 1 = 5$

# 0.3  Multiplying and Factoring Algebraic Expressions

## Multiplying Algebraic Expressions

### Distributive Law

The **distributive law** for real numbers states that

$$a(b \pm c) = ab \pm ac$$
$$(a \pm b)c = ac \pm bc$$

for any real numbers $a$, $b$, and $c$.

#### Quick Examples

1. $2(x-3)$ is *not* equal to $2x - 3$ but is equal to $2x - 2(3) = 2x - 6$.
2. $x(x+1) = x^2 + x$
3. $2x(3x - 4) = 6x^2 - 8x$
4. $(x-4)x^2 = x^3 - 4x^2$
5. $(x+2)(x+3) = (x+2)x + (x+2)3$
   $$= (x^2 + 2x) + (3x + 6) = x^2 + 5x + 6$$
6. $(x+2)(x-3) = (x+2)x - (x+2)3$
   $$= (x^2 + 2x) - (3x + 6) = x^2 - x - 6$$

There is a quicker way of expanding expressions like the last two, called the "FOIL" method (First, Outer, Inner, Last). Consider, for instance, the expression $(x+1)(x-2)$. The FOIL method says: Take the product of the first terms: $x \cdot x = x^2$, the product of the outer terms: $x \cdot (-2) = -2x$, the product of the inner terms: $1 \cdot x = x$, and the product of the last terms: $1 \cdot (-2) = -2$, and then add them all up, getting $x^2 - 2x + x - 2 = x^2 - x - 2$.

### EXAMPLE 1  FOIL

**a.** $(x - 2)(2x + 5) = 2x^2 + 5x - 4x - 10 = 2x^2 + x - 10$

↑    ↑    ↑    ↑<br>First  Outer  Inner  Last

**b.** $(x^2 + 1)(x - 4) = x^3 - 4x^2 + x - 4$

**c.** $(a - b)(a + b) = a^2 + ab - ab - b^2 = a^2 - b^2$

**d.** $(a + b)^2 = (a + b)(a + b) = a^2 + ab + ab + b^2 = a^2 + 2ab + b^2$

**e.** $(a - b)^2 = (a - b)(a - b) = a^2 - ab - ab + b^2 = a^2 - 2ab + b^2$

The last three are particularly important and are worth memorizing.

## Special Formulas

$$(a - b)(a + b) = a^2 - b^2 \qquad \text{Difference of two squares}$$
$$(a + b)^2 = a^2 + 2ab + b^2 \qquad \text{Square of a sum}$$
$$(a - b)^2 = a^2 - 2ab + b^2 \qquad \text{Square of a difference}$$

### Quick Examples

**1.** $(2 - x)(2 + x) = 4 - x^2$

**2.** $(1 + a)(1 - a) = 1 - a^2$

**3.** $(x + 3)^2 = x^2 + 6x + 9$

**4.** $(4 - x)^2 = 16 - 8x + x^2$

Here are some longer examples that require the distributive law.

### EXAMPLE 2  Multiplying Algebraic Expressions

**a.** $(x + 1)(x^2 + 3x - 4) = (x + 1)x^2 + (x + 1)3x - (x + 1)4$

$$= (x^3 + x^2) + (3x^2 + 3x) - (4x + 4)$$
$$= x^3 + 4x^2 - x - 4$$

**b.** $\left(x^2 - \dfrac{1}{x} + 1\right)(2x + 5) = \left(x^2 - \dfrac{1}{x} + 1\right)2x + \left(x^2 - \dfrac{1}{x} + 1\right)5$

$$= (2x^3 - 2 + 2x) + \left(5x^2 - \frac{5}{x} + 5\right)$$
$$= 2x^3 + 5x^2 + 2x + 3 - \frac{5}{x}$$

**c.** $(x - y)(x - y)(x - y) = (x^2 - 2xy + y^2)(x - y)$

$$= (x^2 - 2xy + y^2)x - (x^2 - 2xy + y^2)y$$
$$= (x^3 - 2x^2y + xy^2) - (x^2y - 2xy^2 + y^3)$$
$$= x^3 - 3x^2y + 3xy^2 - y^3$$

## Factoring Algebraic Expressions

We can think of factoring as applying the distributive law in reverse—for example,

$$2x^2 + x = x(2x + 1),$$

which can be checked by using the distributive law. Factoring is an art that you will learn with experience and the help of a few useful techniques.

### Factoring Using a Common Factor

To use this technique, locate a **common factor**—a term that occurs as a factor in each of the expressions being added or subtracted (for example, $x$ is a common factor in $2x^2 + x$, because it is a factor of both $2x^2$ and $x$). Once you have located a common factor, "factor it out" by applying the distributive law.

#### Quick Examples

1. $2x^3 - x^2 + x$ has $x$ as a common factor, so
$$2x^3 - x^2 + x = x(2x^2 - x + 1)$$

2. $2x^2 + 4x$ has $2x$ as a common factor, so
$$2x^2 + 4x = 2x(x + 2)$$

3. $2x^2y + xy^2 - x^2y^2$ has $xy$ as a common factor, so
$$2x^2y + xy^2 - x^2y^2 = xy(2x + y - xy)$$

4. $(x^2 + 1)(x + 2) - (x^2 + 1)(x + 3)$ has $x^2 + 1$ as a common factor, so
$$(x^2 + 1)(x + 2) - (x^2 + 1)(x + 3) = (x^2 + 1)[(x + 2) - (x + 3)]$$
$$= (x^2 + 1)(x + 2 - x - 3)$$
$$= (x^2 + 1)(-1) = -(x^2 + 1)$$

5. $12x(x^2 - 1)^5(x^3 + 1)^6 + 18x^2(x^2 - 1)^6(x^3 + 1)^5$ has $6x(x^2 - 1)^5(x^3 + 1)^5$ as a common factor, so
$$12x(x^2 - 1)^5(x^3 + 1)^6 + 18x^2(x^2 - 1)^6(x^3 + 1)^5$$
$$= 6x(x^2 - 1)^5(x^3 + 1)^5[2(x^3 + 1) + 3x(x^2 - 1)]$$
$$= 6x(x^2 - 1)^5(x^3 + 1)^5(2x^3 + 2 + 3x^3 - 3x)$$
$$= 6x(x^2 - 1)^5(x^3 + 1)^5(5x^3 - 3x + 2)$$

We would also like to be able to reverse calculations such as $(x + 2)(2x - 5) = 2x^2 - x - 10$. That is, starting with the expression $2x^2 - x - 10$, we would like to **factor** it to get the expression $(x + 2)(2x - 5)$. An expression of the form $ax^2 + bx + c$, where $a$, $b$, and $c$ are real numbers, is called a **quadratic** expression in $x$. Thus, given a quadratic expression $ax^2 + bx + c$, we would like to write it in the form $(dx + e)(fx + g)$ for some real numbers $d, e, f$, and $g$. There are some quadratics, such as $x^2 + x + 1$, that cannot be factored in this form at all. Here, we consider only quadratics that do factor, and in such a way that the numbers $d$, $e$, $f$, and $g$ are integers (whole numbers; other cases are discussed in Section 5). The usual technique of factoring such quadratics is a "trial and error" approach.

## Factoring Quadratics by Trial and Error

To factor the quadratic $ax^2 + bx + c$, factor $ax^2$ as $(a_1 x)(a_2 x)$ (with $a_1$ positive) and $c$ as $c_1 c_2$, and then check whether or not $ax^2 + bx + c = (a_1 x \pm c_1)(a_2 x \pm c_2)$. If not, try other factorizations of $ax^2$ and $c$.

### Quick Examples

1. To factor $x^2 - 6x + 5$, first factor $x^2$ as $(x)(x)$, and 5 as $(5)(1)$:

$$(x + 5)(x + 1) = x^2 + 6x + 5. \qquad \text{No good}$$
$$(x - 5)(x - 1) = x^2 - 6x + 5. \qquad \text{Desired factorization}$$

2. To factor $x^2 - 4x - 12$, first factor $x^2$ as $(x)(x)$, and $-12$ as $(1)(-12)$, $(2)(-6)$, or $(3)(-4)$. Trying them one by one gives

$$(x + 1)(x - 12) = x^2 - 11x - 12. \qquad \text{No good}$$
$$(x - 1)(x + 12) = x^2 + 11x - 12. \qquad \text{No good}$$
$$(x + 2)(x - 6) = x^2 - 4x - 12. \qquad \text{Desired factorization}$$

3. To factor $4x^2 - 25$, we can follow the above procedure, or recognize $4x^2 - 25$ as the difference of two squares:

$$4x^2 - 25 = (2x)^2 - 5^2 = (2x - 5)(2x + 5).$$

**Note:** Not all quadratic expressions factor. In Section 5 we look at a test that tells us whether or not a given quadratic factors.

Here are examples requiring either a little more work or a little more thought.

### EXAMPLE 3  Factoring Quadratics

Factor the following: **a.** $4x^2 - 5x - 6$    **b.** $x^4 - 5x^2 + 6$

### Solution

**a.** Possible factorizations of $4x^2$ are $(2x)(2x)$ or $(x)(4x)$. Possible factorizations of $-6$ are $(1)(-6)$, $(2)(-3)$. We now systematically try out all the possibilities until we come up with the correct one.

$$(2x)(2x) \text{ and } (1)(-6): \quad (2x + 1)(2x - 6) = 4x^2 - 10x - 6 \qquad \text{No good}$$
$$(2x)(2x) \text{ and } (2)(-3): \quad (2x + 2)(2x - 3) = 4x^2 - 2x - 6 \qquad \text{No good}$$
$$(x)(4x) \text{ and } (1)(-6): \quad (x + 1)(4x - 6) = 4x^2 - 2x - 6 \qquad \text{No good}$$
$$(x)(4x) \text{ and } (2)(-3): \quad (x + 2)(4x - 3) = 4x^2 + 5x - 6 \qquad \text{Almost!}$$
$$\text{Change signs:} \quad (x - 2)(4x + 3) = 4x^2 - 5x - 6 \qquad \text{Correct}$$

**b.** The expression $x^4 - 5x^2 + 6$ is not a quadratic, you say? Correct. It's a quartic (a fourth degree expression). However, it looks rather like a quadratic. In fact, it is quadratic *in* $x^2$, meaning that it is

$$(x^2)^2 - 5(x^2) + 6 = y^2 - 5y + 6$$

where $y = x^2$. The quadratic $y^2 - 5y + 6$ factors as

$$y^2 - 5y + 6 = (y - 3)(y - 2)$$

so

$$x^4 - 5x^2 + 6 = (x^2 - 3)(x^2 - 2)$$

This is a sometimes useful technique.

Our last example is here to remind you why we should want to factor polynomials in the first place. We shall return to this in Section 5.

---

**EXAMPLE 4  Solving a Quadratic Equation by Factoring**

Solve the equation $3x^2 + 4x - 4 = 0$.

**Solution**  We first factor the left-hand side to get

$$(3x - 2)(x + 2) = 0.$$

Thus, the product of the two quantities $(3x - 2)$ and $(x + 2)$ is zero. Now, if a product of two numbers is zero, one of the two must be zero. In other words, either $3x - 2 = 0$, giving $x = \frac{2}{3}$, or $x + 2 = 0$, giving $x = -2$. Thus, there are two solutions: $x = \frac{2}{3}$ and $x = -2$.

---

# 0.3  EXERCISES

*Expand each expression in Exercises 1–22.*

**1.** $x(4x + 6)$

**2.** $(4y - 2)y$

**3.** $(2x - y)y$

**4.** $x(3x + y)$

**5.** $(x + 1)(x - 3)$

**6.** $(y + 3)(y + 4)$

**7.** $(2y + 3)(y + 5)$

**8.** $(2x - 2)(3x - 4)$

**9.** $(2x - 3)^2$

**10.** $(3x + 1)^2$

**11.** $\left(x + \dfrac{1}{x}\right)^2$

**12.** $\left(y - \dfrac{1}{y}\right)^2$

**13.** $(2x - 3)(2x + 3)$

**14.** $(4 + 2x)(4 - 2x)$

**15.** $\left(y - \dfrac{1}{y}\right)\left(y + \dfrac{1}{y}\right)$

**16.** $(x - x^2)(x + x^2)$

**17.** $(x^2 + x - 1)(2x + 4)$

**18.** $(3x + 1)(2x^2 - x + 1)$

**19.** $(x^2 - 2x + 1)^2$

**20.** $(x + y - xy)^2$

**21.** $(y^3 + 2y^2 + y)(y^2 + 2y - 1)$

**22.** $(x^3 - 2x^2 + 4)(3x^2 - x + 2)$

*In Exercises 23–30, factor each expression and simplify as much as possible.*

**23.** $(x + 1)(x + 2) + (x + 1)(x + 3)$

**24.** $(x + 1)(x + 2)^2 + (x + 1)^2(x + 2)$

**25.** $(x^2 + 1)^5(x + 3)^4 + (x^2 + 1)^6(x + 3)^3$

**26.** $10x(x^2 + 1)^4(x^3 + 1)^5 + 15x^2(x^2 + 1)^5(x^3 + 1)^4$

**27.** $(x^3 + 1)\sqrt{x + 1} - (x^3 + 1)^2\sqrt{x + 1}$

**28.** $(x^2 + 1)\sqrt{x + 1} - \sqrt{(x + 1)^3}$

**29.** $\sqrt{(x + 1)^3} + \sqrt{(x + 1)^5}$

**30.** $(x^2 + 1)\sqrt[3]{(x + 1)^4} - \sqrt[3]{(x + 1)^7}$

*In Exercises 31–48, **(a)** factor the given expression; **(b)** set the expression equal to zero and solve for the unknown (x in the odd-numbered exercises and y in the even-numbered exercises).*

**31.** $2x + 3x^2$

**32.** $y^2 - 4y$

**33.** $6x^3 - 2x^2$

**34.** $3y^3 - 9y^2$

**35.** $x^2 - 8x + 7$

**36.** $y^2 + 6y + 8$

**37.** $x^2 + x - 12$

**38.** $y^2 + y - 6$

**39.** $2x^2 - 3x - 2$

**40.** $3y^2 - 8y - 3$

**41.** $6x^2 + 13x + 6$

**42.** $6y^2 + 17y + 12$

**43.** $12x^2 + x - 6$

**44.** $20y^2 + 7y - 3$

**45.** $x^2 + 4xy + 4y^2$

**46.** $4y^2 - 4xy + x^2$

**47.** $x^4 - 5x^2 + 4$

**48.** $y^4 + 2y^2 - 3$

## 0.4  Rational Expressions

### Rational Expression

A **rational expression** is an algebraic expression of the form $\dfrac{P}{Q}$, where $P$ and $Q$ are simpler expressions (usually polynomials) and the denominator $Q$ is not zero.

**Quick Examples**

1. $\dfrac{x^2 - 3x}{x}$    $P = x^2 - 3x,\ Q = x$

2. $\dfrac{x + \frac{1}{x} + 1}{2x^2 y + 1}$    $P = x + \dfrac{1}{x} + 1,\ Q = 2x^2 y + 1$

3. $3xy - x^2$    $P = 3xy - x^2,\ Q = 1$

### Algebra of Rational Expressions

We manipulate rational expressions in the same way that we manipulate fractions, using the following rules:

| Algebraic Rule | Quick Example |
|---|---|
| **Product:** $\dfrac{P}{Q} \cdot \dfrac{R}{S} = \dfrac{PR}{QS}$ | $\dfrac{x+1}{x} \cdot \dfrac{x-1}{2x+1} = \dfrac{(x+1)(x-1)}{x(2x+1)} = \dfrac{x^2-1}{2x^2+x}$ |
| **Sum:** $\dfrac{P}{Q} + \dfrac{R}{S} = \dfrac{PS + RQ}{QS}$ | $\dfrac{2x-1}{3x+2} + \dfrac{1}{x} = \dfrac{(2x-1)x + 1(3x+2)}{x(3x+2)}$ $= \dfrac{2x^2 + 2x + 2}{3x^2 + 2x}$ |
| **Difference:** $\dfrac{P}{Q} - \dfrac{R}{S} = \dfrac{PS - RQ}{QS}$ | $\dfrac{x}{3x+2} - \dfrac{x-4}{x} = \dfrac{x^2 - (x-4)(3x+2)}{x(3x+2)}$ $= \dfrac{-2x^2 + 10x + 8}{3x^2 + 2x}$ |
| **Reciprocal:** $\dfrac{1}{\left(\frac{P}{Q}\right)} = \dfrac{Q}{P}$ | $\dfrac{1}{\left(\frac{2xy}{3x-1}\right)} = \dfrac{3x-1}{2xy}$ |
| **Quotient:** $\dfrac{\left(\frac{P}{Q}\right)}{\left(\frac{R}{S}\right)} = \dfrac{P}{Q} \cdot \dfrac{S}{R} = \dfrac{PS}{QR}$ | $\dfrac{\left(\frac{x}{x-1}\right)}{\left(\frac{y-1}{y}\right)} = \dfrac{xy}{(x-1)(y-1)} = \dfrac{xy}{xy - x - y + 1}$ |
| **Cancellation:** $\dfrac{P\cancel{R}}{Q\cancel{R}} = \dfrac{P}{Q}$ | $\dfrac{(x-1)(xy+4)}{(x^2y-8)(x-1)} = \dfrac{xy+4}{x^2y-8}$ |

**Caution** Cancellation of summands is *invalid*. For instance,

$$\frac{\cancel{x} + (2xy^2 - y)}{\cancel{x} + 4y} = \frac{(2xy^2 - y)}{4y} \quad ✗ \ \textit{WRONG!} \quad \text{Do } not \text{ cancel a summand.}$$

$$\frac{\cancel{x}(2xy^2 - y)}{4\cancel{x}y} = \frac{(2xy^2 - y)}{4y} \quad ✔ \ \textit{CORRECT} \quad \text{Do cancel a factor.}$$

Here are some examples that require several algebraic operations.

---

**EXAMPLE 1 Simplifying Rational Expressions**

**a.** $\dfrac{\left(\frac{1}{x+y} - \frac{1}{x}\right)}{y} = \dfrac{\left(\frac{x - (x+y)}{x(x+y)}\right)}{y} = \dfrac{\left(\frac{-y}{x(x+y)}\right)}{y} = \dfrac{-y}{xy(x+y)} = -\dfrac{1}{x(x+y)}$

**b.** $\dfrac{(x+1)(x+2)^2 - (x+1)^2(x+2)}{(x+2)^4} = \dfrac{(x+1)(x+2)[(x+2) - (x+1)]}{(x+2)^4}$

$$= \frac{(x+1)(x+2)(x+2-x-1)}{(x+2)^4} = \frac{(x+1)(x+2)}{(x+2)^4} = \frac{x+1}{(x+2)^3}$$

**c.** $\dfrac{2x\sqrt{x+1} - \frac{x^2}{\sqrt{x+1}}}{x+1} = \dfrac{\left(\frac{2x(\sqrt{x+1})^2 - x^2}{\sqrt{x+1}}\right)}{x+1} = \dfrac{2x(x+1) - x^2}{(x+1)\sqrt{x+1}}$

$$= \frac{2x^2 + 2x - x^2}{(x+1)\sqrt{x+1}} = \frac{x^2 + 2x}{\sqrt{(x+1)^3}} = \frac{x(x+2)}{\sqrt{(x+1)^3}}$$

---

## 0.4  EXERCISES

*Rewrite each expression in Exercises 1–16 as a single rational expression, simplified as much as possible.*

**1.** $\dfrac{x-4}{x+1} \cdot \dfrac{2x+1}{x-1}$

**2.** $\dfrac{2x-3}{x-2} \cdot \dfrac{x+3}{x+1}$

**3.** $\dfrac{x-4}{x+1} + \dfrac{2x+1}{x-1}$

**4.** $\dfrac{2x-3}{x-2} + \dfrac{x+3}{x+1}$

**5.** $\dfrac{x^2}{x+1} - \dfrac{x-1}{x+1}$

**6.** $\dfrac{x^2-1}{x-2} - \dfrac{1}{x-1}$

**7.** $\dfrac{1}{\left(\frac{x}{x-1}\right)} + x - 1$

**8.** $\dfrac{2}{\left(\frac{x-2}{x^2}\right)} - \dfrac{1}{x-2}$

**9.** $\dfrac{1}{x}\left[\dfrac{x-3}{xy} + \dfrac{1}{y}\right]$

**10.** $\dfrac{y^2}{x}\left[\dfrac{2x-3}{y} + \dfrac{x}{y}\right]$

**11.** $\dfrac{(x+1)^2(x+2)^3 - (x+1)^3(x+2)^2}{(x+2)^6}$

**12.** $\dfrac{6x(x^2+1)^2(x^3+2)^3 - 9x^2(x^2+1)^3(x^3+2)^2}{(x^3+2)^6}$

**13.** $\dfrac{(x^2-1)\sqrt{x^2+1} - \frac{x^4}{\sqrt{x^2+1}}}{x^2+1}$

**14.** $\dfrac{x\sqrt{x^3-1} - \frac{3x^4}{\sqrt{x^3-1}}}{x^3-1}$

**15.** $\dfrac{\frac{1}{(x+y)^2} - \frac{1}{x^2}}{y}$

**16.** $\dfrac{\frac{1}{(x+y)^3} - \frac{1}{x^3}}{y}$

# 0.5  Solving Polynomial Equations

## Polynomial Equation

A **polynomial equation** in one unknown is an equation that can be written in the form

$$ax^n + bx^{n-1} + \cdots + rx + s = 0$$

where $a, b, \ldots, r$, and $s$ are constants.

We call the largest exponent of $x$ appearing in a nonzero term of a polynomial the **degree** of that polynomial.

### Quick Examples

1. $3x + 1 = 0$ has degree 1 because the largest power of $x$ that occurs is $x = x^1$. Degree 1 equations are called **linear** equations.

2. $x^2 - x - 1 = 0$ has degree 2 because the largest power of $x$ that occurs is $x^2$. Degree 2 equations are also called **quadratic equations**, or just **quadratics**.

3. $x^3 = 2x^2 + 1$ is a degree 3 polynomial (or **cubic**) in disguise. It can be rewritten as $x^3 - 2x^2 - 1 = 0$, which is in the standard form for a degree 3 equation.

4. $x^4 - x = 0$ has degree 4. It is called a **quartic**.

Now comes the question: How do we solve these equations for $x$? This question was asked by mathematicians as early as 1600 BCE. Let's look at these equations one degree at a time.

## Solution of Linear Equations

By definition, a linear equation can be written in the form

$$ax + b = 0. \qquad \text{$a$ and $b$ are fixed numbers with $a \neq 0$.}$$

Solving this is a nice mental exercise: Subtract $b$ from both sides and then divide by $a$, getting $x = -b/a$. Don't bother memorizing this formula; just go ahead and solve linear equations as they arise. If you feel you need practice, see the exercises at the end of the section.

## Solution of Quadratic Equations

By definition, a quadratic equation has the form

$$ax^2 + bx + c = 0. \qquad \text{$a$, $b$, and $c$ are fixed numbers and $a \neq 0$.[3]}$$

---

[3] What happens if $a = 0$?

The solutions of this equation are also called the **roots** of $ax^2 + bx + c$. We're assuming that you saw quadratic equations somewhere in high school but may be a little hazy about the details of their solution. There are two ways of solving these equations—one works sometimes, and the other works every time.

## Solving Quadratic Equations by Factoring (works sometimes)

If we can factor[4] a quadratic equation $ax^2 + bx + c = 0$, we can solve the equation by setting each factor equal to zero.

### Quick Examples

1. $x^2 + 7x + 10 = 0$

    $(x + 5)(x + 2) = 0$             Factor the left-hand side.

    $x + 5 = 0$ or $x + 2 = 0$        If a product is zero, one or both factors is zero.

    Solutions: $x = -5$ and $x = -2$

2. $2x^2 - 5x - 12 = 0$

    $(2x + 3)(x - 4) = 0$             Factor the left-hand side.

    $2x + 3 = 0$ or $x - 4 = 0$

    Solutions: $x = -3/2$ and $x = 4$

## Test for Factoring

The quadratic $ax^2 + bx + c$, with $a$, $b$, and $c$ being integers (whole numbers), factors into an expression of the form $(rx + s)(tx + u)$ with $r$, $s$, $t$, and $u$ integers precisely when the quantity $b^2 - 4ac$ is a perfect square. (That is, it is the square of an integer.) If this happens, we say that the quadratic **factors over the integers**.

### Quick Examples

1. $x^2 + x + 1$ has $a = 1$, $b = 1$, and $c = 1$, so $b^2 - 4ac = -3$, which is not a perfect square. Therefore, this quadratic does not factor over the integers.

2. $2x^2 - 5x - 12$ has $a = 2$, $b = -5$, and $c = -12$, so $b^2 - 4ac = 121$. Because $121 = 11^2$, this quadratic does factor over the integers. (We factored it above.)

## Solving Quadratic Equations with the Quadratic Formula (works every time)

The solutions of the general quadratic $ax^2 + bx + c = 0$ ($a \neq 0$) are given by

$$x = \frac{-b \pm \sqrt{b^2 - 4ac}}{2a}.$$

---

[4] See the section on factoring for a review of how to factor quadratics.

We call the quantity $\Delta = b^2 - 4ac$ the **discriminant** of the quadratic ($\Delta$ is the Greek letter delta), and we have the following general rules:

• If $\Delta$ is positive, there are two distinct real solutions.

• If $\Delta$ is zero, there is only one real solution: $x = -\dfrac{b}{2a}$. (Why?)

• If $\Delta$ is negative, there are no real solutions.

### Quick Examples

1. $2x^2 - 5x - 12 = 0$ has $a = 2$, $b = -5$, and $c = -12$.

$$x = \frac{-b \pm \sqrt{b^2 - 4ac}}{2a} = \frac{5 \pm \sqrt{25 + 96}}{4} = \frac{5 \pm \sqrt{121}}{4} = \frac{5 \pm 11}{4}$$

$$= \frac{16}{4} \text{ or } -\frac{6}{4} = 4 \text{ or } -3/2 \qquad \Delta \text{ is positive in this example.}$$

2. $4x^2 = 12x - 9$ can be rewritten as $4x^2 - 12x + 9 = 0$, which has $a = 4$, $b = -12$, and $c = 9$.

$$x = \frac{-b \pm \sqrt{b^2 - 4ac}}{2a} = \frac{12 \pm \sqrt{144 - 144}}{8} = \frac{12 \pm 0}{8} = \frac{12}{8} = \frac{3}{2}$$

$\Delta$ is zero in this example.

3. $x^2 + 2x - 1 = 0$ has $a = 1$, $b = 2$, and $c = -1$.

$$x = \frac{-b \pm \sqrt{b^2 - 4ac}}{2a} = \frac{-2 \pm \sqrt{8}}{2} = \frac{-2 \pm 2\sqrt{2}}{2} = -1 \pm \sqrt{2}$$

The two solutions are $x = -1 + \sqrt{2} = 0.414\ldots$ and
$x = -1 - \sqrt{2} = -2.414\ldots$ $\qquad \Delta$ is positive in this example.

4. $x^2 + x + 1 = 0$ has $a = 1$, $b = 1$, and $c = 1$. Because $\Delta = -3$ is negative, there are no real solutions. $\qquad \Delta$ is negative in this example.

---

**Q :** *This is all very useful, but where does the quadratic formula come from?*

**A :** To see where it comes from, we will solve a general quadratic equation using "brute force." Start with the general quadratic equation.

$$ax^2 + bx + c = 0.$$

First, divide out the nonzero number $a$ to get

$$x^2 + \frac{bx}{a} + \frac{c}{a} = 0.$$

Now we **complete the square:** Add and subtract the quantity $\dfrac{b^2}{4a^2}$ to get

$$x^2 + \frac{bx}{a} + \frac{b^2}{4a^2} - \frac{b^2}{4a^2} + \frac{c}{a} = 0.$$

We do this to get the first three terms to factor as a perfect square:

$$\left(x + \frac{b}{2a}\right)^2 - \frac{b^2}{4a^2} + \frac{c}{a} = 0.$$

(Check this by multiplying out.) Adding $\dfrac{b^2}{4a^2} - \dfrac{c}{a}$ to both sides gives:

$$\left(x + \frac{b}{2a}\right)^2 = \frac{b^2}{4a^2} - \frac{c}{a} = \frac{b^2 - 4ac}{4a^2}.$$

Taking square roots gives

$$x + \frac{b}{2a} = \frac{\pm\sqrt{b^2 - 4ac}}{2a}.$$

Finally, adding $-\dfrac{b}{2a}$ to both sides yields the result:

$$x = -\frac{b}{2a} + \frac{\pm\sqrt{b^2 - 4ac}}{2a}$$

or

$$x = \frac{-b \pm \sqrt{b^2 - 4ac}}{2a}.$$

## Solution of Cubic Equations

By definition, a cubic equation can be written in the form

$$ax^3 + bx^2 + cx + d = 0. \qquad a, b, c, \text{ and } d \text{ are fixed numbers and } a \neq 0.$$

Now we get into something of a bind. Although there is a perfectly respectable formula for the solutions, it is very complicated and involves the use of complex numbers rather heavily.[5] So we discuss instead a much simpler method that *sometimes* works nicely. Here is the method in a nutshell.

### Solving Cubics by Finding One Factor

Start with a given cubic equation $ax^3 + bx^2 + cx + d = 0$.

**Step 1** By trial and error, find one solution $x = s$. If $a$, $b$, $c$, and $d$ are integers, the only possible *rational* solutions[6] are those of the form $s = \pm(\text{factor of } d)/(\text{factor of } a)$.

**Step 2** It will now be possible to factor the cubic as

$$ax^3 + bx^2 + cx + d = (x - s)(ax^2 + ex + f) = 0$$

To find $ax^2 + ex + f$, divide the cubic by $x - s$, using long division.[7]

**Step 3** The factored equation says that either $x - s = 0$ or $ax^2 + ex + f = 0$. We already know that $s$ is a solution, and now we see that the other solutions are the roots of the quadratic. Note that this quadratic may or may not have any real solutions, as usual.

---

[5]It was when this formula was discovered in the 16th century that complex numbers were first taken seriously. Although we would like to show you the formula, it is too large to fit in this footnote.

[6]There may be *irrational* solutions, however; for example, $x^3 - 2 = 0$ has the single solution $x = \sqrt[3]{2}$.

[7]Alternatively, use "synthetic division," a shortcut that would take us too far afield to describe.

### Quick Example

To solve the cubic $x^3 - x^2 + x - 1 = 0$, we first find a single solution. Here, $a = 1$ and $d = -1$. Because the only factors of $\pm 1$ are $\pm 1$, the only possible rational solutions are $x = \pm 1$. By substitution, we see that $x = 1$ is a solution. Thus, $(x - 1)$ is a factor. Dividing by $(x - 1)$ yields the quotient $(x^2 + 1)$. Thus,

$$x^3 - x^2 + x - 1 = (x - 1)(x^2 + 1) = 0$$

so that either $x - 1 = 0$ or $x^2 + 1 = 0$.

Because the discriminant of the quadratic $x^2 + 1$ is negative, we don't get any real solutions from $x^2 + 1 = 0$, so the only real solution is $x = 1$.

## Possible Outcomes When Solving a Cubic Equation

If you consider all the cases, there are three possible outcomes when solving a cubic equation:

1. One real solution (as in the Quick Example above)
2. Two real solutions (try, for example, $x^3 + x^2 - x - 1 = 0$)
3. Three real solutions (see the next example)

### EXAMPLE 1    Solving a Cubic

Solve the cubic $2x^3 - 3x^2 - 17x + 30 = 0$.

**Solution**  First we look for a single solution. Here, $a = 2$ and $d = 30$. The factors of $a$ are $\pm 1$ and $\pm 2$, and the factors of $d$ are $\pm 1$, $\pm 2$, $\pm 3$, $\pm 5$, $\pm 6$, $\pm 10$, $\pm 15$, and $\pm 30$. This gives us a large number of possible ratios: $\pm 1$, $\pm 2$, $\pm 3$, $\pm 5$, $\pm 6$, $\pm 10$, $\pm 15$, $\pm 30$, $\pm 1/2$, $\pm 3/2$, $\pm 5/2$, $\pm 15/2$. Undaunted, we first try $x = 1$ and $x = -1$, getting nowhere. So we move on to $x = 2$, and we hit the jackpot, because substituting $x = 2$ gives $16 - 12 - 34 + 30 = 0$. Thus, $(x - 2)$ is a factor. Dividing yields the quotient $2x^2 + x - 15$. Here is the calculation:

$$
\begin{array}{r}
2x^2 + x - 15 \\
x - 2 \overline{\smash{\big)}\, 2x^3 - 3x^2 - 17x + 30} \\
\underline{2x^3 - 4x^2} \phantom{ - 17x + 30} \\
x^2 - 17x \phantom{+ 30} \\
\underline{x^2 - 2x} \phantom{+ 30} \\
-15x + 30 \\
\underline{-15x + 30} \\
0.
\end{array}
$$

Thus,

$$2x^3 - 3x^2 - 17x + 30 = (x - 2)(2x^2 + x - 15) = 0.$$

Setting the factors equal to zero gives either $x - 2 = 0$ or $2x^2 + x - 15 = 0$. We could solve the quadratic using the quadratic formula, but, luckily, we notice that it factors as

$$2x^2 + x - 15 = (x + 3)(2x - 5).$$

Thus, the solutions are $x = 2$, $x = -3$ and $x = 5/2$.

## Solution of Higher-Order Polynomial Equations

Logically speaking, our next step should be a discussion of quartics, then quintics (fifth degree equations), and so on forever. Well, we've got to stop somewhere, and cubics may be as good a place as any. On the other hand, since we've gotten so far, we ought to at least tell you what is known about higher order polynomials.

**Quartics**   Just as in the case of cubics, there is a formula to find the solutions of quartics.[8]

**Quintics and Beyond**   All good things must come to an end, we're afraid. It turns out that there is no "quintic formula." In other words, there is no single algebraic formula or collection of algebraic formulas that gives the solutions to all quintics. This question was settled by the Norwegian mathematician Niels Henrik Abel in 1824 after almost 300 years of controversy about this question. (In fact, several notable mathematicians had previously claimed to have devised formulas for solving the quintic, but these were all shot down by other mathematicians—this being one of the favorite pastimes of practitioners of our art.) The same negative answer applies to polynomial equations of degree 6 and higher. It's not that these equations don't have solutions; it's just that they can't be found using algebraic formulas.[9] However, there are certain special classes of polynomial equations that can be solved with algebraic methods. The way of identifying such equations was discovered around 1829 by the French mathematician Évariste Galois.[10]

---

[8]See, for example, *First Course in the Theory of Equations* by L. E. Dickson (New York: Wiley, 1922), or *Modern Algebra* by B. L. van der Waerden (New York: Frederick Ungar, 1953).

[9]What we mean by an "algebraic formula" is a formula in the coefficients using the operations of addition, subtraction, multiplication, division, and the taking of radicals. Mathematicians call the use of such formulas in solving polynomial equations "solution by radicals." If you were a math major, you would eventually go on to study this under the heading of Galois theory.

[10]Both Abel (1802–1829) and Galois (1811–1832) died young. Abel died of tuberculosis at the age of 26, while Galois was killed in a duel at the age of 20.

---

## 0.5   EXERCISES

*Solve the equations in Exercises 1–12 for x (mentally, if possible).*

**1.** $x + 1 = 0$

**2.** $x - 3 = 1$

**3.** $-x + 5 = 0$

**4.** $2x + 4 = 1$

**5.** $4x - 5 = 8$

**6.** $\frac{3}{4}x + 1 = 0$

**7.** $7x + 55 = 98$

**8.** $3x + 1 = x$

**9.** $x + 1 = 2x + 2$

**10.** $x + 1 = 3x + 1$

**11.** $ax + b = c$   $(a \neq 0)$

**12.** $x - 1 = cx + d$   $(c \neq 1)$

*By any method, determine all possible real solutions of each equation in Exercises 13–30. Check your answers by substitution.*

**13.** $2x^2 + 7x - 4 = 0$

**14.** $x^2 + x + 1 = 0$

**15.** $x^2 - x + 1 = 0$

**16.** $2x^2 - 4x + 3 = 0$

**17.** $2x^2 - 5 = 0$

**18.** $3x^2 - 1 = 0$

**19.** $-x^2 - 2x - 1 = 0$

**20.** $2x^2 - x - 3 = 0$

**21.** $\frac{1}{2}x^2 - x - \frac{3}{2} = 0$

**22.** $-\frac{1}{2}x^2 - \frac{1}{2}x + 1 = 0$

**23.** $x^2 - x = 1$

**24.** $16x^2 = -24x - 9$

**25.** $x = 2 - \frac{1}{x}$

**26.** $x + 4 = \frac{1}{x-2}$

**27.** $x^4 - 10x^2 + 9 = 0$

**28.** $x^4 - 2x^2 + 1 = 0$

**29.** $x^4 + x^2 - 1 = 0$

**30.** $x^3 + 2x^2 + x = 0$

*Find all possible real solutions of each equation in Exercises 31–44.*

**31.** $x^3 + 6x^2 + 11x + 6 = 0$

**32.** $x^3 - 6x^2 + 12x - 8 = 0$

**33.** $x^3 + 4x^2 + 4x + 3 = 0$

**34.** $y^3 + 64 = 0$

**35.** $x^3 - 1 = 0$

**36.** $x^3 - 27 = 0$

**37.** $y^3 + 3y^2 + 3y + 2 = 0$

**38.** $y^3 - 2y^2 - 2y - 3 = 0$

**39.** $x^3 - x^2 - 5x + 5 = 0$

**40.** $x^3 - x^2 - 3x + 3 = 0$

**41.** $2x^6 - x^4 - 2x^2 + 1 = 0$

**42.** $3x^6 - x^4 - 12x^2 + 4 = 0$

**43.** $(x^2 + 3x + 2)(x^2 - 5x + 6) = 0$

**44.** $(x^2 - 4x + 4)^2(x^2 + 6x + 5)^3 = 0$

## 0.6 Solving Miscellaneous Equations

Equations often arise in calculus that are not polynomial equations of low degree. Many of these complicated-looking equations can be solved easily if you remember the following, which we used in the previous section:

**Solving an Equation of the Form $P \cdot Q = 0$**

If a product is equal to 0, then at least one of the factors must be 0. That is, if $P \cdot Q = 0$, then either $P = 0$ or $Q = 0$.

**Quick Examples**

**1.** $x^5 - 4x^3 = 0$

$x^3(x^2 - 4) = 0$   Factor the left-hand side.

Either $x^3 = 0$ or $x^2 - 4 = 0$   Either $P = 0$ or $Q = 0$.

$x = 0, 2$ or $-2$.   Solve the individual equations.

**2.** $(x^2 - 1)(x + 2) + (x^2 - 1)(x + 4) = 0$

$(x^2 - 1)[(x + 2) + (x + 4)] = 0$   Factor the left-hand side.

$(x^2 - 1)(2x + 6) = 0$

Either $x^2 - 1 = 0$ or $2x + 6 = 0$   Either $P = 0$ or $Q = 0$.

$x = -3, -1$, or $1$.   Solve the individual equations.

### EXAMPLE 1 Solving by Factoring

Solve $12x(x^2-4)^5(x^2+2)^6 + 12x(x^2-4)^6(x^2+2)^5 = 0$.

**Solution**
Again, we start by factoring the left-hand side:

$$12x(x^2-4)^5(x^2+2)^6 + 12x(x^2-4)^6(x^2+2)^5$$
$$= 12x(x^2-4)^5(x^2+2)^5[(x^2+2)+(x^2-4)]$$
$$= 12x(x^2-4)^5(x^2+2)^5(2x^2-2)$$
$$= 24x(x^2-4)^5(x^2+2)^5(x^2-1).$$

Setting this equal to 0, we get:

$$24x(x^2-4)^5(x^2+2)^5(x^2-1) = 0,$$

which means that at least one of the factors of this product must be zero. Now it certainly cannot be the 24, but it could be the $x$: $x=0$ is one solution. It could also be that

$$(x^2-4)^5 = 0$$

or

$$x^2-4 = 0,$$

which has solutions $x = \pm 2$. Could it be that $(x^2+2)^5 = 0$? If so, then $x^2+2 = 0$, but this is impossible because $x^2+2 \geq 2$, no matter what $x$ is. Finally, it could be that $x^2-1 = 0$, which has solutions $x = \pm 1$. This gives us five solutions to the original equation:

$$x = -2, -1, 0, 1, \text{ or } 2.$$

### EXAMPLE 2 Solving by Factoring

Solve $(x^2-1)(x^2-4) = 10$.

**Solution** Watch out! You may be tempted to say that $x^2-1 = 10$ or $x^2-4 = 10$, but this does not follow. If two numbers multiply to give you 10, what must they be? There are lots of possibilities: 2 and 5, 1 and 10, $-500,000$ and $-0.00002$ are just a few. The fact that the left-hand side is factored is nearly useless to us if we want to solve this equation. What we will have to do is multiply out, bring the 10 over to the left, and hope that we can factor what we get. Here goes:

$$x^4 - 5x^2 + 4 = 10$$
$$x^4 - 5x^2 - 6 = 0$$
$$(x^2-6)(x^2+1) = 0$$

(Here we used a sometimes useful trick that we mentioned in Section 3: We treated $x^2$ like $x$ and $x^4$ like $x^2$, so factoring $x^4 - 5x^2 - 6$ is essentially the same as factoring $x^2 - 5x - 6$.) *Now* we are allowed to say that one of the factors must be 0: $x^2-6 = 0$ has solutions $x = \pm\sqrt{6} = \pm 2.449\ldots$ and $x^2+1 = 0$ has no real solutions. Therefore, we get exactly two solutions, $x = \pm\sqrt{6} = \pm 2.449\ldots$.

To solve equations involving rational expressions, the following rule is very useful.

## Solving an Equation of the Form $P/Q = 0$

If $\dfrac{P}{Q} = 0$, then $P = 0$.

How else could a fraction equal 0? If that is not convincing, multiply both sides by $Q$ (which cannot be 0 if the quotient is defined).

### Quick Example

$$\frac{(x + 1)(x + 2)^2 - (x + 1)^2(x + 2)}{(x + 2)^4} = 0$$

$(x + 1)(x + 2)^2 - (x + 1)^2(x + 2) = 0$    If $\frac{P}{Q} = 0$, then $P = 0$.

$(x + 1)(x + 2)[(x + 2) - (x + 1)] = 0$    Factor.

$(x + 1)(x + 2)(1) = 0$

Either $x + 1 = 0$ or $x + 2 = 0$,

$x = -1$ or $x = -2$

$x = -1$    $x = -2$ does not make sense in the original equation: it makes the denominator 0. So it is not a solution and $x = -1$ is the only solution.

## EXAMPLE 3  Solving a Rational Equation

Solve $1 - \dfrac{1}{x^2} = 0$.

**Solution**  Write 1 as $\frac{1}{1}$, so that we now have a difference of two rational expressions:

$$\frac{1}{1} - \frac{1}{x^2} = 0.$$

To combine these we can put both over a common denominator of $x^2$, which gives

$$\frac{x^2 - 1}{x^2} = 0.$$

Now we can set the numerator, $x^2 - 1$, equal to zero. Thus,

$$x^2 - 1 = 0$$

so

$$(x - 1)(x + 1) = 0,$$

giving $x = \pm 1$.

➡ **Before we go on...**  This equation could also have been solved by writing

$$1 = \frac{1}{x^2}$$

and then multiplying both sides by $x^2$.  ∎

### EXAMPLE 4  Another Rational Equation

Solve $\dfrac{2x-1}{x} + \dfrac{3}{x-2} = 0$.

**Solution**  We *could* first perform the addition on the left and then set the top equal to 0, but here is another approach. Subtracting the second expression from both sides gives

$$\frac{2x-1}{x} = \frac{-3}{x-2}$$

Cross-multiplying [multiplying both sides by both denominators—that is, by $x(x-2)$] now gives

$$(2x-1)(x-2) = -3x$$

so

$$2x^2 - 5x + 2 = -3x.$$

Adding $3x$ to both sides gives the quadratic equation

$$2x^2 - 2x + 1 = 0.$$

The discriminant is $(-2)^2 - 4 \cdot 2 \cdot 1 = -4 < 0$, so we conclude that there is no real solution.

➡ **Before we go on...**  Notice that when we said that $(2x-1)(x-2) = -3x$, we were *not* allowed to conclude that $2x - 1 = -3x$ or $x - 2 = -3x$.  ∎

### EXAMPLE 5  A Rational Equation with Radicals

Solve $\dfrac{\left(2x\sqrt{x+1} - \frac{x^2}{\sqrt{x+1}}\right)}{x+1} = 0$.

**Solution**  Setting the top equal to 0 gives

$$2x\sqrt{x+1} - \frac{x^2}{\sqrt{x+1}} = 0.$$

This still involves fractions. To get rid of the fractions, we could put everything over a common denominator ($\sqrt{x+1}$) and then set the top equal to 0, or we could multiply the whole equation by that common denominator in the first place to clear fractions. If we do the second, we get

$$2x(x+1) - x^2 = 0$$
$$2x^2 + 2x - x^2 = 0$$
$$x^2 + 2x = 0.$$

Factoring,

$$x(x+2) = 0$$

so either $x = 0$ or $x + 2 = 0$, giving us $x = 0$ or $x = -2$. Again, one of these is not really a solution. The problem is that $x = -2$ cannot be substituted into $\sqrt{x+1}$, because we would then have to take the square root of $-1$, and we are not allowing ourselves to do that. Therefore, $x = 0$ is the only solution.

## 0.6  EXERCISES

*Solve the following equations:*

**1.** $x^4 - 3x^3 = 0$

**2.** $x^6 - 9x^4 = 0$

**3.** $x^4 - 4x^2 = -4$

**4.** $x^4 - x^2 = 6$

**5.** $(x + 1)(x + 2) + (x + 1)(x + 3) = 0$

**6.** $(x + 1)(x + 2)^2 + (x + 1)^2(x + 2) = 0$

**7.** $(x^2 + 1)^5(x + 3)^4 + (x^2 + 1)^6(x + 3)^3 = 0$

**8.** $10x(x^2 + 1)^4(x^3 + 1)^5 - 10x^2(x^2 + 1)^5(x^3 + 1)^4 = 0$

**9.** $(x^3 + 1)\sqrt{x + 1} - (x^3 + 1)^2\sqrt{x + 1} = 0$

**10.** $(x^2 + 1)\sqrt{x + 1} - \sqrt{(x + 1)^3} = 0$

**11.** $\sqrt{(x + 1)^3} + \sqrt{(x + 1)^5} = 0$

**12.** $(x^2 + 1)\sqrt[3]{(x + 1)^4} - \sqrt[3]{(x + 1)^7} = 0$

**13.** $(x + 1)^2(2x + 3) - (x + 1)(2x + 3)^2 = 0$

**14.** $(x^2 - 1)^2(x + 2)^3 - (x^2 - 1)^3(x + 2)^2 = 0$

**15.** $\dfrac{(x + 1)^2(x + 2)^3 - (x + 1)^3(x + 2)^2}{(x + 2)^6} = 0$

**16.** $\dfrac{6x(x^2 + 1)^2(x^2 + 2)^4 - 8x(x^2 + 1)^3(x^2 + 2)^3}{(x^2 + 2)^8} = 0$

**17.** $\dfrac{2(x^2 - 1)\sqrt{x^2 + 1} - \frac{x^4}{\sqrt{x^2 + 1}}}{x^2 + 1} = 0$

**18.** $\dfrac{4x\sqrt{x^3 - 1} - \frac{3x^4}{\sqrt{x^3 - 1}}}{x^3 - 1} = 0$

**19.** $x - \dfrac{1}{x} = 0$

**20.** $1 - \dfrac{4}{x^2} = 0$

**21.** $\dfrac{1}{x} - \dfrac{9}{x^3} = 0$

**22.** $\dfrac{1}{x^2} - \dfrac{1}{x + 1} = 0$

**23.** $\dfrac{x - 4}{x + 1} - \dfrac{x}{x - 1} = 0$

**24.** $\dfrac{2x - 3}{x - 1} - \dfrac{2x + 3}{x + 1} = 0$

**25.** $\dfrac{x + 4}{x + 1} + \dfrac{x + 4}{3x} = 0$

**26.** $\dfrac{2x - 3}{x} - \dfrac{2x - 3}{x + 1} = 0$

## 0.7  The Coordinate Plane

**Q**: *Just what is the xy-plane?*

**A**: The *xy*-plane is an infinite flat surface with two perpendicular lines, usually labeled the **x-axis** and **y-axis**. These axes are calibrated as shown in Figure 2. (Notice also how the plane is divided into four **quadrants**.)

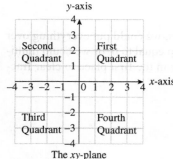

The *xy*-plane

**Figure 2**

Thus, the *xy*-plane is nothing more than a very large—in fact, infinitely large—flat surface. The purpose of the axes is to allow us to locate specific positions, or **points**, on the plane, with the use of **coordinates**. (If Captain Picard wants to have himself beamed to a specific location, he must supply its coordinates, or he's in trouble.)

**Q**: *So how do we use coordinates to locate points?*

**A**: The rule is simple. Each point in the plane has two coordinates, an **x-coordinate** and a **y-coordinate**. These can be determined in two ways:

**1.** The *x*-coordinate measures a point's distance to the right or left of the *y*-axis. It is positive if the point is to the right of the axis, negative if it is to the left of the axis, and 0 if it is on the axis. The *y*-coordinate measures a point's distance above or below the *x*-axis. It is positive if the point is above the axis, negative if it is below the axis, and 0 if it is on the axis. Briefly, the *x*-coordinate tells us the *horizontal* position (distance left or right), and the *y*-coordinate tells us the *vertical* position (height).

2. Given a point $P$, we get its $x$-coordinate by drawing a vertical line from $P$ and seeing where it intersects the $x$-axis. Similarly, we get the $y$-coordinate by extending a horizontal line from $P$ and seeing where it intersects the $y$-axis.

This way of assigning coordinates to points in the plane is often called the system of **Cartesian** coordinates, in honor of the mathematician and philosopher René Descartes (1596–1650), who was the first to use them extensively.

Here are a few examples to help you review coordinates.

---

### EXAMPLE 1 **Coordinates of Points**

**a.** Find the coordinates of the indicated points. (See Figure 3. The grid lines are placed at intervals of one unit.)

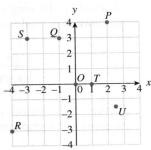

**Figure 3**

**b.** Locate the following points in the $xy$-plane.

$$A(2, 3), B(-4, 2), C(3, -2.5), D(0, -3), E(3.5, 0), F(-2.5, -1.5)$$

#### Solution

**a.** Taking them in alphabetical order, we start with the origin $O$. This point has height zero and is also zero units to the right of the $y$-axis, so its coordinates are $(0, 0)$. Turning to $P$, dropping a vertical line gives $x = 2$ and extending a horizontal line gives $y = 4$. Thus, $P$ has coordinates $(2, 4)$. For practice, determine the coordinates of the remaining points, and check your work against the list that follows:

$$Q(-1, 3), R(-4, -3), S(-3, 3), T(1, 0), U(2.5, -1.5)$$

**b.** In order to locate the given points, we start at the origin $(0, 0)$, and proceed as follows. (See Figure 4.)

To locate $A$, we move 2 units to the right and 3 up, as shown.

To locate $B$, we move $-4$ units to the right (that is, 4 to the *left*) and 2 up, as shown.

To locate $C$, we move 3 units right and 2.5 down.

We locate the remaining points in a similar way.

**Figure 4**

# The Graph of an Equation

One of the more surprising developments of mathematics was the realization that equations, which are algebraic objects, can be represented by graphs, which are geometric objects. The kinds of equations that we have in mind are equations in $x$ and $y$, such as

$$y = 4x - 1, \quad 2x^2 - y = 0, \quad y = 3x^2 + 1, \quad y = \sqrt{x - 1}.$$

The **graph** of an equation in the two variables $x$ and $y$ consists of all points $(x, y)$ in the plane whose coordinates are solutions of the equation.

---

**EXAMPLE 2  Graph of an Equation**

Obtain the graph of the equation $y - x^2 = 0$.

**Solution**  We can solve the equation for $y$ to obtain $y = x^2$. Solutions can then be obtained by choosing values for $x$ and then computing $y$ by squaring the value of $x$, as shown in the following table:

| $x$ | $-3$ | $-2$ | $-1$ | $0$ | $1$ | $2$ | $3$ |
|---|---|---|---|---|---|---|---|
| $y = x^2$ | 9 | 4 | 1 | 0 | 1 | 4 | 9 |

Plotting these points $(x, y)$ gives the following picture (left side of Figure 5), suggesting the graph on the right in Figure 5.

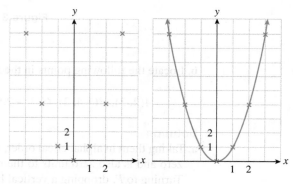

**Figure 5**

---

# Distance

The distance between two points in the $xy$-plane can be expressed as a function of their coordinates, as follows:

## Distance Formula

The distance between the points $P(x_1, y_1)$ and $Q(x_2, y_2)$ is

$$d = \sqrt{(x_2 - x_1)^2 + (y_2 - y_1)^2} = \sqrt{(\Delta x)^2 + (\Delta y)^2}.$$

## Derivation

The distance $d$ is shown in the figure below.

By the Pythagorean theorem applied to the right triangle shown, we get

$$d^2 = (x_2 - x_1)^2 + (y_2 - y_1)^2.$$

Taking square roots ($d$ is a distance, so we take the positive square root), we get the distance formula. Notice that if we switch $x_1$ with $x_2$ or $y_1$ with $y_2$, we get the same result.

### Quick Examples

**1.** The distance between the points $(3, -2)$ and $(-1, 1)$ is

$$d = \sqrt{(-1 - 3)^2 + (1 + 2)^2} = \sqrt{25} = 5.$$

**2.** The distance from $(x, y)$ to the origin $(0, 0)$ is

$$d = \sqrt{(x - 0)^2 + (y - 0)^2} = \sqrt{x^2 + y^2}. \qquad \text{Distance to the origin}$$

The set of all points $(x, y)$ whose distance from the origin $(0, 0)$ is a fixed quantity $r$ is a circle centered at the origin with radius $r$. From the second Quick Example, we get the following equation for the circle centered at the origin with radius $r$:

$$\sqrt{x^2 + y^2} = r. \qquad \text{Distance from the origin} = r.$$

Squaring both sides gives the following equation:

### Equation of the Circle of Radius $r$ Centered at the Origin

$$x^2 + y^2 = r^2$$

### Quick Examples

**1.** The circle of radius 1 centered at the origin has equation $x^2 + y^2 = 1$.

**2.** The circle of radius 2 centered at the origin has equation $x^2 + y^2 = 4$.

## 0.7  EXERCISES

**1.** Referring to the following figure, determine the coordinates of the indicated points as accurately as you can.

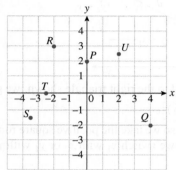

**2.** Referring to the following figure, determine the coordinates of the indicated points as accurately as you can.

**3.** Graph the following points.

$P(4, 4), Q(4, -4), R(3, 0), S(4, 0.5), T(0.5, 2.5),$
$U(-2, 0), V(-4, 4)$

**4.** Graph the following points.

$P(4, -2), Q(2, -4), R(1, -3), S(-4, 2), T(2, -1),$
$U(-2, 0), V(-4, -4)$

*Sketch the graphs of the equations in Exercises 5–12.*

**5.** $x + y = 1$          **6.** $y - x = -1$

**7.** $2y - x^2 = 1$        **8.** $2y + \sqrt{x} = 1$

**9.** $xy = 4$             **10.** $x^2 y = -1$

**11.** $xy = x^2 + 1$       **12.** $xy = 2x^3 + 1$

*In Exercises 13–16, find the distance between the given pairs of points.*

**13.** $(1, -1)$ and $(2, -2)$     **14.** $(1, 0)$ and $(6, 1)$

**15.** $(a, 0)$ and $(0, b)$        **16.** $(a, a)$ and $(b, b)$

**17.** Find the value of $k$ such that $(1, k)$ is equidistant from $(0, 0)$ and $(2, 1)$.

**18.** Find the value of $k$ such that $(k, k)$ is equidistant from $(-1, 0)$ and $(0, 2)$.

**19.** Describe the set of points $(x, y)$ such that $x^2 + y^2 = 9$.

**20.** Describe the set of points $(x, y)$ such that $x^2 + y^2 = 0$.

# 1

# Functions and Applications

**WM** Website

www.WanerMath.com

At the Website you will find:

- Section-by-section tutorials, including game tutorials with randomized quizzes
- A detailed chapter summary
- A true/false quiz
- Additional review exercises
- Graphers, Excel tutorials, and other resources
- The following extra topic:

  New Functions from Old: Scaled and Shifted Functions

## Case Study Modeling Spending on Internet Advertising

You are the new director of *Impact Advertising Inc.*'s Internet division, which has enjoyed a steady 0.25% of the Internet advertising market. You have drawn up an ambitious proposal to expand your division in light of your anticipation that Internet advertising will continue to skyrocket. The VP in charge of Financial Affairs feels that current projections (based on a linear model) do not warrant the level of expansion you propose. **How can you persuade the VP that those projections do not fit the data convincingly?**

Jeff Titcomb/Photographer's Choice / Getty Images

39

# Introduction

To analyze recent trends in spending on Internet advertising and to make reasonable projections, we need a mathematical model of this spending. Where do we start? To apply mathematics to real-world situations like this, we need a good understanding of basic mathematical concepts. Perhaps the most fundamental of these concepts is that of a function: a relationship that shows how one quantity depends on another. Functions may be described numerically and, often, algebraically. They can also be described graphically—a viewpoint that is extremely useful.

The simplest functions—the ones with the simplest formulas and the simplest graphs—are linear functions. Because of their simplicity, they are also among the most useful functions and can often be used to model real-world situations, at least over short periods of time. In discussing linear functions, we will meet the concepts of slope and rate of change, which are the starting point of the mathematics of change.

**algebra Review**

For this chapter, you should be familiar with real numbers and intervals. To review this material, see **Chapter 0**.

In the last section of this chapter, we discuss *simple linear regression*: construction of linear functions that best fit given collections of data. Regression is used extensively in applied mathematics, statistics, and quantitative methods in business. The inclusion of regression utilities in computer spreadsheets like Excel® makes this powerful mathematical tool readily available for anyone to use.

# Functions from the Numerical, Algebraic, and Graphical Viewpoints

The following table gives the approximate number of Facebook users at various times since its establishment early in 2004.[1]

| Year $t$ (Since start of 2004) | 0 | 1 | 2 | 3 | 4 | 5 | 6 |
|---|---|---|---|---|---|---|---|
| Facebook Members $n$ (Millions) | 0 | 1 | 5.5 | 12 | 58 | 150 | 450 |

Let's write $n(0)$ for the number of members (in millions) at time $t = 0$, $n(1)$ for the number at time $t = 1$, and so on (we read $n(0)$ as "$n$ of 0"). Thus, $n(0) = 0$, $n(1) = 1$, $n(2) = 5.5, \ldots, n(6) = 450$. In general, we write $n(t)$ for the number of members (in millions) at time $t$. We call $n$ a **function** of the variable $t$, meaning that for each value of $t$ between 0 and 6, $n$ gives us a single corresponding number $n(t)$ (the number of members at that time).

In general, we think of a function as a way of producing new objects from old ones. The functions we deal with in this text produce new numbers from old numbers. The numbers we have in mind are the *real* numbers, including not only positive and negative integers and fractions but also numbers like $\sqrt{2}$ or $\pi$. (See Chapter 0 for more on real numbers.) For this reason, the functions we use are called **real-valued functions of a real variable**. For example, the function $n$ takes the year since the start of 2004 as input and returns the number of Facebook members as output (Figure 1).

Year
$t$

$n$

$n(t)$
Members

**Figure 1**

[1]Sources: www.facebook.com, www.insidefacebook.com.

The variable $t$ is called the **independent variable**, while $n$ is called the **dependent variable** as its value depends on $t$. A function may be specified in several different ways. Here, we have specified the function $n$ **numerically** by giving the values of the function for a number of values of the independent variable, as in the preceding table.

---

**Q**: *For which values of t does it make sense to ask for n(t)? In other words, for which years t is the function n defined?*

**A**: Because $n(t)$ refers to the number of members from the start of 2004 to the start of 2010, $n(t)$ is defined when $t$ is any number between 0 and 6, that is, when $0 \le t \le 6$. Using interval notation (see Chapter 0), we can say that $n(t)$ is defined when $t$ is in the interval [0, 6].

---

The set of values of the independent variable for which a function is defined is called its **domain** and is a necessary part of the definition of the function. Notice that the preceding table gives the value of $n(t)$ at only some of the infinitely many possible values in the domain [0, 6]. The domain of a function is not always specified explicitly; if no domain is specified for the function $f$, we take the domain to be the largest set of numbers $x$ for which $f(x)$ makes sense. This "largest possible domain" is sometimes called the **natural domain**.

The previous Facebook data can also be represented on a graph by plotting the given pairs of numbers $(t, n(t))$ in the $xy$-plane. (See Figure 2. We have connected successive points by line segments.) In general, the **graph** of a function $f$ consists of all points $(x, f(x))$ in the plane with $x$ in the domain of $f$.

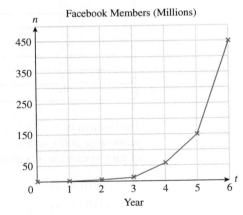

Facebook Members (Millions)

**Figure 2**

*In a graphically defined function, we can never know the $y$-coordinates of points exactly; no matter how accurately a graph is drawn, we can obtain only *approximate* values of the coordinates of points. That is why we have been using the word *estimate* rather than *calculate* and why we say $n(5) \approx 150$ rather than $n(5) = 150$.

In Figure 2 we specified the function $n$ **graphically** by using a graph to display its values. Suppose now that we had only the graph without the table of data. We could use the graph to find approximate values of $n$. For instance, to find $n(5)$ from the graph, we do the following:

**1.** Find the desired value of $t$ at the bottom of the graph ($t = 5$ in this case).

**2.** Estimate the height ($n$-coordinate) of the corresponding point on the graph (around 150 in this case).

Thus, $n(5) \approx 150$ million members.*

In some cases we may be able to use an algebraic formula to calculate the function, and we say that the function is specified **algebraically**. These are not the only ways in which a function can be specified; for instance, it could also be specified **verbally**, as in "Let $n(t)$ be the number of Facebook members, in millions, $t$ years since the start of 2004."[*] Notice that any function can be represented graphically by plotting the points $(x, f(x))$ for a number of values of $x$ in its domain.

Here is a summary of the terms we have just introduced.

[*] Specifying a function verbally in this way is useful for understanding what the function is doing, but it gives no numerical information.

## Functions

A **real-valued function** $f$ **of a real-valued variable** $x$ assigns to each real number $x$ in a specified set of numbers, called the **domain** of $f$, a unique real number $f(x)$, read "$f$ of $x$." The variable $x$ is called the **independent variable**, and $f$ is called the **dependent variable**. A function is usually specified **numerically** using a table of values, **graphically** using a graph, or **algebraically** using a formula. The **graph of a function** consists of all points $(x, f(x))$ in the plane with $x$ in the domain of $f$.

### Quick Examples

1. **A function specified numerically:** Take $c(t)$ to be the world emission of carbon dioxide in year $t$ since 2000, represented by the following table:[2]

| $t$ (Year Since 2000) | $c(t)$ (Billion Metric Tons of $CO_2$) |
|---|---|
| 0 | 24 |
| 5 | 28 |
| 10 | 31 |
| 15 | 33 |
| 20 | 36 |
| 25 | 38 |
| 30 | 41 |

The domain of $c$ is [0, 30], the independent variable is $t$, the number of years since 2000, and the dependent variable is $c$, the world production of carbon dioxide in a given year. Some values of $c$ are:

$c(0) = 24$     24 billion metric tons of $CO_2$ were produced in 2000.

$c(10) = 31$     31 billion metric tons of $CO_2$ were produced in 2010.

$c(30) = 41$     41 billion metric tons of $CO_2$ were projected to be produced in 2030.

[2]Figures for 2015 and later are projections. Source: Energy Information Administration (EIA) (www.eia.doe.gov)

**Graph of c:** Plotting the pairs $(t, c(t))$ gives the following graph:

**2. A function specified graphically:** Take $m(t)$ to be the median U.S. home price in thousands of dollars, $t$ years since 2000, as represented by the following graph:[3]

The domain of $m$ is $[0, 14]$, the independent variable is $t$, the number of years since 2000, and the dependent variable is $m$, the median U.S. home price in thousands of dollars. Some values of $m$ are:

$m(2) \approx 180$    The median home price in 2002 was about $180,000.

$m(10) \approx 210$.    The median home price in 2010 was about $210,000.

**3. A function specified algebraically:** Let $f(x) = \frac{1}{x}$. The function $f$ is specified algebraically. The independent variable is $x$ and the dependent variable is $f$. The natural domain of $f$ consists of all real numbers except zero because $f(x)$ makes sense for all values of $x$ other than $x = 0$. Some specific values of $f$ are

$$f(2) = \frac{1}{2} \qquad f(3) = \frac{1}{3} \qquad f(-1) = \frac{1}{-1} = -1$$

$f(0)$ is not defined because 0 is not in the domain of $f$.

---

[3]Source for data through end of 2010: www.zillow.com/local-info.

4. **The graph of a function:** Let $f(x) = x^2$, with domain the set of all real numbers. To draw the graph of $f$, first choose some convenient values of $x$ in the domain and compute the corresponding $y$-coordinates $f(x)$:

| $x$ | $-3$ | $-2$ | $-1$ | $0$ | $1$ | $2$ | $3$ |
|---|---|---|---|---|---|---|---|
| $f(x) = x^2$ | 9 | 4 | 1 | 0 | 1 | 4 | 9 |

Plotting these points $(x, f(x))$ gives the picture on the left, suggesting the graph on the right.✳

✳ If you plot more points, you will find that they lie on a smooth curve as shown. That is why we did not use line segments to connect the points.

(This particular curve happens to be called a **parabola**, and its lowest point, at the origin, is called its **vertex**.)

## EXAMPLE 1  iPod Sales

The total number of iPods sold by Apple up to the end of year $x$ can be approximated by

$$f(x) = 4x^2 + 16x + 2 \text{ million iPods } (0 \le x \le 6),$$

where $x = 0$ represents 2003.[4]

**a.** What is the domain of $f$? Compute $f(0)$, $f(2)$, $f(4)$, and $f(6)$. What do these answers tell you about iPod sales? Is $f(-1)$ defined?

**b.** Compute $f(a)$, $f(-b)$, $f(a + h)$, and $f(a) + h$ assuming that the quantities $a$, $-b$, and $a + h$ are in the domain of $f$.

**c.** Sketch the graph of $f$. Does the shape of the curve suggest that iPod sales were accelerating or decelerating?

### Solution

**a.** The domain of $f$ is the set of numbers $x$ with $0 \le x \le 6$—that is, the interval $[0, 6]$. If we substitute 0 for $x$ in the formula for $f(x)$, we get

$$f(0) = 4(0)^2 + 16(0) + 2 = 2.$$    By the end of 2003 approximately 2 million iPods had been sold.

---

[4]Source for data: Apple quarterly earnings reports at www.apple.com/investor/.

Similarly,

$$f(2) = 4(2)^2 + 16(2) + 2 = 50$$    By the end of 2005 approximately 50 million iPods had been sold.

$$f(4) = 4(4)^2 + 16(4) + 2 = 130$$    By the end of 2007 approximately 130 million iPods had been sold.

$$f(6) = 4(6)^2 + 16(6) + 2 = 242.$$    By the end of 2009 approximately 242 million iPods had been sold.

As $-1$ is not in the domain of $f$, $f(-1)$ is not defined.

**b.** To find $f(a)$ we substitute $a$ for $x$ in the formula for $f(x)$ to get

$$f(a) = 4a^2 + 16a + 2.$$    Substitute $a$ for $x$.

Similarly,

$$f(-b) = 4(-b)^2 + 16(-b) + 2$$    Substitute $-b$ for $x$.

$$= 4b^2 - 16b + 2$$    $(-b)^2 = b^2$

$$f(a + h) = 4(a + h)^2 + 16(a + h) + 2$$    Substitute $(a + h)$ for $x$.

$$= 4(a^2 + 2ah + h^2) + 16a + 16h + 2$$    Expand.

$$= 4a^2 + 8ah + 4h^2 + 16a + 16h + 2$$

$$f(a) + h = 4a^2 + 16a + 2 + h.$$    Add $h$ to $f(a)$.

Note how we placed parentheses around the quantities at which we evaluated the function. If we tried to do without any of these parentheses we would likely get an error:

Correct expression: $f(a + h) = 4(a + h)^2 + 16(a + h) + 2.$ ✓

*NOT*  $4a + h^2 + 16a + h + 2x$

Also notice the distinction between $f(a + h)$ and $f(a) + h$: To find $f(a + h)$, we replace $x$ by the quantity $(a + h)$; to find $f(a) + h$ we add $h$ to $f(a)$.

**c.** To draw the graph of $f$ we plot points of the form $(x, f(x))$ for several values of $x$ in the domain of $f$. Let us use the values we computed in part (a):

| $x$ | 0 | 2 | 4 | 6 |
|---|---|---|---|---|
| $f(x) = 4x^2 + 16x + 2$ | 2 | 50 | 130 | 242 |

Graphing these points gives the graph shown in Figure 3, suggesting the curve shown on the right.

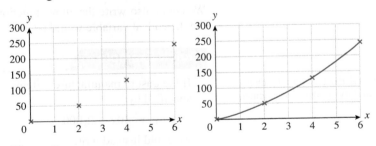

**Figure 3**

The graph becomes more steep as we move from left to right, suggesting that iPod sales were accelerating.

 **using Technology**

See the Technology Guides at the end of the chapter for detailed instructions on how to obtain the table of values and graph in Example 1 using a TI-83/84 Plus or Excel. Here is an outline:

**TI-83/84 Plus**
Table of values:
Y₁=4X^2+16X+2
[2ND] [TABLE].
Graph: [WINDOW];
Xmin = 0, Xmax = 6
[ZOOM] [0].
[More details on page 114.]

**Spreadsheet**
Table of values: Headings $x$ and $f(x)$ in A1–B1; $x$-values 0, 2, 4, 6 in A2–A5.
=4*A2^2+16*A2+2
in B2; copy down through B5.
Graph: Highlight A1 through B5 and insert a Scatter chart. [More details on page 119.]

**Website**
www.WanerMath.com
Go to the Function Evaluator and Grapher under Online Utilities, and enter
4x^2+16x+2
for $y_1$. To obtain a table of values, enter the $x$-values 0, 1, 2, 3 in the Evaluator box, and press "Evaluate" at the top of the box. Graph: Set Xmin = 0, Xmax = 6, and press "Plot Graphs".

➡ **Before we go on...** The following table compares the value of $f$ in Example 1 with the actual sales figures:

| $x$ | 0 | 2 | 4 | 6 |
|---|---|---|---|---|
| $f(x) = 4x^2 + 16x + 2$ | 2 | 50 | 130 | 242 |
| **Actual iPod Sales (Millions)** | 2 | 32 | 141 | 240 |

The actual figures are only stated here for (some) integer values of $x$; for instance, $x = 4$ gives the total sales up to the end of 2007. But what were, for instance, the sales through June of 2008 ($x = 4.5$)? This is where our formula comes in handy: We can use the formula for $f$ to **interpolate**—that is, to find sales at values of $x$ other than those between values that are stated:

$$f(4.5) = 4(4.5)^2 + 16(4.5) + 2 = 155 \text{ million iPods.}$$

We can also use the formula to **extrapolate**—that is, to predict sales at values of $x$ *outside* the domain—say, for $x = 6.5$ (that is, sales through June 2009):

$$f(6.5) = 4(6.5)^2 + 16(6.5) + 2 = 275 \text{ million iPods.}$$

As a general rule, extrapolation is far less reliable than interpolation: Predicting the future from current data is difficult, especially given the vagaries of the marketplace.

We call the algebraic function $f$ an **algebraic model** of iPod sales because it uses an algebraic formula to model—or mathematically represent (approximately)—the annual sales. The particular kind of algebraic model we used is called a **quadratic model**. (See the end of this section for the names of some commonly used models.) ∎

## Functions and Equations

Instead of using the usual "function notation" to specify a function, as in, say,

$$f(x) = 4x^2 + 16x + 2, \qquad \text{Function notation}$$

we could have specified it by an equation by replacing $f(x)$ by $y$:

$$y = 4x^2 + 16x + 2 \qquad \text{Equation notation}$$

(the choice of the letter $y$ is a convention, but any letter will do).

Technically, $y = 4x^2 + 16x + 2$ is an equation and not a function. However, an equation of this type, $y = Expression\ in\ x$, can be thought of as "specifying $y$ as a function of $x$." When we specify a function in this way, the variable $x$ is the independent variable and $y$ is the dependent variable.

We could also write the above function as $f = 4x^2 + 16x + 2$, in which case the dependent variable would be $f$.

### Quick Example

✳ We will discuss cost functions more fully in the next section.

If the cost to manufacture $x$ items is given by the "cost function"✳ $C$ specified by

$$C(x) = 40x + 2{,}000, \qquad \text{Cost function}$$

we could instead write

$$C = 40x + 2{,}000 \qquad \text{Cost equation}$$

and think of $C$, the cost, as a function of $x$.

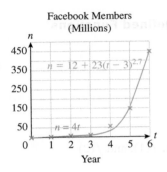

Facebook Members
(Millions)

**Figure 4**

Function notation and equation notation, sometimes using the same letter for the function name and the dependent variable, are often used interchangeably. It is important to be able to switch back and forth between function notation and equation notation, and we shall do so when it is convenient.

Look again at the graph of the number of Facebook users in Figure 2. From year 0 through year 3 the membership appears to increase more-or-less linearly (that is, the graph is almost a straight line), but then curves upward quite sharply from year 3 to year 6. This behavior can be modeled by using two different functions: one for the interval [0, 3] and another for the interval [3, 6] (see Figure 4).

A function specified by two or more different formulas like this is called a **piecewise-defined function**.

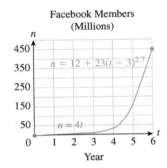

Facebook Members
(Millions)

**Figure 5**

### EXAMPLE 2  A Piecewise-Defined Function: Facebook Membership

The number $n(t)$ of Facebook members can be approximated by the following function of time $t$ in years ($t = 0$ represents January 2004):

$$n(t) = \begin{cases} 4t & \text{if } 0 \le t \le 3 \\ 12 + 23(t-3)^{2.7} & \text{if } 3 < t \le 6 \end{cases} \quad \text{million members.}$$

What was the approximate membership of Facebook in January 2005, January 2007, and June 2009? Sketch the graph of $n$ by plotting several points.

**Solution**  We evaluate the given function at the corresponding values of $t$:

Jan. 2005 ($t = 1$):   $n(1) = 4(1) = 4$       Use the first formula because $0 \le t \le 3$.

Jan. 2007 ($t = 3$):   $n(3) = 4(3) = 12$       Use the first formula because $0 \le t \le 3$.

June 2009 ($t = 5.5$): $n(5.5) = 12 + 23(5.5 - 3)^{2.7} \approx 285$.   Use the second formula because $3 < t \le 6$.

Thus, the number of Facebook members was approximately 4 million in January 2005, 12 million in January 2007, and 285 million in June 2009.

To sketch the graph of $n$ we use a table of rounded values of $n(t)$ (some of which we have already calculated above), plot the points, and connect them to sketch the graph:

| $t$ | 0 | 1 | 2 | 3 | 4 | 5 | 6 |
|---|---|---|---|---|---|---|---|
| $n(t)$ | 0 | 4 | 8 | 12 | 35 | 161 | 459 |

First Formula        Second Formula

The graph (Figure 5) has the following features:

**1.** The first formula (the line) is used for $0 \le t \le 3$.

**2.** The second formula (ascending curve) is used for $3 < t \le 6$.

**3.** The domain is [0, 6], so the graph is cut off at $t = 0$ and $t = 6$.

**4.** The heavy solid dots at the ends indicate the endpoints of the domain.

---

**:::: using Technology**

See the Technology Guides at the end of the chapter for detailed instructions on how to obtain the table of values and graph in Example 2 using a TI-83/84 Plus or Excel. Here is an outline:

**TI-83/84 Plus**
Table of values:
$Y_1 = (X\le 3)*(4X)+(X>3)*$
$(12+23*abs(X-3)^2.7)$
2ND TABLE.
Graph: WINDOW; Xmin = 0,
Xmax = 6; ZOOM 0.
[More details on page 114.]

**Spreadsheet**
Table of values: Headings $t$ and $n(t)$ in A1–B1; $t$-values 0, 1, ..., 6 in A2–A8.
$=(A2<=3)*(4*A2)+(A2>3)*$
$(12+23*abs(A2-3)^2.7)$
in B2; copy down through B8.
Graph: Highlight A1 through B8 and insert a Scatter chart.
[More details on page 120.]

## EXAMPLE 3  More Complicated Piecewise-Defined Functions

Let $f$ be the function specified by

$$f(x) = \begin{cases} -1 & \text{if } -4 \le x < -1 \\ x & \text{if } -1 \le x \le 1 \\ x^2 - 1 & \text{if } 1 < x \le 2 \end{cases}.$$

**a.** What is the domain of $f$? Find $f(-2)$, $f(-1)$, $f(0)$, $f(1)$, and $f(2)$.
**b.** Sketch the graph of $f$.

### Solution

**a.** The domain of $f$ is $[-4, 2]$, because $f(x)$ is specified only when $-4 \le x \le 2$.

| | |
|---|---|
| $f(-2) = -1$ | We used the first formula because $-4 \le x < -1$. |
| $f(-1) = -1$ | We used the second formula because $-1 \le x \le 1$. |
| $f(0) = 0$ | We used the second formula because $-1 \le x \le 1$. |
| $f(1) = 1$ | We used the second formula because $-1 \le x \le 1$. |
| $f(2) = 2^2 - 1 = 3$ | We used the third formula because $1 < x \le 2$. |

**b.** To sketch the graph by hand, we first sketch the three graphs $y = -1$, $y = x$, and $y = x^2 - 1$, and then use the appropriate portion of each (Figure 6).

**Figure 6**

Note that solid dots indicate points on the graph, whereas the open dots indicate points not on the graph. For example, when $x = 1$, the inequalities in the formula tell us that we are to use the middle formula ($x$) rather than the bottom one ($x^2 - 1$). Thus, $f(1) = 1$, not 0, so we place a solid dot at $(1, 1)$ and an open dot at $(1, 0)$.

## Vertical Line Test

Every point in the graph of a function has the form $(x, f(x))$ for some $x$ in the domain of $f$. Because $f$ assigns a *single* value $f(x)$ to each value of $x$ in the domain, it follows that, in the graph of $f$, there should be only one $y$ corresponding to any such value of $x$—namely, $y = f(x)$. In other words, *the graph of a function cannot contain two or more points with the same x-coordinate—that is, two or more points on the same vertical line.* On the other hand, a vertical line at a value of $x$ not in the domain will not contain any points in the graph. This gives us the following rule.

## Vertical-Line Test

For a graph to be the graph of a function, every vertical line must intersect the graph in *at most* one point.

### Quick Examples

As illustrated below, only graph B passes the vertical line test, so only graph B is the graph of a function.

Table 1 lists some common types of functions that are often used to model real world situations.

**Table 1**  A Compendium of Functions and Their Graphs

| *Type of Function* | *Examples* |
|---|---|
| **Linear**<br><br>$f(x) = mx + b$<br>$m, b$ constant<br><br>Graphs of linear functions are straight lines. The quantity $m$ is the **slope** of the line; the quantity $b$ is the **y-intercept** of the line. [See Section 1.3.]<br><br>Technology formulas: | $y = x$  $y = -2x + 2$<br><br>x        −2*x+2 |
| **Quadratic**<br><br>$f(x) = ax^2 + bx + c$<br>$a, b, c$ constant ($a \neq 0$)<br><br>Graphs of quadratic functions are called **parabolas**.<br><br>Technology formulas: | $y = x^2$  $y = -2x^2 + 2x + 4$<br><br>x^2        −2*x^2+2*x+4 |
| **Cubic**<br><br>$f(x) = ax^3 + bx^2 + cx + d$<br>$a, b, c, d$ constant ($a \neq 0$)<br><br>Technology formulas: | $y = x^3$  $y = -x^3 + 3x^2 + 1$<br><br>x^3        −x^3+3*x^2+1 |
| **Polynomial**<br><br>$f(x) = ax^n + bx^{n-1} + \ldots + rx + s$<br>$a, b, \ldots, r, s$ constant<br>(includes all of the above functions)<br><br>Technology formula: | All the above, and<br>$f(x) = x^6 - 2x^5 - 2x^4 + 4x^2$<br><br>x^6−2x^5−2x^4+4x^2 |

**Table 1** (*Continued*)

| *Type of Function* | *Examples* |
|---|---|
| **Exponential**<br><br>$f(x) = Ab^x$<br>$A, b$ constant<br>$(b > 0$ and $b \neq 1)$<br><br>The $y$-coordinate is multiplied by $b$ every time $x$ increases by 1.<br><br>Technology formulas: | $y = 2^x$ $\qquad\qquad$ $y = 4(0.5)^x$<br><br> <br><br>$y$ is doubled every time $\quad$ $y$ is halved every time<br>$x$ increases by 1. $\qquad\qquad$ $x$ increases by 1.<br><br>`2^x` $\qquad\qquad$ `4*0.5^x` |
| **Rational**<br><br>$f(x) = \dfrac{P(x)}{Q(x)};$<br><br>$P(x)$ and $Q(x)$ polynomials<br><br>The graph of $y = 1/x$ is a **hyperbola**. The domain excludes zero because $1/0$ is not defined.<br><br>Technology formulas: | $y = \dfrac{1}{x}$ $\qquad\qquad$ $y = \dfrac{x}{x-1}$<br><br> <br><br>`1/x` $\qquad\qquad$ `x/(x-1)` |
| **Absolute value**<br><br>For $x$ positive or zero, the graph of $y = |x|$ is the same as that of $y = x$. For $x$ negative or zero, it is the same as that of $y = -x$.<br><br>Technology formulas: | $y = |x|$ $\qquad\qquad$ $y = |2x + 2|$<br><br> <br><br>`abs(x)` $\qquad\qquad$ `abs(2*x+2)` |
| **Square Root**<br><br>The domain of $y = \sqrt{x}$ must be restricted to the nonnegative numbers, because the square root of a negative number is not real. Its graph is the top half of a horizontally oriented parabola.<br><br>Technology formulas: | $y = \sqrt{x}$ $\qquad\qquad$ $y = \sqrt{4x - 2}$<br><br> <br><br>`x^0.5 or √(x)` $\quad$ `(4*x-2)^0.5 or`<br>$\qquad\qquad\qquad$ `√(4*x-2)` |

Go to the Website and follow the path

Online Text

→ New Functions from Old: Scaled and Shifted Functions

where you will find complete online interactive text, examples, and exercises on scaling and translating the graph of a function by changing the formula.

Functions and models other than linear ones are called **nonlinear**.

## 1.1   EXERCISES

▼ more advanced    ◆ challenging

**T** indicates exercises that should be solved using technology

*In Exercises 1–4, evaluate or estimate each expression based on the following table.* HINT *[See Quick Example 1 on page 42.]*

| $x$ | $-3$ | $-2$ | $-1$ | $0$ | $1$ | $2$ | $3$ |
|------|------|------|------|-----|-----|-----|------|
| $f(x)$ | 1 | 2 | 4 | 2 | 1 | 0.5 | 0.25 |

**1. a.** $f(0)$    **b.** $f(2)$      **2. a.** $f(-1)$    **b.** $f(1)$

**3. a.** $f(2) - f(-2)$    **b.** $f(-1)f(-2)$    **c.** $-2f(-1)$

**4. a.** $f(1) - f(-1)$    **b.** $f(1)f(-2)$    **c.** $3f(-2)$

*In Exercises 5–8, use the graph of the function f to find approximations of the given values.* HINT *[See Example 1.]*

**5.**

**a.** $f(1)$    **b.** $f(2)$
**c.** $f(3)$    **d.** $f(5)$
**e.** $f(3) - f(2)$

**6.**

**a.** $f(1)$    **b.** $f(2)$
**c.** $f(3)$    **d.** $f(5)$
**e.** $f(3) - f(2)$

**7.**

**a.** $f(-3)$    **b.** $f(0)$
**c.** $f(1)$      **d.** $f(2)$
**e.** $\dfrac{f(2) - f(1)}{2 - 1}$

**8.**

**a.** $f(-2)$    **b.** $f(0)$
**c.** $f(1)$      **d.** $f(3)$
**e.** $\dfrac{f(3) - f(1)}{3 - 1}$

*In Exercises 9–12, say whether or not $f(x)$ is defined for the given values of $x$. If it is defined, give its value.* HINT *[See Quick Example 3 on page 43.]*

**9.** $f(x) = x - \dfrac{1}{x^2}$, with domain $(0, +\infty)$
   **a.** $x = 4$    **b.** $x = 0$    **c.** $x = -1$

**10.** $f(x) = \dfrac{2}{x} - x^2$, with domain $[2, +\infty)$
   **a.** $x = 4$    **b.** $x = 0$    **c.** $x = 1$

**11.** $f(x) = \sqrt{x + 10}$, with domain $[-10, 0)$
   **a.** $x = 0$    **b.** $x = 9$    **c.** $x = -10$

**12.** $f(x) = \sqrt{9 - x^2}$, with domain $(-3, 3)$
   **a.** $x = 0$    **b.** $x = 3$    **c.** $x = -3$

**13.** Given $f(x) = 4x - 3$, find   **a.** $f(-1)$   **b.** $f(0)$
   **c.** $f(1)$    **d.** $f(y)$   **e.** $f(a + b)$ HINT *[See Example 1.]*

**14.** Given $f(x) = -3x + 4$, find
   **a.** $f(-1)$    **b.** $f(0)$    **c.** $f(1)$    **d.** $f(y)$    **e.** $f(a + b)$

**15.** Given $f(x) = x^2 + 2x + 3$, find
   **a.** $f(0)$    **b.** $f(1)$    **c.** $f(-1)$    **d.** $f(-3)$
   **e.** $f(a)$    **f.** $f(x + h)$ HINT *[See Example 1.]*

**16.** Given $g(x) = 2x^2 - x + 1$, find
   **a.** $g(0)$    **b.** $g(-1)$    **c.** $g(r)$    **d.** $g(x + h)$

**17.** Given $g(s) = s^2 + \dfrac{1}{s}$, find
   **a.** $g(1)$    **b.** $g(-1)$    **c.** $g(4)$    **d.** $g(x)$    **e.** $g(s + h)$
   **f.** $g(s + h) - g(s)$

**18.** Given $h(r) = \dfrac{1}{r + 4}$, find
   **a.** $h(0)$    **b.** $h(-3)$    **c.** $h(-5)$    **d.** $h(x^2)$
   **e.** $h(x^2 + 1)$    **f.** $h(x^2) + 1$

*In Exercises 19–24, graph the given functions. Give the technology formula and use technology to check your graph. We suggest that you become familiar with these graphs in addition to those in Table 1.* HINT *[See Quick Example 4 on page 44.]*

**19.** $f(x) = -x^3$ (domain $(-\infty, +\infty)$)

**20.** $f(x) = x^3$ (domain $[0, +\infty)$)

**21.** $f(x) = x^4$ (domain $(-\infty, +\infty)$)

**22.** $f(x) = \sqrt[3]{x}$ (domain $(-\infty, +\infty)$)

**23.** $f(x) = \dfrac{1}{x^2}$   $(x \neq 0)$   **24.** $f(x) = x + \dfrac{1}{x}$   $(x \neq 0)$

*In Exercises 25 and 26, match the functions to the graphs. Using technology to draw the graphs is suggested, but not required.*

**25.** **T** **a.** $f(x) = x$   $(-1 \leq x \leq 1)$
   **b.** $f(x) = -x$   $(-1 \leq x \leq 1)$
   **c.** $f(x) = \sqrt{x}$   $(0 < x < 4)$
   **d.** $f(x) = x + \dfrac{1}{x} - 2$   $(0 < x < 4)$
   **e.** $f(x) = |x|$   $(-1 \leq x \leq 1)$
   **f.** $f(x) = x - 1$   $(-1 \leq x \leq 1)$

(I)                              (II)

(III)                            (IV)

(V)  y

(VI)  y

**26.** a. $f(x) = -x + 4$  $(0 < x \le 4)$
b. $f(x) = 2 - |x|$  $(-2 < x \le 2)$
c. $f(x) = \sqrt{x + 2}$  $(-2 < x \le 2)$
d. $f(x) = -x^2 + 2$  $(-2 < x \le 2)$
e. $f(x) = \dfrac{1}{x} - 1$  $(0 < x \le 4)$
f. $f(x) = x^2 - 1$  $(-2 < x \le 2)$

(I)   y

(II)  y

(III)  y

(IV)  y

(V)  y

(VI)  y

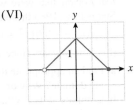

In Exercises 27–30, first give the technology formula for the given function and then use technology to evaluate the function for the given values of x (when defined there).

**27.** $f(x) = 0.1x^2 - 4x + 5$; $x = 0, 1, \ldots, 10$

**28.** $g(x) = 0.4x^2 - 6x - 0.1$; $x = -5, -4, \ldots, 4, 5$

**29.** $h(x) = \dfrac{x^2 - 1}{x^2 + 1}$; $x = 0.5, 1.5, 2.5, \ldots, 10.5$ (Round all answers to four decimal places.)

**30.** $r(x) = \dfrac{2x^2 + 1}{2x^2 - 1}$; $x = -1, 0, 1, \ldots, 9$ (Round all answers to four decimal places.)

In Exercises 31–36, sketch the graph of the given function, evaluate the given expressions, and then use technology to duplicate the graphs. Give the technology formula. HINT [See Example 2.]

**31.** $f(x) = \begin{cases} x & \text{if } -4 \le x < 0 \\ 2 & \text{if } 0 \le x \le 4 \end{cases}$
a. $f(-1)$  b. $f(0)$  c. $f(1)$

**32.** $f(x) = \begin{cases} -1 & \text{if } -4 \le x \le 0 \\ x & \text{if } 0 < x \le 4 \end{cases}$
a. $f(-1)$  b. $f(0)$  c. $f(1)$

**33.** $f(x) = \begin{cases} x^2 & \text{if } -2 < x \le 0 \\ 1/x & \text{if } 0 < x \le 4 \end{cases}$
a. $f(-1)$  b. $f(0)$  c. $f(1)$

**34.** $f(x) = \begin{cases} -x^2 & \text{if } -2 < x \le 0 \\ \sqrt{x} & \text{if } 0 < x < 4 \end{cases}$
a. $f(-1)$  b. $f(0)$  c. $f(1)$

**35.** $f(x) = \begin{cases} x & \text{if } -1 < x \le 0 \\ x + 1 & \text{if } 0 < x \le 2 \\ x & \text{if } 2 < x \le 4 \end{cases}$
a. $f(0)$  b. $f(1)$  c. $f(2)$  d. $f(3)$  HINT [See Example 3.]

**36.** $f(x) = \begin{cases} -x & \text{if } -1 < x < 0 \\ x - 2 & \text{if } 0 \le x \le 2 \\ -x & \text{if } 2 < x \le 4 \end{cases}$
a. $f(0)$  b. $f(1)$  c. $f(2)$  d. $f(3)$

In Exercises 37–40, find and simplify (a) $f(x + h) - f(x)$
(b) $\dfrac{f(x + h) - f(x)}{h}$

**37.** ▼ $f(x) = x^2$         **38.** ▼ $f(x) = 3x - 1$
**39.** ▼ $f(x) = 2 - x^2$     **40.** ▼ $f(x) = x^2 + x$

## APPLICATIONS

**41.** *Petrochemical Sales: Mexico* The following table shows annual petrochemical sales in Mexico by Pemex, Mexico's national oil company, for 2005–2010 ($t = 0$ represents 2005):[5]

| Year $t$ (Year since 2005) | 0 | 1 | 2 | 3 | 4 | 5 |
|---|---|---|---|---|---|---|
| Petrochemical Sales $s$ (Billion metric tons) | 3.7 | 3.8 | 4.0 | 4.1 | 4.0 | 4.1 |

a. Find $s(1)$, $s(4)$, and $s(5)$. Interpret your answers.
b. What is the domain of $s$?
c. Represent $s$ graphically and use your graph to estimate $s(1.5)$. Interpret your answer. HINT [See Quick Example 1 on page 42.]

[5]2010 figure is a projection based on data through May 2010. Source: www.pemex.com (July 2010).

**42. Petrochemical Production: Mexico** The following table shows annual petrochemical production in Mexico by Pemex, Mexico's national oil company, for 2005–2010 ($t = 0$ represents 2005):[6]

| Year $t$ (Year since 2005) | 0 | 1 | 2 | 3 | 4 | 5 |
|---|---|---|---|---|---|---|
| Petrochemical Production $p$ (Billion metric tons) | 10.8 | 11.0 | 11.8 | 12.0 | 12.0 | 13.3 |

**a.** Find $p(0)$, $p(2)$, and $p(4)$. Interpret your answers.
**b.** What is the domain of $p$?
**c.** Represent $p$ graphically and use your graph to estimate $p(2.5)$. Interpret your answer. HINT [See Quick Example 1 on page 42]

**Housing Starts** *Exercises 43–46 refer to the following graph, which shows the number $f(t)$ of housing starts in the U.S. each year from 2000 through 2010 ($t = 0$ represents 2000, and $f(t)$ is the number of housing starts in year $t$ in thousands of units).[7]*

f(t)    Housing starts (Thousands)

**43.** Estimate $f(4)$, $f(5)$, and $f(7.5)$. Interpret your answers.
**44.** Estimate $f(3)$, $f(6)$, and $f(8.5)$. Interpret your answers.
**45.** Which has the larger magnitude: $f(5) - f(0)$ or $f(8) - f(0)$? Interpret the answer.
**46.** Which has the larger magnitude: $f(5) - f(0)$ or $f(9) - f(7)$? Interpret the answer.

**47. Airline Net Income** The following graph shows the approximate annual after-tax net income $P(t)$, in millions of dollars, of Continental Airlines for 2005–2009 ($t = 0$ represents 2005):[8]

Continental Airlines

**a.** Estimate $P(0)$, $P(4)$, and $P(1.5)$ to the nearest 100. Interpret your answers.

**b.** At which of the following values of $t$ is $P(t)$ *increasing* most rapidly: 0, 0.5, 1, 2.5, 3.5? Interpret the result.
**c.** At which of the following values of $t$ is $P(t)$ *decreasing* most rapidly: 0, 0.5, 1, 2.5, 3.5? Interpret the result.

**48. Airline Net Income** The following graph shows the approximate annual after-tax net income $P(t)$, in millions of dollars, of Delta Air Lines for 2005–2009 ($t = 0$ represents 2005):[9]

Delta Air Lines

**a.** Estimate $P(0)$, $P(1)$, and $P(3.5)$ to the nearest 100. Interpret your answers.

**b.** At which of the following values of $t$ is $P(t)$ *increasing* most rapidly: 0, 0.5, 1, 2.5, 3.5? Interpret the result.

**c.** At which of the following values of $t$ is $P(t)$ *decreasing* most rapidly: 0, 0.5, 1, 2.5, 3.5? Interpret the result.

**49. Trade with China** The value of U.S. trade with China from 1994 through 2004 can be approximated by

$$C(t) = 3t^2 - 7t + 50 \text{ billion dollars}$$

($t$ is time in years since 1994).[10]

**a.** Find an appropriate domain of $C$. Is $t \geq 0$ an appropriate domain? Why or why not?
**b.** Compute $C(10)$. What does the answer say about trade with China?

**50. Scientific Research** The number of research articles in *Physical Review* that were written by researchers in the United States from 1983 through 2003 can be approximated by

$$A(t) = -0.01\,t^2 + 0.24t + 3.4 \text{ hundred articles}$$

($t$ is time in years since 1983).[11]

**a.** Find an appropriate domain of $A$. Is $t \leq 20$ an appropriate domain? Why or why not?
**b.** Compute $A(10)$. What does the answer say about the number of research articles?

**51.** ☐ **Acquisition of Language** The percentage $p(t)$ of children who can speak in at least single words by the age of $t$ months can be approximated by the equation[12]

$$p(t) = 100\left(1 - \frac{12{,}200}{t^{4.48}}\right) \quad (t \geq 8.5).$$

[6]*Ibid.*

[7]Sources for data: www.census.gov, www.forecast-chart.com

[8]"Net income" is an accounting term for profit (see Section 1.2). Model is the authors'. Source for data: Company reports.

Unless otherwise noted, all content on this page is © Cengage Learning.

[9]*Ibid.*

[10]Based on a regression by the authors. Source for data: U.S. Census Bureau/*New York Times*, September 23, 2004, p. C1.

[11]Based on a regression by the authors. Source for data: The American Physical Society/*New York Times*, May 3, 2003, p. A1.

[12]The model is the authors' and is based on data presented in the article *The Emergence of Intelligence* by William H. Calvin, *Scientific American*, October, 1994, pp. 101–107.

a. Give a technology formula for $p$.

b. Graph $p$ for $8.5 \le t \le 20$ and $0 \le p \le 100$.

c. Create a table of values of $p$ for $t = 9, 10, \ldots, 20$ (rounding answers to one decimal place).

d. What percentage of children can speak in at least single words by the age of 12 months?

e. By what age are 90% or more children speaking in at least single words?

52. ☐ *Acquisition of Language* The percentage $p(t)$ of children who can speak in sentences of five or more words by the age of $t$ months can be approximated by the equation[13]

$$p(t) = 100 \left( 1 - \frac{5.27 \times 10^{17}}{t^{12}} \right) \quad (t \ge 30).$$

a. Give a technology formula for $p$.

b. Graph $p$ for $30 \le t \le 45$ and $0 \le p \le 100$.

c. Create a table of values of $p$ for $t = 30, 31, \ldots, 40$ (rounding answers to one decimal place).

d. What percentage of children can speak in sentences of five or more words by the age of 36 months?

e. By what age are 75% or more children speaking in sentences of five or more words?

53. ▼ *Processor Speeds* The processor speed, in megahertz (MHz), of Intel processors during the period 1980–2010 could be approximated by the following function of time $t$ in years since the start of 1980:[14]

$$v(t) = \begin{cases} 8(1.22)^t & \text{if } 0 \le t < 16 \\ 400t - 6{,}200 & \text{if } 16 \le t < 25 \\ 3{,}800 & \text{if } 25 \le t \le 30 \end{cases}$$

a. Evaluate $v(10)$, $v(16)$, and $v(28)$ and interpret the results.

b. Write down a technology formula for $v$.

c. ☐ Use technology to sketch the graph of $v$ and to generate a table of values for $v(t)$ with $t = 0, 2, \ldots, 30$. (Round values to two significant digits.)

d. When, to the nearest year, did processor speeds reach 3 gigahertz (1 gigahertz = 1,000 megahertz) according to the model?

54. ▼ *Processor Speeds* The processor speed, in megahertz (MHz), of Intel processors during the period 1970–2000 could be approximated by the following function of time $t$ in years since the start of 1970:[15]

$$v(t) = \begin{cases} 0.12t^2 + 0.04t + 0.2 & \text{if } 0 \le t < 12 \\ 1.1(1.22)^t & \text{if } 12 \le t < 26 \\ 400t - 10{,}200 & \text{if } 26 \le t \le 30 \end{cases}$$

a. Evaluate $v(2)$, $v(12)$, and $v(28)$ and interpret the results.

b. Write down a technology formula for $v$.

c. ☐ Use technology to sketch the graph of $v$ and to generate a table of values for $v(t)$ with $t = 0, 2, \ldots, 30$. (Round values to two significant digits.)

---

[13]*Ibid.*

[14]Based on the fastest processors produced by Intel. Source for data: www.intel.com.

[15]*Ibid.*

d. When, to the nearest year, did processor speeds reach 500 megahertz?

55. ▼ *Income Taxes* The U.S. Federal income tax is a function of taxable income. Write $T(x)$ for the tax owed on a taxable income of $x$ dollars. For tax year 2010, the function $T$ for a single taxpayer was specified as follows:

| If your taxable income was | | | of the |
|---|---|---|---|
| Over... | But not over... | Your tax is | amount over... |
| $0 | 8,375 | ......... 10% | $0 |
| 8,375 | 34,000 | $837.50 + 15% | $8,375 |
| 34,000 | 82,400 | 4,663.25 + 25% | $34,000 |
| 82,400 | 171,850 | 16,763.25 + 28% | $82,400 |
| 171,850 | 373,650 | 41,809.25 + 33% | $171,850 |
| 373,650 | ......... | 108,403.25 + 35% | $373,650 |

a. Represent $T$ as a piecewise-defined function of income $x$. HINT [Each row of the table defines a formula with a condition.]

b. Use your function to compute the tax owed by a single taxpayer on a taxable income of $36,000.

56. ▼ *Income Taxes* Repeat Exercise 55 using the following information for tax year 2009.

| If your taxable income was | | | of the |
|---|---|---|---|
| Over... | But not over... | Your tax is | amount over... |
| $0 | 8,350 | ......... 10% | $0 |
| 8,350 | 33,950 | $835.00 + 15% | $8,350 |
| 33,950 | 82,250 | 4,675.00 + 25% | $33,950 |
| 82,250 | 171,550 | 16,750.00 + 28% | $82,250 |
| 171,550 | 372,950 | 41,754.00 + 33% | $171,550 |
| 372,950 | ......... | 108,216.00 + 35% | $372,950 |

## COMMUNICATION AND REASONING EXERCISES

57. Complete the following sentence: If the market price $m$ of gold varies with time $t$, then the independent variable is ____ and the dependent variable is ____.

58. Complete the following sentence: If weekly profit $P$ is specified as a function of selling price $s$, then the independent variable is ____ and the dependent variable is ____.

59. Complete the following: The function notation for the equation $y = 4x^2 - 2$ is ____.

60. Complete the following: The equation notation for $C(t) = -0.34t^2 + 0.1t$ is ____.

**61.** True or false? Every graphically specified function can also be specified numerically. Explain.

**62.** True or false? Every algebraically specified function can also be specified graphically. Explain.

**63.** True or false? Every numerically specified function with domain [0, 10] can also be specified algebraically. Explain.

**64.** True or false? Every graphically specified function can also be specified algebraically. Explain.

**65.** ▼ True or false? Every function can be specified numerically. Explain.

**66.** ▼ Which supplies more information about a situation: a numerical model or an algebraic model?

**67.** ▼ Why is the following assertion false? "If $f(x) = x^2 - 1$, then $f(x + h) = x^2 + h - 1$."

**68.** ▼ Why is the following assertion false? "If $f(2) = 2$ and $f(4) = 4$, then $f(3) = 3$."

**69.** How do the graphs of two functions differ if they are specified by the same formula but have different domains?

**70.** How do the graphs of two functions $f$ and $g$ differ if $g(x) = f(x) + 10$? (Try an example.)

**71.** ▼ How do the graphs of two functions $f$ and $g$ differ if $g(x) = f(x - 5)$? (Try an example.)

**72.** ▼ How do the graphs of two functions $f$ and $g$ differ if $g(x) = f(-x)$? (Try an example.)

# 1.2 Functions and Models

The functions we used in Examples 1 and 2 in Section 1.1 are **mathematical models** of real-life situations, because they model, or represent, situations in mathematical terms.

## Mathematical Modeling

To mathematically model a situation means to represent it in mathematical terms. The particular representation used is called a **mathematical model** of the situation. Mathematical models do not always represent a situation perfectly or completely. Some (like Example 1 of Section 1.1) represent a situation only approximately, whereas others represent only some aspects of the situation.

### Quick Examples

1. The temperature is now 10°F and increasing by 20° per hour.

   **Model:** $T(t) = 10 + 20t$ ($t$ = time in hours, $T$ = temperature)

2. I invest $1,000 at 5% interest compounded quarterly. Find the value of the investment after $t$ years.

   **Model:** $A(t) = 1,000 \left(1 + \dfrac{0.05}{4}\right)^{4t}$ (This is the compound interest formula we will study in Example 6.)

3. I am fencing a rectangular area whose perimeter is 100 ft. Find the area as a function of the width $x$.

   **Model:** Take $y$ to be the length, so the perimeter is
   $$100 = x + y + x + y = 2(x + y).$$
   This gives
   $$x + y = 50.$$
   Thus the length is $y = 50 - x$, and the area is
   $$A = xy = x(50 - x).$$

**4.** You work 8 hours a day Monday through Friday, 5 hours on Saturday, and have Sunday off. Model the number of hours you work as a function of the day of the week $n$, with $n = 1$ being Sunday.

**Model:** Take $f(n)$ to be the number of hours you work on the $n$th day of the week, so

$$f(n) = \begin{cases} 0 & \text{if } n = 1 \\ 8 & \text{if } 2 \le n \le 6 \\ 5 & \text{if } n = 7 \end{cases}.$$

Note that the domain of $f$ is $\{1, 2, 3, 4, 5, 6, 7\}$—a discrete set rather than a continuous interval of the real line.

**5.** The function

$$f(x) = 4x^2 + 16x + 2 \text{ million iPods sold } (x = \text{years since 2003})$$

in Example 1 of Section 1.1 is a model of iPod sales.

**6.** The function

$$n(t) = \begin{cases} 4t & \text{if } 0 \le t \le 3 \\ 12 + 23(t - 3)^{2.7} & \text{if } 3 < t \le 6 \end{cases} \text{ million members}$$

($t$ = years since January 2004) in Example 2 of Section 1.1 is a model of Facebook membership.

## Types of Models

Quick Examples 1–4 are **analytical models**, obtained by analyzing the situation being modeled, whereas Quick Examples 5 and 6 are **curve-fitting models**, obtained by finding mathematical formulas that approximate observed data. All the models except for Quick Example 4 are **continuous models**, defined by functions whose domains are intervals of the real line, whereas Quick Example 4 is a **discrete model** as its domain is a discrete set, as mentioned above. Discrete models are used extensively in probability and statistics.

## Cost, Revenue, and Profit Models

### EXAMPLE 1  Modeling Cost: Cost Function

As of August 2010, Yellow Cab Chicago's rates amounted to $2.05 on entering the cab plus $1.80 for each mile.[16]

**a.** Find the cost $C$ of an $x$-mile trip.

**b.** Use your answer to calculate the cost of a 40-mile trip.

**c.** What is the cost of the second mile? What is the cost of the tenth mile?

**d.** Graph $C$ as a function of $x$.

---

[16]According to their Web site at www.yellowcabchicago.com.

### Solution

**a.** We are being asked to find how the cost $C$ depends on the length $x$ of the trip, or to find $C$ as a function of $x$. Here is the cost in a few cases:

Cost of a 1-mile trip: $C = 1.80(1) + 2.05 = 3.85$    1 mile at $1.80 per mile plus $2.05

Cost of a 2-mile trip: $C = 1.80(2) + 2.05 = 5.65$    2 miles at $1.80 per mile plus $2.05

Cost of a 3-mile trip: $C = 1.80(3) + 2.05 = 7.45$    3 miles at $1.80 per mile plus $2.05

Do you see the pattern? The cost of an $x$-mile trip is given by the linear function

$$C(x) = 1.80x + 2.05.$$

Notice that the cost function is a sum of two terms: The **variable cost** $1.80x$, which depends on $x$, and the **fixed cost** $2.05$, which is independent of $x$:

$$\text{Cost} = \text{Variable Cost} + \text{Fixed Cost}.$$

The quantity 1.80 by itself is the incremental cost per mile; you might recognize it as the *slope* of the given linear function. In this context we call 1.80 the **marginal cost**. You might recognize the fixed cost 2.05 as the *C-intercept* of the given linear function.

**b.** We can use the formula for the cost function to calculate the cost of a 40-mile trip as

$$C(40) = 1.80(40) + 2.05 = \$74.05.$$

**c.** To calculate the cost of the second mile, we *could* proceed as follows:

Find the cost of a 1-mile trip: $C(1) = 1.80(1) + 2.05 = \$3.85.$

Find the cost of a 2-mile trip: $C(2) = 1.80(2) + 2.05 = \$5.65.$

Therefore, the cost of the second mile is $\$5.65 - \$3.85 = \$1.80.$

But notice that this is just the marginal cost. In fact, the marginal cost is the cost of each additional mile, so we could have done this more simply:

$$\text{Cost of second mile} = \text{Cost of tenth mile} = \text{Marginal cost} = \$1.80.$$

**d.** Figure 7 shows the graph of the cost function, which we can interpret as a *cost vs. miles* graph. The fixed cost is the starting height on the left, while the marginal cost is the slope of the line: It rises 1.80 units per unit of $x$. (See Section 1.3 for a discussion of properties of straight lines.)

Dollars

$C = 1.80x + 2.05$

Miles

**Figure 7**

➡ **Before we go on...** The cost function in Example 1 is an example of an *analytical model:* We derived the form of the cost function from a knowledge of the cost per mile and the fixed cost.

As we discussed in Section 1.1, we can specify the cost function in Example 1 using equation notation:

$$C = 1.80x + 2.05.$$    Equation notation

Here, the independent variable is $x$, and the dependent variable is $C$. (This is the notation we have used in Figure 7. Remember that we will often switch between function and equation notation when it is convenient to do so.) ∎

Here is a summary of some terms we used in Example 1, along with an introduction to some new terms:

## Cost, Revenue, and Profit Functions

A **cost function** specifies the cost $C$ as a function of the number of items $x$. Thus, $C(x)$ is the cost of $x$ items, and has the form

$$\text{Cost} = \text{Variable cost} + \text{Fixed cost}$$

where the variable cost is a function of $x$ and the fixed cost is a constant. A cost function of the form

$$C(x) = mx + b$$

is called a **linear cost function**; the variable cost is $mx$ and the fixed cost is $b$. The slope $m$, the **marginal cost**, measures the incremental cost per item.

The **revenue**, or **net sales**, resulting from one or more business transactions is the total income received. If $R(x)$ is the revenue from selling $x$ items at a price of $m$ each, then $R$ is the linear function $R(x) = mx$ and the selling price $m$ can also be called the **marginal revenue**.

The **profit**, or **net income**, on the other hand, is what remains of the revenue when costs are subtracted. If the profit depends linearly on the number of items, the slope $m$ is called the **marginal profit**. Profit, revenue, and cost are related by the following formula.

$$\text{Profit} = \text{Revenue} - \text{Cost}$$
$$P(x) = R(x) - C(x).^*$$

If the profit is negative, say $-\$500$, we refer to a **loss** (of $500 in this case). To **break even** means to make neither a profit nor a loss. Thus, breakeven occurs when $P = 0$, or

$$R(x) = C(x). \qquad \text{Breakeven}$$

The **break-even point** is the number of items $x$ at which breakeven occurs.

✱ We say that the profit function $P$ is the **difference** between the revenue and cost functions, and express this fact as a formula about functions: $P = R - C$. (We will discuss this further when we talk about the algebra of functions at the end of this section.)

| Quick Example |

If the daily cost (including operating costs) of manufacturing $x$ T-shirts is $C(x) = 8x + 100$, and the revenue obtained by selling $x$ T-shirts is $R(x) = 10x$, then the daily profit resulting from the manufacture and sale of $x$ T-shirts is

$$P(x) = R(x) - C(x) = 10x - (8x + 100) = 2x - 100.$$

Breakeven occurs when $P(x) = 0$, or $x = 50$.

## EXAMPLE 2 Cost, Revenue, and Profit

The annual operating cost of *YSport* Fitness gym is estimated to be

$$C(x) = 100{,}000 + 160x - 0.2x^2 \text{ dollars} \qquad (0 \le x \le 400),$$

where $x$ is the number of members. Annual revenue from membership averages $800 per member. What is the variable cost? What is the fixed cost? What is the profit function? How many members must *YSport* have to make a profit? What will happen if it has fewer members? If it has more?

**Solution**  The variable cost is the part of the cost function that depends on $x$:

$$\text{Variable cost} = 160x - 0.2x^2.$$

The fixed cost is the constant term:

$$\text{Fixed cost} = 100,000.$$

The annual revenue *YSport* obtains from a single member is $800. So, if it has $x$ members, it earns an annual revenue of

$$R(x) = 800x.$$

For the profit, we use the formula

$$P(x) = R(x) - C(x) \qquad \text{Formula for profit}$$
$$= 800x - (100,000 + 160x - 0.2x^2) \qquad \text{Substitute } R(x) \text{ and } C(x).$$
$$= -100,000 + 640x + 0.2x^2.$$

To make a profit, *YSport* needs to do better than break even, so let us find the break-even point: the value of $x$ such that $P(x) = 0$. All we have to do is set $P(x) = 0$ and solve for $x$:

$$-100,000 + 640x + 0.2x^2 = 0.$$

Notice that we have a quadratic equation $ax^2 + bx + c = 0$ with $a = 0.2$, $b = 640$, and $c = -100,000$. Its solution is given by the quadratic formula:

$$x = \frac{-b \pm \sqrt{b^2 - 4ac}}{2a} = \frac{-640 \pm \sqrt{640^2 + 4(0.2)(100,000)}}{2(0.2)}$$
$$\approx \frac{-640 \pm 699.71}{2(0.2)}$$
$$\approx 149.3 \text{ or } -3,349.3.$$

We reject the negative solution (as the domain is $[0, 400]$) and conclude that $x \approx 149.3$ members. To make a profit, should *YSport* have 149 members or 150 members? To decide, take a look at Figure 8, which shows two graphs: On the left we see the graph of revenue and cost, and on the right we see the graph of the profit function.

Cost: $C(x) = 100,000 + 160x - 0.2x^2$
Revenue: $R(x) = 800x$
Breakeven occurs at the point of intersection
of the graphs of revenue and cost.

Profit: $P(x) = -100,000 + 640x + 0.2x^2$
Breakeven occurs when $P(x) = 0$

**Figure 8**

For values of $x$ less than the break-even point of 149.3, $P(x)$ is negative, so the company will have a loss. For values of $x$ greater than the break-even point, $P(x)$ is positive, so the company will make a profit. Thus, *YSport Fitness* needs at least 150 members to make a profit. (Note that we rounded 149.3 up to 150 in this case.)

## Demand and Supply Models

The demand for a commodity usually goes down as its price goes up. It is traditional to use the letter $q$ for the (quantity of) demand, as measured, for example, in sales. Consider the following example.

### EXAMPLE 3  Demand: Private Schools

The demand for private schools in Michigan depends on the tuition cost and can be approximated by

$$q = 77.8p^{-0.11} \text{ thousand students} \qquad (200 \le p \le 2{,}200), \qquad \text{Demand curve}$$

where $p$ is the net tuition cost in dollars.[17]

**a.** Use technology to plot the demand function.

**b.** What is the effect on demand if the tuition cost is increased from \$1,000 to \$1,500?

### Solution

**a.** The demand function is given by $q(p) = 77.8p^{-0.11}$. Its graph is known as a **demand curve** (Figure 9).

**b.** The demand at tuition costs of \$1,000 and \$1,500 is

$$q(1{,}000) = 77.8(1{,}000)^{-0.11} \approx 36.4 \text{ thousand students}$$
$$q(1{,}500) = 77.8(1{,}500)^{-0.11} \approx 34.8 \text{ thousand students}.$$

The change in demand is therefore

$$q(1{,}500) - q(1{,}000) \approx 34.8 - 36.4 = -1.6 \text{ thousand students}.$$

Technology Formula:
y = 77.8x^(-0.11)
**Figure 9**

We have seen that a demand function gives the number of items consumers are willing to buy at a given price, and a higher price generally results in a lower demand. However, as the price rises, suppliers will be more inclined to produce these items (as opposed to spending their time and money on other products), so supply will generally rise. A **supply function** gives $q$, the number of items suppliers are willing to make available for sale*, as a function of $p$, the price per item.

\* Although a bit confusing at first, it is traditional to use the same letter $q$ for the quantity of supply and the quantity of demand, particularly when we want to compare them, as in the next example.

## Demand, Supply, and Equilibrium Price

A **demand equation** or **demand function** expresses demand $q$ (the number of items demanded) as a function of the unit price $p$ (the price per item). A **supply equation** or **supply function** expresses supply $q$ (the number of items a supplier is willing to make available) as a function of the unit price $p$ (the price per item). It is usually the case that demand decreases and supply increases as the unit price increases.

[17]The tuition cost is net cost: tuition minus tax credit. The model is based on data in "The Universal Tuition Tax Credit: A Proposal to Advance Personal Choice in Education," Patrick L. Anderson, Richard McLellan, J.D., Joseph P. Overton, J.D., Gary Wolfram, Ph.D., Mackinac Center for Public Policy, www.mackinac.org/

Demand and supply are said to be in **equilibrium** when demand equals supply. The corresponding values of $p$ and $q$ are called the **equilibrium price** and **equilibrium demand**. To find the equilibrium price, determine the unit price $p$ where the demand and supply curves cross (sometimes we can determine this value analytically by setting demand equal to supply and solving for $p$). To find the equilibrium demand, evaluate the demand (or supply) function at the equilibrium price.

### Quick Example

If the demand for your exclusive T-shirts is $q = -20p + 800$ shirts sold per day and the supply is $q = 10p - 100$ shirts per day, then the equilibrium point is obtained when demand = supply:

$$-20p + 800 = 10p - 100$$
$$30p = 900, \text{ giving } p = \$30.$$

The equilibrium price is therefore \$30 and the equilibrium demand is $q = -20(30) + 800 = 200$ shirts per day. What happens at prices other than the equilibrium price is discussed in Example 4.

**Note** In economics it is customary to plot the independent variable (price) on the vertical axis and the dependent variable (demand or supply) on the horizontal axis, but in this book we follow the usual mathematical convention for all graphs and plot the independent variable on the horizontal axis.

## EXAMPLE 4  Demand, Supply, and Equilibrium Price

Continuing with Example 3, suppose that private school institutions are willing to create private schools to accommodate

$$q = 30.4 + 0.006p \text{ thousand students} \qquad (200 \le p \le 2{,}200) \qquad \text{Supply curve}$$

who pay a net tuition of $p$ dollars.

**a.** Graph the demand curve of Example 3 and the supply curve given here on the same set of axes. Use your graph to estimate, to the nearest \$100, the tuition at which the demand equals the supply. Approximately how many students will be accommodated at that price, known as the **equilibrium price**?

**b.** What happens if the tuition is higher than the equilibrium price? What happens if it is lower?

**c.** Estimate the shortage or surplus of openings at private schools if tuition is set at \$1,200.

## Solution

**a.** Figure 10 shows the graphs of demand $q = 77.8p^{-0.11}$ and supply $q = 30.4 + 0.006p$. (See the margin note for a brief description of how to plot them.)

Demand: $q = 77.8p^{-0.11}$
Supply: $q = 30.4 + 0.006p$

**Figure 10**

The lines cross close to $p = \$1,000$, so we conclude that demand = supply when $p \approx \$1,000$ (to the nearest $100). This is the (approximate) equilibrium tuition price. At that price, we can estimate the demand or supply at around

Demand: $q = 77.8(1,000)^{-0.11} \approx 36.4$

Supply: $q = 30.4 + 0.006(1,000) = 36.4$     Demand = Supply at equilibrium

or 36,400 students.

**b.** Take a look at Figure 11, which shows what happens if schools charge more or less than the equilibrium price.

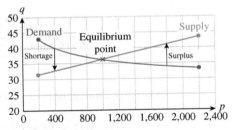

**Figure 11**

If tuition is, say, $1,800, then the supply will be larger than demand and there will be a surplus of available openings at private schools. Similarly, if tuition is less—say $400—then the supply will be less than the demand, and there will be a shortage of available openings.

**c.** The discussion in part (b) shows that if tuition is set at $1,200 there will be a surplus of available openings. To estimate that number, we calculate the projected demand and supply when $p = \$1,200$:

Demand: $q = 77.8(1,200)^{-0.11} \approx 35.7$ thousand seats

Supply: $q = 30.4 + 0.006(1,200) = 37.6$ thousand seats

Surplus = Supply − Demand $\approx 37.6 - 35.7 = 1.9$ thousand seats.

So, there would be a surplus of around 1,900 available seats.

---

### using Technology

See the Technology Guides at the end of the chapter for detailed instructions on how to obtain the table of values and graph in Example 4 using a TI-83/84 Plus or Excel. Here is an outline:

**TI-83/84 Plus**
Graphs:
$Y_1$=77.8*X^(-0.11)
$Y_2$=30.4+0.006*X
2ND TABLE Graph:
Xmin = 200, Xmax = 2200;
ZOOM 0 [More details on page 115.]

**Spreadsheet**
Headings $p$, Demand, Supply in A1–C1; $p$-values 200, 300, ..., 2200 in A2-A22.
=77.8*A2^(-0.11) in B2
=30.4+0.006*A2 in C2
Copy down through C22.
Highlight A1–C22; insert Scatter chart. [More details on page 120.]

**Website**
www.WanerMath.com
Go to the Function Evaluator and Grapher under Online Utilities, and enter
77.8*x^(-0.11) for $y_1$ and
30.4+0.006*x for $y_2$.
Graph: Set Xmin = 200, Xmax = 2200, and press "Plot Graphs".

➡ **Before we go on...** We just saw in Example 4 that if tuition is less than the equilibrium price there will be a shortage. If schools were to raise their tuition toward the equilibrium, they would create and fill more openings and increase revenue, because it is the supply equation—and not the demand equation—that determines what one can sell below the equilibrium price. On the other hand, if they were to charge more than the equilibrium price, they will be left with a possibly costly surplus of unused openings (and will want to lower tuition to reduce the surplus). Prices tend to move toward the equilibrium, so supply tends to equal demand. When supply equals demand, we say that the market **clears**. ■

## Modeling Change over Time

Things around us change with time. Thus, there are many quantities, such as your income or the temperature in Honolulu, that are natural to think of as functions of time. Example 1 on page 44 (on iPod sales) and Example 2 on page 47 (on Facebook membership) in Section 1.1 are models of change over time. Both of those models are curve-fitting models: We used algebraic functions to approximate observed data.

**Note** We usually use the independent variable $t$ to denote time (in seconds, hours, days, years, etc.). If a quantity $q$ changes with time, then we can regard $q$ as a function of $t$. ■

In the next example we are asked to select from among several curve-fitting models for given data.

**⊩ EXAMPLE 5  Model Selection: Sales**

The following table shows annual sales, in billions of dollars, by Nike from 2005 through 2010:[18]

| Year | 2005 | 2006 | 2007 | 2008 | 2009 | 2010 |
|---|---|---|---|---|---|---|
| Sales ($ billion) | 13.5 | 15 | 16.5 | 18.5 | 19 | 19 |

Take $t$ to be the number of years since 2005, and consider the following four models:

(1)  $s(t) = 14 + 1.2t$                           Linear model

(2)  $s(t) = 13 + 2.2t - 0.2t^2$                Quadratic model

(3)  $s(t) = 14(1.07^t)$                          Exponential model

(4)  $s(t) = \dfrac{19.5}{1 + 0.48(1.8^{-t})}$      Logistic model

**a.** Which models fit the data significantly better than the rest?

**b.** Of the models you selected in part (a), which gives the most reasonable prediction for 2013?

---

[18]Figures are rounded. Source: http://invest.nike.com.

## Solution

**a.** The following table shows the original data together with the values, rounded to the nearest 0.5, for all four models:

| $t$ | 0 | 1 | 2 | 3 | 4 | 5 |
|---|---|---|---|---|---|---|
| **Sales ($billion)** | 13.5 | 15 | 16.5 | 18.5 | 19 | 19 |
| **Linear: $s(t) = 14 + 1.2t$**<br>Technology: `14+1.2*x` | 14 | 15 | 16.5 | 17.5 | 19 | 20 |
| **Quadratic: $s(t) = 13 + 2.2t - 0.2t^2$**<br>Technology: `13+2.2*x-0.2*x^2` | 13 | 15 | 16.5 | 18 | 18.5 | 19 |
| **Exponential: $s(t) = 14(1.07^t)$**<br>Technology: `14*1.07^x` | 14 | 15 | 16 | 17 | 18.5 | 19.5 |
| **Logistic: $s(t) = \dfrac{19.5}{1 + 0.48(1.8^{-t})}$**<br>Technology: `19.5/(1+0.48*1.8^(-x))` | 13 | 15.5 | 17 | 18 | 18.5 | 19 |

Notice that all the models give values that seem reasonably close to the actual sales values. However, the quadratic and logistic curves seem to model their behavior more accurately than the others (see Figure 12).

We therefore conclude that the quadratic and logistic models fit the data significantly better than the others.

**b.** Although the quadratic and logistic models both appear to fit the data well, they do not both extrapolate to give reasonable predictions for 2013:

$$\text{Quadratic Model: } s(8) = 13 + 2.2(8) - 0.2(8)^2 = 17.8$$

$$\text{Logistic Model: } s(8) = \frac{19.5}{1 + 0.48(1.8^{-8})} \approx 19.4.$$

Linear: $s(t) = 14 + 1.2t$

Quadratic: $s(t) = 13 + 2.2t - 0.2t^2$

Exponential: $s(t) = 14(1.07^t)$

Logistic: $s(t) = \dfrac{19.5}{1 + 0.48(1.8^{-t})}$

**Figure 12**

Notice that the quadratic model predicts a significant *decline* in sales whereas the logistic model predicts a more reasonable modest increase. This discrepancy can be seen quite dramatically in Figure 13.

**Figure 13**

We now derive an analytical model of change over time based on the idea of **compound interest**. Suppose you invest $500 (the **present value**) in an investment account with an annual yield of 15%, and the interest is reinvested at the end of every year (we say that the interest is **compounded** or **reinvested** once a year). Let $t$ represent the number of years since you made the initial $500 investment. Each year, the investment is worth 115% (or 1.15 times) of its value the previous year. The **future value** $A$ of your investment changes over time $t$, so we think of $A$ as a function of $t$. The following table illustrates how we can calculate the future value for several values of $t$:

| $t$ | 0 | 1 | 2 | 3 |
|---|---|---|---|---|
| **Future Value $A(t)$** | 500 | 575 | 661.25 | 760.44 |
| $A$ | | $500(1.15)$ | $500(1.15)^2$ | $500(1.15)^3$ |

$\times 1.15$     $\times 1.15$     $\times 1.15$

Thus, $A(t) = 500(1.15)^t$. A traditional way to write this formula is

$$A(t) = P(1+r)^t,$$

where $P$ is the present value ($P = 500$) and $r$ is the annual interest rate ($r = 0.15$).

If, instead of compounding the interest once a year, we compound it every three months (four times a year), we would earn one quarter of the interest ($r/4$ of the current investment) every three months. Because this would happen $4t$ times in $t$ years, the formula for the future value becomes

$$A(t) = P\left(1 + \frac{r}{4}\right)^{4t}.$$

## Compound Interest

If an amount (**present value**) $P$ is invested for $t$ years at an annual rate of $r$, and if the interest is compounded (reinvested) $m$ times per year, then the **future value** $A$ is

$$A(t) = P\left(1 + \frac{r}{m}\right)^{mt}.$$

A special case is **interest compounded once a year**:

$$A(t) = P(1+r)^t.$$

> ### Quick Example
>
> If \$2,000 is invested for two and a half years in a mutual fund with an annual yield of 12.6% and the earnings are reinvested each month, then $P = 2,000, r = 0.126, m = 12$, and $t = 2.5$, which gives
>
> $$A(2.5) = 2,000 \left(1 + \frac{0.126}{12}\right)^{12 \times 2.5} \qquad \text{2000*(1+0.126/12)^(12*2.5)}$$
>
> $$= 2,000(1.0105)^{30} = \$2,736.02.$$

## EXAMPLE 6 Compound Interest: Investments

Consider the scenario in the preceding Quick Example: You invest \$2,000 in a mutual fund with an annual yield of 12.6% and the interest is reinvested each month.

**a.** Find the associated exponential model.

**b.** 🖥 Use a table of values to estimate the year during which the value of your investment reaches \$5,000.

**c.** Use a graph to confirm your answer in part (b).

### Solution

**a.** Apply the formula

$$A(t) = P\left(1 + \frac{r}{m}\right)^{mt}$$

with $P = 2,000, r = 0.126$, and $m = 12$. We get

$$A(t) = 2,000\left(1 + \frac{0.126}{12}\right)^{12t}$$

$$= 2,000(1.0105)^{12t}. \qquad \text{2000*(1+0.126/12)^(12*t)}$$

This is the exponential model. (What would happen if we left out the last set of parentheses in the technology formula?)

**b.** We need to find the value of $t$ for which $A(t) = \$5,000$, so we need to solve the equation

$$5,000 = 2,000(1.0105)^{12t}.$$

In Section 9.3 we will learn how to use logarithms to do this algebraically, but we can answer the question now using a graphing calculator, a spreadsheet, or the Function Evaluator and Grapher utility at the Website. Just enter the model and compute the balance at the end of several years. Here are examples of tables obtained using three forms of technology:

 **using Technology**

**TI-83/84 Plus**
Y₁=2000(1+0.126/12)^
(12X)
[2ND] [TABLE]

**Spreadsheet**
Headings *t* and *A* in A1–B1;
*t*-values 0–11 in A2–A13.
=2000*(1+0.126/12)^
(12*A2)
in B2; copy down through B13.

 **Website**
www.WanerMath.com
Go to the Function Evaluator
and Grapher under Online
Utilities, and enter
2000(1+0.126/12)^(12x)
for *y*₁. Scroll down to the
Evaluator, enter the values
0–11 under *x*-values and press
"Evaluate."

TI-83/84 Plus

| | A | B |
|---|---|---|
| 1 | t | A |
| 2 | 0 | $ 2,000.00 |
| 3 | 1 | $ 2,267.07 |
| 4 | 2 | $ 2,569.81 |
| 5 | 3 | $ 2,912.98 |
| 6 | 4 | $ 3,301.97 |
| 7 | 5 | $ 3,742.91 |
| 8 | 6 | $ 4,242.72 |
| 9 | 7 | $ 4,809.29 |
| 10 | 8 | $ 5,451.51 |
| 11 | 9 | $ 6,179.49 |

Excel

| x-Values | $y_1$-Values |
|---|---|
| 3 | 2912.98 |
| 4 | 3301.97 |
| 5 | 3742.91 |
| 6 | 4242.72 |
| 7 | 4809.29 |
| 8 | 5451.51 |

Website

Because the balance first exceeds \$5,000 at $t = 8$ (the end of year 8), your investment has reached \$5,000 during year 8.

c. Figure 14 shows the graph of $A(t) = 2,000(1.0105)^{12t}$ together with the horizontal line $y = 5,000$. The graphs cross between $t = 7$ and $t = 8$, confirming that year 8 is the first year during which the value of the investment reaches \$5,000.

Technology formula:
2000*1.0105^(12*x)

**Figure 14**

The compound interest examples we saw above are instances of **exponential growth:** a quantity whose magnitude is an increasing exponential function of time. The decay of unstable radioactive isotopes provides instances of **exponential decay:** a quantity whose magnitude is a *decreasing* exponential function of time. For example, carbon 14, an unstable isotope of carbon, decays exponentially to nitrogen. Because carbon 14 decay is extremely slow, it has important applications in the dating of fossils.

### EXAMPLE 7 Exponential Decay: Carbon Dating

The amount of carbon 14 remaining in a sample that originally contained $A$ grams is approximately

$$C(t) = A(0.999879)^t,$$

where $t$ is time in years.

**a.** What percentage of the original amount remains after one year? After two years?

**b.** Graph the function $C$ for a sample originally containing 50 g of carbon 14, and use your graph to estimate how long, to the nearest 1,000 years, it takes for half the original carbon 14 to decay.

**c.** A fossilized plant unearthed in an archaeological dig contains 0.50 g of carbon 14 and is known to be 50,000 years old. How much carbon 14 did the plant originally contain?

### Solution

Notice that the given model is exponential as it has the form $f(t) = Ab^t$. (See page 51.)

**a.** At the start of the first year, $t = 0$, so there are
$$C(0) = A(0.999879)^0 = A \text{ grams.}$$
At the end of the first year, $t = 1$, so there are
$$C(1) = A(0.999879)^1 = 0.999879A \text{ grams;}$$

Technology formula:
`50*0.999879^x`

**Figure 15**

that is, 99.9879% of the original amount remains. After the second year, the amount remaining is

$$C(2) = A(0.999879)^2 \approx 0.999758A \text{ grams,}$$

or about 99.9758% of the original sample.

**b.** For a sample originally containing 50 g of carbon 14, $A = 50$, so $C(t) = 50(0.999879)^t$. Its graph is shown in Figure 15. We have also plotted the line $y = 25$ on the same graph. The graphs intersect at the point where the original sample has decayed to 25 g: about $t = 6,000$ years.

**c.** We are given the following information: $C = 0.50$, $A =$ the unknown, and $t = 50,000$. Substituting gives

$$0.50 = A(0.999879)^{50,000}.$$

Solving for $A$ gives

$$A = \frac{0.5}{0.999879^{50,000}} \approx 212 \text{ grams.}$$

Thus, the plant originally contained 212 g of carbon 14.

⇒ **Before we go on...**

The formula we used for $A$ in Example 7(c) has the form

$$A(t) = \frac{C}{0.999879^t},$$

which gives the original amount of carbon 14 $t$ years ago in terms of the amount $C$ that is left now. A similar formula can be used in finance to find the present value, given the future value. ■

## Algebra of Functions

If you look back at some of the functions considered in this section, you will notice that we frequently constructed them by combining simpler or previously constructed functions. For instance:

Quick Example 3 on page 56:    Area = Width × Length: $A(x) = x(50 - x)$

Example 1:    Cost = Variable Cost + Fixed Cost: $C(x) = 1.80x + 2.05$

Quick Example on page 59:    Profit = Revenue − Cost: $P(x) = 10x - (8x + 100)$.

Let us look a little more deeply at each of the above examples:

***Area Example:*** $A(x) =$ Width × Length $= x(50 - x)$:
Think of the width and length as separate functions of $x$:

Width: $W(x) = x$;    Length: $L(x) = 50 - x$

so that

$$A(x) = W(x)L(x). \qquad \text{Area = Width × Length}$$

We say that the area function $A$ is the **product of the functions** $W$ and $L$, and we write

$$A = WL. \qquad \text{\textit{A} is the product of the functions \textit{W} and \textit{L}.}$$

To calculate $A(x)$, we multiply $W(x)$ by $L(x)$.

**Cost Example:** $C(x) = $ Variable Cost $+$ Fixed Cost $= 1.80x + 2.05$:

Think of the variable and fixed costs as separate functions of $x$:

✻ *F* is called a constant function
as its value, 2.05, is the same
for every value of *x*.

Variable Cost: $V(x) = 1.80x$;      Fixed Cost: $F(x) = 2.05$✻

so that

$$C(x) = V(x) + F(x).$$    Cost = Variable Cost + Fixed Cost

We say that the cost function $C$ is the **sum of the functions $V$ and $F$**, and we write

$$C = V + F.$$    $C$ is the sum of the functions $V$ and $F$.

To calculate $C(x)$, we add $V(x)$ to $F(x)$.

**Profit Example:** $P(x) = $ Revenue $-$ Cost $= 10x - (8x + 100)$:

Think of the revenue and cost as separate functions of $x$:

Revenue: $R(x) = 10x$;      Cost: $C(x) = 8x + 100$

so that

$$P(x) = R(x) - C(x).$$    Profit = Revenue − Cost

We say that the profit function $P$ is the **difference between the functions $R$ and $C$**, and we write

$$P = R - C.$$    $P$ is the difference of the functions $R$ and $C$.

To calculate $P(x)$, we subtract $C(x)$ from $R(x)$.

## Algebra of Functions

If $f$ and $g$ are real-valued functions of the real variable $x$, then we define their **sum $s$, difference $d$, product $p$,** and **quotient $q$** as follows:

$s = f + g$ is the function specified by $s(x) = f(x) + g(x)$.

$d = f - g$ is the function specified by $d(x) = f(x) - g(x)$.

$p = fg$ is the function specified by $p(x) = f(x)g(x)$.

$q = \dfrac{f}{g}$ is the function specified by $q(x) = \dfrac{f(x)}{g(x)}$.

Also, if $f$ is as above and $c$ is a constant (real number), then we define the associated **constant multiple $m$ of $f$** by

$m = cf$ is the function specified by $m(x) = cf(x)$.

### Note on Domains

In order for any of the expressions $f(x) + g(x)$, $f(x) - g(x)$, $f(x)g(x)$, or $f(x)/g(x)$ to make sense, $x$ must be simultaneously in the domains of both $f$ and $g$. Further, for the quotient, the denominator $g(x)$ cannot be zero. Thus, we specify the domains of these functions as follows:

**Domain of $f + g, f - g,$ and $fg$:** All real numbers $x$ simultaneously in the domains of $f$ and $g$

**Domain of $f/g$:** All real numbers $x$ simultaneously in the domains of $f$ and $g$ such that $g(x) \neq 0$

**Domain of $cf$:** Same as the domain of $f$

### Quick Examples

1. If $f(x) = x^2 - 1$ and $g(x) = \sqrt{x}$ with domain $[0, +\infty)$, then the sum $s$ of $f$ and $g$ has domain $[0, +\infty)$ and is specified by
$$s(x) = f(x) + g(x) = x^2 - 1 + \sqrt{x}.$$

2. If $f(x) = x^2 - 1$ and $c = 3$, then the associated constant multiple $m$ of $f$ is specified by $m(x) = 3f(x) = 3(x^2 - 1)$.

3. If there are $N = 1,000t$ Mars shuttle passengers in year $t$ who pay a total cost of $C = 40,000 + 800t$ million dollars, then the cost per passenger is given by the quotient of the two functions,

$$\text{Cost per passenger} = q(t) = \frac{C(t)}{N(t)}$$
$$= \frac{40,000 + 800t}{1,000t} \text{ million dollars per passenger.}$$

The largest possible domain of $C/N$ is $(0, +\infty)$, as the quotient is not defined if $t = 0$.

## 1.2 EXERCISES

▼ more advanced   ◆ challenging

T indicates exercises that should be solved using technology

*Exercises 1–8 are based on the following functions:*

$f(x) = x^2 + 1$ *with domain all real numbers*
$g(x) = x - 1$ *with domain all real numbers*
$h(x) = x + 4$ *with domain $x \geq 10$*
$u(x) = \sqrt{x + 10}$ *with domain $[-10, 0)$*
$v(x) = \sqrt{10 - x}$ *with domain $[0, 10]$*

*In each exercise, **a.** write a formula for the indicated function, **b.** give its domain, and **c.** specify its value at the given point $a$, if defined.*

1. $s = f + g; a = -3$
2. $d = g - f; a = -1$
3. $p = gu; a = -6$
4. $p = hv; a = 1$
5. $q = \dfrac{v}{g}; a = 1$
6. $q = \dfrac{g}{v}; a = 1$
7. $m = 5f; a = 1$
8. $m = 3u; a = -1$

## APPLICATIONS

9. *Resources* You now have 200 music files on your hard drive, and this number is increasing by 10 music files each day. Find a mathematical model for this situation. HINT [See Quick Example 1 on page 56.]

10. *Resources* The amount of free space left on your hard drive is now 50 gigabytes (GB) and is decreasing by 5 GB/month. Find a mathematical model for this situation.

11. *Soccer* My rectangular soccer field site has a length equal to twice its width. Find its area in terms of its length $x$. HINT [See Quick Example 3 on page 56.]

12. *Cabbage* My rectangular cabbage patch has a total area of 100 sq. ft. Find its perimeter in terms of the width $x$.

13. *Vegetables* I want to fence in a square vegetable patch. The fencing for the east and west sides costs $4 per foot, and the fencing for the north and south sides costs only $2 per foot. Find the total cost of the fencing as a function of the length of a side $x$.

14. *Orchids* My square orchid garden abuts my house so that the house itself forms the northern boundary. The fencing for the southern boundary costs $4 per foot, and the fencing for the east and west sides costs $2 per foot. Find the total cost of the fencing as a function of the length of a side $x$.

15. *Study* You study math 4 hours a day Sunday through Thursday, and take the rest of the week off. Model the number of hours $h$ you study math as a function of the day of the week $n$ (with $n = 1$ being Sunday). HINT [See Quick Example 4 on page 57.]

16. *Recreation* You spend 5 hours per day on Saturdays and Sundays watching movies, but only two hours per day during the

week. Model the number of hours $h$ you watch movies as a function of the day of the week $n$ (with $n = 1$ being Sunday).

17. **Cost** A piano manufacturer has a daily fixed cost of $1,200 and a marginal cost of $1,500 per piano. Find the cost $C(x)$ of manufacturing $x$ pianos in one day. Use your function to answer the following questions:

   a. On a given day, what is the cost of manufacturing 3 pianos?
   b. What is the cost of manufacturing the 3rd piano that day?
   c. What is the cost of manufacturing the 11th piano that day?
   d. What is the variable cost? What is the fixed cost? What is the marginal cost?
   e. Graph $C$ as a function of $x$. HINT [See Example 1.]

18. **Cost** The cost of renting tuxes for the Choral Society's formal is $20 down, plus $88 per tux. Express the cost $C$ as a function of $x$, the number of tuxedos rented. Use your function to answer the following questions.

   a. What is the cost of renting 2 tuxes?
   b. What is the cost of the 2nd tux?
   c. What is the cost of the 4,098th tux?
   d. What is the variable cost? What is the fixed cost? What is the marginal cost?
   e. Graph $C$ as a function of $x$.

19. **Break-Even Analysis** Your college newspaper, *The Collegiate Investigator*, has fixed production costs of $70 per edition and marginal printing and distribution costs of 40¢ per copy. *The Collegiate Investigator* sells for 50¢ per copy.

   a. Write down the associated cost, revenue, and profit functions. HINT [See Examples 1 and 2.]
   b. What profit (or loss) results from the sale of 500 copies of *The Collegiate Investigator*?
   c. How many copies should be sold in order to break even?

20. **Break-Even Analysis** The Audubon Society at Enormous State University (ESU) is planning its annual fund-raising "Eat-a-thon." The society will charge students 50¢ per serving of pasta. The only expenses the society will incur are the cost of the pasta, estimated at 15¢ per serving, and the $350 cost of renting the facility for the evening.

   a. Write down the associated cost, revenue, and profit functions.
   b. How many servings of pasta must the Audubon Society sell in order to break even?
   c. What profit (or loss) results from the sale of 1,500 servings of pasta?

21. **Break-Even Analysis** Gymnast Clothing manufactures expensive hockey jerseys for sale to college bookstores in runs of up to 200. Its cost (in dollars) for a run of $x$ hockey jerseys is

$$C(x) = 2,000 + 10x + 0.2x^2 \quad (0 \le x \le 200).$$

Gymnast Clothing sells the jerseys at $100 each. Find the revenue and profit functions. How many should Gymnast Clothing manufacture to make a profit? HINT [See Example 2.]

22. **Break-Even Analysis** Gymnast Clothing also manufactures expensive soccer cleats for sale to college bookstores in runs of up to 500. Its cost (in dollars) for a run of $x$ pairs of cleats is

$$C(x) = 3,000 + 8x + 0.1x^2 \quad (0 \le x \le 500).$$

Gymnast Clothing sells the cleats at $120 per pair. Find the revenue and profit functions. How many should Gymnast Clothing manufacture to make a profit?

23. **Break-Even Analysis: School Construction Costs** The cost, in millions of dollars, of building a two-story high school in New York State was estimated to be

$$C(x) = 1.7 + 0.12x - 0.0001x^2 \quad (20 \le x \le 400),$$

where $x$ is the number of thousands of square feet.[19] Suppose that you are contemplating building a for-profit two-story high school and estimate that your total revenue will be $0.1 million dollars per thousand square feet. What is the profit function? What size school should you build in order to break even?

24. **Break-Even Analysis: School Construction Costs** The cost, in millions of dollars, of building a three-story high school in New York State was estimated to be

$$C(x) = 1.7 + 0.14x - 0.0001x^2 \quad (20 \le x \le 400),$$

where $x$ is the number of thousands of square feet.[20] Suppose that you are contemplating building a for-profit three-story high school and estimate that your total revenue will be $0.2 million dollars per thousand square feet. What is the profit function? What size school should you build in order to break even?

25. ▼ **Profit Analysis—Aviation** The hourly operating cost of a Boeing 747-100, which seats up to 405 passengers, is estimated to be $5,132.[21] If an airline charges each passenger a fare of $100 per hour of flight, find the hourly profit $P$ it earns operating a 747-100 as a function of the number of passengers $x$. (Be sure to specify the domain.) What is the least number of passengers it must carry in order to make a profit? HINT [The cost function is constant (Variable cost = 0).]

26. ▼ **Profit Analysis—Aviation** The hourly operating cost of a McDonnell Douglas DC 10-10, which seats up to 295 passengers, is estimated to be $3,885.[22] If an airline charges each passenger a fare of $100 per hour of flight, find the hourly profit $P$ it earns operating a DC10-10 as a function of the number of passengers $x$. (Be sure to specify the domain.) What is the least number of passengers it must carry in order to make a profit? HINT [The cost function is constant (Variable cost = 0).]

27. ▼ **Break-Even Analysis** *(based on a question from a CPA exam)* The Oliver Company plans to market a new product. Based on its market studies, Oliver estimates that it can sell up to 5,500 units in 2005. The selling price will be $2 per unit. Variable costs are estimated to be 40% of total revenue. Fixed costs are estimated to be $6,000 for 2005. How many units should the company sell to break even?

---

[19]The model is the authors'. Source for data: *Project Labor Agreements and Public Construction Cost in New York State,* Paul Bachman and David Tuerck, Beacon Hill Institute at Suffolk University, April 2006, www.beaconhill.org.
[20]*Ibid.*
[21]In 1992. Source: Air Transportation Association of America.
[22]*Ibid.*

**28.** ▼ *Break-Even Analysis (based on a question from a CPA exam)* The Metropolitan Company sells its latest product at a unit price of $5. Variable costs are estimated to be 30% of the total revenue, while fixed costs amount to $7,000 per month. How many units should the company sell per month in order to break even, assuming that it can sell up to 5,000 units per month at the planned price?

**29.** ◆ *Break-Even Analysis (from a CPA exam)* Given the following notations, write a formula for the break-even sales level.

$SP$ = Selling price per unit
$FC$ = Total fixed cost
$VC$ = Variable cost per unit

**30.** ◆ *Break-Even Analysis (based on a question from a CPA exam)* Given the following notation, give a formula for the total fixed cost.

$SP$ = Selling price per unit
$VC$ = Variable cost per unit
$BE$ = Break even sales level in units

**31.** ◆ *Break-Even Analysis—Organized Crime* The organized crime boss and perfume king Butch (Stinky) Rose has daily overheads (bribes to corrupt officials, motel photographers, wages for hit men, explosives, and so on) amounting to $20,000 per day. On the other hand, he has a substantial income from his counterfeit perfume racket: He buys imitation French perfume (Chanel № 22.5) at $20 per gram, pays an additional $30 per 100 grams for transportation, and sells it via his street thugs for $600 per gram. Specify Stinky's profit function, $P(x)$, where $x$ is the quantity (in grams) of perfume he buys and sells, and use your answer to calculate how much perfume should pass through his hands per day in order that he break even.

**32.** ◆ *Break-Even Analysis—Disorganized Crime* Butch (Stinky) Rose's counterfeit Chanel № 22.5 racket has run into difficulties; it seems that the *authentic* Chanel № 22.5 perfume is selling for less than his counterfeit perfume. However, he has managed to reduce his fixed costs to zero, and his overall costs are now $400 per gram plus $30 per gram transportation costs and commission. (The perfume's smell is easily detected by specially trained Chanel Hounds, and this necessitates elaborate packaging measures.) He therefore decides to sell it for $420 per gram in order to undercut the competition. Specify Stinky's profit function, $P(x)$, where $x$ is the quantity (in grams) of perfume he buys and sells, and use your answer to calculate how much perfume should pass through his hands per day in order that he break even. Interpret your answer.

**33.** *Demand for Monorail Service, Las Vegas* The demand for monorail service in Las Vegas can be approximated by

$$q(p) = 64p^{-0.76} \text{ thousand rides per day} \quad (3 \le p \le 5),$$

where $p$ is the cost per ride in dollars.[23]

**a.** Graph the demand function.
**b.** What is the result on demand if the cost per ride is increased from $3.00 to $3.50? HINT [See Example 3.]

**34.** *Demand for Monorail Service, Mars* The demand for monorail service on the Utarek monorail, which links the three urbynes (or districts) of Utarek, Mars, can be approximated by

$$q(p) = 30p^{-0.49} \text{ million rides per day} \quad (3 \le p \le 5),$$

where $p$ is the cost per ride in zonars ($\overline{\overline{\mathbf{Z}}}$).[24]

**a.** Graph the demand function.
**b.** What is the result on demand if the cost per ride is decreased from $\overline{\overline{\mathbf{Z}}}5.00$ to $\overline{\overline{\mathbf{Z}}}3.50$?

**35.** ▼ *Demand: Smart Phones* The worldwide demand for smart phones may be modeled by

$$q(p) = 0.0009p^2 - 0.63p + 245 \text{ million units sold annually} \quad (50 \le p \le 150),$$

where $p$ is the unit price in dollars.[25]

**a.** Graph the demand function.
**b.** Use the demand function to estimate, to the nearest million units, worldwide sales of smart phones if the price is $60.
**c.** Extrapolate the demand function to estimate, to the nearest million units, worldwide sales of smart phones if the price is $20.
**d.** Model the worldwide annual revenue from the sale of smart phones as a function of unit price, and use your model to estimate, to the nearest billion dollars, worldwide annual revenue when the price is set at $60. HINT [Revenue = Price × Quantity = p · q(p)]

**36.** ▼ *Demand: Smart Phones* (See Exercise 35.) Here is another model for worldwide demand for smart phones:

$$q(p) = 700p^{-0.284} \text{ million units sold annually} \quad (50 \le p \le 150),$$

where $p$ is the unit price in dollars.[26]

**a.** Graph the demand function.
**b.** Use the demand function to estimate, to the nearest million units, worldwide sales of smart phones if the price is $70.
**c.** Extrapolate the demand function to estimate, to the nearest million units, worldwide sales of smart phones if the price is $180.
**d.** Model the worldwide annual revenue from the sale of smart phones as a function of unit price, and use your model to estimate, to the nearest billion dollars, worldwide annual revenue when the price is set at $70. HINT [Revenue = Price × Quantity = p · q(p)]

---

[23]Source: *New York Times*, February 10, 2007, p. A9.

[24]The zonar ($\overline{\overline{\mathbf{Z}}}$) is the official currency in the city-state of Utarek, Mars (formerly www.Marsnext.com, a now extinct virtual society).

[25]The model is the authors' based on data available in 2010. Sources for Data: www.businessweek.com, http://techcrunch.com.

[26]*Ibid.*

**37. *Equilibrium Price*** The demand for your hand-made skate-boards, in weekly sales, is $q = -3p + 700$ if the selling price is $\$p$. You are prepared to supply $q = 2p - 500$ per week at the price $\$p$. At what price should you sell your skateboards so that there is neither a shortage nor a surplus? HINT [See Quick Example on page 62.]

**38. *Equilibrium Price*** The demand for your factory-made skateboards, in weekly sales, is $q = -5p + 50$ if the selling price is $\$p$. If you are selling them at that price, you can obtain $q = 3p - 30$ per week from the factory. At what price should you sell your skateboards so that there is neither a shortage nor a surplus?

**39. *Equilibrium Price: Cell Phones*** Worldwide quarterly sales of Nokia® cell phones were approximately $q = -p + 156$ million phones when the wholesale price[27] was $\$p$.

   **a.** If Nokia was prepared to supply $q = 4p - 394$ million phones per quarter at a wholesale price of $\$p$, what would have been the equilibrium price?

   **b.** The actual wholesale price was $\$105$ in the fourth quarter of 2004. Estimate the projected shortage or surplus at that price. HINT [See Quick Example on page 62 and also Example 4.]

**40. *Equilibrium Price: Cell Phones*** Worldwide annual sales of all cell phones were approximately $-10p + 1,600$ million phones when the wholesale price[28] was $\$p$.

   **a.** If manufacturers were prepared to supply $q = 14p - 800$ million phones per year at a wholesale price of $\$p$, what would have been the equilibrium price?

   **b.** The actual wholesale price was projected to be $\$80$ in the fourth quarter of 2008. Estimate the projected shortage or surplus at that price.

**41.** 🖥 ***Equilibrium Price: Las Vegas Monorail Service*** The demand for monorail service in Las Vegas could be approximated by

$$q = 64p^{-0.76} \text{ thousand rides per day},$$

where $p$ was the fare the Las Vegas Monorail Company charges in dollars.[29] Assume the company was prepared to provide service for

$$q = 2.5p + 15.5 \text{ thousand rides per day}$$

at a fare of $\$p$.

   **a.** Graph the demand and supply equations, and use your graph to estimate the equilibrium price (to the nearest 50¢).

   **b.** Estimate, to the nearest 10 rides, the shortage or surplus of monorail service at the December 2005 fare of $\$5$ per ride.

**42.** 🖥 ***Equilibrium Price: Mars Monorail Service*** The demand for monorail service in the three urbynes (or districts) of Utarek, Mars can be approximated by

$$q = 31p^{-0.49} \text{ million rides per day},$$

where $p$ is the fare the Utarek Monorail Cooperative charges in zonars ($\overline{\overline{Z}}$).[30] Assume the cooperative is prepared to provide service for

$$q = 2.5p + 17.5 \text{ million rides per day}$$

at a fare of $\overline{\overline{Z}}p$.

   **a.** Graph the demand and supply equations, and use your graph to estimate the equilibrium price (to the nearest 0.50 zonars).

   **b.** Estimate the shortage or surplus of monorail service at the December 2085 fare of $\overline{\overline{Z}}1$ per ride.

**43.** ▽ ***Toxic Waste Treatment*** The cost of treating waste by removing PCPs goes up rapidly as the quantity of PCPs removed goes up. Here is a possible model:

$$C(q) = 2,000 + 100q^2,$$

where $q$ is the reduction in toxicity (in pounds of PCPs removed per day) and $C(q)$ is the daily cost (in dollars) of this reduction.

   **a.** Find the cost of removing 10 pounds of PCPs per day.

   **b.** Government subsidies for toxic waste cleanup amount to

$$S(q) = 500q,$$

where $q$ is as above and $S(q)$ is the daily dollar subsidy. The *net cost* function is given by $N = C - S$. Give a formula for $N(q)$ and interpret your answer.

   **c.** Find $N(20)$ and interpret your answer.

**44.** ▽ ***Dental Plans*** A company pays for its employees' dental coverage at an annual cost $C$ given by

$$C(q) = 1,000 + 100\sqrt{q},$$

where $q$ is the number of employees covered and $C(q)$ is the annual cost in dollars.

   **a.** If the company has 100 employees, find its annual outlay for dental coverage.

   **b.** Assume that the government subsidizes coverage by an annual dollar amount of

$$S(q) = 200q.$$

The *net cost* function is given by $N = C - S$. Give a formula for $N(q)$ and interpret your answer.

   **c.** Find $N(100)$ and interpret your answer.

---

[27]Source: Embedded.com/Company reports December, 2004.

[28]Wholesale price projections are the authors'. Source for sales prediction: I-Stat/NDR December, 2004.

[29]The model is the authors'. Source for data: *New York Times*, February 10, 2007, p. A9.

[30]The official currency of Utarek, Mars. (See the footnote to Exercise 34.)

**45. *Spending on Corrections in the 1990s*** The following table shows the annual spending by all states in the United States on corrections ($t = 0$ represents the year 1990):[31]

| Year ($t$) | 0 | 2 | 4 | 6 | 7 |
|---|---|---|---|---|---|
| Spending ($ billion) | 16 | 18 | 22 | 28 | 30 |

**a.** Which of the following functions best fits the given data? (Warning: None of them fits exactly, but one fits more closely than the others.) HINT [See Example 5.]

(A) $S(t) = -0.2t^2 + t + 16$
(B) $S(t) = 0.2t^2 + t + 16$
(C) $S(t) = t + 16$

**b.** Use your answer to part (a) to "predict" spending on corrections in 1998, assuming that the trend continued.

**46. *Spending on Corrections in the 1990s*** Repeat Exercise 45, this time choosing from the following functions:

(A) $S(t) = 16 + 2t$
(B) $S(t) = 16 + t + 0.5t^2$
(C) $S(t) = 16 + t - 0.5t^2$

**47. *Soccer Gear*** The East Coast College soccer team is planning to buy new gear for its road trip to California. The cost per shirt depends on the number of shirts the team orders as shown in the following table:

| $x$ (Shirts ordered) | 5 | 25 | 40 | 100 | 125 |
|---|---|---|---|---|---|
| $A(x)$ (Cost/shirt, $) | 22.91 | 21.81 | 21.25 | 21.25 | 22.31 |

**a.** Which of the following functions best models the data?

(A) $A(x) = 0.005x + 20.75$
(B) $A(x) = 0.01x + 20 + \dfrac{25}{x}$
(C) $A(x) = 0.0005x^2 - 0.07x + 23.25$
(D) $A(x) = 25.5(1.08)^{(x-5)}$

**b.** 🔲 Graph the model you chose in part (a) for $10 \le x \le 100$. Use your graph to estimate the lowest cost per shirt and the number of shirts the team should order to obtain the lowest price per shirt.

**48. *Hockey Gear*** The South Coast College hockey team wants to purchase wool hats for its road trip to Alaska. The cost per hat depends on the number of hats the team orders as shown in the following table:

| $x$ (Hats ordered) | 5 | 25 | 40 | 100 | 125 |
|---|---|---|---|---|---|
| $A(x)$ (Cost/hat, $) | 25.50 | 23.50 | 24.63 | 30.25 | 32.70 |

**a.** Which of the following functions best models the data?

(A) $A(x) = 0.05x + 20.75$
(B) $A(x) = 0.1x + 20 + \dfrac{25}{x}$
(C) $A(x) = 0.0008x^2 - 0.07x + 23.25$
(D) $A(x) = 25.5(1.08)^{(x-5)}$

**b.** 🔲 Graph the model you chose in part (a) with $5 \le x \le 30$. Use your graph to estimate the lowest cost per hat and the number of hats the team should order to obtain the lowest price per hat.

***Cost: Hard Drive Storage*** Exercises 49 and 50 are based on the following data showing how the approximate retail cost of a gigabyte of hard drive storage has fallen since 2000:[32]

| $t$ (Year since 2000) | $c(t)$ (Cost per Gigabyte ($)) |
|---|---|
| 0 | 7.5 |
| 2 | 2.5 |
| 4 | 1.2 |
| 6 | 0.6 |
| 8 | 0.2 |
| 10 | 0.1 |
| 12 | 0.06 |

**49.** 🔲 **a.** Graph each of the following models together with the data points above and use your graph to decide which two of the models best fit the data: HINT [See Example 5 and accompanying technology note.]

(A) $c(t) = 6.3(0.67)^t$
(B) $c(t) = 0.093t^2 - 1.6t + 6.7$
(C) $c(t) = 4.75 - 0.50t$
(D) $c(t) = \dfrac{12.8}{t^{1.7} + 1.7}$

**b.** Of the two models you chose in part (a), which predicts the lower price in 2020? What price does that model predict?

**50.** 🔲 **a.** Graph each of the following models together with the data points above and use your graph to decide which three of the models best fit the given data:

(A) $c(t) = \dfrac{15}{1 + 2^t}$
(B) $c(t) = (7.32)0.59^t + 0.10$
(C) $c(t) = 0.00085(t - 9.6)^4$
(D) $c(t) = 7.5 - 0.82t$

**b.** One of the three best-fit models in part (a) gives an unreasonable prediction for the price in 2020. Which is it, what price does it predict, and why is the prediction unreasonable?

---

[31]Data are rounded. Source: National Association of State Budget Officers/*New York Times*, February 28, 1999, p. A1.

[32]2012 price is estimated. Source for data: Historical Notes about the Cost of Hard Drive Storage Space http://ns1758.ca/winch/winchest.html.

**51. *Value of Euro*** The following table shows the approximate value $V$ of one euro in U.S. dollars during three months in 2010. ($t = 1$ represents January, 2010.)[33]

| $t$ (Month) | 1 | 5 | 6 |
|---|---|---|---|
| $V$ (Value in $) | 1.42 | 1.26 | 1.22 |

Which of the following kinds of models would best fit the given data? Explain your choice of model. ($A$, $a$, $b$, $c$, and $m$ are constants.)

(A) Linear: $V(t) = mt + b$
(B) Quadratic: $V(t) = at^2 + bt + c$
(C) Exponential: $V(t) = Ab^t$

**52. *Value of Yen*** The following table shows the approximate value $V$ of one yen in U.S. dollars during three months in 2010. ($t = 1$ represents January, 2010.)[34]

| $t$ (Month) | 1 | 4 | 10 |
|---|---|---|---|
| $V$ (Value in $) | 0.0110 | 0.0107 | 0.0122 |

Which of the following kinds of models would best fit the given data? Explain your choice of model. ($A$, $a$, $b$, $c$, and $m$ are constants.)

(A) Linear: $V(t) = mt + b$
(B) Quadratic: $V(t) = at^2 + bt + c$
(C) Exponential: $V(t) = Ab^t$

**53. *Petrochemical Sales: Mexico*** The following table shows annual petrochemial sales in Mexico by Pemex, Mexico's national oil company, for 2005–2010 ($t = 0$ represents 2005):[35]

| Year $t$ (Year since 2005) | 0 | 2 | 3 | 4 |
|---|---|---|---|---|
| Petrochemical Sales $s$ (Billion metric tons) | 3.7 | 4.0 | 4.1 | 4.0 |

Which of the following kinds of models would best fit the given data? Explain your choice of model. ($A$, $a$, $b$, and $c$ are constants.)

(A) $s(t) = Ab^t$    $(b > 1)$
(B) $s(t) = Ab^t$    $(b < 1)$

(C) $s(t) = at^2 + bt + c$    $(a > 0)$
(D) $s(t) = at^2 + bt + c$    $(a < 0)$

**54. *Petrochemical Production: Mexico*** The following table shows annual petrochemical production in Mexico by Pemex, Mexico's national oil company, for 2005–2010 ($t = 0$ represents 2005):[36]

| Year $t$ (Year since 2005) | 1 | 3 | 5 |
|---|---|---|---|
| Petrochemical Production $p$ (Billion metric tons) | 11.0 | 12.0 | 13.3 |

Which of the following kinds of models would best fit the given data? Explain your choice of model. ($A$, $a$, $b$, $c$, and $m$ are constants.)

(A) $s(t) = Ab^t$    $(b > 1)$
(B) $s(t) = Ab^t$    $(b < 1)$
(C) $s(t) = mt + b$    $(m > 0)$
(D) $s(t) = at^2 + bt + c$    $(a < 0)$

**55. *Investments*** In November 2010, E*Trade Financial was offering only 0.3% interest on its online savings accounts, with interest reinvested monthly.[37] Find the associated exponential model for the value of a $5,000 deposit after $t$ years. Assuming this rate of return continued for seven years, how much would a deposit of $5,000 in November 2010 be worth in November 2017? (Answer to the nearest $1.) HINT [See Quick Example on page 67.]

**56. *Investments*** In November 2010, ING Direct was offering 2.4% interest on its Orange Savings Account, with interest reinvested quarterly.[38] Find the associated exponential model for the value of a $4,000 deposit after $t$ years. Assuming this rate of return continued for eight years, how much would a deposit of $4,000 in November 2010 be worth in November 2018? (Answer to the nearest $1.)

**57.** ☐ *Investments* Refer to Exercise 55. In November of which year will an investment of $5,000 made in November of 2010 first exceed $5,200? HINT [See Example 6.]

**58.** ☐ *Investments* Refer to Exercise 56. In November of which year will an investment of $4,000 made in November of 2010 first exceed $5,200?

**59. *Carbon Dating*** A fossil originally contained 104 grams of carbon 14. Refer to the formula for $C(t)$ in Example 7 and estimate the amount of carbon 14 left in the sample after 10,000 years, 20,000 years, and 30,000 years. HINT [See Example 7.]

---

[33]Source: www.exchange-rates.org.

[34]*Ibid.*

[35]2010 figure is a projection based on data through May 2010. Source: www.pemex.com (July 2010)

[36]*Ibid.*

[37]Interest rate based on annual percentage yield. Source: https://us.etrade.com, November 2010.

[38]Interest rate based on annual percentage yield. Source: http://home.ingdirect.com, November 2010.

**60. *Carbon Dating*** A fossil contains 4.06 grams of carbon 14. Refer to the formula for $A(t)$ at the end of Example 7, and estimate the amount of carbon 14 in the sample 10,000 years, 20,000 years, and 30,000 years ago.

**61. *Carbon Dating*** A fossil contains 4.06 grams of carbon 14. It is estimated that the fossil originally contained 46 grams of carbon 14. By calculating the amount left after 5,000 years, 10,000 years, . . . , 35,000 years, estimate the age of the sample to the nearest 5,000 years. (Refer to the formula for $C(t)$ in Example 7.)

**62. *Carbon Dating*** A fossil contains 2.8 grams of carbon 14. It is estimated that the fossil originally contained 104 grams of carbon 14. By calculating the amount 5,000 years, 10,000 years, . . . , 35,000 years ago, estimate the age of the sample to the nearest 5,000 years. (Refer to the formula for $C(t)$ at the end of Example 7.)

**63. *Radium Decay*** The amount of radium 226 remaining in a sample that originally contained $A$ grams is approximately

$$C(t) = A(0.999567)^t$$

where $t$ is time in years.

**a.** Find, to the nearest whole number, the percentage of radium 226 left in an originally pure sample after 1,000 years, 2,000 years, and 3,000 years.

**b.** Use a graph to estimate, to the nearest 100 years, when one half of a sample of 100 grams will have decayed.

**64. *Iodine Decay*** The amount of iodine 131 remaining in a sample that originally contained $A$ grams is approximately

$$C(t) = A(0.9175)^t$$

where $t$ is time in days.

**a.** Find, to the nearest whole number, the percentage of iodine 131 left in an originally pure sample after 2 days, 4 days, and 6 days.

**b.** Use a graph to estimate, to the nearest day, when one half of a sample of 100 g will have decayed.

## COMMUNICATION AND REASONING EXERCISES

**65.** If the population of the lunar station at Clavius has a population of $P = 200 + 30t$, where $t$ is time in years since the station was established, then the population is increasing by _____ per year.

**66.** My bank balance can be modeled by $B(t) = 5,000 - 200t$ dollars, where $t$ is time in days since I opened the account. The balance on my account is _____ by $200 per day.

**67.** Classify the following model as analytical or curve fitting, and give a reason for your choice: The price of gold was $700 on Monday, $710 on Tuesday, and $700 on Wednesday. Therefore, the price can be modeled by $p(t) = -10t^2 + 20t + 700$ where $t$ is the day since Monday.

**68.** Classify the following model as analytical or curve fitting, and give a reason for your choice: The width of a small animated square on my computer screen is currently 10 mm and is growing by 2 mm per second. Therefore, its area can be modeled by $a(t) = (10 + 2t)^2$ square mm where $t$ is time in seconds.

**69.** Fill in the missing information for the following *analytical model* (answers may vary): _____. Therefore, the cost of downloading a movie can be modeled by $c(t) = 4 - 0.2t$, where $t$ is time in months since January.

**70.** Repeat Exercise 69, but this time regard the given model as a *curve-fitting model*.

**71.** Fill in the blanks: In a linear cost function, the _____ cost is $x$ times the _____ cost.

**72.** Complete the following sentence: In a linear cost function, the marginal cost is the _____.

**73.** ▼ We said on page 61 that the demand for a commodity generally goes down as the price goes up. Assume that the demand for a certain commodity goes up as the price goes up. Is it still possible for there to be an equilibrium price? Explain with the aid of a demand and supply graph.

**74.** ▼ What would happen to the price of a certain commodity if the demand was always greater than the supply? Illustrate with a demand and supply graph.

**75.** You have a set of data points showing the sales of videos on your Web site versus time that are closely approximated by two different mathematical models. Give one criterion that would lead you to choose one over the other. (Answers may vary.)

**76.** Would it ever be reasonable to use a quadratic model $s(t) = at^2 + bt + c$ to predict long-term sales if $a$ is negative? Explain.

**77.** If $f$ and $g$ are functions with $f(x) \geq g(x)$ for every $x$, what can you say about the values of the function $f - g$?

**78.** If $f$ and $g$ are functions with $f(x) > g(x) > 0$ for every $x$, what can you say about the values of the function $\frac{f}{g}$?

**79.** If $f$ is measured in books, and $g$ is measured in people, what are the units of measurement of the function $\frac{f}{g}$?

**80.** If $f$ and $g$ are linear functions, then what can you say about $f - g$?

# 1.3  Linear Functions and Models

Linear functions are among the simplest functions and are perhaps the most useful of all mathematical functions.

## Linear Function

A **linear function** is one that can be written in the form

$$f(x) = mx + b \qquad \text{Function form}$$

or

$$y = mx + b \qquad \text{Equation form}$$

**Quick Example**

$$f(x) = 3x - 1$$

$$y = 3x - 1$$

where $m$ and $b$ are fixed numbers. (The names $m$ and $b$ are traditional.*)

✱ Actually, $c$ is sometimes used instead of $b$. As for $m$, there has even been some research into the question of its origin, but no one knows exactly why the letter $m$ is used.

## Linear Functions from the Numerical and Graphical Point of View

The following table shows values of $y = 3x - 1\,(m = 3, b = -1)$ for some values of $x$:

| $x$ | $-4$ | $-3$ | $-2$ | $-1$ | 0 | 1 | 2 | 3 | 4 |
|---|---|---|---|---|---|---|---|---|---|
| $y$ | $-13$ | $-10$ | $-7$ | $-4$ | $-1$ | 2 | 5 | 8 | 11 |

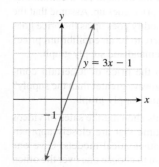

**Figure 16**

Its graph is shown in Figure 16.

Looking first at the table, notice that setting $x = 0$ gives $y = -1$, the value of $b$.

*Numerically, b is the value of y when x = 0.*

On the graph, the corresponding point $(0, -1)$ is the point where the graph crosses the $y$-axis, and we say that $b = -1$ is the **y-intercept** of the graph (Figure 17).

What about $m$? Looking once again at the table, notice that $y$ increases by $m = 3$ units for every increase of 1 unit in $x$. This is caused by the term $3x$ in the formula: for every increase of 1 in $x$ we get an increase of $3 \times 1 = 3$ in $y$.

*Numerically, y increases by m units for every 1-unit increase of x.*

Likewise, for every increase of 2 in $x$ we get an increase of $3 \times 2 = 6$ in $y$. In general, if $x$ increases by some amount, $y$ will increase by three times that amount. We write:

$$\text{Change in } y = 3 \times \text{Change in } x.$$

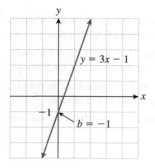

$y$-intercept $= b = -1$
*Graphically, b is the y-intercept of the graph.*

**Figure 17**

## The Change in a Quantity: Delta Notation

If a quantity $q$ changes from $q_1$ to $q_2$, the **change in $q$** is just the difference:

$$\text{Change in } q = \text{Second value} - \text{First value}$$
$$= q_2 - q_1.$$

Mathematicians traditionally use $\Delta$ (delta, the Greek equivalent of the Roman letter $D$) to stand for change, and write the change in $q$ as $\Delta q$.

$$\Delta q = \text{Change in } q = q_2 - q_1$$

### Quick Examples

1. If $x$ is changed from 1 to 3, we write

$$\Delta x = \text{Second value} - \text{First value} = 3 - 1 = 2.$$

2. Looking at our linear function, we see that, when $x$ changes from 1 to 3, $y$ changes from 2 to 8. So,

$$\Delta y = \text{Second value} - \text{First value} = 8 - 2 = 6.$$

Slope $= m = 3$
*Graphically, m is the slope of the graph.*

**Figure 18**

Using delta notation, we can now write, for our linear function,

$$\Delta y = 3\Delta x \qquad \text{Change in } y = 3 \times \text{Change in } x.$$

or

$$\frac{\Delta y}{\Delta x} = 3.$$

Because the value of $y$ increases by exactly 3 units for every increase of 1 unit in $x$, the graph is a straight line rising by 3 units for every 1 unit we go to the right. We say that we have a **rise** of 3 units for each **run** of 1 unit. Because the value of $y$ changes by $\Delta y = 3\Delta x$ units for every change of $\Delta x$ units in $x$, in general we have a rise of $\Delta y = 3\Delta x$ units for each run of $\Delta x$ units (Figure 18). Thus, we have a rise of 6 for a run of 2, a rise of 9 for a run of 3, and so on. So, $m = 3$ is a measure of the steepness of the line; we call $m$ the **slope of the line:**

$$\text{Slope} = m = \frac{\Delta y}{\Delta x} = \frac{\text{Rise}}{\text{Run}}.$$

In general (replace the number 3 by a general number $m$), we can say the following.

### The Roles of *m* and *b* in the Linear Function $f(x) = mx + b$

#### Role of *m*

**Numerically** If $y = mx + b$, then $y$ changes by $m$ units for every 1-unit change in $x$. A change of $\Delta x$ units in $x$ results in a change of $\Delta y = m\Delta x$ units in $y$. Thus,

$$m = \frac{\Delta y}{\Delta x} = \frac{\text{Change in } y}{\text{Change in } x}.$$

**Graphically** $m$ is the slope of the line $y = mx + b$:

$$m = \frac{\Delta y}{\Delta x} = \frac{\text{Rise}}{\text{Run}} = \text{Slope}.$$

For positive $m$, the graph rises $m$ units for every 1-unit move to the right, and rises $\Delta y = m \Delta x$ units for every $\Delta x$ units moved to the right. For negative $m$, the graph drops $|m|$ units for every 1-unit move to the right, and drops $|m|\Delta x$ units for every $\Delta x$ units moved to the right.

Graph of $y = mx + b$

### Role of *b*

**Numerically** When $x = 0$, $y = b$.

**Graphically** $b$ is the $y$-intercept of the line $y = mx + b$.

**Quick Examples**

1. $f(x) = 2x + 1$ has slope $m = 2$ and $y$-intercept $b = 1$. To sketch the graph, we start at the $y$-intercept $b = 1$ on the $y$-axis, and then move 1 unit to the right and up $m = 2$ units to arrive at a second point on the graph. Now connect the two points to obtain the graph on the left.

2. The line $y = -1.5x + 3.5$ has slope $m = -1.5$ and $y$-intercept $b = 3.5$. Because the slope is negative, the graph (above right) goes *down* 1.5 units for every 1 unit it moves to the right.

It helps to be able to picture what different slopes look like, as in Figure 19. Notice that the larger the absolute value of the slope, the steeper is the line.

Slope $\frac{1}{2}$

Slope $-\frac{1}{2}$

Slope 1

Slope $-1$

Slope 2

Slope $-2$

Slope 3

Slope $-3$

**Figure 19**

## using Technology

See the Technology Guides at the end of the chapter for detailed instructions on how to obtain a table with the successive quotients $m = \Delta y / \Delta x$ for the functions $f$ and $g$ in Example 1 using a TI-83/84 Plus or Excel. These tables show at a glance that $f$ is not linear. Here is an outline:

**TI-83/84 Plus**
[STAT] EDIT; Enter values of $x$ and $f(x)$ in lists $L_1$ and $L_2$.
Highlight the heading $L_3$ then enter the following formula (including the quotes)
`"ΔList(L₂)/ΔList(L₁)"`
[More details on page 116.]

**Spreadsheet**
Enter headings $x, f(x), Df/Dx$ in cells A1–C1, and the corresponding values from one of the tables in cells A2–B8. Enter
`=(B3-B2)/(A3-A2)`
in cell C2, and copy down through C8.
[More details on page 122.]

## EXAMPLE 1  Recognizing Linear Data Numerically and Graphically

Which of the following two tables gives the values of a linear function? What is the formula for that function?

| $x$ | 0 | 2 | 4 | 6 | 8 | 10 | 12 |
|---|---|---|---|---|---|---|---|
| $f(x)$ | 3 | −1 | −3 | −6 | −8 | −13 | −15 |

| $x$ | 0 | 2 | 4 | 6 | 8 | 10 | 12 |
|---|---|---|---|---|---|---|---|
| $g(x)$ | 3 | −1 | −5 | −9 | −13 | −17 | −21 |

**Solution**  The function $f$ cannot be linear: If it were, we would have $\Delta f = m \Delta x$ for some fixed number $m$. However, although the change in $x$ between successive entries in the table is $\Delta x = 2$ each time, the change in $f$ is not the same each time. Thus, the ratio $\Delta f / \Delta x$ is not the same for every successive pair of points.

On the other hand, the ratio $\Delta g / \Delta x$ is the same each time, namely,

$$\frac{\Delta g}{\Delta x} = \frac{-4}{2} = -2$$

as we see in the following table:

| $\Delta x$ | | $2-0=2$ | $4-2=2$ | $6-4=2$ | $8-6=2$ | $10-8=2$ | $12-10=2$ |
|---|---|---|---|---|---|---|---|
| $x$ | 0 | 2 | 4 | 6 | 8 | 10 | 12 |
| $g(x)$ | 3 | −1 | −5 | −9 | −13 | −17 | −21 |
| $\Delta g$ | | $-1-3$ $=-4$ | $-5-(-1)$ $=-4$ | $-9-(-5)$ $=-4$ | $-13-(-9)$ $=-4$ | $-17-(-13)$ $=-4$ | $-21-(-17)$ $=-4$ |

Thus, $g$ is linear with slope $m = -2$. By the table, $g(0) = 3$, hence $b = 3$. Thus,

$$g(x) = -2x + 3. \qquad \text{Check that this formula gives the values in the table.}$$

If you graph the points in the tables defining $f$ and $g$ above, it becomes easy to see that $g$ is linear and $f$ is not; the points of $g$ lie on a straight line (with slope $-2$), whereas the points of $f$ do not lie on a straight line (Figure 20).

**Figure 20**

## Finding a Linear Equation from Data

If we happen to know the slope and $y$-intercept of a line, writing down its equation is straightforward. For example, if we know that the slope is 3 and the $y$-intercept is $-1$, then the equation is $y = 3x - 1$. Sadly, the information we are given is seldom so convenient. For instance, we may know the slope and a point other than the $y$-intercept, two points on the line, or other information. We therefore need to know how to use the information we are given to obtain the slope and the intercept.

## Computing the Slope

We can always determine the slope of a line if we are given two (or more) points on the line, because any two points—say $(x_1, y_1)$ and $(x_2, y_2)$—determine the line, and hence its slope. To compute the slope when given two points, recall the formula

$$\text{Slope} = m = \frac{\text{Rise}}{\text{Run}} = \frac{\Delta y}{\Delta x}.$$

**Figure 21**

To find its slope, we need a run $\Delta x$ and corresponding rise $\Delta y$. In Figure 21, we see that we can use $\Delta x = x_2 - x_1$, the change in the $x$-coordinate from the first point to the second, as our run, and $\Delta y = y_2 - y_1$, the change in the $y$-coordinate, as our rise. The resulting formula for computing the slope is given in the box.

## Computing the Slope of a Line

We can compute the slope $m$ of the line through the points $(x_1, y_1)$ and $(x_2, y_2)$ using

$$m = \frac{\Delta y}{\Delta x} = \frac{y_2 - y_1}{x_2 - x_1}.$$

### Quick Examples

1. The slope of the line through $(x_1, y_1) = (1, 3)$ and $(x_2, y_2) = (5, 11)$ is

$$m = \frac{\Delta y}{\Delta x} = \frac{y_2 - y_1}{x_2 - x_1} = \frac{11 - 3}{5 - 1} = \frac{8}{4} = 2.$$

Notice that we can use the points in the reverse order: If we take $(x_1, y_1) = (5, 11)$ and $(x_2, y_2) = (1, 3)$, we obtain the same answer:

$$m = \frac{\Delta y}{\Delta x} = \frac{y_2 - y_1}{x_2 - x_1} = \frac{3 - 11}{1 - 5} = \frac{-8}{-4} = 2.$$

2. The slope of the line through $(x_1, y_1) = (1, 2)$ and $(x_2, y_2) = (2, 1)$ is

$$m = \frac{\Delta y}{\Delta x} = \frac{y_2 - y_1}{x_2 - x_1} = \frac{1 - 2}{2 - 1} = \frac{-1}{1} = -1.$$

3. The slope of the line through $(2, 3)$ and $(-1, 3)$ is

$$m = \frac{\Delta y}{\Delta x} = \frac{y_2 - y_1}{x_2 - x_1} = \frac{3 - 3}{-1 - 2} = \frac{0}{-3} = 0.$$

A line of slope 0 has zero rise, so is a *horizontal* line, as shown in Figure 22.

4. The line through $(3, 2)$ and $(3, -1)$ has slope

$$m = \frac{\Delta y}{\Delta x} = \frac{y_2 - y_1}{x_2 - x_1} = \frac{-1 - 2}{3 - 3} = \frac{-3}{0},$$

which is undefined. The line passing through these points is *vertical*, as shown in Figure 23.

**Figure 22**

*Vertical lines have undefined slope.*

**Figure 23**

## Computing the y-Intercept

Once we know the slope $m$ of a line, and also the coordinates of a point $(x_1, y_1)$, then we can calculate its $y$-intercept $b$ as follows: The equation of the line must be

$$y = mx + b,$$

where $b$ is as yet unknown. To determine $b$ we use the fact that the line must pass through the point $(x_1, y_1)$, so that $(x_1, y_1)$ satisfies the equation $y = mx + b$. In other words,

$$y_1 = mx_1 + b.$$

Solving for $b$ gives

$$b = y_1 - mx_1.$$

In summary:

---

### Computing the y-Intercept of a Line

The $y$-intercept of the line passing through $(x_1, y_1)$ with slope $m$ is

$$b = y_1 - mx_1.$$

**Quick Example**

The line through $(2, 3)$ with slope 4 has

$$b = y_1 - mx_1 = 3 - (4)(2) = -5.$$

Its equation is therefore

$$y = mx + b = 4x - 5.$$

---

### EXAMPLE 2   Finding Linear Equations

Find equations for the following straight lines.

**a.** Through the points $(1, 2)$ and $(3, -1)$
**b.** Through $(2, -2)$ and parallel to the line $3x + 4y = 5$
**c.** Horizontal and through $(-9, 5)$
**d.** Vertical and through $(-9, 5)$

### Solution

**a.** To write down the equation of the line, we need the slope $m$ and the $y$-intercept $b$.

- **Slope**  Because we are given two points on the line, we can use the slope formula:

$$m = \frac{y_2 - y_1}{x_2 - x_1} = \frac{-1 - 2}{3 - 1} = -\frac{3}{2}.$$

**using** Technology

See the Technology Guides at the end of the chapter for detailed instructions on how to obtain the slope and intercept in Example 2(a) using a TI-83/84 Plus or a spreadsheet. Here is an outline:

**TI-83/84 Plus**

STAT EDIT; Enter values of $x$ and $y$ in lists $L_1$ and $L_2$.
Slope: Home screen
$(L_2(2)-L_2(1))/(L_1(2)-L_1(1))\to M$
Intercept: Home screen
$L_2(1)-M*L_1(1)$
[More details on page 117.]

**Spreadsheet**

Enter headings $x$, $y$, $m$, $b$, in cells A1–D1, and the values $(x, y)$ in cells A2–B3. Enter
$=(B3-B2)/(A3-A2)$
in cell C2, and
$=B2-C2*A2$
in cell D2.
[More details on page 122.]

• **Intercept** We now have the slope of the line, $m = -3/2$, and also a point—we have two to choose from, so let us choose $(x_1, y_1) = (1, 2)$. We can now use the formula for the $y$-intercept:

$$b = y_1 - mx_1 = 2 - \left(-\frac{3}{2}\right)(1) = \frac{7}{2}.$$

Thus, the equation of the line is

$$y = -\frac{3}{2}x + \frac{7}{2}. \qquad y = mx + b$$

**b.** Proceeding as before,

• **Slope** We are not given two points on the line, but we are given a parallel line. We use the fact that *parallel lines have the same slope.* (Why?) We can find the slope of $3x + 4y = 5$ by solving for $y$ and then looking at the coefficient of $x$:

$$y = -\frac{3}{4}x + \frac{5}{4} \qquad \text{To find the slope, solve for } y.$$

so the slope is $-3/4$.

• **Intercept** We now have the slope of the line, $m = -3/4$, and also a point $(x_1, y_1) = (2, -2)$. We can now use the formula for the $y$-intercept:

$$b = y_1 - mx_1 = -2 - \left(-\frac{3}{4}\right)(2) = -\frac{1}{2}.$$

Thus, the equation of the line is

$$y = -\frac{3}{4}x - \frac{1}{2}. \qquad y = mx + b$$

**c.** We are given a point: $(-9, 5)$. Furthermore, we are told that the line is horizontal, which tells us that the slope is $m = 0$. Therefore, all that remains is the calculation of the $y$-intercept:

$$b = y_1 - mx_1 = 5 - (0)(-9) = 5$$

so the equation of the line is

$$y = 5. \qquad y = mx + b$$

**d.** We are given a point: $(-9, 5)$. This time, we are told that the line is vertical, which means that the slope is undefined. Thus, we can't express the equation of the line in the form $y = mx + b$. (This formula makes sense only when the slope $m$ of the line is defined.) What can we do? Well, here are some points on the desired line:

$$(-9, 1), (-9, 2), (-9, 3), \ldots$$

so $x = -9$ and $y = $ *anything*. If we simply say that $x = -9$, then these points are all solutions, so the equation is $x = -9$.

## Applications: Linear Models

Using linear functions to describe or approximate relationships in the real world is called **linear modeling**.

**Figure 24**

**Figure 25**

Recall from Section 1.2 that a **cost function** specifies the cost $C$ as a function of the number of items $x$.

## EXAMPLE 3  Linear Cost Function from Data

The manager of the FrozenAir Refrigerator factory notices that on Monday it cost the company a total of $25,000 to build 30 refrigerators and on Tuesday it cost $30,000 to build 40 refrigerators. Find a linear cost function based on this information. What is the daily fixed cost, and what is the marginal cost?

**Solution**  We are seeking the cost $C$ as a linear function of $x$, the number of refrigerators sold:

$$C = mx + b.$$

We are told that $C = 25{,}000$ when $x = 30$, and this amounts to being told that $(30, 25{,}000)$ is a point on the graph of the cost function. Similarly, $(40, 30{,}000)$ is another point on the line (Figure 24).

We can use the two points on the line to construct the linear cost equation:

- **Slope**   $m = \dfrac{C_2 - C_1}{x_2 - x_1} = \dfrac{30{,}000 - 25{,}000}{40 - 30} = 500$   $C$ plays the role of $y$.

- **Intercept** $b = C_1 - mx_1 = 25{,}000 - (500)(30) = 10{,}000$.  We used the point $(x_1, C_1) = (30, 25{,}000)$.

The linear cost function is therefore

$$C(x) = 500x + 10{,}000.$$

Because $m = 500$ and $b = 10{,}000$ the factory's fixed cost is $10,000 each day, and its marginal cost is $500 per refrigerator. (See page 59 in Section 1.2.) These are illustrated in Figure 25.

➡ **Before we go on...**   Recall that, in general, the slope $m$ measures the number of units of change in $y$ per 1-unit change in $x$, so it is measured in units of $y$ per unit of $x$:

Units of Slope = Units of $y$ per unit of $x$.

In Example 3, $y$ is the cost $C$, measured in dollars, and $x$ is the number of items, measured in refrigerators. Hence,

Units of Slope = Units of $y$ per Unit of $x$ = Dollars per refrigerator.

The $y$-intercept $b$, being a value of $y$, is measured in the same units as $y$. In Example 3, $b$ is measured in dollars.  ■

In Section 1.2 we saw that a **demand function** specifies the demand $q$ as a function of the price $p$ per item.

## **using** Technology

To obtain the cost equation for Example 3 with technology, apply the Technology note for Example 2(a) to the given points (30, 25,000) and (40, 30,000) on the graph of the cost equation.

## EXAMPLE 4  Linear Demand Function from Data

You run a small supermarket and must determine how much to charge for Hot'n'Spicy brand baked beans. The following chart shows weekly sales figures (the demand) for Hot'n'Spicy at two different prices.

| | Price/Can | $0.50 | $0.75 |
|---|---|---|---|
| | Demand (cans sold/week) | 400 | 350 |

**a.** Model these data with a linear demand function. (See Example 4 in Section 1.2.)

**b.** How do we interpret the slope and $q$-intercept of the demand function?

### Solution

**a.** Recall that a demand equation—or demand function—expresses demand $q$ (in this case, the number of cans of beans sold per week) as a function of the unit price $p$ (in this case, price per can). We model the demand using the two points we are given: $(0.50, 400)$ and $(0.75, 350)$.

*Slope:*    $m = \dfrac{q_2 - q_1}{p_2 - p_1} = \dfrac{350 - 400}{0.75 - 0.50} = \dfrac{-50}{0.25} = -200$

*Intercept:*  $b = q_1 - mp_1 = 400 - (-200)(0.50) = 500$

So, the demand equation is

$$q = -200p + 500. \qquad q = mp + b$$

**b.** The key to interpreting the slope in a demand equation is to recall (see the "Before we go on" note at the end of Example 3) that we measure the slope in *units of y per unit of x*. Here, $m = -200$, and the units of $m$ are units of $q$ per unit of $p$, or the number of cans sold per dollar change in the price. Because $m$ is negative, we see that the number of cans sold decreases as the price increases. We conclude that the weekly sales will drop by 200 cans per \$1 increase in the price.

To interpret the $q$-intercept, recall that it gives the $q$-coordinate when $p = 0$. Hence it is the number of cans the supermarket can "sell" every week if it were to give them away.*

### Before we go on...

**Q** : *Just how reliable is the linear model used in Example 4?*

**A** : The *actual* demand graph could in principle be obtained by tabulating demand figures for a large number of different prices. If the resulting points were plotted on the $pq$ plane, they would probably suggest a curve and not a straight line. However, if you looked at a small enough portion of any curve, you could closely *approximate* it by a straight line. In other words, *over a small range of values of p, a linear model is accurate.* Linear models of real-world situations are generally reliable only for small ranges of the variables. (This point will come up again in some of the exercises.)

■

The next example illustrates modeling change over time $t$ with a linear function of $t$.

### EXAMPLE 5  Modeling Change Over Time: Growth of Sales

The worldwide market for portable navigation devices was expected to grow from 50 million units in 2007 to around 530 million units in 2015.[39]

---

[39]Sales were expected to grow to more than 500 million in 2015 according to a January 2008 press release by Telematics Research Group. Source: www.telematicsresearch.com.

---

**using** Technology

To obtain the demand equation for Example 4 with technology, apply the Technology note for Example 2(a) to the given points (0.50, 400) and (0.75, 350) on the graph of the demand equation.

---

\* Does this seem realistic? Demand is not always unlimited if items were given away. For instance, campus newspapers are sometimes given away, and yet piles of them are often left untaken. Also see the "Before we go on" discussion at the end of this example.

**a.** Use this information to model annual worldwide sales of portable navigation devices as a linear function of time $t$ in years since 2007. What is the significance of the slope?

**b.** Use the model to predict when annual sales of mobile navigation devices will reach 440 million units.

### Solution

**a.** Since we are interested in worldwide sales $s$ of portable navigation devices as a function of time, we take time $t$ to be the independent coordinate (playing the role of $x$) and the annual sales $s$, in million of units, to be the dependent coordinate (in the role of $y$). Notice that 2007 corresponds to $t = 0$ and 2015 corresponds to $t = 8$, so we are given the coordinates of two points on the graph of sales $s$ as a function of time $t$: $(0, 50)$ and $(8, 530)$. We model the sales using these two points:

$$m = \frac{s_2 - s_1}{t_2 - t_1} = \frac{530 - 50}{8 - 0} = \frac{480}{8} = 60$$

$$b = s_1 - mt_1 = 50 - (60)(0) = 50$$

So,     $s = 60t + 50$ million units.      $s = mt + b$

The slope $m$ is measured in units of $s$ per unit of $t$; that is, millions of devices per year, and is thus the *rate of change of annual sales*. To say that $m = 60$ is to say that annual sales are increasing at a rate of 60 million devices per year.

**b.** Our model of annual sales as a function of time is

$$s = 60t + 50 \text{ million units.}$$

Annual sales of mobile portable devices will reach 440 million when $s = 440$, or

$$440 = 60t + 50$$

Solving for $t$,      $60t = 440 - 50 = 390$

$$t = \frac{390}{60} = 6.5 \text{ years},$$

which is midway through 2013. Thus annual sales are expected to reach 440 million midway through 2013.

**using Technology**

To use technology to obtain $s$ as a function of $t$ in Example 5, apply the Technology note for Example 2(a) to the points $(0, 50)$ and $(8, 530)$ on its graph.

### EXAMPLE 6 Velocity

You are driving down the Ohio Turnpike, watching the mileage markers to stay awake. Measuring time in hours after you see the 20-mile marker, you see the following markers each half hour:

| Time (h)    | 0  | 0.5 | 1  | 1.5 | 2   |
|-------------|----|-----|----|-----|-----|
| Marker (mi) | 20 | 47  | 74 | 101 | 128 |

Find your location $s$ as a function of $t$, the number of hours you have been driving. (The number $s$ is also called your **position** or **displacement**.)

### Solution

**Figure 26**

If we plot the location $s$ versus the time $t$, the five markers listed give us the graph in Figure 26. These points appear to lie along a straight line. We can verify this by calculating how far you traveled in each half hour. In the first half hour, you traveled $47 - 20 = 27$ miles. In the second half hour you traveled $74 - 47 = 27$ miles also. In fact, you traveled exactly 27 miles each half hour. The points we plotted lie on a straight line that rises 27 units for every 0.5 unit we go to the right, for a slope of $27/0.5 = 54$.

To get the equation of that line, notice that we have the $s$-intercept, which is the starting marker of 20. Thus, the equation of $s$ as a function of time $t$ is

$$s(t) = 54t + 20.$$    We used $s$ in place of $y$ and $t$ in place of $x$.

**using Technology**

To use technology to obtain $s$ as a function of $t$ in Example 6, apply the Technology note for Example 2(a) to the points $(0, 20)$ and $(1, 74)$ on its graph.

Notice the significance of the slope: For every hour you travel, you drive a distance of 54 miles. In other words, you are traveling at a constant velocity of 54 mph. We have uncovered a very important principle:

*In the graph of displacement versus time, velocity is given by the slope.*

---

## Linear Change over Time

If a quantity $q$ is a linear function of time $t$,

$$q = mt + b,$$

then the slope $m$ measures the **rate of change** of $q$, and $b$ is the quantity at time $t = 0$, the **initial quantity**. If $q$ represents the position of a moving object, then the rate of change is also called the **velocity**.

### Units of $m$ and $b$

The units of measurement of $m$ are units of $q$ per unit of time; for instance, if $q$ is income in dollars and $t$ is time in years, then the rate of change $m$ is measured in dollars per year.

The units of $b$ are units of $q$; for instance, if $q$ is income in dollars and $t$ is time in years, then $b$ is measured in dollars.

### Quick Example

If the accumulated revenue from sales of your video game software is given by $R = 2,000t + 500$ dollars, where $t$ is time in years from now, then you have earned \$500 in revenue so far, and the accumulated revenue is increasing at a rate of \$2,000 per year.

Examples 3–6 share the following common theme.

---

## General Linear Models

If $y = mx + b$ is a linear model of changing quantities $x$ and $y$, then the slope $m$ is the rate at which $y$ is increasing per unit increase in $x$, and the $y$-intercept $b$ is the value of $y$ that corresponds to $x = 0$.

### Units of $m$ and $b$

The slope $m$ is measured in units of $y$ per unit of $x$, and the intercept $b$ is measured in units of $y$.

**Quick Example**

If the number $n$ of spectators at a soccer game is related to the number $g$ of goals your team has scored so far by the equation $n = 20g + 4$, then you can expect 4 spectators if no goals have been scored and 20 additional spectators per additional goal scored.

**FAQs**

**What to Use as $x$ and $y$, and How to Interpret a Linear Model**

**Q**: *In a problem where I must find a linear relationship between two quantities, which quantity do I use as $x$ and which do I use as $y$?*

**A**: The key is to decide which of the two quantities is the independent variable, and which is the dependent variable. Then use the independent variable as $x$ and the dependent variable as $y$. In other words, $y$ depends on $x$.

Here are examples of phrases that convey this information, usually of the form *Find $y$ [dependent variable] in terms of $x$ [independent variable]:*

- Find the cost in terms of the number of items.    $y = $ Cost, $x = $ # Items
- How does color depend on wavelength?    $y = $ Color, $x = $ Wavelength

If no information is conveyed about which variable is intended to be independent, then you can use whichever is convenient.

**Q**: *How do I interpret a general linear model $y = mx + b$?*

**A**: The key to interpreting a linear model is to remember the units we use to measure $m$ and $b$:

*The slope $m$ is measured in units of $y$ per unit of $x$; the intercept $b$ is measured in units of $y$.*

For instance, if $y = 4.3x + 8.1$ and you know that $x$ is measured in feet and $y$ in kilograms, then you can already say, "$y$ is 8.1 kilograms when $x = 0$ feet, and increases at a rate of 4.3 kilograms per foot" without knowing anything more about the situation!

# 1.3  EXERCISES

▼ more advanced    ◆ challenging
**T** indicates exercises that should be solved using technology

*In Exercises 1–6, a table of values for a linear function is given. Fill in the missing value and calculate $m$ in each case.*

**1.**

| $x$ | −1 | 0 | 1 |
|-----|-----|-----|-----|
| $y$ | 5 | 8 | |

**2.**

| $x$ | −1 | 0 | 1 |
|-----|-----|-----|-----|
| $y$ | −1 | −3 | |

**3.**

| $x$ | 2 | 3 | 5 |
|-----|-----|-----|-----|
| $y$ | −1 | −2 | |

**4.**

| $x$ | 2 | 4 | 5 |
|-----|-----|-----|-----|
| $y$ | −1 | −2 | |

**5.**

| x | −2 | 0 | 2 |
|---|----|---|---|
| y | 4 |   | 10 |

**6.**

| x | 0 | 3 | 6 |
|---|---|---|---|
| y | −1 |   | −5 |

*In Exercises 7–10, first find f(0), if not supplied, and then find the equation of the given linear function.*

**7.**

| x | −2 | 0 | 2 | 4 |
|---|----|---|---|---|
| f(x) | −1 | −2 | −3 | −4 |

**8.**

| x | −6 | −3 | 0 | 3 |
|---|----|----|---|---|
| f(x) | 1 | 2 | 3 | 4 |

**9.**

| x | −4 | −3 | −2 | −1 |
|---|----|----|----|----|
| f(x) | −1 | −2 | −3 | −4 |

**10.**

| x | 1 | 2 | 3 | 4 |
|---|---|---|---|---|
| f(x) | 4 | 6 | 8 | 10 |

*In each of Exercises 11–14, decide which of the two given functions is linear and find its equation.* HINT **[See Example 1.]**

**11.**

| x | 0 | 1 | 2 | 3 | 4 |
|---|---|---|---|---|---|
| f(x) | 6 | 10 | 14 | 18 | 22 |
| g(x) | 8 | 10 | 12 | 16 | 22 |

**12.**

| x | −10 | 0 | 10 | 20 | 30 |
|---|-----|---|----|----|----|
| f(x) | −1.5 | 0 | 1.5 | 2.5 | 3.5 |
| g(x) | −9 | −4 | 1 | 6 | 11 |

**13.**

| x | 0 | 3 | 6 | 10 | 15 |
|---|---|---|---|----|----|
| f(x) | 0 | 3 | 5 | 7 | 9 |
| g(x) | −1 | 5 | 11 | 19 | 29 |

**14.**

| x | 0 | 3 | 5 | 6 | 9 |
|---|---|---|---|---|---|
| f(x) | 2 | 6 | 9 | 12 | 15 |
| g(x) | −1 | 8 | 14 | 17 | 26 |

*In Exercises 15–24, find the slope of the given line, if it is defined.*

**15.** $y = -\frac{3}{2}x - 4$          **16.** $y = \frac{2x}{3} + 4$

**17.** $y = \frac{x+1}{6}$          **18.** $y = -\frac{2x-1}{3}$

**19.** $3x + 1 = 0$          **20.** $8x - 2y = 1$

**21.** $3y + 1 = 0$          **22.** $2x + 3 = 0$

**23.** $4x + 3y = 7$          **24.** $2y + 3 = 0$

*In Exercises 25–38, graph the given equation.* HINT **[See Quick Examples on page 80.]**

**25.** $y = 2x - 1$          **26.** $y = x - 3$

**27.** $y = -\frac{2}{3}x + 2$          **28.** $y = -\frac{1}{2}x + 3$

**29.** $y + \frac{1}{4}x = -4$          **30.** $y - \frac{1}{4}x = -2$

**31.** $7x - 2y = 7$          **32.** $2x - 3y = 1$

**33.** $3x = 8$          **34.** $2x = -7$

**35.** $6y = 9$          **36.** $3y = 4$

**37.** $2x = 3y$          **38.** $3x = -2y$

*In Exercises 39–58, calculate the slope, if defined, of the straight line through the given pair of points. Try to do as many as you can without writing anything down except the answer.* HINT **[See Quick Examples on page 82.]**

**39.** $(0, 0)$ and $(1, 2)$          **40.** $(0, 0)$ and $(-1, 2)$

**41.** $(-1, -2)$ and $(0, 0)$          **42.** $(2, 1)$ and $(0, 0)$

**43.** $(4, 3)$ and $(5, 1)$          **44.** $(4, 3)$ and $(4, 1)$

**45.** $(1, -1)$ and $(1, -2)$          **46.** $(-2, 2)$ and $(-1, -1)$

**47.** $(2, 3.5)$ and $(4, 6.5)$          **48.** $(10, -3.5)$ and $(0, -1.5)$

**49.** $(300, 20.2)$ and $(400, 11.2)$

**50.** $(1, -20.2)$ and $(2, 3.2)$

**51.** $(0, 1)$ and $\left(-\frac{1}{2}, \frac{3}{4}\right)$

**52.** $\left(\frac{1}{2}, 1\right)$ and $\left(-\frac{1}{2}, \frac{3}{4}\right)$

**53.** $(a, b)$ and $(c, d)$ $(a \neq c)$

**54.** $(a, b)$ and $(c, b)$ $(a \neq c)$

**55.** $(a, b)$ and $(a, d)$ $(b \neq d)$

**56.** $(a, b)$ and $(-a, -b)$ $(a \neq 0)$

**57.** $(-a, b)$ and $(a, -b)$ $(a \neq 0)$

**58.** $(a, b)$ and $(b, a)$ $(a \neq b)$

**59.** In the following figure, estimate the slopes of all line segments.

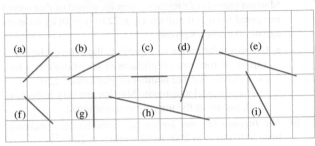

**60.** In the following figure, estimate the slopes of all line segments.

*In Exercises 61–78, find a linear equation whose graph is the straight line with the given properties.* HINT [See Example 2.]

**61.** Through $(1, 3)$ with slope 3

**62.** Through $(2, 1)$ with slope 2

**63.** Through $(1, -\frac{3}{4})$ with slope $\frac{1}{4}$

**64.** Through $(0, -\frac{1}{3})$ with slope $\frac{1}{3}$

**65.** Through $(20, -3.5)$ and increasing at a rate of 10 units of $y$ per unit of $x$

**66.** Through $(3.5, -10)$ and increasing at a rate of 1 unit of $y$ per 2 units of $x$

**67.** Through $(2, -4)$ and $(1, 1)$

**68.** Through $(1, -4)$ and $(-1, -1)$

**69.** Through $(1, -0.75)$ and $(0.5, 0.75)$

**70.** Through $(0.5, -0.75)$ and $(1, -3.75)$

**71.** Through $(6, 6)$ and parallel to the line $x + y = 4$

**72.** Through $(1/3, -1)$ and parallel to the line $3x - 4y = 8$

**73.** Through $(0.5, 5)$ and parallel to the line $4x - 2y = 11$

**74.** Through $(1/3, 0)$ and parallel to the line $6x - 2y = 11$

**75.** ▼ Through $(0, 0)$ and $(p, q)$ $(p \neq 0)$

**76.** ▼ Through $(p, q)$ parallel to $y = rx + s$

**77.** ▼ Through $(p, q)$ and $(r, q)$ $(p \neq r)$

**78.** ▼ Through $(p, q)$ and $(r, s)$ $(p \neq r)$

Unless otherwise noted, all content on this page is © Cengage Learning.

## APPLICATIONS

**79.** *Cost* The RideEm Bicycles factory can produce 100 bicycles in a day at a total cost of $10,500 and it can produce 120 bicycles in a day at a total cost of $11,000. What are the company's daily fixed costs, and what is the marginal cost per bicycle? HINT [See Example 3.]

**80.** *Cost* A soft-drink manufacturer can produce 1,000 cases of soda in a week at a total cost of $6,000, and 1,500 cases of soda at a total cost of $8,500. Find the manufacturer's weekly fixed costs and marginal cost per case of soda.

**81.** *Cost: iPhones* If it costs Apple $1,900 to manufacture 10 iPhone 4s per hour and $3,780 to manufacture 20 per hour at a particular plant,[40] obtain the corresponding linear cost function. What was the cost to manufacture each additional iPhone? Use the cost function to estimate the cost of manufacturing 40 iPhones in an hour.

**82.** *Cost: Kinects* If it costs Microsoft $1,230 to manufacture 8 Kinects per hour for the Xbox 360 and $2,430 to manufacture 16 per hour at a particular plant,[41] obtain the corresponding linear cost function. What was the cost to manufacture each additional Kinect? Use the cost function to estimate the cost of manufacturing 30 Kinects in an hour.

**83.** *Demand* Sales figures show that your company sold 1,960 pen sets each week when they were priced at $1/pen set and 1,800 pen sets each week when they were priced at $5/pen set. What is the linear demand function for your pen sets? HINT [See Example 4.]

**84.** *Demand* A large department store is prepared to buy 3,950 of your tie-dye shower curtains per month for $5 each, but only 3,700 shower curtains per month for $10 each. What is the linear demand function for your tie-dye shower curtains?

**85.** *Demand for Cell Phones* The following table shows worldwide sales of cell phones by Nokia and their average wholesale prices in the first quarters of 2009 and 2010:[42]

| Quarter | Q1 2009 | Q1 2010 |
|---|---|---|
| **Wholesale Price ($)** | 85 | 81 |
| **Sales (millions)** | 103 | 111 |

**a.** Use the data to obtain a linear demand function for (Nokia) cell phones, and use your demand equation to predict sales if Nokia lowered the price further to $75.
**b.** Fill in the blanks: For every ____ increase in price, sales of cell phones decrease by ___ units.

---

[40]Based on marginal cost data at www.isuppli.com.

[41]Based on marginal cost data provided by a "highly-positioned, trusted source." www.develop-online.net.

[42]Data are approximate. Source: www.nokia.com.

**86.** *Demand for Smart Phones* The following table shows worldwide sales of smart phones and their average whole-sale prices in 2009 and 2010:[43]

| Quarter | Q1 2009 | Q1 2010 |
|---|---|---|
| Wholesale Price ($) | 162 | 152 |
| Sales (millions) | 174 | 213 |

**a.** Use the data to obtain a linear demand function for smart phones, and use your demand equation to predict sales to the nearest million phones if the price was set at $155.

**b.** Fill in the blanks: For every ____ increase in price, sales of smart phones decrease by ____ units.

**87.** *Demand for Monorail Service, Las Vegas* In 2005, the Las Vegas monorail charged $3 per ride and had an average ridership of about 28,000 per day. In December 2005 the Las Vegas Monorail Company raised the fare to $5 per ride, and average ridership in 2006 plunged to around 19,000 per day.[44]

**a.** Use the given information to find a linear demand equation.

**b.** Give the units of measurement and interpretation of the slope.

**c.** What would have been the effect on ridership of raising the fare to $6 per ride?

**88.** *Demand for Monorail Service, Mars* The Utarek monorail, which links the three urbynes (or districts) of Utarek, Mars, charged $\overline{\overline{Z}}5$ per ride[45] and sold about 14 million rides per day. When the Utarek City Council lowered the fare to $\overline{\overline{Z}}3$ per ride, the number of rides increased to 18 million per day.

**a.** Use the given information to find a linear demand equation.

**b.** Give the units of measurement and interpretation of the slope.

**c.** What would have been the effect on ridership of raising the fare to $\overline{\overline{Z}}10$ per ride?

**89.** *Pasta Imports in the 90s* During the period 1990–2001, U.S. imports of pasta increased from 290 million pounds in 1990 ($t = 0$) by an average of 40 million pounds/year.[46]

**a.** Use this information to express $y$, the annual U.S. imports of pasta (in millions of pounds), as a linear function of $t$, the number of years since 1990.

**b.** Use your model to estimate U.S. pasta imports in 2005, assuming the import trend continued.

**90.** *Mercury Imports in the 10s* During the period 2210–2220, Martian imports of mercury (from the planet of that name) increased from 550 million kg in 2210 ($t = 0$) by an average of 60 million kg/year.

**a.** Use this information to express $y$, the annual Martian imports of mercury (in millions of kilograms), as a linear function of $t$, the number of years since 2210.

**b.** Use your model to estimate Martian mercury imports in 2230, assuming the import trend continued.

**91.** *Net Income* The net income of Amazon.com increased from $10 billion in 2006 to $25 billion in 2009.[47]

**a.** Use this information to find a linear model for Amazon's net income $N$ (in billions of dollars) as a function of time $t$ in years since 2000.

**b.** Give the units of measurement and interpretation of the slope.

**c.** Use the model from part (a) to estimate the 2008 net income. (The actual 2008 net income was approximately $19 billion.)

**92.** *Operating Expenses* The operating expenses of Amazon.com increased from $2.0 billion in 2006 to $4.4 billion in 2009.[48]

**a.** Use this information to find a linear model for Amazon's operating expenses $E$ (in billions of dollars) as a function of time $t$ in years since 2000.

**b.** Give the units of measurement and interpretation of the slope.

**c.** Use the model from part (a) to estimate the 2007 operating expenses. (The actual 2007 operating expenses were approximately $2.5 billion.)

**93.** *Velocity* The position of a model train, in feet along a railroad track, is given by

$$s(t) = 2.5t + 10$$

after $t$ seconds.

**a.** How fast is the train moving?

**b.** Where is the train after 4 seconds?

**c.** When will it be 25 feet along the track?

**94.** *Velocity* The height of a falling sheet of paper, in feet from the ground, is given by

$$s(t) = -1.8t + 9$$

after $t$ seconds.

**a.** What is the velocity of the sheet of paper?

**b.** How high is it after 4 seconds?

**c.** When will it reach the ground?

**95.** ▼ *Fast Cars* A police car was traveling down Ocean Park-way in a high-speed chase from Jones Beach. It was at Jones Beach at exactly 10 pm ($t = 10$) and was at Oak Beach, 13 miles from Jones Beach, at exactly 10:06 pm.

**a.** How fast was the police car traveling? HINT [See Example 6.]

**b.** How far was the police car from Jones Beach at time $t$?

---

[43]Data are approximate. Sources: www.techcrunch.com, www.businessweek.com.

[44]Source: *New York Times*, February 10, 2007, p. A9.

[45]The zonar ($\overline{\overline{Z}}$) is the official currency in the city-state of Utarek, Mars (formerly www.Marsnext.com, a now extinct virtual society).

[46]Data are rounded. Sources: Department of Commerce/*New York Times*, September 5, 1995, p. D4; International Trade Administration, March 31, 2002, www.ita.doc.gov/.

[47]Recall that "net income" is another term for "profit." Source: www.wikinvest.com.

[48]*Ibid.*

**96.** ▼ *Fast Cars* The car that was being pursued by the police in Exercise 95 was at Jones Beach at exactly 9:54 pm ($t = 9.9$) and passed Oak Beach (13 miles from Jones Beach) at exactly 10:06 pm, where it was overtaken by the police.

**a.** How fast was the car traveling? HINT [See Example 6.]
**b.** How far was the car from Jones Beach at time $t$?

**97.** *Textbook Sizes* The second edition of *Applied Calculus* by Waner and Costenoble was 585 pages long. By the time we got to the fifth edition, the book had grown to 750 pages.

**a.** Use this information to obtain the page length $L$ as a linear function of the edition number $n$.
**b.** What are the units of measurement of the slope? What does the slope tell you about the length of *Applied Calculus*?
**c.** At this rate, by which edition will the book have grown to over 1,500 pages?

**98.** *Textbook Sizes* The second edition of *Finite Mathematics* by Waner and Costenoble was 603 pages long. By the time we got to the fifth edition, the book had grown to 690 pages.

**a.** Use this information to obtain the page length $L$ as a linear function of the edition number $n$.
**b.** What are the units of measurement of the slope? What does the slope tell you about the length of *Finite Mathematics*?
**c.** At this rate, by which edition will the book have grown to over 1,000 pages?

**99.** *Fahrenheit and Celsius* In the Fahrenheit temperature scale, water freezes at 32°F and boils at 212°F. In the Celsius scale, water freezes at 0°C and boils at 100°C. Further, the Fahrenheit temperature $F$ and the Celsius temperature $C$ are related by a linear equation. Find $F$ in terms of $C$. Use your equation to find the Fahrenheit temperatures corresponding to 30°C, 22°C, −10°C, and −14°C, to the nearest degree.

**100.** *Fahrenheit and Celsius* Use the information about Celsius and Fahrenheit given in Exercise 99 to obtain a linear equation for $C$ in terms of $F$, and use your equation to find the Celsius temperatures corresponding to 104°F, 77°F, 14°F, and −40°F, to the nearest degree.

*Airline Net Income Exercises 101 and 102 are based on the following table, which compares the net incomes, in millions of dollars, of American Airlines, Continental Airlines, and Southwest Airlines:*[49]

| Year | 2005 | 2006 | 2007 | 2008 | 2009 | 2010 |
|---|---|---|---|---|---|---|
| American Airlines | −850 | 250 | 450 | −2,100 | −1,450 | −300 |
| Continental Airlines | −50 | 350 | 450 | −600 | −300 | 150 |
| Southwest Airlines | 550 | 500 | 650 | 200 | 100 | 300 |

**101. a.** Use the 2005 and 2009 data to obtain American's net income $A$ as a linear function of Continental's net income $C$. (Use millions of dollars for all units.)

**b.** How far off is your model in estimating American's net income based on Continental's 2007 income?
**c.** What does the slope of the linear function from part (a) suggest about the net income of these two airlines?

**102. a.** Use the 2006 and 2010 data to obtain Continental's net income $C$ as a linear function of Southwest's net income $S$. (Use millions of dollars for all units.)
**b.** How far off is your model in estimating Continental's net income based on Southwest's 2008 income?
**c.** What does the slope of the linear function from part (a) suggest about the net income of these two airlines?

**103.** ▼ *Income* The well-known romance novelist Celestine A. Lafleur (a.k.a. Bertha Snodgrass) has decided to sell the screen rights to her latest book, *Henrietta's Heaving Heart*, to Boxoffice Success Productions for $50,000. In addition, the contract ensures Ms. Lafleur royalties of 5% of the net profits.[50] Express her income $I$ as a function of the net profit $N$, and determine the net profit necessary to bring her an income of $100,000. What is her marginal income (share of each dollar of net profit)?

**104.** ▼ *Income* Due to the enormous success of the movie *Henrietta's Heaving Heart* based on a novel by Celestine A. Lafleur (see Exercise 103), Boxoffice Success Productions decides to film the sequel, *Henrietta, Oh Henrietta*. At this point, Bertha Snodgrass (whose novels now top the best seller lists) feels she is in a position to demand $100,000 for the screen rights and royalties of 8% of the net profits. Express her income $I$ as a function of the net profit $N$ and determine the net profit necessary to bring her an income of $1,000,000. What is her marginal income (share of each dollar of net profit)?

**105.** *Processor Speeds* The processor speed, in megahertz (MHz), of Intel processors during the period 1996–2010 could be approximated by the following function of time $t$ in years since the start of 1990:[51]

$$v(t) = \begin{cases} 400t - 2{,}200 & \text{if } 6 \le t < 15 \\ 3{,}800 & \text{if } 15 \le t \le 20. \end{cases}$$

How fast and in what direction was processor speed changing in 2000?

**106.** *Processor Speeds* The processor speed, in megahertz (MHz), of Intel processors during the period 1970–2000 could be approximated by the following function of time $t$ in years since the start of 1970:[52]

$$v(t) = \begin{cases} 3t & \text{if } 0 \le t < 20 \\ 174t - 3{,}420 & \text{if } 20 \le t \le 30. \end{cases}$$

How fast and in what direction was processor speed changing in 1995?

---

[49]Data are rounded. 2010 net incomes projected based on first three quarters. Source: www.wikinvest.com.

[50]Percentages of net profit are commonly called "monkey points." Few movies ever make a net profit on paper, and anyone with any clout in the business gets a share of the *gross*, not the net.

[51]A rough model based on the fastest processors produced by Intel. Source: www.intel.com.

[52]*Ibid.*

## Superbowl Advertising

*Exercises 107 and 108 are based on the following graph and data from Wikipedia showing the increasing cost of a 30-second television ad during the Super Bowl.*[53]

| Year | 1970 | 1980 | 1990 | 2000 | 2010 |
|---|---|---|---|---|---|
| Cost (Thousands of $) | 78 | 222 | 700 | 1,100 | 2,500 |

**107.** ▼ Take $t$ to be the number of years since 1970 and $y$ to be the cost, in thousands of dollars, of a Super Bowl ad.
  **a.** Model the 1970 and 1990 data with a linear equation.
  **b.** Model the 1990 and 2010 data with a linear equation.
  **c.** Use the results of parts (a) and (b) to obtain a piecewise linear model of the cost of a Super Bowl ad during 1970–2010.
  **d.** Use your model to estimate the cost in 2004. Is your answer in rough agreement with the graph? Explain any discrepancy.

**108.** ▼ Take $t$ to be the number of years since 1980 and $y$ to be the cost, in thousands of dollars, of a Super Bowl ad.
  **a.** Model the 1980 and 2000 data with a linear equation.
  **b.** Model the 2000 and 2010 data with a linear equation.
  **c.** Use the results of parts (a) and (b) to obtain a piecewise linear model of the cost of a Super Bowl ad during 1980–2010.
  **d.** Use your model to estimate the cost in 1992. Is your answer in rough agreement with the graph? Explain any discrepancy.

**109.** ▼ *Employment in Mexico* The number of workers employed in manufacturing jobs in Mexico was 3 million in 1995, rose to 4.1 million in 2000, and then dropped to 3.5 million in 2004.[54] Model this number $N$ as a piecewise-linear function of the time $t$ in years since 1995, and use your model to estimate the number of manufacturing jobs in Mexico in 2002. (Take the units of $N$ to be millions.)

**110.** ▼ *Mortgage Delinquencies* The percentage of borrowers in the highest risk category who were delinquent on their payments decreased from 9.7% in 2001 to 4.3% in 2004 and

then shot up to 10.3% in 2007.[55] Model this percentage $P$ as a piecewise-linear function of the time $t$ in years since 2001, and use your model to estimate the percentage of delinquent borrowers in 2006.

## COMMUNICATION AND REASONING EXERCISES

**111.** How would you test a table of values of $x$ and $y$ to see if it comes from a linear function?

**112.** You have ascertained that a table of values of $x$ and $y$ corresponds to a linear function. How do you find an equation for that linear function?

**113.** To what linear function of $x$ does the linear equation $ax + by = c$ $(b \neq 0)$ correspond? Why did we specify $b \neq 0$?

**114.** Complete the following. The slope of the line with equation $y = mx + b$ is the number of units that ____ increases per unit increase in ____.

**115.** Complete the following. If, in a straight line, $y$ is increasing three times as fast as $x$, then its ____ is ____.

**116.** Suppose that $y$ is decreasing at a rate of 4 units per 3-unit increase of $x$. What can we say about the slope of the linear relationship between $x$ and $y$? What can we say about the intercept?

**117.** If $y$ and $x$ are related by the linear expression $y = mx + b$, how will $y$ change as $x$ changes if $m$ is positive? negative? zero?

**118.** Your friend April tells you that $y = f(x)$ has the property that, whenever $x$ is changed by $\Delta x$, the corresponding change in $y$ is $\Delta y = -\Delta x$. What can you tell her about $f$?

**119.** ▣ ▼ Consider the following worksheet:

| ◇ | A | B | C | D |
|---|---|---|---|---|
| 1 | x | y | m | b |
| 2 | | 1 | 2 | =(B3-B2)/(A3-A2) | =B2-C2*A2 |
| 3 | | 3 | -1 | Slope | Intercept |

What is the effect on the slope of increasing the $y$-coordinate of the second point (the point whose coordinates are in Row 3)? Explain.

**120.** ▣ ▼ Referring to the worksheet in Exercise 119, what is the effect on the slope of increasing the $x$-coordinate of the second point (the point whose coordinates are in row 3)? Explain.

**121.** If $y$ is measured in bootlags[56] and $x$ is measured in zonars[57], and $y = mx + b$, then $m$ is measured in _____ and $b$ is measured in _____.

---

[53]Source: http//en.wikipedia.org/wiki/Super_Bowl_advertising.

[54]Source: *New York Times*, Februrary 18, 2007, p. WK4.

[55]The 2007 figure was projected from data through October 2006.
Source: *New York Times*, Februrary 18, 2007, p. BU9.

[56]An ancient Martian unit of length; one bootlag is the mean distance from a Martian's foreleg to its rearleg.

[57]The official currency of Utarek, Mars. (See the footnote to Exercise 88.)

122. If the slope in a linear relationship is measured in miles per dollar, then the independent variable is measured in _____ and the dependent variable is measured in _____.

123. If a quantity is changing linearly with time, and it increases by 10 units in the first day, what can you say about its behavior in the third day?

124. The quantities $Q$ and $T$ are related by a linear equation of the form

$$Q = mT + b.$$

When $T = 0$, $Q$ is positive, but decreases to a negative quantity when $T$ is 10. What are the signs of $m$ and $b$? Explain your answers.

125. ▼ The velocity of an object is given by $v = 0.1t + 20\,\text{m/sec}$, where $t$ is time in seconds. The object is

(A) moving with fixed speed  (B) accelerating
(C) decelerating  (D) impossible to say from the given information

126. ▼ The position of an object is given by $x = 0.2t - 4$, where $t$ is time in seconds. The object is

(A) moving with fixed speed  (B) accelerating
(C) decelerating  (D) impossible to say from the given information

127. If $f$ and $g$ are linear functions with slope $m$ and $n$ respectively, then what can you say about $f + g$?

128. If $f$ and $g$ are linear functions, then is $\frac{f}{g}$ linear? Explain.

129. Give examples of nonlinear functions $f$ and $g$ whose product is linear.

130. Give examples of nonlinear functions $f$ and $g$ whose quotient is linear (on a suitable domain).

131. ▼ Suppose the cost function is $C(x) = mx + b$ (with $m$ and $b$ positive), the revenue function is $R(x) = kx\ (k > m)$, and the number of items is increased from the break-even quantity. Does this result in a loss, a profit, or is it impossible to say? Explain your answer.

132. ▼ You have been constructing a demand equation, and you obtained a (correct) expression of the form $p = mq + b$, whereas you would have preferred one of the form $q = mp + b$. Should you simply switch $p$ and $q$ in the answer, should you start again from scratch, using $p$ in the role of $x$ and $q$ in the role of $y$, or should you solve your demand equation for $q$? Give reasons for your decision.

## 1.4  Linear Regression

We have seen how to find a linear model given two data points: We find the equation of the line that passes through them. However, we often have more than two data points, and they will rarely all lie on a single straight line, but may often come close to doing so. The problem is to find the line coming *closest* to passing through all of the points.

Suppose, for example, that we are conducting research for a company interested in expanding into Mexico. Of interest to us would be current and projected growth in that country's economy. The following table shows past and projected per capita gross domestic product (GDP)[58] of Mexico for 2000–2014.[59]

| Year $t$ ($t = 0$ represents 2000) | 0 | 2 | 4 | 6 | 8 | 10 | 12 | 14 |
|---|---|---|---|---|---|---|---|---|
| Per Capita GDP $y$ ($1,000) | 9 | 9 | 10 | 11 | 11 | 12 | 13 | 13 |

**Figure 27(a)**

**Figure 27(b)**

A plot of these data suggests a roughly linear growth of the GDP (Figure 27(a)). These points suggest a roughly linear relationship between $t$ and $y$, although they clearly do not all lie on a single straight line. Figure 27(b) shows the points together with several lines, some fitting better than others. Can we precisely measure which lines fit better than others? For instance, which of the two lines labeled as "good" fits in Figure 27(b) models the data more accurately? We begin by considering, for each value of $t$, the difference between the actual GDP (the **observed value**) and the GDP

---

[58]The GDP is a measure of the total market value of all goods and services produced within a country.
[59]Data are approximate and/or projected. Sources: CIA World Factbook/www.indexmundi.com, www.economist.com.

Residual = Observed Value − Predicted Value

**Figure 28**

✱ Why not add the absolute values of the residuals instead? Mathematically, using the squares rather than the absolute values results in a simpler and more elegant solution. Further, using the squares always results in a *single* best-fit line in cases where the x-coordinates are all different, whereas this is not the case if we use absolute values.

predicted by a linear equation (the **predicted value**). The difference between the predicted value and the observed value is called the **residual**.

$$\text{Residual} = \text{Observed Value} - \text{Predicted Value}$$

On the graph, the residuals measure the vertical distances between the (observed) data points and the line (Figure 28) and they tell us how far the linear model is from predicting the actual GDP.

The more accurate our model, the smaller the residuals should be. We can combine all the residuals into a single measure of accuracy by adding their *squares*. (We square the residuals in part to make them all positive.✱) The sum of the squares of the residuals is called the **sum-of-squares error**, **SSE**. Smaller values of SSE indicate more accurate models.

Here are some definitions and formulas for what we have been discussing.

## Observed and Predicted Values

Suppose we are given a collection of data points $(x_1, y_1), \ldots, (x_n, y_n)$. The $n$ quantities $y_1, y_2, \ldots, y_n$ are called the **observed y-values**. If we model these data with a linear equation

$$\hat{y} = mx + b, \qquad \text{$\hat{y}$ stands for "estimated $y$" or "predicted $y$."}$$

then the y-values we get by substituting the given x-values into the equation are called the **predicted y-values**:

$$\hat{y}_1 = mx_1 + b \qquad \text{Substitute $x_1$ for $x$.}$$
$$\hat{y}_2 = mx_2 + b \qquad \text{Substitute $x_2$ for $x$.}$$
$$\ldots$$
$$\hat{y}_n = mx_n + b. \qquad \text{Substitute $x_n$ for $x$.}$$

### Quick Example

Consider the three data points $(0, 2)$, $(2, 5)$, and $(3, 6)$. The observed y-values are $y_1 = 2$, $y_2 = 5$, and $y_3 = 6$. If we model these data with the equation $\hat{y} = x + 2.5$, then the predicted values are:

$$\hat{y}_1 = x_1 + 2.5 = 0 + 2.5 = 2.5$$
$$\hat{y}_2 = x_2 + 2.5 = 2 + 2.5 = 4.5$$
$$\hat{y}_3 = x_3 + 2.5 = 3 + 2.5 = 5.5.$$

## Residuals and Sum-of-Squares Error (SSE)

If we model a collection of data $(x_1, y_1), \ldots, (x_n, y_n)$ with a linear equation $\hat{y} = mx + b$, then the **residuals** are the $n$ quantities (Observed Value − Predicted Value):

$$(y_1 - \hat{y}_1), (y_2 - \hat{y}_2), \ldots, (y_n - \hat{y}_n).$$

The **sum-of-squares error (SSE)** is the sum of the squares of the residuals:

$$\text{SSE} = (y_1 - \hat{y}_1)^2 + (y_2 - \hat{y}_2)^2 + \cdots + (y_n - \hat{y}_n)^2.$$

**Quick Example**

For the data and linear approximation given above, the residuals are:

$$y_1 - \hat{y}_1 = 2 - 2.5 = -0.5$$
$$y_2 - \hat{y}_2 = 5 - 4.5 = 0.5$$
$$y_3 - \hat{y}_3 = 6 - 5.5 = 0.5$$

and so $SSE = (-0.5)^2 + (0.5)^2 + (0.5)^2 = 0.75$.

---

### EXAMPLE 1 Computing SSE

Using the data above on the GDP in Mexico, compute SSE for the linear models $y = 0.5t + 8$ and $y = 0.25t + 9$. Which model is the better fit?

**Solution** We begin by creating a table showing the values of $t$, the observed (given) values of $y$, and the values predicted by the first model.

**using** Technology

See the Technology Guides at the end of the chapter for detailed instructions on how to obtain the tables and graphs in Example 1 using a TI-83/84 Plus or a spreadsheet. Here is an outline:

**TI-83/84 Plus**

STAT EDIT

Values of $t$ in $L_1$, and $y$ in $L_2$.
Predicted $y$: Highlight $L_3$. Enter
`0.5*L₁+8`
Squares of residuals: Highlight $L_4$.
Enter
`(L₂-L₃)^2`
SSE: Home screen `sum(L₄)`
Graph: `Y₁=0.5X+8`
Y = screen: Turn on Plot 1 ZOOM
(STAT) [More details on page 117.]

**Spreadsheet**
Headings $t$, $y$, $y$-hat, Residual^2, $m$, $b$, and SSE in A1–F1.
$t$-values in A2–A9, $y$-values in B2–B9; 0.25 for $m$ and 9 for $b$ in E2–F2
Predicted $y$: `=$E$2*A2+$F$2` in C2 and copy down to C9.
Squares of residuals: `=(B2-C2)^2` in D2 and copy down to D9.
SSE: `=SUM(D2:D9)` in G2
Graph: Highlight A1–C9. Insert a Scatter chart.
[More details on page 122.]

| Year $t$ | Observed $y$ | Predicted $\hat{y} = 0.5t + 8$ |
|---|---|---|
| 0 | 9 | 8 |
| 2 | 9 | 9 |
| 4 | 10 | 10 |
| 6 | 11 | 11 |
| 8 | 11 | 12 |
| 10 | 12 | 13 |
| 12 | 13 | 14 |
| 14 | 13 | 15 |

We now add two new columns for the residuals and their squares.

| Year $t$ | Observed $y$ | Predicted $\hat{y} = 0.5t + 8$ | Residual $y - \hat{y}$ | Residual$^2$ $(y - \hat{y})^2$ |
|---|---|---|---|---|
| 0 | 9 | 8 | $9 - 8 = 1$ | $1^2 = 1$ |
| 2 | 9 | 9 | $9 - 9 = 0$ | $0^2 = 0$ |
| 4 | 10 | 10 | $10 - 10 = 0$ | $0^2 = 0$ |
| 6 | 11 | 11 | $11 - 11 = 0$ | $0^2 = 0$ |
| 8 | 11 | 12 | $11 - 12 = -1$ | $(-1)^2 = 1$ |
| 10 | 12 | 13 | $12 - 13 = -1$ | $(-1)^2 = 1$ |
| 12 | 13 | 14 | $13 - 14 = -1$ | $(-1)^2 = 1$ |
| 14 | 13 | 15 | $13 - 15 = -2$ | $(-2)^2 = 4$ |

SSE, the sum of the squares of the residuals, is then the sum of the entries in the last column,

SSE = 8.

Repeating the process using the second model, $0.25t + 9$, yields the following table:

| Year $t$ | Observed $y$ | Predicted $\hat{y} = 0.25t + 9$ | Residual $y - \hat{y}$ | Residual$^2$ $(y - \hat{y})^2$ |
|---|---|---|---|---|
| 0 | 9 | 9 | $9 - 9 = 0$ | $0^2 = 0$ |
| 2 | 9 | 9.5 | $9 - 9.5 = -0.5$ | $(-0.5)^2 = 0.25$ |
| 4 | 10 | 10 | $10 - 10 = 0$ | $0^2 = 0$ |
| 6 | 11 | 10.5 | $11 - 10.5 = 0.5$ | $0.5^2 = 0.25$ |
| 8 | 11 | 11 | $11 - 11 = 0$ | $0^2 = 0$ |
| 10 | 12 | 11.5 | $12 - 11.5 = 0.5$ | $0.5^2 = 0.25$ |
| 12 | 13 | 12 | $13 - 12 = 1$ | $1^2 = 1$ |
| 14 | 13 | 12.5 | $13 - 12.5 = 0.5$ | $0.5^2 = 0.25$ |

This time, SSE = 2 and so the second model is a better fit.

Figure 29 shows the data points and the two linear models in question.

**Figure 29**

---

➡ **Before we go on...**

**Q** : *It seems clear from the figure that the second model in Example 1 gives a better fit. Why bother to compute SSE to tell me this?*

**A** : The difference between the two models we chose is so great that it is clear from the graphs which is the better fit. However, if we used a third model with $m = 0.25$ and $b = 9.1$, then its graph would be almost indistinguishable from that of the second, but a slightly better fit as measured by SSE = 1.68.

---

Among all possible lines, there ought to be one with the least possible value of SSE—that is, the greatest possible accuracy as a model. The line (and there is only one such line) that minimizes the sum of the squares of the residuals is called the **regression line**, the **least-squares line**, or the **best-fit line**.

To find the regression line, we need a way to find values of $m$ and $b$ that give the smallest possible value of SSE. As an example, let us take the second linear model in the example above. We said in the "Before we go on" discussion that increasing $b$ from 9 to 9.1 had the desirable effect of decreasing SSE from 2 to 1.68. We could then increase $m$ to 0.26, further reducing SSE to 1.328. Imagine this as a kind of game: Alternately alter the values of $m$ and $b$ by small amounts until SSE is as small as you can make it. This works, but is extremely tedious and time-consuming.

Fortunately, there is an algebraic way to find the regression line. Here is the calculation. To justify it rigorously requires calculus of several variables or linear algebra.

---

## Regression Line

The **regression line** (**least squares line, best-fit line**) associated with the points $(x_1, y_1)$, $(x_2, y_2)$, . . . , $(x_n, y_n)$ is the line that gives the minimum SSE. The regression line is

$$y = mx + b,$$

where $m$ and $b$ are computed as follows:

$$m = \frac{n\left(\sum xy\right) - \left(\sum x\right)\left(\sum y\right)}{n\left(\sum x^2\right) - \left(\sum x\right)^2}$$

$$b = \frac{\sum y - m\left(\sum x\right)}{n}$$

$n$ = number of data points.

The quantities $m$ and $b$ are called the **regression coefficients**.

---

Here, "$\sum$" means "the sum of." Thus, for example,

$$\sum x = \text{Sum of the } x\text{-values} = x_1 + x_2 + \cdots + x_n$$
$$\sum xy = \text{Sum of products} = x_1 y_1 + x_2 y_2 + \cdots + x_n y_n$$
$$\sum x^2 = \text{Sum of the squares of the } x\text{-values} = x_1^2 + x_2^2 + \cdots + x_n^2.$$

On the other hand,

$$\left(\sum x\right)^2 = \text{Square of } \sum x = \text{Square of the sum of the } x\text{-values.}$$

---

### EXAMPLE 2  Per Capita Gross Domestic Product in Mexico

In Example 1 we considered the following data on the per capita gross domestic product (GDP) of Mexico:

| Year $x$ ($x = 0$ represents 2000) | 0 | 2 | 4 | 6 | 8 | 10 | 12 | 14 |
|---|---|---|---|---|---|---|---|---|
| Per Capita GDP $y$ ($1,000) | 9 | 9 | 10 | 11 | 11 | 12 | 13 | 13 |

Find the best-fit linear model for these data and use the model to predict the per capita GDP in Mexico in 2016.

**using Technology**

See the Technology Guides at the end of the chapter for detailed instructions on how to obtain the regression line and graph in Example 2 using a TI-83/84 Plus or a spreadsheet. Here is an outline:

**TI-83/84 Plus**
STAT EDIT
Values of $x$ in $L_1$, and $y$ in $L_2$.
Regression equation: STAT CALC
option #4: LinReg(ax+b)
Graph: Y= VARS 5 EQ 1,
then ZOOM 9
[More details on page 118.]

**Spreadsheet**
$x$-values in A2–A9, $y$-values in B2–B9
Graph: Highlight A2–B9. Insert a Scatter Chart.
Regression line: Add a linear trend-line. [More details on page 123.]

**Website**
www.WanerMath.com
The following two utilities will calculate and plot regression lines (link to either from Math Tools for Chapter 1):

Simple Regression

Function Evaluator and Grapher

**Solution** Let's organize our work in the form of a table, where the original data are entered in the first two columns and the bottom row contains the column sums.

| $x$ | $y$ | $xy$ | $x^2$ |
|---|---|---|---|
| 0 | 9 | 0 | 0 |
| 2 | 9 | 18 | 4 |
| 4 | 10 | 40 | 16 |
| 6 | 11 | 66 | 36 |
| 8 | 11 | 88 | 64 |
| 10 | 12 | 120 | 100 |
| 12 | 13 | 156 | 144 |
| 14 | 13 | 182 | 196 |
| $\sum$ **(Sum)** 56 | 88 | 670 | 560 |

Because there are $n = 8$ data points, we get

$$m = \frac{n\left(\sum xy\right) - \left(\sum x\right)\left(\sum y\right)}{n\left(\sum x^2\right) - \left(\sum x\right)^2} = \frac{8(670) - (56)(88)}{8(560) - (56)^2} \approx 0.321$$

and

$$b = \frac{\sum y - m\left(\sum x\right)}{n} \approx \frac{88 - (0.321)(56)}{8} \approx 8.75.$$

So, the regression line is

$$y = 0.321x + 8.75.$$

To predict the per capita GDP in Mexico in 2016 we substitute $x = 16$ and get $y \approx 14$, or $14,000 per capita.

Figure 30 shows the data points and the regression line (which has SSE $\approx 0.643$; a lot lower than in Example 1).

**Figure 30**

## Coefficient of Correlation

If all the data points do not lie on one straight line, we would like to be able to measure how closely they can be approximated by a straight line. Recall that SSE measures the sum of the squares of the deviations from the regression line; therefore it constitutes a measurement of what is called "goodness of fit." (For instance, if SSE = 0, then all the points lie on a straight line.) However, SSE depends on the units we use to measure $y$, and also on the number of data points (the more data points we use, the larger SSE tends to be). Thus, while we can (and do) use SSE to compare the goodness of fit of two lines to the same data, we cannot use it to compare the goodness of fit of one line to one set of data with that of another to a different set of data.

To remove this dependency, statisticians have found a related quantity that can be used to compare the goodness of fit of lines to different sets of data. This quantity, called the **coefficient of correlation** or **correlation coefficient**, and usually denoted $r$, is between $-1$ and $1$. The closer $r$ is to $-1$ or $1$, the better the fit. For an *exact* fit, we would have $r = -1$ (for a line with negative slope) or $r = 1$ (for a line with positive slope). For a bad fit, we would have $r$ close to $0$. Figure 31 shows several collections of data points with least-squares lines and the corresponding values of $r$.

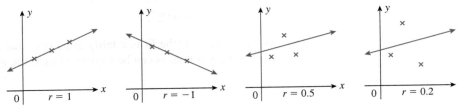

**Figure 31**

## Correlation Coefficient

The coefficient of correlation of the $n$ data points $(x_1, y_1), (x_2, y_2), \ldots, (x_n, y_n)$ is

$$r = \frac{n\left(\sum xy\right) - \left(\sum x\right)\left(\sum y\right)}{\sqrt{n\left(\sum x^2\right) - \left(\sum x\right)^2} \cdot \sqrt{n\left(\sum y^2\right) - \left(\sum y\right)^2}}.$$

It measures how closely the data points $(x_1, y_1), (x_2, y_2), \ldots, (x_n, y_n)$ fit the regression line. (The value $r^2$ is sometimes called the **coefficient of determination**.)

### Interpretation

- If $r$ is positive, the regression line has positive slope; if $r$ is negative, the regression line has negative slope.

- If $r = 1$ or $-1$, then all the data points lie exactly on the regression line; if it is close to $\pm 1$, then all the data points are close to the regression line.

- On the other hand, if $r$ is not close to $\pm 1$, then the data points are not close to the regression line, so the fit is not a good one. As a general rule of thumb, a value of $|r|$ less than around $0.8$ indicates a poor fit of the data to the regression line.

**using Technology**

See the Technology Guides at the end of the chapter for detailed instructions on how to obtain the correlation coefficient in Example 3 using a TI-83/84 Plus or a spreadsheet. Here is an outline:

**TI-83/84 Plus**

2ND CATALOG DiagnosticOn
Then STAT CALC option #4:
LinReg(ax+b) [More details on page 119.]

**Spreadsheet**

Add a trendline and select the option to "Display R-squared value on chart."
[More details and other alternatives on page 124.]

**Website**
www.WanerMath.com
The following two utilities will show regression lines and also $r^2$ (link to either from Math Tools for Chapter 1):
    Simple Regression
    Function Evaluator and Grapher

### EXAMPLE 3 Computing the Coefficient of Correlation

Find the correlation coefficient for the data in Example 2. Is the regression line a good fit?

**Solution** The formula for $r$ requires $\sum x, \sum x^2, \sum xy, \sum y$, and $\sum y^2$. We have all of these except for $\sum y^2$, which we find in a new column as shown.

| $x$ | $y$ | $xy$ | $x^2$ | $y^2$ |
|---|---|---|---|---|
| 0 | 9 | 0 | 0 | 81 |
| 2 | 9 | 18 | 4 | 81 |
| 4 | 10 | 40 | 16 | 100 |
| 6 | 11 | 66 | 36 | 121 |
| 8 | 11 | 88 | 64 | 121 |
| 10 | 12 | 120 | 100 | 144 |
| 12 | 13 | 156 | 144 | 169 |
| 14 | 13 | 182 | 196 | 169 |
| $\sum$ **(Sum)** | 56 | 88 | 670 | 560 | 986 |

Substituting these values into the formula, we get

$$r = \frac{n(\sum xy) - (\sum x)(\sum y)}{\sqrt{n(\sum x^2) - (\sum x)^2} \cdot \sqrt{n(\sum y^2) - (\sum y)^2}}$$

$$= \frac{8(670) - (56)(88)}{\sqrt{8(560) - 56^2} \cdot \sqrt{8(986) - 88^2}}$$

$$\approx 0.982.$$

As $r$ is close to 1, the fit is a fairly good one; that is, the original points lie nearly along a straight line, as can be confirmed from the graph in Example 2.

## 1.4 EXERCISES

▼ more advanced   ◆ challenging
T indicates exercises that should be solved using technology

*In Exercises 1–4, compute the sum-of-squares error (SSE) by hand for the given set of data and linear model.* HINT [See Example 1.]

**1.** $(1, 1), (2, 2), (3, 4); y = x - 1$

**2.** $(0, 1), (1, 1), (2, 2); y = x + 1$

**3.** $(0, -1), (1, 3), (4, 6), (5, 0); y = -x + 2$

**4.** $(2, 4), (6, 8), (8, 12), (10, 0); y = 2x - 8$

T *In Exercises 5–8, use technology to compute the sum-of-squares error (SSE) for the given set of data and linear models. Indicate which linear model gives the better fit.*

**5.** $(1, 1), (2, 2), (3, 4);$   **a.** $y = 1.5x - 1$   **b.** $y = 2x - 1.5$

**6.** $(0, 1), (1, 1), (2, 2);$   **a.** $y = 0.4x + 1.1$   **b.** $y = 0.5x + 0.9$

**7.** $(0, -1), (1, 3), (4, 6), (5, 0);$   **a.** $y = 0.3x + 1.1$
   **b.** $y = 0.4x + 0.9$

**8.** $(2, 4), (6, 8), (8, 12), (10, 0);$   **a.** $y = -0.1x + 7$
   **b.** $y = -0.2x + 6$

*Find the regression line associated with each set of points in Exercises 9–12. Graph the data and the best-fit line. (Round all coefficients to 4 decimal places.)* HINT [See Example 2.]

**9.** $(1, 1), (2, 2), (3, 4)$

**10.** $(0, 1), (1, 1), (2, 2)$

**11.** $(0, -1), (1, 3), (3, 6), (4, 1)$

**12.** $(2, 4), (4, 8), (8, 12), (10, 0)$

*In the next two exercises, use correlation coefficients to determine which of the given sets of data is best fit by its associated regression line and which is fit worst. Is it a perfect fit for any of the data sets?* HINT [See Example 3.]

**13. a.** $\{(1, 3), (2, 4), (5, 6)\}$
   **b.** $\{(0, -1), (2, 1), (3, 4)\}$
   **c.** $\{(4, -3), (5, 5), (0, 0)\}$

**14. a.** $\{(1, 3), (-2, 9), (2, 1)\}$
   **b.** $\{(0, 1), (1, 0), (2, 1)\}$
   **c.** $\{(0, 0), (5, -5), (2, -2.1)\}$

## APPLICATIONS

**15.** *Mobile Broadband Subscriptions* The following table shows the number of mobile broadband subscribers worldwide ($x$ is the number of years since 2000):[60]

| Year $x$ | 2 | 6 | 10 |
|---|---|---|---|
| Subscribers $y$ (Millions) | 0 | 70 | 900 |

Complete the following table and obtain the associated regression line. (Round coefficients to one decimal place.) HINT [See Example 3.]

| $x$ | $y$ | $xy$ | $x^2$ |
|---|---|---|---|
| 2 | 0 | | |
| 6 | 70 | | |
| 10 | 900 | | |
| $\sum$ (Sum) | | | |

Use your regression equation to project the number in 2012.

**16.** *Investment in Gold* Following are approximate values of the Amex Gold BUGS Index ($x = 0$ represents 2000):[61]

| Year $x$ | 0 | 5 | 10 |
|---|---|---|---|
| Index $y$ | 50 | 200 | 500 |

---

[60]Data are rounded (2010 figure is estimated). Source: International Telecommunication Union www.itu.int/ITU-D/ict/statistics/at_glance/KeyTelecom.html.

[61]BUGS stands for "basket of unhedged gold stocks." Figures are approximate. Source: www.google.com/finance.

Complete the following table and obtain the associated regression line. (Round coefficients to the nearest whole number.)
HINT [See Example 3.]

| $x$ | $y$ | $xy$ | $x^2$ |
|---|---|---|---|
| 0 | 50 | | |
| 5 | 200 | | |
| 10 | 500 | | |
| $\sum$ (Sum) | | | |

Use your regression equation to project the index in 2011.

**17. E-Commerce** The following chart shows second quarter total retail e-commerce sales in the U.S. in 2000, 2005, and 2010 ($t = 0$ represents 2000):[62]

| Year $t$ | 0 | 5 | 10 |
|---|---|---|---|
| Sales (\$ billion) $y$ | 6 | 20 | 40 |

Find the regression line (round coefficients to one decimal place) and use it to estimate second quarter retail e-commerce sales in 2006. (The actual figure was approximately \$25 billion.)

**18. Net Sales** The following table show the reported net sales by Nokia in the third quarters of 2008, 2009, and 2010 ($t = 0$ represents the third quarter of 2008):[63]

| Year $t$ | 0 | 1 | 2 |
|---|---|---|---|
| Net Sales (€ billion) $y$ | 12 | 10 | 10 |

Find the regression line (round coefficients to one decimal place) and use it to "predict" Nokia's net income in the third quarter of 2011. (The actual figure was approximately €9.0 billion.)

**19. Oil Recovery** The Texas Bureau of Economic Geology published a study on the economic impact of using carbon dioxide enhanced oil recovery (EOR) technology to extract additional oil from fields that have reached the end of their conventional economic life. The following table gives the approximate number of jobs for the citizens of Texas that would be created at various levels of recovery.[64]

| Percent Recovery (%) | 20 | 40 | 80 | 100 |
|---|---|---|---|---|
| Jobs Created (Millions) | 3 | 6 | 9 | 15 |

Find the regression line and use it to estimate the number of jobs that would be created at a recovery level of 50%.

**20. Oil Recovery** (Refer to Exercise 19.) The following table gives the approximate economic value associated with various levels of oil recovery in Texas.[65]

| Percent Recovery (%) | 10 | 40 | 50 | 80 |
|---|---|---|---|---|
| Economic Value (\$ billions) | 200 | 900 | 1,000 | 2,000 |

Find the regression line and use it to estimate the economic value associated with a recovery level of 70%.

**21. Profit: Amazon.com** The following table shows Amazon.com's approximate net sales (revenue) and net income (profit) in the period 2006–2009:[66]

| Net Sales (\$ billions) | 10 | 15 | 20 | 25 |
|---|---|---|---|---|
| Net Income (\$ millions) | 200 | 500 | 600 | 900 |

**a.** Use this information to find a linear regression model for Amazon's net income $I$ (in millions of dollars) as a function of net sales $S$ (in billions of dollars).
**b.** Give the units of measurement and interpretation of the slope.
**c.** What, according to the model, would Amazon.com need to earn in net sales in order for its net income to be \$1 billion? (Round answer to the nearest \$ billion.)
**d.** Plot the data and regression line. Based on the graph, would you say that the linear model is reasonable? Why?

**22. Operating Expenses: Amazon.com** The following table shows Amazon.com's approximate net sales (revenue) and operating expenses in 2006–2009:[67]

| Net Sales (\$ billions) | 10 | 15 | 20 | 25 |
|---|---|---|---|---|
| Operating Expenses (\$ billions) | 2 | 2.5 | 3 | 4.5 |

**a.** Use this information to find a linear regression model for Amazon's operating expenses $E$ (in billions of dollars) as a function of net sales $S$ (in billions of dollars).
**b.** Give the units of measurement and interpretation of the slope.
**c.** What, according to the model, would Amazon.com need to earn in net sales in order for its operating expenses to be \$5 billion? (Round answer to the nearest \$ billion.)
**d.** Plot the data and regression line. Based on the graph, would you say that the linear model is reasonable? Why?

---

[62]Figures are rounded. Source: U.S. Census Bureau www.census.gov

[63]Data are approximate. Source: www.nokia.com

[64]Source: "CO2–Enhanced Oil Recovery Resource Potential in Texas: Potential Positive Economic Impacts," Texas Bureau of Economic Geology, April 2004, www.rrc.state.tx.us/tepc/CO2-EOR_white_paper.pdf.

[65]Ibid.

[66]Figures are approximate. Source: www.wikinvest.com.

[67]Ibid.

**23.** 🔲 *Textbook Sizes* The following table shows the numbers of pages in previous editions of *Applied Calculus* by Waner and Costenoble:

| Edition $n$ | 2 | 3 | 4 | 5 |
|---|---|---|---|---|
| Number of Pages $L$ | 585 | 656 | 694 | 748 |

**a.** With the edition number as the independent variable, use technology to obtain a regression line and a plot of the points together with the regression line. (Round coefficients to two decimal places.)

**b.** Interpret the slope of the regression line.

**24.** 🔲 *Textbook Sizes* Repeat Exercise 23 using the following corresponding table for *Finite Mathematics* by Waner and Costenoble:

| Edition $n$ | 2 | 3 | 4 | 5 |
|---|---|---|---|---|
| Number of Pages $L$ | 603 | 608 | 676 | 692 |

**25.** 🔲 *Soybean Production: Cerrados* The following table shows soybean production, in millions of tons, in Brazil's *Cerrados* region, as a function of the cultivated area, in millions of acres.[68]

| Area (millions of acres) | 25 | 30 | 32 | 40 | 52 |
|---|---|---|---|---|---|
| Production (millions of tons) | 15 | 25 | 30 | 40 | 60 |

**a.** Use technology to obtain the regression line and to show a plot of the points together with the regression line. (Round coefficients to two decimal places.)

**b.** Interpret the slope of the regression line.

**26.** 🔲 *Soybean Production: U.S.* The following table shows soybean production, in millions of tons, in the U.S. as a function of the cultivated area, in millions of acres.[69]

| Area (millions of acres) | 30 | 42 | 69 | 59 | 74 | 74 |
|---|---|---|---|---|---|---|
| Production (millions of tons) | 20 | 33 | 55 | 57 | 83 | 88 |

**a.** Use technology to obtain the regression line and to show a plot of the points together with the regression line. (Round coefficients to two decimal places.)

**b.** Interpret the slope of the regression line.

**27.** 🔲 *Airline Profits and the Price of Oil* A common perception is that airline profits are strongly correlated with the price of oil. Following are annual net incomes of Continental Airlines together with the approximate price of oil in the period 2005–2010:[70]

---

[68]Source: Brazil Agriculture Ministry/*New York Times*, December 12, 2004, p. N32.

[69]Data are approximate. Source for data: L. David Roper, June 2010, *Crop Production in the World & the United States*, www.roperld.com/science/cropsworld&us.htm.

[70]Figures are rounded and oil prices are inflation adjusted. Sources: www.wikinvest.com, www.inflationdata.com.

| Year | 2005 | 2006 | 2007 | 2008 | 2009 | 2010 |
|---|---|---|---|---|---|---|
| Price of Oil ($ per barrel) | 56 | 63 | 67 | 92 | 54 | 71 |
| Continental Net Income ($ million) | −70 | 370 | 430 | −590 | −280 | 150 |

**a.** Use technology to obtain a regression line showing Continental's net income as a function of the price of oil, and also the coefficient of correlation $r$.

**b.** What does the value of $r$ suggest about the relationship of Continental's net income to the price of oil?

**c.** Support your answer to part (b) with a plot of the data and regression line.

**28.** 🔲 *Airline Profits and the Price of Oil* Repeat Exercise 27 using the following corresponding data for American Airlines:[71]

| Year | 2005 | 2006 | 2007 | 2008 | 2009 | 2010 |
|---|---|---|---|---|---|---|
| Price of Oil ($ per barrel) | 56 | 63 | 67 | 92 | 54 | 71 |
| American Net Income ($ million) | −850 | 250 | 450 | −2,100 | −1,450 | −700 |

🔲 *Doctorates in Mexico* Exercises 29–32 are based on the following table showing the annual number of PhD graduates in Mexico in various fields.[72]

| | Natural Sciences | Engineering | Social Sciences |
|---|---|---|---|
| 1990 | 84 | 8 | 98 |
| 1995 | 107 | 55 | 161 |
| 2000 | 174 | 247 | 222 |
| 2005 | 515 | 371 | 584 |
| 2010 | 979 | 578 | 1,230 |

**29. a.** With $x$ = the number of natural science doctorates and $y$ = the number of engineering doctorates, use technology to obtain the regression equation and graph the associated points and regression line. (Round coefficients to three significant digits.)

**b.** What does the slope tell you about the relationship between the number of natural science doctorates and the number of engineering doctorates?

**c.** Use technology to obtain the coefficient of correlation $r$. Does the value of $r$ suggest a strong correlation between $x$ and $y$?

**d.** Judging from the graph, would you say that a linear relationship between $x$ and $y$ is appropriate? Why?

**30. a.** With $x$ = the number of natural science doctorates and $y$ = the number of social science doctorates, use technology to obtain the regression equation and graph the associated points and regression line. (Round coefficients to three significant digits.)

---

[71]*Ibid.*

[72]2010 data is estimated. Source: Instituto Nacional de Estadística y Geografía www.inegi.org.mx.

**b.** What does the slope tell you about the relationship between the number of natural science doctorates and the number of social science doctorates?

**c.** Use technology to obtain the coefficient of correlation $r$. Does the value of $r$ suggest a strong correlation between $x$ and $y$?

**d.** Judging from the graph, would you say that a linear relationship between $x$ and $y$ is appropriate? Why?

**31. ▼ a.** Use technology to obtain the regression equation and the coefficient of correlation $r$ for the number of natural science doctorates as a function of time $t$ in years since 1990, and graph the associated points and regression line. (Round coefficients to three significant digits.)

**b.** What does the slope tell you about the number of natural science doctorates?

**c.** Judging from the graph, would you say that the number of natural science doctorates is increasing at a faster and faster rate, a slower and slower rate, or neither? Why?

**d.** If $r$ had been equal to 1, could you have drawn the same conclusion as in part (c)? Explain.

**32. ▼ a.** Use technology to obtain the regression equation and the coefficient of correlation $r$ for the number of engineering doctorates as a function of time $t$ in years since 1990, and graph the associated points and regression line. (Round coefficients to three significant digits.)

**b.** What does the slope tell you about the number of engineering doctorates?

**c.** Judging from the graph, would you say that the number of engineering doctorates is increasing at a faster and faster rate, a slower and slower rate, or neither? Why?

**d.** If $r$ had been close to 0, could you have drawn the same conclusion as in part (c)? Explain.

**33. ▯ ▼ *NY City Housing Costs: Downtown*** The following table shows the average price of a two-bedroom apartment in downtown New York City from 1994 to 2004 ($t = 0$ represents 1994).[73]

| Year $t$ | 0 | 2 | 4 | 6 | 8 | 10 |
|---|---|---|---|---|---|---|
| Price $p$ ($ million) | 0.38 | 0.40 | 0.60 | 0.95 | 1.20 | 1.60 |

**a.** Use technology to obtain the linear regression line and correlation coefficient $r$, with all coefficients rounded to two decimal places, and plot the regression line and the given points.

**b.** Does the graph suggest that a non-linear relationship between $t$ and $p$ would be more appropriate than a linear one? Why?

**c.** Use technology to obtain the residuals. What can you say about the residuals in support of the claim in part (b)?

**34. ▯ ▼ *Fiber Optic Connections*** The following table shows the number of fiber optic cable connections to homes in the U.S. from 2000 to 2004 ($t = 0$ represents 2000):[74]

| Year $t$ | 0 | 1 | 2 | 3 | 4 |
|---|---|---|---|---|---|
| Connections $c$ (Thousands) | 0 | 10 | 25 | 65 | 150 |

**a.** Use technology to obtain the linear regression line and correlation coefficient $r$, with all coefficients rounded to two decimal places, and plot the regression line and the given points.

**b.** Does the graph suggest that a non-linear relationship between $t$ and $c$ would be more appropriate than a linear one? Why?

**c.** Use technology to obtain the residuals. What can you say about the residuals in support of the claim in part (b)?

## COMMUNICATION AND REASONING EXERCISES

**35.** Why is the regression line associated with the two points $(a, b)$ and $(c, d)$ the same as the line that passes through both? (Assume that $a \neq c$.)

**36.** What is the smallest possible sum-of-squares error if the given points happen to lie on a straight line? Why?

**37.** If the points $(x_1, y_1), (x_2, y_2), \ldots, (x_n, y_n)$ lie on a straight line, what can you say about the regression line associated with these points?

**38.** If all but one of the points $(x_1, y_1), (x_2, y_2), \ldots, (x_n, y_n)$ lie on a straight line, must the regression line pass through all but one of these points?

**39. ▼** Verify that the regression line for the points $(0, 0)$, $(-a, a)$, and $(a, a)$ has slope 0. What is the value of $r$? (Assume that $a \neq 0$.)

**40. ▼** Verify that the regression line for the points $(0, a)$, $(0, -a)$, and $(a, 0)$ has slope 0. What is the value of $r$? (Assume that $a \neq 0$.)

**41. ▼** Must the regression line pass through at least one of the data points? Illustrate your answer with an example.

**42. ▼** Why must care be taken when using mathematical models to extrapolate?

**43. ▼** Your friend Imogen tells you that if $r$ for a collection of data points is more than 0.9, then the most appropriate relationship between the variables is a linear one. Explain why she is wrong by referring to one of the exercises.

**44. ▼** Your other friend Mervyn tells you that if $r$ for a collection of data points has an absolute value less than 0.8, then the most appropriate relationship between the variables is a quadratic and not a linear one. Explain why *he* is wrong.

---

[73]Data are rounded and 2004 figure is an estimate. Source: Miller Samuel/*New York Times*, March 28, 2004, p. RE 11.

[74]Source: Render, Vanderslice & Associates/*New York Times*, October 11, 2004, p. C1.

## KEY CONCEPTS

  Website www.WanerMath.com
Go to the Website at www.WanerMath
.com to find a comprehensive and
interactive Web-based summary
of Chapter 1.

### 1.1 Functions from the Numerical, Algebraic, and Graphical Viewpoints

Real-valued function $f$ of a real-valued variable $x$, domain  *p. 42*
Independent and dependent variables  *p. 42*
Graph of the function $f$  *p. 42*
Numerically specified function  *p. 42*
Graphically specified function  *p. 42*
Algebraically defined function  *p. 42*
Piecewise-defined function  *p. 47*
Vertical line test  *p. 49*
Common types of algebraic functions and their graphs  *p. 50*

### 1.2 Functions and Models

Mathematical model  *p. 56*
Analytical model  *p. 57*
Curve-fitting model  *p. 57*
Cost, revenue, and profit; marginal cost, revenue, and profit; break-even point  *p. 59*
Demand, supply, and equilibrium price  *pp. 61–62*
Selecting a model  *p. 64*
Compound interest  *p. 66*
Exponential growth and decay  *p. 68*
Algebra of functions (sum, difference, product, quotient)  *p. 70*

### 1.3 Linear Functions and Models

Linear function $f(x) = mx + b$  *p. 78*
Change in $q$: $\Delta q = q_2 - q_1$  *p. 79*
Slope of a line:
$$m = \frac{\Delta y}{\Delta x} = \frac{\text{Change in } y}{\text{Change in } x} \quad p.\ 79$$

Interpretations of $m$  *p. 80*
Interpretation of $b$: $y$-intercept  *p. 80*
Recognizing linear data  *p. 81*
Computing the slope of a line  *p. 82*
Slopes of horizontal and vertical lines  *p. 82*
Computing the $y$-intercept  *p. 83*
Linear modeling  *p. 84*
Linear cost  *p. 85*
Linear demand  *p. 85*
Linear change over time; rate of change; velocity  *p. 88*
General linear models  *p. 88*

### 1.4 Linear Regression

Observed and predicted values  *p. 96*
Residuals and sum-of-squares error (SSE)  *p. 96*
Regression line (least-squares line, best-fit line)  *p. 99*
Correlation coefficient; coefficient of determination  *p. 101*

## REVIEW EXERCISES

In Exercises 1–4, use the graph of the function $f$ to find approximations of the given values.

**1.**

a. $f(-2)$   b. $f(0)$
c. $f(2)$   d. $f(2) - f(-2)$

**2.**

a. $f(-2)$   b. $f(0)$
c. $f(2)$   d. $f(2) - f(-2)$

**3.**

a. $f(-1)$   b. $f(0)$
c. $f(1)$   d. $f(1) - f(-1)$

**4.**

a. $f(-1)$   b. $f(0)$
c. $f(1)$   d. $f(1) - f(-1)$

In Exercises 5–8, graph the given function or equation.

**5.** $y = -2x + 5$

**6.** $2x - 3y = 12$

**7.** $y = \begin{cases} \frac{1}{2}x & \text{if } -1 \le x \le 1 \\ x - 1 & \text{if } 1 < x \le 3 \end{cases}$

**8.** $f(x) = 4x - x^2$ with domain $[0, 4]$

In Exercises 9–14, decide whether the specified values come from a linear, quadratic, exponential, or absolute value function.

**9.**

| $x$ | −2 | 0 | 1 | 2 | 4 |
|---|---|---|---|---|---|
| $f(x)$ | 4 | 2 | 1 | 0 | 2 |

**10.**

| $x$ | −2 | 0 | 1 | 2 | 4 |
|---|---|---|---|---|---|
| $g(x)$ | −5 | −3 | −2 | −1 | 1 |

**11.**

| $x$ | −2 | 0 | 1 | 2 | 4 |
|---|---|---|---|---|---|
| $h(x)$ | 1.5 | 1 | 0.75 | 0.5 | 0 |

**12.**

| $x$ | −2 | 0 | 1 | 2 | 4 |
|---|---|---|---|---|---|
| $k(x)$ | 0.25 | 1 | 2 | 4 | 16 |

**13.**

| $x$ | −2 | 0 | 1 | 2 | 4 |
|---|---|---|---|---|---|
| $u(x)$ | 0 | 4 | 3 | 0 | −12 |

**14.**

| $x$ | $-2$ | 0 | 1 | 2 | 4 |
|---|---|---|---|---|---|
| $w(x)$ | 32 | 8 | 4 | 2 | 0.5 |

*In Exercises 15–22, find the equation of the specified line.*

**15.** Through $(3, 2)$ with slope $-3$

**16.** Through $(-2, 4)$ with slope $-1$

**17.** Through $(1, -3)$ and $(5, 2)$

**18.** Through $(-1, 2)$ and $(1, 0)$

**19.** Through $(1, 2)$ parallel to $x - 2y = 2$

**20.** Through $(-3, 1)$ parallel to $-2x - 4y = 5$

**21.** With slope 4 crossing $2x - 3y = 6$ at its $x$-intercept

**22.** With slope $1/2$ crossing $3x + y = 6$ at its $x$-intercept

*In Exercises 23 and 24, determine which of the given lines better fits the given points.*

**23.** $(-1, 1)$, $(1, 2)$, $(2, 0)$; $y = -x/2 + 1$ or $y = -x/4 + 1$

**24.** $(-2, -1)$, $(-1, 1)$, $(0, 1)$, $(1, 2)$, $(2, 4)$, $(3, 3)$; $y = x + 1$ or $y = x/2 + 1$

*In Exercises 25 and 26, find the line that best fits the given points and compute the correlation coefficient.*

**25.** $(-1, 1)$, $(1, 2)$, $(2, 0)$

**26.** $(-2, -1)$, $(-1, 1)$, $(0, 1)$, $(1, 2)$, $(2, 4)$, $(3, 3)$

## APPLICATIONS: OHaganBooks.com

**27.** *Web Site Traffic* John Sean O'Hagan is CEO of the online bookstore OHaganBooks.com and notices that, since the establishment of the company Web site six years ago ($t = 0$), the number of visitors to the site has grown quite dramatically, as indicated by the following table:

| Year $t$ | 0 | 1 | 2 | 3 | 4 | 5 | 6 |
|---|---|---|---|---|---|---|---|
| Web site Traffic $V(t)$ (visits/day) | 100 | 300 | 1,000 | 3,300 | 10,500 | 33,600 | 107,400 |

    **a.** Graph the function $V$ as a function of time $t$. Which of the following types of function seem to fit the curve best: linear, quadratic, or exponential?

    **b.** Compute the ratios $\dfrac{V(1)}{V(0)}, \dfrac{V(2)}{V(1)}, \ldots,$ and $\dfrac{V(6)}{V(5)}$. What do you notice?

    **c.** Use the result of part (b) to predict Web site traffic next year (to the nearest 100).

**28.** *Publishing Costs* Marjory Maureen Duffin is CEO of publisher Duffin House, a major supplier of paperback titles to OHaganBooks.com. She notices that publishing costs over the past five years have varied considerably as indicated by the following table, which shows the average cost to the company of publishing a paperback novel ($t$ is time in years, and the current year is $t = 5$):

| Year $t$ | 0 | 1 | 2 | 3 | 4 | 5 |
|---|---|---|---|---|---|---|
| Cost $C(t)$ | \$5.42 | \$5.10 | \$5.00 | \$5.12 | \$5.40 | \$5.88 |

    **a.** Graph the function $C$ as a function of time $t$. Which of the following types of function seem to fit the curve best: linear, quadratic, or exponential?

    **b.** Compute the differences $C(1) - C(0)$, $C(2) - C(1), \ldots,$ and $C(5) - C(4)$, rounded to one decimal place. What do you notice?

    **c.** Use the result of part (b) to predict the cost of producing a paperback novel next year.

**29.** *Web Site Stability* John O'Hagan is considering upgrading the Web server equipment at OHaganBooks.com because of frequent crashes. The tech services manager has been monitoring the frequency of crashes as a function of Web site traffic (measured in thousands of visits per day) and has obtained the following model:

$$c(x) = \begin{cases} 0.03x + 2 & \text{if } 0 \le x \le 50 \\ 0.05x + 1 & \text{if } x > 50 \end{cases}$$

where $c(x)$ is the average number of crashes in a day in which there are $x$ thousand visitors.

    **a.** On average, how many times will the Web site crash on a day when there are 10,000 visits? 50,000 visits? 100,000 visits?

    **b.** What does the coefficient 0.03 tell you about the Web site's stability?

    **c.** Last Friday, the Web site went down 8 times. Estimate the number of visits that day.

**30.** *Book Sales* As OHaganBooks.com has grown in popularity, the sales manager has been monitoring book sales as a function of the Web site traffic (measured in thousands of visits per day) and has obtained the following model:

$$s(x) = \begin{cases} 1.55x & \text{if } 0 \le x \le 100 \\ 1.75x - 20 & \text{if } 100 < x \le 250 \end{cases}$$

where $s(x)$ is the average number of books sold in a day in which there are $x$ thousand visitors.

    **a.** On average, how many books per day does the model predict OHaganBooks.com will sell when it has 60,000 visits in a day? 100,000 visits in a day? 160,000 visits in a day?

    **b.** What does the coefficient 1.75 tell you about book sales?

    **c.** According to the model, approximately how many visitors per day will be needed in order to sell an average of 300 books per day?

**31.** *New Users* The number of registered users at OHaganBooks.com has increased substantially over the past few months. The following table shows the number of new users registering each month for the past six months:

| Month $t$ | 1 | 2 | 3 | 4 | 5 | 6 |
|---|---|---|---|---|---|---|
| New Users (thousands) | 12.5 | 37.5 | 62.5 | 72.0 | 74.5 | 75.0 |

**a.** Which of the following models best approximates the data?

**(A)** $n(t) = \dfrac{300}{4 + 100(5^{-t})}$     **(B)** $n(t) = 13.3t + 8.0$

**(C)** $n(t) = -2.3t^2 + 30.0t - 3.3$

**(D)** $n(t) = 7(3^{0.5t})$

**b.** What do each of the above models predict for the number of new users in the next few months: rising, falling, or leveling off?

**32. *Purchases*** OHaganBooks.com has been promoting a number of books published at Duffin House. The following table shows the number of books purchased each month from Duffin House for the past five months:

| Month $t$ | 1 | 2 | 3 | 4 | 5 |
|---|---|---|---|---|---|
| Purchases (books) | 1,330 | 520 | 520 | 1,340 | 2,980 |

**a.** Which of the following models best approximates the data?

**(A)** $n(t) = \dfrac{3,000}{1 + 12(2^{-t})}$     **(B)** $n(t) = \dfrac{2,000}{4.2 - 0.7t}$

**(C)** $n(t) = 300(1.6^t)$

**(D)** $n(t) = 100(4.1t^2 - 20.4t + 29.5)$

**b.** What do each of the above models predict for the number of new users in the next few months: rising, falling, leveling off, or something else?

**33. *Internet Advertising*** Several months ago. John O'Hagan investigated the effect on the popularity of OHaganBooks.com of placing banner ads at well-known Internet portals. The following model was obtained from available data:

$$v(c) = -0.000005c^2 + 0.085c + 1,750 \text{ new visits per day}$$

where $c$ is the monthly expenditure on banner ads.

**a.** John O'Hagan is considering increasing expenditure on banner ads from the current level of $5,000 to $6,000 per month. What will be the resulting effect on Web site popularity?

**b.** According to the model, would the Web site popularity continue to grow at the same rate if he continued to raise expenditure on advertising $1,000 each month? Explain.

**c.** Does this model give a reasonable prediction of traffic at expenditures larger than $8,500 per month? Why?

**34. *Production Costs*** Over at Duffin House, Marjory Duffin is trying to decide on the size of the print runs for the best-selling new fantasy novel *Larry Plotter and the Simplex Method*. The following model shows a calculation of the total cost to produce a million copies of the novel, based on an analysis of setup and storage costs:

$$c(n) = 0.0008n^2 - 72n + 2,000,000 \text{ dollars}$$

where $n$ is the print run size (the number of books printed in each run).

**a.** What would be the effect on cost if the run size was increased from 20,000 to 30,000?

**b.** Would increasing the run size in further steps of 10,000 result in the same changes in the total cost? Explain.

**c.** What approximate run size would you recommend that Marjoy Duffin use for a minimum cost?

**35. *Internet Advertising*** When OHaganBooks.com actually went ahead and increased Internet advertising from $5,000 per month to $6,000 per month (see Exercise 33) it was noticed that the number of new visits increased from an estimated 2,050 per day to 2,100 per day. Use this information to construct a linear model giving the average number $v$ of new visits per day as a function of the monthly advertising expenditure $c$.

**a.** What is the model?

**b.** Based on the model, how many new visits per day could be anticipated if OHaganBooks.com budgets $7,000 per month for Internet advertising?

**c.** The goal is to eventually increase the number of new visits to 2,500 per day. Based on the model, how much should be spent on Internet advertising in order to accomplish this?

**36. *Production Costs*** When Duffin House printed a million copies of *Larry Plotter and the Simplex Method* (see Exercise 34), it used print runs of 20,000, which cost the company $880,000. For the sequel, *Larry Plotter and the Simplex Method, Phase 2* it used print runs of 40,000 which cost the company $550,000. Use this information to construct a linear model giving the production cost $c$ as a function of the run size $n$.

**a.** What is the model?

**b.** Based on the model, what would print runs of 25,000 have cost the company?

**c.** Marjory Duffin has decided to budget $418,000 for production of the next book in the *Simplex Method* series. Based on the model, how large should the print runs be to accomplish this?

**37. *Recreation*** John O'Hagan has just returned from a sales convention at Puerto Vallarta, Mexico where, in order to win a bet he made with Marjory Duffin (Duffin House was also at the convention), he went bungee jumping at a nearby mountain retreat. The bungee cord he used had the property that a person weighing 70 kg would drop a total distance of 74.5 meters, while a 90 kg person would drop 93.5 meters. Express the distance $d$ a jumper drops as a linear function of the jumper's weight $w$. John OHagan dropped 90 m. What was his approximate weight?

**38. *Crickets*** The mountain retreat near Puerto Vallarta was so quiet at night that all one could hear was the chirping of the snowy tree crickets. These crickets behave in a rather interesting way: The rate at which they chirp depends linearly on the temperature. Early in the evening, John O'Hagan counted 140 chirps/minute and noticed that the temperature was 80°F. Later in the evening the temperature dropped to 75°F, and the chirping slowed down to 120 chirps/minute. Express the temperature $T$ as a function of the rate of chirping $r$. The temperature that night dropped to a low of 65°F. At approximately what rate were the crickets chirping at that point?

39. **Break-Even Analysis** OHaganBooks.com has recently decided to start selling music albums online through a service it calls *o'Tunes*.[75] Users pay a fee to download an entire music album. Composer royalties and copyright fees cost an average of $5.50 per album, and the cost of operating and maintaining *o'Tunes* amounts to $500 per week. The company is currently charging customers $9.50 per album.

a. What are the associated (weekly) cost, revenue, and profit functions?

b. How many albums must be sold per week in order to make a profit?

c. If the charge is lowered to $8.00 per album, how many albums must be sold per week in order to make a profit?

40. **Break-Even Analysis** OHaganBooks.com also generates revenue through its *o'Books* e-book service. Author royalties and copyright fees cost the company an average of $4 per novel, and the monthly cost of operating and maintaining the service amounts to $900 per month. The company is currently charging readers $5.50 per novel.

a. What are the associated cost, revenue, and profit functions?

b. How many novels must be sold per month in order to break even?

c. If the charge is lowered to $5.00 per novel, how many books must be sold in order to break even?

41. **Demand and Profit** In order to generate a profit from its new *o'Tunes* service, OHaganBooks.com needs to know how the demand for music albums depends on the price it charges. During the first week of the service, it was charging $7 per album, and sold 500. Raising the price to $9.50 had the effect of lowering demand to 300 albums per week.

a. Use the given data to construct a linear demand equation.

b. Use the demand equation you constructed in part (a) to estimate the demand if the price was raised to $12 per album.

[75]The (highly original) name was suggested to John O'Hagan by Marjory Duffin over cocktails one evening.

c. Using the information on cost given in Exercise 39, determine which of the three prices ($7, $9.50 and $12) would result in the largest weekly profit, and the size of that profit.

42. **Demand and Profit** In order to generate a profit from its *o'Books* e-book service, OHaganBooks.com needs to know how the demand for novels depends on the price it charges. During the first month of the service, it was charging $10 per novel, and sold 350. Lowering the price to $5.50 per novel had the effect of increasing demand to 620 novels per month.

a. Use the given data to construct a linear demand equation.

b. Use the demand equation you constructed in part (a) to estimate the demand if the price was raised to $15 per novel.

c. Using the information on cost given in Exercise 40, determine which of the three prices ($5.50, $10 and $15) would result in the largest profit, and the size of that profit.

43. **Demand** OHaganBooks.com has tried selling music albums on *o'Tunes* at a variety of prices, with the following results:

| Price | $8.00 | $8.50 | $10 | $11.50 |
|---|---|---|---|---|
| Demand (Weekly sales) | 440 | 380 | 250 | 180 |

a. Use the given data to obtain a linear regression model of demand.

b. Use the demand model you constructed in part (a) to estimate the demand if the company charged $10.50 per album. (Round the answer to the nearest album.)

44. **Demand** OHaganBooks.com has tried selling novels through *o'Books* at a variety of prices, with the following results:

| Price | $5.50 | $10 | $11.50 | $12 |
|---|---|---|---|---|
| Demand (Monthly sales) | 620 | 350 | 350 | 300 |

a. Use the given data to obtain a linear regression model of demand.

b. Use the demand model you constructed in part (a) to estimate the demand if the company charged $8 per novel. (Round the answer to the nearest novel.)

Jeff Titcomb/Photographer's Choice / Getty Images

**Case Study**   **Modeling Spending on Internet Advertising**

You are the new director of Impact Advertising Inc.'s Internet division, which has enjoyed a steady 0.25% of the Internet advertising market. You have drawn up an ambitious proposal to expand your division in light of your anticipation that Internet advertising will continue to skyrocket. However, upper management sees things differently and, based on the following email, does not seem likely to approve the budget for your proposal.

TO: JCheddar@impact.com (J. R. Cheddar)
CC: CVODoylePres@impact.com (C. V. O'Doyle, CEO)
FROM: SGLombardoVP@impact.com (S. G. Lombardo, VP Financial Affairs)
SUBJECT: Your Expansion Proposal
DATE: May 30, 2014

Hi John:

Your proposal reflects exactly the kind of ambitious planning and optimism we like to see in our new upper management personnel. Your presentation last week was most impressive, and obviously reflected a great deal of hard work and preparation.

I am in full agreement with you that Internet advertising is on the increase. Indeed, our Market Research department informs me that, based on a regression of the most recently available data, Internet advertising revenue in the United States will continue to grow at a rate of approximately $2.7 billion per year. This translates into approximately $6.75 million in increased revenues per year for Impact, given our 0.25% market share. This rate of expansion is exactly what our planned 2015 budget anticipates. Your proposal, on the other hand, would require a budget of approximately *twice* the 2015 budget allocation, even though your proposal provides no hard evidence to justify this degree of financial backing.

At this stage, therefore, I am sorry to say that I am inclined not to approve the funding for your project, although I would be happy to discuss this further with you. I plan to present my final decision on the 2015 budget at next week's divisional meeting.

Regards, Sylvia

Refusing to admit defeat, you contact the Market Research department and request the details of their projections on Internet advertising. They fax you the following information:[76]

| Year | 2007 | 2008 | 2009 | 2010 | 2011 | 2012 | 2013 | 2014 |
|---|---|---|---|---|---|---|---|---|
| Internet Advertising Revenue ($ Billion) | 21.2 | 23.4 | 22.7 | 25.8 | 28.5 | 32.6 | 36 | 40.5 |

**Regression Model:** $y = 2.744x + 19.233$   ($x$ = time in years since 2007)

**Correlation Coefficient:** $r = 0.970$

Now you see where the VP got that $2.7 billion figure: The slope of the regression equation is close to 2.7, indicating a rate of increase of about $2.7 billion per year. Also, the correlation coefficient is very high—an indication that the linear model fits the data well. In view of this strong evidence, it seems difficult to argue that revenues will increase by significantly more than the projected $2.7 billion per year. To get a

---

[76]The 2011–2014 figures are projections by eMarketer. Source: www.eMarketer.com.

Internet Advertising Revenue
($ billions)

Year since 2007

**Figure 32**

✳ Note that this *r* is *not* the linear correlation coefficient we defined on page 101; what this *r* measures is how closely the *quadratic* regression model fits the data.

Internet Advertising Revenue
($ billions)

Year since 2007

**Figure 33**

† The number of degrees of freedom in a regression model is 1 less than the number of coefficients. For a linear model, it is 1 (there are two coefficients: the slope *m* and the intercept *b*), and for a quadratic model it is 2. For a detailed discussion, consult a text on regression analysis.

better picture of what's going on, you decide to graph the data together with the regression line in your spreadsheet. What you get is shown in Figure 32. You immediately notice that the data points seem to suggest a curve, and not a straight line. Then again, perhaps the suggestion of a curve is an illusion. Thus there are, you surmise, two possible interpretations of the data:

1. (Your first impression)  As a function of time, Internet advertising revenue is nonlinear, and is in fact accelerating (the rate of change is increasing), so a linear model is inappropriate.

2. (Devil's advocate)  Internet advertising revenue *is* a linear function of time; the fact that the points do not lie on the regression line is simply a consequence of random factors that do not reflect a long-term trend, such as world events, mergers and acquisitions, short-term fluctuations in economy or the stock market, etc.

You suspect that the VP will probably opt for the second interpretation and discount the graphical evidence of accelerating growth by claiming that it is an illusion: a "statistical fluctuation." That is, of course, a possibility, but you wonder how likely it really is.

For the sake of comparison, you decide to try a regression based on the simplest nonlinear model you can think of—a quadratic function.

$$y = ax^2 + bx + c$$

Your spreadsheet allows you to fit such a function with a click of the mouse. The result is the following.

$$y = 0.3208x^2 + 0.4982x + 21.479 \quad (x = \text{number of years since 2007})$$
$$r = 0.996 \qquad \text{See Note.}✳$$

Figure 33 shows the graph of the regression function together with the original data.
Aha! The fit is visually far better, and the correlation coefficient is even higher! Further, the quadratic model predicts 2015 revenue as

$$y = 0.3208(8)^2 + 0.4982(8) + 21.479 \approx \$46.0 \text{ billion,}$$

which is $5.5 billion above the 2014 spending figure in the table above. Given Impact Advertising's 0.25% market share, this translates into an increase in revenues of $13.75 million, which is about double the estimate predicted by the linear model!

You quickly draft an email to Lombardo, and are about to click "Send" when you decide, as a precaution, to check with a colleague who is knowledgeable in statistics. He tells you to be cautious: The value of *r* will always tend to increase if you pass from a linear model to a quadratic one because of the increase in "degrees of freedom."† A good way to test whether a quadratic model is more appropriate than a linear one is to compute a statistic called the "*p*-value" associated with the coefficient of $x^2$. A low value of *p* indicates a high degree of confidence that the coefficient of $x^2$ cannot be zero (see below). Notice that if the coefficient of $x^2$ is zero, then you have a linear model.

You can, your colleague explains, obtain the *p*-value using your spreadsheet as follows (the method we describe here works on all the popular spreadsheets, including *Excel, Google Docs,* and *Open Office Calc*).

First, set up the data in columns, with an extra column for the values of $x^2$:

| ◇ | A | B | C |
|---|---|---|---|
| 1 | y | x | x^2 |
| 2 | 21.2 | 0 | 0 |
| 3 | 23.4 | 1 | 1 |
| 4 | 22.7 | 2 | 4 |
| 5 | 25.8 | 3 | 9 |
| 6 | 28.5 | 4 | 16 |
| 7 | 32.6 | 5 | 25 |
| 8 | 36 | 6 | 36 |
| 9 | 40.5 | 7 | 49 |

Then, highlight a vacant $5 \times 3$ block (the block E1:G5 say), type the formula =LINEST(A2:A9,B2:C9,,TRUE), and press Cntl+Shift+Enter (not just Enter!). You will see a table of statistics like the following:

| ◇ | E | F | G |
|---|---|---|---|
| 1 | =LINEST(A2:A9,B2:C9,,TRUE) | | |
| 2 | | | |
| 3 | | | |
| 4 | | | |
| 5 | | | |

↓
Cntl+Shift+Enter

| ◇ | E | F | G |
|---|---|---|---|
| 1 | 0.32083333 | 0.49821429 | 21.4791667 |
| 2 | 0.05808804 | 0.42288732 | 0.63366572 |
| 3 | 0.99157395 | 0.75290706 | #N/A |
| 4 | 294.198816 | 5 | #N/A |
| 5 | 333.544405 | 2.83434524 | #N/A |

(Notice the coefficients of the quadratic model in the first row.) The $p$-value is then obtained by the formula =TDIST(ABS(E1/E2),F4,2), which you can compute in any vacant cell. You should get $p \approx 0.00267$.

---

**Q:** *What does p actually measure?*

**A:** *Roughly speaking, $1 - p \approx 0.997733$ gives the degree of confidence you can have (99.7733%) in asserting that the coefficient of $x^2$ is not zero. (Technically, $p$ is the probability—allowing for random fluctuation in the data—that, if the coefficient of $x^2$ were in fact zero, the ratio E1/E2 could be as large as it is.)*

---

In short, you can go ahead and send your email with almost 100% confidence!

## EXERCISES

*Suppose you are given the following data for the spending on Internet advertising in a hypothetical country in which Impact Advertising also has a 0.25% share of the market.*

| Year | 2010 | 2011 | 2012 | 2013 | 2014 | 2015 | 2016 |
|---|---|---|---|---|---|---|---|
| Spending on Advertising ($ Billion) | 0 | 0.3 | 1.5 | 2.6 | 3.4 | 4.3 | 5.0 |

1. Obtain a linear regression model and the correlation coefficient $r$. (Take $t$ to be time in years since 2010.) According to the model, at what rate is spending on Internet Advertising increasing in this country? How does this translate to annual revenues for Impact Advertising?

2. Use a spreadsheet or other technology to graph the data together with the best-fit line. Does the graph suggest a quadratic model (parabola)?

3. Test your impression in the preceding exercise by using technology to fit a quadratic function and graphing the resulting curve together with the data. Does the graph suggest that the quadratic model is appropriate?

4. Perform a regression analysis using the quadratic model and find the associated $p$-value. What does it tell you about the appropriateness of a quadratic model?

**TECHNOLOGY GUIDE**

## TI-83/84 Plus    Technology Guide

### Section 1.1

**Example 1(a) and (c) (page 44)**    The total number of iPods sold by Apple up to the end of year $x$ can be approximated by $f(x) = 4x^2 + 16x + 2$ million iPods ($0 \le x \le 6$), where $x = 0$ represents 2003. Compute $f(0)$, $f(2)$, $f(4)$, and $f(6)$, and obtain the graph of $f$.

#### Solution with Technology

You can use the Y= screen to enter an algebraically defined function.

1. Enter the function in the Y= screen, as

    $Y_1 = 4X\char`\^2+16X+2$

    or    $Y_1 = 4X^2+16X+2$

(See Chapter 0 for a discussion of technology formulas.)

2. To evaluate $f(0)$, for example, enter $Y_1(0)$ in the Home screen to evaluate the function $Y_1$ at 0. Alternatively, you can use the table feature: After entering the function under $Y_1$, press 2ND TBLSET, and set Indpnt to Ask. (You do this once and for all; it will permit you to specify values for $x$ in the table screen.) Then, press 2ND TABLE, and you will be able to evaluate the function at several values of $x$. Below (top) is a table showing the values requested:

3. To obtain the graph above press WINDOW, set Xmin = 0, Xmax = 6 (the range of $x$-values we are interested in), Ymin = 0, Ymax = 300 (we estimated Ymin and Ymax from the corresponding set of $y$-values in the table) and press GRAPH to obtain the curve. Alternatively, you can avoid having to estimate Ymin and Ymax by pressing ZoomFit ( ZOOM 0 ), which automatically sets Ymin and Ymax to the smallest and greatest values of $y$ in the specified range for $x$.

**Example 2 (page 47)**    The number $n(t)$ of Facebook members can be approximated by the following function of time $t$ in years ($t = 0$ represents January 2004):

$$n(t) = \begin{cases} 4t & \text{if } 0 \le t \le 3 \\ 12 + 23(t - 3)^{2.7} & \text{if } 3 < t \le 6 \end{cases}$$

$$\text{million members.}$$

Obtain a table showing the values $n(t)$ for $t = 0, \ldots, 6$ and also obtain the graph of $n$.

#### Solution with Technology

You can enter a piecewise-defined function using the logical inequality operators $<$, $>$, $\le$, and $\ge$, which are found by pressing 2ND TEST:

1. Enter the function $n$ in the Y= screen as:

    $Y_1=(X\le3)*(4X)+(X>3)*(12+23*abs(X-3)\char`\^2.7)$

    When $x$ is less than or equal to 3, the logical expression (X≤3) evaluates to 1 because it is true, and the expression (X>3) evaluates to 0 because it is false. The value of the function is therefore given by the expression (4X). When $x$ is greater than 3, the expression (X≤3) evaluates to 0 while the expression (X>3) evaluates to 1, so the value of the function is given by the expression (12+23*abs(X-3)^2.7). (The reason we use the abs in the formula is to prevent an error in evaluating $(x - 3)^{2.7}$ when $x < 3$; even though we don't use that formula when $x < 3$, we are in fact evaluating it and multiplying it by zero.)

2. As in Example 1, use the Table feature to compute several values of the function at once by pressing 2ND TABLE.

3. To obtain the graph, we proceed as in Example 1: Press WINDOW, set Xmin = 0, Xmax = 6 (the range of $x$-values we are interested in), Ymin = 0, Ymax = 500 (see the $y$-values in the table) and press GRAPH.

## Section 1.2

**Example 4(a) (page 62)**    The demand and supply curves for private schools in Michigan are $q = 77.8p^{-0.11}$ and $q = 30.4 + 0.006p$ thousand students, respectively ($200 \le p \le 2{,}200$), where $p$ is the net tuition cost in dollars. Graph the demand and supply curves on the same set of axes. Use your graph to estimate, to the nearest $100, the tuition at which the demand equals the supply (equilibrium price). Approximately how many students will be accommodated at that price?

### Solution with Technology

To obtain the graphs of demand and supply:

1. Enter $Y_1=77.8*X^{(-0.11)}$ and $Y_2=30.4+0.006*X$ in the "Y=" screen.
2. Press WINDOW, enter Xmin = 200, Xmax = 2200, Ymin = 0, Ymax = 50 and press GRAPH for the graph shown below:

3. To estimate the equilibrium price, press TRACE and use the arrow keys to follow the curve to the approximate point of intersection (around X = 1008) as shown below.

4. For a more accurate estimate, zoom in by pressing ZOOM and selecting Option 1 ZBox.
5. Move the curser to a point slightly above and to the left of the intersection, press ENTER, and then move the curser to a point slightly below and to the right and press ENTER again to obtain a box.

6. Now press ENTER again for a zoomed-in view of the intersection.
7. You can now use TRACE to obtain the intersection coordinates more accurately: X ≈ 1,000, representing a tuition cost of $1,000. The associated demand is the Y-coordinate: around 36.4 thousand students.

**Example 5(a) (page 64)**    The following table shows annual sales, in billions of dollars, by *Nike* from 2005 through 2010 ($t = 0$ represents 2005):

| $t$ | 0 | 1 | 2 | 3 | 4 | 5 |
|---|---|---|---|---|---|---|
| Sales ($ billion) | 13.5 | 15 | 16.5 | 18.5 | 19 | 19 |

Consider the following four models:

(1) $s(t) = 14 + 1.2t$        Linear model

(2) $s(t) = 13 + 2.2t - 0.2t^2$        Quadratic model

(3) $s(t) = 14(1.07^t)$        Exponential model

(4) $s(t) = \dfrac{19.5}{1 + 0.48(1.8^{-t})}$.        Logistic model

**a.** Which models fit the data significantly better than the rest?

**b.** Of the models you selected in part (a), which gives the most reasonable prediction for 2013?

**TECHNOLOGY GUIDE**

### Solution with Technology

1. First enter the actual revenue data in the stat list editor ([STAT] EDIT) with the values of $t$ in $L_1$, and the values of $s(t)$ in $L_2$.

2. Now go to the Y= window and turn Plot1 on by selecting it and pressing [ENTER]. (You can also turn it on in the [2ND] STAT PLOT screen.) Then press ZoomStat ([ZOOM] [9]) to obtain a plot of the points (above).

3. To see any of the four curves plotted along with the points, enter its formula in the Y= screen (for instance, Y₁=13+2.2x-0.2X² for the second model) and press [GRAPH] (figure on top below).

4. To see the extrapolation of the curve to 2013, just change Xmax to 8 (in the [WINDOW] screen) and press [GRAPH] again (lower figure above).

5. Now change Y₁ to see similar graphs for the remaining curves.

6. When you are done, turn Plot1 off again so that the points you entered do not show up in other graphs.

## Section 1.3

**Example 1 (page 81)**    Which of the following two tables gives the values of a linear function? What is the formula for that function?

| $x$ | 0 | 2 | 4 | 6 | 8 | 10 | 12 |
|-----|---|----|----|----|----|-----|-----|
| $f(x)$ | 3 | $-1$ | $-3$ | $-6$ | $-8$ | $-13$ | $-15$ |

| $x$ | 0 | 2 | 4 | 6 | 8 | 10 | 12 |
|-----|---|----|----|----|----|-----|-----|
| $g(x)$ | 3 | $-1$ | $-5$ | $-9$ | $-13$ | $-17$ | $-21$ |

### Solution with Technology

We can use the "List" feature in the TI-83/84 Plus to automatically compute the successive quotients $m = \Delta y / \Delta x$ for either $f$ or $g$ as follows:

1. Use the stat list editor ([STAT] EDIT) to enter the values of $x$ and $f(x)$ in the first two columns, called $L_1$ and $L_2$, as shown in the screenshot below. (If there is already data in a column you want to use, you can clear it by highlighting the column heading (e.g., $L_1$) using the arrow key, and pressing [CLEAR] [ENTER].)

2. Highlight the heading $L_3$ by using the arrow keys, and enter the following formula (with the quotes, as explained below):

"ΔList(L₂)/ΔList(L₁)"    ΔList is found under [2ND] [LIST] OPS. L₁ is [2ND] [1]

The "ΔList" function computes the differences between successive elements of a list, returning a list with one less element. The formula above then computes the quotients $\Delta y / \Delta x$ in the list $L_3$ as shown in the following screenshot. As you can see in the third column, $f(x)$ is not linear.

**3.** To redo the computation for $g(x)$, all you need to do is edit the values of $L_2$ in the stat list editor. By putting quotes around the formula we used for $L_3$, we told the calculator to remember the formula, so it automatically recalculates the values.

**Example 2(a) (page 83)** Find the equation of the line through the points $(1, 2)$ and $(3, -1)$.

### Solution with Technology

**1.** Enter the coordinates of the given points in the stat list editor (STAT EDIT) with the values of $x$ in $L_1$, and the values of $y$ in $L_2$.

**2.** To compute the slope, enter the following formula in the Home screen:

$$(L_2(2)-L_2(1))/(L_1(2)-L_1(1)) \rightarrow M$$

$L_1$ and $L_2$ are under 2ND LIST and the arrow is STO

**3.** Then, to compute the $y$-intercept, enter

$$L_2(1)-M*L_1(1)$$

## Section 1.4

**Example 1(a) (page 97)** Using the data on the per capita GDP in Mexico given at the beginning of Section 1.4, compute SSE, the sum-of-squares error, for the linear models $y = 0.5t + 8$ and $y = 0.25t + 9$, and graph the data with the given models.

### Solution with Technology

We can use the "List" feature in the TI-83/84 Plus to automate the computation of SSE.

**1.** Use the stat list editor (STAT EDIT) to enter the given data in the lists $L_1$ and $L_2$, as shown in the first screenshot below. (If there is already data in a column you want to use, you can clear it by highlighting the column heading (e.g., $L_1$) using the arrow key, and pressing CLEAR ENTER .)

**2.** To compute the predicted values, highlight the heading $L_3$ using the arrow keys, and enter the following formula for the predicted values (figure on the top below):

$$0.5*L_1+8 \qquad L_1 \text{ is } 2ND \boxed{1}$$

Pressing ENTER again will fill column 3 with the predicted values (below bottom). Note that only seven of the eight data points can be seen on the screen at one time.

**3.** Highlight the heading $L_4$ and enter the following formula (including the quotes):

$$"(L_2-L_3)^2" \qquad \text{Squaring the residuals}$$

4. Pressing $\boxed{\text{ENTER}}$ will fill $L_4$ with the squares of the residuals. (Putting quotes around the formula will allow us to easily check the second model, as we shall see.)

5. To compute SSE, the sum of the entries in $L_4$, go to the home screen and enter $\text{sum}(L_4)$ (see below; "sum" is under $\boxed{\text{2ND}}$ $\boxed{\text{LIST}}$ $\boxed{\text{MATH}}$.)

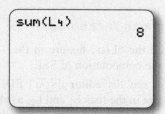

6. To check the second model, go back to the List screen, highlight the heading $L_3$, enter the formula for the second model, $0.25*L_1+9$, and press $\boxed{\text{ENTER}}$. Because we put quotes around the formula for the residuals in $L_4$, the TI-83/84 Plus will remember the formula and automatically recalculate the values (below top). On the home screen we can again calculate $\text{sum}(L_4)$ to get SSE for the second model (below bottom).

The second model gives a much smaller SSE, so is the better fit.

7. You can also use the TI-83/84 Plus to plot both the original data points and the two lines (see below). Turn Plot1 on in the STAT PLOT window, obtained by pressing $\boxed{\text{2ND}}$ $\boxed{\text{STAT PLOT}}$. To show the lines, enter them in the "Y=" screen as usual. To obtain a convenient window showing all the points and the lines, press $\boxed{\text{ZOOM}}$ and choose 9: ZoomStat.

**Example 2 (page 99)**    Use the data on the per capita GDP in Mexico to find the best-fit linear model.

**Solution with Technology**

1. Enter the data in the TI-83/84 Plus using the List feature, putting the $x$-coordinates in $L_1$ and the $y$-coordinates in $L_2$, just as in Example 1.

2. Press $\boxed{\text{STAT}}$, select CALC, and choose #4: LinReg(ax+b). Pressing $\boxed{\text{ENTER}}$ will cause the equation of the regression line to be displayed in the home screen:

So, the regression line is $y \approx 0.321x + 8.75$.

3. To graph the regression line without having to enter it by hand in the "Y=" screen, press $\boxed{\text{Y=}}$, clear the contents of $Y_1$, press $\boxed{\text{VARS}}$, choose #5: Statistics, select EQ, and then choose #1:RegEQ. The regression equation will then be entered under $Y_1$.

4. To simultaneously show the data points, press $\boxed{\text{2ND}}$ $\boxed{\text{STATPLOT}}$ and turn Plot1 on as in Example 1. To obtain a convenient window showing all the points and the line (see below), press $\boxed{\text{ZOOM}}$ and choose #9: ZoomStat.

**Example 3 (page 101)**   Find the correlation coefficient for the data in Example 2.

### Solution with Technology

To find the correlation coefficient using a TI-83/84 Plus you need to tell the calculator to show you the coefficient at the same time that it shows you the regression line.

1. Press [2ND] [CATALOG] and select DiagnosticOn from the list. The command will be pasted to the home screen, and you should then press [ENTER] to execute the command.

2. Once you have done this, the "LinReg(ax+b)" command (see the discussion for Example 2) will show you not only $a$ and $b$, but $r$ and $r^2$ as well:

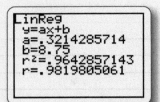

```
LinReg
y=ax+b
a=.3214285714
b=8.75
r²=.9642857143
r=.9819805061
```

---

## SPREADSHEET Technology Guide

### Section 1.1

**Example 1(a) and (c) (page 44)**   The total number of iPods sold by Apple up to the end of year $x$ can be approximated by $f(x) = 4x^2 + 16x + 2$ million iPods ($0 \le x \le 6$), where $x = 0$ represents 2003. Compute $f(0)$, $f(2)$, $f(4)$, and $f(6)$, and obtain the graph of $f$.

### Solution with Technology

To create a table of values of $f$ using a spreadsheet:

1. Set up two columns: one for the values of $x$ and one for the values of $f(x)$. Then enter the sequence of values 0, 2, 4, 6 in the $x$ column as shown.

|   | A | B |
|---|---|---|
| 1 | x | f(x) |
| 2 | 0 | |
| 3 | 2 | |
| 4 | 4 | |
| 5 | 6 | |

2. Now we enter a formula for $f(x)$ in cell B2 (below). The technology formula is `4*x^2+16*x+2`. To use this formula in a spreadsheet, we modify it slightly:

`=4*A2^2+16*A2+2`   Spreadsheet version of tech formula

Notice that we have preceded the Excel formula by an equals sign (=) and replaced each occurrence of $x$ by the name of the cell holding the value of $x$ (cell A2 in this case).

|   | A | B | C |
|---|---|---|---|
| 1 | x | f(x) | |
| 2 | | 0 | =4*A2^2+16*A2+2 |
| 3 | | 2 | |
| 4 | | 4 | |
| 5 | | 6 | |

**Note**  Instead of typing in the name of the cell "A2" each time, you can simply click on the cell A2, and "A2" will be automatically inserted.  ∎

3. Now highlight cell B2 and drag the **fill handle** (the little square at the lower right-hand corner of the selection) down until you reach Row 5 as shown below on the top, to obtain the result shown on the bottom.

|   | A | B | C |
|---|---|---|---|
| 1 | x | f(x) | |
| 2 | | 0 | =4*A2^2+16*A2+2 |
| 3 | | 2 | |
| 4 | | 4 | |
| 5 | | 6 | |

|   | A | B |
|---|---|---|
| 1 | x | f(x) |
| 2 | | 0 | 2 |
| 3 | | 2 | 50 |
| 4 | | 4 | 130 |
| 5 | | 6 | 242 |

4. To graph the data, highlight A1 through B5, and insert a "Scatter chart" (the exact method of doing this depends on the specific version of the spreadsheet program). When choosing the style of the chart,

choose a style that shows points connected by lines (if possible) to obtain a graph something like the following:

**Example 2 (page 47)** The number $n(t)$ of Facebook members can be approximated by the following function of time $t$ in years ($t = 0$ represents January 2004):

$$n(t) = \begin{cases} 4t & \text{if } 0 \leq t \leq 3 \\ 12 + 23(t-3)^{2.7} & \text{if } 3 < t \leq 6 \end{cases} \text{ million members.}$$

Obtain a table showing the values $n(t)$ for $t = 0, \ldots, 6$ and also obtain the graph of $n$.

## Solution with Technology

We can generate a table of values of $n(t)$ for $t = 0, 1, \ldots, 6$ as follows:

1. Set up two columns; one for the values of $t$ and one for the values of $n(t)$, and enter the values $0, 1, \ldots, 6$ in the $t$ column as shown in the first screenshot below:

2. We must now enter the formula for $n$ in cell B2. The following formula defines the function $n$ in Excel:

    ```
    =(x<=3)*(4*x)+(x>3)*(12+23*abs
    (x-3)^2.7)
    ```

When $x$ is less than or equal to 3, the logical expression $(x\leq3)$ evaluates to 1 because it is true, and the expression $(x>3)$ evaluates to 0 because it is false. The value of the function is therefore given by the expression $(4*x)$. When $x$ is greater than 3, the expression $(x\leq3)$ evaluates to 0 while the expression $(x>3)$ evaluates to 1, so the value of the function is given by the expression $(12+23*abs (x-3)^2.7)$. (The reason we use the abs in the formula is to prevent an error in evaluating $(x-3)^{2.7}$ when $x < 3$; even though we don't use that formula

when $x < 3$, we are in fact evaluating it and multiplying it by zero.) We therefore enter the formula

```
=(A2<=3)*(4*A2)+(A2>3)*(12+23
*ABS(A2-3)^2.7)
```

in cell B2 and then copy down to cell B8 (below top) to obtain the result shown on the bottom:

3. To graph the data, highlight A1 through B8, and insert a "Scatter chart" as in Example 1 to obtain the result shown below:

## Section 1.2

**Example 4(a) (page 62)** The demand and supply curves for private schools in Michigan are $q = 77.8p^{-0.11}$ and $q = 30.4 + 0.006p$ thousand students, respectively ($200 \leq p \leq 2{,}200$), where $p$ is the net tuition cost in dollars. Graph the demand and supply curves on the same set of axes. Use your graph to estimate, to the nearest \$100, the tuition at which the demand equals the supply (equilibrium price). Approximately how many students will be accommodated at that price?

## Solution with Technology

To obtain the graphs of demand and supply:

1. Enter the headings $p$, Demand, and Supply in cells A1–C1 and the $p$-values 200, 300, ..., 2,200 in A2–A22.

**2.** Next, enter the formulas for the demand and supply functions in cells B2 and C2.

Demand: `=77.8*A2^(-0.11)`    in cell B2

Supply:  `=30.4+0.006*A2`    in cell C2

 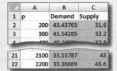

**3.** To graph the data, highlight A1 through C22, and insert a Scatter chart:

**4.** If you place the cursor as close as you can get to the intersection point (or just look at the table of values), you will see that the curves cross close to $p = \$1,000$ (to the nearest $100).

**5.** To more accurately determine where the curves cross, you can narrow down the range of values shown on the $x$-axis by changing the $p$-values to 990, 991, . . . , 1010.

**Example 5(a) (page 64)** The following table shows annual sales, in billions of dollars, by *Nike* from 2005 through 2010 ($t = 0$ represents 2005):

| $t$ | 0 | 1 | 2 | 3 | 4 | 5 |
|---|---|---|---|---|---|---|
| Sales ($ billion) | 13.5 | 15 | 16.5 | 18.5 | 19 | 19 |

Consider the following four models:

(1) $s(t) = 14t + 1.2t$    Linear model

(2) $s(t) = 13 + 2.2t - 0.2t^2$    Quadratic model

(3) $s(t) = 14(1.07^t)$    Exponential model

(4) $s(t) = \dfrac{19.5}{1 + 0.48(1.8^{-t})}$    Logistic model

**a.** Which models fit the data significantly better than the rest?

**b.** Of the models you selected in part (a), which gives the most reasonable prediction for 2013?

## Solution with Technology

**1.** First create a scatter plot of the given data by tabulating the data as shown below, and selecting the Insert tab and choosing a "Scatter" chart:

**2.** In column C use the formula for the model you are interested in seeing; for example, model (2): `=13+2.2*A2-0.2*A2^2`

**3.** To adjust the graph to include the graph of the model you have added, you need to change the graph data from $A$2:$B$7 to $A$2:$C$7 so as to include column C. In Excel you can obtain this by right-clicking on the graph to select "Source Data". In OpenOffice, double-click on the graph and then right-click it to choose "Data Ranges". In Excel, you can also click once on the graph—the effect will be to outline the data you have graphed in columns A and B—and then use the fill handle at the bottom of Column B to extend the selection to Column C as shown:

The graph will now include markers showing the values of both the actual sales and the model you inserted in Column C.

**4.** Right-click on any of the markers corresponding to column B in the graph (in OpenOffice you would first double-click on the graph), select "Format data series" to add lines connecting the points and remove the markers. The effect will be as shown below, with the model represented by a curve and the actual data points represented by dots:

5. To see the extrapolation of the curve to 2013, add the values 6, 7, 8 to Column A. The values of $s(t)$ may automatically be computed in Column C as you type, depending on the spreadsheet. If not, you will need to copy the formula in column C down to C10. (Do not touch Column B, as that contains the observed data up through $t = 5$ only.) Click on the graph, and use the fill handle at the base of Column C to include the new data in the graph:

6. To see the plots for the remaining curves, change the formula in Column B (and don't forget to copy the new formula down to cell C10 when you do so).

## Section 1.3

**Example 1 (page 81)**    Which of the following two tables gives the values of a linear function? What is the formula for that function?

| x | 0 | 2 | 4 | 6 | 8 | 10 | 12 |
|---|---|---|---|---|---|----|----|
| f(x) | 3 | −1 | −3 | −6 | −8 | −13 | −15 |

| x | 0 | 2 | 4 | 6 | 8 | 10 | 12 |
|---|---|---|---|---|---|----|----|
| g(x) | 3 | −1 | −5 | −9 | −13 | −17 | −21 |

### Solution with Technology

1. The following worksheet shows how you can compute the successive quotients $m = \Delta y/\Delta x$, and hence check whether a given set of data shows a linear relationship, in which case all the quotients will be the same. (The shading indicates that the formula is to be copied down only as far as cell C7. Why not cell C8?)

2. Here are the results for both $f$ and $g$.

**Example 2(a) (page 83)**    Find the equation of the line through the points $(1, 2)$ and $(3, -1)$.

### Solution with Technology

1. Enter the $x$- and $y$-coordinates in columns A and B, as shown below on the left.

| | A | B |
|---|---|---|
| 1 | x | y |
| 2 | 1 | 2 |
| 3 | 3 | -1 |

| | A | B | C | D |
|---|---|---|---|---|
| 1 | x | y | m | b |
| 2 | 1 | 2 | =(B3-B2)/(A3-A2) | =B2-C2*A2 |
| 3 | 3 | -1 | Slope | Intercept |

2. Add the headings $m$ and $b$ in C1-D1, and then the formulas for the slope and intercept in C2-D2, as shown above on the right. The result will be as shown below:

| | A | B | C | D |
|---|---|---|---|---|
| 1 | x | y | m | b |
| 2 | 1 | 2 | -1.5 | 3.5 |
| 3 | 3 | -1 | Slope | Intercept |

## Section 1.4

**Example 1(a) (page 97)**    Using the data on the per capita GDP in Mexico given at the beginning of Section 1.4, compute SSE, the sum-of-squares error, for the linear models $y = 0.5t + 8$ and $y = 0.25t + 9$, and graph the data with the given models.

### Solution with Technology

1. Begin by setting up your worksheet with the observed data in two columns, $t$ and $y$, and the predicted data for the first model in the third.

2. Notice that, instead of using the numerical equation for the first model in column C, we used absolute references to the cells containing the slope $m$ and the intercept $b$. This way, we can switch from one linear model to the next by changing only $m$ and $b$ in cells E2 and F2. (We have deliberately left column D empty in anticipation of the next step.)

3. In column D we compute the squares of the residuals using the Excel formula $=(B2-C2)^2$.

4. We now compute SSE in cell F4 by summing the entries in column D:

5. Here is the completed spreadsheet:

6. Changing $m$ to 0.25 and $b$ to 9 gives the sum of squares error for the second model, SSE $= 2$.

7. To plot both the original data points and each of the two lines, use a scatter plot to graph the data in columns A through C in each of the last two worksheets above.

$$y = 0.5t + 8 \qquad\qquad y = 0.25t + 9$$

**Example 2 (page 99)** Use the data on the per capita GDP in Mexico to find the best-fit linear model.

### Solution with Technology

Here are two spreadsheet shortcuts for linear regression; one graphical and one based on a spreadsheet formula:

*Using a Trendline*

1. Start with the original data and insert a scatter plot (below left and right).

2. Insert a "linear trendline", choosing the option to display the equation on the chart. The method for doing so varies from spreadsheet to spreadsheet.[77] In Excel, you can right-click on one of the points in the graph and choose "Add Trendline" (in OpenOffice you would first double-click on the graph). Then, under "Trendline Options", select "Display Equation on chart". The procedure for OpenOffice is almost identical, but you first need to double-click on the graph. The result is shown below.

*Using a Formula*

1. Enter your data as above, and select a block of unused cells two wide and one tall; for example, C2:D2. Then enter the formula

    ```
    =LINEST(B2:B9,A2:A9)
    ```

---

[77]At the time of this writing, Google Docs has no trendline feature for its spreadsheet, so you would need to use the formula method.

**TECHNOLOGY GUIDE**

as shown on the left. Then press Control-Shift-Enter. The result should appear as on the right, with $m$ and $b$ appearing in cells C2 and D2 as shown:

**Example 3 (page 101)**    Find the correlation coefficient for the data in Example 2.

### Solution with Technology

1. When you add a trendline to a chart you can select the option "Display R-squared value on chart" to show the value of $r^2$ on the chart (it is common to examine $r^2$, which takes on values between 0 and 1, instead of $r$).

2. Alternatively, the LINEST function we used above in 2 can be used to display quite a few statistics about a best-fit line, including $r^2$. Instead of selecting a block of cells two wide and one tall as we did in Example 2, we select one two wide and *five* tall. We now enter the requisite LINEST formula with two additional arguments set to "TRUE" as shown, and press Control-Shift-Enter:

The values of $m$ and $b$ appear in cells C2 and D2 as before, and the value of $r^2$ in cell C4. (Among the other numbers shown is SSE in cell D6. For the meanings of the remaining numbers shown, do a Web search for "LINEST"; you will see numerous articles, including many that explain all the terms. A good course in statistics wouldn't hurt, either.)

# 2

# The Mathematics of Finance

**Website**

www.WanerMath.com

At the Website you will find:

- A detailed chapter summary
- A true/false quiz
- A time value of money utility
- Excel tutorials for every section

## Case Study  Adjustable Rate and Subprime Mortgages

Mr. and Mrs. Wong have an appointment tomorrow with you, their investment counselor, to discuss their plan to purchase a $400,000 house in Orlando, Florida. Their combined annual income is $80,000 per year, which they estimate will increase by 4% annually over the foreseeable future, and they are considering three different specialty 30-year mortgages:

**Hybrid:** The interest is fixed at a low introductory rate of 4% for 5 years.

**Interest-Only:** During the first 5 years, the rate is set at 4.2% and no principal is paid.

**Negative Amortization:** During the first 5 years, the rate is set at 4.7% based on a principal of 60% of the purchase price of the home.

**How would you advise them?**

Andy Dean Photography/Shutterstock

125

# Introduction

A knowledge of the mathematics of investments and loans is important not only for business majors but also for everyone who deals with money, which is all of us. This chapter is largely about *interest*: interest paid by an investment, interest paid on a loan, and variations on these.

We focus on three forms of investment: investments that pay simple interest, investments in which interest is compounded, and annuities. An investment that pays *simple interest* periodically gives interest directly to the investor, perhaps in the form of a monthly check. If instead, the interest is reinvested, the interest is *compounded*, and the value of the account grows as the interest is added. An *annuity* is an investment earning compound interest into which periodic payments are made or from which periodic withdrawals are made; in the case of periodic payments, such an investment is more commonly called a *sinking fund*. From the point of view of the lender, a loan is a kind of annuity.

We also look at bonds, the primary financial instrument used by companies and governments to raise money. Although bonds nominally pay simple interest, determining their worth, particularly in the secondary market, requires an annuity calculation.

## 2.1　Simple Interest

You deposit $1,000, called the **principal** or **present value**, into a savings account. The bank pays you 5% interest, in the form of a check, each year. How much interest will you earn each year? Because the bank pays you 5% interest each year, your annual (or yearly) interest will be 5% of $1,000, or $1,000 \times 0.05 = \$50$.

Generalizing this calculation, call the present value $PV$ and the interest rate (expressed as a decimal) $r$. Then $INT$, the annual interest paid to you, is given by*

$$INT = PVr.$$

If the investment is made for a period of $t$ years, then the total interest accumulated is $t$ times this amount, which gives us the following:

* Multiletter variables like $PV$ and $INT$ used here may be unusual in a math textbook but are almost universally used in finance textbooks, calculators (such as the TI-83/84 Plus), and such places as study guides for the finance portion of the Society of Actuaries exams. Just watch out for expressions like $PVr$, which is the product of two things, $PV$ and $r$, not three.

### Simple Interest

The **simple interest** on an investment (or loan) of $PV$ dollars at an annual interest rate of $r$ for a period of $t$ years is

$$INT = PVrt.$$

#### Quick Example

The simple interest over a period of 4 years on a $5,000 investment earning 8% per year is

$$INT = PVrt$$
$$= (5,000)(0.08)(4) = \$1,600.$$

Given your $1,000 investment at 5% simple interest, how much money will you have after 2 years? To find the answer, we need to add the accumulated interest to the principal to get the **future value** $(FV)$ of your deposit.

$$FV = PV + INT = \$1,000 + (1,000)(0.05)(2) = \$1,100$$

In general, we can compute the future value as follows:

$$FV = PV + INT = PV + PVrt = PV(1 + rt).$$

---

## Future Value for Simple Interest

The **future value** of an investment of $PV$ dollars at an annual simple interest rate of $r$ for a period of $t$ years is

$$FV = PV(1 + rt).$$

### Quick Examples

1. The value, at the end of 4 years, of a $5,000 investment earning 8% simple interest per year is

$$FV = PV(1 + rt)$$
$$= 5,000[1 + (0.08)(4)] = \$6,600.$$

2. Writing the future value in Quick Example 1 as a function of time, we get

$$FV = 5,000(1 + 0.08t)$$
$$= 5,000 + 400t,$$

which is a linear function of time $t$. The intercept is $PV = \$5,000$, and the slope is the annual interest, $400 per year.

In general: *Simple interest growth is a linear function of time, with intercept given by the present value and slope given by annual interest.*

---

### EXAMPLE 1 Savings Accounts

In November 2011, the Bank of Montreal was paying 1.30% interest on savings accounts.[1] If the interest is paid as simple interest, find the future value of a $2,000 deposit in 6 years. What is the total interest paid over the period?

---

[1]Source: Canoe Money (http://money.canoe.ca/rates/savings.html)

**Solution**  We use the future value formula:

$$FV = PV(1 + rt)$$

$$= 2{,}000[1 + (0.013)(6)] = 2{,}000[1.078] = \$2{,}156.$$

The total interest paid is given by the simple interest formula:

$$INT = PVrt$$

$$= (2{,}000)(0.013)(6) = \$156.$$

**Note**  To find the interest paid, we could also have computed

$$INT = FV - PV = 2{,}156 - 2{,}000 = \$156. \quad \blacksquare$$

**➡ Before we go on...**  In the preceding example, we could look at the future value as a function of time:

$$FV = 2{,}000(1 + 0.013t) = 2{,}000 + 26t.$$

Thus, the future value is growing linearly at a rate of \$26 per year.  ■

## EXAMPLE 2  **Bridge Loans**

When "trading up," homeowners sometimes have to buy a new house before they sell their old house. One way to cover the costs of the new house until they get the proceeds from selling the old house is to take out a short-term *bridge loan*. Suppose a bank charges 12% simple annual interest on such a loan. How much will be owed at the maturation (the end) of a 90-day bridge loan of \$90,000?

**Solution**  We use the future value formula

$$FV = PV(1 + rt)$$

with $t = 90/365$, the fraction of a year represented by 90 days:

$$FV = 90{,}000[1 + (0.12)(90/365)]$$

$$= \$92{,}663.01.$$

(We will always round our answers to the nearest cent after calculation. Be careful not to round intermediate results.)

**➡ Before we go on...**  Many banks use 360 days for this calculation rather than 365, which makes a "year" for the purposes of the loan slightly shorter than a calendar year. The effect is to increase the amount owed:

$$FV = 90{,}000[1 + (0.12)(90/360)] = \$92{,}700 \qquad\qquad ■$$

One of the primary ways companies and governments raise money is by selling **bonds**. At its most straightforward, a corporate bond promises to pay simple interest, usually twice a year, for a length of time until it **matures**, at which point it returns the original investment to the investor (U.S. Treasury notes and bonds are similar). Things get more complicated when the selling price is negotiable, as we will see later in this chapter.

## EXAMPLE 3  Corporate Bonds

The Megabucks Corporation is issuing 10-year bonds paying an annual rate of 6.5%. If you buy $10,000 worth of bonds, how much interest will you earn every 6 months, and how much interest will you earn over the life of the bonds?

**Solution**  Using the simple interest formula, every 6 months you will receive

$$INT = PVrt$$
$$= (10,000)(0.065)\left(\frac{1}{2}\right) = \$325.$$

Over the 10-year life of the bonds, you will earn

$$INT = PVrt$$
$$= (10,000)(0.065)(10) = \$6,500$$

in interest. So, at the end of 10 years, when your original investment is returned to you, your $10,000 will have turned into $16,500.

We often want to turn an interest calculation around: Rather than starting with the present value and finding the future value, there are times when we know the future value and need to determine the present value. Solving the future value formula for $PV$ gives us the following.

---

## Present Value for Simple Interest

The present value of an investment at an annual simple interest rate of $r$ for a period of $t$ years, with future value $FV$, is

$$PV = \frac{FV}{1 + rt}.$$

### Quick Example

If an investment earns 5% simple interest and will be worth $1,000 in 4 years, then its present value (its initial value) is

$$PV = \frac{FV}{1 + rt}$$
$$= \frac{1,000}{1 + (0.05)(4)} = \$833.33.$$

---

Here is a typical example. U.S. Treasury bills (T-bills) are short-term investments (up to 1 year) that pay you a set amount after a period of time; what you pay to buy a T-bill depends on the interest rate.

---

## EXAMPLE 4  Treasury Bills

A U.S. Treasury bill paying $10,000 after 6 months earns 3.67% simple annual interest. How much did it cost to buy?

**Solution** The future value of the T-bill is $10,000; the price we paid is its present value. We know that

$$FV = \$10,000$$
$$r = 0.0367$$

and

$$t = 0.5.$$

Substituting into the present value formula, we have

$$PV = \frac{10,000}{1 + (0.0367)(0.5)} = 9,819.81$$

so we paid $9,819.81 for the T-bill.

---

➡ **Before we go on...** The simplest way to find the interest earned on the T-bill is by subtraction:

$$INT = FV - PV = 10,000 - 9,819.81 = \$180.19.$$

So, after 6 months we received back $10,000, which is our original investment plus $180.19 in interest. ∎

Here is some additional terminology on Treasury bills:

---

### Treasury Bills (T-Bills): Maturity Value, Discount Rate, and Yield

The **maturity value** of a T-bill is the amount of money it will pay at the end of its life, that is, upon **maturity**.

**Quick Example**

> A 1-year $10,000 T-bill has a maturity value of $10,000, and so will pay you $10,000 after one year.

The cost of a T-bill is generally less than its maturity value.* In other words, a T-bill will generally sell at a *discount*, and the **discount rate** is the *annualized* percentage of this discount; that is, the percentage is adjusted to give an annual percentage. (See Quick Examples 2 and 3.)

**Quick Examples**

1. A 1-year $10,000 T-bill with a discount rate of 5% will sell for 5% less than its maturity value of $10,000, that is, for $9,500.

2. A 6-month $10,000 T-bill with a discount rate of 5% will sell at an actual discount of half of that—2.5% less than its maturity value, or $9,750—because 6 months is half of a year.

3. A 3-month $10,000 T-bill with a discount rate of 5% will sell at an actual discount of a fourth of that: 1.25% less than its maturity value, or $9,875.

The annual **yield** of a T-bill is the simple annual interest rate an investor earns when the T-bill matures, as calculated in the next example.

* An exception occurred during the financial meltdown of 2008, when T-bills were heavily in demand as "safe haven" investments and were sometimes selling at—or even above—their maturity values.

### EXAMPLE 5  Treasury Bills

A T-bill paying $10,000 after 6 months sells at a discount rate of 3.6%. What does it sell for? What is the annual yield?

**Solution**  The (annualized) discount rate is 3.6%; so, for a 6-month bill, the actual discount will be half of that: $3.6\%/2 = 1.8\%$ below its maturity value. This makes the selling price

$$10,000 - (0.018)(10,000) = \$9,820. \qquad \text{Maturity value} - \text{Discount}$$

To find the annual yield, note that the present value of the investment is the price the investor pays, $9,820, and the future value is its maturity value, $10,000 six months later. So,

$$PV = \$9,820 \qquad FV = \$10,000 \qquad t = 0.5$$

and we wish to find the annual interest rate $r$. Substituting in the future value formula, we get

$$FV = PV(1 + rt)$$
$$10,000 = 9,820(1 + 0.5r)$$

so

$$1 + 0.5r = 10,000/9,820$$

and

$$r = (10,000/9,820 - 1)/0.5 \approx 0.0367.$$

Thus, the T-bill is paying 3.67% simple annual interest, so we say that its annual yield is 3.67%.

➡ **Before we go on...** The T-bill in Example 5 is the same one as in Example 4 (with a bit of rounding). The yield and the discount rate are two different ways of telling what the investment pays. One of the Communication and Reasoning Exercises asks you to find a formula for the yield in terms of the discount rate. ■

Fees on loans can also be thought of as a form of interest.

### EXAMPLE 6  Tax Refunds

You are expecting a tax refund of $800. Because it may take up to 6 weeks to get the refund, your tax preparation firm offers, for a fee of $40, to give you an "interest-free" loan of $800 to be paid back with the refund check. If we think of the fee as interest, what simple annual interest rate is the firm actually charging?

**Solution**  If we view the $40 as interest, then the future value of the loan (the value of the loan to the firm, or the total you will pay the firm) is $840. Thus, we have

$$FV = 840$$
$$PV = 800$$
$$t = 6/52 \qquad \text{Using 52 weeks in a year}$$

and we wish to find $r$. Substituting, we get

$$FV = PV(1 + rt)$$

$$840 = 800(1 + 6r/52) = 800 + \frac{4{,}800r}{52}$$

so

$$\frac{4{,}800r}{52} = 840 - 800 = 40$$

$$r = \frac{40 \times 52}{4{,}800} \approx 0.43.$$

In other words, the firm is charging you 43% annual interest! Save your money and wait 6 weeks for your refund.

## 2.1  EXERCISES

▼ more advanced    ◆ challenging

T indicates exercises that should be solved using technology

*In Exercises 1–6, compute the simple interest for the specified period and the future value at the end of the period. Round all answers to the nearest cent.* HINT [See Example 1.]

1. $2,000 is invested for 1 year at 6% per year.

2. $1,000 is invested for 10 years at 4% per year.

3. $20,200 is invested for 6 months at 5% per year.

4. $10,100 is invested for 3 months at 11% per year.

5. You borrow $10,000 for 10 months at 3% per year.

6. You borrow $6,000 for 5 months at 9% per year.

*In Exercises 7–12, find the present value of the given investment.* HINT [See the Quick Example on page 129.]

7. An investment earns 2% per year and is worth $10,000 after 5 years.

8. An investment earns 5% per year and is worth $20,000 after 2 years.

9. An investment earns 7% per year and is worth $1,000 after 6 months.

10. An investment earns 10% per year and is worth $5,000 after 3 months.

11. An investment earns 3% per year and is worth $15,000 after 15 months.

12. An investment earns 6% per year and is worth $30,000 after 20 months.

## APPLICATIONS

*In Exercises 13–30, compute the specified quantity. Round all answers to the nearest month, the nearest cent, or the nearest 0.001%, as appropriate.*

13. **Simple Loans** You take out a 6-month, $5,000 loan at 8% simple interest. How much would you owe at the end of the 6 months? HINT [See Example 2.]

14. **Simple Loans** You take out a 15-month, $10,000 loan at 11% simple interest. How much would you owe at the end of the 15 months? HINT [See Example 2.]

15. **Savings** How much would you have to deposit in an account earning 4.5% simple interest if you wanted to have $1,000 after 6 years? HINT [See Example 4.]

16. **Simple Loans** Your total payment on a 4-year loan, which charged 9.5% simple interest, amounted to $30,360. How much did you originally borrow? HINT [See Example 4.]

17. **Bonds** A 5-year bond costs $1,000 and will pay a total of $250 interest over its lifetime. What is its annual interest rate? HINT [See Example 3.]

18. **Bonds** A 4-year bond costs $10,000 and will pay a total of $2,800 in interest over its lifetime. What is its annual interest rate? HINT [See Example 3.]

19. **Treasury Bills** In December 2008 (in the midst of the financial crisis) a $5,000 6-month T-bill was selling at a discount rate of only 0.25%.[2] What was its simple annual yield? HINT [See Example 5.]

20. **Treasury Bills** In December 2008 (in the midst of the financial crisis) a $15,000 3-month T-bill was selling at a discount rate of only 0.06%.[3] What was its simple annual yield? HINT [See Example 5.]

21. ▼ **Simple Loans** A $4,000 loan, taken now, with a simple interest rate of 8% per year, will require a total repayment of $4,640. When will the loan mature?

22. ▼ **Simple Loans** The simple interest on a $1,000 loan at 8% per year amounted to $640. When did the loan mature?

[2]Discount rate on December 29, 2008. Source: U.S. Treasury (www.ustreas.gov/offices/domestic-finance/debt-management/interest-rate/daily_treas_bill_rates.shtml).
[3]Ibid.

23. ▼ *Treasury Bills* At auction on August 18, 2005, 6-month T-bills were sold at a discount of 3.705%.[4] What was the simple annual yield? HINT [See Example 5.]

24. ▼ *Treasury Bills* At auction on August 18, 2005, 3-month T-bills were sold at a discount of 3.470%.[5] What was the simple annual yield? HINT [See Example 5.]

25. ▼ *Fees* You are expecting a tax refund of $1,000 in 4 weeks. A tax preparer offers you a $1,000 loan for a fee of $50 to be repaid by your refund check when it arrives in 4 weeks. Thinking of the fee as interest, what annual simple interest rate would you be paying on this loan? HINT [See Example 6.]

26. ▼ *Fees* You are expecting a tax refund of $1,500 in 3 weeks. A tax preparer offers you a $1,500 loan for a fee of $60 to be repaid by your refund check when it arrives in 3 weeks. Thinking of the fee as interest, what annual simple interest rate would you be paying on this loan?

27. ▼ *Fees* You take out a 2-year, $5,000 loan at 9% simple annual interest. The lender charges you a $100 fee. Thinking of the fee as additional interest, what is the actual annual interest rate you will pay?

28. ▼ *Fees* You take out a 3-year, $7,000 loan at 8% simple annual interest. The lender charges you a $100 fee. Thinking of the fee as additional interest, what is the actual annual interest rate you will pay?

29. ▼ *Layaway Fees* Layaway plans allow you, for a fee, to pay for an item over a period of time and then receive the item when you finish paying for it. In November 2011, Senator Charles E. Schumer of New York warned that the holiday layaway programs recently reinstated by several popular retailers were, when you took the fees into account, charging interest at a rate significantly higher than the highest credit card rates.[6] Suppose that you bought a $69 item on November 15 on layaway, with the final payment due December 15, and that the retailer charged you a $5 service fee. Thinking of the fee as interest, what simple interest rate would you be paying for this layaway plan?

30. ▼ *Layaway Fees* Referring to Exercise 29, suppose that you bought a $99 item on October 15 on layaway, with the final payment due December 15, and that the retailer charged you a $10 service fee. Thinking of the fee as interest, what simple interest rate would you be paying for this layaway plan?

*Stock Investments* Exercises 31–36 are based on the following chart, which shows monthly figures for Apple Computer, Inc. stock in 2010.[7]

*Marked are the following points on the chart:*

| Jan. 10 | Feb. 10 | Mar. 10 | Apr. 10 | May 10 | June 10 |
|---------|---------|---------|---------|--------|---------|
| 211.98  | 195.46  | 218.95  | 235.97  | 235.86 | 255.96  |
| July 10 | Aug. 10 | Sep. 10 | Oct. 10 | Nov. 10 | Dec. 10 |
| 246.94  | 260.09  | 258.77  | 294.07  | 317.13 | 317.44  |

31. Calculate to the nearest 0.01% your annual percentage return (on a simple interest basis) if you had bought Apple stock in June and sold in December.

32. Calculate to the nearest 0.01% your annual percentage return (on a simple interest basis) if you had bought Apple stock in February and sold in June.

33. ▼ Suppose you bought Apple stock in April. If you later sold at one of the marked dates on the chart, which of those dates would have given you the largest annual return (on a simple interest basis), and what would that return have been?

34. ▼ Suppose you bought Apple stock in May. If you later sold at one of the marked dates on the chart, which of those dates would have given you the largest annual return (on a simple interest basis), and what would that return have been?

35. ▼ Did Apple's stock undergo simple interest change in the period January through May? (Give a reason for your answer.)

36. ▼ If Apple's stock had undergone simple interest change from April to June at the same simple rate as from March to April, what would the price have been in June?

*Population* Exercises 37–42 are based on the following graph, which shows the population of San Diego County from 1950 to 2000.[8]

[4]Source: The Bureau of the Public Debt's Web site: www.publicdebt.treas.gov.

[5]*Ibid.*

[6]Source: November 14, 2011 press release from the office of Senator Charles E. Schumer (http://schumer.senate.gov/Newsroom/releases.cfm)

[7]Source: Yahoo! Finance (http://finance.yahoo.com).

[8]Source: Census Bureau/*New York Times*, April 2, 2002, p. F3.

37. At what annual (simple interest) rate did the population of San Diego County increase from 1950 to 2000?

38. At what annual (simple interest) rate did the population of San Diego County increase from 1950 to 1990?

39. ▼ If you used your answer to Exercise 37 as the annual (simple interest) rate at which the population was growing since 1950, what would you predict the San Diego County population to be in 2010?

40. ▼ If you used your answer to Exercise 38 as the annual (simple interest) rate at which the population was growing since 1950, what would you predict the San Diego County population to be in 2010?

41. ▼ Use your answer to Exercise 37 to give a linear model for the population of San Diego County from 1950 to 2000. Draw the graph of your model over that period of time.

42. ▼ Use your answer to Exercise 38 to give a linear model for the population of San Diego County from 1950 to 2000. Draw the graph of your model over that period of time.

## COMMUNICATION AND REASONING EXERCISES

43. One or more of the following three graphs represents the future value of an investment earning simple interest. Which one(s)? Give the reason for your choice(s).

(A)

(B)

(C)

44. Given that $FV = 5t + 400$, for what interest rate is this the equation of future value (in dollars) as a function of time $t$ (in years)?

45. ▼ *Interpreting the News* You hear the following on your local radio station's business news: "The economy last year grew by 1%. This was the second year in a row in which the economy showed a 1% growth." Because the rate of growth was the same two years in a row, this represents simple interest growth, right? Explain your answer.

46. ▼ *Interpreting the News* You hear the following on your local radio station's business news: "The economy last year grew by 1%. This was the second year in a row in which the economy showed a 1% growth." This means that, in dollar terms, the economy grew more last year than the year before. Why?

47. ▼ Explain why simple interest is not the appropriate way to measure interest on a savings account that pays interest directly into your account.

48. ▼ Suppose that a one-year T-bill sells at a discount rate $d$. Find a formula, in terms of $d$, for the simple annual interest rate $r$ the bill will pay.

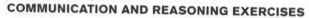

# 2.2   Compound Interest

You deposit $1,000 into a savings account. The bank pays you 5% interest, which it deposits into your account, or **reinvests**, at the end of each year. At the end of 5 years, how much money will you have accumulated? Let us compute the amount you have at the end of each year. At the end of the first year, the bank will pay you simple interest of 5% on your $1,000, which gives you

$$PV(1 + rt) = 1,000(1 + 0.05)$$
$$= \$1,050.$$

At the end of the second year, the bank will pay you another 5% interest, but this time computed on the total in your account, which is $1,050. Thus, you will have a total of

$$1,050(1 + 0.05) = \$1,102.50.$$

If you were being paid simple interest on your original $1,000, you would have only $1,100 at the end of the second year. The extra $2.50 is the interest earned on the $50 interest added to your account at the end of the first year. Having interest earn interest is called **compounding** the interest. We could continue like this until the end of the fifth year, but notice what we are doing: Each year we are multiplying by $1 + 0.05$. So, at the end of 5 years, you will have

$$1,000(1 + 0.05)^5 \approx \$1,276.28.$$

It is interesting to compare this to the amount you would have if the bank paid you simple interest:

$$1,000(1 + 0.05 \times 5) = \$1,250.00.$$

The extra \$26.28 is again the effect of compounding the interest.

Banks often pay interest more often than once a year. Paying interest quarterly (four times per year) or monthly is common. If your bank pays interest monthly, how much will your \$1,000 deposit be worth after 5 years? The bank will not pay you 5% interest every month, but will give you 1/12 of that,* or 5/12% interest each month. Thus, instead of multiplying by $1 + 0.05$ every year, we should multiply by $1 + 0.05/12$ each month. Because there are $5 \times 12 = 60$ months in 5 years, the total amount you will have at the end of 5 years is

$$1,000 \left( 1 + \frac{0.05}{12} \right)^{60} \approx \$1,283.36.$$

Compare this to the \$1,276.28 you would get if the bank paid the interest every year. You earn an extra \$7.08 if the interest is paid monthly because interest gets into your account and starts earning interest earlier. The amount of time between interest payments is called the **compounding period**.

The following table summarizes the results above.

* This is approximate. They will actually give you 31/365 of the 5% at the end of January and so on, but it's simpler and reasonably accurate to call it 1/12.

| Time in Years | Amount with Simple Interest | Amount with Annual Compounding | Amount with Monthly Compounding |
|---|---|---|---|
| 0 | \$1,000 | \$1,000 | \$1,000 |
| 1 | $1,000(1 + 0.05)$ $= \$1,050$ | $1,000(1 + 0.05)$ $= \$1,050$ | $1,000(1 + 0.05/12)^{12}$ $= \$1,051.16$ |
| 2 | $1,000(1 + 0.05 \times 2)$ $= \$1,100$ | $1,000(1 + 0.05)^2$ $= \$1,102.50$ | $1,000(1 + 0.05/12)^{24}$ $= \$1,104.94$ |
| 5 | $1,000(1 + 0.05 \times 5)$ $= \$1,250$ | $1,000(1 + 0.05)^5$ $= \$1,276.28$ | $1,000(1 + 0.05/12)^{60}$ $= \$1,283.36$ |

The preceding calculations generalize easily to give the general formula for future value when interest is compounded.

## Future Value for Compound Interest

The future value of an investment of $PV$ dollars earning interest at an annual rate of $r$ compounded (reinvested) $m$ times per year for a period of $t$ years is

$$FV = PV \left( 1 + \frac{r}{m} \right)^{mt}$$

or

$$FV = PV(1 + i)^n,$$

where $i = r/m$ is the interest paid each compounding period and $n = mt$ is the total number of compounding periods.

## Quick Examples

1. To find the future value after 5 years of a $10,000 investment earning 6% interest, with interest reinvested every month, we set $PV = 10,000$, $r = 0.06$, $m = 12$, and $t = 5$. Thus,

$$FV = PV\left(1 + \frac{r}{m}\right)^{mt} = 10,000\left(1 + \frac{0.06}{12}\right)^{60} \approx \$13,488.50.$$

2. Writing the future value in Quick Example 1 as a function of time, we get

$$FV = 10,000\left(1 + \frac{0.06}{12}\right)^{12t}$$
$$= 10,000(1.005)^{12t},$$

which is an **exponential** function of time $t$.

In general: *Compound interest growth is an exponential function of time.*

### using Technology

All three technologies discussed in this book have built-in mathematics of finance capabilities. See the Technology Guides at the end of the chapter for details on using a TI-83/84 Plus or a spreadsheet to do the calculations in Example 1. [Details: TI-83/84 Plus: page 166, Spreadsheet: page 170]

### Website
www.WanerMath.com
→ On Line Utilities
→ Time Value of Money Utility

This utility is similar to the TVM Solver on the TI-83/84 Plus. To compute the future value, enter the values shown, and press "Compute" next to FV.

(For an explanation of the terms, see the Technology Guide for the TI-83/84 Plus.)

## EXAMPLE 1 Savings Accounts

In November 2011, the Bank of Montreal was paying 1.30% interest on savings accounts.[9] If the interest is compounded quarterly, find the future value of a $2,000 deposit in 6 years. What is the total interest paid over the period?

**Solution** We use the future value formula with $m = 4$:

$$FV = PV\left(1 + \frac{r}{m}\right)^{mt}$$
$$= 2,000\left(1 + \frac{0.0130}{4}\right)^{4 \times 6} \approx \$2,161.97.$$

The total interest paid is

$$INT = FV - PV = 2,161.97 - 2,000 = \$161.97.$$

---

[9]Source: Canoe Money (http://money.canoe.ca/rates/savings.html)

Example 1 illustrates the concept of the **time value of money**: A given amount of money received now will usually be worth a different amount to us than the same amount received some time in the future. In the example above, we can say that $2,000 received now is worth the same as $2,161.97 received 6 years from now, because if we receive $2,000 now, we can turn it into $2,161.97 by the end of 6 years.

We often want to know, for some amount of money in the future, what is the equivalent value at present. As we did for simple interest, we can solve the future value formula for the present value and obtain the following formula.

---

### Present Value for Compound Interest

The present value of an investment earning interest at an annual rate of $r$ compounded $m$ times per year for a period of $t$ years, with future value $FV$, is

$$PV = \frac{FV}{\left(1 + \frac{r}{m}\right)^{mt}}$$

or

$$PV = \frac{FV}{(1+i)^n} = FV(1+i)^{-n},$$

where $i = r/m$ is the interest paid each compounding period and $n = mt$ is the total number of compounding periods.

#### Quick Example

To find the amount we need to invest in an investment earning 12% per year, compounded annually, so that we will have $1 million in 20 years, use $FV = \$1,000,000$, $r = 0.12$, $m = 1$, and $t = 20$:

$$PV = \frac{FV}{\left(1 + \frac{r}{m}\right)^{mt}} = \frac{1,000,000}{(1 + 0.12)^{20}} \approx \$103,666.77.$$

Put another way, $1,000,000 20 years from now is worth only $103,666.77 to us now, if we have a 12% investment available.

---

In the preceding section, we mentioned that a bond pays interest until it reaches maturity, at which point it pays you back an amount called its **maturity value** or **par value**. The two parts, the interest and the maturity value, can be separated and sold and traded by themselves. A **zero coupon bond** is a form of corporate bond that pays no interest during its life but, like U.S. Treasury bills, promises to pay you the maturity value when it reaches maturity. Zero coupon bonds are often created by removing or *stripping* the interest coupons from an ordinary bond, so are also known as **strips**. Zero coupon bonds sell for less than their maturity value, and the return on the investment is the difference between what the investor pays and the maturity value. Although no interest is actually paid, we measure the return on investment by thinking of the interest rate that would make the selling price (the present value) grow to become the maturity value (the future value).*

* The IRS refers to this kind of interest as **original issue discount (OID)** and taxes it as if it were interest actually paid to you each year.

✳ The return investors look for depends on a number of factors, including risk (the chance that the company will go bankrupt and you will lose your investment); the higher the risk, the higher the return. U.S. Treasuries are considered risk free because the federal government has never defaulted on its debts. On the other hand, so-called junk bonds are high-risk investments (below investment grade) and have correspondingly high yields.

 **using Technology**

See the Technology Guides at the end of the chapter for details on using TVM Solver on the TI-83/84 Plus or the built-in finance functions in spreadsheets to do the calculations in Example 2. [Details: TI-83/84 Plus: page 166, Spreadsheet: page 171]

 **Website**
www.WanerMath.com

→ On Line Utilities

→ Time Value of Money Utility

This utility is similar to the TVM Solver on the TI-83/84 Plus. To compute the present value, enter the values shown, and press "Compute" next to PV.

### EXAMPLE 2   Zero Coupon Bonds

Megabucks Corporation is issuing 10-year zero coupon bonds. How much would you pay for bonds with a maturity value of $10,000 if you wish to get a return of 6.5% compounded annually?✳

**Solution**   As we said earlier, we think of a zero coupon bond as if it were an account earning (compound) interest. We are asked to calculate the amount you will pay for the bond—the present value $PV$. We have

$$FV = \$10,000$$
$$r = 0.065$$
$$t = 10$$
$$m = 1.$$

We can now use the present value formula:

$$PV = \frac{FV}{\left(1 + \frac{r}{m}\right)^{mt}}$$

$$PV = \frac{10,000}{\left(1 + \frac{0.065}{1}\right)^{10 \times 1}} \approx \$5,327.26.$$

Thus, you should pay $5,327.26 to get a return of 6.5% annually.

➡ **Before we go on...** Particularly in financial applications, you will hear the word "discounted" in place of "compounded" when discussing present value. Thus, the result of Example 2 might be phrased, "The present value of $10,000 to be received 10 years from now, with an interest rate of 6.5% discounted annually, is $5,327.26." ∎

Time value of money calculations are often done to take into account inflation, which behaves like compound interest. Suppose, for example, that inflation is running at 5% per year. Then prices will increase by 5% each year, so if $PV$ represents the price now, the price one year from now will be 5% higher, or $PV(1 + 0.05)$. The price a year from then will be 5% higher still, or $PV(1 + 0.05)^2$. Thus, the effects of inflation are compounded just as reinvested interest is.

### EXAMPLE 3   Inflation

Inflation in East Avalon is 5% per year. TruVision television sets cost $200 today. How much will a comparable set cost 2 years from now?

**Solution**   To find the price of a television set 2 years from now, we compute the future value of $200 at an inflation rate of 5% compounded yearly:

$$FV = 200(1 + 0.05)^2 = \$220.50.$$

## EXAMPLE 4  Constant Dollars

Inflation in North Avalon is 6% per year. Which is really more expensive, a car costing $20,000 today or one costing $22,000 in 3 years?

**Solution**  We cannot compare the two costs directly because inflation makes $1 today worth more (it buys more) than a dollar 3 years from now. We need the two prices expressed in comparable terms, so we convert to **constant dollars**. We take the car costing $22,000 three years from now and ask what it would cost in today's dollars. In other words, we convert the future value of $22,000 to its present value:

$$PV = FV(1 + i)^{-n}$$
$$= 22,000(1 + 0.06)^{-3}$$
$$\approx \$18,471.62.$$

Thus, the car costing $22,000 in 3 years actually costs less, after adjusting for inflation, than the one costing $20,000 now.

➡ **Before we go on...**  In the presence of inflation, the only way to compare prices at different times is to convert all prices to constant dollars. We pick some fixed time and compute future or present values as appropriate to determine what things would have cost at that time.  ■

There are some other interesting calculations related to compound interest, besides present and future values.

## EXAMPLE 5  Effective Interest Rate

You have just won $1 million in the lottery and are deciding what to do with it during the next year before you move to the South Pacific. Bank Ten offers 10% interest, compounded annually, while Bank Nine offers 9.8% compounded monthly. In which should you deposit your money?

**Solution**  Let's calculate the future value of your $1 million after one year in each of the banks:

Bank Ten:    $FV = 1(1 + 0.10)^1 = \$1.1$ million

Bank Nine:    $FV = 1\left(1 + \dfrac{0.098}{12}\right)^{12} = \$1.1025$ million.

Bank Nine turns out to be better: It will pay you a total of $102,500 in interest over the year, whereas Bank Ten will pay only $100,000 in interest.

Another way of looking at the calculation in Example 5 is that Bank Nine gave you a total of 10.25% interest on your investment over the year. We call 10.25% the **effective interest rate** of the investment (also referred to as the **annual percentage**

**yield**, or **APY** in the banking industry); the stated 9.8% is called the **nominal** interest rate. In general, to best compare two different investments, it is wisest to compare their *effective*—rather than nominal—interest rates.

Notice that we got 10.25% by computing

$$\left(1 + \frac{0.098}{12}\right)^{12} = 1.1025$$

and then subtracting 1 to get 0.1025, or 10.25%. Generalizing, we get the following formula.

## Effective Interest Rate

The effective interest rate $r_{\text{eff}}$ of an investment paying a nominal interest rate of $r_{\text{nom}}$ compounded $m$ times per year is

$$r_{\text{eff}} = \left(1 + \frac{r_{\text{nom}}}{m}\right)^{m} - 1.$$

To compare rates of investments with different compounding periods, always compare the effective interest rates rather than the nominal rates.

### Quick Example

To calculate the effective interest rate of an investment that pays 8% per year, with interest reinvested monthly, set $r_{\text{nom}} = 0.08$ and $m = 12$, to obtain

$$r_{\text{eff}} = \left(1 + \frac{0.08}{12}\right)^{12} - 1 \approx 0.0830, \text{ or } 8.30\%.$$

## EXAMPLE 6  How Long to Invest

You have $5,000 to invest at 6% interest compounded monthly. How long will it take for your investment to grow to $6,000?

**Solution**  If we use the future value formula, we already have the values

$$FV = 6,000$$
$$PV = 5,000$$
$$r = 0.06$$
$$m = 12.$$

Substituting, we get

$$6,000 = 5,000\left(1 + \frac{0.06}{12}\right)^{12t}.$$

**using** Technology

See the Technology Guides at the end of the chapter for details on using TVM Solver on the TI-83/84 Plus or the built-in finance functions in spreadsheets to do the calculations in Example 6. [Details: TI-83/84 Plus: page 167, Spreadsheet: page 171]

 **Website**

www.WanerMath.com

→ On Line Utilities

→ Time Value of Money Utility

This utility is similar to the TVM Solver on the TI-83/84 Plus. To compute the time needed, enter the values shown, and press "Compute" next to *t*.

If you are familiar with logarithms, you can solve explicitly for *t* as follows:

$$\left(1 + \frac{0.06}{12}\right)^{12t} = \frac{6,000}{5,000} = 1.2$$

$$\log\left(1 + \frac{0.06}{12}\right)^{12t} = \log 1.2$$

$$12t \log\left(1 + \frac{0.06}{12}\right) = \log 1.2$$

$$t = \frac{\log 1.2}{12 \log\left(1 + \frac{0.06}{12}\right)} \approx 3.046 \approx 3 \text{ years.}$$

Another approach is to use a bit of trial and error to find the answer. Let's see what the future value is after 2, 3, and 4 years:

$$5,000\left(1 + \frac{0.06}{12}\right)^{12\times2} = 5,635.80 \quad \text{Future value after 2 years}$$

$$5,000\left(1 + \frac{0.06}{12}\right)^{12\times3} = 5,983.40 \quad \text{Future value after 3 years}$$

$$5,000\left(1 + \frac{0.06}{12}\right)^{12\times4} = 6,352.45. \quad \text{Future value after 4 years}$$

From these calculations, it looks as if the answer should be a bit more than 3 years, but is certainly between 3 and 4 years. We could try 3.5 years next and then narrow it down systematically until we have a pretty good approximation of the correct answer. Graphing calculators and spreadsheets give us alternative methods of solution.

---

## FAQs

### Recognizing When to Use Compound Interest and the Meaning of Present Value

**Q**: *How do I distinguish a problem that calls for compound interest from one that calls for simple interest?*

**A**: Study the scenario to ascertain whether the interest is being withdrawn as it is earned or reinvested (deposited back into the account). If the interest is being withdrawn, the problem is calling for simple interest because the interest is not itself earning interest. If it is being reinvested, the problem is calling for compound interest.

**Q**: *How do I distinguish present value from future value in a problem?*

**A**: The present value always refers to the value of an investment before any interest is included (or, in the case of a depreciating investment, before any depreciation takes place). As an example, the future value of a bond is its maturity value. The value of $1 today in constant 2010 dollars is its present value (even though 2010 is in the past).

## 2.2   EXERCISES

▼ more advanced    ◆ challenging

**T** indicates exercises that should be solved using technology

*In Exercises 1–8, calculate, to the nearest cent, the future value of an investment of $10,000 at the stated interest rate after the stated amount of time.* HINT [See Example 1.]

**1.** 3% per year, compounded annually, after 10 years

**2.** 4% per year, compounded annually, after 8 years

**3.** 2.5% per year, compounded quarterly (4 times/year), after 5 years

**4.** 1.5% per year, compounded weekly (52 times/year), after 5 years

**5.** 6.5% per year, compounded daily (assume 365 days/year), after 10 years

**6.** 11.2% per year, compounded monthly, after 12 years

**7.** 0.2% per month, compounded monthly, after 10 years

**8.** 0.45% per month, compounded monthly, after 20 years

*In Exercises 9–14, calculate, to the nearest cent, the present value of an investment that will be worth $1,000 at the stated interest rate after the stated amount of time.* HINT [See Example 2.]

**9.** 10 years, at 5% per year, compounded annually

**10.** 5 years, at 6% per year, compounded annually

**11.** 5 years, at 4.2% per year, compounded weekly (assume 52 weeks per year)

**12.** 10 years, at 5.3% per year, compounded quarterly

**13.** 4 years, depreciating 5% each year

**14.** 5 years, depreciating 4% each year

*In Exercises 15–20, find the effective annual interest rates of the given annual interest rates. Round your answers to the nearest 0.01%.* HINT [See Quick Example on page 140.]

**15.** 5% compounded quarterly

**16.** 5% compounded monthly

**17.** 10% compounded monthly

**18.** 10% compounded daily (assume 365 days per year)

**19.** 10% compounded hourly (assume 365 days per year)

**20.** 10% compounded every minute (assume 365 days per year)

## APPLICATIONS

**21.** *Savings* You deposit $1,000 in an account at the *Lifelong Trust Savings and Loan* that pays 6% interest compounded quarterly. By how much will your deposit have grown after 4 years?

**22.** *Investments* You invest $10,000 in *Rapid Growth Funds*, which appreciate by 2% per year, with yields reinvested

quarterly. By how much will your investment have grown after 5 years?

**23.** *Depreciation* During 2008, the S&P 500 index depreciated by 37.6%.[10] Assuming that this trend had continued, how much would a $3,000 investment in an S&P index fund have been worth after 3 years?

**24.** *Depreciation* During 2008, the NASDAQ Composite Index depreciated by 42%.[11] Assuming that this trend had continued, how much would a $5,000 investment in a NASDAQ Composite Index fund have been worth after 4 years?

**25.** *Bonds* You want to buy a 10-year zero coupon bond with a maturity value of $5,000 and a yield of 5.5% annually. How much will you pay?

**26.** *Bonds* You want to buy a 15-year zero coupon bond with a maturity value of $10,000 and a yield of 6.25% annually. How much will you pay?

**27.** ▼ *Investments* When I was considering what to do with my $10,000 Lottery winnings, my broker suggested I invest half of it in gold, the value of which was growing by 10% per year, and the other half in certificates of deposit (CDs), which were yielding 5% per year, compounded every 6 months. Assuming that these rates are sustained, how much will my investment be worth in 10 years?

**28.** ▼ *Investments* When I was considering what to do with the $10,000 proceeds from my sale of technology stock, my broker suggested I invest half of it in municipal bonds, whose value was growing by 6% per year, and the other half in CDs, which were yielding 3% per year, compounded every 2 months. Assuming that these interest rates are sustained, how much will my investment be worth in 10 years?

**29.** ▼ *Depreciation* During a prolonged recession, property values on Long Island depreciated by 2% every 6 months. If my house cost $200,000 originally, how much was it worth 5 years later?

**30.** ▼ *Depreciation* Stocks in the health industry depreciated by 5.1% in the first 9 months of 1993.[12] Assuming that this trend continued, how much would a $40,000 investment be worth in 9 years? HINT [Nine years corresponds to 12 nine-month periods.]

**31.** ▼ *Retirement Planning* I want to be earning an annual salary of $100,000 when I retire in 15 years. I have been offered a job that guarantees an annual salary increase of 4% per year, and the starting salary is negotiable. What salary should I request in order to meet my goal?

---

[10]Source: http://finance.google.com.

[11]*Ibid.*

[12]Source: *New York Times*, October 9, 1993, p. 37.

32. ▼ *Retirement Planning* I want to be earning an annual salary of $80,000 when I retire in 10 years. I have been offered a job that guarantees an annual salary increase of 5% per year, and the starting salary is negotiable. What salary should I request in order to meet my goal?

33. ▼ *Present Value* Determine the amount of money, to the nearest dollar, you must invest at 6% per year, compounded annually, so that you will be a millionaire in 30 years.

34. ▼ *Present Value* Determine the amount of money, to the nearest dollar, you must invest now at 7% per year, compounded annually, so that you will be a millionaire in 40 years.

35. ▼ *Stocks* Six years ago, I invested some money in *Dracubunny Toy Co.* stock, acting on the advice of a "friend." As things turned out, the value of the stock decreased by 5% every 4 months, and I discovered yesterday (to my horror) that my investment was worth only $297.91. How much did I originally invest?

36. ▼ *Sales* My recent marketing idea, the *Miracle Algae Growing Kit*, has been remarkably successful, with monthly sales growing by 6% every 6 months over the past 5 years. Assuming that I sold 100 kits the first month, how many kits did I sell in the first month of this year?

37. ▼ *Inflation* Inflation has been running 2% per year. A car now costs $30,000. How much would it have cost 5 years ago? HINT [See Example 3.]

38. ▼ *Inflation* (Compare Exercise 37.) Inflation has been running 1% every 6 months. A car now costs $30,000. How much would it have cost 5 years ago?

39. ▼ *Inflation* Housing prices have been rising 6% per year. A house now costs $200,000. What would it have cost 10 years ago?

40. ▼ *Inflation* (Compare Exercise 39.) Housing prices have been rising 0.5% each month. A house now costs $200,000. What would it have cost 10 years ago? HINT [See Example 4.]

41. ▼ *Constant Dollars* Inflation is running 3% per year when you deposit $1,000 in an account earning interest of 5% per year compounded annually. In *constant dollars*, how much money will you have 2 years from now? HINT [First calculate the value of your account in 2 years' time, and then find its present value based on the inflation rate.]

42. ▼ *Constant Dollars* Inflation is running 1% per month when you deposit $10,000 in an account earning 8% compounded monthly. In *constant dollars*, how much money will you have 2 years from now? HINT [See Exercise 41.]

43. ▼ *Investments* You are offered two investments. One promises to earn 12% compounded annually. The other will earn 11.9% compounded monthly. Which is the better investment? HINT [See Example 5.]

44. ▼ *Investments* You are offered three investments. The first promises to earn 15% compounded annually, the second will earn 14.5% compounded quarterly, and the third will earn 14% compounded monthly. Which is the best investment? HINT [See Example 5.]

45. ▼ *History* Legend has it that a band of Lenape Indians known as the "Manhatta" sold Manhattan Island to the Dutch in 1626 for $24. In 2011, the total value of Manhattan real estate was estimated to be $314,119 million.[13] Suppose that the Lenape had instead taken that $24 and invested it at 6.3% compounded annually (a relatively conservative investment goal). Could they then have bought back the island in 2011?

46. ▼ *History* Repeat Exercise 45, assuming that the Lenape had invested the $24 at 6.2% compounded annually.

*Inflation* Exercises 47–54 are based on the following table, which shows the 2008 annual inflation rates in several Latin American countries.[14] Assume that the rates shown continue indefinitely.

| Country | Argentina | Brazil | Bolivia | Nicaragua | Venezuela | Mexico | Uruguay |
|---|---|---|---|---|---|---|---|
| Currency | Peso | Real | Boliviano | Gold cordoba | Bolivar | Peso | Peso |
| Inflation Rate (%) | 9.2 | 6.3 | 15.1 | 13.8 | 25.7 | 5.0 | 8.5 |

47. If an item in Brazil now costs 100 reals, what do you expect it to cost 5 years from now? (Answer to the nearest real.)

48. If an item in Argentina now costs 1,000 pesos, what do you expect it to cost 5 years from now? (Answer to the nearest peso.)

49. If an item in Bolivia will cost 1,000 bolivianos in 10 years, what does it cost now? (Answer to the nearest boliviano.)

50. If an item in Mexico will cost 20,000 pesos in 10 years, what does it cost now? (Answer to the nearest peso.)

51. ▼ You wish to invest 1,000 bolivars in Venezuela at 8% annually, compounded twice a year. Find the value of your investment in 10 years, expressing the answer in constant bolivars. (Answer to the nearest bolivar.)

52. ▼ You wish to invest 1,000 pesos in Uruguay at 8% annually, compounded twice a year. Find the value of your investment in 10 years, expressing the answer in constant pesos. (Answer to the nearest peso.)

53. ▼ Which is the better investment: an investment in Mexico yielding 5.3% per year, compounded annually, or an investment in Nicaragua yielding 14% per year, compounded every 6 months? Support your answer with figures that show the future value of an investment of 1 unit of currency in constant units.

---

[13]Source: FY12 Tentative Assessment Roll summary, New York City Department of Finance, January 14, 2011 (www.nyc.gov/finance)

[14]Sources: www.bloomberg.com, www.csmonitor.com, www.indexmundi.com.

**54.** ▼ Which is the better investment: an investment in Argentina yielding 10% per year, compounded annually, or an investment in Uruguay, yielding 9% per year, compounded every 6 months? Support your answer with figures that show the future value of an investment of 1 unit of currency in constant units.

***Stock Investments*** *Exercises 55–60 are based on the following chart, which shows monthly figures for* Apple Computer, Inc. *stock in 2010.*[15]

*Marked are the following points on the chart:*

| Jan. 10 | Feb. 10 | Mar. 10 | Apr. 10 | May 10 | June 10 |
|---------|---------|---------|---------|--------|---------|
| 211.98  | 195.46  | 218.95  | 235.97  | 235.86 | 255.96  |

| July 10 | Aug. 10 | Sep. 10 | Oct. 10 | Nov. 10 | Dec. 10 |
|---------|---------|---------|---------|---------|---------|
| 246.94  | 260.09  | 258.77  | 294.07  | 317.13  | 317.44  |

**55.** ▼ Calculate to the nearest 0.01% your annual percentage return (assuming annual compounding) if you had bought Apple stock in June and sold in December.

**56.** ▼ Calculate to the nearest 0.01% your annual percentage return (assuming annual compounding) if you had bought Apple stock in February and sold in June.

**57.** ▼ Suppose you bought Apple stock in April. If you later sold at one of the marked dates on the chart, which of those dates would have given you the largest annual return (assuming annual compounding), and what would that return have been?

**58.** ▼ Suppose you bought Apple stock in May. If you later sold at one of the marked dates on the chart, which of those dates would have given you the largest annual return (assuming annual compounding), and what would that return have been?

**59.** ▼ Did Apple's stock undergo compound interest change in the period January through May? (Give a reason for your answer.)

**60.** ▼ If Apple's stock had undergone compound interest change from April to June at the same monthly rate as from March to April, what would the price have been in June?

**61.** ▣ ▼ ***Competing Investments*** I just purchased $5,000 worth of municipal funds that are expected to yield 5.4% per year, compounded every 6 months. My friend has just purchased $6,000 worth of CDs that will earn 4.8% per year,

compounded every 6 months. Determine when, to the nearest year, the value of my investment will be the same as hers, and what this value will be. HINT [You can either graph the values of both investments or make tables of the values of both investments.]

**62.** ▣ ▼ ***Investments*** Determine when, to the nearest year, $3,000 invested at 5% per year, compounded daily, will be worth $10,000.

**63.** ▣ ▼ ***Epidemics*** At the start of 1985, the incidence of AIDS was doubling every 6 months and 40,000 cases had been reported in the United States. Assuming this trend would have continued, determine when, to the nearest tenth of a year, the number of cases would have reached 1 million.

**64.** ▣ ▼ ***Depreciation*** My investment in *Genetic Splicing, Inc.,* is now worth $4,354 and is depreciating by 5% every 6 months. For some reason, I am reluctant to sell the stock and swallow my losses. Determine when, to the nearest year, my investment will drop below $50.

**65.** ◆ ***Bonds*** Once purchased, bonds can be sold in the secondary market. The value of a bond depends on the prevailing interest rates, which vary over time. Suppose that in January 2020, you buy a 30-year zero-coupon U.S. Treasury bond with a maturity value of $100,000 and a yield of 15% annually.
   **a.** How much do you pay for the bond?
   **b.** In January 2037, your bond has 13 years remaining until maturity. Rates on U.S. Treasury bonds of comparable length are now about 4.75%. If you sell your bond to an investor looking for a return of 4.75% annually, how much money do you receive?
   **c.** Using your answers to parts (a) and (b), what was the annual yield (assuming annual compounding) on your 17-year investment?

**66.** ◆ ***Bonds*** Suppose that in January 2040, you buy a 30-year zero-coupon U.S. Treasury bond with a maturity value of $100,000 and a yield of 5% annually.
   **a.** How much do you pay for the bond?
   **b.** Suppose that, 15 years later, interest rates have risen again, to 12%. If you sell your bond to an investor looking for a return of 12%, how much money will you receive?
   **c.** Using your answers to parts (a) and (b), what will be the annual yield (assuming annual compounding) on your 15-year investment?

## COMMUNICATION AND REASONING EXERCISES

**67.** Why is the graph of the future value of a compound interest investment as a function of time not a straight line (assuming a nonzero rate of interest)?

**68.** An investment that earns 10% (compound interest) every year is the same as an investment that earns 5% (compound interest) every 6 months, right?

**69.** ▼ If a bacteria culture is currently 0.01 g and increases in size by 10% each day, then its growth is linear, right?

**70.** ▼ At what point is the future value of a compound interest investment the same as the future value of a simple interest investment at the same annual rate of interest?

[15]Source: Yahoo! Finance (http://finance.yahoo.com).

**71.** ▼ If two equal investments have the same effective interest rate and you graph the future value as a function of time for each of them, are the graphs necessarily the same? Explain your answer.

**72.** ▼ For what kind of compound interest investments is the effective rate the same as the nominal rate? Explain your answer.

**73.** ▼ For what kind of compound interest investments is the effective rate greater than the nominal rate? When is it smaller? Explain your answer.

**74.** ▼ If an investment appreciates by 10% per year for 5 years (compounded annually) and then depreciates by 10% per year (compounded annually) for 5 more years, will it have the same value as it had originally? Explain your answer.

**75.** ▼ You can choose between two investments that mature at different times in the future. If you knew the rate of inflation, how would you decide which is the better investment?

**76.** ▼ If you knew the various inflation rates for the years 2000 through 2011, how would you convert $100 in 2012 dollars to 2000 dollars?

**77.** ⊤ ▼ On the same set of axes, graph the future value of a $100 investment earning 10% per year as a function of time over a 20-year period, compounded once a year, 10 times a year, 100 times a year, 1,000 times a year, and 10,000 times a year. What do you notice?

**78.** ⊤ ▼ By graphing the future value of a $100 investment that is depreciating by 1% each year, convince yourself that, eventually, the future value will be less than $1.

## 2.3  Annuities, Loans, and Bonds

> ✳ Defined-contribution pension plans have largely replaced the defined-benefit pensions that were once the norm in private industry. In a defined-benefit plan, the size of your pension is guaranteed; it is typically a percentage of your final working salary. In a defined-contribution plan, the size of your pension depends on how well your investments do.

A typical defined-contribution pension fund works as follows✳: Every month while you work, you and your employer deposit a certain amount of money in an account. This money earns (compound) interest from the time it is deposited. When you retire, the account continues to earn interest, but you may then start withdrawing money at a rate calculated to reduce the account to zero after some number of years. This account is an example of an **annuity**, an account earning interest into which you make periodic deposits or from which you make periodic withdrawals. In common usage, the term "annuity" is used for an account from which you make withdrawals. There are various terms used for accounts into which you make payments, based on their purpose. Examples include **savings account**, **pension fund**, and **sinking fund**. A sinking fund is generally used by businesses or governments to accumulate money to pay off an anticipated debt, but we'll use the term to refer to any account into which you make periodic payments.

### Sinking Funds

Suppose you make a payment of $100 at the end of every month into an account earning 3.6% interest per year, compounded monthly. This means that your investment is earning $3.6\%/12 = 0.3\%$ per month. We write $i = 0.036/12 = 0.003$. What will be the value of the investment at the end of 2 years (24 months)?

Think of the deposits separately. Each earns interest from the time it is deposited, and the total accumulated after 2 years is the sum of these deposits and the interest they earn. In other words, the accumulated value is the sum of the future values of the deposits, taking into account how long each deposit sits in the account. Figure 1 shows a timeline with the deposits and the contribution of each to the final value. For example, the very last deposit (at the end of month 24) has no time to earn interest, so it contributes only $100. The very first deposit, which earns interest for 23 months, by the future value formula for compound interest contributes $100(1 + 0.003)^{23}$ to the total. Adding together all of the future values gives us the total future value:

$$FV = 100 + 100(1 + 0.003) + 100(1 + 0.003)^2 + \cdots + 100(1 + 0.003)^{23}$$
$$= 100[1 + (1 + 0.003) + (1 + 0.003)^2 + \cdots + (1 + 0.003)^{23}]$$

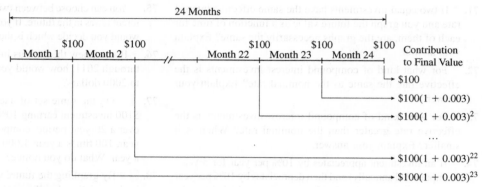

**Figure 1**

✻ It is called a **geometric series**.

† The quickest way to convince yourself that this formula is correct is to multiply out $(x - 1)(1 + x + x^2 + \cdots + x^{n-1})$ and see that you get $x^n - 1$. You should also try substituting some numbers. For example, $1 + 3 + 3^2 = 13 = (3^3 - 1)/(3 - 1)$.

Fortunately, this sort of sum is well-known (to mathematicians, anyway✻) and there is a convenient formula for its value†:

$$1 + x + x^2 + \cdots + x^{n-1} = \frac{x^n - 1}{x - 1}.$$

In our case, with $x = 1 + 0.003$, this formula allows us to calculate the future value:

$$FV = 100\frac{(1 + 0.003)^{24} - 1}{(1 + 0.003) - 1} = 100\frac{(1.003)^{24} - 1}{0.003} \approx \$2,484.65.$$

It is now easy to generalize this calculation.

## Future Value of a Sinking Fund

A **sinking fund** is an account earning compound interest into which you make periodic deposits. Suppose that the account has an annual rate of $r$ compounded $m$ times per year, so that $i = r/m$ is the interest rate per compounding period. If you make a payment of $PMT$ at the end of each period, then the future value after $t$ years, or $n = mt$ periods, will be

$$FV = PMT\frac{(1 + i)^n - 1}{i}.$$

### Quick Example

At the end of each month you deposit $50 into an account earning 2% annual interest compounded monthly. To find the future value after 5 years, we use $i = 0.02/12$ and $n = 12 \times 5 = 60$ compounding periods, so

$$FV = 50\frac{(1 + 0.02/12)^{60} - 1}{0.02/12} = \$3,152.37.$$

▦ **using** Technology

To automate the computations in Example 1 using a graphing calculator or a spreadsheet, see the Technology Guides at the end of the chapter. Outline:

## EXAMPLE 1  Retirement Account

Your retirement account has $5,000 in it and earns 5% interest per year compounded monthly. Every month for the next 10 years you will deposit $100 into the account. How much money will there be in the account at the end of those 10 years?

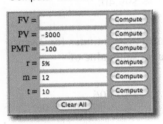

**Solution** This is a sinking fund with $PMT = \$100$, $r = 0.05$, $m = 12$, so $i = 0.05/12$, and $n = 12 \times 10 = 120$. Ignoring for the moment the $5,000 already in the account, your payments have the following future value:

$$FV = PMT\frac{(1+i)^n - 1}{i}$$
$$= 100\frac{(1+0.05/12)^{120} - 1}{0.05/12}$$
$$\approx \$15,528.23.$$

What about the $5,000 that was already in the account? That sits there and earns interest, so we need to find its future value as well, using the compound interest formula:

$$FV = PV(1+i)^n$$
$$= 5,000(1+0.05/12)^{120}$$
$$= \$8,235.05.$$

Hence, the total amount in the account at the end of 10 years will be

$$\$15,528.23 + 8,235.05 = \$23,763.28.$$

Sometimes we know what we want the future value to be and need to determine the payments necessary to achieve that goal. We can simply solve the future value formula for the payment.

### Payment Formula for a Sinking Fund

Suppose that an account has an annual rate of $r$ compounded $m$ times per year, so that $i = r/m$ is the interest rate per compounding period. If you want to accumulate a total of $FV$ in the account after $t$ years, or $n = mt$ periods, by making payments of $PMT$ at the end of each period, then each payment must be

$$PMT = FV\frac{i}{(1+i)^n - 1}.$$

### EXAMPLE 2 Education Fund

Tony and Maria have just had a son, José Phillipe. They establish an account to accumulate money for his college education, in which they would like to have $100,000 after 17 years. If the account pays 4% interest per year compounded quarterly, and they make deposits at the end of every quarter, how large must each deposit be for them to reach their goal?

**Solution** This is a sinking fund with $FV = \$100,000$, $m = 4$, $n = 4 \times 17 = 68$, and $r = 0.04$, so $i = 0.04/4 = 0.01$. From the payment formula, we get

$$PMT = 100,000\frac{0.01}{(1+0.01)^{68} - 1} \approx \$1,033.89.$$

So, Tony and Maria must deposit $1,033.89 every quarter in order to meet their goal.

**Spreadsheet**
=PMT(4/4,17*4,0,100000)
[More details on page 172.]

**Website**
www.WanerMath.com
→ On Line Utilities
→ Time Value of Money Utility
Enter the values shown, and press
"Compute" next to PMT.

## Annuities

Suppose we deposit an amount $PV$ now in an account earning 3.6% interest per year, compounded monthly. Starting 1 month from now, the bank will send us monthly payments of $100. What must $PV$ be so that the account will be drawn down to $0 in exactly 2 years?

As before, we write $i = r/m = 0.036/12 = 0.003$, and we have $PMT = 100$. The first payment of $100 will be made 1 month from now, so its present value is

$$\frac{PMT}{(1+i)^n} = \frac{100}{1+0.003} = 100(1+0.003)^{-1} \approx \$99.70.$$

In other words, that much of the original $PV$ goes toward funding the first payment. The second payment, 2 months from now, has a present value of

$$\frac{PMT}{(1+i)^n} = \frac{100}{(1+0.003)^2} = 100(1+0.003)^{-2} \approx \$99.40.$$

That much of the original $PV$ funds the second payment. This continues for 2 years, at which point we receive the last payment, which has a present value of

$$\frac{PMT}{(1+i)^n} = \frac{100}{(1+0.003)^{24}} = 100(1+0.003)^{-24} \approx \$93.06$$

and that exhausts the account. Figure 2 shows a timeline with the payments and the present value of each.

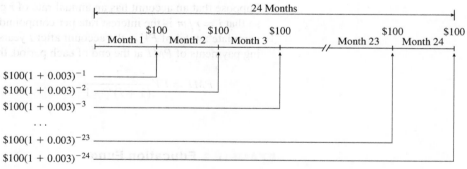

**Figure 2**

Because $PV$ must be the sum of these present values, we get

$$PV = 100(1+0.003)^{-1} + 100(1+0.003)^{-2} + \cdots + 100(1+0.003)^{-24}$$
$$= 100[(1+0.003)^{-1} + (1+0.003)^{-2} + \cdots + (1+0.003)^{-24}].$$

We can again find a simpler formula for this sum:

$$x^{-1} + x^{-2} + \cdots + x^{-n} = \frac{1}{x^n}(x^{n-1} + x^{n-2} + \cdots + 1)$$
$$= \frac{1}{x^n} \cdot \frac{x^n - 1}{x - 1} = \frac{1 - x^{-n}}{x - 1}.$$

So, in our case,

$$PV = 100\frac{1 - (1 + 0.003)^{-24}}{(1 + 0.003) - 1}$$

or

$$PV = 100\frac{1 - (1.003)^{-24}}{0.003} \approx \$2,312.29.$$

If we deposit \$2,312.29 initially and the bank sends us \$100 per month for 2 years, our account will be exhausted at the end of that time.

Generalizing, we get the following formula:

### Present Value of an Annuity

An **annuity** is an account earning compound interest from which periodic withdrawals are made. Suppose that the account has an annual rate of $r$ compounded $m$ times per year, so that $i = r/m$ is the interest rate per compounding period. Suppose also that the account starts with a balance of $PV$. If you receive a payment of $PMT$ at the end of each compounding period, and the account is down to \$0 after $t$ years, or $n = mt$ periods, then

$$PV = PMT\frac{1 - (1 + i)^{-n}}{i}.$$

**Quick Example**

At the end of each month you want to withdraw \$50 from an account earning 2% annual interest compounded monthly. If you want the account to last for 5 years (60 compounding periods), it must have the following amount to begin with:

$$PV = 50\frac{1 - (1 + 0.02/12)^{-60}}{0.02/12} = \$2,852.62.$$

**Note** If you make your withdrawals at the end of each compounding period, as we've discussed so far, you have an **ordinary annuity**. If, instead, you make withdrawals at the beginning of each compounding period, you have an **annuity due**. Because each payment occurs one period earlier, there is one less period in which to earn interest, hence the present value must be larger by a factor of $(1 + i)$ to fund each payment. So, the present value formula for an annuity due is

$$PV = PMT(1 + i)\frac{1 - (1 + i)^{-n}}{i}.$$

In this book, we will concentrate on ordinary annuities. ∎

---

### EXAMPLE 3  Trust Fund

You wish to establish a trust fund from which your niece can withdraw \$2,000 every 6 months for 15 years, at the end of which time she will receive the remaining money in the trust, which you would like to be \$10,000. The trust will be invested at 7% per year compounded every 6 months. How large should the trust be?

---

**using Technology**

To automate the computations in Example 3 using a graphing calculator or a spreadsheet, see the Technology Guides at the end of the chapter. Outline:

**TI-83/84 Plus**

APPS 1:Finance, then
1:TVM Solver
N = 30, I% = 7, PMT = 2000,
FV = 10000, P/Y = 2, C/Y = 2
With cursor on PV line.
ALPHA SOLVE
[More details on page 168.]

**Spreadsheet**

=PV(7/2,15*2,2000,10000)
[More details on page 172.]

**Website**
www.WanerMath.com
→ On Line Utilities
→ Time Value of Money Utility
Enter the values shown, and press "Compute" next to PV.

**Solution** We view this account as having two parts, one funding the semiannual payments and the other funding the $10,000 lump sum at the end. The amount of money necessary to fund the semiannual payments is the present value of an annuity, with $PMT = 2,000$, $r = 0.07$, and $m = 2$, so $i = 0.07/2 = 0.035$, and $n = 2 \times 15 = 30$. Substituting gives

$$PV = 2,000 \frac{1 - (1 + 0.035)^{-30}}{0.035}$$
$$= \$36,784.09.$$

To fund the lump sum of $10,000 after 15 years, we need the present value of $10,000 under compound interest:

$$PV = 10,000(1 + 0.035)^{-30}$$
$$= \$3,562.78.$$

Thus the trust should start with $36,784.09 + 3,562.78 = \$40,346.87$.

Sometimes we know how much money we begin with and for how long we want to make withdrawals. We then want to determine the amount of money we can withdraw each period. For this, we simply solve the present value formula for the payment.

**using Technology**

To automate the computations in Example 4 using a graphing calculator or a spreadsheet, see the Technology Guides at the end of the chapter. Outline:

**TI-83/84 Plus**
APPS 1:Finance, then
1:TVM Solver
N = 16, I% = 4, PV = -100000,
FV = 0, P/Y = 4, C/Y = 4
With cursor on PMT line,
ALPHA SOLVE
[More details on page 168.]

**Spreadsheet**
=PMT(4/4,4*4,-100000,0)
[More details on page 172.]

**Website**
www.WanerMath.com
→ On Line Utilities
→ Time Value of Money Utility

Enter the values shown, and press "Compute" next to PMT.

## Payment Formula for an Ordinary Annuity

Suppose that an account has an annual rate of $r$ compounded $m$ times per year, so that $i = r/m$ is the interest rate per compounding period. Suppose also that the account starts with a balance of $PV$. If you want to receive a payment of $PMT$ at the end of each compounding period, and the account is down to $0 after $t$ years, or $n = mt$ periods, then

$$PMT = PV \frac{i}{1 - (1+i)^{-n}}.$$

### EXAMPLE 4 Education Fund

Tony and Maria (see Example 2), having accumulated $100,000 for José Phillipe's college education, would now like to make quarterly withdrawals over the next 4 years. How much money can they withdraw each quarter in order to draw down the account to zero at the end of the 4 years? (Recall that the account pays 4% interest compounded quarterly.)

**Solution** Now Tony and Maria's account is acting as an annuity with a present value of $100,000. So, $PV = \$100,000$, $r = 0.04$, and $m = 4$, giving $i = 0.04/4 = 0.01$, and $n = 4 \times 4 = 16$. We use the payment formula to get

$$PMT = 100,000 \frac{0.01}{1 - (1 + 0.01)^{-16}} \approx \$6,794.46.$$

So, they can withdraw $6,794.46 each quarter for 4 years, at the end of which time their account balance will be 0.

### EXAMPLE 5  Saving for Retirement

Jane Q. Employee has just started her new job with *Big Conglomerate, Inc.*, and is already looking forward to retirement. BCI offers her as a pension plan an annuity that is guaranteed to earn 6% annual interest compounded monthly. She plans to work for 40 years before retiring and would then like to be able to draw an income of $7,000 per month for 20 years. How much do she and BCI together have to deposit per month into the fund to accomplish this?

**Solution**  Here we have the situation we described at the beginning of the section: a sinking fund accumulating money to be used later as an annuity. We know the desired payment out of the annuity, so we work backward. The first thing we need to do is calculate the present value of the annuity required to make the pension payments. We use the annuity present value formula with $PMT = 7,000$, $i = r/m = 0.06/12 = 0.005$ and $n = 12 \times 20 = 240$.

$$PV = PMT\frac{1 - (1 + i)^{-n}}{i}$$

$$= 7,000\frac{1 - (1 + 0.005)^{-240}}{0.005} \approx \$977,065.40$$

This is the total that must be accumulated in the sinking fund during the 40 years she plans to work. In other words, this is the *future* value, $FV$, of the sinking fund. (Thus, the present value in the first step of our calculation is the future value in the second step.) To determine the payments necessary to accumulate this amount, we use the sinking fund payment formula with $FV = 977,065.40$, $i = 0.005$, and $n = 12 \times 40 = 480$.

$$PMT = FV\frac{i}{(1 + i)^n - 1}$$

$$= 977,065.40\frac{0.005}{1.005^{480} - 1}$$

$$\approx \$490.62$$

So, if she and BCI collectively deposit $490.62 per month into her retirement fund, she can retire with the income she desires.

## Installment Loans

In a typical installment loan, such as a car loan or a home mortgage, we borrow an amount of money and then pay it back with interest by making fixed payments (usually every month) over some number of years. From the point of view of the lender, this is an annuity. Thus, loan calculations are identical to annuity calculations.

### EXAMPLE 6  Home Mortgages

Marc and Mira are buying a house, and have taken out a 30-year, $90,000 mortgage at 8% interest per year. What will their monthly payments be?

**Solution**  From the bank's point of view, a mortgage is an annuity. In this case, the present value is $PV = \$90,000$, $r = 0.08$, $m = 12$, and $n = 12 \times 30 = 360$. To find the payments, we use the payment formula:

$$PMT = 90,000\frac{0.08/12}{1 - (1 + 0.08/12)^{-360}} \approx \$660.39.$$

The word "mortgage" comes from the French for "dead pledge." The process of paying off a loan is called **amortizing** the loan, meaning to kill the debt owed.

## EXAMPLE 7 Amortization Schedule

Continuing Example 6: Mortgage interest is tax deductible, so it is important to know how much of a year's mortgage payments represents interest. How much interest will Marc and Mira pay in the first year of their mortgage?

**Solution** Let us calculate how much of each month's payment is interest and how much goes to reducing the outstanding principal. At the end of the first month Marc and Mira must pay 1 month's interest on $90,000, which is

$$\$90,000 \times \frac{0.08}{12} = \$600.$$

The remainder of their first monthly payment, $660.39 − 600 = $60.39, goes to reducing the principal. Thus, in the second month the outstanding principal is $90,000 − 60.39 = $89,939.61, and part of their second monthly payment will be for the interest on this amount, which is

$$\$89,939.61 \times \frac{0.08}{12} \approx \$599.60.$$

The remaining $660.39 − $599.60 = $60.79 goes to further reduce the principal. If we continue this calculation for the 12 months of the first year, we get the beginning of the mortgage's **amortization schedule**.

**using** Technology

To automate the construction of the amortization schedule in Example 7 using a graphing calculator or a spreadsheet, see the Technology Guides at the end of the chapter. [Details: TI-83/84 Plus: page 169, Spreadsheet: page 173]

| Month | Interest Payment | Payment on Principal | Outstanding Principal |
|-------|-----------------|---------------------|----------------------|
| 0 | | | $90,000.00 |
| 1 | $600.00 | $60.39 | 89,939.61 |
| 2 | 599.60 | 60.79 | 89,878.82 |
| 3 | 599.19 | 61.20 | 89,817.62 |
| 4 | 598.78 | 61.61 | 89,756.01 |
| 5 | 598.37 | 62.02 | 89,693.99 |
| 6 | 597.96 | 62.43 | 89,631.56 |
| 7 | 597.54 | 62.85 | 89,568.71 |
| 8 | 597.12 | 63.27 | 89,505.44 |
| 9 | 596.70 | 63.69 | 89,441.75 |
| 10 | 596.28 | 64.11 | 89,377.64 |
| 11 | 595.85 | 64.54 | 89,313.10 |
| 12 | 595.42 | 64.97 | 89,248.13 |
| **Total** | $7,172.81 | $751.87 | |

As we can see from the totals at the bottom of the columns, Marc and Mira will pay a total of $7,172.81 in interest in the first year.

## Bonds

Suppose that a corporation offers a 10-year bond paying 6.5% with payments every 6 months. As we saw in Example 3 of Section 2.1, this means that if we pay $10,000 for bonds with a maturity value of $10,000, we will receive $6.5/2 = 3.25\%$ of $10,000, or $325, every 6 months for 10 years, at the end of which time the corporation will give us the original $10,000 back. But bonds are rarely sold at their maturity value. Rather, they are auctioned off and sold at a price the bond market determines they are worth.

For example, suppose that bond traders are looking for an investment that has a **rate of return** or **yield** of 7% rather than the stated 6.5% (sometimes called the **coupon interest rate** to distinguish it from the rate of return). How much would they be willing to pay for the bonds above with a maturity value of $10,000? Think of the bonds as an investment that will pay the owner $325 every 6 months for 10 years, and will pay an additional $10,000 on maturity at the end of the 10 years. We can treat the $325 payments as if they come from an annuity and determine how much an investor would pay for such an annuity if it earned 7% compounded semiannually. Separately, we determine the present value of an investment worth $10,000 ten years from now, if it earned 7% compounded semiannually. For the first calculation, we use the annuity present value formula, with $i = 0.07/2$ and $n = 2 \times 10 = 20$.

$$PV = PMT\frac{1 - (1+i)^{-n}}{i}$$
$$= 325\frac{1 - (1+0.07/2)^{-20}}{0.07/2}$$
$$= \$4{,}619.03$$

For the second calculation, we use the present value formula for compound interest:

$$PV = 10{,}000(1 + 0.07/2)^{-20}$$
$$= \$5{,}025.66.$$

Thus, an investor looking for a 7% return will be willing to pay $4,619.03 for the semiannual payments of $325 and $5,025.66 for the $10,000 payment at the end of 10 years, for a total of $4,619.03 + 5,025.66 = \$9,644.69$ for the $10,000 bond.

---

### EXAMPLE 8  Bonds

Suppose that bond traders are looking for only a 6% yield on their investment. How much would they pay per $10,000 for the 10-year bonds above, which have a coupon interest rate of 6.5% and pay interest every 6 months?

**Solution**  We redo the calculation with $r = 0.06$. For the annuity calculation we now get

$$PV = 325\frac{1 - (1+0.06/2)^{-20}}{0.06/2} = \$4{,}835.18.$$

For the compound interest calculation we get

$$PV = 10{,}000(1 + 0.06/2)^{-20} = \$5{,}536.76.$$

Thus, traders would be willing to pay a total of $4,835.18 + 5,536.76 = \$10,371.94$ for bonds with a maturity value of $10,000.

➡ **Before we go on...** Notice how the selling price of the bonds behaves as the desired yield changes. As desired yield goes up, the price of the bonds goes down, and as desired yield goes down, the price of the bonds goes up. When the desired yield equals the coupon interest rate, the selling price will equal the maturity value. Therefore, when the yield is higher than the coupon interest rate, the price of the bond will be below its maturity value, and when the yield is lower than the coupon interest rate, the price will be above the maturity value.

As we've mentioned before, the desired yield depends on many factors, but it generally moves up and down with prevailing interest rates. And interest rates have historically gone up and down cyclically. The effect on the value of bonds can be quite dramatic (see Exercises 65 and 66 in the preceding section). Because bonds can be sold again once bought, someone who buys bonds while interest rates are high and then resells them when interest rates decline can make a healthy profit. ■

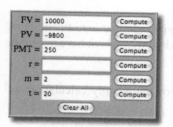

---

## EXAMPLE 9  Rate of Return on a Bond

Suppose that a 5%, 20-year bond sells for $9,800 per $10,000 maturity value. What rate of return will investors get?

**Solution** Assuming the usual semiannual payments, we know the following about the annuity calculation:

$$PMT = 0.05 \times 10,000/2 = 250$$
$$n = 20 \times 2 = 40.$$

What we do not know is $r$ or $i$, the annual or semiannual rate of return, respectively. So, we write

$$PV = 250\frac{1 - (1 + i)^{-40}}{i}.$$

For the compound interest calculation, we know $FV = 10,000$ and $n = 40$ again, so we write

$$PV = 10,000(1 + i)^{-40}.$$

Adding these together should give the selling price of $9,800:

$$250\frac{1 - (1 + i)^{-40}}{i} + 10,000(1 + i)^{-40} = 9,800.$$

This equation cannot be solved for $i$ directly. The best we can do by hand is a sort of trial-and-error approach, substituting a few values for $i$ in the left-hand side of the above equation to get an estimate; we use the fact that, because the selling price is below the maturity value, $r$ must be larger than 0.05, so $i$ must be more than 0.025:

| $i$ | 0.025 | 0.03 | 0.035 |
|---|---|---|---|
| $250\dfrac{1 - (1 + i)^{-40}}{i} + 10,000(1 + i)^{-40}$ | 10,000 | 8,844 | 7,864 |

Since we want the value to be 9,800, we see that the correct answer is somewhere between $i = 0.025$ and $i = 0.03$. Let us try the value midway between 0.025 and 0.03; namely $i = 0.0275$.

| $i$ | 0.025 | 0.0275 | 0.03 |
|---|---|---|---|
| $250\dfrac{1 - (1 + i)^{-40}}{i} + 10,000(1 + i)^{-40}$ | 10,000 | 9,398 | 8,844 |

Now we know that the correct value of $i$ is somewhere between 0.025 and 0.0275, so we can choose for our next estimate of $i$ the number midway between them: 0.02625. We could continue in this fashion to obtain $i$ as accurately as we like. In fact, $i \approx 0.02581$, corresponding to an annual rate of return of approximately 5.162%.

---

## FAQs

### Which Formula to Use

**Q :** *We have retirement accounts, trust funds, loans, bonds, and so on. Some are sinking funds, others are annuities. How do we distinguish among them, so we can tell which formula to use?*

**A :** In general, remember that a sinking fund is an interest-bearing fund into which payments are made, while an annuity is an interest-bearing fund from which money is withdrawn. Here is a list of some of the accounts we have discussed in this section:

- *Retirement Accounts* A retirement account is a sinking fund while payments are being made into the account (prior to retirement) and an annuity while a pension is being withdrawn (after retirement).
- *Education Funds* These are similar to retirement accounts.
- *Trust Funds* A trust fund is an annuity if periodic withdrawals are made.
- *Installment Loans* We think of an installment loan as an investment a bank makes in the lender. In this way, the lender's payments can be viewed as the bank's withdrawals, and so a loan is an annuity.
- *Bonds* A bond pays regular fixed amounts until it matures, at which time it pays its maturity value. We think of the bond as an annuity coupled with a compound interest investment funding the payment of the maturity value. We can then determine its present value based on the current market interest rate.

From a mathematical point of view, sinking funds and annuities are really the same thing. See the Communication and Reasoning Exercises for more about this.

---

## 2.3 EXERCISES

▼ more advanced   ◆ challenging

**T** indicates exercises that should be solved using technology

*Find the amount accumulated in the sinking funds in Exercises 1–6. (Assume end-of-period deposits and compounding at the same intervals as deposits.)* HINT [See Example 1.]

1. $100 deposited monthly for 10 years at 5% per year

2. $150 deposited monthly for 20 years at 3% per year

3. $1,000 deposited quarterly for 20 years at 7% per year

4. $2,000 deposited quarterly for 10 years at 7% per year

5. ▼ $100 deposited monthly for 10 years at 5% per year in an account containing $5,000 at the start

6. ▼ $150 deposited monthly for 20 years at 3% per year in an account containing $10,000 at the start

*Find the periodic payments necessary to accumulate the amounts given in Exercises 7–12 in a sinking fund. (Assume end-of-period deposits and compounding at the same intervals as deposits.)* HINT [See Example 2.]

7. $10,000 in a fund paying 5% per year, with monthly payments for 5 years

8. $20,000 in a fund paying 3% per year, with monthly payments for 10 years

9. $75,000 in a fund paying 6% per year, with quarterly payments for 20 years

10. $100,000 in a fund paying 7% per year, with quarterly payments for 20 years

11. ▼ $20,000 in a fund paying 5% per year, with monthly payments for 5 years, if the fund contains $10,000 at the start

**12.** ▼ $30,000 in a fund paying 3% per year, with monthly payments for 10 years, if the fund contains $10,000 at the start

*Find the present value of the annuity necessary to fund the withdrawals given in Exercises 13–18. (Assume end-of-period withdrawals and compounding at the same intervals as withdrawals.)* HINT [See Example 3.]

**13.** $500 per month for 20 years, if the annuity earns 3% per year

**14.** $1,000 per month for 15 years, if the annuity earns 5% per year

**15.** $1,500 per quarter for 20 years, if the annuity earns 6% per year

**16.** $2,000 per quarter for 20 years, if the annuity earns 4% per year

**17.** ▼ $500 per month for 20 years, if the annuity earns 3% per year and if there is to be $10,000 left in the annuity at the end of the 20 years

**18.** ▼ $1,000 per month for 15 years, if the annuity earns 5% per year and if there is to be $20,000 left in the annuity at the end of the 15 years

*Find the periodic withdrawals for the annuities given in Exercises 19–24. (Assume end-of-period withdrawals and compounding at the same intervals as withdrawals.)* HINT [See Example 4.]

**19.** $100,000 at 3%, paid out monthly for 20 years

**20.** $150,000 at 5%, paid out monthly for 15 years

**21.** $75,000 at 4%, paid out quarterly for 20 years

**22.** $200,000 at 6%, paid out quarterly for 15 years

**23.** ▼ $100,000 at 3%, paid out monthly for 20 years, leaving $10,000 in the account at the end of the 20 years

**24.** ▼ $150,000 at 5%, paid out monthly for 15 years, leaving $20,000 in the account at the end of the 15 years

*Determine the periodic payments on the loans given in Exercises 25–28.* HINT [See Example 6.]

**25.** $10,000 borrowed at 9% for 4 years, with monthly payments

**26.** $20,000 borrowed at 8% for 5 years, with monthly payments

**27.** $100,000 borrowed at 5% for 20 years, with quarterly payments

**28.** $1,000,000 borrowed at 4% for 10 years, with quarterly payments

*Determine the selling price, per $1,000 maturity value, of the bonds[16] in Exercises 29–34. (Assume twice-yearly interest payments.)* HINT [See Example 8.]

**29.** ▼ 10 year, 4.875% bond, with a yield of 4.880%

**30.** ▼ 30 year, 5.375% bond, with a yield of 5.460%

**31.** ▼ 2 year, 3.625% bond, with a yield of 3.705%

**32.** ▼ 5 year, 4.375% bond, with a yield of 4.475%

**33.** ▼ 10 year, 5.5% bond, with a yield of 6.643%

**34.** ▼ 10 year, 6.25% bond, with a yield of 33.409%

⊤ *Determine the yield on the bonds[17] in Exercises 35–40. (Assume twice-yearly interest payments.)* HINT [See Example 9.]

**35.** ▼ 5 year, 3.5% bond, selling for $994.69 per $1,000 maturity value

**36.** ▼ 10 year, 3.375% bond, selling for $991.20 per $1,000 maturity value

**37.** ▼ 2 year, 3% bond, selling for $998.86 per $1,000 maturity value

**38.** ▼ 5 year, 4.625% bond, selling for $998.45 per $1,000 maturity value

**39.** ▼ 5 year, 4.25% bond, selling for $923.81 per $1,000 maturity value

**40.** ▼ 5 year 3.6% bond, selling for $216.96 per $1,000 maturity value

## APPLICATIONS

**41.** *Car Loans* While shopping for a car loan, you get the following offers: Solid Savings & Loan is willing to loan you $10,000 at 9% interest for 4 years. Fifth Federal Bank & Trust will loan you the $10,000 at 7% interest for 3 years. Both require monthly payments. You can afford to pay $250 per month. Which loan, if either, can you take?

**42.** *Business Loans* You need to take out a loan of $20,000 to start up your T-shirt business. You have two possibilities: One bank is offering a 10% loan for 5 years, and another is offering a 9% loan for 4 years. Which will have the lower monthly payments? On which will you end up paying more interest total?

**43.** ▼ *Pensions* Your pension plan is an annuity with a guaranteed return of 3% per year (compounded monthly). You would like to retire with a pension of $5,000 per month for 20 years. If you work 40 years before retiring, how much must you and your employer deposit each month into the fund? HINT [See Example 5.]

**44.** ▼ *Pensions* Meg's pension plan is an annuity with a guaranteed return of 5% per year (compounded quarterly). She would like to retire with a pension of $12,000 per quarter for 25 years. If she works 45 years before retiring, how much money must she and her employer deposit each quarter? HINT [See Example 5.]

**45.** ▼ *Pensions* Your pension plan is an annuity with a guaranteed return of 4% per year (compounded quarterly). You can afford to put $1,200 per quarter into the fund, and you will work for 40 years before retiring. After you retire, you will

---

[16]The first four are actual U.S. Treasury notes and bonds auctioned in 2001 and 2002. The last two are, respectively, a Spanish and a Greek government bond auctioned November 17, 2011, during the Eurozone crisis. Sources: The Bureau of the Public Debt's Web site (www.publicdebt.treas.gov) and The Wall Street Journal's Web site (www.wsj.com).

[17]*Ibid.*

be paid a quarterly pension based on a 25-year payout. How much will you receive each quarter?

46. ▼ *Pensions* Jennifer's pension plan is an annuity with a guaranteed return of 5% per year (compounded monthly). She can afford to put $300 per month into the fund, and she will work for 45 years before retiring. If her pension is then paid out monthly based on a 20-year payout, how much will she receive per month?

*Note: In Exercises 47–52 we suggest the use of the finance functions in the TI-83/84 Plus or amortization tables in a spreadsheet.*

47. ⬛ ▼ *Mortgages* You take out a 15-year mortgage for $50,000, at 8%, to be paid off monthly. Construct an amortization table showing how much you will pay in interest each year, and how much goes toward paying off the principal. HINT [See Example 7.]

48. ⬛ ▼ *Mortgages* You take out a 30-year mortgage for $95,000 at 9.75%, to be paid off monthly. If you sell your house after 15 years, how much will you still owe on the mortgage? HINT [See Example 7.]

49. ⬛ ▼ *Mortgages* This exercise describes a popular kind of mortgage. You take out a $75,000, 30-year mortgage. For the first 5 years the interest rate is held at 5%, but for the remaining 25 years it rises to 9.5%. The payments for the first 5 years are calculated as if the 5% rate were going to remain in effect for all 30 years, and then the payments for the last 25 years are calculated to amortize the debt remaining at the end of the fifth year. What are your monthly payments for the first 5 years, and what are they for the last 25 years?

50. ⬛ ▼ *Adjustable Rate Mortgages* You take out an adjustable rate mortgage for $100,000 for 20 years. For the first 5 years, the rate is 4%. It then rises to 7% for the next 10 years, and then 9% for the last 5 years. What are your monthly payments in the first 5 years, the next 10 years, and the last 5 years? (Assume that each time the rate changes, the payments are recalculated to amortize the remaining debt if the interest rate were to remain constant for the remaining life of the mortgage.)

51. ⬛ ▼ *Refinancing* Your original mortgage was a $96,000, 30-year 9.75% mortgage. After 4 years you refinance the remaining principal for 30 years at 6.875%. What was your original monthly payment? What is your new monthly payment? How much will you save in interest over the course of the loan by refinancing?

52. ⬛ ▼ *Refinancing* Kara and Michael take out a $120,000, 30-year, 10% mortgage. After 3 years they refinance the remaining principal with a 15-year, 6.5% loan. What were their original monthly payments? What is their new monthly payment? How much did they save in interest over the course of the loan by refinancing?

53. ⬛ ▼ *Fees* You take out a 2-year, $5,000 loan at 9% interest with monthly payments. The lender charges you a $100 fee that can be paid off, interest free, in equal monthly installments over the life of the loan. Thinking of the fee as additional interest, what is the actual annual interest rate you will pay?

54. ⬛ ▼ *Fees* You take out a 3-year, $7,000 loan at 8% interest with monthly payments. The lender charges you a $100 fee that can be paid off, interest free, in equal monthly installments over the life of the loan. Thinking of the fee as additional interest, what is the actual annual interest rate you will pay?

55. ⬛ ▼ *Savings* You wish to accumulate $100,000 through monthly payments of $500. If you can earn interest at an annual rate of 4% compounded monthly, how long (to the nearest year) will it take to accomplish your goal?

56. ⬛ ▼ *Retirement* Alonzo plans to retire as soon as he has accumulated $250,000 through quarterly payments of $2,500. If Alonzo invests this money at 5.4% interest, compounded quarterly, when (to the nearest year) can he retire?

57. ⬛ ▼ *Loans* You have a $2,000 credit card debt, and you plan to pay it off through monthly payments of $50. If you are being charged 15% interest per year, how long (to the nearest 0.5 years) will it take you to repay your debt?

58. ⬛ ▼ *Loans* You owe $2,000 on your credit card, which charges you 15% interest. Determine, to the nearest 1¢, the minimum monthly payment that will allow you to eventually repay your debt.

59. ⬛ ▼ *Savings* You are depositing $100 per month in an account that pays 4.5% interest per year (compounded monthly) while your friend Lucinda is depositing $75 per month in an account that earns 6.5% interest per year (compounded monthly). When, to the nearest year, will her balance exceed yours?

60. ⬛ ▼ *Car Leasing* You can lease a $15,000 car for $300 per month. For how long (to the nearest year) should you lease the car so that your monthly payments are lower than if you were purchasing it with an 8%-per-year loan?

## COMMUNICATION AND REASONING EXERCISES

61. Your cousin Simon claims that you have wasted your time studying annuities: If you wish to retire on an income of $1,000 per month for 20 years, you need to save $1,000 per month for 20 years, period. Explain why he is wrong.

62. ▼ Your other cousin Cecilia claims that you will earn more interest by depositing $10,000 through smaller more frequent payments than through larger less frequent payments. Is she correct? Give a reason for your answer.

63. ▼ A real estate broker tells you that doubling the period of a mortgage halves the monthly payments. Is he correct? Support your answer by means of an example.

64. ▼ Another real estate broker tells you that doubling the size of a mortgage doubles the monthly payments. Is he correct? Support your answer by means of an example.

65. ◆ Consider the formula for the future value of a sinking fund with given payments. Show algebraically that the present value of that future value is the same as the present value of the annuity required to fund the same payments.

66. ◆ Give a non-algebraic justification for the result from the preceding exercise.

# CHAPTER 2 REVIEW

## KEY CONCEPTS

 Website www.WanerMath.com

Go to the Website at www.WanerMath
.com to find a comprehensive and
interactive Web-based summary
of Chapter 2.

### 2.1 Simple Interest

Simple interest *p. 126*
Future value *p. 127*
Bond, maturity *p. 128*
Present value *p. 129*
Treasury bills (T-bills); maturity value,
discount rate, yield *p. 130*

### 2.2 Compound Interest

Compound interest *p. 134*
Future value for compound
interest *p. 135*
Present value for compound
interest *p. 137*
Zero coupon bond or strip *p. 137*
Inflation, constant dollars *pp. 138–139*
Effective interest rate, annual
percentage yield (APY) *p. 139*

### 2.3 Annuities, Loans, and Bonds

Annuity, sinking fund *p. 145*
Future value of a sinking fund *p. 146*
Payment formula for a sinking
fund *p. 147*
Present value of an annuity *p. 149*
Ordinary annuity, annuity due *p. 149*
Payment formula for an annuity *p. 150*
Installment loan *p. 151*
Amortization schedule *p. 152*
Bond *p. 153*

## REVIEW EXERCISES

*In each of Exercises 1–6, find the future value of the investment.*

1. $6,000 for 5 years at 4.75% simple annual interest

2. $10,000 for 2.5 years at 5.25% simple annual interest

3. $6,000 for 5 years at 4.75% compounded monthly

4. $10,000 for 2.5 years at 5.25% compounded semiannually

5. $100 deposited at the end of each month for 5 years, at 4.75% interest compounded monthly

6. $2,000 deposited at the end of each half-year for 2.5 years, at 5.25% interest compounded semiannually

*In each of Exercises 7–12, find the present value of the investment.*

7. Worth $6,000 after 5 years at 4.75% simple annual interest

8. Worth $10,000 after 2.5 years at 5.25% simple annual interest

9. Worth $6,000 after 5 years at 4.75% compounded monthly

10. Worth $10,000 after 2.5 years at 5.25% compounded semiannually

11. Funding $100 withdrawals at the end of each month for 5 years, at 4.75% interest compounded monthly

12. Funding $2,000 withdrawals at the end of each half-year for 2.5 years, at 5.25% interest compounded semiannually

*In each of Exercises 13–18, find the amounts indicated.*

13. The monthly deposits necessary to accumulate $12,000 after 5 years in an account earning 4.75% compounded monthly

14. The semiannual deposits necessary to accumulate $20,000 after 2.5 years in an account earning 5.25% compounded semiannually

15. The monthly withdrawals possible over 5 years from an account earning 4.75% compounded monthly and starting with $6,000

16. The semiannual withdrawals possible over 2.5 years from an account earning 5.25% compounded semiannually and starting with $10,000

17. The monthly payments necessary on a 5-year loan of $10,000 at 4.75%

18. The semiannual payments necessary on a 2.5-year loan of $15,000 at 5.25%

19. How much would you pay for a $10,000, 5-year, 6% bond if you want a return of 7%? (Assume that the bond pays interest every 6 months.)

20. How much would you pay for a $10,000, 5-year, 6% bond if you want a return of 5%? (Assume that the bond pays interest every 6 months.)

21. A $10,000, 7-year, 5% bond sells for $9,800. What return does it give you? (Assume that the bond pays interest every 6 months.)

22. A $10,000, 7-year, 5% bond sells for $10,200. What return does it give you? (Assume that the bond pays interest every 6 months.)

*In each of Exercises 23–28, find the time requested, to the nearest 0.1 year.*

23. The time it would take $6,000 to grow to $10,000 at 4.75% simple annual interest

24. The time it would take $10,000 to grow to $15,000 at 5.25% simple annual interest

25. The time it would take $6,000 to grow to $10,000 at 4.75% interest compounded monthly

26. The time it would take $10,000 to grow to $15,000 at 5.25% interest compounded semiannually

27. The time it would take to accumulate $10,000 by depositing $100 at the end of each month in an account earning 4.75% interest compounded monthly

28. The time it would take to accumulate $15,000 by depositing $2,000 at the end of each half-year in an account earning 5.25% compounded semiannually

## APPLICATIONS

*Stock Investments* Exercises 29–34 are based on the following table, which shows some values of ABCromD (ABCD) stock:

| Dec. 2002 | Aug. 2004 | Mar. 2005 | May 2005 | Aug. 2005 | Dec. 2005 |
|-----------|-----------|-----------|----------|-----------|-----------|
| 3.28 | 16.31 | 33.95 | 21.00 | 30.47 | 7.44 |
| Jan. 2007 | Mar. 2008 | Oct. 2008 | Nov. 2009 | Feb. 2010 | Aug. 2010 |
| 12.36 | 7.07 | 11.44 | 33.53 | 44.86 | 45.74 |

29. Marjory Duffin purchased ABCD stock in December 2002 and sold it in August 2010. Calculate her annual return on a simple interest basis to the nearest 0.01%.

30. John O'Hagan purchased ABCD stock in March 2005 and sold it in January 2007. Calculate his annual return on a simple interest basis to the nearest 0.01%.

31. Suppose Marjory Duffin had bought ABCD stock in January 2007. If she had later sold at one of the dates in the table, which of those dates would have given her the largest annual return on a simple interest basis, and what would that return have been?

32. Suppose John O'Hagan had purchased ABCD stock in August 2004. If he had later sold at one of the dates in the table, which of those dates would have given him the largest annual loss on a simple interest basis, and what would that loss have been?

33. Did ABCD stock undergo simple interest increase in the period December 2002 through March 2005? (Give a reason for your answer.)

34. If ABCD stock underwent simple interest increase from February 2010 through August 2010 and into 2011, what would the price have been in December 2011?

35. *Revenue* Total Online revenues at OHaganBooks.com during 1999, its first year of operation, amounted to $150,000. After December, 1999, revenues increased by a steady 20% each year. Track OHaganBooks.com's revenues for the subsequent 5 years, assuming that this rate of growth continued. During which year did the revenue surpass $300,000?

36. *Net Income* Unfortunately, the picture for net income was not so bright: The company lost $20,000 in the fourth quarter of 1999. However, the quarterly loss decreased at an annual rate of 15%. How much did the company lose during the third quarter of 2001?

37. *Stocks* In order to finance anticipated expansion, CEO John O'Hagan is considering making a public offering of OHaganBooks.com shares at $3.00 a share. O'Hagan is not sure how many shares to offer, but would like the shares to reach a total market value of at least $500,000 6 months after the initial offering. He estimates that the value of the stock will double in the first day of trading, and then appreciate at around 8% per month for the first 6 months. How many shares of stock should the company offer?

38. *Stocks* Unfortunately, renewed panic about the Monaco debt crisis cause the U.S. stock market to plunge on the very first day of OHaganBooks.com's initial public offering (IPO) of 600,000 shares at $3.00 per share, and the shares ended the trading day 60% lower. Subsequently, as the Monaco debt crisis worsened, the stock depreciated by 10% per week during the subsequent 5 weeks. What was the total market value of the stocks at the end of 5 weeks?

39. *Loans* OHaganBooks.com is seeking a $250,000 loan to finance its continuing losses. One of the best deals available is offered by Industrial Bank, which offers a 10-year 9.5% loan. What would the monthly payments be for this loan?

40. *Loans* (See Exercise 39.) Expansion Loans offers an 8-year 6.5% loan. What would the monthly payments be for the $250,000 loan from Expansion?

41. *Loans* (See Exercise 39.) OHaganBooks.com can afford to pay only $3,000 per month to service its debt. What, to the nearest dollar, is the largest amount the company can borrow from Industrial Bank?

42. *Loans* (See Exercise 40.) OHaganBooks.com can afford to pay only $3,000 per month to service its debt. What, to the nearest dollar, is the largest amount the company can borrow from Expansion Loans?

43. ⊤ *Loans* (See Exercise 39.) What interest rate would Industrial Bank have to offer in order to meet the company's loan requirements at a price (no more than $3,000 per month) it can afford?

44. ⊤ *Loans* (See Exercise 40.) What interest rate would Expansion Loans have to offer in order to meet the company's loan requirements at a price (no more than $3,000 per month) it can afford?

*Retirement Planning OHaganBooks.com has just introduced a retirement package for the employees. Under the annuity plan operated by Sleepy Hollow, the monthly contribution by the company on behalf of each employee is $800. Each employee can then supplement that amount through payroll deductions. The current rate of return of Sleepy Hollow's retirement fund is 7.3%. Use this information in Exercises 45–52.*

45. Jane Callahan, the Web site developer at OHaganBooks.com, plans to retire in 10 years. She contributes $1,000 per month to the plan (in addition to the company contribution of $800). Currently, there is $50,000 in her retirement annuity. How much (to the nearest dollar) will it be worth when she retires?

46. Percy Egan, the assistant Web site developer at OHaganBooks.com, plans to retire in 8 years. He contributes $950 per month to the plan (in addition to the company contribution of $800). Currently, there is $60,000 in his retirement annuity. How much (to the nearest dollar) will it be worth when he retires?

47. When she retires, how much of Jane Callahan's retirement fund will have resulted from the company contribution? (See Exercise 45. The company did not contribute toward the $50,000 Callahan now has.)

**48.** When he retires, how much of Percy Egan's retirement fund will have resulted from the company contribution? (See Exercise 46. The company began contributing $800 per month to his retirement fund when he was hired 5 years ago. Assume the rate of return from Sleepy Hollow has been unchanged.)

**49.** (See Exercise 45.) Jane Callahan actually wants to retire with $500,000. How much should she contribute each month to the annuity?

**50.** (See Exercise 46.) Percy Egan actually wants to retire with $600,000. How much should he contribute each month to the annuity?

**51.** (See Exercise 45.) On second thought, Callahan wants to be in a position to draw at least $5,000 per month for 30 years after her retirement. She feels she can invest the proceeds of her retirement annuity at 8.7% per year in perpetuity. Given the information in Exercise 45, how much will she need to contribute to the plan starting now?

**52.** (See Exercise 46.) On second thought, Egan wants to be in a position to draw at least $6,000 per month for 25 years after

his retirement. He feels he can invest the proceeds of his retirement annuity at 7.8% per year in perpetuity. Given the information in Exercise 46, how much will he need to contribute to the plan starting now?

*Actually, Jane Callahan is quite pleased with herself; 1 year ago she purchased a $50,000 government bond paying 7.2% per year (with interest paid every 6 months) and maturing in 10 years, and interest rates have come down since then.*

**53.** The current interest rate on 10-year government bonds is 6.3%. If she were to auction the bond at the current interest rate, how much would she get?

**54.** If she holds on to the bond for 6 more months and the interest rate drops to 6%, how much will the bond be worth then?

**55.** ▯ Jane suspects that interest rates will come down further during the next 6 months. If she hopes to auction the bond for $54,000 in 6 months' time, what will the interest rate need to be at that time?

**56.** ▯ If, in 6 months' time, the bond is auctioned for only $52,000, what will the interest rate be at that time?

---

**Case Study**

## Adjustable Rate and Subprime Mortgages

The term **subprime mortgage** refers to mortgages given to home buyers with a heightened perceived risk of default, as when, for instance, the price of the home being purchased is higher than the borrower can reasonably afford. Such loans are typically **adjustable rate** loans, meaning that the lending rate varies through the duration of the loan.* Subprime adjustable rate loans typically start at artificially low "teaser rates" that the borrower can afford, but then increase significantly over the life of the mortgage. The U.S. real estate bubble of 2000–2005 led to a frenzy of subprime lending, the rationale being that a borrower having trouble meeting mortgage payments could either sell the property at a profit or re-finance the loan, or the lending institution could earn a hefty profit by repossessing the property in the event of foreclosure.

Mr. and Mrs. Wong have an appointment tomorrow with you, their investment counselor, to discuss their plan to purchase a $400,000 house in Orlando, Florida. They have saved $20,000 for a down payment, so want to take out a $380,000 mortgage. Their combined annual income is $80,000 per year, which they estimate will increase by 4% annually over the foreseeable future, and they are considering three different specialty 30-year mortgages:

**Hybrid:** The interest is fixed at a low introductory rate of 4% for 5 years, and then adjusts annually to 5% over the U.S. federal funds rate.†

**Interest-Only:** During the first 5 years, the rate is set at 4.2% and no principal is paid. After that time, the mortgage adjusts annually to 5% over the U.S. federal funds rate.

**Negative Amortization:** During the first 5 years, the rate is set at 4.7% based on a principal of 60% of the purchase price of the home, with the result that the balance on the principal actually grows during this period. After that time, the mortgage adjusts annually to 5% over the U.S. federal funds rate.

\* In an adjustable rate mortgage, the payments are recalculated each time the interest rate changes, based on the assumption that the new interest rate will be unchanged for the remaining life of the loan. We say that the loan is **re-amortized** at the new rate.

† The U.S. federal funds rate is the rate banks charge each other for loans and is often used to set rates for other loans. Manipulating this rate is one way the U.S. Federal Reserve regulates the money supply.

Federal Funds Rate History
(Effective Rate 1955–2011)

Source: Board of Governors of the Federal Reserve System (www.federalreserve.gov)

**Figure 3**

You decide that you should create an Excel worksheet that will compute the monthly payments for the three types of loan. Of course, you have no way of predicting what the U.S. federal funds rate will be in over the next 30 years (see Figure 3 for historical values), so you decide to include three scenarios for the federal funds rate in each case:

**Scenario 1:** Federal funds rate is 4.25% in year 6 and then increases by 0.25% per year.

**Scenario 2:** Federal funds rate is steady at 4% during the term of the loan.

**Scenario 3:** Federal funds rate is 15% in year 6 and then decreases by 0.25% per year.

Each worksheet will show month-by-month payments for the specific type of loan. Typically, to be affordable, payments should not exceed 28% of gross monthly income, so you will tabulate that quantity as well.

**Hybrid Loan:** You begin to create your worksheet by estimating 28% of the Wongs' monthly income, assuming a 4% increase each year (the income is computed using the compound interest formula for annual compounding):

| | A | B | C | D | E | F | G | H |
|---|---|---|---|---|---|---|---|---|
| 1 | **Year** | **28% of Monthly Income** | | **Interest Rate** | | | **Monthly Payment** | |
| 2 | | | **Scenario 1** | **Scenario 2** | **Scenario 3** | **Scenario 1** | **Scenario 2** | **Scenario 3** |
| 3 | 1 | =0.28*80000*(1.04)^(A3-1)/12 | | | | | | |
| 4 | 2 | | | | | | | |
| 5 | 3 | | | | | | | |
| 30 | 28 | | | | | | | |
| 31 | 29 | | | | | | | |
| 32 | 30 | | | | | | | |

The next sheet shows the result, as well as the formulas for computing the interest rate in each scenario.

| | A | B | C | D | E | F | G | H |
|---|---|---|---|---|---|---|---|---|
| 1 | Year | 28% of Monthly Income | Interest Rate | | | Monthly Payment | | |
| 2 | | | Scenario 1 | Scenario 2 | Scenario 3 | Scenario 1 | Scenario 2 | Scenario 3 |
| 3 | 1 | $1,866.67 | 4 | 4 | 4 | | | |
| 4 | 2 | $1,941.33 | 4 | 4 | 4 | | | |
| 5 | 3 | $2,018.99 | 4 | 4 | 4 | | | |
| 6 | 4 | $2,099.75 | 4 | 4 | 4 | | | |
| 7 | 5 | $2,183.74 | 4 | 4 | 4 | | | |
| 8 | 6 | $2,271.09 | 9.25 | 9 | 15 | | | |
| 9 | 7 | $2,361.93 | =C8+0.25 | =D8 | =E8-0.25 | | | |
| 10 | 8 | $2,456.41 | | | | | | |
| 30 | 28 | $5,382.29 | | | | | | |
| 31 | 29 | $5,597.58 | | | | | | |
| 32 | 30 | $5,821.48 | | | | | | |

To compute the monthly payment, you decide to use the built-in function PMT, which has the format

$$PMT(i, n, PV, [FV], [type]),$$

where $i$ = interest per period, $n$ = total number of periods of the loan, $PV$ = present value, $FV$ = future value (optional); the *type*, also optional, is 0 or omitted if payments are at the end of each period, and 1 if at the start of each period. The present value will be the outstanding principal owed on the home each time the rate is changed and so that too will need to be known. During the first 5 years we can use as the present value the original cost of the home, but each year thereafter, the loan is re-amortized at the new interest rate, and so the outstanding principal will need to be computed. Although Excel has a built-in function that calculates payment on the principal, it returns only the payment for a single period (month), so without creating a month-by-month amortization table it would be difficult to use this function to track the outstanding principal. On the other hand, the total outstanding principal at any point in time can be computed using the future value formula $FV$. You decide to add three more columns to your Excel worksheet to show the principal outstanding at the start of each year. Here is the spreadsheet with the formulas for the payments and outstanding principal for the first 5 years.

| | A | B | C | D | E | F | G | H | I | J | K |
|---|---|---|---|---|---|---|---|---|---|---|---|
| 1 | Year | 28% of Monthly Income | Interest Rate | | | Monthly Payment | | | Balance on Principal | | |
| 2 | | | Scenario 1 | Scenario 2 | Scenario 3 | Scenario 1 | Scenario 2 | Scenario 3 | Scenario 1 | Scenario 2 | Scenario 3 |
| 3 | 1 | $1,866.67 | 4 | 4 | 4 | =-PMT(C3/1200,360,I$3) | | | $380,000.00 | $380,000.00 | $380,000.00 |
| 4 | 2 | $1,941.33 | 4 | 4 | 4 | | | | =FV(C3/1200,12*$A3,F3,-I$3) | | |
| 5 | 3 | $2,018.99 | 4 | 4 | 4 | | | | | | |
| 6 | 4 | $2,099.75 | 4 | 4 | 4 | | | | | | |
| 7 | 5 | $2,183.74 | 4 | 4 | 4 | | | | | | |
| 8 | 6 | $2,271.09 | 9.25 | 9 | 15 | | | | | | |
| 9 | 7 | $2,361.93 | 9.5 | 9 | 14.75 | | | | | | |
| 10 | 8 | $2,456.41 | 9.75 | 9 | 14.5 | | | | | | |

The two formulas will each be copied across to the adjacent two cells for the other scenarios. A few things to notice: The negative sign before *PMT* converts the negative quantity returned by *PMT* to a positive amount. The dollar sign in I$3 in the *PMT* formula fixes the present value for each year at the original cost of the home for the first 5 years, during which payments are computed as for a fixed rate loan. In the formula for the balance on principal at the start of each year, the number of periods is the total number of months up through the preceding year, and the present value is the same initial price of the home each year during the 5-year fixed rate period.

The next sheet shows the calculated results for the fixed-rate period, and the new formulas to be added for the adjustable rate period starting with the sixth year.

| | C | D | E | F | G | H | I | J | K |
|---|---|---|---|---|---|---|---|---|---|
| 1 | | Interest Rate | | | Monthly Payment | | | Balance on Principal | |
| 2 | Scenario 1 | Scenario 2 | Scenario 3 | Scenario 1 | Scenario 2 | Scenario 3 | Scenario 1 | Scenario 2 | Scenario 3 |
| 3 | 4 | 4 | 4 | $1,814.18 | $1,814.18 | $1,814.18 | $380,000.00 | $380,000.00 | $380,000.00 |
| 4 | 4 | 4 | 4 | $1,814.18 | $1,814.18 | $1,814.18 | $373,308.06 | $373,308.06 | $373,308.06 |
| 5 | 4 | 4 | 4 | $1,814.18 | $1,814.18 | $1,814.18 | $366,343.48 | $366,343.48 | $366,343.48 |
| 6 | 4 | 4 | 4 | $1,814.18 | $1,814.18 | $1,814.18 | $359,095.16 | $359,095.16 | $359,095.16 |
| 7 | 4 | 4 | 4 | $1,814.18 | $1,814.18 | $1,814.18 | $351,551.52 | $351,551.52 | $351,551.52 |
| 8 | 9.25 | 9 | 15 | =-PMT(C8/1200,360-12*$A7,I8) | | | =FV(C7/1200,12,F7,-I7) | | |
| 9 | 9.5 | 9 | 14.75 | | | | | | |
| 10 | 9.75 | 9 | 14.5 | | | | | | |

Notice the changes: The loan is re-amortized each year starting with year 6, and the payment calculation needs to take into account the reduced, remaining lifetime of the loan each time. You now copy these formulas across for the remaining two scenarios, and then copy all six formulas down the remaining rows to complete the calculation. Following is a portion of the complete worksheet showing, in red, those years during which the monthly payment will exceed 28% of the gross monthly income.

| | A | B | C | D | E | F | G | H | I | J | K |
|---|---|---|---|---|---|---|---|---|---|---|---|
| 1 | Year | 28% of Monthly Income | | Interest Rate | | | Monthly Payment | | | Balance on Principal | |
| 2 | | | Scenario 1 | Scenario 2 | Scenario 3 | Scenario 1 | Scenario 2 | Scenario 3 | Scenario 1 | Scenario 2 | Scenario 3 |
| 3 | 1 | $1,866.67 | 4 | 4 | 4 | $1,814.18 | $1,814.18 | $1,814.18 | $380,000.00 | $380,000.00 | $380,000.00 |
| 4 | 2 | $1,941.33 | 4 | 4 | 4 | $1,814.18 | $1,814.18 | $1,814.18 | $373,308.06 | $373,308.06 | $373,308.06 |
| 5 | 3 | $2,018.99 | 4 | 4 | 4 | $1,814.18 | $1,814.18 | $1,814.18 | $366,343.48 | $366,343.48 | $366,343.48 |
| 6 | 4 | $2,099.75 | 4 | 4 | 4 | $1,814.18 | $1,814.18 | $1,814.18 | $359,095.16 | $359,095.16 | $359,095.16 |
| 7 | 5 | $2,183.74 | 4 | 4 | 4 | $1,814.18 | $1,814.18 | $1,814.18 | $351,551.52 | $351,551.52 | $351,551.52 |
| 8 | 6 | $2,271.09 | 9.25 | 9 | 15 | $2,943.39 | $2,884.32 | $4,402.22 | $343,700.55 | $343,700.55 | $343,700.55 |
| 9 | 7 | $2,361.93 | 9.5 | 9 | 14.75 | $3,001.60 | $2,884.32 | $4,336.46 | $340,018.68 | $339,866.12 | $342,337.80 |
| 10 | 8 | $2,456.41 | 9.75 | 9 | 14.5 | $3,058.89 | $2,884.32 | $4,271.84 | $336,135.05 | $335,671.99 | $340,686.40 |
| 11 | 9 | $2,554.66 | 10 | 9 | 14.25 | $3,115.17 | $2,884.32 | $4,208.48 | $332,020.97 | $331,084.43 | $338,694.96 |
| 12 | 10 | $2,656.85 | 10.25 | 9 | 14 | $3,170.39 | $2,884.32 | $4,146.53 | $327,643.98 | $326,066.52 | $336,305.10 |
| 13 | 11 | $2,763.12 | 10.5 | 9 | 13.75 | $3,224.44 | $2,884.32 | $4,086.13 | $322,967.19 | $320,577.89 | $333,450.89 |
| 14 | 12 | $2,873.65 | 10.75 | 9 | 13.5 | $3,277.24 | $2,884.32 | $4,027.41 | $317,948.51 | $314,574.40 | $330,058.36 |
| 15 | 13 | $2,988.59 | 11 | 9 | 13.25 | $3,328.70 | $2,884.32 | $3,970.53 | $312,539.70 | $308,007.73 | $326,045.01 |
| 16 | 14 | $3,108.14 | 11.25 | 9 | 13 | $3,378.71 | $2,884.32 | $3,915.64 | $306,685.30 | $300,825.07 | $321,319.42 |
| 17 | 15 | $3,232.46 | 11.5 | 9 | 12.75 | $3,427.16 | $2,884.32 | $3,862.89 | $300,321.34 | $292,968.62 | $315,780.91 |
| 18 | 16 | $3,361.76 | 11.75 | 9 | 12.5 | $3,473.93 | $2,884.32 | $3,812.43 | $293,373.72 | $284,375.19 | $309,319.29 |
| 19 | 17 | $3,496.23 | 12 | 9 | 12.25 | $3,518.89 | $2,884.32 | $3,764.40 | $285,756.40 | $274,975.63 | $301,814.76 |
| 20 | 18 | $3,636.08 | 12.25 | 9 | 12 | $3,561.90 | $2,884.32 | $3,718.94 | $277,369.15 | $264,694.33 | $293,137.86 |
| 21 | 19 | $3,781.52 | 12.5 | 9 | 11.75 | $3,602.81 | $2,884.32 | $3,676.20 | $268,094.83 | $253,448.57 | $283,149.60 |
| 22 | 20 | $3,932.79 | 12.75 | 9 | 11.5 | $3,641.47 | $2,884.32 | $3,636.32 | $257,796.15 | $241,147.88 | $271,701.73 |

In the third scenario, the Wongs' payments would more than double at the start of the sixth year, and remain above what they can reasonably afford for 13 more years. Even if the Fed rate were to remain at the low rate of 4%, the monthly payments would still jump to above what the Wongs can afford at the start of the sixth year.

**Interest-Only Loan:** Here, the only change in the worksheet constructed previously is the computation of the payments for the first 5 years; because the loan is interest-only during this period, the monthly payment is computed as simple interest at 4.2% on the $380,000 loan for a 30-year period:

$$INT = PVr = 380,000 \times .042/12 = \$1,330.00$$

The formula you could use in the spreadsheet in cell F3 is `=C3/1200*I$3`, and then copy this across and down the entire block of payments for the first 5 years. The rest of the spreadsheet (including the balance on principal) will adjust itself accordingly with the formulas you had for the hybrid loan. Below is a portion of the result, with a lot more red than in the hybrid loan case!

| | A | B | C | D | E | F | G | H | I | J | K |
|---|---|---|---|---|---|---|---|---|---|---|---|
| 1 | Year | 28% of Monthly Income | | Interest Rate | | | Monthly Payment | | | Balance on Principal | |
| 2 | | | Scenario 1 | Scenario 2 | Scenario 3 | Scenario 1 | Scenario 2 | Scenario 3 | Scenario 1 | Scenario 2 | Scenario 3 |
| 3 | 1 | $1,866.67 | 4.2 | 4.2 | 4.2 | $1,330.00 | $1,330.00 | $1,330.00 | $380,000.00 | $380,000.00 | $380,000.00 |
| 4 | 2 | $1,941.33 | 4.2 | 4.2 | 4.2 | $1,330.00 | $1,330.00 | $1,330.00 | $380,000.00 | $380,000.00 | $380,000.00 |
| 5 | 3 | $2,018.99 | 4.2 | 4.2 | 4.2 | $1,330.00 | $1,330.00 | $1,330.00 | $380,000.00 | $380,000.00 | $380,000.00 |
| 6 | 4 | $2,099.75 | 4.2 | 4.2 | 4.2 | $1,330.00 | $1,330.00 | $1,330.00 | $380,000.00 | $380,000.00 | $380,000.00 |
| 7 | 5 | $2,183.74 | 4.2 | 4.2 | 4.2 | $1,330.00 | $1,330.00 | $1,330.00 | $380,000.00 | $380,000.00 | $380,000.00 |
| 8 | 6 | $2,271.09 | 9.25 | 9 | 15 | $3,254.25 | $3,188.95 | $4,867.16 | $380,000.00 | $380,000.00 | $380,000.00 |
| 9 | 7 | $2,361.93 | 9.5 | 9 | 14.75 | $3,318.61 | $3,188.95 | $4,794.45 | $375,929.28 | $375,760.60 | $378,493.33 |
| 10 | 8 | $2,456.41 | 9.75 | 9 | 14.5 | $3,381.95 | $3,188.95 | $4,723.00 | $371,635.48 | $371,123.52 | $376,667.51 |
| 11 | 9 | $2,554.66 | 10 | 9 | 14.25 | $3,444.18 | $3,188.95 | $4,652.96 | $367,086.90 | $366,051.44 | $374,465.75 |
| 12 | 10 | $2,656.85 | 10.25 | 9 | 14 | $3,505.22 | $3,188.95 | $4,584.46 | $362,247.64 | $360,503.57 | $371,823.49 |
| 13 | 11 | $2,763.12 | 10.5 | 9 | 13.75 | $3,564.98 | $3,188.95 | $4,517.68 | $357,076.92 | $354,435.28 | $368,667.84 |
| 14 | 12 | $2,873.65 | 10.75 | 9 | 13.5 | $3,623.37 | $3,188.95 | $4,452.76 | $351,528.19 | $347,797.73 | $364,917.01 |
| 15 | 13 | $2,988.59 | 11 | 9 | 13.25 | $3,680.26 | $3,188.95 | $4,389.88 | $345,548.14 | $340,537.54 | $360,479.80 |
| 16 | 14 | $3,108.14 | 11.25 | 9 | 13 | $3,735.55 | $3,188.95 | $4,329.19 | $339,075.44 | $332,596.29 | $355,255.12 |
| 17 | 15 | $3,232.46 | 11.5 | 9 | 12.75 | $3,789.12 | $3,188.95 | $4,270.87 | $332,039.35 | $323,910.09 | $349,131.66 |
| 18 | 16 | $3,361.76 | 11.75 | 9 | 12.5 | $3,840.82 | $3,188.95 | $4,215.07 | $324,357.97 | $314,409.07 | $341,987.61 |
| 19 | 17 | $3,496.23 | 12 | 9 | 12.25 | $3,890.53 | $3,188.95 | $4,161.97 | $315,936.16 | $304,016.79 | $333,690.51 |
| 20 | 18 | $3,636.08 | 12.25 | 9 | 12 | $3,938.08 | $3,188.95 | $4,111.71 | $306,663.10 | $292,649.65 | $324,097.21 |
| 21 | 19 | $3,781.52 | 12.5 | 9 | 11.75 | $3,983.32 | $3,188.95 | $4,064.46 | $296,409.28 | $280,216.18 | $313,054.05 |
| 22 | 20 | $3,932.79 | 12.75 | 9 | 11.5 | $4,026.06 | $3,188.95 | $4,020.37 | $285,022.93 | $266,616.37 | $300,397.13 |
| 23 | 21 | $4,090.10 | 13 | 9 | 11.25 | $4,066.11 | $3,188.95 | $3,979.57 | $272,325.66 | $251,740.81 | $285,952.80 |
| 24 | 22 | $4,253.70 | 13.25 | 9 | 11 | $4,103.29 | $3,188.95 | $3,942.23 | $258,107.21 | $235,469.81 | $269,538.33 |

In all three scenarios, this type of mortgage is worse for the Wongs than the hybrid loan; in particular, their payments in Scenario 3 would jump to more than double what they can afford at the start of the sixth year.

**Negative Amortization Loan:** Again, the only change in the worksheet is the computation of the payments for the first 5 years. This time, the loan amortizes negatively during the initial 5-year period, so the payment formula in this period is adjusted to reflect this

`=-PMT(C3/1200,360,I$3*0.6)`.

| Year | 28% of Monthly Income | Interest Rate | | | Monthly Payment | | | Balance on Principal | | |
|---|---|---|---|---|---|---|---|---|---|---|
| | | Scenario 1 | Scenario 2 | Scenario 3 | Scenario 1 | Scenario 2 | Scenario 3 | Scenario 1 | Scenario 2 | Scenario 3 |
| 1 | $1,866.67 | 4.7 | 4.7 | 4.7 | $1,182.49 | $1,182.49 | $1,182.49 | $380,000.00 | $380,000.00 | $380,000.00 |
| 2 | $1,941.33 | 4.7 | 4.7 | 4.7 | $1,182.49 | $1,182.49 | $1,182.49 | $383,750.17 | $383,750.17 | $383,750.17 |
| 3 | $2,018.99 | 4.7 | 4.7 | 4.7 | $1,182.49 | $1,182.49 | $1,182.49 | $387,680.45 | $387,680.45 | $387,680.45 |
| 4 | $2,099.75 | 4.7 | 4.7 | 4.7 | $1,182.49 | $1,182.49 | $1,182.49 | $391,799.48 | $391,799.48 | $391,799.48 |
| 5 | $2,183.74 | 4.7 | 4.7 | 4.7 | $1,182.49 | $1,182.49 | $1,182.49 | $396,116.32 | $396,116.32 | $396,116.32 |
| 6 | $2,271.09 | 9.25 | 9 | 15 | $3,431.01 | $3,362.16 | $5,131.53 | $400,640.49 | $400,640.49 | $400,640.49 |
| 7 | $2,361.93 | 9.5 | 9 | 14.75 | $3,498.87 | $3,362.16 | $5,054.87 | $396,348.66 | $396,170.82 | $399,051.98 |
| 8 | $2,456.41 | 9.75 | 9 | 14.5 | $3,565.64 | $3,362.16 | $4,979.54 | $391,821.64 | $391,281.87 | $397,127.00 |
| 9 | $2,554.66 | 10 | 9 | 14.25 | $3,631.26 | $3,362.16 | $4,905.69 | $387,025.99 | $385,934.29 | $394,805.64 |
| 10 | $2,656.85 | 10.25 | 9 | 14 | $3,695.62 | $3,362.16 | $4,833.48 | $381,923.88 | $380,085.08 | $392,019.86 |
| 11 | $2,763.12 | 10.5 | 9 | 13.75 | $3,758.62 | $3,362.16 | $4,763.06 | $376,472.30 | $373,687.17 | $388,692.80 |
| 12 | $2,873.65 | 10.75 | 9 | 13.5 | $3,820.18 | $3,362.16 | $4,694.62 | $370,622.18 | $366,689.09 | $384,738.24 |
| 13 | $2,988.59 | 11 | 9 | 13.25 | $3,880.16 | $3,362.16 | $4,628.32 | $364,317.31 | $359,034.55 | $380,060.01 |
| 14 | $3,108.14 | 11.25 | 9 | 13 | $3,938.46 | $3,362.16 | $4,564.34 | $357,493.03 | $350,661.95 | $374,551.54 |
| 15 | $3,232.46 | 11.5 | 9 | 12.75 | $3,994.93 | $3,362.16 | $4,502.85 | $350,074.77 | $341,503.95 | $368,095.48 |
| 16 | $3,361.76 | 11.75 | 9 | 12.5 | $4,049.45 | $3,362.16 | $4,444.02 | $341,976.15 | $331,486.86 | $360,563.39 |
| 17 | $3,496.23 | 12 | 9 | 12.25 | $4,101.85 | $3,362.16 | $4,388.03 | $333,096.90 | $320,530.10 | $351,815.60 |
| 18 | $3,636.08 | 12.25 | 9 | 12 | $4,151.99 | $3,362.16 | $4,335.05 | $323,320.15 | $308,545.52 | $341,701.22 |
| 19 | $3,781.52 | 12.5 | 9 | 11.75 | $4,199.68 | $3,362.16 | $4,285.23 | $312,509.37 | $295,436.71 | $330,058.23 |
| 20 | $3,932.79 | 12.75 | 9 | 11.5 | $4,244.74 | $3,362.16 | $4,238.74 | $300,504.55 | $281,098.20 | $316,713.83 |
| 21 | $4,090.10 | 13 | 9 | 11.25 | $4,286.97 | $3,362.16 | $4,195.73 | $287,117.60 | $265,414.64 | $301,484.92 |
| 22 | $4,253.70 | 13.25 | 9 | 11 | $4,326.17 | $3,362.16 | $4,156.36 | $272,126.84 | $248,259.85 | $284,178.86 |
| 23 | $4,423.85 | 13.5 | 9 | 10.75 | $4,362.10 | $3,362.16 | $4,120.77 | $255,270.30 | $229,495.82 | $264,594.34 |
| 24 | $4,600.80 | 13.75 | 9 | 10.5 | $4,394.53 | $3,362.16 | $4,089.09 | $236,237.47 | $208,971.60 | $242,522.55 |

Clearly the Wongs should steer clear of this type of loan in order to be able to continue to afford making payments!

In short, it seems unlikely that the Wongs will be able to afford payments on any of the three mortgages in question, and you decide to advise them to either seek a less expensive home or wait until their income has appreciated to enable them to afford a home of this price.

## EXERCISES

1. ▮ In the case of a hybrid loan, what would the federal funds rate have to be in Scenario 2 to ensure that the Wongs can afford to make all payments?

2. ▮ Repeat the preceding exercise in the case of an interest-only loan.

3. ▮ Repeat the preceding exercise in the case of a negative-amortization loan.

4. ▮ What home price, to the nearest $5,000, could the Wongs afford if they took out a hybrid loan, regardless of scenario? HINT [Adjust the original value of the loan on your spreadsheet to obtain the desired result.]

5. ▮ What home price, to the nearest $5,000, could the Wongs afford if they took out a negative-amortization loan, regardless of scenario? HINT [Adjust the original value of the loan on your spreadsheet to obtain the desired result.]

6. ▮ How long would the Wongs need to wait before they could afford to purchase a $400,000 home, assuming that their income continues to increase as above, they still have $20,000 for a down payment, and the mortgage offers remain the same?

TECHNOLOGY GUIDE

## **TI-83/84 Plus**   **Technology Guide**

### Section 2.2

**Example 1 (page 136)**   In November 2011, the Bank of Montreal was paying 1.30% interest on savings accounts. If the interest is compounded quarterly, find the future value of a $2,000 deposit in 6 years. What is the total interest paid over the period?

#### Solution with Technology

We could calculate the future value using the TI-83/84 Plus by entering

        2000(1+0.0130/4)^(4*6)

However, the TI-83/84 Plus has this and other useful calculations built into its TVM (Time Value of Money) Solver.

1. Press APPS then choose item 1:Finance... and then choose item 1:TVM Solver.... This brings up the TVM Solver window.

The second screen shows the values you should enter for this example. The various variables are:

| | |
|---|---|
| N | Number of compounding periods |
| I% | Annual interest rate, as percent, not decimal |
| PV | Negative of present value |
| PMT | Payment per period (0 in this section) |
| FV | Future value |
| P/Y | Payments per year |
| C/Y | Compounding periods per year |
| PMT: | Not used in this section |

Several things to notice:

- *I%* is the *annual* interest rate, corresponding to *r*, not *i*, in the compound interest formula.

- The present value, *PV*, is entered as a negative number. In general, when using the TVM Solver, any amount of money you give to someone else (such as the $2,000 you deposit in the bank) will be a negative number, whereas any amount of money someone gives to you (such as the future value of your deposit, which the bank will give back to you) will be a positive number.

- *PMT* is not used in this example (it will be used in the next section) and should be 0.

- *FV* is the future value, which we shall compute in a moment; it doesn't matter what you enter now.

- *P/Y* and *C/Y* stand for payments per year and compounding periods per year, respectively: They should both be set to the number of compounding periods per year for compound interest problems (setting *P/Y* automatically sets *C/Y* to the same value).

- *PMT*: *END* or *BEGIN* is not used in this example and it doesn't matter which you select.

2. To compute the future value, use the up or down arrow to put the cursor on the *FV* line, then press ALPHA SOLVE.

**Example 2 (page 138)**   Megabucks Corporation is issuing 10-year zero coupon bonds. How much would you pay for bonds with a maturity value of $10,000 if you wish to get a return of 6.5% compounded annually?

**Solution with Technology**

To compute the present value using a TI-83/84 Plus:

1. Enter the numbers shown below (top) in the TVM Solver window.

2. Put the cursor on the PV line, and press [ALPHA] [SOLVE].

Why is the present value given as negative?

**Example 6 (page 141)**   You have $5,000 to invest at 6% interest compounded monthly. How long will it take for your investment to grow to $6,000?

**Solution with Technology**

1. Enter the numbers shown below (top) in the TVM Solver window.

2. Put the cursor on the N line, and press [ALPHA] [SOLVE].

Recall that I% is the annual interest rate, corresponding to $r$ in the formula, but N is the number of compounding

periods, so number of months in this example. Thus, you will need to invest your money for about 36.5 months, or just over 3 years, before it grows to $6,000.

# Section 2.3

**Example 1 (page 146)**   Your retirement account has $5,000 in it and earns 5% interest per year compounded monthly. Every month for the next 10 years, you will deposit $100 into the account. How much money will there be in the account at the end of those 10 years?

**Solution with Technology**

We can use the TVM Solver in the TI-83/84 Plus to calculate future values like these:

1. The TVM Solver allows you to put the $5,000 already in the account as the present value of the account. Following the TI-83/84 Plus's usual convention, set PV to the *negative* of the present value because this is money you paid into the account.

2. Likewise, set PMT to −100 since you are paying $100 each month.

3. Set the number of payment and compounding periods to 12 per year.

4. Set the payments to be made at the end of each period.

5. With the cursor on the FV line, press [ALPHA] [SOLVE] to find the future value.

**Example 2 (page 147)**   Tony and Maria have just had a son, José Phillipe. They establish an account to accumulate money for his college education. They would like to have $100,000 in this account after 17 years. If the account pays 4% interest per year compounded quarterly, and they make deposits at the end of every quarter, how large must each deposit be for them to reach their goal?

## Solution with Technology

**1.** In the TVM Solver in the TI-83/84 Plus, enter the values shown below.

**2.** Solve for *PMT*.

Why is *PMT* negative?

**Example 3 (page 149)**   You wish to establish a trust fund from which your niece can withdraw $2,000 every 6 months for 15 years, at which time she will receive the remaining money in the trust, which you would like to be $10,000. The trust will be invested at 7% per year compounded every 6 months. How large should the trust be?

## Solution with Technology

**1.** In the TVM Solver in the TI-83/84 Plus, enter the values shown below.

**2.** Solve for *PV*.

The payment and future value are positive because you (or your niece) will be receiving these amounts from the investment.

**Note**   We have assumed that your niece receives the withdrawals at the end of each compounding period, so that the trust fund is an ordinary annuity. If, instead, she receives the payments at the beginning of each compounding period, it is an annuity due. You switch between the two types of annuity by changing PMT: END at the bottom to PMT: BEGIN. ■

As mentioned in the text, the present value must be higher to fund payments at the beginning of each period, because the money in the account has less time to earn interest.

**Example 4 (page 150)**   Tony and Maria (see Example 2), having accumulated $100,000 for José Phillipe's college education, would now like to make quarterly withdrawals over the next 4 years. How much money can they withdraw each quarter in order to draw down the account to zero at the end of the 4 years? (Recall that the account pays 4% interest compounded quarterly.)

## Solution with Technology

**1.** Enter the values shown below in the TI-83/84 Plus TVM Solver.

**2.** Solve for *PMT*.

The present value is negative because Tony and Maria do not possess it; the bank does.

**Example 7 (page 152)** Marc and Mira are buying a house and have taken out a 30-year, $90,000 mortgage at 8% interest per year. Mortgage interest is tax deductible, so it is important to know how much of a year's mortgage payments represents interest. How much interest will Marc and Mira pay in the first year of their mortgage?

### Solution with Technology

The TI-83/84 Plus has built-in functions to compute the values in an amortization schedule.

**1.** First, use the TVM Solver to find the monthly payment.

Three functions correspond to the last three columns of the amortization schedule given in the text: ΣInt, ΣPrn, and bal, (found in the Finance menu accessed through APPS). They all require that the values of I%, PV, and PMT be entered or calculated ahead of time; calculating the payment in the TVM Solver in Step 1 accomplishes this.

**2.** Use ΣInt($m, n, 2$) to compute the sum of the interest payments from payment $m$ through payment $n$. For example,

ΣInt(1,12,2)

will return −7,172.81, the total paid in interest in the first year, which answers the question asked in this example. (The last argument, 2, tells the calculator to round all intermediate calculations to two decimal places—that is, the nearest cent—as would the mortgage lender.)

**3.** Use ΣPrn($m, n, 2$) to compute the sum of the payments on principal from payment $m$ through payment $n$. For example,

ΣPrn(1,12,2)

will return −751.87, the total paid on the principal in the first year.

**4.** Finally, bal($n, 2$) finds the balance of the principal outstanding after $n$ payments. For example,

bal(12,2)

will return the value 89,248.13, the balance remaining at the end of one year.

**5.** To construct an amortization schedule as in the text, make sure that FUNC is selected in the MODE window; then enter the functions in the Y= window as shown below.

**6.** Press 2ND TBLSET and enter the values shown here.

**7.** Press 2ND TABLE, to get the table shown here.

The column labeled X gives the month, the column labeled $Y_1$ gives the interest payment for each month, the column labeled $Y_2$ gives the payment on principal for each month, and the column labeled $Y_3$ (use the right arrow button to make it visible) gives the outstanding principal.

**8.** To see later months, use the down arrow. As you can see, some of the values will be rounded in the table, but by selecting a value (as the outstanding principal at the end of the first month is selected in the second screen) you can see its exact value at the bottom of the screen.

**Example 9 (page 154)**    Suppose that a 5%, 20-year bond sells for $9,800 per $10,000 maturity value. What rate of return will investors get?

### Solution with Technology

We can use the TVM Solver in the TI-83/84 Plus to find the interest rate just as we use it to find any other one of the variables.

**1.** Enter the values shown in the TVM Solver window.

**2.** Solve for *I%*. (Recall that *I%* is the annual interest rate, corresponding to $r$ in the formula.)

Thus, at $9,800 per $10,000 maturity value, these bonds yield 5.162% compounded semiannually.

---

## SPREADSHEET    Technology Guide

### Section 2.2

**Example 1 (page 136)**    In November 2011, the Bank of Montreal was paying 1.30% interest on savings accounts. If the interest is compounded quarterly, find the future value of a $2,000 deposit in 6 years. What is the total interest paid over the period?

### Solution with Technology

You can either compute compound interest directly or use financial functions built into your spreadsheet. The following worksheet has more than we need for this example, but will be useful for other examples in this and the next section.

|   | A | B | C | D |
|---|---|---|---|---|
| 1 |  | Entered | Calculated |  |
| 2 | Rate | 1.30% |  |  |
| 3 | Years | 6 |  |  |
| 4 | Payment | $0.00 |  |  |
| 5 | Present Value | -$2,000.00 |  |  |
| 6 | Future Value |  | =FV(B2/B7,B3*B7,B4,B5) |  |
| 7 | Periods per year | 4 |  |  |

For this example the payment amount in B4 should be 0 (we shall use it in the next section).

**1.** Enter the other numbers as shown. As with other technologies, like the TVM Solver in the TI-83/84 Plus calculator, money that you pay to others (such as the $2,000 you deposit in the bank) should be entered as negative, whereas money that is paid to you is positive.

**2.** The formula entered in C6 uses the built-in FV function to calculate the future value based on the entries in column B. This formula has the following format:

$$=FV(i, n, PMT, PV)$$

$i$ = interest per period    We use B2/B7 for the interest.

$n$ = number of periods    We use B3*B7 for the number of periods.

$PMT$ = payment per period    The payment is 0 (cell B4).

$PV$ = present value    The present value is in cell B5.

Instead of using the built-in FV function, we could use

$$=-B5*(1+B2/B7)^{(B3*B7)}$$

based on the future value formula for compound interest. After calculation the result will appear in cell C6.

| | A | B | C | D |
|---|---|---|---|---|
| 1 | | Entered | Calculated | |
| 2 | Rate | 1.30% | | |
| 3 | Years | 6 | | |
| 4 | Payment | $0.00 | | |
| 5 | Present Value | −$2,000.00 | | |
| 6 | Future Value | | $2,161.97 | |
| 7 | Periods per year | 4 | | |

Note that we have formatted the cells B4:C6 as currency with two decimal places. If you change the values in column B, the future value in column C will be automatically recalculated.

**Example 2 (page 138)** Megabucks Corporation is issuing 10-year zero coupon bonds. How much would you pay for bonds with a maturity value of $10,000 if you wish to get a return of 6.5% compounded annually?

### Solution with Technology

You can compute present value in your spreadsheet using the PV worksheet function. The following worksheet is similar to the one in the preceding example, except that we have entered a formula for computing the present value from the entered values.

| | A | B | C | D |
|---|---|---|---|---|
| 1 | | Entered | Calculated | |
| 2 | Rate | 6.50% | | |
| 3 | Years | 10 | | |
| 4 | Payment | $0.00 | | |
| 5 | Present Value | | =PV(B2/B7,B3*B7,B4,B6) | |
| 6 | Future Value | $10,000.00 | | |
| 7 | Periods per year | 1 | | |

The next worksheet shows the calculated value.

| | A | B | C | D |
|---|---|---|---|---|
| 1 | | Entered | Calculated | |
| 2 | Rate | 6.50% | | |
| 3 | Years | 10 | | |
| 4 | Payment | $0.00 | | |
| 5 | Present Value | | −$5,327.26 | |
| 6 | Future Value | $10,000.00 | | |
| 7 | Periods per year | 1 | | |

Why is the present value negative?

**Example 6 (page 141)** You have $5,000 to invest at 6% interest compounded monthly. How long will it take for your investment to grow to $6,000?

### Solution with Technology

You can compute the requisite length of an investment in your spreadsheet using the NPER worksheet function. The following worksheets show the calculation.

| | A | B | C | D |
|---|---|---|---|---|
| 1 | | Entered | Calculated | |
| 2 | Rate | 6.00% | | |
| 3 | Years | | =NPER(B2/B7,B4,B5,B6)/B7 | |
| 4 | Payment | $0.00 | | |
| 5 | Present Value | −$5,000.00 | | |
| 6 | Future Value | $6,000.00 | | |
| 7 | Periods per year | 12 | | |

| | A | B | C | D |
|---|---|---|---|---|
| 1 | | Entered | Calculated | |
| 2 | Rate | 6.00% | | |
| 3 | Years | | 3.04628303 | |
| 4 | Payment | $0.00 | | |
| 5 | Present Value | −$5,000.00 | | |
| 6 | Future Value | $6,000.00 | | |
| 7 | Periods per year | 12 | | |

The NPER function computes the number of compounding periods, months in this case, so we divide by B7, the number of periods per year, to calculate the number of years, which appears as 3.046. So, you need to invest your money for just over 3 years for it to grow to $6,000.

## Section 2.3

**Example 1 (page 146)** Your retirement account has $5,000 in it and earns 5% interest per year compounded monthly. Every month for the next 10 years you will deposit $100 into the account. How much money will there be in the account at the end of those 10 years?

### Solution with Technology

We can use exactly the same worksheet that we used in Example 1 in the preceding section. In fact, we included the "Payment" row in that worksheet just for this purpose.

| | A | B | C | D |
|---|---|---|---|---|
| 1 | | Entered | Calculated | |
| 2 | Rate | 5.00% | | |
| 3 | Years | 10 | | |
| 4 | Payment | −$100.00 | | |
| 5 | Present Value | −$5,000.00 | | |
| 6 | Future Value | | =FV(B2/B7,B3*B7,B4,B5) | |
| 7 | Periods per year | 12 | | |

| | A | B | C | D |
|---|---|---|---|---|
| 1 | | Entered | Calculated | |
| 2 | Rate | 5.00% | | |
| 3 | Years | 10 | | |
| 4 | Payment | −$100.00 | | |
| 5 | Present Value | −$5,000.00 | | |
| 6 | Future Value | | $23,763.28 | |
| 7 | Periods per year | 12 | | |

Note that the FV function allows us to enter, as the last argument, the amount of money already in the account. Following the usual convention, we enter the present value and the payment as *negative*, because these are amounts you pay into the account.

**Example 2 (page 147)**   Tony and Maria have just had a son, José Phillipe. They establish an account to accumulate money for his college education, in which they would like to have $100,000 after 17 years. If the account pays 4% interest per year compounded quarterly, and they make deposits at the end of every quarter, how large must each deposit be for them to reach their goal?

## Solution with Technology

Use the following worksheet, in which the PMT worksheet function is used to calculate the required payments.

| | A | B | C | D |
|---|---|---|---|---|
| 1 | | Entered | Calculated | |
| 2 | Rate | 4.00% | | |
| 3 | Years | 17 | | |
| 4 | Payment | | =PMT(B2/B7,B3*B7,B5,B6) | |
| 5 | Present Value | $0.00 | | |
| 6 | Future Value | $100,000.00 | | |
| 7 | Periods per year | 4 | | |

| | A | B | C | D |
|---|---|---|---|---|
| 1 | | Entered | Calculated | |
| 2 | Rate | 4.00% | | |
| 3 | Years | 17 | | |
| 4 | Payment | | -$1,033.89 | |
| 5 | Present Value | $0.00 | | |
| 6 | Future Value | $100,000.00 | | |
| 7 | Periods per year | 4 | | |

Why is the payment negative?

**Example 3 (page 149)**   You wish to establish a trust fund from which your niece can withdraw $2,000 every 6 months for 15 years, at which time she will receive the remaining money in the trust, which you would like to be $10,000. The trust will be invested at 7% per year compounded every 6 months. How large should the trust be?

## Solution with Technology

You can use the same worksheet as in Example 2 in Section 2.2.

| | A | B | C | D |
|---|---|---|---|---|
| 1 | | Entered | Calculated | |
| 2 | Rate | 7.00% | | |
| 3 | Years | 15 | | |
| 4 | Payment | $2,000.00 | | |
| 5 | Present Value | | =PV(B2/B7,B3*B7,B4,B6) | |
| 6 | Future Value | $10,000.00 | | |
| 7 | Periods per year | 2 | | |

| | A | B | C | D |
|---|---|---|---|---|
| 1 | | Entered | Calculated | |
| 2 | Rate | 7.00% | | |
| 3 | Years | 15 | | |
| 4 | Payment | $2,000.00 | | |
| 5 | Present Value | | -$40,346.87 | |
| 6 | Future Value | $10,000.00 | | |
| 7 | Periods per year | 2 | | |

The payment and future value are positive because you (or your niece) will be receiving these amounts from the investment.

**Note** We have assumed that your niece receives the withdrawals at the end of each compounding period, so that the trust fund is an ordinary annuity. If, instead, she receives the payments at the beginning of each compounding period, it is an annuity due. You switch to an annuity due by adding an optional last argument of 1 to the PV function (and similarly for the other finance functions in spreadsheets).

| | A | B | C | D |
|---|---|---|---|---|
| 1 | | Entered | Calculated | |
| 2 | Rate | 7.00% | | |
| 3 | Years | 15 | | |
| 4 | Payment | $2,000.00 | | |
| 5 | Present Value | | =PV(B2/B7,B3*B7,B4,B6,1) | |
| 6 | Future Value | $10,000.00 | | |
| 7 | Periods per year | 2 | | |

| | A | B | C | D |
|---|---|---|---|---|
| 1 | | Entered | Calculated | |
| 2 | Rate | 7.00% | | |
| 3 | Years | 15 | | |
| 4 | Payment | $2,000.00 | | |
| 5 | Present Value | | -$41,634.32 | |
| 6 | Future Value | $10,000.00 | | |
| 7 | Periods per year | 2 | | |

As mentioned in the text, the present value must be higher to fund payments at the beginning of each period, because the money in the account has less time to earn interest. ∎

**Example 4 (page 150)**   Tony and Maria (see Example 2), having accumulated $100,000 for José Phillipe's college education, would now like to make quarterly withdrawals over the next 4 years. How much money can they withdraw each quarter in order to draw down the account to zero at the end of the four years? (Recall that the account pays 4% interest compounded quarterly.)

## Solution with Technology

You can use the same worksheet as in Example 2.

| | A | B | C | D |
|---|---|---|---|---|
| 1 | | Entered | Calculated | |
| 2 | Rate | 4.00% | | |
| 3 | Years | 4 | | |
| 4 | Payment | | =PMT(B2/B7,B3*B7,B5,B6) | |
| 5 | Present Value | -$100,000.00 | | |
| 6 | Future Value | $0.00 | | |
| 7 | Periods per year | 4 | | |

| | A | B | C | D |
|---|---|---|---|---|
| 1 | | Entered | Calculated | |
| 2 | Rate | 4.00% | | |
| 3 | Years | 4 | | |
| 4 | Payment | | $6,794.46 | |
| 5 | Present Value | -$100,000.00 | | |
| 6 | Future Value | $0.00 | | |
| 7 | Periods per year | 4 | | |

The present value is negative since Tony and Maria do not possess it; the bank does.

**Example 7 (page 152)**   Marc and Mira are buying a house and have taken out a 30-year, $90,000 mortgage at 8% interest per year. Mortgage interest is tax deductible, so it is important to know how much of a year's mortgage payments represents interest. How much interest will Marc and Mira pay in the first year of their mortgage?

## Solution with Technology

We construct an amortization schedule with which we can answer the question.

**1.** Begin with the worksheet below.

| | A | B | C | D | E | F | G | H |
|---|---|---|---|---|---|---|---|---|
| 1 | Month | Interest Payment | Payment on Principal | Outstanding Principal | | | | |
| 2 | 0 | | | $90,000.00 | | Rate | 8% | |
| 3 | | | | | | Years | 30 | |
| 4 | | | | | | Payment | =DOLLAR(-PMT(F2/12,F3*12,D2)) | |

| | A | B | C | D | E | F |
|---|---|---|---|---|---|---|
| 1 | Month | Interest Payment | Payment on Principal | Outstanding Principal | | |
| 2 | 0 | | | $90,000.00 | Rate | 8% |
| 3 | | | | | Years | 30 |
| 4 | | | | | Payment | $660.39 |

Note the formula for the monthly payment:

    =DOLLAR(-PMT(F2/12,F3*12,D2))

The function DOLLAR rounds the payment to the nearest cent, as the bank would.

**2.** Calculate the interest owed at the end of the first month using the formula

    =DOLLAR(D2*F$2/12)

in cell B3.

**3.** The payment on the principal is the remaining part of the payment, so enter

    =F$4-B3

in cell C3.

**4.** Calculate the outstanding principal by subtracting the payment on the principal from the previous outstanding principal, by entering

    =D2-C3

in cell D3.

**5.** Copy the formulas in cells B3, C3, and D3 into the cells below them to continue the table.

| | A | B | C | D | E | F |
|---|---|---|---|---|---|---|
| 1 | Month | Interest Payment | Payment on Principal | Outstanding Principal | | |
| 2 | 0 | | | $90,000.00 | Rate | 8% |
| 3 | 1 | $600.00 | $60.39 | $89,939.61 | Years | 30 |
| 4 | 2 | | | | Payment | $660.39 |
| 5 | 3 | | | | | |
| 6 | 4 | | | | | |
| 7 | 5 | | | | | |
| 8 | 6 | | | | | |
| 9 | 7 | | | | | |
| 10 | 8 | | | | | |
| 11 | 9 | | | | | |
| 12 | 10 | | | | | |
| 13 | 11 | | | | | |
| 14 | 12 | | | | | |

The result should be something like the following:

| | A | B | C | D | E | F |
|---|---|---|---|---|---|---|
| 1 | Month | Interest Payment | Payment on Principal | Outstanding Principal | | |
| 2 | 0 | | | $90,000.00 | Rate | 8% |
| 3 | 1 | $600.00 | $60.39 | $89,939.61 | Years | 30 |
| 4 | 2 | $599.60 | $60.79 | $89,878.82 | Payment | $660.39 |
| 5 | 3 | $599.19 | $61.20 | $89,817.62 | | |
| 6 | 4 | $598.78 | $61.61 | $89,756.01 | | |
| 7 | 5 | $598.37 | $62.02 | $89,693.99 | | |
| 8 | 6 | $597.96 | $62.43 | $89,631.56 | | |
| 9 | 7 | $597.54 | $62.85 | $89,568.71 | | |
| 10 | 8 | $597.12 | $63.27 | $89,505.44 | | |
| 11 | 9 | $596.70 | $63.69 | $89,441.75 | | |
| 12 | 10 | $596.28 | $64.11 | $89,377.64 | | |
| 13 | 11 | $595.85 | $64.54 | $89,313.10 | | |
| 14 | 12 | $595.42 | $64.97 | $89,248.13 | | |

**6.** Adding the calculated interest payments gives us the total interest paid in the first year: $7,172.81.

**Note** Spreadsheets have built-in functions that compute the interest payment (IPMT) or the payment on the principle (PPMT) in a given period. We could also have used the built-in future value function (FV) to calculate the outstanding principal each month. The main problem with using these functions is that, in a sense, they are too accurate. They do not take into account the fact that payments and interest are rounded to the nearest cent. Over time, this rounding causes the actual value of the outstanding principal to differ from what the FV function would tell us. In fact, because the actual payment is rounded slightly upward (to $660.39 from 660.38811...), the principal is reduced slightly faster than necessary and a last payment of $660.39 would be $2.95 larger than needed to clear out the debt. The lender would reduce the last payment by $2.95 for this reason; Marc and Mira will pay only $657.44 for their final payment. This is common: The last payment on an installment loan is usually slightly larger or smaller than the others, to compensate for the rounding of the monthly payment amount. ∎

**Example 9 (page 154)**    Suppose that a 5%, 20-year bond sells for $9,800 per $10,000 maturity value. What rate of return will investors get?

## Solution with Technology

Use the following worksheet, in which the RATE worksheet function is used to calculate the interest rate.

|   | A | B | C | D |
|---|---|---|---|---|
| 1 |   | Entered | Calculated |   |
| 2 | Rate |   | =RATE(B3*B7,B4,B5,B6)*B7 |   |
| 3 | Years | 20 |   |   |
| 4 | Payment | $250.00 |   |   |
| 5 | Present Value | -$9,800.00 |   |   |
| 6 | Future Value | $10,000.00 |   |   |
| 7 | Periods per year | 2 |   |   |

|   | A | B | C | D |
|---|---|---|---|---|
| 1 |   | Entered | Calculated |   |
| 2 | Rate |   | 5.162% |   |
| 3 | Years | 20 |   |   |
| 4 | Payment | $250.00 |   |   |
| 5 | Present Value | -$9,800.00 |   |   |
| 6 | Future Value | $10,000.00 |   |   |
| 7 | Periods per year | 2 |   |   |

# 3

# Systems of Linear Equations and Matrices

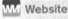

**Website**
www.WanerMath.com
At the Website you will find:

- Section-by-section tutorials, including game tutorials with randomized quizzes

- A detailed chapter summary

- A true/false quiz

- Additional review exercises

- Graphers, Excel tutorials, TI-83/84 Plus programs

- A Web page that pivots and does row operations

- An Excel worksheet that pivots and does row operations

## Case Study  Hybrid Cars—Optimizing the Degree of Hybridization

You are involved in new model development at a major automobile company. The company is planning to introduce two new plug-in hybrid electric vehicles: the subcompact "Green Town Hopper" and the midsize "Electra Supreme," and your department must decide on the degree of hybridization (DOH) for each of these models that will result in the largest reduction in gasoline consumption. The data you have available show the gasoline saving for only three values of the DOH. **How do you estimate the optimal value?**

Oleksiy Maksymenko/Alamy

# Introduction

In Chapter 1 we studied single functions and equations. In this chapter we seek solutions to **systems** of two or more equations. For example, suppose we need to *find two numbers whose sum is* 3 *and whose difference is* 1. In other words, we need to find two numbers $x$ and $y$ such that $x + y = 3$ and $x - y = 1$. The only solution turns out to be $x = 2$ and $y = 1$, a solution you might easily guess. But, how do we know that this is the only solution, and how do we find solutions systematically? When we restrict ourselves to systems of *linear* equations, there is a very elegant method for determining the number of solutions and finding them all. Moreover, as we will see, many real-world applications give rise to just such systems of linear equations.

We begin in Section 3.1 with systems of two linear equations in two unknowns and some of their applications. In Section 3.2 we study a powerful matrix method, called *row reduction*, for solving systems of linear equations in any number of unknowns. In Section 3.3 we look at more applications.

Computers have been used for many years to solve the large systems of equations that arise in the real world. You probably already have access to devices that will do the row operations used in row reduction. Many graphing calculators can do them, as can spreadsheets and various special-purpose applications, including utilities available at the Website. Using such a device or program makes the calculations quicker and helps avoid arithmetic mistakes. Then there are programs (and calculators) into which you simply feed the system of equations and out pop the solutions. We can think of what we do in this chapter as looking inside the "black box" of such a program. More important, we talk about how, starting from a real-world problem, to get the system of equations to solve in the first place. No computer will do this conversion for us yet.

## 3.1 Systems of Two Equations in Two Unknowns

Suppose you have $3 in your pocket to spend on snacks and a drink. If $x$ represents the amount you'll spend on snacks and $y$ represents the amount you'll spend on a drink, you can say that $x + y = 3$. On the other hand, if for some reason you want to spend $1 more on snacks than on your drink, you can also say that $x - y = 1$. These are simple examples of **linear equations in two unknowns**.

### Linear Equations in Two Unknowns

A **linear equation in two unknowns** is an equation that can be written in the form

$$ax + by = c$$

with $a$, $b$, and $c$ being real numbers. The number $a$ is called the **coefficient of $x$** and $b$ is called the **coefficient of $y$**. A **solution** of an equation consists of a pair of numbers: a value for $x$ and a value for $y$ that satisfy the equation.

#### Quick Example

In the linear equation $3x - y = 15$, the coefficients are $a = 3$ and $b = -1$. The point $(x, y) = (5, 0)$ is a solution, because $3(5) - (0) = 15$.

In fact, a single linear equation such as $3x - y = 15$ has infinitely many solutions: We could solve for $y = 3x - 15$ and then, for every value of $x$ we choose, we can get the corresponding value of $y$, giving a solution $(x, y)$. As we saw in Chapter 1, these solutions are the points on a straight line, the *graph* of the equation.

In this section we are concerned with pairs $(x, y)$ that are solutions of *two* linear equations at the same time. For example, $(2, 1)$ is a solution of both of the equations $x + y = 3$ and $x - y = 1$, because substituting $x = 2$ and $y = 1$ into these equations gives $2 + 1 = 3$ (true) and $2 - 1 = 1$ (also true), respectively. So, in the simple example we began with, you could spend \$2 on snacks and \$1 on a drink.

In the following set of examples, you will see how to graphically and algebraically solve a system of two linear equations in two unknowns. Then we'll return to some more interesting applications.

## EXAMPLE 1 Two Ways of Solving a System: Graphically and Algebraically

Find all solutions $(x, y)$ of the following system of two equations:

$$x + y = 3$$
$$x - y = 1.$$

**Solution** We will see how to find the solution(s) in two ways: graphically and algebraically. Remember that a solution is a pair $(x, y)$ that simultaneously satisfies *both* equations.

*Method 1: Graphical* We already know that the solutions of a single linear equation are the points on its graph, which is a straight line. For a point to represent a solution of two linear equations, it must lie simultaneously on both of the corresponding lines. In other words, it must be a point where the two lines cross, or intersect. A look at Figure 1 should convince us that the lines cross only at the point $(2, 1)$, so this is the only possible solution.

*Method 2: Algebraic* In the algebraic approach, we try to combine the equations in such a way as to eliminate one variable. In this case, notice that if we add the left-hand sides of the equations, the terms with $y$ are eliminated. So, we add the first equation to the second (that is, add the left-hand sides and add the right-hand sides*):

$$
\begin{array}{rl}
x + y &= 3 \\
\underline{x - y} &= \underline{1} \\
2x + 0 &= 4 \\
2x &= 4 \\
x &= 2.
\end{array}
$$

Now that we know that $x$ has to be 2, we can substitute back into either equation to find $y$. Choosing the first equation (it doesn't matter which we choose), we have

$$
\begin{array}{rl}
2 + y &= 3 \\
y &= 3 - 2 \\
&= 1.
\end{array}
$$

We have found that the only possible solution is $x = 2$ and $y = 1$, or

$$(x, y) = (2, 1).$$

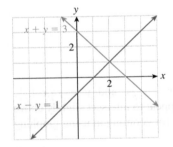

**Figure 1**

✳ We can add these equations because when we add equal amounts to both sides of an equation, the results are equal. That is, if $A = B$ and $C = D$, then $A + C = B + D$.

### using Technology

See the Technology Guides at the end of the chapter for details on the graphical solution of Example 1 using a TI-83/84 Plus or a spreadsheet. Here is an outline:

**TI-83/84 Plus**

$Y_1 = -X+3$  $Y_2 = X-1$

Graph: $\boxed{\text{WINDOW}}$; Xmin $= -4$,

Xmax $= 4$; $\boxed{\text{ZOOM}}$ $\boxed{0}$

Trace to estimate the point of intersection.
[More details on page 225.]

**Spreadsheet**

Headings *x*, *y*1, and *y*2 in A1–C1,
*x*-values −4, 4 in A2, A3
`=-A2+3` and `=A2-1` in B1 and C1;
copy down to B2–C2
Graph the data in columns A–C
with a line-segment scatter plot.
To see a closer view change the values in A2–A3.
[More details on page 227.]

 **Website**

www.WanerMath.com

→ On Line Utilities

→ Function Evaluator and
   Grapher

Enter `-x+3` for $y_1$ and `x-1` for $y_2$.
Set Xmin = −4, Xmax = 4 and
press "Plot Graphs."
Click on the graph and use the trace
arrows below to estimate the point
of intersection.

**Before we go on...** There is another way we could find the solution in Example 1 algebraically: First we solve our two equations for *y* to obtain

$$y = -x + 3$$
$$y = x - 1$$

and then we equate the right-hand sides to get

$$-x + 3 = x - 1,$$

which we solve to find $x = 2$. The value for *y* is then found by substituting $x = 2$ in either equation.

**Q:** *So why don't we solve all systems of equations this way?*

**A:** The elimination method extends more easily to systems with more equations and unknowns. It is the basis for the matrix method of solving systems—a method we discuss in Section 3.2. So, we shall use it exclusively for the rest of this section. ∎

Example 2 illustrates the drawbacks of the graphical method.

### EXAMPLE 2  Solving a System: Algebraically vs. Graphically

Solve the system

$$3x + 5y = 0$$
$$2x + 7y = 1.$$

**Solution**

**Method 1: Graphical** First, solve for *y*, obtaining $y = -\frac{3}{5}x$ and $y = -\frac{2}{7}x + \frac{1}{7}$. Graphing these equations, we get Figure 2. The lines appear to intersect slightly above and to the left of the origin. Redrawing with a finer scale (or zooming in using graphing technology), we can get the graph in Figure 3.

If we look carefully at Figure 3, we see that the graphs intersect near (−0.45, 0.27). Is the point of intersection *exactly* (−0.45, 0.27)? (Substitute these values into the equations to find out.) In fact, it is impossible to find the exact solution of this system graphically, but we now have a ballpark answer that we can use to help check the following algebraic solution.

**Method 2: Algebraic** We first see that adding the equations is not going to eliminate either *x* or *y*. Notice, however, that if we multiply (both sides of) the first equation by 2 and the second by −3, the coefficients of *x* will become 6 and −6. *Then* if we add them, *x* will be eliminated. So we proceed as follows:

$$2(3x + 5y) = 2(0)$$
$$-3(2x + 7y) = -3(1)$$

gives

$$6x + 10y = 0$$
$$-6x - 21y = -3.$$

Adding these equations, we get

$$-11y = -3$$

so that

$$y = \frac{3}{11} = 0.\overline{27}.$$

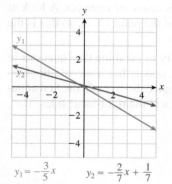

$$y_1 = -\frac{3}{5}x \qquad y_2 = -\frac{2}{7}x + \frac{1}{7}$$

**Figure 2**

Graphs of $y_1$ and $y_2$
with $-1 \le x \le 0$

**Figure 3**

Substituting $y = \dfrac{3}{11}$ in the first equation gives

$$3x + 5\left(\dfrac{3}{11}\right) = 0$$

$$3x = -\dfrac{15}{11}$$

$$x = -\dfrac{5}{11} = -0.\overline{45}.$$

The solution is $(x, y) = \left(-\dfrac{5}{11}, \dfrac{3}{11}\right) = (-0.\overline{45}, 0.\overline{27})$.

Notice that the algebraic method gives us the exact solution that we could not find with the graphical method. Still, we can check that the graph and our algebraic solution agree to the accuracy with which we can read the graph. To be absolutely sure that our answer is correct, we should check it:

$$3\left(-\dfrac{5}{11}\right) + 5\left(\dfrac{3}{11}\right) = -\dfrac{15}{11} + \dfrac{15}{11} = 0 \quad \checkmark$$

$$2\left(-\dfrac{5}{11}\right) + 7\left(\dfrac{3}{11}\right) = -\dfrac{10}{11} + \dfrac{21}{11} = 1. \quad \checkmark$$

Get in the habit of checking your answers.

⮕ **Before we go on...**

**Q**: *In solving the system in Example 2 algebraically, we multiplied (both sides of) the equations by numbers. How does that affect their graphs?*

**A**: Multiplying both sides of an equation by a nonzero number has no effect on its solutions, so the graph (which represents the set of all solutions) is unchanged. ∎

Before doing some more examples, let's summarize what we've said about solving systems of equations.

## Graphical Method for Solving a System of Two Equations in Two Unknowns

Graph both equations on the same graph. (For example, solve each for $y$ to find the slope and $y$-intercept.) A point of intersection gives the solution to the system. To find the point, you may need to adjust the range of $x$-values you use. To find the point accurately you may need to use a smaller range (or zoom in if using technology).

## Algebraic Method for Solving a System of Two Equations in Two Unknowns

Multiply each equation by a nonzero number so that the coefficients of $x$ are the same in absolute value but opposite in sign. Add the two equations to eliminate $x$; this gives an equation in $y$ that we can solve to find its value. Substitute this value of $y$ into one of the original equations to find the value of $x$. (Note that we could eliminate $y$ first instead of $x$ if it's more convenient.)

Sometimes, something appears to go wrong with these methods. The following examples show what can happen.

---

### EXAMPLE 3  Inconsistent System

Solve the system

$$x - 3y = 5$$
$$-2x + 6y = 8.$$

**Solution**  To eliminate $x$, we multiply the first equation by 2 and then add:

$$2x - 6y = 10$$
$$-2x + 6y = 8.$$

Adding gives

$$0 = 18.$$

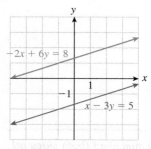

**Figure 4**

But this is absurd! This calculation shows that if we had two numbers $x$ and $y$ that satisfied both equations, it would be true that $0 = 18$. As 0 is *not* equal to 18, there can be no such numbers $x$ and $y$. In other words, *the system has no solutions*, and is called an **inconsistent system**.

In slope-intercept form these lines are $y = \frac{1}{3}x - \frac{5}{3}$ and $y = \frac{1}{3}x + \frac{4}{3}$. Notice that they have the same slope but different $y$-intercepts. This means that they are parallel, but different lines. Plotting them confirms this fact (Figure 4).

Because they are parallel, they do not intersect. A solution must be a point of intersection, so we again conclude that there is no solution.

---

### EXAMPLE 4  Redundant System

Solve the system

$$x + y = 2$$
$$2x + 2y = 4.$$

**Solution**  Multiplying the first equation by $-2$ gives

$$-2x - 2y = -4$$
$$2x + 2y = 4.$$

Adding gives the not-very-enlightening result

$$0 = 0.$$

*Now* what has happened? Looking back at the original system, we note that the second equation is really the first equation in disguise. (It is the first equation multiplied by 2.) Put another way, if we solve both equations for $y$, we find that, in slope-intercept form, both equations become the same:

$$y = -x + 2.$$

The second equation gives us the same information as the first, so we say that this is a **redundant**, or **dependent system**. In other words, we really have only one equation in two unknowns. From Chapter 1, we know that a single linear equation in two unknowns has infinitely many solutions, one for each value of $x$. (Recall that to get the corresponding solution for $y$, we solve the equation for $y$ and substitute the $x$-value.) The entire set of solutions can be summarized as follows:

$x$ is arbitrary

$y = 2 - x$.    Solve the first equation for $y$.

This set of solutions is called the **general solution** because it includes all possible solutions. When we write the general solution this way, we say that we have a **parameterized solution** and that $x$ is the **parameter**.

We can also write the general solution as

$(x, 2 - x)$   $x$ arbitrary.

Different choices of the parameter $x$ lead to different **particular solutions**. For instance, choosing $x = 3$ gives the particular solution

$(x, y) = (3, -1)$.

Because there are infinitely many values of $x$ from which to choose, there are infinitely many solutions.

What does this system of equations look like graphically? The two equations are really the same, so their graphs are identical, each being the line with $x$-intercept 2 and $y$-intercept 2. The "two" lines intersect at every point, so there is a solution for each point on the common line. In other words, we have a "whole line of solutions" (Figure 5).

We could also have solved the first equation for $x$ instead and used $y$ as a parameter, obtaining another form of the general solution:

$(2 - y, y)$   $y$ arbitrary.    Alternate form of the general solution

The line of solutions ($y = 2 - x$) and some particular solutions

**Figure 5**

We summarize the three possible outcomes we have encountered.

## Possible Outcomes for a System of Two Linear Equations in Two Unknowns

1. **A single (or *unique*) solution** This happens when the lines corresponding to the two equations are distinct and not parallel so that they intersect at a single point. (See Example 1.)

2. **No solution** This happens when the two lines are parallel. We say that the system is **inconsistent**. (See Example 3.)

3. **An infinite number of solutions** This occurs when the two equations represent the same straight line, and we say that such a system is **redundant**, or **dependent**. In this case, we can represent the solutions by choosing one variable arbitrarily and solving for the other. (See Example 4.)

In cases 1 and 3, we say that the system of equations is **consistent** because it has at least one solution.

You should think about straight lines and convince yourself that these are the only three possibilities.

# APPLICATIONS

### EXAMPLE 5  Blending

Acme Baby Foods mixes two strengths of apple juice. One quart of Beginner's juice is made from 30 fluid ounces of water and 2 fluid ounces of apple juice concentrate. One quart of Advanced juice is made from 20 fluid ounces of water and 12 fluid ounces of concentrate. Every day Acme has available 30,000 fluid ounces of water and 3,600 fluid ounces of concentrate. If the company wants to use all the water and concentrate, how many quarts of each type of juice should it mix?

**Solution**  In all applications we follow the same general strategy.

**1. *Identify and label the unknowns.*** What are we asked to find? To answer this question, it is common to respond by saying, "The unknowns are Beginner's juice and Advanced juice." Quite frankly, this is a baffling statement. Just what is unknown about juice? We need to be more precise:

The unknowns are (1) the **number of quarts** of Beginner's juice and (2) the **number of quarts** of Advanced juice made each day.

So, we label the unknowns as follows: Let

$x$ = number of quarts of Beginner's juice made each day

$y$ = number of quarts of Advanced juice made each day.

**2. *Use the information given to set up equations in the unknowns.*** This step is trickier, and the strategy varies from problem to problem. Here, the amount of juice the company can make is constrained by the fact that they have limited amounts of water and concentrate. This example shows a kind of application we will often see, and it is helpful in these problems to use a table to record the amounts of the resources used.

|  | Beginner's (x) | Advanced (y) | Available |
|---|---|---|---|
| **Water (fl oz)** | 30 | 20 | 30,000 |
| **Concentrate (fl oz)** | 2 | 12 | 3,600 |

We can now set up an equation for each of the items listed in the left column of the table.

*Water:* We read across the first row. If Acme mixes $x$ quarts of Beginner's juice, each quart using 30 fluid ounces of water, and $y$ quarts of Advanced juice, each using 20 fluid ounces of water, it will use a total of $30x + 20y$ fluid ounces of water. But we are told that the total has to be 30,000 fluid ounces. Thus, $30x + 20y = 30,000$. This is our first equation.

*Concentrate:* We read across the second row. If Acme mixes $x$ quarts of Beginner's juice, each using 2 fluid ounces of concentrate, and $y$ quarts of Advanced juice, each using 12 fluid ounces of concentrate, it will use a total of $2x + 12y$ fluid ounces of concentrate. But we are told that the total has to be 3,600 fluid ounces. Thus, $2x + 12y = 3,600$.

Now we have two equations:

$$30x + 20y = 30,000$$
$$2x + 12y = 3,600.$$

To make the numbers easier to work with, let's divide (both sides of) the first equation by 10 and the second by 2:

$$3x + 2y = 3,000$$
$$x + 6y = 1,800.$$

 **using Technology to Check Your Answer**

 **Website**
www.WanerMath.com

→ Math Tools for Chapter 3

→ Pivot and Gauss-Jordan Tool

To check Example 5 enter the coefficients of $x$ and $y$ and the right-hand sides of the two equations as shown (no commas in numbers!):

| x1 | x2 | x3 |
|---|---|---|
| 30 | 20 | 30000 |
| 2 | 12 | 3600 |

Then press "Reduce Completely". The answer will appear in the last column. (In the next section we will see how this works.)

We can now eliminate $x$ by multiplying the second equation by $-3$ and adding:

$$\begin{array}{r} 3x + 2y = 3,000 \\ -3x - 18y = -5,400 \\ \hline -16y = -2,400. \end{array}$$

So, $y = 2,400/16 = 150$. Substituting this into the equation $x + 6y = 1,800$ gives $x + 900 = 1,800$, and so $x = 900$. The solution is $(x, y) = (900, 150)$. In other words, the company should mix 900 quarts of Beginner's juice and 150 quarts of Advanced juice.

### EXAMPLE 6  Blending

A medieval alchemist's love potion calls for a number of eyes of newt and toes of frog, the total being 20, but with twice as many newt eyes as frog toes. How many of each is required?

**Solution** As in the preceding example, the first step is to identify and label the unknowns. Let

$$x = \text{number of newt eyes}$$
$$y = \text{number of frog toes.}$$

As for the second step—setting up the equations—a table is less appropriate here than in the preceding example. Instead, we translate each phrase of the problem into an equation. The first sentence tells us that the total number of eyes and toes is 20. Thus,

$$x + y = 20.$$

The end of the first sentence gives us more information, but the phrase "twice as many newt eyes as frog toes" is a little tricky: does it mean that $2x = y$ or that $x = 2y$? We can decide which by rewording the statement using the phrases "the *number of* newt eyes," which is $x$, and "the *number of* frog toes," which is $y$. Rephrased, the statement reads:

*The **number of** newt eyes is twice the **number of** frog toes.*

(Notice how the word "twice" is forced into a different place.) With this rephrasing, we can translate directly into algebra:

$$x = 2y.$$

In standard form $(ax + by = c)$, this equation reads

$$x - 2y = 0.$$

Thus, we have the two equations:

$$\begin{array}{r} x + y = 20 \\ x - 2y = 0. \end{array}$$

To eliminate $x$, we multiply the second equation by $-1$ and then add:

$$\begin{array}{r} x + y = 20 \\ -x + 2y = 0. \end{array}$$

We'll leave it to you to finish solving the system and find that $x = 13\frac{1}{3}$ and $y = 6\frac{2}{3}$.

So, the recipe calls for exactly $13\frac{1}{3}$ eyes of newt and $6\frac{2}{3}$ toes of frog. The alchemist needs a very sharp scalpel and a very accurate balance (not to mention a very strong stomach).

We saw in Chapter 1 that the *equilibrium price* of an item (the price at which supply equals demand) and the *break-even point* (the number of items that must be sold to break even) can both be described as the intersection points of two graphs. If the graphs are straight lines, what we need to do to find the intersection is solve a system of two linear equations in two unknowns, as illustrated in the following problem.

### EXAMPLE 7 Equilibrium Price

The demand for refrigerators in West Podunk is given by

$$q = -\frac{p}{10} + 100,$$

where $q$ is the number of refrigerators that the citizens will buy each year if the refrigerators are priced at $p$ dollars each. The supply is

$$q = \frac{p}{20} + 25,$$

where now $q$ is the number of refrigerators the manufacturers will be willing to ship into town each year if they are priced at $p$ dollars each. Find the equilibrium price and the number of refrigerators that will be sold at that price.

**Solution** Figure 6 shows the demand and supply curves. The equilibrium price occurs at the point where these two lines cross, which is where demand equals supply. The graph suggests that the equilibrium price is $500, and zooming in confirms this.

To solve this system algebraically, first write both equations in standard form:

$$\frac{p}{10} + q = 100$$

$$-\frac{p}{20} + q = 25.$$

Supply: $q = \frac{p}{20} + 25$

Demand: $q = -\frac{p}{10} + 100$

**Figure 6**

We can clear fractions and also prepare to eliminate $p$ if we multiply the first equation by 10 and the second by 20:

$$p + 10q = 1{,}000$$

$$-p + 20q = 500,$$

and so:

$$30q = 1{,}500$$

$$q = 50.$$

Substituting this value of $q$ into either equation gives us $p = 500$. Thus, the equilibrium price is $500, and 50 refrigerators will be sold at this price.

We could also have solved this system of equations by setting the two expressions for $q$ (the supply and the demand) equal to each other:

$$-p/10 + 100 = p/20 + 25$$

and then solving for $p$. (See the **Before we go on** discussion at the end of Example 1.)

**FAQs**

**Setting Up the Equations**

**Q:** *Looking through these examples, I notice that in some, we can tabulate the information given and read off the equations (as in Example 5), whereas in others (like Example 6), we have to reword each sentence to turn it into an equation. How do I know what approach to use?*

**A:** There is no hard-and-fast rule, and indeed some applications might call for a bit of each approach. However, it is generally not hard to see when it would be useful to tabulate values: Lists of the numbers of ingredients or components generally lend themselves to tabulation, whereas phrases like "twice as many of these as those" generally require direct translation into equations (after rewording if necessary).

## **3.1 EXERCISES**

▼ more advanced    ◆ challenging
[T] indicates exercises that should be solved using technology

*In Exercises 1–14, find all solutions of the given system of equations and check your answer graphically.* HINT [See Examples 1–4.]

**1.** $x - y = 0$
$x + y = 4$

**2.** $x - y = 0$
$x + y = -6$

**3.** $x + y = 4$
$x - y = 2$

**4.** $2x + y = 2$
$-2x + y = 2$

**5.** $3x - 2y = 6$
$2x - 3y = -6$

**6.** $2x + 3y = 5$
$3x + 2y = 5$

**7.** $0.5x + 0.1y = 0.7$
$0.2x - 0.2y = 0.6$

**8.** $-0.3x + 0.5y = 0.1$
$0.1x - 0.1y = 0.4$

**9.** $\dfrac{x}{3} - \dfrac{y}{2} = 1$
$\dfrac{x}{4} + y = -2$

**10.** $-\dfrac{2x}{3} + \dfrac{y}{2} = -\dfrac{1}{6}$
$\dfrac{x}{4} - y = -\dfrac{3}{4}$

**11.** $2x + 3y = 1$
$-x - \dfrac{3y}{2} = -\dfrac{1}{2}$

**12.** $2x - 3y = 1$
$6x - 9y = 3$

**13.** $2x + 3y = 2$
$-x - \dfrac{3y}{2} = -\dfrac{1}{2}$

**14.** $2x - 3y = 2$
$6x - 9y = 3$

[T] *In Exercises 15–24, use technology to obtain approximate solutions graphically. All solutions should be accurate to one decimal place. (Zoom in for improved accuracy.)*

**15.** $2x + 8y = 10$
$x + y = 5$

**16.** $2x - y = 3$
$x + 3y = 5$

**17.** $3.1x - 4.5y = 6$
$4.5x + 1.1y = 0$

**18.** $0.2x + 4.5y = 1$
$1.5x + 1.1y = 2$

**19.** $10.2x + 14y = 213$
$4.5x + 1.1y = 448$

**20.** $100x + 4.5y = 540$
$1.05x + 1.1y = 0$

**21.** ▼ Find the intersection of the line through $(0, 1)$ and $(4.2, 2)$ and the line through $(2.1, 3)$ and $(5.2, 0)$.

**22.** ▼ Find the intersection of the line through $(2.1, 3)$ and $(4, 2)$ and the line through $(3.2, 2)$ and $(5.1, 3)$.

**23.** ▼ Find the intersection of the line through $(0, 0)$ and $(5.5, 3)$ and the line through $(5, 0)$ and $(0, 6)$.

**24.** ▼ Find the intersection of the line through $(4.3, 0)$ and $(0, 5)$ and the line through $(2.1, 2.2)$ and $(5.2, 1)$.

### **APPLICATIONS**

**25.** *Resource Allocation* You manage an ice cream factory that makes two flavors: Creamy Vanilla and Continental Mocha. Into each quart of Creamy Vanilla go 2 eggs and 3 cups of cream. Into each quart of Continental Mocha go 1 egg and 3 cups of cream. You have in stock 500 eggs and 900 cups of cream. How many quarts of each flavor should you make in order to use up all the eggs and cream? HINT [See Example 5.]

**26.** *Class Scheduling* Enormous State University's Math Department offers two courses: Finite Math and Applied Calculus. Each section of Finite Math has 60 students, and each section of Applied Calculus has 50. The department will offer a total of 110 sections in a semester, and 6,000 students would like to take a math course. How many sections of each course should the department offer in order to fill all sections and accommodate all of the students? HINT [See Example 5.]

**27.** *Nutrition* Gerber Products' Gerber Mixed Cereal for Baby contains, in each serving, 60 calories and 11 grams of carbohydrates. Gerber Mango Tropical Fruit Dessert contains, in each serving, 80 calories and 21 grams of carbohydrates.[1] If you want to provide your child with 200 calories and 43 grams of carbohydrates, how many servings of each should you use?

**28.** *Nutrition* Anthony Altino is mixing food for his young daughter and would like the meal to supply 1 gram of protein and 5 milligrams of iron. He is mixing together cereal, with 0.5 grams of protein and 1 milligram of iron per ounce,

[1]Source: Nutrition information supplied with the products.

and fruit, with 0.2 grams of protein and 2 milligrams of iron per ounce. What mixture will provide the desired nutrition?

29. *Nutrition* One serving of Campbell Soup Company's Campbell's® Pork & Beans contains 5 grams of protein and 21 grams of carbohydrates.[2] A typical slice of white bread provides 2 grams of protein and 11 grams of carbohydrates per slice. The U.S. RDA (Recommended Daily Allowance) is 60 grams of protein each day.[3]

    a. I am planning a meal of "beans on toast" and wish to have it supply one-half of the RDA for protein and 139 grams of carbohydrates. How should I prepare my meal?

    b. Is it possible to have my meal supply the same amount of protein as in part (a) but only 100 g of carbohydrates?

30. *Nutrition* One serving of Campbell Soup Company's Campbell's® Pork & Beans contains 5 grams of protein and 21 grams of carbohydrates.[4] A typical slice of "lite" rye bread contains 4 grams of protein and 12 grams of carbohydrates.

    a. I am planning a meal of "beans on toast" and wish to have it supply one-third of the U.S. RDA for protein (see the preceding exercise) and 80 grams of carbohydrates. How should I prepare my meal?

    b. Is it possible to have my meal supply the same amount of protein as in part (a) but only 60 grams of carbohydrates?

*Protein Supplements Exercises 31–34 are based on the following data on three popular protein supplements. (Figures shown correspond to a single serving.)*[5]

| | Protein (g) | Carbohydrates (g) | Sodium (mg) | Cost ($) |
|---|---|---|---|---|
| **Designer Whey®** (Next) | 18 | 2 | 80 | 0.50 |
| **Muscle Milk®** (Cytosport) | 32 | 16 | 240 | 1.60 |
| **Pure Whey Protein Stack®** (Champion) | 24 | 3 | 100 | 0.60 |

31. You are thinking of combining Designer Whey and Muscle Milk to obtain a 7-day supply that provides exactly 280 grams of protein and 56 grams of carbohydrates. How many servings of each supplement should you combine in order to meet your requirements? What will it cost?

32. You are thinking of combining Muscle Milk and Pure Whey Protein Stack to obtain a supply that provides exactly 640 grams of protein and 3,200 mg of sodium. How many servings of each supplement should you combine in order to meet your requirements? What will it cost?

33. ▼ You have a mixture of Designer Whey and Pure Whey Protein Stack that costs a total of $14 and supplies exactly

540 g of protein. How many grams of carbohydrates does it supply?

34. ▼ You have a mixture of Designer Whey and Muscle Milk that costs a total of $14 and supplies exactly 104 g of carbohydrates. How many grams of protein does it supply?

35. *Investments: Tech Stocks* In July 2011, Apple (AAPL) stock increased from $350 to $400 per share, and Google (GOOG) stock increased from $500 to $600 per share.[6] If you invested a total of $22,000 in these stocks at the beginning of the month and sold them for $26,000 at the end of the month, how many shares of each stock did you buy?

36. *Investments: Energy Stocks* In the two-month period June 1–July 31, 2011, Hess Corp. (HES) stock decreased from $80 to $70 per share, and Exxon Mobil (XOM) stock decreased from $83 to $80 per share.[7] If you invested a total of $19,000 in these stocks at the beginning of June and sold them for $18,100 at the end of July, how many shares of each stock did you buy?

37. ▼ *Investments: Financial Stocks* In August 2011, Bank of Hawaii (BOH) stock cost $45 per share and yielded 4% per year in dividends, while JPMorgan Chase (JPM) stock cost $40 per share and yielded 2.5% per year in dividends.[8] If you invested a total of $25,000 in these stocks and earned $760 in dividends in a year, how many shares of each stock did you purchase? (Assume the dividend rate was unchanged for the year.)

38. ▼ *Investments: Utility Stocks* In April 2011, Consolidated Edison (ED) stock cost $50 per share and yielded 5% per year in dividends, while National Grid (NGG) stock cost $50 per share and yielded 6% per year in dividends.[9] If you invested a total of $45,000 in these stocks and earned $2,400 in dividends in a year, how many shares of each stock did you purchase? (Assume the dividend rate was unchanged for the year.)

39. *Voting* An appropriations bill passes the U.S. House of Representatives with 49 more members voting in favor than against. If all 435 members of the House vote for or against the bill, how many voted in favor and how many voted against?

40. *Voting* The U.S. Senate has 100 members. For a bill to pass with a supermajority, at least twice as many senators must vote in favor of the bill as vote against it. If all 100 senators vote in favor of or against a bill, how many must vote in favor for it to pass with a supermajority?

41. *Intramural Sports* The best sports dorm on campus, Lombardi House, has won a total of 12 games this semester. Some of these games were soccer games, and the others were football games. According to the rules of the university, each win in a soccer game earns the winning house 2 points, whereas each win in a football game earns them 4 points. If the total number of points Lombardi House earned was 38, how many of each type of game did they win?

---

[2]According to the label information on a 16 oz. can.

[3]Recommended Daily Allowance for a person weighing 75 kg (165 lb)

[4]*Ibid.*

[5]Source: Nutritional information supplied by the manufacturers (www.netrition.com). Cost per serving is approximate and varies.

[6]Approximate stock prices at or close to the dates cited. Source: http://finance.google.com.

[7]*Ibid.*

[8]Stock prices and yields are approximate. Source: http://finance.google.com.

[9]*Ibid.*

**42.** *Law* Enormous State University's campus publication, *The Campus Inquirer*, ran a total of 10 exposés five years ago, dealing with alleged recruiting violations by the football team and with theft by the student treasurer of the film society. Each exposé dealing with recruiting violations resulted in a $4 million libel suit, and the treasurer of the film society sued the paper for $3 million as a result of each exposé concerning his alleged theft. Unfortunately for *The Campus Inquirer*, all the lawsuits were successful, and the paper wound up being ordered to pay $37 million in damages. (It closed down shortly thereafter.) How many of each type of exposé did the paper run?

**43.** *Purchasing (from the GMAT)* Elena purchased brand $X$ pens for $4.00 apiece and brand $Y$ pens for $2.80 apiece. If Elena purchased a total of 12 of these pens for $42.00, how many brand $X$ pens did she purchase?

**44.** *Purchasing (based on a question from the GMAT)* Earl is ordering supplies. Yellow paper costs $5.00 per ream while white paper costs $6.50 per ream. He would like to order 100 reams total, and has a budget of $560. How many reams of each color should he order?

**45.** *Equilibrium Price* The demand and supply functions for pet chias are, respectively, $q = -60p + 150$ and $q = 80p - 60$, where $p$ is the price in dollars. At what price should the chias be marked so that there is neither a surplus nor a shortage of chias? HINT [See Example 7.]

**46.** *Equilibrium Price* The demand and supply functions for your college newspaper are, respectively, $q = -10,000p + 2,000$ and $q = 4,000p + 600$, where $p$ is the price in dollars. At what price should the newspapers be sold so that there is neither a surplus nor a shortage of papers? HINT [See Example 7.]

**47.** *Supply and Demand (from the GRE Economics Test)* The demand curve for widgets is given by $D = 85 - 5P$, and the supply curve is given by $S = 25 + 5P$, where $P$ is the price of widgets. When the widget market is in equilibrium, what is the quantity of widgets bought and sold?

**48.** *Supply and Demand (from the GRE Economics Test)* In the market for soybeans, the demand and supply functions are $Q_D = 100 - 10P$ and $Q_S = 20 + 5P$, where $Q_D$ is quantity demanded, $Q_S$ is quantity supplied, and $P$ is price in dollars. If the government sets a price floor of $7, what will be the resulting surplus or shortage?

**49.** *Equilibrium Price* In June 2001, the retail price of a 25-kg bag of cornmeal was $8 in Zambia; by December, the price had risen to $11. The result was that one retailer reported a drop in sales from 15 bags/day to 3 bags/day.[10] Assume that the retailer is prepared to sell 3 bags/day at $8 and 15 bags/day at $11. Find linear demand and supply equations and then compute the retailer's equilibrium price.

**50.** *Equilibrium Price* At the start of December 2001, the retail price of a 25 kg bag of cornmeal was $10 in Zambia, while by the end of the month, the price had fallen to $6.[11] The

result was that one retailer reported an increase in sales from 3 bags/day to 5 bags/day. Assume that the retailer is prepared to sell 18 bags/day at $8 and 12 bags/day at $6. Obtain linear demand and supply equations, and hence compute the retailer's equilibrium price.

**51.** *Pollution* Joe Slo, a college sophomore, neglected to wash his dirty laundry for 6 weeks. By the end of that time, his roommate had had enough and tossed Joe's dirty socks and T-shirts into the trash, counting a total of 44 items. (A pair of dirty socks counts as one item.) The roommate noticed that there were three times as many pairs of dirty socks as T-shirts. How many of each item did he throw out?

**52.** *Diet* The local sushi bar serves 1-ounce pieces of raw salmon (consisting of 50% protein) and $1\frac{1}{4}$-ounce pieces of raw tuna (40% protein). A customer's total intake of protein amounts to $1\frac{1}{2}$ ounces after consuming a total of three pieces. How many of each did the customer consume? (Fractions of pieces are permitted.)

**53.** ▼ *Management (from the GMAT)* A manager has $6,000 budgeted for raises for four full-time and two part-time employees. Each of the full-time employees receives the same raise, which is twice the raise that each of the part-time employees receives. What is the amount of the raise that each full-time employee receives?

**54.** ▼ *Publishing (from the GMAT)* There were 36,000 hardback copies of a certain novel sold before the paperback version was issued. From the time the first paperback copy was sold until the last copy of the novel was sold, nine times as many paperback copies as hardback copies were sold. If a total of 441,000 copies of the novel were sold in all, how many paperback copies were sold?

## COMMUNICATION AND REASONING EXERCISES

**55.** A system of three equations in two unknowns corresponds to three lines in the plane. Describe how these lines might be positioned if the system has a unique solution.

**56.** A system of three equations in two unknowns corresponds to three lines in the plane. Describe several ways that these lines might be positioned if the system has no solutions.

**57.** Both the supply and demand equations for a certain product have negative slope. Can there be an equilibrium price? Explain.

**58.** You are solving a system of equations with $x$ representing the number of rocks and $y$ representing the number of pebbles. The solution is $(200, -10)$. What do you conclude?

**59.** ▼ Referring to Exercise 25, suppose that the solution of the corresponding system of equations was 198.7 gallons of vanilla and 100.89 gallons of mocha. If your factory can produce only whole numbers of gallons, would you recommend rounding the answers to the nearest whole number? Explain.

**60.** ▼ Referring to Exercise 25 but using different data, suppose that the general solution of the corresponding system of equations was $(200 - y, y)$, where $x =$ number of gallons of vanilla, and $y =$ number of gallons of mocha. Your factory can

---

[10]The prices quoted are approximate. (Actual prices varied from retailer to retailer.) Source: *New York Times*, December 24, 2001, p. A4.
[11]*Ibid.*

produce only whole numbers of gallons. There are infinitely many ways of making the ice cream, mixes, right? Explain.

61. ▼ Select one: Multiplying both sides of a linear equation by a nonzero constant results in a linear equation whose graph is

(A) parallel to            (B) the same as
(C) not always parallel to  (D) not the same as
the graph of the original equation.

62. ▼ Select one: If the addition or subtraction of two linear equations results in the equation $3 = 3$, then the graphs of those equations are

(A) equal.                 (B) parallel.
(C) perpendicular.         (D) none of the above.

63. ▼ Select one: If the addition or subtraction of two linear equations results in the equation $0 = 3$, then the graphs of those equations are

(A) equal.                 (B) parallel.
(C) perpendicular.         (D) not parallel.

64. ▼ Select one: If adding two linear equations gives $x = 3$, and subtracting them gives $y = 3$, then the graphs of those equations are

(A) equal.                 (B) parallel.
(C) perpendicular.         (D) not parallel.

65. ▼ Invent an interesting application that leads to a system of two equations in two unknowns with a unique solution.

66. ▼ Invent an interesting application that leads to a system of two equations in two unknowns with no solution.

67. ◆ How likely do you think it is that a "random" system of two equations in two unknowns has a unique solution? Give some justification for your answer.

68. ◆ How likely do you think it is that a "random" system of three equations in two unknowns has a unique solution? Give some justification for your answer.

# 3.2  Using Matrices to Solve Systems of Equations

In this section we describe a systematic method for solving systems of equations that makes solving large systems of equations in any number of unknowns straightforward. Although this method may seem a little cumbersome at first, it will prove *immensely* useful in this and the next several chapters. First, some terminology:

## Linear Equation

A linear equation in the $n$ variables $x_1, x_2, \ldots, x_n$ has the form

$$a_1 x_1 + \cdots + a_n x_n = b. \quad (a_1, a_2, \ldots, a_n, b \text{ constants})$$

The numbers $a_1, a_2, \ldots, a_n$ are called the **coefficients**, and the number $b$ is called the **constant term**, or **right-hand side**.

### Quick Examples

1. $3x - 5y = 0$

   Linear equation in $x$ and $y$
   Coefficients: 3, $-5$ Constant term: 0

2. $x + 2y - z = 6$

   Linear equation in $x, y, z$
   Coefficients: 1, 2, $-1$ Constant term: 6

3. $30x_1 + 18x_2 + x_3 + x_4 = 19$

   Linear equation in $x_1, x_2, x_3, x_4$
   Coefficients: 30, 18, 1, 1 Constant term: 19

**Note**  When the number of variables is small, we will almost always use $x, y, z, \ldots$ (as in Quick Examples 1 and 2) rather than $x_1, x_2, x_3, \ldots$ as the names of the variables. ∎

Notice that a linear equation in any number of unknowns (for example, $2x - y = 3$) is entirely determined by its coefficients and its constant term. In other words, if we were simply given the row of numbers

$$[2 \quad -1 \quad 3]$$

we could easily reconstruct the original linear equation by multiplying the first number by $x$, the second by $y$, and inserting a plus sign and an equals sign, as follows:

$$2 \cdot x + (-1) \cdot y = 3$$

or     $2x - y = 3.$

Similarly, the equation

$$-4x + 2y = 0$$

is represented by the row

$$[-4 \quad 2 \quad 0].$$

and the equation

$$-3y = \frac{1}{4}$$

is represented by

$$\left[ 0 \quad -3 \quad \frac{1}{4} \right].$$

As the last example shows, the first number is always the coefficient of $x$ and the second is the coefficient of $y$. If an $x$ or a $y$ is missing, we write a zero for its coefficient. We shall call such a row the **coefficient row** of an equation.

If we have a system of equations, for example the system

$$2x - \phantom{2}y = 3$$
$$-x + 2y = -4,$$

we can put the coefficient rows together like this:

$$\begin{bmatrix} 2 & -1 & 3 \\ -1 & 2 & -4 \end{bmatrix}.$$

We call this the **augmented matrix** of the system of equations. The term "augmented" means that we have included the right-hand sides 3 and $-4$. We will often drop the word "augmented" and simply refer to the matrix of the system. A **matrix** (plural: **matrices**) is nothing more than a rectangular array of numbers as above.

---

**Matrix, Augmented Matrix**

A **matrix** is a rectangular array of numbers. The **augmented matrix** of a system of linear equations is the matrix whose rows are the coefficient rows of the equations.

**Quick Example**

The augmented matrix of the system

$$x + y = 3$$
$$x - y = 1$$

is $\begin{bmatrix} 1 & 1 & 3 \\ 1 & -1 & 1 \end{bmatrix}.$

We'll be studying matrices in more detail in Chapter 4.

---

**Q** : *What good are coefficient rows and matrices?*

**A** : Think about what we do when we multiply both sides of an equation by a number. For example, consider multiplying both sides of the equation $2x - y = 3$ by $-2$ to get $-4x + 2y = -6$. All we are really doing is multiplying the coefficients and the right-hand side by $-2$. This corresponds to *multiplying the row* $[2 \quad -1 \quad 3]$ *by* $-2$, that is, multiplying every number in the row by $-2$. We shall see that any manipulation we want to do with equations can be done instead with rows, and this fact leads to a method of solving equations that is systematic and generalizes easily to larger systems.

---

Here is the same operation both in the language of equations and the language of rows. (We refer to the equation here as *Equation* 1, or simply $E_1$ for short, and to the row as *Row* 1, or $R_1$.)

| | Equation | Row |
|---|---|---|
| | $E_1$: $\quad 2x - \quad y = 3$ | $[\;2 \quad -1 \quad 3]\quad R_1$ |
| *Multiply by* $-2$: | $(-2)E_1$: $-4x + 2y = -6$ | $[-4 \quad 2 \quad -6]\quad (-2)R_1$ |

*Multiplying both sides of an equation by the number a corresponds to multiplying the coefficient row by a.*

Now look at what we do when we add two equations:

| | Equation | Row |
|---|---|---|
| | $E_1$: $\;2x - \;y = 3$ | $[\;\;2 \quad -1 \quad 3]\quad R_1$ |
| | $E_2$: $-x + 2y = -4$ | $[-1 \quad 2 \quad -4]\quad R_2$ |
| *Add*: | $E_1 + E_2$: $\;x + \;y = -1$ | $[\;\;1 \quad 1 \quad -1]\quad R_1 + R_2$ |

All we are really doing is *adding the corresponding entries in the rows*, or *adding the rows*. In other words,

*Adding two equations corresponds to adding their coefficient rows.*

In short, the manipulations of equations that we saw in the preceding section can be done more easily with rows in a matrix because we don't have to carry $x$, $y$, and other unnecessary notation along with us; $x$ and $y$ can always be inserted at the end if desired.

The manipulations we are talking about are known as **row operations**. In particular, we use three **elementary row operations**.

\* We are using the term *elementary row operations* a little more freely than most books. Some mathematicians insist that $a = 1$ in an operation of Type 2, but the less restrictive version is very useful.

† Multiplying an equation or row by zero gives us the not very surprising result $0 = 0$. In fact, we lose any information that the equation provided, which usually means that the resulting system has more solutions than the original system.

## Elementary Row Operations\*

**Type 1:** Replacing $R_i$ by $a R_i$ (where $a \neq 0$)†
In words: multiplying or dividing a row by a nonzero number.

**Type 2:** Replacing $R_i$ by $a R_i \pm b R_j$ (where $a \neq 0$)
Multiplying a row by a nonzero number and adding or subtracting a multiple of another row.

**using Technology**

See the Technology Guides at the end of the chapter to see how to do row operations using a TI-83/84 Plus or a spreadsheet.

 **Website**
www.WanerMath.com
→ Math Tools for Chapter 3
→ Pivot and Gauss-Jordan Tool

Enter the matrix in columns $x1, x2, x3, \ldots$. To do a row operation, type the instruction(s) next to the row(s) you are changing as shown, and press "Do Row Ops" once.

**Type 3:** Switching the order of the rows
This corresponds to switching the order in which we write the equations; occasionally this will be convenient.

For Types 1 and 2, we write the instruction for the row operation *next to the row we wish to replace.* (See the Quick Examples below.)

---

**Quick Examples**

**Type 1:** $\begin{bmatrix} 1 & 3 & -4 \\ 0 & 4 & 2 \end{bmatrix} 3R_2 \rightarrow \begin{bmatrix} 1 & 3 & -4 \\ 0 & 12 & 6 \end{bmatrix}$    Replace $R_2$ by $3R_2$.

**Type 2:** $\begin{bmatrix} 1 & 3 & -4 \\ 0 & 4 & 2 \end{bmatrix} 4R_1 - 3R_2 \rightarrow \begin{bmatrix} 4 & 0 & -22 \\ 0 & 4 & 2 \end{bmatrix}$    Replace $R_1$ by $4R_1 - 3R_2$.

**Type 3:** $\begin{bmatrix} 1 & 3 & -4 \\ 0 & 4 & 2 \\ 1 & 2 & 3 \end{bmatrix} R_1 \leftrightarrow R_2 \rightarrow \begin{bmatrix} 0 & 4 & 2 \\ 1 & 3 & -4 \\ 1 & 2 & 3 \end{bmatrix}$    Switch $R_1$ and $R_2$.

---

One very important fact about the elementary row operations is that they do not change the solutions of the corresponding system of equations. In other words, the new system of equations that we get by applying any one of these operations will have exactly the same solutions as the original system: It is easy to see that numbers that make the original equations true will also make the new equations true, because each of the elementary row operations corresponds to a valid operation on the original equations. That any solution of the new system is a solution of the old system follows from the fact that these row operations are *invertible:* The effects of a row operation can be reversed by applying another row operation, called its **inverse**. Here are some examples of this invertibility. (Try them out in the above Quick Examples.)

| *Operation* | *Inverse Operation* |
|---|---|
| Replace $R_2$ by $3R_2$. | Replace $R_2$ by $\frac{1}{3}R_2$. |
| Replace $R_1$ by $4R_1 - 3R_2$. | Replace $R_1$ by $\frac{1}{4}R_1 + \frac{3}{4}R_2$. |
| Switch $R_1$ and $R_2$. | Switch $R_1$ and $R_2$. |

Our objective, then, is to use row operations to change the system we are given into one with exactly the same set of solutions in which it is easy to see what the solutions are.

## Solving Systems of Equations by Using Row Operations

Now we put rows to work for us in solving systems of equations. Let's start with a complicated-looking system of equations:

$$-\frac{2x}{3} + \frac{y}{2} = -3$$
$$\frac{x}{4} - y = \frac{11}{4}.$$

We begin by writing the matrix of the system:

$$\begin{bmatrix} -\frac{2}{3} & \frac{1}{2} & -3 \\ \frac{1}{4} & -1 & \frac{11}{4} \end{bmatrix}.$$

Now what do we do with this matrix?

**Step 1** *Clear the fractions and/or decimals (if any) using operations of Type* **1.** To clear the fractions, we multiply the first row by 6 and the second row by 4. We record the operations by writing the symbolic form of an operation next to the row it will change, as follows.

$$\begin{bmatrix} -\frac{2}{3} & \frac{1}{2} & -3 \\ \frac{1}{4} & -1 & \frac{11}{4} \end{bmatrix} \begin{matrix} 6R_1 \\ 4R_2 \end{matrix}$$

By this we mean that we will replace the first row by $6R_1$ and the second by $4R_2$. Doing these operations gives

$$\begin{bmatrix} -4 & 3 & -18 \\ 1 & -4 & 11 \end{bmatrix}.$$

**Step 2** *Designate the first nonzero entry in the first row as the* **pivot.** In this case we designate the entry $-4$ in the first row as the "pivot" by putting a box around it:

$$\begin{bmatrix} \boxed{-4} & 3 & -18 \\ 1 & -4 & 11 \end{bmatrix}. \quad \leftarrow \text{Pivot row}$$
$$\uparrow$$
Pivot column

---

**Q**: What is a "pivot"?

**A**: A **pivot** is an entry in a matrix that is used to "clear a column." (See Step 3.) In this procedure, we will always select the first nonzero entry of a row as our pivot. In Chapter 5, when we study the simplex method, we will select our pivots differently.

---

**Step 3** *Use the pivot to clear its column using operations of Type* **2.** By **clearing a column**, we mean changing the matrix so that the pivot is the only nonzero number in its column. The procedure of clearing a column using a designated pivot is also called **pivoting.**

$$\begin{bmatrix} \boxed{-4} & 3 & -18 \\ 0 & \# & \# \end{bmatrix} \leftarrow \text{Desired row 2 (the "\#"s stand for as yet unknown numbers)}$$
$$\uparrow$$
Cleared pivot column

We want to replace $R_2$ by a row of the form $aR_2 \pm bR_1$ to get a zero in column 1. Moreover—and this will be important when we discuss the simplex method in Chapter 5—*we are going to choose positive values for both a and b.* We need to choose $a$ and $b$ so that we get the desired cancellation. We can do this quite mechanically as follows:

**a.** Write the name of the row you need to change on the left and that of the pivot row on the right.

$$\begin{matrix} R_2 & R_1 \\ \uparrow & \uparrow \end{matrix}$$
Row to change    Pivot row

✻ Thus, the only place a negative sign may appear is between $aR_2$ and $bR_1$ as indicated in the formula $aR_2 \pm bR_1$.

**b.** Focus on the pivot column, $\begin{bmatrix} -4 \\ 1 \end{bmatrix}$. Multiply each row by the *absolute value* of the entry currently in the other. (We are not permitting $a$ or $b$ to be negative.)

$$4R_2 \qquad\qquad 1R_1$$
$$\uparrow \qquad\qquad\quad \uparrow$$
$$\text{From Row 1} \qquad \text{From Row 2}$$

The effect is to make the two entries in the pivot column numerically the same. Sometimes, you can accomplish this by using smaller values of $a$ and $b$.

**c.** If the entries in the pivot column have opposite signs, insert a plus $(+)$. If they have the same sign, insert a minus $(-)$. Here, we get the instruction

$$4R_2 + 1R_1,$$

or simply $4R_2 + R_1$.

**d.** Write the operation next to the row you want to change, and then replace that row using the operation:

$$\begin{bmatrix} \boxed{-4} & 3 & -18 \\ 1 & -4 & 11 \end{bmatrix} 4R_2 + 1R_1 \rightarrow \begin{bmatrix} -4 & 3 & -18 \\ 0 & -13 & 26 \end{bmatrix}.$$

We have cleared the pivot column and completed Step 3.

**Note** In general, the row operation you use should always have the following form[*]:

$$a R_c \qquad \pm \qquad b R_p$$
$$\uparrow \qquad\qquad\qquad \uparrow$$
$$\text{Row to change} \qquad \text{Pivot row}$$

with $a$ and $b$ both positive. ∎

The next step is one that can be performed at any time.

**Simplification Step (Optional)** *If, at any stage of the process, all the numbers in a row are multiples of an integer, divide by that integer*—a Type 1 operation.

This is an optional but extremely helpful step: It makes the numbers smaller and easier to work with. In our case, the entries in $R_2$ are divisible by 13, so we divide that row by 13. (Alternatively, we could divide by $-13$. Try it.)

$$\begin{bmatrix} -4 & 3 & -18 \\ 0 & -13 & 26 \end{bmatrix} \tfrac{1}{13}R_2 \rightarrow \begin{bmatrix} -4 & 3 & -18 \\ 0 & -1 & 2 \end{bmatrix}$$

**Step 4** *Select the first nonzero number in the second row as the pivot, and clear its column.* Here we have combined two steps in one: selecting the new pivot and clearing the column (pivoting). The pivot is shown below, as well as the desired result when the column has been cleared:

$$\begin{bmatrix} -4 & 3 & -18 \\ 0 & \boxed{-1} & 2 \end{bmatrix} \rightarrow \begin{bmatrix} \# & 0 & \# \\ 0 & -1 & 2 \end{bmatrix}. \quad \leftarrow \text{desired row}$$
$$\uparrow \qquad\qquad\qquad \uparrow$$
$$\text{Pivot column} \qquad \text{Cleared pivot column}$$

We now wish to get a 0 in place of the 3 in the pivot column. Let's run once again through the mechanical steps to get the row operation that accomplishes this.

**a.** Write the name of the row you need to change on the left and that of the pivot row on the right:

$$\boldsymbol{R_1} \qquad\qquad\qquad \boldsymbol{R_2}$$
$$\uparrow \qquad\qquad\qquad\quad \uparrow$$
$$\text{Row to change} \qquad \text{Pivot row}$$

[*] We are deviating somewhat from the traditional procedure here. It is traditionally recommended first to divide the pivot row by the pivot, turning the pivot into a 1. This allows us to use $a = 1$, but usually results in fractions. The procedure we use here is easier for hand calculations and, we feel, mathematically more elegant, because it eliminates the need to introduce fractions. See the end of this section for an example done using the traditional procedure.

**b.** Focus on the pivot column, $\begin{bmatrix} 3 \\ -1 \end{bmatrix}$. Multiply each row by the absolute value of the entry currently in the other:

$$1 R_1 \qquad\qquad 3 R_2$$
$$\uparrow \qquad\qquad\qquad \uparrow$$
From Row 2 $\qquad\qquad$ From Row 1

**c.** If the entries in the pivot column have opposite signs, insert a plus $(+)$. If they have the same sign, insert a minus $(-)$. Here, we get the instruction

$$1 R_1 + 3 R_2.$$

**d.** Write the operation next to the row you want to change and then replace that row using the operation.

$$\begin{bmatrix} -4 & 3 & -18 \\ 0 & \boxed{-1} & 2 \end{bmatrix} \begin{matrix} R_1 + 3R_2 \\ {} \end{matrix} \rightarrow \begin{bmatrix} -4 & 0 & -12 \\ 0 & -1 & 2 \end{bmatrix}$$

Now we are essentially done, except for one last step.

**Final Step** *Using operations of Type 1, turn each pivot (the first nonzero entry in each row) into a 1.* We can accomplish this by dividing the first row by $-4$ and multiplying the second row by $-1$:

$$\begin{bmatrix} -4 & 0 & -12 \\ 0 & -1 & 2 \end{bmatrix} \begin{matrix} -\frac{1}{4}R_1 \\ -R_2 \end{matrix} \rightarrow \begin{bmatrix} 1 & 0 & 3 \\ 0 & 1 & -2 \end{bmatrix}.$$

The matrix now has the following nice form:

$$\begin{bmatrix} \boxed{1} & 0 & \# \\ 0 & \boxed{1} & \# \end{bmatrix}.$$

(This is the form we will always obtain with two equations in two unknowns when there is a unique solution.) This form is nice because, when we translate back into equations, we get

$$1x + 0y = 3$$
$$0x + 1y = -2.$$

In other words,

$$x = 3 \text{ and } y = -2$$

and so we have found the solution, which we can also write as $(x, y) = (3, -2)$.

The procedure we've just demonstrated is called **Gauss-Jordan\*** reduction or **row reduction.** It may seem too complicated a way to solve a system of two equations in two unknowns, and it is. However, for systems with more equations and more unknowns, it is very efficient.

In Example 1 below we use row reduction to solve a system of linear equations in *three* unknowns: $x$, $y$, and $z$. Just as for a system in two unknowns, a **solution** of a system in any number of unknowns consists of values for each of the variables that, when substituted, satisfy all of the equations in the system. Again, just as for a system in two unknowns, any system of linear equations in any number of unknowns has either no solution, exactly one solution, or infinitely many solutions. There are no other possibilities.

✱ Gauss-Jordan reduction is named after Carl Friedrich Gauss (1777–1855) and Wilhelm Jordan (1842–1899). Gauss was one of the great mathematicians, making fundamental contributions to number theory, analysis, probability, and statistics, as well as many fields of science. Gauss also made contributions to a method of solving systems of equations that has become known as "Gaussian elimination," even though this method had been described by Isaac Newton in 1707 and, independently, had been known to the Chinese more than 2,000 years ago. (See *Mathematicians of Gaussian Elimination*, Notices of the American Mathematical Society, June/July 2011, for a history of Gaussian elimination.) The method we are showing you here, Gauss-Jordan reduction, is Jordan's variation of Gaussian elimination, first published in 1887.

Solving a system in three unknowns graphically would require the graphing of planes (flat surfaces) in three dimensions. (The graph of a linear equation in three unknowns is a flat surface.) The use of row reduction makes three-dimensional graphing unnecessary.

### EXAMPLE 1  Solving a System by Gauss-Jordan Reduction

Solve the system

$$
\begin{aligned}
x - y + 5z &= -6 \\
3x + 3y - z &= 10 \\
x + 3y + 2z &= 5.
\end{aligned}
$$

**Solution**  The augmented matrix for this system is

$$
\begin{bmatrix}
1 & -1 & 5 & -6 \\
3 & 3 & -1 & 10 \\
1 & 3 & 2 & 5
\end{bmatrix}.
$$

Note that the columns correspond to $x, y, z$, and the right-hand side, respectively. We begin by selecting the pivot in the first row and clearing its column. Remember that clearing the column means that we turn *all* other numbers in the column into zeros. Thus, to clear the column of the first pivot, we need to change two rows, setting up the row operations in exactly the same way as above.

$$
\begin{bmatrix}
\boxed{1} & -1 & 5 & -6 \\
3 & 3 & -1 & 10 \\
1 & 3 & 2 & 5
\end{bmatrix}
\begin{array}{l} \\ R_2 - 3R_1 \\ R_3 - R_1 \end{array} \to
\begin{bmatrix}
1 & -1 & 5 & -6 \\
0 & 6 & -16 & 28 \\
0 & 4 & -3 & 11
\end{bmatrix}
$$

Notice that both row operations have the required form

$$
a R_c \pm b R_1
$$

$$
\begin{array}{ccc}
\uparrow & & \uparrow \\
\text{Row to change} & & \text{Pivot row}
\end{array}
$$

with $a$ and $b$ both positive.

Now we use the optional simplification step to simplify $R_2$:

$$
\begin{bmatrix}
1 & -1 & 5 & -6 \\
0 & 6 & -16 & 28 \\
0 & 4 & -3 & 11
\end{bmatrix}
\begin{array}{l} \\ \frac{1}{2}R_2 \\ \\ \end{array} \to
\begin{bmatrix}
1 & -1 & 5 & -6 \\
0 & 3 & -8 & 14 \\
0 & 4 & -3 & 11
\end{bmatrix}.
$$

Next, we select the pivot in the second row and clear its column:

$$
\begin{bmatrix}
1 & -1 & 5 & -6 \\
0 & \boxed{3} & -8 & 14 \\
0 & 4 & -3 & 11
\end{bmatrix}
\begin{array}{l} 3R_1 + R_2 \\ \\ 3R_3 - 4R_2 \end{array} \to
\begin{bmatrix}
3 & 0 & 7 & -4 \\
0 & 3 & -8 & 14 \\
0 & 0 & 23 & -23
\end{bmatrix}.
$$

$R_1$ and $R_3$ are to be changed.

$R_2$ is the pivot row.

We simplify $R_3$.

$$
\begin{bmatrix}
3 & 0 & 7 & -4 \\
0 & 3 & -8 & 14 \\
0 & 0 & 23 & -23
\end{bmatrix}
\begin{array}{l} \\ \\ \frac{1}{23}R_3 \end{array} \to
\begin{bmatrix}
3 & 0 & 7 & -4 \\
0 & 3 & -8 & 14 \\
0 & 0 & 1 & -1
\end{bmatrix}
$$

**using** Technology

See the Technology Guides at the end of the chapter to see how to use a TI-83/84 Plus or a spreadsheet to solve this system of equations.

**Website**
www.WanerMath.com

Everything

→ Math Tools for Chapter 3

→ Pivot and Gauss-Jordan Tool

Enter the augmented matrix in columns $x1$, $x2$, $x3$, . . . . At each step, type in the row operations exactly as written above next to the rows to which they apply. For example, for the first step type:

R2-3R1 next to Row 2, and

R3-R1 next to Row 3.

Press "Do Row Ops" once, and then "Clear Row Ops" to prepare for the next step.

For the last step, type

(1/3)R1 next to Row 1, and

(1/3)R2 next to Row 2.

Press "Do Row Ops" once.

The utility also does other things, such as automatic pivoting and complete reduction in one step. Use these features to check your work.

Now we select the pivot in the third row and clear its column:

$$\begin{bmatrix} 3 & 0 & 7 & -4 \\ 0 & 3 & -8 & 14 \\ 0 & 0 & \boxed{1} & -1 \end{bmatrix} \begin{matrix} R_1 - 7R_3 \\ R_2 + 8R_3 \\ \ \end{matrix} \rightarrow \begin{bmatrix} 3 & 0 & 0 & 3 \\ 0 & 3 & 0 & 6 \\ 0 & 0 & 1 & -1 \end{bmatrix}.$$

$R_1$ and $R_2$ are to be changed.
$R_3$ is the pivot row.

Finally, we turn all the pivots into 1s:

$$\begin{bmatrix} 3 & 0 & 0 & 3 \\ 0 & 3 & 0 & 6 \\ 0 & 0 & 1 & -1 \end{bmatrix} \begin{matrix} \frac{1}{3}R_1 \\ \frac{1}{3}R_2 \\ \ \end{matrix} \rightarrow \begin{bmatrix} 1 & 0 & 0 & 1 \\ 0 & 1 & 0 & 2 \\ 0 & 0 & 1 & -1 \end{bmatrix}.$$

The matrix is now reduced to a simple form, so we translate back into equations to obtain the solution:

$$x = 1, \ y = 2, \ z = -1, \text{ or } (x, y, z) = (1, 2, -1).$$

Notice the form of the very last matrix in the example:

$$\begin{bmatrix} 1 & 0 & 0 & \# \\ 0 & 1 & 0 & \# \\ 0 & 0 & 1 & \# \end{bmatrix}.$$

The 1s are on the **(main) diagonal** of the matrix; the goal in Gauss-Jordan reduction is to reduce our matrix to this form. If we can do so, then we can easily read off the solution, as we saw in Example 1. However, as we will see in several examples in this section, it is not always possible to achieve this ideal state. After Example 6, we will give a form that is always possible to achieve.

### EXAMPLE 2  Solving a System by Gauss-Jordan Reduction

Solve the system:

$$\begin{aligned} 2x + \ y + 3z &= 1 \\ 4x + 2y + 4z &= 4 \\ x + 2y + \ z &= 4. \end{aligned}$$

#### Solution

$$\begin{bmatrix} \boxed{2} & 1 & 3 & 1 \\ 4 & 2 & 4 & 4 \\ 1 & 2 & 1 & 4 \end{bmatrix} \begin{matrix} \ \\ R_2 - 2R_1 \\ 2R_3 - R_1 \end{matrix} \rightarrow \begin{bmatrix} 2 & 1 & 3 & 1 \\ 0 & 0 & -2 & 2 \\ 0 & 3 & -1 & 7 \end{bmatrix}$$

Now we have a slight problem: The number in the position where we would like to have a pivot—the second column of the second row—is a zero and thus cannot be a pivot. There are two ways out of this problem. One is to move on to the third column and pivot on the $-2$. Another is to switch the order of the second and third rows so that we can use the 3 as a pivot. We will do the latter.

$$\begin{bmatrix} 2 & 1 & 3 & 1 \\ 0 & 0 & -2 & 2 \\ 0 & 3 & -1 & 7 \end{bmatrix} \begin{matrix} \\ \\ R_2 \leftrightarrow R_3 \end{matrix} \rightarrow \begin{bmatrix} 2 & 1 & 3 & 1 \\ 0 & \boxed{3} & -1 & 7 \\ 0 & 0 & -2 & 2 \end{bmatrix} \begin{matrix} 3R_1 - R_2 \\ \\ \\ \end{matrix}$$

$$\rightarrow \begin{bmatrix} 6 & 0 & 10 & -4 \\ 0 & 3 & -1 & 7 \\ 0 & 0 & -2 & 2 \end{bmatrix} \begin{matrix} \\ \\ -\frac{1}{2}R_3 \end{matrix} \rightarrow \begin{bmatrix} 6 & 0 & 10 & -4 \\ 0 & 3 & -1 & 7 \\ 0 & 0 & \boxed{1} & -1 \end{bmatrix} \begin{matrix} R_1 - 10R_3 \\ R_2 + R_3 \\ \\ \end{matrix}$$

$$\rightarrow \begin{bmatrix} 6 & 0 & 0 & 6 \\ 0 & 3 & 0 & 6 \\ 0 & 0 & 1 & -1 \end{bmatrix} \begin{matrix} \frac{1}{6}R_1 \\ \frac{1}{3}R_2 \\ \\ \end{matrix} \rightarrow \begin{bmatrix} 1 & 0 & 0 & 1 \\ 0 & 1 & 0 & 2 \\ 0 & 0 & 1 & -1 \end{bmatrix}$$

Thus, the solution is $(x, y, z) = (1, 2, -1)$, as you can check in the original system.

## EXAMPLE 3  Inconsistent System

Solve the system:

$$\begin{aligned} x + y + z &= 1 \\ 2x - y + z &= 0 \\ 4x + y + 3z &= 3. \end{aligned}$$

**Solution**

$$\begin{bmatrix} \boxed{1} & 1 & 1 & 1 \\ 2 & -1 & 1 & 0 \\ 4 & 1 & 3 & 3 \end{bmatrix} \begin{matrix} \\ R_2 - 2R_1 \\ R_3 - 4R_1 \end{matrix} \rightarrow \begin{bmatrix} 1 & 1 & 1 & 1 \\ 0 & \boxed{-3} & -1 & -2 \\ 0 & -3 & -1 & -1 \end{bmatrix} \begin{matrix} 3R_1 + R_2 \\ \\ R_3 - R_2 \end{matrix}$$

$$\rightarrow \begin{bmatrix} 3 & 0 & 2 & 1 \\ 0 & -3 & -1 & -2 \\ 0 & 0 & 0 & 1 \end{bmatrix}$$

Stop. That last row translates into $0 = 1$, which is nonsense, and so, as in Example 3 in Section 3.1, we can say that this system has no solution. We also say, as we did for systems with only two unknowns, that a system with no solution is **inconsistent**. A system with at least one solution is **consistent**.

➡ **Before we go on...**

**Q** : *How, exactly, does the nonsensical equation $0 = 1$ tell us that there is no solution of the system in Example 3?*

**A** : Here is an argument similar to that in Example 3 in Section 3.1: If there *were* three numbers $x$, $y$, and $z$ satisfying the original system of equations, then manipulating the equations according to the instructions in the row operations above leads us to conclude that $0 = 1$. Because 0 is *not* equal to 1, there can be no such numbers $x$, $y$, and $z$. ∎

## EXAMPLE 4  Infinitely Many Solutions

Solve the system:

$$\begin{aligned} x + y + z &= 1 \\ \frac{1}{4}x - \frac{1}{2}y + \frac{3}{4}z &= 0 \\ x + 7y - 3z &= 3. \end{aligned}$$

**Solution**

$$
\begin{bmatrix} 1 & 1 & 1 & 1 \\ \frac{1}{4} & -\frac{1}{2} & \frac{3}{4} & 0 \\ 1 & 7 & -3 & 3 \end{bmatrix}
\begin{array}{l} \\ 4R_2 \to \\ \\ \end{array}
\begin{bmatrix} \boxed{1} & 1 & 1 & 1 \\ 1 & -2 & 3 & 0 \\ 1 & 7 & -3 & 3 \end{bmatrix}
\begin{array}{l} \\ R_2 - R_1 \\ R_3 - R_1 \end{array}
$$

$$
\to
\begin{bmatrix} 1 & 1 & 1 & 1 \\ 0 & -3 & 2 & -1 \\ 0 & 6 & -4 & 2 \end{bmatrix}
\begin{array}{l} \\ \\ \frac{1}{2}R_3 \end{array}
\to
\begin{bmatrix} 1 & 1 & 1 & 1 \\ 0 & \boxed{-3} & 2 & -1 \\ 0 & 3 & -2 & 1 \end{bmatrix}
\begin{array}{l} 3R_1 + R_2 \\ \\ R_3 + R_2 \end{array}
$$

$$
\to
\begin{bmatrix} 3 & 0 & 5 & 2 \\ 0 & -3 & 2 & -1 \\ 0 & 0 & 0 & 0 \end{bmatrix}
$$

There are no nonzero entries in the third row, so there can be no pivot in the third row. We skip to the final step and turn the pivots we did find into 1s.

$$
\begin{bmatrix} 3 & 0 & 5 & 2 \\ 0 & -3 & 2 & -1 \\ 0 & 0 & 0 & 0 \end{bmatrix}
\begin{array}{l} \frac{1}{3}R_1 \\ -\frac{1}{3}R_2 \\ \\ \end{array}
\to
\begin{bmatrix} 1 & 0 & \frac{5}{3} & \frac{2}{3} \\ 0 & 1 & -\frac{2}{3} & \frac{1}{3} \\ 0 & 0 & 0 & 0 \end{bmatrix}
$$

Now we translate back into equations and obtain:

$$
\begin{aligned}
x \quad + \tfrac{5}{3}z &= \tfrac{2}{3} \\
y - \tfrac{2}{3}z &= \tfrac{1}{3} \\
0 &= 0.
\end{aligned}
$$

But how does this help us find a solution? The last equation doesn't tell us anything useful, so we ignore it. The thing to notice about the other equations is that we can easily solve the first equation for $x$ and the second for $y$, obtaining

$$
\begin{aligned}
x &= \tfrac{2}{3} - \tfrac{5}{3}z \\
y &= \tfrac{1}{3} + \tfrac{2}{3}z.
\end{aligned}
$$

This is the solution! We can choose $z$ to be any number and get corresponding values for $x$ and $y$ from the formulas above. This gives us infinitely many different solutions. Thus, the general solution (see Example 4 in Section 3.1) is

$$
\begin{aligned}
x &= \tfrac{2}{3} - \tfrac{5}{3}z \\
y &= \tfrac{1}{3} + \tfrac{2}{3}z \\
z &\text{ is arbitrary.}
\end{aligned}
$$

General solution

We can also write the general solution as

$$\left(\tfrac{2}{3} - \tfrac{5}{3}z, \ \tfrac{1}{3} + \tfrac{2}{3}z, \ z\right) \ z \text{ arbitrary.}$$

General solution

This general solution has $z$ as the parameter. Specific choices of values for the parameter $z$ give particular solutions. For example, the choice $z = 6$ gives the particular solution

$$
\begin{aligned}
x &= \tfrac{2}{3} - \tfrac{5}{3}(6) = -\tfrac{28}{3} \\
y &= \tfrac{1}{3} + \tfrac{2}{3}(6) = \tfrac{13}{3} \\
z &= 6,
\end{aligned}
$$

Particular solution

while the choice $z = 0$ gives the particular solution $(x, \ y, \ z) = \left(\tfrac{2}{3}, \tfrac{1}{3}, 0\right)$.

Note that, unlike the system given in the preceding example, the system given in this example does have solutions, and is thus *consistent*.

➡ **Before we go on...** Why were there infinitely many solutions to Example 4? The reason is that the third equation was really a combination of the first and second equations to begin with, so we effectively had only two equations in three unknowns.* Choosing a specific value for $z$ (say, $z = 6$) has the effect of supplying the "missing" equation. ∎

\* In fact, you can check that the third equation, $E_3$, is equal to $3E_1 - 8E_2$. Thus, the third equation could have been left out because it conveys no more information than the first two. The process of row reduction always eliminates such a redundancy by creating a row of zeros.

**Q**: *How do we know when there are infinitely many solutions?*

**A**: When there are solutions (we have a consistent system, unlike the one in Example 3), and when the matrix we arrive at by row reduction has fewer pivots than there are unknowns. In Example 4 we had three unknowns but only two pivots.

**Q**: *How do we know which variables to use as parameters in a parameterized solution?*

**A**: The variables to use as parameters are those in the columns without pivots. In Example 4 there were pivots in the $x$ and $y$ columns, but no pivot in the $z$ column, and it was $z$ that we used as a parameter.

### EXAMPLE 5 Four Unknowns

Solve the system:

$$x + 3y + 2z - \phantom{0}w = 6$$
$$2x + 6y + 6z + \phantom{0}3w = 16$$
$$x + 3y - 2z - 11w = -2$$
$$2x + 6y + 8z + \phantom{0}8w = 20.$$

#### Solution

$$
\begin{bmatrix}
\boxed{1} & 3 & 2 & -1 & 6 \\
2 & 6 & 6 & 3 & 16 \\
1 & 3 & -2 & -11 & -2 \\
2 & 6 & 8 & 8 & 20
\end{bmatrix}
\begin{matrix} \\ R_2 - 2R_1 \\ R_3 - R_1 \\ R_4 - 2R_1 \end{matrix}
\rightarrow
\begin{bmatrix}
1 & 3 & 2 & -1 & 6 \\
0 & 0 & 2 & 5 & 4 \\
0 & 0 & -4 & -10 & -8 \\
0 & 0 & 4 & 10 & 8
\end{bmatrix}
$$

There is no pivot available in the second column, so we move on to the third column.

$$
\begin{bmatrix}
1 & 3 & 2 & -1 & 6 \\
0 & 0 & \boxed{2} & 5 & 4 \\
0 & 0 & -4 & -10 & -8 \\
0 & 0 & 4 & 10 & 8
\end{bmatrix}
\begin{matrix} R_1 - R_2 \\ \\ R_3 + 2R_2 \\ R_4 - 2R_2 \end{matrix}
\rightarrow
\begin{bmatrix}
1 & 3 & 0 & -6 & 2 \\
0 & 0 & 2 & 5 & 4 \\
0 & 0 & 0 & 0 & 0 \\
0 & 0 & 0 & 0 & 0
\end{bmatrix}
\begin{matrix} \\ \tfrac{1}{2}R_2 \\ \\ \end{matrix}
$$

$$
\rightarrow
\begin{bmatrix}
1 & 3 & 0 & -6 & 2 \\
0 & 0 & 1 & \tfrac{5}{2} & 2 \\
0 & 0 & 0 & 0 & 0 \\
0 & 0 & 0 & 0 & 0
\end{bmatrix}
$$

Translating back into equations, we get:

$$x + 3y \phantom{00} - 6w = 2$$
$$z + \tfrac{5}{2}w = 2.$$

(We have not written down the equations corresponding to the last two rows, each of which is $0 = 0$.) There are no pivots in the $y$ or $w$ columns, so we use these two variables as parameters. We bring them over to the right-hand sides of the equations above and write the general solution as

$$x = 2 - 3y + 6w$$
$$y \text{ is arbitrary}$$
$$z = 2 - \tfrac{5}{2}w$$
$$w \text{ is arbitrary}$$

or

$$(x, y, z, w) = (2 - 3y + 6w, y, 2 - 5w/2, w) \quad y, w \text{ arbitrary}.$$

➡ **Before we go on...** In Examples 4 and 5, you might have noticed an interesting phenomenon: If at any time in the process, two rows are equal or one is a multiple of the other, then one of those rows (eventually) becomes all zero. ∎

Up to this point, we have always been given as many equations as there are unknowns. However, we shall see in the next section that some applications lead to systems where the number of equations is not the same as the number of unknowns. As the following example illustrates, such systems can be handled the same way as any other.

### EXAMPLE 6  Number of Equations ≠ Number of Unknowns

Solve the system:

$$x + \quad y = 1$$
$$13x - 26y = -11$$
$$26x - 13y = 2.$$

**Solution** We proceed exactly as before and ignore the fact that there is one more equation than unknown.

$$\begin{bmatrix} \boxed{1} & 1 & 1 \\ 13 & -26 & -11 \\ 26 & -13 & 2 \end{bmatrix} \begin{matrix} \\ R_2 - 13R_1 \\ R_3 - 26R_1 \end{matrix} \rightarrow \begin{bmatrix} 1 & 1 & 1 \\ 0 & -39 & -24 \\ 0 & -39 & -24 \end{bmatrix} \begin{matrix} \\ \frac{1}{3}R_2 \\ \frac{1}{3}R_3 \end{matrix}$$

$$\rightarrow \begin{bmatrix} 1 & 1 & 1 \\ 0 & \boxed{-13} & -8 \\ 0 & -13 & -8 \end{bmatrix} \begin{matrix} 13R_1 + R_2 \\ \\ R_3 - R_2 \end{matrix} \rightarrow \begin{bmatrix} 13 & 0 & 5 \\ 0 & -13 & -8 \\ 0 & 0 & 0 \end{bmatrix} \begin{matrix} \frac{1}{13}R_1 \\ -\frac{1}{13}R_2 \\ \\ \end{matrix}$$

$$\rightarrow \begin{bmatrix} 1 & 0 & \frac{5}{13} \\ 0 & 1 & \frac{8}{13} \\ 0 & 0 & 0 \end{bmatrix}$$

Thus, the solution is $(x, y) = \left( \frac{5}{13}, \frac{8}{13} \right)$.

If, instead of a row of zeros, we had obtained, say, [0  0  6] in the last row, we would immediately have concluded that the system was inconsistent.

The fact that we wound up with a row of zeros indicates that one of the equations was actually a combination of the other two; you can check that the third equation can be obtained by multiplying the first equation by 13 and adding the result to the second. Because the third equation therefore tells us nothing that we don't already know from the first two, we call the system of equations **redundant**, or **dependent**. (Compare Example 4 in Section 3.1.)

➡ **Before we go on...** Example 5 above is another example of a redundant system; we could have started with the following smaller system of two equations in four unknowns

$$x + 3y + 2z - \ w = 6$$
$$2x + 6y + 6z + 3w = 16$$

and obtained the same general solution as we did with the larger system. Verify this by solving the smaller system.  ■

The preceding examples illustrated that we cannot always reduce a matrix to the form shown before Example 2, with pivots going all the way down the diagonal. What we *can* always do is reduce a matrix to the following form:

## Reduced Row Echelon Form

A matrix is said to be in **reduced row echelon form** or to be **row-reduced** if it satisfies the following properties.

**P1.** The first nonzero entry in each row (called the **leading entry** of that row) is a 1.

**P2.** The columns of the leading entries are **clear** (i.e., they contain zeros in all positions other than that of the leading entry).

**P3.** The leading entry in each row is to the right of the leading entry in the row above, and any rows of zeros are at the bottom.

### Quick Examples

$$\begin{bmatrix} 1 & 0 & 0 & 2 \\ 0 & 1 & 0 & 4 \\ 0 & 0 & 1 & -3 \end{bmatrix}, \begin{bmatrix} 0 & 1 & -3 \\ 0 & 0 & 0 \end{bmatrix}, \text{ and } \begin{bmatrix} 1 & 0 & 0 & -2 \\ 0 & 0 & 1 & 4 \\ 0 & 0 & 0 & 0 \end{bmatrix} \text{ are row-reduced.}$$

$$\begin{bmatrix} 1 & 1 & 0 & 2 \\ 0 & 1 & 0 & 4 \\ 0 & 0 & 1 & -3 \end{bmatrix}, \begin{bmatrix} 0 & 1 & -3 \\ 0 & 0 & 1 \end{bmatrix}, \text{ and } \begin{bmatrix} 0 & 0 & 1 & 4 \\ 1 & 0 & 0 & -2 \\ 0 & 0 & 0 & 0 \end{bmatrix} \text{ are not row-reduced.}$$

You should check in the examples we did that the final matrices were all in reduced row echelon form.

It is an interesting and useful fact, though not easy to prove, that any two people who start with the same matrix and row-reduce it will reach exactly the same row-reduced matrix, even if they use different row operations.

# The Traditional Gauss-Jordan Method (Optional)

In the version of the Gauss-Jordan method we have presented, we eliminated fractions and decimals in the first step and then worked with integer matrices, partly to make hand computation easier and partly for mathematical elegance. However, complicated fractions and decimals present no difficulty when we use technology. The following example illustrates the more traditional approach to Gauss-Jordan reduction used in many of the computer programs that solve the huge systems of equations that arise in practice.[*]

✳ Actually, for reasons of efficiency and accuracy, the methods used in commercial programs are closer to the method presented above. To learn more, consult a text on numerical methods.

## EXAMPLE 7  Solving a System with the Traditional Gauss-Jordan Method

Solve the following system using the traditional Gauss-Jordan method:

$$2x + y + 3z = 5$$
$$3x + 2y + 4z = 7$$
$$2x + y + 5z = 10.$$

**Solution**  We make two changes in our method. First, there is no need to get rid of decimals (because computers and calculators can handle decimals as easily as they can integers). Second, after selecting a pivot, *divide the pivot row by the pivot value, turning the pivot into a* 1. It is easier to determine the row operations that will clear the pivot column if the pivot is a 1.

If we use technology to solve this system of equations, the sequence of matrices might look like this:

$$
\begin{bmatrix} \boxed{2} & 1 & 3 & 5 \\ 3 & 2 & 4 & 7 \\ 2 & 1 & 5 & 10 \end{bmatrix}
\begin{matrix} \frac{1}{2}R_1 \\ \\ \end{matrix}
\rightarrow
\begin{bmatrix} \boxed{1} & 0.5 & 1.5 & 2.5 \\ 3 & 2 & 4 & 7 \\ 2 & 1 & 5 & 10 \end{bmatrix}
\begin{matrix} \\ R_2 - 3R_1 \\ R_3 - 2R_1 \end{matrix}
$$

$$
\rightarrow
\begin{bmatrix} 1 & 0.5 & 1.5 & 2.5 \\ 0 & \boxed{0.5} & -0.5 & -0.5 \\ 0 & 0 & 2 & 5 \end{bmatrix}
\begin{matrix} \\ 2R_2 \\ \end{matrix}
\rightarrow
\begin{bmatrix} 1 & 0.5 & 1.5 & 2.5 \\ 0 & \boxed{1} & -1 & -1 \\ 0 & 0 & 2 & 5 \end{bmatrix}
\begin{matrix} R_1 - 0.5R_2 \\ \\ \end{matrix}
$$

$$
\rightarrow
\begin{bmatrix} 1 & 0 & 2 & 3 \\ 0 & 1 & -1 & -1 \\ 0 & 0 & \boxed{2} & 5 \end{bmatrix}
\begin{matrix} \\ \\ \frac{1}{2}R_3 \end{matrix}
\rightarrow
\begin{bmatrix} 1 & 0 & 2 & 3 \\ 0 & 1 & -1 & -1 \\ 0 & 0 & \boxed{1} & 2.5 \end{bmatrix}
\begin{matrix} R_1 - 2R_3 \\ R_2 + R_3 \\ \end{matrix}
$$

$$
\rightarrow
\begin{bmatrix} 1 & 0 & 0 & -2 \\ 0 & 1 & 0 & 1.5 \\ 0 & 0 & 1 & 2.5 \end{bmatrix}.
$$

The solution is $(x, y, z) = (-2, 1.5, 2.5)$.

**Q**: *The solution to Example 7 looked quite easy. Why didn't we use the traditional method from the start like the other textbooks?*

**A**: It looked easy because we deliberately chose an example that leads to simple decimals. In all but the most contrived examples, the decimals or fractions involved get very complicated very quickly.

**FAQs**

**Getting Unstuck, Going Round in Circles, and Knowing When to Stop**

**Q** : *Help! I have been doing row operations on this matrix for half an hour. I have filled two pages, and I am getting nowhere. What do I do?*

**A** : Here is a way of keeping track of where you are *at any stage of the process* and also deciding what to do next.

Starting at the top row of your current matrix:

1. Scan along the row until you get to the leading entry: the first nonzero entry.
   If there is none—that is, the row is all zero—go to the next row.
2. Having located the leading entry, scan up and down its *column.* If its column is not clear (that is, it contains other nonzero entries), use your leading entry as a pivot to clear its column as in the examples in this section.
3. Now go to the next row and start again at Step 1.

When you have scanned all the rows and find that all the columns of the leading entries are clear, it means you are done (except possibly for reordering the rows so that the leading entries go from left to right as you read down the matrix, and zero rows are at the bottom).

**Q** : *No good. I have been following these instructions, but every time I try to clear a column, I unclear a column I had already cleared. What is going on?*

**A** : Are you using *leading entries* as pivots? Also, are you *using the pivot* to clear its column? That is, are your row operations all of the following form?

$$aR_c \pm bR_p$$

$\uparrow$     $\uparrow$

Row to change    Pivot row

The instruction next to the row you are changing should involve only that row and the pivot row, even though you might be tempted to use some other row instead.

**Q** : *Must I continue until I get a matrix that has 1s down the leading diagonal and 0s above and below?*

**A** : Not necessarily. You are completely done when your matrix is row-reduced: Each leading entry is a 1, the column of each leading entry is clear, and the leading entries go from left to right. You are done *pivoting* when the column of each leading entry is clear. After that, all that remains is to turn each pivot into a 1 (the "Final Step") and, if necessary, rearrange the rows.

## **3.2** EXERCISES

▼ more advanced ◆ challenging
**T** indicates exercises that should be solved using technology

*In Exercises 1–42, use Gauss-Jordan row reduction to solve the given systems of equation. We suggest doing some by hand, and others using technology.* HINT [See Examples 1–6.]

1. $x + y = 4$
   $x - y = 2$

2. $2x + y = 2$
   $-2x + y = 2$

3. $3x - 2y = 6$
   $2x - 3y = -6$

4. $2x + 3y = 5$
   $3x + 2y = 5$

5. $2x + 3y = 1$
   $-x - \dfrac{3y}{2} = -\dfrac{1}{2}$

6. $2x - 3y = 1$
   $6x - 9y = 3$

**7.** $2x + 3y = 2$

$-x - \dfrac{3y}{2} = -\dfrac{1}{2}$

**8.** $2x - 3y = 2$

$6x - 9y = 3$

**9.** $x + y = 1$

$3x - y = 0$

$x - 3y = -2$

**10.** $x + y = 1$

$3x - 2y = -1$

$5x - y = \dfrac{1}{5}$

**11.** $x + y = 0$

$3x - y = 1$

$x - y = -1$

**12.** $x + 2y = 1$

$3x - 2y = -2$

$5x - y = \dfrac{1}{5}$

**13.** $0.5x + 0.1y = 1.7$

$0.1x - 0.1y = 0.3$

$x + y = \dfrac{11}{3}$

**14.** $-0.3x + 0.5y = 0.1$

$x - y = 4$

$\dfrac{x}{17} + \dfrac{y}{17} = 1$

**15.** $-x + 2y - z = 0$

$-x - y + 2z = 0$

$2x \quad - z = 4$

**16.** $x + 2y = 4$

$y - z = 0$

$x + 3y - 2z = 5$

**17.** $x + y + 6z = -1$

$\dfrac{1}{3}x - \dfrac{1}{3}y + \dfrac{2}{3}z = 1$

$\dfrac{1}{2}x \quad + z = 0$

**18.** $x - \dfrac{1}{2}y \quad = 0$

$\dfrac{1}{3}x + \dfrac{1}{3}y + \dfrac{1}{3}z = 2$

$\dfrac{1}{2}x \quad - \dfrac{1}{2}z = -1$

**19.** $-\dfrac{1}{2}x + y - \dfrac{1}{2}z = 0$

$-\dfrac{1}{2}x - \dfrac{1}{2}y + z = 0$

$x - \dfrac{1}{2}y - \dfrac{1}{2}z = 0$

**20.** $x - \dfrac{1}{2}y \quad = 0$

$\dfrac{1}{2}x - \dfrac{1}{2}z = -1$

$3x - y - z = -2$

**21.** $x + y + 2z = -1$

$2x + 2y + 2z = 2$

$\dfrac{3}{5}x + \dfrac{3}{5}y + \dfrac{3}{5}z = \dfrac{2}{5}$

**22.** $x + y - z = -2$

$x - y - 7z = 0$

$\dfrac{2}{7}x \quad - \dfrac{8}{7}z = 14$

**23.** $-0.5x + 0.5y + 0.5z = 1.5$

$4.2x + 2.1y + 2.1z = 0$

$0.2x \quad + 0.2z = 0$

**24.** $0.25x - 0.5y \quad = 0$

$0.2x + 0.2y - 0.2z = -0.6$

$0.5x - 1.5y + z = 0.5$

**25.** $2x - y + z = 4$

$3x - y + z = 5$

**26.** $3x - y - z = 0$

$x + y + z = 4$

**27.** $0.75x - 0.75y - z = 4$

$x - y + 4z = 0$

**28.** $2x - y + z = 4$

$-x + 0.5y - 0.5z = 1.5$

**29.** ▼ $3x + y - z = 12$

**30.** ▼ $x + y - 3z = 21$

(Yes: One equation in three unknowns!)

**31.** ▼ $x + y + 2z = -1$

$2x + 2y + 2z = 2$

$0.75x + 0.75y + z = 0.25$

$-x \quad - 2z = 21$

**32.** $x + y - z = -2$

$x - y - 7z = 0$

$0.75x - 0.5y + 0.25z = 14$

$x + y + z = 4$

**33.** ▼ $x + y + 5z = 1$

$y + 2z + w = 1$

$x + 3y + 7z + 2w = 2$

$x + y + 5z + w = 1$

**34.** ▼ $x + y \quad + 4w = 1$

$2x - 2y - 3z + 2w = -1$

$4y + 6z + w = 4$

$2x + 4y + 9z \quad = 6$

**35.** ▼ $x + y + 5z \quad = 1$

$y + 2z + w = 1$

$x + y + 5z + w = 1$

$x + 2y + 7z + 2w = 2$

**36.** ▼ $x + y + \quad 4w = 1$

$2x - 2y - 3z + 2w = -1$

$4y + 6z + w = 4$

$3x + 3y + 3z + 7w = 4$

**37.** ▼ $x - 2y + z - 4w = 1$

$x + 3y + 7z + 2w = 2$

$2x + y + 8z - 2w = 3$

**38.** ▼ $x - 3y - 2z - w = 1$

$x + 3y + z + 2w = 2$

$2x \quad - z + w = 3$

**39.** ▼ $x + y + z + u + v = 15$

$y - z + u - v = -2$

$z + u + v = 12$

$u - v = -1$

$v = 5$

**40.** ▼ $x - y + z - u + v = 1$

$y + z + u + v = 2$

$z - u + v = 1$

$u + v = 1$

$v = 1$

**41.** ▼ $x - y + z - u + v = 0$

$y - z + u - v = -2$

$x \quad - 2v = -2$

$2x - y + z - u - 3v = -2$

$4x - y + z - u - 7v = -6$

**42.** ▼ $x + y + z + u + v = 15$

$y + z + u + v = 3$

$x + 2y + 2z + 2u + 2v = 18$

$x - y - z - u - v = 9$

$x - 2y - 2z - 2u - 2v = 6$

⊤ *In Exercises 43–46, use technology to solve the systems of equations. Express all solutions as fractions.*

**43.**
$$x + 2y - z + w = 30$$
$$2x \quad - z + 2w = 30$$
$$x + 3y + 3z - 4w = 2$$
$$2x - 9y \quad + w = 4$$

**44.** $4x - 2y + z + w = 20$
$$3y + 3z - 4w = 2$$
$$2x + 4y \quad - w = 4$$
$$x + 3y + 3z \quad = 2$$

**45.** $x + 2y + 3z + 4w + 5t = 6$
$$2x + 3y + 4z + 5w + t = 5$$
$$3x + 4y + 5z + w + 2t = 4$$
$$4x + 5y + z + 2w + 3t = 3$$
$$5x + y + 2z + 3w + 4t = 2$$

**46.** $x - 2y + 3z - 4w \quad = 0$
$$-2x + 3y - 4z + \quad t = 0$$
$$3x - 4y \quad + w - 2t = 0$$
$$-4x \quad + z - 2w + 3t = 0$$
$$y - 2z + 3w - 4t = 1$$

⊤ *In Exercises 47–50, use technology to solve the systems of equations. Express all solutions as decimals, rounded to one decimal place.*

**47.** $1.6x + 2.4y - 3.2z = 4.4$
$$5.1x - 6.3y + 0.6z = -3.2$$
$$4.2x + 3.5y + 4.9z = 10.1$$

**48.** $2.1x + 0.7y - 1.4z = -2.3$
$$3.5x - 4.2y - 4.9z = 3.3$$
$$1.1x + 2.2y - 3.3z = -10.2$$

**49.** $-0.2x + 0.3y + 0.4z - t = 4.5$
$$2.2x + 1.1y - 4.7z + 2t = 8.3$$
$$9.2y \quad - 1.3t = 0$$
$$3.4x \quad + 0.5z - 3.4t = 0.1$$

**50.** $1.2x - 0.3y + 0.4z - 2t = 4.5$
$$1.9x \quad - 0.5z - 3.4t = 0.2$$
$$12.1y \quad - 1.3t = 0$$
$$3x + 2y - 1.1z \quad = 9$$

## COMMUNICATION AND REASONING EXERCISES

**51.** What is meant by a pivot? What does pivoting do?

**52.** Give instructions to check whether or not a matrix is row-reduced.

**53.** You are row-reducing a matrix and have chosen a $-6$ as a pivot in Row 4. Directly above the pivot, in Row 1, is a 15. What row operation can you use to clear the 15?

**54.** You are row-reducing a matrix and have chosen a $-4$ as a pivot in Row 2. Directly below the pivot, in Row 4, is a $-6$. What row operation can you use to clear the $-6$?

**55.** In the matrix of a system of linear equations, suppose that two of the rows are equal. What can you say about the row-reduced form of the matrix?

**56.** In the matrix of a system of linear equations, suppose that one of the rows is a multiple of another. What can you say about the row-reduced form of the matrix?

**57.** ▼ Your friend Frans tells you that the system of linear equations you are solving cannot have a unique solution because the reduced matrix has a row of zeros. Comment on his claim.

**58.** ▼ Your other friend Hans tells you that because he is solving a consistent system of five linear equations in six unknowns, he will get infinitely many solutions. Comment on his claim.

**59.** ▼ If the reduced matrix of a consistent system of linear equations has five rows, three of which are zero, and five columns, how many parameters does the general solution contain?

**60.** ▼ If the reduced matrix of a consistent system of linear equations has five rows, two of which are zero, and seven columns, how many parameters does the general solution contain?

**61.** ▼ Suppose a system of equations has a unique solution. What must be true of the number of pivots in the reduced matrix of the system? Why?

**62.** ▼ Suppose a system has infinitely many solutions. What must be true of the number of pivots in the reduced matrix of the system? Why?

**63.** ▼ Give an example of a system of three linear equations with the general solution $x = 1$, $y = 1 + z$, $z$ arbitrary. (Check your system by solving it.)

**64.** ▼ Give an example of a system of three linear equations with the general solution $x = y - 1$, $y$ arbitrary, $z = y$. (Check your system by solving it.)

**65.** ◆ A system of linear equations is called **homogeneous** if the right-hand side of each equation in the system is 0. If a homogeneous system has a unique solution, what can you say about that solution? Why?

**66.** ◆ A certain homogeneous system of linear equations in three unknowns (see the preceding exercise) has a solution $(1, 2, 0)$. Is this solution unique? Explain.

**67.** ◆ Can a homogeneous system (see Exercise 65) of linear equations be inconsistent? Explain.

**68.** ◆ Can a non-homogeneous system (see Exercise 65) of linear equations have the zero solution? Explain.

# Applications of Systems of Linear Equations

In the examples and the exercises of this section, we consider scenarios that lead to systems of linear equations in three or more unknowns. Some of these applications will strike you as a little idealized or even contrived compared with the kinds of problems you might encounter in the real world.* One reason is that we will not have tools to handle more realistic versions of these applications until we have studied linear programming in Chapter 5.

**✳ See the discussion at the end of the first example below.**

In each example that follows, we set up the problem as a linear system and then give the solution; the emphasis in this section is on modeling a scenario by a system of linear equations and then interpreting the solution that results, rather than on obtaining the solution. For practice, you should do the row reduction necessary to get the solution.

### EXAMPLE 1  Resource Allocation

The Arctic Juice Company makes three juice blends: PineOrange, using 2 quarts of pineapple juice and 2 quarts of orange juice per gallon; PineKiwi, using 3 quarts of pineapple juice and 1 quart of kiwi juice per gallon; and OrangeKiwi, using 3 quarts of orange juice and 1 quart of kiwi juice per gallon. Each day the company has 800 quarts of pineapple juice, 650 quarts of orange juice, and 350 quarts of kiwi juice available. How many gallons of each blend should it make each day if it wants to use up all of the supplies?

**Solution**  We take the same steps to understand the problem that we took in Section 3.1. The first step is to identify and label the unknowns. Looking at the question asked in the last sentence, we see that we should label the unknowns like this:

$x$ = number of gallons of PineOrange made each day

$y$ = number of gallons of PineKiwi made each day

$z$ = number of gallons of OrangeKiwi made each day.

Next, we can organize the information we are given in a table:

|  | *PineOrange (x)* | *PineKiwi (y)* | *OrangeKiwi (z)* | *Total Available* |
|---|---|---|---|---|
| **Pineapple Juice (qt)** | 2 | 3 | 0 | 800 |
| **Orange Juice (qt)** | 2 | 0 | 3 | 650 |
| **Kiwi Juice (qt)** | 0 | 1 | 1 | 350 |

Notice how we have arranged the table; we have placed headings corresponding to the unknowns along the top, rather than down the side, and we have added a heading for the available totals. This gives us a table that is essentially the matrix of the system of linear equations we are looking for. (However, read the caution in the **Before we go on** section.)

| x1 | x2 | x3 | x4 |
|----|----|----|----|
| 2  | 3  | 0  | 800 |
| 2  | 0  | 3  | 650 |
| 0  | 1  | 1  | 350 |

| | | | |
|----|----|----|----|
| 1  | 0  | 0  | 100 |
| 0  | 1  | 0  | 200 |
| 0  | 0  | 1  | 150 |

Now we read across each row of the table. The fact that we want to use exactly the amount of each juice that is available leads to the following three equations:

$$2x + 3y \qquad = 800$$
$$2x \qquad + 3z = 650$$
$$y + z = 350.$$

The solution of this system is $(x, y, z) = (100, 200, 150)$, so Arctic Juice should make 100 gallons of PineOrange, 200 gallons of PineKiwi, and 150 gallons of OrangeKiwi each day.

➡ **Before we go on...**

*Caution*

We do not recommend relying on the coincidence that the table we created to organize the information in Example 1 happened to be the matrix of the system; it is too easy to set up the table "sideways" and get the wrong matrix. You should always write down the system of equations *and be sure you understand each equation*. For example, the equation $2x + 3y = 800$ in Example 1 indicates that the number of quarts of pineapple juice that will be used ($2x + 3y$) is equal to the amount available (800 quarts). By thinking of the reason for each equation, you can check that you have the correct system. If you have the wrong system of equations to begin with, solving it won't help you.

**Q**: *Just how realistic is the scenario in Example 1?*

**A**: This is a very unrealistic scenario, for several reasons:

1. Isn't it odd that we happened to end up with exactly the same number of equations as unknowns? Real scenarios are rarely so considerate. If there had been four equations, there would in all likelihood have been no solution at all. However, we need to understand these idealized problems before we can tackle the real world.

2. Even if a real-world scenario does give the same number of equations as unknowns, there is still no guarantee that there will be a unique solution consisting of positive values. What, for instance, would we have done in this example if $x$ had turned out to be negative?

3. The requirement that we use exactly all the ingredients would be an unreasonable constraint in real life. When we discuss linear programming, we will be able to substitute the more reasonable constraint that you use no more than is available, and we will add the more reasonable objective that you maximize profit. ∎

### EXAMPLE 2  Aircraft Purchases: Airbus **and** Boeing

A new airline has recently purchased a fleet of Airbus A330-300s, Boeing 767-200ERs, and Boeing Dreamliner 787-9s to meet an estimated demand for 4,800 seats. The A330-300s seat 320 passengers and cost \$200 million each, the 767-200ERs each seat 250 passengers and cost \$125 million each, while the Dreamliner 787-9s seat 275 passengers and cost \$200 million each.[12] The total cost of the fleet, which had twice as many Dreamliners as 767s, was \$3,100 million. How many of each type of aircraft did the company purchase?

---

[12]The prices are approximate 2008 prices. Prices and seating capacities found at the companies' Web sites and in Wikipedia.

**Solution**  We label the unknowns as follows:

$$x = \text{number of Airbus A330-300s}$$
$$y = \text{number of Boeing 767-200ERs}$$
$$z = \text{number of Boeing Dreamliner 787-9s.}$$

We must now set up the equations. We can organize some (but not all) of the given information in a table:

|  | A330-300 | 767-200ER | 787-9 Dreamliner | Total |
|---|---|---|---|---|
| **Capacity** | 320 | 250 | 275 | 4,800 |
| **Cost ($ million)** | 200 | 125 | 200 | 3,100 |

Reading across, we get the equations expressing the facts that the airline needed to seat 4,800 passengers and that it spent $3,100 million:

$$320x + 250y + 275z = 4,800$$
$$200x + 125y + 200z = 3,100.$$

There is an additional piece of information we have not yet used: the airline bought twice as many Dreamliners as 767s. As we said in Section 3.1, it is easiest to translate a statement like this into an equation if we first reword it using the phrase "the number of." Thus, we say: "The number of Dreamliners ordered was twice the number of 767s ordered," or

$$z = 2y$$
$$2y - z = 0.$$

We now have a system of three equations in three unknowns:

$$320x + 250y + 275z = 4,800$$
$$200x + 125y + 200z = 3,100$$
$$2y - \quad z = 0.$$

Solving the system, we get the solution $(x, y, z) = (5, 4, 8)$. Thus, the airline ordered five A330-300s, four 767-200ERs, and eight Dreamliner 787-9s.

---

### EXAMPLE 3  Traffic Flow

Traffic through downtown Urbanville flows through the one-way system shown in Figure 7.

Traffic counting devices installed in the road (shown as boxes) count 200 cars entering town from the west each hour, 150 leaving town on the north each hour, and 50 leaving town on the south each hour.

**Figure 7**

**a.** From this information, is it possible to determine how many cars drive along Allen, Baker, and Coal streets every hour?

**b.** What is the maximum possible traffic flow along Baker Street?

**c.** What is the minimum possible traffic along Allen Street?

**d.** What is the maximum possible traffic flow along Coal Street?

## Solution

**a.** Our unknowns are:

$x =$ number of cars per hour on Allen Street

$y =$ number of cars per hour on Baker Street

$z =$ number of cars per hour on Coal Street.

Assuming that, at each intersection, cars do not fall into a pit or materialize out of thin air, the number of cars entering each intersection has to equal the number exiting. For example, at the intersection of Allen and Baker Streets there are 200 cars entering and $x + y$ cars exiting:

$$\text{Traffic in} = \text{Traffic out}$$
$$200 = x + y.$$

At the intersection of Allen and Coal Streets, we get:

$$\text{Traffic in} = \text{Traffic out}$$
$$x = z + 150$$

and at the intersection of Baker and Coal Streets, we get:

$$\text{Traffic in} = \text{Traffic out}$$
$$y + z = 50.$$

We now have the following system of equations:

$$x + y \qquad = 200$$
$$x \qquad - z = 150$$
$$y + z = 50.$$

If we solve this system using the methods of the preceding section, we find that it has infinitely many solutions. The general solution is:

$$x = z + 150$$
$$y = -z + 50$$
$$z \text{ is arbitrary.}$$

Because we do not have a unique solution, it is *not* possible to determine how many cars drive along Allen, Baker and Coal Streets every hour.

**b.** The traffic flow along Baker Street is measured by $y$. From the general solution,

$$y = -z + 50$$

where $z$ is arbitrary. How arbitrary is $z$? It makes no sense for any of the variables $x$, $y$, or $z$ to be negative in this scenario, so $z \geq 0$. Therefore, the largest possible value $y$ can have is

$$y = -0 + 50 = 50 \text{ cars per hour.}$$

**c.** The traffic flow along Allen Street is measured by $x$. From the general solution,

$$x = z + 150,$$

where $z \geq 0$, as we saw in part (b). Therefore, the smallest possible value $x$ can have is

$$x = 0 + 150 = 150 \text{ cars per hour.}$$

**d.** The traffic flow along Coal Street is measured by $z$. Referring to the general solution, we see that $z$ shows up in the expressions for both $x$ and $y$:

$$x = z + 150$$
$$y = -z + 50.$$

In the first of these equations, there is nothing preventing $z$ from being as big as we like; the larger we make $z$, the larger $x$ becomes. However, the second equation places a limit on how large $z$ can be: If $z > 50$, then $y$ is negative, which is impossible. Therefore, the largest value $z$ can take is 50 cars per hour.

From the discussion above, we see that $z$ is not completely arbitrary: We must have $z \geq 0$ and $z \leq 50$. Thus, $z$ has to satisfy $0 \leq z \leq 50$ for us to get a realistic answer.

---

➡ **Before we go on...**  Here are some questions to think about in Example 3: If you wanted to nail down $x$, $y$, and $z$ to see where the cars are really going, how would you do it with only one more traffic counter? Would it make sense for $z$ to be fractional? What if you interpreted $x$, $y$, and $z$ as *average* numbers of cars per hour over a long period of time?

Traffic flow is only one kind of flow in which we might be interested. Water and electricity flows are others. In each case, to analyze the flow, we use the fact that the amount entering an intersection must equal the amount leaving it.  ∎

---

### EXAMPLE 4  **Transportation**

A car rental company has four locations in the city: Southwest, Northeast, Southeast, and Northwest. The Northwest location has 20 more cars than it needs, and the Northeast location has 15 more cars than it needs. The Southwest location needs 10 more cars than it has, and the Southeast location needs 25 more cars than it has. It costs \$10 (in salary and gas) to have an employee drive a car from Northwest to Southwest. It costs \$20 to drive a car from Northwest to Southeast. It costs \$5 to drive a car from Northeast to Southwest, and it costs \$10 to drive a car from Northeast to Southeast. If the company will spend a total of \$475 rearranging its cars, how many cars will it drive from each of Northwest and Northeast to each of Southwest and Southeast?

### Solution

Figure 8 shows a diagram of this situation. Each arrow represents a route along which the rental company can drive cars. At each location is written the number of extra cars the location has or the number it needs. Along each route is written the cost of driving a car along that route.

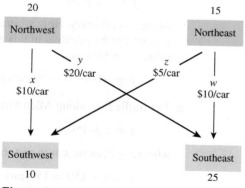

**Figure 8**

The unknowns are the number of cars the company will drive along each route, so we have the following four unknowns, as indicated in the figure:

$x$ = number of cars driven from Northwest to Southwest
$y$ = number of cars driven from Northwest to Southeast
$z$ = number of cars driven from Northeast to Southwest
$w$ = number of cars driven from Northeast to Southeast.

Consider the Northwest location. It has 20 more cars than it needs, so the total number of cars being driven out of Northwest should be 20. This gives us the equation

$$x + y = 20.$$

Similarly, the total number of cars being driven out of Northeast should be 15, so

$$z + w = 15.$$

Considering the number of cars needed at the Southwest and Southeast locations, we get the following two equations as well:

$$x + z = 10$$
$$y + w = 25.$$

There is one more equation that we should write down, the equation that says that the company will spend $475:

$$10x + 20y + 5z + 10w = 475.$$

Thus, we have the following system of five equations in four unknowns:

$$
\begin{aligned}
x + y &= 20 \\
z + w &= 15 \\
x \quad + z \quad &= 10 \\
y \quad + w &= 25 \\
10x + 20y + 5z + 10w &= 475
\end{aligned}
$$

Solving this system, we find that $(x, y, z, w) = (5, 15, 5, 10)$. In words, the company will drive 5 cars from Northwest to Southwest, 15 from Northwest to Southeast, 5 from Northeast to Southwest, and 10 from Northeast to Southeast.

➡ **Before we go on...** A very reasonable question to ask in Example 4 is, Can the company rearrange its cars for less than $475? Even better, what is the least possible cost? In general, a question asking for the optimal cost may require the techniques of linear programming, which we will discuss in Chapter 5. However, in this case we can approach the problem directly. If we remove the equation that says that the total cost is $475 and solve the system consisting of the other four equations, we find that there are infinitely many solutions and that the general solution may be written as

$$x = w - 5$$
$$y = 25 - w$$
$$z = 15 - w$$
$$w \text{ is arbitrary.}$$

This allows us to write the total cost as a function of $w$:

$$\text{Cost} = 10x + 20y + 5z + 10w = 10(w - 5) + 20(25 - w) + 5(15 - w) + 10w$$
$$= 525 - 5w.$$

So, the larger we make $w$, the smaller the total cost will be. The largest we can make $w$ is 15 (why?), and if we do so we get $(x, y, z, w) = (10, 10, 0, 15)$ and a total cost of \$450. ■

## 3.3    EXERCISES

▼ more advanced    ◆ challenging
⊤ indicates exercises that should be solved using technology

### APPLICATIONS

**1. Resource Allocation** You manage an ice cream factory that makes three flavors: Creamy Vanilla, Continental Mocha, and Succulent Strawberry. Into each batch of Creamy Vanilla go 2 eggs, 1 cup of milk, and 2 cups of cream. Into each batch of Continental Mocha go 1 egg, 1 cup of milk, and 2 cups of cream, while into each batch of Succulent Strawberry go 1 egg, 2 cups of milk, and 1 cup of cream. You have in stock 350 eggs, 350 cups of milk, and 400 cups of cream. How many batches of each flavor should you make in order to use up all of your ingredients? HINT [See Example 1.]

**2. Resource Allocation** You own a hamburger franchise and are planning to shut down operations for the day, but you are left with 13 bread rolls, 19 defrosted beef patties, and 15 opened cheese slices. Rather than throw them out, you decide to use them to make burgers that you will sell at a discount. Plain burgers each require 1 beef patty and 1 bread roll, double cheeseburgers each require 2 beef patties, 1 bread roll, and 2 slices of cheese, while regular cheeseburgers each require 1 beef patty, 1 bread roll, and 1 slice of cheese. How many of each should you make? HINT [See Example 1.]

**3. Resource Allocation** Urban Community College is planning to offer courses in Finite Math, Applied Calculus, and Computer Methods. Each section of Finite Math has 40 students and earns the college \$40,000 in revenue. Each section of Applied Calculus has 40 students and earns the college \$60,000, while each section of Computer Methods has 10 students and earns the college \$20,000. Assuming the college wishes to offer a total of six sections, to accommodate 210 students, and to bring in \$260,000 in revenues, how many sections of each course should it offer? HINT [See Example 2.]

**4. Resource Allocation** The Enormous State University History Department offers three courses, Ancient, Medieval, and Modern History, and the chairperson is trying to decide how many sections of each to offer this semester. The department is allowed to offer 45 sections total, there are 5,000 students who would like to take a course, and there are 60 professors

to teach them. Sections of Ancient History have 100 students each, sections of Medieval History hold 50 students each, and sections of Modern History have 200 students each. Modern History sections are taught by a team of 2 professors, while Ancient and Medieval History need only 1 professor per section. How many sections of each course should the chair schedule in order to offer all the sections that they are allowed to, accommodate all of the students, and give one teaching assignment to each professor? HINT [See Example 2.]

**5. Latin Music Sales (Digital)** In 2010, total revenues from digital sales of regional (Mexican/Tejano), pop/rock, and tropical (salsa/merengue/cumbia/bachata) Latin music in the U.S. amounted to \$46 million. Regional music brought in fives times as much as tropical music and \$9 million more than pop/rock music.[13] How much revenue was earned from digital sales in each of the three categories?

**6. Latin Music Sales (Digital)** In 2009, total revenues from digital sales of pop/rock, tropical (salsa/merengue/cumbia/bachata), and urban (reggaeton) Latin music in the U.S. amounted to \$28 million. Tropical music brought in twice as much as urban music, and pop/rock music brought in \$10 million more than the others combined.[14] How much revenue was earned from digital sales in each of the three categories?

**7. Purchasing Aircraft** In Example 2 we saw that Airbus A330-300s seat 320 passengers and cost \$200 million each, Boeing 767-200ERs seat 250 passengers and cost \$125 million each, while Boeing Dreamliner 787-9s seat 275 passengers and cost \$200 million each. You are the purchasing manager of an airline company and have a spending goal of \$2,900 million for the purchase of new aircraft to seat a total of 4,480 passengers. Your company has a policy of supporting U.S. industries, and you have been instructed to buy twice as many Dreamliners as Airbus A330-300s. Given the selection of three aircraft, how many of each should you order?

**8. Purchasing Aircraft** Refer to Exercise 7. René DuFleur has just been appointed the new CEO of your airline, and you

[13]Revenues are based on suggested list prices, and rounded to the nearest \$1 million. Source: Recording Industry Association of America http://riaa.org
[14]Ibid.

have received instructions that the company policy is now to purchase 25% more Airbus A330-300s than Boeing 767-200ERs (that is, 1.25 times as many Airbus A330-300s as 767-200ERs). Further, the desired seating capacity has been revised downward to 2,600 passengers, and the cost target to $1,500 million. Given the selection of three aircraft, how many of each should you order?

9. *Supply* A bagel store orders cream cheese from three suppliers, Cheesy Cream Corp. (CCC), Super Smooth & Sons (SSS), and Bagel's Best Friend Co. (BBF). One month, the total order of cheese came to 100 tons (they do a booming trade). The costs were $80, $50, and $65 per ton from the three suppliers, respectively, with total cost amounting to $5,990. Given that the store ordered the same amount from CCC and BBF, how many tons of cream cheese were ordered from each supplier?

10. *Supply* Refer to Exercise 9. The bagel store's outlay for cream cheese the following month was $2,310, when it purchased a total of 36 tons. Two more tons of cream cheese came from Bagel's Best Friend Co. than from Super Smooth & Sons. How many tons of cream cheese came from each supplier?

11. *Pest Control* Aragorn the Great has boasted to his hordes of followers that many a notorious villain has fallen to his awesome sword: His total of 560 victims consists of evil sorcerers, trolls, and orcs. These he has slain with a total of 620 mighty thrusts of his sword; evil sorcerers and trolls each requiring two thrusts (to the chest) and orcs each requiring one thrust (to the neck). When asked about the number of trolls he has slain, he replies, "I, the mighty Aragorn, despise trolls five times as much as I despise evil sorcerers. Accordingly, five times as many trolls as evil sorcerers have fallen to my sword!" How many of each type of villain has he slain?

12. *Manufacturing Perfume* The Fancy French Perfume Company recently had its secret formula divulged. It turned out that it was using, as the three ingredients, rose oil, oil of fermented prunes, and alcohol. Moreover, each 22-ounce econo-size bottle contained 4 more ounces of alcohol than oil of fermented prunes, while the amount of alcohol was equal to the combined volume of the other two ingredients. How much of each ingredient did it use in an econo-size bottle?[15]
HINT [The answer is the brand-name of a famous eau de cologne.]

13. ▽ *Donations* The Enormous State University Good Works Society recently raised funds for three worthwhile causes: the Math Professors' Benevolent Fund (MPBF), the Society of Computer Nerds (SCN), and the NY Jets. Because the society's members are closet jocks, the society donated twice as much to the NY Jets as to the MPBF, and equal amounts to the first two funds (it is unable to distinguish between mathematicians and nerds). Further, for every $1 it gave to the MPBF, it decided to keep $1 for itself; for every $1 it gave to the SCN, it kept $2, and for every $1 to the Jets, it also kept $2. The treasurer of the Society, Johnny Treasure, was required to itemize all donations for the Dean of Students, but discovered to his consternation that he had lost

the receipts! The only information available to him was that the society's bank account had swelled by $4,200. How much did the society donate to each cause?

14. ▽ *Tenure* Professor Walt is up for tenure, and wishes to submit a portfolio of written student evaluations as evidence of his good teaching. He begins by grouping all the evaluations into four categories: good reviews, bad reviews (a typical one being "GET RID OF WALT! THE MAN CAN'T TEACH!"), mediocre reviews (such as "I suppose he's OK, given the general quality of teaching at this college"), and reviews left blank. When he tallies up the piles, Walt gets a little worried: There are 280 more bad reviews than good ones and only half as many blank reviews as bad ones. The good reviews and blank reviews together total 170. On an impulse, he decides to even up the piles a little by removing 280 of the bad reviews, and this leaves him with a total of 400 reviews of all types. How many of each category of reviews were there originally?

◫ *Airline Costs* Exercises 15 and 16 are based on the following table, which shows the amount spent by four U.S. airlines to fly one available seat one mile in the second quarter of 2011.[16] Set up each system and then solve using technology.
HINT [See the technology note accompanying Example 1.]

| Airline | American | Continental | Delta | Southwest |
|---|---|---|---|---|
| Cost | 13.1¢ | 13.0¢ | 14.4¢ | 12.5¢ |

15. ▽ Suppose that, on a 3,000-mile New York–Los Angeles flight, Continental, American, and Southwest flew a total of 210 empty seats, costing them a total of $81,270. If Continental had three times as many empty seats as American, how many empty seats did each of these three airlines carry on its flight?

16. ▽ Suppose that, on a 2,000-mile Miami–Memphis flight, Continental, Delta, and Southwest flew a total of 200 empty seats, costing them a total of $54,300. If Delta had twice as many empty seats as Southwest, how many empty seats did each of these three airlines carry on its flight?

*Investing: Inverse Mutual Funds* Inverse mutual funds, sometimes referred to as "bear market" or "short" funds, seek to deliver the opposite of the performance of the index or category they track, and can thus be used by traders to bet against the stock market. Exercises 17 and 18 are based on the following table, which shows the performance of three such funds as of August 12, 2011.[17]

| | Year-to-date Loss |
|---|---|
| SHPIX (Short Smallcap Profund) | 6% |
| RYURX (Rydex Inverse S&P 500) | 5% |
| RYIHX (Rydex Inverse High Yield) | 7% |

[15] Most perfumes consist of 10 to 20% perfume oils dissolved in alcohol. This may or may not be reflected in this company's formula.

[16] Costs are rounded to the nearest 0.1¢. Source: Company financial statements, obtained from their respective Web sites. The cost per available seat-mile (CASM) is a widely used operating statistic in the airline industry.

[17] Based on prices at the close of the stock market on August 12, 2011. YTD losses rounded to the nearest percentage point. Source: www.fidelity.com.

**17.** You invested a total of $9,000 in the three funds at the beginning of 2011, including an equal amount in RYURX and RYIHX. Your year-to-date loss from the first two funds amounted to $400. How much did you invest in each of the three funds?

**18.** You invested a total of $6,000 in the three funds at the beginning of 2011, including an equal amount in SHPIX and RYURX. Your total year-to-date loss amounted to $360. How much did you invest in each of the three funds?

*Investing: Lesser-Known Stocks Exercises 19 and 20 are based on the following information about the stocks of* KT Corporation, International Flavors & Fragrances, *and* Quality Systems.[18]

|  | *Price* | *Dividend Yield* |
|---|---|---|
| **KT (KT Corporation)** | $16 | 7% |
| **IFF (International Flavors & Fragrances)** | 56 | 2 |
| **QSII (Quality Systems)** | 80 | 2 |

**19.** ▼ You invested a total of $8,400 in shares of the three stocks at the given prices, and expected to earn $248 in annual dividends. If you purchased a total of 200 shares, how many shares of each stock did you purchase?

**20.** ▼ You invested a total of $11,200 in shares of the three stocks at the given prices, and expected to earn $304 in annual dividends. If you purchased a total of 250 shares, how many shares of each stock did you purchase?

**21.** ▮ *Internet Audience* At the end of 2003, the four companies with the largest number of home Internet users in the United States were Microsoft, Time Warner, Yahoo, and Google, with a combined audience of 284 million users.[19] Taking $x$ to be the Microsoft audience in millions, $y$ the Time Warner audience in millions, $z$ the Yahoo audience in millions, and $u$ the Google audience in millions, it was observed that

$$z - u = 3(x - y) + 6$$
$$x + y = 50 + z + u$$

and  $x - y + z - u = 42.$

How large was the audience of each of the four companies in November 2003?

**22.** ▮ *Internet Audience* At the end of 2003, the four organizations ranking 5 through 8 in home Internet users in the United States were eBay, the U.S. government, Amazon, and Lycos, with a combined audience of 112 million users.[20] Taking $x$ to be the eBay audience in millions, $y$ the U.S. Government audience in millions, $z$ the Amazon audience in millions, and $u$ the Lycos audience in millions, it was observed that

$$y - z = z - u$$
$$x - y = 3(y - u) + 5$$

and  $x - y + z - u = 12.$

How large was the audience of each of the four companies in November 2003?

**23.** ▼ *Market Share: Homeowners Insurance* The three market leaders in homeowners insurance in Missouri are State Farm, American Family Insurance Group, and Allstate. Based on data from 2007, two relationships between the Missouri homeowners insurance percentage market shares are found to be

$$x = 1 + y + z$$
$$z = 16 - 0.2w,$$

where $x$, $y$, $z$, and $w$ are, respectively, the percentages of the market held by State Farm, American Family, Allstate, and other companies.[21] Given that the four groups account for the entire market, obtain a third equation relating $x$, $y$, $z$, and $w$, and solve the associated system of three linear equations to show how the market shares of State Farm, American Family, and Allstate depend on the share held by other companies. Which of the three companies' market share is most impacted by the share held by other companies?

**24.** ▼ *Market Share: Auto Insurance* Repeat the preceding exercise using the following relationships among the auto insurance percentage market shares:

$$x = -40 + 5y + z$$
$$z = 3 - 2y + w.$$

**25.** *Inventory Control* Red Bookstore wants to ship books from its warehouses in Brooklyn and Queens to its stores, one on Long Island and one in Manhattan. Its warehouse in Brooklyn has 1,000 books and its warehouse in Queens has 2,000. Each store orders 1,500 books. It costs $5 to ship each book from Brooklyn to Long Island and $1 to ship each book from Brooklyn to Manhattan. It costs $4 to ship each book from Queens to Long Island and $2 to ship each book from Queens to Manhattan.

**a.** If Red has a transportation budget of $9,000 and is willing to spend all of it, how many books should Red ship from each warehouse to each store in order to fill all the orders?

**b.** Is there a way of doing this for less money?
HINT [See Example 4.]

**26.** *Inventory Control* The Tubular Ride Boogie Board Company has manufacturing plants in Tucson, AZ and Toronto, Ontario. You have been given the job of coordinating distribution of the latest model, the Gladiator, to outlets in Honolulu and Venice Beach. The Tucson plant, when operating at full capacity, can manufacture 620 Gladiator boards per week, while the

[18]Yields rounded to the nearest percentage point and stock prices at the close of the stock market on August 12, 2011 rounded to the nearest $1. Source: www.google.com/finance.

[19]Source: Nielsen/NetRatings www.nielsen-netratings.com/ January 1, 2004.

[20]*Ibid.*

[21]Source: Missouri Dept. of Insurance (www.insurance.mo.gov).

Toronto plant, beset by labor disputes, can produce only 410 boards per week. The outlet in Honolulu orders 500 Gladiator boards per week, while Venice Beach orders 530 boards per week. Transportation costs are as follows:

Tucson to Honolulu: $10 per board; Tucson to Venice Beach: $5 per board.

Toronto to Honolulu: $20 per board; Toronto to Venice Beach: $10 per board.

a. Assuming that you wish to fill all orders and assure full capacity production at both plants, is it possible to meet a total transportation budget of $10,200? If so, how many Gladiator boards are shipped from each manufacturing plant to each distribution outlet?

b. Is there a way of doing this for less money? HINT [See Example 4.]

27. ▼ *Tourism in the 1990s* In the 1990s, significant numbers of tourists traveled from North America and Europe to Australia and South Africa. In 1998, a total of 1,390,000 of these tourists visited Australia, while 1,140,000 of them visited South Africa. Further, 630,000 of them came from North America and 1,900,000 of them came from Europe.[22] (Assume no single tourist visited both destinations or traveled from both North America and Europe.)

a. The given information is not sufficient to determine the number of tourists from each region to each destination. Why?

b. If you were given the additional information that a total of 2,530,000 tourists traveled from these two regions to these two destinations, would you now be able to determine the number of tourists from each region to each destination? If so, what are these numbers?

c. If you were given the additional information that the same number of people from Europe visited South Africa as visited Australia, would you now be able to determine the number of tourists from each region to each destination? If so, what are these numbers?

28. ▼ *Tourism in the 1990s* In the 1990s, significant numbers of tourists traveled from North America and Asia to Australia and South Africa. In 1998, a total of 2,230,000 of these tourists visited Australia, while 390,000 of them visited South Africa. Also, 630,000 of these tourists came from North America, and a total of 2,620,000 tourists traveled from these two regions to these two destinations.[23] (Assume no single tourist visited both destinations or traveled from both North America and Asia.)

a. The given information is not sufficient to determine the number of tourists from each region to each destination. Why?

b. If you were given the additional information that a total of 1,990,000 tourists came from Asia, would you now be

able to determine the number of tourists from each region to each destination? If so, what are these numbers?

c. If you were given the additional information that 200,000 tourists visited South Africa from Asia, would you now be able to determine the number of tourists from each region to each destination? If so, what are these numbers?

29. ▼ *Alcohol* The following table shows some data from a 2000 study on substance use among 10th graders in the United States and Europe.[24]

|  | *Used Alcohol* | *Alcohol-Free* | *Totals* |
|---|---|---|---|
| **U.S.** | $x$ | $y$ | 14,000 |
| **Europe** | $z$ | $w$ | 95,000 |
| **Totals** | 63,550 | 45,450 | |

a. The table leads to a linear system of four equations in four unknowns. What is the system? Does it have a unique solution? What does this indicate about the given and the missing data?

b. ⬜ Given that the number of U.S. 10th graders who were alcohol-free was 50% more than the number who had used alcohol, find the missing data.

30. ▼ *Tobacco* The following table shows some data from the same study cited in Exercise 30.[25]

|  | *Smoked Cigarettes* | *Cigarette-Free* | *Totals* |
|---|---|---|---|
| **U.S.** | $x$ | $y$ | 14,000 |
| **Europe** | $z$ | $w$ | 95,000 |
| **Totals** |  | 70,210 | 109,000 |

a. The table leads to a linear system of four equations in four unknowns. What is the system? Does it have a unique solution? What does this indicate about the missing data?

b. ⬜ Given that 31,510 more European 10th graders smoked cigarettes than U.S. 10th graders, find the missing data.

31. *Traffic Flow* One-way traffic through Enormous State University is shown in the figure, where the numbers indicate daily counts of vehicles.

[22]Figures are rounded to the nearest 10,000. Sources: South African Dept. of Environmental Affairs and Tourism; Australia Tourist Commission/*New York Times*, January 15, 2000, p. C1.

[23]*Ibid.*

[24]"Used Alcohol" indicates consumption of alcohol at least once in the past 30 days. Source: Council of Europe/University of Michigan. "Monitoring the Future"/*New York Times*, February 21, 2001, p. A10.

[25]"Smoked Cigarettes" indicates that at least one cigarette was smoked in the past 30 days. Source: *Ibid.*

a. Is it possible to determine the daily flow of traffic along each of the three streets from the information given? If your answer is *yes*, what is the traffic flow along each street? If your answer is *no*, what additional information would suffice?

b. Is a flow of 60 vehicles per day along Southwest Lane consistent with the information given?

c. What is the minimum traffic flow possible along Northwest Lane consistent with the information given? HINT [See Example 3.]

**32. *Traffic Flow*** The traffic through downtown East Podunk flows through the one-way system shown below.

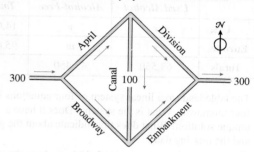

Traffic counters find that 300 vehicles enter town from the west each hour, and 300 leave town toward the east each hour. Also, 100 cars drive down Canal Street each hour.

a. Write down the general solution of the associated system of linear equations. Is it possible to determine the number of vehicles on each street per hour?

b. On which street could you put another traffic counter in order to determine the flow completely?

c. What is the minimum traffic flow along April Street consistent with the information given? HINT [See Example 3.]

**33. *Traffic Flow*** The traffic through downtown Johannesburg follows the one-way system shown below, with traffic movement recorded at incoming and outgoing streets (in cars per minute) as shown.

a. Set up and solve a system of equations to solve for the traffic flow along the middle sections of the four streets.

b. Is there sufficient information to calculate the traffic along the middle section of Jeppe Street? If so, what is it; if not, why not?

c. Given that 400 cars per minute flow down the middle section of Bree Street, how many cars per minute flow down the middle section of Simmons?

d. What is the minimum traffic flow along the middle section of Harrison Street?

e. Is there an upper limit to the possible traffic down Simmons Street consistent with the information given? Explain.

**34. *Traffic Flow*** Town of Hempstead officials were planning to make Hempstead Turnpike and the surrounding streets into one-way streets to prepare for the opening of Hofstra USA. In an experiment, they restricted traffic flow along the streets near Hofstra as shown in the diagram. The numbers show traffic flow per minute and the arrows indicate the direction of traffic.

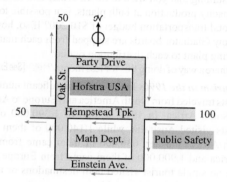

a. Set up and solve the associated traffic flow problem with the following unknowns:
$x$ = traffic per minute along the middle stretch of Oak St.,
$y$ = traffic per minute along the middle stretch of Hempstead Tpk.,
$z$ = traffic per minute along Einstein Ave., and
$u$ = traffic per minute along Party Drive.

b. If 20 cars per minute drive along Party Drive, what is the traffic like along the middle stretch of Oak Street?

c. If 20 vehicles per minute drive along Einstein Ave., and 90 vehicles per minute drive down the middle stretch of Hempstead Tpk., how many cars per minute drive along the middle stretch of Oak Street?

d. If Einstein Avenue is deserted, what is the minimum traffic along the middle stretch of Hempstead Tpk.?

**35. ▼ *Traffic Management*** The Outer Village Town Council has decided to convert its (rather quiet) main street, Broadway, to a one-way street, but is not sure of the direction of most of the traffic. The accompanying diagram illustrates the downtown area of Outer Village, as well as the *net* traffic flow along the intersecting streets (in vehicles per day). (There are no one-way streets; a net traffic flow in a certain direction is defined as the traffic flow in that direction minus the flow in the opposite direction.)

a. Is the given information sufficient to determine the net traffic flow along the three portions of Broadway shown? If your answer is *yes*, give the traffic flow along each stretch. If your answer is *no*, what additional information would suffice? HINT [For the direction of net traffic flow, choose either east or west. If a corresponding value is negative, it indicates net flow in the opposite direction.]

b. Assuming that there is little traffic (less than 160 vehicles per day) east of Fleet Street, in what direction is the net flow of traffic along the remaining stretches of Broadway?

**36.** ▼ *Electric Current* Electric current measures (in **amperes** or **amps**) the flow of electrons through wires. Like traffic flow, the current entering an intersection of wires must equal the current leaving it.[26] Here is an electrical circuit known as a **Wheatstone bridge.**

a. If the currents in two of the wires are 10 amps and 5 amps as shown, determine the currents in the unlabeled wires in terms of suitable parameters.

b. In which wire should you measure the current in order to know all of the currents exactly?

**37.** ▼ *Econometrics* (*from the GRE Economics Test*) This and the next exercise are based on the following simplified model of the determination of the money stock.

$$M = C + D$$
$$C = 0.2D$$
$$R = 0.1D$$
$$H = R + C$$

where

$M$ = money stock
$C$ = currency in circulation
$R$ = bank reserves
$D$ = deposits of the public
$H$ = high-powered money

If the money stock were $120 billion, what would bank reserves have to be?

**38.** ▼ *Econometrics* (*from the GRE Economics Test*) With the model in the previous exercise, if $H$ were equal to $42 billion, what would $M$ equal?

*CAT Scans* CAT (*Computerized Axial Tomographic*) scans are used to map the exact location of interior features of the human body. CAT scan technology is based on the following principles: (1) different components of the human body (water, gray matter, bone, etc.) absorb X-rays to different extents; and (2) to measure the X-ray absorption by a specific region of, say, the brain, it suffices to pass a

number of line-shaped pencil beams of X-rays through the brain at different angles and measure the total absorption for each beam, which is the sum of the absorptions of the regions through which it passes. The accompanying diagram illustrates a simple example. (The number in each region shows its absorption, and the number on each X-ray beam shows the total absorption for that beam.)[27]

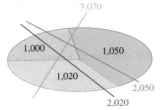

In Exercises 39–44, use the table and the given X-ray absorption diagrams to identify the composition of each of the regions marked by a letter.

| Type | Air | Water | Gray Matter | Tumor | Blood | Bone |
|---|---|---|---|---|---|---|
| Absorption | 0 | 1,000 | 1,020 | 1,030 | 1,050 | 2,000 |

**39.**

**40.**

**41.**

**42.**

---

[26]This is known as **Kirchhoff's current law,** named after Gustav Robert Kirchhoff (1824–1887). Kirchhoff made important contributions to the fields of geometric optics, electromagnetic radiation, and electrical network theory.

[27]Based on a COMAP article *Geometry, New Tools for New Technologies*, by J. Malkevitch, Video Applications Library, COMAP, 1992. The absorptions are actually calibrated on a logarithmic scale. In real applications, the size of the regions is very small, and very large numbers of beams must be used.

**43.** ▼ Identify the composition of site $x$.

(The horizontal and slanted beams each pass through five regions.)

**44.** ▼ Identify the composition of site $x$.

(The horizontal and slanted beams each pass through five regions.)

**45.** ◆ *Voting* In the 75th Congress (1937–1939) the U.S. House of Representatives had 333 Democrats, 89 Republicans, and 13 members of other parties. Suppose that a bill passed the House with 31 more votes in favor than against, with 10 times as many Democrats voting for the bill as Republicans, and with 36 more non-Democrats voting against the bill than for it. If every member voted either for the bill or against it, how many Democrats, how many Republicans, and how many members of other parties voted in favor of the bill?

**46.** ◆ *Voting* In the 75th Congress (1937–1939) there were in the Senate 75 Democrats, 17 Republicans, and 4 members of other parties. Suppose that a bill passed the Senate with 16 more votes in favor than against, with three times as many Democrats voting in favor as non-Democrats voting in favor, and 32 more Democrats voting in favor than Republicans voting in favor. If every member voted either for the bill or against it, how many Democrats, how many Republicans, and how many members of other parties voted in favor of the bill?

**47.** ◆ *Investments* Things have not been going too well here at Accurate Accounting, Inc. since we hired Todd Smiley. He has a tendency to lose important documents, especially around April, when tax returns of our business clients are due. Today Smiley accidentally shredded Colossal Conglomerate Corp.'s investment records. We must therefore reconstruct them based on the information he can gather. Todd recalls that the company earned an $8 million return on investments totaling $65 million last year. After a few frantic telephone calls to sources in Colossal, he learned that Colossal had made investments in four companies last year: X, Y, Z, and W. (For reasons of confidentiality, we are withholding their names.) Investments in company X earned 15% last year, investments in Y depreciated by 20% last year, investments in Z neither appreciated nor depreciated last year, while investments in W earned 20% last year. Smiley was also told that Colossal invested twice as much in company X as in company Z, and three times as much in company W as in company Z. Does Smiley have sufficient information to piece together Colossal's investment portfolio before its tax return is due next week? If so, what does the investment portfolio look like?

**48.** ◆ *Investments* Things are going from bad to worse here at Accurate Accounting, Inc.! Colossal Conglomerate Corp.'s tax return is due tomorrow and the accountant Todd Smiley seems to have no idea how Colossal earned a return of $8 million on a $65 million investment last year. It appears that, although the returns from companies X, Y, Z, and W were as listed in Exercise 47, the rest of the information there was wrong. What Smiley is now being told is that Colossal only invested in companies X, Y, and Z and that the investment in X amounted to $30 million. His sources in Colossal still maintain that twice as much was invested in company X as in company Z. What should Smiley do?

## COMMUNICATION AND REASONING EXERCISES

**49.** Are Exercises 1 and 2 realistic in their expectation of using up all the ingredients? What does your answer have to do with the solution(s) of the associated system of equations?

**50.** Suppose that you obtained a solution for Exercise 3 or 4 consisting of positive values that were not all whole numbers. What would such a solution signify about the situation in the exercise? Should you round these values to the nearest whole numbers?

*In Exercises 51–56, x, y and z represent the weights of the three ingredients X, Y, and Z in a gasoline blend. Say which of the following is represented by a linear equation in x, y, and z, and give a form of the equation when it is:*

**51.** The blend consists of 100 pounds of ingredient $X$.

**52.** The blend is free of ingredient $X$.

**53.** The blend contains 30% ingredient $Y$ by weight.

**54.** The weight of ingredient $X$ is the product of the weights of ingredients $Y$ and $Z$.

**55.** There is at least 30% ingredient $Y$ by weight.

**56.** There is twice as much ingredient $X$ by weight as $Y$ and $Z$ combined.

**57.** Make up an entertaining word problem leading to the following system of equations.

$$
\begin{aligned}
10x + 20y + 10z &= 100 \\
5x + 15y \phantom{{}+10z} &= 50 \\
x + y + z &= 10
\end{aligned}
$$

**58.** Make up an entertaining word problem leading to the following system of equations.

$$
\begin{aligned}
10x + 20y \phantom{{}+ 20w} &= 300 \\
10z + 20w &= 400 \\
20x \phantom{{}+20y} + 10z \phantom{{}+20w} &= 400 \\
10y \phantom{{}+10z} + 20w &= 300
\end{aligned}
$$

## KEY CONCEPTS

**WM** Website www.WanerMath.com
Go to the Website at www.WanerMath
.com to find a comprehensive and
interactive Web-based summary
of Chapter 3.

### 3.1 Systems of Two Equations in Two Unknowns

Linear equation in two unknowns *p. 176*
Coefficient *p. 176*
Solution of an equation in two
unknowns *p. 176*
Graphical method for solving a system
of two linear equations *p. 179*

Algebraic method for solving a system
of two linear equations *p. 179*
Redundant or dependent system *p. 180*
Possible outcomes for a system of two
linear equations *p. 181*
Consistent system *p. 181*

### 3.2 Using Matrices to Solve Systems of Equations

Linear equation (in any number of
unknowns) *p. 188*
Matrix *p. 189*

Augmented matrix of a system of linear
equations *p. 189*
Elementary row operations *p. 190*
Pivot *p. 192*
Clearing a column; pivoting *p. 192*
Gauss-Jordan or row reduction *p. 194*
Reduced row echelon form *p. 201*

### 3.3 Applications of Systems of Linear Equations

Resource allocation *p. 206*
(Traffic) flow *p. 208*
Transportation *p. 210*

## REVIEW EXERCISES

*In each of Exercises 1–6, graph the equations and determine
how many solutions the system has, if any.*

**1.** $x + 2y = 4$
$2x - y = 1$

**2.** $0.2x - 0.1y = 0.3$
$0.2x + 0.2y = 0.4$

**3.** $\frac{1}{2}x - \frac{3}{4}y = 0$
$6x - 9y = 0$

**4.** $2x + 3y = 2$
$-x - 3y/2 = 1/2$

**5.** $x + y = 1$
$2x + y = 0.3$
$3x + 2y = \frac{13}{10}$

**6.** $3x + 0.5y = 0.1$
$6x + y = 0.2$
$\frac{3x}{10} - 0.05y = 0.01$

*Solve each of the systems of linear equations in Exercises 7–18.*

**7.** $x + 2y = 4$
$2x - y = 1$

**8.** $0.2x - 0.1y = 0.3$
$0.2x + 0.2y = 0.4$

**9.** $\frac{1}{2}x - \frac{3}{4}y = 0$
$6x - 9y = 0$

**10.** $2x + 3y = 2$
$-x - 3y/2 = 1/2$

**11.** $x + y = 1$
$2x + y = 0.3$
$3x + 2y = \frac{13}{10}$

**12.** $3x + 0.5y = 0.1$
$6x + y = 0.2$
$\frac{3x}{10} - 0.05y = 0.01$

**13.** $x + 2y = -3$
$x - z = 0$
$x + 3y - 2z = -2$

**14.** $x - y + z = 2$
$7x + y - z = 6$
$x - \frac{1}{2}y + \frac{1}{3}z = 1$
$x + y + z = 6$

**15.** $x - \frac{1}{2}y + z = 0$
$\frac{1}{2}x - \frac{1}{2}z = -1$
$\frac{3}{2}x - \frac{1}{2}y + \frac{1}{2}z = -1$

**16.** $x + y - 2z = -1$
$-2x - 2y + 4z = 2$
$0.75x + 0.75y - 1.5z = -0.75$

**17.** $x = \frac{1}{2}y$
$\frac{1}{2}x = -\frac{1}{2}z + 2$
$z = -3x + y$

**18.** $x - y + z = 1$
$y - z + w = 1$
$x + z - w = 1$
$2x + z = 3$

*Exercises 19–22 are based on the following equation relating
the Fahrenheit and Celsius (or centigrade) temperature scales:*

$$5F - 9C = 160,$$

*where F is the Fahrenheit temperature of an object and C is its
Celsius temperature.*

**19.** What temperature should an object be if its Fahrenheit and
Celsius temperatures are the same?

**20.** What temperature should an object be if its Celsius temperature is half its Fahrenheit temperature?

**21.** Is it possible for the Fahrenheit temperature of an object to
be 1.8 times its Celsius temperature? Explain.

**22.** Is it possible for the Fahrenheit temperature of an object to be
30° more than 1.8 times its Celsius temperature? Explain.

*In Exercises 23–28, let x, y, z, and w represent the population
in millions of four cities A, B, C, and D, respectively. Express
the given statement as an equation in x, y, z, and w. If the
equation is linear, say so and express it in the standard form
$ax + by + cz + dw = k$.*

**23.** The total population of the four cities is 10 million people.

219

**24.** City $A$ has three times as many people as cities $B$ and $C$ combined.

**25.** City $D$ is actually a ghost town; there are no people living in it.

**26.** The population of City $A$ is the sum of the squares of the populations of the other three cities.

**27.** City $C$ has 30% more people than City $B$.

**28.** City $C$ has 30% fewer people than City $B$.

## APPLICATIONS: OHaganBooks.com

*Purchasing You are the buyer for OHaganBooks.com and are considering increasing stocks of romance and horror novels at the new OHaganBooks.com warehouse in Texas. You have offers from two publishers: Duffin House and Higgins Press. Duffin offers a package of 5 horror novels and 5 romance novels for $50, and Higgins offers a package of 5 horror and 11 romance novels for $15. Exercises 29–32 deal with your purchasing options given these offers.*

**29.** How many packages should you purchase from each publisher to get exactly 4,500 horror novels and 6,600 romance novels?

**30.** You want to spend a total of $50,000 on books and have promised to buy twice as many packages from Duffin as from Higgins. How many packages should you purchase from each publisher?

**31.** The accountant tells you that the company can actually afford to spend a total of $90,000 on romance and horror books. She also reminds you that you had signed an agreement to spend twice as much money for books from Duffin as from Higgins. How many packages should you purchase from each publisher?

**32.** Upon revising her records, the accountant now tells you that the company can afford to spend a total of only $60,000 on romance and horror books, and that it is company policy to spend the same amount of money at both publishers. How many packages should you purchase from each publisher?

**33.** *Equilibrium* The demand per year for *Finite Math the OHagan Way* is given by $q = -1,000p + 140,000$, where $p$ is the price per book in dollars. The supply is given by $q = 2,000p + 20,000$. Find the price at which supply and demand balance.

**34.** *Equilibrium* OHaganbooks.com CEO John O'Hagan announces to a stunned audience at the annual board meeting that he is considering expanding into the jumbo jet airline manufacturing business. The demand per year for jumbo jets is given by $q = -2p + 18$ where $p$ is the price per jet in millions of dollars. The supply is given by $q = 3p + 3$. Find the price the envisioned O'Hagan jumbo jet division should charge to balance supply and demand.

**35.** *Feeding Schedules* Billy-Sean O'Hagan is John O'Hagan's son and a freshman in college. Billy's 36-gallon tropical fish tank contains three types of carnivorous creatures: baby sharks, piranhas and squids, and he feeds them three types of delicacies: goldfish, angelfish and butterfly fish. Each baby shark can consume 1 goldfish, 2 angelfish, and 2 butterfly fish per day; each piranha can consume 1 goldfish and 3 butterfly fish per day (the piranhas are rather large as a result of their diet); while each squid can consume 1 goldfish and 1 angelfish per day. After a trip to the local pet store, he was able to feed his creatures to capacity, and noticed that 21 goldfish, 21 angelfish, and 35 butterfly fish were eaten. How many of each type of creature does he have?

**36.** *Resource Allocation* Duffin House is planning its annual Song Festival, when it will serve three kinds of delicacies: granola treats, nutty granola treats, and nuttiest granola treats. The following table shows the ingredients required (in ounces) for a single serving of each delicacy, as well as the total amount of each ingredient available.

|  | Granola | Nutty Granola | Nuttiest Granola | Total Available |
|---|---|---|---|---|
| **Toasted Oats** | 1 | 1 | 5 | 1,500 |
| **Almonds** | 4 | 8 | 8 | 10,000 |
| **Raisins** | 2 | 4 | 8 | 4,000 |

The Song Festival planners at Duffin House would like to use up all the ingredients. Is this possible? If so, how many servings of each kind of delicacy can they make?

**37.** *Web Site Traffic* OHaganBooks.com has two principal competitors: JungleBooks.com and FarmerBooks.com. Combined Web site traffic at the three sites is estimated at 10,000 hits per day. Only 10% of the hits at OHaganBooks.com result in orders, whereas JungleBooks.com and FarmerBooks.com report that 20% of the hits at their sites result in book orders. Together, the three sites process 1,500 book orders per day. FarmerBooks.com appears to be the most successful of the three, and gets as many book orders as the other two combined. What is the traffic (in hits per day) at each of the sites?

**38.** *Sales* As the buyer at OHaganBooks.com you are planning to increase stocks of books about music, and have been monitoring worldwide sales. Last year, worldwide sales of books about rock, rap, and classical music amounted to $5.8 billion. Books on rock music brought in twice as much revenue as books on rap music and they brought in 900% the revenue of books on classical music. How much revenue was earned in each of the three categories of books?

**39.** *Investing in Stocks* Billy-Sean O'Hagan is the treasurer at his college fraternity, which recently earned $12,400 in its annual carwash fundraiser. Billy-Sean decided to invest all the proceeds in the purchase of three computer stocks: HAL, POM, and WELL.

|  | Price per Share | Dividend Yield |
|---|---|---|
| **HAL** | $100 | 0.5% |
| **POM** | $20 | 1.50% |
| **WELL** | $25 | 0% |

If the investment was expected to earn $56 in annual dividends, and he purchased a total of 200 shares, how many shares of each stock did he purchase?

**40. Initial Public Offerings (IPOs)** Duffin House, Higgins Press, and Sickle Publications all went public on the same day recently. John O'Hagan had the opportunity to participate in all three initial public offerings (partly because he and Marjory Duffin are good friends). He made a considerable profit when he sold all of the stock two days later on the open market. The following table shows the purchase price and percentage yield on the investment in each company.

|  | Purchase Price per Share | Yield |
| --- | --- | --- |
| **Duffin House (DHS)** | $8 | 20% |
| **Higgins Press (HPR)** | $10 | 15% |
| **Sickle Publications (SPUB)** | $15 | 15% |

He invested $20,000 in a total of 2,000 shares, and made a $3,400 profit from the transactions. How many shares in each company did he purchase?

**41. Degree Requirements** During his lunch break, John O'Hagan decides to devote some time to assisting his son Billy-Sean, who is having a terrible time coming up with a college course schedule. One reason for this is the very complicated Bulletin of Suburban State University. It reads as follows:

*All candidates for the degree of Bachelor of Science at SSU must take a total of 124 credits from the Sciences, Fine Arts, Liberal Arts, and Mathematics,[28] including an equal number of Science and Fine Arts credits, and twice as many Mathematics credits as Science credits and Fine Arts credits combined, but with Liberal Arts credits exceeding Mathematics credits by exactly one-third of the number of Fine Arts credits.*

What are all the possible degree programs for Billy-Sean?

**42. Degree Requirements** Having finally decided on his degree program, Billy-Sean learns that the Suburban State University Senate (under pressure from the English Department) has revised the Bulletin to include a "Verbal Expression" component in place of the Fine Arts requirement in all programs (including the sciences):

*All candidates for the degree of Bachelor of Science at SSU must take a total of 120 credits from the Liberal Arts, Sciences, Verbal Expression, and Mathematics, including an equal number of Science and Liberal Arts credits, and twice as many Verbal Expression credits as Science credits and Liberal Arts credits combined, but with Liberal Arts credits exceeding Mathematics credits by one quarter of the number of Verbal Expression Credits.*

What are now the possible degree programs for Billy-Sean?

[28]Strictly speaking, mathematics is not a science; it is the Queen of the Sciences, although we like to think of it as the Mother of all Sciences.

**43. Network Traffic** All book orders received at the Order Department at OHaganBooks.com are transmitted through a small computer network to the Shipping Department. The following diagram shows the network (which uses two intermediate computers as routers), together with some of the average daily traffic measured in book orders.

OD = Order department
SD = Shipping department

**a.** Set up a system of linear equations in which the unknowns give the average traffic along the paths labeled $x, y, z, w$, and find the general solution.
**b.** What is the minimum volume of traffic along $y$?
**c.** What is the maximum volume of traffic along $w$?
**d.** If there is no traffic along $z$, find the volume of traffic along all the paths.
**e.** If there is the same volume of traffic along $y$ and $z$, what is the volume of traffic along $w$?

**44. Business Retreats** Marjory Duffin is planning a joint business retreat for Duffin House and OHaganBooks.com at Laguna Surf City, but is concerned about traffic conditions (she feels that too many cars tend to spoil the ambiance of a seaside retreat). She managed to obtain the following map from the Laguna Surf City Engineering Department (all the streets are one-way as indicated). The counters show traffic every five minutes.

**a.** Set up and solve the associated system of linear equations. *Be sure to give the general solution.*
(Take $x$ = traffic along North Beach Way, $y$ = traffic along South Beach Way, $z$ = traffic along Beach View

St., $u$ = traffic along Lagoon Drive and $v$ = traffic along Dock Road.)

**b.** Assuming that all roads are one-way in the directions shown, what, if any, is the maximum possible traffic along Lagoon Drive?

**c.** The Laguna Surf City Traffic Department is considering opening up Beach View Street to two-way traffic, but an environmentalist group is concerned that this will result in increased traffic on Lagoon Drive. What, if any, is the maximum possible traffic along Lagoon Drive assuming that Beach View Street is two-way and the traffic counter readings are as shown?

**45.** *Shipping* On the same day that the sales department at Duffin House received an order for 600 packages from the OHaganBooks.com Texas headquarters, it received an additional order for 200 packages from FantasyBooks.com, based in California. Duffin House has warehouses in New York and Illinois. The Illinois warehouse is closing down

and must clear all 300 packages it has in stock. Shipping costs per package of books are as follows:

New York to Texas: $20    New York to California: $50
Illinois to Texas: $30    Illinois to California: $40

Is it possible to fill both orders and clear the Illinois warehouse at a cost of $22,000? If so, how many packages should be sent from each warehouse to each online bookstore?

**46.** *Transportation Scheduling* Duffin House is about to start a promotional blitz for its new book, *Physics for the Liberal Arts*. The company has 20 salespeople stationed in Chicago and 10 in Denver, and would like to fly 15 to sales fairs at each of Los Angeles and New York. A round-trip plane flight from Chicago to LA costs $200; from Chicago to NY costs $150; from Denver to LA costs $400; and from Denver to NY costs $200. For tax reasons, Duffin House needs to budget exactly $6,500 for the total cost of the plane flights. How many salespeople should the company fly from each of Chicago and Denver to each of LA and NY?

---

**Case Study**

## Hybrid Cars—Optimizing the Degree of Hybridization

You are involved in new model development at a major automobile company. The company is planning to introduce two new plug-in hybrid electric vehicles: the subcompact "Green Town Hopper" and the midsize "Electra Supreme," and your department must decide on the degree of hybridization (DOH) for each of these models that will result in the largest reduction in gasoline consumption. (The DOH of a vehicle is defined as the ratio of electric motor power to the total power, and typically ranges from 10% to 50%. For example, a model with a 20% DOH has an electric motor that delivers 20% of the total power of the vehicle.)

The tables below show the benefit for each of the two models, measured as the estimated reduction in annual gasoline consumption, as well as an estimate of retail cost increment, for various DOH percentages[29]: (The retail cost increment estimate is given by the formula $5,000 + 50(DOH - 10)$ for the Green Town Hopper and $7,000 + 50(DOH - 10)$ for the Electra Supreme.)

**Green Town Hopper**

| DOH (%) | 10 | 20 | 50 |
|---|---|---|---|
| Reduction in Annual Consumption (gals) | 180 | 230 | 200 |
| Retail Cost Increment ($) | 5,000 | 5,500 | 7,000 |

**Electra Supreme**

| DOH (%) | 10 | 20 | 50 |
|---|---|---|---|
| Reduction in Annual Consumption (gals) | 220 | 270 | 260 |
| Retail Cost Increment ($) | 7,000 | 7,500 | 9,000 |

---

[29]The figures are approximate and based on data for two actual vehicles as presented in a 2006 paper entitled *Cost-Benefit Analysis of Plug-In Hybrid Electric Vehicle Technology* by A. Simpson. Source: National Renewable Energy Laboratory, U.S. Department of Energy (www.nrel.gov).

Green Town Hopper

Electra Supreme

**Figure 9**

Notice that increasing the DOH toward 50% results in a decreased benefit. This is due in part to the need to increase the weight of the batteries while keeping the vehicle performance at a desirable level, thus necessitating a more powerful gasoline engine. The *optimum* DOH is the percentage that gives the largest reduction in gasoline consumption, and this is what you need to determine. Since the optimum DOH may not be 20%, you would like to create a mathematical model to compute the reduction $R$ in gas consumption as a function of the DOH $x$. Your first inclination is to try linear equations—that is, an equation of the form

$$R = ax + b \quad (a \text{ and } b \text{ constants}),$$

but you quickly discover that the data simply won't fit, no matter what the choice of the constants. The reason for this can be seen graphically by plotting $R$ versus $x$ (Figure 9). In neither case do the three points lie on a straight line. In fact, the data are not even *close* to being linear. Thus, you will need curves to model these data. After giving the matter further thought, you remember something your mathematics instructor once said: The simplest curve passing through any three points not all on the same line is a parabola. Since you are looking for a simple model of the data, you decide to try a parabola. A general parabola has the equation

$$R = ax^2 + bx + c,$$

where $a$, $b$, and $c$ are constants. The problem now is: What are $a$, $b$, and $c$? You decide to try substituting the values of $R$ and $x$ for the Green Town Hopper into the general equation, and you get the following:

$$x = 10, R = 180 \quad \text{gives} \quad 180 = 100a + 10b + c$$
$$x = 20, R = 230 \quad \text{gives} \quad 230 = 400a + 20b + c$$
$$x = 50, R = 200 \quad \text{gives} \quad 200 = 2{,}500a + 50b + c.$$

Now you notice that you have three linear equations in three unknowns! You solve the system:

$$a = -0.15, b = 9.5, c = 100.$$

Thus your reduction equation for the Green Town Hopper becomes

$$R = -0.15x^2 + 9.5x + 100.$$

Green Town Hopper

Electra Supreme

**Figure 10**

For the Electra Supreme, you get

$$x = 10, R = 220: \quad 220 = 100a + 10b + c$$
$$x = 20, R = 270: \quad 270 = 400a + 20b + c$$
$$x = 50, R = 260: \quad 260 = 2{,}500a + 50b + c.$$

$$a = -0.1\bar{3}, b = 9, c = 143.\bar{3}$$

and so

$$R = -0.1\bar{3}x^2 + 9x + 143.\bar{3}.$$

Figure 10 shows the parabolas superimposed on the data points. You can now estimate a value for the optimal DOH as the value of $x$ that gives the largest benefit $R$.

Recalling that the $x$-coordinate of the vertex of the parabola $y = ax^2 + bx + c$ is $x = -\frac{b}{2a}$, you obtain the following estimates:

Green Town Hopper: Optimal DOH $= -\dfrac{9.5}{2(-0.15)} \approx 31.67\%$

Electra Supreme: Optimal DOH $= -\dfrac{9}{2(-0.1\overline{3})} \approx 33.75\%.$

You can now use the formulas given earlier to estimate the resulting reductions in gasoline consumption and increases in cost:

Green Town Hopper:

Reduction in gasoline consumption $= R \approx -0.15(31.67)^2 + 9.5(31.67) + 100$
$\approx 250.4$ gals/year

Retail Cost Increment $\approx 5,000 + 50(31.67 - 10) \approx \$6,080$

Electra Supreme:

Reduction in gasoline consumption $= R = -0.1\overline{3}(33.75)^2 + 9(33.75) + 143.\overline{3}$
$\approx 295.2$ gals/year

Retail Cost Increment $\approx 7,000 + 50(33.75 - 10) \approx \$8,190.$

You thus submit the following estimates: The optimal degree of hybridization for the Green Town Hopper is about 31.67% and will result in a reduction in gasoline consumption of 250.4 gals/year and a retail cost increment of around $6,080. The optimal degree of hybridization for the Electra Supreme is about 33.75% and will result in a reduction in gasoline consumption of 295.2 gals/year and a retail cost increment of around $8,190.

## EXERCISES

1. Repeat the computations above for the "Earth Suburban" using the following data:

### Earth Suburban

| DOH (%) | 10 | 20 | 50 |
|---|---|---|---|
| Reduction in Annual Consumption (gals) | 240 | 330 | 300 |
| Retail Cost Increment ($) | 9,000 | 9,500 | 11,000 |

Retail cost increment $= 9,000 + 50(DOH - 10)$

2. ⊤ Repeat the analysis for the Green Town Hopper, but this time take $x$ to be the cost increment, in thousands of dollars. (The curve of benefit versus cost is referred to as a *cost-benefit* curve.) What value of DOH corresponds to the optimal cost? What do you notice? Comment on the answer.

3. ⊤ Repeat the analysis for the Electra Supreme, but this time use the optimal values, DOH $= 33.8$, Reduction $= 295.2$ gals/year in place of the 20% data. What do you notice?

4. Find the equation of the parabola that passes through the points $(1, 2)$, $(2, 9)$, and $(3, 19)$.

5. Is there a parabola that passes through the points $(1, 2)$, $(2, 9)$, and $(3, 16)$?

6. Is there a parabola that passes though the points $(1, 2)$, $(2, 9)$, $(3, 19)$, and $(-1, 2)$?

7. ⊤ You submit your recommendations to your manager, and she tells you, "Thank you very much, but we have additional data for the Green Town Hopper: A 30% DOH results in a saving of 255 gals/year. Please resubmit your recommendations taking this into account by tomorrow." HINT [You now have four data points on each graph, so try a general cubic instead: $R = ax^3 + bx^2 + cx + d$. Use a graph to estimate the optimal DOH.]

## TI-83/84 Plus Technology Guide

## Section 3.1

**Example 1 (page 177)** Find all solutions $(x, y)$ of the following system of two equations:

$$x + y = 3$$
$$x - y = 1.$$

### Solution with Technology

You can use a graphing calculator to draw the graphs of the two equations on the same set of axes and to check the solution. First, solve the equations for $y$, obtaining $y = -x + 3$ and $y = x - 1$. On the TI-83/84 Plus:

**1.** Set

   Y₁=-X+3
   Y₂=X-1

   $$Y_1 = -X + 3$$
   $$Y_2 = X - 1$$

**2.** Decide on the range of $x$-values you want to use. As in Figure 1, let us choose the range $[-4, 4]$.[30]

**3.** In the WINDOW menu set Xmin $= -4$ and Xmax $= 4$.

**4.** Press ZOOM and select Zoomfit to set the $y$ range.

You can now zoom in for a more accurate view by choosing a smaller $x$-range that includes the point of intersection, like $[1.5, 2.5]$, and using Zoomfit again. You can also use the trace feature to see the coordinates of points near the point of intersection.

To check that $(2, 1)$ is the correct solution, use the table feature to compare the two values of $y$ corresponding to $x = 2$:

**1.** Press 2ND TABLE.

**2.** Set X $= 2$, and compare the corresponding values of $Y_1$ and $Y_2$; they should each be 1.

**Q:** *How accurate is the answer shown using the trace feature?*

**A:** That depends. We can increase the accuracy up to a point by zooming in on the point of intersection of the two graphs. But there is a limit to this: Most graphing calculators are capable of giving an answer correct to about 13 decimal places. This means, for instance, that, in the eyes of the TI-83/84 Plus, 2.000 000 000 000 1 is exactly the same as 2 (subtracting them yields 0). It follows that if you attempt to use a window so narrow that you need approximately 13 significant digits to distinguish the left and right edges, you will run into accuracy problems.

## Section 3.2

**Row Operations with a TI-83/84 Plus** Start by entering the matrix into [A] using MATRIX EDIT. You can then do row operations on [A] using the following instructions (*row, *row+, and rowSwap are found in the MATRIX MATH menu).

---

[30]How did we come up with this interval? Trial and error. You might need to try several intervals before finding one that gives a graph showing the point of intersection clearly.

**TECHNOLOGY GUIDE**

| Row Operation | TI-83/84 Plus Instruction (Matrix name is [A]) |
|---|---|
| $R_i \to k R_i$ | `*row(k,[A],i)→[A]` |
| **Example:** $R_2 \to 3R_2$ | `*row(3,[A],2)→[A]` |
| $R_i \to R_i + k R_j$ | `*row+(k,[A],j,i)→[A]` |
| **Examples:** $R_1 \to R_1 - 3R_2$ | `*row+(-3,[A],2,1)→[A]` |
| $R_1 \to 4R_1 - 3R_2$ | `*row(4,[A],1)→[A]`<br>`*row+(-3,[A],2,1)→[A]` |
| Swap $R_i$ and $R_j$ | `rowSwap([A],i,j)→[A]` |
| **Example:** Swap $R_1$ and $R_2$ | `rowSwap([A],1,2)→[A]` |

### Example 1 (page 195)   Solve the system:

$$x - y + 5z = -6$$
$$3x + 3y - z = 10$$
$$x + 3y + 2z = 5.$$

### Solution with Technology

1. Begin by entering the matrix into `[A]` using MATRX EDIT. Only three columns can be seen at a time; you can see the rest of the matrix by scrolling left or right.

```
MATRIX[A] 3 ×4
[[1    -1    5      -
 [3     3   -1      -
 [1     3    2      -

1,1=1
```

```
MATRIX[A] 3 ×4
 -1     5    -6  ]
  3    -1    10  ]
  3     2     5  ]

1,4=-6
```

2. Now perform the operations given in Example 1:

```
*row+(-3,[A],1,2
)→[A]
[[1  -1   5   -6]
 [0   6  -16  28]
 [1   3   2    5 ]]
■
```

```
*row+(-1,[A],1,3
)→[A]
[[1  -1   5   -6]
 [0   6  -16  28]
 [0   4  -3   11]]
■
```

```
*row(1/2,[A],2)→
[A]
[[1  -1   5   -6]
 [0   3  -8   14]
 [0   4  -3   11]]
■
```

```
*row(3,[A],1)→[A
]
[[3  -3  15  -18]
 [0   3  -8   14 ]
 [0   4  -3   11 ]]
■
```

```
*row+(1,[A],2,1)
→[A]
[[3   0   7   -4]
 [0   3  -8   14]
 [0   4  -3   11]]
■
```

```
*row(3,[A],3)→[A
]
[[3   0   7   -4]
 [0   3  -8   14]
 [0  12  -9   33]]
■
```

As in Example 1, we can now read the solution from the right-hand column: $x = 1$, $y = 2$, $z = -1$.

**Note** The TI-83/84 Plus has a function, `rref`, that gives the reduced row echelon form of a matrix in one step. (See the text for the definition of reduced row echelon form.) Internally it uses a variation of Gauss-Jordan reduction to do this. ■

## SPREADSHEET Technology Guide

### Section 3.1

**Example 1 (page 177)** Find all solutions $(x, y)$ of the following system of two equations:

$$x + y = 3$$
$$x - y = 1.$$

**Solution with Technology**

You can use a spreadsheet to draw the graphs of the two equations on the same set of axes, and to check the solution.

1. Solve the equations for $y$, obtaining $y = -x + 3$ and $y = x - 1$.

2. To graph these lines we can use the following simple worksheet:

The two values of $x$ give the $x$-coordinates of the two points we will use as endpoints of the lines. (We have—somewhat arbitrarily—chosen the range $[-4, 4]$ for $x$.) The formula for the first line, $y = -x + 3$, is in cell B2, and the formula for the second line, $y = x - 1$, is in C2.

3. Copy these two cells as shown to yield the following result:

4. For the graph, select all nine cells and create a scatter graph with line segments joining the data points. Instruct the spreadsheet to insert a chart and select the "scatter" option. In the same dialogue box, select the option that shows points connected by lines. If you are using Excel, press "Next" to bring up a new dialogue box called "Data Type," where you should make sure that the "Series in Columns" option is selected,

**TECHNOLOGY GUIDE**

telling the program that the *x*- and *y*-coordinates are arranged vertically, down columns. (Other spreadsheet programs have corresponding options you may need to set.) Your graph should appear as shown below.

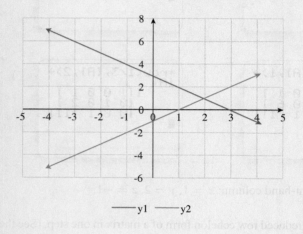

—— y1  —— y2

To zoom in:

1. First decide on a new *x*-range, say, [1, 3].
2. Change the value in cell A2 to 1 and the value in cell A3 to 3, and the spreadsheet will automatically update the *y*-values and the graph.*

To check that (2, 1) is the correct solution:

1. Enter the value 2 in cell A4 in your spreadsheet.
2. Copy the formulas in B3 and C3 down to row 4 to obtain the corresponding values of *y*.

|  | A | B | C |
|---|---|---|---|
| 1 | x | y1 | y2 |
| 2 | -4 | 7 | -5 |
| 3 | 4 | -1 | 3 |
| 4 | 2 | 1 | 1 |

Because the values of *y* agree, we have verified that (2, 1) is a solution of both equations.

## Section 3.2

**Row Operations with a Spreadsheet**   To a spreadsheet, a block of data with one or more rows or columns is an **array**, and a spreadsheet has built in the capability to handle arrays in much the same way as it handles single cells. Consider the following example:

$$\begin{bmatrix} 1 & 3 & -4 \\ 0 & 4 & 2 \end{bmatrix} 3R_2 \rightarrow \begin{bmatrix} 1 & 3 & -4 \\ 0 & 12 & 6 \end{bmatrix}.$$   Replace $R_2$ by $3R_2$.

1. Enter the original matrix in a convenient location, say the cells A1 through C2.
2. To get the first row of the new matrix, which is simply a copy of the first row of the old matrix, decide where you want to place the new matrix, and highlight *the whole row of the new matrix,* say A4:C4.
3. Enter the formula `=A1:C1` (the easiest way to do this is to type "=" and then use the mouse to select cells A1 through C1, i.e., the first row of the old matrix).

Enter formula for new first row.                          Press Control-Shift-Enter.

---

* **NOTE**  Your spreadsheet will set the *y*-range of the graph automatically, and the range it chooses may not always be satisfactory. It may sometimes be necessary to "narrow down" the range of one or both axes (in Excel, double-click on the axis in question).

4. Press **Control-Shift-Enter** (instead of "Enter" alone) and the whole row will be copied.[31] Pressing Control-Shift-Enter tells the spreadsheet that your formula is an *array formula,* one that returns an array rather than a single number. Once entered, the spreadsheet will show an array formula enclosed in "curly braces." (Note that you must also use Control-Shift-Enter to delete any array you create: Select the block you wish to delete and press Delete followed by Control-Shift-Enter.)

5. Similarly, to get the second row, select cells A5 through C5, where the new second row will go, enter the formula =3*A2:C2, and press Control-Shift-Enter. (Again, the easiest way to enter the formula is to type "=3*" and then select the cells A2 through C2 using the mouse.)

Enter formula for new second row.              Press Control-Shift-Enter.

We can perform the following operation in a similar way:

$$\begin{bmatrix} 1 & 3 & -4 \\ 0 & 4 & 2 \end{bmatrix} \xrightarrow{4R_1 - 3R_2} \begin{bmatrix} 4 & 0 & -22 \\ 0 & 4 & 2 \end{bmatrix}.$$

Replace $R_1$ by $4R_1 - 3R_2$.

(To easily enter the formula for $4R_1 - 3R_2$, type "=4*", select the first row A1:C1 using the mouse, type "-3*", and then select the second row A2:C2 using the mouse.)

**Example 1 (page 195)**    Solve the system:

$$\begin{aligned} x - y + 5z &= -6 \\ 3x + 3y - z &= 10 \\ x + 3y + 2z &= 5. \end{aligned}$$

---

[31] Note that on a Mac, Command-Enter and Command-Shift-Enter have the same effect as Control-Shift-Enter.

## Solution with Technology

Here is the complete row reduction as it would appear in a spreadsheet.

**Table 1 (left):**

|    | A | B | C | D | E |
|----|---|---|---|---|---|
| 1  | 1 | -1 | 5 | -6 | |
| 2  | 3 | 3 | -1 | 10 | R2 - 3R1 |
| 3  | 1 | 3 | 2 | 5 | R3 - R1 |
| 4  | | | | | |
| 5  | =A1:D1 | | | | |
| 6  | =A2:D2-3*A1:D1 | | | | |
| 7  | =A3:D3-A1:D1 | | | | |

**Table 1 (right):**

|    | A | B | C | D | E |
|----|---|---|---|---|---|
| 1  | 1 | -1 | 5 | -6 | |
| 2  | 3 | 3 | -1 | 10 | R2 - 3R1 |
| 3  | 1 | 3 | 2 | 5 | R3 - R1 |
| 4  | | | | | |
| 5  | 1 | -1 | 5 | -6 | |
| 6  | 0 | 6 | -16 | 28 | (1/2)R2 |
| 7  | 0 | 4 | -3 | 11 | |
| 8  | | | | | |
| 9  | =A5:D5 | | | | |
| 10 | =(1/2)*A6:D6 | | | | |
| 11 | =A7:D7 | | | | |

**Table 2 (left):**

|    | A | B | C | D | E |
|----|---|---|---|---|---|
| 8  | | | | | |
| 9  | 1 | -1 | 5 | -6 | 3R1 + R2 |
| 10 | 0 | 3 | -8 | 14 | |
| 11 | 0 | 4 | -3 | 11 | 3R3 - 4R2 |
| 12 | | | | | |
| 13 | =3*A9:D9+A10:D10 | | | | |
| 14 | =A10*D10 | | | | |
| 15 | =3*A11:D11-4*A10:D10 | | | | |

**Table 2 (right):**

|    | A | B | C | D | E |
|----|---|---|---|---|---|
| 12 | | | | | |
| 13 | 3 | 0 | 7 | -4 | |
| 14 | 0 | 3 | -8 | 14 | |
| 15 | 0 | 0 | 23 | -23 | (1/23)R3 |
| 16 | | | | | |
| 17 | =A13:D13 | | | | |
| 18 | =A14:D14 | | | | |
| 19 | =(1/23)*A15:D15 | | | | |

**Table 3 (left):**

|    | A | B | C | D | E |
|----|---|---|---|---|---|
| 16 | | | | | |
| 17 | 3 | 0 | 7 | -4 | R1 - 7R3 |
| 18 | 0 | 3 | -8 | 14 | R2 + 8R3 |
| 19 | 0 | 0 | 1 | -1 | |
| 20 | | | | | |
| 21 | =A17:D17-7*A19:D19 | | | | |
| 22 | =A18:D18+8*A19:D19 | | | | |
| 23 | =A19:D19 | | | | |

**Table 3 (right):**

|    | A | B | C | D | E |
|----|---|---|---|---|---|
| 20 | | | | | |
| 21 | 3 | 0 | 0 | 3 | (1/3)R1 |
| 22 | 0 | 3 | 0 | 6 | (1/3)R2 |
| 23 | 0 | 0 | 1 | -1 | |
| 24 | | | | | |
| 25 | =(1/3)*A21:D21 | | | | |
| 26 | =(1/3)*A22:D22 | | | | |
| 27 | =A23:D23 | | | | |

**Table 4:**

|    | A | B | C | D |
|----|---|---|---|---|
| 24 | | | | |
| 25 | 1 | 0 | 0 | 1 |
| 26 | 0 | 1 | 0 | 2 |
| 27 | 0 | 0 | 1 | -1 |

As in Example 1, we can now read the solution from the right hand column: $x = 1, y = 2, z = -1$. What do you notice if you change the entries in cells D1, D2, and D3?

# 4

# Matrix Algebra and Applications

 **Website**

www.WanerMath.com

At the Website you will find:

- Section-by-section tutorials
- A detailed chapter summary
- A true/false quiz
- Additional review exercises
- A matrix algebra utility
- A game theory utility
- The following optional extra sections:

    Determinants

    Using Determinants to Solve Systems: Cramer's Rule

## Case Study Predicting Market Share

You are the sales director at *Selular*, a cellphone provider, and things are not looking good for your company: The recently launched *iClone* competitor is beginning to chip away at *Selular's* market share. Particularly disturbing are rumors of fierce brand loyalty by *iClone* customers, with several bloggers suggesting that *iClone* retains close to 100% of their customers. Worse, you will shortly be presenting a sales report to the board of directors, and the CEO has "suggested" that your report include 2-, 5-, and 10-year projections of *Selular's* market share given the recent impact on the market by *iClone*, and also a "worst-case" scenario projecting what would happen if *iClone* customers are so loyal that none of them ever switch services. You have results from two market surveys, taken one quarter apart, of the major cellphone providers, which shows the current market shares and the percentages of subscribers who switched from one service to another during the quarter. **How should you respond?**

GOH CHAI HIN/AFP/Newscom

231

# Introduction

We used matrices in Chapter 3 simply to organize our work. It is time we examined them as interesting objects in their own right. There is much that we can do with matrices besides row operations: We can add, subtract, multiply, and even, in a sense, "divide" matrices. We use these operations to study game theory and input-output models in this chapter, and Markov chains in a later chapter.

Many calculators, spreadsheets, and other computer programs can do these matrix operations, which is a big help in doing calculations. However, we need to know how these operations are defined to see why they are useful and to understand which to use in any particular application.

## 4.1    Matrix Addition and Scalar Multiplication

Let's start by formally defining what a matrix is and introducing some basic terms.

### Matrix, Dimension, and Entries

An **m × n matrix** $A$ is a rectangular array of real numbers with $m$ rows and $n$ columns. We refer to $m$ and $n$ as the **dimensions** of the matrix. The numbers that appear in the matrix are called its **entries**. We customarily use capital letters $A, B, C, \ldots$ for the names of matrices.

#### Quick Examples

1. $A = \begin{bmatrix} 2 & 0 & 1 \\ 33 & -22 & 0 \end{bmatrix}$ is a $2 \times 3$ matrix because it has two rows and three columns.

2. $B = \begin{bmatrix} 2 & 3 \\ 10 & 44 \\ -1 & 3 \\ 8 & 3 \end{bmatrix}$ is a $4 \times 2$ matrix because it has four rows and two columns.*

The entries of $A$ are 2, 0, 1, 33, −22, and 0. The entries of $B$ are the numbers 2, 3, 10, 44, −1, 3, 8, and 3.

* Remember that the number of rows is given first and the number of columns second. An easy way to remember this is to think of the acronym "RC" for "Row then Column."

### Referring to the Entries of a Matrix

There is a systematic way of referring to particular entries in a matrix. If $i$ and $j$ are numbers, then the entry in the $i$th row and $j$th column of the matrix $A$ is called the **$ij$th entry** of $A$. We usually write this entry as $a_{ij}$ or $A_{ij}$. (If the matrix was called $B$, we would write its $ij$th entry as $b_{ij}$ or $B_{ij}$.) Notice that this follows the "RC" convention: The row number is specified first and the column number second.

---

## Quick Example

With $A = \begin{bmatrix} 2 & 0 & 1 \\ 33 & -22 & 0 \end{bmatrix}$,

$a_{13} = 1$        First row, third column

$a_{21} = 33.$        Second row, first column

---

According to the labeling convention, the entries of the matrix $A$ above are

$$A = \begin{bmatrix} a_{11} & a_{12} & a_{13} \\ a_{21} & a_{22} & a_{23} \end{bmatrix}.$$

In general, the $m \times n$ matrix $A$ has its entries labeled as follows:

$$A = \begin{bmatrix} a_{11} & a_{12} & a_{13} & \cdots & a_{1n} \\ a_{21} & a_{22} & a_{23} & \cdots & a_{2n} \\ \vdots & \vdots & \vdots & \ddots & \vdots \\ a_{m1} & a_{m2} & a_{m3} & \cdots & a_{mn} \end{bmatrix}.$$

We say that two matrices $A$ and $B$ are **equal** if they have the same dimensions and the corresponding entries are equal. Note that a $3 \times 4$ matrix can never equal a $3 \times 5$ matrix because they do not have the same dimensions.

---

### EXAMPLE 1  Matrix Equality

Let $A = \begin{bmatrix} 7 & 9 & x \\ 0 & -1 & y+1 \end{bmatrix}$ and $B = \begin{bmatrix} 7 & 9 & 0 \\ 0 & -1 & 11 \end{bmatrix}$. Find the values of $x$ and $y$ such that $A = B$.

**Solution** For the two matrices to be equal, we must have corresponding entries equal, so

$$x \quad\;\; = 0 \qquad\qquad\qquad a_{13} = b_{13}$$
$$y + 1 = 11 \quad \text{or} \quad y = 10. \qquad a_{23} = b_{23}$$

---

➡ **Before we go on...** Note in Example 1 that the matrix equation

$$\begin{bmatrix} 7 & 9 & x \\ 0 & -1 & y+1 \end{bmatrix} = \begin{bmatrix} 7 & 9 & 0 \\ 0 & -1 & 11 \end{bmatrix}$$

is really six equations in one: $7 = 7$, $9 = 9$, $x = 0$, $0 = 0$, $-1 = -1$, and $y + 1 = 11$. We used only the two that were interesting. ∎

## Row Matrix, Column Matrix, and Square Matrix

A matrix with a single row is called a **row matrix**, or **row vector**. A matrix with a single column is called a **column matrix** or **column vector**. A matrix with the same number of rows as columns is called a **square matrix**.

### Quick Examples

The $1 \times 5$ matrix $C = \begin{bmatrix} 3 & -4 & 0 & 1 & -11 \end{bmatrix}$ is a row matrix.

The $4 \times 1$ matrix $D = \begin{bmatrix} 2 \\ 10 \\ -1 \\ 8 \end{bmatrix}$ is a column matrix.

The $3 \times 3$ matrix $E = \begin{bmatrix} 1 & -2 & 0 \\ 0 & 1 & 4 \\ -4 & 32 & 1 \end{bmatrix}$ is a square matrix.

## Matrix Addition and Subtraction

The first matrix operations we discuss are matrix addition and subtraction. The rules for these operations are simple.

### Matrix Addition and Subtraction

Two matrices can be added (or subtracted) if and only if they have the same dimensions. To add (or subtract) two matrices of the same dimensions, we add (or subtract) the corresponding entries. More formally, if $A$ and $B$ are $m \times n$ matrices, then $A + B$ and $A - B$ are the $m \times n$ matrices whose entries are given by:

$(A + B)_{ij} = A_{ij} + B_{ij}$     $ij$th entry of the sum = sum of the $ij$th entries
$(A - B)_{ij} = A_{ij} - B_{ij}$.     $ij$th entry of the difference = difference of the $ij$th entries

**Visualizing Matrix Addition**

$$\begin{bmatrix} 2 & -3 \\ 1 & 0 \end{bmatrix} + \begin{bmatrix} 1 & 1 \\ -2 & 1 \end{bmatrix} = \begin{bmatrix} 3 & -2 \\ -1 & 1 \end{bmatrix}$$

### Quick Examples

1. $\begin{bmatrix} 2 & -3 \\ 1 & 0 \\ -1 & 3 \end{bmatrix} + \begin{bmatrix} 9 & -5 \\ 0 & 13 \\ -1 & 3 \end{bmatrix} = \begin{bmatrix} 11 & -8 \\ 1 & 13 \\ -2 & 6 \end{bmatrix}$    Corresponding entries added

2. $\begin{bmatrix} 2 & -3 \\ 1 & 0 \\ -1 & 3 \end{bmatrix} - \begin{bmatrix} 9 & -5 \\ 0 & 13 \\ -1 & 3 \end{bmatrix} = \begin{bmatrix} -7 & 2 \\ 1 & -13 \\ 0 & 0 \end{bmatrix}$    Corresponding entries subtracted

**using** Technology

Technology can be used to enter, add, and subtract matrices. Here is an outline (see the Technology Guides at the end of the chapter for additional details on using a TI-83/84 Plus or a spreadsheet):

**TI-83/84 Plus**

Entering a matrix: MATRX ; EDIT
Select a name, ENTER ; type in the entries.
Adding two matrices:
Home Screen: [A]+[B]
(Use MATRX ; NAMES to enter them on the home screen.)
[More details on page 304.]

**Spreadsheet**

Entering a matrix: Type entries in a convenient block of cells.
Adding two matrices: Highlight block where you want the answer to appear.
Type "="; highlight first matrix; type "+"; highlight second matrix; press Control+Shift+Enter
[More details on page 307.]

**Website**

www.WanerMath.com
  Student Home
    → On Line Utilities
    → Matrix Algebra Tool

Enter matrices as shown (use a single letter for the name).

```
Enter your matrices here.
J = [20, 15
10, 12
8, 4]

F = [23, 12,
8, 12
4, 5]
```

To compute their difference, type F-J in the formula box and press "Compute". (You can enter multiple formulas separated by commas in the formula box. For instance, F+J, F-J will compute both the sum and the difference. *Note:* The utility is case sensitive, so be consistent.)

## EXAMPLE 2  Sales

The A-Plus auto parts store chain has two outlets, one in Vancouver and one in Quebec. Among other things, it sells wiper blades, windshield cleaning fluid, and floor mats. The monthly sales of these items at the two stores for two months are given in the following tables:

**January Sales**

|  | Vancouver | Quebec |
|---|---|---|
| **Wiper Blades** | 20 | 15 |
| **Cleaning Fluid (bottles)** | 10 | 12 |
| **Floor Mats** | 8 | 4 |

**February Sales**

|  | Vancouver | Quebec |
|---|---|---|
| **Wiper Blades** | 23 | 12 |
| **Cleaning Fluid (bottles)** | 8 | 12 |
| **Floor Mats** | 4 | 5 |

Use matrix arithmetic to calculate the change in sales of each product in each store from January to February.

**Solution**  The tables suggest two matrices:

$$J = \begin{bmatrix} 20 & 15 \\ 10 & 12 \\ 8 & 4 \end{bmatrix} \quad \text{and} \quad F = \begin{bmatrix} 23 & 12 \\ 8 & 12 \\ 4 & 5 \end{bmatrix}.$$

To compute the change in sales of each product for both stores, we want to subtract corresponding entries in these two matrices. In other words, we want to compute the difference of the two matrices:

$$F - J = \begin{bmatrix} 23 & 12 \\ 8 & 12 \\ 4 & 5 \end{bmatrix} - \begin{bmatrix} 20 & 15 \\ 10 & 12 \\ 8 & 4 \end{bmatrix} = \begin{bmatrix} 3 & -3 \\ -2 & 0 \\ -4 & 1 \end{bmatrix}.$$

Thus, the change in sales of each product is the following:

|  | Vancouver | Quebec |
|---|---|---|
| **Wiper Blades** | 3 | -3 |
| **Cleaning Fluid (bottles)** | -2 | 0 |
| **Floor Mats** | -4 | 1 |

## Scalar Multiplication

A matrix $A$ can be added to itself because the expression $A + A$ is the sum of two matrices that have the same dimensions. When we compute $A + A$, we end up doubling every entry in $A$. So we can think of the expression $2A$ as telling us to *multiply every element in A by* 2.

In general, to multiply a matrix by a number, multiply every entry in the matrix by that number. For example,

$$6 \begin{bmatrix} \frac{5}{2} & -3 \\ 1 & 0 \\ -1 & \frac{5}{6} \end{bmatrix} = \begin{bmatrix} 15 & -18 \\ 6 & 0 \\ -6 & 5 \end{bmatrix}.$$

It is traditional when talking about matrices to call individual numbers **scalars**. For this reason, we call the operation of multiplying a matrix by a number **scalar multiplication**.

 **using** Technology

Technology can be used to compute scalar multiples:

**TI-83/84 Plus**
Enter the matrix [A] using MATRX ;
EDIT
Home Screen: 0.65 [A]

**Spreadsheets**
Enter the matrix *A* in a convenient 3 × 2 block of cells.
Highlight block where you want the answer to appear.
Type =0.65*; highlight matrix *A*; press Control+Shift+Enter.

**Website**
www.WanerMath.com
    Student Home
        → On Line Utilities
        → Matrix Algebra Tool
Enter the matrix *A* as shown.

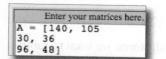

Type 0.65*A in the formula box and press "Compute".

## EXAMPLE 3  Sales

The revenue generated by sales in the Vancouver and Quebec branches of the A-Plus auto parts store (see Example 2) was as follows:

**January Sales in Canadian Dollars**

|  | Vancouver | Quebec |
|---|---|---|
| **Wiper Blades** | 140.00 | 105.00 |
| **Cleaning Fluid** | 30.00 | 36.00 |
| **Floor Mats** | 96.00 | 48.00 |

If the Canadian dollar was worth $0.65 U.S. at the time, compute the revenue in U.S. dollars.

**Solution**  We need to multiply each revenue figure by 0.65. Let *A* be the matrix of revenue figures in Canadian dollars:

$$A = \begin{bmatrix} 140.00 & 105.00 \\ 30.00 & 36.00 \\ 96.00 & 48.00 \end{bmatrix}.$$

The revenue figures in U.S. dollars are then given by the scalar multiple

$$0.65A = 0.65 \begin{bmatrix} 140.00 & 105.00 \\ 30.00 & 36.00 \\ 96.00 & 48.00 \end{bmatrix} = \begin{bmatrix} 91.00 & 68.25 \\ 19.50 & 23.40 \\ 62.40 & 31.20 \end{bmatrix}.$$

In other words, in U.S. dollars, $91 worth of wiper blades was sold in Vancouver, $68.25 worth of wiper blades was sold in Quebec, and so on.

Formally, scalar multiplication is defined as follows:

### Scalar Multiplication

If *A* is an $m \times n$ matrix and *c* is a real number, then $cA$ is the $m \times n$ matrix obtained by multiplying all the entries of *A* by *c*. (We usually use lowercase letters $c, d, e, \ldots$ to denote scalars.) Thus, the $ij$th entry of $cA$ is given by

$$(cA)_{ij} = c(A_{ij}).$$

In words, this rule is: To get the $ij$th entry of $cA$, multiply the $ij$th entry of *A* by *c*.

## EXAMPLE 4  Combining Operations

Let $A = \begin{bmatrix} 2 & -1 & 0 \\ 3 & 5 & -3 \end{bmatrix}$, $B = \begin{bmatrix} 1 & 3 & -1 \\ 5 & -6 & 0 \end{bmatrix}$, and $C = \begin{bmatrix} x & y & w \\ z & t+1 & 3 \end{bmatrix}$.

Evaluate the following: $4A, xB$, and $A + 3C$.

**Solution** First, we find $4A$ by multiplying each entry of $A$ by 4:

$$4A = 4 \begin{bmatrix} 2 & -1 & 0 \\ 3 & 5 & -3 \end{bmatrix} = \begin{bmatrix} 8 & -4 & 0 \\ 12 & 20 & -12 \end{bmatrix}.$$

Similarly, we find $xB$ by multiplying each entry of $B$ by $x$:

$$xB = x \begin{bmatrix} 1 & 3 & -1 \\ 5 & -6 & 0 \end{bmatrix} = \begin{bmatrix} x & 3x & -x \\ 5x & -6x & 0 \end{bmatrix}.$$

We get $A + 3C$ in two steps as follows:

$$A + 3C = \begin{bmatrix} 2 & -1 & 0 \\ 3 & 5 & -3 \end{bmatrix} + 3 \begin{bmatrix} x & y & w \\ z & t+1 & 3 \end{bmatrix}$$

$$= \begin{bmatrix} 2 & -1 & 0 \\ 3 & 5 & -3 \end{bmatrix} + \begin{bmatrix} 3x & 3y & 3w \\ 3z & 3t+3 & 9 \end{bmatrix}$$

$$= \begin{bmatrix} 2+3x & -1+3y & 3w \\ 3+3z & 3t+8 & 6 \end{bmatrix}.$$

Addition and scalar multiplication of matrices have nice properties, reminiscent of the properties of addition and multiplication of real numbers. Before we state them, we need to introduce some more notation.

If $A$ is any matrix, then $-A$ is the matrix $(-1)A$. In other words, $-A$ is $A$ multiplied by the scalar $-1$. This amounts to changing the signs of all the entries in $A$. For example,

$$-\begin{bmatrix} 4 & -2 & 0 \\ 6 & 10 & -6 \end{bmatrix} = \begin{bmatrix} -4 & 2 & 0 \\ -6 & -10 & 6 \end{bmatrix}.$$

For any two matrices $A$ and $B$, $A - B$ is the same as $A + (-B)$. (Why?)

Also, a **zero matrix** is a matrix all of whose entries are zero. Thus, for example, the $2 \times 3$ zero matrix is

$$O = \begin{bmatrix} 0 & 0 & 0 \\ 0 & 0 & 0 \end{bmatrix}.$$

Now we state the most important properties of the operations that we have been talking about:

## Properties of Matrix Addition and Scalar Multiplication

If $A$, $B$, and $C$ are any $m \times n$ matrices and if $O$ is the zero $m \times n$ matrix, then the following hold:

| | |
|---|---|
| $A + (B + C) = (A + B) + C$ | *Associative law* |
| $A + B = B + A$ | *Commutative law* |
| $A + O = O + A = A$ | *Additive identity law* |
| $A + (-A) = O = (-A) + A$ | *Additive inverse law* |
| $c(A + B) = cA + cB$ | *Distributive law* |
| $(c + d)A = cA + dA$ | *Distributive law* |
| $1A = A$ | *Scalar unit* |
| $0A = O$ | *Scalar zero* |

These properties would be obvious if we were talking about addition and multiplication of *numbers*, but here we are talking about addition and multiplication of *matrices*. We are using "+" to mean something new: matrix addition. There is no reason why matrix addition has to obey *all* the properties of addition of numbers. It happens that it does obey many of them, which is why it is convenient to call it *addition* in the first place. This means that we can manipulate equations involving matrices in much the same way that we manipulate equations involving numbers. One word of caution: We haven't yet discussed how to multiply matrices, and it probably isn't what you think. It will turn out that multiplication of matrices does *not* obey all the same properties as multiplication of numbers.

## Transposition

We mention one more operation on matrices:

### Transposition

If $A$ is an $m \times n$ matrix, then its **transpose** is the $n \times m$ matrix obtained by writing its rows as columns, so that the $i$th row of the original matrix becomes the $i$th column of the transpose. We denote the transpose of the matrix $A$ by $A^T$.

### Visualizing Transposition

$$\begin{bmatrix} 2 & -3 \\ 1 & 0 \\ 5 & 1 \end{bmatrix} \qquad \begin{bmatrix} 2 & 1 & 5 \\ -3 & 0 & 1 \end{bmatrix}$$

**using Technology**

Technology can be used to compute transposes like Quick Example 1 on the right:

**TI-83/84 Plus**
Enter the matrix [A] using
MATRX ; EDIT
Home Screen: [A]$^T$
(The symbol $^T$ is found in
MATRX ; MATH.)

**Spreadsheet**
Enter the matrix $A$ in a convenient $3 \times 4$ block of cells (e.g., A1–D3). Highlight a $4 \times 3$ block for the answer (e.g., A5–C8).
Type =TRANSPOSE(A1:D3)
Press Control+Shift+Enter

 **Website**
www.WanerMath.com
  Student Home
    → On Line Utilities
    → Matrix Algebra Tool
Enter the matrix $A$ as shown.

```
Enter your matrices here.
A = [2, 0, 1, 0
33, -22, 0, 5
1, -1, 2, -2]
```

Type A^T in the formula box and press "Compute".

### Quick Examples

**1.** Let $A = \begin{bmatrix} 2 & 0 & 1 & 0 \\ 33 & -22 & 0 & 5 \\ 1 & -1 & 2 & -2 \end{bmatrix}$. Then $A^T = \begin{bmatrix} 2 & 33 & 1 \\ 0 & -22 & -1 \\ 1 & 0 & 2 \\ 0 & 5 & -2 \end{bmatrix}$.

$3 \times 4$ matrix        $4 \times 3$ matrix

**2.** $\begin{bmatrix} -1 & 1 & 2 \end{bmatrix}^T = \begin{bmatrix} -1 \\ 1 \\ 2 \end{bmatrix}$.

$1 \times 3$ matrix    $3 \times 1$ matrix

## Properties of Transposition

If $A$ and $B$ are $m \times n$ matrices, then the following hold:

$$(A + B)^T = A^T + B^T$$
$$(cA)^T = c(A^T)$$
$$(A^T)^T = A.$$

To see why the laws of transposition are true, let us consider the first one: $(A + B)^T = A^T + B^T$. The left-hand side is the transpose of $A + B$, and so is obtained by first adding $A$ and $B$, and then writing the rows as columns. This is the same as first writing the rows of $A$ and $B$ individually as columns before adding, which gives the right-hand side. Similar arguments can be used to establish the other laws of transposition.

## 4.1 EXERCISES

▼ more advanced   ◆ challenging
🔲 indicates exercises that should be solved using technology

*In each of Exercises 1–10, find the dimensions of the given matrix and identify the given entry.*

**1.** $A = [1 \quad 5 \quad 0 \quad \frac{1}{4}]; a_{13}$    **2.** $B = [44 \quad 55]; b_{12}$

**3.** $C = \begin{bmatrix} \frac{5}{2} \\ 1 \\ -2 \\ 8 \end{bmatrix}; C_{11}$    **4.** $D = \begin{bmatrix} 15 & -18 \\ 6 & 0 \\ -6 & 5 \\ 48 & 18 \end{bmatrix}; d_{31}$

**5.** $E = \begin{bmatrix} e_{11} & e_{12} & e_{13} & \cdots & e_{1q} \\ e_{21} & e_{22} & e_{23} & \cdots & e_{2q} \\ \vdots & \vdots & \vdots & \ddots & \vdots \\ e_{p1} & e_{p2} & e_{p3} & \cdots & e_{pq} \end{bmatrix}; E_{22}$

**6.** $A = \begin{bmatrix} 2 & -1 & 0 \\ 3 & 5 & -3 \end{bmatrix}; A_{21}$    **7.** $B = \begin{bmatrix} 1 & 3 \\ 5 & -6 \end{bmatrix}; b_{12}$

**8.** $C = \begin{bmatrix} x & y & w & e \\ z & t+1 & 3 & 0 \end{bmatrix}; C_{23}$

**9.** $D = [d_1 \quad d_2 \quad \cdots \quad d_n]; D_{1r}$ (any $r$)

**10.** $E = [d \quad d \quad d \quad d]; E_{1r}$ (any $r$)

**11.** Solve for $x, y, z,$ and $w$. HINT [See Example 1.]
$$\begin{bmatrix} x+y & y+z \\ z+w & w \end{bmatrix} = \begin{bmatrix} 3 & 5 \\ 7 & 4 \end{bmatrix}$$

**12.** Solve for $x, y, z,$ and $w$. HINT [See Example 1.]
$$\begin{bmatrix} x-y & x-z \\ y-w & w \end{bmatrix} = \begin{bmatrix} 0 & 0 \\ 0 & 6 \end{bmatrix}$$

*In Exercises 13–20, evaluate the given expression. Take*
$$A = \begin{bmatrix} 0 & -1 \\ 1 & 0 \\ -1 & 2 \end{bmatrix}, B = \begin{bmatrix} 0.25 & -1 \\ 0 & 0.5 \\ -1 & 3 \end{bmatrix}, \text{ and}$$
$$C = \begin{bmatrix} 1 & -1 \\ 1 & 1 \\ -1 & -1 \end{bmatrix}.$$
HINT [See Example 4 and Quick Examples page 234.]

**13.** $A + B$    **14.** $A - C$
**15.** $A + B - C$    **16.** $12B$
**17.** $2A - C$    **18.** $2A + 0.5C$
**19.** $2A^T$    **20.** $A^T + 3C^T$

*In Exercises 21–28, evaluate the given expression. Take*
$$A = \begin{bmatrix} 1 & -1 & 0 \\ 0 & 2 & -1 \end{bmatrix}, B = \begin{bmatrix} 3 & 0 & -1 \\ 5 & -1 & 1 \end{bmatrix}, \text{ and}$$
$$C = \begin{bmatrix} x & 1 & w \\ z & r & 4 \end{bmatrix}.$$
HINT [See Example 4 and Quick Examples page 234.]

**21.** $A + B$    **22.** $B - C$
**23.** $A - B + C$    **24.** $\frac{1}{2}B$
**25.** $2A - B$    **26.** $2A - 4C$
**27.** $3B^T$    **28.** $2A^T - C^T$

🔲 *Use technology in Exercises 29–36.*
$$\text{Let } A = \begin{bmatrix} 1.5 & -2.35 & 5.6 \\ 44.2 & 0 & 12.2 \end{bmatrix}, B = \begin{bmatrix} 1.4 & 7.8 \\ 5.4 & 0 \\ 5.6 & 6.6 \end{bmatrix},$$
$$C = \begin{bmatrix} 10 & 20 & 30 \\ -10 & -20 & -30 \end{bmatrix}.$$

**29.** $A - C$    **30.** $C - A$
**31.** $1.1B$    **32.** $-0.2B$
**33.** $A^T + 4.2B$    **34.** $(A + 2.3C)^T$
**35.** $(2.1A - 2.3C)^T$    **36.** $(A - C)^T - B$

## APPLICATIONS

**37.** *Sales* The following table shows the number of Mac computers, iPods, and iPhones sold, in millions of units, in the first quarter (Q1) of 2009, and the changes over the previous year in each of 2010 Q1 and 2011 Q1.[1]

|  | Macs | iPods | iPhones |
|---|---|---|---|
| **Q1 2009** | 2.5 | 22.7 | 4.4 |
| **Change in 2010** | 0.8 | -1.7 | 4.3 |
| **Change in 2011** | 0.8 | -1.6 | 7.5 |

Use matrix algebra to find the sales of Macs, iPods, and iPhones in Q1 2010 and Q1 2011. HINT [See Example 2.]

[1] Apple's first quarter ends at the end of December on the preceding year. Source for data: Apple company reports (www.apple.com/investor).

**38. Home Prices** The following table shows median home prices, in thousands of dollars, in four regions of the U.S. in December 2010, as well as the month-over-month changes for the subsequent two months.[2]

|  | Northeast | Midwest | South | West |
|---|---|---|---|---|
| **Dec. 2010** | 238 | 140 | 149 | 205 |
| **Change in Jan. 2011** | −2 | −13 | −13 | −14 |
| **Change in Feb. 2011** | −6 | −5 | −1 | −1 |

Use matrix algebra to find the median home price in each region in January and February 2011. HINT [See Example 2.]

**39. Inventory** The *Left Coast Bookstore* chain has two stores, one in San Francisco and one in Los Angeles. It stocks three kinds of book: hardcover, softcover, and plastic (for infants). At the beginning of January, the central computer showed the following books in stock:

|  | Hard | Soft | Plastic |
|---|---|---|---|
| **San Francisco** | 1,000 | 2,000 | 5,000 |
| **Los Angeles** | 1,000 | 5,000 | 2,000 |

Suppose its sales in January were as follows: 700 hardcover books, 1,300 softcover books, and 2,000 plastic books sold in San Francisco, and 400 hardcover, 300 softcover, and 500 plastic books sold in Los Angeles. Write these sales figures in the form of a matrix, and then show how matrix algebra can be used to compute the inventory remaining in each store at the end of January.

**40. Inventory** The *Left Coast Bookstore* chain discussed in Exercise 39 actually maintained the same sales figures for the first 6 months of the year. Each month, the chain restocked the stores from its warehouse by shipping 600 hardcover, 1,500 softcover, and 1,500 plastic books to San Francisco and 500 hardcover, 500 softcover, and 500 plastic books to Los Angeles.

**a.** Use matrix operations to determine the total sales over the 6 months, broken down by store and type of book.

**b.** Use matrix operations to determine the inventory in each store at the end of June.

**41. Profit** Annual revenues and production costs at *Luddington's Wellington Boots & Co.* are shown in the following spreadsheet.

|  | A | B | C | D |
|---|---|---|---|---|
| 1 | Revenue | | | |
| 2 | | 2004 | 2005 | 2006 |
| 3 | Full Boots | $10,000 | $9,000 | $11,000 |
| 4 | Half Boots | $8,000 | $7,200 | $8,800 |
| 5 | Sandals | $4,000 | $5,000 | $6,000 |
| 6 | | | | |
| 7 | Production Costs | | | |
| 8 | | 2004 | 2005 | 2006 |
| 9 | Full Boots | $2,000 | $1,800 | $2,200 |
| 10 | Half Boots | $2,400 | $1,440 | $1,760 |
| 11 | Sandals | $1,200 | $1,500 | $2,000 |

Use matrix algebra to compute the profits from each sector each year.

**42. Revenue** The following spreadsheet gives annual production costs and profits at *Gauss-Jordan Sneakers, Inc.*

|  | A | B | C | D |
|---|---|---|---|---|
| 1 | Production Costs | | | |
| 2 | | 2004 | 2005 | 2006 |
| 3 | Gauss Grip | $1,800 | $2,200 | $2,400 |
| 4 | Air Gauss | $1,400 | $1,700 | $1,200 |
| 5 | Gauss Gel | $1,500 | $2,000 | $1,300 |
| 6 | | | | |
| 7 | Profit | | | |
| 8 | | 2004 | 2005 | 2006 |
| 9 | Gauss Grip | $10,000 | $14,000 | $16,000 |
| 10 | Air Gauss | $8,000 | $12,000 | $14,000 |
| 11 | Gauss Gel | $9,000 | $14,000 | $12,000 |

Use matrix algebra to compute the revenues from each sector each year.

**43. Population Movement** In 2000 the U.S. population, broken down by regions, was 53.6 million in the Northeast, 64.4 million in the Midwest, 100.2 million in the South, and 63.2 million in the West.[3] In 2010 the population was 55.3 million in the Northeast, 66.9 million in the Midwest, 114.6 million in the South, and 71.9 million in the West. Set up the population figures for each year as a row vector, and then show how to use matrix operations to find the net increase or decrease of population in each region from 2000 to 2010. Assuming the same population growth from 2010 to 2020 as from 2000 to 2010, use matrix operations to predict the population in each region in 2020.

**44. Population Movement** In 1990 the U.S. population, broken down by regions, was 50.8 million in the Northeast, 59.7 million in the Midwest, 85.4 million in the South, and 52.8 million in the West.[4] Between 1990 and 2000, the population in the Northeast grew by 2.8 million, the population in the Midwest grew by 4.7 million, the population in the South grew by 14.8 million, and the population in the West grew by 10.4 million. Set up the population figures for 1990 and the growth figures for the decade as row vectors. Assuming the same population growth from 2000 to 2010 as from 1990 to 2000, use matrix operations to estimate the population in each region in 2010. Compare the predicted 2010 population in the Northeast with the actual population given in the preceding exercise.

*Foreclosures Exercises 45–48 are based on the following table, which shows the numbers of foreclosures in three states in April through August of 2011.[5]*

|  | April | May | June | July | Aug. |
|---|---|---|---|---|---|
| **California** | 55,900 | 51,900 | 54,100 | 56,200 | 59,400 |
| **Florida** | 19,600 | 19,200 | 23,800 | 22,400 | 23,600 |
| **Texas** | 8,800 | 9,100 | 9,300 | 10,600 | 10,100 |

---

[2]Source for data: Census Bureau (www.census.gov).

[3]Source: U.S. Census Bureau http://2010.census.gov/2010census/data/apportionment-pop-text.php.

[4]*Ibid.*

[5]Figures are rounded. Source: www.realtytrac.com.

**45.** Use matrix algebra to determine the total number of foreclosures in each of the given months.

**46.** Use matrix algebra to determine the total number of foreclosures in each of the given states during the entire period shown.

**47.** Use matrix algebra to determine in which month the difference between the number of foreclosures in California and in Florida was greatest.

**48.** Use matrix algebra to determine in which region the difference between the number of foreclosures in April and August was greatest.

**49.** ▼ *Inventory* Microbucks Computer Company makes two computers, the Pomegranate II and the Pomegranate Classic, at two different factories. The Pom II requires 2 processor chips, 16 memory chips, and 20 vacuum tubes, while the Pom Classic requires 1 processor chip, 4 memory chips, and 40 vacuum tubes. Microbucks has in stock at the beginning of the year 500 processor chips, 5,000 memory chips, and 10,000 vacuum tubes at the Pom II factory, and 200 processor chips, 2,000 memory chips, and 20,000 vacuum tubes at the Pom Classic factory. It manufactures 50 Pom IIs and 50 Pom Classics each month.

   **a.** Find the company's inventory of parts after 2 months, using matrix operations.

   **b.** When (if ever) will the company run out of one of the parts?

**50.** ▼ *Inventory* Microbucks Computer Company, besides having the stock mentioned in Exercise 49, gets shipments of parts every month in the amounts of 100 processor chips, 1,000 memory chips, and 3,000 vacuum tubes at the Pom II factory, and 50 processor chips, 1,000 memory chips, and 2,000 vacuum tubes at the Pom Classic factory.

   **a.** What will the company's inventory of parts be after 6 months?

   **b.** When (if ever) will the company run out of one of the parts?

**51.** ▼ *Tourism in the 1990s* The following table gives the number of people (in thousands) who visited Australia and South Africa in 1998[6]:

| | To | Australia | South Africa |
|---|---|---|---|
| *From* | **North America** | 440 | 190 |
| | **Europe** | 950 | 950 |
| | **Asia** | 1,790 | 200 |

It was predicted that in 2008, 20,000 fewer people from North America would visit Australia and 40,000 more would visit South Africa, 50,000 more people from Europe would visit each of Australia and South Africa, and 100,000 more people from Asia would visit South Africa, but there would be no change in the number visiting Australia.

   **a.** Represent the changes predicted in 2008 in the form of a matrix, and use matrix algebra to predict the number

of visitors from the three regions to Australia and South Africa in 2008.

   **b.** Take $A$ to be the $3 \times 2$ matrix whose entries are the 1998 tourism figures and take $B$ to be the $3 \times 2$ matrix whose entries are the 2008 tourism figures. Give a formula (in terms of $A$ and $B$) that predicts the average of the numbers of visitors from the three regions to Australia and South Africa in 1998 and 2008. Compute its value.

**52.** ▼ *Tourism in the 1990s* Referring to the 1998 tourism figures given in the preceding exercise, assume that the following (fictitious) figures represent the corresponding numbers from 1988.

| | To | Australia | South Africa |
|---|---|---|---|
| *From* | **North America** | 500 | 100 |
| | **Europe** | 900 | 800 |
| | **Asia** | 1,400 | 50 |

Take $A$ to be the $3 \times 2$ matrix whose entries are the 1998 tourism figures and take $B$ to be the $3 \times 2$ matrix whose entries are the 1988 tourism figures.

   **a.** Compute the matrix $A - B$. What does this matrix represent?

   **b.** Assuming that the changes in tourism over 1988–1998 are repeated in 1998–2008, give a formula (in terms of $A$ and $B$) that predicts the number of visitors from the three regions to Australia and South Africa in 2008.

## COMMUNICATION AND REASONING EXERCISES

**53.** Is it possible for a $2 \times 3$ matrix to equal a $3 \times 2$ matrix? Explain.

**54.** If $A$ and $B$ are $2 \times 3$ matrices and $A = B$, what can you say about $A - B$? Explain.

**55.** What does it mean when we say that $(A + B)_{ij} = A_{ij} + B_{ij}$?

**56.** What does it mean when we say that $(cA)_{ij} = c(A_{ij})$?

**57.** What would a $5 \times 5$ matrix $A$ look like if $A_{ii} = 0$ for every $i$?

**58.** What would a matrix $A$ look like if $A_{ij} = 0$ whenever $i \neq j$?

**59.** ▼ Give a formula for the $ij$th entry of the transpose of a matrix $A$.

**60.** ▼ A matrix is **symmetric** if it is equal to its transpose. Give an example of **a.** a nonzero $2 \times 2$ symmetric matrix and **b.** a nonzero $3 \times 3$ symmetric matrix.

**61.** ▼ A matrix is **skew-symmetric** or **antisymmetric** if it is equal to the negative of its transpose. Give an example of **a.** a nonzero $2 \times 2$ skew-symmetric matrix and **b.** a nonzero $3 \times 3$ skew-symmetric matrix.

**62.** ▼ Referring to Exercises 60 and 61, what can be said about a matrix that is both symmetric and skew-symmetric?

**63.** ▼ Why is matrix addition associative?

**64.** Describe a scenario (possibly based on one of the preceding examples or exercises) in which you might wish to compute $A - 2B$ for certain matrices $A$ and $B$.

**65.** Describe a scenario (possibly based on one of the preceding examples or exercises) in which you might wish to compute $A + B - C$ for certain matrices $A$, $B$, and $C$.

---

[6]Figures are rounded to the nearest 10,000. Sources: South African Dept. of Environmental Affairs and Tourism; Australia Tourist Commission/*New York Times*, January 15, 2000, p. C1.

## **4.2** **Matrix Multiplication**

Suppose we download 3 movies at \$10 each and 5 Chopin albums at \$8 each. We calculate our total cost by computing the products' price × quantity and adding:

$$\text{Cost} = 10 \times 3 + 8 \times 5 = \$70.$$

Let us instead put the prices in a row vector

$$P = [10 \quad 8] \qquad \text{The price matrix}$$

and the quantities purchased in a column vector,

$$Q = \begin{bmatrix} 3 \\ 5 \end{bmatrix}. \qquad \text{The quantity matrix}$$

---

**Q**: *Why a row and a column instead of, say, two rows?*

**A**: It's rather a long story, but mathematicians found that it works best this way . . .

---

Because $P$ represents the prices of the items we are purchasing and $Q$ represents the quantities, it would be useful if the product $PQ$ represented the total cost, a *single number* (which we can think of as a $1 \times 1$ matrix). For this to work, $PQ$ should be calculated the same way we calculated the total cost:

$$PQ = [10 \quad 8] \begin{bmatrix} 3 \\ 5 \end{bmatrix} = [10 \times 3 + 8 \times 5] = [70].$$

Notice that we obtain the answer by multiplying each entry in $P$ (going from left to right) by the corresponding entry in $Q$ (going from top to bottom) and then adding the results.

---

### The Product *Row* × *Column*

The **product** $AB$ of a row matrix $A$ and a column matrix $B$ is a $1 \times 1$ matrix. The length of the row in $A$ must match the length of the column in $B$ for the product to be defined. To find the product, multiply each entry in $A$ (going from left to right) by the corresponding entry in $B$ (going from top to bottom) and then add the results.

### Visualizing Matrix Multiplication

$$\begin{bmatrix} 2 \\ 10 \\ -1 \end{bmatrix}$$

$$[2 \quad 4 \quad 1]$$

| | |
|---|---|
| $2 \times 2 \quad = \quad 4$ | Product of first entries $= 4$ |
| $4 \times 10 \quad = \quad 40$ | Product of second entries $= 40$ |
| $1 \times (-1) = \underline{-1}$ | Product of third entries $= -1$ |
| $\phantom{1 \times (-1) =} 43$ | Sum of products $= 43$ |

### Quick Examples

1. $[2 \quad 1]\begin{bmatrix} -3 \\ 1 \end{bmatrix} = [2 \times (-3) + 1 \times 1] = [-6 + 1] = [-5]$

2. $[2 \quad 4 \quad 1]\begin{bmatrix} 2 \\ 10 \\ -1 \end{bmatrix} = [2 \times 2 + 4 \times 10 + 1 \times (-1)] = [4 + 40 + (-1)]$

$$= [43]$$

---

**Notes**

1. In the discussion so far, *the row is on the left and the column is on the right* (RC again). (Later we will consider products where the column matrix is on the left and the row matrix is on the right.)

2. The row size has to match the column size. This means that, if we have a $1 \times 3$ row on the left, then the column on the right must be $3 \times 1$ in order for the product to make sense. For example, the product

$$[a \quad b \quad c]\begin{bmatrix} x \\ y \end{bmatrix}$$

is not defined.  ∎

**EXAMPLE 1  Revenue**

The A-Plus auto parts store mentioned in examples in the previous section had the following sales in its Vancouver store:

|  | Vancouver |
|---|---|
| **Wiper Blades** | 20 |
| **Cleaning Fluid (bottles)** | 10 |
| **Floor Mats** | 8 |

The store sells wiper blades for $7.00 each, cleaning fluid for $3.00 per bottle, and floor mats for $12.00 each. Use matrix multiplication to find the total revenue generated by sales of these items.

**Solution**  We need to multiply each sales figure by the corresponding price and then add the resulting revenue figures. We represent the sales by a column vector, as suggested by the table:

$$Q = \begin{bmatrix} 20 \\ 10 \\ 8 \end{bmatrix}.$$

We put the selling prices in a row vector:

$$P = [7.00 \quad 3.00 \quad 12.00].$$

We can now compute the total revenue as the product

$$R = PQ = [7.00 \quad 3.00 \quad 12.00]\begin{bmatrix} 20 \\ 10 \\ 8 \end{bmatrix}$$

$$= [140.00 + 30.00 + 96.00] = [266.00].$$

So, the sale of these items generated a total revenue of $266.00.

**Note**  We could also have written the quantity sold as a row vector (which would be $Q^T$) and the prices as a column vector (which would be $P^T$) and then multiplied them in the opposite order ($Q^T P^T$). Try this.  ∎

### EXAMPLE 2 **Relationship with Linear Equations**

**a.** Represent the matrix equation

$$[2 \quad -4 \quad 1] \begin{bmatrix} x \\ y \\ z \end{bmatrix} = [5]$$

as an ordinary equation.

**b.** Represent the linear equation $3x + y - z + 2w = 8$ as a matrix equation.

**Solution**

**a.** If we perform the multiplication on the left, we get the $1 \times 1$ matrix $[2x - 4y + z]$. Thus the equation may be rewritten as

$$[2x - 4y + z] = [5]. \quad \text{\small $1 \times 1$ matrix on the left $= 1 \times 1$ matrix on the right}$$

Saying that these two $1 \times 1$ matrices are equal means that their entries are equal, so we get the equation

$$2x - 4y + z = 5.$$

**b.** This is the reverse of part (a):

$$[3 \quad 1 \quad -1 \quad 2] \begin{bmatrix} x \\ y \\ z \\ w \end{bmatrix} = [8].$$

➡ **Before we go on...** The row matrix $[3 \quad 1 \quad -1 \quad 2]$ in Example 2 is the row of **coefficients** of the original equation. (See Section 3.1.) ∎

Now to the general case of matrix multiplication:

## The Product of Two Matrices: General Case

In general, for matrices $A$ and $B$, we can take the product $AB$ only if the number of columns of $A$ equals the number of rows of $B$ (so that we can multiply the rows of $A$ by the columns of $B$ as above). The product $AB$ is then obtained by taking its $ij$th entry to be

$$ij\text{th entry of } AB = \text{Row } i \text{ of } A \times \text{Column } j \text{ of } B. \quad \text{\small As defined above}$$

### Quick Examples

($R$ stands for row; $C$ stands for column.)

$$1. \; R_1 \to [2 \quad 0 \quad -1 \quad 3] \begin{matrix} C_1 & C_2 & C_3 \\ \downarrow & \downarrow & \downarrow \end{matrix} \begin{bmatrix} 1 & 1 & -8 \\ 1 & -6 & 0 \\ 0 & 5 & 2 \\ -3 & 8 & 1 \end{bmatrix} = [R_1 \times C_1 \quad R_1 \times C_2 \quad R_1 \times C_3]$$

$$= [-7 \quad 21 \quad -15]$$

$$2. \; \begin{matrix} R_1 \to \\ R_2 \to \end{matrix} \begin{bmatrix} 1 & -1 \\ 0 & 2 \end{bmatrix} \begin{matrix} C_1 & C_2 \\ \downarrow & \downarrow \end{matrix} \begin{bmatrix} 3 & 0 \\ 5 & -1 \end{bmatrix} = \begin{bmatrix} R_1 \times C_1 & R_1 \times C_2 \\ R_2 \times C_1 & R_2 \times C_2 \end{bmatrix} = \begin{bmatrix} -2 & 1 \\ 10 & -2 \end{bmatrix}$$

In matrix multiplication we always take

Rows on the left × Columns on the right.

Look at the dimensions in the two Quick Examples on the previous page.

$$(1 \times 4)\overset{\text{Match}}{(4} \times 3) \to 1 \times 3 \qquad (2 \times 2)\overset{\text{Match}}{(2} \times 2) \to 2 \times 2$$

The fact that the number of columns in the left-hand matrix equals the number of rows in the right-hand matrix amounts to saying that the middle two numbers must match as above. If we "cancel" the middle matching numbers, we are left with the dimensions of the product.

Before continuing with examples, we state the rule for matrix multiplication formally.

## Multiplication of Matrices: Formal Definition

If $A$ is an $m \times n$ matrix and $B$ is an $n \times k$ matrix, then the product $AB$ is the $m \times k$ matrix whose $ij$th entry is the product

$$\text{Row } i \text{ of } A \times \text{Column } j \text{ of } B$$

$$(AB)_{ij} = [a_{i1} \; a_{i2} \; a_{i3} \ldots a_{in}] \begin{bmatrix} b_{1j} \\ b_{2j} \\ b_{3j} \\ \vdots \\ b_{nj} \end{bmatrix} = a_{i1}b_{1j} + a_{i2}b_{2j} + a_{i3}b_{3j} + \cdots + a_{in}b_{nj}.$$

### EXAMPLE 3  Matrix Product

Calculate:

**a.** $\begin{bmatrix} 2 & 0 & -1 & 3 \\ 1 & -1 & 2 & -2 \end{bmatrix} \begin{bmatrix} 1 & 1 & -8 \\ 1 & 0 & 0 \\ 0 & 5 & 2 \\ -2 & 8 & -1 \end{bmatrix}$    **b.** $\begin{bmatrix} -3 \\ 1 \end{bmatrix} [2 \quad 1]$

### Solution

**a.** Before we start the calculation, we check that the dimensions of the matrices match up.

$$\underset{2 \times 4}{} \overset{\text{Match}}{\diagup \diagdown} \underset{4 \times 3}{}$$

$$\begin{bmatrix} 2 & 0 & -1 & 3 \\ 1 & -1 & 2 & -2 \end{bmatrix} \begin{bmatrix} 1 & 1 & -8 \\ 1 & 0 & 0 \\ 0 & 5 & 2 \\ -2 & 8 & -1 \end{bmatrix}$$

Technology can be used to multiply the matrices in Example 3. Here is an outline for part (a) (see the Technology Guides at the end of the chapter for additional details on using a TI-83/84 Plus or a spreadsheet):

**TI-83/84 Plus**
Enter the matrices $A$ and $B$ using MATRX; EDIT
Home Screen: [A]*[B]
[More details on page 304.]

**Spreadsheet**
Enter the matrices in convenient blocks of cells (e.g., A1–D2 and F1–H4).
Highlight a $2 \times 3$ block for the answer.
Type =MMULT(A1:D2,F1:H4)
Press Control+Shift+Enter
[More details on page 308.]

**Website**
www.WanerMath.com
    Student Home
        → On Line Utilities
        → Matrix Algebra Tool
Enter matrices as shown:

```
Enter your matrices here.
A = [2, 0, -1, 3
1, -1, 2, -2]

B = [1, 1, -8
1, 0, 0
0, 5, 2
-2, 8, -1]
```

Type A*B in the formula box and press "Compute". If you try to multiply two matrices whose product is not defined, you will get an error alert box telling you that.

The product of the two matrices is defined, and the product will be a $2 \times 3$ matrix (we remove the matching 4s: $(2 \times 4)(4 \times 3) \to 2 \times 3$). To calculate the product, we follow the previous prescription:

$$
\begin{array}{c}
R_1 \to \\ R_2 \to
\end{array}
\begin{bmatrix} 2 & 0 & -1 & 3 \\ 1 & -1 & 2 & -2 \end{bmatrix}
\begin{array}{c}
\begin{matrix} C_1 & C_2 & C_3 \\ \downarrow & \downarrow & \downarrow \end{matrix} \\
\begin{bmatrix} 1 & 1 & -8 \\ 1 & 0 & 0 \\ 0 & 5 & 2 \\ -2 & 8 & -1 \end{bmatrix}
\end{array}
= \begin{bmatrix} R_1 \times C_1 & R_1 \times C_2 & R_1 \times C_3 \\ R_2 \times C_1 & R_2 \times C_2 & R_2 \times C_3 \end{bmatrix}
$$

$$
= \begin{bmatrix} -4 & 21 & -21 \\ 4 & -5 & -2 \end{bmatrix}.
$$

**b.** The dimensions of the two matrices given are $2 \times 1$ and $1 \times 2$. Because the 1s match, the product is defined, and the result will be a $2 \times 2$ matrix.

$$
\begin{array}{c}
R_1 \to \\ R_2 \to
\end{array}
\begin{bmatrix} -3 \\ 1 \end{bmatrix}
\begin{matrix} C_1 & C_2 \\ \downarrow & \downarrow \end{matrix}
[2 \quad 1]
= \begin{bmatrix} R_1 \times C_1 & R_1 \times C_2 \\ R_2 \times C_1 & R_2 \times C_2 \end{bmatrix}
= \begin{bmatrix} -6 & -3 \\ 2 & 1 \end{bmatrix}.
$$

**Note** In part (a) we *cannot* multiply the matrices in the opposite order—the dimensions do not match. We say simply that the product in the opposite order is **not defined**. In part (b) we *can* multiply the matrices in the opposite order, but we would get a $1 \times 1$ matrix if we did so. Thus, order is important when multiplying matrices. In general, if $AB$ is defined, then $BA$ need not even be defined. If $BA$ is also defined, it may not have the same dimensions as $AB$. And even if $AB$ and $BA$ have the same dimensions, they may have different entries. (See the next example.) ∎

---

**EXAMPLE 4    AB versus BA**

Let $A = \begin{bmatrix} 1 & -1 \\ 0 & 2 \end{bmatrix}$ and $B = \begin{bmatrix} 3 & 0 \\ 5 & -1 \end{bmatrix}$. Find $AB$ and $BA$.

**Solution** Note first that $A$ and $B$ are both $2 \times 2$ matrices, so the products $AB$ and $BA$ are both defined and are both $2 \times 2$ matrices—unlike the case in Example 3(b). We first calculate $AB$:

$$
AB = \begin{bmatrix} 1 & -1 \\ 0 & 2 \end{bmatrix} \begin{bmatrix} 3 & 0 \\ 5 & -1 \end{bmatrix} = \begin{bmatrix} -2 & 1 \\ 10 & -2 \end{bmatrix}.
$$

Now let's calculate $BA$:

$$
BA = \begin{bmatrix} 3 & 0 \\ 5 & -1 \end{bmatrix} \begin{bmatrix} 1 & -1 \\ 0 & 2 \end{bmatrix} = \begin{bmatrix} 3 & -3 \\ 5 & -7 \end{bmatrix}.
$$

Notice that $BA$ has no resemblance to $AB$! Thus, we have discovered that, even for square matrices:

*Matrix multiplication is not commutative.*

In other words, $AB \neq BA$ in general, even when $AB$ and $BA$ both exist and have the same dimensions. (There are instances when $AB = BA$ for particular matrices $A$ and $B$, but this is an exception, not the rule.)

### EXAMPLE 5 Revenue

January sales at the A-Plus auto parts stores in Vancouver and Quebec are given in the following table.

|  | Vancouver | Quebec |
|---|---|---|
| Wiper Blades | 20 | 15 |
| Cleaning Fluid (bottles) | 10 | 12 |
| Floor Mats | 8 | 4 |

The usual selling prices for these items are $7.00 each for wiper blades, $3.00 per bottle for cleaning fluid, and $12.00 each for floor mats. The discount prices for A-Plus Club members are $6.00 each for wiper blades, $2.00 per bottle for cleaning fluid, and $10.00 each for floor mats. Use matrix multiplication to compute the total revenue at each store, assuming first that all items were sold at the usual prices, and then that they were all sold at the discount prices.

**Solution** We can do all of the requested calculations at once with a single matrix multiplication. Consider the following two labeled matrices.

$$Q = \begin{array}{c} \\ \textbf{Wb} \\ \textbf{Cf} \\ \textbf{Fm} \end{array} \begin{array}{cc} \textbf{V} & \textbf{Q} \\ \left[\begin{array}{cc} 20 & 15 \\ 10 & 12 \\ 8 & 4 \end{array}\right] \end{array}$$

$$P = \begin{array}{c} \\ \textbf{Usual} \\ \textbf{Discount} \end{array} \begin{array}{ccc} \textbf{Wb} & \textbf{Cf} & \textbf{Fm} \\ \left[\begin{array}{ccc} 7.00 & 3.00 & 12.00 \\ 6.00 & 2.00 & 10.00 \end{array}\right] \end{array}$$

The first matrix records the quantities sold, while the second records the sales prices under the two assumptions. To compute the revenue at both stores under the two different assumptions, we calculate $R = PQ$.

$$R = PQ = \begin{bmatrix} 7.00 & 3.00 & 12.00 \\ 6.00 & 2.00 & 10.00 \end{bmatrix} \begin{bmatrix} 20 & 15 \\ 10 & 12 \\ 8 & 4 \end{bmatrix}$$

$$= \begin{bmatrix} 266.00 & 189.00 \\ 220.00 & 154.00 \end{bmatrix}$$

We can label this matrix as follows:

$$R = \begin{array}{c} \\ \textbf{Usual} \\ \textbf{Discount} \end{array} \begin{array}{cc} \textbf{V} & \textbf{Q} \\ \left[\begin{array}{cc} 266.00 & 189.00 \\ 220.00 & 154.00 \end{array}\right] \end{array}.$$

In other words, if the items were sold at the usual price, then Vancouver had a revenue of $266 while Quebec had a revenue of $189, and so on.

➡ **Before we go on...** In Example 5 we were able to calculate $PQ$ because the dimensions matched correctly: $(2 \times 3)(3 \times 2) \to 2 \times 2$. We could also have multiplied them in the opposite order and gotten a $3 \times 3$ matrix. Would the product $QP$ be meaningful? In an application like this, not only do the dimensions have to match, but also the *labels* have to match for the result to be meaningful. The labels on the three columns of $P$ are the parts that were sold, and these are also the labels on the three rows of $Q$. Therefore, we can "cancel labels" at the same time that we cancel the dimensions in the product. However, the labels on the two columns of $Q$ do not match the labels on the two rows of $P$, and there is no useful interpretation of the product $QP$ in this situation.  ∎

There are very special square matrices of every size: $1 \times 1$, $2 \times 2$, $3 \times 3$, and so on, called the **identity** matrices.

## Identity Matrix

The $n \times n$ identity matrix $I$ is the matrix with 1s down the **main diagonal** (the diagonal starting at the top left) and 0s everywhere else. In symbols,

$$I_{ii} = 1, \quad \text{and}$$
$$I_{ij} = 0 \quad \text{if } i \neq j.$$

### Quick Examples

**1.** $1 \times 1$ identity matrix $\quad I = [1]$

**2.** $2 \times 2$ identity matrix $\quad I = \begin{bmatrix} 1 & 0 \\ 0 & 1 \end{bmatrix}$

**3.** $3 \times 3$ identity matrix $\quad I = \begin{bmatrix} 1 & 0 & 0 \\ 0 & 1 & 0 \\ 0 & 0 & 1 \end{bmatrix}$

**4.** $4 \times 4$ identity matrix $\quad I = \begin{bmatrix} 1 & 0 & 0 & 0 \\ 0 & 1 & 0 & 0 \\ 0 & 0 & 1 & 0 \\ 0 & 0 & 0 & 1 \end{bmatrix}$

**Note** Identity matrices are always square matrices, meaning that they have the same number of rows as columns. There is no such thing, for example, as the "$2 \times 4$ identity matrix."

The next example shows why $I$ is interesting.  ∎

---

### EXAMPLE 6    Identity Matrix

Evaluate the products $AI$ and $IA$, where $A = \begin{bmatrix} a & b & c \\ d & e & f \\ g & h & i \end{bmatrix}$ and $I$ is the $3 \times 3$ identity matrix.

### Solution

First notice that $A$ is arbitrary; it could be any $3 \times 3$ matrix.

$$AI = \begin{bmatrix} a & b & c \\ d & e & f \\ g & h & i \end{bmatrix} \begin{bmatrix} 1 & 0 & 0 \\ 0 & 1 & 0 \\ 0 & 0 & 1 \end{bmatrix} = \begin{bmatrix} a & b & c \\ d & e & f \\ g & h & i \end{bmatrix}$$

and

$$IA = \begin{bmatrix} 1 & 0 & 0 \\ 0 & 1 & 0 \\ 0 & 0 & 1 \end{bmatrix} \begin{bmatrix} a & b & c \\ d & e & f \\ g & h & i \end{bmatrix} = \begin{bmatrix} a & b & c \\ d & e & f \\ g & h & i \end{bmatrix}.$$

In both cases, the answer is the matrix $A$ we started with. In symbols,

$$AI = A$$

and

$$IA = A$$

no matter which $3 \times 3$ matrix $A$ you start with. Now this should remind you of a familiar fact from arithmetic:

$$a \cdot 1 = a$$

and

$$1 \cdot a = a.$$

That is why we call the matrix $I$ the $3 \times 3$ *identity* matrix, because it appears to play the same role for $3 \times 3$ matrices that the identity 1 does for numbers.

➡ **Before we go on...** Try a similar calculation using $2 \times 2$ matrices: Let $A = \begin{bmatrix} a & b \\ c & d \end{bmatrix}$, let $I$ be the $2 \times 2$ identity matrix, and check that $AI = IA = A$. In fact, the equation

$$AI = IA = A$$

works for square matrices of every dimension. It is also interesting to notice that $AI = A$ if $I$ is the $2 \times 2$ identity matrix and $A$ is any $3 \times 2$ matrix (try one). In fact, if $I$ is any identity matrix, then $AI = A$ whenever the product is defined, and $IA = A$ whenever this product is defined. ■

We can now add to the list of properties we gave for matrix arithmetic at the end of Section 4.1 by writing down properties of matrix multiplication. In stating these properties, we shall assume that all matrix products we write are defined—that is, that the matrices have correctly matching dimensions. The first eight properties are the ones we've already seen; the rest are new.

## Properties of Matrix Addition and Multiplication

If $A$, $B$, and $C$ are matrices, if $O$ is a zero matrix, and if $I$ is an identity matrix, then the following hold:

| | |
|---|---|
| $A + (B + C) = (A + B) + C$ | *Additive associative law* |
| $A + B = B + A$ | *Additive commutative law* |
| $A + O = O + A = A$ | *Additive identity law* |
| $A + (-A) = O = (-A) + A$ | *Additive inverse law* |
| $c(A + B) = cA + cB$ | *Distributive law* |
| $(c + d)A = cA + dA$ | *Distributive law* |
| $1A = A$ | *Scalar unit* |

$$
\begin{array}{ll}
0A = O & \textit{Scalar zero} \\
A(BC) = (AB)C & \textit{Multiplicative associative law} \\
c(AB) = (cA)B & \textit{Multiplicative associative law} \\
c(dA) = (cd)A & \textit{Multiplicative associative law} \\
AI = IA = A & \textit{Multiplicative identity law} \\
A(B + C) = AB + AC & \textit{Distributive law} \\
(A + B)C = AC + BC & \textit{Distributive law} \\
OA = AO = O & \textit{Multiplication by zero matrix}
\end{array}
$$

Note that we have not included a multiplicative commutative law for matrices, because the equation $AB = BA$ does not hold in general. In other words, matrix multiplication is *not* exactly like multiplication of numbers. (You have to be a little careful because it is easy to apply the commutative law without realizing it.)

We should also say a bit more about transposition. Transposition and multiplication have an interesting relationship. We write down the properties of transposition again, adding one new one.

## Properties of Transposition

$$
(A + B)^T = A^T + B^T
$$
$$
(cA)^T = c(A^T)
$$
$$
(AB)^T = B^T A^T
$$

Notice the change in order in the last one. The order is crucial.

### Quick Examples

1. $\left( \begin{bmatrix} 1 & -1 \\ 0 & 2 \end{bmatrix} \begin{bmatrix} 3 & 0 \\ 5 & -1 \end{bmatrix} \right)^T = \begin{bmatrix} -2 & 1 \\ 10 & -2 \end{bmatrix}^T = \begin{bmatrix} -2 & 10 \\ 1 & -2 \end{bmatrix}$   $(AB)^T$

2. $\begin{bmatrix} 3 & 0 \\ 5 & -1 \end{bmatrix}^T \begin{bmatrix} 1 & -1 \\ 0 & 2 \end{bmatrix}^T = \begin{bmatrix} 3 & 5 \\ 0 & -1 \end{bmatrix} \begin{bmatrix} 1 & 0 \\ -1 & 2 \end{bmatrix} = \begin{bmatrix} -2 & 10 \\ 1 & -2 \end{bmatrix}$   $B^T A^T$

3. $\begin{bmatrix} 1 & -1 \\ 0 & 2 \end{bmatrix}^T \begin{bmatrix} 3 & 0 \\ 5 & -1 \end{bmatrix}^T = \begin{bmatrix} 1 & 0 \\ -1 & 2 \end{bmatrix} \begin{bmatrix} 3 & 5 \\ 0 & -1 \end{bmatrix} = \begin{bmatrix} 3 & 5 \\ -3 & -7 \end{bmatrix}$   $A^T B^T$

These properties give you a glimpse of the field of mathematics known as **abstract algebra**. Algebraists study operations like these that resemble the operations on numbers but differ in some way, such as the lack of commutativity for multiplication seen here.

We end this section with more on the relationship between linear equations and matrix equations, which is one of the important applications of matrix multiplication.

---

### EXAMPLE 7 Matrix Form of a System of Linear Equations

**a.** If

$$
A = \begin{bmatrix} 1 & -2 & 3 \\ 2 & 0 & -1 \\ -3 & 1 & 1 \end{bmatrix}, X = \begin{bmatrix} x \\ y \\ z \end{bmatrix}, \text{ and } B = \begin{bmatrix} 3 \\ -1 \\ 0 \end{bmatrix},
$$

rewrite the matrix equation $AX = B$ as a system of linear equations.

**b.** Express the following system of equations as a matrix equation of the form $AX = B$:

$$2x + y = 3$$
$$4x - y = -1.$$

**Solution**

**a.** The matrix equation $AX = B$ is

$$\begin{bmatrix} 1 & -2 & 3 \\ 2 & 0 & -1 \\ -3 & 1 & 1 \end{bmatrix} \begin{bmatrix} x \\ y \\ z \end{bmatrix} = \begin{bmatrix} 3 \\ -1 \\ 0 \end{bmatrix}.$$

As in Example 2(a), we first evaluate the left-hand side and then set it equal to the right-hand side.

$$\begin{bmatrix} 1 & -2 & 3 \\ 2 & 0 & -1 \\ -3 & 1 & 1 \end{bmatrix} \begin{bmatrix} x \\ y \\ z \end{bmatrix} = \begin{bmatrix} x - 2y + 3z \\ 2x - z \\ -3x + y + z \end{bmatrix}$$

$$\begin{bmatrix} x - 2y + 3z \\ 2x - z \\ -3x + y + z \end{bmatrix} = \begin{bmatrix} 3 \\ -1 \\ 0 \end{bmatrix}$$

Because these two matrices are equal, their corresponding entries must be equal.

$$x - 2y + 3z = 3$$
$$2x \quad\quad - z = -1$$
$$-3x + y + z = 0$$

In other words, the matrix equation $AX = B$ is equivalent to this system of linear equations. Notice that the coefficients of the left-hand sides of these equations are the entries of the matrix $A$. We call $A$ the **coefficient matrix** of the system of equations. The entries of $X$ are the unknowns and the entries of $B$ are the right-hand sides $3, -1$, and $0$.

**b.** As we saw in part (a), the coefficient matrix $A$ has entries equal to the coefficients of the left-hand sides of the equations. Thus,

$$A = \begin{bmatrix} 2 & 1 \\ 4 & -1 \end{bmatrix}.$$

$X$ is the column matrix consisting of the unknowns, while $B$ is the column matrix consisting of the right-hand sides of the equations, so

$$X = \begin{bmatrix} x \\ y \end{bmatrix} \quad \text{and} \quad B = \begin{bmatrix} 3 \\ -1 \end{bmatrix}.$$

The system of equations can be rewritten as the matrix equation $AX = B$ with this $A$, $X$, and $B$.

This translation of systems of linear equations into matrix equations is really the first step in the method of solving linear equations discussed in Chapter 3. There we worked with the **augmented matrix** of the system, which is simply $A$ with $B$ adjoined as an extra column.

**Q:** *When we write a system of equations as $AX = B$, couldn't we solve for the unknown $X$ by dividing both sides by $A$?*

**A:** If we interpret division as multiplication by the inverse (for example, $2 \div 3 = 2 \times 3^{-1}$), we shall see in the next section that *certain* systems of the form $AX = B$ can be solved in this way, by multiplying both sides by $A^{-1}$. We first need to discuss what we mean by $A^{-1}$ and how to calculate it.

## 4.2  EXERCISES

▼ more advanced   ◆ challenging

**T** indicates exercises that should be solved using technology

In Exercises 1–28, compute the products. Some of these may be undefined. Exercises marked **T** should be done using technology. The others should be done two ways: by hand and by using technology where possible. HINT [See Example 3.]

**1.** $\begin{bmatrix} 1 & 3 & -1 \end{bmatrix} \begin{bmatrix} 9 \\ 1 \\ -1 \end{bmatrix}$

**2.** $\begin{bmatrix} 4 & 0 & -1 \end{bmatrix} \begin{bmatrix} -4 \\ 1 \\ 8 \end{bmatrix}$

**3.** $\begin{bmatrix} -1 & \frac{1}{2} \end{bmatrix} \begin{bmatrix} -\frac{1}{3} \\ 1 \end{bmatrix}$

**4.** $\begin{bmatrix} -1 & 1 \end{bmatrix} \begin{bmatrix} \frac{3}{4} \\ \frac{1}{4} \end{bmatrix}$

**5.** $\begin{bmatrix} 0 & -2 & 1 \end{bmatrix} \begin{bmatrix} x \\ y \\ z \end{bmatrix}$

**6.** $\begin{bmatrix} 4 & -1 & 1 \end{bmatrix} \begin{bmatrix} -x \\ x \\ y \end{bmatrix}$

**7.** $\begin{bmatrix} 1 & 3 & 2 \end{bmatrix} \begin{bmatrix} 1 \\ 1 \\ -1 \end{bmatrix}$

**8.** $\begin{bmatrix} 3 & 2 \end{bmatrix} \begin{bmatrix} 1 & -2 \end{bmatrix}$

**9.** $\begin{bmatrix} -1 & 1 \end{bmatrix} \begin{bmatrix} -3 & 1 & 4 & 3 \\ 0 & 1 & -2 & 1 \end{bmatrix}$

**10.** $\begin{bmatrix} 2 & -1 \end{bmatrix} \begin{bmatrix} -3 & 1 & 4 & 3 \\ 4 & 0 & 1 & 3 \end{bmatrix}$

**11.** $\begin{bmatrix} 1 & -1 & 2 & 3 \end{bmatrix} \begin{bmatrix} -1 & 2 & 0 \\ 2 & -1 & 0 \\ 0 & 5 & 2 \\ -1 & 8 & 1 \end{bmatrix}$

**12.** $\begin{bmatrix} 0 & 1 & -1 & 2 \end{bmatrix} \begin{bmatrix} 1 & -2 & 1 \\ 0 & 1 & 3 \\ 6 & 0 & 2 \\ -1 & -2 & 11 \end{bmatrix}$

**13.** $\begin{bmatrix} 1 & 0 & -1 \\ 1 & 1 & 2 \end{bmatrix} \begin{bmatrix} 0 & 1 & -1 \\ 1 & 0 & 1 \\ 4 & 8 & 0 \end{bmatrix}$

**14.** $\begin{bmatrix} 0 & 1 & -1 \\ 3 & 1 & -1 \end{bmatrix} \begin{bmatrix} 1 & 1 \\ 4 & 2 \\ 0 & 1 \end{bmatrix}$

**15.** $\begin{bmatrix} 1 & 0 \\ 1 & -1 \end{bmatrix} \begin{bmatrix} 0 & 1 \\ 0 & 1 \end{bmatrix}$

**16.** $\begin{bmatrix} 1 & -1 \\ 1 & -1 \end{bmatrix} \begin{bmatrix} 3 & -3 \\ 5 & -7 \end{bmatrix}$

**17.** $\begin{bmatrix} 0 & 1 \\ 0 & 1 \end{bmatrix} \begin{bmatrix} 1 & 0 \\ 1 & -1 \end{bmatrix}$

**18.** $\begin{bmatrix} 3 & -3 \\ 5 & -7 \end{bmatrix} \begin{bmatrix} 1 & -1 \\ 1 & -1 \end{bmatrix}$

**19.** $\begin{bmatrix} 1 & -1 \\ 1 & -1 \end{bmatrix} \begin{bmatrix} 2 & 3 \\ 2 & 3 \end{bmatrix}$

**20.** $\begin{bmatrix} 0 & 1 \\ 1 & 0 \end{bmatrix} \begin{bmatrix} 3 & -3 \\ 2 & -1 \end{bmatrix}$

**21.** $\begin{bmatrix} 1 & -1 \\ -1 & 1 \end{bmatrix} \begin{bmatrix} 2 & 3 \\ 2 & 3 \\ 1 & 1 \end{bmatrix}$

**22.** $\begin{bmatrix} 0 & 1 & -1 \\ 0 & -1 & 1 \end{bmatrix} \begin{bmatrix} 3 & -3 \\ 2 & -1 \end{bmatrix}$

**23.** $\begin{bmatrix} 1 & 0 & -1 \\ 2 & -2 & 1 \\ 0 & 0 & 1 \end{bmatrix} \begin{bmatrix} 1 & -1 & 4 \\ 1 & 1 & 0 \\ 0 & 4 & 1 \end{bmatrix}$

**24.** $\begin{bmatrix} 1 & 2 & 0 \\ 4 & -1 & 1 \\ 1 & 0 & 1 \end{bmatrix} \begin{bmatrix} 1 & 2 & -4 \\ 4 & 1 & 0 \\ 0 & -2 & 1 \end{bmatrix}$

**25.** $\begin{bmatrix} 1 & 0 & 1 & 0 \\ -1 & 1 & 0 & 1 \\ -2 & 0 & 1 & 4 \\ 0 & -1 & 0 & 1 \end{bmatrix} \begin{bmatrix} 1 \\ -3 \\ 2 \\ 0 \end{bmatrix}$

**26.** $\begin{bmatrix} 1 & 1 & -7 & 0 \\ -1 & 0 & 2 & 4 \\ -1 & 0 & -2 & 1 \\ 1 & -1 & 1 & 1 \end{bmatrix} \begin{bmatrix} 1 \\ -3 \\ 2 \\ 1 \end{bmatrix}$

**27. T** $\begin{bmatrix} 1.1 & 2.3 & 3.4 & -1.2 \\ 3.4 & 4.4 & 2.3 & 1.1 \\ 2.3 & 0 & -2.2 & 1.1 \\ 1.2 & 1.3 & 1.1 & 1.1 \end{bmatrix} \begin{bmatrix} -2.1 & 0 & -3.3 \\ -3.4 & -4.8 & -4.2 \\ 3.4 & 5.6 & 1 \\ 1 & 2.2 & 9.8 \end{bmatrix}$

**28. T** $\begin{bmatrix} 1.2 & 2.3 & 3.4 & 4.5 \\ 3.3 & 4.4 & 5.5 & 6.6 \\ 2.3 & -4.3 & -2.2 & 1.1 \\ 2.2 & -1.2 & -1 & 1.1 \end{bmatrix} \begin{bmatrix} 9.8 & 1 & -1.1 \\ 8.8 & 2 & -2.2 \\ 7.7 & 3 & -3.3 \\ 6.6 & 4 & -4.4 \end{bmatrix}$

**29.** Find[7] $A^2 = A \cdot A$, $A^3 = A \cdot A \cdot A$, $A^4$, and $A^{100}$, given that

$$A = \begin{bmatrix} 0 & 1 & 1 & 1 \\ 0 & 0 & 1 & 1 \\ 0 & 0 & 0 & 1 \\ 0 & 0 & 0 & 0 \end{bmatrix}.$$

**30.** Repeat the preceding exercise with $A = \begin{bmatrix} 0 & 2 & 0 & -1 \\ 0 & 0 & 2 & 0 \\ 0 & 0 & 0 & 2 \\ 0 & 0 & 0 & 0 \end{bmatrix}.$

*Exercises 31–38 should be done two ways: by hand and by using technology where possible.*

Let

$$A = \begin{bmatrix} 0 & -1 & 0 & 1 \\ 10 & 0 & 1 & 0 \end{bmatrix}, \quad B = \begin{bmatrix} 0 & -1 \\ 1 & 1 \\ -1 & 3 \\ 5 & 0 \end{bmatrix}, \quad C = \begin{bmatrix} 1 & -1 \\ 1 & 1 \\ 1 & 1 \\ 1 & 1 \end{bmatrix}.$$

Evaluate:

**31.** $AB$    **32.** $AC$    **33.** $A(B-C)$    **34.** $(B-C)A$

Let $A = \begin{bmatrix} 1 & -1 \\ 0 & 2 \\ 0 & -2 \end{bmatrix}$, $B = \begin{bmatrix} 3 & 0 & -1 \\ 5 & -1 & 1 \end{bmatrix}$, $C = \begin{bmatrix} x & 1 & w \\ z & r & 4 \end{bmatrix}.$

Evaluate:

**35.** $AB$    **36.** $AC$    **37.** $A(B+C)$    **38.** $(B+C)A$

*In Exercises 39–44, calculate (a)* $P^2 = P \cdot P$ *(b)* $P^4 = P^2 \cdot P^2$ *and (c)* $P^8$. *(Round all entries to four decimal places.)* *(d) Without computing it explicitly, find* $P^{1,000}$.

**39.** ▼ $P = \begin{bmatrix} 0.2 & 0.8 \\ 0.2 & 0.8 \end{bmatrix}$    **40.** ▼ $P = \begin{bmatrix} 0.1 & 0.1 \\ 0.9 & 0.9 \end{bmatrix}$

**41.** ▼ $P = \begin{bmatrix} 0.1 & 0.9 \\ 0 & 1 \end{bmatrix}$    **42.** ▼ $P = \begin{bmatrix} 1 & 0 \\ 0.8 & 0.2 \end{bmatrix}$

**43.** ▼ $P = \begin{bmatrix} 0.3 & 0.3 & 0.4 \\ 0.3 & 0.3 & 0.4 \\ 0.3 & 0.3 & 0.4 \end{bmatrix}$

What do you notice about the rows of $P$? Compare with Exercise 39 and state a general result about square matrices that these two exercises seem to suggest.

**44.** ▼ $P = \begin{bmatrix} -0.3 & -0.3 & -0.3 \\ 0.9 & 0.9 & 0.9 \\ 0.4 & 0.4 & 0.4 \end{bmatrix}$

What do you notice about the columns of $P$? Compare with Exercise 40 and state a general result about square matrices that these two exercises seem to suggest.

*In Exercises 45–48, translate the given matrix equations into systems of linear equations.* HINT [See Example 7.]

**45.** $\begin{bmatrix} 2 & -1 & 4 \\ -4 & \frac{3}{4} & \frac{1}{3} \\ -3 & 0 & 0 \end{bmatrix} \begin{bmatrix} x \\ y \\ z \end{bmatrix} = \begin{bmatrix} 3 \\ -1 \\ 0 \end{bmatrix}$

**46.** $\begin{bmatrix} 1 & -1 & 4 \\ -\frac{1}{3} & -3 & \frac{1}{3} \\ 3 & 0 & 1 \end{bmatrix} \begin{bmatrix} x \\ y \\ z \end{bmatrix} = \begin{bmatrix} -3 \\ -1 \\ 2 \end{bmatrix}$

**47.** $\begin{bmatrix} 1 & -1 & 0 & 1 \\ 1 & 1 & 2 & 4 \end{bmatrix} \begin{bmatrix} x \\ y \\ z \\ w \end{bmatrix} = \begin{bmatrix} -1 \\ 2 \end{bmatrix}$

**48.** $\begin{bmatrix} 0 & 1 & 6 & 1 \\ 1 & -5 & 0 & 0 \end{bmatrix} \begin{bmatrix} x \\ y \\ z \\ w \end{bmatrix} = \begin{bmatrix} -2 \\ 9 \end{bmatrix}$

*In Exercises 49–52, translate the given systems of equations into matrix form.* HINT [See Example 7.]

**49.** $x - y = 4$
    $2x - y = 0$

**50.** $2x + y = 7$
    $-x \quad = 9$

**51.** $x + y - z = 8$
    $2x + y + z = 4$
    $\dfrac{3x}{4} + \dfrac{z}{2} = 1$

**52.** $x + y + 2z = -2$
    $4x + 2y - z = -8$
    $\dfrac{x}{2} - \dfrac{y}{3} \quad = 4$

## APPLICATIONS

**53.** *Revenue* Your T-shirt operation is doing a booming trade. Last week you sold 50 tie-dye shirts for $15 each, 40 Suburban State University Crew shirts for $10 each, and 30 Lacrosse T-shirts for $12 each. Use matrix operations to calculate your total revenue for the week. HINT [See Example 1.]

**54.** *Revenue* Karen Sandberg, your competitor in Suburban State U's T-shirt market, has apparently been undercutting your prices and outperforming you in sales. Last week she sold 100 tie-dye shirts for $10 each, 50 (low quality) Crew shirts at $5 apiece, and 70 Lacrosse T-shirts for $8 each. Use matrix operations to calculate her total revenue for the week. HINT [See Example 1.]

**55.** *Real Estate* The following table shows the cost of 1,000 square feet of "business executive"-grade residential real estate in three cities at the end of 2011[8] together with the quantity your development company intends to purchase in each city.

| | London | New York | Mumbai |
|---|---|---|---|
| **Cost ($ millions)** | 1.15 | 0.80 | 0.50 |
| **Quantity Purchased (thousands of square feet)** | 10 | 20 | 10 |

Use matrix multiplication to estimate the total cost of the purchase.

---

[7]$A \cdot A \cdot A$ is $A(A \cdot A)$, or the equivalent $(A \cdot A)A$ by the associative law. Similarly, $A \cdot A \cdot A \cdot A = A(A \cdot A \cdot A) = (A \cdot A \cdot A)A = (A \cdot A)(A \cdot A)$; it doesn't matter where we place parentheses.

[8]Source: *Insights, World Cities Review* World Press, Savallis, Autumn 2011.

56. **Real Estate** Repeat Exercise 55 using the following table for Hong Kong, Paris, and Tokyo:[9]

|  | Hong Kong | Paris | Tokyo |
|---|---|---|---|
| Cost ($ millions) | 2.10 | 1.15 | 1.30 |
| Quantity Purchased (thousands of square feet) | 10 | 30 | 5 |

Use matrix multiplication to estimate the total cost of the purchase.

57. **Revenue** Recall the *Left Coast Bookstore* chain from the preceding section. In January, it sold 700 hardcover books, 1,300 softcover books, and 2,000 plastic books in San Francisco; it sold 400 hardcover, 300 softcover, and 500 plastic books in Los Angeles. Now, hardcover books sell for $30 each, softcover books sell for $10 each, and plastic books sell for $15 each. Write a column matrix with the price data and show how matrix multiplication (using the sales and price data matrices) may be used to compute the total revenue at the two stores. HINT [See Example 5.]

58. **Profit** Refer back to Exercise 57, and now suppose that each hardcover book costs the stores $10, each softcover book costs $5, and each plastic book costs $10. Use matrix operations to compute the total *profit* at each store in January. HINT [See Example 5.]

**Income** Exercises 59–62 are based on the following spreadsheet, which shows the projected 2020 and 2030 U.S. male and female population in various age groups, as well as per capita incomes.[10]

|  | A | B | C | D | E | F |
|---|---|---|---|---|---|---|
| 1 |  |  | 2020 Population | | 2030 population | |
| 2 | Age | Mean Income ($1000) | Female (Millions) | Male (Millions) | Female (Millions) | Male (Millions) |
| 3 | 15 to 24 | 14 | 23 | 24 | 25 | 26 |
| 4 | 25 to 44 | 42 | 43 | 43 | 45 | 46 |
| 5 | 45 to 64 | 48 | 43 | 41 | 42 | 41 |
| 6 | 65 to 84 | 29 | 31 | 20 | 40 | 31 |

59. Use matrix algebra to estimate the total income for females in 2020. (Round the answer to two significant digits.)

60. Use matrix algebra to estimate the total income for males in 2030. (Round the answer to two significant digits.)

61. Give a single matrix formula that expresses the difference in total income between males and females in 2020, and compute its value, rounded to two significant digits.

62. Give a single matrix formula that expresses the total income in 2030, and compute its value, rounded to two significant digits.

63. **Cheese Production** The total amount of cheese, in billions of pounds, produced in the 13 western and 12 north central U.S. states in 1999 and 2000 was as follows:[11]

|  | 1999 | 2000 |
|---|---|---|
| Western States | 2.7 | 3.0 |
| North Central States | 3.9 | 4.0 |

Thinking of this table as a (labeled) $2 \times 2$ matrix $P$, compute the matrix product $[-1 \quad 1]P$. What does this product represent?

64. **Milk Production** The total amount of milk, in billions of pounds, produced in the 13 western and 12 north central U.S. states in 1999 and 2000 was as follows.[12]

|  | 1999 | 2000 |
|---|---|---|
| Western States | 56 | 60 |
| North Central States | 57 | 59 |

Thinking of this table as a (labeled) $2 \times 2$ matrix $P$, compute the matrix product $P \begin{bmatrix} -1 \\ 1 \end{bmatrix}$. What does this product represent?

**Foreclosures** Exercises 65–70 are based on the following table, which shows the numbers of foreclosures in three states during three months of 2011.[13]

|  | June | July | Aug. |
|---|---|---|---|
| California | 54,100 | 56,200 | 59,400 |
| Florida | 23,800 | 22,400 | 23,600 |
| Texas | 9,300 | 10,600 | 10,100 |

65. Each month, your law firm handles 10% of all foreclosures in California, 5% of all foreclosures in Florida, and 20% of all foreclosures in Texas. Use matrix multiplication to compute the total number of foreclosures handled by your firm in each of the months shown.

66. Your law firm handled 10% of all foreclosures in each state in June, 30% of all foreclosures in July, and 20% of all foreclosures in August 2011. Use matrix multiplication to compute the total number of foreclosures handled by your firm in each of the states shown.

67. Let $A$ be the $3 \times 3$ matrix whose entries are the figures in the table, and let $B = [1 \quad 1 \quad 0]$. What does the matrix $BA$ represent?

68. Let $A$ be the $3 \times 3$ matrix whose entries are the figures in the table, and let $B = [1 \quad 1 \quad 0]^T$. What does the matrix $AB$ represent?

[9]*Ibid.*

[10]The population figures are Census Bureau estimates, and the income figures are 2007 mean per capita incomes. All figures are approximate. Source: U.S. Census Bureau (www.census.gov).

[11]Figures are approximate. Source: Department of Agriculture/*New York Times*, June 28, 2001, p. C1.

[12]*Ibid.*

[13]Figures are rounded. Source: www.realtytrac.com.

69. ▼ Write a matrix product whose computation gives the total number by which the combined foreclosures for all three months in California and Texas exceeded the foreclosures in Florida. Calculate the product.

70. ▼ Write a matrix product whose computation gives the total number by which combined foreclosures in August exceeded foreclosures in June. Calculate the product.

71. ▼ *Costs Microbucks Computer Co.* makes two computers, the Pomegranate II and the Pomegranate Classic. The Pom II requires 2 processor chips, 16 memory chips, and 20 vacuum tubes, while the Pom Classic requires 1 processor chip, 4 memory chips, and 40 vacuum tubes. There are two companies that can supply these parts: Motorel can supply them at $100 per processor chip, $50 per memory chip, and $10 per vacuum tube, while Intola can supply them at $150 per processor chip, $40 per memory chip, and $15 per vacuum tube. Write down all of this data in two matrices, one showing the parts required for each model computer, and the other showing the prices for each part from each supplier. Then show how matrix multiplication allows you to compute the total cost for parts for each model when parts are bought from either supplier.

72. ▼ *Profits* Refer back to Exercise 71. It actually costs Motorel only $25 to make each processor chip, $10 for each memory chip, and $5 for each vacuum tube. It costs Intola $50 per processor chip, $10 per memory chip, and $7 per vacuum tube. Use matrix operations to find the total profit Motorel and Intola would make on each model.

73. ▼ *Tourism in the 1990s* The following table gives the number of people (in thousands) who visited Australia and South Africa in 1998.[14]

| | To | Australia | South Africa |
|---|---|---|---|
| *From* | **North America** | 440 | 190 |
| | **Europe** | 950 | 950 |
| | **Asia** | 1,790 | 200 |

You estimate that 5% of all visitors to Australia and 4% of all visitors to South Africa decide to settle there permanently. Take $A$ to be the $3 \times 2$ matrix whose entries are the 1998 tourism figures in the above table and take

$$B = \begin{bmatrix} 0.05 \\ 0.04 \end{bmatrix} \quad \text{and} \quad C = \begin{bmatrix} 0.05 & 0 \\ 0 & 0.04 \end{bmatrix}.$$

Compute the products $AB$ and $AC$. What do the entries in these matrices represent?

74. ▼ *Tourism in the 1990s* Referring to the tourism figures in the preceding exercise, you estimate that from 1998 to 2018, tourism from North America to each of Australia and South Africa will have increased by 20%, tourism from Europe by 30%, and tourism from Asia by 10%. Take $A$ to be the $3 \times 2$ matrix whose entries are the 1998 tourism figures and take

$$B = [1.2 \quad 1.3 \quad 1.1] \quad \text{and} \quad C = \begin{bmatrix} 1.2 & 0 & 0 \\ 0 & 1.3 & 0 \\ 0 & 0 & 1.1 \end{bmatrix}.$$

Compute the products $BA$ and $CA$. What do the entries in these matrices represent?

75. 🔟▼ *Population Movement* In 2008, the population of the U.S., broken down by regions, was 54.1 million in the Northeast, 65.7 million in the Midwest, 112.9 million in the South, and 70.3 million in the West. The table below shows the population movement during the period 2008–2009. (Thus, 99.23% of the population in the Northeast stayed there, while 0.16% of the population in the Northeast moved to the Midwest, and so on.)[15]

| | To | Northeast | Midwest | South | West |
|---|---|---|---|---|---|
| *From* | **Northeast** | 0.9923 | 0.0016 | 0.0042 | 0.0019 |
| | **Midwest** | 0.0018 | 0.9896 | 0.0047 | 0.0039 |
| | **South** | 0.0056 | 0.0059 | 0.9827 | 0.0058 |
| | **West** | 0.0024 | 0.0033 | 0.0044 | 0.9899 |

Set up the 2008 population figures as a row vector. Then use matrix multiplication to estimate the population in each region in 2009. (Round all answers to the nearest 0.1 million.)

76. 🔟▼ *Population Movement* Assume that the percentages given in the preceding exercise also describe the population movements from 2009 to 2010. Use two matrix multiplications to estimate from the data in the preceding exercise the population in each region in 2010.

## COMMUNICATION AND REASONING EXERCISES

77. Give an example of two matrices $A$ and $B$ such that $AB$ is defined but $BA$ is not defined.

78. Give an example of two matrices $A$ and $B$ of different dimensions such that both $AB$ and $BA$ are defined.

79. Compare addition and multiplication of $1 \times 1$ matrices to the arithmetic of numbers.

80. In comparing the algebra of $1 \times 1$ matrices, as discussed so far, to the algebra of real numbers (see Exercise 79), what important difference do you find?

81. Comment on the following claim: Every matrix equation represents a system of equations.

[14]Figures are rounded to the nearest 10,000. Sources: South African Dept. of Environmental Affairs and Tourism; Australia Tourist Commission/*New York Times*, January 15, 2000, p. C1.

[15]Note that this exercise ignores migration into or out of the country. Source: U.S. Census Bureau, Current Population Survey, 2009 Annual Social and Economic Supplement.

82. When is it true that both $AB$ and $BA$ are defined, even though neither $A$ nor $B$ is a square matrix?

83. ▼ Find a scenario in which it would be useful to "multiply" two row vectors according to the rule

$$[a \quad b \quad c][d \quad e \quad f] = [ad \quad be \quad cf].$$

84. ▼ Make up an application whose solution reads as follows.

$$\text{"Total revenue} = [10 \quad 100 \quad 30] \begin{bmatrix} 10 & 0 & 3 \\ 1 & 2 & 0 \\ 0 & 1 & 40 \end{bmatrix}."$$

85. ▼ What happens in a spreadsheet if, instead of using the function MMULT, you use "ordinary multiplication" as shown here?

86. ▼ Define the *naïve product* $A \square B$ of two $m \times n$ matrices $A$ and $B$ by

$$(A \square B)_{ij} = A_{ij} B_{ij}.$$

(This is how someone who has never seen matrix multiplication before might think to multiply matrices.) Referring to Example 1 in this section, compute and comment on the meaning of $P \square (Q^T)$.

## 4.3  Matrix Inversion

Now that we've discussed matrix addition, subtraction, and multiplication, you may well be wondering about matrix *division*. In the realm of real numbers, division can be thought of as a form of multiplication: Dividing 3 by 7 is the same as multiplying 3 by 1/7, the inverse of 7. In symbols, $3 \div 7 = 3 \times (1/7)$, or $3 \times 7^{-1}$. In order to imitate division of real numbers in the realm of matrices, we need to discuss the multiplicative **inverse**, $A^{-1}$, of a matrix $A$.

**Note** Because multiplication of real numbers is commutative, we can write, for example, $\frac{3}{7}$ as either $3 \times 7^{-1}$ or $7^{-1} \times 3$. In the realm of matrices, multiplication is not commutative, so from now on we shall *never* talk about "division" of matrices (by "$\frac{B}{A}$" should we mean $A^{-1}B$ or $BA^{-1}$?). ■

Before we try to find the inverse of a matrix, we must first know exactly what we *mean* by the inverse. Recall that the inverse of a number $a$ is the number, often written $a^{-1}$, with the property that $a^{-1} \cdot a = a \cdot a^{-1} = 1$. For example, the inverse of 76 is the number $76^{-1} = 1/76$, because $(1/76) \cdot 76 = 76 \cdot (1/76) = 1$. This is the number calculated by the $x^{-1}$ button found on most calculators. Not all numbers have an inverse. For example—and this is the only example—the number 0 has no inverse, because you cannot get 1 by multiplying 0 by anything.

The inverse of a matrix is defined similarly. To make life easier, we shall restrict attention to **square** matrices, matrices that have the same number of rows as columns.*

\* Nonsquare matrices *cannot* have inverses in the sense that we are talking about here. This is not a trivial fact to prove.

### Inverse of a Matrix

The **inverse** of an $n \times n$ matrix $A$ is that $n \times n$ matrix $A^{-1}$ which, when multiplied by $A$ on either side, yields the $n \times n$ identity matrix $I$. Thus,

$$AA^{-1} = A^{-1}A = I.$$

If $A$ has an inverse, it is said to be **invertible**. Otherwise, it is said to be **singular**.

**Quick Examples**

1. The inverse of the $1 \times 1$ matrix $[3]$ is $[1/3]$, because $[3][1/3] = [1] = [1/3][3]$.

2. The inverse of the $n \times n$ identity matrix $I$ is $I$ itself, because $I \times I = I$. Thus, $I^{-1} = I$.

3. The inverse of the $2 \times 2$ matrix $A = \begin{bmatrix} 1 & -1 \\ -1 & -1 \end{bmatrix}$ is $A^{-1} = \begin{bmatrix} \frac{1}{2} & -\frac{1}{2} \\ -\frac{1}{2} & -\frac{1}{2} \end{bmatrix}$,

   because $\begin{bmatrix} 1 & -1 \\ -1 & -1 \end{bmatrix}\begin{bmatrix} \frac{1}{2} & -\frac{1}{2} \\ -\frac{1}{2} & -\frac{1}{2} \end{bmatrix} = \begin{bmatrix} 1 & 0 \\ 0 & 1 \end{bmatrix}$    $AA^{-1} = I$

   and    $\begin{bmatrix} \frac{1}{2} & -\frac{1}{2} \\ -\frac{1}{2} & -\frac{1}{2} \end{bmatrix}\begin{bmatrix} 1 & -1 \\ -1 & -1 \end{bmatrix} = \begin{bmatrix} 1 & 0 \\ 0 & 1 \end{bmatrix}$.    $A^{-1}A = I$

**Notes**

1. It is possible to show that if $A$ and $B$ are square matrices with $AB = I$, then it must also be true that $BA = I$. In other words, once we have checked that $AB = I$, we know that $B$ is the inverse of $A$. The second check, that $BA = I$, is unnecessary.

2. If $B$ is the inverse of $A$, then we can also say that $A$ is the inverse of $B$ (why?). Thus, we sometimes refer to such a pair of matrices as an **inverse pair** of matrices. ■

---

**EXAMPLE 1  Singular Matrix**

Can $A = \begin{bmatrix} 1 & 1 \\ 0 & 0 \end{bmatrix}$ have an inverse?

**Solution** No. To see why not, notice that both entries in the second row of $AB$ will be 0, no matter what $B$ is. So $AB$ cannot equal $I$, no matter what $B$ is. Hence, $A$ is singular.

---

➡ **Before we go on...** If you think about it, you can write down many similar examples of singular matrices. There is only one number with no multiplicative inverse (0), but there are many matrices having no inverses. ■

## Finding the Inverse of a Square Matrix

**Q** : *In the box, it was stated that the inverse of* $\begin{bmatrix} 1 & -1 \\ -1 & -1 \end{bmatrix}$ *is* $\begin{bmatrix} \frac{1}{2} & -\frac{1}{2} \\ -\frac{1}{2} & -\frac{1}{2} \end{bmatrix}$. *How was that obtained?*

**A** : We can think of the problem of finding $A^{-1}$ as a problem of finding four unknowns, the four unknown entries of $A^{-1}$:

$$A^{-1} = \begin{bmatrix} x & y \\ z & w \end{bmatrix}.$$

These unknowns must satisfy the equation $AA^{-1} = I$, or

$$\begin{bmatrix} 1 & -1 \\ -1 & -1 \end{bmatrix}\begin{bmatrix} x & y \\ z & w \end{bmatrix} = \begin{bmatrix} 1 & 0 \\ 0 & 1 \end{bmatrix}.$$

If we were to try to find the first column of $A^{-1}$, consisting of $x$ and $z$, we would have to solve

$$\begin{bmatrix} 1 & -1 \\ -1 & -1 \end{bmatrix} \begin{bmatrix} x \\ z \end{bmatrix} = \begin{bmatrix} 1 \\ 0 \end{bmatrix}$$

or

$$x - z = 1$$
$$-x - z = 0.$$

To solve this system by Gauss-Jordan reduction, we would row-reduce the augmented matrix, which is $A$ with the column $\begin{bmatrix} 1 \\ 0 \end{bmatrix}$ adjoined.

$$\begin{bmatrix} 1 & -1 & | & 1 \\ -1 & -1 & | & 0 \end{bmatrix} \rightarrow \begin{bmatrix} 1 & 0 & | & x \\ 0 & 1 & | & z \end{bmatrix}$$

To find the second column of $A^{-1}$ we would similarly row-reduce the augmented matrix obtained by tacking on to $A$ the second column of the identity matrix.

$$\begin{bmatrix} 1 & -1 & | & 0 \\ -1 & -1 & | & 1 \end{bmatrix} \rightarrow \begin{bmatrix} 1 & 0 & | & y \\ 0 & 1 & | & w \end{bmatrix}$$

The row operations used in doing these two reductions would be exactly the same. We could do both reductions simultaneously by "doubly augmenting" $A$, putting both columns of the identity matrix to the right of $A$.

$$\begin{bmatrix} 1 & -1 & | & 1 & 0 \\ -1 & -1 & | & 0 & 1 \end{bmatrix} \rightarrow \begin{bmatrix} 1 & 0 & | & x & y \\ 0 & 1 & | & z & w \end{bmatrix}$$

We carry out this reduction in the following example.

---

### EXAMPLE 2 Computing Matrix Inverse

Find the inverse of each matrix.

**a.** $P = \begin{bmatrix} 1 & -1 \\ -1 & -1 \end{bmatrix}$     **b.** $Q = \begin{bmatrix} 1 & 0 & 1 \\ 2 & -2 & -1 \\ 3 & 0 & 0 \end{bmatrix}$

### Solution

**a.** As described above, we put the matrix $P$ on the left and the identity matrix $I$ on the right to get a $2 \times 4$ matrix.

$$\begin{bmatrix} 1 & -1 & | & 1 & 0 \\ -1 & -1 & | & 0 & 1 \end{bmatrix}$$
$$\quad P \qquad\qquad I$$

We now row-reduce the whole matrix:

$$\begin{bmatrix} 1 & -1 & | & 1 & 0 \\ -1 & -1 & | & 0 & 1 \end{bmatrix} \begin{matrix} \\ R_2 + R_1 \end{matrix} \rightarrow \begin{bmatrix} 1 & -1 & | & 1 & 0 \\ 0 & -2 & | & 1 & 1 \end{bmatrix} \begin{matrix} 2R_1 - R_2 \\ \end{matrix} \rightarrow$$

$$\begin{bmatrix} 2 & 0 & | & 1 & -1 \\ 0 & -2 & | & 1 & 1 \end{bmatrix} \begin{matrix} \frac{1}{2}R_1 \\ -\frac{1}{2}R_2 \end{matrix} \rightarrow \begin{bmatrix} 1 & 0 & | & \frac{1}{2} & -\frac{1}{2} \\ 0 & 1 & | & -\frac{1}{2} & -\frac{1}{2} \end{bmatrix}.$$
$$\qquad\qquad\qquad\qquad\qquad\qquad\qquad\qquad I \qquad\qquad P^{-1}$$

We have now solved the systems of linear equations that define the entries of $P^{-1}$. Thus,

$$P^{-1} = \begin{bmatrix} \frac{1}{2} & -\frac{1}{2} \\ -\frac{1}{2} & -\frac{1}{2} \end{bmatrix}.$$

**b.** The procedure to find the inverse of a 3 × 3 matrix (or larger) is just the same as for a 2 × 2 matrix. We place $Q$ on the left and the identity matrix (now 3 × 3) on the right, and reduce.

$$
\begin{array}{c} Q \qquad\qquad I \\ \left[\begin{array}{ccc|ccc} 1 & 0 & 1 & 1 & 0 & 0 \\ 2 & -2 & -1 & 0 & 1 & 0 \\ 3 & 0 & 0 & 0 & 0 & 1 \end{array}\right] \begin{array}{c} \\ R_2 - 2R_1 \\ R_3 - 3R_1 \end{array} \to \end{array}
\left[\begin{array}{ccc|ccc} 1 & 0 & 1 & 1 & 0 & 0 \\ 0 & -2 & -3 & -2 & 1 & 0 \\ 0 & 0 & -3 & -3 & 0 & 1 \end{array}\right] \begin{array}{c} 3R_1 + R_3 \\ R_2 - R_3 \\ \end{array} \to
$$

$$
\left[\begin{array}{ccc|ccc} 3 & 0 & 0 & 0 & 0 & 1 \\ 0 & -2 & 0 & 1 & 1 & -1 \\ 0 & 0 & -3 & -3 & 0 & 1 \end{array}\right] \begin{array}{c} \frac{1}{3}R_1 \\ -\frac{1}{2}R_2 \\ -\frac{1}{3}R_3 \end{array} \to
\begin{array}{c} \left[\begin{array}{ccc|ccc} 1 & 0 & 0 & 0 & 0 & \frac{1}{3} \\ 0 & 1 & 0 & -\frac{1}{2} & -\frac{1}{2} & \frac{1}{2} \\ 0 & 0 & 1 & 1 & 0 & -\frac{1}{3} \end{array}\right]. \\ I \qquad\qquad Q^{-1} \end{array}
$$

Thus,

$$
Q^{-1} = \begin{bmatrix} 0 & 0 & \frac{1}{3} \\ -\frac{1}{2} & -\frac{1}{2} & \frac{1}{2} \\ 1 & 0 & -\frac{1}{3} \end{bmatrix}.
$$

We have already checked that $P^{-1}$ is the inverse of $P$. You should also check that $Q^{-1}$ is the inverse of $Q$.

The method we used in Example 2 can be summarized as follows:

### Inverting an *n* × *n* Matrix

In order to determine whether an $n \times n$ matrix $A$ is invertible or not, and to find $A^{-1}$ if it does exist, follow this procedure:

1. Write down the $n \times 2n$ matrix $[A\,|\,I]$ (this is $A$ with the $n \times n$ identity matrix set next to it).

2. Row-reduce $[A\,|\,I]$.

3. If the reduced form is $[I\,|\,B]$ (i.e., has the identity matrix in the left part), then $A$ is invertible and $B = A^{-1}$. If you cannot obtain $I$ in the left part, then $A$ is singular. (See Example 3.)

Although there is a general formula for the inverse of a matrix, it is not a simple one. In fact, using the formula for anything larger than a 3 × 3 matrix is so inefficient that the row-reduction procedure is the method of choice even for computers. However, the general formula is very simple for the special case of 2 × 2 matrices:

### Formula for the Inverse of a 2 × 2 Matrix

The inverse of a 2 × 2 matrix is

$$
\begin{bmatrix} a & b \\ c & d \end{bmatrix}^{-1} = \frac{1}{ad - bc} \begin{bmatrix} d & -b \\ -c & a \end{bmatrix}, \quad \text{provided } ad - bc \neq 0.
$$

If the quantity $ad - bc$ is zero, then the matrix is singular (noninvertible). The quantity $ad - bc$ is called the **determinant** of the matrix $\begin{bmatrix} a & b \\ c & d \end{bmatrix}$.

**Quick Examples**

1. $\begin{bmatrix} 1 & 2 \\ 3 & 4 \end{bmatrix}^{-1} = \dfrac{1}{(1)(4)-(2)(3)} \begin{bmatrix} 4 & -2 \\ -3 & 1 \end{bmatrix} = -\dfrac{1}{2}\begin{bmatrix} 4 & -2 \\ -3 & 1 \end{bmatrix}$

   $= \begin{bmatrix} -2 & 1 \\ \frac{3}{2} & -\frac{1}{2} \end{bmatrix}$

2. $\begin{bmatrix} 1 & -1 \\ 2 & -2 \end{bmatrix}$ has determinant $ad - bc = (1)(-2) - (-1)(2) = 0$ and so

   is singular.

The formula for the inverse of a $2 \times 2$ matrix can be obtained using the technique of row reduction. (See the Communication and Reasoning Exercises at the end of the section.)

As we have mentioned earlier, not every square matrix has an inverse, as we see in the next example.

### EXAMPLE 3  Singular 3 × 3 Matrix

Find the inverse of the matrix $S = \begin{bmatrix} 1 & 1 & 2 \\ -2 & 0 & 4 \\ 3 & 1 & -2 \end{bmatrix}$, if it exists.

**Solution**  We proceed as before.

$$
\begin{matrix} S & & & & I \end{matrix}
$$

$$
\left[\begin{array}{ccc|ccc} 1 & 1 & 2 & 1 & 0 & 0 \\ -2 & 0 & 4 & 0 & 1 & 0 \\ 3 & 1 & -2 & 0 & 0 & 1 \end{array}\right] \begin{matrix} \\ R_2 + 2R_1 \\ R_3 - 3R_1 \end{matrix} \rightarrow \left[\begin{array}{ccc|ccc} 1 & 1 & 2 & 1 & 0 & 0 \\ 0 & 2 & 8 & 2 & 1 & 0 \\ 0 & -2 & -8 & -3 & 0 & 1 \end{array}\right] \begin{matrix} 2R_1 - R_2 \\ \\ R_3 + R_2 \end{matrix}
$$

$$
\rightarrow \left[\begin{array}{ccc|ccc} 2 & 0 & -4 & 0 & -1 & 0 \\ 0 & 2 & 8 & 2 & 1 & 0 \\ 0 & 0 & 0 & -1 & 1 & 1 \end{array}\right]
$$

We stopped here, even though the reduction is incomplete, because there is *no hope* of getting the identity on the left-hand side. Completing the row reduction will not change the three zeros in the bottom row. So what is wrong? Nothing. As in Example 1, we have here a singular matrix. Any square matrix that, after row reduction, winds up with a row of zeros is singular. (See Exercise 77.)

➡ **Before we go on...** In practice, deciding whether a given matrix is invertible or singular is easy: Simply try to find its inverse. If the process works, then the matrix is invertible, and we get its inverse. If the process fails, then the matrix is singular. If you try to invert a singular matrix using a spreadsheet, calculator, or computer program, you should get an error. Sometimes, instead of an error, you will get a spurious answer due to round-off errors in the device. ∎

## Using the Inverse to Solve a System of *n* Linear Equations in *n* Unknowns

Having used systems of equations and row reduction to find matrix inverses, we will now use matrix inverses to solve systems of equations. Recall that, at the end of the previous section, we saw that a system of linear equations could be written in the form

$$AX = B,$$

where $A$ is the coefficient matrix, $X$ is the column matrix of unknowns, and $B$ is the column matrix of right-hand sides. Now suppose that there are as many unknowns as equations, so that $A$ is a square matrix, and suppose that $A$ is invertible. The object is to solve for the matrix $X$ of unknowns, so we multiply both sides of the equation by the inverse $A^{-1}$ of $A$, getting

$$A^{-1}AX = A^{-1}B.$$

Notice that we put $A^{-1}$ on the left on both sides of the equation. Order matters when multiplying matrices, so we have to be careful to do the same thing to both sides of the equation. But now $A^{-1}A = I$, so we can rewrite the last equation as

$$IX = A^{-1}B.$$

Also, $IX = X$ ($I$ being the identity matrix), so we really have

$$X = A^{-1}B,$$

and we have solved for $X$!

Moreover, we have shown that, if $A$ is invertible and $AX = B$, then the only *possible* solution is $X = A^{-1}B$. We should check that $A^{-1}B$ is actually a solution by substituting back into the original equation.

$$AX = A(A^{-1}B) = (AA^{-1})B = IB = B$$

Thus, $X = A^{-1}B$ is a solution and is the only solution. Therefore, if $A$ is invertible, $AX = B$ has exactly one solution.

On the other hand, if $AX = B$ has no solutions or has infinitely many solutions, we can conclude that $A$ is not invertible (why?). To summarize:

## Solving the Matrix Equation $AX = B$

If $A$ is an invertible matrix, then the matrix equation $AX = B$ has the unique solution

$$X = A^{-1}B.$$

### Quick Example

The system of linear equations

$$\begin{aligned} 2x \quad\;\;\; + z &= 9 \\ 2x + y - z &= 6 \\ 3x + y - z &= 9 \end{aligned}$$

can be written as $AX = B$, where

$$A = \begin{bmatrix} 2 & 0 & 1 \\ 2 & 1 & -1 \\ 3 & 1 & -1 \end{bmatrix}, X = \begin{bmatrix} x \\ y \\ z \end{bmatrix} \quad \text{and} \quad B = \begin{bmatrix} 9 \\ 6 \\ 9 \end{bmatrix}.$$

The matrix $A$ is invertible with inverse

$$A^{-1} = \begin{bmatrix} 0 & -1 & 1 \\ 1 & 5 & -4 \\ 1 & 2 & -2 \end{bmatrix}. \qquad \text{You should check this.}$$

Thus,

$$X = A^{-1}B = \begin{bmatrix} 0 & -1 & 1 \\ 1 & 5 & -4 \\ 1 & 2 & -2 \end{bmatrix} \begin{bmatrix} 9 \\ 6 \\ 9 \end{bmatrix} = \begin{bmatrix} 3 \\ 3 \\ 3 \end{bmatrix}$$

so that $(x, y, z) = (3, 3, 3)$ is the (unique) solution to the system.

## using Technology

Here is an outline on the use of technology to do the calculations in Example 4 (see the Technology Guides at the end of the chapter for additional details on using a TI-83/84 Plus or a spreadsheet):

### TI-83/84 Plus

Enter the four matrices `[A]`, `[B]`, `[C]`, `[D]` using MATRX; EDIT
Home Screen: `[A]⁻¹*[B]`, `[A]⁻¹*[C]`, `[A]⁻¹*[D]`
[More details on page 305.]

### Spreadsheet

Enter the matrices A and B in convenient blocks of cells (e.g., A1–C3, E1–E3).
Highlight a 3 × 1 block for the answer.
Type `=MMULT(MINVERSE(A1:C3), E1:E3)`
Press Control+Shift+Enter.
[More details on page 309.]

### Website
www.WanerMath.com
  Student Website
    → On Line Utilities
    → Matrix Algebra Tool

Enter `A^-1*B`, `A^-1*C`, and `A^-1*D` to compute these products after entering the matrices A, B, C, and D as shown earlier.

## EXAMPLE 4 Solving Systems of Equations Using an Inverse

Solve the following three systems of equations.

**a.** 
$$2x \qquad + z = 1$$
$$2x + y - z = 1$$
$$3x + y - z = 1$$

**b.** 
$$2x \qquad + z = 0$$
$$2x + y - z = 1$$
$$3x + y - z = 2$$

**c.** 
$$2x \qquad + z = 0$$
$$2x + y - z = 0$$
$$3x + y - z = 0$$

**Solution** We *could* go ahead and row-reduce all three augmented matrices as we did in Chapter 3, but this would require a lot of work. Notice that the coefficients are the same in all three systems. In other words, we can write the three systems in matrix form as

**a.** $AX = B$  **b.** $AX = C$  **c.** $AX = D$

where the matrix $A$ is the same in all three cases:

$$A = \begin{bmatrix} 2 & 0 & 1 \\ 2 & 1 & -1 \\ 3 & 1 & -1 \end{bmatrix}.$$

Now the solutions to these systems are

**a.** $X = A^{-1}B$  **b.** $X = A^{-1}C$  **c.** $X = A^{-1}D$

so the main work is the calculation of the single matrix $A^{-1}$, which we have already noted (Quick Example above) is

$$A^{-1} = \begin{bmatrix} 0 & -1 & 1 \\ 1 & 5 & -4 \\ 1 & 2 & -2 \end{bmatrix}.$$

Thus, the three solutions are:

**a.** $X = A^{-1}B = \begin{bmatrix} 0 & -1 & 1 \\ 1 & 5 & -4 \\ 1 & 2 & -2 \end{bmatrix}\begin{bmatrix} 1 \\ 1 \\ 1 \end{bmatrix} = \begin{bmatrix} 0 \\ 2 \\ 1 \end{bmatrix}$

**b.** $X = A^{-1}C = \begin{bmatrix} 0 & -1 & 1 \\ 1 & 5 & -4 \\ 1 & 2 & -2 \end{bmatrix}\begin{bmatrix} 0 \\ 1 \\ 2 \end{bmatrix} = \begin{bmatrix} 1 \\ -3 \\ -2 \end{bmatrix}$

**c.** $X = A^{-1}D = \begin{bmatrix} 0 & -1 & 1 \\ 1 & 5 & -4 \\ 1 & 2 & -2 \end{bmatrix}\begin{bmatrix} 0 \\ 0 \\ 0 \end{bmatrix} = \begin{bmatrix} 0 \\ 0 \\ 0 \end{bmatrix}$

➡ **Before we go on...** We have been speaking of *the* inverse of a matrix $A$. Is there only one? It is not hard to prove that a matrix $A$ cannot have more than one inverse: If $B$ and $C$ were both inverses of $A$, then

$$B = BI \qquad \text{Property of the identity}$$
$$= B(AC) \qquad \text{Because } C \text{ is an inverse of } A$$
$$= (BA)C \qquad \text{Associative law}$$
$$= IC \qquad \text{Because } B \text{ is an inverse of } A$$
$$= C. \qquad \text{Property of the identity}$$

In other words, if $B$ and $C$ were both inverses of $A$, then $B$ and $C$ would have to be equal. ■

**FAQs**

**Which Method to Use in Solving a System**

**Q :** *Now we have two methods to solve a system of linear equations AX = B: (1) Compute X = $A^{-1}B$, or (2) row-reduce the augmented matrix. Which is the better method?*

**A :** Each method has its advantages and disadvantages. Method (1), as we have seen, is very efficient when you must solve several systems of equations with the same coefficients, but it works only when the coefficient matrix is *square* (meaning that you have the same number of equations as unknowns) *and invertible* (meaning that there is a unique solution). The row-reduction method will work for all systems. Moreover, for all but the smallest systems, the most efficient way to find $A^{-1}$ is to use row reduction. Thus, in practice, the two methods are essentially the same when both apply.

# 4.3 EXERCISES

▼ more advanced   ◆ challenging
**T** indicates exercises that should be solved using technology

*In Exercises 1–6, determine whether or not the given pairs of matrices are inverse pairs.* HINT [See Quick Examples page 257.]

**1.** $A = \begin{bmatrix} 0 & 1 \\ 1 & 0 \end{bmatrix}, B = \begin{bmatrix} 0 & 1 \\ 1 & 0 \end{bmatrix}$

**2.** $A = \begin{bmatrix} 2 & 0 \\ 0 & 3 \end{bmatrix}, B = \begin{bmatrix} \frac{1}{2} & 0 \\ 0 & \frac{1}{2} \end{bmatrix}$

**3.** $A = \begin{bmatrix} 2 & 1 & 1 \\ 0 & 1 & 1 \\ 0 & 0 & 1 \end{bmatrix}, B = \begin{bmatrix} \frac{1}{2} & -\frac{1}{2} & 0 \\ 0 & 1 & -1 \\ 0 & 0 & 1 \end{bmatrix}$

**4.** $A = \begin{bmatrix} 1 & 1 & 1 \\ 0 & 1 & 1 \\ 0 & 0 & 1 \end{bmatrix}, B = \begin{bmatrix} 1 & -1 & 0 \\ 0 & 1 & -1 \\ 0 & 0 & 1 \end{bmatrix}$

**5.** $A = \begin{bmatrix} a & 0 & 0 \\ 0 & b & 0 \\ 0 & 0 & 0 \end{bmatrix}, B = \begin{bmatrix} a^{-1} & 0 & 0 \\ 0 & b^{-1} & 0 \\ 0 & 0 & 0 \end{bmatrix}$ $(a, b \neq 0)$

**6.** $A = \begin{bmatrix} a & 0 & 0 \\ 0 & b & 0 \\ 0 & 0 & c \end{bmatrix}, B = \begin{bmatrix} a^{-1} & 0 & 0 \\ 0 & b^{-1} & 0 \\ 0 & 0 & c^{-1} \end{bmatrix}$ $(a, b, c \neq 0)$

*In Exercises 7–26, use row reduction to find the inverses of the given matrices if they exist, and check your answers by multiplication.* HINT [See Example 2.]

**7.** $\begin{bmatrix} 1 & 1 \\ 2 & 1 \end{bmatrix}$   **8.** $\begin{bmatrix} 0 & 1 \\ 1 & 1 \end{bmatrix}$   **9.** $\begin{bmatrix} 0 & 1 \\ 1 & 0 \end{bmatrix}$   **10.** $\begin{bmatrix} 4 & 0 \\ 0 & 2 \end{bmatrix}$

**11.** $\begin{bmatrix} 2 & 1 \\ 1 & 1 \end{bmatrix}$   **12.** $\begin{bmatrix} 3 & 0 \\ 0 & \frac{1}{2} \end{bmatrix}$   **13.** $\begin{bmatrix} 2 & 1 \\ 4 & 2 \end{bmatrix}$   **14.** $\begin{bmatrix} 1 & 1 \\ 6 & 6 \end{bmatrix}$

**15.** $\begin{bmatrix} 1 & 1 & 1 \\ 0 & 1 & 1 \\ 0 & 0 & 1 \end{bmatrix}$   **16.** $\begin{bmatrix} 1 & 2 & 3 \\ 0 & 1 & 2 \\ 0 & 0 & 1 \end{bmatrix}$   **17.** $\begin{bmatrix} 1 & 1 & 1 \\ 1 & 0 & 2 \\ 1 & -1 & 1 \end{bmatrix}$

**18.** $\begin{bmatrix} 1 & 2 & 3 \\ 0 & 2 & 3 \\ 1 & 0 & 1 \end{bmatrix}$   **19.** $\begin{bmatrix} 1 & 1 & 1 \\ 1 & -1 & 0 \\ 1 & 2 & 3 \end{bmatrix}$   **20.** $\begin{bmatrix} 1 & -1 & 3 \\ 0 & 1 & 3 \\ 1 & 1 & 1 \end{bmatrix}$

**21.** $\begin{bmatrix} 1 & 1 & 1 \\ 1 & 0 & 1 \\ 1 & -1 & 1 \end{bmatrix}$   **22.** $\begin{bmatrix} 1 & 1 & 1 \\ 0 & 1 & 1 \\ 1 & 0 & 0 \end{bmatrix}$

**23.** $\begin{bmatrix} 1 & 0 & 1 & 0 \\ -1 & 1 & 0 & 1 \\ -1 & 0 & 0 & 1 \\ 0 & -1 & 0 & 1 \end{bmatrix}$   **24.** $\begin{bmatrix} 0 & 1 & 1 & 0 \\ -1 & 1 & 1 & 1 \\ -1 & 1 & 0 & 1 \\ 0 & -1 & 0 & 1 \end{bmatrix}$

**25.** $\begin{bmatrix} 1 & 2 & 3 & 4 \\ 0 & 1 & 2 & 3 \\ 0 & 0 & 1 & 2 \\ 0 & 0 & 0 & 1 \end{bmatrix}$   **26.** $\begin{bmatrix} 0 & 0 & 0 & 1 \\ 0 & 0 & 1 & 0 \\ 0 & 1 & 0 & 0 \\ 1 & 0 & 0 & 0 \end{bmatrix}$

*In Exercises 27–34, compute the determinant of the given matrix. If the determinant is nonzero, use the formula for inverting a 2 × 2 matrix to calculate the inverse of the given matrix.* HINT [See Quick Examples page 260.]

**27.** $\begin{bmatrix} 1 & 1 \\ 1 & -1 \end{bmatrix}$   **28.** $\begin{bmatrix} 4 & 1 \\ 0 & 2 \end{bmatrix}$   **29.** $\begin{bmatrix} 1 & 2 \\ 3 & 4 \end{bmatrix}$

**30.** $\begin{bmatrix} 1 & 0 \\ 0 & 1 \end{bmatrix}$   **31.** $\begin{bmatrix} \frac{1}{6} & -\frac{1}{6} \\ 0 & \frac{1}{6} \end{bmatrix}$   **32.** $\begin{bmatrix} 2 & 1 \\ 4 & 2 \end{bmatrix}$

**33.** $\begin{bmatrix} 1 & 0 \\ \frac{3}{4} & 0 \end{bmatrix}$   **34.** $\begin{bmatrix} 1 & 1 \\ 1 & 1 \end{bmatrix}$

**T** *In Exercises 35–42, use technology to find the inverse of the given matrix (when it exists). Round all entries in your answer to two decimal places. [Caution: Because of rounding errors, technology sometimes produces an "inverse" of a singular matrix. These often can be recognized by their huge entries.]*

35. $\begin{bmatrix} 1.1 & 1.2 \\ 1.3 & -1 \end{bmatrix}$

36. $\begin{bmatrix} 0.1 & -3.2 \\ 0.1 & -1.5 \end{bmatrix}$

37. $\begin{bmatrix} 3.56 & 1.23 \\ -1.01 & 0 \end{bmatrix}$

38. $\begin{bmatrix} 9.09 & -5.01 \\ 1.01 & 2.20 \end{bmatrix}$

39. $\begin{bmatrix} 1.1 & 3.1 & 2.4 \\ 1.7 & 2.4 & 2.3 \\ 0.6 & -0.7 & -0.1 \end{bmatrix}$

40. $\begin{bmatrix} 2.1 & 2.4 & 3.5 \\ 6.1 & -0.1 & 2.3 \\ -0.3 & -1.2 & 0.1 \end{bmatrix}$

41. $\begin{bmatrix} 0.01 & 0.32 & 0 & 0.04 \\ -0.01 & 0 & 0 & 0.34 \\ 0 & 0.32 & -0.23 & 0.23 \\ 0 & 0.41 & 0 & 0.01 \end{bmatrix}$

42. $\begin{bmatrix} 0.01 & 0.32 & 0 & 0.04 \\ -0.01 & 0 & 0 & 0.34 \\ 0 & 0.32 & -0.23 & 0.23 \\ 0.01 & 0.96 & -0.23 & 0.65 \end{bmatrix}$

*In Exercises 43–48, use matrix inversion to solve the given systems of linear equations. (You previously solved similar systems using row reduction in Chapter 3.)*
HINT [See Quick Example page 261.]

43. $x + y = 4$
    $x - y = 1$

44. $2x + y = 2$
    $2x - 3y = 2$

45. $\dfrac{x}{3} + \dfrac{y}{2} = 0$
    $\dfrac{x}{2} + y = -1$

46. $\dfrac{2x}{3} - \dfrac{y}{2} = \dfrac{1}{6}$
    $\dfrac{x}{2} - \dfrac{y}{2} = -1$

47. $-x + 2y - z = 0$
    $-x - y + 2z = 0$
    $2x \quad - z = 6$

48. $x + 2y \quad = 4$
    $\quad y - z = 0$
    $x + 3y - 2z = 5$

*In Exercises 49 and 50, use matrix inversion to solve each collection of systems of linear equations.* HINT [See Example 4.]

49. **a.** $-x - 4y + 2z = 4$
    $x + 2y - z = 3$
    $x + y - z = 8$

    **b.** $-x - 4y + 2z = 0$
    $x + 2y - z = 3$
    $x + y - z = 2$

    **c.** $-x - 4y + 2z = 0$
    $x + 2y - z = 0$
    $x + y - z = 0$

50. **a.** $-x - 4y + 2z = 8$
    $x \quad - z = 3$
    $x + y - z = 8$

    **b.** $-x - 4y + 2z = 8$
    $x \quad - z = 3$
    $x + y - z = 2$

    **c.** $-x - 4y + 2z = 0$
    $x \quad - z = 0$
    $x + y - z = 0$

## APPLICATIONS

*Some of the following exercises are similar or identical to exercises and examples in Chapter 3. Use matrix inverses to find the solutions. We suggest you invert some of the matrices by hand, and others using technology.*

51. **Nutrition** One serving of Campbell Soup Company's Campbell's® Pork & Beans contains 5 grams of protein and 21 grams of carbohydrates.[16] A typical slice of "lite" rye bread contains 4 grams of protein and 12 grams of carbohydrates.

    **a.** I am planning a meal of "beans-on-toast" and I want it to supply 20 grams of protein and 80 grams of carbohydrates. How should I prepare my meal?

    **b.** If I require $A$ grams of protein and $B$ grams of carbohydrates, give a formula that tells me how many slices of bread and how many servings of Pork & Beans to use.

52. **Nutrition** According to the nutritional information on a package of General Mills' Honey Nut Cheerios® brand cereal, each 1-ounce serving of Cheerios contains 3 grams protein and 24 grams carbohydrates.[17] Each half-cup serving of enriched skim milk contains 4 grams protein and 6 grams carbohydrates.

    **a.** I am planning a meal of cereal and milk and I want it to supply 26 grams of protein and 78 grams of carbohydrates. How should I prepare my meal?

    **b.** If I require $A$ grams of protein and $B$ grams of carbohydrates, give a formula that tells me how many servings of milk and Cheerios to use.

53. **Resource Allocation** You manage an ice cream factory that makes three flavors: Creamy Vanilla, Continental Mocha, and Succulent Strawberry. Into each batch of Creamy Vanilla go two eggs, one cup of milk, and two cups of cream. Into each batch of Continental Mocha go one egg, one cup of milk, and two cups of cream. Into each batch of Succulent Strawberry go one egg, two cups of milk, and one cup of cream. Your stocks of eggs, milk, and cream vary from day to day. How many batches of each flavor should you make in order to use up all of your ingredients if you have the following amounts in stock?

    **a.** 350 eggs, 350 cups of milk, and 400 cups of cream
    **b.** 400 eggs, 500 cups of milk, and 400 cups of cream
    **c.** $A$ eggs, $B$ cups of milk, and $C$ cups of cream

54. **Resource Allocation** The *Arctic Juice Company* makes three juice blends: PineOrange, using 2 quarts of pineapple juice and 2 quarts of orange juice per gallon; PineKiwi, using 3 quarts of pineapple juice and 1 quart of kiwi juice per gallon; and OrangeKiwi, using 3 quarts of orange juice and 1 quart of kiwi juice per gallon. The amount of each kind of juice the company has on hand varies from day to day. How many gallons of each blend can it make on a day with the following stocks?

---

[16] According to the label information on a 16-oz. can.

[17] Actually, it is 23 grams carbohydrates. We made it 24 grams to simplify the calculation.

a. 800 quarts of pineapple juice, 650 quarts of orange juice, 350 quarts of kiwi juice.

b. 650 quarts of pineapple juice, 800 quarts of orange juice, 350 quarts of kiwi juice.

c. A quarts of pineapple juice, B quarts of orange juice, C quarts of kiwi juice.

*Investing: Inverse Mutual Funds Inverse mutual funds, sometimes referred to as "bear market" or "short" funds, are designed to deliver the opposite of the performance of the index or category they track, and so can be used by traders to bet against the stock market. Exercises 55–56 are based on the following table, which shows the performance of three such funds as of August 12, 2011.*[18]

|  | Year-to-Date Loss |
|---|---|
| **SHPIX (Short Smallcap Profund)** | 6% |
| **RYURX (Rydex Inverse S&P 500)** | 5% |
| **RYIHX (Rydex Inverse High Yield)** | 7% |

**55.** You invested a total of $9,000 in the three funds at the beginning of 2011, including an equal amount in RYURX and RYIHX. Your year-to-date loss from the first two funds amounted to $400. How much did you invest in each of the three funds?

**56.** You invested a total of $6,000 in the three funds at the beginning of 2011, including an equal amount in SHPIX and RYURX. Your total year-to-date loss amounted to $360. How much did you invest in each of the three funds?

*Investing: Lesser-Known Stocks Exercises 57–58 are based on the following information about the stocks of* KT Corporation, International Flavors & Fragrances, *and* Quality Systems.[19]

|  | Price | Dividend Yield |
|---|---|---|
| **KT (KT Corporation)** | $16 | 7% |
| **IFF (International Flavors & Fragrances)** | 56 | 2 |
| **QSII (Quality Systems)** | 80 | 2 |

**57.** ▼ You invested a total of $8,400 in shares of the three stocks at the given prices and expected to earn $248 in annual dividends. If you purchased a total of 200 shares, how many shares of each stock did you purchase?

**58.** ▼ You invested a total of $11,200 in shares of the three stocks at the given prices and expected to earn $304 in annual dividends. If you purchased a total of 250 shares, how many shares of each stock did you purchase?

---

[18]Based on prices at the close of the stock market on August 12, 2011. YTD losses rounded to the nearest percentage point. Source: www.fidelity.com.

[19]Yields rounded to the nearest percentage point and stock prices at the close of the stock market on August 12, 2011 rounded to the nearest $1. Source: www.google.com/finance.

**59.** ▣ ▼ *Population Movement* In 2009, the population of the U.S., broken down by regions, was 54.6 million in the Northeast, 66.0 million in the Midwest, 111.8 million in the South, and 70.6 million in the West. The table below shows the population movement during the period 2008–2009. (Thus, 99.23% of the population in the Northeast stayed there, while 0.16% of the population in the Northeast moved to the Midwest, and so on.)[20]

| | To | Northeast | Midwest | South | West |
|---|---|---|---|---|---|
| *From* | **Northest** | 0.9923 | 0.0016 | 0.0042 | 0.0019 |
| | **Midwest** | 0.0018 | 0.9896 | 0.0047 | 0.0039 |
| | **South** | 0.0056 | 0.0059 | 0.9827 | 0.0058 |
| | **West** | 0.0024 | 0.0033 | 0.0044 | 0.9899 |

Set up the 2009 population figures as a row vector. Use matrix inversion and multiplication to estimate the population in each region in 2008. (Round all answers to the nearest 0.1 million.)

**60.** ▼ ▣ *Population Movement* Assume that the percentages given in the preceding exercise also describe the population movements from 2007 to 2008. Use two matrix multiplications to estimate from the data in the preceding exercise the population in each region in 2007.

**61.** ◆ ▣ *Rotations* If a point $(x, y)$ in the plane is rotated counterclockwise about the origin through an angle of $45°$, its new coordinates $(x', y')$ are given by

$$\begin{bmatrix} x' \\ y' \end{bmatrix} = R \begin{bmatrix} x \\ y \end{bmatrix}$$

where $R$ is the $2 \times 2$ matrix $\begin{bmatrix} a & -a \\ a & a \end{bmatrix}$ and $a = \sqrt{1/2} \approx 0.7071$.

a. If the point $(2, 3)$ is rotated counterclockwise through an angle of $45°$, what are its (approximate) new coordinates?

b. Multiplication by what matrix would result in a counterclockwise rotation of $90°$? $135°$? (Express the matrices in terms of $R$.) HINT [Think of a rotation through $90°$ as two successive rotations through $45°$.]

c. Multiplication by what matrix would result in a *clockwise* rotation of $45°$?

**62.** ◆ ▣ *Rotations* If a point $(x, y)$ in the plane is rotated counterclockwise about the origin through an angle of $60°$, its new coordinates are given by

$$\begin{bmatrix} x' \\ y' \end{bmatrix} = S \begin{bmatrix} x \\ y \end{bmatrix}$$

where $S$ is the $2 \times 2$ matrix $\begin{bmatrix} a & -b \\ b & a \end{bmatrix}$ and $a = 1/2$ and $b = \sqrt{3/4} \approx 0.8660$.

---

[20]Note that this exercise ignores migration into or out of the country. Source: U.S. Census Bureau, Current Population Survey, 2009 Annual Social and Economic Supplement.

**a.** If the point (2, 3) is rotated counterclockwise through an angle of 60°, what are its (approximate) new coordinates?

**b.** Referring to Exercise 61, multiplication by what matrix would result in a counterclockwise rotation of 105°? (Express the matrices in terms of $S$ and the matrix $R$ from Exercise 61.) HINT [Think of a rotation through 105° as a rotation through 60° followed by one through 45°.]

**c.** Multiplication by what matrix would result in a *clockwise* rotation of 60°?

📝 *Encryption Matrices are commonly used to encrypt data. Here is a simple form such an encryption can take. First, we represent each letter in the alphabet by a number, so let us take <space> = 0, A = 1, B = 2, and so on. Thus, for example, "ABORT MISSION" becomes*

[1  2  15  18  20  0  13  9  19  19  9  15  14]

*To encrypt this coded phrase, we use an invertible matrix of any size with integer entries. For instance, let us take A to be the 2 × 2 matrix $\begin{bmatrix} 1 & 2 \\ 3 & 4 \end{bmatrix}$. We can first arrange the coded sequence of numbers in the form of a matrix with two rows (using zero in the last place if we have an odd number of characters) and then multiply on the left by A.*

Encrypted Matrix $= \begin{bmatrix} 1 & 2 \\ 3 & 4 \end{bmatrix} \begin{bmatrix} 1 & 15 & 20 & 13 & 19 & 9 & 14 \\ 2 & 18 & 0 & 9 & 19 & 15 & 0 \end{bmatrix}$

$= \begin{bmatrix} 5 & 51 & 20 & 31 & 57 & 39 & 14 \\ 11 & 117 & 60 & 75 & 133 & 87 & 42 \end{bmatrix}$,

*which we can also write as*

[5  11  51  117  20  60  31  75  57  133  39  87  14  42].

*To decipher the encoded message, multiply the encrypted matrix by $A^{-1}$. Exercises 63–66 use the above matrix A for encoding and decoding.*

**63.** ▼ Use the matrix $A$ to encode the phrase "GO TO PLAN B".

**64.** ▼ Use the matrix $A$ to encode the phrase "ABANDON SHIP".

**65.** ▼ Decode the following message, which was encrypted using the matrix $A$.

[33  69  54  126  11  27  20  60  29  59  65  149  41  87]

**66.** ▼ Decode the following message, which was encrypted using the matrix $A$.

[59  141  43  101  7  21  29  59  65  149  41  87]

## COMMUNICATION AND REASONING EXERCISES

**67.** Multiple Choice: If $A$ and $B$ are square matrices with $AB = I$ and $BA = I$, then

**(A)** $B$ is the inverse of $A$.

**(B)** $A$ and $B$ must be equal.

**(C)** $A$ and $B$ must both be singular.

**(D)** At least one of $A$ and $B$ is singular.

**68.** Multiple Choice: If $A$ is a square matrix with $A^3 = I$, then

**(A)** $A$ must be the identity matrix.

**(B)** $A$ is invertible.

**(C)** $A$ is singular.

**(D)** $A$ is both invertible and singular.

**69.** What can you say about the inverse of a $2 \times 2$ matrix of the form $\begin{bmatrix} a & b \\ a & b \end{bmatrix}$?

**70.** If you think of numbers as $1 \times 1$ matrices, which numbers are invertible $1 \times 1$ matrices?

**71.** ▼ Use matrix multiplication to check that the inverse of a general $2 \times 2$ matrix is given by

$\begin{bmatrix} a & b \\ c & d \end{bmatrix}^{-1} = \dfrac{1}{ad - bc} \begin{bmatrix} d & -b \\ -c & a \end{bmatrix}$ (provided $ad - bc \neq 0$).

**72.** ◆ Derive the formula in Exercise 71 using row reduction. (Assume that $ad - bc \neq 0$.)

**73.** ▼ A **diagonal** matrix $D$ has the following form.

$$D = \begin{bmatrix} d_1 & 0 & 0 & \ldots & 0 \\ 0 & d_2 & 0 & \ldots & 0 \\ 0 & 0 & d_3 & \ldots & 0 \\ \vdots & \vdots & \vdots & \ddots & \vdots \\ 0 & 0 & 0 & \ldots & d_n \end{bmatrix}$$

When is $D$ singular? Why?

**74.** ▼ If a square matrix $A$ row-reduces to the identity matrix, must it be invertible? If so, say why, and if not, give an example of such a (singular) matrix.

**75.** ▼ If $A$ and $B$ are invertible, check that $B^{-1}A^{-1}$ is the inverse of $AB$.

**76.** ▼ Solve the matrix equation $A(B + CX) = D$ for $X$. (You may assume that $A$ and $C$ are invertible square matrices.)

**77.** ◆ In Example 3 we said that, if a square matrix $A$ row-reduces to a matrix with a row of zeros, then it is singular. Why?

**78.** ◆ Your friend has two square matrices $A$ and $B$, neither of them the zero matrix, with the property that $AB$ is the zero matrix. You immediately tell him that neither $A$ nor $B$ can possibly be invertible. How can you be so sure?

# 4.4 Game Theory

It frequently happens that you are faced with making a decision or choosing a best strategy from several possible choices. For instance, you might need to decide whether to invest in stocks or bonds, whether to cut prices of the product you sell, or what offensive play to use in a football game. In these examples, the result depends on something you cannot control. In the first case, your success depends on the future behavior of the economy. In the second case, it depends in part on whether your competitors also cut prices, and in the third case, it depends on the defensive strategy chosen by the opposing team.

We can model situations like these using **game theory**. We represent the various options and payoffs in a matrix and can then calculate the best single strategy or combination of strategies using matrix algebra and other techniques.

Game theory is very new compared with most of the mathematics you learn. It was invented in the 1920s by the noted mathematicians Emile Borel (1871–1956) and John von Neumann (1903–1957). Game theory's connection with linear programming was discovered even more recently, in 1947, by von Neumann, and further advances were made by the mathematician John Nash\* (1928–), for which he received the 1994 Nobel Prize for Economics.

## The Payoff Matrix and Expected Payoff

We have probably all played the simple game "Rock, Paper, Scissors" at some time in our lives. It goes as follows: There are two players—let us call them A and B—and at each turn, both players produce, by a gesture of the hand, either paper, a pair of scissors, or a rock. Rock beats scissors (since a rock can crush scissors), but is beaten by paper (since a rock can be covered by paper), while scissors beat paper (since scissors can cut paper). The round is a draw if both A and B show the same item. We could turn this into a betting game if, at each turn, we require the loser to pay the winner 1¢. For instance, if A shows a rock and B shows paper, then A pays B 1¢.

Rock, Paper, Scissors is an example of a **two-person zero-sum game**. It is called a zero-sum game because each player's loss is equal to the other player's gain.† We can represent this game by a matrix, called the **payoff matrix**.

$$
\begin{array}{c}
 \\
\mathbf{A}
\end{array}
\begin{array}{c}
\\
\begin{array}{ccc} r & p & s \end{array} \\
\begin{array}{c} r \\ p \\ s \end{array}
\left[\begin{array}{ccc}
0 & -1 & 1 \\
1 & 0 & -1 \\
-1 & 1 & 0
\end{array}\right]
\end{array}
\quad \text{or just} \quad
P = \left[\begin{array}{ccc}
0 & -1 & 1 \\
1 & 0 & -1 \\
-1 & 1 & 0
\end{array}\right]
$$

if we choose to omit the labels. In the payoff matrix, Player A's options, or **moves**, are listed on the left, while Player B's options are listed on top. We think of A as playing the rows and B as playing the columns. Positive numbers indicate a win for the row player, while negative numbers indicate a loss for the row player. Thus, for example, the $p, s$ entry represents the outcome if A plays $p$ (paper) and B plays $s$ (scissors). In this event, B wins, and the –1 entry there indicates that A loses 1¢. (If that entry were –2 instead, it would have meant that A loses 2¢.)

\* Nash's turbulent life is the subject of the biography *A Beautiful Mind* by Sylvia Nasar (Simon & Schuster, 1998). The 2001 Academy Award-winning movie of the same title is a somewhat fictionalized account.

† An example of a *nonzero-sum game* would be one in which the government taxed the earnings of the winner. In that case the winner's gain would be less than the loser's loss.

## Two-Person Zero-Sum Game

A **two-person zero-sum game** is one in which one player's loss equals the other's gain. We assume that the outcome is determined by each player's choice from among a fixed, finite set of moves. If Player A has $m$ moves to choose from and Player B has $n$, we can represent the game using the **payoff matrix**, the $m \times n$ matrix showing the result of each possible pair of choices of moves.

In each round of the game, the way a player chooses a move is called a **strategy**. A player using a **pure strategy** makes the same move each round of the game. For example, if a player in the above game chooses to play scissors at each turn, then that player is using the pure strategy $s$. A player using a **mixed strategy** chooses each move a certain percentage of the time in a random fashion; for instance, Player A might choose to play $p$ 50% of the time, and each of $s$ and $r$ 25% of the time.

Our ultimate goal is to be able to determine which strategy is best for each player to use. To do that, we need to know how to evaluate strategies. The fundamental calculation we need is that of the **expected payoff** resulting from a pair of strategies. Let's look at a simple example.

### EXAMPLE 1  Expected Payoff

Consider the following game:

$$\begin{array}{cc} & \mathbf{B} \\ & \begin{array}{cc} a & \;\; b \end{array} \\ \mathbf{A} \;\; \begin{array}{c} p \\ q \end{array} & \begin{bmatrix} 3 & -1 \\ -2 & 3 \end{bmatrix}. \end{array}$$

Player A decides to pick moves at random, choosing to play $p$ 75% of the time and $q$ 25% of the time. Player B also picks moves at random, choosing $a$ 20% of the time and $b$ 80% of the time. On average, how much does A expect to win or lose?

**Solution**  Suppose they play the game 100 times. Each time they play there are four possible outcomes:

*Case 1: A plays p, B plays a.*
Because A plays $p$ only 75% of the time and B plays $a$ only 20% of the time, we expect this case to occur $0.75 \times 0.20 = 0.15$, or 15% of the time, or 15 times out of 100. Each time this happens, A gains 3 points, so we get a contribution of $15 \times 3 = 45$ points to A's total winnings.

*Case 2: A plays p, B plays b.*
Because A plays $p$ only 75% of the time and B plays $b$ only 80% of the time, we expect this case to occur $0.75 \times 0.80 = 0.60$, or 60 times out of 100. Each time this happens, A loses 1 point, so we get a contribution of $60 \times -1 = -60$ to A's total winnings.

*Case 3: A plays q, B plays a.*
This case occurs $0.25 \times 0.20 = 0.05$, or 5 out of 100 times, with a loss of 2 points to A each time, giving a contribution of $5 \times -2 = -10$ to A's total winnings.

*Case 4: A plays q, B plays b.*
This case occurs $0.25 \times 0.80 = 0.20$, or 20 out of 100 times, with a gain of 3 points to A each time, giving a contribution of $20 \times 3 = 60$ to A's total winnings.

Summing to get A's total winnings and then dividing by the number of times the game is played gives the average value of

$$(45 - 60 - 10 + 60)/100 = 0.35$$

so that $A$ can expect to win an average of 0.35 points per play of the game. We call 0.35 the **expected payoff** of the game resulting from these particular strategies for A and B.

This calculation was somewhat tedious and it would only get worse if A and B had many moves to choose from. There is a far more convenient way of doing exactly the same calculation, using matrix multiplication: We start by representing the player's strategies as matrices. For reasons to become clear in a moment, we record A's strategy as a row matrix:

$$R = [0.75 \quad 0.25].$$

We record B's strategy as a column matrix:

$$C = \begin{bmatrix} 0.20 \\ 0.80 \end{bmatrix}.$$

(We will sometimes write column vectors using transpose notation, writing, for example, $[0.20 \quad 0.80]^T$ for the column above, to save space.) Now: *The expected payoff is the matrix product RPC, where P is the payoff matrix!*

$$\text{Expected payoff} = RPC = [0.75 \quad 0.25] \begin{bmatrix} 3 & -1 \\ -2 & 3 \end{bmatrix} \begin{bmatrix} 0.20 \\ 0.80 \end{bmatrix}$$

$$= [1.75 \quad 0] \begin{bmatrix} 0.20 \\ 0.80 \end{bmatrix} = [0.35].$$

Why does this work? Write out the arithmetic involved in the matrix product $RPC$ to see what we calculated:

$$[0.75 \times 3 + 0.25 \times (-2)] \times 0.20 + [0.75 \times (-1) + 0.25 \times 3] \times 0.80$$
$$= 0.75 \times 3 \times 0.20 + 0.25 \times (-2) \times 0.20 + 0.75 \times (-1) \times 0.80 + 0.25 \times 3 \times 0.80$$
$$= \quad \text{Case 1} \quad + \quad \text{Case 3} \quad + \quad \text{Case 2} \quad + \quad \text{Case 4.}$$

So, the matrix product does all at once the various cases we considered above.

The use of technology becomes indispensable when we need to do several calculations or when the matrices involved are big. See the technology note accompanying Example 3 in Section 4.2 on page 246 for instructions on multiplying matrices using a TI-83/84 Plus, a spreadsheet, and the Matrix Algebra Tool at the Website.

**using Technology**

To summarize what we just saw:

## The Expected Payoff resulting from Mixed Strategies R and C

The **expected payoff of a game resulting from given mixed strategies** is the average payoff that occurs if the game is played a large number of times with the row and column players using the given strategies.

To compute the expected payoff resulting from mixed strategies $R$ and $C$:

**1.** Write the row player's mixed strategy as a row matrix $R$.

**2.** Write the column player's mixed strategy as a column matrix $C$.

**3.** Calculate the product $RPC$, where $P$ is the payoff matrix. This product is a $1 \times 1$ matrix whose entry is the expected payoff $e$.

---

**Quick Example**

Consider "Rock, Paper, Scissors."

$$\mathbf{A} \; \begin{matrix} \\ r \\ p \\ s \end{matrix} \overset{\begin{matrix} & \mathbf{B} & \\ r & p & s \end{matrix}}{\begin{bmatrix} 0 & -1 & 1 \\ 1 & 0 & -1 \\ -1 & 1 & 0 \end{bmatrix}}$$

Suppose that the row player plays *rock* half the time and each of the other two strategies a quarter of the time, and the column player always plays *paper*. We write

$$R = \begin{bmatrix} \frac{1}{2} & \frac{1}{4} & \frac{1}{4} \end{bmatrix}, \quad C = \begin{bmatrix} 0 \\ 1 \\ 0 \end{bmatrix}.$$

So,

$$e = RPC = \begin{bmatrix} \frac{1}{2} & \frac{1}{4} & \frac{1}{4} \end{bmatrix} \begin{bmatrix} 0 & -1 & 1 \\ 1 & 0 & -1 \\ -1 & 1 & 0 \end{bmatrix} \begin{bmatrix} 0 \\ 1 \\ 0 \end{bmatrix}$$

$$= \begin{bmatrix} \frac{1}{2} & \frac{1}{4} & \frac{1}{4} \end{bmatrix} \begin{bmatrix} -1 \\ 0 \\ 1 \end{bmatrix} = -\frac{1}{4}.$$

Thus, player A can expect to lose an average of once every four plays.

---

## Solving a Game

Now that we know how to evaluate particular strategies, we want to find the *best* strategy. The next example takes us another step toward that goal.

---

### EXAMPLE 2    Television Ratings Wars

Commercial TV station RTV and cultural station CTV are competing for viewers in the Tuesday prime-time 9–10 PM time slot. RTV is trying to decide whether to show a sitcom, a docudrama, a reality show, or a movie, while CTV is thinking about either a nature documentary, a symphony concert, a ballet, or an opera. A television rating company estimates the payoffs for the various alternatives as follows. (Each point indicates a shift of 1,000 viewers from one channel to the other; thus, for instance, −2 indicates a shift of 2,000 viewers from RTV to CTV.)

|  |  | CTV | | | |
|---|---|---|---|---|---|
|  |  | **Nature Doc.** | **Symphony** | **Ballet** | **Opera** |
|  | **Sitcom** | 2 | 1 | −2 | 2 |
| **RTV** | **Docudrama** | −1 | 1 | −1 | 2 |
|  | **Reality Show** | −2 | 0 | 0 | 1 |
|  | **Movie** | 3 | 1 | −1 | 1 |

**a.** If RTV notices that CTV is showing nature documentaries half the time and symphonies the other half, what would RTV's best strategy be, and how many viewers would it gain if it followed this strategy?

**b.** If, on the other hand, CTV notices that RTV is showing docudramas half the time and reality shows the other half, what would CTV's best strategy be, and how many viewers would it gain or lose if it followed this strategy?

**Solution**

**a.** We are given the matrix of the game, $P$, in the table above, and we are given CTV's strategy $C = [0.50 \ \ 0.50 \ \ 0 \ \ 0]^T$. We are not given RTV's strategy $R$. To say that RTV is looking for its best strategy is to say that it wants the resulting expected payoff $e = RPC$ to be as high as possible. So, we take $R = [x \ \ y \ \ z \ \ t]$ and look for values for $x$, $y$, $z$, and $t$ that make $RPC$ as high as possible. First, we calculate $e$ in terms of these unknowns:

$$e = RPC = \begin{bmatrix} x & y & z & t \end{bmatrix} \begin{bmatrix} 2 & 1 & -2 & 2 \\ -1 & 1 & -1 & 2 \\ -2 & 0 & 0 & 1 \\ 3 & 1 & -1 & 1 \end{bmatrix} \begin{bmatrix} 0.50 \\ 0.50 \\ 0 \\ 0 \end{bmatrix}$$

$$= \begin{bmatrix} x & y & z & t \end{bmatrix} \begin{bmatrix} 1.5 \\ 0 \\ -1 \\ 2 \end{bmatrix} = 1.5x - z + 2t.$$

Now, the unknowns $x$, $y$, $z$, and $t$ must be nonnegative and add up to 1 (why?). Because $t$ has the largest coefficient, 2, we'll get the best result by making it as large as possible, namely, $t = 1$, leaving $x = y = z = 0$. Thus, RTV's best strategy is $R = [0 \ \ 0 \ \ 0 \ \ 1]$. In other words, RTV should use the pure strategy of showing a movie every Tuesday evening. If it does so, the expected payoff will be

$$e = 1.5(0) - 0 + 2(1) = 2$$

so RTV can expect to gain 2,000 viewers.

**b.** Here, we are given $R = [0 \ \ 0.50 \ \ 0.50 \ \ 0]$ and are not given CTV's strategy $C$, so this time we take $C = [x \ \ y \ \ z \ \ t]^T$ and calculate the resulting expected payoff $e$:

$$e = RPC = \begin{bmatrix} 0 & 0.50 & 0.50 & 0 \end{bmatrix} \begin{bmatrix} 2 & 1 & -2 & 2 \\ -1 & 1 & -1 & 2 \\ -2 & 0 & 0 & 1 \\ 3 & 1 & -1 & 1 \end{bmatrix} \begin{bmatrix} x \\ y \\ z \\ t \end{bmatrix}$$

$$= \begin{bmatrix} -1.5 & 0.5 & -0.5 & 1.5 \end{bmatrix} \begin{bmatrix} x \\ y \\ z \\ t \end{bmatrix} = -1.5x + 0.5y - 0.5z + 1.5t.$$

Now, CTV wants $e$ to be as *low* as possible (why?). Because $x$ has the largest negative coefficient, CTV would like it to be as large as possible: $x = 1$, so the rest of the unknowns must be zero. Thus, CTV's best strategy is $C = [1 \ \ 0 \ \ 0 \ \ 0]^T$; that is, show a nature documentary every night. If it does so, the expected payoff will be

$$e = -1.5(1) + 0.5(0) - 0.5(0) + 1.5(0) = -1.5.$$

So CTV can expect to gain 1,500 viewers.

This example illustrates the fact that, no matter what mixed strategy one player selects, the other player can choose an appropriate *pure* counterstrategy in order to maximize its gain. How does this affect what decisions you should make as one of the players? If you were on the board of directors of RTV, you might reason as follows: Since for every mixed strategy you try, CTV can find a best counterstrategy (as in part (b)), it is in your company's best interest to select a mixed strategy that *minimizes* the effect of CTV's best counterstrategy. This is called the **minimax criterion**.

## Minimax Criterion

A player using the **minimax criterion** chooses a strategy that, among all possible strategies, minimizes the effect of the other player's best counterstrategy. That is, an optimal (best) strategy according to the minimax criterion is one that minimizes the maximum damage the opponent can cause.

This criterion assumes that your opponent is determined to win. More precisely, it assumes the following.

## Fundamental Principle of Game Theory

Each player tries to use its best possible strategy, and assumes that the other player is doing the same.

This principle is not always followed by every player. For example, one of the players may be nature and may choose its move at random, with no particular purpose in mind. In such a case, criteria other than the minimax criterion may be more appropriate. For example, there is the "maximax" criterion, which maximizes the maximum possible payoff (also known as the "reckless" strategy), or the criterion that seeks to minimize "regret" (the difference between the payoff you get and the payoff you *would have gotten* if you had known beforehand what was going to happen).\* But, we shall assume here the fundamental principle and try to find optimal strategies under the minimax criterion.

Finding the optimal strategy is called **solving the game**. In general, solving a game can be done using linear programming, as we shall see in the next chapter. However, we can solve $2 \times 2$ games "by hand," as we shall see in the next example. First, we notice that some large games can be reduced to smaller games.

Consider the game in the preceding example, which had the following matrix:

$$P = \begin{bmatrix} 2 & 1 & -2 & 2 \\ -1 & 1 & -1 & 2 \\ -2 & 0 & 0 & 1 \\ 3 & 1 & -1 & 1 \end{bmatrix}.$$

Compare the second and third columns through the eyes of the column player, CTV. Every payoff in the third column is as good as or better, from CTV's point of view, than the corresponding entry in the second column. Thus, no matter what RTV does, CTV will do better showing a ballet (third column) than a symphony (second column). We say that the third column **dominates** the second column. As far as CTV is concerned, we might as well forget about symphonies entirely, so we remove the second column. Similarly, the third column dominates the fourth,

✴ See *Location in Space: Theoretical Perspectives in Economic Geography*, 3rd Edition, by Peter Dicken and Peter E. Lloyd, HarperCollins Publishers, 1990, p. 276.

so we can remove the fourth column, too. This gives us a smaller game to work with:

$$P = \begin{bmatrix} 2 & -2 \\ -1 & -1 \\ -2 & 0 \\ 3 & -1 \end{bmatrix}.$$

Now compare the first and last rows. Every payoff in the last row is larger than the corresponding payoff in the first row, so the last row is always better to RTV. Again, we say that the last row dominates the first row, and we can discard the first row. Similarly, the last row dominates the second row, so we discard the second row as well. This reduces us to the following game:

$$P = \begin{bmatrix} -2 & 0 \\ 3 & -1 \end{bmatrix}.$$

In this matrix, neither row dominates the other and neither column dominates the other. So, this is as far as we can go with this line of argument. We call this **reduction by dominance**.

## Reduction by Dominance

One *row* **dominates** another if every entry in the former is greater than or equal to the corresponding entry in the latter. Put another way, one row dominates another if it is always at least as good for the row player.

One *column* dominates another if every entry in the former is less than or equal to the corresponding entry in the latter. Put another way, one column dominates another if it is always at least as good for the column player.

### Procedure for Reducing by Dominance:
1. Check whether there is any row in the (remaining) matrix that is dominated by another row. Remove all dominated rows.

2. Check whether there is any column in the (remaining) matrix that is dominated by another column. Remove all dominated columns.

3. Repeat steps 1 and 2 until there are no dominated rows or columns.

Let us now go back to the "television ratings wars" example and see how we can solve a game using the minimax criterion once we are down to a 2 × 2 payoff matrix.

EXAMPLE 3 **Solving a 2 × 2 Game**

Continuing the preceding example:
**a.** Find the optimal strategy for RTV.
**b.** Find the optimal strategy for CTV.
**c.** Find the expected payoff of the game if RTV and CTV use their optimal strategies.

**Solution** As in the text, we begin by reducing the game by dominance, which brings us down to the following $2 \times 2$ game:

|  |  | CTV | |
|---|---|---|---|
|  |  | **Nature Doc.** | **Ballet** |
| **RTV** | **Reality Show** | $-2$ | $0$ |
|  | **Movie** | $3$ | $-1$ |

**a.** Now let's find RTV's optimal strategy. Because we don't yet know what it is, we write down a general strategy:

$$R = [x \quad y].$$

Because $x + y = 1$, we can replace $y$ by $1 - x$:

$$R = [x \quad 1 - x].$$

We know that CTV's best counterstrategy to $R$ will be a pure strategy (see the discussion after Example 2), so let's compute the expected payoff that results from each of CTV's possible pure strategies:

$$e = [x \quad 1 - x]\begin{bmatrix} -2 & 0 \\ 3 & -1 \end{bmatrix}\begin{bmatrix} 1 \\ 0 \end{bmatrix}$$

$$= (-2)x + 3(1 - x) = -5x + 3$$

$$f = [x \quad 1 - x]\begin{bmatrix} -2 & 0 \\ 3 & -1 \end{bmatrix}\begin{bmatrix} 0 \\ 1 \end{bmatrix}$$

$$= 0x - (1 - x) = x - 1.$$

**Figure 1**

Because both $e$ and $f$ depend on $x$, we can graph them as in Figure 1.

If, for instance, RTV happened to choose $x = 0.5$, then the expected payoffs resulting from CTV's two pure strategies are $e = -5(1/2) + 3 = 1/2$ and $f = 1/2 - 1 = -1/2$. The worst outcome for RTV is the lower of the two, $f$, and this will be true wherever the graph of $f$ is below the graph of $e$. On the other hand, if RTV chose $x = 1$, the graph of $e$ would be lower and the worst possible expected value would be $e = -5(1) + 3 = -2$. Since RTV can choose $x$ to be any value between 0 and 1, the worst possible outcomes are those shown by the colored portion of the graph in Figure 2.

**Figure 2**

Because RTV is trying to make the worst possible outcome as large as possible (that is, to minimize damages), it is seeking the point on the colored portion of the graph that is highest. This is the intersection point of the two lines. To calculate its coordinates, it's easiest to equate the two functions of $x$:

$$-5x + 3 = x - 1,$$

$$-6x = -4,$$

or

$$x = \frac{2}{3}.$$

The $e$ (or $f$) coordinate is then obtained by substituting $x = 2/3$ into the expression for $e$ (or $f$), giving:

$$e = -5\left(\frac{2}{3}\right) + 3$$

$$= -\frac{1}{3}.$$

We conclude that RTV's best strategy is to take $x = 2/3$, giving an expected value of $-1/3$. In other words, RTV's optimal mixed strategy is:

$$R = \begin{bmatrix} \frac{2}{3} & \frac{1}{3} \end{bmatrix}.$$

Going back to the original game, RTV should show reality shows 2/3 of the time and movies 1/3 of the time. It should not bother showing any sitcoms or docudramas. It expects to lose, on average, 333 viewers to CTV, but all of its other options are worse.

**b.** To find CTV's optimal strategy, we must reverse roles and start by writing its unknown strategy as follows:

$$C = \begin{bmatrix} x \\ 1 - x \end{bmatrix}.$$

We calculate the expected payoffs for the two pure row strategies:

$$e = \begin{bmatrix} 1 & 0 \end{bmatrix} \begin{bmatrix} -2 & 0 \\ 3 & -1 \end{bmatrix} \begin{bmatrix} x \\ 1 - x \end{bmatrix}$$

$$= -2x$$

and

$$f = \begin{bmatrix} 0 & 1 \end{bmatrix} \begin{bmatrix} -2 & 0 \\ 3 & -1 \end{bmatrix} \begin{bmatrix} x \\ 1 - x \end{bmatrix}$$

$$= 3x - (1 - x)$$

$$= 4x - 1.$$

**Figure 3**

As with the row player, we know that the column player's best strategy will correspond to the intersection of the graphs of $e$ and $f$ (Figure 3). (Why is the upper edge colored, rather than the lower edge?) The graphs intersect when

$$-2x = 4x - 1$$

or

$$x = \frac{1}{6}.$$

The corresponding value of $e$ (or $f$) is

$$e = -2\left(\frac{1}{6}\right) = -\frac{1}{3}.$$

Thus, CTV's optimal mixed strategy is $\begin{bmatrix} \frac{1}{6} & \frac{5}{6} \end{bmatrix}^T$ and the expected payoff is $-1/3$. So, CTV should show nature documentaries 1/6 of the time and ballets 5/6 of the time. It should not bother to show symphonies or operas. It expects to gain, on average, 333 viewers from RTV.

**c.** We can now calculate the expected payoff as usual, using the optimal strategies we found in parts (a) and (b).

$$e = RPC$$

$$= \begin{bmatrix} \frac{2}{3} & \frac{1}{3} \end{bmatrix} \begin{bmatrix} -2 & 0 \\ 3 & -1 \end{bmatrix} \begin{bmatrix} \frac{1}{6} \\ \frac{5}{6} \end{bmatrix}$$

$$= -\frac{1}{3}$$

➡ **Before we go on...** In Example 3, it is no accident that the expected payoff resulting from the optimal strategies equals the expected payoff we found in (a) and (b). If we call the expected payoff resulting from the optimal strategies the **expected value of the game**, the row player's optimal strategy guarantees an expected payoff no smaller than the expected value, while the column player's optimal strategy guarantees an expected payoff no larger. Together they force the average payoff to be the expected value of the game. ∎

### Expected Value of a Game

The **expected value of a game** is its expected payoff that results when the row and column players use their optimal (minimax) strategies. By using its optimal strategy, the row player guarantees an expected payoff no lower than the expected value of the game, no matter what the column player does. Similarly, by using its optimal strategy, the column player guarantees an expected payoff no higher than the expected value of the game, no matter what the row player does.

---

**Q** : *What about games that don't reduce to 2 × 2 matrices? Are these solved in a similar way?*

**A** : The method illustrated in the previous example cannot easily be generalized to solve bigger games (i.e., games that cannot be reduced to 2 × 2 matrices); solving even a 2 × 3 game using this approach would require us to consider graphs in three dimensions. To be able to solve games of arbitrary size, we need to wait until the next chapter (Section 5.5), where we describe a method for solving a game, using the simplex method, that works for all payoff matrices.

---

## Strictly Determined Games

Although we haven't yet discussed how to solve general $m \times n$ games, there are certain kinds of games that can be solved quite simply regardless of size, as illustrated by the following example.

---

**EXAMPLE 4  Strictly Determined Game**

Solve the following game:

$$\mathbf{A} \begin{array}{c} \\ s \\ t \\ u \end{array} \overset{\overset{\textstyle \mathbf{B}}{p \quad q \quad r}}{\begin{bmatrix} -4 & -3 & 3 \\ 2 & -1 & -2 \\ 1 & 0 & 2 \end{bmatrix}}.$$

**Solution** If we look carefully at this matrix, we see that no row dominates another and no column dominates another, so we can't reduce it. Nor do we know how to solve a 3 × 3 game, so it looks as if we're stuck. However, there is a way to understand this particular game. With the minimax criterion in mind, let's begin by considering the worst possible outcomes for the row player for each possible move. We do this by circling the smallest payoff in each row, the **row minima**:

$$
\begin{array}{c}
 & & \mathbf{B} \\
 & \begin{array}{ccc} p & q & r \end{array} & \text{Row minima} \\
\mathbf{A}\begin{array}{c} s \\ t \\ u \end{array}
\left[\begin{array}{ccc}
\boxed{-4} & -3 & 3 \\
2 & -1 & \boxed{-2} \\
1 & \boxed{0} & 2
\end{array}\right]
&
\begin{array}{c} -4 \\ -2 \\ 0 \leftarrow \text{(largest)} \end{array}
\end{array}
$$

So, for example, if A plays move $s$, the worst possible outcome is to lose 4. Player A takes the least risk by using move $u$, which has the largest row minimum.

We do the same thing for the column player, remembering that smaller payoffs are better for B and larger payoffs worse. We draw a box around the largest payoff in each column, the **column maxima**:

$$
\begin{array}{c}
 & \mathbf{B} \\
 & \begin{array}{ccc} p & q & r \end{array} \\
\mathbf{A}\begin{array}{c} s \\ t \\ u \end{array}
\left[\begin{array}{ccc}
-4 & -3 & \boxed{3} \\
\boxed{2} & -1 & -2 \\
1 & \boxed{0} & 2
\end{array}\right]
\end{array}
$$

$$
\begin{array}{cccc}
\text{Column maxima} & 2 & 0 & 3 \\
 & & \uparrow & \\
 & & \text{(smallest)} &
\end{array}
$$

Player B takes the least risk by using move $q$, which has the smallest column maximum.

Now put the circles and boxes together:

$$
\begin{array}{c}
 & \mathbf{B} \\
 & \begin{array}{ccc} p & q & r \end{array} \\
\mathbf{A}\begin{array}{c} s \\ t \\ u \end{array}
\left[\begin{array}{ccc}
\boxed{-4} & -3 & \boxed{3} \\
\boxed{2} & -1 & \boxed{-2} \\
1 & \boxed{\boxed{0}} & 2
\end{array}\right]
\end{array}
$$

Notice that the $uq$ entry is both circled and boxed: It is both a row minimum and a column maximum. We call such an entry a **saddle point.**

Now we claim that the optimal strategy for A is to always play $u$ while the optimal strategy for B is to always play $q$. By playing $u$, A guarantees that the payoff will be 0 or higher, no matter what B does, so the expected value of the game has to be *at least* 0. On the other hand, by playing $q$, B guarantees that the payoff will be 0 or less, so the expected value of the game has to be *no more than* 0. Combining these facts, we conclude that the expected value of the game must be exactly 0, and A and B have no strategies that could do any better for them than the pure strategies $u$ and $q$.

➡ **Before we go on...** You should consider what happens in an example like the television rating wars game of Example 3. In that game, the largest row minimum is $-1$ while the smallest column maximum is 0; there is no saddle point. The row player can force a payoff of at least $-1$ by playing a pure strategy (always showing movies, for example), but can do better, forcing an expected payoff of $-1/3$, by playing a mixed strategy as we saw in Example 3. Similarly, the column player can force the payoff to be 0 or less with a pure strategy, but can do better, forcing an expected payoff of $-1/3$, with a mixed strategy. Only when there is a saddle point will pure strategies be optimal. ∎

## Strictly Determined Game

A **saddle point** is a payoff $v$ that is simultaneously a row minimum and a column maximum (both boxed and circled in our approach). If a game has a saddle point, the corresponding row and column strategies are the optimal ones, the expected value of the game is the payoff $v$, and we say that the game is **strictly determined**.

## FAQs

### Solving a Game

**Q**: *We've seen several ways of trying to solve a game. What should I do and in what order?*

**A**: Here are the steps you should take when trying to solve a game:

1. Reduce by dominance. This should always be your first step.
2. If you were able to reduce to a 1 × 1 game, you're done. The optimal strategies are the corresponding pure strategies, as they dominate all the others.
3. Look for a saddle point in the reduced game. If it has one, the game is strictly determined, and the corresponding pure strategies are optimal.
4. If your reduced game is 2 × 2 and has no saddle point, use the method of Example 3 to find the optimal mixed strategies.
5. If your reduced game is larger than 2 × 2 and has no saddle point, you have to use linear programming to solve it, but that will have to wait until the following chapter.

## 4.4 EXERCISES

▼ more advanced ◆ challenging
T indicates exercises that should be solved using technology

*In Exercises 1–4, calculate the expected payoff of the game with payoff matrix*

$$P = \begin{bmatrix} 2 & 0 & -1 & 2 \\ -1 & 0 & 0 & -2 \\ -2 & 0 & 0 & 1 \\ 3 & 1 & -1 & 1 \end{bmatrix}$$

*using the mixed strategies supplied.* HINT [See Example 1.]

1. $R = [0 \quad 1 \quad 0 \quad 0], C = [1 \quad 0 \quad 0 \quad 0]^T$
2. $R = [0 \quad 0 \quad 0 \quad 1], C = [0 \quad 1 \quad 0 \quad 0]^T$
3. $R = [0.5 \quad 0.5 \quad 0 \quad 0], C = [0 \quad 0 \quad 0.5 \quad 0.5]^T$
4. $R = [0 \quad 0.5 \quad 0 \quad 0.5], C = [0.5 \quad 0.5 \quad 0 \quad 0]^T$

*In Exercises 5–8, either a mixed column or row strategy is given. In each case, use*

$$P = \begin{bmatrix} 0 & -1 & 5 \\ 2 & -2 & 4 \\ 0 & 3 & 0 \\ 1 & 0 & -5 \end{bmatrix}$$

*and find the optimal pure strategy (or strategies) the other player should use. Express the answer as a row or column matrix. Also determine the resulting expected payoff.* HINT [See Example 2.]

5. $C = [0.25 \quad 0.75 \quad 0]^T$
6. $C = \left[\frac{1}{3} \quad \frac{1}{3} \quad \frac{1}{3}\right]^T$
7. $R = \left[\frac{1}{2} \quad 0 \quad \frac{1}{4} \quad \frac{1}{4}\right]$
8. $R = [0.8 \quad 0.2 \quad 0 \quad 0]$

*Reduce the payoff matrices in Exercises 9–14 by dominance.*

9.
$$\begin{array}{c} \\ a \\ b \end{array} \begin{array}{cc} \begin{array}{ccc} p & q & r \end{array} \\ \begin{bmatrix} 1 & 1 & 10 \\ 2 & 3 & -4 \end{bmatrix} \end{array}$$

10.
$$\begin{array}{c} \\ a \\ b \end{array} \begin{array}{cc} \begin{array}{ccc} p & q & r \end{array} \\ \begin{bmatrix} 2 & 0 & 10 \\ 15 & -4 & -5 \end{bmatrix} \end{array}$$

11.
$$\begin{array}{c} \\ 1 \\ 2 \\ 3 \end{array} \begin{array}{cc} \begin{array}{ccc} a & b & c \end{array} \\ \begin{bmatrix} 2 & -4 & -9 \\ -1 & -2 & -3 \\ 5 & 0 & -1 \end{bmatrix} \end{array}$$

12.
$$\begin{array}{c} \\ 1 \\ 2 \\ 3 \end{array} \begin{array}{cc} \begin{array}{ccc} a & b & c \end{array} \\ \begin{bmatrix} 0 & -1 & -5 \\ -3 & -10 & 10 \\ 2 & 3 & -4 \end{bmatrix} \end{array}$$

**13.**

$$\begin{array}{c} \\ p \\ q \\ r \\ s \end{array} \begin{array}{ccc} a & b & c \\ \left[\begin{array}{ccc} 1 & -1 & -5 \\ 4 & 0 & 2 \\ 3 & -3 & 10 \\ 3 & -5 & -4 \end{array}\right] \end{array}$$

**14.**

$$\begin{array}{c} \\ p \\ q \\ r \\ s \end{array} \begin{array}{ccc} a & b & c \\ \left[\begin{array}{ccc} 2 & -4 & 9 \\ 1 & 1 & 0 \\ -1 & -2 & -3 \\ 1 & 1 & -1 \end{array}\right] \end{array}$$

*In Exercises 15–20, decide whether the game is strictly determined. If it is, give the players' optimal pure strategies and the value of the game.* HINT [See Example 4.]

**15.**

$$\begin{array}{c} \\ a \\ b \end{array} \begin{array}{cc} p & q \\ \left[\begin{array}{cc} 1 & 1 \\ 2 & -4 \end{array}\right] \end{array}$$

**16.**

$$\begin{array}{c} \\ a \\ b \end{array} \begin{array}{cc} p & q \\ \left[\begin{array}{cc} -1 & 2 \\ 10 & -1 \end{array}\right] \end{array}$$

**17.**

$$\begin{array}{c} \\ a \\ b \end{array} \begin{array}{ccc} p & q & r \\ \left[\begin{array}{ccc} 2 & 0 & -2 \\ -1 & 3 & 0 \end{array}\right] \end{array}$$

**18.**

$$\begin{array}{c} \\ a \\ b \end{array} \begin{array}{ccc} p & q & r \\ \left[\begin{array}{ccc} -2 & 1 & -3 \\ -2 & 3 & -2 \end{array}\right] \end{array}$$

**19.**

$$\begin{array}{c} \\ P \\ Q \\ R \\ S \end{array} \begin{array}{ccc} a & b & c \\ \left[\begin{array}{ccc} 1 & -1 & -5 \\ 4 & -4 & 2 \\ 3 & -3 & -10 \\ 5 & -5 & -4 \end{array}\right] \end{array}$$

**20.**

$$\begin{array}{c} \\ P \\ Q \\ R \\ S \end{array} \begin{array}{ccc} a & b & c \\ \left[\begin{array}{ccc} -2 & -4 & 9 \\ 1 & 1 & 0 \\ -1 & -2 & -3 \\ 1 & 1 & -1 \end{array}\right] \end{array}$$

*In Exercises 21–24, find:* **(a)** *the optimal mixed row strategy;* **(b)** *the optimal mixed column strategy, and* **(c)** *the expected value of the game in the event that each player uses his or her optimal mixed strategy.* HINT [See Example 3.]

**21.** $P = \begin{bmatrix} -1 & 2 \\ 0 & -1 \end{bmatrix}$      **22.** $P = \begin{bmatrix} -1 & 0 \\ 1 & -1 \end{bmatrix}$

**23.** $P = \begin{bmatrix} -1 & -2 \\ -2 & 1 \end{bmatrix}$      **24.** $P = \begin{bmatrix} -2 & -1 \\ -1 & -3 \end{bmatrix}$

## APPLICATIONS

*Set up the payoff matrix in each of Exercises 25–32.*

**25. Games to Pass the Time** You and your friend have come up with the following simple game to pass the time: at each round, you simultaneously call "heads" or "tails." If you have both called the same thing, your friend wins 1 point; if your calls differ, you win 1 point.

**26. Games to Pass the Time** Bored with the game in Exercise 25, you decide to use the following variation instead: If you both call "heads" your friend wins 2 points; if you both call "tails" your friend wins 1 point; if your calls differ, then you win 2 points if you called "heads," and 1 point if you called "tails."

**27. War Games** You are deciding whether to invade France, Sweden, or Norway, and your opponent is simultaneously deciding which of these three countries to defend. If you invade a country that your opponent is defending, you will be defeated (payoff: $-1$), but if you invade a country your opponent is not defending, you will be successful (payoff: $+1$).

**28. War Games** You must decide whether to attack your opponent by sea or air, and your opponent must simultaneously decide whether to mount an all-out air defense, an all-out coastal defense (against an attack from the sea), or a combined air and coastal defense. If there is no defense for your mode of attack, you win 100 points. If your attack is met by a shared air and coastal defense, you win 50 points. If your attack is met by an all-out defense, you lose 200 points.

**29.** ▼ **Marketing** Your fast-food outlet, *Burger Queen*, has obtained a license to open branches in three closely situated South African cities: Brakpan, Nigel, and Springs. Your market surveys show that Brakpan and Nigel each provide a potential market of 2,000 burgers a day, while Springs provides a potential market of 1,000 burgers per day. Your company can finance an outlet in only one of those cities. Your main competitor, Burger Princess, has also obtained licenses for these cities, and is similarly planning to open only one outlet. If you both happen to locate at the same city, you will share the total business from all three cities equally, but if you locate in different cities, you will each get all the business in the cities in which you have located, plus half the business in the third city. The payoff is the number of burgers you will sell per day minus the number of burgers your competitor will sell per day.

**30.** ▼ **Marketing** Repeat Exercise 29, given that the potential sales markets in the three cities are: Brakpan, 2,500 per day; Nigel, 1,500 per day; Springs, 1,200 per day.

**31.** ▼ **Betting** When you bet on a racehorse with odds of *m–n*, you stand to win *m* dollars for every bet of *n* dollars if your horse wins; for instance, if the horse you bet is running at 5–2 and wins, you will win $5 for every $2 you bet. (Thus a $2 bet will return $7.) Here are some actual odds from a 1992 race at Belmont Park, NY.[21] The favorite at 5–2 was Pleasant Tap. The second choice was Thunder Rumble at 7–2, while the third choice was Strike the Gold at 4–1. Assume you are making a $10 bet on one of these horses. The payoffs are your winnings. (If your horse does not win, you lose your entire bet. Of course, it is possible for none of your horses to win.)

**32.** ▼ **Betting** Referring to Exercise 31, suppose that just before the race there has been frantic betting on Thunder Rumble with the result that the odds have dropped to 2–5. The odds on the other two horses remain unchanged.

**33. Retail Discount Wars** Just one week after your *Abercrom B* men's fashion outlet has opened at a new location opposite *Burger Prince* in the Mall, your rival, *Abercrom A*, opens up

---

[21] Source: *New York Times*, September 18, 1992, p. B14.

directly across from you. You have been informed that Abercrom A is about to launch either a 30% off everything sale or a 50% off everything sale. You, on the other hand, have decided to either *increase* prices (to make your store seem more exclusive) or do absolutely nothing. You construct the following payoff matrix, where the payoffs represent the number of customers your outlet can expect to gain from Abercrom A:

$$\begin{array}{c}\phantom{xx}\\ \text{Abercrom B}\end{array}\begin{array}{cc}\phantom{x} & \text{Abercrom A}\\ & \begin{array}{cc}\text{30\% Off} & \text{50\% Off}\end{array}\\ \begin{array}{c}\text{Do Nothing}\\ \text{Increase Prices}\end{array} & \begin{bmatrix}-60 & -40\\ 30 & -50\end{bmatrix}.\end{array}$$

There is a 20% chance that Abercrom A will opt for the "30% off" sale and an 80% chance that they will opt for the "50% Off" sale. Your sense from upper management at Abercrom B is that there is a 50% chance you will be given the go-ahead to raise prices. What is the expected resulting effect on your customer base?

**34.** *More Retail Discount Wars* Your *Abercrom B* men's fashion outlet has a 30% chance of launching an expensive new line of used auto-mechanic dungarees (complete with grease stains) and a 70% chance of staying instead with its traditional torn military-style dungarees. Your rival across from you in the mall, *Abercrom A*, appears to be deciding between a line of torn gym shirts and a more daring line of "empty shirts" (that is, empty shirt boxes). Your corporate spies reveal that there is a 20% chance that Abercrom A will opt for the empty shirt option. The following payoff matrix gives the number of customers your outlet can expect to gain from Abercrom A in each situation:

$$\begin{array}{c}\phantom{xx}\\ \text{Abercrom B}\end{array}\begin{array}{cc}\phantom{x} & \text{Abercrom A}\\ & \begin{array}{cc}\text{Torn Shirts} & \text{Empty Shirts}\end{array}\\ \begin{array}{c}\text{Mechanics}\\ \text{Military}\end{array} & \begin{bmatrix}10 & -40\\ -30 & 50\end{bmatrix}.\end{array}$$

What is the expected resulting effect on your customer base?

**35.** *Factory Location*[22] A manufacturer of electrical machinery is located in a cramped, though low-rent, factory close to the center of a large city. The firm needs to expand, and it could do so in one of three ways: (1) Remain where it is and install new equipment, (2) move to a suburban site in the same city, or (3) relocate in a different part of the country where labor is cheaper. Its decision will be influenced by the fact that one of the following will happen: (I) The government may introduce a program of equipment grants, (II) a new suburban highway may be built, or (III) the government may institute a policy of financial help to companies who move into regions of high unemployment. The value to the company of each combination is given in the following payoff matrix.

| | | Government's Options | |
|---|---|---|---|
| | **I** | **II** | **III** |
| **1** | 200 | 150 | 140 |
| **2** | 130 | 220 | 130 |
| **3** | 110 | 110 | 220 |

**Manufacturer's Options** (rows 1, 2, 3)

If the manufacturer judges that there is a 20% probability that the government will go with option I, a 50% probability that they will go with option II, and a 30% probability that they will go with option III, what is the manufacturer's best option?

**36.** *Crop Choice*[23] A farmer has a choice of growing wheat, barley, or rice. Her success will depend on the weather, which could be dry, average, or wet. Her payoff matrix is as follows.

| | | Weather | |
|---|---|---|---|
| | **Dry** | **Average** | **Wet** |
| **Wheat** | 20 | 20 | 10 |
| **Barley** | 10 | 15 | 20 |
| **Rice** | 10 | 20 | 20 |

**Crop Choices** (rows Wheat, Barley, Rice)

If the probability that the weather will be dry is 10%, the probability that it will be average is 60%, and the probability that it will be wet is 30%, what is the farmer's best choice of crop?

**37.** *Study Techniques* Your mathematics test is tomorrow and will cover the following topics: game theory, linear programming, and matrix algebra. You have decided to do an "all-nighter" and must determine how to allocate your eight hours of study time among the three topics. If you were to spend the entire eight hours on any one of these topics (thus using a pure strategy), you feel confident that you would earn a 90% score on that portion of the test but would not do so well on the other topics. You have come up with the following table, where the entries are your expected scores. (The fact that linear programming and matrix algebra are used in game theory is reflected in these numbers.)

| | | Test | |
|---|---|---|---|
| **Your Strategies** | **Game Theory** | **Linear Programming** | **Matrix Algebra** |
| **Game Theory** | 90 | 70 | 70 |
| **Linear Programming** | 40 | 90 | 40 |
| **Matrix Algebra** | 60 | 40 | 90 |

You have been told that the test will be weighted as follows: game theory: 25%; linear programming: 50%; matrix algebra: 25%.

[22] Adapted from an example in *Location in Space: Theoretical Perspectives in Economic Geography* by P. Dicken and P.E. Lloyd (Harper & Row, 1990).

[23] *Ibid.*

**a.** If you spend 25% of the night on game theory, 50% on linear programming, and 25% on matrix algebra, what score do you expect to get on the test?

**b.** Is it possible to improve on this by altering your study schedule? If so, what is the highest score you can expect on the test?

**c.** If your study schedule is according to part (a) and your teacher decides to forget her promises about how the test will be weighted and instead base it all on a single topic, which topic would be worst for you, and what score could you expect on the test?

**38.** *Study Techniques* Your friend Joe has been spending all of his time on fraternity activities and thus knows absolutely nothing about any of the three topics on tomorrow's math test. (See Exercise 37.) Because you are recognized as an expert on the use of game theory to solve study problems, he has turned to you for advice as to how to spend his "all-nighter." As the following table shows, his situation is not so rosy. (Since he knows no linear programming or matrix algebra, the table shows, for instance, that studying game theory all night will not be much use in preparing him for this topic.)

|  | Test | | |
|---|---|---|---|
| **Joe's Strategies** | **Game Theory** | **Linear Programming** | **Matrix Algebra** |
| **Game Theory** | 30 | 0 | 20 |
| **Linear Programming** | 0 | 70 | 0 |
| **Matrix Algebra** | 0 | 0 | 70 |

Assuming that the test will be weighted as described in Exercise 37, what are the answers to parts (a), (b), and (c) as they apply to Joe?

**39.** ▼*Staff Cutbacks* Frank Tempest manages a large snowplow service in Manhattan, Kansas, and is alarmed by the recent weather trends; there have been no significant snowfalls in recent years. He is therefore contemplating laying off some of his workers, but is unsure about whether to lay off 5, 10, or 15 of his 50 workers. Being very methodical, he estimates his annual net profits based on four possible annual snowfall figures: 0 inches, 20 inches, 40 inches, and 60 inches. (He takes into account the fact that, if he is running a small operation in the face of a large annual snowfall, he will lose business to his competitors because he will be unable to discount on volume.)

|  | **0 Inches** | **20 Inches** | **40 Inches** | **60 Inches** |
|---|---|---|---|---|
| **5 Laid Off** | −$500,000 | −$200,000 | $10,000 | $200,000 |
| **10 Laid Off** | −$200,000 | $0 | $0 | $0 |
| **15 Laid Off** | −$100,000 | $10,000 | −$200,000 | −$300,000 |

**a.** During the past 10 years, the region has had 0 inches twice, 20 inches twice, 40 inches three times, and 60 inches three times. Based on this information, how many workers should Tempest lay off, and how much would it cost him?

**b.** There is a 50% chance that Tempest will lay off 5 workers and a 50% chance that he will lay off 15 workers. What is the worst thing Nature can do to him in terms of snowfall? How much would it cost him?

**c.** The Gods of Chaos (who control the weather) know that Tempest is planning to use the strategy in part (a), and are determined to hurt Tempest as much as possible. Tempest, being somewhat paranoid, suspects it too. What should he do?

**40.** ▼ *Textbook Writing* You are writing a college-level textbook on finite mathematics and are trying to come up with the best combination of word problems. Over the years, you have accumulated a collection of amusing problems, serious applications, long complicated problems, and "generic" problems.[24] Before your book is published, it must be scrutinized by several reviewers, who, it seems, are never satisfied with the mix you use. You estimate that there are three kinds of reviewers: the "no-nonsense" types, who prefer applications and generic problems; the "dead serious" types, who feel that a college-level text should be contain little or no humor and lots of long complicated problems; and the "laid-back" types, who believe that learning best takes place in a light-hearted atmosphere bordering on anarchy. You have drawn up the following chart, where the payoffs represent the reactions of reviewers on a scale of −10 (ballistic) to +10 (ecstatic):

|  | Reviewers | | |
|---|---|---|---|
|  | **No-Nonsense** | **Dead Serious** | **Laid-Back** |
| **Amusing** | −5 | −10 | 10 |
| **Serious** | 5 | 3 | 0 |
| **Long** | −5 | 5 | 3 |
| **Generic** | 5 | 3 | −10 |

(You)

**a.** Your first draft of the book contained no generic problems, and equal numbers of the other categories. If half the reviewers of your book were "dead serious" and the rest were equally divided between the "no-nonsense" and "laid-back" types, what score would you expect?

**b.** In your second draft of the book, you tried to balance the content by including some generic problems and eliminating several amusing ones, and wound up with a mix of which one eighth were amusing, one quarter were serious, three eighths were long, and a quarter were generic. What kind of reviewer would be *least* impressed by this mix?

**c.** What kind of reviewer would be *most* impressed by the mix in your second draft?

---

[24]Of the following type: "A certain company has three processing plants: A, B, and C, each of which uses three processes: $P_1$, $P_2$, and $P_3$. Process $P_1$ uses 100 units of chemical $C_1$, 50 units of $C_2$, . . ." and so on.

**41. *Price Wars*** Computer Electronics, Inc. (CE) and the *Gigantic Computer Store* (GCS) are planning to discount the price they charge for the HAL Laptop Computer, of which they are the only distributors. Because Computer Electronics provides a free warranty service, they can generally afford to charge more. A market survey provides the following data on the gains to CE's market share that will result from different pricing decisions:

| | | GCS | | |
|---|---|---|---|---|
| | | $900 | $1,000 | $1,200 |
| CE | $1,000 | 15% | 60% | 80% |
| | $1,200 | 15% | 60% | 60% |
| | $1,300 | 10% | 20% | 40% |

**a.** Use reduction by dominance to determine how much each company should charge. What is the effect on CE's market share?

**b.** CE is aware that GCS is planning to use reduction by dominance to determine its pricing policy, and wants its market share to be as large as possible. What effect, if any, would the information about GCS have on CE's best strategy?

**42. *More Price Wars*** (Refer to Exercise 41.) A new market survey results in the following revised data:

| | | GCS | | |
|---|---|---|---|---|
| | | $900 | $1,000 | $1,200 |
| CE | $1,000 | 20% | 60% | 60% |
| | $1,200 | 15% | 60% | 60% |
| | $1,300 | 10% | 20% | 40% |

**a.** Use reduction by dominance to determine how much each company should charge. What is the effect on CE's market share?

**b.** In general, why do price wars tend to force prices down?

**43. *Wrestling Tournaments*** City Community College (CCC) plans to host Midtown Military Academy (MMA) for a wrestling tournament. Each school has three wrestlers in the 190 lb. weight class: CCC has Pablo, Sal, and Edison, while MMA has Carlos, Marcus, and Noto. Pablo can beat Carlos and Marcus, Marcus can beat Edison and Sal, Noto can beat Edison, while the other combinations will result in an even match. Set up a payoff matrix, and use reduction by dominance to decide which wrestler each team should choose as their champion. Does one school have an advantage over the other?

**44. *Wrestling Tournaments*** (Refer to Exercise 43.) One day before the wrestling tournament discussed in Exercise 43, Pablo sustains a hamstring injury and is replaced by Hans, who (unfortunately for CCC) can be beaten by both Carlos and Marcus. Set up the payoff matrix, and use reduction by dominance to decide which wrestler each team should

choose as their champion. Does one school have an advantage over the other?

**45. *The Battle of Rabaul-Lae***[25] In the Second World War, during the struggle for New Guinea, intelligence reports revealed that the Japanese were planning to move a troop and supply convoy from the port of Rabaul at the Eastern tip of New Britain to Lae, which lies just west of New Britain on New Guinea. It could either travel via a northern route, which was plagued by poor visibility, or by a southern route, where the visibility was clear. General Kenney, who was the commander of the Allied Air Forces in the area, had the choice of concentrating reconnaissance aircraft on one route or the other and bombing the Japanese convoy once it was sighted. Kenney's staff drafted the following outcomes for his choices, where the payoffs are estimated days of bombing time:

| | | Japanese Commander's Strategies | |
|---|---|---|---|
| | | Northern Route | Southern Route |
| Kenney's Strategies | Northern Route | 2 | 2 |
| | Southern Route | 1 | 3 |

What would you have recommended to General Kenney? What would you have recommended to the Japanese Commander?[26] How much bombing time results if these recommendations are followed?

**46. *The Battle of Rabaul-Lae*** Referring to Exercise 45, suppose that General Kenney had a third alternative: splitting his reconnaissance aircraft between the two routes, which would result in the following estimates:

| | | Japanese Commander's Strategies | |
|---|---|---|---|
| | | Northern Route | Southern Route |
| Kenney's Strategies | Northern Route | 2 | 2 |
| | Split Reconnaissance | 1.5 | 2.5 |
| | Southern Route | 1 | 3 |

What would you have recommended to General Kenney? What would you have recommended to the Japanese Commander? How much bombing time results if these recommendations are followed?

---

[25] As discussed in *Games and Decisions* by R. D. Luce and H. Raiffa Section 11.3 (New York: Wiley, 1957). This is based on an article in the *Journal of the Operations Research Society of America* 2 (1954) 365–385.

[26] The correct answers to these two questions correspond to the actual decisions both commanders made.

47. ▼ *The Prisoner's Dilemma* Slim Lefty and Joe Rap have been arrested for grand theft auto, having been caught red-handed driving away in a stolen 2012 Porsche. Although the police have more than enough evidence to convict them both, they feel that a confession would simplify the work of the prosecution. They decide to interrogate the prisoners separately. Slim and Joe are both told of the following plea-bargaining arrangement: If both confess, they will each receive a 2-year sentence; if neither confesses, they will both receive 5-year sentences, and if only one confesses (and thus squeals on the other), he will receive a suspended sentence, while the other will receive a 10-year sentence. What should Slim do?

48. ▼ *More Prisoners' Dilemmas* Jane Good and Prudence Brown have been arrested for robbery, but the police lack sufficient evidence for a conviction, and so decide to interrogate them separately in the hope of extracting a confession. Both Jane and Prudence are told the following: If they both confess, they will each receive a 5-year sentence; if neither confesses, they will be released; if one confesses, she will receive a suspended sentence, while the other will receive a 10-year sentence. What should Jane do?

49. ▼ *Campaign Strategies*[27] Florida and Ohio are "swing states" that have a large bounty of electoral votes and are therefore highly valued by presidential campaign strategists. Suppose it is now the weekend before Election Day 2012, and each candidate (Romney and Obama) can visit only one more state. Further, to win the election, Romney needs to win both of these states. Currently Romney has a 40% chance of winning Ohio and a 60% chance of winning Florida. Therefore, he has a $0.40 \times 0.60 = 0.24$, or 24% chance of winning the election. Assume that each candidate can increase his probability of winning a state by 10% if he, and not his opponent, visits that state. If both candidates visit the same state, there is no effect.

   a. Set up a payoff matrix with Romney as the row player and Obama as the column player, where the payoff for a specific set of circumstances is the probability (expressed as a percentage) that Romney will win both states.
   b. Where should each candidate visit under the circumstances?

50. ▼ *Campaign Strategies* Repeat Exercise 49, this time assuming that Romney has an 80% chance of winning Ohio and a 90% chance of winning Florida.

51. *Advertising* The *Softex Shampoo Company* is considering how to split its advertising budget between ads on two radio stations: WISH and WASH. Its main competitor, Splash Shampoo, Inc. has found out about this and is considering countering Softex's ads with its own, on the same radio stations. (Proposed jingle: *Softex, Shmoftex; Splash with Splash*) Softex has calculated that, were it to devote its entire advertising budget to ads on WISH, it would increase revenues in the coming month by $100,000 in the event that Splish was running all its ads on the less popular WASH, but would lose $20,000 in revenues if Splish ran its ads on WISH. If, on the other hand, it devoted its entire budget to WASH ads, it would neither increase nor decrease revenues in the event that Splish was running all its ads on the more popular WISH, and would in fact lose $20,000 in revenues if Splish ran its ads on WASH. What should Softex do, and what effect will this have on revenues?

52. *Labor Negotiations* The management team of the *Abstract Concrete Company* is negotiating a 3-year contract with the labor unions at one of its plants and is trying to decide on its offer for a salary increase. If it offers a 5% increase and the unions accept the offer, Abstract Concrete will gain $20 million in projected profits in the coming year, but if labor rejects the offer, the management team predicts that it will be forced to increase the offer to the union demand of 15%, thus halving the projected profits. If Abstract Concrete offers a 15% increase, the company will earn $10 million in profits over the coming year if the unions accept. If the unions reject, they will probably go out on strike (because management has set 15% as its upper limit) and management has decided that it can then in fact gain $12 million in profits by selling out the defunct plant in retaliation. What intermediate percentage should the company offer, and what profit should it project?

## COMMUNICATION AND REASONING EXERCISES

53. Why is a saddle point called a "saddle point"?

54. Can the payoff in a saddle point ever be larger than all other payoffs in a game? Explain.

55. ◆ One day, while browsing through an old *Statistical Abstract of the United States,* you came across the following data, which show the number of females employed (in thousands) in various categories according to their educational attainment.[28]

| | Managerial/ Professional | Technical/Sales/ Administrative | Service | Precision Production | Operators/ Fabricators |
|---|---|---|---|---|---|
| Less Than 4 Years High School | 260 | 1,080 | 2,020 | 260 | 1,400 |
| 4 Years of High School Only | 2,430 | 9,510 | 3,600 | 570 | 2,130 |
| 1 to 3 Years of College | 2,690 | 5,080 | 1,080 | 160 | 350 |
| At Least 4 Years of College | 7,210 | 2,760 | 380 | 70 | 110 |

Because you had been studying game theory that day, the first thing you did was to search for a saddle point. Having

---

[27]Based on *Game Theory for Swingers: What states should the candidates visit before Election Day?* by Jordan Ellenberg. Source: www.slate.com.

[28]Source: *Statistical Abstract of the United States 1991* (111th Ed.) U.S. Department of Commerce, Economics and Statistics Administration, and Bureau of the Census.

found one, you conclude that, as a female, your best strategy in the job market is to forget about a college career. Find the flaw in this reasoning.

56. ◆Exercises 37 and 38 seem to suggest that studying a single topic prior to an exam is better than studying all the topics in that exam. Comment on this discrepancy between the game theory result and common sense.

57. ◆Explain what is wrong with a decision to play the mixed strategy [0.5  0.5] by alternating the two strategies: Play the first strategy on the odd-numbered moves and the second strategy on the even-numbered moves. Illustrate your argument by devising a game in which your best strategy is [0.5  0.5].

58. ◆Describe a situation in which a both a mixed strategy and a pure strategy are equally effective.

# 4.5  Input-Output Models

In this section we look at an application of matrix algebra developed by Wassily Leontief (1906–1999) in the middle of the twentieth century. In 1973, he won the Nobel Prize in Economics for this work. The application involves analyzing national and regional economies by looking at how various parts of the economy interrelate. We'll work out some of the details by looking at a simple scenario.

First, we can think of the economy of a country or a region as being composed of various **sectors**, or groups of one or more industries. Typical sectors are the manufacturing sector, the utilities sector, and the agricultural sector. To introduce the basic concepts, we shall consider two specific sectors: the coal-mining sector (Sector 1) and the electric utilities sector (Sector 2). Both produce a commodity: The coal-mining sector produces coal, and the electric utilities sector produces electricity. We measure these products by their dollar value. By **one unit** of a product, we mean $1 worth of that product.

Here is the scenario.

1. To produce one unit ($1 worth) of coal, assume that the coal-mining sector uses 50¢ worth of coal (to power mining machinery, say) and 10¢ worth of electricity.

2. To produce one unit ($1 worth) of electricity, assume that the electric utilities sector uses 25¢ worth of coal and 25¢ worth of electricity.

These are *internal* usage figures. In addition to this, assume that there is an *external* demand (from the rest of the economy) of 7,000 units ($7,000 worth) of coal and 14,000 units ($14,000 worth) of electricity over a specific time period (one year, say). Our basic question is: How much should each of the two sectors supply in order to meet both internal and external demand?

The key to answering this question is to set up equations of the form:

Total supply = Total demand.

The unknowns, the values we are seeking, are

$x_1$ = the total supply (in units) from Sector 1 (coal) and
$x_2$ = the total supply (in units) from Sector 2 (electricity).

Our equations then take the following form:

Total supply from Sector 1 = Total demand for Sector 1 products

$$x_1 = 0.50x_1 \qquad + \qquad 0.25x_2 \qquad + \qquad 7,000$$
$$\uparrow \qquad\qquad\qquad \uparrow \qquad\qquad\qquad \uparrow$$

Coal required by Sector 1     Coal required by Sector 2     External demand for coal

Total supply from Sector 2 = Total demand for Sector 2 products

$$x_2 = \underset{\uparrow}{0.10x_1} + \underset{\uparrow}{0.25x_2} + \underset{\uparrow}{14{,}000.}$$

Electricity required by Sector 1     Electricity required by Sector 2     External demand for electricity

This is a system of two linear equations in two unknowns:

$$x_1 = 0.50x_1 + 0.25x_2 + 7{,}000$$
$$x_2 = 0.10x_1 + 0.25x_2 + 14{,}000.$$

We can rewrite this system of equations in matrix form as follows:

$$\underset{\text{Production}}{\underbrace{\begin{bmatrix} x_1 \\ x_2 \end{bmatrix}}} = \underset{\text{Internal demand}}{\underbrace{\begin{bmatrix} 0.50 & 0.25 \\ 0.10 & 0.25 \end{bmatrix} \begin{bmatrix} x_1 \\ x_2 \end{bmatrix}}} + \underset{\text{External demand}}{\underbrace{\begin{bmatrix} 7{,}000 \\ 14{,}000 \end{bmatrix}}}.$$

In symbols,

$$X = AX + D.$$

Here,

$$X = \begin{bmatrix} x_1 \\ x_2 \end{bmatrix}$$

is called the **production vector**. Its entries are the amounts produced by the two sectors. The matrix

$$D = \begin{bmatrix} 7{,}000 \\ 14{,}000 \end{bmatrix}$$

is called the **external demand** vector, and

$$A = \begin{bmatrix} 0.50 & 0.25 \\ 0.10 & 0.25 \end{bmatrix} \qquad \text{Organization: } \begin{bmatrix} 1 \to 1 & 1 \to 2 \\ 2 \to 1 & 2 \to 2 \end{bmatrix}$$

is called the **technology matrix**. The entries of the technology matrix have the following meanings:

$a_{11}$ = units of Sector 1 needed to produce one unit of Sector 1
$a_{12}$ = units of Sector 1 needed to produce one unit of Sector 2
$a_{21}$ = units of Sector 2 needed to produce one unit of Sector 1
$a_{22}$ = units of Sector 2 needed to produce one unit of Sector 2.

Now that we have the matrix equation

$$X = AX + D$$

we can solve it as follows. First, subtract $AX$ from both sides:

$$X - AX = D.$$

Because $X = IX$, where $I$ is the $2 \times 2$ identity matrix, we can rewrite this as

$$IX - AX = D.$$

Now factor out $X$:

$$(I - A)X = D.$$

If we multiply both sides by the inverse of $(I - A)$, we get the solution

$$X = (I - A)^{-1}D.$$

## Input-Output Model

In an input-output model, an economy (or part of one) is divided into $n$ **sectors**. We then record the $n \times n$ **technology matrix** $A$, whose $ij$th entry is the number of units from Sector $i$ used in producing one unit from Sector $j$ (in symbols, "$i \rightarrow j$"). To meet an **external demand** of $D$, the economy must produce $X$, where $X$ is the **production vector**. These are related by the equations

$$X = AX + D$$

or

$$X = (I - A)^{-1}D. \qquad \text{Provided } (I - A) \text{ is invertible}$$

### Quick Example

In the previous scenario, $A = \begin{bmatrix} 0.50 & 0.25 \\ 0.10 & 0.25 \end{bmatrix}$, $X = \begin{bmatrix} x_1 \\ x_2 \end{bmatrix}$, and $D = \begin{bmatrix} 7{,}000 \\ 14{,}000 \end{bmatrix}$.

The solution is

$$X = (I - A)^{-1}D$$

$$\begin{bmatrix} x_1 \\ x_2 \end{bmatrix} = \left( \begin{bmatrix} 1 & 0 \\ 0 & 1 \end{bmatrix} - \begin{bmatrix} 0.50 & 0.25 \\ 0.10 & 0.25 \end{bmatrix} \right)^{-1} \begin{bmatrix} 7{,}000 \\ 14{,}000 \end{bmatrix} \qquad \text{Calculate } I - A.$$

$$= \begin{bmatrix} 0.50 & -0.25 \\ -0.10 & 0.75 \end{bmatrix}^{-1} \begin{bmatrix} 7{,}000 \\ 14{,}000 \end{bmatrix} \qquad \text{Calculate } (I - A)^{-1}.$$

$$= \begin{bmatrix} \frac{15}{7} & \frac{5}{7} \\ \frac{2}{7} & \frac{10}{7} \end{bmatrix} \begin{bmatrix} 7{,}000 \\ 14{,}000 \end{bmatrix}$$

$$= \begin{bmatrix} 25{,}000 \\ 22{,}000 \end{bmatrix}.$$

In other words, to meet the demand, the economy must produce $25,000 worth of coal and $22,000 worth of electricity.

The next example uses actual data from the U.S. economy (we have rounded the figures to make the computations less complicated). It is rare to find input-output data already packaged for you as a technology matrix. Instead, the data commonly found in statistical sources come in the form of "input-output tables," from which we will have to construct the technology matrix.

### EXAMPLE 1   Petroleum and Natural Gas

Consider two sectors of the U.S. economy: crude petroleum and natural gas (*crude*) and petroleum refining and related industries (*refining*). According to government figures,[29] in 1998 the crude sector used $27,000 million worth of its own products and $750 million worth of the products of the refining sector to produce $87,000 million worth of goods (crude oil and natural gas). The refining sector in the same

---

[29]The data have been rounded to two significant digits. Source: *Survey of Current Business*, December, 2001, U.S. Department of Commerce. The *Survey of Current Business* and the input-output tables themselves are available at the Web site of the Department of Commerce's Bureau of Economic Analysis (www.bea.gov).

year used $59,000 million worth of the products of the crude sector and $15,000 million worth of its own products to produce $140,000 million worth of goods (refined oil and the like). What was the technology matrix for these two sectors? What was left over from each of these sectors for use by other parts of the economy or for export?

**Solution**  First, for convenience, we record the given data in the form of a table, called the **input-output table**. (All figures are in millions of dollars.)

|  |  | Crude | Refining |
|---|---|---|---|
| *From* | **Crude** | 27,000 | 59,000 |
|  | **Refining** | 750 | 15,000 |
|  | **Total Output** | 87,000 | 140,000 |

The entries in the top portion are arranged in the same way as those of the technology matrix: The *ij*th entry represents the number of units of Sector *i* that went to Sector *j*. Thus, for instance, the 59,000 million entry in the 1, 2 position represents the number of units of Sector 1, crude, that were used by Sector 2, refining. ("From the side, to the top.")

We now construct the technology matrix. The technology matrix has entries $a_{ij}$ = units of Sector *i* used to produce *one* unit of Sector *j*. Thus,

$a_{11}$ = units of crude to produce one unit of crude. We are told that 27,000 million units of crude were used to produce 87,000 million units of crude. Thus, to produce *one* unit of crude, $27,000/87,000 \approx 0.31$ units of crude were used, and so $a_{11} \approx 0.31$. (We have rounded this value to two significant digits; further digits are not reliable due to rounding of the original data.)

$a_{12}$ = units of crude to produce one unit of refined:
$$a_{12} = 59,000/140,000 \approx 0.42$$

$a_{21}$ = units of refined to produce one unit of crude:
$$a_{21} = 750/87,000 \approx 0.0086$$

$a_{22}$ = units of refined to produce one unit of refined:
$$a_{22} = 15,000/140,000 \approx 0.11.$$

This gives the technology matrix

$$A = \begin{bmatrix} 0.31 & 0.42 \\ 0.0086 & 0.11 \end{bmatrix}.$$   Technology matrix

In short *we obtained the technology matrix from the input-output table by dividing the Sector 1 column by the Sector 1 total, and the Sector 2 column by the Sector 2 total.*

Now we also know the total output from each sector, so *we have already been given the production vector:*

$$X = \begin{bmatrix} 87,000 \\ 140,000 \end{bmatrix}.$$   Production vector

What we are asked for is the external demand vector *D*, the amount available for the outside economy. To find *D*, we use the equation

$$X = AX + D,$$   Relationship of *X*, *A*, and *D*

where, this time, we are given $A$ and $X$, and must solve for $D$. Solving for $D$ gives

$$D = X - AX$$

* Why?

$$= \begin{bmatrix} 87,000 \\ 140,000 \end{bmatrix} - \begin{bmatrix} 0.31 & 0.42 \\ 0.0086 & 0.11 \end{bmatrix} \begin{bmatrix} 87,000 \\ 140,000 \end{bmatrix}$$

$$\approx \begin{bmatrix} 87,000 \\ 140,000 \end{bmatrix} - \begin{bmatrix} 86,000 \\ 16,000 \end{bmatrix} = \begin{bmatrix} 1,000 \\ 124,000 \end{bmatrix}.$$

We rounded to 2 digits.*

The first number, $\$1,000$ million, is the amount produced by the crude sector that is available to be used by other parts of the economy or to be exported. (In fact, because something has to happen to all that crude petroleum and natural gas, this is the amount actually used or exported, where use can include stockpiling.) The second number, $\$124,000$ million, represents the amount produced by the refining sector that is available to be used by other parts of the economy or to be exported.

Note that we could have calculated $D$ more simply from the input-output table. The internal use of units from the crude sector was the sum of the outputs from that sector:

**using Technology**

See the Technology Guides at the end of the chapter to see how to compute the technology matrix and the external demand vector in Example 1 using a TI-83/84 Plus or a spreadsheet.

$$27,000 + 59,000 = 86,000.$$

Because 87,000 units were actually produced by the sector, that left a surplus of $87,000 - 86,000 = 1,000$ units for export. We could compute the surplus from the refining sector similarly. (The two calculations actually come out slightly different, because we rounded the intermediate results.) The calculation in Example 2 below cannot be done as trivially, however.

## Input-Output Table

National economic data are often given in the form of an **input-output table**. The $ij$th entry in the top portion of the table is the number of units that go from Sector $i$ to Sector $j$. The "Total outputs" are the total numbers of units produced by each sector. We obtain the technology matrix from the input-output table by dividing the Sector 1 column by the Sector 1 total, the Sector 2 column by the Sector 2 total, and so on.

### Quick Example

Input-Output Table:

† The production of skateboards required skateboards due to the fact that skateboard workers tend to commute to work on (what else?) skateboards!

| | To | Skateboards | Wood |
|---|---|---|---|
| *From* | **Skateboards** | 20,000† | 0 |
| | **Wood** | 100,000 | 500,000 |
| | **Total Output** | 200,000 | 5,000,000 |

Technology Matrix:

$$A = \begin{bmatrix} \dfrac{20,000}{200,000} & \dfrac{0}{5,000,000} \\ \dfrac{100,000}{200,000} & \dfrac{500,000}{5,000,000} \end{bmatrix} = \begin{bmatrix} 0.1 & 0 \\ 0.5 & 0.1 \end{bmatrix}.$$

## EXAMPLE 2  Rising Demand

Suppose that external demand for refined petroleum rises to $200,000 million, but the demand for crude remains $1,000 million (as in Example 1). How do the production levels of the two sectors considered in Example 1 have to change?

**Solution**  We are being told that now

$$D = \begin{bmatrix} 1,000 \\ 200,000 \end{bmatrix}$$

and we are asked to find $X$. Remember that we can calculate $X$ from the formula

$$X = (I - A)^{-1}D.$$

Now

$$I - A = \begin{bmatrix} 1 & 0 \\ 0 & 1 \end{bmatrix} - \begin{bmatrix} 0.31 & 0.42 \\ 0.0086 & 0.11 \end{bmatrix} = \begin{bmatrix} 0.69 & -0.42 \\ -0.0086 & 0.89 \end{bmatrix}.$$

We take the inverse using our favorite technique and find that, to four significant digits,*

$$(I - A)^{-1} \approx \begin{bmatrix} 1.458 & 0.6880 \\ 0.01409 & 1.130 \end{bmatrix}.$$

* Because $A$ is accurate to two digits, we should use more than two significant digits in intermediate calculations so as not to lose additional accuracy. We must, of course, round the final answer to two digits.

Now we can compute $X$:

$$X = (I - A)^{-1}D = \begin{bmatrix} 1.458 & 0.6880 \\ 0.01409 & 1.130 \end{bmatrix} \begin{bmatrix} 1,000 \\ 200,000 \end{bmatrix} \approx \begin{bmatrix} 140,000 \\ 230,000 \end{bmatrix}.$$

(As in Example 1, we have rounded all the entries in the answer to two significant digits.) Comparing this vector to the production vector used in Example 1, we see that production in the crude sector has to increase from $87,000 million to $140,000 million, while production in the refining sector has to increase from $140,000 million to $230,000 million.

---

**Note**  Using the matrix $(I - A)^{-1}$, we have a slightly different way of solving Example 2. We are asking for the effect on production of a *change* in the final demand of 0 for crude and $200,000 - 124,000 = $76,000 million for refined products. If we multiply $(I - A)^{-1}$ by the matrix representing this *change,* we obtain

$$\begin{bmatrix} 1.458 & 0.6880 \\ 0.01409 & 1.130 \end{bmatrix} \begin{bmatrix} 0 \\ 76,000 \end{bmatrix} \approx \begin{bmatrix} 53,000 \\ 90,000 \end{bmatrix}.$$

$(I - A)^{-1} \times$ Change in Demand $=$ Change in Production

We see the changes required in production: an increase of $53,000 million in the crude sector and an increase of $90,000 million in the refining sector.

Notice that the increase in external demand for the products of the refining sector requires the crude sector to increase production as well, even though there is no increase in the *external* demand for its products. The reason is that, in order to increase production, the refining sector needs to use more crude oil, so that the *internal* demand for crude oil goes up. The inverse matrix $(I - A)^{-1}$ takes these **indirect effects** into account in a nice way.

By replacing the $76,000 by $1 in the computation we just did, we see that a $1 increase in external demand for refined products will require an increase in production of $0.6880 in the crude sector, as well as an increase in production of $1.130 in the refining sector. This is how we interpret the entries in $(I - A)^{-1}$, and this is why it is useful to look at this matrix inverse rather than just solve $(I - A)X = D$ for $X$ using, say, Gauss-Jordan reduction. Looking at $(I - A)^{-1}$, we can also find the effects of an increase of $1 in external demand for crude: an increase in production of $1.458 in the crude sector and an increase of $0.01409 in the refining sector.

Here are some questions to think about: Why are the diagonal entries of $(I - A)^{-1}$ (slightly) larger than 1? Why is the entry in the lower left so small compared to the others? ■

## Interpreting $(I - A)^{-1}$: Indirect Effects

If $A$ is the technology matrix, then the $ij$th entry of $(I - A)^{-1}$ is the change in the number of units Sector $i$ must produce in order to meet a one-unit increase in external demand for Sector $j$ products. To meet a rising external demand, the necessary change in production for each sector is given by

$$\text{Change in production} = (I - A)^{-1}D^+,$$

where $D^+$ is the change in external demand.

### Quick Example

Take Sector 1 to be skateboards, and Sector 2 to be wood, and assume that

$$(I - A)^{-1} = \begin{bmatrix} 1.1 & 0 \\ 0.6 & 1.1 \end{bmatrix}.$$

Then

$a_{11} = 1.1 =$ number of additional units of skateboards that must be produced to meet a one-unit increase in the demand for skateboards (Why is this number larger than 1?)

$a_{12} = 0 =$ number of additional units of skateboards that must be produced to meet a one-unit increase in the demand for wood (Why is this number 0?)

$a_{21} = 0.6 =$ number of additional units of wood that must be produced to meet a one-unit increase in the demand for skateboards

$a_{22} = 1.1 =$ number of additional units of wood that must be produced to meet a one-unit increase in the demand for wood.

To meet an increase in external demand of 100 skateboards and 400 units of wood, the necessary change in production is

$$(I - A)^{-1}D^+ = \begin{bmatrix} 1.1 & 0 \\ 0.6 & 1.1 \end{bmatrix} \begin{bmatrix} 100 \\ 400 \end{bmatrix} = \begin{bmatrix} 110 \\ 500 \end{bmatrix}$$

so 110 additional skateboards and 500 additional units of wood will need to be produced.

In the preceding examples, we used only two sectors of the economy. The data used in Examples 1 and 2 were taken from an input-output table published by the U.S. Department of Commerce, in which the whole U.S. economy was broken down into 85 sectors. This in turn was a simplified version of a model in which the economy was broken into about 500 sectors. Obviously, computers are required to make a realistic input-output analysis possible. Many governments collect and publish input-output data as part of their national planning. The United Nations collects these data and publishes collections of national statistics. The United Nations also has a useful set of links to government statistics at the following URL:

www.un.org/Depts/unsd/sd_natstat.htm

### EXAMPLE 3  Kenya Economy

Consider four sectors of the economy of Kenya[30]: (1) the traditional economy, (2) agriculture, (3) manufacture of metal products and machinery, and (4) wholesale and retail trade. The input-output table for these four sectors for 1976 looks like this (all numbers are thousands of K£):

| | To | 1 | 2 | 3 | 4 |
|---|---|---|---|---|---|
| From | 1 | 8,600 | 0 | 0 | 0 |
| | 2 | 0 | 20,000 | 24 | 0 |
| | 3 | 1,500 | 530 | 15,000 | 660 |
| | 4 | 810 | 8,500 | 5,800 | 2,900 |
| **Total Output** | | 87,000 | 530,000 | 110,000 | 180,000 |

Suppose that external demand for agriculture increased by K£50,000,000 and that external demand for metal products and machinery increased by K£10,000,000. How would production in these four sectors have to change to meet this rising demand?

**Solution**  To find the change in production necessary to meet the rising demand, we need to use the formula

$$\text{Change in production} = (I - A)^{-1} D^{+},$$

where $A$ is the technology matrix and $D^{+}$ is the change in demand:

$$D^{+} = \begin{bmatrix} 0 \\ 50{,}000 \\ 10{,}000 \\ 0 \end{bmatrix}.$$

With entries shown rounded to two significant digits, the matrix $A$ is

$$A = \begin{bmatrix} 0.099 & 0 & 0 & 0 \\ 0 & 0.038 & 0.00022 & 0 \\ 0.017 & 0.001 & 0.14 & 0.0037 \\ 0.0093 & 0.016 & 0.053 & 0.016 \end{bmatrix}.$$

Entries shown are rounded to two significant digits.

---

[30]Figures are rounded. Source: *Input-Output Tables for Kenya 1976*, Central Bureau of Statistics of the Ministry of Economic Planning and Community Affairs, Kenya.

**using Technology**

Technology can be used to do the computations in Example 3. Here is an outline for part (a) (see the Technology Guides at the end of the chapter for additional details on using a TI-83/84 Plus or a spreadsheet):

**TI-83/84 Plus**
Enter the matrices $A$ and $D$ using MATRX ; EDIT.
(For $A$ type the entries as quotients: Column entry/Column total.) Home Screen:
(identity(4)-[A])$^{-1}$[D]
[More details on page 306.]

**Spreadsheet**
Obtain the technology matrix as in Example 1.
Insert the 4×4 identity matrix and then use MINVERSE and MMULT to compute $(I - A)^{-1} D$.
[More details on page 310.]

**Website**
www.WanerMath.com
Student Website
→ On Line Utilities
→ Matrix Algebra Tool
Enter the matrices $A$ and $D$. (For $A$ type the entries as quotients: Column entry/Column total.)
Type (I-A)^(-1)*D in the formula box and press "Compute".

The next calculation is best done using technology:

$$\text{Change in production} = (I - A)^{-1}D^+ = \begin{bmatrix} 0 \\ 52{,}000 \\ 12{,}000 \\ 1{,}500 \end{bmatrix}. \quad \begin{array}{l} \text{Entries shown are rounded} \\ \text{to two significant digits.} \end{array}$$

Looking at this result, we see that the changes in external demand will leave the traditional economy unaffected, production in agriculture will rise by K£52 million, production in the manufacture of metal products and machinery will rise by K£12 million, and activity in wholesale and retail trade will rise by K£1.5 million.

➡ **Before we go on...** Can you see why the traditional economy was unaffected in Example 3? Although it takes inputs from other parts of the economy, it is not itself an input to any other part. In other words, there is no intermediate demand for the products of the traditional economy coming from any other part of the economy, and so an increase in production in any other sector of the economy will require no increase from the traditional economy. On the other hand, the wholesale and retail trade sector does provide input to the agriculture and manufacturing sectors, so increases in those sectors do require an increase in the trade sector.

One more point: If you calculate $(I - A)^{-1}$ you will notice how small the off-diagonal entries are. This says that increases in each sector have relatively small effects on the other sectors. We say that these sectors are **loosely coupled**. Regional economies, where many products are destined to be shipped out to the rest of the country, tend to show this phenomenon even more strongly. Notice in Example 2 that those two sectors are **strongly coupled**, because a rise in demand for refined products requires a comparable rise in the production of crude. ∎

## 4.5  EXERCISES

▼ more advanced   ◆ challenging
🅣 indicates exercises that should be solved using technology

**1.** Let $A$ be the technology matrix $A = \begin{bmatrix} 0.2 & 0.05 \\ 0.8 & 0.01 \end{bmatrix}$, where Sector 1 is paper and Sector 2 is wood. Fill in the missing quantities.

**a.** ___ units of wood are needed to produce one unit of paper.

**b.** ___ units of paper are used in the production of one unit of paper.

**c.** The production of each unit of wood requires the use of ___ units of paper.

**2.** Let $A$ be the technology matrix $A = \begin{bmatrix} 0.01 & 0.001 \\ 0.2 & 0.004 \end{bmatrix}$, where Sector 1 is computer chips and Sector 2 is silicon. Fill in the missing quantities.

**a.** ___ units of silicon are required in the production of one unit of silicon.

**b.** ___ units of computer chips are used in the production of one unit of silicon.

**c.** The production of each unit of computer chips requires the use of ___ units of silicon.

**3.** Each unit of television news requires 0.2 units of television news and 0.5 units of radio news. Each unit of radio news requires 0.1 units of television news and no radio news. With Sector 1 as television news and Sector 2 as radio news, set up the technology matrix $A$.

**4.** Production of one unit of cologne requires no cologne and 0.5 units of perfume. Into one unit of perfume go 0.1 units of cologne and 0.3 units of perfume. With Sector 1 as cologne and Sector 2 as perfume, set up the technology matrix $A$.

*In each of Exercises 5–12, you are given a technology matrix A and an external demand vector D. Find the corresponding production vector X.* HINT [See Quick Example page 286.]

**5.** $A = \begin{bmatrix} 0.5 & 0.4 \\ 0 & 0.5 \end{bmatrix}$, $D = \begin{bmatrix} 10{,}000 \\ 20{,}000 \end{bmatrix}$

**6.** $A = \begin{bmatrix} 0.5 & 0.4 \\ 0 & 0.5 \end{bmatrix}$, $D = \begin{bmatrix} 20{,}000 \\ 10{,}000 \end{bmatrix}$

**7.** $A = \begin{bmatrix} 0.1 & 0.4 \\ 0.2 & 0.5 \end{bmatrix}$, $D = \begin{bmatrix} 25{,}000 \\ 15{,}000 \end{bmatrix}$

**8.** $A = \begin{bmatrix} 0.1 & 0.2 \\ 0.4 & 0.5 \end{bmatrix}$, $D = \begin{bmatrix} 24{,}000 \\ 14{,}000 \end{bmatrix}$

9. $A = \begin{bmatrix} 0.5 & 0.1 & 0 \\ 0 & 0.5 & 0.1 \\ 0 & 0 & 0.5 \end{bmatrix}$, $D = \begin{bmatrix} 1,000 \\ 1,000 \\ 2,000 \end{bmatrix}$

10. $A = \begin{bmatrix} 0.5 & 0.1 & 0 \\ 0 & 0.5 & 0.1 \\ 0 & 0 & 0.5 \end{bmatrix}$, $D = \begin{bmatrix} 3,000 \\ 3,800 \\ 2,000 \end{bmatrix}$

11. $A = \begin{bmatrix} 0.2 & 0.2 & 0 \\ 0.2 & 0.4 & 0.2 \\ 0 & 0.2 & 0.2 \end{bmatrix}$, $D = \begin{bmatrix} 16,000 \\ 8,000 \\ 8,000 \end{bmatrix}$

12. $A = \begin{bmatrix} 0.2 & 0.2 & 0.2 \\ 0.2 & 0.4 & 0.2 \\ 0.2 & 0.2 & 0.2 \end{bmatrix}$, $D = \begin{bmatrix} 7,000 \\ 14,000 \\ 7,000 \end{bmatrix}$

13. Given $A = \begin{bmatrix} 0.1 & 0.4 \\ 0.2 & 0.5 \end{bmatrix}$, find the changes in production required to meet an increase in demand of 50 units of Sector 1 products and 30 units of Sector 2 products.

14. Given $A = \begin{bmatrix} 0.5 & 0.4 \\ 0 & 0.5 \end{bmatrix}$, find the changes in production required to meet an increase in demand of 20 units of Sector 1 products and 10 units of Sector 2 products.

15. Let $(I - A)^{-1} = \begin{bmatrix} 1.5 & 0.1 & 0 \\ 0.2 & 1.2 & 0.1 \\ 0.1 & 0.7 & 1.6 \end{bmatrix}$ and assume that the external demand for the products in Sector 1 increases by 1 unit. By how many units should each sector increase production? What do the columns of the matrix $(I - A)^{-1}$ tell you? HINT [See Quick Example on page 290.]

16. Let $(I - A)^{-1} = \begin{bmatrix} 1.5 & 0.1 & 0 \\ 0.1 & 1.1 & 0.1 \\ 0 & 0 & 1.3 \end{bmatrix}$, and assume that the external demand for the products in each of the sectors increases by 1 unit. By how many units should each sector increase production? HINT [See Quick Example on page 290.]

*In Exercises 17 and 18, obtain the technology matrix from the given input-output table.* HINT [See Example 1.]

17.

| To | A | B | C |
|---|---|---|---|
| From **A** | 1,000 | 2,000 | 3,000 |
| **B** | 0 | 4,000 | 0 |
| **C** | 0 | 1,000 | 3,000 |
| **Total Output** | 5,000 | 5,000 | 6,000 |

18.

| To | A | B | C |
|---|---|---|---|
| From **A** | 0 | 100 | 300 |
| **B** | 500 | 400 | 300 |
| **C** | 0 | 0 | 600 |
| **Total Output** | 1,000 | 2,000 | 3,000 |

## APPLICATIONS

19. *Campus Food* The two campus cafeterias, the Main Dining Room and Bits & Bytes, typically use each other's food in doing business on campus. One weekend, the input-output table was as follows.[31]

| To | Main DR | Bits & Bytes |
|---|---|---|
| *From* **Main DR** | $10,000 | $20,000 |
| **Bits & Bytes** | 5,000 | 0 |
| **Total Output** | 50,000 | 40,000 |

Given that the demand for food on campus last weekend was $45,000 from the Main Dining Room and $30,000 from Bits & Bytes, how much did the two cafeterias have to produce to meet the demand last weekend? HINT [See Example 1.]

20. *Plagiarism* Two student groups at Enormous State University, the Choral Society and the Football Club, maintain files of term papers that they write and offer to students for research purposes. Some of these papers they use themselves in generating more papers. In order to avoid suspicion of plagiarism by faculty members (who seem to have astute memories), each paper is given to students or used by the clubs only once (no copies are kept). The number of papers that were used in the production of new papers last year is shown in the following input-output table:

| To | Choral Soc. | Football Club |
|---|---|---|
| *From* **Choral Soc.** | 20 | 10 |
| **Football Club** | 10 | 30 |
| **Total Output** | 100 | 200 |

Given that 270 Choral Society papers and 810 Football Club papers will be used by students outside of these two clubs next year, how many new papers do the two clubs need to write?

21. ▣ *Communication Equipment* Two sectors of the U.S. economy are (1) audio, video, and communication equipment and (2) electronic components and accessories. In 1998, the input-output table involving these two sectors was as follows (all figures are in millions of dollars):[32]

| To | Equipment | Components |
|---|---|---|
| *From* **Equipment** | 6,000 | 500 |
| **Components** | 24,000 | 30,000 |
| **Total Output** | 90,000 | 140,000 |

Determine the production levels necessary in these two sectors to meet an external demand for $80,000 million of communication equipment and $90,000 million of electronic components. Round answers to two significant digits.

---

[31]For some reason, the Main Dining Room consumes a lot of its own food!

[32]The data have been rounded. Source: *Survey of Current Business*, December, 2001, U.S. Department of Commerce.

**22.** ▢ *Wood and Paper* Two sectors of the U.S. economy are (1) lumber and wood products and (2) paper and allied products. In 1998 the input-output table involving these two sectors was as follows (all figures are in millions of dollars).[33]

|  | *To* | **Wood** | **Paper** |
|---|---|---|---|
| *From* | **Wood** | 36,000 | 7,000 |
|  | **Paper** | 100 | 17,000 |
|  | **Total Output** | 120,000 | 120,000 |

If external demand for lumber and wood products rises by $10,000 million and external demand for paper and allied products rises by $20,000 million, what increase in output of these two sectors is necessary? Round answers to two significant digits.

**23.** *Australia Economy* Two sectors of the Australian economy are (1) textiles and (2) clothing and footwear. The 1977 input-output table[34] involving these two sectors results in the following value for $(I - A)^{-1}$:

$$(I - A)^{-1} = \begin{bmatrix} 1.228 & 0.182 \\ 0.006 & 1.1676 \end{bmatrix}.$$

Complete the following sentences.

**a.** ____ additional dollars worth of clothing and footwear must be produced to meet a $1 increase in the demand for textiles.

**b.** 0.182 additional dollars worth of ____ must be produced to meet a one-dollar increase in the demand for ____.

**24.** *Australia Economy* Two sectors of the Australian economy are (1) community services and (2) recreation services. The 1978–79 input-output table[35] involving these two sectors results in the following value for $(I - A)^{-1}$:

$$(I - A)^{-1} = \begin{bmatrix} 1.0066 & 0.00576 \\ 0.00496 & 1.04206 \end{bmatrix}.$$

Complete the following sentences.

**a.** 0.00496 additional dollars worth of ____ must be produced to meet a $1 increase in the demand for ____.

**b.** ____ additional dollars worth of community services must be produced to meet a one-dollar increase in the demand for community services.

▢ *Exercises 25–28 require the use of technology.*

**25.** ▢ *United States Input-Output Table* Four sectors of the U.S. economy are (1) livestock and livestock products,

(2) other agricultural products, (3) forestry and fishery products, and (4) agricultural, forestry, and fishery services. In 1977 the input-output table involving these four sectors was as follows (all figures are in millions of dollars):[36]

|  | *To* | **1** | **2** | **3** | **4** |
|---|---|---|---|---|---|
| *From* | **1** | 11,937 | 9 | 109 | 855 |
|  | **2** | 26,649 | 4,285 | 0 | 4,744 |
|  | **3** | 0 | 0 | 439 | 61 |
|  | **4** | 5,423 | 10,952 | 3,002 | 216 |
|  | **Total Output** | 97,795 | 120,594 | 14,642 | 47,473 |

Determine how these four sectors would react to an increase in demand for livestock (Sector 1) of $1,000 million, how they would react to an increase in demand for other agricultural products (Sector 2) of $1,000 million, and so on.

**26.** ▢ *United States Input-Output Table* Four sectors of the U.S. economy are (1) motor vehicles, (2) truck and bus bodies, trailers, and motor vehicle parts, (3) aircraft and parts, and (4) other transportation equipment. In 1998 the input-output table involving these four sectors was (all figures in millions of dollars):[37]

|  | *To* | **1** | **2** | **3** | **4** |
|---|---|---|---|---|---|
| *From* | **1** | 75 | 1,092 | 0 | 1,207 |
|  | **2** | 64,858 | 13,081 | 7 | 1,070 |
|  | **3** | 0 | 0 | 21,782 | 0 |
|  | **4** | 0 | 0 | 0 | 1,375 |
|  | **Total Output** | 230,676 | 135,108 | 129,376 | 44,133 |

Determine how these four sectors would react to an increase in demand for motor vehicles (Sector 1) of $1,000 million, how they would react to an increase in demand for truck and bus bodies (Sector 2) of $1,000 million, and so on.

**27.** ▽ *Australia Input-Output Table* Four sectors of the Australian economy are (1) agriculture, (2) forestry, fishing, and hunting, (3) meat and milk products, and (4) other food products. In 1978–79 the input-output table involving these four sectors was as follows (all figures are in millions of Australian dollars).[38]

---

[33]The data have been rounded. Source: *Survey of Current Business,* December, 2001, U.S. Department of Commerce.

[34]Source: *Australian National Accounts and Input-Output Tables 1978–1979,* Australian Bureau of Statistics.

[35]*Ibid.*

[36]Source: *Survey of Current Business,* December 2001, U.S. Department of Commerce.

[37]*Ibid.*

[38]Source: *Australian National Accounts and Input-Output Tables 1978–1979,* Australian Bureau of Statistics.

| | To | 1 | 2 | 3 | 4 |
|---|---|---|---|---|---|
| *From* | 1 | 678.4 | 3.7 | 3,341.5 | 1,023.5 |
| | 2 | 15.5 | 6.9 | 17.1 | 124.5 |
| | 3 | 47.3 | 4.3 | 893.1 | 145.8 |
| | 4 | 312.5 | 22.1 | 83.2 | 693.5 |
| **Total Output** | | 9,401.3 | 685.8 | 6,997.3 | 4,818.3 |

**a.** How much additional production by the meat and milk sector is necessary to accommodate a $100 increase in the demand for agriculture?

**b.** Which sector requires the most of its own product in order to meet a $1 increase in external demand for that product?

**28.** ▼ *Australia Input-Output Table* Four sectors of the Australian economy are (1) petroleum and coal products, (2) nonmetallic mineral products, (3) basic metals and products, and (4) fabricated metal products. In 1978–79 the input-output table involving these four sectors was as follows (all figures are in millions of Australian dollars).[39]

| | To | 1 | 2 | 3 | 4 |
|---|---|---|---|---|---|
| *From* | 1 | 174.1 | 30.5 | 120.3 | 14.2 |
| | 2 | 0 | 190.1 | 55.8 | 12.6 |
| | 3 | 2.1 | 40.2 | 1,418.7 | 1,242.0 |
| | 4 | 0.1 | 7.3 | 40.4 | 326.0 |
| **Total Output** | | 3,278.0 | 2,188.8 | 6,541.7 | 4,065.8 |

[39]Source: *Australian National Accounts and Input-Output Tables 1978–1979,* Australian Bureau of Statistics.

**a.** How much additional production by the petroleum and coal products sector is necessary to accommodate a $1,000 increase in the demand for fabricated metal products?

**b.** Which sector requires the most of the product of some other sector in order to meet a $1 increase in external demand for that product?

## COMMUNICATION AND REASONING EXERCISES

**29.** What would it mean if the technology matrix $A$ were the zero matrix?

**30.** Can an external demand be met by an economy whose technology matrix $A$ is the identity matrix? Explain.

**31.** ▼ What would it mean if the total output figure for a particular sector of an input-output table were equal to the sum of the figures in the row for that sector?

**32.** ▼ What would it mean if the total output figure for a particular sector of an input-output table were less than the sum of the figures in the row for that sector?

**33.** ▼ What does it mean if an entry in the matrix $(I - A)^{-1}$ is zero?

**34.** ▼ Why do we expect the diagonal entries in the matrix $(I - A)^{-1}$ to be slightly larger than 1?

**35.** ▼ Why do we expect the off-diagonal entries of $(I - A)^{-1}$ to be less than 1?

**36.** ▼ Why do we expect all the entries of $(I - A)^{-1}$ to be nonnegative?

## KEY CONCEPTS

Website www.WanerMath.com
Go to the Website at www.WanerMath
  .com to find a comprehensive and
  interactive Web-based summary
  of Chapter 4.

### 4.1 Matrix Addition and Scalar Multiplication

$m \times n$ matrix, dimensions, entries
  *p. 232*
Referring to the entries of a matrix
  *p. 232*
Matrix equality *p. 233*
Row, column, and square matrices
  *p. 234*
Addition and subtraction of matrices
  *p. 234*
Scalar multiplication *p. 236*
Properties of matrix addition and scalar
  multiplication *p. 237*
The transpose of a matrix *p. 238*
Properties of transposition *p. 238*

### 4.2 Matrix Multiplication

Multiplying a row by a column
  *p. 242*
Linear equation as a matrix
  equation *p. 244*

The product of two matrices: general
  case *p. 244*
Identity matrix *p. 248*
Properties of matrix addition and
  multiplication *p. 249*
Properties of transposition and
  multiplication *p. 250*
A system of linear equations can be
  written as a single matrix equation
  *p. 250*

### 4.3 Matrix Inversion

The inverse of a matrix, singular matrix
  *p. 256*
Procedure for finding the inverse of a
  matrix *p. 256*
Formula for the inverse of a $2 \times 2$
  matrix; determinant of a $2 \times 2$
  matrix *p. 257*
Using an inverse matrix to solve a
  system of equations *p. 261*

### 4.4 Game Theory

Two-person zero-sum game, payoff
  matrix *p. 268*
A strategy specifies how a player
  chooses a move *p. 268*
The expected payoff of a game for given
  mixed strategies $R$ and $C$ *p. 269*

An optimal strategy, according to the
  minimax criterion, is one that minimizes the maximum damage your
  opponent can cause you. *p. 272*
The Fundamental Principle of Game
  Theory *p. 272*
Procedure for reducing by dominance
  *p. 273*
Procedure for solving a $2 \times 2$ game
  *p. 273*
The expected value of a game is its
  expected payoff when the players use
  their optimal strategies *p. 276*
A strictly determined game is one with
  a saddle point *p. 278*
Steps to follow in solving a game
  *p. 278*

### 4.5 Input-Output Models

An input-output model divides an
  economy into sectors. The technology matrix records the interactions of
  these sectors and allows us to relate
  external demand to the production
  vector. *p. 286*
Procedure for finding a technology
  matrix from an input-output table
  *p. 288*
The entries of $(I - A)^{-1}$ *p. 290*

## REVIEW EXERCISES

*For Exercises 1–10, let*

$$A = \begin{bmatrix} 1 & 2 & 3 \\ 4 & 5 & 6 \end{bmatrix}, \ B = \begin{bmatrix} 1 & -1 \\ 0 & 1 \end{bmatrix},$$

$$C = \begin{bmatrix} -1 & 0 \\ 1 & 1 \\ 0 & 1 \end{bmatrix}, \text{ and } D = \begin{bmatrix} -3 & -2 & -1 \\ 1 & 2 & 3 \end{bmatrix}.$$

*Determine whether each expression is defined, and if it is, evaluate it.*

**1.** $A + B$

**2.** $A - D$

**3.** $2A^T + C$

**4.** $AB$

**5.** $A^T B$

**6.** $A^2$

**7.** $B^2$

**8.** $B^3$

**9.** $AC + B$

**10.** $CD + B$

*In Exercises 11–16, find the inverse of the given matrix or determine that the matrix is singular.*

**11.** $\begin{bmatrix} 1 & -1 \\ 0 & 1 \end{bmatrix}$

**12.** $\begin{bmatrix} 1 & 2 \\ 0 & 0 \end{bmatrix}$

**13.** $\begin{bmatrix} 1 & 2 & 3 \\ 0 & 4 & 1 \\ 0 & 0 & 1 \end{bmatrix}$

**14.** $\begin{bmatrix} 1 & 2 & 3 & 4 \\ 1 & 3 & 4 & 2 \\ 0 & 1 & 2 & 3 \\ 0 & 0 & 1 & 2 \end{bmatrix}$

**15.** $\begin{bmatrix} 1 & 2 & 3 & 4 \\ 2 & 3 & 3 & 3 \\ 0 & 1 & 2 & 3 \\ 0 & 0 & 1 & 2 \end{bmatrix}$

**16.** $\begin{bmatrix} 0 & 1 & 0 & 0 \\ 1 & 0 & 0 & 0 \\ 0 & 0 & 0 & 1 \\ 0 & 0 & 1 & 0 \end{bmatrix}$

*In Exercises 17–20, write the given system of linear equations as a matrix equation, and solve by inverting the coefficient matrix.*

**17.** $x + 2y = 0$
$3x + 4y = 2$

**18.** $x + y + z = 3$
$y + 2z = 4$
$y - z = 1$

**19.**
$$\begin{aligned} x + y + z &= 2 \\ x + 2y + z &= 3 \\ x + y + 2z &= 1 \end{aligned}$$

**20.**
$$\begin{aligned} x + y \quad\quad &= 0 \\ y + z \quad\ &= 1 \\ z + w &= 0 \\ x \quad\quad\ - w &= 3 \end{aligned}$$

*In each of Exercises 21–24, solve the game with the given payoff matrix and give the expected value of the game.*

**21.** $P = \begin{bmatrix} 2 & 1 & 3 & 2 \\ -1 & 0 & -2 & 1 \\ 2 & 0 & 1 & 3 \end{bmatrix}$

**22.** $P = \begin{bmatrix} 3 & -3 & -2 \\ -1 & 3 & 0 \\ 2 & 2 & 1 \end{bmatrix}$

**23.** $P = \begin{bmatrix} -1 & -3 & -2 \\ -1 & 3 & 0 \\ 3 & 3 & -1 \end{bmatrix}$

**24.** $P = \begin{bmatrix} 1 & 4 & 3 & 3 \\ 0 & -1 & 2 & 3 \\ 2 & 0 & -1 & 2 \end{bmatrix}$

*In each of Exercises 25–28, find the production vector X corresponding to the given technology matrix A and external demand vector D.*

**25.** $A = \begin{bmatrix} 0.3 & 0.1 \\ 0 & 0.3 \end{bmatrix}$, $D = \begin{bmatrix} 700 \\ 490 \end{bmatrix}$

**26.** $A = \begin{bmatrix} 0.7 & 0.1 \\ 0.1 & 0.7 \end{bmatrix}$, $D = \begin{bmatrix} 1,000 \\ 2,000 \end{bmatrix}$

**27.** $A = \begin{bmatrix} 0.2 & 0.2 & 0.2 \\ 0 & 0.2 & 0.2 \\ 0 & 0 & 0.2 \end{bmatrix}$, $D = \begin{bmatrix} 32,000 \\ 16,000 \\ 8,000 \end{bmatrix}$

**28.** $A = \begin{bmatrix} 0.5 & 0.1 & 0 \\ 0.1 & 0.5 & 0.1 \\ 0 & 0.1 & 0.5 \end{bmatrix}$, $D = \begin{bmatrix} 23,000 \\ 46,000 \\ 23,000 \end{bmatrix}$

## APPLICATIONS: OHaganBooks.com

*It is now July 1 and online sales of romance, science fiction, and horror novels at OHaganBooks.com were disappointingly slow over the past month. Exercises 29–34 are based on the following tables:*

*Inventory of books in stock on June 1 at the OHaganBooks.com warehouses in Texas and Nevada:*

**Books in Stock (June 1)**

|  | Romance | Sci Fi | Horror |
|---|---|---|---|
| **Texas** | 2,500 | 4,000 | 3,000 |
| **Nevada** | 1,500 | 3,000 | 1,000 |

*Online sales during June:*

**June Sales**

|  | Romance | Sci Fi | Horror |
|---|---|---|---|
| **Texas** | 300 | 500 | 100 |
| **Nevada** | 100 | 600 | 200 |

*New books purchased each month:*

**Monthly Purchases**

|  | Romance | Sci Fi | Horror |
|---|---|---|---|
| **Texas** | 400 | 400 | 300 |
| **Nevada** | 200 | 400 | 300 |

**July Sales (Projected)**

|  | Romance | Sci Fi | Horror |
|---|---|---|---|
| **Texas** | 280 | 550 | 100 |
| **Nevada** | 50 | 500 | 120 |

**29.** *Inventory* Use matrix algebra to compute the inventory at each warehouse at the end of June.

**30.** *Inventory* Use matrix algebra to compute the change in inventory at each warehouse during June.

**31.** *Inventory* Assuming that sales continue at the level projected for July for the next few months, write down a matrix equation showing the inventory $N$ at each warehouse $x$ months after July 1. How many months from now will OHaganBooks.com run out of Sci Fi novels at the Nevada warehouse?

**32.** *Inventory* Assuming that sales continue at the level projected for July for the next few months, write down a matrix equation showing the change in inventory $N$ at each warehouse $x$ months after July 1. How many months from now will OHaganBooks.com have 1,000 more horror novels in stock in Texas than currently?

**33.** *Revenue* It is now the end of July and OHaganBooks.com's e-commerce manager bursts into the CEO's office. "I thought you might want to know, John, that our sales figures are exactly what I projected a month ago. Is that good market analysis or what?" OHaganBooks.com has charged an average of \$5 for romance novels, \$6 for science fiction novels, and \$5.50 for horror novels. Use the projected July sales figures from above and matrix arithmetic to compute the total revenue OHaganBooks.com earned at each warehouse in July.

**34.** *Cost* OHaganBooks.com pays an average of \$2 for romance novels, \$3.50 for science fiction novels, and \$1.50 for horror novels. Use this information together with the monthly purchasing information to compute the monthly purchasing cost.

*Acting on a "tip" from Marjory Duffin, John O'Hagan decided that his company should invest a significant sum in shares of Duffin House Publishers (DHP) and Duffin Subprime Ventures (DSV). Exercises 35–38 are based on the following table, which shows what information John was able to piece together later, after some of the records had been deleted by an angry student intern.*

| Date | Number of Shares: DHP | Price per Share: DHP | Number of Shares: DSV | Price per Share: DSV |
|---|---|---|---|---|
| **July 1** | ? | \$20 | ? | \$10 |
| **August 1** | ? | \$10 | ? | \$20 |
| **September 1** | ? | \$5 | ? | \$40 |
| **Total** | 5,000 | | 7,000 | |

**35. *Investments*** Over the three months shown, the company invested a total of $50,000 in DHP stock, and, on August 15, was paid dividends of 10¢ per share held on that date, for a total of $300. Use matrix inversion to determine how many shares of DHP OHaganBooks.com purchased on each of the three dates shown.

**36. *Investments*** Over the three months shown, the company invested a total of $150,000 in DSV stock, and, on July 15, was paid dividends of 20¢ per share held on that date, for a total of $600. Use matrix inversion to determine how many shares of DSV OHaganBooks.com purchased on each of the three dates shown.

**37. *Investments*** (Refer to Exercise 35.) On October 1, the shares of DHP purchased on July 1 were sold at $3 per share. The remaining shares were sold one month later at $1 per share. Use matrix algebra to determine the total loss (taking into account the dividends paid on August 15) incurred as a result of the Duffin stock debacle.

**38. *Investments*** (Refer to Exercise 36.) On September 15, DSV announced a two-for-one stock split, so that each share originally purchased was converted into two shares. On October 1, the company paid an additional special dividend of 10¢ per share. On October 15, OHaganBooks.com sold 3,000 shares at $20 per share. One week later the subprime market crashed, Duffin Subprime Ventures declared bankruptcy (after awarding its fund manager a $10 million bonus), and DSV stock became worthless. Use matrix algebra to determine the total loss (taking into account the dividends paid on July 15) incurred as a result of the Duffin stock debacle.

*OHaganBooks.com has two main competitors: JungleBooks.com and FarmerBooks.com, and no other competitors of any significance on the horizon. Exercises 39–42 are based on the following table, which shows the movement of customers during July.[40] (Thus, for instance, the first row tells us that 80% of OHaganBooks.com's customers remained loyal, 10% of them went to JungleBooks.com, and the remaining 10% went to FarmerBooks.com.)*

|  | To OHagan | To Jungle | To Farmer |
|---|---|---|---|
| **From OHagan** | 0.8 | 0.1 | 0.1 |
| **From Jungle** | 0.4 | 0.6 | 0 |
| **From Farmer** | 0.2 | 0 | 0.8 |

*At the beginning of July, OHaganBooks.com had an estimated 2,000 customers, while its two competitors had 4,000 each.*

**39. *Competition*** Set up the July 1 customer numbers in a row matrix, and use matrix arithmetic to estimate the number of customers each company has at the end of July.

**40. *Competition*** Assuming the July trends continue in August, predict the number of customers each company will have at the end of August.

---
[40]By a "customer" of one of the three e-commerce sites, we mean someone who purchases more at that site than at either of the two competitors.

**41. *Competition*** Assuming the July trends continue in August, why is it not possible for a customer of FarmerBooks.com on July 1 to have ended up as a JungleBooks.com customer two months later without having ever been an OHaganBooks.com customer?

**42. *Competition*** Name one or more important factors that the model we have used does not take into account.

*Publisher Marjory Duffin reveals that JungleBooks may be launching a promotional scheme in which it will offer either two books for the price of one, or three books for the price of two (Marjory can't quite seem to remember which, and is not certain whether they will go with the scheme at all). John O'Hagan's marketing advisers Flood and O'Lara seem to have different ideas as to how to respond. Flood suggests that the company counter by offering three books for the price of one, while O'Lara suggests that it offer instead a free copy of the* Finite Mathematics Student Solutions Manual *with every purchase. After a careful analysis, O'Hagan comes up with the following payoff matrix, where the payoffs represent the number of customers, in thousands, he expects to gain from JungleBooks.*

|  | JungleBooks | | |
|---|---|---|---|
|  | No Promo | 2 for Price of 1 | 3 for Price of 2 |
| **No Promo** | 0 | −60 | −40 |
| **O'Hagan  3 for Price of 1** | 30 | 20 | 10 |
| **Finite Math** | 20 | 0 | 15 |

*Use the above information in Exercises 43–48.*

**43. *Competition*** After a very expensive dinner at an exclusive restaurant, Marjory suddenly "remembers" that the JungleBooks CEO mentioned to her (at a less expensive restaurant) that there is only a 20% chance JungleBooks will launch a "2 for the price of 1" promotion, and a 40% chance that it will launch a "3 for the price of 2" promotion. What should OHaganBooks.com do in view of this information, and what will the expected effect be on its customer base?

**44. *Competition*** JungleBooks CEO François Dubois has been told by someone with personal ties to OHaganBooks.com staff that OHaganBooks is in fact 80% certain to opt for the "3 for the price of 1" option and will certainly go with one of the two possible promos. What should JungleBooks.com do in view of this information, and what will the expected effect be on its customer base?

**45. *Corporate Spies*** One of John O'Hagan's trusted marketing advisers has, without knowing it, accidentally sent him a copy of the following e-mail:

> To: René, JungleBooks Marketing Department
>
> From: O'Lara
>
> Hey René somehow the CEO here says he has learned that there is a 20% chance that you will opt for the "2 for the price of 1" promotion, and a 40% chance that you will opt for the "3 for the price of 2" promotion. Thought you might want to know. So when do I get my "commission"? —Jim

What will each company do in view of this new information, and what will the expected effect be on its customer base?

**46. *More Corporate Spies*** The next day, everything changes: OHaganBooks.com's mole at JungleBooks, Davíde DuPont, is found unconscious in the coffee room at JungleBooks.com headquarters, clutching in his hand the following correspondence he had apparently received moments earlier:

> To: Davíde
> From: John O'Hagan
> Subject: Re: Urgent Information
> Davíde: This information is much appreciated and definitely changes my plans—J
>
> >
> >To: John O
> >From: Davíde
> >Subject: Urgent Information
> >Thought you might want to know that JungleBooks
> >thinks you are 80% likely to opt for 3 for 1 and 20%
> >likely to opt for Finite Math.
> >—D

What will each company do in view of this information, and what will the expected effect be on its customer base?

**47. *Things Unravel*** John O'Hagan is about to go with the option chosen in Exercise 45 when he hears word about an exposé in the *Publisher Enquirer* on corporate spying in the two companies. Each company now knows that no information about the other's intentions can be trusted. Now what should OHaganBooks.com do, and how many customers should it expect to gain or lose?

**48. *Competition*** It is now apparent as a result of the *Publisher Enquirer* exposé that, not only can each company make no assumptions about the strategies the other might be using, but the payoff matrix they have been using is wrong: A crack investigative reporter at the *Enquirer* publishes the following revised matrix:

$$P = \begin{bmatrix} 0 & -60 & -40 \\ 30 & 20 & 10 \\ 20 & 0 & 20 \end{bmatrix}.$$

*Now* what should JungleBooks.com do, and how many customers should it expect to gain or lose?

*Some of the books sold by OHaganBooks.com are printed at Bruno Mills, Inc., a combined paper mill and printing company. Exercises 49–52 are based on the following typical monthly input-output table for Bruno Mills' paper and book printing sectors.*

|  | *To* | **Paper** | **Books** |
|---|---|---|---|
| *From* | **Paper** | $20,000 | $50,000 |
|  | **Books** | 2,000 | 5,000 |
|  | **Total Output** | 200,000 | 100,000 |

**49. *Production*** Find the technology matrix for Bruno Mills' paper and book printing sectors.

**50. *Production*** Compute $(I - A)^{-1}$. What is the significance of the (1, 2)-entry?

**51. *Production*** Approximately $1,700 worth of the books sold each month by OHaganBooks.com are printed at Bruno Mills, Inc., and OHaganBooks.com uses approximately $170 worth of Bruno Mills' paper products each month. What is the total value of paper and books that must be produced by Bruno Mills, Inc. in order to meet demand from OHaganBooks.com?

**52. *Production*** Currently, Bruno Mills, Inc. has a monthly capacity of $500,000 of paper products and $200,000 of books. What level of external demand would cause Bruno to meet the capacity for both products?

## Case Study    Projecting Market Share

You are the sales director at *Selular*, a cellphone provider, and things are not looking good for your company: The recently launched *iClone* competitor is beginning to chip away at Selular's market share. Particularly disturbing are rumors of fierce brand loyalty by iClone customers, with several bloggers suggesting that iClone retains close to 100% of their customers. Worse, you will shortly be presenting a sales report to the board of directors, and the CEO has "suggested" that your report include 2-, 5-, and 10-year projections of Selular's market share given the recent impact on the market by iClone, and also a "worst-case" scenario projecting what would happen if iClone customers are so loyal that none of them ever switch services.

The sales department has conducted two market surveys, taken one quarter apart, of the major cellphone providers, which are *iClone, Selular, AB&C,* and some

✱ If you go on to study probability theory in Chapter 7, you will see how this scenario can be interpreted as a Markov system, and you will revisit the analysis below in that context.

The percentages in the figure and market shares below reflect actual data for several cellphone services in the United States during a single quarter of 2003. (See the exercises for Section 7.7.)

smaller companies lumped together as "Other," and has given you the data shown in Figure 4, which shows the percentages of subscribers who switched from one service to another during the quarter.✱

**Figure 4**

For example, by the end of the quarter 0.5% of iClone's customers had switched to Selular, 0.5% to AB&C, and 0.9% to Other. The current market shares are as follows: iClone: 29.7%, Selular: 19.3%, AB&C: 18.1%, Other: 32.9%.

"This is simple," you tell yourself. "Since I know the current market shares and percentage movements over one quarter, I can easily calculate the market shares next quarter (assuming the percentages that switch services remain the same), then repeat the calculation for the following quarter, and so on, until I get the long-term prediction I am seeking." So you begin your calculations by computing the market shares next quarter. First you note that, since a total of $0.5 + 0.5 + 0.9 = 1.9\%$ of iClone users switched to other brands, the rest, 98.1%, stayed with iClone. Similarly, 97.2% of Selular users, 97.3% of AB&C, and 98.1% of Other stayed with their respective brands. Then you compute:

iClone share after one quarter

$$= 98.1\% \text{ of iClone} + 1.0\% \text{ of Selular} + 1.0\% \text{ of AB\&C} + 0.8\% \text{ of Other}$$

$$= (0.981)(0.297) + (0.010)(0.193) + (0.010)(0.181) + (0.008)(0.329)$$

$$= 0.297729, \text{ or } 29.7729\%.$$

Similarly,

Selular share:

$$= (0.005)(0.297) + (0.972)(0.193) + (0.006)(0.181) + (0.006)(0.329)$$

$$= 0.192141$$

AB&C share:

$$= (0.005)(0.297) + (0.006)(0.193) + (0.973)(0.181) + (0.005)(0.329)$$

$$= 0.180401$$

Other share:

$$= (0.009)(0.297) + (0.012)(0.193) + (0.011)(0.181) + (0.981)(0.329)$$

$$= 0.329729.$$

So, you project the market shares after one quarter to be: iClone: 29.7729%, Selular: 19.2141%, AB&C: 18.0401%, Other: 32.9729%. You now begin to realize that repeating this kind of calculation for the large number of quarters required for long-term projections will be tedious. You call in your student intern (who happens to be a mathematics major) to see if she can help. After taking one look at the calculations, she makes the observation that all you have really done is compute the product of two matrices:

$$[0.297729 \quad 0.192141 \quad 0.180401 \quad 0.329729]$$

$$= [0.297 \quad 0.193 \quad 0.181 \quad 0.329] \begin{bmatrix} .981 & .005 & .005 & .009 \\ .010 & .972 & .006 & .012 \\ .010 & .006 & .973 & .011 \\ .008 & .006 & .005 & .981 \end{bmatrix}$$

Market shares after one quarter = Market shares at start of quarter $\times A$

The $4 \times 4$ matrix $A$ is organized as follows:

|  |  | *To* | | | |
|---|---|---|---|---|---|
|  |  | **iClone** | **Selular** | **AB&C** | **Other** |
|  | **iClone** | .981 | .005 | .005 | .009 |
| *From* | **Selular** | .010 | .972 | .006 | .012 |
|  | **AB&C** | .010 | .006 | .973 | .011 |
|  | **Other** | .008 | .006 | .005 | .981 |

Since the market shares one quarter later can be obtained from the shares at the start of the quarter, you realize that you can now obtain the shares *two* quarters later by multiplying the result by $A$—and at this point you start using technology (such as the Matrix Algebra Tool at www.WanerMath.com) to continue the calculation:

Market shares after two quarters = Market shares after one quarter $\times A$

$$= [0.297729 \quad 0.192141 \quad 0.180401 \quad 0.329729] \begin{bmatrix} .981 & .005 & .005 & .009 \\ .010 & .972 & .006 & .012 \\ .010 & .006 & .973 & .011 \\ .008 & .006 & .005 & .981 \end{bmatrix}$$

$$\approx [0.298435 \quad 0.191310 \quad 0.179820 \quad 0.330434].$$

Although the use of matrices has simplified your work, continually multiplying the result by $A$ over and over again to get the market shares for successive months is still tedious. Would it not be possible to get, say, the market share after 10 years (40 quarters) with a single calculation? To explore this, you decide to use symbols for the various market shares:

$m_0$ = Starting market shares = $[0.297 \quad 0.193 \quad 0.181 \quad 0.329]$

$m_1$ = Market shares after 1 quarter
$\quad = [0.297729 \quad 0.192141 \quad 0.180401 \quad 0.329729]$

$m_2$ = Market shares after 2 quarters

. . .

$m_n$ = Market share after $n$ quarters.

You then rewrite the relationships above as

$m_1 = m_0 A$    Shares after one quarter = Shares at start of quarter $\times A$
$m_2 = m_1 A$    Shares after two quarters = Shares after one quarter $\times A$

On an impulse, you substitute the first equation in the second:

$$m_2 = m_1 A = (m_0 A) A = m_0 A^2.$$

Continuing,

$$m_3 = m_2 A = (m_0 A^2) A = m_0 A^3$$

$$\ldots$$

$$m_n = m_0 A^n,$$

which is exactly the formula you need! You can now obtain the 2-, 5-, and 10-year projections each in a single step (with the aid of technology):

2-year projection: $m_8 = m_0 A^8$

$$= [0.297 \quad 0.193 \quad 0.181 \quad 0.329] \begin{bmatrix} .981 & .005 & .005 & .009 \\ .010 & .972 & .006 & .012 \\ .010 & .006 & .973 & .011 \\ .008 & .006 & .005 & .981 \end{bmatrix}^8$$

$$\approx [0.302237 \quad 0.186875 \quad 0.176691 \quad 0.334198]$$

5-year projection: $m_{20} = m_0 A^{20} \approx [0.307992 \quad 0.180290 \quad 0.171938 \quad 0.33978]$

10-year projection: $m_{40} = m_0 A^{40} \approx [0.313857 \quad 0.173809 \quad 0.167074 \quad 0.34526].$

In particular, Selular's market shares are projected to decline slightly: 2-year projection: 18.7%, 5-year projection: 18.0%, 10-year projection: 17.4%.

You now move on to the "worst-case" scenario, which you represent by assuming 100% loyalty by iClone users with the remaining percentages staying the same (Figure 5).

**Figure 5**

The matrix $A$ corresponding to this diagram is $\begin{bmatrix} 1 & 0 & 0 & 0 \\ .010 & .972 & .006 & .012 \\ .010 & .006 & .973 & .011 \\ .008 & .006 & .005 & .981 \end{bmatrix}$, and

you find:

2-year projection: $m_8 = m_0 A^8 \approx [0.346335 \quad 0.175352 \quad 0.165200 \quad 0.313113]$

5-year projection: $m_{20} = m_0 A^{20} \approx [0.413747 \quad 0.152940 \quad 0.144776 \quad 0.288536]$

10-year projection: $m_{40} = m_0 A^{40} \approx [0.510719 \quad 0.123553 \quad 0.117449 \quad 0.248279].$

Thus, in the worst-case scenario, Selular's market shares are projected to decline more rapidly: 2-year projection: 17.5%, 5-year projection: 15.3%, 10-year projection: 12.4%.

## EXERCISES

1. Project Selular's market share 20 and 30 years from now based on the original data shown in Figure 4. (Round all figures to the nearest 0.1%.)

2. Using the original data, what can you say about each company's share in 60 and 80 years (assuming current trends continue)? (Round all figures to the nearest 0.1%.)

3. Obtain a sequence of projections several hundreds of years into the future using the scenarios in both Figure 4 and Figure 5. What do you notice?

4. Compute a sequence of larger and larger powers of the matrix $A$ for both scenarios with all figures rounded to four decimal places. What do you notice?

5. Suppose the trends noted in the market survey were in place for several quarters *before* the current quarter. Using the scenario in Figure 4, determine what the companies' market shares were one quarter before the present and one year before the present (to the nearest 0.1%). Do the same for the scenario in Figure 5.

6. If $A$ is the matrix from the scenario in Figure 4, compute $A^{-1}$. In light of the preceding exercise, what do the entries in $A^{-1}$ mean?

**TECHNOLOGY GUIDE**

## TI-83/84 Plus    Technology Guide

### Section 4.1

**Example 2 (page 235)**   The A-Plus auto parts store chain has two outlets, one in Vancouver and one in Quebec. Among other things, it sells wiper blades, windshield cleaning fluid, and floor mats. The monthly sales of these items at the two stores for two months are given in the following tables:

**January Sales**

|                        | Vancouver | Quebec |
|------------------------|-----------|--------|
| **Wiper Blades**       | 20        | 15     |
| **Cleaning Fluid (bottles)** | 10  | 12     |
| **Floor Mats**         | 8         | 4      |

**February Sales**

|                        | Vancouver | Quebec |
|------------------------|-----------|--------|
| **Wiper Blades**       | 23        | 12     |
| **Cleaning Fluid (bottles)** | 8   | 12     |
| **Floor Mats**         | 4         | 5      |

Use matrix arithmetic to calculate the change in sales of each product in each store from January to February.

#### Solution with Technology

On the TI-83/84 Plus, matrices are referred to as [A], [B], and so on through [J]. To enter a matrix, press MATRX to bring up the matrix menu, select EDIT, select a matrix, and press ENTER. Then enter the dimensions of the matrix followed by its entries. When you want to use a matrix, press MATRX, select the matrix and press ENTER.

On the TI-83/84 Plus, adding matrices is similar to adding numbers. The sum of the matrices [A] and [B] is [A] + [B]; their difference, of course, is [A] − [B]. As in the text, for this example,

1. Create two matrices, [J] and [F].
2. Compute their difference, [F] − [J] using [F] − [J] → [D].

Note that we have stored the difference in the matrix [D] in case we need it for later use.

### Section 4.2

#### Example 3(a) (page 245)     Calculate

$$\begin{bmatrix} 2 & 0 & -1 & 3 \\ 1 & -1 & 2 & -2 \end{bmatrix} \begin{bmatrix} 1 & 1 & -8 \\ 1 & 0 & 0 \\ 0 & 5 & 2 \\ -2 & 8 & -1 \end{bmatrix}.$$

#### Solution with Technology

On the TI-83/84 Plus, the format for multiplying matrices is the same as for multiplying numbers: [A][B] or [A]*[B] will give the product. We enter the matrices and then multiply them. (Note that, while editing, you can see only three columns of [A] at a time.)

Note that if you try to multiply two matrices whose product is not defined, you will get the error "DIM MISMATCH" (dimension mismatch).

**Example 6 (page 248)—Identity Matrix**    On the TI-83/84 Plus, the function identity(n) (in the MATRX MATH menu) returns the $n \times n$ identity matrix.

## Section 4.3

### Example 2(b) (page 259)

Find the inverse of

$$Q = \begin{bmatrix} 1 & 0 & 1 \\ 2 & -2 & -1 \\ 3 & 0 & 0 \end{bmatrix}.$$

### Solution with Technology

On a TI-83/84 Plus, you can invert the square matrix [A] by entering [A] $X^{-1}$ ENTER.

(Note that you can use the right and left arrow keys to scroll the answer, which cannot be shown on the screen all at once.) You could also use the calculator to help you go through the row reduction, as described in Chapter 2.

### Example 4 (page 262)

Solve the following three systems of equations.

**a.**
$2x \quad + z = 1$
$2x + y - z = 1$
$3x + y - z = 1$

**b.**
$2x \quad + z = 0$
$2x + y - z = 1$
$3x + y - z = 2$

**c.**
$2x \quad + z = 0$
$2x + y - z = 0$
$3x + y - z = 0$

### Solution with Technology

**1.** Enter the four matrices $A$, $B$, $C$, and $D$:

**2.** Compute the solutions $A^{-1}B$, $A^{-1}C$, and $A^{-1}D$:

## Section 4.5

**Example 1 (page 286)**    Recall that the input-output table in Example 1 looks like this:

| | To | Crude | Refining |
|---|---|---|---|
| *From* | Crude | 27,000 | 59,000 |
| | Refining | 750 | 15,000 |
| | Total Output | 87,000 | 140,000 |

What was the technology matrix for these two sectors? What was left over from each of these sectors for use by other parts of the economy or for export?

## Solution with Technology

There are several ways to use these data to create the technology matrix in your TI-83/84 Plus. For small matrices like this, the most straightforward is to use the matrix editor, where you can give each entry as the appropriate quotient:

and so on. Once we have the technology matrix [A] and the production vector [B] (remember that we can't use [X] as a matrix name), we can calculate the external demand vector:

Of course, when interpreting these numbers, we must remember to round to two significant digits, because our original data were accurate to only that many digits.

Here is an alternative way to calculate the technology matrix that may be better for examples with more sectors.

1. Begin by entering the columns of the input-output table as lists in the list editor (STAT EDIT) (see below left).

2. We now want to divide each column by the total output of its sector and assemble the results into a matrix. We can do this using the List ▶ matr function (under the MATRX MATH menu) shown on the right above.

**Example 3 (page 291)** Consider four sectors of the economy of Kenya[41]: (1) the traditional economy, (2) agriculture, (3) manufacture of metal products and machinery, and (4) wholesale and retail trade. The input-output table for these four sectors for 1976 looks like this (all numbers are thousands of K£):

| | *To* | 1 | 2 | 3 | 4 |
|---|---|---|---|---|---|
| *From* | 1 | 8,600 | 0 | 0 | 0 |
| | 2 | 0 | 20,000 | 24 | 0 |
| | 3 | 1,500 | 530 | 15,000 | 660 |
| | 4 | 810 | 8,500 | 5,800 | 2,900 |
| **Total Output** | | 87,000 | 530,000 | 110,000 | 180,000 |

Suppose that external demand for agriculture increased by K£50,000,000 and that external demand for metal products and machinery increased by K£10,000,000. How would production in these four sectors have to change to meet this rising demand?

## Solution with Technology

1. Enter the technology matrices $A$ as [A] and $D^+$ as [D] using one of the techniques above:

2. You can then compute the change in production with the formula: `(identity(4)-[A])`$^{-1}$ `[D]`.

---

[41]Figures are rounded. Source: *Input-Output Tables for Kenya 1976,* Central Bureau of Statistics of the Ministry of Economic Planning and Community Affairs, Kenya.

## SPREADSHEET Technology Guide

### Section 4.1

**Example 2 (page 235)** The A-Plus auto parts store chain has two outlets, one in Vancouver and one in Quebec. Among other things, it sells wiper blades, windshield cleaning fluid, and floor mats. The monthly sales of these items at the two stores for two months are given in the following tables:

**January Sales**

|                          | Vancouver | Quebec |
|--------------------------|-----------|--------|
| Wiper Blades             | 20        | 15     |
| Cleaning Fluid (bottles) | 10        | 12     |
| Floor Mats               | 8         | 4      |

**February Sales**

|                          | Vancouver | Quebec |
|--------------------------|-----------|--------|
| Wiper Blades             | 23        | 12     |
| Cleaning Fluid (bottles) | 8         | 12     |
| Floor Mats               | 4         | 5      |

Use matrix arithmetic to calculate the change in sales of each product in each store from January to February.

### Solution with Technology

To enter a matrix in a spreadsheet, we put its entries in any convenient block of cells. For example, the matrix $A$ in the first Quick Example of this section might look like this:

|   | A  | B   | C |
|---|----|-----|---|
| 1 | 2  | 0   | 1 |
| 2 | 33 | -22 | 0 |

Spreadsheets refer to such blocks of data as **arrays**, which it can handle in much the same way as it handles single cells of data. For instance, when typing a formula, just as clicking on a cell creates a reference to that cell, selecting a whole array of cells will create a reference to that array. An array is referred to using an **array range** consisting of the top-left and bottom-right cell coordinates, separated by a colon. For example, the array range A1:C2 refers to the 2 × 3 matrix above, with top-left corner A1 and bottom-right corner C2.

**1.** To add or subtract two matrices in a spreadsheet, first input their entries in two separate arrays in the spreadsheet (we've also added labels as in the previous tables, which you might do if you wanted to save the spreadsheet for later use):

|    | A                        | B         | C      |
|----|--------------------------|-----------|--------|
| 1  | January Sales            |           |        |
| 2  |                          | Vancouver | Quebec |
| 3  | wiper blades             | 20        | 15     |
| 4  | cleaning fluid (bottles) | 10        | 12     |
| 5  | floor mats               | 8         | 4      |
| 6  |                          |           |        |
| 7  | February Sales           |           |        |
| 8  |                          | Vancouver | Quebec |
| 9  | wiper blades             | 23        | 12     |
| 10 | cleaning fluid (bottles) | 8         | 12     |
| 11 | floor mats               | 4         | 5      |

**2.** Select (highlight) a block of the same size (3 × 2 in this case) where you would like the answer, $F - J$, to appear, enter the formula =B9:C11-B3:C5, and then type Control+Shift+Enter. The easiest way to do this is as follows:

- Highlight cells B15:C17.  Where you want the answer to appear
- Type "=".
- Highlight the matrix $F$.  Cells B9 through C11
- Type "-".
- Highlight the matrix $J$.  Cells B3 through C5
- Press Control+Shift+Enter.  Not just Enter

|    | A                        | B              | C      |
|----|--------------------------|----------------|--------|
| 1  | January Sales            |                |        |
| 2  |                          | Vancouver      | Quebec |
| 3  | wiper blades             | 20             | 15     |
| 4  | cleaning fluid (bottles) | 10             | 12     |
| 5  | floor mats               | 8              | 4      |
| 6  |                          |                |        |
| 7  | February Sales           |                |        |
| 8  |                          | Vancouver      | Quebec |
| 9  | wiper blades             | 23             | 12     |
| 10 | cleaning fluid (bottles) | 8              | 12     |
| 11 | floor mats               | 4              | 5      |
| 12 |                          |                |        |
| 13 | Change in Sales          |                |        |
| 14 |                          | Vancouver      | Quebec |
| 15 | wiper blades             | =B9:C11-B3:C5  |        |
| 16 | cleaning fluid (bottles) |                |        |
| 17 | floor mats               |                |        |

Typing Control+Shift+Enter (instead of Enter) tells the spreadsheet that your formula is an *array formula*, one that returns a matrix rather than a single number.[42] Once entered, the formula bar will show the formula you entered enclosed in "curly braces," indicating that it is an array formula. Note that you must use Control+Shift+Enter to delete any array you create: Select the block you wish to delete and press Delete followed by Control+Shift+Enter.

## Section 4.2

**Example 3(a) (page 245)**    Calculate the product

$$\begin{bmatrix} 2 & 0 & -1 & 3 \\ 1 & -1 & 2 & -2 \end{bmatrix} \begin{bmatrix} 1 & 1 & -8 \\ 1 & 0 & 0 \\ 0 & 5 & 2 \\ -2 & 8 & -1 \end{bmatrix}.$$

**Solution with Technology**

In a spreadsheet, the function we use for matrix multiplication is MMULT. (Ordinary multiplication, *, will *not* work.)

1. Enter the two matrices as shown in the spreadsheet and highlight a block where you want the answer to appear. (Note that it should have the correct dimensions for the product: 2 × 3.)

2. Enter the formula =MMULT(A1:D2,F1:H4) (using the mouse to avoid typing the array ranges if you like) and press Control+Shift+Enter. The product will appear in the region you highlighted. If you try to multiply two matrices whose product is not defined, you will get the error "#VALUE!".

**Example 6—Identity Matrix (page 248)**    There is no spreadsheet function that returns an identity matrix. If you need a small identity matrix, it's simplest to just

enter the 1s and 0s by hand. If you need a large identity matrix, here is one way to get it quickly.

1. Say we want a 4 × 4 identity matrix in the cells B1:E4. Enter the following formula in cell B1:

```
=IF(ROW(B1)-ROW($B$1)
=COLUMN(B1)-COLUMN($B$1),1,0)
```

| | A | B | C | D | E |
|---|---|---|---|---|---|
| 1 | | =IF(ROW(B1)-ROW($B$1)=COLUMN(B1)-COLUMN($B$1),1,0) | | | |
| 2 | | | | | |
| 3 | | | | | |
| 4 | | | | | |

2. Press Enter, then copy cell B1 to cells B1:E4. The formula will return 1s along the diagonal of the matrix and 0s elsewhere, giving you the identity matrix. Why does this formula work?

| | A | B | C | D | E |
|---|---|---|---|---|---|
| 1 | | 1 | 0 | 0 | 0 |
| 2 | | 0 | 1 | 0 | 0 |
| 3 | | 0 | 0 | 1 | 0 |
| 4 | | 0 | 0 | 0 | 1 |

## Section 4.3

**Example 2(b) (page 259)**    Find the inverse of

$$Q = \begin{bmatrix} 1 & 0 & 1 \\ 2 & -2 & -1 \\ 3 & 0 & 0 \end{bmatrix}.$$

**Solution with Technology**

In a spreadsheet, the function MINVERSE computes the inverse of a matrix.

1. Enter Q somewhere convenient, for example, in cells A1:C3.

2. Choose the block where you would like the inverse to appear, highlight the whole block.

3. Enter the formula =MINVERSE(A1:C3) and press Control+Shift+Enter.

---

[42]Note that on a Mac, Command-Enter has the same effect as Control+Shift+Enter.

The inverse will appear in the region you highlighted. (To convert the answer to fractions, format the cells as fractions.)

If a matrix is singular, a spreadsheet will register an error by showing #NUM! in each cell.

| | A | B | C | D | E | F | G |
|---|---|---|---|---|---|---|---|
| 1 | 1 | 0 | 1 | | 0 | 0 | 0.3333333 |
| 2 | 2 | -2 | -1 | | -0.5 | -0.5 | 0.5 |
| 3 | 3 | 0 | 0 | | 1 | 0 | -0.3333333 |

Although the spreadsheet appears to invert the matrix in one step, it is going through the procedure in the text or some variation of it to find the inverse. Of course, you could also use the spreadsheet to help you go through the row reduction, just as in Chapter 3.

**Example 4 (page 262)** Solve the following three systems of equations.

**a.** $2x \quad + z = 1$
$2x + y - z = 1$
$3x + y - z = 1$

**b.** $2x \quad + z = 0$
$2x + y - z = 1$
$3x + y - z = 2$

**c.** $2x \quad + z = 0$
$2x + y - z = 0$
$3x + y - z = 0$

**Solution with Technology**

Spreadsheets instantly update calculated results every time the contents of a cell are changed. We can take advantage of this to solve the three systems of equations given above using the same worksheet as follows.

**1.** Enter the matrices $A$ and $B$ from the matrix equation $AX = B$.

**2.** Select a $3 \times 1$ block of cells for the matrix $X$.

**3.** The Excel formula we can use to calculate $X$ is:

$$=MMULT(MINVERSE(A1:C3),E1:E3) \qquad A^{-1}B$$

(As usual, use the mouse to select the ranges for $A$ and $B$ while typing the formula, and don't forget to press Control+Shift+Enter.) Having obtained the solution to

part (a), you can now simply modify the entries in Column E to see the solutions for parts (b) and (c).

**Note** Your spreadsheet for part (a) may look like this:

| | A | B | C | D | E |
|---|---|---|---|---|---|
| 1 | 2 | 0 | 1 | | 1 |
| 2 | 2 | 1 | -1 | | 1 |
| 3 | 3 | 1 | -1 | | 1 |
| 4 | | A | | | B |
| 5 | | | | | |
| 6 | 1.11022E-16 | | | | |
| 7 | 2 | | | | |
| 8 | 1 | | | | |
| 9 | $A^{-1}B$ | | | | |

What is that strange number doing in cell A6? "E-16" represents "$\times 10^{-16}$", so the entry is really

$$1.11022 \times 10^{-16} = 0.000\,000\,000\,000\,000\,111022 \approx 0.$$

Mathematically, it is supposed to be *exactly* zero (see the solution to part (a) in the text) but Excel made a small error in computing the inverse of $A$, resulting in this spurious value. Note, however, that it is accurate (agrees with zero) to 15 decimal places! In practice, when we see numbers arise in matrix calculations that are far smaller than all the other entries, we can usually assume they are supposed to be zero. ∎

## Section 4.5

**Example 1 (page 286)** Recall that the input-output table in Example 1 looks like this:

| | To | Crude | Refining |
|---|---|---|---|
| *From* **Crude** | | 27,000 | 59,000 |
| **Refining** | | 750 | 15,000 |
| **Total Output** | | 87,000 | 140,000 |

What was the technology matrix for these two sectors? What was left over from each of these sectors for use by other parts of the economy or for export?

**Solution with Technology**

**1.** Enter the input-output table in a spreadsheet:

| | A | B | C | D |
|---|---|---|---|---|
| 1 | | | To | |
| 2 | | | Crude | Refining |
| 3 | From | Crude | 27000 | 59000 |
| 4 | | Refining | 750 | 15000 |
| 5 | | Total Output | 87000 | 140000 |

**2.** To obtain the technology matrix, we divide each column by the total output of its sector:

| | A | B | C | D |
|---|---|---|---|---|
| 1 | | | To | |
| 2 | | | Crude | Refining |
| 3 | From | Crude | 27000 | 59000 |
| 4 | | Refining | 750 | 15000 |
| 5 | | Total Output | 87000 | 140000 |
| 6 | | | | |
| 7 | | | =C3/C$5 | |
| 8 | | | | |

| | A | B | C | D |
|---|---|---|---|---|
| 1 | | | To | |
| 2 | | | Crude | Refining |
| 3 | From | Crude | 27000 | 59000 |
| 4 | | Refining | 750 | 15000 |
| 5 | | Total Output | 87000 | 140000 |
| 6 | | | | |
| 7 | | | 0.3103448 | 0.4214286 |
| 8 | | | 0.0086207 | 0.1071429 |

The formula =C3/C$5 is copied into the shaded $2 \times 2$ block shown above. (The $ sign in front of the 5 forces the program to always divide by the total in Row 5 even when the formula is copied from Row 7 to Row 8.) The result is the technology matrix shown in the bottom screenshot above.

**3.** Using the techniques discussed in the second section, we can now compute $D = X - AX$ to find the demand vector.

**Example 3 (page 291)** Consider four sectors of the economy of Kenya[43]: (1) the traditional economy, (2) agriculture, (3) manufacture of metal products and machinery, and (4) wholesale and retail trade. The input-output table for these four sectors for 1976 looks like this (all numbers are thousands of K£):

| | To | 1 | 2 | 3 | 4 |
|---|---|---|---|---|---|
| From | 1 | 8,600 | 0 | 0 | 0 |
| | 2 | 0 | 20,000 | 24 | 0 |
| | 3 | 1,500 | 530 | 15,000 | 660 |
| | 4 | 810 | 8,500 | 5,800 | 2,900 |
| Total Output | | 87,000 | 530,000 | 110,000 | 180,000 |

[43]Figures are rounded. Source: *Input-Output Tables for Kenya 1976*, Central Bureau of Statistics of the Ministry of Economic Planning and Community Affairs, Kenya.

Unless otherwise noted, all content on this page is © Cengage Learning.

Suppose that external demand for agriculture increased by K£50,000,000 and that external demand for metal products and machinery increased by K£10,000,000. How would production in these four sectors have to change to meet this rising demand?

### Solution with Technology

**1.** Enter the input-output table in the spreadsheet.

**2.** Compute the technology matrix by dividing each column by the column total.

**3.** Insert the identity matrix $I$ in preparation for the next step.

| | A | B | C | D | E | F | G | H | I |
|---|---|---|---|---|---|---|---|---|---|
| 1 | 8600 | 0 | 0 | 0 | | | | | |
| 2 | 0 | 20000 | 24 | 0 | | | | | |
| 3 | 1500 | 530 | 15000 | 660 | | | | | |
| 4 | 810 | 8500 | 5800 | 2900 | | | | | |
| 5 | 87000 | 530000 | 110000 | 180000 | Totals | | | | |
| 6 | | | | | | | | | |
| 7 | | Technology Matrix | | | | | | Identity Matrix | |
| 8 | =A1/A$5 | | | | | | 1 | 0 | 0 | 0 |
| 9 | | | | | | | 0 | 1 | 0 | 0 |
| 10 | | | | | | | 0 | 0 | 1 | 0 |
| 11 | | | | | | | 0 | 0 | 0 | 1 |

**4.** To see how each sector reacts to rising external demand, you must calculate the inverse matrix $(I - A)^{-1}$, as shown below. (Remember to use Control+Shift+Enter each time.)

| | A | B | C | D | E | F | G | H | I |
|---|---|---|---|---|---|---|---|---|---|
| 1 | 8600 | 0 | 0 | 0 | | | | | |
| 2 | 0 | 20000 | 24 | 0 | | | | | |
| 3 | 1500 | 530 | 15000 | 660 | | | | | |
| 4 | 810 | 8500 | 5800 | 2900 | | | | | |
| 5 | 87000 | 530000 | 110000 | 180000 | Totals | | | | |
| 6 | | | | | | | | | |
| 7 | | Technology Matrix | | | | | | Identity Matrix | |
| 8 | 0.0988505 | 0 | 0 | 0 | | | 1 | 0 | 0 | 0 |
| 9 | 0 | 0.0377358 | 0.0002182 | 0 | | | 0 | 1 | 0 | 0 |
| 10 | 0.0172414 | 0.001 | 0.1363636 | 0.0036667 | | | 0 | 0 | 1 | 0 |
| 11 | 0.0093103 | 0.0160377 | 0.0527273 | 0.0161111 | | | 0 | 0 | 0 | 1 |
| 12 | | | | | | | | | |
| 13 | | (I − A) | | | | | (I - A)⁻¹ | |
| 14 | =F8:I11-A8:D11 | | | | | | =MINVERSE(A14:D17) | |
| 15 | | | | | | | | | |
| 16 | | | | | | | | | |
| 17 | | | | | | | | | |

**5.** To compute $(I - A)^{-1}D^+$, enter $D^+$ as a column and use the MMULT operation. (See Example 3 in Section 4.2.)

| | A | B | C | D | E | F | G | H | I |
|---|---|---|---|---|---|---|---|---|---|
| 13 | | (I − A) | | | | | (I - A)⁻¹ | |
| 14 | 0.9011494 | 0 | 0 | 0 | | 1.1096939 | 0 | 0 | 0 |
| 15 | 0 | 0.9622642 | -0.0002182 | 0 | | 5.034E-06 | 1.039216 | 0.0002626 | 9.786E-07 |
| 16 | -0.0172414 | -0.001 | 0.8636364 | -0.0036667 | | 0.0222032 | 0.0012755 | 1.1581586 | 0.0043161 |
| 17 | -0.0093103 | -0.0160377 | -0.0527273 | 0.9838889 | | 0.0116908 | 0.0170079 | 0.0620708 | 1.0166082 |
| 18 | | | | | | | | | |
| 19 | D | | (I - A)⁻¹D | | | | | | |
| 20 | 0 | | 0 | | | | | | |
| 21 | 50000 | | 51963.425 | | | | | | |
| 22 | 10000 | | 11645.361 | | | | | | |
| 23 | 0 | | 1471.105 | | | | | | |

**Note** Here is one of the beauties of spreadsheet programs: Once you are done with the calculation, you can use the spreadsheet as a template for any $4 \times 4$ input-output table by just changing the entries of the input-output matrix and/or $D^+$. The rest of the computation will then be done automatically as the spreadsheet is updated. In other words, you can use it to do your homework! ∎

# 9

# Nonlinear Functions and Models

**WW** Website

www.WanerMath.com

At the Website you will find:

- Section-by-section tutorials, including game tutorials with randomized quizzes

- A detailed chapter summary

- A true/false quiz

- Additional review exercises

- Graphers, Excel tutorials, and other resources

- The following extra topics:

  Inverse functions

  Using and deriving algebraic properties of logarithms

## Case Study  Checking up on Malthus

In 1798 Thomas R. Malthus (1766–1834) published an influential pamphlet, later expanded into a book, titled *An Essay on the Principle of Population as It Affects the Future Improvement of Society*. One of his main contentions was that population grows geometrically (exponentially), while the supply of resources such as food grows only arithmetically (linearly). Some 200+ years later, you have been asked to check the validity of Malthus's contention. **How do you go about doing so?**

Robert Nickelsberg/Getty Images

619

# Introduction

To see if Malthus was right, we need to see if the data fit the models (linear and exponential) that he suggested or if other models would be better. We saw in Chapter 1 how to fit a linear model. In this chapter we discuss how to construct models that use various *nonlinear* functions.

The nonlinear functions we consider in this chapter are the *quadratic* functions, the simplest nonlinear functions; the *exponential* functions, essential for discussing many kinds of growth and decay, including the growth (and decay) of money in finance and the initial growth of an epidemic; the *logarithmic* functions, needed to fully understand the exponential functions; and the *logistic* functions, used to model growth with an upper limit, such as the spread of an epidemic.

**algebra Review**
For this chapter, you should be familiar with the algebra reviewed in **Chapter 0, Section 2.**

## 9.1    Quadratic Functions and Models

In Chapter 1 we studied linear functions. Linear functions are useful, but in real-life applications, they are often accurate for only a limited range of values of the variables. The relationship between two quantities is often best modeled by a curved line rather than a straight line. The simplest function with a graph that is not a straight line is a *quadratic* function.

### Quadratic Function

A **quadratic function** of the variable $x$ is a function that can be written in the form

$$f(x) = ax^2 + bx + c \qquad \text{Function form}$$

or

$$y = ax^2 + bx + c \qquad \text{Equation form}$$

where $a$, $b$, and $c$ are fixed numbers (with $a \neq 0$).

### Quick Examples

1. $f(x) = 3x^2 - 2x + 1$        $a = 3, b = -2, c = 1$
2. $g(x) = -x^2$        $a = -1, b = 0, c = 0$
3. $R(p) = -5{,}600p^2 + 14{,}000p$        $a = -5{,}600, b = 14{,}000, c = 0$

✳ We shall not fully justify the formula for the vertex and the axis of symmetry until we have studied some calculus, although it is possible to do so with just algebra.

Every quadratic function $f(x) = ax^2 + bx + c$ ($a \neq 0$) has a **parabola** as its graph. Following is a summary of some features of parabolas that we can use to sketch the graph of any quadratic function.✳

# Features of a Parabola

The graph of $f(x) = ax^2 + bx + c \ (a \neq 0)$ is a **parabola**. If $a > 0$ the parabola opens upward (concave up) and if $a < 0$ it opens downward (concave down):

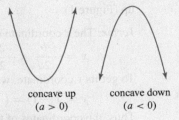

concave up            concave down
$(a > 0)$                $(a < 0)$

## Vertex, Intercepts, and Symmetry

**Vertex**   The vertex is the highest or lowest point of the parabola (see the figure above). Its $x$-coordinate is $-\dfrac{b}{2a}$. Its $y$-coordinate is $f\left(-\dfrac{b}{2a}\right)$.

**$x$-Intercepts** (if any)   These occur when $f(x) = 0$; that is, when

$$ax^2 + bx + c = 0.$$

Solve this equation for $x$ by either factoring or using the quadratic formula. The $x$-intercepts are

$$x = \frac{-b \pm \sqrt{b^2 - 4ac}}{2a}.$$

If the **discriminant** $b^2 - 4ac$ is positive, there are two $x$-intercepts. If it is zero, there is a single $x$-intercept (at the vertex). If it is negative, there are no $x$-intercepts (so the parabola doesn't touch the $x$-axis at all).

**$y$-Intercept**   This occurs when $x = 0$, so

$$y = a(0)^2 + b(0) + c = c.$$

**Symmetry**   The parabola is symmetric with respect to the vertical line through the vertex, which is the line $x = -\dfrac{b}{2a}$.

Note that the $x$-intercepts can also be written as

$$x = -\frac{b}{2a} \pm \frac{\sqrt{b^2 - 4ac}}{2a},$$

making it clear that they are located symmetrically on either side of the line $x = -b/(2a)$. This partially justifies the claim that the whole parabola is symmetric with respect to this line.

**Figure 1**

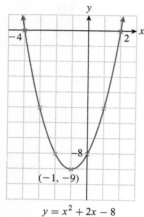

$$y = x^2 + 2x - 8$$

**Figure 2**

### EXAMPLE 1  Sketching the Graph of a Quadratic Function

Sketch the graph of $f(x) = x^2 + 2x - 8$ by hand.

**Solution**  Here, $a = 1$, $b = 2$, and $c = -8$. Because $a > 0$, the parabola is concave up (Figure 1).

*Vertex:* The $x$ coordinate of the vertex is

$$x = -\frac{b}{2a} = -\frac{2}{2} = -1.$$

To get its $y$ coordinate, we substitute the value of $x$ back into $f(x)$ to get

$$y = f(-1) = (-1)^2 + 2(-1) - 8 = 1 - 2 - 8 = -9.$$

Thus, the coordinates of the vertex are $(-1, -9)$.

*$x$-Intercepts:* To calculate the $x$-intercepts (if any), we solve the equation

$$x^2 + 2x - 8 = 0.$$

Luckily, this equation factors as $(x + 4)(x - 2) = 0$. Thus, the solutions are $x = -4$ and $x = 2$, so these values are the $x$-intercepts. (We could also have used the quadratic formula here.)

*$y$-Intercept:* The $y$-intercept is given by $c = -8$.

*Symmetry:* The graph is symmetric around the vertical line $x = -1$.

Now we can sketch the curve as in Figure 2. (As we see in the figure, it is helpful to plot additional points by using the equation $y = x^2 + 2x - 8$, and to use symmetry to obtain others.)

### EXAMPLE 2  One *x*-Intercept and No *x*-Intercepts

Sketch the graph of each quadratic function, showing the location of the vertex and intercepts.

**a.** $f(x) = 4x^2 - 12x + 9$

**b.** $g(x) = -\frac{1}{2}x^2 + 4x - 12$

**Solution**

**a.** We have $a = 4$, $b = -12$, and $c = 9$. Because $a > 0$, this parabola is concave up.

*Vertex:*   $x = -\dfrac{b}{2a} = \dfrac{12}{8} = \dfrac{3}{2}$      *x* coordinate of vertex

$$y = f\left(\frac{3}{2}\right) = 4\left(\frac{3}{2}\right)^2 - 12\left(\frac{3}{2}\right) + 9 = 0 \qquad y \text{ coordinate of vertex}$$

Thus, the vertex is at the point $(3/2, 0)$.

*x-Intercepts:*   $4x^2 - 12x + 9 = 0$

$$(2x - 3)^2 = 0$$

The only solution is $2x - 3 = 0$, or $x = 3/2$. Note that this coincides with the vertex, which lies on the $x$-axis.

*y-Intercept:*   $c = 9$

*Symmetry:* The graph is symmetric around the vertical line $x = 3/2$.

The graph is the narrow parabola shown in Figure 3. (As we remarked in Example 1, plotting additional points and using symmetry helps us obtain an accurate sketch.)

$$y = 4x^2 - 12x + 9$$

**Figure 3**

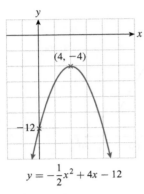

$$y = -\frac{1}{2}x^2 + 4x - 12$$

**Figure 4**

To automate the computations in Example 2 using a graphing calculator or a spreadsheet, see the Technology Guides at the end of the chapter. Outline:

**TI-83/84 Plus**
Y₁=AX^2+BX+C
4→A:-12→B:9→C
WINDOW Xmin=0, Xmax=3
ZOOM 0
[More details on page 680.]

**Spreadsheet**
Enter *x* values in column A.
Compute the corresponding *y* values in column B.
Graph the data in columns A and B.
[More details on page 682.]

**Website**
www.WanerMath.com
In the Function Evaluator and Grapher, enter 4x^2-12x+9 for y₁. For a table of values, enter the various *x*-values in the Evaluator box, and press "Evaluate".

**b.** Here, $a = -1/2$, $b = 4$, and $c = -12$. Because $a < 0$, the parabola is concave down. The vertex has $x$ coordinate $-b/(2a) = 4$, with corresponding $y$ coordinate $f(4) = -\frac{1}{2}(4)^2 + 4(4) - 12 = -4$. Thus, the vertex is at $(4, -4)$.

For the $x$-intercepts, we must solve $-\frac{1}{2}x^2 + 4x - 12 = 0$. If we try to use the quadratic formula, we discover that the discriminant is $b^2 - 4ac = 16 - 24 = -8$. Because the discriminant is negative, there are no solutions of the equation, so there are no $x$-intercepts.

The $y$-intercept is given by $c = -12$, and the graph is symmetric around the vertical line $x = 4$.

Because there are no $x$-intercepts, the graph lies entirely below the $x$-axis, as shown in Figure 4. (Again, you should plot additional points and use symmetry to ensure that your sketch is accurate.)

## APPLICATIONS

Recall that the **revenue** resulting from one or more business transactions is the total payment received. Thus, if $q$ units of some item are sold at $p$ dollars per unit, the revenue resulting from the sale is

$$\text{revenue} = \text{price} \times \text{quantity}$$
$$R = pq.$$

### EXAMPLE 3 Demand and Revenue

*Alien Publications, Inc.* predicts that the demand equation for the sale of its latest illustrated sci-fi novel *Episode 93: Yoda vs. Alien* is

$$q = -2{,}000p + 150{,}000$$

where $q$ is the number of books it can sell each year at a price of $\$p$ per book. What price should Alien Publications, Inc., charge to obtain the maximum annual revenue?

**Solution**  The total revenue depends on the price, as follows:

$$\begin{aligned} R &= pq && \text{Formula for revenue} \\ &= p(-2{,}000p + 150{,}000) && \text{Substitute for } q \text{ from demand equation.} \\ &= -2{,}000p^2 + 150{,}000p. && \text{Simplify.} \end{aligned}$$

We are after the price $p$ that gives the maximum possible revenue. Notice that what we have is a quadratic function of the form $R(p) = ap^2 + bp + c$, where $a = -2{,}000$, $b = 150{,}000$, and $c = 0$. Because $a$ is negative, the graph of the function is a parabola, concave down, so its vertex is its highest point (Figure 5). The $p$ coordinate of the vertex is

$$p = -\frac{b}{2a} = -\frac{150{,}000}{-4{,}000} = 37.5.$$

This value of $p$ gives the highest point on the graph and thus gives the largest value of $R(p)$. We may conclude that Alien Publications, Inc., should charge $\$37.50$ per book to maximize its annual revenue.

**Figure 5**

➡ **Before we go on...** You might ask what the maximum annual revenue is for the publisher in Example 3. Because $R(p)$ gives us the revenue at a price of $\$p$, the answer is $R(37.5) = -2,000(37.5)^2 + 150,000(37.5) = 2,812,500$. In other words, the company will earn total annual revenues from this book amounting to $\$2,812,500$. ∎

### EXAMPLE 4 Demand, Revenue, and Profit

As the operator of *YSport Fitness* gym, you calculate your demand equation to be

$$q = -0.06p + 84,$$

where $q$ is the number of members in the club and $p$ is the annual membership fee you charge.

**a.** Your annual operating costs are a fixed cost of $\$20,000$ per year plus a variable cost of $\$20$ per member. Find the annual revenue and profit as functions of the membership price $p$.

**b.** At what price should you set the annual membership fee to obtain the maximum revenue? What is the maximum possible revenue?

**c.** At what price should you set the annual membership fee to obtain the maximum profit? What is the maximum possible profit? What is the corresponding revenue?

### Solution

**a.** The annual revenue is given by

$$R = pq \qquad \text{Formula for revenue}$$
$$= p(-0.06p + 84) \qquad \text{Substitute for } q \text{ from demand equation.}$$
$$= -0.06p^2 + 84p. \qquad \text{Simplify.}$$

The annual cost $C$ is given by

$$C = 20,000 + 20q. \qquad \text{\$20,000 plus \$20 per member}$$

However, this is a function of $q$, and not $p$. To express $C$ as a function of $p$ we substitute for $q$ using the demand equation $q = -0.06p + 84$:

$$C = 20,000 + 20(-0.06p + 84)$$
$$= 20,000 - 1.2p + 1,680$$
$$= -1.2p + 21,680.$$

$R = -0.06p^2 + 84p$

**Figure 6**

$P = -0.06p^2 + 85.2p - 21,680$

**Figure 7**

Thus, the profit function is:

$$P = R - C \qquad \text{Formula for profit}$$
$$= -0.06p^2 + 84p - (-1.2p + 21,680) \qquad \text{Substitute for revenue and cost.}$$
$$= -0.06p^2 + 85.2p - 21,680.$$

**b.** From part (a) the revenue function is given by

$$R = -0.06p^2 + 84p.$$

This is a quadratic function ($a = -0.06, b = 84, c = 0$) whose graph is a concave-down parabola (Figure 6). The maximum revenue corresponds to the highest point of the graph: the vertex, of which the $p$ coordinate is

$$p = -\frac{b}{2a} = -\frac{84}{2(-0.06)} \approx \$700.$$

This is the membership fee you should charge for the maximum revenue. The corresponding maximum revenue is given by the $y$ coordinate of the vertex in Figure 6:

$$R(700) = -0.06(700)^2 + 84(700) = \$29,400.$$

**c.** From part (a), the profit function is given by

$$P = -0.06p^2 + 85.2p - 21,680.$$

Like the revenue function, the profit function is quadratic ($a = -0.06, b = 85.2, c = -21,680$). Figure 7 shows both the revenue and profit functions. The maximum profit corresponds to the vertex, whose $p$ coordinate is

$$p = -\frac{b}{2a} = -\frac{85.2}{2(-0.06)} \approx \$710.$$

This is the membership fee you should charge for the maximum profit. The corresponding maximum profit is given by the $y$ coordinate of the vertex of the profit curve in Figure 7:

$$P(710) = -0.06(710)^2 + 85.2(710) - 21,680 = \$8,566.$$

The corresponding revenue is

$$R(710) = -0.06(710)^2 + 84(710) = \$29,394,$$

slightly less than the maximum possible revenue of \$29,400.

➡ **Before we go on...** The result of part (c) of Example 4 tells us that the vertex of the profit curve in Figure 7 is slightly to the right of the vertex in the revenue curve. However, the difference is tiny compared to the scale of the graphs, so the graphs appear to be parallel. ■

Q: *Charging $710 membership brings in less revenue than charging $700. So why charge $710?*

A: A membership fee of $700 does bring in slightly larger revenue than a fee of $710, but it also brings in a slightly larger membership, which in turn raises the operating expense and has the effect of *lowering* the profit slightly (to $8,560). In other words, the slightly higher fee, while bringing in less revenue, also lowers the cost, and the net result is a larger profit.

## Fitting a Quadratic Function to Data: Quadratic Regression

In Section 1.4 we saw how to fit a regression line to a collection of data points. Here, we see how to use technology to obtain the **quadratic regression curve** associated with a set of points. The quadratic regression curve is the quadratic curve $y = ax^2 + bx + c$ that best fits the data points in the sense that the associated sum-of-squares error (SSE—see Section 1.4) is a minimum. Although there are algebraic methods for obtaining the quadratic regression curve, it is normal to use technology to do this.

### EXAMPLE 5  Carbon Dioxide Concentration

The following table shows the annual mean carbon dioxide concentration measured at Mauna Loa Observatory in Hawaii, in parts per million, every 10 years from 1960 through 2010 ($t = 0$ represents 1960).[1]

| Year $t$ | 0 | 10 | 20 | 30 | 40 | 50 |
|---|---|---|---|---|---|---|
| PPM $CO_2$ $C$ | 317 | 326 | 339 | 354 | 369 | 390 |

**a.** Is a linear model appropriate for these data?

**b.** Find the quadratic model

$$C(t) = at^2 + bt + c$$

that best fits the data.

### Solution

**a.** To see whether a linear model is appropriate, we plot the data points and the regression line using one of the methods of Example 2 in Section 1.4 (Figure 8).

**Figure 8**

---

[1]Figures are approximate. Source: U.S. Department of Commerce/National Oceanic and Atmospheric Administration (NOAA) Earth System Research Laboratory, data downloaded from www.esrl.noaa.gov/gmd/ccgg/trends/ on March 13, 2011.

From the graph, we can see that the given data suggest a curve and not a straight line: The observed points are above the regression line at the ends but below in the middle. (We would expect the data points from a linear relation to fall randomly above and below the regression line.)

**b.** The quadratic model that best fits the data is the quadratic regression model. As with linear regression, there are algebraic formulas to compute $a$, $b$, and $c$, but they are rather involved. However, we exploit the fact that these formulas are built into graphing calculators, spreadsheets, and other technology and obtain the regression curve using technology (see Figure 9):

$$C(t) = 0.012t^2 + 0.85t + 320.$$    Coefficients rounded to two significant digits

Notice from the graphs that the quadratic regression model appears to give a better fit than the linear regression model. This impression is supported by the values of SSE: For the linear regression model, SSE $\approx$ 58, while for the quadratic regression model, SSE is much smaller, approximately 2.6, indicating a much better fit.

**Figure 9**

## 9.1 EXERCISES

▼ more advanced   ◆ challenging
⊤ indicates exercises that should be solved using technology

*In Exercises 1–10, sketch the graphs of the quadratic functions, indicating the coordinates of the vertex, the y-intercept, and the x-intercepts (if any).* HINT [See Example 1.]

**1.** $f(x) = x^2 + 3x + 2$
**2.** $f(x) = -x^2 - x$
**3.** $f(x) = -x^2 + 4x - 4$
**4.** $f(x) = x^2 + 2x + 1$
**5.** $f(x) = -x^2 - 40x + 500$
**6.** $f(x) = x^2 - 10x - 600$
**7.** $f(x) = x^2 + x - 1$
**8.** $f(x) = x^2 + \sqrt{2}x + 1$
**9.** $f(x) = x^2 + 1$
**10.** $f(x) = -x^2 + 5$

*In Exercises 11–14, for each demand equation, express the total revenue R as a function of the price p per item, sketch the graph of the resulting function, and determine the price p that maximizes total revenue in each case.* HINT [See Example 3.]

**11.** $q = -4p + 100$
**12.** $q = -3p + 300$
**13.** $q = -2p + 400$
**14.** $q = -5p + 1,200$

⊤ *In Exercises 15–18, use technology to find the quadratic regression curve through the given points. (Round all coefficients to four decimal places.)* HINT [See Example 5.]

**15.** $\{(1, 2), (3, 5), (4, 3), (5, 1)\}$
**16.** $\{(-1, 2), (-3, 5), (-4, 3), (-5, 1)\}$
**17.** $\{(-1, 2), (-3, 5), (-4, 3)\}$
**18.** $\{(2, 5), (3, 5), (5, 3)\}$

### APPLICATIONS

**19.** *World Military Expenditure* The following chart shows total military and arms trade expenditure from 1990 to 2008 ($t = 0$ represents 1990).[2]

_____
[2]Approximate figures in constant 2005 dollars. The 2008 figure is an estimate, based on the increase in U.S. military expenditure.

World military expenditure

Source: www.globalissues.org/Geopolitics/ArmsTrade/Spending.asp

**a.** If you want to model the expenditure figures with a function of the form

$$f(t) = at^2 + bt + c,$$

would you expect the coefficient $a$ to be positive or negative? Why? HINT [See "Features of a Parabola," page 621.]

**b.** Which of the following models best approximates the data given? (Try to answer this without actually computing values.)

**(A)** $f(t) = 5t^2 - 80t - 1{,}150$
**(B)** $f(t) = -5t^2 - 80t + 1{,}150$
**(C)** $f(t) = 5t^2 - 80t + 1{,}150$
**(D)** $f(t) = -5t^2 - 80t - 1{,}150$

**c.** What is the nearest year that would correspond to the vertex of the graph of the correct model from part (b)? What is the danger of extrapolating the data in either direction?

**20. Education Expenditure** The following chart shows the percentage of the U.S. Discretionary Budget allocated to education from 2003 to 2009 ($t = 3$ represents the start of 2003).

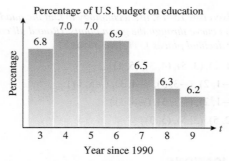

Percentage of U.S. budget on education

Source: www.globalissues.org/Geopolitics/ArmsTrade/Spending.asp

**a.** If you want to model the percentage figures with a function of the form

$$f(t) = at^2 + bt + c,$$

would you expect the coefficient $a$ to be positive or negative? Why? HINT [See "Features of a Parabola," page 621.]

**b.** Which of the following models best approximates the data given? (Try to answer this without actually computing values)

**(A)** $f(t) = 0.04t^2 + 0.3t - 6$
**(B)** $f(t) = -0.04t^2 + 0.3t + 6$
**(C)** $f(t) = 0.04t^2 + 0.3t + 6$
**(D)** $f(t) = -0.04t^2 + 0.3t - 6$

**c.** What is the nearest year that would correspond to the vertex of the graph of the correct model from part (b)? What is the danger of extrapolating the data in either direction?

**21. Oil Imports from Mexico** Daily oil imports to the United States from Mexico can be approximated by

$$I(t) = -0.015t^2 + 0.1t + 1.4$$
$$\text{million barrels/day } (0 \le t \le 8)$$

where $t$ is time in years since the start of 2000.[3] According to the model, in what year were oil imports to the United States greatest? How many barrels per day were imported that year? HINT [See Example 1.]

**22. Oil Production in Mexico** Daily oil production by Pemex, Mexico's national oil company, for 2001–2009 can be approximated by

$$P(t) = -0.022t^2 + 0.2t + 2.9$$
$$\text{million barrels/day } (1 \le t \le 9)$$

where $t$ is time in years since the start of 2000.[4] According to the model, in what year was oil production by Pemex greatest? How many barrels per day were produced that year?

**23. Net Income** The annual net income of General Electric for the period 2005–2010 could be approximated by

$$P(t) = -2.0t^2 + 6.6t + 16 \text{ billion dollars } (0 \le t \le 5)$$

where $t$ is time in years since 2005.[5] According to the model, in what year in this period was General Electric's net income highest? What was its net income that year? Would you trust this model to continue to be valid long past this period? Why?

**24. Net Income** The annual net income of General Electric for the period 2007–2012 could be approximated by

$$P(t) = 3.0t^2 - 24t + 59 \text{ billion dollars } (2 \le t \le 7)$$

where $t$ is time in years since 2005.[6] According to the model, in what year in this period was General Electric's net income lowest? What was its net income that year? Would you trust this model to continue to be valid long past this period? Why?

---

[3] Source: Energy Information Administration/Pemex/www.eia.gov

[4] Figures are approximate, and 2008–2009 figures are projections by the Department of Energy. Source: Energy Information Administration: Pemex.

[5] Source for data: www.wikinvest.com/

[6] *Ibid.* 2011 net income is projected.

**25.** *Revenue* The market research department of the *Better Baby Buggy Co.* predicts that the demand equation for its buggies is given by $q = -0.5p + 140$, where $q$ is the number of buggies it can sell in a month if the price is \$$p$ per buggy. At what price should it sell the buggies to get the largest revenue? What is the largest monthly revenue? HINT [See Example 3.]

**26.** *Revenue* The *Better Baby Buggy Co.* has just come out with a new model, the Turbo. The market research department predicts that the demand equation for Turbos is given by $q = -2p + 320$, where $q$ is the number of buggies it can sell in a month if the price is \$$p$ per buggy. At what price should it sell the buggies to get the largest revenue? What is the largest monthly revenue?

**27.** *Revenue* *Pack-Em-In Real Estate* is building a new housing development. The more houses it builds, the less people will be willing to pay, due to the crowding and smaller lot sizes. In fact, if it builds 40 houses in this particular development, it can sell them for \$200,000 each, but if it builds 60 houses, it will only be able to get \$160,000 each. Obtain a linear demand equation and hence determine how many houses Pack-Em-In should build to get the largest revenue. What is the largest possible revenue? HINT [See Example 3.]

**28.** *Revenue* *Pack-Em-In* has another development in the works. If it builds 50 houses in this development, it will be able to sell them at \$190,000 each, but if it builds 70 houses, it will get only \$170,000 each. Obtain a linear demand equation and hence determine how many houses it should build to get the largest revenue. What is the largest possible revenue?

**29.** ▼ *Revenue from Monorail Service, Las Vegas* In 2005, the Las Vegas monorail charged \$3 per ride and had an average ridership of about 28,000 per day. In December 2005 the Las Vegas Monorail Company raised the fare to \$5 per ride, and average ridership in 2006 plunged to around 19,000 per day.[7]

  **a.** Use the given information to find a linear demand equation.

  **b.** Find the price the company should have charged to maximize revenue from ridership. What is the corresponding daily revenue?

  **c.** The Las Vegas Monorail Company would have needed \$44.9 million in revenues from ridership to break even in 2006. Would it have been possible to break even in 2006 by charging a suitable price?

**30.** ▼ *Revenue from Monorail Service, Mars* The Utarek monorail, which links the three urbynes (or districts) of Utarek, Mars, charged $\overline{\overline{Z}}$5 per ride[8] and sold about 14 million rides per day. When the Utarek City Council lowered the fare to $\overline{\overline{Z}}$3 per ride, the number of rides increased to 18 million per day.

  **a.** Use the given information to find a linear demand equation.

**b.** Find the price the City Council should have charged to maximize revenue from ridership. What is the corresponding daily revenue?

  **c.** The City Council would have needed to raise $\overline{\overline{Z}}$48 billion in revenues from ridership each Martian year (670 days[9]) to finance the new Mars organism research lab. Would this have been possible by charging a suitable price?

**31.** *Web Site Profit* You operate a gaming Web site, www.mudbeast.net, where users must pay a small fee to log on. When you charged \$2 the demand was 280 log-ons per month. When you lowered the price to \$1.50, the demand increased to 560 log-ons per month.

  **a.** Construct a linear demand function for your Web site and hence obtain the monthly revenue $R$ as a function of the log-on fee $x$.

  **b.** Your Internet provider charges you a monthly fee of \$30 to maintain your site. Express your monthly profit $P$ as a function of the log-on fee $x$, and hence determine the log-on fee you should charge to obtain the largest possible monthly profit. What is the largest possible monthly profit? HINT [See Example 4.]

**32.** *T-Shirt Profit* Two fraternities, Sig Ep and Ep Sig, plan to raise money jointly to benefit homeless people on Long Island. They will sell *Yoda vs. Alien* T-shirts in the student center, but are not sure how much to charge. Sig Ep treasurer Augustus recalls that they once sold 400 shirts in a week at \$8 per shirt, but Ep Sig treasurer Julius has solid research indicating that it is possible to sell 600 per week at \$4 per shirt.

  **a.** Based on this information, construct a linear demand equation for Yoda vs. Alien T-shirts, and hence obtain the weekly revenue $R$ as a function of the unit price $x$.

  **b.** The university administration charges the fraternities a weekly fee of \$500 for use of the Student Center. Write down the monthly profit $P$ as a function of the unit price $x$, and hence determine how much the fraternities should charge to obtain the largest possible weekly profit. What is the largest possible weekly profit? HINT [See Example 4.]

**33.** *Web Site Profit* The latest demand equation for your gaming Web site, www.mudbeast.net, is given by

$$q = -400x + 1{,}200$$

where $q$ is the number of users who log on per month and $x$ is the log-on fee you charge. Your Internet provider bills you as follows:

| Site maintenance fee: | \$20 per month |
| High-volume access fee: | 50¢ per log-on |

Find the monthly cost as a function of the log-on fee $x$. Hence, find the monthly profit as a function of $x$ and determine the log-on fee you should charge to obtain the largest possible monthly profit. What is the largest possible monthly profit?

---

[7]Source: *New York Times*, Februrary 10, 2007, p. A9.

[8]The zonar ($\overline{\overline{Z}}$) is the official currency in the city-state of Utarek, Mars (formerly www.Marsnext.com, a now extinct virtual society).

[9]As measured in Mars days. The actual length of a Mars year is about 670.55 Mars days, so frequent leap years are designated by the Mars Planetary Authority to adjust.

**34. *T-Shirt Profit*** The latest demand equation for your *Yoda vs. Alien* T-shirts is given by

$$q = -40x + 600$$

where $q$ is the number of shirts you can sell in one week if you charge $\$x$ per shirt. The Student Council charges you $\$400$ per week for use of their facilities, and the T-shirts cost you $\$5$ each. Find the weekly cost as a function of the unit price $x$. Hence, find the weekly profit as a function of $x$ and determine the unit price you should charge to obtain the largest possible weekly profit. What is the largest possible weekly profit?

**35. ▼*Nightclub Management*** You have just opened a new nightclub, *Russ' Techno Pitstop*, but are unsure of how high to set the cover charge (entrance fee). One week you charged $\$10$ per guest and averaged 300 guests per night. The next week you charged $\$15$ per guest and averaged 250 guests per night.

  **a.** Find a linear demand equation showing the number of guests $q$ per night as a function of the cover charge $p$.

  **b.** Find the nightly revenue $R$ as a function of the cover charge $p$.

  **c.** The club will provide two free nonalcoholic drinks for each guest, costing the club $\$3$ per head. In addition, the nightly overheads (rent, salaries, dancers, DJ, etc.) amount to $\$3,000$. Find the cost $C$ as a function of the cover charge $p$.

  **d.** Now find the profit in terms of the cover charge $p$, and hence determine the entrance fee you should charge for a maximum profit.

**36. ▼*Television Advertising*** As sales manager for *Montevideo Productions, Inc.,* you are planning to review the prices you charge clients for television advertisement development. You currently charge each client an hourly development fee of $\$2,500$. With this pricing structure, the demand, measured by the number of contracts Montevideo signs per month, is 15 contracts. This is down 5 contracts from the figure last year, when your company charged only $\$2,000$.

  **a.** Construct a linear demand equation giving the number of contracts $q$ as a function of the hourly fee $p$ Montevideo charges for development.

  **b.** On average, Montevideo bills for 50 hours of production time on each contract. Give a formula for the total revenue obtained by charging $\$p$ per hour.

  **c.** The costs to Montevideo Productions are estimated as follows:

      Fixed costs:      $\$120,000$ per month
      Variable costs:  $\$80,000$ per contract

  Express Montevideo Productions' monthly cost **(i)** as a function of the number $q$ of contracts and **(ii)** as a function of the hourly production charge $p$.

  **d.** Express Montevideo Productions' monthly profit as a function of the hourly development fee $p$ and hence find the price it should charge to maximize the profit.

**37.** ◱ ***World Military Expenditure*** The following table shows total military and arms trade expenditure in 1994, 1998, and 2006. (See Exercise 19; $t = 4$ represents 1994.)[10]

| Year $t$ | 4 | 8 | 16 |
|---|---|---|---|
| Military Expenditure ($ billion) | 900 | 800 | 1,200 |

Find a quadratic model for these data, and use your model to estimate world military expenditure in 2008. Compare your answer with the actual figure shown in Exercise 19. HINT [See Example 5.]

**38.** ◱ ***Education Expenditure*** The following table shows the percentage of the U.S. Discretionary Budget allocated to education in 2003, 2005, and 2009. (See Exercise 20; $t = 3$ represents the start of 2003.)[11]

| Year $t$ | 3 | 5 | 9 |
|---|---|---|---|
| Percentage | 6.8 | 7 | 6.2 |

Find a quadratic model for these data, and use your model to estimate the percentage of the U.S. Discretionary Budget allocated to education in 2008. Compare your answer with the actual figure shown in Exercise 20.

**39.** ◱ ***iPod Sales*** The following table shows Apple iPod sales from the last quarter of 2009 through the last quarter of 2010.[12]

| Quarter $t$ | 4 | 5 | 6 | 7 | 8 |
|---|---|---|---|---|---|
| iPod Sales (millions) | 21.0 | 10.9 | 9.4 | 9.1 | 19.5 |

  **a.** Find a quadratic regression model for these data. (Round coefficients to two significant digits.) Graph the model together with the data.

  **b.** What does the model predict for iPod sales in the first quarter of 2011 ($t = 9$), to the nearest million? Comment on the answer.

**40.** ◱ ***iPod Sales*** The following table shows Apple iPod sales from the second quarter of 2009 through the second quarter of 2010.[13]

| Quarter $t$ | 2 | 3 | 4 | 5 | 6 |
|---|---|---|---|---|---|
| iPod Sales (millions) | 10.2 | 10.2 | 21.0 | 10.9 | 9.4 |

---

[10] Approximate figures in constant 2005 dollars. The 2008 figure is an estimate, based on the increase in U.S. military expenditure. Source: www.globalissues.org/Geopolitics/ArmsTrade/Spending.asp.

[11] Source: www.globalissues.org/Geopolitics/ArmsTrade/Spending.asp.

[12] Source: Apple quarterly press releases (www.apple.com/investor/).

[13] *Ibid.*

**a.** Find a quadratic regression model for these data. (Round coefficients to two significant digits.) Graph the model together with the data.

**b.** What does the model predict for iPod sales in the third quarter of 2010 ($t = 7$), to the nearest million? Comment on the answer.

## COMMUNICATION AND REASONING EXERCISES

**41.** What can you say about the graph of $f(x) = ax^2 + bx + c$ if $a = 0$?

**42.** What can you say about the graph of $f(x) = ax^2 + bx + c$ if $c = 0$?

**43. Multiple choice:** Following is the graph of $f(x) = ax^2 + bx + c$:

**(A)** $a$ is positive and $c$ is positive.
**(B)** $a$ is negative and $c$ is positive.
**(C)** $a$ is positive and $c$ is negative.
**(D)** $a$ is negative and $c$ is negative.

**44. Multiple choice:** Following is the graph of $f(x) = ax^2 + bx + c$:

**(A)** $a$ is positive and $c$ is positive.
**(B)** $a$ is negative and $c$ is positive.
**(C)** $a$ is positive and $c$ is negative.
**(D)** $a$ is negative and $c$ is negative.

**45.** ▼ Refer to the graph of $f(x) = ax^2 + bx + c$ in Exercise 43. Is $b$ positive or negative? Why?

**46.** ▼ Refer to the graph of $f(x) = ax^2 + bx + c$ in Exercise 44. Is $b$ positive or negative? Why?

**47.** Suppose the graph of revenue as a function of unit price is a parabola that is concave down. What is the significance of the coordinates of the vertex, the $x$-intercepts, and the $y$-intercept?

**48.** Suppose the height of a stone thrown vertically upward is given by a quadratic function of time. What is the significance of the coordinates of the vertex, the (possible) $x$-intercepts, and the $y$-intercept?

**49.** How might you tell, roughly, whether a set of data should be modeled by a quadratic rather than by a linear equation?

**50.** A member of your study group tells you that, because the following set of data does not suggest a straight line, the data are best modeled by a quadratic.

| $x$ | 0 | 2 | 4 | 6 | 8 |
|-----|---|---|---|---|---|
| $y$ | 1 | 2 | 1 | 0 | 1 |

Comment on her suggestion.

**51.** Is a quadratic model useful for long-term prediction of sales of an item? Why?

**52.** Of what use is a quadratic model, if not for long-term prediction?

**53.** ▼ Explain why, if demand is a linear function of unit price $p$ (with negative slope), then there must be a *single value of p* that results in the maximum revenue.

**54.** ▼ Explain why, if the average cost of a commodity is given by $y = 0.1x^2 - 4x - 2$, where $x$ is the number of units sold, there is a single choice of $x$ that results in the lowest possible average cost.

**55.** ▼ If the revenue function for a particular commodity is $R(p) = -50p^2 + 60p$, what is the (linear) demand function? Give a reason for your answer.

**56.** ▼ If the revenue function for a particular commodity is $R(p) = -50p^2 + 60p + 50$, can the demand function be linear? What is the associated demand function?

# 9.2 Exponential Functions and Models

The quadratic functions we discussed in Section 9.1 can be used to model many nonlinear situations. However, exponential functions give better models in some applications, including population growth, radioactive decay, the growth or depreciation of financial investments, and many other phenomena. (We already saw some of these applications in Section 1.2.)

To work effectively with exponential functions, we need to know the laws of exponents. The following list, taken from the algebra review in Chapter 0, gives the laws of exponents we will be using.

## The Laws of Exponents

If $b$ and $c$ are positive and $x$ and $y$ are any real numbers, then the following laws hold:

| Law | Quick Examples |
|---|---|
| 1. $b^x b^y = b^{x+y}$ | $2^3 2^2 = 2^5 = 32$ $\qquad$ $2^{3-x} = 2^3 2^{-x}$ |
| 2. $\dfrac{b^x}{b^y} = b^{x-y}$ | $\dfrac{4^3}{4^2} = 4^{3-2} = 4^1 = 4$ $\qquad$ $3^{x-2} = \dfrac{3^x}{3^2} = \dfrac{3^x}{9}$ |
| 3. $\dfrac{1}{b^x} = b^{-x}$ | $9^{-0.5} = \dfrac{1}{9^{0.5}} = \dfrac{1}{3}$ $\qquad$ $2^{-x} = \dfrac{1}{2^x}$ |
| 4. $b^0 = 1$ | $(3.3)^0 = 1$ $\qquad$ $x^0 = 1$ if $x \neq 0$ |
| 5. $(b^x)^y = b^{xy}$ | $(2^3)^2 = 2^6 = 64$ $\qquad$ $\left(\dfrac{1}{2}\right)^x = (2^{-1})^x = 2^{-x}$ |
| 6. $(bc)^x = b^x c^x$ | $(4 \cdot 2)^2 = 4^2 2^2 = 64$ $\qquad$ $10^x = 5^x 2^x$ |
| 7. $\left(\dfrac{b}{c}\right)^x = \dfrac{b^x}{c^x}$ | $\left(\dfrac{4}{3}\right)^2 = \dfrac{4^2}{3^2} = \dfrac{16}{9}$ $\qquad$ $\left(\dfrac{1}{2}\right)^x = \dfrac{1^x}{2^x} = \dfrac{1}{2^x}$ |

Here are the functions we will study in this section.

## Exponential Function

An **exponential function** has the form

$$f(x) = Ab^x, \qquad \text{Technology: A*b\textasciicircum x}$$

where $A$ and $b$ are constants with $A \neq 0$ and $b$ positive and not equal to 1. We call $b$ the **base** of the exponential function.

### Quick Examples

1. $f(x) = 2^x$ $\qquad$ $A = 1, b = 2$; Technology: 2\textasciicircum x

   $f(1) = 2^1 = 2$ $\qquad$ 2\textasciicircum 1

   $f(-3) = 2^{-3} = \dfrac{1}{8}$ $\qquad$ 2\textasciicircum (-3)

   $f(0) = 2^0 = 1$ $\qquad$ 2\textasciicircum 0

2. $g(x) = 20(3^x)$ $\qquad$ $A = 20, b = 3$; Technology: 20*3\textasciicircum x

   $g(2) = 20(3^2) = 20(9) = 180$ $\qquad$ 20*3\textasciicircum 2

   $g(-1) = 20(3^{-1}) = 20\left(\dfrac{1}{3}\right) = 6\dfrac{2}{3}$ $\qquad$ 20*3\textasciicircum (-1)

**3.** $h(x) = 2^{-x} = \left(\dfrac{1}{2}\right)^x$          $A = 1, b = \frac{1}{2};$ Technology: 2^(-x)
                                                          or (1/2)^x

$h(1) = 2^{-1} = \dfrac{1}{2}$          2^(-1) or (1/2)^1

$h(2) = 2^{-2} = \dfrac{1}{4}$          2^(-2) or (1/2)^2

**4.** $k(x) = 3 \cdot 2^{-4x} = 3(2^{-4})^x$          $A = 3, b = 2^{-4};$ Technology: 3*2^(-4*x)

$k(-2) = 3 \cdot 2^{-4(-2)}$          3*2^(-4*(-2))

$= 3 \cdot 2^8 = 3 \cdot 256 = 768$

## Exponential Functions from the Numerical and Graphical Points of View

The following table shows values of $f(x) = 3(2^x)$ for some values of $x$ ($A = 3$, $b = 2$):

| $x$ | $-3$ | $-2$ | $-1$ | 0 | 1 | 2 | 3 |
|---|---|---|---|---|---|---|---|
| $f(x)$ | $\frac{3}{8}$ | $\frac{3}{4}$ | $\frac{3}{2}$ | 3 | 6 | 12 | 24 |

The graph of $f$ is shown in Figure 10.

Notice that the $y$-intercept is $A = 3$ (obtained by setting $x = 0$). In general:

*In the graph of $f(x) = Ab^x$, $A$ is the $y$-intercept, or the value of $y$ when $x = 0$.*

What about $b$? Notice from the table that the value of $y$ is multiplied by $b = 2$ for every increase of 1 in $x$. If we decrease $x$ by 1, the $y$ coordinate gets *divided* by $b = 2$.

*The value of $y$ is multiplied by $b$ for every one-unit increase of $x$.*

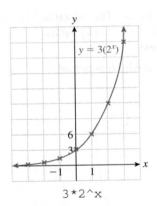

$3*2$^$x$

**Figure 10**

| $x$ | $-3$ | $-2$ | $-1$ | 0 | 1 | 2 | 3 |
|---|---|---|---|---|---|---|---|
| $f(x)$ | $\frac{3}{8}$ | $\frac{3}{4}$ | $\frac{3}{2}$ | 3 | 6 | 12 | 24 |

Multiply by 2.

On the graph, if we move one unit to the right from any point on the curve, the $y$ coordinate doubles. Thus, the curve becomes dramatically steeper as the value of $x$ increases. This phenomenon is called **exponential growth**. (See Section 1.2.)

## Exponential Function Numerically and Graphically

For the exponential function $f(x) = Ab^x$:

### Role of *A*
$f(0) = A$, so $A$ is the $y$-intercept of the graph of $f$.

### Role of *b*
If $x$ increases by 1, $f(x)$ is multiplied by $b$.
If $x$ increases by 2, $f(x)$ is multiplied by $b^2$.
$\vdots$
If $x$ increases by $\Delta x$, $f(x)$ is multiplied by $b^{\Delta x}$.

If $x$ increases by 1, $y$ is multiplied by $b$.

## Quick Examples

**1.** $f_1(x) = 2^x$, $f_2(x) = \left(\dfrac{1}{2}\right)^x = 2^{-x}$

|   | A | B | C |
|---|---|---|---|
| 1 | x | 2^x | 2^(-x) |
| 2 | -3 | 1/8 | 8 |
| 3 | -2 | 1/4 | 4 |
| 4 | -1 | 1/2 | 2 |
| 5 | 0 | 1 | 1 |
| 6 | 1 | 2 | 1/2 |
| 7 | 2 | 4 | 1/4 |
| 8 | 3 | 8 | 1/8 |

Technology: 2^x;  2^(-x)

When $x$ increases by 1, $f_2(x)$ is multiplied by $\frac{1}{2}$. The function $f_1(x) = 2^x$ illustrates exponential growth, while $f_2(x) = \left(\frac{1}{2}\right)^x$ illustrates the opposite phenomenon: **exponential decay**.

**2.** $f_1(x) = 2^x$, $f_2(x) = 3^x$, $f_3(x) = 1^x$ (Can you see why $f_3$ is not an exponential function?)

|   | A | B | C | D |
|---|---|---|---|---|
| 1 | x | 2^x | 3^x | 1^x |
| 2 | -3 | 1/8 | 1/27 | 1 |
| 3 | -2 | 1/4 | 1/9 | 1 |
| 4 | -1 | 1/2 | 1/3 | 1 |
| 5 | 0 | 1 | 1 | 1 |
| 6 | 1 | 2 | 3 | 1 |
| 7 | 2 | 4 | 9 | 1 |
| 8 | 3 | 8 | 27 | 1 |

If $x$ increases by 1, $3^x$ is multiplied by 3. Note also that all the graphs pass through $(0, 1)$. (Why?)

## EXAMPLE 1   Recognizing Exponential Data Numerically and Graphically

Some of the values of two functions, $f$ and $g$, are given in the following table:

| $x$ | $-2$ | $-1$ | 0 | 1 | 2 |
|---|---|---|---|---|---|
| $f(x)$ | $-7$ | $-3$ | 1 | 5 | 9 |
| $g(x)$ | $\frac{2}{9}$ | $\frac{2}{3}$ | 2 | 6 | 18 |

One of these functions is linear, and the other is exponential. Which is which?

x g(x)    o f(x)

**Figure 11**

**Solution** Remember that a linear function increases (or decreases) by the same amount every time $x$ increases by 1. The values of $f$ behave this way: Every time $x$ increases by 1, the value of $f(x)$ increases by 4. Therefore, $f$ is a linear function with a *slope* of 4. Because $f(0) = 1$, we see that

$$f(x) = 4x + 1$$

is a linear formula that fits the data.

On the other hand, every time $x$ increases by 1, the value of $g(x)$ is *multiplied* by 3. Because $g(0) = 2$, we find that

$$g(x) = 2(3^x)$$

is an exponential function fitting the data.

We can visualize the two functions $f$ and $g$ by plotting the data points (Figure 11). The data points for $f(x)$ clearly lie along a straight line, whereas the points for $g(x)$ lie along a curve. The $y$ coordinate of each point for $g(x)$ is 3 times the $y$ coordinate of the preceding point, demonstrating that the curve is an exponential one.

In Section 1.3, we discussed a method for calculating the equation of the line that passes through two given points. In the following example, we show a method for calculating the equation of the exponential curve through two given points.

### EXAMPLE 2 Finding the Exponential Curve through Two Points

Find an equation of the exponential curve through $(1, 6.3)$ and $(4, 170.1)$.

**Solution** We want an equation of the form

$$y = Ab^x \quad (b > 0).$$

Substituting the coordinates of the given points, we get

$$6.3 = Ab^1 \qquad \text{Substitute } (1, 6.3).$$
$$170.1 = Ab^4. \qquad \text{Substitute } (4, 170.1).$$

If we now divide the second equation by the first, we get

$$\frac{170.1}{6.3} = \frac{Ab^4}{Ab} = b^3$$
$$b^3 = 27$$
$$b = 27^{1/3} \qquad \text{Take reciprocal power of both sides.}$$
$$b = 3.$$

Now that we have $b$, we can substitute its value into the first equation to obtain

$$6.3 = 3A \qquad \text{Substitute } b = 3 \text{ into the equation } 6.3 = Ab^1.$$
$$A = \frac{6.3}{3} = 2.1.$$

We have both constants, $A = 2.1$ and $b = 3$, so the model is

$$y = 2.1(3^x).$$

Example 6 will show how to use technology to fit an exponential function to two or more data points.

## APPLICATIONS

Recall some terminology we mentioned earlier: A quantity $y$ experiences **exponential growth** if $y = Ab^t$ with $b > 1$. (Here we use $t$ for the independent variable, thinking of time.) It experiences **exponential decay** if $y = Ab^t$ with $0 < b < 1$. We already saw several examples of exponential growth and decay in Section 1.2.

---

### EXAMPLE 3  Exponential Growth and Decay

**a. Compound Interest** (See Section 1.2 Example 6.) If $2,000 is invested in a mutual fund with an annual yield of 12.6% and the earnings are reinvested each month, then the future value after $t$ years is

$$A(t) = P\left(1 + \frac{r}{m}\right)^{mt} = 2{,}000\left(1 + \frac{0.126}{12}\right)^{12t} = 2{,}000(1.0105)^{12t},$$

which can be written as $2{,}000(1.0105^{12})^t$, so $A = 2{,}000$ and $b = 1.0105^{12}$. This is an example of exponential growth, because $b > 1$.

**b. Carbon Decay** (See Section 1.2 Example 7.) The amount of carbon 14 remaining in a sample that originally contained $A$ grams is approximately

$$C(t) = A(0.999879)^t.$$

This is an instance of exponential decay, because $b < 1$.

---

→ **Before we go on...**   Refer again to part (a). In Example 6(b) of Section 1.2 we showed how to use technology to answer questions such as the following: "When, to the nearest year, will the value of your investment reach $5,000?"  ∎

The next example shows an application to public health.

---

### EXAMPLE 4  Exponential Growth: Epidemics

In the early stages of the AIDS epidemic during the 1980s, the number of cases in the United States was increasing by about 50% every 6 months. By the start of 1983, there were approximately 1,600 AIDS cases in the United States.[14]

**a.** Assuming an exponential growth model, find a function that predicts the number of people infected $t$ years after the start of 1983.

**b.** Use the model to estimate the number of people infected by October 1, 1986, and also by the end of that year.

### Solution

**a.** One way of finding the desired exponential function is to reason as follows: At time $t = 0$ (January 1, 1983), the number of people infected was 1,600, so $A = 1{,}600$. Every 6 months, the number of cases increased to 150% of the number 6 months earlier—that is, to 1.50 times that number. Each year, it therefore increased to $(1.50)^2 = 2.25$ times the number one year earlier. Hence, after $t$ years, we need to multiply the original 1,600 by $2.25^t$, so the model is

$$y = 1{,}600(2.25^t)\ \text{cases}.$$

---

[14]Data based on regression of the 1982–1986 figures. Source for data: Centers for Disease Control and Prevention. HIV/AIDS Surveillance Report, 2000;12 (No. 2).

Alternatively, if we wish to use the method of Example 2, we need two data points. We are given one point: (0, 1,600). Because $y$ increased by 50% every 6 months, 6 months later it reached $1{,}600 + 800 = 2{,}400$ ($t = 0.5$). This information gives a second point: (0.5, 2,400). We can now apply the method in Example 2 to find the model above.

**b.** October 1, 1986, corresponds to $t = 3.75$ (because October 1 is 9 months, or $9/12 = 0.75$ of a year after January 1). Substituting this value of $t$ in the model gives

$$y = 1{,}600(2.25^{3.75}) \approx 33{,}481 \text{ cases} \qquad \text{1600*2.25^3.75}$$

By the end of 1986, the model predicts that

$$y = 1{,}600(2.25^4) = 41{,}006 \text{ cases.}$$

(The actual number of cases was around 41,700.)

---

➡ **Before we go on...** Increasing the number of cases by 50% every 6 months couldn't continue for very long and this is borne out by observations. If increasing by 50% every 6 months did continue, then by January 2003 ($t = 20$), the number of infected people would have been

$$1{,}600(2.25^{20}) \approx 17{,}700{,}000{,}000$$

a number that is more than 50 times the size of the U.S. population! Thus, although the exponential model is fairly reliable in the early stages of the epidemic, it is unreliable for predicting long-term trends.  ■

Epidemiologists use more sophisticated models to measure the spread of epidemics, and these models predict a leveling-off phenomenon as the number of cases becomes a significant part of the total population. We discuss such a model, the **logistic function**, in Section 9.4.

## The Number e and More Applications

In nature we find examples of growth that occurs *continuously*, as though "interest" is being added more often than every second or fraction of a second. To model this, we need to see what happens to the compound interest formula of Section 1.2 as we let $m$ (the number of times interest is added per year) become extremely large. Something very interesting does happen: We end up with a more compact and elegant formula than we began with. To see why, let's look at a very simple situation.

Suppose we invest \$1 in the bank for 1 year at 100% interest, compounded $m$ times per year. If $m = 1$, then 100% interest is added every year, and so our money doubles at the end of the year. In general, the accumulated capital at the end of the year is

$$A = 1\left(1 + \frac{1}{m}\right)^m = \left(1 + \frac{1}{m}\right)^m. \qquad \text{(1+1/m)^m}$$

Now, we are interested in what $A$ becomes for large values of $m$. On the left is a spreadsheet showing the quantity $\left(1 + \frac{1}{m}\right)^m$ for larger and larger values of $m$.

Something interesting *does* seem to be happening! The numbers appear to be getting closer and closer to a specific value. In mathematical terminology, we say that the numbers **converge** to a fixed number, $2.71828\ldots$, called the **limiting value**[*] of the quantities $\left(1 + \frac{1}{m}\right)^m$. This number, called $e$, is one of the most important in mathematics. The number $e$ is irrational, just as the more familiar number $\pi$ is, so we cannot write down its exact numerical value. To 20 decimal places,

$$e = 2.71828182845904523536\ldots.$$

| | A | B |
|---|---|---|
| | m | (1+1/m)^m |
| 1 | | |
| 2 | 1 | 2 |
| 3 | 10 | 2.59374246 |
| 4 | 100 | 2.704813829 |
| 5 | 1000 | 2.716923932 |
| 6 | 10000 | 2.718145927 |
| 7 | 100000 | 2.718268237 |
| 8 | 1000000 | 2.718280469 |
| 9 | 10000000 | 2.718281694 |
| 10 | 100000000 | 2.718281786 |
| 11 | 1000000000 | 2.718282031 |

[*] See Chapter 10 for more on limits.

We now say that, if $1 is invested for 1 year at 100% interest **compounded continuously**, the accumulated money at the end of that year will amount to $$e = \$2.72$$ (to the nearest cent). But what about the following more general question?

**Q :** *What about a more general scenario: If we invest an amount $P for t years at an interest rate of r, compounded continuously, what will be the accumulated amount A at the end of that period?*

**A :** In the special case above (*P*, *t*, and *r* all equal 1), we took the compound interest formula and let *m* get larger and larger. We do the same more generally, after a little preliminary work with the algebra of exponentials.

$$A = P\left(1 + \frac{r}{m}\right)^{mt}$$

$$= P\left(1 + \frac{1}{(m/r)}\right)^{mt} \qquad \text{Substituting } \tfrac{r}{m} = \tfrac{1}{(m/r)}$$

$$= P\left(1 + \frac{1}{(m/r)}\right)^{(m/r)rt} \qquad \text{Substituting } m = \left(\tfrac{m}{r}\right)r$$

$$= P\left[\left(1 + \frac{1}{(m/r)}\right)^{(m/r)}\right]^{rt} \qquad \text{Using the rule } a^{bc} = (a^b)^c$$

For continuous compounding of interest, we let *m*, and hence *m/r*, get very large. This affects only the term in brackets, which converges to *e*, and we get the formula

$$A = Pe^{rt}.$$

**Q :** *How do I obtain powers of e or e itself on a TI-83/84 Plus or in a spreadsheet?*

**A :** On the TI-83/84 Plus, enter $e^x$ as `e^(x)`, where `e^(` can be obtained by pressing [2ND] [LN] . To obtain the number *e* on the TI-83/84 Plus, enter `e^(1)`. Spreadsheets have a built-in function called `EXP`; `EXP(x)` gives the value of $e^x$. To obtain the number *e* in a spreadsheet, enter `= EXP(1)`.

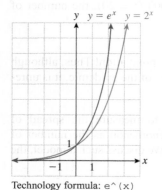

$y$  $y = e^x$  $y = 2^x$

Technology formula: `e^(x)`
or `EXP(x)`

**Figure 12**

Figure 12 shows the graph of $y = e^x$ with that of $y = 2^x$ for comparison.

## The Number e and Continuous Compounding

The number *e* is the limiting value of the quantities $\left(1 + \frac{1}{m}\right)^m$ as *m* gets larger and larger, and has the value 2.71828182845904523536...

If $P is invested at an annual interest rate *r* compounded continuously, the accumulated amount after *t* years is

$$A(t) = Pe^{rt}. \qquad \texttt{P*e\^{}(r*t) or P*EXP(r*t)}$$

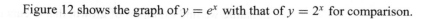

### Quick Examples

1. If $100 is invested in an account that bears 15% interest compounded continuously, at the end of 10 years the investment will be worth

$$A(10) = 100e^{(0.15)(10)} = \$448.17. \qquad \texttt{100*e\^{}(0.15*10) or}$$
$$\texttt{100*EXP(0.15*10)}$$

2. If $1 is invested in an account that bears 100% interest compounded continuously, at the end of *x* years the investment will be worth

$$A(x) = e^x \text{dollars.}$$

### EXAMPLE 5    Continuous Compounding

**a.** You invest $10,000 at *Fastrack Savings & Loan,* which pays 6% compounded continuously. Express the balance in your account as a function of the number of years $t$ and calculate the amount of money you will have after 5 years.

**b.** Your friend has just invested $20,000 in *Constant Growth Funds,* whose stocks are continuously *declining* at a rate of 6% per year. How much will her investment be worth in 5 years?

**c.** 🖩 During which year will the value of your investment first exceed that of your friend?

### Solution

**a.** We use the continuous growth formula with $P = 10,000$, $r = 0.06$, and $t$ variable, getting

$$A(t) = Pe^{rt} = 10,000e^{0.06t}.$$

In 5 years,

$$A(5) = 10,000e^{0.06(5)}$$
$$= 10,000e^{0.3}$$
$$\approx \$13,498.59.$$

**b.** Because the investment is depreciating, we use a negative value for $r$ and take $P = 20,000$, $r = -0.06$, and $t = 5$, getting

$$A(t) = Pe^{rt} = 20,000e^{-0.06t}$$
$$A(5) = 20,000e^{-0.06(5)}$$
$$= 20,000e^{-0.3}$$
$$\approx \$14,816.36.$$

**c.** We can answer the question now using a graphing calculator, a spreadsheet, or the Function Evaluator and Grapher tool at the Website. Just enter the exponential models of parts (a) and (b) and create tables to compute the values at the end of several years:

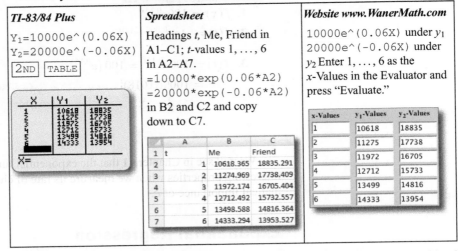

| TI-83/84 Plus | Spreadsheet | Website www.WanerMath.com |
|---|---|---|

From the table, we see that the value of your investment overtakes that of your friend after $t = 5$ (the end of year 5) and before $t = 6$ (the end of year 6). Thus your investment first exceeds that of your friend sometime during year 6.

➡ **Before we go on...**

**Q :** *How does continuous compounding compare with monthly compounding?*

**A :** To repeat the calculation in part (a) of Example 5 using monthly compounding instead of continuous compounding, we use the compound interest formula with $P = 10{,}000$, $r = 0.06$, $m = 12$, and $t = 5$ and find

$$A(5) = 10{,}000(1 + 0.06/12)^{60} \approx \$13{,}488.50.$$

Thus, continuous compounding earns you approximately \$10 more than monthly compounding on a 5-year, \$10,000 investment. This is little to get excited about.

If we write the continuous compounding formula $A(t) = Pe^{rt}$ as $A(t) = P(e^r)^t$, we see that $A(t)$ is an exponential function of $t$, where the base is $b = e^r$, so we have really not introduced a new kind of function. In fact, exponential functions are often written in this way:

### Exponential Functions: Alternative Form

We can write any exponential function in the following alternative form:

$$f(x) = Ae^{rx}$$

where $A$ and $r$ are constants. If $r$ is positive, $f$ models exponential growth; if $r$ is negative, $f$ models exponential decay.

#### Quick Examples

1. $f(x) = 100e^{0.15x}$      Exponential growth $A = 100, r = 0.15$
2. $f(t) = Ae^{-0.000\,121\,01t}$      Exponential decay of carbon 14; $r = -0.000\,121\,01$
3. $f(t) = 100e^{0.15t} = 100\left(e^{0.15}\right)^t$
   $= 100(1.1618)^t$      Converting $Ae^{rt}$ to the form $Ab^t$

We will see in Chapter 11 that the exponential function with base $e$ exhibits some interesting properties when we measure its rate of change, and this is the real mathematical importance of $e$.

## Exponential Regression

Starting with a set of data that suggests an exponential curve, we can use technology to compute the exponential regression curve in much the same way as we did for the quadratic regression curve in Example 5 of Section 9.1.

**Figure 13**

**using Technology**

See the Technology Guides at the end of the chapter for detailed instructions on how to obtain the regression curve and graph for Example 6 using a graphing calculator or spreadsheet. Outline:

**TI-83/84 Plus**

STAT EDIT values of $t$ in $L_1$ and values of C in $L_2$
Regression curve: STAT
CALC option #0 ExpReg ENTER
Graph: Y= VARS 5 EQ 1,
then ZOOM 9
[More details on page 681.]

**Spreadsheet**

$t$- and C-values in Columns A and B, graph these data. Regression curve: Add exponential Trendline with option to show equation.
[More details on page 684.]

 **Website**

www.WanerMath.com
In the Simple Regression utility, enter the data in the x and y columns and press "y=a(b^x)".

## EXAMPLE 6 ☐ Exponential Regression: Health Expenditures

The following table shows annual expenditure on health in the U.S. from 1980 through 2009 ($t = 0$ represents 1980).[15]

| Year $t$ | 0 | 5 | 10 | 15 | 20 | 25 | 29 |
|---|---|---|---|---|---|---|---|
| Expenditure ($ billion) | 256 | 444 | 724 | 1,030 | 1,380 | 2,020 | 2,490 |

**a.** Find the exponential regression model

$$C(t) = Ab^t$$

for the annual expenditure.

**b.** Use the regression model to estimate the expenditure in 2002 ($t = 22$; the actual expenditure was approximately $1,640 billion).

## Solution

**a.** We use technology to obtain the exponential regression curve (See Figure 13):

$$C(t) \approx 296(1.08)^t. \qquad \text{Coefficients rounded}$$

**b.** Using the model $C(t) \approx 296(1.08)^t$ we find that

$$C(22) \approx 296(1.08)^{22} \approx \$1,609 \text{ billion,}$$

which is close to the actual number of about $1,640 billion.

➡ **Before we go on...** We said in the preceding section that the regression curve gives the smallest value of the sum-of-squares error, SSE (the sum of the squares of the residuals). However, exponential regression as computed via technology generally minimizes the sum of the squares of the residuals of the *logarithms* (logarithms are discussed in the next section). Using logarithms allows one easily to convert an exponential function into a linear one and then use linear regression formulas. However, in Section 9.4, we will discuss a way of using Excel's Solver to minimize SSE directly, which allows us to find the best-fit exponential curve directly without the need for devices to simplify the mathematics. If we do this, we obtain a very different equation:

$$C(t) \approx 353(1.07)^t.$$

If you plot this function, you will notice that it seems to fit the data more closely than the regression curve. ■

**FAQs**

### When to Use an Exponential Model for Data Points, and when to Use e in Your Model

**Q** : *Given a set of data points that appear to be curving upward, how can I tell whether to use a quadratic model or an exponential model?*

[15]Data are rounded. Source: U.S. Department of Health & Human Services/Centers for Medicare & Medicaid Services, National Health Expenditure Data, downloaded April 2011 from www.cms.gov.

**A** : Here are some things to look for:

- Do the data values appear to double at regular intervals? (For example, do the values approximately double every 5 units?) If so, then an exponential model is appropriate. If it takes longer and longer to double, then a quadratic model may be more appropriate.
- Do the values first decrease to a low point and then increase? If so, then a quadratic model is more appropriate.

It is also helpful to use technology to graph both the regression quadratic and exponential curves and to visually inspect the graphs to determine which gives the closest fit to the data.

**Q** : *We have two ways of writing exponential functions: $f(x) = Ab^x$ and $f(x) = Ae^{rx}$. How do we know which one to use?*

**A** : The two forms are equivalent, and it is always possible to convert from one form to the other.* So, use whichever form seems to be convenient for a particular situation. For instance, $f(t) = A(3^t)$ conveniently models exponential growth that is tripling every unit of time, whereas $f(t) = Ae^{0.06t}$ conveniently models an investment with continuous compounding at 6%.

\* Quick Example 3 on page 640 shows how to convert $Ae^{rx}$ to $Ab^x$. Conversion from $Ab^x$ to $Ae^{rx}$ involves logarithms: $r = \ln b$.

## 9.2  EXERCISES

▼ more advanced   ◆ challenging

**T** indicates exercises that should be solved using technology

*For each function in Exercises 1–12, compute the missing values in the following table and supply a valid technology formula for the given function:* HINT [See Quick Examples on page 632.]

| x | −3 | −2 | −1 | 0 | 1 | 2 | 3 |
|---|----|----|----|---|---|---|---|
| f(x) | | | | | | | |

1. $f(x) = 4^x$
2. $f(x) = 3^x$
3. $f(x) = 3^{-x}$
4. $f(x) = 4^{-x}$
5. $g(x) = 2(2^x)$
6. $g(x) = 2(3^x)$
7. $h(x) = -3(2^{-x})$
8. $h(x) = -2(3^{-x})$
9. $r(x) = 2^x - 1$
10. $r(x) = 2^{-x} + 1$
11. $s(x) = 2^{x-1}$
12. $s(x) = 2^{1-x}$

*Using a chart of values, graph each of the functions in Exercises 13–18. (Use $-3 \leq x \leq 3$.)*

13. $f(x) = 3^{-x}$
14. $f(x) = 4^{-x}$
15. $g(x) = 2(2^x)$
16. $g(x) = 2(3^x)$
17. $h(x) = -3(2^{-x})$
18. $h(x) = -2(3^{-x})$

*In Exercises 19–24, the values of two functions, f and g, are given in a table. One, both, or neither of them may be exponential. Decide which, if any, are exponential, and give the exponential models for those that are.* HINT [See Example 1.]

19.
| x | −2 | −1 | 0 | 1 | 2 |
|---|----|----|---|---|---|
| f(x) | 0.5 | 1.5 | 4.5 | 13.5 | 40.5 |
| g(x) | 8 | 4 | 2 | 1 | $\frac{1}{2}$ |

20.
| x | −2 | −1 | 0 | 1 | 2 |
|---|----|----|---|---|---|
| f(x) | $\frac{1}{2}$ | 1 | 2 | 4 | 8 |
| g(x) | 3 | 0 | −1 | 0 | 3 |

21.
| x | −2 | −1 | 0 | 1 | 2 |
|---|----|----|---|---|---|
| f(x) | 22.5 | 7.5 | 2.5 | 7.5 | 22.5 |
| g(x) | 0.3 | 0.9 | 2.7 | 8.1 | 16.2 |

22.
| x | −2 | −1 | 0 | 1 | 2 |
|---|----|----|---|---|---|
| f(x) | 0.3 | 0.9 | 2.7 | 8.1 | 24.3 |
| g(x) | 3 | 1.5 | 0.75 | 0.375 | 0.1875 |

23.
| x | −2 | −1 | 0 | 1 | 2 |
|---|----|----|---|---|---|
| f(x) | 100 | 200 | 400 | 600 | 800 |
| g(x) | 100 | 20 | 4 | 0.8 | 0.16 |

24.
| x | −2 | −1 | 0 | 1 | 2 |
|---|----|----|---|---|---|
| f(x) | 0.8 | 0.2 | 0.1 | 0.05 | 0.025 |
| g(x) | 80 | 40 | 20 | 10 | 2 |

⊤ *For each function in Exercises 25–30, supply a valid technology formula and then use technology to compute the missing values in the following table:* HINT [See Quick Examples on page 632.]

| $x$ | $-3$ | $-2$ | $-1$ | 0 | 1 | 2 | 3 |
|---|---|---|---|---|---|---|---|
| $f(x)$ | | | | | | | |

**25.** $f(x) = e^{-2x}$      **26.** $g(x) = e^{x/5}$

**27.** $h(x) = 1.01(2.02^{-4x})$      **28.** $h(x) = 3.42(3^{-x/5})$

**29.** $r(x) = 50\left(1 + \dfrac{1}{3.2}\right)^{2x}$

**30.** $r(x) = 0.043\left(4.5 - \dfrac{5}{1.2}\right)^{-x}$

*In Exercises 31–38, supply a valid technology formula for the given function.*

**31.** $2^{x-1}$    **32.** $2^{-4x}$    **33.** $\dfrac{2}{1 - 2^{-4x}}$    **34.** $\dfrac{2^{3-x}}{1 - 2^x}$

**35.** $\dfrac{(3+x)^{3x}}{x+1}$    **36.** $\dfrac{20.3^{3x}}{1 + 20.3^{2x}}$    **37.** $2e^{(1+x)/x}$    **38.** $\dfrac{2e^{2/x}}{x}$

⊤ *On the same set of axes, use technology to graph the pairs of functions in Exercises 39–46 with $-3 \leq x \leq 3$. Identify which graph corresponds to which function.* HINT [See Quick Examples on page 633.]

**39.** $f_1(x) = 1.6^x$, $f_2(x) = 1.8^x$

**40.** $f_1(x) = 2.2^x$, $f_2(x) = 2.5^x$

**41.** $f_1(x) = 300(1.1^x)$, $f_2(x) = 300(1.1^{2x})$

**42.** $f_1(x) = 100(1.01^{2x})$, $f_2(x) = 100(1.01^{3x})$

**43.** $f_1(x) = 2.5^{1.02x}$, $f_2(x) = e^{1.02x}$

**44.** $f_1(x) = 2.5^{-1.02x}$, $f_2(x) = e^{-1.02x}$

**45.** $f_1(x) = 1,000(1.045^{-3x})$, $f_2(x) = 1,000(1.045^{3x})$

**46.** $f_1(x) = 1,202(1.034^{-3x})$, $f_2(x) = 1,202(1.034^{3x})$

*For Exercises 47–54, model the data using an exponential function $f(x) = Ab^x$.* HINT [See Example 1.]

**47.**

| $x$ | 0 | 1 | 2 |
|---|---|---|---|
| $f(x)$ | 500 | 250 | 125 |

**48.**

| $x$ | 0 | 1 | 2 |
|---|---|---|---|
| $f(x)$ | 500 | 1,000 | 2,000 |

**49.**

| $x$ | 0 | 1 | 2 |
|---|---|---|---|
| $f(x)$ | 10 | 30 | 90 |

**50.**

| $x$ | 0 | 1 | 2 |
|---|---|---|---|
| $f(x)$ | 90 | 30 | 10 |

**51.**

| $x$ | 0 | 1 | 2 |
|---|---|---|---|
| $f(x)$ | 500 | 225 | 101.25 |

**52.**

| $x$ | 0 | 1 | 2 |
|---|---|---|---|
| $f(x)$ | 5 | 3 | 1.8 |

**53.**

| $x$ | 1 | 2 |
|---|---|---|
| $f(x)$ | $-110$ | $-121$ |

**54.**

| $x$ | 1 | 2 |
|---|---|---|
| $f(x)$ | $-41$ | $-42.025$ |

*Find equations for exponential functions that pass through the pairs of points given in Exercises 55–62. (Round all coefficients to 4 decimal places when necessary.)* HINT [See Example 2.]

**55.** Through $(2, 36)$ and $(4, 324)$

**56.** Through $(2, -4)$ and $(4, -16)$

**57.** Through $(-2, -25)$ and $(1, -0.2)$

**58.** Through $(1, 1.2)$ and $(3, 0.108)$

**59.** Through $(1, 3)$ and $(3, 6)$    **60.** Through $(1, 2)$ and $(4, 6)$

**61.** Through $(2, 3)$ and $(6, 2)$    **62.** Through $(-1, 2)$ and $(3, 1)$

*Obtain exponential functions in the form $f(t) = Ae^{rt}$ in Exercises 63–66.* HINT [See Example 5.]

**63.** $f(t)$ is the value after $t$ years of a \$5,000 investment earning 10% interest compounded continuously.

**64.** $f(t)$ is the value after $t$ years of a \$2,000 investment earning 5.3% interest compounded continuously.

**65.** $f(t)$ is the value after $t$ years of a \$1,000 investment depreciating continuously at an annual rate of 6.3%.

**66.** $f(t)$ is the value after $t$ years of a \$10,000 investment depreciating continuously at an annual rate of 60%.

⊤ *In Exercises 67–70, use technology to find the exponential regression function through the given points. (Round all coefficients to 4 decimal places.)* HINT [See Example 6.]

**67.** $\{(1, 2), (3, 5), (4, 9), (5, 20)\}$

**68.** $\{(-1, 2), (-3, 5), (-4, 9), (-5, 20)\}$

**69.** $\{(-1, 10), (-3, 5), (-4, 3)\}$

**70.** $\{(3, 3), (4, 5), (5, 10)\}$

## APPLICATIONS

**71.** *Aspirin* Soon after taking an aspirin, a patient has absorbed 300 mg of the drug. After 2 hours, only 75 mg remain. Find an exponential model for the amount of aspirin in the bloodstream after $t$ hours, and use your model to find the amount of aspirin in the bloodstream after 5 hours. HINT [See Example 2.]

**72.** *Alcohol* After a large number of drinks, a person has a blood alcohol level of 200 mg/dL (milligrams per deciliter). If the amount of alcohol in the blood decays exponentially, and after 2 hours, 112.5 mg/dL remain, find an exponential model for the person's blood alcohol level, and use your model to estimate the person's blood alcohol level after 4 hours. HINT [See Example 2.]

**73.** *Freon Production* The production of ozone-layer-damaging Freon 22 (chlorodifluoromethane) in developing countries rose from 200 tons in 2004 to a projected 590 tons in 2010.[16]

---

[16] Figures are approximate. Source: Lampert Kuijpers (Panel of the Montreal Protocol), National Bureau of Statistics in China, via CEIC Data/*New York Times*, February 23, 2007, p. C1.

**a.** Use this information to find both a linear model and an exponential model for the amount $F$ of Freon 22 (in tons) as a function of time $t$ in years since 2000. (Round all coefficients to three significant digits.) HINT [See Example 2.] Which of these models would you judge to be more appropriate to the data shown below?

| t (year since 2000) | 0 | 2 | 4 | 6 | 8 | 10 |
|---|---|---|---|---|---|---|
| F (tons of Freon 22) | 100 | 140 | 200 | 270 | 400 | 590 |

**b.** Use the better of the two models from part (a) to predict the 2008 figure and compare it with the projected figure above.

**74. Revenue** The annual revenue of Amazon.com rose from approximately $10.7 billion in 2006 to $34.2 billion in 2010.[17]

**a.** Use this information to find both a linear model and an exponential model for Amazon.com's annual revenue $I$ (in billions of dollars) as a function of time $t$ in years since 2000. (Round all coefficients to three significant digits.) HINT [See Example 2.] Which of these models would you judge to be more appropriate to the data shown below?

| t (Year since 2000) | 6 | 7 | 8 | 9 | 10 |
|---|---|---|---|---|---|
| I ($ billions) | 10.7 | 14.8 | 19.2 | 24.5 | 34.2 |

**b.** Use the better of the two models from part (a) to predict the 2008 figure and compare it with the actual figure above.

**75. ▼U.S. Population** The U.S. population was 180 million in 1960 and 309 million in 2010.[18]

**a.** Use these data to give an exponential growth model showing the U.S. population $P$ as a function of time $t$ in years since 1960. Round coefficients to 6 significant digits. HINT [See Example 2.]

**b.** By experimenting, determine the smallest number of significant digits to which you should round the coefficients in part (a) in order to obtain the correct 2010 population figure accurate to 3 significant digits.

**c.** Using the model in part (a), predict the population in 2020.

**76. ▼World Population** World population was estimated at 2.56 billion people in 1950 and 6.91 billion people in 2011.[19]

**a.** Use these data to give an exponential growth model showing the world population $P$ as a function of time $t$ in years since 1950. Round coefficients to 6 significant digits. HINT [See Example 2.]

**b.** By experimenting, determine the smallest number of significant digits to which you should round the coefficients in part (a) in order to obtain the correct 2011 population figure to 3 significant digits.

**c.** Assuming the exponential growth model from part (a), estimate the world population in the year 1000. Comment on your answer.

**77. ▼Frogs** Frogs have been breeding like flies at the Enormous State University (ESU) campus! Each year, the pledge class of the Epsilon Delta fraternity is instructed to tag all the frogs residing on the ESU campus. Two years ago they managed to tag all 50,000 of them (with little Epsilon Delta Fraternity tags). This year's pledge class discovered that last year's tags had all fallen off, and they wound up tagging a total of 75,000 frogs.

**a.** Find an exponential model for the frog population.

**b.** Assuming exponential population growth, and that all this year's tags have fallen off, how many tags should Epsilon Delta order for next year's pledge class?

**78. ▼Flies** Flies in Suffolk County have been breeding like frogs! Three years ago the Health Commission caught 4,000 flies in a trap in 1 hour. This year it caught 7,000 flies in 1 hour.

**a.** Find an exponential model for the fly population.

**b.** Assuming exponential population growth, how many flies should the commission expect to catch next year?

**79. Bacteria** A bacteria culture starts with 1,000 bacteria and doubles in size every 3 hours. Find an exponential model for the size of the culture as a function of time $t$ in hours and use the model to predict how many bacteria there will be after 2 days. HINT [See Example 4.]

**80. Bacteria** A bacteria culture starts with 1,000 bacteria. Two hours later there are 1,500 bacteria. Find an exponential model for the size of the culture as a function of time $t$ in hours, and use the model to predict how many bacteria there will be after 2 days. HINT [See Example 4.]

**81. SARS** In the early stages of the deadly SARS (Severe Acute Respiratory Syndrome) epidemic in 2003, the number of cases was increasing by about 18% each day.[20] On March 17, 2003 (the first day for which statistics were reported by the World Health Organization), there were 167 cases. Find an exponential model that predicts the number of cases $t$ days after March 17, 2003, and use it to estimate the number of cases on March 31, 2003. (The actual reported number of cases was 1,662.)

**82. SARS** A few weeks into the deadly SARS (Severe Acute Respiratory Syndrome) epidemic in 2003, the number of cases was increasing by about 4% each day.[21] On April 1,

[17]Source for data: www.wikinvest.com/

[18]Figures are rounded to 3 significant digits. Source: U.S. Census Bureau, www.census.gov.

[19]Ibid.

[20]Source: World Health Organization, www.who.int.

[21]Ibid.

2003 there were 1,804 cases. Find an exponential model that predicts the number of cases $t$ days after April 1, 2003, and use it to estimate the number of cases on April 30, 2003. (The actual reported number of cases was 5,663.)

**83. Investments** In November 2010, **E*TRADE Financial** was offering only 0.3% interest on its online savings accounts, with interest reinvested monthly.[22] Find the associated exponential model for the value of a $5,000 deposit after $t$ years. Assuming this rate of return continued for seven years, how much would a deposit of $5,000 in November 2010 be worth in November 2017? (Answer to the nearest $1.) HINT [See Example 3; you saw this exercise before in Section 1.2.]

**84. Investments** In November 2010, **ING Direct** was offering 2.4% interest on its Orange Savings Account, with interest reinvested quarterly.[23] Find the associated exponential model for the value of a $4,000 deposit after $t$ years. Assuming this rate of return continued for 8 years, how much would a deposit of $4,000 in November 2010 be worth in November 2018? (Answer to the nearest $1.) HINT [See Example 3; you saw this exercise before in Section 1.2.]

**85.** ▣ **Investments** Refer to Exercise 83. In November of which year will an investment of $5,000 made in November of 2010 first exceed $5,200? HINT [See Example 5; you saw this exercise before in Section 1.2.]

**86.** ▣ **Investments** Refer to Exercise 84. In November of which year will an investment of $4,000 made in November of 2010 first exceed $5,200? HINT [See Example 5; you saw this exercise before in Section 1.2.]

**87. Investments** *Rock Solid Bank & Trust* is offering a CD (certificate of deposit) that pays 4% compounded continuously. How much interest would a $1,000 deposit earn over 10 years? HINT [See Example 5.]

**88. Savings** *FlybynightSavings.com* is offering a savings account that pays 31% interest compounded continuously. How much interest would a deposit of $2,000 earn over 10 years?

**89. Home Sales** Sales of existing homes in the U.S. rose continuously over the period 2008–2011 at the rate of 3.2% per year from 4.9 million in 2008.[24] Write down a formula that predicts sales of existing homes $t$ years after 2008. Use your model to estimate, to the nearest 0.1 million, sales of existing homes in 2010 and 2012.

**90. Home Prices** The median selling price of an existing home in the U.S. declined continuously over the period 2008–2011 at the rate of 7.6% per year from approximately $198 thousand in 2008.[25] Write down a formula that predicts the median selling price of an existing home $t$ years after 2008. Use your model to estimate, to the nearest $1,000, the median selling price of an existing home in 2011 and 2013.

**91. Global Warming** The most abundant greenhouse gas is carbon dioxide. According to figures from the Intergovernmental Panel on Climate Change (IPCC), the amount of carbon dioxide in the atmosphere (in parts of volume per million) can be approximated by

$$C(t) \approx 280e^{0.00119t} \text{ parts per million}$$

where $t$ is time in years since 1750.[26]
**a.** Use the model to estimate the amount of carbon dioxide in the atmosphere in 1950, 2000, 2050, and 2100.
**b.** According to the model, when, to the nearest decade, will the level surpass 390 parts per million?

**92. Global Warming** Another greenhouse gas is methane. According to figures from the Intergovernmental Panel on Climate Change (IPCC), the amount of methane in the atmosphere (in parts of volume per billion) can be approximated by

$$C(t) \approx 715e^{0.00356t} \text{ parts per billion}$$

where $t$ is time in years since 1750.[27]
**a.** Use the model to estimate the amount of methane in the atmosphere in 1950, 2000, 2050, and 2100. (Round your answers to the nearest 10 parts per billion.)
**b.** According to the model, when, to the nearest decade, will the level surpass 2,000 parts per billion?

**93.** ▣ **New York City Housing Costs: Downtown** The following table shows the average price of a two-bedroom apartment in downtown New York City during the real estate boom from 1994 to 2004.[28]

| $t$ | 0 (1994) | 2 | 4 | 6 | 8 | 10 (2004) |
|---|---|---|---|---|---|---|
| Price ($ million) | 0.38 | 0.40 | 0.60 | 0.95 | 1.20 | 1.60 |

**a.** Use exponential regression to model the price $P(t)$ as a function of time $t$ since 1994. Include a sketch of the points and the regression curve. (Round the coefficients to 3 decimal places.) HINT [See Example 6.]
**b.** Extrapolate your model to estimate the cost of a two-bedroom downtown apartment in 2005.

**94.** ▣ **New York City Housing Costs: Uptown** The following table shows the average price of a two-bedroom apartment in uptown New York City during the real estate boom from 1994 to 2004.[29]

| $t$ | 0 (1994) | 2 | 4 | 6 | 8 | 10 (2004) |
|---|---|---|---|---|---|---|
| Price ($ million) | 0.18 | 0.18 | 0.19 | 0.2 | 0.35 | 0.4 |

[22] Interest rate based on annual percentage yield. Source: us.etrade.com, November 2010.

[23] Interest rate based on annual percentage yield. Source: home.ingdirect.com, November 2010.

[24] Source: National Association of Realtors, www.realtor.org.

[25] *Ibid.*

[26] Authors' exponential model based on the 1750 and 2005 figures. Source for data: IPCC Fourth Assessment Report: Climate Change 2007, www.ipcc.ch.

[27] *Ibid.*

[28] Data are rounded and 2004 figure is an estimate. Source: Miller Samuel/*New York Times*, March 28, 2004, p. RE 11.

[29] *Ibid.*

a. Use exponential regression to model the price $P(t)$ as a function of time $t$ since 1994. Include a sketch of the points and the regression curve. (Round the coefficients to 3 decimal places.)

b. Extrapolate your model to estimate the cost of a two-bedroom uptown apartment in 2005.

**95. ▯ *Facebook*** The following table gives the approximate numbers of Facebook members at various times early in its history.[30]

| Year $t$ (Since start of 2005) | 0 | 0.5 | 1 | 1.5 | 2 | 2.5 | 3 | 3.5 |
|---|---|---|---|---|---|---|---|---|
| Facebook Members $n$ (millions) | 1 | 2 | 5.5 | 7 | 12 | 30 | 58 | 80 |

a. Use exponential regression to model Facebook membership as a function of time in years since the start of 2005, and graph the data points and regression curve. (Round coefficients to 3 decimal places.)

b. Fill in the missing quantity: According to the model, Facebook membership each year was _____ times that of the year before.

c. Use your model to estimate Facebook membership in early 2009 to the nearest million.

**96. ▯ *Freon Production*** The following table shows Freon 22 production in developing countries in various years since 2000.[31]

| $t$ (Year since 2000) | 0 | 2 | 4 | 6 | 8 | 10 |
|---|---|---|---|---|---|---|
| $F$ (Tons of Freon 22) | 100 | 140 | 200 | 270 | 400 | 590 |

a. Use exponential regression to model Freon production as a function of time in years since 2000, and graph the data points and regression curve. (Round coefficients to 3 decimal places.)

b. Fill in the missing quantity: According to the model, Freon production each year was _____ times that of the year before.

c. Use your model to estimate freon production in 2009 to the nearest ton.

## COMMUNICATION AND REASONING EXERCISES

**97.** Which of the following three functions will be largest for large values of $x$?

**(A)** $f(x) = x^2$    **(B)** $r(x) = 2^x$    **(C)** $h(x) = x^{10}$

**98.** Which of the following three functions will be smallest for large values of $x$?

**(A)** $f(x) = x^{-2}$    **(B)** $r(x) = 2^{-x}$    **(C)** $h(x) = x^{-10}$

**99.** What limitations apply to using an exponential function to model growth in real-life situations? Illustrate your answer with an example.

**100.** Explain in words why 5% per year compounded continuously yields more interest than 5% per year compounded monthly.

**101.** ▽ The following commentary and graph appeared in politicalcalculations.blogspot.com on August 30, 2005:[32]

One of the neater blogs I've recently encountered is The Real Returns, which offers a wealth of investing, market and economic data. Earlier this month, The Real Returns posted data related to the recent history of U.S. median house prices over the period from 1963 to 2004. The original source of the housing data is the U.S. Census Bureau.

Well, that kind of data deserves some curve-fitting and a calculator to estimate what the future U.S. median house price might be, so Political Calculations has extracted the data from 1973 onward to create the following chart:

U.S. median house prices since 1973

$y = 41152e^{0.0551x}$

Comment on the article and graph. HINT [See Exercise 90.]

**102.** ▽ Refer to Exercise 101. Of what possible predictive use, then, is the kind of exponential model given by the blogger in the article referred to?

**103.** ▽ Describe two real-life situations in which a linear model would be more appropriate than an exponential model, and two situations in which an exponential model would be more appropriate than a linear model.

**104.** ▽ Describe a real-life situation in which a quadratic model would be more appropriate than an exponential model and one in which an exponential model would be more appropriate than a quadratic model.

**105.** How would you check whether data points of the form $(1, y_1), (2, y_2), (3, y_3)$ lie on an exponential curve?

[30] Sources: www.facebook.com, www.insidehighered.com. (Some data are interpolated.)

[31] Figures are approximate. Source: Lampert Kuijpers (Panel of the Montreal Protocol), National Bureau of Statistics in China, via CEIC Data/*New York Times*, February 23, 2007, p. C1.

[32] The graph was re-created by the authors using the blog author's data source.
Source for article: politicalcalculations.blogspot.com/2005/08/projecting-us-median-housing-prices.html.
Source for data: therealreturns.blogspot.com/2005_08_01_archive.html.

**106.** ▼ You are told that the points $(1, y_1)$, $(2, y_2)$, $(3, y_3)$ lie on an exponential curve. Express $y_3$ in terms of $y_1$ and $y_2$.

**107.** ▼ Your local banker tells you that the reason his bank doesn't compound interest continuously is that it would be too demanding of computer resources because the computer would need to spend a great deal of time keeping all accounts updated. Comment on his reasoning.

**108.** ▼ Your other local banker tells you that the reason *her* bank doesn't offer continuously compounded interest is that it is equivalent to offering a fractionally higher interest rate compounded daily. Comment on her reasoning.

## 9.3 Logarithmic Functions and Models

Logarithms were invented by John Napier (1550–1617) in the late sixteenth century as a means of aiding calculation. His invention made possible the prodigious hand calculations of astronomer Johannes Kepler (1571–1630), who was the first to describe accurately the orbits and the motions of the planets. Today, computers and calculators have done away with that use of logarithms, but many other uses remain. In particular, the logarithm is used to model real-world phenomena in numerous fields, including physics, finance, and economics.

From the equation

$$2^3 = 8$$

we can see that the power to which we need to raise 2 in order to get 8 is 3. We abbreviate the phrase "the power to which we need to raise 2 in order to get 8" as "$\log_2 8$." Thus, another way of writing the equation $2^3 = 8$ is

$$\log_2 8 = 3.\qquad \text{\small The power to which we need to raise 2 in order to get 8 is 3.}$$

This is read "the base 2 logarithm of 8 is 3" or "the log, base 2, of 8 is 3."

Here is the general definition.

---

### Base *b* Logarithm

The **base *b* logarithm of *x***, $\log_b x$, is the power to which we need to raise *b* in order to get *x*. Symbolically,

$$\log_b x = y \qquad\qquad \text{means} \qquad\qquad b^y = x.$$

*Logarithmic form*                                      *Exponential form*

#### Quick Examples

1. The following table lists some exponential equations and their equivalent logarithmic forms:

| Exponential Form | $10^3 = 1000$ | $4^2 = 16$ | $3^3 = 27$ | $5^1 = 5$ | $7^0 = 1$ | $4^{-2} = \dfrac{1}{16}$ | $25^{1/2} = 5$ |
|---|---|---|---|---|---|---|---|
| Logarithmic Form | $\log_{10} 1000 = 3$ | $\log_4 16 = 2$ | $\log_3 27 = 3$ | $\log_5 5 = 1$ | $\log_7 1 = 0$ | $\log_4 \dfrac{1}{16} = -2$ | $\log_{25} 5 = \dfrac{1}{2}$ |

2. $\log_3 9 =$ the power to which we need to raise 3 in order to get 9. Because $3^2 = 9$, this power is 2, so $\log_3 9 = 2$.

3. $\log_{10} 10{,}000 =$ the power to which we need to raise 10 in order to get 10,000. Because $10^4 = 10{,}000$, this power is 4, so $\log_{10} 10{,}000 = 4$.

4. $\log_3 \frac{1}{27}$ is the power to which we need to raise 3 in order to get $\frac{1}{27}$. Because $3^{-3} = \frac{1}{27}$ this power is $-3$, so $\log_3 \frac{1}{27} = -3$.

5. $\log_b 1 = 0$ for every positive number $b$ other than 1 because $b^0 = 1$.

**Note** The number $\log_b x$ is defined only if $b$ and $x$ are both positive and $b \neq 1$. Thus, it is impossible to compute, say, $\log_3(-9)$ (because there is no power of 3 that equals $-9$), or $\log_1(2)$ (because there is no power of 1 that equals 2). ∎

Logarithms with base 10 and base $e$ are frequently used, so they have special names and notations.

## Common Logarithm, Natural Logarithm

The following are standard abbreviations.

                                                                    **TI-83/84 Plus & Spreadsheet Formula**

**Base 10:** $\log_{10} x = \log x$     *Common Logarithm*     `log(x)`

**Base $e$:** $\log_e x = \ln x$     *Natural Logarithm*     `ln(x)`

### Quick Examples

| **Logarithmic Form** | **Exponential Form** |
|---|---|
| 1. $\log 10{,}000 = 4$ | $10^4 = 10{,}000$ |
| 2. $\log 10 = 1$ | $10^1 = 10$ |
| 3. $\log \dfrac{1}{10{,}000} = -4$ | $10^{-4} = \dfrac{1}{10{,}000}$ |
| 4. $\ln e = 1$ | $e^1 = e$ |
| 5. $\ln 1 = 0$ | $e^0 = 1$ |
| 6. $\ln 2 = 0.69314718\ldots$ | $e^{0.69314718\ldots} = 2$ |

Some technologies (such as calculators) do not permit direct calculation of logarithms other than common and natural logarithms. To compute logarithms with other bases with these technologies, we can use the following formula:

## Change-of-Base Formula

$$\log_b a = \frac{\log a}{\log b} = \frac{\ln a}{\ln b} \qquad \text{Change-of-base formula}^*$$

*Here is a quick explanation of why this formula works: To calculate $\log_b a$, we ask, "to what power must we raise $b$ to get $a$?" To check the formula, we try using $\log a/\log b$ as the exponent.

$$b^{\frac{\log a}{\log b}} = (10^{\log b})^{\frac{\log a}{\log b}}$$
$$\text{(because } b = 10^{\log b})$$
$$= 10^{\log a} = a$$

so this exponent works!

### Quick Examples

1. $\log_{11} 9 = \dfrac{\log 9}{\log 11} \approx 0.91631$     `log(9)/log(11)`

2. $\log_{11} 9 = \dfrac{\ln 9}{\ln 11} \approx 0.91631$     `ln(9)/ln(11)`

3. $\log_{3.2}\left(\dfrac{1.42}{3.4}\right) \approx -0.75065$     `log(1.42/3.4)/log(3.2)`

**Using Technology to Compute Logarithms**

To compute $\log_b x$ using technology, use the following formulas:

| | | |
|---|---|---|
| TI-83/84 Plus | `log(x)/log(b)` | Example: $\log_2(16)$ is `log(16)/log(2)` |
| Spreadsheet: | `=LOG(x,b)` | Example: $\log_2(16)$ is `= LOG(16,2)` |

One important use of logarithms is to solve equations in which the unknown is in the exponent.

---

**EXAMPLE 1  Solving Equations with Unknowns in the Exponent**

Solve the following equations.

**a.** $5^{-x} = 125$          **b.** $3^{2x-1} = 6$          **c.** $100(1.005)^{3x} = 200$

**Solution**

**a.** Write the given equation $5^{-x} = 125$ in logarithmic form:

$$-x = \log_5 125$$

This gives $x = -\log_5 125 = -3$.

**b.** In logarithmic form, $3^{2x-1} = 6$ becomes

$$2x - 1 = \log_3 6$$

$$2x = 1 + \log_3 6$$

giving     $x = \dfrac{1 + \log_3 6}{2} \approx \dfrac{1 + 1.6309}{2} \approx 1.3155.$

**c.** We cannot write the given equation, $100(1.005)^{3x} = 200$, directly in logarithmic form. We must first divide both sides by 100:

$$1.005^{3x} = \frac{200}{100} = 2$$

$$3x = \log_{1.005} 2$$

$$x = \frac{\log_{1.005} 2}{3} \approx \frac{138.9757}{3} \approx 46.3252.$$

---

Now that we know what logarithms are, we can talk about functions based on logarithms:

**Logarithmic Function**

A **logarithmic function** has the form

$$f(x) = \log_b x + C \qquad \text{(}b \text{ and } C \text{ are constants with } b > 0, b \neq 1\text{)}$$

or, alternatively,

$$f(x) = A \ln x + C. \qquad \text{(}A, C \text{ constants with } A \neq 0\text{)}$$

**Quick Examples**

1. $f(x) = \log x$
2. $g(x) = \ln x - 5$
3. $h(x) = \log_2 x + 1$
4. $k(x) = 3.2 \ln x + 7.2$

**Q:** *What is the difference between the two forms of the logarithmic function?*

**A:** None, really—they're equivalent: We can start with an equation in the first form and use the change-of-base formula to rewrite it:

$$f(x) = \log_b x + C$$
$$= \frac{\ln x}{\ln b} + C \qquad \text{Change-of-base formula}$$
$$= \left(\frac{1}{\ln b}\right) \ln x + C.$$

Our function now has the form $f(x) = A \ln x + C$, where $A = 1/\ln b$. We can go the other way as well, to rewrite $A \ln x + C$ in the form $\log_b x + C$.

### EXAMPLE 2 Graphs of Logarithmic Functions

**a.** Sketch the graph of $f(x) = \log_2 x$ by hand.

**b.** Use technology to compare the graph in part (a) with the graphs of $\log_b x$ for $b = 1/4,\ 1/2,$ and $4$.

### Solution

**Figure 14**

**a.** To sketch the graph of $f(x) = \log_2 x$ by hand, we begin with a table of values. Because $\log_2 x$ is not defined when $x = 0$, we choose several values of $x$ close to zero and also some larger values, all chosen so that their logarithms are easy to compute:

| $x$ | $\frac{1}{8}$ | $\frac{1}{4}$ | $\frac{1}{2}$ | 1 | 2 | 4 | 8 |
|---|---|---|---|---|---|---|---|
| $f(x) = \log_2 x$ | $-3$ | $-2$ | $-1$ | 0 | 1 | 2 | 3 |

Graphing these points and joining them by a smooth curve gives us Figure 14.

**b.** We enter the logarithmic functions in graphing utilities as follows (note the use of the change-of-base formula in the TI-83/84 Plus version):

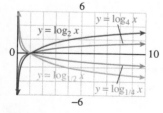

**Figure 15**

**TI-83/84 Plus**

```
Y₁=log(X)/log(0.25)
Y₂=log(X)/log(0.5)
Y₃=log(X)/log(2)
Y₄=log(X)/log(4)
```

**Spreadsheet**

```
=LOG(x,0.25)
=LOG(x,0.5)
=LOG(x,2)
=LOG(x,4)
```

Figure 15 shows the resulting graphs.

➡ **Before we go on...** Notice that the graphs of the logarithmic functions in Example 2 all pass through the point $(1, 0)$. (Why?) Notice further that the graphs of the logarithmic functions with bases less than 1 are upside-down versions of the others. Finally, how are these graphs related to the graphs of exponential functions? ■

Below are some important algebraic properties of logarithms we shall use throughout the rest of this section.

**W** **Website**
www.WanerMath.com
Follow the path
  Chapter 9
  → Using and Deriving
    Algebraic Properties of
    Logarithms
to find a list of logarithmic identities and a discussion on where they come from.
Follow the path
  Chapter 9
  → Inverse Functions
for a general discussion of inverse functions, including further discussion of the relationship between logarithmic and exponential functions.

## Logarithm Identities

The following identities hold for all positive bases $a \neq 1$ and $b \neq 1$, all positive numbers $x$ and $y$, and every real number $r$. These identities follow from the laws of exponents.

**Identity**

1. $\log_b(xy) = \log_b x + \log_b y$
2. $\log_b\left(\dfrac{x}{y}\right) = \log_b x - \log_b y$
3. $\log_b(x^r) = r \log_b x$
4. $\log_b b = 1$; $\log_b 1 = 0$
5. $\log_b\left(\dfrac{1}{x}\right) = -\log_b x$
6. $\log_b x = \dfrac{\log_a x}{\log_a b}$

**Quick Examples**

$\log_2 16 = \log_2 8 + \log_2 2$

$\log_2\left(\dfrac{5}{3}\right) = \log_2 5 - \log_2 3$

$\log_2(6^5) = 5\log_2 6$

$\log_2 2 = 1$; $\ln e = 1$; $\log_{11} 1 = 0$

$\log_2\left(\dfrac{1}{3}\right) = -\log_2 3$

$\log_2 5 = \dfrac{\log_{10} 5}{\log_{10} 2} = \dfrac{\log 5}{\log 2}$

### Relationship with Exponential Functions

The following two identities demonstrate that the operations of taking the base $b$ logarithm and raising $b$ to a power are *inverse* to each other.

**Identity**

1. $\log_b(b^x) = x$

  *In words:* The power to which you raise $b$ in order to get $b^x$ is $x$. (!)

2. $b^{\log_b x} = x$

  *In words:* Raising $b$ to the power to which it must be raised to get $x$ yields $x$. (!)

**Quick Examples**

$\log_2(2^7) = 7$

$5^{\log_5 8} = 8$

# APPLICATIONS

EXAMPLE 3 **Investments: How Long?**

Global bonds sold by Mexico are yielding an average of 2.51% per year.[33] At that interest rate, how long will it take a $1,000 investment to be worth $1,200 if the interest is compounded monthly?

---

[33] In 2011 (Bonds maturing 03/03/2015). Source: www.bloomberg.com.

**Solution**   Substituting $A = 1,200$, $P = 1,000$, $r = 0.0251$, and $m = 12$ in the compound interest equation gives

$$A(t) = P\left(1 + \frac{r}{m}\right)^{mt}$$

$$1,200 = 1,000\left(1 + \frac{0.0251}{12}\right)^{12t}$$

$$\approx 1,000(1.002092)^{12t}$$

and we must solve for $t$. We first divide both sides by 1,000, getting an equation in exponential form:

$$1.2 = 1.002092^{12t}.$$

In logarithmic form, this becomes

$$12t = \log_{1.002092}(1.2).$$

We can now solve for $t$:

$$t = \frac{\log_{1.002092}(1.2)}{12} \approx 7.3 \text{ years.}$$    `log(1.2)/(log(1.002092)*12)`

Thus, it will take approximately 7.3 years for a \$1,000 investment to be worth \$1,200.

➡ **Before we go on...**   We can use the logarithm identities to solve the equation

$$1.2 = 1.002092^{12t}$$

that arose in Example 3 (and also more general equations with unknowns in the exponent) by taking the natural logarithm of both sides:

$$\ln 1.2 = \ln(1.002092^{12t})$$
$$= 12t \ln 1.002092.$$    By Identity 3

We can now solve this for $t$ to get

$$t = \frac{\ln 1.2}{12 \ln 1.002092},$$

which, by the change-of-base formula, is equivalent to the answer we got in Example 3.

■

### EXAMPLE 4   Half-Life

**a.** The weight of carbon 14 that remains in a sample that originally contained $A$ grams is given by

$$C(t) = A(0.999879)^t$$

where $t$ is time in years. Find the **half-life**, the time it takes half of the carbon 14 in a sample to decay.

**b.** Repeat part (a) using the following alternative form of the exponential model in part (a):

$$C(t) = Ae^{-0.000\,121\,01t} \quad \text{See Quick Examples, page 640.}$$

**c.** Another radioactive material has a half-life of 7,000 years. Find an exponential decay model in the form

$$R(t) = Ae^{-kt}$$

for the amount of undecayed material remaining. (The constant $k$ is called the **decay constant**.)

**d.** How long will it take for 99.95% of the substance in a sample of the material in part (c) to decay?

### Solution

**a.** We want to find the value of $t$ for which $C(t) =$ the weight of undecayed carbon 14 left = half the original weight = $0.5A$. Substituting, we get

$$0.5A = A(0.999879)^t.$$

Dividing both sides by $A$ gives

$$0.5 = 0.999879^t \qquad\qquad \text{Exponential form}$$

$$t = \log_{0.999879} 0.5 \approx 5,728 \text{ years.} \qquad\qquad \text{Logarithmic form}$$

**b.** This is similar to part (a): We want to solve the equation

$$0.5A = Ae^{-0.000\,121\,01t}$$

for $t$. Dividing both sides by $A$ gives

$$0.5 = e^{-0.000\,121\,01t}.$$

Taking the natural logarithm of both sides gives

$$\ln(0.5) = \ln(e^{-0.000\,121\,01t}) = -0.000\,121\,01t \qquad \text{Identity 3: } \ln(e^a) = a \ln e = a$$

$$t = \frac{\ln(0.5)}{-0.000\,121\,01} \approx 5,728 \text{ years,}$$

as we obtained in part (a).

**c.** This time we are given the half-life, which we can use to find the exponential model $R(t) = Ae^{-kt}$. At time $t = 0$, the amount of radioactive material is

$$R(0) = Ae^0 = A.$$

Because half of the sample decays in 7,000 years, this sample will decay to $0.5A$ grams in 7,000 years ($t = 7,000$). Substituting this information gives

$$0.5A = Ae^{-k(7,000)}.$$

Canceling $A$ and taking natural logarithms (again using Identity 3) gives

$$\ln(0.5) = -7,000k$$

so the decay constant $k$ is

$$k = -\frac{\ln(0.5)}{7,000} \approx 0.000\,099\,021.$$

Therefore, the model is

$$R(t) = Ae^{-0.000\,099\,021t}.$$

d. If 99.95% of the substance in a sample has decayed, then the amount of undecayed material left is 0.05% of the original amount, or $0.0005A$. We have

$$0.0005A = Ae^{-0.000\,099\,021t}$$

$$0.0005 = e^{-0.000\,099\,021t}$$

$$\ln(0.0005) = -0.000\,099\,021t$$

$$t = \frac{\ln(0.0005)}{-0.000\,099\,021} \approx 76{,}760 \text{ years.}$$

➡ **Before we go on...**

**Q**: *In parts (a) and (b) of Example 4 we were given two different forms of the model for carbon 14 decay. How do we convert an exponential function in one form to the other?*

**A**: We have already seen (See Quick Example 3 on page 640) how to convert from the form $f(t) = Ae^{rt}$ in part (b) to the form $f(t) = Ab^t$ in part (a). To go the other way, start with the model in part (a), and equate it to the desired form:

$$C(t) = A(0.999\,879)^t = Ae^{rt}.$$

To solve for $r$, cancel the $A$s and take the natural logarithm of both sides:

$$t\ln(0.999\,879) = rt\ln e = rt$$

so $r = \ln(0.999\,879) \approx -0.000\,121\,007,$

giving

$$C(t) = Ae^{-0.000\,121\,01t}$$

as in part (b).

■

We can use the work we did in parts (b) and (c) of the above example to obtain a formula for the decay constant in an exponential decay model for any radioactive substance when we know its half-life. Write the half-life as $t_h$. Then the calculation in part (b) gives

$$k = -\frac{\ln(0.5)}{t_h} = \frac{\ln 2}{t_h}. \qquad -\ln(0.5) = -\ln\left(\frac{1}{2}\right) = \ln 2$$

Multiplying both sides by $t_h$ gives us the relationship $t_h k = \ln 2$.

### Exponential Decay Model and Half-Life

An **exponential decay function** has the form

$$Q(t) = Q_0 e^{-kt}. \qquad Q_0, k \text{ both positive}$$

$Q_0$ represents the value of $Q$ at time $t = 0$, and $k$ is the **decay constant**. The decay constant $k$ and half-life $t_h$ for $Q$ are related by

$$t_h k = \ln 2.$$

> **Quick Examples**
>
> 1. $Q(t) = Q_0 e^{-0.000\,121\,01t}$ is the decay function for carbon 14 (see Example 4b).
> 2. If $t_h = 10$ years, then $10k = \ln 2$, so $k = \dfrac{\ln 2}{10} \approx 0.06931$ and the decay model is
> $$Q(t) = Q_0 e^{-0.06931t}.$$
> 3. If $k = 0.0123$, then $t_h(0.0123) = \ln 2$, so the half-life is
> $$t_h = \frac{\ln 2}{0.0123} \approx 56.35 \text{ years.}$$

We can repeat the analysis above for exponential growth models:

> **Exponential Growth Model and Doubling Time**
>
> An **exponential growth function** has the form
> $$Q(t) = Q_0 e^{kt}. \qquad Q_0, k \text{ both positive}$$
> $Q_0$ represents the value of $Q$ at time $t = 0$, and $k$ is the **growth constant**. The growth constant $k$ and doubling time $t_d$ for $Q$ are related by
> $$t_d k = \ln 2.$$
>
> **Quick Examples**
>
> 1. $P(t) = 1,000e^{0.05t}$    $\$1,000$ invested at 5% annually with interest compounded continuously
> 2. If $t_d = 10$ years, then $10k = \ln 2$, so $k = \dfrac{\ln 2}{10} \approx 0.06931$ and the growth model is
> $$Q(t) = Q_0 e^{0.06931t}.$$
> 3. If $k = 0.0123$, then $t_d(0.0123) = \ln 2$, so the doubling time is
> $$t_d = \frac{\ln 2}{0.0123} \approx 56.35 \text{ years.}$$

**using Technology**

To obtain the regression curve and graph for Example 5 using a graphing calculator or a spreadsheet, see the Technology Guides at the end of the chapter. Outline:

**TI-83/84 Plus**

[STAT] EDIT values of $t$ in $L_1$ and values of $S$ in $L_2$

Regression curve: [STAT]

CALC option #9 LnReg [ENTER]

Graph: [Y=] [VARS] [5] EQ [1], then [ZOOM] [9]

[More details on page 681.]

**Spreadsheet**

$t$- and $S$-values in Columns A and B
Graph: Highlight data and insert a Scatter chart.
Regression curve: Right-click a dat-apoint and add logarithmic Trendline with option to show equation on chart.
[More details on page 684.]

**Website**

www.WanerMath.com
In the Function Evaluator and Grapher, enter the data as shown, press "Examples" until the logarithmic model $1*ln(x)+$2 shows in the first box, and press "Fit Curve".

## Logarithmic Regression

If we start with a set of data that suggests a logarithmic curve we can, by repeating the methods from previous sections, use technology to find the logarithmic regression curve $y = \log_b x + C$ approximating the data.

### EXAMPLE 5 ⬛ Research & Development

The following table shows the total spent on research and development by universities and colleges in the U.S., in billions of dollars, for the period 1998–2008 ($t$ is the number of years since 1990).[34]

---

[34]2008 data is preliminary. Source: National Science Foundation, Division of Science Resources Statistics. 2010. *National Patterns of R&D Resources: 2008 Data Update*. NSF 10-314. Arlington, VA. www.nsf.gov/statistics/nsf10314/.

| Year $t$ | 8 | 9 | 10 | 11 | 12 | 13 | 14 | 15 | 16 | 17 | 18 |
|---|---|---|---|---|---|---|---|---|---|---|---|
| Spending ($ billions) | 27 | 29 | 31 | 33 | 36 | 38 | 39 | 40 | 40 | 41 | 42 |

Find the best-fit logarithmic model of the form

$$S(t) = A \ln t + C$$

and use the model to project total spending on research by universities and colleges in 2012, assuming the trend continues.

**Solution**  We use technology to get the following regression model:

$$S(t) = 19.3 \ln t - 12.8.$$    Coefficients rounded

Because 2012 is represented by $t = 22$, we have

$$S(22) = 19.3 \ln(22) - 12.8 \approx 47.$$    Why did we round the result to two significant digits?

So, research and development spending by universities and colleges is projected to be around $47 billion in 2012.

**Figure 16**

➡ **Before we go on...**  The model in Example 5 seems to give reasonable estimates when we extrapolate forward, but extrapolating backward is quite another matter: The logarithm curve drops sharply to the left of the given range and becomes negative for small values of $t$ (Figure 16). ∎

# 9.3  EXERCISES

▼ more advanced    ◆ challenging
🔲 indicates exercises that should be solved using technology

*In Exercises 1–4, complete the given tables.* HINT [See Quick Examples on page 647.]

**1.**

| Exponential Form | $10^4 = 10,000$ | $4^2 = 16$ | $3^3 = 27$ | $5^1 = 5$ | $7^0 = 1$ | $4^{-2} = \frac{1}{16}$ |
|---|---|---|---|---|---|---|
| Logarithmic Form | | | | | | |

**2.**

| Exponential Form | $4^3 = 64$ | $10^{-1} = 0.1$ | $2^8 = 256$ | $5^0 = 1$ | $(0.5)^2 = 0.25$ | $6^{-2} = \frac{1}{36}$ |
|---|---|---|---|---|---|---|
| Logarithmic Form | | | | | | |

**3.**

| Exponential Form | | | | | | |
|---|---|---|---|---|---|---|
| Logarithmic Form | $\log_{0.5} 0.25 = 2$ | $\log_5 1 = 0$ | $\log_{10} 0.1 = -1$ | $\log_4 64 = 3$ | $\log_2 256 = 8$ | $\log_2 \frac{1}{4} = -2$ |

**4.**

| Exponential Form | | | | | | |
|---|---|---|---|---|---|---|
| Logarithmic Form | $\log_5 5 = 1$ | $\log_4 \frac{1}{16} = -2$ | $\log_4 16 = 2$ | $\log_{10} 10,000 = 4$ | $\log_3 27 = 3$ | $\log_7 1 = 0$ |

*In Exercises 5–12, use logarithms to solve the given equation. (Round answers to 4 decimal places.)* HINT [See Example 1.]

**5.** $3^x = 5$

**6.** $4^x = 3$

**7.** $5^{-2x} = 40$

**8.** $6^{3x+1} = 30$

**9.** $4.16e^x = 2$

**10.** $5.3(10^x) = 2$

**11.** $5(1.06^{2x+1}) = 11$

**12.** $4(1.5^{2x-1}) = 8$

*In Exercises 13–18, graph the given function.* HINT [See Example 2.]

**13.** $f(x) = \log_4 x$

**14.** $f(x) = \log_5 x$

**15.** $f(x) = \log_4(x - 1)$

**16.** $f(x) = \log_5(x + 1)$

**17.** $f(x) = \log_{1/4} x$

**18.** $f(x) = \log_{1/5} x$

*In Exercises 19–22 find the associated exponential decay or growth model.* HINT [See Quick Examples on page 655.]

**19.** $Q = 1,000$ when $t = 0$; half-life $= 1$

**20.** $Q = 2,000$ when $t = 0$; half-life $= 5$

**21.** $Q = 1,000$ when $t = 0$; doubling time $= 2$

**22.** $Q = 2,000$ when $t = 0$; doubling time $= 5$

*In Exercises 23–26 find the associated half-life or doubling time.* HINT [See Quick Examples on page 655.]

**23.** $Q = 1,000e^{0.5t}$

**24.** $Q = 1,000e^{-0.025t}$

**25.** $Q = Q_0 e^{-4t}$

**26.** $Q = Q_0 e^t$

*In Exercises 27–32 convert the given exponential function to the form indicated. Round all coefficients to 4 significant digits.* HINT [See Example 4 "Before we go on."]

**27.** $f(x) = 4e^{2x}$; $f(x) = Ab^x$

**28.** $f(x) = 2.1e^{-0.1x}$; $f(x) = Ab^x$

**29.** $f(t) = 2.1(1.001)^t$; $f(t) = Q_0 e^{kt}$

**30.** $f(t) = 23.4(0.991)^t$; $f(t) = Q_0 e^{-kt}$

**31.** $f(t) = 10(0.987)^t$; $f(t) = Q_0 e^{-kt}$

**32.** $f(t) = 2.3(2.2)^t$; $f(t) = Q_0 e^{kt}$

## APPLICATIONS

**33.** *Investments* How long will it take a $500 investment to be worth $700 if it is continuously compounded at 10% per year? (Give the answer to two decimal places.) HINT [See Example 3.]

**34.** *Investments* How long will it take a $500 investment to be worth $700 if it is continuously compounded at 15% per year? (Give the answer to two decimal places.) HINT [See Example 3.]

**35.** *Investments* How long, to the nearest year, will it take an investment to triple if it is continuously compounded at 10% per year? HINT [See Example 3.]

**36.** *Investments* How long, to the nearest year, will it take me to become a millionaire if I invest $1,000 at 10% interest compounded continuously? HINT [See Example 3.]

**37.** *Investments* I would like my investment to double in value every 3 years. At what rate of interest would I need to invest it, assuming the interest is compounded continuously? HINT [See Quick Examples page 655.]

**38.** *Depreciation* My investment in OHaganBooks.com stocks is losing half its value every 2 years. Find and interpret the associated decay rate. HINT [See Quick Examples page 655.]

**39.** *Carbon Dating* The amount of carbon 14 remaining in a sample that originally contained $A$ grams is given by

$$C(t) = A(0.999879)^t$$

where $t$ is time in years. If tests on a fossilized skull reveal that 99.95% of the carbon 14 has decayed, how old, to the nearest 1,000 years, is the skull? HINT [See Example 4.]

**40.** *Carbon Dating* Refer back to Exercise 39. How old, to the nearest 1,000 years, is a fossil in which only 30% of the carbon 14 has decayed? HINT [See Example 4.]

***Long-Term Investments*** *Exercises 41–48 are based on the following table, which lists interest rates on long-term investments (based on 10-year government bonds) in several countries in 2011.*[35] HINT [See Example 3.]

| Country | U.S. | Japan | Germany | Australia | Brazil |
|---------|------|-------|---------|-----------|--------|
| Yield | 3.4% | 1.2% | 3.3% | 5.5% | 12.7% |

**41.** Assuming that you invest $10,000 in the U.S., how long (to the nearest year) must you wait before your investment is worth $15,000 if the interest is compounded annually?

**42.** Assuming that you invest $10,000 in Japan, how long (to the nearest year) must you wait before your investment is worth $15,000 if the interest is compounded annually?

**43.** If you invest $10,400 in Germany and the interest is compounded monthly, when, to the nearest month, will your investment be worth $20,000?

**44.** If you invest $10,400 in the U.S., and the interest is compounded monthly, when, to the nearest month, will your investment be worth $20,000?

**45.** How long, to the nearest year, will it take an investment in Australia to double its value if the interest is compounded every 6 months?

**46.** How long, to the nearest year, will it take an investment in Brazil to double its value if the interest is compounded every 6 months?

**47.** If the interest on a long-term U.S. investment is compounded continuously, how long will it take the value of an investment to double? (Give the answer correct to 2 decimal places.)

---

[35] Approximate interest rates based on 10-year government bonds. Source: www.bloomberg.com.

**48.** If the interest on a long-term Australia investment is compounded continuously, how long will it take the value of an investment to double? (Give an answer correct to 2 decimal places.)

**49.** *Half-Life* The amount of radium 226 remaining in a sample that originally contained $A$ grams is approximately

$$C(t) = A(0.999\ 567)^t$$

where $t$ is time in years. Find the half-life to the nearest 100 years. HINT [See Example 4a.]

**50.** *Half-Life* The amount of iodine 131 remaining in a sample that originally contained $A$ grams is approximately

$$C(t) = A(0.9175)^t$$

where $t$ is time in days. Find the half-life to 2 decimal places. HINT [See Example 4a.]

**51.** *Automobiles* The rate of auto thefts triples every 6 months.

**a.** Determine, to 2 decimal places, the base $b$ for an exponential model $y = Ab^t$ of the rate of auto thefts as a function of time in months.

**b.** Find the doubling time to the nearest tenth of a month. HINT [(a) See Section 9.2 Example 2. (b) See Quick Examples page 655.]

**52.** *Televisions* The rate of television thefts is doubling every 4 months.

**a.** Determine, to 2 decimal places, the base $b$ for an exponential model $y = Ab^t$ of the rate of television thefts as a function of time in months.

**b.** Find the tripling time to the nearest tenth of a month. HINT [(a) See Section 9.2 Example 2. (b) See Quick Examples page 655.]

**53.** *Half-Life* The half-life of cobalt 60 is 5 years.

**a.** Obtain an exponential decay model for cobalt 60 in the form $Q(t) = Q_0 e^{-kt}$. (Round coefficients to 3 significant digits.)

**b.** Use your model to predict, to the nearest year, the time it takes one third of a sample of cobalt 60 to decay.

**54.** *Half-Life* The half-life of strontium 90 is 28 years.

**a.** Obtain an exponential decay model for strontium 90 in the form $Q(t) = Q_0 e^{-kt}$. (Round coefficients to 3 significant digits.)

**b.** Use your model to predict, to the nearest year, the time it takes three fifths of a sample of strontium 90 to decay.

**55.** *Radioactive Decay* Uranium 235 is used as fuel for some nuclear reactors. It has a half-life of 710 million years. How long will it take 10 grams of uranium 235 to decay to 1 gram? (Round your answer to 3 significant digits.)

**56.** *Radioactive Decay* Plutonium 239 is used as fuel for some nuclear reactors, and also as the fissionable material in atomic bombs. It has a half-life of 24,400 years. How long would it take 10 grams of plutonium 239 to decay to 1 gram? (Round your answer to 3 significant digits.)

**57.** ▼*Aspirin* Soon after taking an aspirin, a patient has absorbed 300 mg of the drug. If the amount of aspirin in the bloodstream decays exponentially, with half being removed every 2 hours, find, to the nearest 0.1 hour, the time it will take for the amount of aspirin in the bloodstream to decrease to 100 mg.

**58.** ▼*Alcohol* After a large number of drinks, a person has a blood alcohol level of 200 mg/dL (milligrams per deciliter). If the amount of alcohol in the blood decays exponentially, with one fourth being removed every hour, find the time it will take for the person's blood alcohol level to decrease to 80 mg/dL. HINT [See Example 4.]

**59.** ▼*Radioactive Decay* You are trying to determine the half-life of a new radioactive element you have isolated. You start with 1 gram, and 2 days later you determine that it has decayed down to 0.7 grams. What is its half-life? (Round your answer to 3 significant digits.) HINT [First find an exponential model, then see Example 4.]

**60.** ▼*Radioactive Decay* You have just isolated a new radioactive element. If you can determine its half-life, you will win the Nobel Prize in physics. You purify a sample of 2 grams. One of your colleagues steals half of it, and 3 days later you find that 0.1 grams of the radioactive material is still left. What is the half-life? (Round your answer to 3 significant digits.) HINT [First find an exponential model, then see Example 4.]

**61.** 🔢 *Population Aging* The following table shows the percentage of U.S. residents over the age of 65 in 1950, 1960, . . . , 2010 ($t$ is time in years since 1900):[36]

| t (Year since 1900) | 50 | 60 | 70 | 80 | 90 | 100 | 110 |
|---|---|---|---|---|---|---|---|
| P (% over 65) | 8.2 | 9.2 | 9.9 | 11.3 | 12.6 | 12.6 | 13 |

**a.** Find the logarithmic regression model of the form $P(t) = A \ln t + C$. (Round the coefficients to 4 significant digits). HINT [See Example 5.]

**b.** In 1940, 6.9% of the population was over 65. To how many significant digits does the model reflect this figure?

**c.** Which of the following is correct? The model, if extrapolated into the indefinite future, predicts that

**(A)** The percentage of U.S. residents over the age of 65 will increase without bound.

**(B)** The percentage of U.S. residents over the age of 65 will level off at around 14.2%.

**(C)** The percentage of U.S. residents over the age of 65 will eventually decrease.

**62.** 🔢 *Population Aging* The following table shows the percentage of U.S. residents over the age of 85 in 1950, 1960, . . . , 2010 ($t$ is time in years since 1900):[37]

| t (Year since 1900) | 50 | 60 | 70 | 80 | 90 | 100 | 110 |
|---|---|---|---|---|---|---|---|
| P (% over 85) | 0.4 | 0.5 | 0.7 | 1 | 1.2 | 1.6 | 1.9 |

**a.** Find the logarithmic regression model of the form $P(t) = A \ln t + C$. (Round the coefficients to 4 significant digits). HINT [See Example 5.]

---

[36]Source: U.S. Census Bureau.

[37]*Ibid.*

**b.** In 2020, 2.1% of the population is projected to be over 85. To how many significant digits does the model reflect this figure?

**c.** Which of the following is correct? If you increase $A$ by 0.1 and decrease $C$ by 0.1 in the logarithmic model, then

  **(A)** The new model predicts eventually lower percentages.
  **(B)** The long-term prediction is essentially the same.
  **(C)** The new model predicts eventually higher percentages.

**63.** ☐ *Research & Development: Industry* The following table shows the total spent on research and development by industry in the United States, in billions of dollars, for the period 1998–2008 ($t$ is the year since 1990).[38]

| Year $t$ | 8 | 9 | 10 | 11 | 12 | 13 | 14 | 15 | 16 | 17 | 18 |
|---|---|---|---|---|---|---|---|---|---|---|---|
| Spending ($ billions) | 150 | 165 | 183 | 181 | 170 | 172 | 172 | 181 | 191 | 203 | 215 |

Find the logarithmic regression model of the form $S(t) = A \ln t + C$ with coefficients $A$ and $C$ rounded to 2 decimal places. Also obtain a graph showing the data points and the regression curve. In which direction is it more reasonable to extrapolate the model? Why?

**64.** ☐ *Research & Development: Federal* The following table shows the total spent on research and development by the federal government in the United States, in billions of dollars, for the period 1998–2008 ($t$ is the year since 1990).[39]

| Year $t$ | 8 | 9 | 10 | 11 | 12 | 13 | 14 | 15 | 16 | 17 | 18 |
|---|---|---|---|---|---|---|---|---|---|---|---|
| Spending ($ billions) | 18 | 18 | 18 | 20 | 21 | 21 | 21 | 22 | 22 | 21 | 22 |

Find the logarithmic regression model of the form $S(t) = A \ln t + C$ with coefficients $A$ and $C$ rounded to 2 decimal places. Also obtain a graph showing the data points and the regression curve. In which direction is it more reasonable to extrapolate the model? Why?

**65.** ▼ *Richter Scale* The **Richter scale** is used to measure the intensity of earthquakes. The Richter scale rating of an earthquake is given by the formula

$$R = \frac{2}{3}(\log E - 11.8)$$

where $E$ is the energy released by the earthquake (measured in ergs[40]).

**a.** The San Francisco earthquake of 1906 is estimated to have registered $R = 7.9$ on the Richter scale. How many ergs of energy were released?

**b.** The Japan earthquake of 2011 registered 9.0 on the Richter scale. Compare the two: The energy released in the 1906 earthquake was what percentage of the energy released in the 2011 quake?

**c.** Solve the equation given above for $E$ in terms of $R$.

**d.** Use the result of part (c) to show that if two earthquakes registering $R_1$ and $R_2$ on the Richter scale release $E_1$ and $E_2$ ergs of energy, respectively, then

$$\frac{E_2}{E_1} = 10^{1.5(R_2 - R_1)}.$$

**e.** Fill in the blank: If one earthquake registers 2 points more on the Richter scale than another, then it releases ____ times the amount of energy.

**66.** ▼ *Sound Intensity* The loudness of a sound is measured in **decibels**. The decibel level of a sound is given by the formula

$$D = 10 \, \log \frac{I}{I_0},$$

where $D$ is the decibel level (dB), $I$ is its intensity in watts per square meter (W/m$^2$), and $I_0 = 10^{-12}$ W/m$^2$ is the intensity of a barely audible "threshold" sound. A sound intensity of 90 dB or greater causes damage to the average human ear.

**a.** Find the decibel levels of each of the following, rounding to the nearest decibel:

| | |
|---|---|
| Whisper: | $115 \times 10^{-12}$ W/m$^2$ |
| TV (average volume from 10 feet): | $320 \times 10^{-7}$ W/m$^2$ |
| Loud music: | $900 \times 10^{-3}$ W/m$^2$ |
| Jet aircraft (from 500 feet): | $100$ W/m$^2$ |

**b.** Which of the sounds above damages the average human ear?

**c.** Solve the given equation to express $I$ in terms of $D$.

**d.** Use the answer to part (c) to show that if two sounds of intensity $I_1$ and $I_2$ register decibel levels of $D_1$ and $D_2$, respectively, then

$$\frac{I_2}{I_1} = 10^{0.1(D_2 - D_1)}.$$

**e.** Fill in the blank: If one sound registers one decibel more than another, then it is ____ times as intense.

**67.** ▼ *Sound Intensity* The decibel level of a TV set decreases with the distance from the set according to the formula

$$D = 10 \log \left( \frac{320 \times 10^7}{r^2} \right)$$

---

[38]Excludes Federal funding; 2008 data is preliminary. Source: National Science Foundation, Division of Science Resources Statistics. 2010. *National Patterns of R&D Resources: 2008 Data Update.* NSF 10-314. Arlington, VA. www.nsf.gov/statistics/nsf10314/.

[39]Excludes Federal funding to industry and nonprofit organizations; 2008 data is preliminary. Source: National Science Foundation, Division of Science Resources Statistics. 2010. *National Patterns of R&D Resources: 2008 Data Update.* NSF 10-314. Arlington, VA. www.nsf.gov/statistics/nsf10314/.

---

[40]An erg is a unit of energy. One erg is the amount of energy it takes to move a mass of one gram one centimeter in one second. The term "Richter scale" is used loosely to refer to several ways of measuring earthquake magnitudes, calibrated to agree where they overlap.

where $D$ is the decibel level (dB) and $r$ is the distance from the TV set in feet.

**a.** Find the decibel level (to the nearest decibel) at distances of 10, 20, and 50 feet.

**b.** Express $D$ in the form $D = A + B \log r$ for suitable constants $A$ and $B$. (Round $A$ and $D$ to 2 significant digits.)

**c.** How far must a listener be from a TV so that the decibel level drops to 0 dB? (Round the answer to two significant digits.)

**68.** ▼ *Acidity* The acidity of a solution is measured by its pH, which is given by the formula

$$pH = -\log(H^+)$$

where $H^+$ measures the concentration of hydrogen ions in moles per liter.[41] The pH of pure water is 7. A solution is referred to as *acidic* if its pH is below 7 and as *basic* if its pH is above 7.

**a.** Calculate the pH of each of the following substances.

| | |
|---|---|
| Blood: | $3.9 \times 10^{-8}$ moles/liter |
| Milk: | $4.0 \times 10^{-7}$ moles/liter |
| Soap solution: | $1.0 \times 10^{-11}$ moles/liter |
| Black coffee: | $1.2 \times 10^{-7}$ moles/liter |

**b.** How many moles of hydrogen ions are contained in a liter of acid rain that has a pH of 5.0?

**c.** Complete the following sentence: If the pH of a solution increases by 1.0, then the concentration of hydrogen ions _____.

## COMMUNICATION AND REASONING EXERCISES

**69.** On the same set of axes, graph $y = \ln x$, $y = A \ln x$, and $y = A \ln x + C$ for various choices of *positive A* and *C*. What is the effect on the graph of $y = \ln x$ of multiplying by $A$? What is the effect of then adding $C$?

---

[41] A mole corresponds to about $6.0 \times 10^{23}$ hydrogen ions. (This number is known as Avogadro's number.)

**70.** On the same set of axes, graph $y = -\ln x$, $y = A \ln x$, and $y = A \ln x + C$ for various choices of *negative A* and *C*. What is the effect on the graph of $y = \ln x$ of multiplying by $A$? What is the effect of then adding $C$?

**71.** Why is the logarithm of a negative number not defined?

**72.** Of what use are logarithms, now that they are no longer needed to perform complex calculations?

**73.** Your company's market share is undergoing steady growth. Explain why a logarithmic function is *not* appropriate for long-term future prediction of your market share.

**74.** Your company's market share is undergoing steady growth. Explain why a logarithmic function is *not* appropriate for long-term backward extrapolation of your market share.

**75.** If $y = 4^x$, then $x = $ _____.

**76.** If $y = \log_6 x$, then $x = $ _____.

**77.** Simplify: $2^{\log_2 8}$.

**78.** Simplify: $e^{\ln x}$.

**79.** Simplify: $\ln(e^x)$.

**80.** Simplify: $\ln\sqrt{a}$.

**81.** ▼ If a town's population is increasing exponentially with time, how is time increasing with population? Explain.

**82.** ▼ If a town's population is increasing logarithmically with time, how is time increasing with population? Explain.

**83.** ▼ If two quantities $Q_1$ and $Q_2$ are logarithmic functions of time $t$, show that their sum, $Q_1 + Q_2$, is also a logarithmic function of time $t$.

**84.** 🖥 ▼ In Exercise 83 we saw that the sum of two logarithmic functions is a logarithmic function. In Exercises 63 and 64 you modeled research and development expenditure by industry and government. Now do a logarithmic regression on the sum of the two sets of figures. Does the result coincide with the sum of the two individual regression models? What does your answer tell you about the sum of logarithmic regression models?

---

## 9.4 Logistic Functions and Models

Figure 17 shows wired broadband penetration in the United States as a function of time $t$ in years ($t = 0$ represents 2000).[42]

The left-hand part of the curve in Figure 17, from $t = 2$ to, say, $t = 6$, looks roughly like exponential growth: $P$ behaves (roughly) like an exponential function, with the $y$-coordinates growing by a factor of around 1.5 per year. Then, as the market starts to become saturated, the growth of $P$ slows and its value approaches a "saturation" point that appears to be around 30%. **Logistic** functions have just this

---

[42] Broadband penetration is the number of broadband installations divided by the total population. Source for data: Organisation for Economic Co-operation and Development (OECD) Directorate for Science, Technology, and Industry, table of Historical Penetration Rates, June 2010, downloaded April 2011 from www.oecd.org/sti/ict/broadband.

**Figure 17**

kind of behavior, growing exponentially at first and then leveling off. In addition to modeling the demand for a new technology or product, logistic functions are often used in epidemic and population modeling. In an epidemic, the number of infected people often grows exponentially at first and then slows when a significant proportion of the entire susceptible population is infected and the epidemic has "run its course." Similarly, populations may grow exponentially at first and then slow as they approach the capacity of the available resources.

## Logistic Function

A **logistic function** has the form

$$f(x) = \frac{N}{1 + Ab^{-x}}$$

for nonzero constants $N$, $A$, and $b$ ($A$ and $b$ positive and $b \neq 1$).

### Quick Example

$N = 6$, $A = 2$, $b = 1.1$ gives $f(x) = \dfrac{6}{1 + 2(1.1^{-x})}$    `6/(1+2*1.1^-x)`

$f(0) = \dfrac{6}{1+2} = 2$    The $y$-intercept is $N/(1+A)$.

$f(1{,}000) = \dfrac{6}{1 + 2(1.1^{-1{,}000})} \approx \dfrac{6}{1+0} = 6 = N$    When $x$ is large, $f(x) \approx N$.

## Graph of a Logistic Function

$b > 1$

$0 < b < 1$

$$y = \frac{N}{1 + Ab^{-x}}$$

**Properties of the Logistic Curve** $y = \dfrac{N}{1 + Ab^{-x}}$

- The graph is an S-shaped curve sandwiched between the horizontal lines $y = 0$ and $y = N$. $N$ is called the **limiting value** of the logistic curve.
- If $b > 1$ the graph rises; if $b < 1$, the graph falls.
- The $y$-intercept is $\dfrac{N}{1 + A}$.
- The curve is steepest when $t = \dfrac{\ln A}{\ln b}$.    We will see why in Chapter 12.

**Note** If we write $b^{-x}$ as $e^{-kx}$ (so that $k = \ln b$), we get the following alternative form of the logistic function:

$$f(x) = \frac{N}{1 + Ae^{-kx}}.$$    ∎

---

**Q**: *How does the constant b affect the graph?*

**A**: To understand the role of $b$, we first rewrite the logistic function by multiplying top and bottom by $b^x$:

$$f(x) = \frac{N}{1 + Ab^{-x}}$$

$$= \frac{Nb^x}{(1 + Ab^{-x})b^x}$$

$$= \frac{Nb^x}{b^x + A}    \qquad \text{Because } b^{-x}b^x = 1$$

For values of $x$ close to 0, the quantity $b^x$ is close to 1, so the denominator is approximately $1 + A$, giving

$$f(x) \approx \frac{Nb^x}{1 + A} = \left(\frac{N}{1 + A}\right)b^x.$$

In other words, $f(x)$ is approximately exponential with base $b$ for values of $x$ close to 0. Put another way, if $x$ represents time, then initially the logistic function behaves like an exponential function.

---

To summarize:

**Logistic Function for Small $x$ and the Role of $b$**

For small values of $x$, we have

$$\frac{N}{1 + Ab^{-x}} \approx \left(\frac{N}{1 + A}\right)b^x.$$

Thus, for small $x$, the logistic function grows approximately exponentially with base $b$.

**Quick Example**

Let

$$f(x) = \frac{50}{1 + 24(3^{-x})}.$$    $N = 50, A = 24, b = 3$

Then

$$f(x) \approx \left( \frac{50}{1 + 24} \right)(3^x) = 2(3^x)$$

for small values of $x$. The following figure compares their graphs:

The upper curve is the exponential curve.

## Modeling with the Logistic Function

### EXAMPLE 1 Epidemics

A flu epidemic is spreading through the U.S. population. An estimated 150 million people are susceptible to this particular strain, and it is predicted that all susceptible people will eventually become infected. There are 10,000 people already infected, and the number is doubling every 2 weeks. Use a logistic function to model the number of people infected. Hence predict when, to the nearest week, 1 million people will be infected.

**Solution**    Let $t$ be time in weeks, and let $P(t)$ be the total number of people infected at time $t$. We want to express $P$ as a logistic function of $t$, so that

$$P(t) = \frac{N}{1 + Ab^{-t}}.$$

We are told that, in the long run, 150 million people will be infected, so that

$$N = 150,000,000.$$    Limiting value of $P$

At the current time ($t = 0$), 10,000 people are infected, so

$$10,000 = \frac{N}{1 + A} = \frac{150,000,000}{1 + A}.$$    Value of $P$ when $t = 0$

Solving for $A$ gives

$$10,000(1 + A) = 150,000,000$$
$$1 + A = 15,000$$
$$A = 14,999.$$

**Figure 18**

What about $b$? At the beginning of the epidemic ($t$ near 0), $P$ is growing approximately exponentially, doubling every 2 weeks. Using the technique of Section 9.2, we find that the exponential curve passing through the points (0, 10,000) and (2, 20,000) is

$$y = 10,000(\sqrt{2})^t$$

giving us $b = \sqrt{2}$. Now that we have the constants $N$, $A$, and $b$, we can write down the logistic model:

$$P(t) = \frac{150,000,000}{1 + 14,999(\sqrt{2})^{-t}}.$$

The graph of this function is shown in Figure 18.

Now we tackle the question of prediction: When will 1 million people be infected? In other words: When is $P(t) = 1,000,000$?

$$1,000,000 = \frac{150,000,000}{1 + 14,999(\sqrt{2})^{-t}}$$

$$1,000,000[1 + 14,999(\sqrt{2})^{-t}] = 150,000,000$$

$$1 + 14,999(\sqrt{2})^{-t} = 150$$

$$14,999(\sqrt{2})^{-t} = 149$$

$$(\sqrt{2})^{-t} = \frac{149}{14,999}$$

$$-t = \log_{\sqrt{2}}\left(\frac{149}{14,999}\right) \approx -13.31 \qquad \text{Logarithmic form}$$

Thus, 1 million people will be infected by about the thirteenth week.

➡ **Before we go on...** We said earlier that the logistic curve is steepest when $t = \dfrac{\ln A}{\ln b}$. In Example 1, this occurs when $t = \dfrac{\ln 14,999}{\ln \sqrt{2}} \approx 28$ weeks into the epidemic. At that time, the number of cases is growing most rapidly (look at the apparent slope of the graph at the corresponding point). ∎

## Logistic Regression

Let's go back to the data on broadband penetration in the United States with which we began this section and try to determine the long-term percentage of broadband penetration. In order to be able to make predictions such as this, we require a model for the data, so we will need to do some form of regression.

**Figure 19**

### EXAMPLE 2    ⓘ Broadband Penetration

Here are the data graphed in Figure 17:

| Year ($t$) | 2 | 3 | 4 | 5 | 6 | 7 | 8 | 9 | 10 |
|---|---|---|---|---|---|---|---|---|---|
| Penetration (%) ($P$) | 5.5 | 7.9 | 10.9 | 14.2 | 18.2 | 21.9 | 23.9 | 26.5 | 27.1 |

**using** Technology

See the Technology Guide at the end of the chapter for detailed instructions on how to obtain the regression curve and graph for Example 2 using a graphing calculator or spreadsheet. Outline:

**TI-83/84 Plus**
[STAT] EDIT values of $t$ in $L_1$ and values of $P$ in $L_2$
Regression curve: [STAT]
CALC option #B Logistic [ENTER]
Graph: [Y=] [VARS] [5] EQ [1],
then [ZOOM] [9] [More details on page 682.]

**Spreadsheet**
Use the Solver Add-in to obtain the best-fit logistic curve. [More details on page 684.]

 **Website**
www.WanerMath.com
In the Function Evaluator and Grapher, enter the data as shown, press "Examples" until the logistic model `$1/(1+$2*$3^(-x))` shows in the first box, and press "Fit Curve".

Find a logistic regression curve of the form

$$P(t) = \frac{N}{1 + Ab^{-t}}.$$

In the long term, what percentage of broadband penetration in the United States does the model predict?

**Solution** We can use technology to obtain the following regression model:

$$P(t) \approx \frac{29.842}{1 + 12.502(1.642)^{-t}}. \qquad \text{Coefficients rounded to 3 decimal places}$$

Its graph and the original data are shown in Figure 19. Because $N = 29.842$, this model predicts that, in the long term, the percentage of broadband penetration in the United States will be 29.842%, or about 30%.

➡ **Before we go on...** Logistic regression programs generally estimate all three constants $N$, $A$, and $b$ for a model $y = \dfrac{N}{1 + Ab^{-x}}$. However, there are times, as in Example 1, when we already know the limiting value $N$ and require estimates of only $A$ and $b$. In such cases, we can use technology like Excel Solver to find $A$ and $b$ for the best-fit curve with $N$ fixed. Alternatively, we can use exponential regression to compute estimates of $A$ and $b$ as follows: First rewrite the logistic equation as

$$\frac{N}{y} = 1 + Ab^{-x},$$

so that

$$\frac{N}{y} - 1 = Ab^{-x} = A(b^{-1})^x.$$

This equation gives $N/y - 1$ as an exponential function of $x$. Thus, if we do exponential regression using the data points $(x, N/y - 1)$, we can obtain estimates for $A$ and $b^{-1}$ (and hence $b$). This is done in Exercises 35 and 36.

It is important to note that the resulting curve is not the best-fit curve (in the sense of minimizing SSE; see the "Before we go on" discussion on page 641 after Example 6 in Section 9.2) and will be thus be different from that obtained using the method in Example 2. ■

# 9.4 EXERCISES

▼ more advanced ◆ challenging
🔢 indicates exercises that should be solved using technology

*In Exercises 1–6, find N, A, and b, give a technology formula for the given function, and use technology to sketch its graph for the given range of values of x.* HINT [See Quick Examples on page 661.]

**1.** $f(x) = \dfrac{7}{1 + 6(2^{-x})}$;   [0, 10]

**2.** $g(x) = \dfrac{4}{1 + 0.333(4^{-x})}$;   [0, 2]

**3.** $f(x) = \dfrac{10}{1 + 4(0.3^{-x})}$;   [−5, 5]

**4.** $g(x) = \dfrac{100}{1 + 5(0.5^{-x})}$;   [−5, 5]

**5.** $h(x) = \dfrac{2}{0.5 + 3.5(1.5^{-x})}$;   [0, 15]
(First divide top and bottom by 0.5.)

**6.** $k(x) = \dfrac{17}{2 + 6.5(1.05^{-x})}$; $\quad [0, 100]$
(First divide top and bottom by 2.)

*In Exercises 7–10, find the logistic function f with the given properties.* HINT [See Example 1.]

**7.** $f(0) = 10$, $f$ has limiting value 200, and for small values of $x$, $f$ is approximately exponential and doubles with every increase of 1 in $x$.

**8.** $f(0) = 1$, $f$ has limiting value 10, and for small values of $x$, $f$ is approximately exponential and grows by 50% with every increase of 1 in $x$.

**9.** $f$ has limiting value 6 and passes through $(0, 3)$ and $(1, 4)$.
HINT [First find *A*, then substitute.]

**10.** $f$ has limiting value 4 and passes through $(0, 1)$ and $(1, 2)$.
HINT [First find *A*, then substitute.]

*In Exercises 11–16, choose the logistic function that best approximates the given curve.*

**11.**

**(A)** $f(x) = \dfrac{6}{1 + 0.5(3^{-x})}$

**(B)** $f(x) = \dfrac{9}{1 + 3.5(2^{-x})}$

**(C)** $f(x) = \dfrac{9}{1 + 0.5(1.01)^{-x}}$

**12.**

**(A)** $f(x) = \dfrac{8}{1 + 7(2)^{-x}}$    **(B)** $f(x) = \dfrac{8}{1 + 3(2)^{-x}}$

**(C)** $f(x) = \dfrac{6}{1 + 11(5)^{-x}}$

**13.**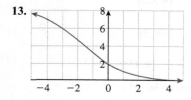

**(A)** $f(x) = \dfrac{8}{1 + 7(0.5)^{-x}}$    **(B)** $f(x) = \dfrac{8}{1 + 3(0.5)^{-x}}$

**(C)** $f(x) = \dfrac{8}{1 + 3(2)^{-x}}$

**14.**

**(A)** $f(x) = \dfrac{10}{1 + 3(1.01)^{-x}}$    **(B)** $f(x) = \dfrac{8}{1 + 7(0.1)^{-x}}$

**(C)** $f(x) = \dfrac{10}{1 + 3(0.1)^{-x}}$

**15.**

**(A)** $f(x) = \dfrac{18}{2 + 7(5)^{-x}}$    **(B)** $f(x) = \dfrac{18}{2 + 3(1.1)^{-x}}$

**(C)** $f(x) = \dfrac{18}{2 + 7(1.1)^{-x}}$

**16.**

**(A)** $f(x) = \dfrac{14}{2 + 5(15)^{-x}}$    **(B)** $f(x) = \dfrac{14}{1 + 13(1.05)^{-x}}$

**(C)** $f(x) = \dfrac{14}{2 + 5(1.05)^{-x}}$

⬛ *In Exercises 17–20, use technology to find a logistic regression curve* $y = \dfrac{N}{1 + Ab^{-x}}$ *approximating the given data. Draw a graph showing the data points and regression curve. (Round b to 3 significant digits and A and N to 2 significant digits.)* HINT [See Example 2.]

**17.**

| x | 0 | 20 | 40 | 60 | 80 | 100 |
|---|---|----|----|----|----|-----|
| y | 2.1 | 3.6 | 5.0 | 6.1 | 6.8 | 6.9 |

**18.**

| x | 0 | 30 | 60 | 90 | 120 | 150 |
|---|---|----|----|----|-----|-----|
| y | 2.8 | 5.8 | 7.9 | 9.4 | 9.7 | 9.9 |

**19.**

| x | 0 | 20 | 40 | 60 | 80 | 100 |
|---|---|----|----|----|----|-----|
| y | 30.1 | 11.6 | 3.8 | 1.2 | 0.4 | 0.1 |

**20.**

| x | 0 | 30 | 60 | 90 | 120 | 150 |
|---|---|----|----|----|-----|-----|
| y | 30.1 | 20 | 12 | 7.2 | 3.8 | 2.4 |

## APPLICATIONS

**21.** *Subprime Mortgages during the Housing Bubble* The following graph shows the approximate percentage of mortgages issued in the United States during the real-estate run-up in 2000–2008 that were subprime (normally classified as risky) as well as the logistic regression curve:[43]

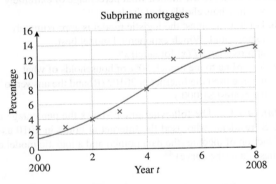

Subprime mortgages

**a.** Which of the following logistic functions best approximates the curve? (*t* is the number of years since the start of 2000.) Try to determine the correct model without actually computing data points. HINT [See Properties of the Logistic Curve on page 662.]

**(A)** $A(t) = \dfrac{15.0}{1 + 8.6(1.8)^{-t}}$

**(B)** $A(t) = \dfrac{2.0}{1 + 6.8(0.8)^{-t}}$

**(C)** $A(t) = \dfrac{2.0}{1 + 6.8(1.8)^{-t}}$

**(D)** $A(t) = \dfrac{15.0}{1 + 8.6(0.8)^{-t}}$

**b.** According to the model you selected, during which year was the percentage growing fastest? HINT [See the "Before we go on" discussion after Example 1.]

---

[43]2008 figure is an estimate. Sources: Mortgage Bankers Association, UBS.

**22.** *Subprime Mortgage Debt during the Housing Bubble* The following graph shows the approximate value of subprime (normally classified as risky) mortgage debt outstanding in the United States during the real-estate run-up in 2000–2008 as well as the logistic regression curve:[44]

Subprime debt outstanding

**a.** Which of the following logistic functions best approximates the curve? (*t* is the number of years since the start of 2000.) Try to determine the correct model without actually computing data points. HINT [See Properties of the Logistic Curve on page 662.]

**(A)** $A(t) = \dfrac{1,850}{1 + 5.36(1.8)^{-t}}$

**(B)** $A(t) = \dfrac{1,350}{1 + 4.2(1.7)^{-t}}$

**(C)** $A(t) = \dfrac{1,020}{1 + 5.3(1.8)^{-t}}$

**(D)** $A(t) = \dfrac{1,300}{1 + 4.2(0.9)^{-t}}$

**b.** According to the model you selected, during which year was outstanding debt growing fastest? HINT [See the "Before we go on" discussion after Example 1.]

**23.** *Scientific Research* The following graph shows the number of research articles in the prominent journal *Physical Review* that were written by researchers in Europe during 1983–2003 (*t* = 0 represents 1983).[45]

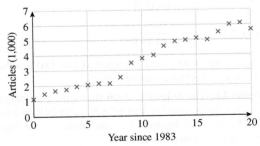

**a.** Which of the following logistic functions best models the data? (*t* is the number of years since 1983.) Try to determine the correct model without actually computing data points.

---

[44]2008–2009 figures are estimates. Source: www.data360.org/dataset .aspx?Data_Set_Id=9549.

[45]Source: The American Physical Society/*New York Times* May 3, 2003, p. A1.

**(A)** $A(t) = \dfrac{7.0}{1 + 5.4(1.2)^{-t}}$

**(B)** $A(t) = \dfrac{4.0}{1 + 3.4(1.2)^{-t}}$

**(C)** $A(t) = \dfrac{4.0}{1 + 3.4(0.8)^{-t}}$

**(D)** $A(t) = \dfrac{7.0}{1 + 5.4(6.2)^{-t}}$

**b.** According to the model you selected, at what percentage was the number of articles growing around 1985?

**24. Scientific Research** The following graph shows the percentage, above 25%, of research articles in the prominent journal *Physical Review* that were written by researchers in the United States during 1983–2003 ($t = 0$ represents 1983).[46]

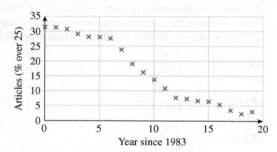

**a.** Which of the following logistic functions best models the data? ($t$ is the number of years since 1983 and $P$ is the actual percentage.) Try to determine the correct model without actually computing data points.

**(A)** $P(t) = \dfrac{36}{1 + 0.06(0.02)^{-t}}$

**(B)** $P(t) = \dfrac{12}{1 + 0.06(1.7)^{-t}}$

**(C)** $P(t) = \dfrac{12}{1 + 0.06(0.7)^{-t}}$

**(D)** $P(t) = \dfrac{36}{1 + 0.06(0.7)^{-t}}$

**b.** According to the model you selected, how fast was the value of $P$ declining around 1985?

**25. Internet Use** The following graph shows the percentage of U.S. households using the Internet at home in 2010 as a function of household income (the data points) and a logistic model of these data (the curve).[47]

The logistic model is

$$P(x) = \frac{95.0}{1 + 2.78(1.064)^{-x}} \text{ percent}$$

where $x$ is the household income in thousands of dollars.

**a.** According to the model, what percentage of extremely wealthy households used the Internet?

**b.** For low incomes, the logistic model is approximately exponential. Which exponential model best approximates $P(x)$ for small $x$?

**c.** According to the model, 50% of households of what income used the Internet in 2010? (Round the answer to the nearest $1,000.)

**26. Internet Use** The following graph shows the percentage of U.S. residents who used the Internet at home in 2010 as a function of income (the data points) and a logistic model of these data (the curve).[48]

The logistic model is given by

$$P(x) = \frac{86.2}{1 + 2.49(1.054)^{-x}} \text{ percent}$$

where $x$ is the household income in thousands of dollars.

[46]Source: The American Physical Society/*New York Times* May 3, 2003, p. A1.

[47]Income levels are midpoints of income brackets. Source: *Current Population Survey (CPS) Internet Use 2010*, National Telecommunications and Information Administration, www.ntia.doc.gov, January 2011.

[48]*Ibid.*

**a.** According to the model, what percentage of extremely wealthy people used the Internet at home?

**b.** For low incomes, the logistic model is approximately exponential. Which exponential model best approximates $P(x)$ for small $x$?

**c.** According to the model, 50% of individuals with what household income used the Internet at home in 2010? (Round the answer to the nearest $1,000.)

**27. *Epidemics*** There are currently 1,000 cases of Venusian flu in a total susceptible population of 10,000 and the number of cases is increasing by 25% each day. Find a logistic model for the number of cases of Venusian flu and use your model to predict the number of flu cases a week from now. HINT [See Example 1.]

**28. *Epidemics*** Last year's epidemic of Martian flu began with a single case in a total susceptible population of 10,000. The number of cases was increasing initially by 40% per day. Find a logistic model for the number of cases of Martian flu and use your model to predict the number of flu cases 3 weeks into the epidemic. HINT [See Example 1.]

**29. *Sales*** You have sold 100 "I ♥ Calculus" T-shirts and sales appear to be doubling every 5 days. You estimate the total market for "I ♥ Calculus" T-shirts to be 3,000. Give a logistic model for your sales and use it to predict, to the nearest day, when you will have sold 700 T-shirts.

**30. *Sales*** In Russia the average consumer drank two servings of Coca-Cola® in 1993. This amount appeared to be increasing exponentially with a doubling time of 2 years.[49] Given a long-range market saturation estimate of 100 servings per year, find a logistic model for the consumption of Coca-Cola in Russia and use your model to predict when, to the nearest year, the average consumption reached 50 servings per year.

**31.** ▣ *Scientific Research* The following chart shows some the data shown in the graph in Exercise 23 ($t = 0$ represents 1983).[50]

| Year $t$ | 0 | 5 | 10 | 15 | 20 |
|---|---|---|---|---|---|
| Research Articles $A$ (1,000) | 1.2 | 2.1 | 3.8 | 5.1 | 5.7 |

**a.** What is the logistic regression model for the data? (Round all coefficients to 2 significant digits.) At what value does the model predict that the number of research articles will level off? HINT [See Example 2.]

**b.** According to the model, how many *Physical Review* articles were published by U.S. researchers in 2000 ($t = 17$)? (The actual number was about 5,500 articles.)

**32.** ▣ *Scientific Research* The following chart shows some the data shown in the graph in Exercise 24 ($t = 0$ represents 1983).[51]

| Year $t$ | 0 | 5 | 10 | 15 | 20 |
|---|---|---|---|---|---|
| Percentage $P$ (over 25) | 36 | 28 | 16 | 7 | 3 |

**a.** What is the logistic regression model for the data? (Round all coefficients to 2 significant digits.) HINT [See Example 2.]

**b.** According to the model, what percentage of *Physical Review* articles were published by researchers in the United States in 2000 ($t = 17$)? (The actual figure was about 30.1%.)

**33.** ▣ *College Basketball: Men* The following table shows the number of NCAA men's college basketball teams in the U.S. for various years since 1990.[52]

| $t$ (Year since 1990) | 0 | 5 | 10 | 11 | 12 | 13 | 14 |
|---|---|---|---|---|---|---|---|
| Teams | 767 | 868 | 932 | 937 | 936 | 967 | 981 |
| $t$ (Year since 1990) | 15 | 16 | 17 | 18 | 19 | 20 | |
| Teams | 983 | 984 | 982 | 1,017 | 1,017 | 1,011 | |

**a.** What is the logistic regression model for the data? (Round all coefficients to 3 significant digits.) At what value does the model predict that the number of basketball teams will level off?

**b.** According to the model, for what value of $t$ is the regression curve steepest? Interpret the answer.

**c.** Interpret the coefficient $b$ in the context of the number of men's basketball teams.

**34.** ▣ *College Basketball: Women* The following table shows the number of NCAA women's college basketball teams in the U.S. for various years since 1990.[53]

| $t$ (Year since 1990) | 0 | 5 | 10 | 11 | 12 | 13 | 14 |
|---|---|---|---|---|---|---|---|
| Teams | 782 | 864 | 956 | 958 | 975 | 1,009 | 1,008 |
| $t$ (Year since 1990) | 15 | 16 | 17 | 18 | 19 | 20 | |
| Teams | 1,036 | 1,018 | 1,003 | 1,013 | 1,032 | 1,036 | |

**a.** What is the logistic regression model for the data? (Round all coefficients to 3 significant digits.) At what value does the model predict that the number of basketball teams will level off?

---

[49]The doubling time is based on retail sales of Coca-Cola products in Russia. Sales in 1993 were double those in 1991, and were expected to double again by 1995. Source: *New York Times*, September 26, 1994, p. D2.

[50]Source: The American Physical Society/*New York Times* May 3, 2003, p. A1.

[51]*Ibid.*

[52]2010 figure is an estimate. Source: www.census.gov.

[53]*Ibid.*

**b.** According to the model, for what value of $t$ is the regression curve steepest? Interpret the answer.

**c.** Interpret the coefficient $b$ in the context of the number of women's basketball teams.

⬛ *Exercises 35 and 36 are based on the discussion following Example 2. If the limiting value N is known, then*

$$\frac{N}{y} - 1 = A(b^{-1})^x$$

*and so $N/y - 1$ is an exponential function of x. In Exercises 35 and 36, use the given value of N and the data points $(x, N/y - 1)$ to obtain A and b, and hence a logistic model.*

**35.** ⬛ ◆ *Population: Puerto Rico* The following table and graph show the population of Puerto Rico in thousands from 1950 to 2025.[54]

| $t$ (year since 1950) | 0 | 10 | 20 | 30 | 40 | 50 |
|---|---|---|---|---|---|---|
| Population (thousands) | 2,220 | 2,360 | 2,720 | 3,210 | 3,540 | 3,820 |
| $t$ (year since 1950) | 55 | 60 | 65 | 70 | 75 | |
| Population (thousands) | 3,910 | 3,990 | 4,050 | 4,080 | 4,100 | |

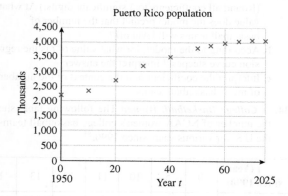

Puerto Rico population

Take $t$ to be the number of years since 1950, and find a logistic model based on the assumption that, eventually, the population of Puerto Rico will grow to 4.5 million. (Round coefficients to 4 decimal places.) In what year does your model predict the population of Puerto Rico will first reach 4.0 million?

**36.** ⬛ ◆ *Population: Virgin Islands* The following table and graph show the population of the Virgin Islands in thousands from 1950 to 2025.[55]

| $t$ (year since 1950) | 0 | 10 | 20 | 30 | 40 | 50 |
|---|---|---|---|---|---|---|
| Population (thousands) | 27 | 33 | 63 | 98 | 104 | 106 |
| $t$ (year since 1950) | 55 | 60 | 65 | 70 | 75 | |
| Population (thousands) | 108 | 108 | 107 | 107 | 108 | |

Virgin Islands population

Take $t$ to be the number of years since 1950, and find a logistic model based on the assumption that, eventually, the population of the Virgin Islands will grow to 110,000. (Round coefficients to 4 decimal places.) In what year does your model predict the population of the Virgin Islands first reached 80,000?

## COMMUNICATION AND REASONING EXERCISES

**37.** Logistic functions are commonly used to model the spread of epidemics. Given this fact, explain why a logistic function is also useful to model the spread of a new technology.

**38.** Why is a logistic function more appropriate than an exponential function for modeling the spread of an epidemic?

**39.** Give one practical use for logistic regression.

**40.** Refer to an exercise or example in this section to find a scenario in which a logistic model may not be a good predictor of long-term behavior.

**41.** What happens to the function $P(t) = \dfrac{N}{1 + Ab^{-t}}$ if we replace $b^{-t}$ by $b^t$ when $b > 1$? When $b < 1$?

**42.** ▼ What happens to the function $P(t) = \dfrac{N}{1 + Ab^{-t}}$ if $A = 0$? If $A < 0$?

**43.** ▼ We said that the logistic curve $y = \dfrac{N}{1 + Ab^{-t}}$ is steepest when $t = \dfrac{\ln A}{\ln b}$. Show that the corresponding value of $y$ is $N/2$. HINT [Use the fact that $\dfrac{\ln A}{\ln b} = \log_b A$.]

**44.** ▼ We said that the logistic curve $y = \dfrac{N}{1 + Ab^{-t}}$ is steepest when $t = \dfrac{\ln A}{\ln b}$. For which values of $A$ and $b$ is this value of $t$ positive, zero, and negative?

---

[54]Figures from 2010 on are U.S. census projections. Source: The 2008 Statistical Abstract, www.census.gov.

[55]Ibid.

# CHAPTER 9 REVIEW

## KEY CONCEPTS

Website www.WanerMath.com
Go to the Website at www.WanerMath
.com to find a comprehensive and
interactive Web-based summary
of Chapter 9.

### 9.1 Quadratic Functions and Models

A **quadratic function** has the form
$f(x) = ax^2 + bx + c.$ *p. 620*
The graph of $f(x) = ax^2 + bx + c$
$(a \neq 0)$ is a **parabola.** *p. 621*
The *x*-coordinate of the **vertex** is $-\frac{b}{2a}$.
The *y*-coordinate is $f\left(-\frac{b}{2a}\right)$. *p. 621*
**x-intercepts** (if any) occur at
$x = \dfrac{-b \pm \sqrt{b^2 - 4ac}}{2a}$. *p. 621*
The **y-intercept** occurs at $y = c.$ *p. 621*
The parabola is **symmetric** with respect to
the vertical line through the vertex. *p. 621*
Sketching the graph of a quadratic
function *p. 622*
Application to maximizing revenue *p. 624*
Application to maximizing profit *p. 624*
Finding the quadratic regression curve
*p. 626*

### 9.2 Exponential Functions and Models

An **exponential function** has the form
$f(x) = Ab^x.$ *p. 632*

Roles of the constants *A* and *b* in an
exponential function $f(x) = Ab^x$
*p. 633*
Recognizing exponential data *p. 634*
Finding the exponential curve through
two points *p. 635*
Application to compound interest
*p. 636*
Application to exponential decay
(carbon dating) *p. 636*
Application to exponential growth
(epidemics) *p. 636*
The number *e* and continuous
compounding *p. 638*
Alternative form of an exponential
function: $f(x) = Ae^{rx}$ *p. 640*
Finding the exponential regression
curve *p. 640*

### 9.3 Logarithmic Functions and Models

The **base *b* logarithm of *x*:** $y = \log_b x$
means $b^y = x$ *p. 647*
**Common logarithm,** $\log x = \log_{10} x$,
and **natural logarithm,**
$\ln x = \log_e x$ *p. 648*
Change-of-base formula *p. 648*
Solving equations with unknowns in
the exponent *p. 649*

A **logarithmic function** has the form
$f(x) = \log_b x + C$ or
$f(x) = A \ln x + C.$ *p. 649*
Graphs of logarithmic functions *p. 650*
Logarithm identities *p. 651*
Application to investments
(How long?) *p. 651*
Application to half-life *p. 652*
Exponential decay models and half-life
*p. 654*
Exponential growth models and
doubling time *p. 655*
Finding the logarithmic regression
curve *p. 655*

### 9.4 Logistic Functions and Models

A **logistic function** has the form
$f(x) = \dfrac{N}{1 + Ab^{-x}}.$ *p. 661*
Properties of the logistic curve, point
where curve is steepest *p. 662*
Logistic function for small *x*, the role
of *b* *p. 662*
Application to epidemics *p. 663*
Finding the logistic regression curve
*p. 664*

## REVIEW EXERCISES

*Sketch the graph of the quadratic functions in Exercises 1 and 2, indicating the coordinates of the vertex, the y-intercept, and the x-intercepts (if any).*

**1.** $f(x) = x^2 + 2x - 3$

**2.** $f(x) = -x^2 - x - 1$

*In Exercises 3 and 4, the values of two functions, f and g, are given in a table. One, both, or neither of them may be exponential. Decide which, if any, are exponential, and give the exponential models for those that are.*

**3.**

| $x$ | $-2$ | $-1$ | $0$ | $1$ | $2$ |
|---|---|---|---|---|---|
| $f(x)$ | 20 | 10 | 5 | 2.5 | 1.25 |
| $g(x)$ | 8 | 4 | 2 | 1 | 0 |

**4.**

| $x$ | $-2$ | $-1$ | $0$ | $1$ | $2$ |
|---|---|---|---|---|---|
| $f(x)$ | 8 | 6 | 4 | 2 | 1 |
| $g(x)$ | $\frac{3}{4}$ | $\frac{3}{2}$ | 3 | 6 | 12 |

*In Exercises 5 and 6, graph the given pairs of functions on the same set of axes with $-3 \leq x \leq 3$.*

**5.** $f(x) = \frac{1}{2}(3^x); \ g(x) = \frac{1}{2}(3^{-x})$

**6.** $f(x) = 2(4^x); \ g(x) = 2(4^{-x})$

**T** *On the same set of axes, use technology to graph the pairs of functions in Exercises 7 and 8 for the given range of x. Identify which graph corresponds to which function.*

**7.** $f(x) = e^x; \ g(x) = e^{0.8x}; \ -3 \leq x \leq 3$

**8.** $f(x) = 2(1.01)^x; \ g(x) = 2(0.99)^x; \ -100 \leq x \leq 100$

*In Exercises 9–14, compute the indicated quantity.*

**9.** A \$3,000 investment earns 3% interest, compounded monthly. Find its value after 5 years.

**10.** A \$10,000 investment earns 2.5% interest, compounded quarterly. Find its value after 10 years.

**11.** An investment earns 3% interest, compounded monthly, and is worth \$5,000 after 10 years. Find its initial value.

**12.** An investment earns 2.5% interest, compounded quarterly, and is worth $10,000 after 10 years. Find its initial value.

**13.** A $3,000 investment earns 3% interest, compounded continuously. Find its value after 5 years.

**14.** A $10,000 investment earns 2.5% interest, compounded continuously. Find its value after 10 years.

*In Exercises 15–18, find a formula of the form $f(x) = Ab^x$ using the given information.*

**15.** $f(0) = 4.5$; the value of $f$ triples for every half-unit increase in $x$.

**16.** $f(0) = 5$; the value of $f$ decreases by 75% for every 1-unit increase in $x$.

**17.** $f(1) = 2$, $f(3) = 18$.

**18.** $f(1) = 10$, $f(3) = 5$.

*In Exercises 19–22, use logarithms to solve the given equation for x.*

**19.** $3^{-2x} = 4$        **20.** $2^{2x^2-1} = 2$

**21.** $300(10^{3x}) = 315$    **22.** $P(1+i)^{mx} = A$

*On the same set of axes, graph the pairs of functions in Exercises 23 and 24.*

**23.** $f(x) = \log_3 x$; $g(x) = \log_{(1/3)} x$

**24.** $f(x) = \log x$; $g(x) = \log_{(1/10)} x$

*In Exercises 25–28, use the given information to find an exponential model of the form $Q = Q_0 e^{-kt}$ or $Q = Q_0 e^{kt}$, as appropriate. Round all coefficients to 3 significant digits when rounding is necessary.*

**25.** $Q$ is the amount of radioactive substance with a half-life of 100 years in a sample originally containing 5 g ($t$ is time in years).

**26.** $Q$ is the number of cats on an island whose cat population was originally 10,000 but is being cut in half every 5 years ($t$ is time in years).

**27.** $Q$ is the diameter (in cm) of a circular patch of mold on your roommate's damp towel you have been monitoring with morbid fascination. You measured the patch at 2.5 cm across 4 days ago, and have observed that it is doubling in diameter every 2 days ($t$ is time in days).

**28.** $Q$ is the population of cats on another island whose cat population was originally 10,000 but is doubling every 15 months ($t$ is time in months).

*In Exercises 29–32, find the time required, to the nearest 0.1 year, for the investment to reach the desired goal.*

**29.** $2,000 invested at 4%, compounded monthly; goal: $3,000

**30.** $2,000 invested at 6.75%, compounded daily; goal: $3,000

**31.** $2,000 invested at 3.75%, compounded continuously; goal: $3,000

**32.** $1,000 invested at 100%, compounded quarterly; goal: $1,200

*In Exercises 33–36, find equations for the logistic functions of x with the stated properties.*

**33.** Through (0, 100), initially increasing by 50% per unit of $x$, and limiting value 900.

**34.** Initially exponential of the form $y = 5(1.1)^x$ with limiting value 25.

**35.** Passing through (0, 5) and decreasing from a limiting value of 20 to 0 at a rate of 20% per unit of $x$ when $x$ is near 0.

**36.** Initially exponential of the form $y = 2(0.8)^x$ with a value close to 10 when $x = -60$.

## APPLICATIONS: OHaganBooks.com

**37. Web Site Traffic** The daily traffic ("hits per day") at OHaganBooks.com apparently depends on the monthly expenditure on Internet advertising. The following model is based on information collected over the past few months:

$$h = -0.000005c^2 + 0.085c + 1,750.$$

Here, $h$ is the average number of hits per day at OHaganBooks.com, and $c$ is the monthly advertising expenditure.

**a.** According to the model, what monthly advertising expenditure will result in the largest volume of traffic at OHaganBooks.com? What is that volume?

**b.** In addition to predicting a maximum volume of traffic, the model predicts that the traffic will eventually drop to zero if the advertising expenditure is increased too far. What expenditure (to the nearest dollar) results in no Web site traffic?

**c.** What feature of the formula for this quadratic model indicates that it will predict an eventual decline in traffic as advertising expenditure increases?

**38. Broadband Access** Pablo Pelogrande, a new summer intern at OHaganBooks.com in 2013, was asked by John O'Hagan to research the extent of broadband access in the United States. Pelogrande found some very old data online on broadband access from the start of 2001 to the end of 2003 and used it to construct the following quadratic model of the growth rate of broadband access:

$$n(t) = 2t^2 - 6t + 12 \text{ million new American adults}$$
with broadband per year

($t$ is time in years; $t = 0$ represents the start 2000).[56]

**a.** What is the appropriate domain of $n$?

**b.** According to the model, when was the growth rate at a minimum?

**c.** Does the model predict a zero growth rate at any particular time? If so, when?

**d.** What feature of the formula for this quadratic model indicates that the growth rate eventually increases?

**e.** Does the fact that $n(t)$ decreases for $t \leq 1.5$ suggest that the number of broadband users actually declined before June 2001? Explain.

---

[56]Based on data for 2001–2003. Source for data: Pew Internet and American Life Project data memos dated May 18, 2003 and April 19, 2004, downloaded from www.pewinternet.org.

**f.** Pelogande extrapolated the model in order to estimate the growth rate at the beginning of 2013 and 2014. What did he find? Comment on the answer.

**39. Revenue and Profit** Some time ago, a consultant formulated the following linear model of demand for online novels:

$$q = -60p + 950$$

where $q$ is the monthly demand for OHaganBooks.com's online novels at a price of $p$ dollars per novel.

**a.** Use this model to express the monthly revenue as a function of the unit price $p$, and hence determine the price you should charge for a maximum monthly revenue.

**b.** Author royalties and copyright fees cost the company an average of $4 per novel, and the monthly cost of operating and maintaining the online publishing service amounts to $900 per month. Express the monthly profit $P$ as a function of the unit price $p$, and hence determine the unit price you should charge for a maximum monthly profit. What is the resulting profit (or loss)?

**40. Revenue and Profit** Billy-Sean O'Hagan is John O'Hagan's son and a freshman in college. He notices that the demand for the college newspaper was 2,000 copies each week when the paper was given away free of charge, but dropped to 1,000 each week when the college started charging 10¢/copy.

**a.** Write down the associated linear demand function.

**b.** Use your demand function to express the revenue as a function of the unit price $p$, and hence determine the price the college should charge for a maximum revenue. At that price, what is the revenue from sales of one edition of the newspaper?

**c.** It costs the college 4¢ to produce each copy of the paper, plus an additional fixed cost of $200. Express the profit $P$ as a function of the unit price $p$, and hence determine the unit price the college should charge for a maximum monthly profit (or minimum loss). What is the resulting profit (or loss)?

**41. Lobsters** Marjory Duffin, CEO of Duffin House, is particularly fond of having steamed lobster at working lunches with executives from OHaganBooks.com and is therefore alarmed by the fact that the yearly lobster harvest from New York's Long Island Sound has been decreasing dramatically since 1997. Indeed, the size of the annual harvest can be approximated by

$$n(t) = 9.1(0.81^t) \text{ million pounds}$$

where $t$ is time in years since 1997.[57]

**a.** The model tells us that the harvest was _____ million pounds in 1997 and decreasing by ____% each year.

**b.** What does the model predict for the 2013 harvest?

**42. Stock Prices** In the period immediately following its initial public offering (IPO), OHaganBooks.com's stock was doubling in value every 3 hours. If you bought $10,000 worth of the stock when it was first offered, how much was your stock worth after 8 hours?

[57]Authors' regression model. Source for data: Long Island Sound Study, data downloaded May 2011 from longislandsoundstudy.net/2010/07/lobster-landings/.

**43. Lobsters** (See Exercise 41.) Marjory Duffin has just left John O'Hagan, CEO of OHaganBooks.com, a frantic phone message to the effect that this year's lobster harvest from New York's Long Island Sound is predicted to dip below 200,000 pounds, making that planned lobster working lunch more urgent than ever. What year is it?

**44. Stock Prices** We saw in Exercise 42 that OHaganBooks.com's stock was doubling in value every 3 hours, following its IPO. If you bought $10,000 worth of the stock when it was first offered, how long from the initial offering did it take your investment to reach $50,000?

**45. Lobsters** We saw in Exercise 41 that the Long Island Sound lobster harvest was given by $n(t) = 9.1(0.81^t)$ million pounds $t$ years after 1997. However, in 2010, thanks to the efforts of Duffin House, Inc. it turned around and started increasing by 24% each year.[58] What, to the nearest 10,000 pounds, was the actual size of the harvest in 2013?

**46. Stock Prices** We saw in Exercise 42 that OHaganBooks.com's stock was doubling in value every 3 hours, following its IPO. After 10 hours of trading, the stock turns around and starts losing one third of its value every 4 hours. How long (from the initial offering) will it be before your stock is once again worth $10,000?

**47.** ⊞ **Lobsters** The following chart shows some of the data that went into the model in Exercise 41:

Annual Lobster Harvest from Long Island Sound

Use these data to obtain an exponential regression curve of the form $n(t) = Ab^t$, with $t = 0$ corresponding to 1997 and coefficients rounded to 2 significant digits.

**48.** ⊞ **Stock Prices** The actual stock price of OHaganBooks.com in the hours following its IPO is shown in the following chart:

OHaganBooks.com stock price

[58]This claim, like Duffin House, is fiction.

Use the data to obtain an exponential regression curve of the form $P(t) = Ab^t$, with $t = 0$ the time in hours since the IPO and coefficients rounded to 3 significant digits. At the end of which hour will the stock price first be above $10?

49. **Hardware Life** *(Based on a question from the GRE economics exam)* To estimate the rate at which new computer hard drives will have to be retired, OHaganBooks.com uses the "survivor curve":

$$L_x = L_0 e^{-x/t}$$

where

$L_x$ = number of surviving hard drives at age $x$
$L_0$ = number of hard drives initially
$t$ = average life in years.

All of the following are implied by the curve *except:*

(A) Some of the equipment is retired during the first year of service.

(B) Some equipment survives three average lives.

(C) More than half the equipment survives the average life.

(D) Increasing the average life of equipment by using more durable materials would increase the number surviving at every age.

(E) The number of survivors never reaches zero.

50. **Sales** OHaganBooks.com modeled its weekly sales over a period of time with the function

$$s(t) = 6{,}050 + \frac{4{,}470}{1 + 14\,(1.73^{-t})}$$

as shown in the following graph ($t$ is measured in weeks):

a. As time goes on, it appears that weekly sales are leveling off. At what value are they leveling off?

b. When did weekly sales rise above 10,000?

c. When, to the nearest week, were sales rising most rapidly?

---

**Case Study**

Robert Nickelsberg/Getty Images

## Checking up on Malthus

In 1798 Thomas R. Malthus (1766–1834) published an influential pamphlet, later expanded into a book, titled *An Essay on the Principle of Population As It Affects the Future Improvement of Society.* One of his main contentions was that population grows geometrically (exponentially) while the supply of resources such as food grows only arithmetically (linearly). This led him to the pessimistic conclusion that population would always reach the limits of subsistence and precipitate famine, war, and ill health, unless population could be checked by other means. He advocated "moral restraint," which includes the pattern of late marriage common in Western Europe at the time and now common in most developed countries and which leads to a lower reproduction rate.

Two hundred years later, you have been asked to check the validity of Malthus's contention. That population grows geometrically, at least over short periods of time, is commonly assumed. That resources grow linearly is more questionable. You decide to check the actual production of a common crop, wheat, in the United States. Agricultural statistics like these are available from the U.S. government on the Internet, through the U.S. Department of Agriculture's National Agricultural Statistics Service (NASS). As of 2011, this service was available at www.nass.usda.gov. Looking through this site, you locate data on the annual production of all wheat in the United States from 1900 through 2010.

**Website**
www.WanerMath.com
To download an Excel sheet with the data used in the case study, go to Everything, scroll down to the case study for Chapter 9, and click on "Wheat Production Data (Excel)".

| Year | 1900 | 1901 | . . . | 2009 | 2010 |
|---|---|---|---|---|---|
| **Thousands of Bushels** | 599,315 | 762,546 | . . . | 2,218,061 | 2,208,391 |

Graphing these data (using Excel, for example), you obtain the graph in Figure 20.

**Figure 20**

This does not look very linear, particularly in the last half of the twentieth century, but you continue checking the mathematics. Using Excel's built-in linear regression capabilities, you find that the line that best fits these data, shown in Figure 21, has $r^2 = 0.8039$. (Recall the discussion of the correlation coefficient $r$ in Section 1.4. A similar statistic is available for other types of regression as well.)

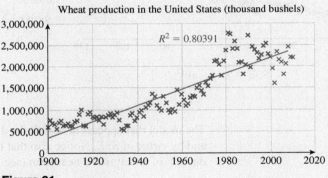

**Figure 21**

✳ Recall that the residuals are defined as $y_{Observed} - y_{Predicted}$ (see Section 1.4) and are the vertical distances between the observed data points and the regression line.

Although that is a fairly high correlation, you notice that the residuals✳ are not distributed randomly: The actual wheat production starts out higher than the line, dips below the line from about 1920 to about 1970, then rises above the line, and finally appears to dip below the line around 2002. This behavior seems to suggest a logistic curve or perhaps a cubic curve. On the other hand, it is also possible that the apparent dip at the end of the data is not statistically significant—it could be nothing more than a transitory fluctuation in the wheat production industry—so perhaps we should also consider models that do not bend downward, like exponential and quadratic models.

Following is a comparison of the four proposed models (with coefficients rounded to 3 significant digits). For the independent variable, we used $t = $ time in years since 1900. SSE is the sum-of-squares error.

**Quadratic**

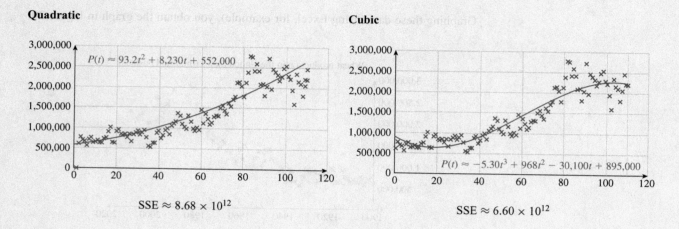

SSE $\approx 8.68 \times 10^{12}$

**Cubic**

SSE $\approx 6.60 \times 10^{12}$

**Exponential**

SSE $\approx 9.34 \times 10^{12}$

**Logistic**

SSE $\approx 8.26 \times 10^{12}$

The model that appears to best fit the data seems to be the cubic model; both visually and by virtue of SSE. Notice also that the cubic model predicts a *decrease* in the production of wheat in the near term (see Figure 22).

**Figure 22**

So you prepare a report that documents your findings and concludes that things are even worse than Malthus predicted, at least as far as wheat production in the United States is concerned: The supply is deceasing while the population is still increasing more-or-less exponentially. (See Exercise 75 in Section 9.2.)

You are about to hit "Send," which will dispatch copies of your report to a significant number of people on whom the success of your career depends, when you notice something strange about the pattern of data in Figure 22: The observed data points appear to hug the regression curve quite closely for small values of $t$, but appear to become more and more scattered as the value of $t$ increases. In the language of residuals, the residuals are small for small values of $t$ but then tend to get larger with increasing $t$. Figure 23 shows a plot of the residuals that shows this trend even more clearly.

This reminds you vaguely of something that came up in your college business statistics course, so you consult the textbook from that class that (fortunately) you still own and discover that a pattern of residuals with increasing magnitude suggests that, instead of modeling $y$ versus $t$ directly, you instead model $\ln y$ versus $t$. (The residuals for large values of $t$ will then be scaled down by the logarithm.)

Figure 24 shows the resulting plot together with the regression line (what we call the "linear transformed model").

*t* Residual plot

**Figure 23**

Natural log of wheat production versus time

$y = 0.0139x + 13.26$
$R^2 = 0.84951$

$\ln(y)$

SSE $\approx 3.89$
**Linear Transformed Model**

**Figure 24**

Notice that this time, the regression patterns no longer suggest an obvious curve. Further, they no longer appear to grow with increasing $t$. Although SSE is dramatically lower than the values for the earlier models, the contrast is a false one; the units of $y$ are now different, and comparing SSE with that of the earlier models is like comparing apples and oranges. While SSE depends on the units of measurement used, the coefficient of determination $r^2$ discussed in Section 1.4 is independent of the units used. A similar statistic is available for other types of regression as well, as well as something called "adjusted $r^2$."

The value of $r^2$ for the transformed model is approximately 0.850, while $r^2$ for the cubic model* is about 0.861, which is fairly close.

\* The "adjusted $r^2$" from statistics that corrects for model size.

**Q** : *If the cubic model and the linear transformed model have similar values of $r^2$, how do I decide which is more appropriate?*

**A** : The cubic model, if extrapolated, predicts unrealistically that the production of wheat will plunge in the near future, but the linear transformed model sees the recent drop-off as just one of several market fluctuations that show up in the residuals. You should therefore favor the more reasonable linear transformed model.

**Q** : *The linear transformed model gives us ln y versus t. What does it say about y versus t?*

**A** : Accurately write down the equation of the transformed linear model, being careful to replace $y$ by $\ln y$:

$$\ln y = 0.0139t + 13.26.$$

Rewriting this in exponential form gives

$$y = e^{0.0139t + 13.26}$$
$$= e^{13.26} e^{0.0139t}$$
$$\approx 574{,}000 e^{0.0139t}, \qquad \text{Coefficients rounded to 3 digits}$$

which is exactly the exponential model we found earlier! (In fact, using the natural logarithm transformation is the standard method of computing the regression exponential curve.)

**Q** : *What of the logistic model; should that not be the most realistic?*

**A** : The logistic model seems as though it *ought* to be the most appropriate, because wheat production cannot reasonably be expected to continue increasing exponentially forever; eventually resource limitations must lead to a leveling off of wheat production. Such a leveling off, if it occurred before the population started to level off, would seem to vindicate Malthus's pessimistic predictions. However, the logistic regression model has the same problem as the cubic model: It is trying to interpret the recent large fluctuations in the data as evidence of leveling off, but we really do not yet have significant evidence that that is occurring. Wheat production—even if it is logistic—appears still in the early (exponential) stage of growth. In general, for a logistic model to be reliable in its prediction of the leveling-off value $N$, we would need to see significant evidence of leveling off in the data. (See, however, Exercise 2 following.)

You now conclude that wheat production for the past 100 years is better described as increasing exponentially than linearly, contradicting Malthus, and moreover that it shows no sign of leveling off as yet.

### EXERCISES

1. Use the wheat production data starting at 1950 to construct the exponential regression model in two ways: directly, and using a linear transformed model as above. (Round coefficients to 3 digits.) Compare the growth constant $k$ of your model with that of the exponential model based on the data from 1900 on. How would you interpret the difference?

2. Compute the least-squares logistic model for the data in the preceding exercise. (Round coefficients to 3 significant digits.) At what level does it predict that wheat production will level off? (Note on using Excel Solver for logistic regression: Before running Solver, press Options in the Solver window and turn "Automatic Scaling" on. This adjusts the algorithm for the fact that the constants $A$, $N$, and $b$ have vastly different orders of magnitude.) Give

two graphs: one showing the data with the exponential regression model, and the other showing the data with the logistic regression model. Which model gives a better fit visually? Justify your observation by computing SSE directly for both models. Comment on your answer in terms of Malthus's assertions.

3. Find the production figures for another common crop grown in the United States. Compare the linear, quadratic, cubic, exponential, and logistic models. What can you conclude?

4. Below are the census figures for the population of the U.S. (in thousands) from 1820 to 2010.[59] Compare the linear, quadratic, and exponential models. What can you conclude?

| Year | 1820 | 1830 | 1840 | 1850 | 1860 | 1870 | 1880 | 1890 | 1900 | 1910 |
|---|---|---|---|---|---|---|---|---|---|---|
| Population (1000s) | 9,638 | 12,861 | 17,063 | 23,192 | 31,443 | 38,558 | 50,189 | 62,980 | 76,212 | 92,228 |
| Year | 1920 | 1930 | 1940 | 1950 | 1960 | 1970 | 1980 | 1990 | 2000 | 2010 |
| Population (1000s) | 106,022 | 123,203 | 132,165 | 151,326 | 179,323 | 203,302 | 226,542 | 248,710 | 281,422 | 308,746 |

[59] Source: Bureau of the Census, U.S. Department of Commerce.

**TECHNOLOGY GUIDE**

**TI-83/84 Plus**    **Technology Guide**

## Section 9.1

**Example 2 (page 622)** Sketch the graph of each quadratic function, showing the location of the vertex and intercepts.

**a.** $f(x) = 4x^2 - 12x + 9$    **b.** $g(x) = -\frac{1}{2}x^2 + 4x - 12$

### Solution with Technology

We will do part (a).

1. Start by storing the coefficients $a$, $b$, $c$ using

    $4 \to A: -12 \to B: 9 \to C$

    STO> gives the arrow    ALPHA  .  gives the colon

2. Save your quadratic as $Y_1$ using the $Y =$ screen:

    $Y_1 = AX^2 + BX + C$

3. To obtain the $x$-coordinate of the vertex, enter its formula as shown below on the left.

4. The $y$-coordinate of the vertex can be obtained from the Table screen by entering $x = 1.5$ as shown above on the right. (If you can't enter values of $x$, press 2ND  TBLSET , and set Indpnt to Ask.) From the table, we see that the vertex is at the point (1.5, 0).

5. To obtain the $x$-intercepts, enter the quadratic formula on the home screen as shown:

Because both intercepts agree, we conclude that the graph intersects the $x$-axis on a single point (at the vertex).

6. To graph the function, we need to select good values for Xmin and Xmax. In general, we would like our graph to show the vertex as well as all the intercepts. To see the vertex, make sure that its $x$ coordinate (1.5) is between Xmin and Xmax. To see the $x$-intercepts, make sure that they are also between Xmin and

Xmax. To see the $y$-intercept, make sure that $x = 0$ is between Xmin and Xmax. Thus, to see everything, choose Xmin and Xmax so that the interval [xMin, xMax] contains the $x$ coordinate of the vertex, the $x$-intercepts, and 0. For this example, we can choose an interval like $[-1, 3]$.

7. Once xMin and xMax are chosen, you can obtain convenient values of yMin and yMax by pressing ZOOM and selecting the option ZoomFit. (Make sure that your quadratic equation is entered in the $Y =$ screen before doing this!)

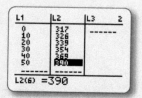

**Example 5 (page 626)** The following table shows the annual mean carbon dioxide concentration measured at Mauna Loa Observatory in Hawaii, in parts per million, every 10 years from 1960 through 2010 ($t = 0$ represents 1960).

| Year $t$ | 0 | 10 | 20 | 30 | 40 | 50 |
|---|---|---|---|---|---|---|
| PPM $CO_2$ $C$ | 317 | 326 | 339 | 354 | 369 | 390 |

Find the quadratic model

$$C(t) = at^2 + bt + c$$

that best fits the data.

### Solution with Technology

1. Using STAT EDIT enter the data with the $x$-coordinates (values of $t$) in $L_1$ and the $y$-coordinates (values of $C$) in $L_2$, just as in Section 1.4:

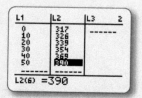

**2.** Press STAT , select CALC, and choose option #5 QuadReg. Pressing ENTER gives the quadratic regression curve in the home screen:

$$y \approx 0.01214x^2 + 0.8471x + 316.9.$$  Coefficients rounded to 4 decimal places

**3.** Now go to the Y= window and turn Plot1 on by selecting it and pressing ENTER . (You can also turn it on in the 2ND STAT PLOT screen.)

**4.** Next, enter the regression equation in the Y= screen by pressing Y= , clearing out whatever function is there, and pressing VARS 5 and selecting EQ option #1: RegEq as shown below left.

**5.** To obtain a convenient window showing all the points and the lines, press ZOOM and choose option #9: ZoomStat as shown above on the right.

**Note** When you are done viewing the graph, it is a good idea to turn Plot1 off again to avoid errors in graphing or data points showing up in your other graphs. ∎

## Section 9.2

**Example 6(a) (page 641)** The following table shows annual expenditure on health in the U.S. from 1980 through 2009 ($t = 0$ represents 1980).

| Year $t$ | 0 | 5 | 10 | 15 | 20 | 25 | 29 |
|---|---|---|---|---|---|---|---|
| Expenditure ($ billion) | 256 | 444 | 724 | 1,030 | 1,380 | 2,020 | 2,490 |

Find the exponential regression model

$$C(t) = Ab^t$$

for the annual expenditure.

### Solution with Technology

This is very similar to Example 5 in Section 9.1 (see the Technology Guide for Section 9.1):

**1.** Use STAT EDIT to enter the table of values.

**2.** Press STAT , select CALC, and choose option #0 ExpReg. Pressing ENTER gives the exponential regression curve in the home screen:

$$C(t) \approx 296.25(1.0798)^t.$$  Coefficients rounded

**3.** To graph the points and regression line in the same window, turn Plot1 on (see the Technology Guide for Example 5 in Section 9.1) and enter the regression equation in the Y= screen by pressing Y= , clearing out whatever function is there, and pressing VARS 5 and selecting EQ option #1: RegEq. Then press ZOOM and choose option #9: ZoomStat to see the graph.

**Note** When you are done viewing the graph, it is a good idea to turn Plot1 off again to avoid errors in graphing or data points showing up in your other graphs. ∎

## Section 9.3

**Example 5 (page 655)** The following table shows the total spent on research and development by universities and colleges in the U.S., in billions of dollars, for the period 1998–2008 ($t$ is the year since 1990).

| Year $t$ | 8 | 9 | 10 | 11 | 12 | 13 |
|---|---|---|---|---|---|---|
| Spending ($ billions) | 27 | 29 | 31 | 33 | 36 | 38 |
| Year $t$ | 14 | 15 | 16 | 17 | 18 | |
| Spending ($ billions) | 39 | 40 | 40 | 41 | 42 | |

Find the best-fit logarithmic model of the form

$$S(t) = A \ln t + C$$

and use the model to project total spending on research by universities and colleges in 2012, assuming the trend continues.

## Solution with Technology

This is very similar to Example 5 in Section 9.1 (see the Technology Guide for Section 9.1):

1. Use [STAT] EDIT to enter the table of values.
2. Press [STAT], select CALC, and choose option #9 LnReg. Pressing [ENTER] gives the logarithmic regression curve in the home screen:

$$S(t) \approx 19.25 \ln t - 12.78.$$  Coefficients rounded

3. To graph the points and regression line in the same window, turn Plot1 on (see the Technology Guide for Example 5 in Section 9.1) and enter the regression equation in the Y= screen by pressing [Y=], clearing out whatever function is there, and pressing [VARS] [5] and selecting EQ option #1: RegEq. Then press [ZOOM] and choose option #9: ZoomStat to see the graph.

## Section 9.4

**Example 2 (page 664)**   The following table shows wired broadband penetration in the United States as a function of time $t$ in years ($t = 0$ represents 2000).

| Year ($t$) | 2 | 3 | 4 | 5 | 6 | 7 | 8 | 9 | 10 |
|---|---|---|---|---|---|---|---|---|---|
| Penetration (%) ($P$) | 5.5 | 7.9 | 10.9 | 14.2 | 18.2 | 21.9 | 23.9 | 26.5 | 27.1 |

Find a logistic regression curve of the form

$$P(t) = \frac{N}{1 + Ab^{-t}}.$$

## Solution with Technology

This is very similar to Example 5 in Section 9.1 (see the Technology Guide for Section 9.1):

1. Use [STAT] EDIT to enter the table of values.
2. Press [STAT], select CALC, and choose option #B Logistic. Pressing [ENTER] gives the logistic regression curve in the home screen:

$$P(t) \approx \frac{29.842}{1 + 12.502e^{-0.49592t}}.$$  Coefficients rounded

This is not exactly the form we are seeking, but we can convert it to that form by writing

$$e^{-0.49592t} = (e^{0.49592})^{-t} \approx 1.642^{-t}$$

so

$$P(t) \approx \frac{29.842}{1 + 12.502(1.642)^{-t}}.$$

3. To graph the points and regression line in the same window, turn Plot1 on (see the Technology Guide for Example 5 in Section 9.1) and enter the regression equation in the Y= screen by pressing [Y=], clearing out whatever function is there, and pressing [VARS] [5] and selecting EQ option #1: RegEq. Then press [ZOOM] and choose option #9: ZoomStat to see the graph.

## SPREADSHEET Technology Guide

## Section 9.1

**Example 2 (page 622)**   Sketch the graph of each quadratic function, showing the location of the vertex and intercepts.

a. $f(x) = 4x^2 - 12x + 9$

b. $g(x) = -\dfrac{1}{2}x^2 + 4x - 12$

## Solution with Technology

We can set up a worksheet so that all we have to enter are the coefficients $a$, $b$, and $c$, and a range of $x$-values for the graph. Here is a possible layout that will plot 101 points using the coefficients for part (a).

**1.** First, we compute the $x$ coordinates:

**2.** To add the $y$ coordinates, we use the technology formula

$$a*x^2+b*x+c$$

replacing $a$, $b$, and $c$ with (absolute) references to the cells containing their values.

**3.** Graphing the data in columns A and B gives the graph shown here:

$$y = 4x^2 - 12x + 9$$

**4.** We can go further and compute the exact coordinates of the vertex and intercepts:

The completed sheet should look like this:

We can now save this sheet as a template to handle all quadratic functions. For instance, to do part (b), we just change the values of $a$, $b$, and $c$ in column D to $a = -1/2$, $b = 4$, and $c = -12$.

**Example 5 (page 626)**  The following table shows the annual mean carbon dioxide concentration measured at Mauna Loa Observatory in Hawaii, in parts per million, every 10 years from 1960 through 2010 ($t = 0$ represents 1960).

| Year $t$ | 0 | 10 | 20 | 30 | 40 | 50 |
|---|---|---|---|---|---|---|
| PPM $CO_2$ $C$ | 317 | 326 | 339 | 354 | 369 | 390 |

Find the quadratic model

$$C(t) = at^2 + bt + c$$

that best fits the data.

### Solution with Technology

As in Section 1.4, Example 2, we start with a scatter plot of the original data, and add a trendline:

**1.** Start with the original data and a "Scatter plot" (see Section 1.2 Example 5).

**2.** Add a quadratic trend line. (As of the time of this writing, among the common spreadsheets only Excel has the ability to add a polynomial trendline.) Right-click on any data point in the chart and select "Add Trendline," then select a "Polynomial" type of order 2 and check the option to "Display Equation on chart."

## Section 9.2

**Example 6(a) (page 641)**   The following table shows annual expenditure on health in the U.S. from 1980 through 2009 ($t = 0$ represents 1980).

| Year $t$ | 0 | 5 | 10 | 15 | 20 | 25 | 29 |
|---|---|---|---|---|---|---|---|
| Expenditure ($ billion) | 256 | 444 | 724 | 1,030 | 1,380 | 2,020 | 2,490 |

Find the exponential regression model

$$C(t) = Ab^t$$

for the annual expenditure.

### Solution with Technology

This is very similar to Example 5 in Section 9.1 (see the Technology Guide for Section 9.1):

1. Start with a "Scatter plot" of the observed data.

2. Add an exponential trendline:[60] The details vary from spreadsheet to spreadsheet—in OpenOffice, first double-click on the graph. Right-click on any data point in the chart and select "Add Trendline," then select an "Exponential" type and check the option to "Display Equation on chart."

Notice that the regression curve is given in the form $Ae^{kt}$ rather than $Ab^t$. To transform it, write

$$296.25e^{0.0768t} = 296.25(e^{0.0768})^t$$
$$\approx 296.25(1.0798)^t. \qquad e^{0.0768} \approx 1.0798$$

## Section 9.3

**Example 5 (page 655)**   The following table shows the total spent on research and development by universities and colleges in the U.S., in billions of dollars, for the period 1998–2008 ($t$ is the year since 1990).

---

| Year $t$ | 8 | 9 | 10 | 11 | 12 | 13 |
|---|---|---|---|---|---|---|
| Spending ($ billions) | 27 | 29 | 31 | 33 | 36 | 38 |
| Year $t$ | 14 | 15 | 16 | 17 | 18 | |
| Spending ($ billions) | 39 | 40 | 40 | 41 | 42 | |

Find the best-fit logarithmic model of the form

$$S(t) = A \ln t + C$$

and use the model to project total spending on research by universities and colleges in 2012, assuming the trend continues.

### Solution with Technology

This is very similar to Example 5 in Section 9.1 (see the Technology Guide for Section 9.1): We start, as usual, with a "Scatter plot" of the observed data and add a Logarithmic trendline. Here is the result:

## Section 9.4

**Example 2 (page 664)**   The following table shows wired broadband penetration in the United States as a function of time $t$ in years ($t = 0$ represents 2000).

| Year ($t$) | 2 | 3 | 4 | 5 | 6 | 7 | 8 | 9 | 10 |
|---|---|---|---|---|---|---|---|---|---|
| Penetration (%) ($P$) | 5.5 | 7.9 | 10.9 | 14.2 | 18.2 | 21.9 | 23.9 | 26.5 | 27.1 |

Find a logistic regression curve of the form

$$P(t) = \frac{N}{1 + Ab^{-t}}.$$

### Solution with Technology

At the time of this writing, available spreadsheets did not have a built-in logistic regression calculation, so we use an alternative method that works for any type of regression curve. The Solver included with Windows versions of Excel and some Mac versions can find logistic regression curves while the Solver included with some other spreadsheets is not yet capable of this, so the instructions here are specific to Excel.

---

[60] At the time of this writing, Google Docs has no trendline feature for its spreadsheet.

1. First use rough estimates for $N$, $A$, and $b$, and compute the sum-of-squares error (SSE; see Section 1.4) directly:

Cells E2:G2 contain our initial rough estimates of $N$, $A$, and $b$. For $N$, we used 30 (notice that the $y$-coordinates do appear to level off around 30). For $A$, we used the fact that the $y$-intercept is $N/(1 + A)$ and the $y$-intercept appears to be approximately 3. In other words,

$$3 = \frac{30}{1 + A}.$$

Because a very rough estimate is all we are after, using $A = 10$ will do just fine. For $b$, we chose 1.5 as the values of $P$ appear to be increasing by around 50% per year initially (again, this is rough).

2. Cell C2 contains the formula for $P(t)$, and the square of the resulting residual is computed in D2.

3. Cell F6 will contain SSE. The completed spreadsheet should look like this:

The best-fit curve will result from values of $N$, $A$, and $b$ that give a minimum value for SSE. We shall use Excel's "Solver," found in the "Analysis" group on the "Data" tab. (If "Solver" does not appear in the Analysis group, you will have to install the Solver Add-in using the Excel Options dialogue.) The Solver dialogue box with the necessary fields completed to solve the problem looks like this:

- The Target Cell refers to the cell that contains SSE.
- "Min" is selected because we are minimizing SSE.
- "Changing Cells" are obtained by selecting the cells that contain the current values of $N$, $A$, and $b$.

4. When you have filled in the values for the three items above, press "Solve" and tell Solver to Keep Solver Solution when done. You will find $N \approx 29.842$, $A \approx 12.501$, and $b \approx 1.642$ so that

$$P(t) \approx \frac{29.842}{1 + 12.501(1.642)^{-t}}.$$

If you use a scatter plot to graph the data in columns A, B and C, you will obtain the following graph:

# 10

# Introduction to the Derivative

 **Website**
www.WanerMath.com
At the Website you will find:

- Section-by-section tutorials, including game tutorials with randomized quizzes

- A detailed chapter summary

- A true/false quiz

- Additional review exercises

- Graphers, Excel tutorials, and other resources

- The following extra topics:

    Sketching the graph of the derivative

    Continuity and differentiability

## Case Study  Reducing Sulfur Emissions

The Environmental Protection Agency (EPA) wants to formulate a policy that will encourage utilities to reduce sulfur emissions. Its goal is to reduce annual emissions of sulfur dioxide by a total of 10 million tons from the current level of 25 million tons by imposing a fixed charge for every ton of sulfur released into the environment per year. The EPA has some data showing the marginal cost to utilities of reducing sulfur emissions. As a consultant to the EPA, you must determine the amount to be charged per ton of sulfur emissions in light of these data.

Norbert Schaefer/CORBIS

# Introduction

In the world around us, everything is changing. The mathematics of change is largely about the rate of change: how fast and in which direction the change is occurring. Is the Dow Jones average going up, and if so, how fast? If I raise my prices, how many customers will I lose? If I launch this missile, how fast will it be traveling after two seconds, how high will it go, and where will it come down?

We have already discussed the concept of rate of change for linear functions (straight lines), where the slope measures the rate of change. But this works only because a straight line maintains a constant rate of change along its whole length. Other functions rise faster here than there—or rise in one place and fall in another—so that the rate of change varies along the graph. The first achievement of calculus is to provide a systematic and straightforward way of calculating (hence the name) these rates of change. To describe a changing world, we need a language of change, and that is what calculus is.

The history of calculus is an interesting story of personalities, intellectual movements, and controversy. Credit for its invention is given to two mathematicians: Isaac Newton (1642–1727) and Gottfried Leibniz (1646–1716). Newton, an English mathematician and scientist, developed calculus first, probably in the 1660s. We say "probably" because, for various reasons, he did not publish his ideas until much later. This allowed Leibniz, a German mathematician and philosopher, to publish his own version of calculus first, in 1684. Fifteen years later, stirred up by nationalist fervor in England and on the continent, controversy erupted over who should get the credit for the invention of calculus. The debate got so heated that the Royal Society (of which Newton and Leibniz were both members) set up a commission to investigate the question. The commission decided in favor of Newton, who happened to be president of the society at the time. The consensus today is that both mathematicians deserve credit because they came to the same conclusions working independently. This is not really surprising: Both built on well-known work of other people, and it was almost inevitable that someone would put it all together at about that time.

**algebra Review**

For this chapter, you should be familiar with the algebra reviewed in **Chapter 0, Section 2.**

---

## 10.1 Limits: Numerical and Graphical Viewpoints

Rates of change are calculated by derivatives, but an important part of the definition of the derivative is something called a **limit**. Arguably, much of mathematics since the eighteenth century has revolved around understanding, refining, and exploiting the idea of the limit. The basic idea is easy, but getting the technicalities right is not.

### Evaluating Limits Numerically

Start with a very simple example: Look at the function $f(x) = 2 + x$ and ask: What happens to $f(x)$ as $x$ approaches 3? The following table shows the value of $f(x)$ for values of $x$ close to and on either side of 3:

*x* approaching 3 from the left →     ← *x* approaching 3 from the right

| $x$ | 2.9 | 2.99 | 2.999 | 2.9999 | 3 | 3.0001 | 3.001 | 3.01 | 3.1 |
|---|---|---|---|---|---|---|---|---|---|
| $f(x) = 2 + x$ | 4.9 | 4.99 | 4.999 | 4.9999 | | 5.0001 | 5.001 | 5.01 | 5.1 |

We have left the entry under 3 blank to emphasize that when calculating the limit of $f(x)$ as $x$ *approaches* 3, we are not interested in its value when $x$ *equals* 3.

Notice from the table that the closer $x$ gets to 3 from either side, the closer $f(x)$ gets to 5. We write this as

$$\lim_{x \to 3} f(x) = 5.$$    The limit of $f(x)$, as $x$ approaches 3, equals 5.

---

**Q** : *Why all the fuss? Can't we simply substitute $x = 3$ and avoid having to use a table?*

**A** : This happens to work for *some* functions, but not for *all* functions. The following example illustrates this point.

---

### EXAMPLE 1  Estimating a Limit Numerically

Use a table to estimate the following limits:

**a.** $\lim\limits_{x \to 2} \dfrac{x^3 - 8}{x - 2}$    **b.** $\lim\limits_{x \to 0} \dfrac{e^{2x} - 1}{x}$

### Solution

> \* However, if you factor $x^3 - 8$, you will find that $f(x)$ can be simplified to a function that *is* defined at $x = 2$. This point will be discussed (and this example redone) in Section 10.3. The function in part (b) cannot be simplified by factoring.

**a.** We cannot simply substitute $x = 2$, because the function $f(x) = \dfrac{x^3 - 8}{x - 2}$ is not defined at $x = 2$. (Why?)\* Instead, we use a table of values as we did above, with $x$ approaching 2 from both sides.

$x$ approaching 2 from the left →    ← $x$ approaching 2 from the right

| $x$ | 1.9 | 1.99 | 1.999 | 1.9999 | 2 | 2.0001 | 2.001 | 2.01 | 2.1 |
|---|---|---|---|---|---|---|---|---|---|
| $f(x) = \dfrac{x^3 - 8}{x - 2}$ | 11.41 | 11.9401 | 11.9940 | 11.9994 | | 12.0006 | 12.0060 | 12.0601 | 12.61 |

We notice that as $x$ approaches 2 from either side, $f(x)$ appears to be approaching 12. This suggests that the limit is 12, and we write

$$\lim_{x \to 2} \frac{x^3 - 8}{x - 2} = 12.$$

> **using Technology**
>
> To automate the computations in Example 1 using a graphing calculator or a spreadsheet, see the Technology Guides at the end of the chapter. Outline for part (a):
>
> **TI-83/84 Plus**
> Home screen: Y₁=(X^3−8)/(X−2)
> 2ND TBLSET Indpnt
> set to Ask
> 2ND TABLE Enter some values of $x$ from the example:
> 1.9, 1.99, 1.999 . . .
> [More details on page 778.]

**b.** The function $g(x) = \dfrac{e^{2x} - 1}{x}$ is not defined at $x = 0$ (nor can it even be simplified to one which *is* defined at $x = 0$). In the following table, we allow $x$ to approach 0 from both sides:

$x$ approaching 0 from the left →    ← $x$ approaching 0 from the right

| $x$ | −0.1 | −0.01 | −0.001 | −0.0001 | 0 | 0.0001 | 0.001 | 0.01 | 0.1 |
|---|---|---|---|---|---|---|---|---|---|
| $g(x) = \dfrac{e^{2x} - 1}{x}$ | 1.8127 | 1.9801 | 1.9980 | 1.9998 | | 2.0002 | 2.0020 | 2.0201 | 2.2140 |

The table suggests that $\lim\limits_{x \to 0} \dfrac{e^{2x} - 1}{x} = 2$.

**Spreadsheet**
Headings $x$, $f(x)$ in A1–B1 and again in C1–D1.
In A2–A5 enter 1.9, 1.99, 1.999, 1.9999.
In C1–C5 enter 2.1, 2.01, 2.001, 2.0001. Enter
`=(A2^3-8)/(A2-2)`
in B2 and copy down to B5.
Copy and paste the same formula in D2–D5.
[More details on page 780.]

 **Website**
www.WanerMath.com
In the Function Evaluator and Grapher, enter `(x^3-8)/(x-2)` for $y_1$. For a table of values, enter the various $x$-values in the Evaluator box, and press "Evaluate".

➡ **Before we go on...** Although the table *suggests* that the limit in Example 1 part (b) is 2, it by no means establishes that fact conclusively. It is *conceivable* (though not in fact the case here) that putting $x = 0.000000087$ could result in, say, $g(x) = 426$. Using a table can only *suggest* a value for the limit. In the next two sections we shall discuss algebraic techniques to allow us to actually *calculate* limits. ∎

Before we continue, let us make a more formal definition.

## Definition of a Limit

If $f(x)$ approaches the number $L$ as $x$ approaches (but is not equal to) $a$ from both sides, then we say that $f(x)$ **approaches $L$ as $x \to a$** ("$x$ approaches $a$") or that the **limit** of $f(x)$ as $x \to a$ is $L$. More precisely, *we can make $f(x)$ be as close to $L$ as we like by choosing any $x$ sufficiently close to (but not equal to) $a$ on either side*. We write

$$\lim_{x \to a} f(x) = L$$

or

$$f(x) \to L \text{ as } x \to a.$$

If $f(x)$ *fails* to approach *a single fixed number* as $x$ approaches $a$ from both sides, then we say that $f(x)$ **has no limit** as $x \to a$, or

$$\lim_{x \to a} f(x) \text{ does not exist.}$$

### Quick Examples

1. $\displaystyle\lim_{x \to 3}(2 + x) = 5$ — See discussion before Example 1.

2. $\displaystyle\lim_{x \to -2}(3x) = -6$ — As $x$ approaches $-2$, $3x$ approaches $-6$.

3. $\displaystyle\lim_{x \to 0}(x^2 - 2x + 1)$ exists. — In fact, the limit is 1.

4. $\displaystyle\lim_{x \to 5}\frac{1}{x} = \frac{1}{5}$ — As $x$ approaches 5, $\frac{1}{x}$ approaches $\frac{1}{5}$.

5. $\displaystyle\lim_{x \to 2}\frac{x^3 - 8}{x - 2} = 12$ — See Example 1. (We cannot just put $x = 2$ here.)

(For examples where the limit does not exist, see Example 2.)

### Notes

1. It is important that $f(x)$ approach the same number as $x$ approaches $a$ from *both sides*. For instance, if $f(x)$ approaches 5 for $x = 1.9$, $1.99$, $1.999$, ... , but approaches 4 for $x = 2.1$, $2.01$, $2.001$, ... , then the limit as $x \to 2$ does not exist. (See Example 2 for such a situation.)

2. It may happen that $f(x)$ does not approach any fixed number at all as $x \to a$ from either side. In this case, we also say that the limit does not exist.

3. If $a$ happens to be an endpoint of the domain of $f$, then the function is only defined on one side of $a$, and so the limit as $x \to a$ does not exist. For example, the natural domain of $f(x) = \sqrt{x}$ is $[0, +\infty)$, so the limit of $f(x)$ as $x \to 0$ does not exist. The appropriate kind of limit to consider in such a case is a **one-sided limit** (see Example 2). ∎

The next example gives instances in which a stated limit does not exist.

### EXAMPLE 2 Limits Do Not Always Exist

Do the following limits exist?

**a.** $\lim_{x \to 0} \dfrac{1}{x^2}$     **b.** $\lim_{x \to 0} \dfrac{|x|}{x}$     **c.** $\lim_{x \to 2} \dfrac{1}{x-2}$     **d.** $\lim_{x \to 1} \sqrt{x-1}$

#### Solution

**a.** Here is a table of values for $f(x) = \dfrac{1}{x^2}$, with $x$ approaching 0 from both sides.

$x$ approaching 0 from the left → ← $x$ approaching 0 from the right

| $x$ | −0.1 | −0.01 | −0.001 | −0.0001 | 0 | 0.0001 | 0.001 | 0.01 | 0.1 |
|---|---|---|---|---|---|---|---|---|---|
| $f(x) = \dfrac{1}{x^2}$ | 100 | 10,000 | 1,000,000 | 100,000,000 | | 100,000,000 | 1,000,000 | 10,000 | 100 |

The table shows that as $x$ gets closer to zero on either side, $f(x)$ gets larger and larger **without bound**—that is, if you name any number, no matter how large, $f(x)$ will be even larger than that if $x$ is sufficiently close to 0. Because $f(x)$ is not approaching any real number, we conclude that $\lim_{x \to 0} \dfrac{1}{x^2}$ does not exist. Because $f(x)$ is becoming arbitrarily large, we also say that $\lim_{x \to 0} \dfrac{1}{x^2}$ **diverges to** $+\infty$, or just

$$\lim_{x \to 0} \frac{1}{x^2} = +\infty.$$

**Note** This is not meant to imply that the limit exists; the symbol $+\infty$ does not represent any real number. We write $\lim_{x \to a} f(x) = +\infty$ to indicate two things: (1) The limit does not exist and (2) the function gets large without bound as $x$ approaches $a$. ■

**b.** Here is a table of values for $f(x) = \dfrac{|x|}{x}$, with $x$ approaching 0 from both sides.

$x$ approaching 0 from the left → ← $x$ approaching 0 from the right

| $x$ | −0.1 | −0.01 | −0.001 | −0.0001 | 0 | 0.0001 | 0.001 | 0.01 | 0.1 |
|---|---|---|---|---|---|---|---|---|---|
| $f(x) = \dfrac{|x|}{x}$ | −1 | −1 | −1 | −1 | | 1 | 1 | 1 | 1 |

The table shows that $f(x)$ does not approach the same limit as $x$ approaches 0 from both sides. There appear to be two *different* limits: the limit as we approach 0 from the left and the limit as we approach from the right. We write

$$\lim_{x \to 0^-} f(x) = -1$$

read as "the limit as $x$ approaches 0 from the left (or from below) is −1" and

$$\lim_{x \to 0^+} f(x) = 1$$

read as "the limit as $x$ approaches 0 from the right (or from above) is 1." These are called the **one-sided limits** of $f(x)$. In order for $f$ to have a **two-sided limit**, the two one-sided limits must be equal. Because they are not, we conclude that $\lim_{x \to 0} f(x)$ does not exist.

**c.** Near $x = 2$, we have the following table of values for $f(x) = \dfrac{1}{x-2}$:

| | $x$ approaching 2 from the left → | | | | | ← $x$ approaching 2 from the right | | | |
|---|---|---|---|---|---|---|---|---|---|
| $x$ | 1.9 | 1.99 | 1.999 | 1.9999 | 2 | 2.0001 | 2.001 | 2.01 | 2.1 |
| $f(x) = \dfrac{1}{x-2}$ | $-10$ | $-100$ | $-1{,}000$ | $-10{,}000$ | | 10,000 | 1,000 | 100 | 10 |

Because $f(x)$ is approaching no (single) real number as $x \to 2$, we see that $\displaystyle\lim_{x \to 2} \dfrac{1}{x-2}$ does not exist. Notice also that $\dfrac{1}{x-2}$ diverges to $+\infty$ as $x \to 2$ from the positive side (right half of the table) and to $-\infty$ as $x \to 2$ from the left (left half of the table). In other words,

$$\lim_{x \to 2^-} \frac{1}{x-2} = -\infty$$

$$\lim_{x \to 2^+} \frac{1}{x-2} = +\infty$$

$$\lim_{x \to 2} \frac{1}{x-2} \text{ does not exist.}$$

**d.** The natural domain of $f(x) = \sqrt{x-1}$ is $[1, +\infty)$, as $f(x)$ is defined only when $x \geq 1$. Thus we cannot evaluate $f(x)$ if $x$ is to the left of 1. This means that

$$\lim_{x \to 1^-} \sqrt{x-1} \text{ does not exist,}$$

and our table looks like this:

| | | ← $x$ approaching 1 from the right | | | | |
|---|---|---|---|---|---|---|
| $x$ | 1 | 1.00001 | 1.0001 | 1.001 | 1.01 | 1.1 |
| $f(x) = \sqrt{x-1}$ | | 0.0032 | 0.0100 | 0.0316 | 0.1000 | 0.3162 |

The values suggest that

$$\lim_{x \to 1^+} \sqrt{x-1} = 0.$$

In fact, we can obtain this limit by substituting $x = 1$ in the formula for $f(x)$ (see the comments after the example).

---

**Q** : *In Example 2(d) (and also in some of the Quick Examples before that) we could find a limit of an algebraically specified function by simply substituting the value of x in the formula for f(x). Does this always work?*

**A** : Short answer: Yes, when it makes sense. If the function is specified by a *single* algebraic formula and if $x = a$ is in the domain of $f$, then the limit can be obtained by substituting. We will say more about this when we discuss the algebraic approach to limits in Section 10.3. Remember, however, that, by definition, the limit of a function as $x \to a$ has nothing to do with its value at $x = a$, but rather its values for $x$ close to $a$.

**Q** : *If f(x) is undefined when x = a, then the limit does not exist, right?*

**A** : Wrong. If *f(a)* is not defined, then the limit may or may not exist. Example 1 shows instances where the limit *does* exist, and Example 2 shows instances where it does not. Again, the limit of a function as *x → a* has nothing to do with its value at *x = a*, but rather its values for *x close to a*.

In another useful kind of limit, we let *x* approach either $+\infty$ or $-\infty$, by which we mean that we let *x* get arbitrarily large or let *x* become an arbitrarily large negative number. The next example illustrates this.

### EXAMPLE 3  Limits at Infinity

Use a table to estimate: **a.** $\lim\limits_{x \to +\infty} \dfrac{2x^2 - 4x}{x^2 - 1}$ and **b.** $\lim\limits_{x \to -\infty} \dfrac{2x^2 - 4x}{x^2 - 1}$.

### Solution

**a.** By saying that *x* is "approaching $+\infty$," we mean that *x* is getting larger and larger without bound, so we make the following table:

*x* approaching $+\infty \to$

| $x$ | 10 | 100 | 1,000 | 10,000 | 100,000 |
|---|---|---|---|---|---|
| $f(x) = \dfrac{2x^2 - 4x}{x^2 - 1}$ | 1.6162 | 1.9602 | 1.9960 | 1.9996 | 2.0000 |

(Note that we are only approaching $+\infty$ from the left because we can hardly approach it from the right!) What seems to be happening is that $f(x)$ is approaching 2. Thus we write

$$\lim_{x \to +\infty} f(x) = 2.$$

**b.** Here, *x* is approaching $-\infty$, so we make a similar table, this time with *x* assuming negative values of greater and greater magnitude (read this table from right to left):

$\leftarrow$ *x* approaching $-\infty$

| $x$ | $-100,000$ | $-10,000$ | $-1,000$ | $-100$ | $-10$ |
|---|---|---|---|---|---|
| $f(x) = \dfrac{2x^2 - 4x}{x^2 - 1}$ | 2.0000 | 2.0004 | 2.0040 | 2.0402 | 2.4242 |

Once again, $f(x)$ is approaching 2. Thus, $\lim_{x \to -\infty} f(x) = 2$.

## Estimating Limits Graphically

We can often estimate a limit from a graph, as the next example shows.

### EXAMPLE 4  Estimating Limits Graphically

The graph of a function *f* is shown in Figure 1. (Recall that the solid dots indicate points on the graph, and the hollow dots indicate points not on the graph.) From the graph, analyze the following limits.

**a.** $\lim\limits_{x \to -2} f(x)$    **b.** $\lim\limits_{x \to 0} f(x)$    **c.** $\lim\limits_{x \to 1} f(x)$    **d.** $\lim\limits_{x \to +\infty} f(x)$

**Figure 1**

**Solution** Since we are given only a graph of $f$, we must analyze these limits graphically.

**a.** Imagine that Figure 1 was drawn on a graphing calculator equipped with a trace feature that allows us to move a cursor along the graph and see the coordinates as we go. To simulate this, place a pencil point on the graph to the left of $x = -2$, and move it along the curve so that the $x$-coordinate approaches $-2$. (See Figure 2.) We evaluate the limit numerically by noting the behavior of the $y$-coordinates.*

However, we can see directly from the graph that the $y$-coordinate approaches 2. Similarly, if we place our pencil point to the right of $x = -2$ and move it to the left, the $y$ coordinate will approach 2 from that side as well (Figure 3). Therefore, as $x$ approaches $-2$ from either side, $f(x)$ approaches 2, so

$$\lim_{x \to -2} f(x) = 2.$$

*For a visual animation of this process, look at the online tutorial for this section at the Website.

**Figure 2**          **Figure 3**

**b.** This time we move our pencil point toward $x = 0$. Referring to Figure 4, if we start from the left of $x = 0$ and approach 0 (by moving right), the $y$-coordinate approaches $-1$. However, if we start from the right of $x = 0$ and approach 0 (by moving left), the $y$-coordinate approaches 3. Thus (see Example 2),

$$\lim_{x \to 0^-} f(x) = -1$$

and

$$\lim_{x \to 0^+} f(x) = 3.$$

Because these limits are not equal, we conclude that

$$\lim_{x \to 0} f(x) \text{ does not exist.}$$

In this case there is a "break" in the graph at $x = 0$, and we say that the function is **discontinuous** at $x = 0$ (see Section 10.2).

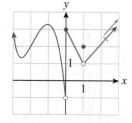

**Figure 4**          **Figure 5**          **Figure 6**

**c.** Once more we think about a pencil point moving along the graph with the $x$-coordinate this time approaching $x = 1$ from the left and from the right (Figure 5).

As the $x$-coordinate of the point approaches 1 from either side, the $y$-coordinate approaches 1 also. Therefore,

$$\lim_{x \to 1} f(x) = 1.$$

**d.** For this limit, $x$ is supposed to approach infinity. We think about a pencil point moving along the graph further and further to the right as shown in Figure 6.

As the $x$-coordinate gets larger, the $y$-coordinate also gets larger and larger without bound. Thus, $f(x)$ diverges to $+\infty$:

$$\lim_{x \to +\infty} f(x) = +\infty.$$

Similarly,

$$\lim_{x \to -\infty} f(x) = +\infty.$$

➡ **Before we go on...** In Example 4(c) $\lim_{x \to 1} f(x) = 1$ but $f(1) = 2$ (why?). Thus, $\lim_{x \to 1} f(x) \neq f(1)$. In other words, the limit of $f(x)$ as $x$ *approaches* 1 is not the same as the value of $f$ at $x = 1$. Always keep in mind that when we evaluate a limit as $x \to a$, *we do not care about the value of the function at $x = a$*. We only care about the value of $f(x)$ as $x$ *approaches* $a$. In other words, $f(a)$ may or may not equal $\lim_{x \to a} f(x)$. ∎

Here is a summary of the graphical method we used in Example 4, together with some additional information:

### Evaluating Limits Graphically

To decide whether $\lim_{x \to a} f(x)$ exists and to find its value if it does:

1. Draw the graph of $f(x)$ by hand or with graphing technology.
2. Position your pencil point (or the Trace cursor) on a point of the graph to the right of $x = a$.
3. Move the point *along the graph* toward $x = a$ from the right and read the $y$-coordinate as you go. The value the $y$-coordinate approaches (if any) is the limit $\lim_{x \to a^+} f(x)$.
4. Repeat Steps 2 and 3, this time starting from a point on the graph to the left of $x = a$, and approaching $x = a$ along the graph from the left. The value the $y$-coordinate approaches (if any) is $\lim_{x \to a^-} f(x)$.
5. If the left and right limits both exist and have the same value $L$, then $\lim_{x \to a} f(x) = L$. Otherwise, the limit does not exist. The value $f(a)$ has no relevance whatsoever.
6. To evaluate $\lim_{x \to +\infty} f(x)$, move the pencil point toward the far right of the graph and estimate the value the $y$-coordinate approaches (if any). For $\lim_{x \to -\infty} f(x)$, move the pencil point toward the far left.
7. If $x = a$ happens to be an endpoint of the domain of $f$, then only a one-sided limit is possible at $x = a$. For instance, if the domain is $(-\infty, 4]$, then $\lim_{x \to 4^-} f(x)$ may exist, but neither $\lim_{x \to 4^+} f(x)$ nor $\lim_{x \to 4} f(x)$ exists.

In the next example we use both the numerical and graphical approaches.

### EXAMPLE 5  Infinite Limit

Does $\lim\limits_{x \to 0^+} \dfrac{1}{x}$ exist?

### Solution

**Numerical Method** Because we are asked for only the right-hand limit, we need only list values of $x$ approaching 0 from the right.

$\leftarrow x$ approaching 0 from the right

| $x$ | 0 | 0.0001 | 0.001 | 0.01 | 0.1 |
|---|---|---|---|---|---|
| $f(x) = \dfrac{1}{x}$ | | 10,000 | 1,000 | 100 | 10 |

What seems to be happening as $x$ approaches 0 from the right is that $f(x)$ is increasing without bound, as in Example 4(d). That is, if you name any number, no matter how large, $f(x)$ will be even larger than that if $x$ is sufficiently close to zero. Thus, the limit diverges to $+\infty$, so

$$\lim_{x \to 0^+} \frac{1}{x} = +\infty$$

**Figure 7**

**Graphical Method** Recall that the graph of $f(x) = \dfrac{1}{x}$ is the standard hyperbola shown in Figure 7. The figure also shows the pencil point moving so that its $x$-coordinate approaches 0 from the right. Because the point moves along the graph, it is forced to go higher and higher. In other words, its $y$-coordinate becomes larger and larger, approaching $+\infty$. Thus, we conclude that

$$\lim_{x \to 0^+} \frac{1}{x} = +\infty.$$

➡ **Before we go on...** In Example 5 you should check that

$$\lim_{x \to 0^-} \frac{1}{x} = -\infty. \qquad \frac{1}{x} \text{ diverges to } -\infty \text{ as } x \to 0^-.$$

Also, check that

$$\lim_{x \to +\infty} \frac{1}{x} = \lim_{x \to -\infty} \frac{1}{x} = 0. \ \blacksquare$$

## APPLICATION

### EXAMPLE 6  Broadband Penetration

Wired broadband penetration in the United States can be modeled by

$$P(t) = \frac{29.842}{1 + 12.502(1.642)^{-t}} \quad (t \geq 0),$$

where $t$ is time in years since 2000.[1]

---

[1]See Example 2 in Section 9.4. Broadband penetration is the number of broadband installations divided by the total population. Source for data: Organization for Economic Cooperation and Development (OECD) Directorate for Science, Technology, and Industry, table of Historical Penetration Rates, June 2010, downloaded April 2011 from www.oecd.org/sti /ict/broadband.

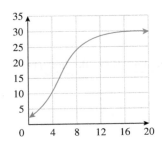

**Figure 8**

**a.** Estimate $\lim_{t \to +\infty} P(t)$ and interpret the answer.

**b.** Estimate $\lim_{t \to 0^+} P(t)$ and interpret the answer.

**Solution**

**a.** Figure 8 shows a plot of $P(t)$ for $0 \le t \le 20$. Using either the numerical or the graphical approach, we find

$$\lim_{t \to +\infty} P(t) = \lim_{t \to +\infty} \frac{29.842}{1 + 12.502(1.642)^{-t}} \approx 30.$$

(The actual limit is 29.842. Why?) Thus, in the long term (as $t$ gets larger and larger), broadband penetration in the United States is expected to approach 30%; that is, the number of installations is expected to approach 30% of the total population.

**b.** The limit here is

$$\lim_{t \to 0^+} P(t) = \lim_{t \to 0^+} \frac{29.842}{1 + 12.502(1.642)^{-t}} \approx 2.21.$$

(Notice that in this case, we can simply put $t = 0$ to evaluate this limit.) Thus, the closer $t$ gets to 0 (representing 2000) from the right, the closer $P(t)$ gets to 2.21%, meaning that, in 2000, broadband penetration was about 2.2% of the population.

---

### FAQs

**Determining when a Limit Does or Does Not Exist**

**Q**: *If I substitute $x = a$ in the formula for a function and find that the function is defined there, it means that $\lim_{x \to a} f(x)$ exists and equals f(a), right?*

**A**: Correct, provided the function is specified by a *single algebraic formula* and is not, say, piecewise-defined. We shall say more about this in the next two sections.

**Q**: *If I substitute $x = a$ in the formula for a function and find that the function is not defined there, it means that $\lim_{x \to a} f(x)$ does not exist, right?*

**A**: Wrong. The limit may still exist, as in Example 1, or may not exist, as in Example 2. In general, whether or not $\lim_{x \to a} f(x)$ exists has nothing to do with f(a), but rather the values of f when x is *very close to, but not equal to a.*

**Q**: *Is there a quick and easy way of telling from a graph whether $\lim_{x \to a} f(x)$ exists?*

**A**: Yes. If you cover up the portion of the graph corresponding to $x = a$ and it appears as though the visible part of the graph could be made into a continuous curve by filling in a suitable point at $x = a$, then the limit exists. (The "suitable point" need not be (a, f(a)).) Otherwise, it does not. Try this method with the curves in Example 4.

## 10.1    EXERCISES

▼ more advanced     ◆ challenging
**T** indicates exercises that should be solved using technology

*Estimate the limits in Exercises 1–20 numerically.*
HINT [See Examples 1–3.]

**1.** $\lim\limits_{x\to 0}\dfrac{x^2}{x+1}$

**2.** $\lim\limits_{x\to 0}\dfrac{x-3}{x-1}$

**3.** $\lim\limits_{x\to 2}\dfrac{x^2-4}{x-2}$

**4.** $\lim\limits_{x\to 2}\dfrac{x^2-1}{x-2}$

**5.** $\lim\limits_{x\to -1}\dfrac{x^2+1}{x+1}$

**6.** $\lim\limits_{x\to -1}\dfrac{x^2+2x+1}{x+1}$

**7.** $\lim\limits_{x\to +\infty}\dfrac{3x^2+10x-1}{2x^2-5x}$

**8.** $\lim\limits_{x\to +\infty}\dfrac{6x^2+5x+100}{3x^2-9}$

**9.** $\lim\limits_{x\to -\infty}\dfrac{x^5-1{,}000x^4}{2x^5+10{,}000}$

**10.** $\lim\limits_{x\to -\infty}\dfrac{x^6+3{,}000x^3+1{,}000{,}000}{2x^6+1{,}000x^3}$

**11.** $\lim\limits_{x\to +\infty}\dfrac{10x^2+300x+1}{5x+2}$

**12.** $\lim\limits_{x\to +\infty}\dfrac{2x^4+20x^3}{1{,}000x^6+6}$

**13.** $\lim\limits_{x\to +\infty}\dfrac{10x^2+300x+1}{5x^3+2}$

**14.** $\lim\limits_{x\to +\infty}\dfrac{2x^4+20x^3}{1{,}000x^3+6}$

**15.** $\lim\limits_{x\to 2}e^{x-2}$

**16.** $\lim\limits_{x\to +\infty}e^{-x}$

**17.** $\lim\limits_{x\to +\infty}xe^{-x}$

**18.** $\lim\limits_{x\to -\infty}xe^{x}$

**19.** $\lim\limits_{x\to -\infty}(x^{10}+2x^5+1)e^{x}$     **20.** $\lim\limits_{x\to +\infty}(x^{50}+x^{30}+1)e^{-x}$

*In each of Exercises 21–34, the graph of f is given. Use the graph to compute the quantities asked for.* HINT [See Examples 4–5.]

**21. a.** $\lim\limits_{x\to 1}f(x)$     **b.** $\lim\limits_{x\to -1}f(x)$

**22. a.** $\lim\limits_{x\to -1}f(x)$  **b.** $\lim\limits_{x\to 1}f(x)$

**23. a.** $\lim\limits_{x\to 0^-}f(x)$  **b.** $\lim\limits_{x\to 2}f(x)$  **c.** $\lim\limits_{x\to -\infty}f(x)$  **d.** $\lim\limits_{x\to +\infty}f(x)$

**24. a.** $\lim\limits_{x\to -1}f(x)$  **b.** $\lim\limits_{x\to 1}f(x)$  **c.** $\lim\limits_{x\to +\infty}f(x)$  **d.** $\lim\limits_{x\to -\infty}f(x)$

**25. a.** $\lim\limits_{x\to 2}f(x)$     **b.** $\lim\limits_{x\to 0^+}f(x)$     **c.** $\lim\limits_{x\to 0^-}f(x)$

**d.** $\lim\limits_{x\to 0}f(x)$     **e.** $f(0)$     **f.** $\lim\limits_{x\to -\infty}f(x)$

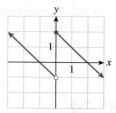

**26. a.** $\lim\limits_{x\to 3}f(x)$     **b.** $\lim\limits_{x\to 1^+}f(x)$     **c.** $\lim\limits_{x\to 1^-}f(x)$

**d.** $\lim\limits_{x\to 1}f(x)$     **e.** $f(1)$     **f.** $\lim\limits_{x\to +\infty}f(x)$

**27. a.** $\lim\limits_{x\to -2}f(x)$     **b.** $\lim\limits_{x\to -1^+}f(x)$     **c.** $\lim\limits_{x\to -1^-}f(x)$

**d.** $\lim\limits_{x\to -1}f(x)$     **e.** $f(-1)$     **f.** $\lim\limits_{x\to +\infty}f(x)$

**28. a.** $\lim\limits_{x \to -1} f(x)$    **b.** $\lim\limits_{x \to 0^+} f(x)$    **c.** $\lim\limits_{x \to 0^-} f(x)$

    **d.** $\lim\limits_{x \to 0} f(x)$    **e.** $f(0)$    **f.** $\lim\limits_{x \to -\infty} f(x)$

**29. a.** $\lim\limits_{x \to -1^+} f(x)$   **b.** $\lim\limits_{x \to -1^-} f(x)$   **c.** $\lim\limits_{x \to -1} f(x)$   **d.** $f(-1)$

**30. a.** $\lim\limits_{x \to 0^+} f(x)$   **b.** $\lim\limits_{x \to 0^-} f(x)$   **c.** $\lim\limits_{x \to 0} f(x)$   **d.** $f(0)$

**31. a.** $\lim\limits_{x \to -1} f(x)$    **b.** $\lim\limits_{x \to 0^+} f(x)$    **c.** $\lim\limits_{x \to 0^-} f(x)$

    **d.** $\lim\limits_{x \to 0} f(x)$    **e.** $f(0)$    **f.** $\lim\limits_{x \to +\infty} f(x)$

**32. a.** $\lim\limits_{x \to 1} f(x)$    **b.** $\lim\limits_{x \to 0^+} f(x)$    **c.** $\lim\limits_{x \to 0^-} f(x)$

    **d.** $\lim\limits_{x \to 0} f(x)$    **e.** $f(0)$    **f.** $\lim\limits_{x \to -\infty} f(x)$

**33. a.** $\lim\limits_{x \to -1} f(x)$    **b.** $\lim\limits_{x \to 0^+} f(x)$    **c.** $\lim\limits_{x \to 0^-} f(x)$

    **d.** $\lim\limits_{x \to 0} f(x)$    **e.** $f(0)$    **f.** $f(-1)$

**34. a.** $\lim\limits_{x \to 0^-} f(x)$    **b.** $\lim\limits_{x \to 1^+} f(x)$    **c.** $\lim\limits_{x \to 0} f(x)$

    **d.** $\lim\limits_{x \to 1} f(x)$    **e.** $f(0)$    **f.** $f(1)$

## APPLICATIONS

**35. *Economic Growth*** The value of sold goods in Mexico can be approximated by

$$v(t) = 210 - 62e^{-0.05t} \text{ trillion pesos per month } (t \geq 0),$$

where $t$ is time in months since January 2005.[2] Numerically estimate $\lim_{t \to +\infty} v(t)$ and interpret the answer. HINT [See Example 6.]

**36. *Economic Growth*** The number of housing starts for single family homes in the U.S. can be approximated by

$$n(t) = 400 + 1{,}420e^{-0.48t} \text{ thousand housing starts per year} \\ (t \geq 0),$$

where $t$ is time in years since January 2005.[3] Numerically estimate $\lim_{t \to +\infty} n(t)$ and interpret the answer.

**37. *Revenue*** The annual revenue of Amazon.com for the period 2006–2010 could be approximated by

$$P(t) = 870t^2 + 2{,}200t + 11{,}000 \text{ million dollars } (0 \leq t \leq 4),$$

where $t$ is time in years since 2006.[4] If one extrapolates the function and numerically estimates $\lim_{t \to +\infty} P(t)$, what does the answer suggest about Amazon's revenue?

---

[2]Source: Instituto Nacional de Estadística y Geografía (INEGI), www.inegi.org.mx.

[3]Source for data: www.census.gov.

[4]Source for data: www.wikinvest.com.

**38. *Net Income*** The annual net income of General Electric for the period 2005–2010 could be approximated by

$$P(t) = -2.0t^2 + 6.6t + 16 \text{ billion dollars} \quad (0 \le t \le 5),$$

where $t$ is time in years since 2005.[5] If one extrapolates the function and numerically estimates $\lim_{t \to +\infty} P(t)$, what does the answer suggest about General Electric's net income?

**39. *Scientific Research*** The number of research articles per year, in thousands, in the prominent journal *Physical Review* written by researchers in Europe can be modeled by

$$A(t) = \frac{7.0}{1 + 5.4(1.2)^{-t}},$$

where $t$ is time in years ($t = 0$ represents 1983).[6] Numerically estimate $\lim_{t \to +\infty} A(t)$ and interpret the answer.

**40. *Scientific Research*** The percentage of research articles in the prominent journal *Physical Review* written by researchers in the United States can be modeled by

$$A(t) = 25 + \frac{36}{1 + 0.6(0.7)^{-t}},$$

where $t$ is time in years ($t = 0$ represents 1983).[7] Numerically estimate $\lim_{t \to +\infty} A(t)$ and interpret the answer.

**41. *SAT Scores by Income*** The following bar graph shows U.S. math SAT scores as a function of household income:[8]

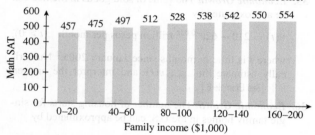

These data can be modeled by

$$S(x) = 573 - 133(0.987)^x,$$

where $S(x)$ is the average math SAT score of students whose household income is $x$ thousand dollars per year. Numerically estimate $\lim_{x \to +\infty} S(x)$ and interpret the answer.

**42. *SAT Scores by Income*** The following bar graph shows U.S. critical reading SAT scores as a function of household income:[9]

These data can be modeled by

$$S(x) = 550 - 136(0.985)^x,$$

where $S(x)$ is the average critical reading SAT score of students whose household income is $x$ thousand dollars per year. Numerically estimate $\lim_{x \to +\infty} S(x)$ and interpret the answer.

**43. *Flash Crash*** The graph shows a rough representation of what happened to the Russel 1000 Growth Index Fund (IWF) stock price on the day of the U.S. stock market crash at 2:45 pm on May 6, 2010 (the "Flash Crash"; $t$ is the time of the day in hours, and $r(t)$ is the price of the stock in dollars).[10]

**a.** Compute the following (if a limit does not exist, say why):

$$\lim_{t \to 14.75^-} r(t), \quad \lim_{t \to 14.75^+} r(t), \quad \lim_{t \to 14.75} r(t), \quad r(14.75).$$

**b.** What do the answers to part (a) tell you about the IWF stock price?

**44. *Flash Crash*** The graph shows a rough representation of the (aggregate) market depth[11] of the stocks comprising the S&P 500 on the day of the U.S. stock market crash at 2:45 pm on May 6, 2010 (the "Flash Crash"; $t$ is the time of the day in hours, and $m(t)$ is the market depth in millions of shares).

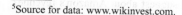

[5]Source for data: www.wikinvest.com.

[6]Based on data from 1983 to 2003. Source: The American Physical Society/*New York Times*, May 3, 2003, p. A1.

[7]*Ibid.*

[8]2009 data. Source: College Board/*New York Times* http://economix .blogs.nytimes.com.

[9]*Ibid.*

[10]The actual graph can be seen at http://seekingalpha.com.

[11]The market depth of a stock is a measure of its ability to withstand relatively large market orders, and is measured in orders to buy or sell the given stock. Source for data on graph: *Findings Regarding the Market Events of May 6, 2010*, U.S. Commodity Futures Trading Commission, U.S. Securities & Exchange Commission.

**a.** Compute the following (if a limit does not exist, say why):
$$\lim_{t \to 14.75^-} m(t), \quad \lim_{t \to 14.75^+} m(t), \quad \lim_{t \to 14.75} m(t), \quad m(14.75).$$

**b.** What do the answers to part (a) tell you about the market depth?

**45. Home Prices** The following graph shows the values of the home price index[12] for 2000–2010 together with a mathematical model $I$ extrapolating the data.

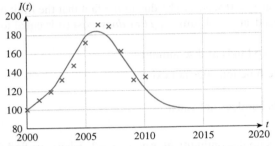

Estimate and interpret $\lim_{t \to +\infty} I(t)$.

**46. Home Prices: Optimist Projection** The following graph shows the values of the home price index[13] for 2000–2010 together with another mathematical model $I$ extrapolating the data.

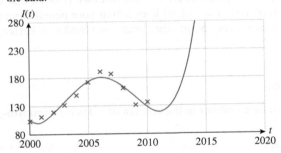

Estimate and interpret $\lim_{t \to +\infty} I(t)$.

**47. Electric Rates** The cost of electricity in Portland, Oregon, for residential customers increased suddenly on October 1, 2001, from around $0.06 to around $0.08 per kilowatt hour.[14] Let $C(t)$ be this cost at time $t$, and take $t = 1$ to represent October 1, 2001. What does the given information tell you about $\lim_{t \to 1} C(t)$?

**48. Airline Stocks** Prior to the September 11, 2001 attacks, United Airlines stock was trading at around $35 per share. Immediately following the attacks, the share price dropped by $15.[15] Let $U(t)$ be this cost at time $t$, and take $t = 11$ to represent September 11, 2001. What does the given information tell you about $\lim_{t \to 11} U(t)$?

---

[12]The index is the Standard & Poor/Case-Shiller Home Price Index. Source for data: www.standardandpoors.com.

[13]*Ibid.*

[14]Source: Portland General Electric/*New York Times*, February 2, 2002, p. C1.

[15]Stock prices are approximate.

**Foreign Trade** Annual U.S. imports from China in the years 1996 through 2003 can be approximated by
$$I(t) = t^2 + 3.5t + 50 \qquad (1 \le t \le 9)$$
billion dollars, where $t$ represents time in years since 1995. Annual U.S. exports to China in the same years can be approximated by
$$E(t) = 0.4t^2 - 1.6t + 14$$
billion dollars.[16] Exercises 49 and 50 are based on these models.

**49.** ▼ Assuming the trends shown in the above models continued indefinitely, numerically estimate
$$\lim_{t \to +\infty} I(t) \text{ and } \lim_{t \to +\infty} \frac{I(t)}{E(t)},$$
interpret your answers, and comment on the results.

**50.** ▼ Repeat Exercise 49, this time calculating
$$\lim_{t \to +\infty} E(t) \text{ and } \lim_{t \to +\infty} \frac{E(t)}{I(t)}.$$

## COMMUNICATION AND REASONING EXERCISES

**51.** Describe the method of evaluating limits numerically. Give at least one disadvantage of this method.

**52.** Describe the method of evaluating limits graphically. Give at least one disadvantage of this method.

**53.** Your friend Dion, a business student, claims that the study of limits that do not exist is completely unrealistic and has nothing to do with the world of business. Give two examples from the world of business that might convince him that he is wrong.

**54.** Your other friend Fiona claims that the study of limits is a complete farce; all you ever need to do to find the limit as $x$ approaches $a$ is substitute $x = a$. Give two examples that show she is wrong.

**55.** ▼ What is wrong with the following statement? "Because $f(a)$ is not defined, $\lim_{x \to a} f(x)$ does not exist." Illustrate your claim with an example.

**56.** ▼ What is wrong with the following statement? "Because $f(a)$ is defined, $\lim_{x \to a} f(x)$ exists." Illustrate your claim with an example.

**57.** ◆ Give an example of a function $f$ with $\lim_{x \to 1} f(x) = f(2)$.

**58.** ◆ If $S(t)$ represents the size of the universe in billions of light years at time $t$ years since the big bang and $\lim_{t \to +\infty} S(t) = 130,000$, is it possible that the universe will continue to expand forever?

**59.** ◆ Investigate $\lim_{x \to +\infty} x^n e^{-x}$ for some large values of $n$. What do you find? What do you think is the value of $\lim_{x \to +\infty} p(x)e^{-x}$ if $p(x)$ is any polynomial?

**60.** ◆ Investigate $\lim_{x \to -\infty} x^n e^x$ for some large values of $n$. What do you find? What do you think is the value of $\lim_{x \to -\infty} p(x)e^x$ if $p(x)$ is any polynomial?

---

[16]Based on quadratic regression using data from the U.S. Census Bureau Foreign Trade Division Web site www.census.gov/foreign-trade/sitc1/ as of December 2004.

**Figure 9**

# 10.2  Limits and Continuity

In the last section we saw examples of graphs that had various kinds of "breaks" or "jumps." For instance, in Example 4 we looked at the graph in Figure 9. This graph appears to have breaks, or **discontinuities**, at $x = 0$ and at $x = 1$. At $x = 0$ we saw that $\lim_{x \to 0} f(x)$ does not exist because the left- and right-hand limits are not the same. Thus, the discontinuity at $x = 0$ seems to be due to the fact that the limit does not exist there. On the other hand, at $x = 1$, $\lim_{x \to 1} f(x)$ *does* exist (it is equal to 1), but is not equal to $f(1) = 2$.

Thus, we have identified two kinds of discontinuity:

**1.** Points where the limit of the function does not exist.    $x = 0$ in Figure 9 because $\lim_{x \to 0} f(x)$ does not exist.

**2.** Points where the limit exists but does not equal the value of the function.    $x = 1$ in Figure 9 because $\lim_{x \to 1} f(x) = 1 \neq f(1)$

On the other hand, there is no discontinuity at, say, $x = -2$, where we find that $\lim_{x \to -2} f(x)$ exists and equals 2 and $f(-2)$ is also equal to 2. In other words,

$$\lim_{x \to -2} f(x) = 2 = f(-2).$$

The point $x = -2$ is an example of a point where $f$ is **continuous**. (Notice that you can draw the portion of the graph near $x = -2$ without lifting your pencil from the paper.) Similarly, $f$ is continuous at *every* point other than $x = 0$ and $x = 1$. Here is the mathematical definition.

## Continuous Function

Let $f$ be a function and let $a$ be a number in the domain of $f$. Then $f$ is **continuous at $a$** if

**a.** $\lim_{x \to a} f(x)$ exists, and

**b.** $\lim_{x \to a} f(x) = f(a)$.

If the number $a$ is an endpoint of the domain, we will understand that the limit is the left or right limit, as appropriate.

The function $f$ is said to be **continuous on its domain** if it is continuous at each point in its domain.

If $f$ is not continuous at a particular $a$ in its domain, we say that $f$ is **discontinuous** at $a$ or that $f$ has a **discontinuity** at $a$. Thus, a discontinuity can occur at $x = a$ if either

**a.** $\lim_{x \to a} f(x)$ does not exist, or

**b.** $\lim_{x \to a} f(x)$ exists but is not equal to $f(a)$.

### Quick Examples

**1.** The function shown in Figure 9 is continuous at $x = -1$ and $x = 2$. It is discontinuous at $x = 0$ and $x = 1$, and so is not continuous on its domain.

**2.** The function $f(x) = x^2$ is continuous on its domain. (Think of its graph, which contains no breaks.)

**Figure 10**

3. The function shown in Figure 10 is continuous on its domain. In particular, it is continuous at the left endpoint $x = -1$ of its domain, because $\lim_{x \to -1^+} f(x) = 1 = f(-1)$ (recall that we use only a one-sided limit at an endpoint of the domain).

4. The function $f$ whose graph is shown on the left in the following figure is continuous on its domain. (Although the graph breaks at $x = 2$, that is not a point of its domain.) The function $g$ whose graph is shown on the right is not continuous on its domain because it has a discontinuity at $x = 2$. (Here, $x = 2$ is a point of the domain of $g$.)

$y = f(x)$: Continuous on its domain

$y = g(x)$: Not continuous on its domain

**Note** Continuity and discontinuity of a function are defined only for points in a function's domain; a function cannot be continuous at a point not in its domain, and it cannot be discontinuous there either. So, if $a$ is not in the domain of $f$—that is, if $f(a)$ is not defined—then it is meaningless to talk about whether $f$ is continuous or discontinuous at $a$.  ■

---

**EXAMPLE 1  Continuous and Discontinuous Functions**

Which of the following functions are continuous on their domains?

**a.** $h(x) = \begin{cases} x + 3 & \text{if } x \leq 1 \\ 5 - x & \text{if } x > 1 \end{cases}$    **b.** $k(x) = \begin{cases} x + 3 & \text{if } x \leq 1 \\ 1 - x & \text{if } x > 1 \end{cases}$

**c.** $f(x) = \dfrac{1}{x}$    **d.** $g(x) = \begin{cases} \dfrac{1}{x} & \text{if } x \neq 0 \\ 0 & \text{if } x = 0 \end{cases}$

**Solution**

**a** and **b.** The graphs of $h$ and $k$ are shown in Figure 11.

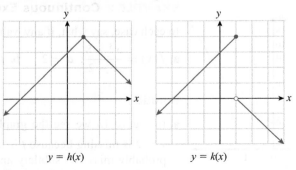

$y = h(x)$          $y = k(x)$

**Figure 11**

Even though the graph of $h$ is made up of two different line segments, it is continuous at every point of its domain, including $x = 1$ because

$$\lim_{x \to 1} h(x) = 4 = h(1).$$

On the other hand, $x = 1$ is also in the domain of $k$, but $\lim_{x \to 1} k(x)$ does not exist. Thus, $k$ is discontinuous at $x = 1$ and thus not continuous on its domain.

**c** and **d.** The graphs of $f$ and $g$ are shown in Figure 12.

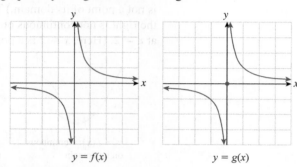

**Figure 12**

The domain of $f$ consists of all real numbers except 0 and $f$ is continuous at all such numbers. (Notice that 0 is not in the domain of $f$, so the question of continuity at 0 does not arise.) Thus, $f$ is continuous on its domain.

The function $g$, on the other hand, has its domain expanded to include 0, so we now need to check whether $g$ is continuous at 0. From the graph, it is easy to see that $g$ is discontinuous there because $\lim_{x \to 0} g(x)$ does not exist. Thus, $g$ is not continuous on its domain because it is discontinuous at 0.

➡ **Before we go on...**

**Q** : *Wait a minute! How can a function like* $f(x) = 1/x$ *be continuous when its graph has a break in it?*

**A** : We are not claiming that $f$ is continuous *at every real number*. What we are saying is that $f$ is continuous *on its domain;* the break in the graph occurs at a point not in the domain of $f$. In other words, $f$ is continuous on the set of all nonzero real numbers; it is not continuous on the set of *all* real numbers because it is not even defined on that set.

■

**EXAMPLE 2  Continuous Except at a Point**

In each case, say what, if any, value of $f(a)$ would make $f$ continuous at $a$.

**a.** $f(x) = \dfrac{x^3 - 8}{x - 2};\ a = 2$    **b.** $f(x) = \dfrac{e^{2x} - 1}{x};\ a = 0$    **c.** $f(x) = \dfrac{|x|}{x};\ a = 0$

**Solution**

**a.** In Figure 13 we see the graph of $f(x) = \dfrac{x^3 - 8}{x - 2}$. The point corresponding to $x = 2$ is missing because $f$ is not (yet) defined there. (Your graphing utility will probably miss this subtlety and render a continuous curve. See the technology note in the margin.) To turn $f$ into a function that is continuous at $x = 2$, we need

**Figure 13**

to "fill in the gap" so as to obtain a continuous curve. Since the graph suggests that the missing point is $(2, 12)$, let us define $f(2) = 12$.

Does $f$ now become continuous if we take $f(2) = 12$? From the graph, or Example 1(a) of Section 10.1,

$$\lim_{x \to 2} f(x) = \lim_{x \to 2} \frac{x^3 - 8}{x - 2} = 12,$$

which is now equal to $f(2)$. Thus, $\lim_{x \to 2} f(x) = f(2)$, showing that $f$ is now continuous at $x = 2$.

**b.** In Example 1(b) of the preceding section, we saw that

$$\lim_{x \to 0} f(x) = \lim_{x \to 0} \frac{e^{2x} - 1}{x} = 2,$$

and so, as in part (a), we must define $f(0) = 2$. This is confirmed by the graph, shown in Figure 14.

**c.** We considered the function $f(x) = |x|/x$ in Example 2 in Section 10.1. Its graph is shown in Figure 15.

**Figure 14**

**Figure 15**

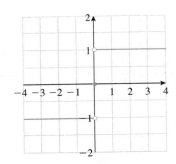

**Figure 16**

**using** Technology

It is instructive to see how technology handles the functions in Example 2. Here are the technology formulas that will work for the TI-83/84 Plus, spreadsheets, and Website function evaluator and grapher. (In spreadsheets, replace x by a cell reference and insert an equal sign in front of the formula.)
**a.** (x^3-8)/(x-2)
**b.** (e^(2x)-1)/x

Spreadsheet:
=(exp(2*A2)-1)/A2
**c.** abs(x)/x
In each case, compare the graph rendered by technology with the corresponding figure in Example 2.

Now we encounter a problem: No matter how we try to fill in the gap at $x = 0$, the result will be a discontinuous function. For example, setting $f(0) = 0$ will result in the discontinuous function shown in Figure 16. We conclude that it is impossible to assign any value to $f(0)$ to turn $f$ into a function that is continuous at $x = 0$.

We can also see this result algebraically: In Example 2 of Section 10.1, we saw that $\lim_{x \to 0} \frac{|x|}{x}$ does not exist. Thus, the resulting function will fail to be continuous at 0, no matter how we define $f(0)$.

A function not defined at an isolated point is said to have a **singularity** at that point. The function in part (a) of Example 2 has a singularity at $x = 2$, and the functions in parts (b) and (c) have singularities at $x = 0$. The functions in parts (a) and (b) have **removable singularities** because we can make these functions continuous at $x = a$ by properly defining $f(a)$. The function in part (c) has an **essential singularity** because we cannot make $f$ continuous at $x = a$ just by defining $f(a)$ properly.

## 10.2 EXERCISES

▼ more advanced    ◆ challenging

**T** indicates exercises that should be solved using technology

*In Exercises 1–14, the graph of a function f is given. Determine whether f is continuous on its domain. If it is not continuous on its domain, say why.* HINT [See Quick Examples page 702.]

**1.**

**2.**

**3.**

**4.**

**5.**

**6.**

**7.**

**8.**

**9.**

**10.**

**11.**

**12.**

**13.**

**14.**

*In Exercises 15 and 16, identify which (if any) of the given graphs represent functions continuous on their domains.* HINT [See Quick Examples page 702.]

**15. (A)**

**(B)**

**(C)**

**(D)**

**(E)**

**(F)**

**16. (A)**

**(B)**

**(C)**

**(D)**

**(E)**

**(F)**

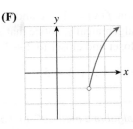

In Exercises 17–24, use a graph of $f$ or some other method to determine what, if any, value to assign to $f(a)$ to make $f$ continuous at $x = a$. HINT [See Example 2.]

**17.** $f(x) = \dfrac{x^2 - 2x + 1}{x - 1}$; $a = 1$

**18.** $f(x) = \dfrac{x^2 + 3x + 2}{x + 1}$; $a = -1$

**19.** $f(x) = \dfrac{x}{3x^2 - x}$; $a = 0$

**20.** $f(x) = \dfrac{x^2 - 3x}{x + 4}$; $a = -4$

**21.** $f(x) = \dfrac{3}{3x^2 - x}$; $a = 0$

**22.** $f(x) = \dfrac{x - 1}{x^3 - 1}$; $a = 1$

**23.** $f(x) = \dfrac{1 - e^x}{x}$; $a = 0$

**24.** $f(x) = \dfrac{1 + e^x}{1 - e^x}$; $a = 0$

In Exercises 25–34, use a graph to determine whether the given function is continuous on its domain. If it is not continuous on its domain, list the points of discontinuity. HINT [See Example 1.]

**25.** $f(x) = |x|$

**26.** $f(x) = \dfrac{|x|}{x}$

**27.** $g(x) = \dfrac{1}{x^2 - 1}$

**28.** $g(x) = \dfrac{x - 1}{x + 2}$

**29.** $f(x) = \begin{cases} x + 2 & \text{if } x < 0 \\ 2x - 1 & \text{if } x \geq 0 \end{cases}$

**30.** $f(x) = \begin{cases} 1 - x & \text{if } x \leq 1 \\ x - 1 & \text{if } x > 1 \end{cases}$

**31.** $h(x) = \begin{cases} \dfrac{|x|}{x} & \text{if } x \neq 0 \\ 0 & \text{if } x = 0 \end{cases}$

**32.** $h(x) = \begin{cases} \dfrac{1}{x^2} & \text{if } x \neq 0 \\ 2 & \text{if } x = 0 \end{cases}$

**33.** $g(x) = \begin{cases} x + 2 & \text{if } x < 0 \\ 2x + 2 & \text{if } x \geq 0 \end{cases}$

**34.** $g(x) = \begin{cases} 1 - x & \text{if } x \leq 1 \\ x + 1 & \text{if } x > 1 \end{cases}$

**COMMUNICATION AND REASONING EXERCISES**

**35.** If a function is continuous on its domain, is it continuous at every real number? Explain.

**36.** True or false? The graph of a function that is continuous on its domain is a continuous curve with no breaks in it. Explain your answer.

**37.** True or false? The graph of a function that is continuous at every real number is a continuous curve with no breaks in it. Explain your answer.

**38.** True or false? If the graph of a function is a continuous curve with no breaks in it, then the function is continuous on its domain. Explain your answer.

**39.** ▼ Give a formula for a function that is continuous on its domain but whose graph consists of three distinct curves.

**40.** ▼ Give a formula for a function that is not continuous at $x = -1$ but is not discontinuous there either.

**41.** ▼ Draw the graph of a function that is discontinuous at every integer.

**42.** ▼ Draw the graph of a function that is continuous on its domain but whose graph has a break at every integer.

**43.** ▼ Describe a real-life scenario in the stock market that can be modeled by a discontinuous function.

**44.** ▼ Describe a real-life scenario in your room that can be modeled by a discontinuous function.

# 10.3 Limits and Continuity: Algebraic Viewpoint

Although numerical and graphical estimation of limits is effective, the estimates these methods yield may not be perfectly accurate. The algebraic method, when it can be used, will always yield an exact answer. Moreover, algebraic analysis of a function often enables us to take a function apart and see "what makes it tick."

Let's start with the function $f(x) = 2 + x$ and ask: What happens to $f(x)$ as $x$ approaches 3? To answer this algebraically, notice that as $x$ gets closer and closer to 3, the quantity $2 + x$ must get closer and closer to $2 + 3 = 5$. Hence,

$$\lim_{x \to 3} f(x) = \lim_{x \to 3}(2 + x) = 2 + 3 = 5$$

---

**Q** : *Is that all there is to the algebraic method? Just substitute $x = a$?*

**A** : Under certain circumstances: Notice that by substituting $x = 3$ we *evaluated the function at $x = 3$*. In other words, we relied on the fact that

$$\lim_{x \to 3} f(x) = f(3).$$

In Section 10.2 we said that a function satisfying this equation is *continuous* at $x = 3$.

---

Thus,

*If we know that the function $f$ is continuous at a point $a$, we can compute $\lim_{x \to a} f(x)$ by simply substituting $x = a$ into $f(x)$.*

To use this fact, we need to know how to recognize continuous functions when we see them. Geometrically, they are easy to spot: A function is continuous at $x = a$ if its graph has no break at $x = a$. Algebraically, a large class of functions are known to be continuous on their domains—those, roughly speaking, that are *specified by a single formula*.

We can be more precise: A **closed-form function** is any function that can be obtained by combining constants, powers of $x$, exponential functions, radicals, logarithms, absolute values, trigonometric functions (and some other functions we do not encounter in this text) into a *single* mathematical formula by means of the usual arithmetic operations and composition of functions. (They can be as complicated as we like.)

### Closed-Form Functions

A function is **written in closed form** if it is specified by combining constants, powers of $x$, exponential functions, radicals, logarithms, absolute values, trigonometric functions (and some other functions we do not encounter in this text) into a *single* mathematical formula by means of the usual arithmetic operations and composition of functions. A **closed-form function** is any function that can be written in closed form.

#### Quick Examples

1. $3x^2 - |x| + 1, \dfrac{\sqrt{x^2 - 1}}{6x - 1}, e^{-\frac{4x^2 - 1}{x}}$, and $\sqrt{\log_3(x^2 - 1)}$ are written in closed form, so they are all closed-form functions.

2. $f(x) = \begin{cases} -1 & \text{if } x \leq -1 \\ x^2 + x & \text{if } -1 < x \leq 1 \\ 2 - x & \text{if } 1 < x \leq 2 \end{cases}$ 
is not written in closed-form because $f(x)$ is not expressed by a *single* mathematical formula.*

---

\* It is possible to rewrite some piecewise-defined functions in closed form (using a single formula), but not this particular function, so $f(x)$ is not a closed-form function.

What is so special about closed-form functions is the following theorem.

### Theorem 10.1  Continuity of Closed-Form Functions

*Every closed-form function is continuous on its domain.* Thus, if $f$ is a closed-form function and $f(a)$ is defined, then $\lim_{x \to a} f(x)$ exists, and equals $f(a)$. (When $a$ is an endpoint of the domain of $f$, we understand the limit to be the appropriate one-sided limit.)

### Quick Example

$f(x) = 1/x$ is a closed-form function, and its natural domain consists of all real numbers except 0. Thus, $f$ is continuous at every nonzero real number. That is,

$$\lim_{x \to a} \frac{1}{x} = \frac{1}{a}$$

provided $a \neq 0$.

Mathematics majors spend a great deal of time studying the proof of this theorem. We ask you to accept it without proof.

### EXAMPLE 1  Limit of a Closed-Form Function

Evaluate the following limits algebraically:

**a.** $\lim\limits_{x \to 1} \dfrac{x^3 - 8}{x - 2}$      **b.** $\lim\limits_{x \to 2} \dfrac{x^3 - 8}{x - 2}$.

### Solution

**a.** First, notice that $(x^3 - 8)/(x - 2)$ is a closed-form function because it is specified by a single algebraic formula. Also, $x = 1$ is in the domain of this function. Therefore the theorem applies, and

$$\lim_{x \to 1} \frac{x^3 - 8}{x - 2} = \frac{1^3 - 8}{1 - 2} = 7.$$

**b.** Although $(x^3 - 8)/(x - 2)$ is a closed-form function, $x = 2$ is not in its domain. Thus, the theorem does not apply and we cannot obtain the limit by substitution. However—and this is the key to finding such limits—*some preliminary algebraic simplification will allow us to obtain a closed-form function with $x = 2$ in its domain.* To do this, notice first that the numerator can be factored as

$$x^3 - 8 = (x - 2)(x^2 + 2x + 4).$$

Thus,

$$\frac{x^3 - 8}{x - 2} = \frac{(x - 2)(x^2 + 2x + 4)}{x - 2} = x^2 + 2x + 4.$$

Once we have canceled the offending $(x - 2)$ in the denominator, we are left with a closed-form function *with 2 in its domain*. Thus,

$$\lim_{x \to 2} \frac{x^3 - 8}{x - 2} = \lim_{x \to 2}(x^2 + 2x + 4)$$

$$= 2^2 + 2(2) + 4 = 12. \quad \text{Substitute } x = 2.$$

This confirms the answer we found numerically in Example 1 in Section 10.1.

➡ **Before we go on...** Notice that in Example 1(b) before simplification, the substitution $x = 2$ yields

$$\frac{x^3 - 8}{x - 2} = \frac{8 - 8}{2 - 2} = \frac{0}{0}.$$

Worse than the fact that 0/0 is undefined, it also conveys absolutely no information as to what the limit might be. (The limit turned out to be 12!) We therefore call the expression 0/0 an **indeterminate form**. Once simplified, the function became $x^2 + 2x + 4$, which, upon the substitution $x = 2$, yielded 12—no longer an indeterminate form. In general, we have the following rule of thumb:

*If the substitution $x = a$ yields the indeterminate form 0/0, try simplifying by the method in Example 1.*

We will say more about indeterminate forms in Example 2.  ∎

---

**Q**: *There is something suspicious about Example 1(b). If 2 was not in the domain before simplifying but was in the domain after simplifying, we must have changed the function, right?*

**A**: Correct. In fact, when we said that

$$\frac{x^3 - 8}{x - 2} = x^2 + 2x + 4$$

Domain excludes 2    Domain includes 2

we were lying a little bit. What we really meant is that these two expressions are equal *where both are defined*. The functions $(x^3 - 8)/(x - 2)$ and $x^2 + 2x + 4$ are different functions. The difference is that $x = 2$ is not in the domain of $(x^3 - 8)/(x - 2)$ and is in the domain of $x^2 + 2x + 4$. Since $\lim_{x \to 2} f(x)$ explicitly *ignores* any value that $f$ may have at 2, this does not affect the limit. From the point of view of the limit at 2, these functions *are* equal. In general we have the following rule.

---

## Functions with Equal Limits

If $f(x) = g(x)$ for all $x$ except possibly $x = a$, then

$$\lim_{x \to a} f(x) = \lim_{x \to a} g(x).$$

### Quick Example

$$\frac{x^2 - 1}{x - 1} = x + 1 \text{ for all } x \text{ except } x = 1. \quad \text{Write } \frac{x^2 - 1}{x - 1} \text{ as } \frac{(x + 1)(x - 1)}{x - 1}$$

$$\text{and cancel the } (x - 1).$$

Therefore,

$$\lim_{x \to 1} \frac{x^2 - 1}{x - 1} = \lim_{x \to 1}(x + 1) = 1 + 1 = 2.$$

**Q** : *How do we find* $\lim_{x \to a} f(x)$ *when* $x = a$ *is not in the domain of the function* $f$ *and we cannot simplify the given function to make* $a$ *a point of the domain?*

**A** : In such a case, it might be necessary to analyze the function by some other method, such as numerically or graphically. However, if we do not obtain the indeterminate form 0/0 upon substitution, we can often say what the limit is, as the following example shows.

---

**EXAMPLE 2  Limit of a Closed-Form Function at a Point Not in Its Domain: The Determinate Form $k/0$**

Evaluate the following limits, if they exist:

**a.** $\lim_{x \to 1^+} \dfrac{x^2 - 4x + 1}{x - 1}$
**b.** $\lim_{x \to 1} \dfrac{x^2 - 4x + 1}{x - 1}$
**c.** $\lim_{x \to 1} \dfrac{x^2 - 4x + 1}{x^2 - 2x + 1}$

**Solution**

**a.** Although the function $f(x) = \dfrac{x^2 - 4x + 1}{x - 1}$ is a closed-form function, $x = 1$ is not in its domain. Notice that substituting $x = 1$ gives

$$\frac{x^2 - 4x + 1}{x - 1} = \frac{1^2 - 4 + 1}{1 - 1} = \frac{-2}{0} \qquad \text{The determinate form } \frac{k}{0}$$

which, although not defined, conveys important information to us: As $x$ gets closer and closer to 1, the numerator approaches $-2$ and the denominator gets closer and closer to 0. Now, if we divide a number close to $-2$ by a number close to zero, we get a number of large absolute value; for instance,

$$\frac{-2.1}{0.0001} = -21{,}000 \qquad \text{and} \qquad \frac{-2.1}{-0.0001} = 21{,}000$$

$$\frac{-2.01}{0.00001} = -201{,}000 \qquad \text{and} \qquad \frac{-2.01}{-0.00001} = 201{,}000.$$

(Compare Example 5 in Section 10.1.) In our limit for part (a), $x$ is approaching 1 from the right, so the denominator $x - 1$ is positive (as $x$ is to the right of 1). Thus we have the scenario illustrated previously on the left, and we can conclude that

$$\lim_{x \to 1^+} \frac{x^2 - 4x + 1}{x - 1} = -\infty. \qquad \text{Think of this as } \frac{-2}{0^+} = -\infty.$$

**b.** This time, $x$ could be approaching 1 from either side. We already have, from part (a)

$$\lim_{x \to 1^+} \frac{x^2 - 4x + 1}{x - 1} = -\infty.$$

The same reasoning we used in part (a) gives

$$\lim_{x \to 1^-} \frac{x^2 - 4x + 1}{x - 1} = +\infty \qquad \text{Think of this as } \frac{-2}{0^-} = +\infty.$$

because now the denominator is negative and still approaching zero while the numerator still approaches $-2$ and therefore is also negative. (See the numerical calculations above on the right.) Because the left and right limits do not agree, we conclude that

$$\lim_{x \to 1} \frac{x^2 - 4x + 1}{x - 1} \text{ does not exist.}$$

**c.** First notice that the denominator factors:

$$\lim_{x \to 1} \frac{x^2 - 4x + 1}{x^2 - 2x + 1} = \lim_{x \to 1} \frac{x^2 - 4x + 1}{(x-1)^2}.$$

As $x$ approaches 1, the numerator approaches $-2$ as before, and the denominator approaches 0. However, this time, the denominator $(x-1)^2$, being a square, is $\geq 0$, regardless of from which side $x$ is approaching 1. Thus, the entire function is negative as $x$ approaches 1, and

$$\lim_{x \to 1} \frac{x^2 - 4x + 1}{(x-1)^2} = -\infty. \qquad \frac{-2}{0^+} = -\infty$$

---

➡ **Before we go on...** In general, the determinate forms $\dfrac{k}{0^+}$ and $\dfrac{k}{0^-}$ will always yield $\pm\infty$, with the sign depending on the sign of the overall expression as $x \to a$. (When we write the form $\dfrac{k}{0}$ we always mean $k \neq 0$.) This and other determinate forms are discussed further after Example 4.

Figure 17 shows the graphs of $\dfrac{x^2 - 4x + 1}{x - 1}$ and $\dfrac{x^2 - 4x + 1}{(x-1)^2}$ from Example 2. You should check that results we obtained above agree with a geometric analysis of these graphs near $x = 1$.

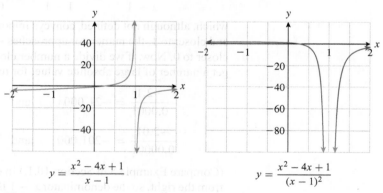

$$y = \frac{x^2 - 4x + 1}{x - 1} \qquad\qquad y = \frac{x^2 - 4x + 1}{(x-1)^2}$$

**Figure 17**

We can also use algebraic techniques to analyze functions that are not given in closed form.

---

### EXAMPLE 3  Functions Not Written in Closed Form

For which values of $x$ are the following piecewise defined functions continuous?

**a.** $f(x) = \begin{cases} x^2 + 2 & \text{if } x < 1 \\ 2x - 1 & \text{if } x \geq 1 \end{cases}$    **b.** $g(x) = \begin{cases} x^2 - x + 1 & \text{if } x \leq 0 \\ 1 - x & \text{if } 0 < x \leq 1 \\ x - 3 & \text{if } x > 1 \end{cases}$

**Solution**

**a.** The function $f(x)$ is given in closed form over the intervals $(-\infty, 1)$ and $[1, +\infty)$. At $x = 1$, $f(x)$ suddenly switches from one closed-form formula to another, so

$x = 1$ is the only place where there is a potential problem with continuity. To investigate the continuity of $f(x)$ at $x = 1$, let's calculate the limit there:

$$\lim_{x \to 1^-} f(x) = \lim_{x \to 1^-} (x^2 + 2) \qquad f(x) = x^2 + 2 \text{ for } x < 1.$$
$$= (1)^2 + 2 = 3 \qquad x^2 + 2 \text{ is closed-form.}$$

$$\lim_{x \to 1^+} f(x) = \lim_{x \to 1^+} (2x - 1) \qquad f(x) = 2x - 1 \text{ for } x > 1.$$
$$= 2(1) - 1 = 1. \qquad 2x - 1 \text{ is closed-form.}$$

Because the left and right limits are different, $\lim_{x \to 1} f(x)$ does not exist, and so $f(x)$ is discontinuous at $x = 1$.

**b.** The only potential points of discontinuity for $g(x)$ occur at $x = 0$ and $x = 1$:

$$\lim_{x \to 0^-} g(x) = \lim_{x \to 0^-} (x^2 - x + 1) = 1$$
$$\lim_{x \to 0^+} g(x) = \lim_{x \to 0^+} (1 - x) = 1.$$

Thus, $\lim_{x \to 0} g(x) = 1$. Further, $g(0) = 0^2 - 0 + 1 = 1$ from the formula, and so

$$\lim_{x \to 0} g(x) = g(0),$$

which shows that $g(x)$ is continuous at $x = 0$. At $x = 1$ we have

$$\lim_{x \to 1^-} g(x) = \lim_{x \to 1^-} (1 - x) = 0$$
$$\lim_{x \to 1^+} g(x) = \lim_{x \to 1^+} (x - 3) = -2$$

so that $\lim_{x \to 1} g(x)$ does not exist. Thus, $g(x)$ is discontinuous at $x = 1$. We conclude that $g(x)$ is continuous at every real number $x$ except $x = 1$.

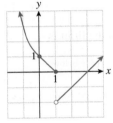

$y = g(x)$

**Figure 18**

➡ **Before we go on...** Figure 18 shows the graph of $g$ from Example 3(b). Notice how the discontinuity at $x = 1$ shows up as a break in the graph, whereas at $x = 0$ the two pieces "fit together" at the point $(0, 1)$. ■

## Limits at Infinity

Let's look once again at some limits similar to those in Examples 3 and 6 in Section 10.1.

### EXAMPLE 4 Limits at Infinity

Compute the following limits, if they exist:

**a.** $\lim_{x \to +\infty} \dfrac{2x^2 - 4x}{x^2 - 1}$ 
**b.** $\lim_{x \to -\infty} \dfrac{2x^2 - 4x}{x^2 - 1}$

**c.** $\lim_{x \to +\infty} \dfrac{-x^3 - 4x}{2x^2 - 1}$ 
**d.** $\lim_{x \to +\infty} \dfrac{2x^2 - 4x}{5x^3 - 3x + 5}$

**e.** $\lim_{t \to +\infty} (e^{0.1t} - 20)$ 
**f.** $\lim_{t \to +\infty} \dfrac{80}{1 + 2.2(3.68)^{-t}}$

**Solution  a** and **b.** While calculating the values for the tables used in Example 3 in Section 10.1, you might have noticed that the highest power of $x$ in both the numerator and denominator dominated the calculations. For instance, when $x = 100,000$, the term $2x^2$ in the numerator has the value of 20,000,000,000, whereas the term $4x$ has the comparatively insignificant value of 400,000. Similarly, the term $x^2$ in the denominator overwhelms the term $-1$. In other words, for large values of $x$ (or negative values with large magnitude),

$$\frac{2x^2 - 4x}{x^2 - 1} \approx \frac{2x^2}{x^2} \qquad \text{Use only the highest powers top and bottom.}$$

$$= 2.$$

Therefore,

$$\lim_{x \to \pm\infty} \frac{2x^2 - 4x}{x^2 - 1} = \lim_{x \to \pm\infty} \frac{2x^2}{x^2}$$

$$= \lim_{x \to \pm\infty} 2 = 2.$$

The procedure of using only the highest powers of $x$ to compute the limit is stated formally and justified after this example.

**c.** Applying the previous technique of looking only at highest powers gives

$$\lim_{x \to +\infty} \frac{-x^3 - 4x}{2x^2 - 1} = \lim_{x \to +\infty} \frac{-x^3}{2x^2} \qquad \text{Use only the highest powers top and bottom.}$$

$$= \lim_{x \to +\infty} \frac{-x}{2}. \qquad \text{Simplify.}$$

As $x$ gets large, $-x/2$ gets large in magnitude but negative, so the limit is

$$\lim_{x \to +\infty} \frac{-x}{2} = -\infty. \qquad \frac{-\infty}{2} = -\infty \ \ \text{(See below.)}$$

**d.** $\displaystyle \lim_{x \to +\infty} \frac{2x^2 - 4x}{5x^3 - 3x + 5} = \lim_{x \to +\infty} \frac{2x^2}{5x^3}$ \qquad Use only the highest powers top and bottom.

$$= \lim_{x \to +\infty} \frac{2}{5x}.$$

As $x$ gets large, $2/(5x)$ gets close to zero, so the limit is

$$\lim_{x \to +\infty} \frac{2}{5x} = 0. \qquad \frac{2}{\infty} = 0 \ \text{(See below.)}$$

**e.** Here we do not have a ratio of polynomials. However, we know that, as $t$ becomes large and positive, so does $e^{0.1t}$, and hence also $e^{0.1t} - 20$. Thus,

$$\lim_{t \to +\infty} (e^{0.1t} - 20) = +\infty \qquad e^{+\infty} = +\infty \ \text{(See below.)}$$

**f.** As $t \to +\infty$, the term $(3.68)^{-t} = \dfrac{1}{3.68^t}$ in the denominator, being 1 divided by a very large number, approaches zero. Hence the denominator $1 + 2.2(3.68)^{-t}$ approaches $1 + 2.2(0) = 1$ as $t \to +\infty$. Thus,

$$\lim_{t \to +\infty} \frac{80}{1 + 2.2(3.68)^{-t}} = \frac{80}{1 + 2.2(0)} = 80 \qquad (3.68)^{-\infty} = 0 \ \text{(See below.)}$$

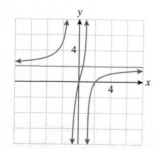

**Figure 19**

➡ **Before we go on...** Let's now look at the graph of the function $\dfrac{2x^2 - 4x}{x^2 - 1}$ in parts (a) and (b) of Example 4. We say that the graph of $f$ has a **horizontal asymptote** at $y = 2$ because of the limits we have just calculated. This means that the graph approaches the horizontal line $y = 2$ far to the right or left (in this case, to both the right and left). Figure 19 shows the graph of $f$ together with the line $y = 2$.

The graph reveals some additional interesting information: as $x \to 1^+$, $f(x) \to -\infty$, and as $x \to 1^-$, $f(x) \to +\infty$. Thus,

$$\lim_{x \to 1} f(x) \text{ does not exist.}$$

See if you can determine what happens as $x \to -1$.

If you graph the functions in parts (d) and (f) of Example 4, you will again see a horizontal asymptote. Do the limits in parts (c) and (e) show horizontal asymptotes? ■

It is worthwhile looking again at what we did in each of the limits in Example 4:

**a** and **b.** We saw that $\dfrac{2x^2 - 4x}{x^2 - 1} \approx \dfrac{2x^2}{x^2}$, and then we canceled the $x^2$. Notice that, before we cancel, letting $x$ approach $\pm\infty$ in the numerator and denominator yields the ratio $\dfrac{\infty}{\infty}$, which, like $\dfrac{0}{0}$, is another *indeterminate form*, and indicates to us that further work is needed—in this case cancellation—before we can write down the limit.

**c.** We obtained $\dfrac{-x^3 - 4x}{2x^2 - 1} \approx \dfrac{-x^3}{2x^2}$, which results in another indeterminate form, $\dfrac{-\infty}{\infty}$, as $x \to +\infty$. Cancellation of the $x^2$ gave us $\dfrac{-x}{2}$, resulting in the *determinate* form $\dfrac{-\infty}{2} = -\infty$ (a very large number divided by 2 is again a very large number).

**d.** Here, $\dfrac{2x^2 - 4x}{5x^3 - 3x + 5} \approx \dfrac{2x^2}{5x^3} = \dfrac{2}{5x}$, and the cancellation step turns the indeterminate form $\dfrac{\infty}{\infty}$ into the determinate form $\dfrac{2}{\infty} = 0$ (dividing 2 by a very large number yields a very small number).

**e.** We reasoned that $e$ raised to a large positive number is large and positive. Putting $t = +\infty$ gives us the determinate form $e^{+\infty} = +\infty$.

**f.** Here we reasoned that 3.68 raised to a large *negative* number is close to zero. Putting $t = -\infty$ gives us the determinate form $3.68^{-\infty} = 1/3.68^{+\infty} = 1/\infty = 0$ (see (d)).

In parts (a)–(d) of Example 4, $f(x)$ was a **rational function**: a quotient of polynomial functions. We calculated the limit of $f(x)$ at $\pm\infty$ by ignoring all powers of $x$ in both the numerator and denominator except for the largest. Following is a theorem that justifies this procedure.

**Theorem 10.2  Evaluating the Limit of a Rational Function at $\pm\infty$**

If $f(x)$ has the form

$$f(x) = \frac{c_n x^n + c_{n-1} x^{n-1} + \cdots + c_1 x + c_0}{d_m x^m + d_{m-1} x^{m-1} + \cdots + d_1 x + d_0}$$

with the $c_i$ and $d_i$ constants ($c_n \neq 0$ and $d_m \neq 0$), then we can calculate the limit of $f(x)$ as $x \to \pm\infty$ by ignoring all powers of $x$ except the highest in both the numerator and denominator. Thus,

$$\lim_{x \to \pm\infty} f(x) = \lim_{x \to \pm\infty} \frac{c_n x^n}{d_m x^m}.$$

> ### Quick Examples
>
> (See Example 4.)
>
> **1.** $\lim\limits_{x \to +\infty} \dfrac{2x^2 - 4x}{x^2 - 1} = \lim\limits_{x \to +\infty} \dfrac{2x^2}{x^2} = \lim\limits_{x \to +\infty} 2 = 2$
>
> **2.** $\lim\limits_{x \to +\infty} \dfrac{-x^3 - 4x}{2x^2 - 1} = \lim\limits_{x \to +\infty} \dfrac{-x^3}{2x^2} = \lim\limits_{x \to +\infty} \dfrac{-x}{2} = -\infty$
>
> **3.** $\lim\limits_{x \to +\infty} \dfrac{2x^2 - 4x}{5x^3 - 3x + 5} = \lim\limits_{x \to +\infty} \dfrac{2x^2}{5x^3} = \lim\limits_{x \to +\infty} \dfrac{2}{5x} = 0$

**Proof** Our function $f(x)$ is a polynomial of degree $n$ divided by a polynomial of degree $m$. If $n$ happens to be larger than $m$, then dividing the top and bottom by the largest power $x^n$ of $x$ gives

$$f(x) = \frac{c_n x^n + c_{n-1} x^{n-1} + \cdots + c_1 x + c_0}{d_m x^m + d_{m-1} x^{m-1} + \cdots + d_1 x + d_0}$$

$$= \frac{c_n x^n / x^n + c_{n-1} x^{n-1} / x^n + \cdots + c_1 x / x^n + c_0 / x^n}{d_m x^m / x^n + d_{m-1} x^{m-1} / x^n + \cdots + d_1 x / x^n + d_0 / x^n}.$$

Canceling powers of $x$ in each term and remembering that $n > m$ leaves us with

$$f(x) = \frac{c_n + c_{n-1}/x + \cdots + c_1 / x^{n-1} + c_0 / x^n}{d_m / x^{n-m} + d_{m-1} / x^{n-m+1} + \cdots + d_1 / x^{n-1} + d_0 / x^n}.$$

As $x \to \pm\infty$, all the terms shown in red approach 0, so we can ignore them in taking the limit. (The first term in the denominator happens to approach 0 as well, but we retain it for convenience.) Thus,

$$\lim_{x \to \pm\infty} f(x) = \lim_{x \to \pm\infty} \frac{c_n}{d_m / x^{n-m}} = \lim_{x \to \pm\infty} \frac{c_n x^n}{d_m x^m},$$

as required. The cases when $n$ is smaller than $m$ and $m = n$ are proved similarly by dividing top and bottom by the largest power of $x$ in each case.

## Some Determinate and Indeterminate Forms

The following table brings these ideas together with our observations in Example 2.

> ### Some Determinate and Indeterminate Forms
>
> $\dfrac{0}{0}$ and $\pm\dfrac{\infty}{\infty}$ are **indeterminate**; evaluating limits in which these arise requires simplification or further analysis.[†]
>
> The following are **determinate** forms for any nonzero number $k$:
>
> $\dfrac{k}{0^{\pm}} = \pm\infty$        $\dfrac{k}{\text{Small}} = \text{Big*}$ (See Example 2.)
>
> $k(\pm\infty) = \pm\infty$              $k \times \text{Big} = \text{Big*}$
>
> $k \pm \infty = \pm\infty$              $k \pm \text{Big} = \pm\text{Big}$

[†] Some other indeterminate forms are: $\pm\infty \cdot 0$, $\infty - \infty$, and $1^{\infty}$. (These are not discussed in this text, but see the Communication and Reasoning exercises for this section.)

$$\pm\frac{\infty}{k} = \pm\infty \qquad\qquad \frac{\text{Big}}{k} = \text{Big*}$$

$$\pm\frac{k}{\infty} = 0 \qquad\qquad \frac{k}{\text{Big}} = \text{Small}$$

and, if $k > 1$, then

$$k^{+\infty} = +\infty \qquad\qquad k^{\text{Big positive}} = \text{Big}$$

$$k^{-\infty} = 0. \qquad\qquad k^{\text{Big negative}} = \text{Small}$$

*The sign gets switched in these forms if $k$ is negative.

## Quick Examples

**1.** $\displaystyle\lim_{x\to 0}\frac{60}{2x^2} = +\infty$ $\qquad\qquad$ $\dfrac{k}{0^+} = +\infty$

**2.** $\displaystyle\lim_{x\to -1^-}\frac{2x-6}{x+1} = +\infty$ $\qquad\qquad$ $\dfrac{-8}{0^-} = +\infty$

**3.** $\displaystyle\lim_{x\to -\infty} 3x - 5 = -\infty$ $\qquad\qquad$ $3(-\infty) - 5 = -\infty - 5 = -\infty$

**4.** $\displaystyle\lim_{x\to +\infty}\frac{2x}{60} = +\infty$ $\qquad\qquad$ $\dfrac{2(\infty)}{60} = \infty$

**5.** $\displaystyle\lim_{x\to -\infty}\frac{60}{2x} = 0$ $\qquad\qquad$ $\dfrac{60}{2(-\infty)} = 0$

**6.** $\displaystyle\lim_{x\to +\infty}\frac{60x}{2x} = 30$ $\qquad\qquad$ $\dfrac{\infty}{\infty}$ is indeterminate but we can cancel.

**7.** $\displaystyle\lim_{x\to -\infty}\frac{60}{e^x - 1} = \frac{60}{0-1} = -60$ $\qquad$ $e^{-\infty} = 0$

## FAQs

### Strategy for Evaluating Limits Algebraically

**Q** : Is there a systematic way to evaluate a limit $\lim_{x\to a} f(x)$ algebraically?

**A** : The following approach is often successful:

**Case 1:** $a$ Is a Finite Number (Not $\pm\infty$)

1. Decide whether $f$ is a closed-form function. If it is not, then find the left and right limits at the values of $x$ where the function changes from one formula to another.

2. If $f$ is a closed-form function, try substituting $x = a$ in the formula for $f(x)$. Then one of the following three things may happen:

   $f(a)$ is defined. Then $\lim_{x\to a} f(x) = f(a)$.

   $f(a)$ is not defined and has the indeterminate form 0/0. Try to simplify the expression for $f$ to cancel one of the terms that gives 0.

   $f(a)$ is not defined and has one of the determinate forms listed above in the above table. Use the table to determine the limit as in the Quick Examples.

> **Case 2:** $a = \pm\infty$
>
> Remember that we can use the determinate forms $k^{+\infty} = \infty$ and $k^{-\infty} = 0$ if $k > 1$. Further, if the given function is a polynomial or ratio of polynomials, use the technique of Example 4: Focus only on the highest powers of $x$ and then simplify to obtain either a number $L$, in which case the limit exists and equals $L$, or one of the determinate forms $\pm\infty/k = \pm\infty$ or $\pm k/\infty = 0$.

There is another technique for evaluating certain difficult limits, called *l'Hospital's rule,* but this uses derivatives, so we'll have to wait to discuss it until Section 11.1.

# 10.3   EXERCISES

▼ more advanced    ◆ challenging

🔳 indicates exercises that should be solved using technology

*In Exercises 1–4, complete the given sentence.*

1. The closed-form function $f(x) = \dfrac{1}{x-1}$ is continuous for all $x$ except _____. HINT [See Quick Example on page 709.]

2. The closed-form function $f(x) = \dfrac{1}{x^2-1}$ is continuous for all $x$ except _____. HINT [See Quick Example on page 709.]

3. The closed-form function $f(x) = \sqrt{x+1}$ has $x = 3$ in its domain. Therefore, $\lim_{x\to3}\sqrt{x+1} = $ ____. HINT [See Example 1.]

4. The closed-form function $f(x) = \sqrt{x-1}$ has $x = 10$ in its domain. Therefore, $\lim_{x\to10}\sqrt{x-1} = $ ____. HINT [See Example 1.]

*In Exercises 5–20, determine whether the given limit leads to a determinate or indeterminate form. Evaluate the limit if it exists, or say why if not.* HINT [See Example 2 and Quick Examples on page 717.]

5. $\lim_{x\to0}\dfrac{60}{x^4}$

6. $\lim_{x\to0}\dfrac{2x^2}{x^2}$

7. $\lim_{x\to0}\dfrac{x^3-1}{x^3}$

8. $\lim_{x\to0}\dfrac{-2}{x^2}$

9. $\lim_{x\to-\infty}(-x^2+5)$

10. $\lim_{x\to0}\dfrac{2x^2+4}{x}$

11. $\lim_{x\to+\infty}4^{-x}$

12. $\lim_{x\to+\infty}\dfrac{60+e^{-x}}{2-e^{-x}}$

13. $\lim_{x\to0}\dfrac{-x^3}{3x^3}$

14. $\lim_{x\to-\infty}3x^2+6$

15. $\lim_{x\to-\infty}\dfrac{-x^3}{3x^6}$

16. $\lim_{x\to+\infty}\dfrac{-x^6}{3x^3}$

17. $\lim_{x\to-\infty}\dfrac{4}{-x+2}$

18. $\lim_{x\to-\infty}e^x$

19. $\lim_{x\to-\infty}\dfrac{60}{e^x-1}$

20. $\lim_{x\to-\infty}\dfrac{2}{2x^2+3}$

*Calculate the limits in Exercises 21–74 algebraically. If a limit does not exist, say why.*

21. $\lim_{x\to0}(x+1)$

    HINT [See Example 1(a).]

22. $\lim_{x\to0}(2x-4)$

    HINT [See Example 1(a).]

23. $\lim_{x\to2}\dfrac{2+x}{x}$

24. $\lim_{x\to-1}\dfrac{4x^2+1}{x}$

25. $\lim_{x\to-1}\dfrac{x+1}{x}$

26. $\lim_{x\to4}(x+\sqrt{x})$

27. $\lim_{x\to8}(x-\sqrt[3]{x})$

28. $\lim_{x\to1}\dfrac{x-2}{x+1}$

29. $\lim_{h\to1}(h^2+2h+1)$

30. $\lim_{h\to0}(h^3-4)$

31. $\lim_{h\to3}2$

32. $\lim_{h\to0}-5$

33. $\lim_{h\to0}\dfrac{h^2}{h+h^2}$

    HINT [See Example 1(b).]

34. $\lim_{h\to0}\dfrac{h^2+h}{h^2+2h}$

    HINT [See Example 1(b).]

35. $\lim_{x\to1}\dfrac{x^2-2x+1}{x^2-x}$

36. $\lim_{x\to-1}\dfrac{x^2+3x+2}{x^2+x}$

37. $\lim_{x\to2}\dfrac{x^3-8}{x-2}$

38. $\lim_{x\to-2}\dfrac{x^3+8}{x^2+3x+2}$

39. $\lim_{x\to0^+}\dfrac{1}{x^2}$   HINT [See Example 2.]

40. $\lim_{x\to0^+}\dfrac{1}{x^2-x}$   HINT [See Example 2.]

41. $\lim_{x\to-1}\dfrac{x^2+1}{x+1}$

42. $\lim_{x\to-1^-}\dfrac{x^2+1}{x+1}$

43. $\lim_{x\to-2^+}\dfrac{x^2+8}{x^2+3x+2}$

44. $\lim_{x\to-1}\dfrac{x^2+3x}{x^2+x}$

**45.** $\lim\limits_{x \to -2} \dfrac{x^2 + 8}{x^2 + 3x + 2}$

**46.** $\lim\limits_{x \to -1} \dfrac{x^2 + 3x}{x^2 + 2x + 1}$

**47.** $\lim\limits_{x \to 2} \dfrac{x^2 + 8}{x^2 - 4x + 4}$

**48.** $\lim\limits_{x \to -1} \dfrac{x^2 + 3x}{x^2 + 3x + 2}$

**49.** ▼ $\lim\limits_{x \to 2^+} \dfrac{x - 2}{\sqrt{x - 2}}$

**50.** ▼ $\lim\limits_{x \to 3^-} \dfrac{\sqrt{3 - x}}{3 - x}$

**51.** ▼ $\lim\limits_{x \to 9} \dfrac{\sqrt{x} - 3}{x - 9}$

**52.** ▼ $\lim\limits_{x \to 4} \dfrac{x - 4}{\sqrt{x} - 2}$

**53.** $\lim\limits_{x \to +\infty} \dfrac{3x^2 + 10x - 1}{2x^2 - 5x}$  HINT [See Example 4.]

**54.** $\lim\limits_{x \to +\infty} \dfrac{6x^2 + 5x + 100}{3x^2 - 9}$  HINT [See Example 4.]

**55.** $\lim\limits_{x \to +\infty} \dfrac{x^5 - 1{,}000x^4}{2x^5 + 10{,}000}$

**56.** $\lim\limits_{x \to +\infty} \dfrac{x^6 + 3{,}000x^3 + 1{,}000{,}000}{2x^6 + 1{,}000x^3}$

**57.** $\lim\limits_{x \to +\infty} \dfrac{10x^2 + 300x + 1}{5x + 2}$

**58.** $\lim\limits_{x \to +\infty} \dfrac{2x^4 + 20x^3}{1{,}000x^3 + 6}$

**59.** $\lim\limits_{x \to -\infty} \dfrac{3x^2 + 10x - 1}{2x^2 - 5x}$

**60.** $\lim\limits_{x \to -\infty} \dfrac{6x^2 + 5x + 100}{3x^2 - 9}$

**61.** $\lim\limits_{x \to -\infty} \dfrac{x^5 - 1{,}000x^4}{2x^5 + 10{,}000}$

**62.** $\lim\limits_{x \to -\infty} \dfrac{x^6 + 3000x^3 + 1{,}000{,}000}{2x^6 + 1{,}000x^3}$

**63.** $\lim\limits_{x \to -\infty} \dfrac{10x^2 + 300x + 1}{5x + 2}$

**64.** $\lim\limits_{x \to -\infty} \dfrac{2x^4 + 20x^3}{1{,}000x^3 + 6}$

**65.** $\lim\limits_{x \to -\infty} \dfrac{10x^2 + 300x + 1}{5x^3 + 2}$

**66.** $\lim\limits_{x \to -\infty} \dfrac{2x^4 + 20x^3}{1{,}000x^6 + 6}$

**67.** $\lim\limits_{x \to +\infty} (4e^{-3x} + 12)$

**68.** $\lim\limits_{x \to +\infty} \dfrac{2}{5 - 5.3e^{-3x}}$

**69.** $\lim\limits_{t \to +\infty} \dfrac{2}{5 - 5.3(3^{3t})}$

**70.** $\lim\limits_{t \to +\infty} (4.1 - 2e^{3t})$

**71.** $\lim\limits_{t \to +\infty} \dfrac{2^{3t}}{1 + 5.3e^{-t}}$

**72.** $\lim\limits_{x \to -\infty} \dfrac{4.2}{2 - 3^{2x}}$

**73.** $\lim\limits_{x \to -\infty} \dfrac{-3^{2x}}{2 + e^x}$

**74.** $\lim\limits_{x \to +\infty} \dfrac{2^{-3x}}{1 + 5.3e^{-x}}$

*In each of Exercises 75–82, find all points of discontinuity of the given function.* HINT [See Example 3.]

**75.** $f(x) = \begin{cases} x + 2 & \text{if } x < 0 \\ 2x - 1 & \text{if } x \geq 0 \end{cases}$

**76.** $g(x) = \begin{cases} 1 - x & \text{if } x \leq 1 \\ x - 1 & \text{if } x > 1 \end{cases}$

**77.** $g(x) = \begin{cases} x + 2 & \text{if } x < 0 \\ 2x + 2 & \text{if } 0 \leq x < 2 \\ x^2 + 2 & \text{if } x \geq 2 \end{cases}$

**78.** $f(x) = \begin{cases} 1 - x & \text{if } x \leq 1 \\ x + 2 & \text{if } 1 < x < 3 \\ x^2 - 4 & \text{if } x \geq 3 \end{cases}$

**79.** ▼ $h(x) = \begin{cases} x + 2 & \text{if } x < 0 \\ 0 & \text{if } x = 0 \\ 2x + 2 & \text{if } x > 0 \end{cases}$

**80.** ▼ $h(x) = \begin{cases} 1 - x & \text{if } x < 1 \\ 1 & \text{if } x = 1 \\ x + 2 & \text{if } x > 1 \end{cases}$

**81.** ▼ $f(x) = \begin{cases} 1/x & \text{if } x < 0 \\ x & \text{if } 0 \leq x \leq 2 \\ 2^{x-1} & \text{if } x > 2 \end{cases}$

**82.** ▼ $f(x) = \begin{cases} x^3 + 2 & \text{if } x \leq -1 \\ x^2 & \text{if } -1 < x < 0 \\ x & \text{if } x \geq 0 \end{cases}$

## APPLICATIONS

**83.** *Processor Speeds* The processor speeds, in megahertz (MHz), of Intel processors during the period 1996–2010 can be approximated by the following function of time $t$ in years since the start of 1990:[17]

$$v(t) = \begin{cases} 400t - 2{,}200 & \text{if } 6 \leq t < 15 \\ 3{,}800 & \text{if } 15 \leq t \leq 20. \end{cases}$$

**a.** Compute $\lim_{t \to 15^-} v(t)$ and $\lim_{t \to 15^+} v(t)$ and interpret each answer. HINT [See Example 3.]

**b.** Is the function $v$ continuous at $t = 15$? According to the model, was there any abrupt change in processor speeds during the period 1996–2010?

**84.** *Processor Speeds* The processor speeds, in megahertz (MHz), of Intel processors during the period 1970–2000 can be approximated by the following function of time $t$ in years since the start of 1970:[18]

$$v(t) = \begin{cases} 3t & \text{if } 0 \leq t < 20 \\ 174t - 3{,}420 & \text{if } 20 \leq t \leq 30. \end{cases}$$

---

[17]A rough model based on the fastest processors produced by Intel. Source for data: www.intel.com.

[18]*Ibid.*

**a.** Compute $\lim_{t \to 20^-} v(t)$ and $\lim_{t \to 20^+} v(t)$ and interpret each answer.

**b.** Is the function $v$ continuous at $t = 20$? According to the model, was there any abrupt change in processor speeds during the period 1970–2000?

**85. *Movie Advertising*** Movie expenditures, in billions of dollars, on advertising in newspapers from 1995 to 2004 can be approximated by

$$f(t) = \begin{cases} 0.04t + 0.33 & \text{if } t \le 4 \\ -0.01t + 1.2 & \text{if } t > 4 \end{cases}$$

where $t$ is time in years since 1995.[19]

**a.** Compute $\lim_{t \to 4^-} f(t)$ and $\lim_{t \to 4^+} f(t)$, and interpret each answer. HINT [See Example 3.]

**b.** Is the function $f$ continuous at $t = 4$? What does the answer tell you about movie advertising expenditures?

**86. *Movie Advertising*** The percentage of movie advertising as a share of newspapers' total advertising revenue from 1995 to 2004 can be approximated by

$$p(t) = \begin{cases} -0.07t + 6.0 & \text{if } t \le 4 \\ 0.3t + 17.0 & \text{if } t > 4 \end{cases}$$

where $t$ is time in years since 1995.[20]

**a.** Compute $\lim_{t \to 4^-} p(t)$ and $\lim_{t \to 4^+} p(t)$, and interpret each answer. HINT [See Example 3.]

**b.** Is the function $p$ continuous at $t = 4$? What does the answer tell you about newspaper revenues?

**87. *Law Enforcement in the 1980s and 1990s*** The cost of fighting crime in the United States increased significantly during the period 1982–1999. Total spending on police and courts can be approximated, respectively, by[21]

$$P(t) = 1.745t + 29.84 \text{ billion dollars} \quad (2 \le t \le 19)$$
$$C(t) = 1.097t + 10.65 \text{ billion dollars} \quad (2 \le t \le 19),$$

where $t$ is time in years since 1980. Compute $\lim_{t \to +\infty} \dfrac{P(t)}{C(t)}$ to two decimal places and interpret the result. HINT [See Example 4.]

**88. *Law Enforcement in the 1980s and 1990s*** Refer to Exercise 87. Total spending on police, courts, and prisons in

the period 1982–1999 could be approximated, respectively, by[22]

$$P(t) = 1.745t + 29.84 \text{ billion dollars} \quad (2 \le t \le 19)$$
$$C(t) = 1.097t + 10.65 \text{ billion dollars} \quad (2 \le t \le 19)$$
$$J(t) = 1.919t + 12.36 \text{ billion dollars} \quad (2 \le t \le 19),$$

where $t$ is time in years since 1980. Compute $\lim_{t \to +\infty} \dfrac{P(t)}{P(t) + C(t) + J(t)}$ to two decimal places and interpret the result. HINT [See Example 4.]

**89. *SAT Scores by Income*** The following bar graph shows U.S. math SAT scores as a function of household income:[23]

These data can be modeled by

$$S(x) = 573 - 33e^{-0.0131x},$$

where $S(x)$ is the average math SAT score of students whose household income is $x$ thousand dollars per year. Calculate $\lim_{x \to +\infty} S(x)$ and interpret the answer.

**90. *SAT Scores by Income*** The following bar graph shows U.S. critical reading SAT scores as a function of household income:[24]

These data can be modeled by

$$S(x) = 550 - 136e^{-0.0151x},$$

where $S(x)$ is the average critical reading SAT score of students whose household income is $x$ thousand dollars per year. Calculate $\lim_{x \to +\infty} S(x)$ and interpret the answer.

---

[19]Model by the authors. Source for data: Newspaper Association of America Business Analysis and Research/*New York Times*, May 16, 2005.

[20]*Ibid.*

[21]Spending is adjusted for inflation and shown in 1999 dollars. Models are based on a linear regression. Source for data: Bureau of Justice Statistics/*New York Times*, February 11, 2002, p. A14.

[22]*Ibid.*

[23]2009 data. Source: College Board/*New York Times* http://economix .blogs.nytimes.com.

[24]*Ibid.*

*Foreign Trade Annual U.S. imports from China in the years
1996 through 2003 could be approximated by*

$$I(t) = t^2 + 3.5t + 50 \quad (1 \le t \le 9)$$

*billion dollars, where t represents time in years since 1995.
Annual U.S. exports to China in the same years could be
approximated by*

$$E(t) = 0.4t^2 - 1.6t + 14 \quad (0 \le t \le 10)$$

*billion dollars.[25] Exercises 91 and 92 are based on these models.*

91. Assuming that the trends shown in the above models
continue indefinitely, calculate the limits

$$\lim_{t \to +\infty} I(t) \text{ and } \lim_{t \to +\infty} \frac{I(t)}{E(t)}$$

algebraically, interpret your answers, and comment on the
results. HINT [See Example 4.]

92. Repeat Exercise 91, this time calculating

$$\lim_{t \to +\infty} E(t) \text{ and } \lim_{t \to +\infty} \frac{E(t)}{I(t)} \quad \text{HINT [See Example 4.]}$$

93. *Acquisition of Language* The percentage $p(t)$ of children
who can speak in at least single words by the age of $t$ months
can be approximated by the equation[26]

$$p(t) = 100 \left( \frac{1 - 12,200}{t^{4.48}} \right) \quad (t \ge 8.5).$$

Calculate $\lim_{t \to +\infty} p(t)$ and interpret the result. HINT [See
Example 4.]

94. *Acquisition of Language* The percentage $q(t)$ of children
who can speak in sentences of five or more words by the age
of $t$ months can be approximated by the equation[27]

$$q(t) = 100 \left( 1 - \frac{5.27 \times 10^{17}}{t^{12}} \right) \quad (t \ge 30).$$

If $p$ is the function referred to in the preceding exercise,
calculate $\lim_{t \to +\infty} [p(t) - q(t)]$ and interpret the result.
HINT [See Example 4.]

## COMMUNICATION AND REASONING EXERCISES

95. Describe the algebraic method of evaluating limits as discussed in this section and give at least one disadvantage of
this method.

96. What is a closed-form function? What can we say about
such functions?

---

[25]Based on quadratic regression using data from the U.S. Census Bureau
Foreign Trade Division Web site www.census.gov/foreign-trade/sitc1/
as of December 2004.

[26]The model is the authors' and is based on data presented in the
article *The Emergence of Intelligence* by William H. Calvin, *Scientific
American,* October, 1994, pp. 101–107.

[27]*Ibid.*

97. Why was the following marked wrong? What is the correct
answer?

$$\lim_{x \to 3} \frac{x^3 - 27}{x - 3} = \frac{0}{0} \text{ undefined} \qquad \text{✗ WRONG!}$$

98. Why was the following marked wrong? What is the correct
answer?

$$\lim_{x \to 1^-} \frac{x - 1}{x^2 - 2x + 1} = \frac{0}{0} = 0 \qquad \text{✗ WRONG!}$$

99. ▼ Your friend Karin tells you that $f(x) = 1/(x - 2)^2$ cannot be a closed-form function because it is not continuous
at $x = 2$. Comment on her assertion.

100. ▼ Give an example of a function $f$ specified by means of
algebraic formulas such that the domain of $f$ consists of all
real numbers and $f$ is not continuous at $x = 2$. Is $f$ a closed-
form function?

101. Give examples of two limits that lead to two different indeterminate forms, but where both limits exist.

102. Give examples of two limits: one that leads to a determinate form and another that leads to an indeterminate form,
but where neither limit exists.

103. ▼ (Compare Exercise 59 in Section 10.1.) Which indetermi-
nate form results from $\lim_{x \to +\infty} \dfrac{p(x)}{e^x}$ if $p(x)$ is a polynomial? Numerically or graphically estimate these limits for
various polynomials $p(x)$. What does this suggest about
limits that result in $\dfrac{p(\infty)}{e^\infty}$?

104. ▼ (Compare Exercise 60 in Section 10.1.) Which indeterminate form results from $\lim_{x \to -\infty} p(x)e^x$ if $p(x)$ is a polynomial?
What does this suggest about the limits that result in
$p(\infty)e^{-\infty}$?

105. ▼ What is wrong with the following statement? If $f(x)$ is
specified algebraically and $f(a)$ is defined, then
$\lim_{x \to a} f(x)$ exists and equals $f(a)$. How can it be corrected?

106. ▼ What is wrong with the following statement? If $f(x)$ is
specified algebraically and $f(a)$ is not defined, then
$\lim_{x \to a} f(x)$ does not exist.

107. ▼ Give the formula for a function that is continuous everywhere except at two points.

108. ▼ Give the formula for a function that is continuous everywhere except at three points.

109. ◆ *The Indeterminate Form* $\infty - \infty$ An indeterminate form
not mentioned in Section 10.3 is $\infty - \infty$. Give examples of
three limits that lead to this indeterminate form, and where
the first limit exists and equals 5, where the second limit
diverges to $+\infty$, and where the third exists and equals $-5$.

110. ◆ *The Indeterminate Form* $1^\infty$ An indeterminate form not
mentioned in Section 10.3 is $1^\infty$. Give examples of three
limits that lead to this indeterminate form, and where the
first limit exists and equals 1, where the second limit exists
and equals $e$, and where the third diverges to $+\infty$. HINT [For
the third, consider modifying the second.]

# 10.4 Average Rate of Change

Calculus is the mathematics of change, inspired largely by observation of continuously changing quantities around us in the real world. As an example, the Consumer Price Index (CPI) $C$ increased from 211 points in January 2009 to 220 points in January 2011.[28] As we saw in Chapter 1, the **change** in this index can be measured as the difference:

$$\Delta C = \text{Second value} - \text{First value} = 220 - 211 = 9 \text{ points}.$$

(The fact that the CPI increased is reflected in the positive sign of the change.) The kind of question we will concentrate on is *how fast* the CPI was changing. Because $C$ increased by 9 points in 2 years, we say it averaged a $9/2 = 4.5$ point rise each year. (It actually rose 5 points the first year and 4 the second, giving an average rise of 4.5 points each year.)

Alternatively, we might want to measure this rate in points per month rather than points per year. Because $C$ increased by 9 points in 24 months, it increased at an average rate of $9/24 = 0.375$ points per month.

In both cases, we obtained the average rate of change by dividing the change by the corresponding length of time:

$$\text{Average rate of change} = \frac{\text{Change in } C}{\text{Change in time}} = \frac{9}{2} = 4.50 \text{ points per year}$$

$$\text{Average rate of change} = \frac{\text{Change in } C}{\text{Change in time}} = \frac{9}{24} = 0.375 \text{ points per month}.$$

## Average Rate of Change of a Function Numerically and Graphically

### EXAMPLE 1 Standard and Poor's 500

The following table lists the approximate value of Standard and Poor's 500 stock market index (S&P) during the period 2005–2011[29] ($t = 5$ represents 2005):

| $t$ (year) | 5 | 6 | 7 | 8 | 9 | 10 | 11 |
|---|---|---|---|---|---|---|---|
| $S(t)$ (points) | 1,200 | 1,300 | 1,400 | 1,400 | 900 | 1,150 | 1,300 |

**a.** What was the average rate of change in the S&P over the 2-year period 2005–2007 (the period $5 \leq t \leq 7$ or $[5, 7]$ in interval notation); over the 4-year period 2005–2009 (the period $5 \leq t \leq 9$ or $[5, 9]$); and over the period $[6, 11]$?

**b.** Graph the values shown in the table. How are the rates of change reflected in the graph?

---

[28]Figures are approximate. Source: Bureau of Labor Statistics www.bls.gov.

[29]The values are approximate values at the start of the given year. Source: http://finance.google.com.

## Solution

**a.** During the 2-year period [5, 7], the S&P changed as follows:

| | |
|---|---|
| Start of the period ($t = 5$): | $S(5) = 1,200$ |
| End of the period ($t = 7$): | $S(7) = 1,400$ |

Change during the period [5, 7]:  $S(7) - S(5) = 200$

Thus, the S&P increased by 200 points in 2 years, giving an average rate of change of $200/2 = 100$ points per year. We can write the calculation this way:

$$\text{Average rate of change of } S = \frac{\text{Change in } S}{\text{Change in } t}$$

$$= \frac{\Delta S}{\Delta t}$$

$$= \frac{S(7) - S(5)}{7 - 5}$$

$$= \frac{1,400 - 1,200}{7 - 5} = \frac{200}{2} = 100 \text{ points per year.}$$

**Interpreting the result:** During the period [5, 7] (that is, 2005–2007), the S&P increased at an average rate of 100 points per year.

Similarly, the average rate of change during the period [5, 9] was

$$\text{Average rate of change of } S = \frac{\Delta S}{\Delta t} = \frac{S(9) - S(5)}{9 - 5} = \frac{900 - 1,200}{9 - 5}$$

$$= \frac{-300}{4} = -75 \text{ points per year.}$$

**Interpreting the result:** During the period [5, 9] (that is, 2005–2009), the S&P *decreased* at an average rate of 75 points per year.

Finally, during the period [6, 11], the average change was

$$\text{Average rate of change of } S = \frac{\Delta S}{\Delta t} = \frac{S(11) - S(6)}{11 - 6} = \frac{1,300 - 1,300}{11 - 6}$$

$$= \frac{0}{5} = 0 \text{ points per year.}$$

**Interpreting the result:** During the period [6, 11] the average rate of change of the S&P was zero points per year (even though its value did fluctuate during that period).

**b.** In Chapter 1, we saw that the rate of change of a quantity that changes linearly with time is measured by the slope of its graph. However, the S&P index does not change linearly with time. Figure 20 shows the data plotted two different ways: (a) as a bar chart and (b) as a piecewise linear graph. Bar charts are more commonly used in the media, but Figure 20(b) on the right illustrates the changing index more clearly.

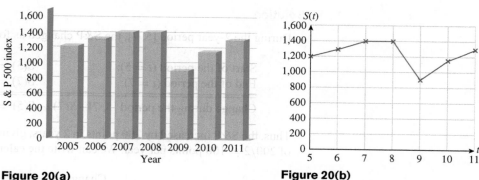

**Figure 20(a)**                     **Figure 20(b)**

We saw in part (a) that the average rate of change of $S$ over the interval $[5, 9]$ is the ratio

$$\text{Average rate of change of } S = \frac{\Delta S}{\Delta t} = \frac{S(9) - S(5)}{9 - 5} = -75 \text{ points per year.}$$

Notice that this rate of change is also the slope of the line through $P$ and $Q$ shown in Figure 21, and we can estimate this slope directly from the graph as shown.

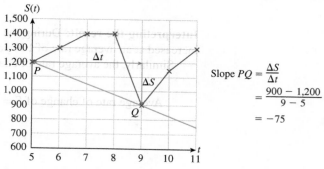

Slope $PQ = \dfrac{\Delta S}{\Delta t}$

$= \dfrac{900 - 1{,}200}{9 - 5}$

$= -75$

**Figure 21**

**Average Rate of Change as Slope:** The average rate of change of the S&P over the interval $[5, 9]$ is the slope of the line passing through the points on the graph where $t = 5$ and $t = 9$.

Similarly, the average rates of change of the S&P over the intervals $[5, 7]$ and $[6, 11]$ are the slopes of the lines through pairs of corresponding points.

Here is the formal definition of the average rate of change of a function over an interval.

## Change and Average Rate of Change of *f* over [*a*, *b*]: Difference Quotient

The **change** in $f(x)$ over the interval $[a, b]$ is

Change in $f = \Delta f$

$= \text{Second value} - \text{First value}$

$= f(b) - f(a)$.

The **average rate of change** of $f(x)$ over the interval $[a, b]$ is

$$\text{Average rate of change of } f = \frac{\text{Change in } f}{\text{Change in } x}$$

$$= \frac{\Delta f}{\Delta x} = \frac{f(b) - f(a)}{b - a}$$

$$= \text{Slope of line through points } P \text{ and } Q$$
$$\quad (\text{see figure}).$$

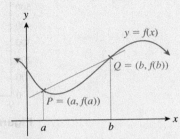

Average rate of change = Slope of $PQ$

We also call this average rate of change the **difference quotient** of $f$ over the interval $[a, b]$. (It is the *quotient* of the *differences* $f(b) - f(a)$ and $b - a$.) A line through two points of a graph like $P$ and $Q$ is called a **secant line** of the graph.

## Units

The units of the change $\Delta f$ in $f$ are the units of $f(x)$.
The units of the average rate of change of $f$ are units of $f(x)$ per unit of $x$.*

* The average rate of change is a slope, and so it is measured in the same units as the slope: units of $y$ (or $f(x)$) per unit of $x$.

### Quick Example

If $f(3) = -1$ billion dollars, $f(5) = 0.5$ billion dollars, and $x$ is measured in years, then the change and average rate of change of $f$ over the interval $[3, 5]$ are given by

$$\text{Change in } f = f(5) - f(3) = 0.5 - (-1) = 1.5 \text{ billion dollars}$$

$$\text{Average rate of change} = \frac{f(5) - f(3)}{5 - 3} = \frac{0.5 - (-1)}{2}$$

$$= 0.75 \text{ billion dollars/year.}$$

## Alternative Formula: Average Rate of Change of *f* over [*a, a + h*]

(Replace $b$ in the formula for the average rate of change by $a + h$.) The average rate of change of $f$ over the interval $[a, a + h]$ is

$$\text{Average rate of change of } f = \frac{f(a + h) - f(a)}{h}. \qquad \text{Replace } b \text{ by } a + h.$$

In Example 1 we saw that the average rate of change of a quantity can be estimated directly from a graph. Here is an example that further illustrates the graphical approach.

### EXAMPLE 2 **Freon 22 Production**

Figure 22 shows the number of tons $f(t)$ of ozone-layer-damaging Freon 22 (chlorodi-fluoromethane) produced annually in developing countries for the period 2000–2010 ($t$ is time in years, and $t = 0$ represents 2000).[30]

**Figure 22**

**a.** Use the graph to estimate the average rate of change of $f(t)$ with respect to $t$ over the interval [4, 8] and interpret the result.

**b.** Over which 2-year period(s) was the average rate of change of Freon 22 production the greatest?

**c.** Multiple choice: For the period of time under consideration, Freon 22 production was

**(A)** increasing at a faster and faster rate.
**(B)** increasing at a slower and slower rate.
**(C)** decreasing at a faster and faster rate.
**(D)** decreasing at a slower and slower rate.

### Solution

**a.** The average rate of change of $f$ over the interval [4, 8] is given by the slope of the line through the points $P$ and $Q$ shown in Figure 23.

**Figure 23**

From the figure,

$$\text{Average rate of change of } f = \frac{\Delta f}{\Delta t} = \text{slope } PQ \approx \frac{400 - 200}{8 - 4} = \frac{200}{4} = 50.$$

Thus, the rate of change of $f$ over the interval [4, 8] is approximately 50.

[30]Figures from 2007 on were projected. Source: *New York Times*, February 23, 2007, p. C1.

**Q**: How do we interpret the result?

**A**: A clue is given by the units of the average rate of change: units of *f* per unit of *t*. The units of *f* are tons of Freon 22 and the units of *t* are years. Thus, the average rate of change of *f* is measured in tons of Freon 22 per year, and we can now interpret the result as follows:

**Interpreting the average rate of change:** Annual production of Freon 22 was increasing at an average rate of 50 tons of Freon 22 per year from 2004 to 2008.

**b.** The rates of change of Freon 22 production over successive 2-year periods are given by the slopes of the individual line segments that make up the graph in Figure 22. Thus, the greatest average rate of change in a single 2-year period corresponds to the segment(s) with the largest slope. If you look at the figure, you will notice that the segment corresponding to [8, 10] is the steepest. Thus, the average rate of change of Freon 22 production was largest over the 2-year period from 2008 to 2010.

**c.** Looking again at the figure, notice that the graph rises as we go from left to right; that is, the value of the function (Freon 22 production) is increasing with increasing *t*. At the same time, the fact that the curve bends up (is concave up) with increasing *t* tells us that the successive slopes get steeper, and so the average rates of change increase as well (Choice (A)).

➡ **Before we go on...** Notice in Example 2 that we do not get exact answers from a graph; the best we can do is *estimate* the rates of change: Was the exact answer to part (a) closer to 49 or 51? Two people can reasonably disagree about results read from a graph, and you should bear this in mind when you check the answers to the exercises. ∎

Perhaps the most sophisticated way to compute the average rate of change of a quantity is through the use of a mathematical formula or model for the quantity in question.

## Average Rate of Change of a Function Using Algebraic Data

### EXAMPLE 3 Average Rate of Change from a Formula

You are a commodities trader and you monitor the price of gold on the spot market very closely during an active morning. Suppose you find that the price of an ounce of gold can be approximated by the function

$$G(t) = 5t^2 - 85t + 1{,}762 \qquad (7.5 \le t \le 10.5),$$

where *t* is time in hours. (See Figure 24. $t = 8$ represents 8:00 am.)

Looking at the graph on the right, we can see that the price of gold was falling at the beginning of the time period, but by $t = 8.5$ the fall had slowed to a stop, whereupon the market turned around and the price began to rise more and more rapidly toward the end of the period. What was the average rate of change of the price of gold over the $1\frac{1}{2}$-hour period starting at 8:00 am (the interval [8, 9.5] on the *t*-axis)?

**using** Technology

See the Technology Guides at the end of the chapter for detailed instructions on how to calculate the average rate of change of the function in Example 3 using a TI-83/84 Plus or a spreadsheet. Here is an outline:

**TI-83/84 Plus**
Y₁=5X^2-85X+1762
Home screen: (Y₁(9.5)-
Y₁(8))/(9.5-8)
[More details on page 778.]

**Spreadsheet**
Headings t, G(t), Rate of Change in A1–C1
t-values 8, 9.5 in A2–A3
=5*A2^2-85*A2+1762 in B2, copied down to B3
=(B3-B2)/(A3-A2) in C2.
[More details on page 780.]

Source: www.kitco.com

**Figure 24**

$$G(t) = 5t^2 - 85t + 1,762$$

**Solution** We have

$$\text{Average rate of change of } G \text{ over } [8, 9.5] = \frac{\Delta G}{\Delta t} = \frac{G(9.5) - G(8)}{9.5 - 8}.$$

From the formula for $G(t)$, we find

$$G(9.5) = 5(9.5)^2 - 85(9.5) + 1,762 = 1,405.75$$
$$G(8) = 5(8)^2 - 85(8) + 1,762 = 1,402.$$

Thus, the average rate of change of $G$ is given by

$$\frac{G(9.5) - G(8)}{9.5 - 8} = \frac{1,405.75 - 1,402}{1.5} = \frac{3.75}{1.5} = \$2.50 \text{ per hour.}$$

In other words, the price of gold increased at an average rate of $2.50 per hour over the $1\frac{1}{2}$-hour period.

**EXAMPLE 4** ▥ **Rates of Change over Shorter Intervals**

Continuing with Example 3, use technology to compute the average rate of change of

$$G(t) = 5t^2 - 85t + 1,762 \qquad (7.5 \leq t \leq 10.5)$$

over the intervals $[8, 8 + h]$, where $h = 1, 0.1, 0.01, 0.001,$ and $0.0001$. What do the answers tell you about the price of gold?

**Solution**

We use the "alternative" formula

$$\text{Average rate of change of } G \text{ over } [a, a + h] = \frac{G(a + h) - G(a)}{h}$$

so

$$\text{Average rate of change of } G \text{ over } [8, 8 + h] = \frac{G(8 + h) - G(8)}{h}.$$

**using** Technology

Example 4 is the kind of example where the use of technology can make a huge difference. See the Technology Guides at the end of the chapter to find out how to do the above computations almost effortlessly using a TI-83/84 Plus or a spreadsheet. Here is an outline:

**TI-83/84 Plus**

$Y_1=5X^2-85X+1762$
Home screen:
$(Y_1(8+1)-Y_1(8))/1$
$(Y_1(8+0.1)-Y_1(8))/0.1$
$(Y_1(8+0.01)-Y_1(8))/0.01$
etc.
[More details on page 778.]

**Spreadsheet**

Headings $a$, $h$, $t$, $G(t)$, Rate of Change in A1–E1
8 in A2, 1 in B2,
$=$A2 in C2, $=$A2+B2 in C3
$=5*$C2$^2-85*$C2$+1762$ in D2; copy down to D3
$=$(D3-D2)/(C3-C2) in E2
[More details on page 780.]

Let us calculate this average rate of change for some of the values of $h$ listed:

$h = 1$:    $G(8 + h) = G(8 + 1) = G(9) = 5(9)^2 - 85(9) + 1{,}762 = 1{,}402$

$\qquad G(8) = 5(8)^2 - 85(8) + 1{,}762 = 1{,}402$

$\qquad$ Average rate of change of $G = \dfrac{G(9) - G(8)}{1} = \dfrac{1{,}402 - 1{,}402}{1} = 0$

$h = 0.1$:   $G(8 + h) = G(8 + 0.1) = G(8.1) = 5(8.1)^2 - 85(8.1) + 1{,}762$

$\qquad = 1{,}401.55$

$\qquad G(8) = 5(8)^2 - 85(8) + 1{,}762 = 1{,}402$

$\qquad$ Average rate of change of $G = \dfrac{G(8.1) - G(8)}{0.1} = \dfrac{1{,}401.55 - 1{,}402}{0.1} = \dfrac{-0.45}{0.1}$

$\qquad = -4.5$

$h = 0.01$:  $G(8 + h) = G(8 + 0.01) = G(8.01) = 5(8.01)^2 - 85(8.01) + 1{,}762$

$\qquad = 1{,}401.9505$

$\qquad G(8) = 5(8)^2 - 85(8) + 1{,}762 = 1{,}402$

$\qquad$ Average rate of change of $G = \dfrac{G(8.01) - G(8)}{0.01} = \dfrac{1{,}401.9505 - 1{,}402}{0.01} = \dfrac{-0.0495}{0.01}$

$\qquad = -4.95$

Continuing in this way, we get the values in the following table:

| $h$ | 1 | 0.1 | 0.01 | 0.001 | 0.0001 |
|---|---|---|---|---|---|
| Ave. Rate of Change $\dfrac{G(8 + h) - G(8)}{h}$ | 0 | $-4.5$ | $-4.95$ | $-4.995$ | $-4.9995$ |

Each value is an average rate of change of $G$. For example, the value corresponding to $h = 0.01$ is $-4.95$, which tells us:

*Over the interval* [8, 8.01] *the price of gold was decreasing at an average rate of* $4.95 *per hour.*

In other words, during the first one hundredth of an hour (or 36 seconds) starting at $t = $ 8:00 am, the price of gold was decreasing at an average rate of $4.95 per hour. Put another way, in those 36 seconds, the price of gold decreased at a rate that, if continued, would have produced a decrease of $4.95 in the price of gold during the next hour. We will return to this example at the beginning of Section 10.5.

---

**FAQs**

**Recognizing When and How to Compute the Average Rate of Change and How to Interpret the Answer**

**Q** : *How do I know, by looking at the wording of a problem, that it is asking for an average rate of change?*

**A** : If a problem does not ask for an average rate of change directly, it might do so indirectly, as in "On average, how fast is quantity $q$ increasing?"

**Q** : *If I know that a problem calls for computing an average rate of change, how should I compute it? By hand or using technology?*

**A** : All the computations can be done by hand, but when hand calculations are not called for, using technology might save time.

**Q** : *Lots of problems ask us to "interpret" the answer. How do I do that for questions involving average rates of change?*

**A** : The *units* of the average rate of change are often the key to interpreting the results:

*The units of the average rate of change of $f(x)$ are units of $f(x)$ per unit of $x$.*

Thus, for instance, if $f(x)$ is the cost, in dollars, of a trip of $x$ miles in length, and the average rate of change of $f$ is calculated to be 10, then the units of the average rate of change are dollars per mile, and so we can interpret the answer by saying that the cost of a trip rises an average of $10 for each additional mile.

---

## **10.4**  EXERCISES

▼ more advanced     ◆ challenging
**T** indicates exercises that should be solved using technology

*In Exercises 1–18, calculate the average rate of change of the given function over the given interval. Where appropriate, specify the units of measurement.* HINT *[See Example 1.]*

**1.**

| $x$ | 0 | 1 | 2 | 3 |
|---|---|---|---|---|
| $f(x)$ | 3 | 5 | 2 | −1 |

Interval: $[1, 3]$

**2.**

| $x$ | 0 | 1 | 2 | 3 |
|---|---|---|---|---|
| $f(x)$ | −1 | 3 | 2 | 1 |

Interval: $[0, 2]$

**3.**

| $x$ | −3 | −2 | −1 | 0 |
|---|---|---|---|---|
| $f(x)$ | −2.1 | 0 | −1.5 | 0 |

Interval: $[-3, -1]$

**4.**

| $x$ | −2 | −1 | 0 | 1 |
|---|---|---|---|---|
| $f(x)$ | −1.5 | −0.5 | 4 | 6.5 |

Interval: $[-1, 1]$

**5.**

| $t$ (months) | 2 | 4 | 6 |
|---|---|---|---|
| $R(t)$ ($ millions) | 20.2 | 24.3 | 20.1 |

Interval: $[2, 6]$

**6.**

| $x$ (kilos) | 1 | 2 | 3 |
|---|---|---|---|
| $C(x)$ (£) | 2.20 | 3.30 | 4.00 |

Interval: $[1, 3]$

**7.**

| $p$ ($) | 5.00 | 5.50 | 6.00 |
|---|---|---|---|
| $q(p)$ (items) | 400 | 300 | 150 |

Interval: $[5, 5.5]$

**8.**

| $t$ (hours) | 0 | 0.1 | 0.2 |
|---|---|---|---|
| $D(t)$ (miles) | 0 | 3 | 6 |

Interval: $[0.1, 0.2]$

**9.** Apple Computer Stock Price ($)

Interval: [2, 5]
HINT [See Example 2.]

**10.**   Cisco Systems Stock Price ($)

Interval: [1, 5]
HINT [See Example 2.]

**11.** Unemployment (%)

Budget deficit (% of GNP)

Interval: [0, 4]

**12.** Inflation (%)

Budget deficit (% of GNP)

Interval: [0, 4]

**13.** $f(x) = x^2 - 3$; [1, 3] HINT [See Example 3.]

**14.** $f(x) = 2x^2 + 4$; [−1, 2] HINT [See Example 3.]

**15.** $f(x) = 2x + 4$; [−2, 0]

**16.** $f(x) = \dfrac{1}{x}$; [1, 4]

**17.** $f(x) = \dfrac{x^2}{2} + \dfrac{1}{x}$; [2, 3]    **18.** $f(x) = 3x^2 - \dfrac{x}{2}$; [3, 4]

*In Exercises 19–24, calculate the average rate of change of the given function f over the intervals [a, a + h] where h = 1, 0.1, 0.01, 0.001, and 0.0001. (Technology is recommended for the cases h = 0.01, 0.001, and 0.0001.)* HINT [See Example 4.]

**19.** $f(x) = 2x^2$; $a = 0$      **20.** $f(x) = \dfrac{x^2}{2}$; $a = 1$

**21.** $f(x) = \dfrac{1}{x}$; $a = 2$      **22.** $f(x) = \dfrac{2}{x}$; $a = 1$

**23.** $f(x) = x^2 + 2x$; $a = 3$    **24.** $f(x) = 3x^2 - 2x$; $a = 0$

## APPLICATIONS

**25.** *World Military Expenditure* The following table shows total military and arms trade expenditure in 2000, 2005, and 2010 ($t = 0$ represents 2000):[31]

| Year $t$ | 0 | 5 | 10 |
|---|---|---|---|
| Military Expenditure $C(t)$ ($ billion) | 1,100 | 1,300 | 1,600 |

Compute and interpret the average rate of change of $C(t)$ **(a)** over the period 2005–2010 (that is, [5, 10]), and **(b)** over the period [0, 10]. Be sure to state the units of measurement. HINT [See Example 1.]

**26.** *Education Expenditure* The following table shows the percentage of the U.S. Discretionary Budget allocated to education in 2003, 2005, and 2009 ($t = 0$ represents 2000):[32]

| Year $t$ | 3 | 5 | 9 |
|---|---|---|---|
| Percentage $P(t)$ | 6.8 | 7 | 6.2 |

Compute and interpret the average rate of change of $P(t)$ **(a)** over the period 2003–2009 (that is, [3, 9]), and **(b)** over the period [5, 9]. Be sure to state the units of measurement. HINT [See Example 1.]

**27.** *Oil Production in Mexico* The following table shows approximate daily oil production by Pemex, Mexico's national oil company, for 2001–2009 ($t = 1$ represents the start of 2001):[33]

| $t$ (year since 2000) | 1 | 2 | 3 | 4 | 5 | 6 | 7 | 8 | 9 |
|---|---|---|---|---|---|---|---|---|---|
| $P(t)$ (million barrels) | 3.1 | 3.3 | 3.4 | 3.4 | 3.4 | 3.3 | 3.2 | 3.1 | 3.0 |

---

[31] Source: www.globalissues.org/Geopolitics/ArmsTrade/Spending.asp.
[32] *Ibid.*
[33] Figures are approximate, and 2008–2009 figures are projections by the Department of Energy. Source: Energy Information Administration/Pemex (www.eia.doe.gov).

**a.** Compute the average rate of change of $P(t)$ over the period 2002–2007. Interpret the result. HINT [See Example 1.]

**b.** Which of the following is true? From 2001 to 2008, the one-year average rate of change of oil production by Pemex

   **(A)** increased in value.
   **(B)** decreased in value.
   **(C)** never increased in value.
   **(D)** never decreased in value.
   HINT [See Example 2.]

28. ***Oil Imports from Mexico*** The following table shows U.S. daily oil imports from Mexico, for 2001–2009 ($t = 1$ represents the start of 2001):[34]

| $t$ (year since 2000) | 1 | 2 | 3 | 4 | 5 | 6 | 7 | 8 | 9 |
|---|---|---|---|---|---|---|---|---|---|
| $I(t)$ (million barrels) | 1.4 | 1.35 | 1.5 | 1.55 | 1.6 | 1.5 | 1.5 | 1.5 | 1.2 |

**a.** Use the data in the table to compute the average rate of change of $I(t)$ over the period 2001–2006. Interpret the result.

**b.** Which of the following is true? From 2002 to 2006, the one-year average rate of change of oil imports from Mexico

   **(A)** increased in value.
   **(B)** decreased in value.
   **(C)** never increased in value.
   **(D)** never decreased in value.

29. ***Subprime Mortgages during the Housing Crisis*** The following graph shows the approximate percentage $P(t)$ of mortgages issued in the U.S. that were subprime (normally classified as risky):[35]

Subprime mortgages

**a.** Use the graph to estimate, to one decimal place, the average rate of change of $P(t)$ with respect to $t$ over the interval [0, 6] and interpret the result.

**b.** Over which 2-year period(s) was the average rate of change of $P(t)$ the greatest? HINT [See Example 2.]

30. ***Subprime Mortgage Debt during the Housing Crisis*** The following graph shows the approximate value $V(t)$ of subprime (normally classified as risky) mortgage debt outstanding in the United States:[36]

Subprime debt outstanding

**a.** Use the graph to estimate, to one decimal place, the average rate of change of $V(t)$ with respect to $t$ over the interval [2, 6] and interpret the result.

**b.** Over which 2-year period(s) was the average rate of change of $V(t)$ the least? HINT [See Example 2.]

31. ***Immigration to Ireland*** The following graph shows the approximate number (in thousands) of people who immigrated to Ireland during the period 2006–2010 ($t$ is time in years since 2000):[37]

During which 2-year interval(s) was the magnitude of the average rate of change of $I(t)$ **(a)** greatest **(b)** least? Interpret your answers by referring to the rates of change.

32. ***Emigration from Ireland*** The following graph shows the approximate number (in thousands) of people who emigrated from Ireland during the period 2006–2010.[38]

---

[34]Figures are approximate, and 2008–2009 figures are projections by the Department of Energy. Source: Energy Information Administration/Pemex (www.eia.doe.gov).

[35]Sources: Mortgage Bankers Association, UBS.

[36]Source: Data 360 www.data360.org.

[37]Sources: Ireland Central Statistic Office/Data 360 www.data360.org.

[38]*Ibid.*

During which 2-year interval(s) was the magnitude of the average rate of change of $E(t)$ (**a**) greatest (**b**) least? Interpret your answers by referring to the rates of change.

**33.** ▼ *Physics Research in the U.S.* The following table shows the number of research articles in the journal *Physical Review* authored by U.S researchers during the period 1993–2003 ($t = 3$ represents 1993):[39]

| $t$ (year since 1990) | 3 | 5 | 7 | 9 | 11 | 13 |
|---|---|---|---|---|---|---|
| $N(t)$ (articles, thousands) | 5.1 | 4.6 | 4.3 | 4.3 | 4.5 | 4.2 |

**a.** Find the interval(s) over which the average rate of change of $N$ was the most negative. What was that rate of change? Interpret your answer.

**b.** The **percentage change of $N$ over the interval $[a, b]$** is defined to be

$$\text{Percentage change of } N = \frac{\text{Change in } N}{\text{First value}} = \frac{N(b) - N(a)}{N(a)}.$$

Compute the percentage change of $N$ over the interval $[3, 13]$ and also the average rate of change. Interpret your answers.

**34.** ▼ *Physics Research in Europe* The following table shows the number of research articles in the journal *Physical Review* authored by researchers in Europe during the period 1993–2003 ($t = 3$ represents 1993):[40]

| $t$ (year since 1990) | 3 | 5 | 7 | 9 | 11 | 13 |
|---|---|---|---|---|---|---|
| $N(t)$ (articles, thousands) | 3.8 | 4.6 | 5.0 | 5.0 | 6.0 | 5.7 |

**a.** Find the interval(s) over which the average rate of change of $N$ was the most positive. What was that rate of change? Interpret your answer.

**b.** The **percentage change of $N$ over the interval $[a, b]$** is defined to be

$$\text{Percentage change of } N = \frac{\text{Change in } N}{\text{First value}} = \frac{N(b) - N(a)}{N(a)}.$$

Compute the percentage change of $N$ over the interval [7, 13] and also the average rate of change. Interpret the answers.

**35.** *College Basketball: Men* The following chart shows the number of NCAA men's college basketball teams in the United States during the period 2000–2010:[41]

**a.** On average, how fast was the number of men's college basketball teams growing over the 4-year period beginning in 2002?

**b.** By inspecting the chart, determine whether the 3-year average rates of change increased or decreased beginning in 2005. HINT [See Example 2.]

**36.** *College Basketball: Women* The following chart shows the number of NCAA women's college basketball teams in the United States during the period 2000–2010:[42]

**a.** On average, how fast was the number of women's college basketball teams growing over the 4-year period beginning in 2004?

**b.** By inspecting the graph, find the 3-year period over which the average rate of change was largest.

[39] Source: The Americal Physical Society/*New York Times*, May 3, 2003, p. A1.
[40] *Ibid.*

[41] 2010 figure is an estimate. Source: www.census.gov.
[42] *Ibid.*

**37. *Funding for the Arts*** State governments in the United States spend a total of between $1 and $2 per person on the arts and culture each year. The following chart shows the data for 2002–2010, together with the regression line:[43]

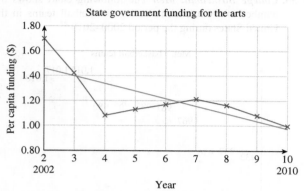

State government funding for the arts

**a.** Over the period [2, 6] the average rate of change of state government funding for the arts was

(A) less than  (B) greater than  (C) approximately equal to

the rate predicted by the regression line.

**b.** Over the period [3, 10] the average rate of change of state government funding for the arts was

(A) less than  (B) greater than  (C) approximately equal to

the rate predicted by the regression line.

**c.** Over the period [4, 8] the average rate of change of state government funding for the arts was

(A) less than  (B) greater than  (C) approximately equal to

the rate predicted by the regression line.

**d.** Estimate, to two significant digits, the average rate of change of per capita state government funding for the arts over the period [2, 10]. (Be careful to state the units of measurement.) How does it compare to the slope of the regression line?

**38. *Funding for the Arts*** The U.S. federal government spends a total of between $6 and $7 per person on the arts and culture each year. The following chart shows the data for 2002–2010, together with the regression line:[44]

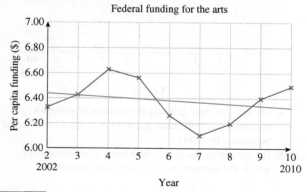

Federal funding for the arts

**a.** Over the period [4, 10] the average rate of change of federal government funding for the arts was

(A) less than  (B) greater than  (C) approximately equal to

the rate predicted by the regression line.

**b.** Over the period [2, 7] the average rate of change of federal government funding for the arts was

(A) less than  (B) greater than  (C) approximately equal to

the rate predicted by the regression line.

**c.** Over the period [3, 10] the average rate of change of federal government funding for the arts was

(A) less than  (B) greater than  (C) approximately equal to

the rate predicted by the regression line.

**d.** Estimate, to one significant digit, the average rate of change of per capita federal government funding for the arts over the period [2, 10]. (Be careful to state the units of measurement.) How does it compare to the slope of the regression line?

**39.** ▼ ***Market Volatility during the Dot-com Boom*** A volatility index generally measures the extent to which a market undergoes sudden changes in value. The volatility of the S&P 500 (as measured by one such index) was decreasing at an average rate of 0.2 points per year during 1991–1995, and was increasing at an average rate of about 0.3 points per year during 1995–1999. In 1995, the volatility of the S&P was 1.1.[45] Use this information to give a rough sketch of the volatility of the S&P 500 as a function of time, showing its values in 1991 and 1999.

**40.** ▼ ***Market Volatility during the Dot-com Boom*** The volatility (see the preceding exercise) of the NASDAQ had an average rate of change of 0 points per year during 1992–1995, and increased at an average rate of 0.2 points per year during 1995–1998. In 1995, the volatility of the NASDAQ was 1.1.[46] Use this information to give a rough sketch of the volatility of the NASDAQ as a function of time.

**41. *Market Index*** Joe Downs runs a small investment company from his basement. Every week he publishes a report on the success of his investments, including the progress of the "Joe Downs Index." At the end of one particularly memorable week, he reported that the index for that week had the value $I(t) = 1{,}000 + 1{,}500t - 800t^2 + 100t^3$ points, where $t$ represents the number of business days into the week; $t$ ranges from 0 at the beginning of the week to 5 at the end of the week. The graph of $I$ is shown below.

$I$ (Joe Downs Index)

---

[43] Figures are in constant 2008 dollars, and the 2010 figure is the authors' estimate. Source: *Americans for the Arts* www.artsusa.org.
[44] *Ibid.*

---

[45] Source for data: Sanford C. Bernstein Company/*New York Times*, March 24, 2000, p. C1.
[46] *Ibid.*

On average, how fast, and in what direction, was the index changing over the first two business days (the interval [0, 2])? HINT [See Example 3.]

42. **Market Index** Refer to the Joe Downs Index in the preceding exercise. On average, how fast, and in which direction, was the index changing over the last three business days (the interval [2, 5])? HINT [See Example 3.]

43. **Crude Oil Prices** The price per barrel of crude oil in constant 2008 dollars can be approximated by

$$P(t) = 0.45t^2 - 12t + 105 \text{ dollars} \quad (0 \le t \le 28),$$

where $t$ is time in years since the start of 1980.[47]

a. What, in constant 2008 dollars, was the average rate of change of the price of oil from the start of 1981 ($t = 1$) to the start of 2006 ($t = 26$)? HINT [See Example 3.]

b. Your answer to part (a) is quite small. Can you conclude that the price of oil hardly changed at all over the 25-year period 1981 to 2006? Explain.

44. **Median Home Price** The median home price in the U.S. over the period 2003–2011 can be approximated by

$$P(t) = -5t^2 + 75t - 30 \text{ thousand dollars} \quad (3 \le t \le 11),$$

where $t$ is time in years since the start of 2000.[48]

a. What was the average rate of change of the median home price from the start of 2007 to the start of 2009? HINT [See Example 3.]

b. What, if anything, does your answer to part (a) say about the median home price in 2008? Explain.

45. **SARS** In the early stages of the deadly SARS (Severe Acute Respiratory Syndrome) epidemic in 2003, the number of reported cases could be approximated by

$$A(t) = 167(1.18)^t \quad (0 \le t \le 20)$$

$t$ days after March 17, 2003 (the first day for which statistics were reported by the World Health Organization).

a. What was the average rate of change of $A(t)$ from March 17 to March 23? Interpret the result.

b. Which of the following is true? For the first 20 days of the epidemic, the number of reported cases
   (A) increased at a faster and faster rate
   (B) increased at a slower and slower rate
   (C) decreased at a faster and faster rate
   (D) decreased at a slower and slower rate HINT [See Example 2.]

46. **SARS** A few weeks into the deadly SARS (Severe Acute Respiratory Syndrome) epidemic in 2003, the number of reported cases could be approximated by

$$A(t) = 1,804(1.04)^t \quad (0 \le t \le 30)$$

$t$ days after April 1, 2003.

a. What was the average rate of change of $A(t)$ from April 19 ($t = 18$) to April 29? Interpret the result.

b. Which of the following is true? During the 30-day period beginning April 1, the number of reported cases
   (A) increased at a faster and faster rate
   (B) increased at a slower and slower rate
   (C) decreased at a faster and faster rate
   (D) decreased at a slower and slower rate
      HINT [See Example 2.]

47. ▼ **Ecology** Increasing numbers of manatees ("sea sirens") have been killed by boats off the Florida coast. The following graph shows the relationship between the number of boats registered in Florida and the number of manatees killed each year:

The regression curve shown is given by

$$f(x) = 3.55x^2 - 30.2x + 81 \text{ manatees} \quad (4.5 < x < 8.5),$$

where $x$ is the number of boats (in hundreds of thousands) registered in Florida in a particular year, and $f(x)$ is the number of manatees killed by boats in Florida that year.[49]

a. Compute the average rate of change of $f$ over the intervals [5, 6] and [7, 8].

b. What does the answer to part (a) tell you about the manatee deaths per boat?

48. ▼ **Ecology** Refer to Exercise 47.

a. Compute the average rate of change of $f$ over the intervals [5, 7] and [6, 8].

b. Had we used a linear model instead of a quadratic one, how would the two answers in part (a) be related to each other?

49. ▣ ▼ **SAT Scores by Income** The math SAT score of a high school graduate can be approximated by

$$S(x) = 573 - 133(0.987)^x \text{ points on the math SAT test,}$$

where $x$ is the household income of the student in thousands of dollars per year.[50]

a. Use technology to complete the following table, which shows the average rate of change of $S$ over successive intervals of length 40. (Round all answers to two decimal places.) HINT [See Example 4.]

[47]Source for data: www.inflationdata.com.

[48]Source for data: www.zillow.com.

[49]Regression model is based on data from 1976 to 2000. Sources for data: Florida Department of Highway Safety & Motor Vehicles, Florida Marine Institute/New York Times, February 12, 2002, p. F4.

[50]The model is the authors'. Source for data: College Board/New York Times http://economix.blogs.nytimes.com.

| Interval | [0, 40] | [40, 80] | [80, 120] | [120, 160] | [160, 200] |
|---|---|---|---|---|---|
| Average Rate of change of S | | | | | |

b. Interpret your answer for the interval [40, 80], being sure to indicate the direction of change and the units of measurement.

c. Multiple choice: As her household income rises, a student's SAT score
   (A) increases.
   (B) decreases.
   (C) increases, then decreases.
   (D) decreases, then increases.

d. Multiple choice: As the household income increases, the effect on a student's SAT score is
   (A) more pronounced.
   (B) less pronounced.

**50.** ☐ ▼ *SAT Scores by Income* Repeat Exercise 49 using the following model for the critical reading SAT score of a high school graduate:

$$S(x) = 550 - 136(0.985)^x \text{ points on the critical reading SAT test,}$$

where $x$ is the household income of the student in thousands of dollars per year.[51]

## COMMUNICATION AND REASONING EXERCISES

**51.** Describe three ways we have used to determine the average rate of change of $f$ over an interval $[a, b]$. Which of the three ways is *least* precise? Explain.

**52.** If $f$ is a linear function of $x$ with slope $m$, what is its average rate of change over any interval $[a, b]$?

**53.** Is the average rate of change of a function over $[a, b]$ affected by the values of the function between $a$ and $b$? Explain.

**54.** If the average rate of change of a function over $[a, b]$ is zero, this means that the function has not changed over $[a, b]$, right?

**55.** Sketch the graph of a function whose average rate of change over $[0, 3]$ is negative but whose average rate of change over $[1, 3]$ is positive.

**56.** Sketch the graph of a function whose average rate of change over $[0, 2]$ is positive but whose average rate of change over $[0, 1]$ is negative.

**57.** ▼ If the rate of change of quantity A is 2 units of quantity A per unit of quantity B, and the rate of change of quantity B is 3 units of quantity B per unit of quantity C, what is the rate of change of quantity A with respect to quantity C?

**58.** ▼ If the rate of change of quantity A is 2 units of quantity A per unit of quantity B, what is the rate of change of quantity B with respect to quantity A?

---

[51]The model is the authors'. Source for data: College Board/*New York Times* http://economix.blogs.nytimes.com.

**59.** ▼ A certain function $f$ has the property that its average rate of change over the interval $[1, 1+h]$ (for positive $h$) increases as $h$ decreases. Which of the following graphs could be the graph of $f$?

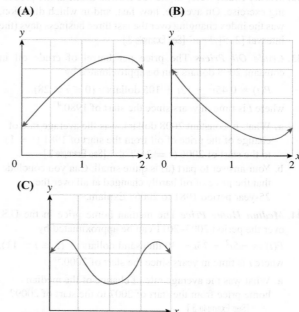

(A)    (B)

(C)

**60.** ▼ A certain function $f$ has the property that its average rate of change over the interval $[1, 1 + h]$ (for positive $h$) decreases as $h$ decreases. Which of the following graphs could be the graph of $f$?

(A)    (B)

(C)

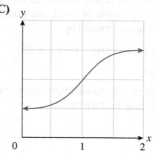

**61.** ▼ Is it possible for a company's revenue to have a negative 3-year average rate of growth, but a positive average rate of growth in 2 of the 3 years? (If not, explain; if so, illustrate with an example.)

**62.** ▼ Is it possible for a company's revenue to have a larger 2-year average rate of change than either of the 1-year average rates of change? (If not, explain why with the aid of a graph; if so, illustrate with an example.)

**63.** ◆ The average rate of change of $f$ over $[1, 3]$ is

**(A)** always equal to    **(B)** never equal to
**(C)** sometimes equal to

the average of its average rates of change over $[1, 2]$ and $[2, 3]$.

**64.** ◆ The average rate of change of $f$ over $[1, 4]$ is

**(A)** always equal to    **(B)** never equal to
**(C)** sometimes equal to

the average of its average rates of change over $[1, 2]$, $[2, 3]$, and $[3, 4]$.

# 10.5 Derivatives: Numerical and Graphical Viewpoints

In Example 4 of Section 10.4, we looked at the average rate of change of the function $G(t) = 5t^2 - 85t + 1,762$ approximating the price of gold on the spot market over smaller and smaller intervals of time. We obtained the following table showing the average rates of change of $G$ over the intervals $[8, 8 + h]$ for successively smaller values of $h$:

$h$ getting smaller; interval $[8, 8 + h]$ getting smaller →

| $h$ | 1 | 0.1 | 0.01 | 0.001 | 0.0001 |
|---|---|---|---|---|---|
| **Ave. Rate of Change over $[8, 8 + h]$** | 0 | −4.5 | −4.95 | −4.995 | −4.9995 |

Rate of change approaching −$5 per hour →

The average rates of change of the price of gold over smaller and smaller periods of time, starting at the instant $t = 8$ (8:00 am), appear to be getting closer and closer to −$5 per hour. As we look at these shrinking periods of time, we are getting closer to looking at what happens at the *instant* $t = 8$. So it seems reasonable to say that the average rates of change are approaching the **instantaneous rate of change** at $t = 8$, which the table suggests is −$5 per hour. This is how fast the price of gold was changing *exactly* at 8:00 am.

> *At* $t = 8$, *the instantaneous rate of change of* $G(t)$ *is* −5.

We express this fact mathematically by writing $G'(8) = -5$ (which we read as "$G$ prime of 8 equals −5"). Thus

> $G'(8) = -5$ *means that, at* $t = 8$, *the instantaneous rate of change of* $G(t)$ *is* −5.

The process of letting $h$ get smaller and smaller is called taking the **limit** as $h$ approaches 0 (as you recognize if you've done the sections on limits). As in the sections on limits, we write $h \to 0$ as shorthand for "$h$ approaches 0." Thus, taking the limit of the average rates of change as $h \to 0$ gives us the instantaneous rate of change.

**Q:** *All these intervals [8, 8 + h] are intervals to the right of 8. What about small intervals to the left of 8, such as [7.9, 8]?*

**A:** We can compute the average rate of change of our function for such intervals by choosing $h$ to be negative ($h = -0.1, -0.01$, etc.) and using the same difference quotient formula we used for positive $h$:

$$\text{Average rate of change of } G \text{ over } [8 + h, 8] = \frac{G(8) - G(8 + h)}{8 - (8 + h)}$$

Here are the results we get using negative $h$:

$h$ getting closer to 0; interval $[8 + h, 8]$ getting smaller $\rightarrow$

| $h$ | $-1$ | $-0.1$ | $-0.01$ | $-0.001$ | $-0.0001$ |
|---|---|---|---|---|---|
| **Ave. Rate of Change over $[8 + h, 8]$** | $-10$ | $-5.5$ | $-5.05$ | $-5.005$ | $-5.0005$ |

Rate of change approaching $-\$5$ per hour $\rightarrow$

Notice that the average rates of change are again getting closer and closer to $-5$ as $h$ approaches 0, suggesting once again that the instantaneous rate of change is $-\$5$ per hour.

## Instantaneous Rate of Change of $f(x)$ at $x = a$: Derivative

The **instantaneous rate of change** of $f(x)$ at $x = a$ is defined as

$$f'(a) = \lim_{h \to 0} \frac{f(a + h) - f(a)}{h}.$$

$f$ prime of $a$ equals the limit, as $h$ approaches 0, of the ratio $\frac{f(a + h) - f(a)}{h}$.

The quantity $f'(a)$ is also called the **derivative of $f(x)$ at $x = a$**. Finding the derivative of $f$ is called **differentiating $f$**.

### Units

The units of $f'(a)$ are the same as the units of the average rate of change: units of $f$ per unit of $x$.

### Quick Examples

1. If $f(x) = 5x^2 - 85x + 1{,}762$, then the two tables above suggest that

$$f'(8) = \lim_{h \to 0} \frac{f(8 + h) - f(8)}{h} = -5.$$

2. If $f(t)$ is the number of insects in your dorm room at time $t$ hours, and you know that $f(3) = 5$ and $f'(3) = 8$, this means that, at time $t = 3$ hours, there are 5 insects in your room, and this number is growing at an instantaneous rate of 8 insects per hour.

**IMPORTANT NOTES**

1. Sections 10.1–10.3 discuss limits in some detail. If you have not (yet) covered those sections, you can trust to your intuition.

2. The formula for the derivative tells us that the instantaneous rate of change is the limit of the average rates of change $[f(a + h) - f(a)]/h$ over smaller and smaller intervals. Thus, the value of $f'(a)$ can be approximated by computing the average rate of change for smaller and smaller values of $h$, both positive and negative.

3. If $a$ happens to be an endpoint of the domain of $f$, then $f'(a)$ does not exist, as then $[f(a + h) - f(a)]/h$ only has a one-sided limit as $h \to 0$, and so

$$\lim_{h \to 0} \frac{f(a + h) - f(a)}{h} \text{ does not exist.}^*$$

* One could define the derivative at an endpoint by instead using the associated one-sided limit; for instance, if $a$ is a left endpoint of the domain of $f$, then one could define

$$f'(a) = \lim_{h \to 0^+} \frac{f(a + h) - f(a)}{h}$$

as we did in previous editions of this book. However, in this edition we have decided to follow the usual convention and say that the derivative at an endpoint does not exist.

4. In this section we will only *approximate* derivatives. In Section 10.6 we will begin to see how we find the *exact* values of derivatives.

5. $f'(a)$ is a number we can calculate, or at least approximate, for various values of $a$, as we have done in the earlier example. Since $f'(a)$ depends on the value of $a$, we can think of $f'$ as *a function of a*. (We return to this idea at the end of this section.) An old name for $f'$ is "the function *derived from f*," which has been shortened to the *derivative* of $f$.

6. It is because $f'$ is a function that we sometimes refer to $f'(a)$ as "the derivative of $f$ evaluated at $a$," or the "derivative of $f(x)$ evaluated at $x = a$."

It may happen that the average rates of change $[f(a + h) - f(a)]/h$ do not approach any fixed number at all as $h$ approaches zero, or that they approach one number on the intervals using positive $h$, and another on those using negative $h$. If this happens, $\lim_{h \to 0}[f(a + h) - f(a)]/h$ does not exist, and we say that $f$ is **not differentiable** at $x = a$, or $f'(a)$ **does not exist**. When the limit *does* exist, we say that $f$ is **differentiable** at the point $x = a$, or $f'(a)$ **exists**. It is comforting to know that all polynomials and exponential functions are differentiable at *every* point. On the other hand, certain functions are not differentiable. Examples are $f(x) = |x|$ and $f(x) = x^{1/3}$, neither of which is differentiable at $x = 0$. (See Section 11.1.)

---

**EXAMPLE 1 Instantaneous Rate of Change: Numerically and Graphically**

The air temperature one spring morning, $t$ hours after 7:00 am, was given by the function $f(t) = 50 + 0.1t^4$ degrees Fahrenheit ($0 \le t \le 4$).

**a.** How fast was the temperature rising at 9:00 am?

**b.** How is the instantaneous rate of change of temperature at 9:00 am reflected in the graph of temperature vs. time?

**Solution**

**a.** We are being asked to find the instantaneous rate of change of the temperature at $t = 2$, so we need to find $f'(2)$. To do this we examine the average rates of change

$$\frac{f(2 + h) - f(2)}{h} \qquad \text{Average rate of change} = \text{difference quotient}$$

for values of $h$ approaching 0. Calculating the average rate of change over $[2, 2 + h]$ for $h = 1$, 0.1, 0.01, 0.001, and 0.0001 we get the following values (rounded to four decimal places):*

| $h$ | 1 | 0.1 | 0.01 | 0.001 | 0.0001 |
|---|---|---|---|---|---|
| Average Rate of Change Over $[2, 2 + h]$ | 6.5 | 3.4481 | 3.2241 | 3.2024 | 3.2002 |

Here are the values we get using negative values of $h$:

| $h$ | −1 | −0.1 | −0.01 | −0.001 | −0.0001 |
|---|---|---|---|---|---|
| Average Rate of Change Over $[2 + h, 2]$ | 1.5 | 2.9679 | 3.1761 | 3.1976 | 3.1998 |

The average rates of change are clearly approaching the number 3.2, so we can say that $f'(2) = 3.2$. Thus, at 9:00 in the morning, the temperature was rising at the rate of 3.2 degrees per hour.

**b.** We saw in Section 10.4 that the average rate of change of $f$ over an interval is the slope of the secant line through the corresponding points on the graph of $f$. Figure 25 illustrates this for the intervals $[2, 2 + h]$ with $h = 1$, 0.5, and 0.1.

**Figure 25**

All three secant lines pass though the point $(2, f(2)) = (2, 51.6)$ on the graph of $f$. Each of them passes through a second point on the curve (the second point is different for each secant line) and this second point gets closer and closer to $(2, 51.6)$ as $h$ gets closer to 0. What seems to be happening is that the secant lines are getting closer and closer to a line that just touches the curve at $(2, 51.6)$: the **tangent line** at $(2, 51.6)$, shown in Figure 26.

**Figure 26**

**Q**: What is the slope of this tangent line?

**A**: Because the slopes of the secant lines are getting closer and closer to 3.2, and because the secant lines are approaching the tangent line, the tangent line must have slope 3.2. In other words,

*At the point on the graph where x = 2, the slope of the tangent line is f'(2).*

**Q**: What is the difference between f(2) and f'(2)?

**A**: An important question. Briefly, *f*(2) is the *value of f* when *t* = 2, while *f'*(2) is the *rate at which f is changing* when *t* = 2. Here,

$$f(2) = 50 + 0.1(2)^4 = 51.6 \text{ degrees.}$$

Thus, at 9:00 am (*t* = 2), the temperature was 51.6 degrees. On the other hand,

$$f'(2) = 3.2 \text{ degrees per hour.} \qquad \text{Units of slope are units of } f \text{ per unit of } t.$$

This means that, at 9:00 am (*t* = 2), the temperature was increasing at a rate of 3.2 degrees per hour.

---

Because we have been talking about tangent lines, we should say more about what they *are*. A tangent line to a *circle* is a line that touches the circle in just one point. A tangent line gives the circle "a glancing blow," as shown in Figure 27.

For a smooth curve other than a circle, a tangent line may touch the curve at more than one point, or pass through it (Figure 28).

Tangent line to the circle at *P*

**Figure 27**

Tangent line at *P* intersects graph at *Q*

Tangent line at *P* passes through curve at *P*

**Figure 28**

However, all tangent lines have the following interesting property in common: If we focus on a small portion of the curve very close to the point *P*—in other words, if we "zoom in" to the graph near the point *P*—the curve will appear almost straight, and almost indistinguishable from the tangent line (Figure 29).

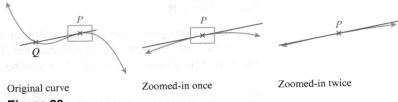

Original curve

Zoomed-in once

Zoomed-in twice

**Figure 29**

You can check this property by zooming in on the curve shown in Figures 25 and 26 in the previous example near the point where *x* = 2.

## Secant and Tangent Lines

The *slope of the secant line* through the points on the graph of $f$ where $x = a$ and $x = a + h$ is given by the average rate of change, or difference quotient,

$$m_{\text{sec}} = \text{slope of secant} = \text{average rate of change} = \frac{f(a + h) - f(a)}{h}.$$

The *slope of the tangent line* through the point on the graph of $f$ where $x = a$ is given by the instantaneous rate of change, or derivative

$$m_{\text{tan}} = \text{slope of tangent} = \text{instantaneous rate of change} = \text{derivative}$$

$$= f'(a) = \lim_{h \to 0} \frac{f(a + h) - f(a)}{h},$$

assuming the limit exists.

Secant line through $(a, f(a))$ and $(a + h, f(a + h))$

Tangent line at $(a, f(a))$

### Quick Example

In the following graph, the tangent line at the point where $x = 2$ has slope 3. Therefore, the derivative at $x = 2$ is 3. That is, $f'(2) = 3$.

**Note** It might happen that the tangent line is vertical at some point or does not exist at all. These are the cases in which $f$ is not differentiable at the given point. (See Section 10.6.) ■

We can now give a more precise definition of what we mean by the tangent line to a point $P$ on the graph of $f$ at a given point: The **tangent line** to the graph of $f$ at the point $P(a, f(a))$ is the straight line passing through $P$ with slope $f'(a)$.

## Quick Approximation of the Derivative

**Q:** *Do we always need to make tables of difference quotients as above in order to calculate an approximate value for the derivative? That seems like a large amount of work just to get an approximation.*

**A**: We can usually *approximate* the value of the derivative by using a single, small value of *h*. In the example above, the value $h = 0.0001$ would have given a pretty good approximation. The problems with using a fixed value of *h* are that (1) we do not get an *exact* answer, only an *approximation* of the derivative, and (2) how good an approximation it is depends on the function we're differentiating.* However, with most of the functions we'll be considering, setting $h = 0.0001$ does give us a good approximation.

* In fact, no matter how small the value we decide to use for *h*, it is possible to craft a function *f* for which the difference quotient at *a* is not even close to $f'(a)$.

---

### Calculating a Quick Approximation of the Derivative

We can calculate an approximate value of $f'(a)$ by using the formula

$$f'(a) \approx \frac{f(a + h) - f(a)}{h} .$$    Rate of change over $[a, a + h]$

with a small value of *h*. The value $h = 0.0001$ works for most examples we encounter (students of numerical methods study the question of exactly how accurate this approximation is).

### Alternative Formula: The Balanced Difference Quotient

The following alternative formula, which measures the rate of change of *f* over the interval $[a - h, a + h]$, often gives a more accurate result, and is the one used in many calculators:

$$f'(a) \approx \frac{f(a + h) - f(a - h)}{2h} .$$    Rate of change over $[a - h, a + h]$

---

**Note** For the quick approximations to be valid, the function *f* must be differentiable; that is, $f'(a)$ must exist. ■

---

### EXAMPLE 2 Quick Approximation of the Derivative

**a.** Calculate an approximate value of $f'(1.5)$ if $f(x) = x^2 - 4x$.

**b.** Find the equation of the tangent line at the point on the graph where $x = 1.5$.

### Solution

**a.** We shall compute both the ordinary difference quotient and the balanced difference quotient.

**Ordinary Difference Quotient:** Using $h = 0.0001$, the ordinary difference quotient is:

$$f'(1.5) \approx \frac{f(1.5 + 0.0001) - f(1.5)}{0.0001}$$    Ordinary difference quotient

$$= \frac{f(1.5001) - f(1.5)}{0.0001}$$

$$= \frac{(1.5001^2 - 4 \times 1.5001) - (1.5^2 - 4 \times 1.5)}{0.0001} = -0.9999.$$

This answer is accurate to 0.0001; in fact, $f'(1.5) = -1$.

Graphically, we can picture this approximation as follows: Zoom in on the curve using the window $1.5 \le x \le 1.5001$ and measure the slope of the secant line joining both ends of the curve segment. Figure 30 shows close-up views of the curve and tangent line near the point $P$ in which we are interested, the third view being the zoomed-in view used for this approximation.

**Figure 30**

Slope $PQ = -0.9999$

Notice that in the third window the tangent line and curve are indistinguishable. Also, the point $P$ in which we are interested is on the left edge of the window.

**Balanced Difference Quotient:** For the balanced difference quotient, we get

$$f'(1.5) \approx \frac{f(1.5 + 0.0001) - f(1.5 - 0.0001)}{2(0.0001)}$$    Balanced difference quotient

$$= \frac{f(1.5001) - f(1.4999)}{0.0002}$$

$$= \frac{(1.5001^2 - 4 \times 1.5001) - (1.4999^2 - 4 \times 1.4999)}{0.0002} = -1.$$

✱ The balanced difference quotient always gives the exact derivative for a quadratic function.

This balanced difference quotient gives the exact answer in this case!✱ Graphically, it is as though we have zoomed in using a window that puts the point $P$ in the *center* of the screen (Figure 31) rather than at the left edge.

Slope $RQ = -1$

**Figure 31**

See the Technology Guides at the end of the chapter to find out how to calculate the quick approximations to the derivative in Example 2 using a TI-83/84 Plus or a spreadsheet. Here is an outline:

**TI-83/84 Plus**
$Y_1$=X^2-4*X
Home screen:
  $(Y_1(1.5001)-Y_1(1.5))/$
  0.0001
  $(Y_1(1.5001)-$
  $Y_1(1.4999))/0.0002$
[More details on page 779.]

**Spreadsheet**
Headings a, h, x, f(x), Diff Quotient, Balanced Diff Quotient in A1–F1
  1.5 in A2, 0.0001 in B2,
  =A2-B2 in C2, =A2 in C3,
  =A2+B2 in C4
  =C2^2-4*C2 in D2; copy down to D4
  =(D3-D2)/(C3-C2) in E2
  =(D4-D2)/(C4-C2) in E3
[More details on page 781.]

**b.** We find the equation of the tangent line from a point on the line and its slope, as we did in Chapter 1:

• *Point* $(1.5, f(1.5)) = (1.5, -3.75)$.
• *Slope* $m = f'(1.5) = -1$.    Slope of the tangent line = derivative.

The equation is

$$y = mx + b,$$

where $m = -1$ and $b = y_1 - mx_1 = -3.75 - (-1)(1.5) = -2.25$. Thus, the equation of the tangent line is

$$y = -x - 2.25.$$

**Q**: *Why can't we simply put* $h = 0.000\,000\,000\,000\,000\,000\,000\,01$ *for an incredibly accurate approximation to the instantaneous rate of change and be done with it?*

**A**: This approach would certainly work if you were patient enough to do the (thankless) calculation by hand! However, doing it with the help of technology—even an ordinary calculator—will cause problems: The issue is that calculators and spreadsheets represent numbers with a maximum number of significant digits (15 in the case of Excel). As the value of $h$ gets smaller, the value of $f(a+h)$ gets closer and closer to the value of $f(a)$. For example, if $f(x) = 50 + 0.1x^4$, Excel might compute

$$f(2 + 0.000\,000\,000\,000\,1) - f(2)$$

$$= 51.600\,000\,000\,000\,3 - 51.6 \qquad \text{Rounded to 15 digits}$$

$$= 0.000\,000\,000\,000\,3$$

and the corresponding difference quotient would be 3, not 3.2 as it should be. If $h$ gets even smaller, Excel will not be able to distinguish between $f(a+h)$ and $f(a)$ at all, in which case it will compute 0 for the rate of change. This loss in accuracy when subtracting two very close numbers is called **subtractive error**.

Thus, there is a trade-off in lowering the value of $h$: Smaller values of $h$ yield *mathematically* more accurate approximations of the derivative, but if $h$ gets too small, subtractive error becomes a problem and decreases the accuracy of computations that use technology.

## Leibniz $d$ Notation

We introduced the notation $f'(x)$ for the derivative of $f$ at $x$, but there is another interesting notation. We have written the average rate of change as

$$\text{Average rate of change} = \frac{\Delta f}{\Delta x}. \qquad \frac{\text{Change in } f}{\text{Change in } x}$$

As we use smaller and smaller values for $\Delta x$, we approach the instantaneous rate of change, or derivative, for which we also have the notation $df/dx$, due to Leibniz:

$$\text{Instantaneous rate of change} = \lim_{\Delta x \to 0} \frac{\Delta f}{\Delta x} = \frac{df}{dx}.$$

That is, $df/dx$ is just another notation for $f'(x)$. Do not think of $df/dx$ as an actual quotient of two numbers: Remember that we only use an actual quotient $\Delta f/\Delta x$ to *approximate* the value of $df/dx$.

In Example 3, we apply the quick approximation method of estimating the derivative.

---

### EXAMPLE 3  **Velocity**

* Eric's claim is difficult to believe; 100 ft/s corresponds to around 68 mph, and professional pitchers can throw *forward* at about 100 mph.

My friend Eric, an enthusiastic baseball player, claims he can "probably" throw a ball upward at a speed of 100 feet per second (ft/s).* Our physicist friends tell us that its height $s$ (in feet) $t$ seconds later would be $s = 100t - 16t^2$. Find its average velocity over the interval [2, 3] and its instantaneous velocity exactly 2 seconds after Eric throws it.

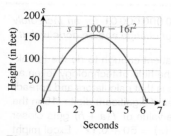

**Figure 32**

**Solution**    The graph of the ball's height as a function of time is shown in Figure 32. Asking for the velocity is really asking for the rate of change of height with respect to time. (Why?) Consider average velocity first. To compute the **average velocity** of the ball from time 2 to time 3, we first compute the change in height:

$$\Delta s = s(3) - s(2) = 156 - 136 = 20 \text{ ft}.$$

Since it rises 20 feet in $\Delta t = 1$ second, we use the defining formula *speed = distance/time* to get the average velocity:

$$\text{Average velocity} = \frac{\Delta s}{\Delta t} = \frac{20}{1} = 20 \text{ ft/sec}.$$

from time $t = 2$ to $t = 3$. This is just the difference quotient, so

*The average velocity is the average rate of change of height.*

To get the **instantaneous velocity** at $t = 2$, we find the instantaneous rate of change of height. In other words, we need to calculate the derivative $ds/dt$ at $t = 2$. Using the balanced quick approximation described earlier, we get

$$\frac{ds}{dt} \approx \frac{s(2 + 0.0001) - s(2 - 0.0001)}{2(0.0001)}$$

$$= \frac{s(2.0001) - s(1.9999)}{0.0002}$$

$$= \frac{100(2.0001) - 16(2.0001)^2 - (100(1.9999) - 16(1.9999)^2)}{0.0002}$$

$$= 36 \text{ ft/sec}.$$

In fact, this happens to be the exact answer; the instantaneous velocity at $t = 2$ is exactly 36 ft/sec. (Try an even smaller value of $h$ to persuade yourself.)

➡ **Before we go on...**    If we repeat the calculation in Example 3 at time $t = 5$, we get

$$\frac{ds}{dt} = -60 \text{ ft/sec}.$$

The negative sign tells us that the ball is *falling* at a rate of 60 feet per second at time $t = 5$. (How does the fact that it is falling at $t = 5$ show up on the graph?) ∎

The preceding example gives another interpretation of the derivative.

## Average and Instantaneous Velocity

For an object moving in a straight line with position $s(t)$ at time $t$, the **average velocity** from time $t$ to time $t + h$ is the average rate of change of position with respect to time:

$$v_{ave} = \frac{s(t + h) - s(t)}{h} = \frac{\Delta s}{\Delta t}. \qquad \begin{array}{l}\text{Average velocity} = \\ \text{Average rate of change of position}\end{array}$$

The **instantaneous velocity** at time $t$ is

$$v = \lim_{h \to 0} \frac{s(t + h) - s(t)}{h} = \frac{ds}{dt}. \qquad \begin{array}{l}\text{Instantaneous velocity} = \\ \text{Instantaneous rate of change of position}\end{array}$$

In other words, *instantaneous velocity is the derivative of position with respect to time.*

Here is one last comment on Leibniz notation. In Example 3, we could have written the velocity either as $s'$ or as $ds/dt$, as we chose to do. To write the answer to the question, that the velocity at $t = 2$ sec was 36 ft/sec, we can write either

$$s'(2) = 36$$

or

$$\left.\frac{ds}{dt}\right|_{t=2} = 36.$$

The notation "$|_{t=2}$" is read "evaluated at $t = 2$." Similarly, if $y = f(x)$, we can write the instantaneous rate of change of $f$ at $x = 5$ in either functional notation as

$$f'(5) \qquad \text{The derivative of } f, \text{ evaluated at } x = 5$$

or in Leibniz notation as

$$\left.\frac{dy}{dx}\right|_{x=5}. \qquad \text{The derivative of } y, \text{ evaluated at } x = 5$$

The latter notation is obviously more cumbersome than the functional notation $f'(5)$, but the notation $dy/dx$ has compensating advantages. You should practice using both notations.

## The Derivative Function

The derivative $f'(x)$ is a number we can calculate, or at least approximate, for various values of $x$. Because $f'(x)$ depends on the value of $x$, we may think of $f'$ as a function of $x$. This function is the **derivative function**.

### Derivative Function

If $f$ is a function, its **derivative function** $f'$ is the function whose value $f'(x)$ is the derivative of $f$ at $x$. Its domain is the set of all $x$ at which $f$ is differentiable. Equivalently, $f'$ associates to each $x$ the slope of the tangent to the graph of the function $f$ at $x$, or the rate of change of $f$ at $x$. The formula for the derivative function is

$$f'(x) = \lim_{h \to 0} \frac{f(x+h) - f(x)}{h}. \qquad \text{Derivative function}$$

### Quick Examples

1. Let $f(x) = 3x - 1$. The graph of $f$ is a straight line that has slope 3 everywhere. In other words, $f'(x) = 3$ for every choice of $x$; that is, $f'$ is a constant function.

| **Original Function** $f$ | **Derivative Function** $f'$ |
|---|---|
| $f(x) = 3x - 1$ | $f'(x) = 3$ |

* This method is discussed in detail on the Website at

Online Text → Sketching the Graph of the Derivative.

2. Given the graph of a function $f$, we can get a rough sketch of the graph of $f'$ by estimating the slope of the tangent to the graph of $f$ at several points, as illustrated below.*

**Original Function $f$**
$$y = f(x)$$

**Derivative Function $f'$**
$$y = f'(x)$$

For $x$ between $-2$ and $0$, the graph of $f$ is linear with slope $-2$. As $x$ increases from $0$ to $2$, the slope increases from $-2$ to $2$. For $x$ larger than $2$, the graph of $f$ is linear with slope $2$. (Notice that, when $x = 1$, the graph of $f$ has a horizontal tangent, so $f'(1) = 0$.)

3. Look again at the graph on the left in Quick Example 2. When $x < 1$ the derivative $f'(x)$ is negative, so the graph has negative slope and $f$ is **decreasing**; its values are going down as $x$ increases. When $x > 1$ the derivative $f'(x)$ is positive, so the graph has positive slope and $f$ is **increasing**; its values are going up as $x$ increases.

$f$ decreasing for $x < 1$
$f$ increasing for $x > 1$

$f'$ negative for $x < 1$
$f'$ positive for $x > 1$

The following example shows how we can use technology to graph the (approximate) derivative of a function, where it exists.

### EXAMPLE 4 Graphing the Derivative with Technology

Use technology to graph the derivative of $f(x) = -2x^2 + 6x + 5$ for values of $x$ starting at $-5$.

**Solution** The TI-83/84 Plus has a built-in function that approximates the derivative, and we can use it to graph the derivative of a function. In a spreadsheet, we need to create the approximation using one of the quick approximation formulas and we can then graph a table of its values. See the technology note in the margin on the next page to find out how to graph the derivative (Figure 33) using the Website graphing utility, the TI-83/84 Plus, and a spreadsheet.

Graph of $f$

Graph of $f'$

**Figure 33**

**using Technology**

See the Technology Guides at the end of the chapter to find out how to obtain a table of values of and graph the derivative in Example 4 using a TI-83/84 Plus or a spreadsheet. Here is an outline:

**TI-83/84 Plus**
```
Y₁=-2X^2+6X+5
Y₂=nDeriv(Y₁,X,X)
```
[More details on page 779.]

**Spreadsheet**
Value of *h* in E2
Values of *x* from A2 down increasing by *h*
`-2*A2^2+6*A2+5` from B2 down
`=(B3-B2)/$E$2` from C2 down
Insert scatter chart using columns A and C [More details on page 781.]

 **Website**
www.WanerMath.com
**Web grapher:**

Online Utilities→ Function Evaluator and Grapher

Enter
`deriv(-2*x^2+6*x+5)` for
$y_1$. Alternatively, enter
`-2*x^2+6*x+5` for $y_1$ and
`deriv(y1)` for $y_2$.

**Excel grapher:**
Student Home→Online Utilities→Excel First and Second Derivative Graphing Utility
Function: `-2*x^2+6*x+5`

We said that $f'$ records the slope of (the tangent line to) the function $f$ at each point. Notice that the graph of $f'$ confirms that the slope of the graph of $f$ is decreasing as $x$ increases from $-5$ to $5$. Note also that the graph of $f$ reaches a high point at $x = 1.5$ (the vertex of the parabola). At that point, the slope of the tangent is zero; that is, $f'(1.5) = 0$, as we see in the graph of $f'$.

### EXAMPLE 5 ◻ An Application: Broadband Penetration

Wired broadband penetration in the United States can be modeled by

$$P(t) = \frac{29.842}{1 + 12.502(1.642)^{-t}} \quad (0 \le t \le 12),$$

where $t$ is time in years since 2000.[52] Graph both $P$ and its derivative, and determine when broadband penetration was growing most rapidly.

**Solution** Using one of the methods in Example 4, we obtain the graphs shown in Figure 34.

Graph of $P$

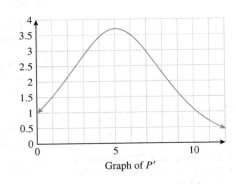
Graph of $P'$

**Figure 34**

---

[52]Broadband penetration is the number of broadband installations divided by the total population. Source for data: Organisation for Economic Co-operation and Development (OECD) Directorate for Science, Technology, and Industry, table of Historical Penetration Rates, June 2010, downloaded April 2011 from www.oecd.org/sti/ict/broadband.

From the graph on the right, we see that $P'$ reaches a peak somewhere near $t = 5$ (the beginning of 2005). Recalling that $P'$ measures the *slope* of the graph of $P$, we can conclude that the graph of $P$ is steepest near $t = 5$, indicating that, according to the model, broadband penetration was growing most rapidly at the start of 2005. Notice that this is not so easy to see directly on the graph of $P$.

To determine the point of maximum growth more accurately, we can zoom in on the graph of $P'$ using the range $4.5 \leq t \leq 5.5$ (Figure 35).

Graph of $P'$

**Figure 35**

We can now see that $P'$ reaches its highest point around $t = 5.1$, so we conclude that broadband penetration was growing most rapidly in early 2005.

➡ **Before we go on...** Besides helping us to determine the point of maximum growth, the graph of $P'$ in Example 5 gives us a great deal of additional information. As just one example, in Figure 35 we can see that the maximum value of $P'$ is about 3.7, indicating that broadband penetration grew at a fastest rate of about 3.7 percentage points per year. ∎

## FAQs

### Recognizing When and How to Compute the Instantaneous Rate of Change

**Q:** *How do I know, by looking at the wording of a problem, that it is asking for an instantaneous rate of change?*

**A:** If a problem does not ask for an instantaneous rate of change directly, it might do so indirectly, as in "How fast is quantity q increasing?" or "Find the rate of increase of q."

**Q:** *If I know that a problem calls for estimating an instantaneous rate of change, how should I estimate it: with a table showing smaller and smaller values of h, or by using a quick approximation?*

**A:** For most practical purposes, a quick approximation is accurate enough. Use a table showing smaller and smaller values of h when you would like to check the accuracy.

**Q:** *Which should I use in computing a quick approximation: the balanced difference quotient or the ordinary difference quotient?*

**A:** In general, the balanced difference quotient gives a more accurate answer.

## 10.5 EXERCISES

▼ more advanced   ◆ challenging
⊤ indicates exercises that should be solved using technology

*In Exercises 1–4, estimate the derivative from the table of average rates of change.* HINT *[See discussion at the beginning of the section.]*

**1.**

| h | 1 | 0.1 | 0.01 | 0.001 | 0.0001 |
|---|---|---|---|---|---|
| **Average Rate of Change of f Over [5, 5 + h]** | 12 | 6.4 | 6.04 | 6.004 | 6.0004 |
| h | −1 | −0.1 | −0.01 | −0.001 | −0.0001 |
| **Average Rate of Change of f Over [5 + h, 5]** | 3 | 5.6 | 5.96 | 5.996 | 5.9996 |

Estimate $f'(5)$.

**2.**

| h | 1 | 0.1 | 0.01 | 0.001 | 0.0001 |
|---|---|---|---|---|---|
| **Average Rate of Change of g Over [7, 7 + h]** | 4 | 4.8 | 4.98 | 4.998 | 4.9998 |
| h | −1 | −0.1 | −0.01 | −0.001 | −0.0001 |
| **Average Rate of Change of g Over [7 + h, 7]** | 5 | 5.3 | 5.03 | 5.003 | 5.0003 |

Estimate $g'(7)$.

**3.**

| h | 1 | 0.1 | 0.01 | 0.001 | 0.0001 |
|---|---|---|---|---|---|
| **Average Rate of Change of r Over [−6, −6 + h]** | −5.4 | −5.498 | −5.4998 | −5.499982 | −5.49999822 |
| h | −1 | −0.1 | −0.01 | −0.001 | −0.0001 |
| **Average Rate of Change of r Over [−6 + h, −6]** | −7.52 | −6.13 | −5.5014 | −5.5000144 | −5.500001444 |

Estimate $r'(-6)$.

**4.**

| h | 1 | 0.1 | 0.01 | 0.001 | 0.0001 |
|---|---|---|---|---|---|
| **Average Rate of Change of s Over [0, h]** | −2.52 | −1.13 | −0.6014 | −0.6000144 | −0.600001444 |
| h | −1 | −0.1 | −0.01 | −0.001 | −0.0001 |
| **Average Rate of Change of s Over [h, 0]** | −0.4 | −0.598 | −0.5998 | −0.599982 | −0.59999822 |

Estimate $s'(0)$.

*Consider the functions in Exercises 5–8 as representing the value of an ounce of palladium in U.S. dollars as a function of the time t in days.[53] Find the average rates of change of R(t)*

over the time intervals $[t, t + h]$, where t is as indicated and $h = 1, 0.1,$ and $0.01$ days. Hence, estimate the instantaneous rate of change of R at time t, specifying the units of measurement. (Use smaller values of h to check your estimates.) HINT [See Example 1.]

**5.** $R(t) = 60 + 50t - t^2; t = 5$

**6.** $R(t) = 60t - 2t^2; t = 3$

**7.** $R(t) = 270 + 20t^3; t = 1$

**8.** $R(t) = 200 + 50t - t^3; t = 2$

*Each of the functions in Exercises 9–12 gives the cost to manufacture x items. Find the average cost per unit of manufacturing h more items (i.e., the average rate of change of the total cost) at a production level of x, where x is as indicated and $h = 10$ and 1. Hence, estimate the instantaneous rate of change of the total cost at the given production level x, specifying the units of measurement. (Use smaller values of h to check your estimates.)* HINT [See Example 1.]

**9.** $C(x) = 10{,}000 + 5x - \dfrac{x^2}{10{,}000}; x = 1{,}000$

**10.** $C(x) = 20{,}000 + 7x - \dfrac{x^2}{20{,}000}; x = 10{,}000$

**11.** $C(x) = 15{,}000 + 100x + \dfrac{1{,}000}{x}; x = 100$

**12.** $C(x) = 20{,}000 + 50x + \dfrac{10{,}000}{x}; x = 100$

*In Exercises 13–16, the graph of a function is shown together with the tangent line at a point P. Estimate the derivative of f at the corresponding x value.* HINT [See Quick Example page 742.]

**13.**

**14.**

**15.**

**16.**
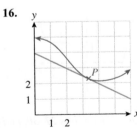

---

[53]Palladium was trading at around $290 in August 2008.

*In each of the graphs given in Exercises 17–22, say at which labeled point the slope of the tangent is **(a)** greatest and **(b)** least (in the sense that −7 is less than 1).* HINT [See Quick Example page 742.]

**17.**

**18.**

**19.**

**20.**

**21.**

**22.**

*In each of Exercises 23–26, three slopes are given. For each slope, determine at which of the labeled points on the graph the tangent line has that slope.*

**23. a.** 0   **b.** 4   **c.** −1

**24. a.** 0   **b.** 1   **c.** −1

**25. a.** 0   **b.** 3   **c.** −3

**26. a.** 0   **b.** 3   **c.** 1

*In each of Exercises 27–30, find the approximate coordinates of all points (if any) where the slope of the tangent is: **(a)** 0, **(b)** 1, **(c)** −1.* HINT [See Quick Example page 742.]

**27.**

**28.**

**29.**

**30.**

**31.** Complete the following: The tangent to the graph of the function $f$ at the point where $x = a$ is the line passing through the point ____ with slope ____ .

**32.** Complete the following: The difference quotient for $f$ at the point where $x = a$ gives the slope of the ____ line that passes through ____ .

**33.** Which is correct? The derivative function assigns to each value $x$

 **(A)** the average rate of change of $f$ at $x$.
 **(B)** the slope of the tangent to the graph of $f$ at $(x, f(x))$.
 **(C)** the rate at which $f$ is changing over the interval $[x, x + h]$ for $h = 0.0001$.
 **(D)** the balanced difference quotient $[f(x + h) − f(x − h)]/(2h)$ for $h \approx 0.0001$.

**34.** Which is correct? The derivative function $f'(x)$ tells us
 **(A)** the slope of the tangent line at each of the points $(x, f(x))$.
 **(B)** the approximate slope of the tangent line at each of the points $(x, f(x))$.
 **(C)** the slope of the secant line through $(x, f(x))$ and $(x + h, f(x + h))$ for $h = 0.0001$.
 **(D)** the slope of a certain secant line through each of the points $(x, f(x))$.

**35.** ▼ Let $f$ have the graph shown.

 **a.** The average rate of change of $f$ over the interval $[2, 4]$ is
  **(A)** greater than $f'(2)$.   **(B)** less than $f'(2)$.
  **(C)** approximately equal to $f'(2)$.

 **b.** The average rate of change of $f$ over the interval $[−1, 1]$ is
  **(A)** greater than $f'(0)$.   **(B)** less than $f'(0)$.
  **(C)** approximately equal to $f'(0)$.

**c.** Over the interval [0, 2], the instantaneous rate of change of $f$ is
  **(A)** increasing.    **(B)** decreasing.    **(C)** neither.

**d.** Over the interval [0, 4], the instantaneous rate of change of $f$ is
  **(A)** increasing, then decreasing.
  **(B)** decreasing, then increasing.
  **(C)** always increasing.
  **(D)** always decreasing.

**e.** When $x = 4$, $f(x)$ is
  **(A)** approximately 0, and increasing at a rate of about 0.7 units per unit of $x$.
  **(B)** approximately 0, and decreasing at a rate of about 0.7 units per unit of $x$.
  **(C)** approximately 0.7, and increasing at a rate of about 1 unit per unit of $x$.
  **(D)** approximately 0.7, and increasing at a rate of about 3 units per unit of $x$.

**36.** ▼ A function $f$ has the following graph.

**a.** The average rate of change of $f$ over [0, 200] is
  **(A)** greater than
  **(B)** less than
  **(C)** approximately equal to
  the instantaneous rate of change at $x = 100$.

**b.** The average rate of change of $f$ over [0, 200] is
  **(A)** greater than
  **(B)** less than
  **(C)** approximately equal to
  the instantaneous rate of change at $x = 150$.

**c.** Over the interval [0, 50] the instantaneous rate of change of $f$ is
  **(A)** increasing, then decreasing.
  **(B)** decreasing, then increasing.
  **(C)** always increasing.
  **(D)** always decreasing.

**d.** On the interval [0, 200], the instantaneous rate of change of $f$ is
  **(A)** always positive.    **(B)** always negative.
  **(C)** negative, positive, and then negative.

**e.** $f'(100)$ is
  **(A)** greater than $f'(25)$.    **(B)** less than $f'(25)$.
  **(C)** approximately equal to $f'(25)$.

*In Exercises 37–40, use a quick approximation to estimate the derivative of the given function at the indicated point.* HINT [See Example 2(a).]

**37.** $f(x) = 1 - 2x$;  $x = 2$    **38.** $f(x) = \dfrac{x}{3} - 1$;  $x = -3$

**39.** $f(x) = \dfrac{x^2}{4} - \dfrac{x^3}{3}$; $x = -1$    **40.** $f(x) = \dfrac{x^2}{2} + \dfrac{x}{4}$; $x = 2$

*In Exercises 41–48, estimate the indicated derivative by any method.* HINT [See Example 2.]

**41.** $g(t) = \dfrac{1}{t^5}$; estimate $g'(1)$

**42.** $s(t) = \dfrac{1}{t^3}$; estimate $s'(-2)$

**43.** $y = 4x^2$; estimate $\left.\dfrac{dy}{dx}\right|_{x=2}$

**44.** $y = 1 - x^2$; estimate $\left.\dfrac{dy}{dx}\right|_{x=-1}$

**45.** $s = 4t + t^2$; estimate $\left.\dfrac{ds}{dt}\right|_{t=-2}$

**46.** $s = t - t^2$; estimate $\left.\dfrac{ds}{dt}\right|_{t=2}$

**47.** $R = \dfrac{1}{p}$; estimate $\left.\dfrac{dR}{dp}\right|_{p=20}$

**48.** $R = \sqrt{p}$; estimate $\left.\dfrac{dR}{dp}\right|_{p=400}$

*In Exercises 49–54, **(a)** use any method to estimate the slope of the tangent to the graph of the given function at the point with the given x-coordinate, and **(b)** find an equation of the tangent line in part (a). In each case, sketch the curve together with the appropriate tangent line.* HINT [See Example 2(b).]

**49.** $f(x) = x^3$; $x = -1$

**50.** $f(x) = x^2$; $x = 0$

**51.** $f(x) = x + \dfrac{1}{x}$; $x = 2$

**52.** $f(x) = \dfrac{1}{x^2}$; $x = 1$

**53.** $f(x) = \sqrt{x}$; $x = 4$

**54.** $f(x) = 2x + 4$; $x = -1$

*In each of Exercises 55–58, estimate the given quantity.*

**55.** $f(x) = e^x$; estimate $f'(0)$

**56.** $f(x) = 2e^x$; estimate $f'(1)$

**57.** $f(x) = \ln x$; estimate $f'(1)$

**58.** $f(x) = \ln x$; estimate $f'(2)$

*In Exercises 59–64, match the graph of f to the graph of f′ (the graphs of f′ are shown after Exercise 64).*

**59.** ▼

**60.** ▼

**61.** ▼

**62.** ▼

**63.** ▼

**64.** ▼

*Graphs of derivatives for Exercises 59–64:*

(A)

(B)

(C)

(D)

(E)

(F)

*In Exercises 65–68, the graph of a function is given. For which x in the range shown is the function increasing? For which x is the function decreasing?* HINT [See Quick Example 3 page 748.]

**65.**

**66.**

**67.**

**68.**

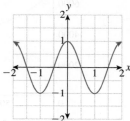

*In Exercises 69–72, the graph of the derivative of a function is given. For which x is the (original) function increasing? For which x is the (original) function decreasing?* HINT [See Quick Example 3 page 748.]

**69.**

**70.**

**71.**

**72.**

In Exercises 73 and 74, use technology to graph the derivative of the given function for the given range of values of x. Then use your graph to estimate all values of x (if any) where the tangent line to the graph of the given function is horizontal. Round answers to one decimal place. HINT [See Example 4.]

**73.** $f(x) = x^4 + 2x^3 - 1; \quad -2 \le x \le 1$

**74.** $f(x) = -x^3 - 3x^2 - 1; \quad -3 \le x \le 1$

In Exercises 75 and 76, use the method of Example 4 to list approximate values of f'(x) for x in the given range. Graph f(x) together with f'(x) for x in the given range.

**75.** $f(x) = \dfrac{x+2}{x-3}; \quad 4 \le x \le 5$

**76.** $f(x) = \dfrac{10x}{x-2}; \quad 2.5 \le x \le 3$

## APPLICATIONS

**77. Demand** Suppose the demand for a new brand of sneakers is given by

$$q = \frac{5,000,000}{p}$$

where $p$ is the price per pair of sneakers, in dollars, and $q$ is the number of pairs of sneakers that can be sold at price $p$. Find $q(100)$ and estimate $q'(100)$. Interpret your answers. HINT [See Example 1.]

**78. Demand** Suppose the demand for an old brand of TV is given by

$$q = \frac{100,000}{p+10}$$

where $p$ is the price per TV set, in dollars, and $q$ is the number of TV sets that can be sold at price $p$. Find $q(190)$ and estimate $q'(190)$. Interpret your answers. HINT [See Example 1.]

**79. Oil Imports from Mexico** The following graph shows approximate daily oil imports to the U.S. from Mexico.[54] Also shown is the tangent line at the point corresponding to year 2005.

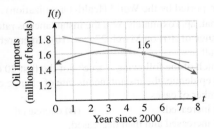

**a.** Estimate the slope of the tangent line shown on the graph. What does the graph tell you about oil imports from Mexico in 2005? HINT [Identify two points on the tangent line. Then see Quick Example page 742.]

**b.** According to the graph, is the rate of change of oil imports from Mexico increasing, decreasing, or increasing then decreasing? Why?

**80. Oil Production in Mexico** The following graph shows approximate daily oil production by **Pemex**, Mexico's national oil company.[55] Also shown is the tangent line at the point corresponding to year 2003.

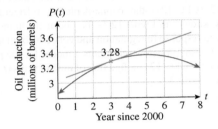

**a.** Estimate the slope of the tangent line shown on the graph. What does the graph tell you about oil production by Pemex in 2003? HINT [Identify two points on the tangent line. Then see Quick Example page 742.]

**b.** According to the graph, is the rate of change of oil production by Pemex increasing or decreasing over the range [0, 4]? Why?

**81. ▼ Prison Population** The following curve is a model of the total U.S. prison population as a function of time in years ($t = 0$ represents 2000).[56]

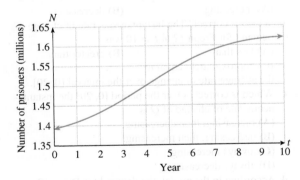

**a.** Which is correct? Over the period [5, 10] the instantaneous rate of change of $N$ is
    **(A)** increasing.        **(B)** decreasing.

**b.** Which is correct? The instantaneous rate of change of prison population at $t = 4$ was
    **(A)** less than        **(B)** greater than
    **(C)** approximately equal to
    the average rate of change over the interval [0, 10].

[54]Figures are approximate, and 2008–2009 figures are projections by the Department of Energy. Source: Energy Information Administration/Pemex (www.eia.doe.gov).

[55]Ibid.

[56]Source: Bureau of Justice Statistics http://bjs.ojp.usdoj.gov.

**c.** Which is correct? Over the period [0, 10] the instantaneous rate of change of $N$ is
  **(A)** increasing, then decreasing.
  **(B)** decreasing, then increasing.
  **(C)** always increasing.
  **(D)** always decreasing.

**d.** According to the model, the U.S. prison population was increasing fastest around what year?

**e.** Roughly estimate the instantaneous rate of change of $N$ at $t = 4$ by using a balanced difference quotient with $h = 1.5$. Interpret the result.

**82.** ▼ *Demand for Freon 12* The demand for chlorofluorocarbon-12 (CFC-12)—the ozone-depleting refrigerant commonly known as Freon 12[57]—has been declining significantly in response to regulation and concern about the ozone layer. The graph below represents a model for the projected demand for CFC-12 as a function of time in years ($t = 0$ represents 1990).[58]

**a.** Which is correct? Over the period [12, 20] the instantaneous rate of change of $Q$ is
  **(A)** increasing.          **(B)** decreasing.

**b.** Which is correct? The instantaneous rate of change of demand for Freon 12 at $t = 10$ was
  **(A)** less than          **(B)** greater than
  **(C)** approximately equal to
  the average rate of change over the interval [0, 20].

**c.** Which is correct? Over the period [0, 20] the instantaneous rate of change of $Q$ is
  **(A)** increasing, then decreasing.
  **(B)** decreasing, then increasing.
  **(C)** always increasing.
  **(D)** always decreasing.

**d.** According to the model, the demand for Freon 12 was decreasing most rapidly around what year?

**e.** Roughly estimate the instantaneous rate of change of $Q$ at $t = 13$ by using a balanced difference quotient with $h = 5$. Interpret the result.

---

[57]The name given to it by DuPont. Freon 12 (dichlorodifluoromethane) is distinct from Freon 22 (chlorodifluoromethane, also registered by DuPont; see p. 232).

[58]Source for data: The Automobile Consulting Group (*New York Times*, December 26, 1993, p. F23). The exact figures were not given, and the chart is a reasonable facsimile of the chart that appeared in *New York Times*.

**83.** *Velocity* If a stone is dropped from a height of 400 feet, its height after $t$ seconds is given by $s = 400 - 16t^2$.
  **a.** Find its average velocity over the period [2, 4].
  **b.** Estimate its instantaneous velocity at time $t = 4$.
  HINT [See Example 3.]

**84.** *Velocity* If a stone is thrown down at 120 ft/s from a height of 1,000 feet, its height after $t$ seconds is given by $s = 1,000 - 120t - 16t^2$.
  **a.** Find its average velocity over the period [1, 3].
  **b.** Estimate its instantaneous velocity at time $t = 3$.
  HINT [See Example 3.]

**85.** *Crude Oil Prices* The price per barrel of crude oil in constant 2008 dollars can be approximated by
$$P(t) = 0.45t^2 - 12t + 105 \text{ dollars} \quad (0 \le t \le 28),$$
where $t$ is time in years since the start of 1980.[59]
  **a.** Compute the average rate of change of $P(t)$ over the interval [0, 28], and interpret your answer. HINT [See Section 10.4 Example 3.]
  **b.** Estimate the instantaneous rate of change of $P(t)$ at $t = 0$, and interpret your answer. HINT [See Example 2(a).]
  **c.** The answers to part (a) and part (b) have opposite signs. What does this indicate about the price of oil?

**86.** *Median Home Prices* The median home price in the U.S. over the period 2003–2011 can be approximated by
$$P(t) = -5t^2 + 75t - 30 \text{ thousand dollars} \quad (3 \le t \le 11),$$
where $t$ is time in years since the start of 2000.[60]
  **a.** Compute the average rate of change of $P(t)$ over the interval [5, 9], and interpret your answer. HINT [See Section 10.4 Example 3.]
  **b.** Estimate the instantaneous rate of change of $P(t)$ at $t = 5$, and interpret your answer. HINT [See Example 2(a).]
  **c.** The answer to part (b) has larger absolute value than the answer to part (a). What does this indicate about the median home price?

**87.** *SARS* In the early stages of the deadly SARS (Severe Acute Respiratory Syndrome) epidemic in 2003, the number of reported cases could be approximated by
$$A(t) = 167(1.18)^t \quad (0 \le t \le 20)$$
$t$ days after March 17, 2003 (the first day in which statistics were reported by the World Health Organization).
  **a.** What, approximately, was the instantaneous rate of change of $A(t)$ on March 27 ($t = 10$)? Interpret the result.
  **b.** Which of the following is true? For the first 20 days of the epidemic, the instantaneous rate of change of the number of cases
  **(A)** increased.          **(B)** decreased.
  **(C)** increased and then decreased.
  **(D)** decreased and then increased.

---

[59]Source for data: www.inflationdata.com.
[60]Source for data: www.zillow.com.

**88. *SARS*** A few weeks into the deadly SARS (Severe Acute Respiratory Syndrome) epidemic in 2003, the number of reported cases could be approximated by

$$A(t) = 1{,}804(1.04)^t \quad (0 \le t \le 30)$$

$t$ days after April 1, 2003.

a. What, approximately, was the instantaneous rate of change of $A(t)$ on April 21 ($t = 20$)? Interpret the result.
b. Which of the following is true? During April, the instantaneous rate of change of the number of cases
 **(A)** increased.  **(B)** decreased.
 **(C)** increased and then decreased.
 **(D)** decreased and then increased.

**89. *Sales*** Weekly sales of a new brand of sneakers are given by

$$S(t) = 200 - 150e^{-t/10}$$

pairs sold per week, where $t$ is the number of weeks since the introduction of the brand. Estimate $S(5)$ and $\left.\dfrac{dS}{dt}\right|_{t=5}$ and interpret your answers.

**90. *Sales*** Weekly sales of an old brand of TV are given by

$$S(t) = 100e^{-t/5}$$

sets per week, where $t$ is the number of weeks after the introduction of a competing brand. Estimate $S(5)$ and $\left.\dfrac{dS}{dt}\right|_{t=5}$ and interpret your answers.

**91. *Early Internet Services*** On January 1, 1996, America Online was the biggest online service provider, with 4.5 million subscribers, and was adding new subscribers at a rate of 60,000 per week.[61] If $A(t)$ is the number of America Online subscribers $t$ weeks after January 1, 1996, what do the given data tell you about values of the function $A$ and its derivative? HINT [See Quick Example 2 on page 738.]

**92. *Early Internet Services*** On January 1, 1996, Prodigy was the third-biggest online service provider, with 1.6 million subscribers, but was losing subscribers.[62] If $P(t)$ is the number of Prodigy subscribers $t$ weeks after January 1, 1996, what do the given data tell you about values of the function $P$ and its derivative? HINT [See Quick Example 2 on page 738.]

**93.** ▼ ***Learning to Speak*** Let $p(t)$ represent the percentage of children who are able to speak at the age of $t$ months.

a. It is found that $p(10) = 60$ and $\left.\dfrac{dp}{dt}\right|_{t=10} = 18.2$. What does this mean?[63] HINT [See Quick Example 2 on page 738.]

b. As $t$ increases, what happens to $p$ and $\dfrac{dp}{dt}$?

**94.** ▼ ***Learning to Read*** Let $p(t)$ represent the percentage of children in your class who learned to read at the age of $t$ years.

a. Assuming that everyone in your class could read by the age of 7, what does this tell you about $p(7)$ and $\left.\dfrac{dp}{dt}\right|_{t=7}$?
HINT [See Quick Example 2 on page 738.]
b. Assuming that 25.0% of the people in your class could read by the age of 5, and that 25.3% of them could read by the age of 5 years and one month, estimate $\left.\dfrac{dp}{dt}\right|_{t=5}$. Remember to give its units.

**95. *Subprime Mortgages during the Housing Crisis*** (Compare Exercise 29 of Section 10.4.) The percentage of mortgages issued in the United States during the period 2000–2009 that were subprime (normally classified as risky) can be approximated by

$$A(t) = \frac{15}{1 + 8.6(1.8)^{-t}} \quad (0 \le t \le 9),$$

where $t$ is the number of years since the start of 2000.[64]

a. Estimate $A(6)$ and $A'(6)$. (Round answers to two significant digits.) What do the answers tell you about subprime mortgages?
b. ▯ Graph the extrapolated function and its derivative for $0 \le t \le 16$ and use your graphs to describe how the derivative behaves as $t$ becomes large. (Express this behavior in terms of limits if you have studied the sections on limits.) What does this tell you about subprime mortgages? HINT [See Example 5.]

**96. *Subprime Mortgage Debt during the Housing Crisis*** (Compare Exercise 30 of Section 10.4.) The value of subprime (normally classified as risky) mortgage debt outstanding in the U.S. during the period 2000–2009 can be approximated by

$$A(t) = \frac{1{,}350}{1 + 4.2(1.7)^{-t}} \quad \text{billion dollars} \quad (0 \le t \le 9),$$

where $t$ is the number of years since the start of 2000.[65]

a. Estimate $A(7)$ and $A'(7)$. (Round answers to three significant digits.) What do the answers tell you about subprime mortgages?
b. ▯ Graph the function and its derivative and use your graphs to estimate when, to the nearest year, $A'(t)$ is greatest. What does this tell you about subprime mortgages? HINT [See Example 5.]

**97.** ▯ ▼ ***Embryo Development*** The oxygen consumption of a turkey embryo increases from the time the egg is laid through the time the turkey chick hatches. In a brush turkey, the oxygen consumption (in milliliters per hour) can be approximated by

$$c(t) = -0.0012t^3 + 0.12t^2 - 1.83t + 3.97 \quad (20 \le t \le 50),$$

---

[61]Source: Information and Interactive Services Report/*New York Times*, January 2, 1996, p. C14.
[62]*Ibid.*
[63]Based on data presented in the article *The Emergence of Intelligence* by William H. Calvin, *Scientific American*, October, 1994, pp. 101–107.

[64]Sources: Mortgage Bankers Association, UBS.
[65]Source: Data 360 www.data360.org.

where $t$ is the time (in days) since the egg was laid.[66] (An egg will typically hatch at around $t = 50$.) Use technology to graph $c'(t)$ and use your graph to answer the following questions. HINT [See Example 5.]

**a.** Over the interval $[20, 32]$ the derivative $c'$ is
    **(A)** increasing, then decreasing.
    **(B)** decreasing, then increasing.
    **(C)** decreasing.        **(D)** increasing.

**b.** When, to the nearest day, is the oxygen consumption increasing at the fastest rate?

**c.** When, to the nearest day, is the oxygen consumption increasing at the slowest rate?

**98.** ☐ ▼ *Embryo Development* The oxygen consumption of a bird embryo increases from the time the egg is laid through the time the chick hatches. In a typical galliform bird, the oxygen consumption (in milliliters per hour) can be approximated by

$$c(t) = -0.0027t^3 + 0.14t^2 - 0.89t + 0.15 \quad (8 \le t \le 30),$$

where $t$ is the time (in days) since the egg was laid.[67] (An egg will typically hatch at around $t = 28$.) Use technology to graph $c'(t)$ and use your graph to answer the following questions. HINT [See Example 5.]

**a.** Over the interval $[8, 30]$ the derivative $c'$ is
    **(A)** increasing, then decreasing.
    **(B)** decreasing, then increasing.
    **(C)** decreasing.        **(D)** increasing.

**b.** When, to the nearest day, is the oxygen consumption increasing the fastest?

**c.** When, to the nearest day, is the oxygen consumption increasing at the slowest rate?

*The next two exercises are applications of Einstein's Special Theory of Relativity and relate to objects that are moving extremely fast. In science fiction terminology, a speed of warp 1 is the speed of light—about $3 \times 10^8$ meters per second. (Thus, for instance, a speed of warp 0.8 corresponds to 80% of the speed of light—about $2.4 \times 10^8$ meters per second.)*

**99.** ◆ *Lorentz Contraction* According to Einstein's Special Theory of Relativity, a moving object appears to get shorter to a stationary observer as its speed approaches the speed of light. If a spaceship that has a length of 100 meters at rest travels at a speed of warp $p$, its length in meters, as measured by a stationary observer, is given by

$$L(p) = 100\sqrt{1 - p^2}$$

with domain $[0, 1)$. Estimate $L(0.95)$ and $L'(0.95)$. What do these figures tell you?

**100.** ◆ *Time Dilation* Another prediction of Einstein's Special Theory of Relativity is that, to a stationary observer, clocks (as well as all biological processes) in a moving object appear to go more and more slowly as the speed of the object approaches that of light. If a spaceship travels at a speed of warp $p$, the time it takes for an onboard clock to register one second, as measured by a stationary observer, will be given by

$$T(p) = \frac{1}{\sqrt{1 - p^2}} \text{ seconds}$$

with domain $[0, 1)$. Estimate $T(0.95)$ and $T'(0.95)$. What do these figures tell you?

## COMMUNICATION AND REASONING EXERCISES

**101.** Explain why we cannot put $h = 0$ in the approximation

$$f'(x) \approx \frac{f(x + h) - f(x)}{h}$$

for the derivative of $f$.

**102.** The balanced difference quotient

$$f'(a) \approx \frac{f(a + 0.0001) - f(a - 0.0001)}{0.0002}$$

is the average rate of change of $f$ on what interval?

**103.** Let $H(t)$ represent the number of Handbook members in millions $t$ years after its inception in 2020. It is found that $H(10) = 50$ and $H'(10) = -6$. This means that, in 2030 (Multiple Choice):

    **(A)** There were 6 million members and this number was decreasing at a rate of 50 million per year.
    **(B)** There were –6 million members and this number was increasing at a rate of 50 million per year.
    **(C)** Membership had dropped by 6 million since the previous year, but was now increasing at a rate of 50 million per year.
    **(D)** There were 50 million members and this number was decreasing at a rate of 6 million per year.
    **(E)** There were 50 million members and membership had dropped by 6 million since the previous year.

**104.** Let $F(t)$ represent the net earnings of Footbook Inc. in millions of dollars $t$ years after its inception in 3020. It is found that $F(100) = -10$ and $F'(100) = 60$. This means that, in 3120 (Multiple Choice):

    **(A)** Footbook lost $10 million but its net earnings were increasing at a rate of $60 million per year.
    **(B)** Footbook earned $60 million but its earnings were decreasing at a rate of $10 million per year.
    **(C)** Footbook's net earnings had increased by $60 million since the year before, but it still lost $10 million.

---

[66] The model approximates graphical data published in the article *The Brush Turkey* by Roger S. *Seymour, Scientific American*, December, 1991, pp. 108–114.

[67] *Ibid.*

**(D)** Footbook earned $10 million but its net earnings were decreasing at a rate of $60 million per year.

**(E)** Footbook's net earnings had decreased by $10 million since the year before, but it still earned $60 million.

**105.** It is now eight months since the Garden City lacrosse team won the national championship, and sales of team paraphernalia, while still increasing, have been leveling off. What does this tell you about the derivative of the sales curve?

**106.** Having been soundly defeated in the national lacrosse championships, Brakpan High has been faced with decreasing sales of its team paraphernalia. However, sales, while still decreasing, appear to be bottoming out. What does this tell you about the derivative of the sales curve?

**107.** ▼ Company A's profits are given by $P(0) = \$1$ million and $P'(0) = -\$1$ million/month. Company B's profits are given by $P(0) = -\$1$ million and $P'(0) = \$1$ million/month. In which company would you rather invest? Why?

**108.** ▼ Company C's profits are given by $P(0) = \$1$ million and $P'(0) = \$0.5$ million/month. Company D's profits are given by $P(0) = \$0.5$ million and $P'(0) = \$1$ million/month. In which company would you rather invest? Why?

**109.** ▼ During the one-month period starting last January 1, your company's profits increased at an average rate of change of $4 million per month. On January 1, profits were increasing at an instantaneous rate of $5 million per month. Which of the following graphs could represent your company's profits? Why?

**(A)**

**(B)**

**(C)**

**110.** ▼ During the one-month period starting last January 1, your company's sales increased at an average rate of change of $3,000 per month. On January 1, sales were changing at an instantaneous rate of –$1,000 per month. Which of the following graphs could represent your company's sales? Why?

**(A)**

**(B)**

**(C)**

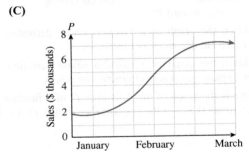

**111.** ▼ If the derivative of $f$ is zero at a point, what do you know about the graph of $f$ near that point?

**112.** ▼ Sketch the graph of a function whose derivative never exceeds 1.

**113.** ▼ Sketch the graph of a function whose derivative exceeds 1 at every point.

**114.** ▼ Sketch the graph of a function whose derivative is exactly 1 at every point.

**115.** ▼ Use the difference quotient to explain the fact that if $f$ is a linear function, then the average rate of change over any interval equals the instantaneous rate of change at any point.

**116.** ▼ Give a numerical explanation of the fact that if $f$ is a linear function, then the average rate of change over any interval equals the instantaneous rate of change at any point.

**117.** ◆ Consider the following values of the function $f$ from Exercise 1.

| $h$ | 0.1 | 0.01 | 0.001 | 0.0001 |
|---|---|---|---|---|
| Average Rate of Change of $f$ over $[5, 5+h]$ | 6.4 | 6.04 | 6.004 | 6.0004 |
| $h$ | $-0.1$ | $-0.01$ | $-0.001$ | $-0.0001$ |
| Average Rate of Change of $f$ over $[5+h, 5]$ | 5.6 | 5.96 | 5.996 | 5.9996 |

Does the table suggests that the instantaneous rate of change of $f$ is

(A) increasing          (B) decreasing

as $x$ increases toward 5?

**118.** ◆ Consider the following values of the function $g$ from Exercise 2.

| $h$ | 0.1 | 0.01 | 0.001 | 0.0001 |
|---|---|---|---|---|
| Average Rate of Change of $g$ over $[7, 7+h]$ | 4.8 | 4.98 | 4.998 | 4.9998 |
| $h$ | $-0.1$ | $-0.01$ | $-0.001$ | $-0.0001$ |
| Average Rate of Change of $g$ over $[7+h, 7]$ | 5.3 | 5.03 | 5.003 | 5.0003 |

Does the table suggest that the instantaneous rate of change of $g$ is

(A) increasing          (B) decreasing

as $x$ increases toward 7?

**119.** ▼ Sketch the graph of a function whose derivative is never zero but decreases as $x$ increases.

**120.** ▼ Sketch the graph of a function whose derivative is never negative but is zero at exactly two points.

**121.** ◆ Here is the graph of the derivative $f'$ of a function $f$. Give a rough sketch of the graph of $f$, given that $f(0) = 0$.

**122.** ◆ Here is the graph of the derivative $f'$ of a function $f$. Give a rough sketch of the graph of $f$, given that $f(0) = 0$.

**123.** ◆ Professor Talker of the physics department drove a 60-mile stretch of road in exactly one hour. The speed limit along that stretch was 55 miles per hour. Which of the following must be correct?

(A) He exceeded the speed limit at no point of the journey.
(B) He exceeded the speed limit at some point of the journey.
(C) He exceeded the speed limit throughout the journey.
(D) He traveled slower than the speed limit at some point of the journey.

**124.** ◆ Professor Silent, another physics professor, drove a 50-mile stretch of road in exactly one hour. The speed limit along that stretch was 55 miles per hour. Which of the following must be correct?

(A) She exceeded the speed limit at no point of the journey.
(B) She exceeded the speed limit at some point of the journey.
(C) She traveled slower than the speed limit throughout the journey.
(D) She traveled slower than the speed limit at some point of the journey.

**125.** ◆ Draw the graph of a function $f$ with the property that the balanced difference quotient gives a more accurate approximation of $f'(1)$ than the ordinary difference quotient.

**126.** ◆ Draw the graph of a function $f$ with the property that the balanced difference quotient gives a less accurate approximation of $f'(1)$ than the ordinary difference quotient.

# 10.6  Derivatives: Algebraic Viewpoint

In Section 10.5 we saw how to estimate the derivative of a function using numerical and graphical approaches. In this section we use an algebraic approach that will give us the *exact value* of the derivative, rather than just an approximation, when the function is specified algebraically.

This algebraic approach is quite straightforward: Instead of subtracting numbers to estimate the average rate of change over smaller and smaller intervals, we subtract algebraic expressions. Our starting point is the definition of the derivative in terms of the difference quotient:

$$f'(a) = \lim_{h \to 0} \frac{f(a+h) - f(a)}{h}.$$

---

### EXAMPLE 1 Calculating the Derivative at a Point Algebraically

Let $f(x) = x^2$. Use the definition of the derivative to compute $f'(3)$ algebraically.

**Solution** Substituting $a = 3$ into the definition of the derivative, we get:

$$f'(3) = \lim_{h \to 0} \frac{f(3+h) - f(3)}{h}$$    Formula for the derivative

$$= \lim_{h \to 0} \frac{\overbrace{(3+h)^2}^{f(3+h)} - \overbrace{3^2}^{f(3)}}{h}$$    Substitute for $f(3)$ and $f(3+h)$.

$$= \lim_{h \to 0} \frac{(9 + 6h + h^2) - 9}{h}$$    Expand $(3+h)^2$.

$$= \lim_{h \to 0} \frac{6h + h^2}{h}$$    Cancel the 9.

$$= \lim_{h \to 0} \frac{h(6+h)}{h}$$    Factor out $h$.

$$= \lim_{h \to 0} (6 + h).$$    Cancel the $h$.

Now we let $h$ approach 0. As $h$ gets closer and closer to 0, the sum $6 + h$ clearly gets closer and closer to $6 + 0 = 6$. Thus,

$$f'(3) = \lim_{h \to 0} (6 + h) = 6.$$    As $h \to 0$, $(6 + h) \to 6$

(Calculations of limits like this are discussed and justified more fully in Sections 10.2 and 10.3.)

---

➡ **Before we go on...** We did the following calculation in Example 1: If $f(x) = x^2$, then $f'(3) = 6$. In other words, the tangent to the graph of $y = x^2$ at the point $(3, 9)$ has slope 6 (Figure 36). ■

There is nothing very special about $a = 3$ in Example 1. Let's try to compute $f'(x)$ for general $x$.

---

### EXAMPLE 2 Calculating the Derivative Function Algebraically

Let $f(x) = x^2$.

**a.** Use the definition of the derivative to compute $f'(x)$ algebraically.
**b.** Use the answer to evaluate $f'(3)$.

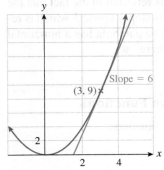

**Figure 36**

**Solution**

**a.** Once again, our starting point is the definition of the derivative in terms of the difference quotient:

$$f'(x) = \lim_{h \to 0} \frac{f(x+h) - f(x)}{h}$$    Formula for the derivative

$$= \lim_{h \to 0} \frac{\overbrace{(x+h)^2}^{f(x+h)} - \overbrace{x^2}^{f(x)}}{h}$$    Substitute for $f(x)$ and $f(x+h)$.

$$= \lim_{h \to 0} \frac{(x^2 + 2xh + h^2) - x^2}{h}$$    Expand $(x+h)^2$.

$$= \lim_{h \to 0} \frac{2xh + h^2}{h}$$    Cancel the $x^2$.

$$= \lim_{h \to 0} \frac{h(2x + h)}{h}$$    Factor out $h$.

$$= \lim_{h \to 0} (2x + h).$$    Cancel the $h$.

Now we let $h$ approach 0. As $h$ gets closer and closer to 0, the sum $2x + h$ clearly gets closer and closer to $2x + 0 = 2x$. Thus,

$$f'(x) = \lim_{h \to 0} (2x + h) = 2x.$$

This is the derivative function.

**b.** Now that we have a *formula* for the derivative of $f$, we can obtain $f'(a)$ for any value of $a$ we choose by simply evaluating $f'$ there. For instance,

$$f'(3) = 2(3) = 6$$

as we saw in Example 1.

---

➡ **Before we go on...** The graphs of $f(x) = x^2$ and $f'(x) = 2x$ from Example 2 are familiar. Their graphs are shown in Figure 37.

When $x < 0$, the parabola slopes downward, which is reflected in the fact that the derivative $2x$ is negative there. When $x > 0$, the parabola slopes upward, which is reflected in the fact that the derivative is positive there. The parabola has a horizontal tangent line at $x = 0$, reflected in the fact that $2x = 0$ there. ■

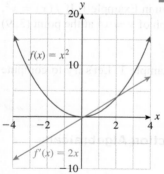

**Figure 37**

---

### EXAMPLE 3   More Computations of Derivative Functions

Compute the derivative $f'(x)$ for each of the following functions:

**a.** $f(x) = x^3$    **b.** $f(x) = 2x^2 - x$    **c.** $f(x) = \dfrac{1}{x}$

**Solution**

**a.** $f'(x) = \lim_{h\to 0} \dfrac{f(x+h) - f(x)}{h}$  Derivative formula

$= \lim_{h\to 0} \dfrac{\overbrace{(x+h)^3}^{f(x+h)} - \overbrace{x^3}^{f(x)}}{h}$  Substitute for $f(x)$ and $f(x+h)$.

$= \lim_{h\to 0} \dfrac{(x^3 + 3x^2h + 3xh^2 + h^3) - x^3}{h}$  Expand $(x+h)^3$.

$= \lim_{h\to 0} \dfrac{3x^2h + 3xh^2 + h^3}{h}$  Cancel the $x^3$.

$= \lim_{h\to 0} \dfrac{h(3x^2 + 3xh + h^2)}{h}$  Factor out $h$.

$= \lim_{h\to 0} (3x^2 + 3xh + h^2)$  Cancel the $h$.

$= 3x^2.$  Let $h$ approach 0.

**b.** $f'(x) = \lim_{h\to 0} \dfrac{f(x+h) - f(x)}{h}$  Derivative formula

$= \lim_{h\to 0} \dfrac{\overbrace{(2(x+h)^2 - (x+h))}^{f(x+h)} - \overbrace{(2x^2 - x)}^{f(x)}}{h}$  Substitute for $f(x)$ and $f(x+h)$.

$= \lim_{h\to 0} \dfrac{(2x^2 + 4xh + 2h^2 - x - h) - (2x^2 - x)}{h}$  Expand.

$= \lim_{h\to 0} \dfrac{4xh + 2h^2 - h}{h}$  Cancel the $2x^2$ and $x$.

$= \lim_{h\to 0} \dfrac{h(4x + 2h - 1)}{h}$  Factor out $h$.

$= \lim_{h\to 0} (4x + 2h - 1)$  Cancel the $h$.

$= 4x - 1.$  Let $h$ approach 0.

**c.** $f'(x) = \lim_{h\to 0} \dfrac{f(x+h) - f(x)}{h}$  Derivative formula

$= \lim_{h\to 0} \dfrac{\left[\overbrace{\dfrac{1}{x+h}}^{f(x+h)} - \overbrace{\dfrac{1}{x}}^{f(x)}\right]}{h}$  Substitute for $f(x)$ and $f(x+h)$.

$= \lim_{h\to 0} \dfrac{\left[\dfrac{x - (x+h)}{(x+h)x}\right]}{h}$  Subtract the fractions.

$= \lim_{h\to 0} \dfrac{1}{h}\left[\dfrac{x - (x+h)}{(x+h)x}\right]$  Dividing by $h$ = Multiplying by $1/h$.

$$= \lim_{h \to 0} \left[ \frac{-h}{h(x+h)x} \right] \qquad \text{Simplify.}$$

$$= \lim_{h \to 0} \left[ \frac{-1}{(x+h)x} \right] \qquad \text{Cancel the } h.$$

$$= \frac{-1}{x^2} \qquad \text{Let } h \text{ approach 0.}$$

In Example 4, we redo Example 3 of Section 10.5, this time getting an exact, rather than approximate, answer.

### EXAMPLE 4  Velocity

My friend Eric, an enthusiastic baseball player, claims he can "probably" throw a ball upward at a speed of 100 feet per second (ft/sec). Our physicist friends tell us that its height $s$ (in feet) $t$ seconds later would be $s(t) = 100t - 16t^2$. Find the ball's instantaneous velocity function and its velocity exactly 2 seconds after Eric throws it.

**Solution** The instantaneous velocity function is the derivative $ds/dt$, which we calculate as follows:

$$\frac{ds}{dt} = \lim_{h \to 0} \frac{s(t+h) - s(t)}{h}.$$

Let us compute $s(t+h)$ and $s(t)$ separately:

$$s(t) = 100t - 16t^2$$
$$s(t+h) = 100(t+h) - 16(t+h)^2$$
$$= 100t + 100h - 16(t^2 + 2th + h^2)$$
$$= 100t + 100h - 16t^2 - 32th - 16h^2.$$

Therefore,

$$\frac{ds}{dt} = \lim_{h \to 0} \frac{s(t+h) - s(t)}{h}$$
$$= \lim_{h \to 0} \frac{100t + 100h - 16t^2 - 32th - 16h^2 - (100t - 16t^2)}{h}$$
$$= \lim_{h \to 0} \frac{100h - 32th - 16h^2}{h}$$
$$= \lim_{h \to 0} \frac{h(100 - 32t - 16h)}{h}$$
$$= \lim_{h \to 0} (100 - 32t - 16h)$$
$$= 100 - 32t \text{ ft/sec.}$$

Thus, the velocity exactly 2 seconds after Eric throws it is

$$\left. \frac{ds}{dt} \right|_{t=2} = 100 - 32(2) = 36 \text{ ft/sec.}$$

This verifies the accuracy of the approximation we made in Section 10.5.

➡ **Before we go on...** From the derivative function in Example 4, we can now describe the behavior of the velocity of the ball: Immediately on release ($t = 0$) the ball is traveling at 100 feet per second upward. The ball then slows down; precisely, it loses 32 feet per second of speed every second. When, exactly, does the velocity become zero and what happens after that? ■

---

**Q** : *Do we always have to calculate the limit of the difference quotient to find a formula for the derivative function?*

**A** : As it turns out, no. In Section 11.1 we will start to look at shortcuts for finding derivatives that allow us to bypass the definition of the derivative in many cases.

---

## A Function Not Differentiable at a Point

Recall from Section 10.5 that a function is **differentiable** at a point $a$ if $f'(a)$ exists; that is, if the difference quotient $[f(a + h) - f(a)]/h$ approaches a fixed value as $h$ approaches 0. In Section 10.5, we mentioned that the function $f(x) = |x|$ is not differentiable at $x = 0$. In Example 5, we find out why.

---

### EXAMPLE 5　A Function Not Differentiable at 0

Numerically, graphically, and algebraically investigate the differentiability of the function $f(x) = |x|$ at the points **(a)** $x = 1$ and **(b)** $x = 0$.

#### Solution

**a.** We compute

$$f'(1) = \lim_{h \to 0} \frac{f(1 + h) - f(1)}{h}$$

$$= \lim_{h \to 0} \frac{|1 + h| - 1}{h}.$$

Numerically, we can make tables of the values of the average rate of change $(|1 + h| - 1)/h$ for $h$ positive or negative and approaching 0:

| $h$ | 1 | 0.1 | 0.01 | 0.001 | 0.0001 |
|---|---|---|---|---|---|
| **Average Rate of Change Over [1, 1 + h]** | 1 | 1 | 1 | 1 | 1 |

| $h$ | $-1$ | $-0.1$ | $-0.01$ | $-0.001$ | $-0.0001$ |
|---|---|---|---|---|---|
| **Average Rate of Change Over [1 + h, 1]** | 1 | 1 | 1 | 1 | 1 |

From these tables it appears that $f'(1)$ is equal to 1. We can verify that algebraically: For $h$ that is sufficiently small, $1 + h$ is positive (even if $h$ is negative) and so

$$f'(1) = \lim_{h \to 0} \frac{1 + h - 1}{h}$$

$$= \lim_{h \to 0} \frac{h}{h} \qquad \text{Cancel the 1s.}$$

$$= \lim_{h \to 0} 1 \qquad \text{Cancel the } h.$$

$$= 1.$$

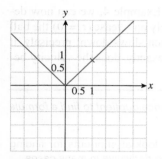

**Figure 38**

Graphically, we are seeing the fact that the tangent line at the point $(1, 1)$ has slope 1 because the graph is a straight line with slope 1 near that point (Figure 38).

**b.** $f'(0) = \lim\limits_{h \to 0} \dfrac{f(0 + h) - f(0)}{h}$

$\qquad\quad = \lim\limits_{h \to 0} \dfrac{|0 + h| - 0}{h}$

$\qquad\quad = \lim\limits_{h \to 0} \dfrac{|h|}{h}$

If we make tables of values in this case we get the following:

| $h$ | 1 | 0.1 | 0.01 | 0.001 | 0.0001 |
|---|---|---|---|---|---|
| **Average Rate of Change over $[0, 0 + h]$** | 1 | 1 | 1 | 1 | 1 |

| $h$ | $-1$ | $-0.1$ | $-0.01$ | $-0.001$ | $-0.0001$ |
|---|---|---|---|---|---|
| **Average Rate of Change over $[0 + h, 0]$** | $-1$ | $-1$ | $-1$ | $-1$ | $-1$ |

For the limit and hence the derivative $f'(0)$ to exist, the average rates of change should approach the same number for both positive and negative $h$. Because they do not, $f$ is not differentiable at $x = 0$. We can verify this conclusion algebraically: If $h$ is positive, then $|h| = h$, and so the ratio $|h|/h$ is 1, regardless of how small $h$ is. Thus, according to the values of the difference quotients with $h > 0$, the limit should be 1. On the other hand if $h$ is negative, then $|h| = -h$ (positive) and so $|h|/h = -1$, meaning that the limit should be $-1$. Because the limit cannot be both $-1$ and 1 (it must be a single number for the derivative to exist), we conclude that $f'(0)$ does not exist.

To see what is happening graphically, take a look at Figure 39, which shows zoomed-in views of the graph of $f$ near $x = 0$.

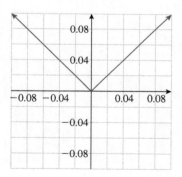

**Figure 39**

No matter what scale we use to view the graph, it has a sharp corner at $x = 0$ and hence has no tangent line there. Since there is no tangent line at $x = 0$, the function is not differentiable there.

➡ **Before we go on...** Notice that $|x| = \begin{cases} -x & \text{if } x < 0 \\ x & \text{if } x \geq 0 \end{cases}$ is an example of a piecewise-linear function whose graph comes to a point at $x = 0$. In general, if $f(x)$ is any piecewise linear function whose graph comes to a point at $x = a$, it will be nondifferentiable at $x = a$ for the same reason that $|x|$ fails to be differentiable at $x = 0$.

If we repeat the computation in Example 5(a) using any nonzero value for $a$ in place of 1, we see that $f$ is differentiable there as well. If $a$ is positive, we find that $f'(a) = 1$ and, if $a$ is negative, $f'(a) = -1$. In other words, the derivative function is

$$f'(x) = \begin{cases} -1 & \text{if } x < 0 \\ 1 & \text{if } x > 0 \end{cases}.$$

Immediately to the left of $x = 0$, we see that $f'(x) = -1$, immediately to the right, $f'(x) = 1$; and when $x = 0$, $f'(x)$ is not defined. ■

---

**Q**: *So does that mean there is no single formula for the derivative of $|x|$?*

**A**: Actually, there is a convenient formula. Consider the ratio $\dfrac{|x|}{x}$. If $x$ is positive, then $|x| = x$, so $\dfrac{|x|}{x} = \dfrac{x}{x} = 1$. On the other hand, if $x$ is negative, then $|x| = -x$, so $\dfrac{|x|}{x} = \dfrac{-x}{x} = -1$. In other words,

$$\frac{|x|}{x} = \begin{cases} -1 & \text{if } x < 0 \\ 1 & \text{if } x > 0 \end{cases},$$

which is exactly the formula we obtained for $f'(x)$. We have therefore obtained a convenient closed-form formula for the derivative of $|x|$!

---

### Derivative of $|x|$

If $f(x) = |x|$, then $f'(x) = \dfrac{|x|}{x}$.

Note that $|x|/x$ is not defined if $x = 0$, reflecting the fact that $f'(x)$ does not exist when $x = 0$.

We will use the above formula extensively in the next chapter.

### FAQ

**Computing Derivatives Algebraically**

**Q**: *The algebraic computation of $f'(x)$ seems to require a number of steps. How do I remember what to do, and when?*

**A**: If you examine the computations in the examples above, you will find the following pattern:

1. Write out the formula for $f'(x)$, as the limit of the difference quotient, then substitute $f(x + h)$ and $f(x)$.
2. Expand and simplify the *numerator* of the expression, but not the denominator.
3. After simplifying the numerator, factor out an $h$ to cancel with the $h$ in the denominator. If $h$ does not factor out of the numerator, you might have made an error. (A frequent error is a wrong sign.)
4. After canceling the $h$, you should be able to see what the limit is by letting $h \to 0$.

## 10.6  EXERCISES

▼ more advanced    ◆ challenging

Ⓣ indicates exercises that should be solved using technology

*In Exercises 1–14, compute $f'(a)$ algebraically for the given value of a.* HINT [See Example 1.]

1. $f(x) = x^2 + 1$; $a = 2$

2. $f(x) = x^2 - 3$; $a = 1$

3. $f(x) = 3x - 4$; $a = -1$

4. $f(x) = -2x + 4$; $a = -1$

5. $f(x) = 3x^2 + x$; $a = 1$

6. $f(x) = 2x^2 + x$; $a = -2$

7. $f(x) = 2x - x^2$; $a = -1$

8. $f(x) = -x - x^2$; $a = 0$

9. $f(x) = x^3 + 2x$; $a = 2$

10. $f(x) = x - 2x^3$; $a = 1$

11. $f(x) = \dfrac{-1}{x}$; $a = 1$ HINT [See Example 3.]

12. $f(x) = \dfrac{2}{x}$; $a = 5$ HINT [See Example 3.]

13. ▼ $f(x) = mx + b$; $a = 43$

14. ▼ $f(x) = \dfrac{x}{k} - b\ (k \neq 0)$; $a = 12$

*In Exercises 15–28, compute the derivative function $f'(x)$ algebraically. (Notice that the functions are the same as those in Exercises 1–14.)* HINT [See Examples 2 and 3.]

15. $f(x) = x^2 + 1$

16. $f(x) = x^2 - 3$

17. $f(x) = 3x - 4$

18. $f(x) = -2x + 4$

19. $f(x) = 3x^2 + x$

20. $f(x) = 2x^2 + x$

21. $f(x) = 2x - x^2$

22. $f(x) = -x - x^2$

23. $f(x) = x^3 + 2x$

24. $f(x) = x - 2x^3$

25. ▼ $f(x) = \dfrac{-1}{x}$

26. ▼ $f(x) = \dfrac{2}{x}$

27. ▼ $f(x) = mx + b$

28. ▼ $f(x) = \dfrac{x}{k} - b\ (k \neq 0)$

*In Exercises 29–38, compute the indicated derivative.*

29. $R(t) = -0.3t^2$; $R'(2)$

30. $S(t) = 1.4t^2$; $S'(-1)$

31. $U(t) = 5.1t^2 + 5.1$; $U'(3)$

32. $U(t) = -1.3t^2 + 1.1$; $U'(4)$

33. $U(t) = -1.3t^2 - 4.5t$; $U'(1)$

34. $U(t) = 5.1t^2 - 1.1t$; $U'(1)$

35. $L(r) = 4.25r - 5.01$; $L'(1.2)$

36. $L(r) = -1.02r + 5.7$; $L'(3.1)$

37. ▼ $q(p) = \dfrac{2.4}{p} + 3.1$; $q'(2)$

38. ▼ $q(p) = \dfrac{1}{0.5p} - 3.1$; $q'(2)$

*In Exercises 39–44, find the equation of the tangent to the graph at the indicated point.* HINT [Compute the derivative algebraically; then see Example 2(b) in Section 10.5.]

39. ▼ $f(x) = x^2 - 3$; $a = 2$    40. ▼ $f(x) = x^2 + 1$; $a = 2$

41. ▼ $f(x) = -2x - 4$; $a = 3$    42. ▼ $f(x) = 3x + 1$; $a = 1$

43. ▼ $f(x) = x^2 - x$; $a = -1$    44. ▼ $f(x) = x^2 + x$; $a = -1$

## APPLICATIONS

45. **Velocity** If a stone is dropped from a height of 400 feet, its height after $t$ seconds is given by $s = 400 - 16t^2$. Find its instantaneous velocity function and its velocity at time $t = 4$. HINT [See Example 4.]

46. **Velocity** If a stone is thrown down at 120 feet per second from a height of 1,000 feet, its height after $t$ seconds is given by $s = 1,000 - 120t - 16t^2$. Find its instantaneous velocity function and its velocity at time $t = 3$. HINT [See Example 4.]

47. **Oil Imports from Mexico** Daily oil imports to the United States from Mexico can be approximated by

$$I(t) = -0.015t^2 + 0.1t + 1.4 \text{ million barrels} \quad (0 \leq t \leq 8),$$

where $t$ is time in years since the start of 2000.[68] Find the derivative function $\dfrac{dI}{dt}$. At what rate were oil imports changing at the start of 2007 $(t = 7)$? HINT [See Example 4.]

48. **Oil Production in Mexico** Daily oil production by Pemex, Mexico's national oil company, can be approximated by

$$P(t) = -0.022t^2 + 0.2t + 2.9 \text{ million barrels} \quad (1 \leq t \leq 9),$$

where $t$ is time in years since the start of 2000.[69] Find the derivative function $\dfrac{dP}{dt}$. At what rate was oil production changing at the start of 2004 $(t = 4)$? HINT [See Example 4.]

49. **Bottled Water Sales** The following chart shows the amount of bottled water sold in the United States for the period 2000–2010:[70]

---

[68] Source for data: Energy Information Administration/Pemex (www.eia.doe.gov).

[69] *Ibid.*

[70] The 2010 figure is an estimate. Source: Beverage Marketing Corporation/www.bottledwater.org.

Bottled water sales in the U.S.

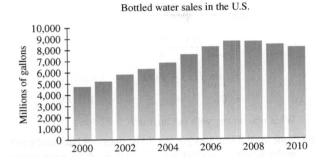

The function

$$R(t) = -45t^2 + 900t + 4{,}200 \text{ million gallons} \quad (0 \le t \le 10)$$

gives a good approximation, where $t$ is time in years since 2000. Find the derivative function $R'(t)$. According to the model, how fast were annual sales of bottled water changing in 2005?

**50. Bottled Water Sales** The following chart shows annual per capita sales of bottled water in the United States for the period 2000–2010:[71]

Per capita bottled water sales in the U.S.

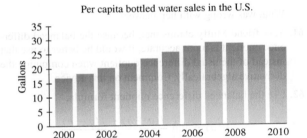

The function

$$R(t) = -0.18t^2 + 3t + 15 \text{ gallons} \quad (0 \le t \le 10)$$

gives a good approximation, where $t$ is time in years since 2000. Find the derivative function $R'(t)$. According to the model, how fast were per capita sales of bottled water changing in 2009?

**51. ▼ Ecology** Increasing numbers of manatees ("sea sirens") have been killed by boats off the Florida coast. The following graph shows the relationship between the number of boats registered in Florida and the number of manatees killed each year.

Boats (100,000)

[71] The 2010 figure is an estimate. Source: Beverage Marketing Corporation/www.bottledwater.org.

The regression curve shown is given by

$$f(x) = 3.55x^2 - 30.2x + 81 \text{ manatee deaths}$$
$$(4.5 \le x \le 8.5),$$

where $x$ is the number of boats (hundreds of thousands) registered in Florida in a particular year and $f(x)$ is the number of manatees killed by boats in Florida that year.[72] Compute and interpret $f'(8)$.

**52. ▼ SAT Scores by Income** The following graph shows U.S. math SAT scores as a function of parents' income level.[73]

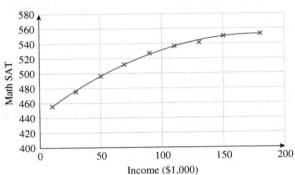

Income ($1,000)

The regression curve shown is given by

$$f(x) = -0.0034x^2 + 1.2x + 444 \quad (10 \le x \le 180),$$

where $f(x)$ is the average math SAT score of a student whose parents earn $x$ thousand dollars per year. Compute and interpret $f'(30)$.

**53. ▼ Television Advertising** The cost, in thousands of dollars, of a 30-second television ad during the Super Bowl in the years 1970 to 2010 can be approximated by the following piecewise linear function ($t = 0$ represents 1970):[74]

$$C(t) = \begin{cases} 31.1t + 78 & \text{if } 0 \le t < 20 \\ 90t - 1{,}100 & \text{if } 20 \le t \le 40 \end{cases}.$$

**a.** Is $C$ a continuous function of $t$? Why? HINT [See Example 4 of Section 10.3.]

**b.** Is $C$ a differentiable function of $t$? Compute $\lim_{t \to 20^-} C'(t)$ and $\lim_{t \to 20^+} C'(t)$ and interpret the results. HINT [See *Before we go on...* after Example 5.]

**54. ▼ Television Advertising** (Compare Exercise 53.) The cost, in thousands of dollars, of a 30-second television ad during the Super Bowl in the years 1980 to 2010 can be approximated by the following piecewise linear function ($t = 0$ represents 1980):[75]

$$C(t) = \begin{cases} 43.9t + 222 & \text{if } 0 \le t \le 20 \\ 140t - 1{,}700 & \text{if } 20 < t \le 30 \end{cases}.$$

[72] Regression model is based on data from 1976 to 2000. Sources for data: Florida Department of Highway Safety & Motor Vehicles, Florida Marine Institute/*New York Times*, February 12, 2002, p. F4.

[73] Regression model is based on 2009 data. Source: College Board/*New York Times* http://economix.blogs.nytimes.com.

[74] Source: http://en.wikipedia.org/wiki/Super_Bowl_advertising.

[75] *Ibid.*

**a.** Is $C$ a continuous function of $t$? Why? HINT [See Example 4 of Section 10.3.]

**b.** Is $C$ a differentiable function of $t$? Compute $\lim_{t\to 20^-} C'(t)$ and $\lim_{t\to 20^+} C'(t)$ and interpret the results. HINT [See *Before we go on...* after Example 5.]

## COMMUNICATION AND REASONING EXERCISES

**55.** Of the three methods (numerical, graphical, algebraic) we can use to estimate the derivative of a function at a given value of $x$, which is always the most accurate? Explain.

**56.** Explain why we cannot put $h = 0$ in the formula

$$f'(a) = \lim_{h\to 0} \frac{f(a+h) - f(a)}{h}$$

for the derivative of $f$.

**57.** You just got your derivatives test back and you can't understand why that teacher of yours deducted so many points for what you thought was your best work:

$$\lim_{h\to 0} \frac{f(x+h) - f(x)}{h}$$
$$= \lim_{h\to 0} \frac{f(x) + h - f(x)}{h}$$
$$= \lim_{h\to 0} \frac{h}{h} \qquad \text{Canceled the } f(x)$$
$$= 1. \qquad\qquad \textbf{✗ } WRONG \quad -10$$

What was wrong with your answer?

**58.** Your friend just got his derivatives test back and can't understand why that teacher of his deducted so many points for the following:

$$\lim_{h\to 0} \frac{f(x+h) - f(x)}{h}$$
$$= \lim_{h\to 0} \frac{f(x) + f(h) - f(x)}{h}$$
$$= \lim_{h\to 0} \frac{f(h)}{h} \qquad \text{Canceled the } f(x)$$
$$= \lim_{h\to 0} \frac{f(\cancel{h})}{\cancel{h}} \qquad \text{Now cancel the } h.$$
$$= f. \qquad\qquad \textbf{✗ } WRONG \quad -50$$

What was wrong with his answer?

**59.** Your other friend just got her derivatives test back and can't understand why that teacher of hers took off so many points for the following:

$$\lim_{h\to 0} \frac{f(x+h) - f(x)}{h}$$
$$= \lim_{h\to 0} \frac{f(x+h) - f(x)}{\cancel{h}} \qquad \text{Now cancel the } h.$$
$$= \lim_{h\to 0} f(x) - f(x) \qquad \text{Cancel the } f(x)$$
$$= 0. \qquad\qquad \textbf{✗ } WRONG \quad -15$$

What was wrong with her answer?

**60.** Your third friend just got her derivatives test back and can't understand why that teacher of hers took off so many points for the following:

$$\lim_{h\to 0} \frac{f(x+h) - f(x)}{h}$$
$$= \lim_{h\to 0} \frac{f(x) + h - f(x)}{h}$$
$$= \lim_{h\to 0} \frac{f(x) + h - f(x)}{\cancel{h}} \qquad \text{Now cancel the } h.$$
$$= \lim_{h\to 0} f(x) - f(x) \qquad \text{Cancel the } f(x)$$
$$= 0. \qquad\qquad \textbf{✗ } WRONG \quad -25$$

What was wrong with her answer?

**61.** Your friend Muffy claims that, because the balanced difference quotient is more accurate, it would be better to use that instead of the usual difference quotient when computing the derivative algebraically. Comment on this advice.

**62.** Use the balanced difference quotient formula,

$$f'(a) = \lim_{h\to 0} \frac{f(a+h) - f(a-h)}{2h},$$

to compute $f'(3)$ when $f(x) = x^2$. What do you find?

**63.** ▼ A certain function $f$ has the property that $f'(a)$ does not exist. How is that reflected in the attempt to compute $f'(a)$ algebraically?

**64.** ▼ One cannot put $h = 0$ in the formula

$$f'(a) = \lim_{h\to 0} \frac{f(a+h) - f(a)}{h}$$

for the derivative of $f$. (See Exercise 56.) However, in the last step of each of the computations in the text, we are effectively setting $h = 0$ when taking the limit. What is going on here?

## KEY CONCEPTS

### 10.1 Limits: Numerical and Graphical Viewpoints

$\lim_{x \to a} f(x) = L$ means that $f(x)$ approaches $L$ as $x$ approaches $a$ p. 690
What it means for a limit to exist p. 691
Limits at infinity p. 693
Estimating limits graphically p. 693
Interpreting limits in real-world situations p. 696

### 10.2 Limits and Continuity

$f$ is continuous at $a$ if $\lim_{x \to a} f(x)$ exists and $\lim_{x \to a} f(x) = f(a)$ p. 702
Discontinuous, continuous on domain p. 702
Determining whether a given function is continuous p. 703

### 10.3 Limits and Continuity: Algebraic Viewpoint

Closed-form function p. 708
Limits of closed form functions p. 709

Simplifying to obtain limits p. 709
The indeterminate form 0/0 p. 710
The determinate form k/0 p. 711
Limits of piecewise-defined functions p. 712
Limits at infinity p. 713
Determinate and indeterminate forms p. 716

### 10.4 Average Rate of Change

Average rate of change of $f(x)$ over
$[a, b]$: $\dfrac{\Delta f}{\Delta x} = \dfrac{f(b) - f(a)}{b - a}$ p. 724
Average rate of change as slope of the secant line p. 725
Computing the average rate of change from a graph p. 726
Computing the average rate of change from a formula p. 727
Computing the average rate of change over short intervals $[a, a + h]$ p. 728

### 10.5 Derivatives: Numerical and Graphical Viewpoints

Instantaneous rate of change of $f(x)$ (derivative of $f$ at $a$);
$f'(a) = \lim_{h \to 0} \dfrac{f(a + h) - f(a)}{h}$ p. 738

The derivative as slope of the tangent line p. 742
Quick approximation of the derivative p. 743
$\dfrac{d}{dx}$ Notation p. 745
The derivative as velocity p. 745
Average and instantaneous velocity p. 746
The derivative function p. 747
Graphing the derivative function with technology p. 748

### 10.6 Derivatives: Algebraic Viewpoint

Derivative at the point $x = a$:
$f'(a) = \lim_{h \to 0} \dfrac{f(a + h) - f(a)}{h}$ p. 761
Derivative function:
$f'(x) = \lim_{h \to 0} \dfrac{f(x + h) - f(x)}{h}$ p. 762
Examples of the computation of $f'(x)$ p. 763
$f(x) = |x|$ is not differentiable at $x = 0$ p. 765

## REVIEW EXERCISES

**T** indicates exercises that must be solved using technology

Numerically *estimate whether the limits in Exercises 1–4 exist. If a limit does exist, give its approximate value.*

**1.** $\lim_{x \to 3} \dfrac{x^2 - x - 6}{x - 3}$

**2.** $\lim_{x \to 3} \dfrac{x^2 - 2x - 6}{x - 3}$

**3.** $\lim_{x \to -1} \dfrac{|x + 1|}{x^2 - x - 2}$

**4.** $\lim_{x \to -1} \dfrac{|x + 1|}{x^2 + x - 2}$

*In Exercises 5 and 6, the graph of a function f is shown. Graphically determine whether the given limits exist. If a limit does exist, give its approximate value.*

**5.**

**a.** $\lim_{x \to 0} f(x)$ **b.** $\lim_{x \to 1} f(x)$
**c.** $\lim_{x \to 2} f(x)$

**6.**

**a.** $\lim_{x \to 0} f(x)$ **b.** $\lim_{x \to -2} f(x)$
**c.** $\lim_{x \to 2} f(x)$

*Calculate the limits in Exercises 7–30 algebraically. If a limit does not exist, say why.*

**7.** $\lim_{x \to -2} \dfrac{x^2}{x - 3}$

**8.** $\lim_{x \to 3} \dfrac{x^2 - 9}{2x - 6}$

**9.** $\lim_{x \to -2} \dfrac{x^2 - 4}{x^3 + 2x^2}$

**10.** $\lim_{x \to -1} \dfrac{x^2 - 9}{2x - 6}$

**11.** $\lim_{x \to 0} \dfrac{x}{2x^2 - x}$

**12.** $\lim_{x \to 1} \dfrac{x^2 - 9}{x - 1}$

**13.** $\lim_{x \to -1} \dfrac{x^2 + 3x}{x^2 - x - 2}$

**14.** $\lim_{x \to -1^+} \dfrac{x^2 + 1}{x^2 + 3x + 2}$

**15.** $\lim_{x \to 8} \dfrac{x^2 - 6x - 16}{x^2 - 9x + 8}$

**16.** $\lim_{x \to 4} \dfrac{x^2 + 3x}{x^2 - 8x + 16}$

**17.** $\lim_{x \to 4} \dfrac{x^2 + 8}{x^2 - 2x - 8}$

**18.** $\lim_{x \to 6} \dfrac{x^2 - 5x - 6}{x^2 - 36}$

**19.** $\lim_{x \to 1/2} \dfrac{x^2 + 8}{4x^2 - 4x + 1}$

**20.** $\lim_{x \to 1/2} \dfrac{x^2 + 3x}{2x^2 + 3x - 1}$

**21.** $\lim_{x \to +\infty} \dfrac{10x^2 + 300x + 1}{5x^3 + 2}$

**22.** $\lim_{x \to +\infty} \dfrac{2x^4 + 20x^3}{1,000x^6 + 6}$

**23.** $\lim_{x \to -\infty} \dfrac{x^2 - x - 6}{x - 3}$

**24.** $\lim_{x \to +\infty} \dfrac{x^2 - x - 6}{4x^2 - 3}$

**25.** $\lim\limits_{t\to+\infty} \dfrac{-5}{5 + 5.3(3^{2t})}$   **26.** $\lim\limits_{t\to+\infty}\left(3 + \dfrac{2}{e^{4t}}\right)$

**27.** $\lim\limits_{x\to+\infty} \dfrac{2}{5 + 4e^{-3x}}$   **28.** $\lim\limits_{x\to+\infty}(4e^{3x} + 12)$

**29.** $\lim\limits_{t\to+\infty} \dfrac{1 + 2^{-3t}}{1 + 5.3e^{-t}}$   **30.** $\lim\limits_{x\to-\infty} \dfrac{8 + 0.5^{x}}{2 - 3^{2x}}$

*In Exercises 31–34, find the average rate of change of the given function over the interval* $[a, a + h]$ *for* $h = 1, 0.01,$ *and* $0.001$. *(Round answers to four decimal places.) Then estimate the slope of the tangent line to the graph of the function at a.*

**31.** $f(x) = \frac{1}{x+1}$; $a = 0$   **32.** $f(x) = x^x$; $a = 2$

**33.** $f(x) = e^{2x}$; $a = 0$   **34.** $f(x) = \ln(2x)$; $a = 1$

*In Exercises 35–38, you are given the graph of a function with four points marked. Determine at which (if any) of these points the derivative of the function is:* **(a)** $-1$, **(b)** $0$, **(c)** $1$, *and* **(d)** $2$.

**35.**

**36.**

**37.**

**38.**

**39.** Let $f$ have the graph shown.

Select the correct answer.

**a.** The average rate of change of $f$ over the interval $[0, 2]$ is
   **(A)** greater than $f'(0)$.   **(B)** less than $f'(0)$.
   **(C)** approximately equal to $f'(0)$.
**b.** The average rate of change of $f$ over the interval $[-1, 1]$ is
   **(A)** greater than $f'(0)$.   **(B)** less than $f'(0)$.
   **(C)** approximately equal to $f'(0)$.
**c.** Over the interval $[0, 2]$, the instantaneous rate of change of $f$ is
   **(A)** increasing.   **(B)** decreasing.
   **(C)** neither increasing nor decreasing.
**d.** Over the interval $[-2, 2]$, the instantaneous rate of change of $f$ is
   **(A)** increasing, then decreasing.
   **(B)** decreasing, then increasing.
   **(C)** approximately constant.
**e.** When $x = 2$, $f(x)$ is
   **(A)** approximately 1 and increasing at a rate of about 2.5 units per unit of $x$.
   **(B)** approximately 1.2 and increasing at a rate of about 1 unit per unit of $x$.
   **(C)** approximately 2.5 and increasing at a rate of about 0.5 units per unit of $x$.
   **(D)** approximately 2.5 and increasing at a rate of about 2.5 units per unit of $x$.

**40.** Let $f$ have the graph shown.

Select the correct answer.

**a.** The average rate of change of $f$ over the interval $[0, 1]$ is
   **(A)** greater than $f'(0)$.   **(B)** less than $f'(0)$.
   **(C)** approximately equal to $f'(0)$.
**b.** The average rate of change of $f$ over the interval $[0, 2]$ is
   **(A)** greater than $f'(1)$.   **(B)** less than $f'(1)$.
   **(C)** approximately equal to $f'(1)$.
**c.** Over the interval $[-2, 0]$, the instantaneous rate of change of $f$ is
   **(A)** increasing.   **(B)** decreasing.
   **(C)** neither increasing nor decreasing.
**d.** Over the interval $[-2, 2]$, the instantaneous rate of change of $f$ is
   **(A)** increasing, then decreasing.
   **(B)** decreasing, then increasing.
   **(C)** approximately constant.
**e.** When $x = 0$, $f(x)$ is
   **(A)** approximately 0 and increasing at a rate of about 1.5 units per unit of $x$.
   **(B)** approximately 0 and decreasing at a rate of about 1.5 units per unit of $x$.
   **(C)** approximately 1.5 and neither increasing nor decreasing.
   **(D)** approximately 0 and neither increasing nor decreasing.

*In Exercises 41–44, use the definition of the derivative to calculate the derivative of each of the given functions algebraically.*

**41.** $f(x) = x^2 + x$   **42.** $f(x) = 3x^2 - x + 1$

**43.** $f(x) = 1 - \dfrac{2}{x}$   **44.** $f(x) = \dfrac{1}{x} + 1$

**T** *In Exercises 45–48, use technology to graph the derivative of the given function. In each case, choose a range of x-values and y-values that shows the interesting features of the graph.*

**45.** $f(x) = 10x^5 + \dfrac{1}{2}x^4 - x + 2$

46. $f(x) = \dfrac{10}{x^5} + \dfrac{1}{2x^4} - \dfrac{1}{x} + 2$

47. $f(x) = 3x^3 + 3\sqrt[3]{x}$

48. $f(x) = \dfrac{2}{x^{2.1}} - \dfrac{x^{0.1}}{2}$

## APPLICATIONS: OHaganBooks.com

49. **Stock Investments** OHaganBooks.com CEO John O'Hagan has terrible luck with stocks. The following graph shows the value of Fly-By-Night Airlines stock that he bought acting on a "hot tip" from Marjory Duffin (CEO of Duffin House publishers and a close business associate):

Fly-by-night stock

a. Compute $P(3)$, $\lim_{t \to 3^-} P(t)$ and $\lim_{t \to 3^+} P(t)$. Does $\lim_{t \to 3} P(t)$ exist? Interpret your answers in terms of Fly-By-Night stock.

b. Is $P$ continuous at $t = 6$? Is $P$ differentiable at $t = 6$? Interpret your answers in terms of Fly-By-Night stock.

50. **Stock Investments** John O'Hagan's golf partner Juan Robles seems to have had better luck with his investment in Gapple Gomputer Inc. stocks as shown in the following graph:

Gapple Inc. Stock

a. Compute $P(6)$, $\lim_{t \to 6^-} P(t)$ and $\lim_{t \to 6^+} P(t)$. Does $\lim_{t \to 6} P(t)$ exist? Interpret your answers in terms of Gapple stock.

b. Is $P$ continuous at $t = 3$? Is $P$ differentiable at $t = 3$? Interpret your answers in terms of Gapple stock.

51. **Real Estate** Marjory Duffin has persuaded John O'Hagan to consider investing a portion of OHaganBooks.com profits in real estate, now that the real estate market seems to have bottomed out. A real estate broker friend of hers

emailed her the following (somewhat optimistic) graph from brokersadvocacy.com:[76]

Home price index

Here, $p(t)$ is the home price percentage over the 2003 level.

a. Assuming the trend shown in the graph were to continue indefinitely, estimate $\lim_{t \to 3} p(t)$ and $\lim_{t \to +\infty} p(t)$ and interpret the results.

b. Estimate $\lim_{t \to +\infty} p'(t)$ and interpret the result.

52. **Advertising Costs** OHaganBooks.com has (on further advice from Marjory Duffin) mounted an aggressive online marketing strategy. The following graph shows the weekly cost of this campaign for the six-week period since the start of July ($t$ is time in weeks):

a. Assuming the trend shown in the graph were to continue indefinitely, estimate $\lim_{t \to 2} C(t)$ and $\lim_{t \to +\infty} C(t)$ and interpret the results.

b. Estimate $\lim_{t \to +\infty} C'(t)$ and interpret the result.

53. **Sales** Since the start of July, OHaganBooks.com has seen its weekly sales increase, as shown in the following table:

| Week | 1 | 2 | 3 | 4 | 5 | 6 |
|---|---|---|---|---|---|---|
| Sales (books) | 6,500 | 7,000 | 7,200 | 7,800 | 8,500 | 9,000 |

a. What was the average rate of increase of weekly sales over this entire period?

b. During which 1-week interval(s) did the rate of increase of sales exceed the average rate?

c. During which 2-week interval(s) did the weekly sales rise at the highest average rate, and what was that average rate?

---

[76] Authors' note: As of August 2012, brokersadvocacy.com is unregistered.

**54. *Rising Sea Level*** Marjory Duffin recently purchased a beachfront condominium in New York and is now in a panic, having just seen some disturbing figures about rising sea levels (sea levels as measured in New York relative to the 1900 level).[77]

| Year since 1900 | 0 | 25 | 50 | 75 | 100 | 125 |
|---|---|---|---|---|---|---|
| Sea Level (mm) | 0 | 60 | 140 | 240 | 310 | 390 |

a. What was the average rate of increase of the sea level over this entire period?

b. During which 25-year interval(s) did the rate of increase of the sea level exceed the average rate?

c. Marjory Duffin's condominium is about 2 meters above sea level. Using the average rate of change from part (a), estimate how long she has before the sea rises to her condominium.

**55. *Real Estate*** The following graph (see Exercise 51) shows the home price index chart emailed to Marjory Duffin by a real estate broker:

Home price index

Use the graph to answer the following questions:

a. What was the average rate of change of the index over the 10-year period beginning 2004?

b. What was the average rate of change of the index over the period [3, 10]?

c. Which of the following is correct? Over the period [4, 6],

(A) The rate of change of the index increased.

(B) The rate of change of the index increased and then decreased.

(C) The rate of change of the index decreased.

(D) The rate of change of the index decreased and then increased.

**56. *Advertising Costs*** The following graph (see Exercise 52) shows the weekly cost of OHaganBooks.com's online ad campaign for the 6-week period since the start of July ($t$ is time in weeks).

[77] The 2025 level is a projection. Source: New England Integrated Science & Assessment (www.neisa.unh.edu/Climate/index.html).

Unless otherwise noted, all content on this page is © Cengage Learning.

Use the graph to answer the following questions:

a. What was the average rate of change of cost over the entire six-week period?

b. What was the average rate of change of cost over the period [2, 6]?

c. Which of the following is correct? Over the period [2, 6],

(A) The rate of change of cost increased and the cost increased.

(B) The rate of change of cost decreased and the cost increased.

(C) The rate of change of cost increased and the cost decreased.

(D) The rate of change of cost decreased and the cost decreased.

**57. *Sales*** OHaganBooks.com fits the curve

$$w(t) = 36t^2 + 250t + 6{,}240 \quad (0 \le t \le 6)$$

to its weekly sales figures from Exercise 53, as shown in the following graph:

a. Compute the derivative function $w'(t)$.

b. According to the model, what was the rate of increase of sales at the beginning of the second week ($t = 1$)?

c. If we extrapolate the model, what would be the rate of increase of weekly sales at the beginning of the 8th week ($t = 7$)?

**58. *Sea Levels*** Marjory Duffin fit the curve

$$s(t) = 0.002t^2 + 3t - 6.4 \quad (0 \le t \le 125)$$

to her sea level figures from Exercise 54, as shown in the following graph:

a. Compute the derivative function $s'(t)$.

b. According to the model, what was the rate of increase of the sea level in 2000 ($t = 100$)?

c. If we extrapolate the model, what would be the rate of increase of the sea level in 2100 ($t = 200$)?

## Case Study | Reducing Sulfur Emissions

The Environmental Protection Agency (EPA) wishes to formulate a policy that will encourage utilities to reduce sulfur emissions. Its goal is to reduce annual emissions of sulfur dioxide by a total of 10 million tons from the current level of 25 million tons by imposing a fixed charge for every ton of sulfur released into the environment per year. As a consultant to the EPA, you must determine the amount to be charged per ton of sulfur emissions.

You would like first to know the cost to the utility industry of reducing sulfur emissions. In other words, you would like to have a cost function of the form

$$C(q) = \text{Cost of removing } q \text{ tons of sulfur dioxide.}$$

Unfortunately, you do not have such a function handy. You do, however, have the following data, which show the *marginal* cost (that is, the *rate of change* of cost) to the utility industry of reducing sulfur emissions at several levels of reduction.[78]

**Figure 40**

| Reduction (tons) $q$ | 8,000,000 | 10,000,000 | 12,000,000 |
|---|---|---|---|
| Marginal Cost ($ per ton) $C'(q)$ | 270 | 360 | 779 |

The table tells you that $C'(8,000,000) = \$270$ per ton, $C'(10,000,000) = \$360$ per ton, and $C'(12,000,000) = \$779$ per ton. Recalling that $C'(q)$ is the slope of the tangent to the graph of the cost function, you can see from the table that this slope is positive and increasing as $q$ increases, so the graph of the cost function has the general shape shown in Figure 40.

Notice that the slope (additional cost) is increasing as you move to the right, so the utility industry has no cost incentive to reduce emissions further, as it costs the industry significantly more per ton for each additional ton of sulfur it removes. What you would like—if the goal of reducing total emissions by 10 million tons is to be reached—is that, somehow, the imposition of a fixed charge for every ton of sulfur dioxide released will *alter* the form of the cost curve so that it has the general shape shown in Figure 41. In this ideal curve, the cost $D$ to utilities is lowest at a reduction level of 10 million tons, so if the utilities act to minimize cost, they can be expected to reduce emissions by 10 million tons, which is exactly the EPA goal! From the graph, you can see that the tangent line to the curve at the point where $q = 10$ million tons is horizontal, and thus has zero slope: $D'(10,000,000) = \$0$ per ton. Further, the slope $D'(q)$ is negative for values of $q$ to the left of 10 million tons and positive for values to the right.

**Figure 41**

So, how much should the EPA charge per ton of sulfur released into the environment? Suppose the EPA charges $\$k$ per ton, so that

$$\text{Emission charge to utilities} = k \times \text{Sulfur emissions.}$$

It is your job to calculate $k$. Because you are working with $q$ as the independent variable, you decide that it would be best to formulate the emission charge as a function of $q$. However, $q$ represents the amount by which sulfur emissions have been reduced from the original 25 million tons; that is, the amount by which sulfur emissions are *lower than* the original 25 million tons:

$$q = 25,000,000 - \text{Sulfur emissions.}$$

---

[78] These figures were produced in a computerized study of reducing sulfur emissions from the 1980 level by the given amounts. Source: Congress of the United States, Congressional Budget Office, *Curbing Acid Rain: Cost, Budget and Coal Market Effects* (Washington, DC: Government Printing Office, 1986): xx, xxii, 23, 80.

Thus, the total annual emission charge to the utilities is

$$k \times \text{Sulfur emissions} = k(25{,}000{,}000 - q) = 25{,}000{,}000k - kq.$$

This results in a total cost to the utilities of

$$\text{Total cost} = \text{Cost of reducing emissions} + \text{Emission charge}$$
$$D(q) = C(q) + 25{,}000{,}000k - kq.$$

You now recall from calculus that the derivative of a sum of two functions is the sum of their derivatives (you will see why in Section 11.1*), so the derivative of $D$ is given by

$$D'(q) = \text{Derivative of } C + \text{Derivative of } (25{,}000{,}000k - kq).$$

The function $y = 25{,}000{,}000k - kq$ is a linear function of $q$ with slope $-k$ and intercept $25{,}000{,}000k$. Thus its derivative is just its slope: $-k$. Therefore:

$$D'(q) = C'(q) - k.$$

Remember that you want

$$D'(10{,}000{,}000) = 0.$$

Thus,

$$C'(10{,}000{,}000) - k = 0.$$

Referring to the table, you see that

$$360 - k = 0$$

or

$$k = \$360 \text{ per ton.}$$

In other words, all you need to do is set the emission charge at $k = \$360$ per ton of sulfur emitted. Further, to ensure that the resulting curve will have the general shape shown in Figure 40, you would like to have $D'(q)$ negative for $q < 10{,}000{,}000$ and positive for $q > 10{,}000{,}000$. To check this, write

$$D'(q) = C'(q) - k$$
$$= C'(q) - 360$$

and refer to the table to obtain

$$D'(8{,}000{,}000) = 270 - 360 = -90 < 0 \checkmark$$

and

$$D'(12{,}000{,}000) = 779 - 360 = 419 > 0 \checkmark$$

Thus, based on the given data, the resulting curve will have the shape you require. You therefore inform the EPA that an annual emissions charge of \$360 per ton of sulfur released into the environment will create the desired incentive: to reduce sulfur emissions by 10 million tons per year.

One week later, you are informed that this charge would be unrealistic because the utilities cannot possibly afford such a cost. You are asked whether there is an alternative plan that accomplishes the 10-million-ton reduction goal and yet is cheaper to the utilities by \$5 billion per year. You then look at your expression for the emission charge

$$25{,}000{,}000k - kq$$

* This statement makes intuitive sense: For instance, if $C$ is changing at a rate of 3 units per second and $D$ is changing at a rate of 2 units per second, then their sum is changing at a rate of $3 + 2 = 5$ units per second.

and notice that, if you decrease this amount by \$5 billion, the derivative will not change at all because it will still have the same slope (only the intercept is affected). Thus, you propose the following revised formula for the emission charge:

$$25,000,000k - kq - 5,000,000,000$$
$$= 25,000,000(360) - 360q - 5,000,000,000$$
$$= 4,000,000,000 - 360q.$$

At the expected reduction level of 10 million tons, the total amount paid by the utilities will then be

$$4,000,000,000 - 360(10,000,000) = \$400,000,000.$$

Thus, your revised proposal is the following: Impose an annual emissions charge of \$360 per ton of sulfur released into the environment and hand back \$5 billion in the form of subsidies. The effect of this policy will be to cause the utilities industry to reduce sulfur emissions by 10 million tons per year and will result in \$400 million in annual revenues to the government.

Notice that this policy also provides an incentive for the utilities to search for cheaper ways to reduce emissions. For instance, if they lowered costs to the point where they could achieve a reduction level of 12 million tons, they would have a total emission charge of

$$4,000,000,000 - 360(12,000,000) = -\$320,000,000.$$

The fact that this is negative means that the government would be paying the utilities industry \$320 million more in annual subsidies than the industry is paying in per ton emission charges.

### EXERCISES

1. Excluding subsidies, what should the annual emission charge be if the goal is to reduce sulfur emissions by 8 million tons?
2. Excluding subsidies, what should the annual emission charge be if the goal is to reduce sulfur emissions by 12 million tons?
3. What is the *marginal emission charge* (derivative of emission charge) in your revised proposal (as stated before the exercise set)? What is the relationship between the marginal cost of reducing sulfur emissions before emissions charges are implemented and the marginal emission charge, at the optimal reduction under your revised proposal?
4. We said that the revised policy provided an incentive for utilities to find cheaper ways to reduce emissions. How would $C(q)$ have to change to make 12 million tons the optimum reduction?
5. What change in $C(q)$ would make 8 million tons the optimum reduction?
6. If the scenario in Exercise 5 took place, what would the EPA have to do in order to make 10 million tons the optimal reduction once again?
7. Due to intense lobbying by the utility industry, you are asked to revise the proposed policy so that the utility industry will pay no charge if sulfur emissions are reduced by the desired 10 million tons. How can you accomplish this?
8. Suppose that instead of imposing a fixed charge per ton of emission, you decide to use a sliding scale, so that the total charge to the industry for annual emissions of $x$ tons will be $\$kx^2$ for some $k$. What must $k$ be to again make 10 million tons the optimum reduction? HINT [The derivative of $kx^2$ is $2kx$.]

**TECHNOLOGY GUIDE**

## TI-83/84 Plus    Technology Guide

### Section 10.1

**Example 1 (page 689)**    Use a table to estimate the following limits.

**a.** $\lim_{x \to 2} \dfrac{x^3 - 8}{x - 2}$    **b.** $\lim_{x \to 0} \dfrac{e^{2x} - 1}{x}$

**Solution with Technology**

On the TI-83/84 Plus, use the table feature to automate these computations as follows:

1. Define $Y_1=(X^3-8)/(X-2)$ for part (a) or $Y_1=(e^{(2X)-1})/X$ for part (b).

2. Press 2ND TABLE to list its values for the given values of $x$. (If the calculator does not allow you to enter values of $x$, press 2ND TBLSET and set Indpnt to Ask).

    Here is the table showing some of the values for part (a):

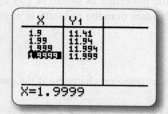

3. For part (b) use $Y_1=(e^{(2X)-1})/X$ and use values of $x$ approaching 0 from either side.

### Section 10.4

**Example 3 (page 727)**    The price of an ounce of gold can be approximated by the function

$$G(t) = 5t^2 - 85t + 1{,}762 \quad (7.5 \le t \le 10.5)$$

where $t$ is time in hours. ($t = 8$ represents 8:00 am.) What was the average rate of change of the price of gold over the $1\frac{1}{2}$-hour period starting at 8:00 am (the interval $[8, 9.5]$ on the $t$-axis)?

**Solution with Technology**

On the TI-83/84 Plus:

1. Enter the function $G$ as $Y_1$ (using X for $t$):
$$Y_1=5X^2-85X+1762$$

2. Now find the average rate of change over $[8, 9.5]$ by evaluating the following on the home screen:
$$(Y_1(9.5)-Y_1(8))/(9.5-8)$$

As shown on the screen, the average rate is of change is 2.5.

**Example 4 (page 728)**    Continuing with Example 3, use technology to compute the average rate of change of

$$G(t) = 5t^2 - 85t + 1{,}762 \quad (7.5 \le t \le 10.5)$$

over the intervals $[8, 8 + h]$, where $h = 1$, 0.1, 0.01, 0.001, and 0.0001.

**Solution with Technology**

1. As in Example 4, enter the function $G$ as $Y_1$ (using X for $t$):
$$Y_1=5X^2-85X+1762$$

2. Now find the average rate of change for $h = 1$ by evaluating, on the home screen,
$$(Y_1(8+1)-Y_1(8))/1,$$
which gives 0.

3. To evaluate for $h = 0.1$, recall the expression using 2ND ENTER and then change the 1, both places it occurs, to 0.1, getting
$$(Y_1(8+0.1)-Y_1(8))/0.1,$$
which gives $-4.95$.

4. Continuing, we can evaluate the average rate of change for all the desired values of $h$:

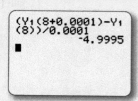

## Section 10.5

**Example 2 (page 743)**   Calculate an approximate value of $f'(1.5)$ if $f(x) = x^2 - 4x$, and then find the equation of the tangent line at the point on the graph where $x = 1.5$.

### Solution with Technology

**1.** In the TI-83/84 Plus, enter the function $f$ as $Y_1$

$$Y_1 = X^2 - 4*X$$

**2.** Go to the home screen to compute the approximations:

$(Y_1(1.5001) - Y_1(1.5))/0.0001$

Usual difference quotient

$(Y_1(1.5001) - Y_1(1.4999))/0.0002$

Balanced difference quotient

From the display on the right, we find that the difference quotient quick approximation is $-0.9999$ and the balanced difference quotient quick approximation is $-1$, which is in fact the exact value of $f'(1.5)$. See the discussion in the text for the calculation of the equation of the tangent line.

**Example 4 (page 748)**   Use technology to graph the derivative of $f(x) = -2x^2 + 6x + 5$ for values of $x$ in starting at $-5$.

### Solution with Technology

On the TI-83/84 Plus, the easiest way to obtain quick approximations of the derivative of a given function is to use the built-in nDeriv function, which calculates balanced difference quotients.

**1.** On the $Y=$ screen, first enter the function:

$$Y_1 = -2X^2 + 6X + 5$$

**2.** Then set

$Y_2 = \text{nDeriv}(Y_1, X, X)$     For nDeriv press [MATH] [8]

which is the TI-83/84 Plus's approximation of $f'(x)$ (see figure on the left below). Alternatively, we can enter the balanced difference quotient directly:

$Y_2 = (Y_1(X+0.001) - Y_1(X-0.001))/0.002$

(The TI-83/84 Plus uses $h = 0.001$ by default in the balanced difference quotient when calculating nDeriv, but this can be changed by giving a value of $h$ as a fourth argument, like nDeriv($Y_1, X, X, 0.0001$).) To see a table of approximate values of the derivative, we press [2ND] [TABLE] and choose a collection of values for $x$ (shown on the right below):

Here, $Y_1$ shows the value of $f$ and $Y_2$ shows the values of $f'$.

To graph the function or its derivative, we can graph $Y_1$ or $Y_2$ in a window showing the given domain $[-5, 5]$:

Graph of $f$                    Graph of $f'$

**TECHNOLOGY GUIDE**

## SPREADSHEET   Technology Guide

## Section 10.1

**Example 1 (page 689)**   Use a table to estimate the following limits.

**a.** $\lim\limits_{x\to 2}\dfrac{x^3-8}{x-2}$   **b.** $\lim\limits_{x\to 0}\dfrac{e^{2x}-1}{x}$

### Solution with Technology

**1.** Set up your spreadsheet to duplicate the table in part (a) as follows:

| | A | B | C | D |
|---|---|---|---|---|
| 1 | x | f(x) | x | f(x) |
| 2 | 1.9 | =(A2^3-8)/(A2-2) | 2.1 | |
| 3 | 1.99 | | 2.01 | |
| 4 | 1.999 | | 2.001 | |
| 5 | 1.9999 | | 2.0001 | |

↓

| | A | B | C | D |
|---|---|---|---|---|
| 1 | x | f(x) | x | f(x) |
| 2 | 1.9 | 11.41 | 2.1 | 12.61 |
| 3 | 1.99 | 11.9401 | 2.01 | 12.0601 |
| 4 | 1.999 | 11.994001 | 2.001 | 12.006001 |
| 5 | 1.9999 | 11.99940001 | 2.0001 | 12.00060001 |

(The formula in cell B2 is copied to columns B and D as indicated by the shading.) The values of $f(x)$ will be calculated in columns B and D.

**2.** For part (b), use the formula $=(\text{EXP}(2*A2)-1)/A2$ in cell B2 and, in columns A and C, use values of $x$ approaching 0 from either side.

## Section 10.4

**Example 3 (page 727)**   The price of an ounce of gold can be approximated by the function

$$G(t)=5t^2+85t+1{,}762 \quad (7.5\le t\le 10.5)$$

where $t$ is time in hours. ($t=8$ represents 8:00 am.) What was the average rate of change of the price of gold over the $1\frac{1}{2}$-hour period starting at 8:00 am (the interval $[8, 9.5]$ on the $t$-axis)?

### Solution with Technology

To use a spreadsheet to compute the average rate of change of $G$:

**1.** Start with two columns, one for values of $t$ and one for values of $G(t)$, which you enter using the formula for $G$:

$$=5*A2^2-85*A2+1762$$

| | A | B |
|---|---|---|
| 1 | t | G(t) |
| 2 | 8 | =5*A2^2-85*A2+1762 |
| 3 | 9.5 | |

**2.** Next, calculate the average rate of change as shown here:

| | A | B | C | D |
|---|---|---|---|---|
| 1 | t | G(t) | | |
| 2 | 8 | 1402 | Rate of change over [8, 9.5]: | |
| 3 | 9.5 | 1405.75 | =(B3-B2)/(A3-A2) | |

↓

| | A | B | C | D |
|---|---|---|---|---|
| 1 | t | G(t) | | |
| 2 | 8 | 1402 | Rate of change over [8, 9.5]: | |
| 3 | 9.5 | 1405.75 | 2.5 | |

In Example 4, we describe another, more versatile Excel template for computing rates of change.

**Example 4 (page 728)**   Continuing with Example 3, use technology to compute the average rate of change of

$$G(t)=5t^2+85t+1{,}762 \quad (7.5\le t\le 10.5)$$

over the intervals $[8, 8+h]$, where $h=1$, 0.1, 0.01, 0.001, and 0.0001.

### Solution with Technology

The template we can use to compute the rates of change is an extension of what we used in Example 3:

**1.** Column C contains the values $t=a$ and $t=a+h$ we are using for the independent variable.

**2.** The formula in cell E2 is the average-rate-of-change formula $\Delta G/\Delta t$. Entering the different values $h=1$, 0.1, 0.01, 0.001, and 0.0001 in cell B2 gives the results shown in Example 4.

## Section 10.5

**Example 2 (page 743)** Calculate an approximate value of $f'(1.5)$ if $f(x) = x^2 - 4x$, and then find the equation of the tangent line at the point on the graph where $x = 1.5$.

### Solution with Technology

You can compute both the difference quotient and the balanced difference quotient approximations in a spreadsheet using the following extension of the worksheet in Example 4 in Section 10.4:

|  | A | B | C | D | E | F |
|---|---|---|---|---|---|---|
| 1 | a | h | x | f(x) | Diff Quotients | Balanced Diff Quotient |
| 2 | 1.5 | 0.0001 | =A2-B2 | =C2^2-4*C2 | =(D3-D2)/(C3-C2) | =(D4-D2)/(C4-C2) |
| 3 |  |  | =A2 |  |  |  |
| 4 |  |  | =A2+B2 |  |  |  |

Notice that we get two difference quotients in column E. The first uses $h = -0.0001$ while the second uses $h = 0.0001$ and is the one we use for our quick approximation. The balanced quotient is their average (column F). The results are as follows.

|  | A | B | C | D | E | F |
|---|---|---|---|---|---|---|
| 1 | a | h | x | f(x) | Diff Quotients | Balanced Diff Quotient |
| 2 | 1.5 | 0.0001 | 1.4999 | -3.7499 | -1.0001 | -1 |
| 3 |  |  | 1.5 | -3.75 | -0.9999 |  |
| 4 |  |  | 1.5001 | -3.7501 |  |  |

From the results shown above, we find that the difference quotient quick approximation is $-0.9999$ and that the balanced difference quotient quick approximation is $-1$, which is in fact the exact value of $f'(1.5)$. See the discussion in the text for the calculation of the equation of the tangent line.

**Example 4 (page 748)** Use technology to graph the derivative of $f(x) = -2x^2 + 6x + 5$ for values of $x$ starting at $-5$.

### Solution with Technology

**1.** Start with a table of values for the function $f$:

**2.** Next, compute approximate derivatives in Column C:

You cannot paste the difference quotient formula into cell C102. (Why?) Notice that this worksheet uses the ordinary difference quotients, $[f(x + h) - f(x)]/h$. If you prefer, you can use balanced difference quotients $[f(x + h) - f(x - h)]/(2h)$, in which case cells C2 and C102 would both have to be left blank.

We now graph the function and the derivative on different graphs as follows:

**1.** First, graph the function $f$ in the usual way, using Columns A and B.

**2.** Make a copy of this graph and click on it once. Columns A and B should be outlined, indicating that these are the columns used in the graph.

**3.** By dragging from the center of the bottom edge of the box, move the Column B box over to Column C as shown:

|  | A | B | C |
|---|---|---|---|
| 96 | 4.4 | -7.32 | -11.8 |
| 97 | 4.5 | -8.5 | -12.2 |
| 98 | 4.6 | -9.72 | -12.6 |
| 99 | 4.7 | -10.98 | -13 |
| 100 | 4.8 | -12.28 | -13.4 |
| 101 | 4.9 | -13.62 | -13.8 |
| 102 | 5 | -15 |  |

↓

|  | A | B | C |
|---|---|---|---|
| 96 | 4.4 | -7.32 | -11.8 |
| 97 | 4.5 | -8.5 | -12.2 |
| 98 | 4.6 | -9.72 | -12.6 |
| 99 | 4.7 | -10.98 | -13 |
| 100 | 4.8 | -12.28 | -13.4 |
| 101 | 4.9 | -13.62 | -13.8 |
| 102 | 5 | -15 |  |

The graph will then show the derivative (Columns A and C):

Graph of $f$

Graph of $f'$

# 11

# Techniques of Differentiation with Applications

**Website**
www.WanerMath.com
At the Website you will find:

- Section-by-section tutorials, including game tutorials with randomized quizzes
- A detailed chapter summary
- A true/false quiz
- Additional review exercises
- Graphers, Excel tutorials, and other resources
- The following extra topic:
   Linear Approximation and Error Estimation

## Case Study  Projecting Market Growth

You are on the board of directors at Fullcourt Academic Press. The sales director of the high school division has just burst into your office with a proposal for an expansion strategy based on the assumption that the number of graduates from private high schools in the U.S. will grow at a rate of at least 4,000 per year through the year 2015. Because the figures actually appear to be leveling off, you are suspicious about this estimate. You would like to devise a model that predicts this trend before tomorrow's scheduled board meeting. **How do you go about doing this?**

Yuri Arcurs/Shutterstock

783

# Introduction

In Chapter 10 we studied the concept of the derivative of a function, and we saw some of the applications for which derivatives are useful. However, computing the derivative of a function algebraically seemed to be a time-consuming process, forcing us to restrict attention to fairly simply functions.

In this chapter we develop shortcut techniques that will allow us to write down the derivative of a function directly without having to calculate any limit. These techniques will also enable us to differentiate any closed-form function—that is, any function, no matter how complicated, that can be specified by a formula involving powers, radicals, absolute values, exponents, and logarithms. (In Chapter 16 we will discuss how to add trigonometric functions to this list.) We also show how to find the derivatives of functions that are only specified *implicitly*—that is, functions for which we are not given an explicit formula for $y$ in terms of $x$ but only an equation relating $x$ and $y$.

**algebra Review**

For this chapter, you should be familiar with the algebra reviewed in **Chapter 0, Sections 3 and 4.**

## 11.1  Derivatives of Powers, Sums, and Constant Multiples

Up to this point we have approximated derivatives using difference quotients, and we have done exact calculations using the definition of the derivative as the limit of a difference quotient. In general, we would prefer to have an exact calculation, and it is also very useful to have a formula for the derivative function when we can find one. However, the calculation of a derivative as a limit is often tedious, so it would be nice to have a quicker method. We discuss the first of the shortcut rules in this section. By the end of this chapter, we will be able to find fairly quickly the derivative of almost any function we can write.

### Shortcut Formula: The Power Rule

If you look at Examples 2 and 3 in Section 10.6, you may notice a pattern:

$$f(x) = x^2 \quad \Rightarrow \quad f'(x) = 2x$$
$$f(x) = x^3 \quad \Rightarrow \quad f'(x) = 3x^2.$$

This pattern generalizes to any power of $x$:

---

**Theorem 11.1  The Power Rule**

If $n$ is any constant and $f(x) = x^n$, then

$$f'(x) = nx^{n-1}.$$

**Quick Examples**

1. If $f(x) = x^2$, then $f'(x) = 2x^1 = 2x$.
2. If $f(x) = x^3$, then $f'(x) = 3x^2$.
3. If $f(x) = x$, rewrite as $f(x) = x^1$, so $f'(x) = 1x^0 = 1$.
4. If $f(x) = 1$, rewrite as $f(x) = x^0$, so $f'(x) = 0x^{-1} = 0$.

**Website**
www.WanerMath.com
At the Website you can find a proof
of the power rule by following:

Everything
→ Chapter 11
→ Proof of the Power Rule

✳ See Section 0.2 in the Precalculus Review to brush up on negative and fractional exponents. Pay particular attention to rational, radical, and exponent forms.

The proof of the power rule involves first studying the case when $n$ is a positive integer, and then studying the cases of other types of exponents (negative integer, rational number, irrational number). You can find a proof at the Website.

### EXAMPLE 1  Using the Power Rule for Negative and Fractional Exponents

Calculate the derivatives of the following:

**a.** $f(x) = \dfrac{1}{x}$    **b.** $f(x) = \dfrac{1}{x^2}$    **c.** $f(x) = \sqrt{x}$

**Solution**

**a.** Rewrite✳ as $f(x) = x^{-1}$. Then $f'(x) = (-1)x^{-2} = -\dfrac{1}{x^2}$.

**b.** Rewrite as $f(x) = x^{-2}$. Then $f'(x) = (-2)x^{-3} = -\dfrac{2}{x^3}$.

**c.** Rewrite as $f(x) = x^{0.5}$. Then $f'(x) = 0.5x^{-0.5} = \dfrac{0.5}{x^{0.5}}$. Alternatively, rewrite

$f(x)$ as $x^{1/2}$, so that $f'(x) = \dfrac{1}{2}x^{-1/2} = \dfrac{1}{2x^{1/2}} = \dfrac{1}{2\sqrt{x}}$.

By rewriting the given functions in Example 1 before taking derivatives, we converted them from **rational** or **radical form** (as in, say, $\dfrac{1}{x^2}$ and $\sqrt{x}$) to **exponent form** (as in $x^{-2}$ and $x^{0.5}$; see the Precalculus Review, Section 0.2) to enable us to use the power rule. (See the Caution below.)

**Caution**

We cannot apply the power rule to terms in the denominators or under square roots. For example:

**1.** The derivative of $\dfrac{1}{x^2}$ is **NOT** $\dfrac{1}{2x}$; it is $-\dfrac{2}{x^3}$.    See Example 1(b).

**2.** The derivative of $\sqrt{x^3}$ is **NOT** $\sqrt{3x^2}$; it is $1.5x^{0.5}$.    Rewrite $\sqrt{x^3}$ as $x^{3/2}$ or $x^{1.5}$ and apply the power rule.

**Table 1**  Table of Derivative Formulas

| $f(x)$ | $f'(x)$ |
|---|---|
| 1 | 0 |
| $x$ | 1 |
| $x^2$ | $2x$ |
| $x^3$ | $3x^2$ |
| $x^n$ | $nx^{n-1}$ |
| $\dfrac{1}{x}$ | $-\dfrac{1}{x^2}$ |
| $\dfrac{1}{x^2}$ | $-\dfrac{2}{x^3}$ |
| $\sqrt{x}$ | $\dfrac{1}{2\sqrt{x}}$ |

Some of the derivatives in Example 1 are very useful to remember, so we summarize them in Table 1. We suggest that you add to this table as you learn more derivatives. It is *extremely* helpful to remember the derivatives of common functions such as $1/x$ and $\sqrt{x}$, even though they can be obtained by using the power rule as in the above example.

## Another Notation: Differential Notation

Here is a useful notation based on the "*d*-notation" we discussed in Section 10.5. **Differential notation** is based on an abbreviation for the phrase "the derivative with respect to $x$." For example, we learned that if $f(x) = x^3$, then $f'(x) = 3x^2$. When we say "$f'(x) = 3x^2$," we mean the following:

The derivative of $x^3$ with respect to $x$ equals $3x^2$.

You may wonder why we sneaked in the words "with respect to $x$." All this means is that the variable of the function is $x$, and not any other variable.* Because we use the phrase "the derivative with respect to $x$" often, we use the following abbreviation.

## Differential Notation; Differentiation

$\dfrac{d}{dx}$ means "the derivative with respect to $x$."

Thus, $\dfrac{d}{dx}[f(x)]$ is the same thing as $f'(x)$, the derivative of $f(x)$ with respect to $x$. If $y$ is a function of $x$, then the derivative of $y$ with respect to $x$ is

$$\frac{d}{dx}(y) \qquad \text{or, more compactly,} \qquad \frac{dy}{dx}.$$

To **differentiate** a function $f(x)$ with respect to $x$ means to take its derivative with respect to $x$.

### Quick Examples

| In Words | Formula |
|---|---|
| **1.** The derivative with respect to $x$ of $x^3$ is $3x^2$. | $\dfrac{d}{dx}(x^3) = 3x^2$ |
| **2.** The derivative with respect to $t$ of $\dfrac{1}{t}$ is $-\dfrac{1}{t^2}$. | $\dfrac{d}{dt}\left(\dfrac{1}{t}\right) = -\dfrac{1}{t^2}$ |
| **3.** If $y = x^4$, then $\dfrac{dy}{dx} = 4x^3$. | |
| **4.** If $u = \dfrac{1}{t^2}$, then $\dfrac{du}{dt} = -\dfrac{2}{t^3}$. | |

### Notes

**1.** $\dfrac{dy}{dx}$ is Leibniz's notation for the derivative we discussed in Section 10.5. (See the discussion before Example 3 there.)

**2.** Leibniz notation illustrates units nicely: Units of $\dfrac{dy}{dx}$ are units of $y$ per unit of $x$.

**3.** We can (and often do!) use different kind of brackets or parentheses in Liebniz notation; for instance, $\dfrac{d}{dx}[x^3]$, $\dfrac{d}{dx}(x^3)$, and $\dfrac{d}{dx}\{x^3\}$ all mean the same thing (and equal $3x^2$). ■

## The Rules for Sums and Constant Multiples

We can now find the derivatives of more complicated functions, such as polynomials, using the following rules. If $f$ and $g$ are functions and if $c$ is a constant, we saw in Section 1.2 how to obtain the **sum**, $f + g$, **difference**, $f - g$, and **constant multiple**, $cf$.

## Theorem 11.2 Derivatives of Sums, Differences, and Constant Multiples

If $f$ and $g$ are any two differentiable functions, and if $c$ is any constant, then the sum, $f + g$, the difference, $f - g$, and the constant multiple, $cf$, are differentiable, and

$$[f \pm g]'(x) = f'(x) \pm g'(x) \qquad \text{Sum Rule}$$

$$[cf]'(x) = cf'(x). \qquad \text{Constant Multiple Rule}$$

### In Words:

- The derivative of a sum is the sum of the derivatives, and the derivative of a difference is the difference of the derivatives.
- The derivative of $c$ times a function is $c$ times the derivative of the function.

### Differential Notation:

$$\frac{d}{dx}[f(x) \pm g(x)] = \frac{d}{dx}f(x) \pm \frac{d}{dx}g(x)$$

$$\frac{d}{dx}[cf(x)] = c\frac{d}{dx}f(x)$$

### Quick Examples

1. $\dfrac{d}{dx}(x^2 - x^4) = \dfrac{d}{dx}(x^2) - \dfrac{d}{dx}(x^4) = 2x - 4x^3$

2. $\dfrac{d}{dx}(7x^3) = 7\dfrac{d}{dx}(x^3) = 7(3x^2) = 21x^2$

   In other words, we multiply the coefficient (7) by the exponent (3), and then decrease the exponent by 1.

3. $\dfrac{d}{dx}(12x) = 12\dfrac{d}{dx}(x) = 12(1) = 12$

   In other words, the derivative of a constant times $x$ is that constant.

4. $\dfrac{d}{dx}(-x^{0.5}) = \dfrac{d}{dx}[(-1)x^{0.5}] = (-1)\dfrac{d}{dx}(x^{0.5}) = (-1)(0.5)x^{-0.5}$

   $= -0.5x^{-0.5}$

5. $\dfrac{d}{dx}(12) = \dfrac{d}{dx}[12(1)] = 12\dfrac{d}{dx}(1) = 12(0) = 0.$

   In other words, the derivative of a constant is zero.

6. If my company earns twice as much (annual) revenue as yours and the derivative of your revenue function is the curve on the left, then the derivative of my revenue function is the curve on the right.

**7.** Suppose that a company's revenue $R$ and cost $C$ are changing with time. Then so is the profit, $P(t) = R(t) - C(t)$, and the rate of change of the profit is

$$P'(t) = R'(t) - C'(t).$$

In words: *The derivative of the profit is the derivative of revenue minus the derivative of cost.*

## Proof of the Sum Rule

By the definition of the derivative of a function,

$$\frac{d}{dx}[f(x) + g(x)] = \lim_{h \to 0} \frac{[f(x + h) + g(x + h)] - [f(x) + g(x)]}{h}$$

$$= \lim_{h \to 0} \frac{[f(x + h) - f(x)] + [g(x + h) - g(x)]}{h}$$

$$= \lim_{h \to 0} \left[ \frac{f(x + h) - f(x)}{h} + \frac{g(x + h) - g(x)}{h} \right]$$

$$= \lim_{h \to 0} \frac{f(x + h) - f(x)}{h} + \lim_{h \to 0} \frac{g(x + h) - g(x)}{h}$$

$$= \frac{d}{dx}[f(x)] + \frac{d}{dx}[g(x)].$$

The next-to-last step uses a property of limits: The limit of a sum is the sum of the limits. Think about why this should be true. The last step uses the definition of the derivative again (and the fact that the functions are differentiable).

The proof of the rule for constant multiples is similar.

---

**EXAMPLE 2  Combining the Sum and Constant Multiple Rules, and Dealing with *x* in the Denominator**

Find the derivatives of the following:

**a.** $f(x) = 3x^2 + 2x - 4$          **b.** $f(x) = \dfrac{2x}{3} - \dfrac{6}{x} + \dfrac{2}{3x^{0.2}} - \dfrac{x^4}{2}$

**c.** $f(x) = \dfrac{|x|}{4} + \dfrac{1}{2\sqrt{x}}$

## Solution

**a.** $\dfrac{d}{dx}(3x^2 + 2x - 4) = \dfrac{d}{dx}(3x^2) + \dfrac{d}{dx}(2x - 4)$      Rule for sums

$$= \dfrac{d}{dx}(3x^2) + \dfrac{d}{dx}(2x) - \dfrac{d}{dx}(4)$$      Rule for differences

$$= 3(2x) + 2(1) - 0$$      See Quick Example 2.

$$= 6x + 2$$

**b.** Notice that $f$ has $x$ and powers of $x$ in the denominator. We deal with these terms the same way we did in Example 1, by rewriting them in exponent form (that is, in the form constant $\times$ power of $x$; see Section 0.2 in the Precalculus Review):

$$f(x) = \dfrac{2x}{3} - \dfrac{6}{x} + \dfrac{2}{3x^{0.2}} - \dfrac{x^4}{2}$$      Rational form

$$= \dfrac{2}{3}x - 6x^{-1} + \dfrac{2}{3}x^{-0.2} - \dfrac{1}{2}x^4.$$      Exponent form

We are now ready to take the derivative:

$$f'(x) = \dfrac{2}{3}(1) - 6(-1)x^{-2} + \dfrac{2}{3}(-0.2)x^{-1.2} - \dfrac{1}{2}(4x^3)$$

$$= \dfrac{2}{3} + 6x^{-2} - \dfrac{0.4}{3}x^{-1.2} - 2x^3$$      Exponent form

$$= \dfrac{2}{3} + \dfrac{6}{x^2} - \dfrac{0.4}{3x^{1.2}} - 2x^3.$$      Rational form

**c.** Rewrite $f(x)$ using exponent form as follows:

$$f(x) = \dfrac{|x|}{4} + \dfrac{1}{2\sqrt{x}}$$      Rational form

$$= \dfrac{1}{4}|x| + \dfrac{1}{2}x^{-1/2}.$$      Exponent form

Now recall from the end of Section 10.6 that the derivative of $|x|$ is $\dfrac{|x|}{x}$. Thus,

$$f'(x) = \dfrac{1}{4}\dfrac{|x|}{x} + \dfrac{1}{2}\left(\dfrac{-1}{2}x^{-3/2}\right)$$

$$= \dfrac{|x|}{4x} - \dfrac{1}{4}x^{-3/2}$$      Simplify

$$= \dfrac{|x|}{4x} - \dfrac{1}{4x^{3/2}}.$$      Rational form

Notice that in Example 2(a) we had three terms in the expression for $f(x)$, not just two. By applying the rule for sums and differences twice, we saw that the derivative of a sum or difference of three terms is the sum or difference of the derivatives of the terms. (One of those terms had zero derivative, so the final answer had only two terms.) In fact, the derivative of a sum or difference of any number of terms is the sum or difference of the derivatives of the terms. Put another way, to take the derivative of a sum or difference of any number of terms, we take derivatives term by term.

**Note** Nothing forces us to use only $x$ as the independent variable when taking derivatives (although it is traditional to give $x$ preference). For instance, part (a) in Example 2 can be rewritten as

$$\frac{d}{dt}(3t^2 + 2t - 4) = 6t + 2 \qquad \frac{d}{dt} \text{ means "derivative with respect to } t\text{."}$$

or

$$\frac{d}{du}(3u^2 + 2u - 4) = 6u + 2. \qquad \frac{d}{du} \text{ means "derivative with respect to } u\text{."} \quad \blacksquare$$

In the previous examples, we saw instances of the following important facts. (Think about these graphically to see why they must be true.)

### The Derivative of a Constant Times $x$ and the Derivative of a Constant

If $c$ is any constant, then:

**Rule**

$$\frac{d}{dx}(cx) = c$$

$$\frac{d}{dx}(c) = 0$$

**Quick Examples**

$$\frac{d}{dx}(6x) = 6 \qquad \frac{d}{dx}(-x) = -1$$

$$\frac{d}{dx}(5) = 0 \qquad \frac{d}{dx}(\pi) = 0$$

In Section 10.5 we pointed out that the derivative of a function cannot exist at an endpoint of its domain, as the defining limit does not exist there (see the "Important Notes" after the definition of instantaneous rate of change). Thus, for instance, $f(x) = \sqrt{x}$ and $g(x) = x^{1/4}$ are not differentiable at the endpoint $x = 0$ of their domains. In Example 5 of Section 10.6 we saw that $h(x) = |x|$ also fails to be differentiable at $x = 0$, even though $x = 0$ is not an endpoint of its domain (the domain of $h$ is the set of all real numbers). In the next example we see how to spot the non-differentiability at a point of this and other functions simply by looking at the formulas for their derivatives.

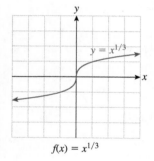

$f(x) = x^{1/3}$

---

### EXAMPLE 3   Functions Not Differentiable at a Point

Find the natural domains of the derivatives of $f(x) = x^{1/3}$, $g(x) = x^{2/3}$, and $h(x) = |x|$.

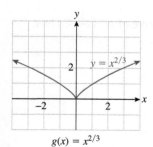

$g(x) = x^{2/3}$

**Solution** Let's look at the derivatives of the three functions given:

$$f(x) = x^{1/3}, \text{ so } \quad f'(x) = \frac{1}{3}x^{-2/3} = \frac{1}{3x^{2/3}}.$$

$$g(x) = x^{2/3}, \text{ so } \quad g'(x) = \frac{2}{3}x^{-1/3} = \frac{2}{3x^{1/3}}.$$

$$h(x) = |x|, \text{ so } \quad h'(x) = \frac{|x|}{x}.$$

The derivatives of all three functions are defined only for nonzero values of $x$, and their natural domains consist of all real numbers except 0. Thus, the derivatives $f'$, $g'$, and $h'$ do not exist at $x = 0$. In other words, these functions are not differentiable at $x = 0$. If we look at Figure 1 we notice why these functions fail to be differentiable at $x = 0$: The graph of $f$ has a vertical tangent line at 0. Because a vertical line

$h(x) = |x|$

**Figure 1**

**using Technology**

If you try to graph the function $f(x) = x^{2/3}$ using the format

$$X^{\wedge}(2/3)$$

you may get only the right-hand portion of the graph of $g$ in Figure 1 because graphing utilities are (often) not programmed to raise negative numbers to fractional exponents. (However, many will handle $X^{\wedge}(1/3)$ correctly, as a special case they recognize.) To avoid this difficulty, you can take advantage of the identity

$$x^{2/3} = (x^2)^{1/3}$$

so that it is always a nonnegative number that is being raised to a fractional exponent. Thus, use the format

$$(X^{\wedge}2)^{\wedge}(1/3)$$

to obtain both portions of the graph.

has undefined slope, the derivative is undefined at that point. The graphs of $g$ and $h$ come to a sharp point at 0, where it is not meaningful to speak about the slope of the tangent line; therefore, the derivatives of $g$ and $h$ are not defined there. (In the case of $g$, where the sharp point is called a *cusp,* a vertical tangent line would seem appropriate, but as in the case of $f$, its slope is undefined.)

You can also detect this nondifferentiability by computing some difference quotients numerically, as we did for $h$ in Section 10.6.

## APPLICATION

### EXAMPLE 4 Gold Price

You are a commodities trader and you monitor the price of gold on the spot market very closely during an active morning. Suppose you find that the price of an ounce of gold can be approximated by the function

$$G(t) = 5t^2 - 85t + 1{,}762 \quad (7.5 \le t \le 10.5),$$

where $t$ is time in hours. (See Figure 2. $t = 8$ represents 8:00 am.)

Source: www.kitco.com

$$G(t) = 5t^2 - 85t + 1{,}762$$

**Figure 2**

**a.** According to the model, how fast was the price of gold changing at 8:00 am?

**b.** According to the model, the price of gold

(A) increased at a faster and faster rate
(B) increased at a slower and slower rate
(C) decreased at a faster and faster rate
(D) decreased at a slower and slower rate

between 7:30 and 8:30 am.

### Solution

**a.** Differentiating the given function with respect to $t$ gives

$$G'(t) = 10t - 85.$$

Because 8:00 am corresponds to $t = 8$, we obtain

$$G'(8) = 10(8) - 85 = -5.$$

The units of the derivative are dollars per hour, so we conclude that, at 8:00 am, the price of gold was dropping at a rate of $5 per hour.

**b.** From the graph, we can see that, between 7:30 and 8:30 am (the interval [7.5, 8.5]), the price of gold was decreasing. Also from the graph, we see that the slope of the tangent becomes less and less negative as $t$ increases, so the price of gold is decreasing at a slower and slower rate (choice (D)).

We can also see this algebraically from the derivative, $G'(t) = 10t - 85$: For values of $t$ less than 8.5, $G'(t)$ is negative; that is, the rate of change of $G$ is negative, so the price of gold is decreasing. Further, as $t$ increases, $G'(t)$ becomes less and less negative, so the price of gold is decreasing at a slower and slower rate, confirming that choice (D) is the correct one.

# An Application to Limits: L'Hospital's Rule

The limits that caused us some trouble in Sections 10.1–10.3 are those of the form $\lim_{x \to a} f(x)$ in which substituting $x = a$ gave us an indeterminate form, such as

$$\lim_{x \to 2} \frac{x^3 - 8}{x - 2} \qquad \text{Substituting } x = 2 \text{ yields } \tfrac{0}{0}.$$

$$\lim_{x \to +\infty} \frac{2x - 4}{x - 1}. \qquad \text{Substituting } x = +\infty \text{ yields } \tfrac{\infty}{\infty}.$$

* Guillaume François Antoine, Marquis de l'Hospital (1661–1704) wrote the first textbook on calculus, *Analyse des infiniment petits pour l'intelligence des lignes courbes,* in 1692. The rule now known as l'Hospital's rule appeared first in this book.

L'Hospital's rule* gives us an alternate way of computing limits such as these without the need to do any preliminary simplification. It also allows us to compute some limits for which algebraic simplification does not work.

## Theorem 11.3  L'Hospital's Rule

If $f$ and $g$ are two differentiable functions such that substituting $x = a$ in the expression $\dfrac{f(x)}{g(x)}$ gives the indeterminate form $\dfrac{0}{0}$ or $\dfrac{\infty}{\infty}$, then

$$\lim_{x \to a} \frac{f(x)}{g(x)} = \lim_{x \to a} \frac{f'(x)}{g'(x)}.$$

That is, we can replace $f(x)$ and $g(x)$ with their *derivatives* and try again to take the limit.

### Quick Examples

**1.** Substituting $x = 2$ in $\dfrac{x^3 - 8}{x - 2}$ yields $\dfrac{0}{0}$. Therefore, l'Hospital's rule applies and

$$\lim_{x \to 2} \frac{x^3 - 8}{x - 2} = \lim_{x \to 2} \frac{3x^2}{1} = \frac{3(2)^2}{1} = 12.$$

**2.** Substituting $x = +\infty$ in $\dfrac{2x - 4}{x - 1}$ yields $\dfrac{\infty}{\infty}$. Therefore, l'Hospital's rule applies and

$$\lim_{x \to +\infty} \frac{2x - 4}{x - 1} = \lim_{x \to +\infty} \frac{2}{1} = 2.$$

† A proof of l'Hospital's rule can be found in most advanced calculus textbooks.

The proof of l'Hospital's rule is beyond the scope of this text.†

## EXAMPLE 5 Applying L'Hospital's Rule

Check whether l'Hospital's rule applies to each of the following limits. If it does, use it to evaluate the limit. Otherwise, use some other method to evaluate the limit.

**a.** $\displaystyle\lim_{x\to 1}\frac{x^2-2x+1}{4x^3-3x^2-6x+5}$

**b.** $\displaystyle\lim_{x\to+\infty}\frac{2x^2-4x}{5x^3-3x+5}$

**c.** $\displaystyle\lim_{x\to 1}\frac{x-1}{x^3-3x^2+3x-1}$

**d.** $\displaystyle\lim_{x\to 1}\frac{x}{x^3-3x^2+3x-1}$

**Solution**

**a.** Setting $x=1$ yields

$$\frac{1-2+1}{4-3-6+5}=\frac{0}{0}.$$

Therefore, l'Hospital's rule applies and

$$\lim_{x\to 1}\frac{x^2-2x+1}{4x^3-3x^2-6x+5}=\lim_{x\to 1}\frac{2x-2}{12x^2-6x-6}.$$

We are left with a closed-form function. However, we cannot substitute $x=1$ to find the limit because the function $(2x-2)/(12x^2-6x-6)$ is still not defined at $x=1$. In fact, if we set $x=1$, we again get $0/0$. Thus, l'Hospital's rule applies again, and

$$\lim_{x\to 1}\frac{2x-2}{12x^2-6x-6}=\lim_{x\to 1}\frac{2}{24x-6}.$$

Once again we have a closed-form function, but this time it is defined when $x=1$, giving

$$\frac{2}{24-6}=\frac{1}{9}.$$

Thus,

$$\lim_{x\to 1}\frac{x^2-2x+1}{4x^3-3x^2-6x+5}=\frac{1}{9}.$$

**b.** Setting $x=+\infty$ yields $\dfrac{\infty}{\infty}$, so

$$\lim_{x\to+\infty}\frac{2x^2-4x}{5x^3-3x+5}=\lim_{x\to+\infty}\frac{4x-4}{15x^2-3}.$$

Setting $x=+\infty$ again yields $\dfrac{\infty}{\infty}$, so we can apply the rule again to obtain

$$\lim_{x\to+\infty}\frac{4x-4}{15x^2-3}=\lim_{x\to+\infty}\frac{4}{30x}.$$

Note that we cannot apply l'Hospital's rule a third time because setting $x=+\infty$ yields the *determinate* form $4/\infty=0$ (see the discussion at the end of Section 10.3). Thus, the limit is 0.

**c.** Setting $x=1$ yields $0/0$ so, by l'Hospital's rule,

$$\lim_{x\to 1}\frac{x-1}{x^3-3x^2+3x-1}=\lim_{x\to 1}\frac{1}{3x^2-6x+3}.$$

We are left with a closed-form function that is still not defined at $x=1$. Further, l'Hospital's rule no longer applies because putting $x=1$ yields the determinate form $1/0$. To investigate this limit, we refer to the discussion at the end of Section 10.3 and find

$$\lim_{x \to 1} \frac{1}{3x^2 - 6x + 3} = \lim_{x \to 1} \frac{1}{3(x-1)^2} = +\infty. \qquad \frac{1}{0^+} = +\infty$$

**d.** Setting $x = 1$ in the expression yields the determinate form $1/0$, so l'Hospital's rule does not apply here. Using the methods of Section 10.3 again, we find that the limit does not exist.

---

**FAQs**

**Using the Rules and Recognizing when a Function Is Not Differentiable**

**Q**: I would *like* to say that the derivative of $5x^2 - 8x + 4$ is just $10x - 8$ without having to go through all that stuff about derivatives of sums and constant multiples. Can I simply forget about all the rules and write down the answer?

**A**: We developed the rules for sums and constant multiples precisely for that reason: so that we could simply write down a derivative without having to think about it too hard. So, you are perfectly justified in simply writing down the derivative without going through the rules, but bear in mind that what you are really doing is applying the power rule, the rule for sums, and the rule for multiples over and over.

**Q**: Is there a way of telling from its formula whether a function $f$ is not differentiable at a point?

**A**: Here are some indicators to look for in the formula for $f$:

- The absolute value of some expression; $f$ may not be differentiable at points where that expression is zero.

  **Example:** $f(x) = 3x^2 - |x - 4|$ is not differentiable at $x = 4$.

- A fractional power smaller than 1 of some expression; $f$ may not be differentiable at points where that expression is zero.

  **Example:** $f(x) = (x^2 - 16)^{2/3}$ is not differentiable at $x = \pm 4$.

---

## 11.1  EXERCISES

▼ more advanced    ◆ challenging
**T** indicates exercises that should be solved using technology

*In Exercises 1–10, use the shortcut rules to **mentally** calculate the derivative of the given function.* HINT [See Examples 1 and 2.]

**1.** $f(x) = x^5$                **2.** $f(x) = x^4$

**3.** $f(x) = 2x^{-2}$            **4.** $f(x) = 3x^{-1}$

**5.** $f(x) = -x^{0.25}$          **6.** $f(x) = -x^{-0.5}$

**7.** $f(x) = 2x^4 + 3x^3 - 1$    **8.** $f(x) = -x^3 - 3x^2 - 1$

**9.** $f(x) = -x + \dfrac{1}{x} + 1$    **10.** $f(x) = \dfrac{1}{x} + \dfrac{1}{x^2}$

*In Exercises 11–16, obtain the derivative dy/dx and state the rules that you use.* HINT [See Example 2.]

**11.** $y = 10$                   **12.** $y = x^3$

**13.** $y = x^2 + x$              **14.** $y = x - 5$

**15.** $y = 4x^3 + 2x - 1$        **16.** $y = 4x^{-1} - 2x - 10$

*In Exercises 17–40, find the derivative of each function.* HINT [See Examples 1 and 2.]

**17.** $f(x) = x^2 - 3x + 5$      **18.** $f(x) = 3x^3 - 2x^2 + x$

**19.** $f(x) = x + x^{0.5}$       **20.** $f(x) = x^{0.5} + 2x^{-0.5}$

**21.** $g(x) = x^{-2} - 3x^{-1} - 2$

**22.** $g(x) = 2x^{-1} + 4x^{-2}$

**23.** $g(x) = \dfrac{1}{x} - \dfrac{1}{x^2}$

**24.** $g(x) = \dfrac{1}{x^2} + \dfrac{1}{x^3}$

**25.** $h(x) = \dfrac{2}{x^{0.4}}$

**26.** $h(x) = -\dfrac{1}{2x^{0.2}}$

**27.** $h(x) = \dfrac{1}{x^2} + \dfrac{2}{x^3}$

**28.** $h(x) = \dfrac{2}{x} - \dfrac{2}{x^3} + \dfrac{1}{x^4}$

**29.** $r(x) = \dfrac{2}{3x} - \dfrac{1}{2x^{0.1}}$

**30.** $r(x) = \dfrac{4}{3x^2} + \dfrac{1}{x^{3.2}}$

**31.** $r(x) = \dfrac{2x}{3} - \dfrac{x^{0.1}}{2} + \dfrac{4}{3x^{1.1}} - 2$

**32.** $r(x) = \dfrac{4x^2}{3} + \dfrac{x^{3.2}}{6} - \dfrac{2}{3x^2} + 4$

**33.** $t(x) = |x| + \dfrac{1}{x}$

**34.** $t(x) = 3|x| - \sqrt{x}$

**35.** $s(x) = \sqrt{x} + \dfrac{1}{\sqrt{x}}$

**36.** $s(x) = x + \dfrac{7}{\sqrt{x}}$

HINT [For Exercises 37–38, first expand the given function.]

**37.** ▼ $s(x) = x\left(x^2 - \dfrac{1}{x}\right)$

**38.** ▼ $s(x) = x^{-1}\left(x - \dfrac{2}{x}\right)$

HINT [For Exercises 39–40, first rewrite the given function.]

**39.** ▼ $t(x) = \dfrac{x^2 - 2x^3}{x}$

**40.** ▼ $t(x) = \dfrac{2x + x^2}{x}$

*In Exercises 41–46, evaluate the given expression.*

**41.** $\dfrac{d}{dx}(2x^{1.3} - x^{-1.2})$

**42.** $\dfrac{d}{dx}(2x^{4.3} + x^{0.6})$

**43.** ▼ $\dfrac{d}{dx}[1.2(x - |x|)]$

**44.** ▼ $\dfrac{d}{dx}[4(x^2 + 3|x|)]$

**45.** ▼ $\dfrac{d}{dt}(at^3 - 4at)$;    (*a* constant)

**46.** ▼ $\dfrac{d}{dt}(at^2 + bt + c)$;    (*a, b, c* constant)

*In Exercises 47–52, find the indicated derivative.*

**47.** $y = \dfrac{x^{10.3}}{2} + 99x^{-1}; \dfrac{dy}{dx}$

**48.** $y = \dfrac{x^{1.2}}{3} - \dfrac{x^{0.9}}{2}; \dfrac{dy}{dx}$

**49.** $s = 2.3 + \dfrac{2.1}{t^{1.1}} - \dfrac{t^{0.6}}{2}; \dfrac{ds}{dt}$

**50.** $s = \dfrac{2}{t^{1.1}} + t^{-1.2}; \dfrac{ds}{dt}$

**51.** ▼ $V = \dfrac{4}{3}\pi r^3; \dfrac{dV}{dr}$

**52.** ▼ $A = 4\pi r^2; \dfrac{dA}{dr}$

*In Exercises 53–58, find the slope of the tangent to the graph of the given function at the indicated point.* HINT [Recall that the slope of the tangent to the graph of f at $x = a$ is $f'(a)$.]

**53.** $f(x) = x^3; (-1, -1)$

**54.** $g(x) = x^4; (-2, 16)$

**55.** $f(x) = 1 - 2x; (2, -3)$

**56.** $f(x) = \dfrac{x}{3} - 1; (-3, -2)$

**57.** $g(t) = \dfrac{1}{t^5}; (1, 1)$

**58.** $s(t) = \dfrac{1}{t^3}; \left(-2, -\dfrac{1}{8}\right)$

*In Exercises 59–64, find the equation of the tangent line to the graph of the given function at the point with the indicated x-coordinate. In each case, sketch the curve together with the appropriate tangent line.*

**59.** ▼ $f(x) = x^3; x = -1$

**60.** ▼ $f(x) = x^2; x = 0$

**61.** ▼ $f(x) = x + \dfrac{1}{x}; x = 2$

**62.** ▼ $f(x) = \dfrac{1}{x^2}; x = 1$

**63.** ▼ $f(x) = \sqrt{x}; x = 4$

**64.** ▼ $f(x) = 2x + 4; x = -1$

*In Exercises 65–70, find all values of x (if any) where the tangent line to the graph of the given equation is horizontal.* HINT [The tangent line is horizontal when its slope is zero.]

**65.** ▼ $y = 2x^2 + 3x - 1$

**66.** ▼ $y = -3x^2 - x$

**67.** ▼ $y = 2x + 8$

**68.** ▼ $y = -x + 1$

**69.** ▼ $y = x + \dfrac{1}{x}$

**70.** ▼ $y = x - \sqrt{x}$

**71.** ◆ Write out the proof that $\dfrac{d}{dx}(x^4) = 4x^3$.

**72.** ◆ Write out the proof that $\dfrac{d}{dx}(x^5) = 5x^4$.

*In Exercises 73–76, determine whether f is differentiable at the given point. If $f'(a)$ exists, give its value.* HINT [See Example 3.]

**73.** $f(x) = x - x^{1/3}$    **a.** $a = 1$    **b.** $a = 0$

**74.** $f(x) = 2x + x^{4/3}$    **a.** $a = 8$    **b.** $a = 0$

**75.** $f(x) = x^{5/4} - 1$    **a.** $a = 16$    **b.** $a = 0$

**76.** $f(x) = x^{1/5} + 5$    **a.** $a = 1$    **b.** $a = 0$

*In Exercises 77–88 say whether l'Hospital's rule applies. If is does, use it to evaluate the given limit. If not, use some other method.*

**77.** $\lim\limits_{x \to 1} \dfrac{x^2 - 2x + 1}{x^2 - x}$

**78.** $\lim\limits_{x \to -1} \dfrac{x^2 + 3x + 2}{x^2 + x}$

**79.** $\lim\limits_{x \to 2} \dfrac{x^3 - 8}{x - 2}$

**80.** $\lim\limits_{x \to 0} \dfrac{x^3 + 8}{x^2 + 3x + 2}$

**81.** $\lim\limits_{x \to 1} \dfrac{x^2 + 3x + 2}{x^2 + x}$

**82.** $\lim\limits_{x \to -2} \dfrac{x^3 + 8}{x^2 + 3x + 2}$

**83.** $\lim\limits_{x \to -\infty} \dfrac{3x^2 + 10x - 1}{2x^2 - 5x}$

**84.** $\lim\limits_{x \to -\infty} \dfrac{6x^2 + 5x + 100}{3x^2 - 9}$

**85.** $\lim\limits_{x \to -\infty} \dfrac{10x^2 + 300x + 1}{5x + 2}$

**86.** $\lim\limits_{x \to -\infty} \dfrac{2x^4 + 20x^3}{1,000x^3 + 6}$

**87.** $\displaystyle\lim_{x\to-\infty}\frac{x^3-100}{2x^2+500}$   **88.** $\displaystyle\lim_{x\to-\infty}\frac{x^2+30x}{2x^6+10x}$

## APPLICATIONS

**89. *Crude Oil Prices*** The price per barrel of crude oil in constant 2008 dollars can be approximated by

$$P(t) = 0.45t^2 - 12t + 105 \text{ dollars} \quad (0 \le t \le 28),$$

where $t$ is time in years since the start of 1980.[1] Find $P'(t)$ and $P'(20)$. What does the answer tell you about the price of crude oil? HINT [See Example 2.]

**90. *Median Home Prices*** The median home price in the United States over the period 2003–2011 can be approximated by

$$P(t) = -5t^2 + 75t - 30 \text{ thousand dollars} \quad (3 \le t \le 11),$$

where $t$ is time in years since the start of 2000.[2] Find $P'(t)$ and $P'(6)$. What does the answer tell you about home prices? HINT [See Example 2.]

**91. *Food versus Education*** The following equation shows the approximate relationship between the percentage $y$ of total personal consumption spent on food and the corresponding percentage $x$ spent on education.[3]

$$y = \frac{35}{x^{0.35}} \text{ percentage points} \quad (6.5 \le x \le 17.5).$$

According to the model, spending on food is decreasing at a rate of _____ percentage points per one percentage point increase in spending on education when 10% of total consumption is spent on education. (Answer should be rounded to two significant digits.) HINT [See Example 2(b).]

**92. *Food versus Recreation*** The following equation shows the approximate relationship between the percentage $y$ of total personal consumption spent on food and the corresponding percentage $x$ spent on recreation.[4]

$$y = \frac{33}{x^{0.63}} \text{ percentage points} \quad (6.5 \le x \le 17.5).$$

According to the model, spending on food is decreasing at a rate of _____ percentage points per one percentage point increase in spending on recreation when 3% of total consumption is spent on recreation. (Answer should be rounded to two significant digits.) HINT [See Example 2(b).]

**93. *Velocity*** If a stone is dropped from a height of 400 feet, its height $s$ after $t$ seconds is given by $s(t) = 400 - 16t^2$, with $s$ in feet.

**a.** Compute $s'(t)$ and hence find its velocity at times $t = 0, 1, 2, 3,$ and 4 seconds.

**b.** When does it reach the ground, and how fast is it traveling when it hits the ground? HINT [It reaches the ground when $s(t) = 0.$]

---

[1] Source for data: www.inflationdata.com.

[2] Source for data: www.zillow.com.

[3] Model based on historical and projected data from 1908–2010. Sources: Historical data, Bureau of Economic Analysis; projected data, Bureau of Labor Statistics/*New York Times*, December 1, 2003, p. C2.

[4] *Ibid.*

**94. *Velocity*** If a stone is thrown down at 120 ft/s from a height of 1,000 feet, its height $s$ after $t$ seconds is given by $s(t) = 1,000 - 120t - 16t^2$, with $s$ in feet.

**a.** Compute $s'(t)$ and hence find its velocity at times $t = 0, 1, 2, 3,$ and 4 seconds.

**b.** When does it reach the ground, and how fast is it traveling when it hits the ground? HINT [It reaches the ground when $s(t) = 0.$]

**95. *GE Net Income 2005–2009*** The annual net income of General Electric for the period 2005–2009 could be approximated by[5]

$$P(t) = -2.0t^2 + 6.6t + 16 \text{ billion dollars} \quad (0 \le t \le 4),$$

where $t$ is time in years since 2005.

**a.** Compute $P'(t)$. How fast was GE's annual net income changing in 2008? (Be careful to give correct units of measurement.)

**b.** According to the model, GE's annual net income

**(A)** increased at a faster and faster rate

**(B)** increased at a slower and slower rate

**(C)** decreased at a faster and faster rate

**(D)** decreased at a slower and slower rate

during the first year and a half shown (the interval $[0, 1.5]$). Justify your answer in two ways: geometrically, reasoning entirely from the graph; and algebraically, reasoning from the derivative of $P$. HINT [See Example 4.]

**96. *GE Net Income 2007–2011*** The annual net income of General Electric for the period 2007–2011 could be approximated by[6]

$$P(t) = 3t^2 - 24t + 59 \text{ billion dollars} \quad (2 \le t \le 6),$$

where $t$ is time in years since 2005.

---

[5] Source for data: www.wikinvest.com.

[6] 2011 value estimated. Source: *Ibid.*

**a.** Compute $P'(t)$. How fast was GE's annual net income changing in 2010? (Be careful to give correct units of measurement.)

**b.** According to the model, GE's annual net income

   **(A)** increased at a faster and faster rate
   **(B)** increased at a slower and slower rate
   **(C)** decreased at a faster and faster rate
   **(D)** decreased at a slower and slower rate

   during the first two years shown (the interval [2, 4]). Justify your answer in two ways: geometrically, reasoning entirely from the graph; and algebraically, reasoning from the derivative of $P$. HINT [See Example 4.]

**97. *Ecology*** Increasing numbers of manatees ("sea sirens") have been killed by boats off the Florida coast. The following graph shows the relationship between the number of boats registered in Florida and the number of manatees killed each year.

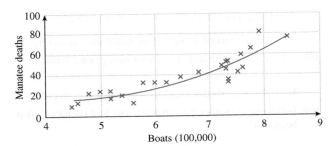

The regression curve shown is given by

$$f(x) = 3.55x^2 - 30.2x + 81 \quad (4.5 \leq x \leq 8.5),$$

where $x$ is the number of boats (hundreds of thousands) registered in Florida in a particular year and $f(x)$ is the number of manatees killed by boats in Florida that year.[7]

**a.** Find $f'(x)$, and use your formula to compute $f'(8)$, stating its units of measurement. What does the answer say about manatee deaths?

**b.** Is $f'(x)$ increasing or decreasing with increasing $x$? Interpret the answer. HINT [See Example 4.]

**98. *SAT Scores by Income*** The following graph shows U.S. math SAT scores as a function of parents' income level.[8]

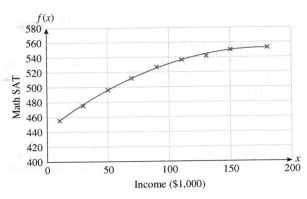

The regression curve shown is given by

$$f(x) = -0.0034x^2 + 1.2x + 444 \quad (10 \leq x \leq 180),$$

where $f(x)$ is the average math SAT score of a student whose parents earn $x$ thousand dollars per year.

**a.** Find $f'(x)$, and use your formula to compute $f'(100)$, stating its units of measurement. What does the answer say about math SAT scores?

**b.** Does $f'(x)$ increase or decrease with increasing $x$? What does your answer say about math SAT scores? HINT [See Example 4.]

**99.** ▼ ***Market Share: Smart Phones*** The following graph shows the approximate market shares, in percentage points, of Apple's *iPhone* and Google's *Android*-based smart phones from January 2009 to May 2010 ($t$ is time in months, and $t = 0$ represents January, 2009):[9]

Market share (%)

Let $I(t)$ be the iPhone market share at time $t$, and let $A(t)$ be the Android market share at time $t$.

**a.** What does the function $I - A$ measure? What does its derivative $(I - A)'$ measure?

**b.** The graph suggests that, on the interval [6, 16], $I - A$ is

   **(A)** increasing.
   **(B)** decreasing.
   **(C)** increasing, then decreasing.
   **(D)** decreasing, then increasing.

---

[7]Regression model is based on data from 1976 to 2000. Sources for data: Florida Department of Highway Safety & Motor Vehicles, Florida Marine Institute/*New York Times*, February 12, 2002, p. F4.

[8]Regression model is based on 2009 data. Source: College Board/ *New York Times* http://economix.blogs.nytimes.com.

[9]Source for data: Quantcast www.quantcast.com.

**c.** The two market shares are approximated by
iPhone: $I(t) = -0.5t + 73$
Android: $A(t) = 0.1t^2 - t + 10$.

Compute $(I - A)'$, stating its units of measurement. On the interval $[0, 16]$, $(I - A)'$ is

**(A)** positive.
**(B)** negative.
**(C)** positive, then negative.
**(D)** negative, then positive.

How is this behavior reflected in the graph, and what does it mean about the market shares of iPhone and Android?

**d.** Compute $(I - A)'(4)$. Interpret your answer.

**100.** ▼ *Market Share: Smart Phones* The following graph shows the approximate market shares, in percentage points, of Research In Motion's *BlackBerry* smartphones and Google's *Android*-based smartphones from July 2009 to May 2010 ($t$ is time in months, and $t = 0$ represents January 2009):[10]

Market share (%)

Let $B(t)$ be the BlackBerry market share at time $t$, and let $A(t)$ be the Android market share at time $t$.

**a.** What does the function $B - A$ measure? What does its derivative $(B - A)'$ measure?

**b.** The graph suggests that, on the interval $[6, 16]$, $B - A$ is

**(A)** increasing.
**(B)** decreasing.
**(C)** increasing, then decreasing.
**(D)** decreasing, then increasing.

**c.** The two market shares are approximated by
BlackBerry: $B(t) = -0.6t + 21.6$
Android: $A(t) = 0.1t^2 - t + 10$.

Compute $(B - A)'$, stating its units of measurement. On the interval $[6, 16]$, $(B - A)'$ is

**(A)** positive.
**(B)** negative.
**(C)** positive, then negative.
**(D)** negative, then positive.

How is this reflected in the graph, and what does it mean about the market shares of BlackBerry and Android?

**d.** Compute $(B - A)'(10)$. Interpret your answer.

---

[10]Source for data: Quantcast www.quantcast.com.

**COMMUNICATION AND REASONING EXERCISES**

**101.** What instructions would you give to a fellow student who wanted to accurately graph the tangent line to the curve $y = 3x^2$ at the point $(-1, 3)$?

**102.** What instructions would you give to a fellow student who wanted to accurately graph a line at right angles to the curve $y = 4/x$ at the point where $x = 0.5$?

**103.** Consider $f(x) = x^2$ and $g(x) = 2x^2$. How do the slopes of the tangent lines of $f$ and $g$ at the same $x$ compare?

**104.** Consider $f(x) = x^3$ and $g(x) = x^3 + 3$. How do the slopes of the tangent lines of $f$ and $g$ compare?

**105.** Suppose $g(x) = -f(x)$. How do the derivatives of $f$ and $g$ compare?

**106.** Suppose $g(x) = f(x) - 50$. How do the derivatives of $f$ and $g$ compare?

**107.** Following is an excerpt from your best friend's graded homework:

$$3x^4 + 11x^5 = 12x^3 + 55x^4. \quad ✗ WRONG \quad -8$$

Why was it marked wrong? How would you correct it?

**108.** Following is an excerpt from your own graded homework:

$$x^n = nx^{n-1}. \quad ✗ WRONG \quad -10$$

Why was it marked wrong? How would you correct it?

**109.** Following is another excerpt from your best friend's graded homework:

$$y = \frac{1}{2x} = 2x^{-1}, \text{ so } \frac{dy}{dx} = -2x^{-2}. \quad ✗ WRONG \quad -5$$

Why was it marked wrong? How would you correct it?

**110.** Following is an excerpt from your second best friend's graded homework:

$$f(x) = \frac{3}{4x^2}; f'(x) = \frac{3}{8x}. \quad ✗ WRONG \quad -10$$

Why was it marked wrong? How would you correct it?

**111.** Following is an excerpt from your worst enemy's graded homework:

$$f(x) = 4x^2; f'(x) = (0)(2x) = 0. \quad ✗ WRONG \quad -6$$

Why was it marked wrong? How would you correct it?

**112.** Following is an excerpt from your second worst enemy's graded homework:

$$f(x) = \frac{3}{4x}; f'(x) = \frac{0}{4} = 0. \quad ✗ WRONG \quad -10$$

Why was it marked wrong? How would you correct it?

**113.** One of the questions in your last calculus test was "**Question 1(a)** Give the definition of the derivative of a function $f$." Following is your answer and the grade you received:

$$nx^{n-1}. \quad ✗ WRONG \quad -10$$

Why was it marked wrong? What is the correct answer?

**114.** ▼ How would you respond to an acquaintance who says, "I finally understand what the derivative is: It is $nx^{n-1}$! Why weren't we taught that in the first place instead of the difficult way using limits?"

**115.** ▼ Sketch the graph of a function whose derivative is undefined at exactly two points but that has a tangent line at all but one point.

**116.** ▼ Sketch the graph of a function that has a tangent line at each of its points, but whose derivative is undefined at exactly two points.

## 11.2  A First Application: Marginal Analysis

In Chapter 1, we considered linear *cost functions* of the form $C(x) = mx + b$, where $C$ is the total cost, $x$ is the number of items, and $m$ and $b$ are constants. The slope $m$ is the *marginal cost*. It measures the *cost of one more item*. Notice that the derivative of $C(x) = mx + b$ is $C'(x) = m$. In other words, for a linear cost function, *the marginal cost is the derivative of the cost function.*

In general, we make the following definition.

---

### Marginal Cost

Recall from Section 1.2 that a **cost function** $C$ specifies the total cost as a function of the number of items $x$, so that $C(x)$ is the total cost of $x$ items. The **marginal cost function** is the derivative $C'$ of the cost function $C$. Thus, $C'(x)$ measures the rate of change of cost with respect to $x$.

### Units
The units of marginal cost are units of cost (dollars, say) per item.

### Interpretation
*We interpret $C'(x)$ as the approximate cost of one more item.*[*]

✳ See Example 1.

#### Quick Example

If $C(x) = 400x + 1,000$ dollars, then the marginal cost function is $C'(x) = \$400$ per item (a constant).

---

### EXAMPLE 1  Marginal Cost

Suppose that the cost in dollars to manufacture portable music players is given by

$$C(x) = 150,000 + 20x - 0.0001x^2$$

† The term $0.0001x^2$ may reflect a cost saving for high levels of production, such as a bulk discount in the cost of electronic components.

where $x$ is the number of music players manufactured.[†] Find the marginal cost function $C'$ and use it to estimate the cost of manufacturing the 50,001st music player.

**Solution** Since

$$C(x) = 150,000 + 20x - 0.0001x^2$$

the marginal cost function is

$$C'(x) = 20 - 0.0002x.$$

The units of $C'(x)$ are units of $C$ (dollars) per unit of $x$ (music players). Thus, $C'(x)$ is measured in dollars per music player.

The cost of the 50,001st music player is the amount by which the total cost would rise if we increased production from 50,000 music players to 50,001. Thus, we need to know the rate at which the total cost rises as we increase production. This rate of change is measured by the derivative, or marginal cost, which we just computed. At $x = 50,000$, we get

$$C'(50,000) = 20 - 0.0002(50,000) = \$10 \text{ per music player.}$$

In other words, we estimate that the 50,001st music player will cost approximately $10.

---

➡ **Before we go on...** In Example 1, the marginal cost is really only an *approximation* to the cost of the 50,001st music player:

$$C'(50,000) \approx \frac{C(50,001) - C(50,000)}{1} \qquad \text{Set } h = 1 \text{ in the definition of the derivative.}$$

$$= C(50,001) - C(50,000)$$

$$= \text{cost of the 50,001st music player}$$

The exact cost of the 50,001st music player is

$$C(50,001) - C(50,000) = [150,000 + 20(50,001) - 0.0001(50,001)^2]$$

$$- [150,000 + 20(50,000) - 0.0001(50,000)^2]$$

$$= \$9.9999$$

So, the marginal cost is a good approximation to the actual cost.

Graphically, we are using the tangent line to approximate the cost function near a production level of 50,000. Figure 3 shows the graph of the cost function together with the tangent line at $x = 50,000$. Notice that the tangent line is essentially indistinguishable from the graph of the function for some distance on either side of 50,000.

$$C(x) = 150,000 + 20x - 0.0001x^2$$

**Figure 3**

**Notes**

1. In general, the difference quotient $[C(x + h) - C(x)]/h$ gives the **average cost per item** to produce $h$ more items at a current production level of $x$ items. (Why?)

2. Notice that $C'(x)$ is much easier to calculate than $[C(x + h) - C(x)]/h$. (Try it.) ∎

We can extend the idea of marginal cost to include other functions we discussed in Section 1.2, like revenue and profit:

---

## Marginal Revenue and Profit

Recall that a **revenue** or **profit function** specifies the total revenue $R$ or profit $P$ as a function of the number of items $x$. The derivatives, $R'$ and $P'$, of these functions are called the **marginal revenue** and **marginal profit** functions. They measure the rate of change of revenue and profit with respect to the number of items.

### Units
The units of marginal revenue and profit are the same as those of marginal cost: dollars (or euros, pesos, etc.) per item.

### Interpretation
*We interpret $R'(x)$ and $P'(x)$ as the approximate revenue and profit from the sale of one more item.*

---

### EXAMPLE 2 Marginal Revenue and Profit

You operate an *iPod* refurbishing service (a typical refurbished iPod might have a custom color case with blinking lights and a personalized logo). The cost to refurbish $x$ iPods in a month is calculated to be

$$C(x) = 0.25x^2 + 40x + 1{,}000 \text{ dollars.}$$

You charge customers $80 per iPod for the work.

**a.** Calculate the marginal revenue and profit functions. Interpret the results.

**b.** Compute the revenue and profit, and also the marginal revenue and profit, if you have refurbished 20 units this month. Interpret the results.

**c.** For which value of $x$ is the marginal profit zero? Interpret your answer.

### Solution

**a.** We first calculate the revenue and profit functions:

$$R(x) = 80x \qquad \text{Revenue = Price} \times \text{Quantity}$$
$$P(x) = R(x) - C(x) \qquad \text{Profit = Revenue } - \text{ Cost}$$
$$= 80x - (0.25x^2 + 40x + 1{,}000)$$
$$P(x) = -0.25x^2 + 40x - 1{,}000.$$

The marginal revenue and profit functions are then the derivatives:

Marginal revenue $= R'(x) = 80$

Marginal profit $= P'(x) = -0.5x + 40.$

**Interpretation:** $R'(x)$ gives the approximate revenue from the refurbishing of one more item, and $P'(x)$ gives the approximate profit from the refurbishing of one more item. Thus, if $x$ iPods have been refurbished in a month, you will earn a revenue of $80 and make a profit of approximately $(-0.5x + 40)$ if you refurbish one more that month.

Notice that the marginal revenue is a constant, so you earn the same revenue ($80) for each iPod you refurbish. However, the marginal profit, $(-0.5x + 40)$, decreases as $x$ increases, so your additional profit is about 50¢ less for each additional iPod you refurbish.

**b.** From part (a), the revenue, profit, marginal revenue, and marginal profit functions are

$$R(x) = 80x$$
$$P(x) = -0.25x^2 + 40x - 1,000$$
$$R'(x) = 80$$
$$P'(x) = -0.5x + 40$$

Because you have refurbished $x = 20$ iPods this month, $x = 20$, so

| | |
|---|---|
| $R(20) = 80(20) = \$1,600$ | Total revenue from 20 iPods |
| $P(20) = -0.25(20)^2 + 40(20) - 1,000 = -\$300$ | Total profit from 20 iPods |
| $R'(20) = \$80$ per unit | Approximate revenue from the 21st iPod |
| $P'(20) = -0.5(20) + 40 = \$30$ per unit | Approximate profit from the 21st iPod |

**Interpretation:** If you refurbish 20 iPods in a month, you will earn a total revenue of $160 and a profit of –$300 (indicating a loss of $300). Refurbishing one more iPod that month will earn you an additional revenue of $80 and an additional profit of about $30.

**c.** The marginal profit is zero when $P'(x) = 0$:

$$-0.5x + 40 = 0$$
$$x = \frac{40}{0.5} = 80 \text{ iPods}$$

Thus, if you refurbish 80 iPods in a month, refurbishing one more will get you (approximately) zero additional profit. To understand this further, let us take a look at the graph of the profit function, shown in Figure 4. Notice that the graph is a parabola (the profit function is quadratic) with vertex at the point $x = 80$, where $P'(x) = 0$, so the profit is a maximum at this value of $x$.

$P(x) = -0.25x^2 + 40x - 1,000$

**Figure 4**

➡ **Before we go on...** In general, setting $P'(x) = 0$ and solving for $x$ will always give the exact values of $x$ for which the profit peaks as in Figure 4, assuming there is such a value. We recommend that you graph the profit function to check whether the profit is indeed a maximum at such a point. ∎

### EXAMPLE 3 Marginal Product

A consultant determines that *Precision Manufacturers'* annual profit (in dollars) is given by

$$P(n) = -200,000 + 400,000n - 4,600n^2 - 10n^3 \qquad (10 \le n \le 50),$$

where $n$ is the number of assembly-line workers it employs.

**a.** Compute $P'(n)$. $P'(n)$ is called the **marginal product** at the employment level of $n$ assembly-line workers. What are its units?

**b.** Calculate $P(20)$ and $P'(20)$, and interpret the results.

**c.** Precision Manufacturers currently employs 20 assembly-line workers and is considering laying off some of them. What advice would you give the company's management?

**Solution**

**a.** Taking the derivative gives

$$P'(n) = 400{,}000 - 9{,}200n - 30n^2.$$

The units of $P'(n)$ are profit (in dollars) per worker.

**b.** Substituting into the formula for $P(n)$, we get

$$P(20) = -200{,}000 + 400{,}000(20) - 4{,}600(20)^2 - 10(20)^3 = \$5{,}880{,}000.$$

Thus, Precision Manufacturers will make an annual profit of $5,880,000 if it employs 20 assembly-line workers. On the other hand,

$$P'(20) = 400{,}000 - 9{,}200(20) - 30(20)^2 = \$204{,}000/\text{worker}.$$

Thus, at an employment level of 20 assembly-line workers, annual profit is increasing at a rate of $204,000 per additional worker. In other words, if the company were to employ one more assembly-line worker, its annual profit would increase by approximately $204,000.

**c.** Because the marginal product is positive, profits will increase if the company increases the number of workers and will decrease if it decreases the number of workers, so your advice would be to hire additional assembly-line workers. Downsizing the assembly-line workforce would reduce its annual profits.

$P(n)$

Figure 5

➡ **Before we go on...** In Example 3, it would be interesting for Precision Manufacturers to ascertain how many additional assembly-line workers it should hire to obtain the *maximum* annual profit. Taking our cue from Example 2, we suspect that such a value of $n$ would correspond to a point where $P'(n) = 0$. Figure 5 shows the graph of $P$, and on it we see that the highest point of the graph is indeed a point where the tangent line is horizontal; that is, $P'(n) = 0$, and occurs somewhere between $n = 35$ and 40. To compute this value of $n$ more accurately, set $P'(n) = 0$ and solve for $n$:

$$P'(n) = 400{,}000 - 9{,}200n - 30n^2 = 0 \quad \text{or} \quad 40{,}000 - 920n - 3n^2 = 0.$$

We can now obtain $n$ using the quadratic formula:

$$n = \frac{-b \pm \sqrt{b^2 - 4ac}}{2a} = \frac{920 \pm \sqrt{920^2 - 4(-3)(40{,}000)}}{2(-3)}$$

$$= \frac{920 \pm \sqrt{1{,}326{,}400}}{-6} \approx -345.3 \text{ or } 38.6.$$

The only meaningful solution is the positive one, $n \approx 38.6$ workers, and we conclude that the company should employ between 38 and 39 assembly-line workers for a maximum profit. To see which gives the larger profit, 38 or 39, we check:

$$P(38) = \$7{,}808{,}880$$

while

$$P(39) = \$7{,}810{,}210.$$

This tells us that the company should employ 39 assembly-line workers for a maximum profit. Thus, instead of laying off any of its 20 assembly-line workers, the company should hire 19 additional assembly-line workers for a total of 39. ∎

## Average Cost

---

### EXAMPLE 4  Average Cost

Suppose the cost in dollars to manufacture portable music players is given by

$$C(x) = 150,000 + 20x - 0.0001x^2$$

where $x$ is the number of music players manufactured. (This is the cost equation we saw in Example 1.)

**a.** Find the average cost per music player if 50,000 music players are manufactured.

**b.** Find a formula for the average cost per music player if $x$ music players are manufactured. This function of $x$ is called the **average cost function, $\bar{C}(x)$.**

### Solution

**a.** The total cost of manufacturing 50,000 music players is given by

$$C(50,000) = 150,000 + 20(50,000) - 0.0001(50,000)^2$$
$$= \$900,000.$$

Because 50,000 music players cost a total of $900,000 to manufacture, the average cost of manufacturing one music player is this total cost divided by 50,000:

$$\bar{C}(50,000) = \frac{900,000}{50,000} = \$18.00 \text{ per music player.}$$

Thus, if 50,000 music players are manufactured, each music player costs the manufacturer an average of $18.00 to manufacture.

**b.** If we replace 50,000 by $x$, we get the general formula for the average cost of manufacturing $x$ music players:

$$\bar{C}(x) = \frac{C(x)}{x}$$

$$= \frac{1}{x}(150,000 + 20x - 0.0001x^2)$$

$$= \frac{150,000}{x} + 20 - 0.0001x. \qquad \text{Average cost function}$$

---

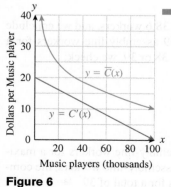

Figure 6

➡ **Before we go on...** Average cost and marginal cost convey different but related information. The average cost $\bar{C}(50,000) = \$18$ that we calculated in Example 4 is the cost per item of manufacturing the first 50,000 music players, whereas the marginal cost $C'(50,000) = \$10$ that we calculated in Example 1 gives the (approximate) cost of manufacturing the *next* music player. Thus, according to our calculations, the first 50,000 music players cost an average of $18 to manufacture, but it costs only about $10 to manufacture the next one. Note that the marginal cost at a production level of 50,000 music players is lower than the average cost. This means that the average cost to manufacture CDs is going down with increasing volume. (Think about why.)

Figure 6 shows the graphs of average and marginal cost. Notice how the decreasing marginal cost seems to pull the average cost down with it. ∎

To summarize:

---

## Average Cost

Given a cost function $C$, the **average cost** of the first $x$ items is given by

$$\bar{C}(x) = \frac{C(x)}{x}.$$

The average cost is distinct from the **marginal cost** $C'(x)$, which tells us the approximate cost of the *next* item.

### Quick Example

For the cost function $C(x) = 20x + 100$ dollars

Marginal Cost $= C'(x) = \$20$ per additional item.

Average Cost $= \bar{C}(x) = \dfrac{C(x)}{x} = \dfrac{20x + 100}{x} = \$(20 + 100/x)$ per item.

---

## 11.2  EXERCISES

▼ more advanced    ◆ challenging
▣ indicates exercises that should be solved using technology

*In Exercises 1–4, for each cost function, find the marginal cost at the given production level x, and state the units of measurement. (All costs are in dollars.)* HINT [See Example 1.]

**1.** $C(x) = 10,000 + 5x - 0.0001x^2$; $x = 1,000$

**2.** $C(x) = 20,000 + 7x - 0.00005x^2$; $x = 10,000$

**3.** $C(x) = 15,000 + 100x + \dfrac{1,000}{x}$; $x = 100$

**4.** $C(x) = 20,000 + 50x + \dfrac{10,000}{x}$; $x = 100$

*In Exercises 5 and 6, find the marginal cost, marginal revenue, and marginal profit functions, and find all values of x for which the marginal profit is zero. Interpret your answer.* HINT [See Example 2.]

**5.** $C(x) = 4x$;    $R(x) = 8x - 0.001x^2$

**6.** $C(x) = 5x^2$;    $R(x) = x^3 + 7x + 10$

**7.** ▼ A certain cost function has the following graph:

**a.** The associated marginal cost is

**(A)** increasing, then decreasing.
**(B)** decreasing, then increasing.
**(C)** always increasing.
**(D)** always decreasing.

**b.** The marginal cost is least at approximately

**(A)** $x = 0$.   **(B)** $x = 50$.   **(C)** $x = 100$.   **(D)** $x = 150$.

**c.** The cost of 50 items is

**(A)** approximately \$20, and increasing at a rate of about \$3,000 per item.
**(B)** approximately \$0.50, and increasing at a rate of about \$3,000 per item.
**(C)** approximately \$3,000, and increasing at a rate of about \$20 per item.
**(D)** approximately \$3,000, and increasing at a rate of about \$0.50 per item.

**8.** ▼ A certain cost function has the following graph:

**a.** The associated marginal cost is

(A) increasing, then decreasing.
(B) decreasing, then increasing.
(C) always increasing.
(D) always decreasing.

**b.** When $x = 100$, the marginal cost is

(A) greater than the average cost.
(B) less than the average cost.
(C) approximately equal to the average cost.

**c.** The cost of 150 items is

(A) approximately $4,400, and increasing at a rate of about $40 per item.
(B) approximately $40, and increasing at a rate of about $4,400 per item.
(C) approximately $4,400, and increasing at a rate of about $1 per item.
(D) approximately $1, and increasing at a rate of about $4,400 per item.

## APPLICATIONS

**9. *Advertising Costs*** The cost, in thousands of dollars, of airing $x$ television commercials during a Super Bowl game is given by[11]

$$C(x) = 150 + 2,500x - 0.02x^2.$$

**a.** Find the marginal cost function and use it to estimate how fast the cost is increasing when $x = 4$. Compare this with the exact cost of airing the fifth commercial. HINT [See Example 1.]

**b.** Find the average cost function $\bar{C}$, and evaluate $\bar{C}(4)$. What does the answer tell you? HINT [See Example 4.]

**10. *Marginal Cost and Average Cost*** The cost of producing $x$ teddy bears per day at the *Cuddly Companion Co.* is calculated by the company's marketing staff to be given by the formula

$$C(x) = 100 + 40x - 0.001x^2.$$

**a.** Find the marginal cost function and use it to estimate how fast the cost is going up at a production level of 100 teddy bears. Compare this with the exact cost of producing the 101st teddy bear. HINT [See Example 1.]

**b.** Find the average cost function $\bar{C}$, and evaluate $\bar{C}(100)$. What does the answer tell you? HINT [See Example 4.]

**11. *Marginal Revenue and Profit*** Your college newspaper, *The Collegiate Investigator*, sells for 90¢ per copy. The cost of producing $x$ copies of an edition is given by

$$C(x) = 70 + 0.10x + 0.001x^2 \text{ dollars.}$$

**a.** Calculate the marginal revenue and profit functions. HINT [See Example 2.]

**b.** Compute the revenue and profit, and also the marginal revenue and profit, if you have produced and sold 500 copies of the latest edition. Interpret the results.

**c.** For which value of $x$ is the marginal profit zero? Interpret your answer.

**12. *Marginal Revenue and Profit*** The Audubon Society at Enormous State University (ESU) is planning its annual fund-raising "Eatathon." The society will charge students $1.10 per serving of pasta. The society estimates that the total cost of producing $x$ servings of pasta at the event will be

$$C(x) = 350 + 0.10x + 0.002x^2 \text{ dollars.}$$

**a.** Calculate the marginal revenue and profit functions. HINT [See Example 2.]

**b.** Compute the revenue and profit, and also the marginal revenue and profit, if you have produced and sold 200 servings of pasta. Interpret the results.

**c.** For which value of $x$ is the marginal profit zero? Interpret your answer.

**13. *Marginal Profit*** Suppose $P(x)$ represents the profit in dollars on the sale of $x$ DVDs. If $P(1,000) = 3,000$ and $P'(1,000) = -3$, what do these values tell you about the profit?

**14. *Marginal Loss*** An automobile retailer calculates that its loss in dollars on the sale of *Type M* cars is given by $L(50) = 5,000$ and $L'(50) = -200$, where $L(x)$ represents the loss on the sale of $x$ Type M cars. What do these values tell you about losses?

**15. *Marginal Profit*** Your monthly profit (in dollars) from selling magazines is given by

$$P = 5x + \sqrt{x}$$

where $x$ is the number of magazines you sell in a month. If you are currently selling 50 magazines per month, find your profit and your marginal profit. Interpret your answers.

**16. *Marginal Profit*** Your monthly profit (in dollars) from your newspaper route is given by

$$P = 2n - \sqrt{n}$$

where $n$ is the number of subscribers on your route. If you currently have 100 subscribers, find your profit and your marginal profit. Interpret your answers.

**17. ▼ *Marginal Revenue: Pricing Tuna*** Assume that the demand equation for tuna in a small coastal town is given by

$$p = \frac{20,000}{q^{1.5}} \qquad (200 \le q \le 800),$$

where $p$ is the price (in dollars) per pound of tuna, and $q$ is the number of pounds of tuna that can be sold at the price $p$ in one month.[12]

---

[11]The cost of a 30-second ad during the 2010 Super Bowl game was about $2.5 million. This explains the coefficient of $x$ in the cost function. Source: http://en.wikipedia.org/wiki/Super_Bowl_advertising.

[12]Notice that here we have specified $p$ as a function of $q$, and not the other way around as we did in Section 1.2. Economists frequently specify demand curves this way.

**a.** Calculate the price that the town's fishery should charge for tuna in order to produce a demand of 400 pounds of tuna per month.

**b.** Calculate the monthly revenue $R$ as a function of the number of pounds of tuna $q$.

**c.** Calculate the revenue and marginal revenue (derivative of the revenue with respect to $q$) at a demand level of 400 pounds per month, and interpret the results.

**d.** If the town fishery's monthly tuna catch amounted to 400 pounds of tuna, and the price is at the level in part (a), would you recommend that the fishery raise or lower the price of tuna in order to increase its revenue?

**18.** ▼ *Marginal Revenue: Pricing Tuna* Repeat Exercise 17, assuming a demand equation of

$$p = \frac{60}{q^{0.5}} \qquad (200 \le q \le 800).$$

**19.** *Marginal Product* A car wash firm calculates that its daily profit (in dollars) depends on the number $n$ of workers it employs according to the formula

$$P = 400n - 0.5n^2.$$

Calculate the marginal product at an employment level of 50 workers, and interpret the result. HINT [See Example 3.]

**20.** *Marginal Product* Repeat the preceding exercise using the formula

$$P = -100n + 25n^2 - 0.005n^4.$$

HINT [See Example 3.]

**21.** *Average and Marginal Cost* The daily cost to manufacture generic trinkets for gullible tourists is given by the cost function

$$C(x) = -0.001x^2 + 0.3x + 500 \text{ dollars}$$

where $x$ is the number of trinkets.

**a.** As $x$ increases, the marginal cost
 **(A)** increases.  **(B)** decreases.  **(C)** increases, then decreases.  **(D)** decreases, then increases.

**b.** As $x$ increases, the average cost
 **(A)** increases.  **(B)** decreases.  **(C)** increases, then decreases.  **(D)** decreases, then increases.

**c.** The marginal cost is
 **(A)** greater than  **(B)** equal to  **(C)** less than
 the average cost when $x = 100$. HINT [See Example 4.]

**22.** *Average and Marginal Cost* Repeat Exercise 21, using the following cost function for imitation oil paintings ($x$ is the number of "oil paintings" manufactured):

$$C(x) = 0.1x^2 - 3.5x + 500 \text{ dollars}.$$

HINT [See Example 4.]

**23.** *Advertising Cost* Your company is planning to air a number of television commercials during the ABC Television Network's presentation of the Academy Awards. ABC is charg- ing your company $1.6 million per 30-second spot.[13] Addi- tional fixed costs (development and personnel costs) amount to $500,000, and the network has agreed to provide a discount of $10,000\sqrt{x}$ for $x$ television spots.

**a.** Write down the cost function $C$, marginal cost function $C'$, and average cost function $\bar{C}$.

**b.** Compute $C'(3)$ and $\bar{C}(3)$. (Round all answers to three significant digits.) Use these two answers to say whether the average cost is increasing or decreasing as $x$ increases.

**24.** *Housing Costs* The cost $C$ of building a house is related to the number $k$ of carpenters used and the number $x$ of electri- cians used by the formula[14]

$$C = 15,000 + 50k^2 + 60x^2.$$

**a.** Assuming that 10 carpenters are currently being used, find the cost function $C$, marginal cost function $C'$, and average cost function $\bar{C}$, all as functions of $x$.

**b.** Use the functions you obtained in part (a) to compute $C'(15)$ and $\bar{C}(15)$. Use these two answers to say whether the average cost is increasing or decreasing as the number of electricians increases.

**25.** ▼ *Emission Control* The cost of controlling emissions at a firm rises rapidly as the amount of emissions reduced in- creases. Here is a possible model:

$$C(q) = 4,000 + 100q^2$$

where $q$ is the reduction in emissions (in pounds of pollutant per day) and $C$ is the daily cost (in dollars) of this reduction.

**a.** If a firm is currently reducing its emissions by 10 pounds each day, what is the marginal cost of reducing emissions further?

**b.** Government clean-air subsidies to the firm are based on the formula

$$S(q) = 500q$$

where $q$ is again the reduction in emissions (in pounds per day) and $S$ is the subsidy (in dollars). At what reduc- tion level does the marginal cost surpass the marginal subsidy?

**c.** Calculate the net cost function, $N(q) = C(q) - S(q)$, given the cost function and subsidy above, and find the value of $q$ that gives the lowest net cost. What is this lowest net cost? Compare your answer to that for part (b) and comment on what you find.

**26.** ▼ *Taxation Schemes* In order to raise revenues during the recent recession, the governor of your state proposed the following taxation formula:

$$T(i) = 0.001i^{0.5},$$

---

[13]ABC charged an average of $1.6 million for a 30-second spot during the 2005 Academy Awards presentation. Source: CNN/Reuters, www.cnn.com, February 9, 2005.

[14]Based on an exercise in *Introduction to Mathematical Economics* by A. L. Ostrosky, Jr., and J. V. Koch (Waveland Press, Prospect Heights, Illinois, 1979).

where $i$ represents total annual income earned by an individual in dollars and $T(i)$ is the income tax rate as a percentage of total annual income. (Thus, for example, an income of $50,000 per year would be taxed at about 22%, while an income of double that amount would be taxed at about 32%.)[15]

a. Calculate the after-tax (net) income $N(i)$ an individual can expect to earn as a function of income $i$.

b. Calculate an individual's marginal after-tax income at income levels of $100,000 and $500,000.

c. At what income does an individual's marginal after-tax income become negative? What is the after-tax income at that level, and what happens at higher income levels?

d. What do you suspect is the most anyone can earn after taxes? (See NOTE at the bottom of this page.)

27. ▼ *Fuel Economy* Your Porsche's gas mileage (in miles per gallon) is given as a function $M(x)$ of speed $x$ in miles per hour. It is found that

$$M'(x) = \frac{3{,}600x^{-2} - 1}{(3{,}600x^{-1} + x)^2}.$$

Estimate $M'(10)$, $M'(60)$, and $M'(70)$. What do the answers tell you about your car?

28. ▼ *Marginal Revenue* The estimated marginal revenue for sales of ESU soccer team T-shirts is given by

$$R'(p) = \frac{(8 - 2p)e^{-p^2 + 8p}}{10{,}000{,}000}$$

where $p$ is the price (in dollars) that the soccer players charge for each shirt. Estimate $R'(3)$, $R'(4)$, and $R'(5)$. What do the answers tell you?

29. ◆ *Marginal Cost* (from the GRE Economics Test) In a multiplant firm in which the different plants have different and continuous cost schedules, if costs of production for a given output level are to be minimized, which of the following is essential?

(A) Marginal costs must equal marginal revenue.

(B) Average variable costs must be the same in all plants.

(C) Marginal costs must be the same in all plants.

(D) Total costs must be the same in all plants.

(E) Output per worker per hour must be the same in all plants.

30. ◆ *Study Time* (from the GRE economics test) A student has a fixed number of hours to devote to study and is certain of the relationship between hours of study and the final grade for each course. Grades are given on a numerical scale (0 to 100), and each course is counted equally in computing the grade average. In order to maximize his or her grade average, the student should allocate these hours to different courses so that

(A) the grade in each course is the same.

(B) the marginal product of an hour's study (in terms of final grade) in each course is zero.

(C) the marginal product of an hour's study (in terms of final grade) in each course is equal, although not necessarily equal to zero.

(D) the average product of an hour's study (in terms of final grade) in each course is equal.

(E) the number of hours spent in study for each course is equal.

31. ◆ *Marginal Product* (from the GRE Economics Test) Assume that the marginal product of an additional senior professor is 50% higher than the marginal product of an additional junior professor and that junior professors are paid one half the amount that senior professors receive. With a fixed overall budget, a university that wishes to maximize its quantity of output from professors should do which of the following?

(A) Hire equal numbers of senior professors and junior professors.

(B) Hire more senior professors and junior professors.

(C) Hire more senior professors and discharge junior professors.

(D) Discharge senior professors and hire more junior professors.

(E) Discharge all senior professors and half of the junior professors.

32. ◆ *Marginal Product* (based on a question from the GRE Economics Test) Assume that the marginal product of an additional senior professor is twice the marginal product of an additional junior professor and that junior professors are paid two thirds the amount that senior professors receive. With a fixed overall budget, a university that wishes to maximize its quantity of output from professors should do which of the following?

(A) Hire equal numbers of senior professors and junior professors.

(B) Hire more senior professors and junior professors.

(C) Hire more senior professors and discharge junior professors.

(D) Discharge senior professors and hire more junior professors.

(E) Discharge all senior professors and half of the junior professors.

## COMMUNICATION AND REASONING EXERCISES

33. The marginal cost of producing the 1,001st item is

(A) equal to
(B) approximately equal to
(C) always slightly greater than
(D) always slightly less than

the actual cost of producing the 1,001st item.

34. For the cost function $C(x) = mx + b$, the marginal cost of producing the 1,001st item is,

(A) equal to
(B) approximately equal to
(C) always slightly greater than
(D) always slightly less than

the actual cost of producing the 1,001st item.

---

[15]This model has the following interesting feature: An income of $1 million per year would be taxed at 100%, leaving the individual penniless!

35. What is a cost function? Carefully explain the difference between *average cost* and *marginal cost* in terms of **(a)** their mathematical definition, **(b)** graphs, and **(c)** interpretation.

36. The cost function for your grand piano manufacturing plant has the property that $\bar{C}(1,000) = \$3,000$ per unit and $C'(1,000) = \$2,500$ per unit. Will the average cost increase or decrease if your company manufactures a slightly larger number of pianos? Explain your reasoning.

37. Give an example of a cost function for which the marginal cost function is the same as the average cost function.

38. Give an example of a cost function for which the marginal cost function is always less than the average cost function.

39. If the average cost to manufacture one grand piano increases as the production level increases, which is greater, the marginal cost or the average cost?

40. If your analysis of a manufacturing company yielded positive marginal profit but negative profit at the company's current production levels, what would you advise the company to do?

41. ▼ If the marginal cost is decreasing, is the average cost necessarily decreasing? Explain.

42. ▼ If the average cost is decreasing, is the marginal cost necessarily decreasing? Explain.

43. ◆ If a company's marginal average cost is zero at the current production level, positive for a slightly higher production level, and negative for a slightly lower production level, what should you advise the company to do?

44. ◆ The **acceleration** of cost is defined as the derivative of the marginal cost function: that is, the derivative of the derivative—or *second derivative*—of the cost function. What are the units of acceleration of cost, and how does one interpret this measure?

# 11.3 The Product and Quotient Rules

We know how to find the derivatives of functions that are sums of powers, such as polynomials. In general, if a function is a sum or difference of functions whose derivatives we know, then we know how to find its derivative. But what about *products and quotients* of functions whose derivatives we know? For instance, how do we calculate the derivative of something like $x^2/(x+1)$? The derivative of $x^2/(x+1)$ is not, as one might suspect, $2x/1 = 2x$. That calculation is based on an assumption that the derivative of a quotient is the quotient of the derivatives. But it is easy to see that this assumption is false: For instance, the derivative of $1/x$ is not $0/1 = 0$, but $-1/x^2$. Similarly, the derivative of a product is not the product of the derivatives: For instance, the derivative of $x = 1 \cdot x$ is not $0 \cdot 1 = 0$, but 1.

To identify the correct method of computing the derivatives of products and quotients, let's look at a simple example. We know that the daily revenue resulting from the sale of $q$ items per day at a price of $p$ dollars per item is given by the product, $R = pq$ dollars. Suppose you are currently selling wall posters on campus. At this time your daily sales are 50 posters, and sales are increasing at a rate of 4 per day. Furthermore, you are currently charging \$10 per poster, and you are also raising the price at a rate of \$2 per day. Let's use this information to estimate how fast your daily revenue is increasing. In other words, let us estimate the rate of change, $dR/dt$, of the revenue $R$.

There are two contributions to the rate of change of daily revenue: the increase in daily sales and the increase in the unit price. We have

$\dfrac{dR}{dt}$ due to increasing price:   \$2 per day $\times$ 50 posters $= \$100$ per day

$\dfrac{dR}{dt}$ due to increasing sales:   \$10 per poster $\times$ 4 posters per day $= \$40$ per day.

Thus, we estimate the daily revenue to be increasing at a rate of $100 + $40 = $140 per day. Let us translate what we have said into symbols:

$$\frac{dR}{dt} \text{ due to increasing price: } \frac{dp}{dt} \times q$$

$$\frac{dR}{dt} \text{ due to increasing sales: } p \times \frac{dq}{dt}.$$

Thus, the rate of change of revenue is given by

$$\frac{dR}{dt} = \frac{dp}{dt}q + p\frac{dq}{dt}.$$

Because $R = pq$, we have discovered the following rule for differentiating a product:

$$\frac{d}{dt}(pq) = \frac{dp}{dt}q + p\frac{dq}{dt}.$$

The derivative of a product is the derivative of the first times the second, plus the first times the derivative of the second.

This rule and a similar rule for differentiating quotients are given next, and also a discussion of how these results are proved rigorously.

## Product Rule

If $f$ and $g$ are differentiable functions of $x$, then so is their product $fg$, and

$$\frac{d}{dx}[f(x)g(x)] = f'(x)g(x) + f(x)g'(x).$$

### Product Rule in Words
*The derivative of a product is the derivative of the first times the second, plus the first times the derivative of the second.*

### Quick Example

Let $f(x) = x^2$ and $g(x) = 3x - 1$. Because $f$ and $g$ are both differentiable functions of $x$, so is their product $fg$, and its derivative is

$$\frac{d}{dx}[x^2(3x-1)] = 2x \cdot (3x-1) + x^2 \cdot (3).$$

| Derivative of first | Second | First | Derivative of second |

## Quotient Rule

If $f$ and $g$ are differentiable functions of $x$, then so is their quotient $f/g$, and

$$\frac{d}{dx}\left(\frac{f(x)}{g(x)}\right) = \frac{f'(x)g(x) - f(x)g'(x)}{[g(x)]^2}.$$

provided $g(x) \neq 0$.*

✱ If $g(x)$ is zero, then the quotient $f(x)/g(x)$ is not defined in the first place.

### Quotient Rule in Words
*The derivative of a quotient is the derivative of the top times the bottom, minus the top times the derivative of the bottom, all over the bottom squared.*

**Quick Example**

Let $f(x) = x^3$ and $g(x) = x^2 - 1$. Because $f$ and $g$ are both differentiable functions of $x$, so is their quotient $f/g$, and its derivative is

Derivative of top    Bottom   Top   Derivative of bottom
↓        ↓    ↓    ↓

$$\frac{d}{dx}\left(\frac{x^3}{x^2 - 1}\right) = \frac{3x^2(x^2 - 1) - x^3 \cdot 2x}{(x^2 - 1)^2},$$

↑
Bottom squared

provided $x \neq 1$ or $-1$.

## Notes

1. Don't try to remember the rules by the symbols we have used, but remember them in words. (The slogans are easy to remember, even if the terms are not precise.)

2. One more time: *The derivative of a product is* NOT *the product of the derivatives, and the derivative of a quotient is* NOT *the quotient of the derivatives.* To find the derivative of a product, you must use the product rule, and to find the derivative of a quotient, you must use the quotient rule.* ■

✳ Leibniz made this mistake at first, too, so you would be in good company if you forgot to use the product or quotient rule.

---

**Q**: *Wait a minute! The expression $2x^3$ is a product, and we already know that its derivative is $6x^2$. Where did we use the product rule?*

**A**: To differentiate functions such as $2x^3$, we have used the rule from Section 11.1:

*The derivative of c times a function is c times the derivative of the function.*

However, the product rule gives us the same result:

Derivative of first  Second    First   Derivative of second
↓    ↓        ↓    ↓

$$\frac{d}{dx}(2x^3) = (0)(x^3) \quad + \quad (2)(3x^2) = 6x^2 \qquad \text{Product rule}$$

$$\frac{d}{dx}(2x^3) = (2)(3x^2) = 6x^2 \qquad \begin{array}{l}\text{Derivative of a constant}\\\text{times a function}\end{array}$$

---

We do not recommend that you use the product rule to differentiate functions such as $2x^3$; continue to use the simpler rule when one of the factors is a constant.

## Derivation of the Product Rule

Before we look at more examples of using the product and quotient rules, let's see why the product rule is true. To calculate the derivative of the product $f(x)g(x)$ of two differentiable functions, we go back to the definition of the derivative:

$$\frac{d}{dx}[f(x)g(x)] = \lim_{h \to 0} \frac{f(x + h)g(x + h) - f(x)g(x)}{h}.$$

We now rewrite this expression so that we can evaluate the limit: Notice that the numerator reflects a simultaneous change in $f$ [from $f(x)$ to $f(x + h)$] and $g$ [from

$g(x)$ to $g(x + h)$]. To separate the two effects, we add and subtract a quantity in the numerator that reflects a change in only one of the functions:

$$\frac{d}{dx}[f(x)g(x)] = \lim_{h \to 0} \frac{f(x + h)g(x + h) - f(x)g(x)}{h}$$

$$= \lim_{h \to 0} \frac{f(x + h)g(x + h) - f(x)g(x + h) + f(x)g(x + h) - f(x)g(x)}{h}$$

We subtracted and added the quantity*
$f(x)g(x + h)$.

$$= \lim_{h \to 0} \frac{[f(x + h) - f(x)] \, g(x + h) + f(x)[g(x + h) - g(x)]}{h}$$

Common factors

$$= \lim_{h \to 0} \left( \frac{f(x + h) - f(x)}{h} \right) g(x + h) + \lim_{h \to 0} f(x) \left( \frac{g(x + h) - g(x)}{h} \right)$$

Limit of sum

$$= \lim_{h \to 0} \left( \frac{f(x + h) - f(x)}{h} \right) \lim_{h \to 0} g(x + h) + \lim_{h \to 0} f(x) \lim_{h \to 0} \left( \frac{g(x + h) - g(x)}{h} \right)$$

Limit of product

Now we already know the following four limits:

**\*** Adding an appropriate form of zero is an age-old mathematical ploy.

$$\lim_{h \to 0} \frac{f(x + h) - f(x)}{h} = f'(x)$$   Definition of derivative of $f$; $f$ is differentiable.

$$\lim_{h \to 0} \frac{g(x + h) - g(x)}{h} = g'(x)$$   Definition of derivative of $g$; $g$ is differentiable.

$$\lim_{h \to 0} g(x + h) = g(x)$$   If $g$ is differentiable, it must be continuous.†

**†** For a proof of the fact that, if $g$ is differentiable, it must be continuous, go to the Website and follow the path

$$\lim_{h \to 0} f(x) = f(x)$$   Limit of a constant

Putting these limits into the one we're calculating, we get

Everything
→ Chapter 11
→ Continuity and
   Differentiability

$$\frac{d}{dx}[f(x)g(x)] = f'(x)g(x) + f(x)g'(x)$$

which is the product rule.

### EXAMPLE 1   Using the Product Rule

Compute the following derivatives.

**a.** $\dfrac{d}{dx}[(x^{3.2} + 1)(1 - x)]$    Simplify the answer.

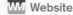 **Website**
www.WanerMath.com
The quotient rule can be proved in a very similar way. Go to the Website and follow the path

**b.** $\dfrac{d}{dx}[(x + 1)(x^2 + 1)(x^3 + 1)]$    Do not expand the answer.

**c.** $\dfrac{d}{dx}\left[ \dfrac{x|x|}{2} \right]$

Everything
→ Chapter 11
→ Proof of Quotient Rule

### Solution

**a.** We can do the calculation in two ways.

| | Derivative of first | Second | First | Derivative of second |
|---|---|---|---|---|
| | ↓ | ↓ | ↓ | ↓ |

**Using the Product Rule:**   $\dfrac{d}{dx}[(x^{3.2} + 1)(1 - x)] = (3.2x^{2.2})(1 - x) + (x^{3.2} + 1)(-1)$

$$= 3.2x^{2.2} - 3.2x^{3.2} - x^{3.2} - 1$$   Expand the answer.

$$= -4.2x^{3.2} + 3.2x^{2.2} - 1$$

**Not Using the Product Rule:** First, expand the given expression.

$$(x^{3.2} + 1)(1 - x) = -x^{4.2} + x^{3.2} - x + 1$$

Thus,

$$\frac{d}{dx}[(x^{3.2} + 1)(1 - x)] = \frac{d}{dx}(-x^{4.2} + x^{3.2} - x + 1)$$
$$= -4.2x^{3.2} + 3.2x^{2.2} - 1$$

In this example the product rule saves us little or no work, but in later sections we shall see examples that can be done in no other way. Learn how to use the product rule now!

**b.** Here we have a product of *three* functions, not just two. We can find the derivative by using the product rule twice:

$$\frac{d}{dx}[(x + 1)(x^2 + 1)(x^3 + 1)]$$
$$= \frac{d}{dx}(x + 1) \cdot [(x^2 + 1)(x^3 + 1)] + (x + 1) \cdot \frac{d}{dx}[(x^2 + 1)(x^3 + 1)]$$
$$= (1)(x^2 + 1)(x^3 + 1) + (x + 1)[(2x)(x^3 + 1) + (x^2 + 1)(3x^2)]$$
$$= (1)(x^2 + 1)(x^3 + 1) + (x + 1)(2x)(x^3 + 1) + (x + 1)(x^2 + 1)(3x^2)$$

We can see here a more general product rule:

$$(fgh)' = f'gh + fg'h + fgh'$$

Notice that every factor has a chance to contribute to the rate of change of the product. There are similar formulas for products of four or more functions.

**c.** First write $\dfrac{x|x|}{2}$ as $\dfrac{1}{2}x|x|$.

$$\frac{d}{dx}\left[\frac{1}{2}x|x|\right] = \frac{1}{2}\frac{d}{dx}[x|x|] \qquad \text{Constant multiple rule}$$
$$= \frac{1}{2}\left((1) \cdot |x| + x \cdot \frac{|x|}{x}\right) \qquad \text{Recall that } \frac{d}{dx}|x| = \frac{|x|}{x}.$$
$$= \frac{1}{2}(|x| + |x|) \qquad \text{Cancel the } x.$$
$$= \frac{1}{2}(2|x|) = |x| \qquad \text{See the note.}*$$

\* Notice that we have found a function whose derivative is $|x|$; namely $x|x|/2$. Notice also that the derivation we gave assumes that $x \neq 0$ because we divided by $x$ in the third step. However, one can verify, using the definition of the derivative as a limit, that $x|x|/2$ is differentiable at $x = 0$ as well, and that its derivative at $x = 0$ is 0, implying that the formula $\dfrac{d}{dx}(x|x|/2) = |x|$ is valid for all values of $x$, including 0.

### EXAMPLE 2 Using the Quotient Rule

Compute the derivatives **a.** $\dfrac{d}{dx}\left[\dfrac{1 - 3.2x^{-0.1}}{x + 1}\right]$   **b.** $\dfrac{d}{dx}\left[\dfrac{(x + 1)(x + 2)}{x - 1}\right]$

**Solution**

Derivative of top   Bottom          Top   Derivative of bottom
        ↓              ↓              ↓          ↓

**a.** $\dfrac{d}{dx}\left[\dfrac{1 - 3.2x^{-0.1}}{x + 1}\right] = \dfrac{(0.32x^{-1.1})(x + 1) - (1 - 3.2x^{-0.1})(1)}{(x + 1)^2}$
                                                            ↑
                                                    Bottom squared

$$= \frac{0.32x^{-0.1} + 0.32x^{-1.1} - 1 + 3.2x^{-0.1}}{(x+1)^2} \qquad \text{Expand the numerator.}$$

$$= \frac{3.52x^{-0.1} + 0.32x^{-1.1} - 1}{(x+1)^2}$$

**b.** Here we have both a product and a quotient. Which rule do we use, the product or the quotient rule? Here is a way to decide. Think about how we would calculate, step by step, the value of $(x+1)(x+2)/(x-1)$ for a specific value of $x$—say $x = 11$. Here is how we would probably do it:

**1.** Calculate $(x+1)(x+2) = (11+1)(11+2) = 156$.

**2.** Calculate $x - 1 = 11 - 1 = 10$.

**3.** Divide 156 by 10 to get 15.6.

Now ask: *What was the last operation we performed?* The last operation we performed was division, so we can regard the whole expression as a *quotient*—that is, as $(x+1)(x+2)$ *divided by* $(x-1)$. Therefore, we should use the quotient rule.

The first thing the quotient rule tells us to do is to take the derivative of the numerator. Now, the numerator is a product, so we must use the product rule to take its derivative. Here is the calculation:

$$\frac{d}{dx}\left[\frac{(x+1)(x+2)}{x-1}\right] = \frac{\overbrace{[(1)(x+2) + (x+1)(1)]}^{\text{Derivative of top}}\overbrace{(x-1)}^{\text{Bottom}} - \overbrace{[(x+1)(x+2)]}^{\text{Top}}\overbrace{(1)}^{\substack{\text{Derivative} \\ \text{of bottom}}}}{\underset{\text{Bottom squared}}{(x-1)^2}}$$

$$= \frac{(2x+3)(x-1) - (x+1)(x+2)}{(x-1)^2}$$

$$= \frac{x^2 - 2x - 5}{(x-1)^2}$$

What is important is to determine the *order of operations* and, in particular, to determine the last operation to be performed. Pretending to do an actual calculation reminds us of the order of operations; we call this technique the **calculation thought experiment**.

➡ **Before we go on...** We used the quotient rule in Example 2 because the function was a quotient; we used the product rule to calculate the derivative of the numerator because the numerator was a product. Get used to this: Differentiation rules usually must be used in combination.

Here is another way we could have done this problem: Our calculation thought experiment could have taken the following form.

**1.** Calculate $(x+1)/(x-1) = (11+1)/(11-1) = 1.2$.

**2.** Calculate $x + 2 = 11 + 2 = 13$.

**3.** Multiply 1.2 by 13 to get 15.6.

We would have then regarded the expression as a *product*—the product of the factors $(x+1)/(x-1)$ and $(x+2)$—and used the product rule instead. We can't escape the quotient rule, however: We need to use it to take the derivative of the first factor, $(x+1)/(x-1)$. Try this approach for practice and check that you get the same answer. ∎

## Calculation Thought Experiment

The **calculation thought experiment** is a technique to determine whether to treat an algebraic expression as a product, quotient, sum, or difference. Given an expression, consider the steps you would use in computing its value. If the last operation is multiplication, treat the expression as a product; if the last operation is division, treat the expression as a quotient; and so on.

### Quick Examples

1. $(3x^2 - 4)(2x + 1)$ can be computed by first calculating the expressions in parentheses and then multiplying. Because the last step is multiplication, we can treat the expression as a product.

2. $\dfrac{2x - 1}{x}$ can be computed by first calculating the numerator and denominator and then dividing one by the other. Because the last step is division, we can treat the expression as a quotient.

3. $x^2 + (4x - 1)(x + 2)$ can be computed by first calculating $x^2$, then calculating the product $(4x - 1)(x + 2)$, and finally adding the two answers. Thus, we can treat the expression as a sum.

4. $(3x^2 - 1)^5$ can be computed by first calculating the expression in parentheses and then raising the answer to the fifth power. Thus, we can treat the expression as a power. (We shall see how to differentiate powers of expressions in Section 11.4.)

5. The expression $(x + 1)(x + 2)/(x - 1)$ can be treated as either a quotient or a product: We can write it as a quotient: $\dfrac{(x + 1)(x + 2)}{x - 1}$ or as a product: $(x + 1)\left(\dfrac{x + 2}{x - 1}\right)$. (See Example 2(b).)

---

### EXAMPLE 3 Using the Calculation Thought Experiment

Find $\dfrac{d}{dx}\left[6x^2 + 5\left(\dfrac{x}{x - 1}\right)\right]$.

**Solution** The calculation thought experiment tells us that the expression we are asked to differentiate can be treated as a *sum*. Because the derivative of a sum is the sum of the derivatives, we get

$$\frac{d}{dx}\left[6x^2 + 5\left(\frac{x}{x - 1}\right)\right] = \frac{d}{dx}(6x^2) + \frac{d}{dx}\left[5\left(\frac{x}{x - 1}\right)\right].$$

In other words, we must take the derivatives of $6x^2$ and $5\left(\dfrac{x}{x-1}\right)$ separately and then add the answers. The derivative of $6x^2$ is $12x$. There are two ways of taking the derivative of $5\left(\dfrac{x}{x-1}\right)$: We could either first multiply the expression $\left(\dfrac{x}{x-1}\right)$ by 5 to get $\left(\dfrac{5x}{x-1}\right)$ and then take its derivative using the quotient rule, or we could pull the 5 out, as we do next.

$$\frac{d}{dx}\left[6x^2 + 5\left(\frac{x}{x-1}\right)\right] = \frac{d}{dx}(6x^2) + \frac{d}{dx}\left[5\left(\frac{x}{x-1}\right)\right] \qquad \text{Derivative of sum}$$

$$= 12x + 5\frac{d}{dx}\left(\frac{x}{x-1}\right) \qquad \text{Constant} \times \text{Function}$$

$$= 12x + 5\left(\frac{(1)(x-1) - (x)(1)}{(x-1)^2}\right) \qquad \text{Quotient rule}$$

$$= 12x + 5\left(\frac{-1}{(x-1)^2}\right)$$

$$= 12x - \frac{5}{(x-1)^2}$$

## APPLICATIONS

In the next example, we return to a scenario similar to the one discussed at the start of this section.

### EXAMPLE 4 Applying the Product and Quotient Rules: Revenue and Average Cost

Sales of your newly launched miniature wall posters for college dorms, *iMiniPosters,* are really taking off. (Those old-fashioned large wall posters no longer fit in today's "downsized" college dorm rooms.) Monthly sales to students at the start of this year were 1,500 iMiniPosters, and since that time, sales have been increasing by 300 posters each month, even though the price you charge has also been going up.

**a.** The price you charge for iMiniPosters is given by

$$p(t) = 10 + 0.05t^2 \text{ dollars per poster,}$$

where $t$ is time in months since the start of January of this year. Find a formula for the monthly revenue, and then compute its rate of change at the beginning of March.

**b.** The number of students who purchase iMiniPosters in a month is given by

$$n(t) = 800 + 0.2t,$$

where $t$ is as in part (a). Find a formula for the average number of posters each student buys, and hence estimate the rate at which this number was growing at the beginning of March.

### Solution

**a.** To compute monthly revenue as a function of time $t$, we use

$$R(t) = p(t)q(t). \qquad \text{Revenue} = \text{Price} \times \text{Quantity}$$

We already have a formula for $p(t)$. The function $q(t)$ measures sales, which were 1,500 posters/month at time $t = 0$, and were rising by 300 per month:

$$q(t) = 1,500 + 300t.$$

Therefore, the formula for revenue is

$$R(t) = p(t)q(t)$$
$$R(t) = (10 + 0.05t^2)(1,500 + 300t).$$

Rather than expand this expression, we shall leave it as a product so that we can use the product rule in computing its rate of change:

$$R'(t) = p'(t)q(t) + p(t)q'(t)$$
$$= [0.10t][1,500 + 300t] + [10 + 0.05t^2][300].$$

Because the beginning of March corresponds to $t = 2$, we have

$$R'(2) = [0.10(2)][1,500 + 300(2)] + [10 + 0.05(2)^2][300]$$
$$= (0.2)(2,100) + (10.2)(300) = \$3,480 \text{ per month.}$$

Therefore, your monthly revenue was increasing at a rate of $3,480 per month at the beginning of March.

**b.** The average number of posters sold to each student is

$$k(t) = \frac{\text{Number of posters}}{\text{Number of students}} = \frac{q(t)}{n(t)} = \frac{1,500 + 300t}{800 + 0.2t}.$$

The rate of change of $k(t)$ is computed with the quotient rule:

$$k'(t) = \frac{q'(t)n(t) - q(t)n'(t)}{n(t)^2}$$
$$= \frac{(300)(800 + 0.2t) - (1,500 + 300t)(0.2)}{(800 + 0.2t)^2}$$

so that

$$k'(2) = \frac{(300)[800 + 0.2(2)] - [1,500 + 300(2)](0.2)}{[800 + 0.2(2)]^2}$$
$$= \frac{(300)(800.4) - (2,100)(0.2)}{800.4^2} \approx 0.37 \text{ posters/student per month.}$$

Therefore, the average number of posters sold to each student was increasing at a rate of about 0.37 posters/student per month.

# 11.3  EXERCISES

▼ more advanced   ◆ challenging

◧ indicates exercises that should be solved using technology

*In Exercises 1–12:*

**a.** *Calculate the derivative of the given function without using either the product or quotient rule.*

**b.** *Use the product or quotient rule to find the derivative. Check that you obtain the same answer.* HINT *[See Quick Examples on pages 810–811.]*

**1.** $f(x) = 3x$

**2.** $f(x) = 2x^2$

**3.** $g(x) = x \cdot x^2$

**4.** $g(x) = x \cdot x$

**5.** $h(x) = x(x + 3)$

**6.** $h(x) = x(1 + 2x)$

**7.** $r(x) = 100x^{2.1}$

**8.** $r(x) = 0.2x^{-1}$

**9.** $s(x) = \dfrac{2}{x}$

**10.** $t(x) = \dfrac{x}{3}$

**11.** $u(x) = \dfrac{x^2}{3}$

**12.** $s(x) = \dfrac{3}{x^2}$

*Calculate* $\dfrac{dy}{dx}$ *in Exercises 13–28. Simplify your answer.*
HINT [See Examples 1 and 2.]

**13.** $y = 3x(4x^2 - 1)$       **14.** $y = 3x^2(2x + 1)$

**15.** $y = x^3(1 - x^2)$       **16.** $y = x^5(1 - x)$

**17.** $y = (2x + 3)^2$       **18.** $y = (4x - 1)^2$

**19.** $y = \dfrac{4x}{5x - 2}$       **20.** $y = \dfrac{3x}{-3x + 2}$

**21.** $y = \dfrac{2x + 4}{3x - 1}$       **22.** $y = \dfrac{3x - 9}{2x + 4}$

**23.** $y = \dfrac{|x|}{x}$       **24.** $y = \dfrac{x}{|x|}$

**25.** $y = \dfrac{|x|}{x^2}$       **26.** $y = \dfrac{x^2}{|x|}$

**27.** $y = x\sqrt{x}$       **28.** $y = x^2\sqrt{x}$

*Calculate* $\dfrac{dy}{dx}$ *in Exercises 29–56. You need not expand your answers.* HINT [See Examples 1 and 2.]

**29.** $y = (x + 1)(x^2 - 1)$

**30.** $y = (4x^2 + x)(x - x^2)$

**31.** $y = (2x^{0.5} + 4x - 5)(x - x^{-1})$

**32.** $y = (x^{0.7} - 4x - 5)(x^{-1} + x^{-2})$

**33.** $y = (2x^2 - 4x + 1)^2$

**34.** $y = (2x^{0.5} - x^2)^2$

**35.** $y = \left(\dfrac{x}{3.2} + \dfrac{3.2}{x}\right)(x^2 + 1)$

**36.** $y = \left(\dfrac{x^{2.1}}{7} + \dfrac{2}{x^{2.1}}\right)(7x - 1)$

**37.** $y = x^2(2x + 3)(7x + 2)$ HINT [See Example 1(b).]

**38.** $y = x(x^2 - 3)(2x^2 + 1)$ HINT [See Example 1(b).]

**39.** $y = (5.3x - 1)(1 - x^{2.1})(x^{-2.3} - 3.4)$

**40.** $y = (1.1x + 4)(x^{2.1} - x)(3.4 - x^{-2.1})$

**41.** ▼$y = (\sqrt{x} + 1)\left(\sqrt{x} + \dfrac{1}{x^2}\right)$

**42.** ▼$y = (4x^2 - \sqrt{x})\left(\sqrt{x} - \dfrac{2}{x^2}\right)$

**43.** $y = \dfrac{2x^2 + 4x + 1}{3x - 1}$       **44.** $y = \dfrac{3x^2 - 9x + 11}{2x + 4}$

**45.** $y = \dfrac{x^2 - 4x + 1}{x^2 + x + 1}$       **46.** $y = \dfrac{x^2 + 9x - 1}{x^2 + 2x - 1}$

**47.** $y = \dfrac{x^{0.23} - 5.7x}{1 - x^{-2.9}}$       **48.** $y = \dfrac{8.43x^{-0.1} - 0.5x^{-1}}{3.2 + x^{2.9}}$

**49.** ▼$y = \dfrac{\sqrt{x} + 1}{\sqrt{x} - 1}$       **50.** ▼$y = \dfrac{\sqrt{x} - 1}{\sqrt{x} + 1}$

**51.** ▼$y = \dfrac{\left(\dfrac{1}{x} + \dfrac{1}{x^2}\right)}{x + x^2}$       **52.** ▼$y = \dfrac{\left(1 - \dfrac{1}{x^2}\right)}{x^2 - 1}$

**53.** $y = \dfrac{(x + 3)(x + 1)}{3x - 1}$ HINT [See Example 2(b).]

**54.** $y = \dfrac{x}{(x - 5)(x - 4)}$ HINT [See Example 2(b).]

**55.** $y = \dfrac{(x + 3)(x + 1)(x + 2)}{3x - 1}$

**56.** $y = \dfrac{3x - 1}{(x - 5)(x - 4)(x - 1)}$

*In Exercises 57–62, compute the indicated derivatives.*

**57.** $\dfrac{d}{dx}[(x^2 + x)(x^2 - x)]$

**58.** $\dfrac{d}{dx}[(x^2 + x^3)(x + 1)]$

**59.** $\dfrac{d}{dx}[(x^3 + 2x)(x^2 - x)]\Big|_{x=2}$

**60.** $\dfrac{d}{dx}[(x^2 + x)(x^2 - x)]\Big|_{x=1}$

**61.** $\dfrac{d}{dt}[(t^2 - t^{0.5})(t^{0.5} + t^{-0.5})]\Big|_{t=1}$

**62.** $\dfrac{d}{dt}[(t^2 + t^{0.5})(t^{0.5} - t^{-0.5})]\Big|_{t=1}$

*In Exercises 63–70, use the calculation thought experiment to say whether the expression is written as a sum, difference, scalar multiple, product, or quotient. Then use the appropriate rules to find its derivative.* HINT [See Quick Examples on page 815 and Example 3.]

**63.** $y = x^4 - (x^2 + 120)(4x - 1)$

**64.** $y = x^4 - \dfrac{x^2 + 120}{4x - 1}$

**65.** $y = x + 1 + 2\left(\dfrac{x}{x + 1}\right)$

**66.** $y = (x + 2) - 4(x^2 - x)\left(x + \dfrac{1}{x}\right)$

(Do not simplify the answer.)

**67.** $y = (x + 2)\left(\dfrac{x}{x + 1}\right)$     (Do not simplify the answer.)

**68.** $y = \dfrac{(x + 2)x}{x + 1}$     (Do not simplify the answer.)

**69.** $y = (x + 1)(x - 2) - 2\left(\dfrac{x}{x + 1}\right)$

**70.** $y = \dfrac{x + 2}{x + 1} + (x + 1)(x - 2)$

*In Exercises 71–76, find the equation of the line tangent to the graph of the given function at the point with the indicated x-coordinate.*

**71.** $f(x) = (x^2 + 1)(x^3 + x); x = 1$

**72.** $f(x) = (x^{0.5} + 1)(x^2 + x); x = 1$

**73.** $f(x) = \dfrac{x+1}{x+2}; x = 0$    **74.** $f(x) = \dfrac{\sqrt{x}+1}{\sqrt{x}+2}; x = 4$

**75.** $f(x) = \dfrac{x^2+1}{x}; x = -1$    **76.** $f(x) = \dfrac{x}{x^2+1}; x = 1$

## APPLICATIONS

**77.** *Revenue* The monthly sales of *Sunny Electronics'* new sound system are given by $q(t) = 2{,}000t - 100t^2$ units per month, $t$ months after its introduction. The price Sunny charges is $p(t) = 1{,}000 - t^2$ dollars per sound system, $t$ months after introduction. Find the rate of change of monthly sales, the rate of change of the price, and the rate of change of monthly revenue 5 months after the introduction of the sound system. Interpret your answers. HINT [See Example 4(a).]

**78.** *Revenue* The monthly sales of *Sunny Electronics'* new *iSun* walkman is given by $q(t) = 2{,}000t - 100t^2$ units per month, $t$ months after its introduction. The price Sunny charges is $p(t) = 100 - t^2$ dollars per iSun, $t$ months after introduction. Find the rate of change of monthly sales, the rate of change of the price, and the rate of change of monthly revenue 6 months after the introduction of the iSun. Interpret your answers. HINT [See Example 4(a).]

**79.** *Saudi Oil Revenues* The price of crude oil during the period 2000–2010 can be approximated by

$$P(t) = 6t + 18 \text{ dollars per barrel } (0 \le t \le 10)$$

in year $t$, where $t = 0$ represents 2000. Saudi Arabia's crude oil production over the same period can be approximated by[16]

$$Q(t) = -0.036t^2 + 0.62t + 8 \text{ million barrels per day.}$$
$$(0 \le t \le 10).$$

Use these models to estimate Saudi Arabia's daily oil revenue and also its rate of change in 2008. (Round your answers to the nearest $1 million.)

**80.** *Russian Oil Revenues* The price of crude oil during the period 2000–2010 can be approximated by

$$P(t) = 6t + 18 \text{ dollars per barrels } (0 \le t \le 10)$$

in year $t$, where $t = 0$ represents 2000. Russia's crude oil production over the same period can be approximated by[17]

$$Q(t) = -0.08t^2 + 1.2t + 5.5 \text{ million barrels per day}$$
$$(0 \le t \le 10).$$

Use these models to estimate Russia's daily oil revenue and also its rate of change in 2005. (Round your answers to the nearest $1 million.)

**81.** *Revenue* Dorothy Wagner is currently selling 20 "I ♥ Calculus" T-shirts per day, but sales are dropping at a rate of 3 per day. She is currently charging $7 per T-shirt, but to compensate for dwindling sales, she is increasing the unit price by $1 per day. How fast, and in what direction, is her daily revenue currently changing?

**82.** *Pricing Policy* Let us turn Exercise 81 around a little: Dorothy Wagner is currently selling 20 "I ♥ Calculus" T-shirts per day, but sales are dropping at a rate of 3 per day. She is currently charging $7 per T-shirt, and she wishes to increase her daily revenue by $10 per day. At what rate should she increase the unit price to accomplish this (assuming that the price increase does not affect sales)?

**83.** *Bus Travel* *Thoroughbred Bus Company* finds that its monthly costs for one particular year were given by $C(t) = 10{,}000 + t^2$ dollars after $t$ months. After $t$ months the company had $P(t) = 1{,}000 + t^2$ passengers per month. How fast is its cost per passenger changing after 6 months? HINT [See Example 4(b).]

**84.** *Bus Travel* *Thoroughbred Bus Company* finds that its monthly costs for one particular year were given by $C(t) = 100 + t^2$ dollars after $t$ months. After $t$ months, the company had $P(t) = 1{,}000 + t^2$ passengers per month. How fast is its cost per passenger changing after 6 months? HINT [See Example 4(b).]

**85.** *Fuel Economy* Your muscle car's gas mileage (in miles per gallon) is given as a function $M(x)$ of speed $x$ in mph, where

$$M(x) = \frac{3{,}000}{x + 3{,}600x^{-1}}.$$

Calculate $M'(x)$, and then $M'(10)$, $M'(60)$, and $M'(70)$. What do the answers tell you about your car?

**86.** *Fuel Economy* Your used Chevy's gas mileage (in miles per gallon) is given as a function $M(x)$ of speed $x$ in mph, where

$$M(x) = \frac{4{,}000}{x + 3{,}025x^{-1}}.$$

Calculate $M'(x)$ and hence determine *the sign* of each of the following: $M'(40)$, $M'(55)$, and $M'(60)$. Interpret your results.

**87.** ▼ *Oil Imports from Mexico* Daily oil production in Mexico and daily U.S. oil imports from Mexico during 2005–2009 could be approximated by

$$P(t) = 3.9 - 0.10t \text{ million barrels } (5 \le t \le 9)$$
$$I(t) = 2.1 - 0.11t \text{ million barrels } (5 \le t \le 9),$$

where $t$ is time in years since the start of 2000.[18]

**a.** What are represented by the functions $P(t) - I(t)$ and $I(t)/P(t)$?

---

[16]Sources for data: Oil price: InflationData.com www.inflationdata.com, Production: Energy Bulletin www.energybulletin.net.

[17]*Ibid.*

[18]Source for data: Energy Information Administration (www.eia.doe.gov)/Pemex.

**b.** Compute $\dfrac{d}{dt}\left[\dfrac{I(t)}{P(t)}\right]\Bigg|_{t=8}$ to two significant digits. What does the answer tell you about oil imports from Mexico?

88. ▼ *Oil Imports from Mexico* Daily oil production in Mexico and daily U.S. oil imports from Mexico during 2000–2004 could be approximated by

$$P(t) = 3.0 + 0.13t \text{ million barrels} \quad (0 \leq t \leq 4)$$

$$I(t) = 1.4 + 0.06t \text{ million barrels} \quad (0 \leq t \leq 4),$$

where $t$ is time in years since the start of 2000.[19]

**a.** What are represented by the functions $P(t) - I(t)$ and $I(t)/P(t)$?

**b.** Compute $\dfrac{d}{dt}\left[\dfrac{I(t)}{P(t)}\right]\Bigg|_{t=3}$ to two significant digits. What does the answer tell you about oil imports from Mexico?

89. ▼ *Military Spending* The annual cost per active-duty armed service member in the United States increased from $80,000 in 1995 to a projected $120,000 in 2007. In 1995, there were 1.5 million armed service personnel, and this number was projected to decrease to 1.4 million in 2003.[20] Use linear models for annual cost and personnel to estimate, to the nearest $10 million, the rate of change of total military personnel costs in 2002.

90. ▼ *Military Spending in the 1990s* The annual cost per active-duty armed service member in the United States increased from $80,000 in 1995 to $90,000 in 2000. In 1990, there were 2 million armed service personnel and this number decreased to 1.5 million in 2000.[21] Use linear models for annual cost and personnel to estimate, to the nearest $10 million, the rate of change of total military personnel costs in 1995.

91. ▼ *Biology—Reproduction* The Verhulst model for population growth specifies the reproductive rate of an organism as a function of the total population according to the following formula:

$$R(p) = \frac{r}{1 + kp}$$

where $p$ is the total population in thousands of organisms, $r$ and $k$ are constants that depend on the particular circumstances and the organism being studied, and $R(p)$ is the reproduction rate in thousands of organisms per hour.[22] If $k = 0.125$ and $r = 45$, find $R'(p)$ and then $R'(4)$. Interpret the result.

92. ▼ *Biology—Reproduction* Another model, the predator satiation model for population growth, specifies that the reproductive rate of an organism as a function of the total population varies according to the following formula:

$$R(p) = \frac{rp}{1 + kp}$$

where $p$ is the total population in thousands of organisms, $r$ and $k$ are constants that depend on the particular circumstances and the organism being studied, and $R(p)$ is the reproduction rate in new organisms per hour.[23] Given that $k = 0.2$ and $r = 0.08$, find $R'(p)$ and $R'(2)$. Interpret the result.

93. ▼ *Embryo Development* Bird embryos consume oxygen from the time the egg is laid through the time the chick hatches. For a typical galliform bird egg, the oxygen consumption (in milliliters) $t$ days after the egg was laid can be approximated by[24]

$$C(t) = -0.016t^4 + 1.1t^3 - 11t^2 + 3.6t \quad (15 \leq t \leq 30).$$

(An egg will usually hatch at around $t = 28$.) Suppose that at time $t = 0$ you have a collection of 30 newly laid eggs and that the number of eggs decreases linearly to zero at time $t = 30$ days. How fast is the total oxygen consumption of your collection of embryos changing after 25 days? (Round your answers to two significant digits.) Comment on the result. HINT [Total oxygen consumption = Oxygen consumption per egg × Number of eggs.]

94. ▼ *Embryo Development* Turkey embryos consume oxygen from the time the egg is laid through the time the chick hatches. For a brush turkey, the oxygen consumption (in milliliters) $t$ days after the egg was laid can be approximated by[25]

$$C(t) = -0.0071t^4 + 0.95t^3 - 22t^2 + 95t \quad (25 \leq t \leq 50).$$

(An egg will typically hatch at around $t = 50$.) Suppose that at time $t = 0$ you have a collection of 100 newly laid eggs and that the number of eggs decreases linearly to zero at time $t = 50$ days. How fast is the total oxygen consumption of your collection of embryos changing after 40 days? (Round your answer to two significant digits.) Interpret the result. HINT [Total oxygen consumption = Oxygen consumption per egg × Number of eggs.]

---

[19]Source for data: Energy Information Administration (www.eia.doe.gov)/Pemex.

[20]Annual costs are adjusted for inflation. Sources: Department of Defense, Stephen Daggett, military analyst, Congressional Research Service/*New York Times*, April 19, 2002, p. A21.

[21]*Ibid.*

[22]Source: *Mathematics in Medicine and the Life Sciences* by F. C. Hoppensteadt and C. S. Peskin (Springer-Verlag, New York, 1992) pp. 20–22.

[23]*Ibid.*

[24]The model is derived from graphical data published in the article "The Brush Turkey" by Roger S. Seymour, *Scientific American*, December, 1991, pp. 108–114.

[25]*Ibid.*

## COMMUNICATION AND REASONING EXERCISES

**95.** If $f$ and $g$ are functions of time, and at time $t = 3$, $f$ equals 5 and is rising at a rate of 2 units per second, and $g$ equals 4 and is rising at a rate of 5 units per second, then the product $fg$ equals _____ and is rising at a rate of _____ units per second.

**96.** If $f$ and $g$ are functions of time, and at time $t = 2$, $f$ equals 3 and is rising at a rate of 4 units per second, and $g$ equals 5 and is rising at a rate of 6 units per second, then $fg$ equals _____ and is rising at a rate of _____ units per second.

**97.** If $f$ and $g$ are functions of time, and at time $t = 3$, $f$ equals 5 and is rising at a rate of 2 units per second, and $g$ equals 4 and is rising at a rate of 5 units per second, then $f/g$ equals _____ and is changing at a rate of _____ units per second.

**98.** If $f$ and $g$ are functions of time, and at time $t = 2$, $f$ equals 3 and is rising at a rate of 4 units per second, and g equals 5 and is rising at a rate of 6 units per second, then $f/g$ equals _____ and is changing at a rate of _____ units per second.

**99.** You have come across the following in a newspaper article: "Revenues of HAL Home Heating Oil Inc. are rising by \$4.2 million per year. This is due to an annual increase of 70¢ per gallon in the price HAL charges for heating oil and an increase in sales of 6 million gallons of oil per year." Comment on this analysis.

**100.** Your friend says that because average cost is obtained by dividing the cost function by the number of units $x$, it follows that the derivative of average cost is the same as marginal cost because the derivative of $x$ is 1. Comment on this analysis.

**101.** ▼Find a demand function $q(p)$ such that, at a price per item of $p = \$100$, revenue will rise if the price per item is increased.

**102.** ▼What must be true about a demand function $q(p)$ so that, at a price per item of $p = \$100$, revenue will decrease if the price per item is increased?

**103.** ▼You and I are both selling a steady 20 T-shirts per day. The price I am getting for my T-shirts is increasing twice as fast as yours, but your T-shirts are currently selling for twice the price of mine. Whose revenue is increasing faster: yours, mine, or neither? Explain.

**104.** ▼You and I are both selling T-shirts for a steady \$20 per shirt. Sales of my T-shirts are increasing at twice the rate of yours, but you are currently selling twice as many as I am. Whose revenue is increasing faster: yours, mine, or neither? Explain.

**105.** ◆*Marginal Product (from the GRE Economics Test)* Which of the following statements about average product and marginal product is correct?

**(A)** If average product is decreasing, marginal product must be less than average product.

**(B)** If average product is increasing, marginal product must be increasing.

**(C)** If marginal product is decreasing, average product must be less than marginal product.

**(D)** If marginal product is increasing, average product must be decreasing.

**(E)** If marginal product is constant over some range, average product must be constant over that range.

**106.** ◆*Marginal Cost (based on a question from the GRE Economics Test)* Which of the following statements about average cost and marginal cost is correct?

**(A)** If average cost is increasing, marginal cost must be increasing.

**(B)** If average cost is increasing, marginal cost must be decreasing.

**(C)** If average cost is increasing, marginal cost must be more than average cost.

**(D)** If marginal cost is increasing, average cost must be increasing.

**(E)** If marginal cost is increasing, average cost must be larger than marginal cost.

# 11.4 The Chain Rule

We can now find the derivatives of expressions involving powers of $x$ combined using addition, subtraction, multiplication, and division, but we still cannot take the derivative of an expression like $(3x + 1)^{0.5}$. For this we need one more rule. The function $h(x) = (3x + 1)^{0.5}$ is not a sum, difference, product, or quotient. To find out what it is, we can use the calculation thought experiment and think about the last operation we would perform in calculating $h(x)$.

**1.** Calculate $3x + 1$.

**2.** Take the 0.5 power (square root) of the answer.

The last operation is "take the 0.5 power." We do not yet have a rule for finding the derivative of the 0.5 power of a quantity other than $x$.

There is a way to build $h(x) = (3x + 1)^{0.5}$ out of two simpler functions: $u(x) = 3x + 1$ (the function that corresponds to the first step in the calculation above) and $f(x) = x^{0.5}$ (the function that corresponds to the second step):

$$\begin{aligned} h(x) &= (3x + 1)^{0.5} \\ &= [u(x)]^{0.5} \qquad u(x) = 3x + 1 \\ &= f(u(x)). \qquad f(x) = x^{0.5} \end{aligned}$$

We say that $h$ is the **composite** of $f$ and $u$. We read $f(u(x))$ as "$f$ of $u$ of $x$."

To compute $h(1)$, say, we first compute $3 \cdot 1 + 1 = 4$ and then take the square root of 4, giving $h(1) = 2$. To compute $f(u(1))$ we follow exactly the same steps: First compute $u(1) = 4$ and then $f(u(1)) = f(4) = 2$. We always compute $f(u(x))$ from the inside out: Given $x$, first compute $u(x)$ and then $f(u(x))$.

Now, $f$ and $u$ are functions *whose derivatives we know*. The *chain rule* allows us to use our knowledge of the derivatives of $f$ and $u$ to find the derivative of $f(u(x))$. For the purposes of stating the rule, let us avoid some of the nested parentheses by abbreviating $u(x)$ as $u$. Thus, we write $f(u)$ instead of $f(u(x))$ and remember that $u$ is a function of $x$.

---

## Chain Rule

If $f$ is a differentiable function of $u$ and $u$ is a differentiable function of $x$, then the composite $f(u)$ is a differentiable function of $x$, and

$$\frac{d}{dx}[f(u)] = f'(u)\frac{du}{dx}. \qquad \text{Chain rule}$$

**In words** *The derivative of f(quantity) is the derivative of f, evaluated at that quantity, times the derivative of the quantity.*

---

### Quick Examples

In the Quick Examples that follow, $u$, "the quantity," is some (unspecified) differentiable function of $x$.

1. Take $f(u) = u^2$. Then

$$\frac{d}{dx}(u^2) = 2u\frac{du}{dx}. \qquad \text{Because } f'(u) = 2u$$

*The derivative of a quantity squared is two times the quantity, times the derivative of the quantity.*

2. Take $f(u) = u^{0.5}$. Then

$$\frac{d}{dx}(u^{0.5}) = 0.5u^{-0.5}\frac{du}{dx}. \qquad \text{Because } f'(u) = 0.5u^{-0.5}$$

*The derivative of a quantity raised to the 0.5 is 0.5 times the quantity raised to the −0.5, times the derivative of the quantity.*

---

As the quick examples illustrate, for every power of a function $u$ whose derivative we know, we now get a "generalized" differentiation rule. The following table gives more examples.

| Original Rule | Generalized Rule | In Words |
|---|---|---|
| $\dfrac{d}{dx}(x^2) = 2x$ | $\dfrac{d}{dx}(u^2) = 2u\dfrac{du}{dx}$ | The derivative of a quantity squared is twice the quantity, times the derivative of the quantity. |
| $\dfrac{d}{dx}(x^3) = 3x^2$ | $\dfrac{d}{dx}(u^3) = 3u^2\dfrac{du}{dx}$ | The derivative of a quantity cubed is 3 times the quantity squared, times the derivative of the quantity. |
| $\dfrac{d}{dx}\left(\dfrac{1}{x}\right) = -\dfrac{1}{x^2}$ | $\dfrac{d}{dx}\left(\dfrac{1}{u}\right) = -\dfrac{1}{u^2}\dfrac{du}{dx}$ | The derivative of 1 over a quantity is negative 1 over the quantity squared, times the derivative of the quantity. |

| Power Rule | Generalized Power Rule | In Words |
|---|---|---|
| $\dfrac{d}{dx}(x^n) = nx^{n-1}$ | $\dfrac{d}{dx}(u^n) = nu^{n-1}\dfrac{du}{dx}$ | The derivative of a quantity raised to the n is n times the quantity raised to the $n-1$, times the derivative of the quantity. |
| $\dfrac{d}{dx}|x| = \dfrac{|x|}{x}$ | $\dfrac{d}{dx}|u| = \dfrac{|u|}{u}\dfrac{du}{dx}$ | The derivative of the absolute value of a quantity is the absolute value of the quantity divided by the quantity, times the derivative of the quantity. |

To motivate the chain rule, let us see why it is true in the special case when $f(u) = u^3$, where the chain rule tells us that

$$\frac{d}{dx}(u^3) = 3u^2\frac{du}{dx}.\qquad \text{Generalized power rule with } n = 3$$

But we could have done this using the product rule instead:

$$\frac{d}{dx}(u^3) = \frac{d}{dx}(u \cdot u \cdot u) = \frac{du}{dx}u \cdot u + u\frac{du}{dx}u + u \cdot u\frac{du}{dx} = 3u^2\frac{du}{dx},$$

which gives us the same result. A similar argument works for $f(u) = u^n$, where $n = 2, 3, 4, \ldots$ . We can then use the quotient rule and the chain rule for positive powers to verify the generalized power rule for *negative* powers as well. For the case of a general differentiable function $f$, the proof of the chain rule is beyond the scope of this book, but you can find one on the Website by following the path

Website → Everything → Chapter 11 → Proof of Chain Rule.

---

### EXAMPLE 1 Using the Chain Rule

Compute the following derivatives.

**a.** $\dfrac{d}{dx}[(2x^2 + x)^3]$    **b.** $\dfrac{d}{dx}[(x^3 + x)^{100}]$    **c.** $\dfrac{d}{dx}\sqrt{3x + 1}$    **d.** $\dfrac{d}{dx}|4x^2 - x|$

### Solution

**a.** Using the calculation thought experiment, we see that the last operation we would perform in calculating $(2x^2 + x)^3$ is that of *cubing*. Thus we think of $(2x^2 + x)^3$ as *a quantity cubed*. There are two similar methods we can use to calculate its derivative.

**Method 1: Using the formula**  We think of $(2x^2 + x)^3$ as $u^3$, where $u = 2x^2 + x$. By the formula,

$$\frac{d}{dx}(u^3) = 3u^2\frac{du}{dx}.$$    Generalized power rule

Now substitute for $u$:

$$\frac{d}{dx}[(2x^2 + x)^3] = 3(2x^2 + x)^2\frac{d}{dx}(2x^2 + x)$$

$$= 3(2x^2 + x)^2(4x + 1)$$

**Method 2: Using the verbal form**  If we prefer to use the verbal form, we get:

*The derivative of $(2x^2 + x)$ cubed is three times $(2x^2 + x)$ squared, times the derivative of $(2x^2 + x)$.*

In symbols,

$$\frac{d}{dx}[(2x^2 + x)^3] = 3(2x^2 + x)^2(4x + 1),$$

as we obtained above.

**b.** First, the calculation thought experiment: If we were computing $(x^3 + x)^{100}$, the last operation we would perform is *raising a quantity to the power* 100. Thus we are dealing with *a quantity raised to the power* 100, and so we must again use the generalized power rule. According to the verbal form of the generalized power rule, the derivative of a quantity raised to the power 100 is 100 times that quantity to the power 99, times the derivative of that quantity. In symbols,

$$\frac{d}{dx}[(x^3 + x)^{100}] = 100(x^3 + x)^{99}(3x^2 + 1).$$

**c.** We first rewrite the expression $\sqrt{3x + 1}$ as $(3x + 1)^{0.5}$ and then use the generalized power rule as in parts (a) and (b):

*The derivative of a quantity raised to the 0.5 is 0.5 times the quantity raised to the −0.5, times the derivative of the quantity.*

Thus,

$$\frac{d}{dx}[(3x + 1)^{0.5}] = 0.5(3x + 1)^{-0.5} \cdot 3 = 1.5(3x + 1)^{-0.5}.$$

**d.** The calculation thought experiment tells us that $|4x^2 - x|$ is the absolute value of a quantity, so we use the generalized rule for absolute values (above):

$$\frac{d}{dx}|u| = \frac{|u|}{u}\frac{du}{dx}, \text{ or, in words,}$$

*The derivative of the absolute value of a quantity is the absolute value of the quantity divided by the quantity, times the derivative of the quantity.*

Thus,

$$\frac{d}{dx}|4x^2 - x| = \frac{|4x^2 - x|}{4x^2 - x} \cdot (8x - 1). \qquad \frac{d}{dx}|u| = \frac{|u|}{u}\frac{du}{dx}$$

➡ **Before we go on...**  The following are examples of common errors in solving Example 1(b):

"$\frac{d}{dx}[(x^3 + x)^{100}] = 100(3x^2 + 1)^{99}$"     ✗  *WRONG!*

"$\frac{d}{dx}[(x^3 + x)^{100}] = 100(x^3 + x)^{99}.$"     ✗  *WRONG!*

Remember that the generalized power rule says that the derivative of a quantity to the power 100 is 100 times *that same quantity* raised to the power 99, *times the derivative of that quantity.* ∎

---

**Q**: *It seems that there are now two formulas for the derivative of an nth power:*

1. $\dfrac{d}{dx}[x^n] = nx^{n-1}$

2. $\dfrac{d}{dx}[u^n] = nu^{n-1}\dfrac{du}{dx}$.

*Which one do I use?*

**A**: Formula 1 is actually a special case of Formula 2: Formula 1 is the original power rule, which applies only to a power of *x*. For instance, it applies to $x^{10}$, but it does not apply to $(2x + 1)^{10}$ because the quantity that is being raised to a power is not *x*. Formula 2 applies to a power of any *function of x*, such as $(2x + 1)^{10}$. It can even be used in place of the original power rule. For example, if we take $u = x$ in Formula 2, we obtain

$$\frac{d}{dx}[x^n] = nx^{n-1}\frac{dx}{dx}$$

$$= nx^{n-1}. \qquad \text{The derivative of } x \text{ with respect to } x \text{ is 1.}$$

Thus, the generalized power rule really *is* a generalization of the original power rule, as its name suggests.

---

### EXAMPLE 2   More Examples Using the Chain Rule

Find:   **a.** $\dfrac{d}{dx}[(2x^5 + x^2 - 20)^{-2/3}]$    **b.** $\dfrac{d}{dx}\left[\dfrac{1}{\sqrt{x+2}}\right]$    **c.** $\dfrac{d}{dx}\left[\dfrac{1}{x^2 + x}\right]$

**Solution**  Each of the given functions is, or can be rewritten as, a power of a function whose derivative we know. Thus, we can use the method of Example 1.

**a.** $\dfrac{d}{dx}[(2x^5 + x^2 - 20)^{-2/3}] = -\dfrac{2}{3}(2x^5 + x^2 - 20)^{-5/3}(10x^4 + 2x)$

**b.** $\dfrac{d}{dx}\left[\dfrac{1}{\sqrt{x+2}}\right] = \dfrac{d}{dx}(x+2)^{-1/2} = -\dfrac{1}{2}(x+2)^{-3/2} \cdot 1 = -\dfrac{1}{2(x+2)^{3/2}}$

**c.** $\dfrac{d}{dx}\left[\dfrac{1}{x^2 + x}\right] = \dfrac{d}{dx}(x^2 + x)^{-1} = -(x^2 + x)^{-2}(2x + 1) = -\dfrac{2x + 1}{(x^2 + x)^2}$

---

➡ **Before we go on...**  In Example 2(c), we could have used the quotient rule instead of the generalized power rule. We can think of the quantity $1/(x^2 + x)$ in two different ways using the calculation thought experiment:

**1.** As 1 divided by something—in other words, as a quotient

**2.** As something raised to the $-1$ power

Of course, we get the same derivative using either approach. ∎

We now look at some more complicated examples.

## EXAMPLE 3  Harder Examples Using the Chain Rule

Find $\dfrac{dy}{dx}$ in each case.    **a.** $y = [(x+1)^{-2.5} + 3x]^{-3}$    **b.** $y = (x+10)^3 \sqrt{1-x^2}$

### Solution

**a.** The calculation thought experiment tells us that the last operation we would perform in calculating $y$ is raising the quantity $[(x+1)^{-2.5} + 3x]$ to the power $-3$. Thus, we use the generalized power rule.

$$\frac{dy}{dx} = -3[(x+1)^{-2.5} + 3x]^{-4} \frac{d}{dx}[(x+1)^{-2.5} + 3x]$$

We are not yet done; we must still find the derivative of $(x+1)^{-2.5} + 3x$. Finding the derivative of a complicated function in several steps helps to keep the problem manageable. Continuing, we have

$$\frac{dy}{dx} = -3[(x+1)^{-2.5} + 3x]^{-4} \frac{d}{dx}[(x+1)^{-2.5} + 3x]$$

$$= -3[(x+1)^{-2.5} + 3x]^{-4} \left( \frac{d}{dx}[(x+1)^{-2.5}] + \frac{d}{dx}(3x) \right). \quad \text{Derivative of a sum}$$

Now we have two derivatives left to calculate. The second of these we know to be 3, and the first is the derivative of a quantity raised to the $-2.5$ power. Thus

$$\frac{dy}{dx} = -3[(x+1)^{-2.5} + 3x]^{-4}[-2.5(x+1)^{-3.5} \cdot 1 + 3].$$

**b.** The expression $(x+10)^3 \sqrt{1-x^2}$ is a product, so we use the product rule:

$$\frac{d}{dx}[(x+10)^3 \sqrt{1-x^2}] = \left( \frac{d}{dx}[(x+10)^3] \right) \sqrt{1-x^2} + (x+10)^3 \left( \frac{d}{dx} \sqrt{1-x^2} \right)$$

$$= 3(x+10)^2 \sqrt{1-x^2} + (x+10)^3 \frac{1}{2\sqrt{1-x^2}}(-2x)$$

$$= 3(x+10)^2 \sqrt{1-x^2} - \frac{x(x+10)^3}{\sqrt{1-x^2}}.$$

## APPLICATIONS

The next example is a new way of looking at Example 3 from Section 11.2.

## EXAMPLE 4  Marginal Product

A consultant determines that *Precision Manufacturers'* annual profit (in dollars) is given by

$$P = -200,000 + 4,000q - 0.46q^2 - 0.00001q^3,$$

where $q$ is the number of surgical lasers it sells each year. The consultant also informs Precision that the number of surgical lasers it can manufacture each year depends on the number $n$ of assembly-line workers it employs according to the equation

$$q = 100n. \quad \text{Each worker contributes 100 lasers per year.}$$

Use the chain rule to find the marginal product $\dfrac{dP}{dn}$.

**Solution** We could calculate the marginal product by substituting the expression for $q$ in the expression for $P$ to obtain $P$ as a function of $n$ (as given in Example 3 from Section 11.2) and then finding $dP/dn$. Alternatively—and this will simplify the calculation—we can use the chain rule. To see how the chain rule applies, notice that $P$ is a function of $q$, where $q$ in turn is given as a function of $n$. By the chain rule,

$$\frac{dP}{dn} = P'(q)\frac{dq}{dn} \qquad \text{Chain rule}$$

$$= \frac{dP}{dq}\frac{dq}{dn}. \qquad \text{Notice how the "quantities" } dq \text{ appear to cancel.}$$

Now we compute

$$\frac{dP}{dq} = 4{,}000 - 0.92q - 0.00003q^2$$

and    $$\frac{dq}{dn} = 100.$$

Substituting into the equation for $\dfrac{dP}{dn}$ gives

$$\frac{dP}{dn} = (4{,}000 - 0.92q - 0.00003q^2)(100)$$

$$= 400{,}000 - 92q - 0.003q^2.$$

Notice that the answer has $q$ as a variable. We can express $dP/dn$ as a function of $n$ by substituting $100n$ for $q$:

$$\frac{dP}{dn} = 400{,}000 - 92(100n) - 0.003(100n)^2$$

$$= 400{,}000 - 9{,}200n - 30n^2.$$

The equation

$$\frac{dP}{dn} = \frac{dP}{dq}\frac{dq}{dn}$$

in the example above is an appealing way of writing the chain rule because it suggests that the "quantities" $dq$ cancel. In general, we can write the chain rule as follows.

## Chain Rule in Differential Notation

If $y$ is a differentiable function of $u$, and $u$ is a differentiable function of $x$, then

$$\frac{dy}{dx} = \frac{dy}{du}\frac{du}{dx}. \qquad \text{The terms } du \text{ cancel.}$$

Notice how the units of measurement also cancel:

$$\frac{\text{Units of } y}{\text{Units of } x} = \frac{\text{Units of } y}{\text{Units of } u}\frac{\text{Units of } u}{\text{Units of } x}.$$

**Quick Examples**

1. If $y = u^3$, where $u = 4x + 1$, then
$$\frac{dy}{dx} = \frac{dy}{du}\frac{du}{dx} = 3u^2 \cdot 4 = 12u^2 = 12(4x+1)^2.$$

2. If $q = 43p^2$, where $p$ (and hence $q$ also) is a differentiable function of $t$, then
$$\frac{dq}{dt} = \frac{dq}{dp}\frac{dp}{dt}$$
$$= 86p\frac{dp}{dt}. \qquad p \text{ is not specified, so we leave } dp/dt \text{ as is.}$$

## EXAMPLE 5  Marginal Revenue

Suppose a company's weekly revenue $R$ is given as a function of the unit price $p$, and $p$ in turn is given as a function of weekly sales $q$ (by means of a demand equation). If

$$\left.\frac{dR}{dp}\right|_{q=1,000} = \$40 \text{ per } \$1 \text{ increase in price}$$

and

$$\left.\frac{dp}{dq}\right|_{q=1,000} = -\$20 \text{ per additional item sold per week}$$

find the marginal revenue when sales are 1,000 items per week.

**Solution**  The marginal revenue is $\dfrac{dR}{dq}$. By the chain rule, we have

$$\frac{dR}{dq} = \frac{dR}{dp}\frac{dp}{dq}. \qquad \text{Units: Revenue per item}$$
$$= \text{Revenue per } \$1 \text{ price increase} \times \text{price increase per additional item.}$$

Because we are interested in the marginal revenue at a demand level of 1,000 items per week, we have

$$\left.\frac{dR}{dq}\right|_{q=1,000} = (40)(-20) = -\$800 \text{ per additional item sold.}$$

Thus, if the price is lowered to increase the demand from 1,000 to 1,001 items per week, the weekly revenue will drop by approximately $800.

Look again at the way the terms "$du$" appeared to cancel in the differential formula $\dfrac{dy}{dx} = \dfrac{dy}{du}\dfrac{du}{dx}$. In fact, the chain rule tells us more:

## Manipulating Derivatives in Differential Notation

* The notion of "thinking of $x$ as a function of $y$" will be made more precise in Section 11.6.

**1.** Suppose $y$ is a function of $x$. Then, thinking of $x$ as a function of $y$ (as, for instance, when we can solve for $x$)* we have

$$\frac{dx}{dy} = \frac{1}{\left(\dfrac{dy}{dx}\right)}, \text{ provided } \frac{dy}{dx} \neq 0. \quad \text{Notice again how } \frac{dy}{dx} \text{ behaves like a fraction.}$$

### Quick Example

In the demand equation $q = -0.2p - 8$, we have $\dfrac{dq}{dp} = -0.2$. Therefore,

$$\frac{dp}{dq} = \frac{1}{\left(\dfrac{dq}{dp}\right)} = \frac{1}{-0.2} = -5.$$

**2.** Suppose $x$ and $y$ are functions of $t$. Then, thinking of $y$ as a function of $x$ (as, for instance, when we can solve for $t$ as a function of $x$, and hence obtain $y$ as a function of $x$), we have

$$\frac{dy}{dx} = \frac{dy/dt}{dx/dt}. \quad \text{The terms } dt \text{ appear to cancel.}$$

### Quick Example

If $x = 3 - 0.2t$ and $y = 6 + 6t$, then

$$\frac{dy}{dx} = \frac{dy/dt}{dx/dt} = \frac{6}{-0.2} = -30.$$

To see why the above formulas work, notice that the second formula,

$$\frac{dy}{dx} = \frac{\left(\dfrac{dy}{dt}\right)}{\left(\dfrac{dx}{dt}\right)}$$

can be written as

$$\frac{dy}{dx}\frac{dx}{dt} = \frac{dy}{dt}, \quad \text{Multiply both sides by } \frac{dx}{dt}.$$

which is just the differential form of the chain rule. For the first formula, use the second formula with $y$ playing the role of $t$:

$$\frac{dy}{dx} = \frac{dy/dy}{dx/dy}$$

$$= \frac{1}{dx/dy}. \quad \frac{dy}{dy} = \frac{d}{dy}[y] = 1$$

**FAQs**

## Using the Chain Rule

**Q:** *How do I decide whether or not to use the chain rule when taking a derivative?*

**A:** Use the calculation thought experiment (Section 11.3): Given an expression, consider the steps you would use in computing its value.

- If the last step is *raising a quantity to a power,* as in $\left(\dfrac{x^2-1}{x+4}\right)^4$, then the first step to use is the chain rule (in the form of the generalized power rule):

$$\frac{d}{dx}\left(\frac{x^2-1}{x+4}\right)^4 = 4\left(\frac{x^2-1}{x+4}\right)^3 \frac{d}{dx}\left(\frac{x^2-1}{x+4}\right).$$

Then use the appropriate rules to finish the computation. You may need to again use the calculation thought experiment to decide on the next step (here the quotient rule):

$$= 4\left(\frac{x^2-1}{x+4}\right)^3 \frac{(2x)(x+4)-(x^2-1)(1)}{(x+4)^2}.$$

- If the last step is *division,* as in $\dfrac{(x^2-1)}{(3x+4)^4}$, then the first step to use is the quotient rule:

$$\frac{d}{dx}\frac{(x^2-1)}{(3x+4)^4} = \frac{(2x)(3x+4)^4-(x^2-1)\dfrac{d}{dx}(3x+4)^4}{(3x+4)^8}.$$

Then use the appropriate rules to finish the computation (here the chain rule):

$$= \frac{(2x)(3x+4)^4-(x^2-1)[4(3x+4)^3(3)]}{(3x+4)^8}.$$

- If the last step is *multiplication, addition, subtraction, or multiplication by a constant,* then the first rule to use is the product rule, or the rule for sums, differences, or constant multiples as appropriate.

**Q:** *Every time I compute a derivative, I leave something out. How do I make sure I am really done when taking the derivative of a complicated-looking expression?*

**A:** Until you are an expert at taking derivatives, the key is to use one rule at a time and write out each step, rather than trying to compute the derivative in a single step.

To illustrate this, try computing the derivative of $(x+10)^3\sqrt{1-x^2}$ in Example 3(b) in two ways: First try to compute it in a single step, and then compute it by writing out each step as shown in the example. How do your results compare? For more practice, try Exercises 87 and 88 following.

## 11.4 EXERCISES

▼ more advanced    ◆ challenging
Ⓣ indicates exercises that should be solved using technology

*Calculate the derivatives of the functions in Exercises 1–50.*
HINT [See Example 1.]

**1.** $f(x) = (2x + 1)^2$

**2.** $f(x) = (3x - 1)^2$

**3.** $f(x) = (x - 1)^{-1}$

**4.** $f(x) = (2x - 1)^{-2}$

**5.** $f(x) = (2 - x)^{-2}$

**6.** $f(x) = (1 - x)^{-1}$

**7.** $f(x) = (2x + 1)^{0.5}$

**8.** $f(x) = (-x + 2)^{1.5}$

**9.** $f(x) = \dfrac{1}{3x - 1}$

**10.** $f(x) = \dfrac{1}{(x + 1)^2}$

**11.** $f(x) = (x^2 + 2x)^4$

**12.** $f(x) = (x^3 - x)^3$

**13.** $f(x) = (2x^2 - 2)^{-1}$

**14.** $f(x) = (2x^3 + x)^{-2}$

**15.** $g(x) = (x^2 - 3x - 1)^{-5}$

**16.** $g(x) = (2x^2 + x + 1)^{-3}$

**17.** $h(x) = \dfrac{1}{(x^2 + 1)^3}$

**18.** $h(x) = \dfrac{1}{(x^2 + x + 1)^2}$

HINT [See Example 2.]    HINT [See Example 2.]

**19.** $r(x) = (0.1x^2 - 4.2x + 9.5)^{1.5}$

**20.** $r(x) = (0.1x - 4.2x^{-1})^{0.5}$

**21.** $r(s) = (s^2 - s^{0.5})^4$

**22.** $r(s) = (2s + s^{0.5})^{-1}$

**23.** $f(x) = \sqrt{1 - x^2}$

**24.** $f(x) = \sqrt{x + x^2}$

**25.** $f(x) = |3x - 6|$

**26.** $f(x) = |-5x + 1|$

HINT [See Example 1(d).]    HINT [See Example 1(d).]

**27.** $f(x) = |-x^3 + 5x|$

**28.** $f(x) = |x - x^4|$

**29.** $h(x) = 2[(x + 1)(x^2 - 1)]^{-1/2}$   HINT [See Example 3.]

**30.** $h(x) = 3[(2x - 1)(x - 1)]^{-1/3}$   HINT [See Example 3.]

**31.** $h(x) = (3.1x - 2)^2 - \dfrac{1}{(3.1x - 2)^2}$

**32.** $h(x) = \left[3.1x^2 - 2 - \dfrac{1}{3.1x - 2}\right]^2$

**33.** $f(x) = [(6.4x - 1)^2 + (5.4x - 2)^3]^2$

**34.** $f(x) = (6.4x - 3)^{-2} + (4.3x - 1)^{-2}$

**35.** $f(x) = (x^2 - 3x)^{-2}(1 - x^2)^{0.5}$

**36.** $f(x) = (3x^2 + x)(1 - x^2)^{0.5}$

**37.** $s(x) = \left(\dfrac{2x + 4}{3x - 1}\right)^2$

**38.** $s(x) = \left(\dfrac{3x - 9}{2x + 4}\right)^3$

**39.** $g(z) = \left(\dfrac{z}{1 + z^2}\right)^3$

**40.** $g(z) = \left(\dfrac{z^2}{1 + z}\right)^2$

**41.** $f(x) = [(1 + 2x)^4 - (1 - x)^2]^3$

**42.** $f(x) = [(3x - 1)^2 + (1 - x)^5]^2$

**43.** $f(x) = (3x - 1)|3x - 1|$

**44.** $f(x) = |(x - 3)^{1/3}|$

**45.** $f(x) = |x - (2x - 3)^{1/2}|$

**46.** $f(x) = (3 - |3x - 1|)^{-2}$

**47.** ▼ $r(x) = (\sqrt{2x + 1} - x^2)^{-1}$

**48.** ▼ $r(x) = (\sqrt{x + 1} + \sqrt{x})^3$

**49.** ▼ $f(x) = (1 + (1 + (1 + 2x)^3)^3)^3$

**50.** ▼ $f(x) = 2x + (2x + (2x + 1)^3)^3$

*Find the indicated derivatives in Exercises 51–58. In each case, the independent variable is a (unspecified) differentiable function of t.* HINT [See Quick Example 2 on page 828.]

**51.** $y = x^{100} + 99x^{-1}$. Find $\dfrac{dy}{dt}$.

**52.** $y = x^{0.5}(1 + x)$. Find $\dfrac{dy}{dt}$.

**53.** $s = \dfrac{1}{r^3} + r^{0.5}$. Find $\dfrac{ds}{dt}$.

**54.** $s = r + r^{-1}$. Find $\dfrac{ds}{dt}$.

**55.** $V = \dfrac{4}{3}\pi r^3$. Find $\dfrac{dV}{dt}$.

**56.** $A = 4\pi r^2$. Find $\dfrac{dA}{dt}$.

**57.** ▼ $y = x^3 + \dfrac{1}{x}$, $x = 2$ when $t = 1$, $\left.\dfrac{dx}{dt}\right|_{t=1} = -1$

Find $\left.\dfrac{dy}{dt}\right|_{t=1}$.

**58.** ▼ $y = \sqrt{x} + \dfrac{1}{\sqrt{x}}$, $x = 9$ when $t = 1$, $\left.\dfrac{dx}{dt}\right|_{t=1} = -1$

Find $\left.\dfrac{dy}{dt}\right|_{t=1}$.

*In Exercises 59–64, compute the indicated derivative using the chain rule.* HINT [See Quick Examples on page 829.]

**59.** $y = 3x - 2$; $\dfrac{dx}{dy}$

**60.** $y = 8x + 4$; $\dfrac{dx}{dy}$

**61.** $x = 2 + 3t$, $y = -5t$; $\dfrac{dy}{dx}$

**62.** $x = 1 - t/2$, $y = 4t - 1$; $\dfrac{dy}{dx}$

**63.** $y = 3x^2 - 2x$; $\left.\dfrac{dx}{dy}\right|_{x=1}$

**64.** $y = 3x - \dfrac{2}{x}$; $\left.\dfrac{dx}{dy}\right|_{x=2}$

## APPLICATIONS

**65. *Marginal Product*** *Paramount Electronics* has an annual profit given by

$$P = -100,000 + 5,000q - 0.25q^2 \text{ dollars,}$$

where $q$ is the number of laptop computers it sells each year. The number of laptop computers it can make and sell each year depends on the number $n$ of electrical engineers Paramount employs, according to the equation

$$q = 30n + 0.01n^2.$$

Use the chain rule to find $\left.\dfrac{dP}{dn}\right|_{n=10}$ and interpret the result. HINT [See Example 3.]

66. **Marginal Product** Refer back to Exercise 65. The average profit $\bar{P}$ per computer is given by dividing the total profit $P$ by $q$:

$$\bar{P} = -\frac{100,000}{q} + 5,000 - 0.25q \text{ dollars.}$$

Determine the **marginal average product**, $d\bar{P}/dn$ at an employee level of 10 engineers. Interpret the result. HINT [See Example 3.]

67. **Food versus Education** The percentage $y$ (of total personal consumption) an individual spends on food is approximately

$$y = 35x^{-0.25} \text{ percentage points} \quad (6.5 \le x \le 17.5),$$

where $x$ is the percentage the individual spends on education.[26] An individual finds that she is spending

$$x = 7 + 0.2t$$

percent of her personal consumption on education, where $t$ is time in months since January 1. Use direct substitution to express the percentage $y$ as a function of time $t$ (do not simplify the expression) and then use the chain rule to estimate how fast the percentage she spends on food is changing on November 1. Be sure to specify the units.

68. **Food versus Recreation** The percentage $y$ (of total personal consumption) an individual spends on food is approximately

$$y = 33x^{-0.63} \text{ percentage points} \quad (2.5 \le x \le 4.5),$$

where $x$ is the percentage the individual spends on recreation.[27] A college student finds that he is spending

$$x = 3.5 + 0.1t$$

percent of his personal consumption on recreation, where $t$ is time in months since January 1. Use direct substitution to express the percentage $y$ as a function of time $t$ (do not simplify the expression) and then use the chain rule to estimate how fast the percentage he spends on food is changing on November 1. Be sure to specify the units.

69. **Marginal Revenue** The weekly revenue from the sale of rubies at *Royal Ruby Retailers* (RRR) is increasing at a rate of \$40 per \$1 increase in price, and the price is decreasing at a rate of \$0.75 per additional ruby sold. What is the marginal revenue? (Be sure to state the units of measurement.) Interpret the result. HINT [See Example 5.]

70. **Marginal Revenue** The weekly revenue from the sale of emeralds at *Eduardo's Emerald Emporium* (EEE) is decreasing at a rate of €500 per €1 increase in price, and the price is decreasing at a rate of €0.45 per additional emerald sold. What is the marginal revenue? (Be sure to state the units of measurement.) Interpret the result. HINT [See Example 5.]

71. **Crime Statistics** The murder rate in large cities (over 1 million residents) can be related to that in smaller cities (500,000–1,000,000 residents) by the following linear model:[28]

$$y = 1.5x - 1.9 \quad (15 \le x \le 25),$$

where $y$ is the murder rate (in murders per 100,000 residents each year) in large cities and $x$ is the murder rate in smaller cities. During the period 1991–1998, the murder rate in small cities was decreasing at an average rate of 2 murders per 100,000 residents each year. Use the chain rule to estimate how fast the murder rate was changing in larger cities during that period. (Show how you used the chain rule in your answer.)

72. **Crime Statistics** Following is a quadratic model relating the murder rates described in the preceding exercise:

$$y = 0.1x^2 - 3x + 39 \quad (15 \le x \le 25).$$

In 1996, the murder rate in smaller cities was approximately 22 murders per 100,000 residents each year and was decreasing at a rate of approximately 2.5 murders per 100,000 residents each year. Use the chain rule to estimate how fast the murder rate was changing for large cities. (Show how you used the chain rule in your answer.)

73. **Existing Home Sales** The following graph shows the approximate value of home prices and existing home sales in 2006–2010 as a percentage change from 2003, together with quadratic approximations.[29]

Home prices and sales of existing homes

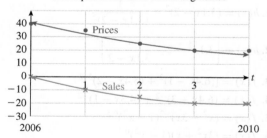

The quadratic approximations are given by

$$\text{Home Prices: } P(t) = t^2 - 10t + 41 \quad (0 \le t \le 4)$$
$$\text{Existing Home Sales: } S(t) = 1.5t^2 - 11t \quad (0 \le t \le 4),$$

---

[26]Model based on historical and projected data from 1908–2010. Sources: Historical data, Bureau of Economic Analysis; projected data, Bureau of Labor Statistics/*New York Times*, December 1, 2003, p. C2.
[27]Ibid.

[28]The model is a linear regression model. Source for data: Federal Bureau of Investigation, Supplementary Homicide Reports/*New York Times*, May 29, 2000, p. A12.
[29]Sources: Standard & Poors/Bloomberg Financial Markets/*New York Times*, September 29, 2007, p. C3. Projection is the authors'.

where $t$ is time in years since the start of 2006. Use the chain rule to estimate $\left.\dfrac{dS}{dP}\right|_{t=2}$. What does the answer tell you about home sales and prices? HINT [See the second Quick Example on page 829.]

74. **Existing Home Sales Leading to the Financial Crisis** The following graph shows the approximate value of home prices and existing home sales in 2004–2007 (the 3 years prior to the 2008 economic crisis) as a percentage change from 2003, together with quadratic approximations.[30]

Home prices and sales of existing homes

The quadratic approximations are given by

Home Prices:    $P(t) = -6t^2 + 27t + 10 \quad (0 \le t \le 3)$

Existing Home Sales:  $S(t) = -4t^2 + 4t + 11 \quad (0 \le t \le 3)$,

where $t$ is time in years since the start of 2004. Use the chain rule to estimate $\left.\dfrac{dS}{dP}\right|_{t=2}$. What does the answer tell you about home sales and prices? HINT [See the second Quick Example on page 829.]

75. ▼ **Pollution** An offshore oil well is leaking oil and creating a circular oil slick. If the radius of the slick is growing at a rate of 2 miles/hour, find the rate at which the area is increasing when the radius is 3 miles. (The area of a disc of radius $r$ is $A = \pi r^2$.) HINT [See Quick Example 2 on page 828.]

76. ▼ **Mold** A mold culture in a dorm refrigerator is circular and growing. The radius is growing at a rate of 0.3 cm/day. How fast is the area growing when the culture is 4 centimeters in radius? (The area of a disc of radius $r$ is $A = \pi r^2$.) HINT [See Quick Example 2 on page 828.]

77. ▼ **Budget Overruns** The Pentagon is planning to build a new, spherical satellite. As is typical in these cases, the specifications keep changing, so that the size of the satellite keeps growing. In fact, the radius of the planned satellite is growing 0.5 feet per week. Its cost will be $1,000 per cubic foot. At the point when the plans call for a satellite 10 feet in radius, how fast is the cost growing? (The volume of a solid sphere of radius $r$ is $V = \frac{4}{3}\pi r^3$.)

78. ▼ **Soap Bubbles** The soap bubble I am blowing has a radius that is growing at a rate of 4 cm/sec. How fast is the surface area growing when the radius is 10 cm? (The surface area of a sphere of radius $r$ is $S = 4\pi r^2$.)

79. 🔲 ▼ **Revenue Growth** The demand for the Cyberpunk II arcade video game is modeled by the logistic curve

$$q(t) = \frac{10{,}000}{1 + 0.5e^{-0.4t}}$$

where $q(t)$ is the total number of units sold $t$ months after its introduction.

**a.** Use technology to estimate $q'(4)$.

**b.** Assume that the manufacturers of Cyberpunk II sell each unit for $800. What is the company's marginal revenue $dR/dq$?

**c.** Use the chain rule to estimate the rate at which revenue is growing 4 months after the introduction of the video game.

80. 🔲 ▼ **Information Highway** The amount of information transmitted each month in the early years of the Internet (1988 to 1994) can be modeled by the equation

$$q(t) = \frac{2e^{0.69t}}{3 + 1.5e^{-0.4t}} \qquad (0 \le t \le 6),$$

where $q$ is the amount of information transmitted each month in billions of data packets and $t$ is the number of years since the start of 1988.[31]

**a.** Use technology to estimate $q'(2)$.

**b.** Assume that it costs $5 to transmit a million packets of data. What is the marginal cost $C'(q)$?

**c.** How fast was the cost increasing at the start of 1990?

**Money Stock** *Exercises 81–84 are based on the following demand function for money (taken from a question on the GRE Economics Test):*

$$M_d = 2 \times y^{0.6} \times r^{-0.3} \times p,$$

*where*

$M_d =$ *demand for nominal money balances (money stock)*
$y =$ *real income*
$r =$ *an index of interest rates*
$p =$ *an index of prices*

*These exercises also use the idea of **percentage rate of growth**:*

$$\text{Percentage Rate of Growth of } M = \frac{\text{Rate of Growth of } M}{M}$$

$$= \frac{dM/dt}{M}.$$

81. ◆ (from the GRE Economics Test) If the interest rate and price level are to remain constant while real income grows at 5 percent per year, the money stock must grow at what percent per year?

82. ◆ (from the GRE Economics Test) If real income and price level are to remain constant while the interest rate grows at 5 percent per year, the money stock must change by what percent per year?

---

[30]Sources: Standard & Poors /Bloomberg Financial Markets/*New York Times*, September 29, 2007, p. C3. Projection is the authors'.

[31]This is the authors' model, based on figures published in *New York Times*, Nov. 3, 1993.

**83.** ◆ (from the GRE Economics Test) If the interest rate is to remain constant while real income grows at 5 percent per year and the price level rises at 5 percent per year, the money stock must grow at what percent per year?

**84.** ◆ (from the GRE Economics Test) If real income grows by 5 percent per year, the interest rate grows by 2 percent per year, and the price level drops by 3 percent per year, the money stock must change by what percent per year?

## COMMUNICATION AND REASONING EXERCISES

**85.** Complete the following: The derivative of 1 over a glob is −1 over . . . .

**86.** Complete the following: The derivative of the square root of a glob is 1 over . . . .

**87.** Say why the following was marked wrong and give the correct answer.

$$\frac{d}{dx}[(3x^3 - x)^3] = 3(9x^2 - 1)^2 \quad \text{✗ } WRONG!$$

**88.** Say why the following was marked wrong and give the correct answer.

$$\frac{d}{dx}\left[\left(\frac{3x^2 - 1}{2x - 2}\right)^3\right] = 3\left(\frac{3x^2 - 1}{2x - 2}\right)^2\left(\frac{6x}{2}\right) \quad \text{✗ } WRONG!$$

**89.** Name two major errors in the following graded test question and give the correct answer.

$$\frac{d}{dx}\left[\left(\frac{3x^2 - 1}{2x - 2}\right)^3\right] = 3\left(\frac{6x}{2}\right)^2 \quad \text{✗ } WRONG! \ SEE \ ME!$$

**90.** Name two major errors in the following graded test question and give the correct answer.

$$\frac{d}{dx}[(3x^3 - x)(2x + 1)]^4 = 4[(9x^2 - 1)(2)]^3 \quad \text{✗ } WRONG! \ SEE \ ME!$$

**91.** ▼ Formulate a simple procedure for deciding whether to apply first the chain rule, the product rule, or the quotient rule when finding the derivative of a function.

**92.** ▼ Give an example of a function $f$ with the property that calculating $f'(x)$ requires use of the following rules in the given order: (1) the chain rule, (2) the quotient rule, and (3) the chain rule.

**93.** ◆ Give an example of a function $f$ with the property that calculating $f'(x)$ requires use of the chain rule five times in succession.

**94.** ◆ What can you say about the composite of two linear functions and what can you say about its derivative?

---

## 11.5  Derivatives of Logarithmic and Exponential Functions

At this point, we know how to take the derivative of any algebraic expression in $x$ (involving powers, radicals, and so on). We now turn to the derivatives of logarithmic and exponential functions.

### Derivative of the Natural Logarithm

$$\frac{d}{dx}[\ln x] = \frac{1}{x} \qquad \text{Recall that } \ln x = \log_e x.$$

#### Quick Examples

**1.** $\dfrac{d}{dx}[3 \ln x] = 3 \cdot \dfrac{1}{x} = \dfrac{3}{x}$    Derivative of a constant times a function

**2.** $\dfrac{d}{dx}[x \ln x] = 1 \cdot \ln x + x \cdot \dfrac{1}{x}$    Product rule, because $x \ln x$ is a product

$= \ln x + 1.$

The above simple formula works only for the natural logarithm (the logarithm with base $e$). For logarithms with bases other than $e$, we have the following:

### Derivative of the Logarithm with Base $b$

$$\frac{d}{dx}[\log_b x] = \frac{1}{x \ln b}$$    Notice that, if $b = e$, we get the same formula as previously.

#### Quick Examples

1. $\dfrac{d}{dx}[\log_3 x] = \dfrac{1}{x \ln 3} \approx \dfrac{1}{1.0986x}$

2. $\dfrac{d}{dx}[\log_2(x^4)] = \dfrac{d}{dx}(4 \log_2 x)$    We used the logarithm identity $\log_b(x^r) = r \log_b x$.

   $$= 4 \cdot \frac{1}{x \ln 2} \approx \frac{4}{0.6931x}$$

## Derivation of the formulas $\dfrac{d}{dx}[\ln x] = \dfrac{1}{x}$ and $\dfrac{d}{dx}[\log_b x] = \dfrac{1}{x \ln b}$

To compute $\dfrac{d}{dx}[\ln x]$, we need to use the definition of the derivative. We also use properties of the logarithm to help evaluate the limit.

$$\frac{d}{dx}[\ln x] = \lim_{h \to 0} \frac{\ln(x+h) - \ln x}{h}$$    Definition of the derivative

$$= \lim_{h \to 0} \frac{1}{h}[\ln(x+h) - \ln x]$$    Algebra

$$= \lim_{h \to 0} \frac{1}{h} \ln\left(\frac{x+h}{x}\right)$$    Properties of the logarithm

$$= \lim_{h \to 0} \frac{1}{h} \ln\left(1 + \frac{h}{x}\right)$$    Algebra

$$= \lim_{h \to 0} \ln\left(1 + \frac{h}{x}\right)^{1/h}$$    Properties of the logarithm

which we rewrite as

$$\lim_{h \to 0} \ln\left[\left(1 + \frac{1}{(x/h)}\right)^{x/h}\right]^{1/x}.$$

As $h \to 0^+$, the quantity $x/h$ gets large and positive, and so the quantity in brackets approaches $e$ (see the definition of $e$ in Section 9.2), which leaves us with

$$\ln[e]^{1/x} = \frac{1}{x} \ln e = \frac{1}{x}$$

**\*** We actually used the fact that the logarithm function is continuous when we took the limit.

**†** Here is an outline of the argument for negative $h$. Because $x$ must be positive for $\ln x$ to be defined, we find that $x/h \to -\infty$ as $h \to 0^-$, and so we must consider the quantity $(1 + 1/m)^m$ for large *negative* $m$. It turns out the limit is still $e$ (check it numerically!) and so the computation above still works.

which is the derivative we are after.**\*** What about the limit as $h \to 0^-$? We will glide over that case and leave it for the interested reader to pursue.**†**

The rule for the derivative of $\log_b x$ follows from the fact that $\log_b x = \ln x / \ln b$.

If we were to take the derivative of the natural logarithm of a *quantity* (a function of $x$), rather than just $x$, we would need to use the chain rule:

## Derivatives of Logarithms of Functions

| Original Rule | Generalized Rule | In Words |
|---|---|---|
| $\dfrac{d}{dx}[\ln x] = \dfrac{1}{x}$ | $\dfrac{d}{dx}[\ln u] = \dfrac{1}{u}\dfrac{du}{dx}$ | The derivative of the natural logarithm of a quantity is 1 over that quantity, times the derivative of that quantity. |
| $\dfrac{d}{dx}[\log_b x] = \dfrac{1}{x \ln b}$ | $\dfrac{d}{dx}[\log_b u] = \dfrac{1}{u \ln b}\dfrac{du}{dx}$ | The derivative of the log to base $b$ of a quantity is 1 over the product of $\ln b$ and that quantity, times the derivative of that quantity. |

### Quick Examples

**§** If we were to evaluate $\ln(x^2 + 1)$, the last operation we would perform would be to take the natural logarithm of a quantity. Thus, the calculation thought experiment tells us that we are dealing with $\ln$ *of a quantity*, and so we need the generalized logarithm rule as stated above.

**1.** $\dfrac{d}{dx}\ln[x^2 + 1] = \dfrac{1}{x^2 + 1}\dfrac{d}{dx}(x^2 + 1)$       $u = x^2 + 1$ (See the margin note.**§**)

$\qquad\qquad\qquad = \dfrac{1}{x^2 + 1}(2x) = \dfrac{2x}{x^2 + 1}$

**2.** $\dfrac{d}{dx}\log_2[x^3 + x] = \dfrac{1}{(x^3 + x)\ln 2}\dfrac{d}{dx}(x^3 + x)$   $u = x^3 + x$

$\qquad\qquad\qquad = \dfrac{1}{(x^3 + x)\ln 2}(3x^2 + 1) = \dfrac{3x^2 + 1}{(x^3 + x)\ln 2}$

## EXAMPLE 1 **Derivative of Logarithmic Function**

Compute the following derivatives:

**a.** $\dfrac{d}{dx}[\ln\sqrt{x + 1}]$     **b.** $\dfrac{d}{dx}[\ln[(1 + x)(2 - x)]]$   **c.** $\dfrac{d}{dx}[\ln|x|]$

**Solution**

**a.** The calculation thought experiment tells us that we have the natural logarithm of a quantity, so

$\dfrac{d}{dx}[\ln\sqrt{x + 1}] = \dfrac{1}{\sqrt{x + 1}}\dfrac{d}{dx}\sqrt{x + 1}$     $\dfrac{d}{dx}\ln u = \dfrac{1}{u}\dfrac{du}{dx}$

$\qquad\qquad\qquad = \dfrac{1}{\sqrt{x + 1}} \cdot \dfrac{1}{2\sqrt{x + 1}}$     $\dfrac{d}{dx}\sqrt{u} = \dfrac{1}{2\sqrt{u}}\dfrac{du}{dx}$

$\qquad\qquad\qquad = \dfrac{1}{2(x + 1)}.$

---

**Q** : *What happened to the square root?*

**A** : As with many problems involving logarithms, we could have done this one differently and much more easily if we had simplified the expression $\ln \sqrt{x+1}$ using the properties of logarithms *before* differentiating. Doing this, we get the following:

---

***Part (a) redone by simplifying first:***

$$\ln\sqrt{x+1} = \ln(x+1)^{1/2} = \frac{1}{2}\ln(x+1). \qquad \text{Simplify the logarithm first.}$$

Thus,

$$\frac{d}{dx}[\ln\sqrt{x+1}] = \frac{d}{dx}\left[\frac{1}{2}\ln(x+1)\right]$$

$$= \frac{1}{2}\left[\frac{1}{x+1}\right]\cdot 1 = \frac{1}{2(x+1)}.$$

A *lot* easier!

**b.** This time, we simplify the expression $\ln[(1+x)(2-x)]$ before taking the derivative:

$$\ln[(1+x)(2-x)] = \ln(1+x) + \ln(2-x). \qquad \text{Simplify the logarithm first.}$$

Thus,

$$\frac{d}{dx}[\ln[(1+x)(2-x)]] = \frac{d}{dx}[\ln[(1+x)]] + \frac{d}{dx}[\ln[(2-x)]]$$

$$= \frac{1}{1+x} - \frac{1}{2-x}. \qquad \frac{d}{dx}\ln u = \frac{1}{u}\frac{du}{dx}$$

For practice, try doing this calculation without simplifying first. What other differentiation rule do you need to use?

**c.** Before we start, we note that $\ln x$ is defined only for positive values of $x$, so its domain is the set of positive real numbers. The domain of $\ln |x|$, on the other hand, is the set of *all* nonzero real numbers. For example, $\ln |-2| = \ln 2 \approx 0.6931$. For this reason, $\ln |x|$ often turns out to be more useful than the ordinary logarithm function.

$$\frac{d}{dx}[\ln |x|] = \frac{1}{|x|}\frac{d}{dx}|x| \qquad \frac{d}{dx}\ln u = \frac{1}{u}\frac{du}{dx}$$

$$= \frac{1}{|x|}\frac{|x|}{x} \qquad \text{Recall that } \frac{d}{dx}|x| = \frac{|x|}{x}.$$

$$= \frac{1}{x}$$

➡ **Before we go on...** Figure 7(a) shows the graphs of $y = \ln |x|$ and $y = 1/x$. Figure 7(b) shows the graphs of $y = \ln |x|$ and $y = 1/|x|$. You should be able to see from these graphs why the derivative of $\ln |x|$ is $1/x$ and not $1/|x|$.

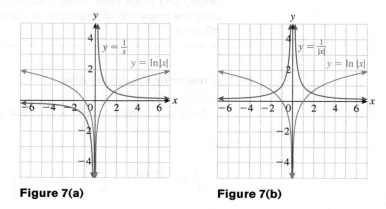

**Figure 7(a)**                    **Figure 7(b)**    ■

This last example, in conjunction with the chain rule, gives us the following formulas.

### Derivative of Logarithms of Absolute Values

| Original Rule | Generalized Rule | In Words |
|---|---|---|
| $\dfrac{d}{dx}[\ln |x|] = \dfrac{1}{x}$ | $\dfrac{d}{dx}[\ln |u|] = \dfrac{1}{u}\dfrac{du}{dx}$ | The derivative of the natural logarithm of the absolute value of a quantity is 1 over that quantity, times the derivative of that quantity. |
| $\dfrac{d}{dx}[\log_b |x|] = \dfrac{1}{x \ln b}$ | $\dfrac{d}{dx}[\log_b |u|] = \dfrac{1}{u \ln b}\dfrac{du}{dx}$ | The derivative of the log to base b of the absolute value of a quantity is 1 over the product of ln b and that quantity, times the derivative of that quantity. |

**Note:** Compare the above formulas with those on page 836. They tell us that we can simply ignore the absolute values in $\ln |u|$ or $\log_b |u|$ when taking the derivative.

### Quick Examples

1. $\dfrac{d}{dx}[\ln |x^2 - 1|] = \dfrac{1}{x^2 - 1}\dfrac{d}{dx}(x^2 - 1)$ $\qquad u = x^2 - 1$

$= \dfrac{1}{x^2 - 1}(2x) = \dfrac{2x}{x^2 - 1}$

**2.** $\dfrac{d}{dx}[\log_2|x^3 + x|] = \dfrac{1}{(x^3 + x)\ln 2}\dfrac{d}{dx}(x^3 + x)$    $u = x^3 + x$

$$= \dfrac{1}{(x^3 + x)\ln 2}(3x^2 + 1) = \dfrac{3x^2 + 1}{(x^3 + x)\ln 2}$$

We now turn to the derivatives of *exponential* functions—that is, functions of the form $f(x) = b^x$. We begin by showing how *not* to differentiate them.

**Caution** The derivative of $b^x$ is *not* $x b^{x-1}$. The power rule applies only to *constant* exponents. In this case the exponent is decidedly *not* constant, and so the power rule does not apply.

The following shows the correct way of differentiating $b^x$, beginning with a special case.

## Derivative of $e^x$

$$\dfrac{d}{dx}[e^x] = e^x$$

### Quick Examples

**1.** $\dfrac{d}{dx}[3e^x] = 3\dfrac{d}{dx}[e^x] = 3e^x$    Constant multiple rule

**2.** $\dfrac{d}{dx}\left[\dfrac{e^x}{x}\right] = \dfrac{e^x x - e^x(1)}{x^2}$    Quotient rule

$$= \dfrac{e^x(x - 1)}{x^2}$$

* There is another—very simple—function that is its own derivative. What is it?

Thus, $e^x$ has the amazing property that its derivative is itself!* For bases other than $e$, we have the following generalization:

## Derivative of $b^x$

If $b$ is any positive number, then

$$\dfrac{d}{dx}[b^x] = b^x \ln b.$$

Note that if $b = e$, we obtain the previous formula.

### Quick Example

$$\dfrac{d}{dx}[3^x] = 3^x \ln 3$$

# Derivation of the Formula $\dfrac{d}{dx}[e^x] = e^x$

✻ This shortcut is an example of a technique called *logarithmic differentiation*, which is occasionally useful. We will see it again in the next section.

To find the derivative of $e^x$, we use a shortcut.✻ Write $g(x) = e^x$. Then

$$\ln g(x) = x.$$

Take the derivative of both sides of this equation to get

$$\frac{g'(x)}{g(x)} = 1$$

or

$$g'(x) = g(x) = e^x.$$

In other words, the exponential function with base $e$ is its own derivative. The rule for exponential functions with other bases follows from the equality $b^x = e^{x\ln b}$ (why?) and the chain rule. (Try it.)

If we were to take the derivative of $e$ raised to a *quantity*, not just $x$, we would need to use the chain rule, as follows.

## Derivatives of Exponentials of Functions

| Original Rule | Generalized Rule | In Words |
|---|---|---|
| $\dfrac{d}{dx}[e^x] = e^x$ | $\dfrac{d}{dx}[e^u] = e^u \dfrac{du}{dx}$ | The derivative of $e$ raised to a quantity is $e$ raised to that quantity, times the derivative of that quantity. |
| $\dfrac{d}{dx}[b^x] = b^x \ln b$ | $\dfrac{d}{dx}[b^u] = b^u \ln b \dfrac{du}{dx}$ | The derivative of $b$ raised to a quantity is $b$ raised to that quantity, times $\ln b$, times the derivative of that quantity. |

### Quick Examples

† The calculation thought experiment tells us that we have $e$ raised to a quantity.

1. $\dfrac{d}{dx}\left[e^{x^2+1}\right] = e^{x^2+1}\dfrac{d}{dx}[x^2+1]$    $u = x^2 + 1$ (See margin note.†)

   $= e^{x^2+1}(2x) = 2x\,e^{x^2+1}$

2. $\dfrac{d}{dx}[2^{3x}] = 2^{3x}\ln 2 \dfrac{d}{dx}[3x]$    $u = 3x$

   $= 2^{3x}(\ln 2)(3) = (3\ln 2)2^{3x}$

3. $\dfrac{d}{dt}[30e^{1.02t}] = 30e^{1.02t}(1.02) = 30.6e^{1.02t}$    $u = 1.02t$

4. If \$1,000 is invested in an account earning 5% per year compounded continuously, then the rate of change of the account balance after $t$ years is

$$\frac{d}{dt}[1{,}000e^{0.05t}] = 1{,}000(0.05)e^{0.05t} = 50e^{0.05t} \text{ dollars/year.}$$

## APPLICATIONS

### EXAMPLE 2 Epidemics

In the early stages of the AIDS epidemic during the 1980s, the number of cases in the United States was increasing by about 50% every 6 months. By the start of 1983, there were approximately 1,600 AIDS cases in the United States.[32] Had this trend continued, how many new cases per year would have been occurring by the start of 1993?

**Solution** To find the answer, we must first model this exponential growth using the methods of Chapter 9. Referring to Example 4 in Section 9.2, we find that $t$ years after the start of 1983 the number of cases is

$$A = 1,600(2.25^t).$$

We are asking for the number of new cases each year. In other words, we want the rate of change, $dA/dt$:

$$\frac{dA}{dt} = 1,600(2.25)^t \ln 2.25 \text{ cases per year.}$$

At the start of 1993, $t = 10$, so the number of new cases per year is

$$\left.\frac{dA}{dt}\right|_{r=10} = 1,600(2.25)^{10} \ln 2.25 \approx 4,300,000 \text{ cases per year.}$$

➡ **Before we go on...** In Example 2, the figure for the number of new cases per year is so large because we assumed that exponential growth—the 50% increase every 6 months—would continue. A more realistic model for the spread of a disease is the logistic model. (See Section 9.4, as well as the next example.) ∎

### EXAMPLE 3 Sales Growth

The sales of the *Cyberpunk II* video game can be modeled by the logistic curve

$$q(t) = \frac{10,000}{1 + 0.5e^{-0.4t}}$$

where $q(t)$ is the total number of units sold $t$ months after its introduction. How fast is the game selling 2 years after its introduction?

**Solution** We are asked for $q'(24)$. We can find the derivative of $q(t)$ using the quotient rule, or we can first write

$$q(t) = 10,000(1 + 0.5e^{-0.4t})^{-1}$$

and then use the generalized power rule:

$$q'(t) = -10,000(1 + 0.5e^{-0.4t})^{-2}(0.5e^{-0.4t})(-0.4)$$
$$= \frac{2,000e^{-0.4t}}{(1 + 0.5e^{-0.4t})^2}.$$

[32] Data based on regression of 1982–1986 figures. Source for data: Centers for Disease Control and Prevention. HIV/AIDS Surveillance Report, 2000;12 (No. 2).

**Figure 8**

**Figure 9**

✱ We can also say this using limits:
$$\lim_{t \to +\infty} q(t) = 10,000.$$

Thus,

$$q'(24) = \frac{2,000e^{-0.4(24)}}{(1 + 0.5e^{-0.4(24)})^2} \approx 0.135 \text{ units per month.}$$

So, after 2 years, sales are quite slow.

➡ **Before we go on...** We can check the answer in Example 3 graphically. If we plot the total sales curve for $0 \le t \le 30$ and $6,000 \le q \le 10,000$, on a TI-83/84 Plus, for example, we get the graph shown in Figure 8. Notice that total sales level off at about 10,000 units.✱ We computed $q'(24)$, which is the slope of the curve at the point with $t$-coordinate 24. If we zoom in to the portion of the curve near $t = 24$, we obtain the graph shown in Figure 9, with $23 \le t \le 25$ and $9,999 \le q \le 10,000$. The curve is almost linear in this range. If we use the two endpoints of this segment of the curve, $(23, 9,999.4948)$ and $(25, 9,999.7730)$, we can approximate the derivative as

$$\frac{9,999.7730 - 9,999.4948}{25 - 23} = 0.1391$$

which is accurate to two decimal places. ∎

# 11.5 EXERCISES

▼ more advanced   ◆ challenging
Ⓣ indicates exercises that should be solved using technology

*Find the derivatives of the functions in Exercises 1–76.*
HINT [See Quick Examples on page 836.]

**1.** $f(x) = \ln(x - 1)$

**2.** $f(x) = \ln(x + 3)$

**3.** $f(x) = \log_2 x$

**4.** $f(x) = \log_3 x$

**5.** $g(x) = \ln|x^2 + 3|$

**6.** $g(x) = \ln|2x - 4|$

**7.** $h(x) = e^{x+3}$
HINT [See Quick Examples on page 840.]

**8.** $h(x) = e^{x^2}$
HINT [See Quick Examples on page 840.]

**9.** $f(x) = e^{-x}$

**10.** $f(x) = e^{1-x}$

**11.** $g(x) = 4^x$

**12.** $g(x) = 5^x$

**13.** $h(x) = 2^{x^2-1}$

**14.** $h(x) = 3^{x^2-x}$

**15.** $f(x) = x \ln x$

**16.** $f(x) = 3 \ln x$

**17.** $f(x) = (x^2 + 1) \ln x$

**18.** $f(x) = (4x^2 - x) \ln x$

**19.** $f(x) = (x^2 + 1)^5 \ln x$

**20.** $f(x) = (x + 1)^{0.5} \ln x$

**21.** $g(x) = \ln|3x - 1|$

**22.** $g(x) = \ln|5 - 9x|$

**23.** $g(x) = \ln|2x^2 + 1|$

**24.** $g(x) = \ln|x^2 - x|$

**25.** $g(x) = \ln(x^2 - 2.1x^{0.3})$

**26.** $g(x) = \ln(x - 3.1x^{-1})$

**27.** $h(x) = \ln[(-2x + 1)(x + 1)]$   HINT [See Example 1b.]

**28.** $h(x) = \ln[(3x + 1)(-x + 1)]$   HINT [See Example 1b.]

**29.** $h(x) = \ln\left(\dfrac{3x + 1}{4x - 2}\right)$

**30.** $h(x) = \ln\left(\dfrac{9x}{4x - 2}\right)$

**31.** $r(x) = \ln\left|\dfrac{(x + 1)(x - 3)}{-2x - 9}\right|$

**32.** $r(x) = \ln\left|\dfrac{-x + 1}{(3x - 4)(x - 9)}\right|$

**33.** $s(x) = \ln(4x - 2)^{1.3}$
HINT [See Example 1a.]

**34.** $s(x) = \ln(x - 8)^{-2}$
HINT [See Example 1a.]

**35.** $s(x) = \ln\left|\dfrac{(x + 1)^2}{(3x - 4)^3(x - 9)}\right|$

**36.** $s(x) = \ln\left|\dfrac{(x + 1)^2(x - 3)^4}{2x + 9}\right|$

**37.** $h(x) = \log_2(x + 1)$

**38.** $h(x) = \log_3(x^2 + x)$

**39.** $r(t) = \log_3(t + 1/t)$

**40.** $r(t) = \log_3\left(t + \sqrt{t}\right)$

**41.** $f(x) = (\ln|x|)^2$

**42.** $f(x) = \dfrac{1}{\ln|x|}$

**43.** $r(x) = \ln(x^2) - [\ln(x - 1)]^2$

**44.** $r(x) = (\ln(x^2))^2$

**45.** $f(x) = xe^x$

**46.** $f(x) = 2e^x - x^2e^x$

**47.** $r(x) = \ln(x + 1) + 3x^3e^x$

**48.** $r(x) = \ln|x + e^x|$

**49.** $f(x) = e^x \ln|x|$

**50.** $f(x) = e^x \log_2|x|$

**51.** $f(x) = e^{2x+1}$

**52.** $f(x) = e^{4x-5}$

**53.** $h(x) = e^{x^2-x+1}$

**54.** $h(x) = e^{2x^2-x+1/x}$

**55.** $s(x) = x^2e^{2x-1}$

**56.** $s(x) = \dfrac{e^{4x-1}}{x^3 - 1}$

**57.** $r(x) = (e^{2x-1})^2$

**58.** $r(x) = (e^{2x^2})^3$

**59.** $t(x) = 3^{2x-4}$

**60.** $t(x) = 4^{-x+5}$

**61.** $v(x) = 3^{2x+1} + e^{3x+1}$

**62.** $v(x) = e^{2x}4^{2x}$

63. $u(x) = \dfrac{3^{x^2}}{x^2 + 1}$

64. $u(x) = (x^2 + 1)4^{x^2-1}$

65. $g(x) = \dfrac{e^x + e^{-x}}{e^x - e^{-x}}$

66. $g(x) = \dfrac{1}{e^x + e^{-x}}$

67. ▼ $g(x) = e^{3x-1}e^{x-2}e^x$

68. ▼ $g(x) = e^{-x+3}e^{2x-1}e^{-x+11}$

69. ▼ $f(x) = \dfrac{1}{x \ln x}$

70. ▼ $f(x) = \dfrac{e^{-x}}{xe^x}$

71. ▼ $f(x) = [\ln(e^x)]^2 - \ln[(e^x)^2]$

72. ▼ $f(x) = e^{\ln x} - e^{2\ln(x^2)}$

73. ▼ $f(x) = \ln|\ln x|$

74. ▼ $f(x) = \ln|\ln|\ln x||$

75. ▼ $s(x) = \ln\sqrt{\ln x}$

76. ▼ $s(x) = \sqrt{\ln(\ln x)}$

*Find the equations of the straight lines described in Exercises 77–82. Use graphing technology to check your answers by plotting the given curve together with the tangent line.*

77. Tangent to $y = e^x \log_2 x$ at the point $(1, 0)$

78. Tangent to $y = e^x + e^{-x}$ at the point $(0, 2)$

79. Tangent to $y = \ln\sqrt{2x + 1}$ at the point where $x = 0$

80. Tangent to $y = \ln\sqrt{2x^2 + 1}$ at the point where $x = 1$

81. At right angles to $y = e^{x^2}$ at the point where $x = 1$

82. At right angles to $y = \log_2(3x + 1)$ at the point where $x = 1$

## APPLICATIONS

83. **Research and Development: Industry** The total spent on research and development by industry in the United States during 1995–2007 can be approximated by

$$S(t) = 57.5 \ln t + 31 \text{ billion dollars} \quad (5 \le t \le 17),$$

where $t$ is the year since 1990.[33] What was the total spent in 2000 ($t = 10$) and how fast was it increasing? HINT [See Quick Examples on page 834.]

84. **Research and Development: Federal** The total spent on research and development by the federal government in the United States during 1995–2007 can be approximated by

$$S(t) = 7.4 \ln t + 3 \text{ billion dollars} \quad (5 \le t \le 17),$$

where $t$ is the year since 1990.[34] What was the total spent in 2005 ($t = 15$) and how fast was it increasing? HINT [See Quick Examples on page 834.]

---

[33]Spending is in constant 2000 dollars. Source for data through 2006: National Science Foundation, Division of Science Resources Statistics, National Patterns of R&D Resources (www.nsf.gov/statistics) August 2008.

[34]Federal funding excluding grants to industry and nonprofit organizations. Spending is in constant 2000 dollars. Source for data through 2006: National Science Foundation, Division of Science Resources Statistics, National Patterns of R&D Resources (www.nsf.gov/statistics) August 2008.

85. **Research and Development: Industry** The function $S(t)$ in Exercise 83 can also be written (approximately) as

$$S(t) = 57.5 \ln (1.71t + 17.1) \text{ billion dollars}$$
$$(-5 \le t \le 7),$$

where this time $t$ is the year since 2000. Use this alternative formula to estimate the amount spent in 2000 and its rate of change, and check your answers by comparing them with those in Exercise 83.

86. **Research and Development: Federal** The function $S(t)$ in Exercise 84 can also be written (approximately) as

$$S(t) = 7.4 \ln (1.5t + 15) \text{ billion dollars}$$
$$(-5 \le t \le 7),$$

where this time $t$ is the year since 2000. Use this alternative formula to estimate the amount spent in 2005 and its rate of change, and check your answers by comparing them with those in Exercise 84.

87. ▼ **Carbon Dating** The age in years of a specimen that originally contained 10g of carbon 14 is given by

$$y = \log_{0.999879}(0.1x),$$

where $x$ is the amount of carbon 14 it currently contains. Compute $\left.\dfrac{dy}{dx}\right|_{x=5}$ and interpret your answer. HINT [For the calculation, see Quick Examples on page 836.]

88. ▼ **Iodine Dating** The age in years of a specimen that originally contained 10g of iodine 131 is given by

$$y = \log_{0.999567}(0.1x),$$

where $x$ is the amount of iodine 131 it currently contains. Compute $\left.\dfrac{dy}{dx}\right|_{x=8}$ and interpret your answer. HINT [For the calculation, see Quick Examples on page 836.]

89. **New York City Housing Costs: Downtown** The average price of a two-bedroom apartment in downtown New York City during the real estate boom from 1994 to 2004 can be approximated by

$$p(t) = 0.33e^{0.16t} \text{ million dollars} \quad (0 \le t \le 10),$$

where $t$ is time in years ($t = 0$ represents 1994).[35] What was the average price of a two-bedroom apartment in downtown New York City in 2003, and how fast was it increasing? (Round your answers to two significant digits.) HINT [See Quick Example 3 on page 840.]

90. **New York City Housing Costs: Uptown** The average price of a two-bedroom apartment in uptown New York City during the real estate boom from 1994 to 2004 can be approximated by

$$p(t) = 0.14e^{0.10t} \text{ million dollars} \quad (0 \le t \le 10),$$

where $t$ is time in years ($t = 0$ represents 1994).[36] What was the average price of a two-bedroom apartment in uptown

---

[35]Model is based on a exponential regression. Source for data: Miller Samuel/*New York Times*, March 28, 2004, p. RE 11.

[36]*Ibid.*

New York City in 2002, and how fast was it increasing? (Round your answers to two significant digits.) HINT [See Quick Example 3 on page 840.]

**91. Big Brother** The following chart shows the total number of wiretaps authorized each year by U.S. state and federal courts from 1990 to 2009 ($t = 0$ represents 1990):[37]

These data can be approximated with the model

$$N(t) = 820e^{0.051t} \quad (0 \le t \le 19).$$

**a.** Find $N(15)$ and $N'(15)$. Be sure to state the units of measurement. To how many significant digits should we round the answers? Why?

**b.** The number of people whose communications are intercepted averages around 100 per wiretap order. What does the answer to part (a) tell you about the number of people whose communications were intercepted?[38]

**c.** According to the model, the number of wiretaps orders each year (choose one)

  **(A)** increased at a linear rate
  **(B)** decreased at a quadratic rate
  **(C)** increased at an exponential rate
  **(D)** increased at a logarithmic rate

over the period shown.

**92. Big Brother** The following chart shows the total number of wiretaps authorized each year by U.S. state courts from 1990 to 2009 ($t = 0$ represents 1990):[39]

These data can be approximated with the model

$$N(t) = 430e^{0.065t} \quad (0 \le t \le 19).$$

**a.** Find $N(10)$ and $N'(10)$. Be sure to state the units of measurement. To how many significant digits should we round the answers? Why?

**b.** The number of people whose communications are intercepted averages around 100 per wiretap order. What does the answer to part (a) tell you about the number of people whose communications were intercepted?[40]

**c.** According to the model, the number of wiretaps orders each year (choose one)

  **(A)** increased at a linear rate
  **(B)** decreased at a quadratic rate
  **(C)** increased at an exponential rate
  **(D)** increased at a logarithmic rate

over the period shown.

**93. Investments** If $10,000 is invested in a savings account offering 4% per year, compounded continuously, how fast is the balance growing after 3 years?

**94. Investments** If $20,000 is invested in a savings account offering 3.5% per year, compounded continuously, how fast is the balance growing after 3 years?

**95. Investments** If $10,000 is invested in a savings account offering 4% per year, compounded semiannually, how fast is the balance growing after 3 years?

**96. Investments** If $20,000 is invested in a savings account offering 3.5% per year, compounded semiannually, how fast is the balance growing after 3 years?

**97. SARS** In the early stages of the deadly SARS (Severe Acute Respiratory Syndrome) epidemic in 2003, the number of cases was increasing by about 18% each day.[41] On March 17, 2003 (the first day for which statistics were reported by the World Health Organization) there were 167 cases. Find an exponential model that predicts the number of people infected $t$ days after March 17, 2003, and use it to estimate how fast the epidemic was spreading on March 31, 2003. (Round your answer to the nearest whole number of new cases per day.) HINT [See Example 2.]

**98. SARS** A few weeks into the deadly SARS (Severe Acute Respiratory Syndrome) epidemic in 2003, the number of cases was increasing by about 4% each day.[42] On April 1, 2003 there were 1,804 cases. Find an exponential model that predicts the number $A(t)$ of people infected $t$ days after April 1, 2003, and use it to estimate how fast the epidemic was spreading on April 30, 2003. (Round your answer to the nearest whole number of new cases per day.) HINT [See Example 2.]

---

[37]Source for data: Wiretap Reports, Administrative Office of the United States Courts www.uscourts.gov/Statistics/WiretapReports.

[38]Assume there is no significant overlap between the people whose communications are intercepted in different wiretap orders.

[39]See Footnote 37.

---

[40]See Footnote 38.

[41]World Health Organization (www.who.int).

[42]*Ibid.*

**99.** ▼ *SAT Scores by Income* The following bar graph shows U.S. math SAT scores as a function of household income:[43]

Family income ($1,000)

**a.** Which of the following best models the data ($C$ is a constant)?

(A) $S(x) = C - 133e^{-0.0131x}$
(B) $S(x) = C + 133e^{-0.0131x}$
(C) $S(x) = C + 133e^{0.0131x}$
(D) $S(x) = C - 133e^{0.0131x}$

($S(x)$ is the average math SAT score of students whose household income is $x$ thousand dollars per year.)

**b.** Use $S'(x)$ to predict how a student's math SAT score is affected by a $1,000 increase in parents' income for a student whose parents earn $45,000.

**c.** Does $S'(x)$ increase or decrease as $x$ increases? Interpret your answer.

**100.** *SAT Scores by Income* The following bar graph shows U.S. critical reading SAT scores as a function of household income:[44]

Family income ($1,000)

**a.** Which of the following best models the data ($C$ is a constant)?

(A) $S(x) = C + \dfrac{1}{136e^{0.015x}}$

(B) $S(x) = C - 136e^{0.015x}$

(C) $S(x) = C - \dfrac{136}{e^{0.015x}}$

(D) $S(x) = C - \dfrac{e^{0.015x}}{136}$

($S(x)$ is the average critical reading SAT score of students whose household income is $x$ thousand dollars per year.)

**b.** Use $S'(x)$ to predict how a student's critical reading SAT score is affected by a $1,000 increase in parents' income for a student whose parents earn $45,000.

**c.** Does $S'(x)$ increase or decrease as $x$ increases? Interpret your answer.

**101.** ▼ *Demographics: Average Age and Fertility* The following graph shows a plot of average age of a population versus fertility rate (the average number of children each woman has in her lifetime) in the United States and Europe over the period 1950–2005.[45]

Fertility rate

The equation of the accompanying curve is

$$a = 28.5 + 120(0.172)^x \quad (1.4 \le x \le 3.7),$$

where $a$ is the average age (in years) of the population and $x$ is the fertility rate.

**a.** Compute $a'(2)$. What does the answer tell you about average age and fertility rates?

**b.** Use the answer to part (a) to estimate how much the fertility rate would need to increase from a level of 2 children per woman to lower the average age of a population by about 1 year.

**102.** ▼ *Demographics: Average Age and Fertility* The following graph shows a plot of average age of a population versus fertility rate (the average number of children each woman has in her lifetime) in Europe over the period 1950–2005.[46]

Fertility rate

The equation of the accompanying curve is

$$g = 27.6 + 128(0.181)^x \quad (1.4 \le x \le 3.7),$$

where $g$ is the average age (in years) of the population and $x$ is the fertility rate.

**a.** Compute $g'(2.5)$. What does the answer tell you about average age and fertility rates?

---

[43]2009 data. Source: College Board/*New York Times* http://economix.blogs.nytimes.com.

[44]*Ibid.*

[45]The separate data for Europe and the United States are collected in the same graph. 2005 figures are estimates. Source: United Nations World Population Division/*New York Times*, June 29, 2003, p. 3.

[46]All European countries including the Russian Federation. 2005 figures are estimates. Source: United Nations World Population Division/*New York Times*, June 29, 2003, p. 3.

**b.** Referring to the model that combines the data for Europe and the United States in Exercise 101, which population's average age is affected more by a changing fertility rate at the level of 2.5 children per woman?

**103. *Epidemics*** A flu epidemic described in Example 1 in Section 9.4 approximately followed the curve

$$P = \frac{150}{1 + 15,000e^{-0.35t}} \text{ million people,}$$

where $P$ is the number of people infected and $t$ is the number of weeks after the start of the epidemic. How fast is the epidemic growing (that is, how many new cases are there each week) after 20 weeks? After 30 weeks? After 40 weeks? (Round your answers to two significant digits.) HINT [See Example 3.]

**104. *Epidemics*** Another epidemic follows the curve

$$P = \frac{200}{1 + 20,000e^{-0.549t}} \text{ million people,}$$

where $t$ is in years. How fast is the epidemic growing after 10 years? After 20 years? After 30 years? (Round your answers to two significant digits.) HINT [See Example 3.]

**105. *Subprime Mortgages during the Housing Bubble*** During the real estate run-up in 2000–2008, the percentage of mortgages issued in the U.S. that were subprime (normally classified as risky) could be approximated by

$$A(t) = \frac{15.0}{1 + 8.6e^{-0.59t}} \text{ percent } \quad (0 \le t \le 8)$$

$t$ years after the start of 2000.[47]

Subprime mortgages

How fast, to the nearest 0.1%, was the percentage increasing at the start of 2003? How would you check that the answer is approximately correct by looking at the graph? HINT [See Example 3.]

**106. *Subprime Mortgage Debt during the Housing Bubble*** During the real estate run-up in 2000–2008, the value of

subprime (normally classified as risky) mortgage debt outstanding in the U.S. was approximately

$$A(t) = \frac{1,350}{1 + 4.2e^{-0.53t}} \text{ billion dollars } \quad (0 \le t \le 8)$$

$t$ years after the start of 2000.[48]

Subprime debt outstanding

How fast, to the nearest $1 billion, was subprime mortgage debt increasing at the start of 2005? How would you check that the answer is approximately correct by looking at the graph? HINT [See Example 3.]

**107. *Subprime Mortgages during the Housing Bubble*** (Compare Exercise 105.) During the real estate run-up in 2000–2008, the percentage of mortgages issued in the U.S. that were subprime (normally classified as risky) could be approximated by

$$A(t) = \frac{15.0}{1 + 8.6(1.8)^{-t}} \text{ percent } \quad (0 \le t \le 8)$$

$t$ years after the start of 2000.[49]
**a.** How fast, to the nearest 0.1%, was the percentage increasing at the start of 2003?
**b.** Compute $\lim_{t \to +\infty} A(t)$ and $\lim_{t \to +\infty} A'(t)$. What do the answers tell you about subprime mortgages?

**108. *Subprime Mortgage Debt during the Housing Bubble*** (Compare Exercise 106.) During the real estate run-up in 2000–2008, the value of subprime (normally classified as risky) mortgage debt outstanding in the U.S. could be approximated by

$$A(t) = \frac{1,350}{1 + 4.2(1.7)^{-t}} \text{ billion dollars } \quad (0 \le t \le 8)$$

$t$ years after the start of 2000.[50]
**a.** How fast, to the nearest $1 billion, was subprime mortgage debt increasing at the start of 2005?
**b.** Compute $\lim_{t \to +\infty} A(t)$ and $\lim_{t \to +\infty} A'(t)$. What do the answers tell you about subprime mortgages?

---

[47]2009 figure is an estimate. Sources: Mortgage Bankers Association, UBS.
[48]2008–2009 figures are estimates. Source: www.data360.org/dataset .aspx? Data_Set_Id=9549.
[49]*Ibid.*
[50]*Ibid.*

**109.** ▼ *Population Growth* The population of Lower Anchovia was 4,000,000 at the start of 2010 and was doubling every 10 years. How fast was it growing per year at the start of 2010? (Round your answer to three significant digits.) HINT [Use the method of Example 2 in Section 9.2 to obtain an exponential model for the population.]

**110.** ▼ *Population Growth* The population of Upper Anchovia was 3,000,000 at the start of 2011 and doubling every 7 years. How fast was it growing per year at the start of 2011? (Round your answer to three significant digits.) HINT [Use the method of Example 2 in Section 9.2 to obtain an exponential model for the population.]

**111.** ▼ *Radioactive Decay* Plutonium 239 has a half-life of 24,400 years. How fast is a lump of 10 grams decaying after 100 years?

**112.** ▼ *Radioactive Decay* Carbon 14 has a half-life of 5,730 years. How fast is a lump of 20 grams decaying after 100 years?

**113.** ◆ *Cellphone Revenues* The number of cellphone subscribers in China for the period 2000–2005 was projected to follow the equation[51]

$$N(t) = 39t + 68 \text{ million subscribers}$$

in year $t$ ($t = 0$ represents 2000). The average annual revenue per cellphone user was $350 in 2000. Assuming that, due to competition, the revenue per cellphone user decreases continuously at an annual rate of 10%, give a formula for the annual revenue in year $t$. Hence, project the annual revenue and its rate of change in 2002. Round all answers to the nearest billion dollars or billion dollars per year.

**114.** ◆ *Cellphone Revenues* The annual revenue for cellphone use in China for the period 2000–2005 was projected to follow the equation[52]

$$R(t) = 14t + 24 \text{ billion dollars}$$

in year $t$ ($t = 0$ represents 2000). At the same time, there were approximately 68 million subscribers in 2000. Assuming that the number of subscribers increases continuously at an annual rate of 10%, give a formula for the annual revenue per subscriber in year $t$. Hence, project to the nearest dollar the annual revenue per subscriber and its rate of change in 2002. (Be careful with units!)

---

[51]Based on a regression of projected figures (coefficients are rounded). Source: Intrinsic Technology/*New York Times*, Nov. 24, 2000, p. C1.

[52]Not allowing for discounting due to increased competition. Source: Ibid.

## COMMUNICATION AND REASONING EXERCISES

**115.** Complete the following: The derivative of $e$ raised to a glob is . . . .

**116.** Complete the following: The derivative of the natural logarithm of a glob is . . . .

**117.** Complete the following: The derivative of 2 raised to a glob is . . . .

**118.** Complete the following: The derivative of the base 2 logarithm of a glob is . . . .

**119.** What is wrong with the following?

$$\frac{d}{dx} \ln|3x + 1| = \frac{3}{|3x + 1|} \qquad ✗ \quad WRONG!$$

**120.** What is wrong with the following?

$$\frac{d}{dx} 2^{2x} = (2)2^{2x} \qquad ✗ \quad WRONG!$$

**121.** What is wrong with the following?

$$\frac{d}{dx} 3^{2x} = (2x)3^{2x-1} \qquad ✗ \quad WRONG!$$

**122.** What is wrong with the following?

$$\frac{d}{dx} \ln(3x^2 - 1) = \frac{1}{6x} \qquad ✗ \quad WRONG!$$

**123.** ▼ The number $N$ of music downloads on campus is growing exponentially with time. Can $N'(t)$ grow linearly with time? Explain.

**124.** ▼ The number $N$ of graphing calculators sold on campus is decaying exponentially with time. Can $N'(t)$ grow with time? Explain.

*The **percentage rate of change** or **fractional rate of change** of a function is defined to be the ratio $f'(x)/f(x)$. (It is customary to express this as a percentage when speaking about percentage rate of change.)*

**125.** ◆ Show that the fractional rate of change of the exponential function $e^{kx}$ is equal to $k$, which is often called its **fractional growth rate**.

**126.** ◆ Show that the fractional rate of change of $f(x)$ is the rate of change of $\ln(f(x))$.

**127.** ◆ Let $A(t)$ represent a quantity growing exponentially. Show that the percentage rate of change, $A'(t)/A(t)$, is constant.

**128.** ◆ Let $A(t)$ be the amount of money in an account that pays interest that is compounded some number of times per year. Show that the percentage rate of growth, $A'(t)/A(t)$, is constant. What might this constant represent?

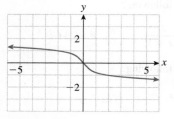

**Figure 10**

# 11.6 Implicit Differentiation

Consider the equation $y^5 + y + x = 0$, whose graph is shown in Figure 10.

How did we obtain this graph? We did not solve for $y$ as a function of $x$; that is impossible. In fact, we solved for $x$ in terms of $y$ to find points to plot. Nonetheless, the graph in Figure 10 is the graph of a function because it passes the vertical line test: Every vertical line crosses the graph no more than once, so for each value of $x$ there is no more than one corresponding value of $y$. Because we cannot solve for $y$ explicitly in terms of $x$, we say that the equation $y^5 + y + x = 0$ determines $y$ as an **implicit function** of $x$.

Now, suppose we want to find the slope of the tangent line to this curve at, say, the point $(2, -1)$ (which, you should check, is a point on the curve). In the following example we find, surprisingly, that it is possible to obtain a formula for $dy/dx$ without having to first solve the equation for $y$.

---

**EXAMPLE 1   Implicit Differentiation**

Find $\dfrac{dy}{dx}$, given that $y^5 + y + x = 0$.

**Solution**  We use the chain rule and a little cleverness. Think of $y$ as a function of $x$ and take the derivative with respect to $x$ of both sides of the equation:

$$y^5 + y + x = 0 \qquad \text{Original equation}$$

$$\frac{d}{dx}[y^5 + y + x] = \frac{d}{dx}[0] \qquad \text{Derivative with respect to } x \text{ of both sides}$$

$$\frac{d}{dx}[y^5] + \frac{d}{dx}[y] + \frac{d}{dx}[x] = 0. \qquad \text{Derivative rules}$$

Now we must be careful. The derivative *with respect to x* of $y^5$ is *not* $5y^4$. Rather, because $y$ is a function of $x$, we must use the chain rule, which tells us that

$$\frac{d}{dx}[y^5] = 5y^4 \frac{dy}{dx}.$$

Thus, we get

$$5y^4 \frac{dy}{dx} + \frac{dy}{dx} + 1 = 0.$$

We want to find $dy/dx$, so we *solve for it*:

$$(5y^4 + 1)\frac{dy}{dx} = -1 \qquad \text{Isolate } dy/dx \text{ on one side.}$$

$$\frac{dy}{dx} = -\frac{1}{5y^4 + 1}. \qquad \text{Divide both sides by } 5y^4 + 1.$$

➡ **Before we go on...** Note that we should not expect to obtain $dy/dx$ as an explicit function of $x$ if $y$ was not an explicit function of $x$ to begin with. For example, the formula we found for $dy/dx$ in Example 1 is not a function of $x$ because there is a $y$ in it. However, the result is still useful because we can evaluate the derivative at any point on the graph. For instance, at the point $(2, -1)$ on the graph, we get

$$\frac{dy}{dx} = -\frac{1}{5y^4 + 1} = -\frac{1}{5(-1)^4 + 1} = -\frac{1}{6}.$$

Thus, the slope of the tangent line to the curve $y^5 + y + x = 0$ at the point $(2, -1)$ is $-1/6$. Figure 11 shows the graph and this tangent line. ∎

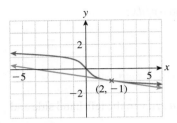

**Figure 11**

This procedure we just used—differentiating an equation to find $dy/dx$ without first solving the equation for $y$—is called **implicit differentiation**.

In Example 1 we were given an equation in $x$ and $y$ that determined $y$ as an (implicit) function of $x$, even though we could not solve for $y$. But an equation in $x$ and $y$ need not always determine $y$ as a function of $x$. Consider, for example, the equation

$$2x^2 + y^2 = 2.$$

Solving for $y$ yields $y = \pm\sqrt{2 - 2x^2}$. The $\pm$ sign reminds us that for some values of $x$ there are two corresponding values for $y$. We can graph this equation by superimposing the graphs of

$$y = \sqrt{2 - 2x^2} \quad \text{and} \quad y = -\sqrt{2 - 2x^2}.$$

The graph, an *ellipse*, is shown in Figure 12.

The graph of $y = \sqrt{2 - 2x^2}$ constitutes the top half of the ellipse, and the graph of $y = -\sqrt{2 - 2x^2}$ constitutes the bottom half.

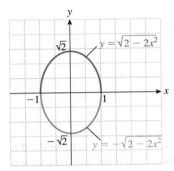

**Figure 12**

---

### EXAMPLE 2 Slope of Tangent Line

Refer to Figure 12. Find the slope of the tangent line to the ellipse $2x^2 + y^2 = 2$ at the point $(1/\sqrt{2}, 1)$.

**Solution**  Because $(1/\sqrt{2}, 1)$ is on the top half of the ellipse in Figure 12, we *could* differentiate the function $y = \sqrt{2 - 2x^2}$ to obtain the result, but it is actually easier to apply implicit differentiation to the original equation.

$$2x^2 + y^2 = 2 \qquad \text{Original equation}$$

$$\frac{d}{dx}[2x^2 + y^2] = \frac{d}{dx}[2] \qquad \text{Derivative with respect to } x \text{ of both sides}$$

$$4x + 2y\frac{dy}{dx} = 0$$

$$2y\frac{dy}{dx} = -4x \qquad \text{Solve for } dy/dx.$$

$$\frac{dy}{dx} = -\frac{4x}{2y} = -\frac{2x}{y}$$

To find the slope at $(1/\sqrt{2}, 1)$ we now substitute for $x$ and $y$:

$$\frac{dy}{dx}\bigg|_{(1/\sqrt{2},1)} = -\frac{2/\sqrt{2}}{1} = -\sqrt{2}.$$

Thus, the slope of the tangent to the ellipse at the point $(1/\sqrt{2}, 1)$ is $-\sqrt{2} \approx -1.414$.

---

### EXAMPLE 3 Tangent Line for an Implicit Function

Find the equation of the tangent line to the curve $\ln y = xy$ at the point where $y = 1$.

**Solution** First, we use implicit differentiation to find $dy/dx$:

$$\frac{d}{dx}[\ln y] = \frac{d}{dx}[xy] \qquad \text{Take } d/dx \text{ of both sides.}$$

$$\frac{1}{y}\frac{dy}{dx} = (1)y + x\frac{dy}{dx}. \qquad \text{Chain rule on left, product rule on right}$$

To solve for $dy/dx$, we bring all the terms containing $dy/dx$ to the left-hand side and all terms not containing it to the right-hand side:

$$\frac{1}{y}\frac{dy}{dx} - x\frac{dy}{dx} = y \qquad \text{Bring the terms with } dy/dx \text{ to the left.}$$

$$\frac{dy}{dx}\left(\frac{1}{y} - x\right) = y \qquad \text{Factor out } dy/dx.$$

$$\frac{dy}{dx}\left(\frac{1-xy}{y}\right) = y$$

$$\frac{dy}{dx} = y\left(\frac{y}{1-xy}\right) = \frac{y^2}{1-xy}. \qquad \text{Solve for } dy/dx.$$

The derivative gives the slope of the tangent line, so we want to evaluate the derivative at the point where $y = 1$. However, the formula for $dy/dx$ requires values for both $x$ and $y$. We get the value of $x$ by substituting $y = 1$ in the original equation:

$$\ln y = xy$$

$$\ln 1 = x \cdot 1$$

But $\ln 1 = 0$, and so $x = 0$ for this point. Thus,

$$\left.\frac{dy}{dx}\right|_{(0,1)} = \frac{1^2}{1-(0)(1)} = 1.$$

Therefore, the tangent line is the line through $(x, y) = (0, 1)$ with slope 1, which is

$$y = x + 1.$$

➡ **Before we go on...** Example 3 presents an instance of an implicit function in which it is simply not possible to solve for $y$. Try it. ∎

Sometimes, it is easiest to differentiate a complicated function of $x$ by first taking the logarithm and then using implicit differentiation—a technique called **logarithmic differentiation**.

**EXAMPLE 4  Logarithmic Differentiation**

Find $\dfrac{d}{dx}\left[\dfrac{(x+1)^{10}(x^2+1)^{11}}{(x^3+1)^{12}}\right]$ without using the product or quotient rules.

**Solution** Write

$$y = \frac{(x+1)^{10}(x^2+1)^{11}}{(x^3+1)^{12}}$$

and then take the natural logarithm of both sides:

$$\ln y = \ln\left[\frac{(x+1)^{10}(x^2+1)^{11}}{(x^3+1)^{12}}\right].$$

We can use properties of the logarithm to simplify the right-hand side:

$$\ln y = \ln(x+1)^{10} + \ln(x^2+1)^{11} - \ln(x^3+1)^{12}$$

$$= 10\ln(x+1) + 11\ln(x^2+1) - 12\ln(x^3+1).$$

Now we can find $\dfrac{dy}{dx}$ using implicit differentiation:

$$\frac{1}{y}\frac{dy}{dx} = \frac{10}{x+1} + \frac{22x}{x^2+1} - \frac{36x^2}{x^3+1} \qquad \text{Take } d/dx \text{ of both sides.}$$

$$\frac{dy}{dx} = y\left(\frac{10}{x+1} + \frac{22x}{x^2+1} - \frac{36x^2}{x^3+1}\right) \qquad \text{Solve for } dy/dx.$$

$$= \frac{(x+1)^{10}(x^2+1)^{11}}{(x^3+1)^{12}}\left(\frac{10}{x+1} + \frac{22x}{x^2+1} - \frac{36x^2}{x^3+1}\right). \qquad \text{Substitute for } y.$$

➡ **Before we go on...** Redo Example 4 using the product and quotient rules (and the chain rule) instead of logarithmic differentiation and compare the answers. Compare also the amount of work involved in both methods. ■

## APPLICATION

Productivity usually depends on both labor and capital. Suppose, for example, you are managing a surfboard manufacturing company. You can measure its productivity by counting the number of surfboards the company makes each year. As a measure of labor, you can use the number of employees, and as a measure of capital you can use its operating budget. The so-called *Cobb-Douglas* model uses a function of the form:

$$P = Kx^a y^{1-a}, \qquad \text{Cobb-Douglas model for productivity}$$

where $P$ stands for the number of surfboards made each year, $x$ is the number of employees, and $y$ is the operating budget. The numbers $K$ and $a$ are constants that depend on the particular situation studied, with $a$ between 0 and 1.

### EXAMPLE 5 Cobb-Douglas Production Function

The surfboard company you own has the Cobb-Douglas production function

$$P = x^{0.3}y^{0.7}$$

where $P$ is the number of surfboards it produces per year, $x$ is the number of employees, and $y$ is the daily operating budget (in dollars). Assume that the production level $P$ is constant.

**a.** Find $\dfrac{dy}{dx}$.

**b.** Evaluate this derivative at $x = 30$ and $y = 10,000$, and interpret the answer.

#### Solution

**a.** We are given the equation $P = x^{0.3}y^{0.7}$, in which $P$ is constant. We find $\dfrac{dy}{dx}$ by implicit differentiation.

$$0 = \frac{d}{dx}[x^{0.3}y^{0.7}] \qquad d/dx \text{ of both sides}$$

$$0 = 0.3x^{-0.7}y^{0.7} + x^{0.3}(0.7)y^{-0.3}\frac{dy}{dx}$$    Product and chain rules

$$-0.7x^{0.3}y^{-0.3}\frac{dy}{dx} = 0.3x^{-0.7}y^{0.7}$$    Bring term with $dy/dx$ to left.

$$\frac{dy}{dx} = -\frac{0.3x^{-0.7}y^{0.7}}{0.7x^{0.3}y^{-0.3}}$$    Solve for $dy/dx$.

$$= -\frac{3y}{7x}.$$    Simplify.

**b.** Evaluating this derivative at $x = 30$ and $y = 10,000$ gives

$$\frac{dy}{dx}\bigg|_{x=30,\ y=10,000} = -\frac{3(10,000)}{7(30)} \approx -143.$$

To interpret this result, first look at the units of the derivative: We recall that the units of $dy/dx$ are units of $y$ per unit of $x$. Because $y$ is the daily budget, its units are dollars; because $x$ is the number of employees, its units are employees. Thus,

$$\frac{dy}{dx}\bigg|_{x=30,\ y=10,000} \approx -\$143 \text{ per employee}.$$

Next, recall that $dy/dx$ measures the rate of change of $y$ as $x$ changes. Because the answer is negative, the daily budget to maintain production at the fixed level is decreasing by approximately $143 per additional employee at an employment level of 30 employees and a daily operating budget of $10,000. In other words, increasing the workforce by one worker will result in a savings of approximately $143 per day. Roughly speaking, *a new employee is worth $143 per day* at the current levels of employment and production.

## 11.6    EXERCISES

▼ more advanced    ◆ challenging
🅣 indicates exercises that should be solved using technology

*In Exercises 1–10, find dy/dx, using implicit differentiation. In each case, compare your answer with the result obtained by first solving for y as a function of x and then taking the derivative.* HINT [See Example 1.]

**1.** $2x + 3y = 7$

**2.** $4x - 5y = 9$

**3.** $x^2 - 2y = 6$

**4.** $3y + x^2 = 5$

**5.** $2x + 3y = xy$

**6.** $x - y = xy$

**7.** $e^x y = 1$

**8.** $e^x y - y = 2$

**9.** $y \ln x + y = 2$

**10.** $\dfrac{\ln x}{y} = 2 - x$

*In Exercises 11–30, find the indicated derivative using implicit differentiation.* HINT [See Example 1.]

**11.** $x^2 + y^2 = 5;\ \dfrac{dy}{dx}$

**12.** $2x^2 - y^2 = 4;\ \dfrac{dy}{dx}$

**13.** $x^2 y - y^2 = 4;\ \dfrac{dy}{dx}$

**14.** $xy^2 - y = x;\ \dfrac{dy}{dx}$

**15.** $3xy - \dfrac{y}{3} = \dfrac{2}{x};\ \dfrac{dy}{dx}$

**16.** $\dfrac{xy}{2} - y^2 = 3;\ \dfrac{dy}{dx}$

**17.** $x^2 - 3y^2 = 8;\ \dfrac{dx}{dy}$

**18.** $(xy)^2 + y^2 = 8;\ \dfrac{dx}{dy}$

**19.** $p^2 - pq = 5p^2 q^2;\ \dfrac{dp}{dq}$

**20.** $q^2 - pq = 5p^2 q^2;\ \dfrac{dp}{dq}$

**21.** $xe^y - ye^x = 1;\ \dfrac{dy}{dx}$

**22.** $x^2 e^y - y^2 = e^x;\ \dfrac{dy}{dx}$

**23.** ▼ $e^{st} = s^2;\ \dfrac{ds}{dt}$

**24.** ▼ $e^{s^2 t} - st = 1;\ \dfrac{ds}{dt}$

**25.** ▼ $\dfrac{e^x}{y^2} = 1 + e^y;\ \dfrac{dy}{dx}$

**26.** ▼ $\dfrac{x}{e^y} + xy = 9y;\ \dfrac{dy}{dx}$

**27.** ▼ $\ln(y^2 - y) + x = y;\ \dfrac{dy}{dx}$

**28.** ▼ $\ln(xy) - x \ln y = y;\ \dfrac{dy}{dx}$

**29.** ▼ $\ln(xy + y^2) = e^y;\ \dfrac{dy}{dx}$

**30.** ▼ $\ln(1 + e^{xy}) = y;\ \dfrac{dy}{dx}$

*In Exercises 31–42, use implicit differentiation to find* **(a)** *the slope of the tangent line, and* **(b)** *the equation of the tangent line at the indicated point on the graph. (Round answers to four decimal places as needed.) If only the x-coordinate is given, you must also find the y-coordinate.*
HINT [See Examples 2, 3.]

**31.** $4x^2 + 2y^2 = 12, (1, -2)$

**32.** $3x^2 - y^2 = 11, (-2, 1)$

**33.** $2x^2 - y^2 = xy, (-1, 2)$

**34.** $2x^2 + xy = 3y^2, (-1, -1)$

**35.** $x^2y - y^2 + x = 1, (1, 0)$

**36.** $(xy)^2 + xy - x = 8, (-8, 0)$

**37.** $xy - 2000 = y, x = 2$

**38.** $x^2 - 10xy = 200, x = 10$

**39.** ▽ $\ln(x + y) - x = 3x^2, x = 0$

**40.** ▽ $\ln(x - y) + 1 = 3x^2, x = 0$

**41.** ▽ $e^{xy} - x = 4x, x = 3$

**42.** ▽ $e^{-xy} + 2x = 1, x = -1$

*In Exercises 43–52, use logarithmic differentiation to find* $dy/dx$. *Do not simplify the result.* HINT [See Example 4.]

**43.** $y = \dfrac{2x + 1}{4x - 2}$

**44.** $y = (3x + 2)(8x - 5)$

**45.** $y = \dfrac{(3x + 1)^2}{4x(2x - 1)^3}$

**46.** $y = \dfrac{x^2(3x + 1)^2}{(2x - 1)^3}$

**47.** $y = (8x - 1)^{1/3}(x - 1)$

**48.** $y = \dfrac{(3x + 2)^{2/3}}{3x - 1}$

**49.** $y = (x^3 + x)\sqrt{x^3 + 2}$

**50.** $y = \sqrt{\dfrac{x - 1}{x^2 + 2}}$

**51.** ▽ $y = x^x$

**52.** ▽ $y = x^{-x}$

## APPLICATIONS

**53. *Productivity*** The number of CDs per hour that *Snappy Hardware* can manufacture at its plant is given by

$$P = x^{0.6}y^{0.4},$$

where $x$ is the number of workers at the plant and $y$ is the monthly budget (in dollars). Assume $P$ is constant, and compute $\dfrac{dy}{dx}$ when $x = 100$ and $y = 200,000$. Interpret the result.
HINT [See Example 5.]

**54. *Productivity*** The number of cellphone accessory kits (neon lights, matching covers, and earpods) per day that *USA Cellular Makeover Inc,* can manufacture at its plant in Cambodia is given by

$$P = x^{0.5}y^{0.5},$$

where $x$ is the number of workers at the plant and $y$ is the monthly budget (in dollars). Assume $P$ is constant, and compute $\dfrac{dy}{dx}$ when $x = 200$ and $y = 100,000$. Interpret the result. HINT [See Example 5.]

**55. *Demand*** The demand equation for soccer tournament T-shirts is

$$xy - 2,000 = y,$$

where $y$ is the number of T-shirts the Enormous State University soccer team can sell at a price of \$$x$ per shirt. Find $\dfrac{dy}{dx}\Big|_{x=5}$, and interpret the result.

**56. *Cost Equations*** The cost $y$ (in cents) of producing $x$ gallons of *Ectoplasm* hair gel is given by the cost equation

$$y^2 - 10xy = 200.$$

Evaluate $\dfrac{dy}{dx}$ at $x = 1$ and interpret the result.

**57. *Housing Costs*[53]** The cost $C$ (in dollars) of building a house is related to the number $k$ of carpenters used and the number $e$ of electricians used by the formula

$$C = 15,000 + 50k^2 + 60e^2.$$

If the cost of the house is fixed at \$200,000, find $\dfrac{dk}{de}\Big|_{e=15}$ and interpret your result.

**58. *Employment*** An employment research company estimates that the value of a recent MBA graduate to an accounting company is

$$V = 3e^2 + 5g^3,$$

where $V$ is the value of the graduate, $e$ is the number of years of prior business experience, and $g$ is the graduate school grade-point average. If $V$ is fixed at 200, find $\dfrac{de}{dg}$ when $g = 3.0$ and interpret the result.

**59.** ▽ ***Grades*[54]** A productivity formula for a student's performance on a difficult English examination is

$$g = 4tx - 0.2t^2 - 10x^2 \quad (t < 30),$$

where $g$ is the score the student can expect to obtain, $t$ is the number of hours of study for the examination, and $x$ is the student's grade-point average.

**a.** For how long should a student with a 3.0 grade-point average study in order to score 80 on the examination?

**b.** Find $\dfrac{dt}{dx}$ for a student who earns a score of 80, evaluate it when $x = 3.0$, and interpret the result.

**60.** ▽ ***Grades*** Repeat the preceding exercise using the following productivity formula for a basket-weaving examination:

$$g = 10tx - 0.2t^2 - 10x^2 \quad (t < 10).$$

Comment on the result.

---

[53]Based on an exercise in *Introduction to Mathematical Economics* by A. L. Ostrosky Jr., and J. V. Koch (Waveland Press, Springfield, Illinois, 1979).

[54]*Ibid.*

*Exercises 61 and 62 are based on the following demand function for money (taken from a question on the GRE Economics Test):*

$$M_d = (2) \times (y)^{0.6} \times (r)^{-0.3} \times (p)$$

*where*

$M_d = $ *demand for nominal money balances (money stock)*

$y = $ *real income*

$r = $ *an index of interest rates*

$p = $ *an index of prices.*

**61. ◆ *Money Stock*** If real income grows while the money stock and the price level remain constant, the interest rate must change at what rate? (First find $dr/dy$, then $dr/dt$; your answers will be expressed in terms of $r$, $y$, and $\dfrac{dy}{dt}$.)

**62. ◆ *Money Stock*** If real income grows while the money stock and the interest rate remain constant, the price level must change at what rate?

## COMMUNICATION AND REASONING EXERCISES

**63.** Fill in the missing terms: The equation $x = y^3 + y - 3$ specifies ___ as a function of ___, and ___ as an implicit function of ___.

**64.** Fill in the missing terms: When $x \neq 0$ in the equation $xy = x^3 + 4$, it is possible to specify ___ as a function of ___. However, ___ is only an implicit function of ___.

**65.** ▼ Use logarithmic differentiation to give another proof of the product rule.

**66.** ▼ Use logarithmic differentiation to give a proof of the quotient rule.

**67.** ▼ If $y$ is given explicitly as a function of $x$ by an equation $y = f(x)$, compare finding $dy/dx$ by implicit differentiation to finding it explicitly in the usual way.

**68.** ▼ Explain why one should not expect $dy/dx$ to be a function of $x$ if $y$ is not a function of $x$.

**69.** ◆ If $y$ is a function of $x$ and $dy/dx \neq 0$ at some point, regard $x$ as an implicit function of $y$ and use implicit differentiation to obtain the equation

$$\frac{dx}{dy} = \frac{1}{dy/dx}.$$

**70.** ◆ If you are given an equation in $x$ and $y$ such that $dy/dx$ is a function of $x$ only, what can you say about the graph of the equation?

## KEY CONCEPTS

### 11.1 Derivatives of Powers, Sums, and Constant Multiples

Power Rule: If $n$ is any constant and $f(x) = x^n$, then $f'(x) = nx^{n-1}$. *p. 784*

Using the power rule for negative and fractional exponents *p. 785*

Sums, differences, and constant multiples *p. 787*

Combining the rules *p. 788*

$\frac{d}{dx}(cx) = c, \quad \frac{d}{dx}(c) = 0$ *p. 790*

$f(x) = x^{1/3}, g(x) = x^{2/3}$, and $h(x) = |x|$ are not differentiable at $x = 0$. *p. 790*

L'Hospital's rule *p. 792*

### 11.2 A First Application: Marginal Analysis

Marginal cost function $C'(x)$ *p. 799*

Marginal revenue and profit functions $R'(x)$ and $P'(x)$ *p. 801*

What it means when the marginal profit is zero *p. 802*

Marginal product *p. 802*

Average cost of the first $x$ items:
$\bar{C}(x) = \frac{C(x)}{x}$ *p. 805*

### 11.3 The Product and Quotient Rules

Product rule: $\frac{d}{dx}[f(x)g(x)] =$
$f'(x)g(x) + f(x)g'(x)$ *p. 810*

Quotient rule: $\frac{d}{dx}\left[\frac{f(x)}{g(x)}\right] =$
$\frac{f'(x)g(x) - f(x)g'(x)}{[g(x)]^2}$ *p. 810*

Using the product rule *p. 812*
Using the quotient rule *p. 813*
Calculation thought experiment *p. 815*
Application to revenue and average cost *p. 816*

### 11.4 The Chain Rule

Chain rule: $\frac{d}{dx}[f(u)] = f'(u)\frac{du}{dx}$ *p. 822*

Generalized power rule:
$\frac{d}{dx}[u^n] = nu^{n-1}\frac{du}{dx}$ *p. 823*

Using the chain rule *p. 823*
Application to marginal product *p. 826*
Chain rule in differential notation:
$\frac{dy}{dx} = \frac{dy}{du}\frac{du}{dx}$ *p. 827*

Manipulating derivatives in differential notation *p. 829*

### 11.5 Derivatives of Logarithmic and Exponential Functions

Derivative of the natural logarithm:
$\frac{d}{dx}[\ln x] = \frac{1}{x}$ *p. 834*

Derivative of logarithm with base $b$:
$\frac{d}{dx}[\log_b x] = \frac{1}{x \ln b}$ *p. 835*

Derivatives of logarithms of functions:
$\frac{d}{dx}[\ln u] = \frac{1}{u}\frac{du}{dx}$
$\frac{d}{dx}[\log_b u] = \frac{1}{u \ln b}\frac{du}{dx}$ *p. 836*

Derivatives of logarithms of absolute values:
$\frac{d}{dx}[\ln|x|] = \frac{1}{x} \quad \frac{d}{dx}[\ln|u|] = \frac{1}{u}\frac{du}{dx}$
$\frac{d}{dx}[\log_b|x|] = \frac{1}{x \ln b}$
$\frac{d}{dx}[\log_b|u|] = \frac{1}{u \ln b}\frac{du}{dx}$ *p. 838*

Derivative of $e^x$: $\frac{d}{dx}[e^x] = e^x$ *p. 839*

Derivative of $b^x$: $\frac{d}{dx}[b^x] = b^x \ln b$ *p. 839*

Derivatives of exponential functions *p. 840*
Application to epidemics *p. 841*
Application to sales growth (logistic function) *p. 841*

### 11.6 Implicit Differentiation

Implicit function of $x$ *p. 848*
Implicit differentiation *p. 848*
Using implicit differentiation *p. 849*
Finding a tangent line *p. 849*
Logarithmic differentiation *p. 850*

## REVIEW EXERCISES

*In Exercises 1–26, find the derivative of the given function.*

**1.** $f(x) = 10x^5 + \frac{1}{2}x^4 - x + 2$

**2.** $f(x) = \frac{10}{x^5} + \frac{1}{2x^4} - \frac{1}{x} + 2$

**3.** $f(x) = 3x^3 + 3\sqrt[3]{x}$

**4.** $f(x) = \frac{2}{x^{2.1}} - \frac{x^{0.1}}{2}$

**5.** $f(x) = x + \frac{1}{x^2}$

**6.** $f(x) = 2x - \frac{1}{x}$

**7.** $f(x) = \frac{4}{3x} - \frac{2}{x^{0.1}} + \frac{x^{1.1}}{3.2} - 4$

**8.** $f(x) = \frac{4}{x} + \frac{x}{4} - |x|$

**9.** $f(x) = e^x(x^2 - 1)$

**10.** $f(x) = \frac{x^2 + 1}{x^2 - 1}$

**11.** $f(x) = \frac{|x| + 1}{3x^2 + 1}$

**12.** $f(x) = (|x| + x)(2 - 3x^2)$

**13.** $f(x) = (4x - 1)^{-1}$

**14.** $f(x) = (x + 7)^{-2}$

**15.** $f(x) = (x^2 - 1)^{10}$

**16.** $f(x) = \frac{1}{(x^2 - 1)^{10}}$

**17.** $f(x) = [2 + (x + 1)^{-0.1}]^{4.3}$

**18.** $f(x) = [(x + 1)^{0.1} - 4x]^{-5.1}$

**19.** $f(x) = e^x(x^2 + 1)^{10}$

20. $f(x) = \left[\dfrac{x-1}{3x+1}\right]^3$

21. $f(x) = \dfrac{3^x}{x-1}$

22. $f(x) = 4^{-x}(x+1)$

23. $f(x) = e^{x^2-1}$

24. $f(x) = (x^2+1)e^{x^2-1}$

25. $f(x) = \ln(x^2-1)$

26. $f(x) = \dfrac{\ln(x^2-1)}{x^2-1}$

*In Exercises 27–34, find all values of x (if any) where the tangent line to the graph of the given equation is horizontal.*

27. $y = -3x^2 + 7x - 1$

28. $y = 5x^2 - 2x + 1$

29. $y = \dfrac{x}{2} + \dfrac{2}{x}$

30. $y = \dfrac{x^2}{2} - \dfrac{8}{x^2}$

31. $y = x - e^{2x-1}$

32. $y = e^{x^2}$

33. $y = \dfrac{x}{x+1}$

34. $y = \sqrt{x}(x-1)$

*In Exercises 35–40, find $dy/dx$ for the given equation.*

35. $x^2 - y^2 = x$

36. $2xy + y^2 = y$

37. $e^{xy} + xy = 2$

38. $\ln\left(\dfrac{y}{x}\right) = y$

39. $y = \dfrac{(2x-1)^4(3x+4)}{(x+1)(3x-1)^3}$

40. $y = x^{x-1}3^x$

*In Exercises 41–46, find the equation of the tangent line to the graph of the given equation at the specified point.*

41. $y = (x^2-3x)^{-2}$; $x = 1$

42. $y = (2x^2-3)^{-3}$; $x = -1$

43. $y = x^2 e^{-x}$; $x = -1$

44. $y = \dfrac{x}{1+e^x}$; $x = 0$

45. $xy - y^2 = x^2 - 3$; $(-1, 1)$

46. $\ln(xy) + y^2 = 1$; $(-1, -1)$

## APPLICATIONS: OHaganBooks.com

47. **Sales** OHaganBooks.com fits the cubic curve

$$w(t) = -3.7t^3 + 74.6t^2 + 135.5t + 6{,}300$$

to its weekly sales figures (see Chapter 10 Review Exercise 57; $t$ is time in weeks), as shown in the following graph:

**a.** According to the cubic model, what was the rate of increase of sales at the beginning of the second week ($t = 1$)? (Round your answer to the nearest unit.)

**b.** If we extrapolate the model, what would be the rate of increase of weekly sales at the beginning of the eighth week ($t = 7$)?

**c.** Graph the function $w$ for $0 \le t \le 20$. Would it be realistic to use the function to predict sales through week 20? Why?

**d.** By examining the graph, say why the choice of a quadratic model would result in radically different long-term predictions of sales.

48. **Rising Sea Level** Marjory Duffin is still toying with various models to fit to the New York sea level figures she had seen after purchasing a beachfront condominium in New York (see Chapter 10 Review Exercise 58). Following is a cubic curve she obtained using regression:

$$L(t) = -0.0001t^3 + 0.02t^2 + 2.2t \text{ mm}$$

($t$ is time in years since 1900). The curve and data are shown in the following graph:

Sea Level Change since 1900

**a.** According to the cubic model, what was the rate at which the sea level was rising in 2000 ($t = 100$)? (Round your answer to two significant digits.)

**b.** If we extrapolate the model, what would be the rate at which the sea level is rising in 2025 ($t = 125$)?

**c.** Graph the function $L$ for $0 \le t \le 200$. Why is it not realistic to use the function to predict the sea level through 2100?

**d.** James Stewart, a summer intern at Duffin House Publishers, differs. As he puts it, "The cubic curve came from doing regression on the actual data, and thus reflects the actual trend of the data. We can't argue against reality!" Comment on this assertion.

49. **Cost** As OHaganBooks.com's sales increase, so do its costs. If we take into account volume discounts from suppliers and shippers, the weekly cost of selling $x$ books is

$$C(x) = -0.00002x^2 + 3.2x + 5{,}400 \text{ dollars.}$$

**a.** What is the marginal cost at a sales level of 8,000 books per week?

**b.** What is the average cost per book at a sales level of 8,000 books per week?

**c.** What is the marginal average cost ($d\bar{C}/dx$) at a sales level of 8,000 books per week?

**d.** Interpret the results of parts (a)–(c).

**50. Cost** OHaganBooks.com has been experiencing a run of bad luck with its summer college intern program in association with PCU (Party Central University), begun as a result of a suggestion by Marjory Duffin over dinner one evening. The frequent errors in filling orders, charges from movie download sites and dating sites, and beverages spilled on computer equipment have resulted in an estimated weekly cost to the company of

$$C(x) = 25x^2 - 5.2x + 4,000 \text{ dollars},$$

where $x$ is the number of college interns employed.

a. What is the marginal cost at a level of 10 interns?
b. What is the average cost per intern at a level of 10 interns?
c. What is the marginal average cost at a level of 10 interns?
d. Interpret the results of parts (a)–(c).

**51. Revenue** At the moment, OHaganBooks.com is selling 1,000 books per week and its sales are rising at a rate of 200 books per week. Also, it is now selling all its books for $20 each, but its price is dropping at a rate of $1 per week.

a. At what rate is OHaganBooks.com's weekly revenue rising or falling?
b. John O'Hagan would like to see the company's weekly revenue increase at a rate of $5,000 per week. At what rate would sales have to have been increasing to accomplish that goal, assuming all the other information is as given above?

**52. Revenue** Due to ongoing problems with its large college intern program in association with PCU (see Exercise 50), OHaganBooks.com has arranged to transfer its interns to its competitor JungleBooks.com (whose headquarters happens to be across the road) for a small fee. At the moment, it is transferring 5 students per week, and this number is rising at a rate of 4 students per week. Also, it is now charging JungleBooks $400 per intern, but this amount is decreasing at a rate of $20 per week.

a. At what rate is OHaganBooks.com's weekly revenue from this transaction rising or falling?
b. Flush with success of the transfer program, John O'Hagan would like to see the company's resulting revenue increase at a rate of $3,900 per week. At what rate would the transfer of interns have to increase to accomplish that goal, assuming all the other information is as given above?

**53. Percentage Rate of Change of Revenue** The percentage rate of change of a quantity $Q$ is $Q'/Q$. Why is the percentage rate of change of revenue always equal to the sum of the percentage rates of change of unit price and weekly sales?

**54. P/E Ratios** At the beginning of last week, OHaganBooks.com stock was selling for $100 per share, rising at a rate of $50 per year. Its earnings amounted to $1 per share, rising at a rate of $0.10 per year. At what rate was its price-to-earnings (P/E) ratio, the ratio of its stock price to its earnings per share, rising or falling?

**55. P/E Ratios** Refer to Exercise 54. Jay Campbell, who recently invested in OHaganBooks.com stock, would have liked to see the P/E ratio increase at a rate of 100 points per year. How fast would the stock have to have been rising, assuming all the other information is as given in Exercise 54?

**56. Percentage Rate of Change of P/E Ratios** Refer to Exercise 54. The percentage rate of change of a quantity $Q$ is $Q'/Q$. Why is the percentage rate of change of P/E always equal to the percentage rate of change of unit price minus the percentage rate of change of earnings?

**57. Sales** OHaganBooks.com decided that the cubic curve in Exercise 47 was not suitable for extrapolation, so instead it tried

$$s(t) = 6,000 + \frac{4,500}{1 + e^{-0.55(t-4.8)}}$$

as shown in the following graph:

Week

a. Compute $s'(t)$ and use the answer to estimate the rate of increase of weekly sales at the beginning of the seventh week ($t = 6$). (Round your answer to the nearest unit.)
b. Compute $\lim_{t \to +\infty} s'(t)$ and interpret the answer.

**58. Rising Sea Level** Upon some reflection, Marjory Duffin decided that the curve in Exercise 48 was not suitable for extrapolation, so instead she tried

$$L(t) = \frac{418}{1 + 17.2e^{-0.041t}} \qquad (0 \le t \le 125)$$

($t$ is time in years since 1900) as shown in the following graph:

Year

a. Compute $L'(t)$ and use the answer to estimate the rate at which the sea level was rising in 2000 ($t = 100$). (Round your answer to two decimal places.)
b. Compute $\lim_{t \to +\infty} L'(t)$ and interpret the answer.

**59. *Web Site Activity*** The number of "hits" on OHaganBooks .com's Web site was 1,000 per day at the beginning of the year, and was growing at a rate of 5% per week. If this growth rate continued for the whole year (52 weeks), find the rate of increase (in hits per day per week) at the end of the year.

**60. *Web Site Activity*** The number of "hits" on ShadyDownload .net during the summer intern program at OHaganBooks .com was 100 per day at the beginning of the intern program, and was growing at a rate of 15% per day. If this growth rate continued for the duration of the whole summer intern program (85 days), find the rate of increase (in hits per day per day) at the end of the program.

**61. *Demand and Revenue*** The price $p$ that OHaganBooks.com charges for its latest leather-bound gift edition of *The Complete Larry Potter* is related to the demand $q$ in weekly sales by the equation

$$250pq + q^2 = 13,500,000.$$

Suppose the price is set at $50, which would make the demand 1,000 copies per week.

**a.** Using implicit differentiation, compute the rate of change of demand with respect to price, and interpret the result. (Round the answer to two decimal places.)

**b.** Use the result of part (a) to compute the rate of change of revenue with respect to price. Should the price be raised or lowered to increase revenue?

**62. *Demand and Revenue*** The price $p$ that OHaganBooks.com charges for its latest leather-bound gift edition of *Lord of the Fields* is related to the demand $q$ in weekly sales by the equation

$$100pq + q^2 = 5,000,000.$$

Suppose the price is set at $40, which would make the demand 1,000 copies per week.

**a.** Using implicit differentiation, compute the rate of change of demand with respect to price, and interpret the result. (Round the answer to two decimal places.)

**b.** Use the result of part (a) to compute the rate of change of revenue with respect to price. Should the price be raised or lowered to increase revenue?

---

**Case Study**    **Projecting Market Growth**

You are on the board of directors at *Fullcourt Academic Press,* a major textbook supplier to private schools, and various expansion strategies will be discussed at tomorrow's board meeting. TJM, the sales director of the high school division, has just burst into your office with his last-minute proposal based on data showing the number of private high school graduates in the U.S. each year over the past 18 years:[55]

| Year | 1995 | 1996 | 1997 | 1998 | 1999 | 2000 | 2001 | 2002 | 2003 |
|---|---|---|---|---|---|---|---|---|---|
| **Graduates (thousands)** | 245 | 254 | 265 | 273 | 279 | 279 | 285 | 296 | 301 |
| **Year** | 2004 | 2005 | 2006 | 2007 | 2008 | 2009 | 2010 | 2011 | 2012 |
| **Graduates (thousands)** | 307 | 307 | 307 | 314 | 314 | 315 | 315 | 316 | 316 |

He is asserts that, despite the unspectacular numbers in the past few years, the long-term trend appears to support a basic premise of his proposal for an expansion strategy: that the number of high school seniors in private schools in the U.S. will be growing at a rate of about 4,000 per year through 2015. He points out that the rate of increase predicted by the regression line is approximately 4,080 students per year, supporting his premise.

In order to decide whether to support TJM's proposal at tomorrow's board meeting, you would like to first determine whether the linear regression prediction of around 4,000 students per year is reasonable, especially in view of the more recent figures. You open your spreadsheet and graph the data with the regression line (Figure 13). The data suggest that the number of graduates began to "level off" (in the language of calculus, the *derivative appears to be decreasing*) toward the end of the

Thousands of graduates

$t$ = Time in years since 1995
Regression Line: $y = 4.08t + 259$

**Figure 13**

---

[55]Data through 2011 are National Center for Educational Statistics actual and projected data as of April 2010. Source: National Center for Educational Statistics http://nces.ed.gov/.

**Figure 14**

period. Moreover, you recall reading somewhere that the numbers of students in the lower grades have also begun to level off, so it is safe to predict that the slowing of growth in the senior class will continue over the next few years, contrary to what TJM has claimed. In order to make a meaningful prediction, you would really need some precise data about numbers in the lower grades, but the meeting is tomorrow and you would like a quick and easy way of extrapolating the data by "extending the curve to the right."

It would certainly be helpful if you had a mathematical model of the data in Figure 13 that you could use to project the current trend. But what kind of model should you use? The linear model is no good because it does not show any change in the derivative (the derivative of a linear function is constant). In addition, best-fit polynomial and exponential functions do not accurately reflect the leveling off, as you realize after trying to fit a few of them (Figure 14).

You then recall that a logistic curve can model the leveling-off property you desire, and so you try a model of the form

$$N(t) = \frac{M}{1 + Ab^{-t}}.$$

Figure 15 shows the best-fit logistic curve, which has a sum-of-squares error (SSE) of around 109. (See Section 1.4 or any of the regression examples in Chapter 9 for a discussion of SSE.)

$$N(t) = \frac{323.9}{1 + 0.3234(1.186)^{-t}} \qquad \text{SSE} \approx 109$$

Its graph shows the leveling off and also gives more reasonable long-term predictions.

**Figure 15**

The rate of increase of high school students—pertinent to TJM's report—is given by the derivative, $N'(t)$:

$$N(t) = \frac{M}{1 + Ab^{-t}}$$

$$N'(t) = -\frac{M}{(1 + Ab^{-t})^2}\frac{d}{dt}[1 + Ab^{-t}]$$

$$= \frac{MAb^{-t}\ln b}{(1 + Ab^{-t})^2}.$$

The rate of increase in the number of high school students in 2015 ($t = 20$) is given by

$$N'(20) = \frac{(323.9)(0.3234)(1.186)^{-20}\ln 1.186}{(1 + 0.3234(1.186)^{-20})^2}$$

$$\approx 0.577 \text{ thousand students per year,}$$

or about 580 students per year—far less than the optimistic estimate of 4,000 in the proposal! Thus, TJM's prediction is suspect and further research will have to be done before the board can even consider the proposal.

To reassure yourself, you decide to look for another kind of S-shaped model as a backup. After flipping through a calculus book, you stumble across a function that is slightly more general than the one you have:

※ To find a detailed discussion of scaled and shifted functions, visit the Website and follow

Chapter 1
→ New Functions from Old: Scaled and Shifted Functions.

$$N(t) = \frac{M}{1 + Ab^{-t}} + C. \qquad \text{Shifted Logistic Curve}^{*}$$

The added term $C$ has the effect of shifting the graph up $C$ units. Turning once again to your calculus book (see the discussion of logistic regression in Section 9.4), you see that a best-fit curve is one that minimizes the sum-of-squares error, and you find the best-fit curve by again using the utility on the Website (this time, with the model $1/(1+2*3^(-x))+4$ and initial guess 323.9, 0.3234, 1.186, 0; that is, keeping the current values and setting $c = 0$). You obtain the model

$$N(t) = \frac{135.5}{1 + 1.192(1.268)^{-t}} + 184.6. \qquad \text{SSE} \approx 100$$

The value of SSE has decreased only slightly, and, as seen in Figure 16, the shifted logistic curve seems almost identical to the unshifted curve, but does seem to level off slightly faster (compare the portions of the two curves on the extreme right).

**Figure 16**

You decide to use the shifted model to obtain another estimate of the projected rate of change in 2015. As the two models differ by a constant, their derivatives are given by the same formula, so you compute

$$N'(20) = \frac{(135.5)(1.192)(1.268)^{-20} \ln 1.268}{(1 + 1.192(1.268)^{-20})^2}$$

$$\approx 0.325 \text{ thousand students per year,}$$

or about 325 students per year, even less than the prediction of the logistic model.

---

**Q:** *Why do the two models give very different predictions of the rate of change in 2015?*

**A:** The long-term prediction in any logistic model is highly sensitive to small changes in the data and/or the model. This is one reason why using regression curve-fitting models to make long-term projections can be a risky undertaking.

**Q :** *Then what is the point of using any model to project in the first place?*

**A :** Projections are always tricky as we cannot foresee the future. But a *good* model is not merely one that seems to fit the data well, but rather a model whose structure is based on the situation being modeled. For instance, a *good* model of student graduation rates should take into account such factors as the birth rate, current school populations at all levels, and the relative popularity of private schools as opposed to public schools. It is by using models of this kind that the National Center for Educational Statistics is able to make the projections shown in the data above.

---

## EXERCISES

1. In 1994 there were 246,000 private high school graduates. What do the two logistic models (unshifted and shifted) "predict" for 1994? (Round answer to the nearest 1,000.) Which gives the better prediction?

2. What is the long-term prediction of each of the two models? (Round answer to the nearest 1,000.)

3. Find $\lim_{t \to +\infty} N'(t)$ for both models, and interpret the results.

4. ☐ You receive a last-minute memo from TJM to the effect that, sorry, the 2011 and 2012 figures are not accurate. Use technology to re-estimate $M$, $A$, $b$, and $C$ for the shifted logistic model in the absence of this data and obtain new estimates for the 2011 and 2012 data. What does the new model predict the rate of change in the number of high school seniors will be in 2015?

5. ☐ *Another Model* Using the original data, find the best-fit shifted logistic curve of the form

$$N(t) = c + b \frac{a(t-m)}{1 + a|t-m|}. \qquad (a, b, c, m \text{ constant})$$

Its graph is shown below:

(Use the model `$1+$2*$3*(x-$4)/(1+$3*abs(x-$4))` and start with the following values: $a = 0.05$, $b = 160$, $c = 250$, $m = 5$; that is, input `250, 160, 0.05, 5` in the "Guess" field.) Graph the data together with the model. What is SSE? Is the model as accurate a fit as the model used in the text? What does this model predict will be the growth rate of the number of high school graduates in 2015? Comment on the answer. (Round the coefficients in the model and all answers to four decimal places.)

6. ☐ *Demand for Freon* The demand for chlorofluorocarbon-12 (CFC-12)—the ozone-depleting refrigerant commonly known as Freon 12 (the name given to it by Du Pont)—has

been declining significantly in response to regulation and concern about the ozone layer. The chart below shows the projected demand for CFC-12 for the period 1994–2005.[56]

a. Use technology to obtain the best-fit equation of the form

$$N(t) = c + b\frac{a(t - m)}{1 + a|t - m|}, \quad (a, b, c, m \text{ constant})$$

where $t$ is the number of years since 1990. Use your function to estimate the total demand for CFC-12 from the start of the year 2000 to the start of 2010. [Start with the following values: $a = 1$, $b = -25$, $c = 35$, and $m = 10$, and round your answers to four decimal places.]

b. According to your model, how fast is the demand for Freon declining in 2000?

---

[56]Source: The Automobile Consulting Group (*New York Times*, December 26, 1993, p. F23). The exact figures were not given, and the chart is a reasonable facsimile of the chart that appeared in *New York Times*.

# 12

# Further Applications of the Derivative

**Website**
www.WanerMath.com
At the Website you will find:

- Section-by-section tutorials, including game tutorials with randomized quizzes
- A detailed chapter summary
- A true/false quiz
- Additional review exercises
- Graphers, Excel tutorials, and other resources
- The following extra topic:
  Linear Approximation and Error Estimation

## Case Study Production Lot Size Management

Your publishing company is planning the production of its latest best seller, which it predicts will sell 100,000 copies each month over the coming year. The book will be printed in several batches of the same number, evenly spaced throughout the year. Each print run has a setup cost of $5,000, a single book costs $1 to produce, and monthly storage costs for books awaiting shipment average 1¢ per book. **To meet the anticipated demand at minimum total cost to your company, how many printing runs should you plan?**

SERDAR/Alamy

863

# Introduction

In this chapter we begin to see the power of calculus as an optimization tool. In Chapter 9 we saw how to price an item in order to get the largest revenue when the demand function is linear. Using calculus, we can handle nonlinear functions, which are much more general. In Section 12.1 we show how calculus can be used to solve the problem of finding the values of a variable that lead to a maximum or minimum value of a given function. In Section 12.2 we show how this helps us in various real-world applications.

Another theme in this chapter is that calculus can help us to draw and understand the graph of a function. By the time you have completed the material in Section 12.1, you will be able to locate and sketch some of the important features of a graph, such as where it rises and where it falls. In Section 12.3 we look at the *second derivative,* the derivative of the derivative function, and what it tells us about how the graph *curves.* We also see how the second derivative is used to model the notion of *acceleration.* In Section 12.4 we put a number of ideas together that help to explain what you see in a graph (drawn, for example, using graphing technology) and to locate its most important points.

We also include sections on related rates and elasticity of demand. The first of these (Section 12.5) examines further the concept of the derivative as a rate of change. The second (Section 12.6) returns to the problem of optimizing revenue based on the demand equation, looking at it in a new way that leads to an important idea in economics—elasticity.

**algebra Review**
For this chapter, you should be familiar with the algebra reviewed in **Chapter 0, sections 5 and 6.**

## 12.1  Maxima and Minima

Figure 1 shows the graph of a function $f$ whose domain is the closed interval $[a, b]$. A mathematician sees lots of interesting things going on here. There are hills and valleys, and even a small chasm (called a *cusp*) near the center. For many purposes, the important features of this curve are the highs and lows. Suppose, for example, you know that the price of the stock of a certain company will follow this graph during the course of a week. Although you would certainly make a handsome profit if you bought at time $a$ and sold at time $b$, your best strategy would be to follow the old adage to "buy low and sell high," buying at all the lows and selling at all the highs.

Figure 2 shows the graph once again with the highs and lows marked. Mathematicians have names for these points: the highs (at the $x$-values $p$, $r$, and $b$) are referred to as **relative maxima**, and the lows (at the $x$-values $a$, $q$, and $s$) are referred to as **relative minima**. Collectively, these highs and lows are referred to as **relative extrema**. (A point of language: The singular forms of the plurals *minima, maxima,* and *extrema* are *minimum, maximum,* and *extremum.*)

**Figure 1**

**Figure 2**

Why do we refer to these points as relative extrema? Take a look at the point corresponding to $x = r$. It is the highest point of the graph *compared to other points nearby.* If you were an extremely nearsighted mountaineer standing at the point

**Figure 3**

where $x = r$, you would *think* that you were at the highest point of the graph, not being able to see the distant peaks at $x = p$ and $x = b$.

Let's translate into mathematical terms. We are talking about the heights of various points on the curve. The height of the curve at $x = r$ is $f(r)$, so we are saying that $f(r)$ is greater than or equal to $f(x)$ for every $x$ near $r$. In other words, $f(r)$ *is the greatest value that* $f(x)$ *has for all choices of* $x$ *between* $r - h$ *and* $r + h$ for some (possibly small) $h$. (See Figure 3.)

We can phrase the formal definition as follows.

## Relative Extrema

$f$ has a **relative maximum** at $x = r$ if there is some interval $(r - h, r + h)$ (even a very small one) for which $f(r) \geq f(x)$ for all $x$ in $(r - h, r + h)$ for which $f(x)$ is defined.

$f$ has a **relative minimum** at $x = r$ if there is some interval $(r - h, r + h)$ (even a very small one) for which $f(r) \leq f(x)$ for all $x$ in $(r - h, r + h)$ for which $f(x)$ is defined.

### Quick Examples

In Figure 2, $f$ has the following relative extrema:

1. Relative maxima at $p$ and $r$.
2. A relative maximum at $b$. (See Figure 4.) Note that $f(x)$ is not defined for $x > b$. However, $f(b) \geq f(x)$ for every $x$ in the interval $(b - h, b + h)$ *for which* $f(x)$ *is defined*—that is, for every $x$ in $(b - h, b]$.*
3. Relative minima at $a$, $q$, and $s$.

**Figure 4**

* Our definition of relative extremum allows $f$ to have a relative extremum at an endpoint of its domain; the definitions used in some books do not. In view of examples like the stock market investing strategy mentioned above, we find it more useful to allow endpoints as relative extrema.

Looking carefully at Figure 2, we can see that the lowest point on the whole graph is where $x = s$ and the highest point is where $x = b$. This means that $f(s)$ is the least value of $f$ on the whole domain of $f$ (the interval $[a, b]$) and $f(b)$ is the greatest value. We call these the *absolute* minimum and maximum.

## Absolute Extrema

$f$ has an **absolute maximum** at $x = r$ if $f(r) \geq f(x)$ for every $x$ in the domain of $f$.

$f$ has an **absolute minimum** at $x = r$ if $f(r) \leq f(x)$ for every $x$ in the domain of $f$.

### Quick Examples

1. In Figure 2, $f$ has an absolute maximum at $b$ and an absolute minimum at $s$.
2. If $f(x) = x^2$, then $f(x) \geq f(0)$ for every real number $x$. Therefore, $f(x) = x^2$ has an absolute minimum at $x = 0$. (See Figure 5.)
3. Generalizing (2), every quadratic function $f(x) = ax^2 + bx + c$ has an absolute extremum at its vertex $x = -b/(2a)$; an absolute minimum if $a > 0$, and an absolute maximum if $a < 0$.

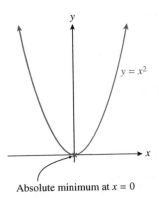

Absolute minimum at $x = 0$

**Figure 5**

**Note** If $f$ has an absolute extremum at $x = r$, then it automatically satisfies the requirement for a *relative* extremum there as well; take $h = 1$ (or any other value) in the definition of relative extremum. Thus, absolute extrema are special types of relative extrema. ∎

Absolute maxima at $x = a$ and $x = b$

**Figure 6**

**Figure 7**

Some graphs have no absolute extrema at all (think of the graph of $y = x$), while others might have an absolute minimum but no absolute maximum (like $y = x^2$), or vice versa. When $f$ does have an absolute maximum, there is only one absolute maximum *value* of $f$, but this value may occur at different values of $x$, and similarly for absolute minima. (See Figure 6.)

**Q :** *At how many different values of $x$ can $f$ take on its absolute maximum value?*

**A :** An extreme case is that of a constant function; because we use $\geq$ in the definition of absolute maximum, a constant function has an absolute maximum (and minimum) at every point in its domain.

Now, how do we go about locating extrema? In many cases we can get a good idea by using graphing technology to zoom in on a maximum or minimum and approximate its coordinates. However, calculus gives us a way to find the exact locations of the extrema and at the same time to understand why the graph of a function behaves the way it does. In fact, it is often best to combine the powers of graphing technology with those of calculus, as we shall see.

In Figure 7 we see the graph from Figure 1 once more, but we have labeled each extreme point as one of three types. Notice that two extrema occur at endpoints and the others at **interior points;** that is, points other than endpoints. Let us look first at the extrema occurring at interior points: At the points labeled "Stationary," the tangent lines to the graph are horizontal, and so have slope 0, so $f'$ (which gives the slope) is 0. Any time $f'(x) = 0$, we say that $f$ has a **stationary point** at $x$ because the rate of change of $f$ is zero there. We call an extremum that occurs at a stationary point a **stationary extremum.** In general, to find the exact location of each stationary point, we need to solve the equation $f'(x) = 0$. Note that stationary points are always interior points, as $f'$ is not defined at endpoints of the domain.

There is a relative minimum in Figure 7 at $x = q$, but there is no horizontal tangent there. In fact, there is no tangent line at all; $f'(q)$ is not defined. (Recall a similar situation with the graph of $f(x) = |x|$ at $x = 0$.) When $f'(x)$ does not exist for some interior point $x$ in the domain of $f$, we say that $f$ has a **singular point** at $x$. We shall call an extremum that occurs at a singular point a **singular extremum**. The points that are either stationary or singular we call collectively the **critical points** of $f$.

The remaining two extrema are at the **endpoints** of the domain (remember that we do allow relative extrema at endpoints). As we see in the figure, they are (almost) always either relative maxima or relative minima.

We bring all the above information together in Figure 8:

**Figure 8**

**Q**: *Are there any other types of relative extrema?*

**A**: No; relative extrema of a function always occur at critical points or endpoints. (A rigorous proof is beyond the scope of this book.)*

* Here is an outline of the argument. Suppose $f$ has a maximum, say, at $x = a$, at some interior point of its domain. Then either $f$ is differentiable there, or it is not. If it is not, then we have a singular point. If $f$ is differentiable at $x = a$, then consider the slope of the secant line through the points where $x = a$ and $x = a + h$ for small positive $h$. Because $f$ has a maximum at $x = a$, it is falling (or level) to the right of $x = a$, and so the slope of this secant line must be $\leq 0$. Thus, we must have $f'(a) \leq 0$ in the limit as $h \to 0$. On the other hand, if $h$ is small and *negative*, then the corresponding secant line must have slope $\geq 0$ because $f$ is also falling (or level) as we move left from $x = a$, and so $f'(a) \geq 0$. Because $f'(a)$ is both $\geq 0$ and $\leq 0$, it must be zero, and so we have a stationary point at $x = a$.

## Locating Candidates for Extrema

If $f$ is a real-valued function, then its extrema occur among the following types of points:

1. **Stationary Points:** $f$ has a stationary point at $x$ if $x$ is in the interior of the domain and $f'(x) = 0$. To locate stationary points, set $f'(x) = 0$ and solve for $x$.

2. **Singular Points:** $f$ has a singular point at $x$ if $x$ is in the interior of the domain and $f'(x)$ is not defined. To locate singular points, find values of $x$ where $f'(x)$ is *not* defined, but $f(x)$ *is* defined.

3. **Endpoints:** These are the endpoints, if any, of the domain. Recall that closed intervals contain endpoints, but open intervals do not. If the domain of $f$ is an open interval or the whole real line, then there are no endpoints.

Once we have a candidate for an extremum of $f$, we find the corresponding point $(x, y)$ on the graph of $f$ using $y = f(x)$.

### Quick Examples

1. **Stationary Points:** Let $f(x) = x^3 - 12x$. Then to locate the stationary points, set $f'(x) = 0$ and solve for $x$. This gives $3x^2 - 12 = 0$, so $f$ has stationary points at $x = \pm 2$. The corresponding points on the graph are $(-2, f(-2)) = (-2, 16)$ and $(2, f(2)) = (2, -16)$.

2. **Singular Points:** Let $f(x) = 3(x - 1)^{1/3}$. Then $f'(x) = (x - 1)^{-2/3} = 1/(x - 1)^{2/3}$. $f'(1)$ is not defined, although $f(1)$ *is* defined. Thus, the (only) singular point occurs at $x = 1$. The corresponding point on the graph is $(1, f(1)) = (1, 0)$.

3. **Endpoints:** Let $f(x) = 1/x$, with domain $(-\infty, 0) \cup [1, +\infty)$. Then the only endpoint in the domain of $f$ occurs at $x = 1$. The corresponding point on the graph is $(1, 1)$. The natural domain of $1/x$, on the other hand, has no endpoints.

Remember, though, that the three types of points we identify above are only *candidates* for extrema. It is quite possible, as we shall see, to have a stationary point or a singular point that is neither a maximum nor a minimum. (It is also possible for an endpoint to be neither a maximum nor a minimum, but only in functions whose graphs are rather bizarre—see Exercise 65.)

Now let's look at some examples of finding maxima and minima. In all of these examples, we will use the following procedure: First, we find the derivative, which we examine to find the stationary points and singular points. Next, we make a table listing the $x$-coordinates of the critical points and endpoints, together with their $y$-coordinates. We use this table to make a rough sketch of the graph. From the table and rough sketch, we usually have enough data to be able to say where the extreme points are and what kind they are.

### EXAMPLE 1  **Maxima and Minima**

Find the relative and absolute maxima and minima of

$$f(x) = x^2 - 2x$$

on the interval $[0, 4]$.

**Solution**  We first calculate $f'(x) = 2x - 2$. We use this derivative to locate the critical points (stationary and singular points).

**Stationary Points**  To locate the stationary points, we solve the equation $f'(x) = 0$, or

$$2x - 2 = 0,$$

getting $x = 1$. The domain of the function is $[0, 4]$, so $x = 1$ is in the interior of the domain. Thus, the only candidate for a stationary relative extremum occurs when $x = 1$.

**Singular Points**  We look for interior points where the derivative is not defined. However, the derivative is $2x - 2$, which is defined for every $x$. Thus, there are no singular points and hence no candidates for singular relative extrema.

**Endpoints**  The domain is $[0, 4]$, so the endpoints occur when $x = 0$ and $x = 4$.

We record these values of $x$ in a table, together with the corresponding $y$-coordinates (values of $f$):

| $x$ | 0 | 1 | 4 |
|---|---|---|---|
| $f(x) = x^2 - 2x$ | 0 | $-1$ | 8 |

**Figure 9**

This gives us three points on the graph, $(0, 0)$, $(1, -1)$, and $(4, 8)$, which we plot in Figure 9. We remind ourselves that the point $(1, -1)$ is a stationary point of the graph by drawing in a part of the horizontal tangent line. Connecting these points must give us a graph something like that in Figure 10.

From Figure 10 we can see that $f$ has the following extrema:

| $x$ | $y = x^2 - 2x$ | *Classification* |
|---|---|---|
| 0 | 0 | Relative maximum (endpoint) |
| 1 | $-1$ | Absolute minimum (stationary point) |
| 4 | 8 | Absolute maximum (endpoint) |

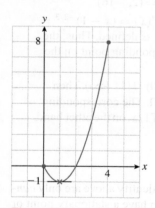

**Figure 10**

➡ **Before we go on...**  A little terminology: If the point $(a, b)$ on the graph of $f$ represents a maximum (or minimum) of $f$, we will sometimes say that *f* **has a maximum (or minimum) value of *b* at *x* = *a***, or simply that *f* **has a maximum (or minimum) at** $(a, b)$. Thus, in the above example, we could have said the following:

- "$f$ has a relative maximum value of 0 at $x = 0$," or "$f$ has a relative maximum at $(0, 0)$."
- "$f$ has an absolute minimum value of $-1$ at $x = 1$," or "$f$ has an absolute minimum at $(1, -1)$."
- "$f$ has an absolute maximum value of 8 at $x = 4$," or "$f$ has an absolute maximum at $(4, 8)$."

**Figure 11**

* Why "first" derivative test? To distinguish it from a test based on the **second derivative** of a function, which we shall discuss in Section 12.3.

**Q**: *How can we be sure that the graph in Example 1 doesn't look like Figure 11?*

**A**: If it did, there would be another critical point somewhere between $x = 1$ and $x = 4$. But we already know that there aren't any other critical points. The table we made listed all of the possible extrema; there can be no more.

## First Derivative Test

The **first derivative test**\* gives another, very systematic, way of checking whether a critical point is a maximum or minimum. To motivate the first derivative test, consider again the critical point $x = 1$ in Example 1. If we look at some values of $f'(x)$ to the left and right of the critical point, we obtain the information shown in the following table:

|  | Point to the Left | Critical Point | Point to the Right |
|---|---|---|---|
| $x$ | 0.5 | 1 | 2 |
| $f'(x) = 2x - 2$ | $-1$ | 0 | 2 |
| Direction of Graph | ↘ | → | ↗ |

At $x = 0.5$ (to the left of the critical point) we see that $f'(0.5) = -1 < 0$, so the graph has negative slope and $f$ is decreasing. We note this with the downward pointing arrow. At $x = 2$ (to the right of the critical point), we find $f'(2) = 2 > 0$, so the graph has positive slope and $f$ is increasing. In fact, because $f'(x) = 0$ only at $x = 1$, we know that $f'(x) < 0$ for all $x$ in $(0, 1)$, and we can say that $f$ is decreasing on the interval $(0, 1)$. Similarly, $f$ is increasing on $(1, 4)$.

So, starting at $x = 0$, the graph of $f$ goes down until we reach $x = 1$ and then it goes back up, telling us that $x = 1$ must be a minimum. Notice how the minimum is suggested by the arrows to the left and right.

### First Derivative Test for Extrema

Suppose that $c$ is a critical point of the continuous function $f$, and that its derivative is defined for $x$ close to, and on both sides of, $x = c$. Then, determine the sign of the derivative to the left and right of $x = c$.

1. If $f'(x)$ is positive to the left of $x = c$ and negative to the right, then $f$ has a maximum at $x = c$.

2. If $f'(x)$ is negative to the left of $x = c$ and positive to the right, then $f$ has a minimum at $x = c$.

3. If $f'(x)$ has the same sign on both sides of $x = c$, then $f$ has neither a maximum nor a minimum at $x = c$.

#### Quick Examples

1. In Example 1 above, we saw that $f(x) = x^2 - 2x$ has a critical point at $x = 1$ with $f'(x)$ negative to the left of $x = 1$ and positive to the right (see the table). Therefore, $f$ has a minimum at $x = 1$.

**2.** Here is a graph showing a function $f$ with a singular point at $x = 1$:

The graph gives us the information shown in the table:

| | Point to the Left | Critical Point | Point to the Right |
|---|---|---|---|
| $x$ | 0.5 | 1 | 1.5 |
| $f'(x)$ | + | Undefined | − |
| Direction of Graph | ↗ | | ↘ |

Since $f'(x)$ is positive to the left of $x = 1$ and negative to the right, we see that $f$ has a maximum at $x = 1$. (Notice again how this is suggested by the direction of the arrows.)

## EXAMPLE 2 Unbounded Interval

Find all extrema of $f(x) = 3x^4 - 4x^3$ on $[-1, \infty)$.

**Solution** We first calculate $f'(x) = 12x^3 - 12x^2$.

*Stationary points* We solve the equation $f'(x) = 0$, which is

$$12x^3 - 12x^2 = 0 \text{ or}$$
$$12x^2(x - 1) = 0.$$

There are two solutions, $x = 0$ and $x = 1$, and both are in the domain. These are our candidates for the $x$-coordinates of stationary extrema.

*Singular points* There are no points where $f'(x)$ is not defined, so there are no singular points.

*Endpoints* The domain is $[-1, \infty)$, so there is one endpoint, at $x = -1$.

We record these points in a table with the corresponding $y$-coordinates:

| $x$ | −1 | 0 | 1 |
|---|---|---|---|
| $f(x) = 3x^4 - 4x^3$ | 7 | 0 | −1 |

We will illustrate three methods we can use to determine which are minima, which are maxima, and which are neither:

**1.** Plot these points and sketch the graph by hand.

**2.** Use the first derivative test.

**3.** Use technology to help us.

Use the method you find most convenient.

**(a)**  **(b)**

**Figure 12**

**Figure 13**

**Figure 14**

***Using a Hand Plot:*** If we plot these points by hand, we obtain Figure 12(a), which suggests Figure 12(b).

We can't be sure what happens to the right of $x = 1$. Does the curve go up, or does it go down? To find out, let's plot a "test point" to the right of $x = 1$. Choosing $x = 2$, we obtain $y = 3(2)^4 - 4(2)^3 = 16$, so $(2, 16)$ is another point on the graph. Thus, it must turn upward to the right of $x = 1$, as shown in Figure 13.

From the graph, we find that $f$ has the following extrema:

A relative (endpoint) maximum at $(-1, 7)$

An absolute (stationary) minimum at $(1, -1)$

***Using the First Derivative Test:*** List the critical and endpoints in a table, and add additional points as necessary so that each critical point has a noncritical point on either side. Then compute the derivative at each of these points, and draw an arrow to indicate the direction of the graph.

|  | Endpoint |  | Critical Point |  | Critical Point |  |
|---|---|---|---|---|---|---|
| $x$ | $-1$ | $-0.5$ | $0$ | $0.5$ | $1$ | $2$ |
| $f'(x) = 12x^3 - 12x^2$ | $-24$ | $-1.5$ | $0$ | $-1.5$ | $0$ | $48$ |
| **Direction of Graph** |  | $\searrow$ | $\rightarrow$ | $\searrow$ | $\rightarrow$ | $\nearrow$ |

Notice that the arrows now suggest the shape of the curve in Figure 13. The first derivative test tells us that the function has a relative maximum at $x = -1$, neither a maximum nor a minimum at $x = 0$, and a relative minimum at $x = 1$. Deciding which of these extrema are absolute and which are relative requires us to compute $y$-coordinates and plot the corresponding points on the graph by hand, as we did in the first method.

### 📟 using Technology

If we use technology to show the graph, we should choose the viewing window so that it contains the three interesting points we found: $x = -1$, $x = 0$, and $x = 1$. Again, we can't be sure yet what happens to the right of $x = 1$; does the graph go up or down from that point? If we set the viewing window to an interval of $[-1, 2]$ for $x$ and $[-2, 8]$ for $y$, we will leave enough room to the right of $x = 1$ and below $y = -1$ to see what the graph will do. The result will be something like Figure 14.

Now we can tell what happens to the right of $x = 1$: the function increases. We know that it cannot later decrease again because if it did, there would have to be another critical point where it turns around, and we found that there are no other critical points. ∎

➡ **Before we go on...** Notice that the stationary point at $x = 0$ in Example 2 is neither a relative maximum nor a relative minimum. It is simply a place where the graph of $f$ flattens out for a moment before it continues to fall. Notice also that $f$ has no absolute maximum because $f(x)$ increases without bound as $x$ gets large. ∎

### EXAMPLE 3 Singular Point

Find all extrema of $f(t) = t^{2/3}$ on $[-1, 1]$.

**Solution** First, $f'(t) = \dfrac{2}{3}t^{-1/3}$.

*Stationary points* We need to solve

$$\frac{2}{3}t^{-1/3} = 0.$$

We can rewrite this equation without the negative exponent:

$$\frac{2}{3t^{1/3}} = 0.$$

Now, the only way that a fraction can equal 0 is if the numerator is 0, so this fraction can never equal 0. Thus, there are no stationary points.

***Singular Points***   The derivative

$$f'(t) = \frac{2}{3t^{1/3}}$$

is not defined for $t = 0$. However, 0 is in the interior of the domain of $f$ (although $f'$ is not defined at $t = 0$, $f$ itself is). Thus, $f$ has a singular point at $t = 0$.

***Endpoints***   There are two endpoints, $-1$ and $1$.

We now put these three points in a table with the corresponding $y$-coordinates:

| $t$ | $-1$ | $0$ | $1$ |
|---|---|---|---|
| $f(t)$ | $1$ | $0$ | $1$ |

***Using a Hand Plot:*** The derivative, $f'(t) = 2/(3t^{1/3})$, is not defined at the singular point $t = 0$. To help us sketch the graph, let's use limits to investigate what happens to the derivative as we approach 0 from either side:

$$\lim_{t \to 0^-} f'(t) = \lim_{t \to 0^-} \frac{2}{3t^{1/3}} = -\infty$$

$$\lim_{t \to 0^+} f'(t) = \lim_{t \to 0^+} \frac{2}{3t^{1/3}} = +\infty.$$

**Figure 15**

Thus, the graph decreases very steeply, approaching $t = 0$ from the left, and then rises very steeply as it leaves to the right. It would make sense to say that the tangent line at $x = 0$ is vertical, as seen in Figure 15.

From this graph, we find the following extrema for $f$:

An absolute (endpoint) maximum at $(-1, 1)$

An absolute (singular) minimum at $(0, 0)$

An absolute (endpoint) maximum at $(1, 1)$.

Notice that the absolute maximum value of $f$ is achieved at two values of $t$: $t = -1$ and $t = 1$.

***First Derivative Test:*** Here is the corresponding table for the first derivative test.

| $t$ | $-0.5$ | $0$ | $0.5$ |
|---|---|---|---|
| $f'(t) = \dfrac{2}{3t^{1/3}}$ | $-\dfrac{2}{3(0.5)^{1/3}}$ | Undefined | $\dfrac{2}{3(0.5)^{1/3}}$ |
| **Direction of Graph** | ↘ | ↕ | ↗ |

(We drew a vertical arrow at $t = 0$ to indicate a vertical tangent.) Again, notice how the arrows suggest the shape of the curve in Figure 15, and the first derivative test confirms that we have a minimum at $x = 0$.

### using Technology

Because there is only one critical point, at $t = 0$, it is clear from this table that $f$ must decrease from $t = -1$ to $t = 0$ and then increase from $t = 0$ to $t = 1$. To graph $f$ using technology, choose a viewing window with an interval of $[-1, 1]$ for $t$ and $[0, 1]$ for $y$. The result will be something like Figure 15.✱ ■

In Examples 1 and 3, we could have found the absolute maxima and minima without doing any graphing. In Example 1, after finding the critical points and endpoints, we created the following table:

| $x$ | 0 | 1 | 4 |
|---|---|---|---|
| $f(x)$ | 0 | −1 | 8 |

From this table we can see that $f$ must decrease from its value of 0 at $x = 0$ to −1 at $x = 1$, and then increase to 8 at $x = 4$. The value of 8 must be the largest value it takes on, and the value of −1 must be the smallest, on the interval $[0, 4]$. Similarly, in Example 3 we created the following table:

| $t$ | −1 | 0 | 1 |
|---|---|---|---|
| $f(t)$ | 1 | 0 | 1 |

From this table we can see that the largest value of $f$ on the interval $[-1, 1]$ is 1 and the smallest value is 0. We are taking advantage of the following fact, the proof of which uses some deep and beautiful mathematics (alas, beyond the scope of this book):

## Extreme Value Theorem

If $f$ is *continuous* on a *closed interval* $[a, b]$, then it will have an absolute maximum and an absolute minimum value on that interval. Each absolute extremum must occur at either an endpoint or a critical point. Therefore, the absolute maximum is the largest value in a table of the values of $f$ at the endpoints and critical points, and the absolute minimum is the smallest value.

### Quick Example

The function $f(x) = 3x - x^3$ on the interval $[0, 2]$ has one critical point at $x = 1$. The values of $f$ at the critical point and the endpoints of the interval are given in the following table:

| | Endpoint | Critical point | Endpoint |
|---|---|---|---|
| $x$ | 0 | 1 | 2 |
| $f(x)$ | 0 | 2 | −2 |

From this table we can say that the absolute maximum value of $f$ on $[0, 2]$ is 2, which occurs at $x = 1$, and the absolute minimum value of $f$ is −2, which occurs at $x = 2$.

As we can see in Example 2 and the following examples, if the domain is not a closed interval, then $f$ may not have an absolute maximum and minimum, and a table of values as above is of little help in determining whether it does.

### EXAMPLE 4  Domain Not a Closed Interval

Find all extrema of $f(x) = x + \dfrac{1}{x}$.

**Solution**  Because no domain is specified, we take the domain to be as large as possible. The function is not defined at $x = 0$ but is at all other points, so we take its domain to be $(-\infty, 0) \cup (0, +\infty)$. We calculate

$$f'(x) = 1 - \frac{1}{x^2}.$$

**Stationary Points**  Setting $f'(x) = 0$, we solve

$$1 - \frac{1}{x^2} = 0$$

to find $x = \pm 1$. Calculating the corresponding values of $f$, we get the two stationary points $(1, 2)$ and $(-1, -2)$.

**Singular Points**  The only value of $x$ for which $f'(x)$ is not defined is $x = 0$, but then $f$ is not defined there either, so there are no singular points in the domain.

**Endpoints**  The domain, $(-\infty, 0) \cup (0, +\infty)$, has no endpoints.

From this scant information, it is hard to tell what $f$ does. If we are sketching the graph by hand, or using the first derivative test, we will need to plot additional "test points" to the left and right of the stationary points $x = \pm 1$.

---

 **using Technology**

For the technology approach, let's choose a viewing window with an interval of $[-3, 3]$ for $x$ and $[-4, 4]$ for $y$, which should leave plenty of room to see how $f$ behaves near the stationary points. The result is something like Figure 16.

From this graph we can see that $f$ has:

A relative (stationary) maximum at $(-1, -2)$

A relative (stationary) minimum at $(1, 2)$

Curiously, the relative maximum is lower than the relative minimum! Notice also that, because of the break in the graph at $x = 0$, the graph did not need to rise to get from $(-1, -2)$ to $(1, 2)$.  ∎

---

**Figure 16**

So far we have been solving the equation $f'(x) = 0$ to obtain our candidates for stationary extrema. However, it is often not easy—or even possible—to solve equations analytically. In the next example, we show a way around this problem by using graphing technology.

---

### EXAMPLE 5  ▣ Finding Approximate Extrema Using Technology

Graph the function $f(x) = (x - 1)^{2/3} - \dfrac{x^2}{2}$ with domain $[-2, +\infty)$. Also graph its derivative and hence locate and classify all extrema of $f$, with coordinates accurate to two decimal places.

**Solution** In Example 4 of Section 10.5, we saw how to draw the graphs of $f$ and $f'$ using technology. Note that the technology formula to use for the graph of $f$ is

```
((x-1)^2)^(1/3)-0.5*x^2
```

instead of

```
(x-1)^(2/3)-0.5*x^2
```

(Why?)

Figure 17 shows the resulting graphs of $f$ and $f'$.

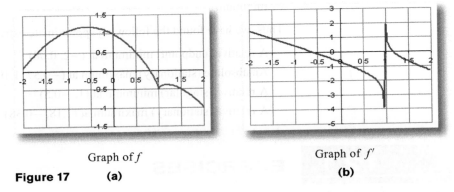

Graph of $f$            Graph of $f'$

**Figure 17**    **(a)**                   **(b)**

If we extend Xmax beyond $x = 2$, we find that the graph continues downward, apparently without any further interesting behavior.

***Stationary Points*** The graph of $f$ shows two stationary points, both maxima, at around $x = -0.6$ and $x = 1.2$. Notice that the graph of $f'$ is zero at these points. Moreover, it is easier to locate these values accurately on the graph of $f'$ because it is easier to pinpoint where a graph crosses the $x$-axis than to locate a stationary point. Zooming in to the stationary point at $x \approx -0.6$ results in Figure 18.

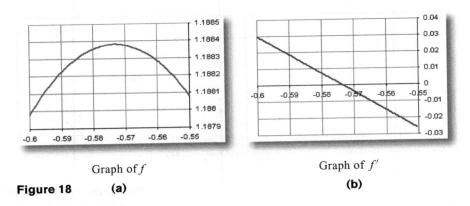

Graph of $f$            Graph of $f'$

**Figure 18**    **(a)**                   **(b)**

From the graph of $f$, we can see that the stationary point is somewhere between $-0.58$ and $-0.57$. The graph of $f'$ shows more clearly that the zero of $f'$, hence the stationary point of $f$ lies somewhat closer to $-0.57$ than to $-0.58$. Thus, the stationary point occurs at $x \approx -0.57$, rounded to two decimal places.

In a similar way, we find the second stationary point at $x \approx 1.18$.

***Singular Points*** Going back to Figure 17, we notice what appears to be a cusp (singular point) at the relative minimum around $x = 1$, and this is confirmed by a glance at the graph of $f'$, which seems to take a sudden jump at that value. Zooming in closer suggests that the singular point occurs at exactly $x = 1$. In fact, we can calculate

$$f'(x) = \frac{2}{3(x-1)^{1/3}} - x.$$

From this formula we see clearly that $f'(x)$ is defined everywhere except at $x = 1$.

***Endpoints*** The only endpoint in the domain is $x = -2$, which gives a relative minimum.

Thus, we have found the following approximate extrema for $f$:

A relative (endpoint) minimum at $(-2, 0.08)$

An absolute (stationary) maximum at $(-0.57, 1.19)$

A relative (singular) minimum at $(1, -0.5)$

A relative (stationary) maximum at $(1.18, -0.38)$.

---

## 12.1  EXERCISES

▼ more advanced    ◆ challenging

Ⓣ indicates exercises that should be solved using technology

*In Exercises 1–12, locate and classify all extrema in each graph. (By classifying the extrema, we mean listing whether each extremum is a relative or absolute maximum or minimum.) Also, locate any stationary points or singular points that are not relative extrema.* HINT [See Figure 8.]

**1.**

**2.**

**3.**

**4.**

**5.**

**6.**

**7.**

**8.**

**9.**

**10.**

**11.**

**12.**

*Find the exact location of all the relative and absolute extrema of each function in Exercises 13–44.* HINT [See Example 1.]

**13.** $f(x) = x^2 - 4x + 1$ with domain $[0, 3]$

**14.** $f(x) = 2x^2 - 2x + 3$ with domain $[0, 3]$

**15.** $g(x) = x^3 - 12x$ with domain $[-4, 4]$

**16.** $g(x) = 2x^3 - 6x + 3$ with domain $[-2, 2]$

**17.** $f(t) = t^3 + t$ with domain $[-2, 2]$

**18.** $f(t) = -2t^3 - 3t$ with domain $[-1, 1]$

**19.** $h(t) = 2t^3 + 3t^2$ with domain $[-2, +\infty)$ HINT [See Example 2.]

**20.** $h(t) = t^3 - 3t^2$ with domain $[-1, +\infty)$ HINT [See Example 2.]

**21.** $f(x) = x^4 - 4x^3$ with domain $[-1, +\infty)$

**22.** $f(x) = 3x^4 - 2x^3$ with domain $[-1, +\infty)$

**23.** $g(t) = \frac{1}{4}t^4 - \frac{2}{3}t^3 + \frac{1}{2}t^2$ with domain $(-\infty, +\infty)$

**24.** $g(t) = 3t^4 - 16t^3 + 24t^2 + 1$ with domain $(-\infty, +\infty)$

**25.** $h(x) = (x - 1)^{2/3}$ with domain $[0, 2]$ HINT [See Example 3.]

**26.** $h(x) = (x + 1)^{2/5}$ with domain $[-2, 0]$ HINT [See Example 3.]

**27.** $k(x) = \frac{2x}{3} + (x + 1)^{2/3}$ with domain $(-\infty, 0]$

**28.** $k(x) = \frac{2x}{5} - (x - 1)^{2/5}$ with domain $[0, +\infty)$

**29.** ▼ $f(t) = \frac{t^2 + 1}{t^2 - 1}; -2 \le t \le 2, t \ne \pm 1$

**30.** ▼ $f(t) = \frac{t^2 - 1}{t^2 + 1}$ with domain $[-2, 2]$

**31.** ▼ $f(x) = \sqrt{x}(x - 1); x \ge 0$

**32.** ▼ $f(x) = \sqrt{x}(x + 1); x \ge 0$

**33.** ▼ $g(x) = x^2 - 4\sqrt{x}$

**34.** ▼ $g(x) = \frac{1}{x} - \frac{1}{x^2}$

**35.** ▼ $g(x) = \frac{x^3}{x^2 + 3}$

**36.** ▼ $g(x) = \frac{x^3}{x^2 - 3}$

**37.** ▼ $f(x) = x - \ln x$ with domain $(0, +\infty)$

**38.** ▼ $f(x) = x - \ln x^2$ with domain $(0, +\infty)$

**39.** ▼ $g(t) = e^t - t$ with domain $[-1, 1]$

**40.** ▼ $g(t) = e^{-t^2}$ with domain $(-\infty, +\infty)$

**41.** ▼ $f(x) = \frac{2x^2 - 24}{x + 4}$

**42.** ▼ $f(x) = \frac{x - 4}{x^2 + 20}$

**43.** ▼ $f(x) = xe^{1-x^2}$

**44.** ▼ $f(x) = x \ln x$ with domain $(0, +\infty)$

*In Exercises 45–48, use graphing technology and the method in Example 5 to find the x-coordinates of the critical points, accurate to two decimal places. Find all relative and absolute maxima and minima.* HINT [See Example 5.]

**45.** ▮ $y = x^2 + \frac{1}{x - 2}$ with domain $(-3, 2) \cup (2, 6)$

**46.** ▮ $y = x^2 - 10(x - 1)^{2/3}$ with domain $(-4, 4)$

**47.** ▮ $f(x) = (x - 5)^2(x + 4)(x - 2)$ with domain $[-5, 6]$

**48.** ▮ $f(x) = (x + 3)^2(x - 2)^2$ with domain $[-5, 5]$

*In Exercises 49–56, the graph of the derivative of a function f is shown. Determine the x-coordinates of all stationary and singular points of f, and classify each as a relative maximum, relative minimum, or neither. (Assume that f(x) is defined and continuous everywhere in $[-3, 3]$.)* HINT [See Example 5.]

**49.** ▼

**50.** ▼

**51.** ▼

**52.** ▼

**53.** ▼

**54.** ▼

**55.** ▼

**56.** ▼

**COMMUNICATION AND REASONING EXERCISES**

**57.** Draw the graph of a function $f$ with domain the set of all real numbers, such that $f$ is not linear and has no relative extrema.

**58.** Draw the graph of a function $g$ with domain the set of all real numbers, such that $g$ has a relative maximum and minimum but no absolute extrema.

**59.** Draw the graph of a function that has stationary and singular points but no relative extrema.

**60.** Draw the graph of a function that has relative, not absolute, maxima and minima, but has no stationary or singular points.

**61.** If a stationary point is not a relative maximum, then must it be a relative minimum? Explain your answer.

**62.** If one endpoint is a relative maximum, must the other be a relative minimum? Explain your answer.

**63.** ▼ We said that if $f$ is continuous on a closed interval $[a, b]$, then it will have an absolute maximum and an absolute minimum. Draw the graph of a function with domain $[0, 1]$ having an absolute maximum but no absolute minimum.

**64.** ▼ Refer to Exercise 63. Draw the graph of a function with domain $[0, 1]$ having no absolute extrema.

**65.** ☐ ▼ Must endpoints always be extrema? Consider the following function (based on the trigonometric sine function—see Chapter 16 for a discussion of its properties):

$$f(x) = \begin{cases} x \sin\left(\dfrac{1}{x}\right) & \text{if } x > 0 \\ 0 & \text{if } x = 0 \end{cases}.$$    Technology formula: x*sin(1/x)

Graph this function using the technology formula above for $0 \le x \le h$, choosing smaller and smaller values of $h$, and decide whether $f$ has a either a relative maximum or relative

minimum at the endpoint $x = 0$. Explain your answer. (Note: Very few graphers can draw this curve accurately; the grapher on the Website does a good job (you can increase the number of points to plot for more beautiful results), the grapher that comes with Mac computers is probably among the best, while the TI-83/84 Plus is probably among the worst.)

**66.** ☐ ▼ Refer to the preceding exercise, and consider the function

$$f(x) = \begin{cases} x^2 \sin\left(\dfrac{1}{x}\right) & \text{if } x \ne 0 \\ 0 & \text{if } x = 0 \end{cases}.$$    Technology formula: x^2*sin(1/x)

Graph this function using the technology formula above for $-h \le x \le h$, choosing smaller and smaller values of $h$, and decide **(a)** whether $x = 0$ is a stationary point, and **(b)** whether $f$ has either a relative maximum or a relative minimum at $x = 0$. Explain your answers. HINT [For part (a), use technology to estimate the derivative at $x = 0$.]

---

## 12.2  Applications of Maxima and Minima

In many applications we would like to find the largest or smallest possible value of some quantity—for instance, the greatest possible profit or the lowest cost. We call this the *optimal* (best) value. In this section we consider several such examples and use calculus to find the optimal value in each.

In all applications the first step is to translate a written description into a mathematical problem. In the problems we look at in this section, there are *unknowns* that we are asked to find, there is an expression involving those unknowns that must be made as large or as small as possible—the **objective function**—and there may be **constraints**—equations or inequalities relating the variables.*

✳ If you have studied linear programming, you will notice a similarity here, but unlike the situation in linear programming, neither the objective function nor the constraints need be linear.

### EXAMPLE 1  Minimizing Average Cost

*Gymnast Clothing* manufactures expensive hockey jerseys for sale to college bookstores in runs of up to 500. Its cost (in dollars) for a run of $x$ hockey jerseys is

$$C(x) = 2{,}000 + 10x + 0.2x^2.$$

How many jerseys should Gymnast produce per run in order to minimize average cost?[†]

**Solution**  Here is the procedure we will follow to solve problems like this.

**1. Identify the unknown(s).** There is one unknown: $x$, the number of hockey jerseys Gymnast should produce per run. (We know this because the question is, How many jerseys . . . ?)

**2. Identify the objective function.** The objective function is the quantity that must be made as small (in this case) as possible. In this example it is the average cost, which is given by

$$\bar{C}(x) = \frac{C(x)}{x} = \frac{2{,}000 + 10x + 0.2x^2}{x}$$

$$= \frac{2{,}000}{x} + 10 + 0.2x \text{ dollars/jersey.}$$

[†] Why don't we seek to minimize total cost? The answer would be uninteresting; to minimize total cost, we would make *no* jerseys at all. Minimizing the average cost is a more practical objective.

**3. *Identify the constraints (if any).*** At most 500 jerseys can be manufactured in a run. Also, $\bar{C}(0)$ is not defined. Thus, $x$ is constrained by

$$0 < x \le 500.$$

Put another way, the domain of the objective function $\bar{C}(x)$ is (0, 500].

**4. *State and solve the resulting optimization problem.*** Our optimization problem is:

$$\text{Minimize } \bar{C}(x) = \frac{2{,}000}{x} + 10 + 0.2x \qquad \text{Objective function}$$

$$\text{subject to } 0 < x \le 500. \qquad\qquad \text{Constraint}$$

We now solve this problem as in Section 12.1. We first calculate

$$\bar{C}'(x) = -\frac{2{,}000}{x^2} + 0.2.$$

We solve $\bar{C}'(x) = 0$ to find $x = \pm 100$. We reject $x = -100$ because $-100$ is not in the domain of $\bar{C}$ (and makes no sense), so we have one stationary point, at $x = 100$. There, the average cost is $\bar{C}(100) = \$50$ per jersey.

The only point at which the formula for $\bar{C}'$ is not defined is $x = 0$, but that is not in the domain of $\bar{C}$, so we have no singular points. We have one endpoint in the domain, at $x = 500$. There, the average cost is $\bar{C}(500) = \$114$.

**Figure 19**

---

▦ **using Technology**

Let's plot $\bar{C}$ in a viewing window with the intervals [0, 500] for $x$ and [0, 150] for $y$, which will show the whole domain and the two interesting points we've found so far. The result is Figure 19.

From the graph of $\bar{C}$, we can see that the stationary point at $x = 100$ gives the absolute minimum. We can therefore say that Gymnast Clothing should produce 100 jerseys per run, for a lowest possible average cost of \$50 per jersey. ∎

---

### EXAMPLE 2 Maximizing Area

Slim wants to build a rectangular enclosure for his pet rabbit, Killer, against the side of his house, as shown in Figure 20. He has bought 100 feet of fencing. What are the dimensions of the largest area that he can enclose?

**Figure 20**

**Solution**

**1. *Identify the unknown(s).*** To identify the unknown(s), we look at the question: What are the *dimensions* of the largest area he can enclose? Thus, the unknowns are the dimensions of the fence. We call these $x$ and $y$, as shown in Figure 21.

**2. *Identify the objective function.*** We look for what it is that we are trying to maximize (or minimize). The phrase "largest area" tells us that our object is to *maximize the area*, which is the product of length and width, so our objective function is

$$A = xy, \text{ where } A \text{ is the area of the enclosure.}$$

**Figure 21**

**3. *Identify the constraints (if any).*** What stops Slim from making the area as large as he wants? He has only 100 feet of fencing to work with. Looking again at Figure 21, we see that the sum of the lengths of the three sides must equal 100, so

$$x + 2y = 100.$$

One more point: Because $x$ and $y$ represent the lengths of the sides of the enclosure, neither can be a negative number.

**4. *State and solve the resulting optimization problem.*** Our mathematical problem is:

Maximize $A = xy$                                Objective function

subject to $x + 2y = 100$, $x \geq 0$, and $y \geq 0$.        Constraints

We know how to find maxima and minima of a function of one variable, but $A$ appears to depend on two variables. We can remedy this by using a constraint to express one variable in terms of the other. Let's take the constraint $x + 2y = 100$ and solve for $x$ in terms of $y$:

$$x = 100 - 2y.$$

Substituting into the objective function gives

$$A = xy = (100 - 2y)y = 100y - 2y^2$$

and we have eliminated $x$ from the objective function. What about the inequalities? One says that $x \geq 0$, but we want to eliminate $x$ from this as well. We substitute for $x$ again, getting

$$100 - 2y \geq 0.$$

Solving this inequality for $y$ gives $y \leq 50$. The second inequality says that $y \geq 0$. Now, we can restate our problem with $x$ eliminated:

Maximize $A(y) = 100y - 2y^2$ subject to $0 \leq y \leq 50$.

We now proceed with our usual method of solving such problems. We calculate $A'(y) = 100 - 4y$. Solving $100 - 4y = 0$, we get one stationary point at $y = 25$. There, $A(25) = 1,250$. There are no points at which $A'(y)$ is not defined, so there are no singular points. We have two endpoints, at $y = 0$ and $y = 50$. The corresponding areas are $A(0) = 0$ and $A(50) = 0$. We record the three points we found in a table:

| $y$ | 0 | 25 | 50 |
|---|---|---|---|
| $A(y)$ | 0 | 1,250 | 0 |

It's clear now how $A$ must behave: It increases from 0 at $y = 0$ to 1,250 at $y = 25$ and then decreases back to 0 at $y = 50$. Thus, the largest possible value of $A$ is 1,250 square feet, which occurs when $y = 25$. To completely answer the question that was asked, we need to know the corresponding value of $x$. We have $x = 100 - 2y$, so $x = 50$ when $y = 25$. Thus, Slim should build his enclosure 50 feet across and 25 feet deep (with the "missing" 50-foot side being formed by part of the house).

➡ **Before we go on...** Notice that the problem in Example 2 came down to finding the absolute maximum value of $A$ on the closed and bounded interval $[0, 50]$. As we noted in the preceding section, the table of values of $A$ at its critical points and the endpoints of the interval gives us enough information to find the absolute maximum. ∎

Let's stop for a moment and summarize the steps we've taken in these two examples.

## Solving an Optimization Problem

1. **Identify the unknown(s), possibly with the aid of a diagram.** These are usually the quantities asked for in the problem.

2. **Identify the objective function.** This is the quantity you are asked to maximize or minimize. You should name it explicitly, as in "Let $S =$ surface area."

3. **Identify the constraint(s).** These can be equations relating variables or inequalities expressing limitations on the values of variables.

4. **State the optimization problem.** This will have the form "Maximize [minimize] the objective function subject to the constraint(s)."

5. **Eliminate extra variables.** If the objective function depends on several variables, solve the constraint equations to express all variables in terms of one particular variable. Substitute these expressions into the objective function to rewrite it as a function of a single variable. In short, if there is only one constraint equation:

    *Solve the constraint for one of the unknowns and substitute into the objective.*

    Also substitute the expressions into any inequality constraints to help determine the domain of the objective function.

6. **Find the absolute maximum (or minimum) of the objective function.** Use the techniques of the preceding section.

Now for some further examples.

---

### EXAMPLE 3 Maximizing Revenue

*Cozy Carriage Company* builds baby strollers. Using market research, the company estimates that if it sets the price of a stroller at $p$ dollars, then it can sell $q = 300,000 - 10p^2$ strollers per year.* What price will bring in the greatest annual revenue?

* This equation is, of course, the demand equation for the baby strollers. However, coming up with a suitable demand equation in real life is hard, to say the least. In this regard, the very entertaining and also insightful article, *Camels and Rubber Duckies* by Joel Spolsky at www.joelonsoftware.com/articles/CamelsandRubberDuckies.html is a must-read.

**Solution** The question we are asked identifies our main unknown, the price $p$. However, there is another quantity that we do not know, $q$, the number of strollers the company will sell per year. The question also identifies the objective function, revenue, which is

$$R = pq.$$

Including the equality constraint given to us, that $q = 300,000 - 10p^2$, and the "reality" inequality constraints $p \geq 0$ and $q \geq 0$, we can write our problem as

Maximize $R = pq$ subject to $q = 300,000 - 10p^2$, $p \geq 0$, and $q \geq 0$.

We are given $q$ in terms of $p$, so let's substitute to eliminate $q$:

$$R = pq = p(300,000 - 10p^2) = 300,000p - 10p^3.$$

Substituting in the inequality $q \geq 0$, we get

$$300,000 - 10p^2 \geq 0.$$

Thus, $p^2 \leq 30,000$, which gives $-100\sqrt{3} \leq p \leq 100\sqrt{3}$. When we combine this with $p \geq 0$, we get the following restatement of our problem:

Maximize $R(p) = 300,000p - 10p^3$ such that $0 \leq p \leq 100\sqrt{3}$.

We solve this problem in much the same way we did the preceding one. We calculate $R'(p) = 300,000 - 30p^2$. Setting $300,000 - 30p^2 = 0$, we find one stationary point at $p = 100$. There are no singular points and we have the endpoints $p = 0$ and $p = 100\sqrt{3}$. Putting these points in a table and computing the corresponding values of $R$, we get the following:

| $p$ | 0 | 100 | $100\sqrt{3}$ |
|-----|---|-----|---------------|
| $R(p)$ | 0 | 20,000,000 | 0 |

Thus, Cozy Carriage should price its strollers at $100 each, which will bring in the largest possible revenue of $20,000,000.

**Figure 22**

## EXAMPLE 4   Optimizing Resources

*The Metal Can Company* has an order to make cylindrical cans with a volume of 250 cubic centimeters. What should be the dimensions of the cans in order to use the least amount of metal in their production?

**Solution**  We are asked to find the dimensions of the cans. It is traditional to take as the dimensions of a cylinder the height $h$ and the radius of the base $r$, as in Figure 22. We are also asked to minimize the amount of metal used in the can, which is the area of the surface of the cylinder. We can look up the formula or figure it out ourselves: Imagine removing the circular top and bottom and then cutting vertically and flattening out the hollow cylinder to get a rectangle, as shown in Figure 23.

**Figure 23**

Our objective function is the (total) surface area $S$ of the can. The area of each disc is $\pi r^2$, while the area of the rectangular piece is $2\pi r h$. Thus, our objective function is

$$S = 2\pi r^2 + 2\pi r h.$$

As usual, there is a constraint: The volume must be exactly 250 cubic centimeters. The formula for the volume of a cylinder is $V = \pi r^2 h$, so

$$\pi r^2 h = 250.$$

It is easiest to solve this constraint for $h$ in terms of $r$:

$$h = \frac{250}{\pi r^2}.$$

Substituting in the objective function, we get

$$S = 2\pi r^2 + 2\pi r \frac{250}{\pi r^2} = 2\pi r^2 + \frac{500}{r}.$$

Now $r$ cannot be negative or 0, but it can become very large (a very wide but very short can could have the right volume). We therefore take the domain of $S(r)$ to be $(0, +\infty)$, so our mathematical problem is as follows:

$$\text{Minimize } S(r) = 2\pi r^2 + \frac{500}{r} \text{ subject to } r > 0.$$

Now we calculate

$$S'(r) = 4\pi r - \frac{500}{r^2}.$$

To find stationary points, we set this equal to 0 and solve:

$$4\pi r - \frac{500}{r^2} = 0$$

$$4\pi r = \frac{500}{r^2}$$

$$4\pi r^3 = 500$$

$$r^3 = \frac{125}{\pi}.$$

So

$$r = \sqrt[3]{\frac{125}{\pi}} = \frac{5}{\sqrt[3]{\pi}} \approx 3.41.$$

The corresponding surface area is approximately $S(3.41) \approx 220$. There are no singular points or endpoints in the domain.

**Figure 24**

▦ **using** Technology

To see how $S$ behaves near the one stationary point, let's graph it in a viewing window with interval $[0, 5]$ for $r$ and $[0, 300]$ for $S$. The result is Figure 24.

From the graph we can clearly see that the smallest surface area occurs at the stationary point at $r \approx 3.41$. The height of the can will be

$$h = \frac{250}{\pi r^2} \approx 6.83.$$

Thus, the can that uses the least amount of metal has a height of approximately 6.83 centimeters and a radius of approximately 3.41 centimeters. Such a can will use approximately 220 square centimeters of metal.

➡ **Before we go on...** We obtained the value of $r$ in Example 4 by solving the equation

$$4\pi r = \frac{500}{r^2}.$$

This time, let us do things differently: Divide both sides by $4\pi$ to obtain

$$r = \frac{500}{4\pi r^2} = \frac{125}{\pi r^2}$$

and compare what we got with the expression for $h$:

$$h = \frac{250}{\pi r^2},$$

which we see is exactly twice the expression for $r$. Put another way, the height is exactly equal to the diameter so that the can looks square when viewed from the side. Have you ever seen cans with that shape? Why do you think most cans do not have this shape? ∎

---

### EXAMPLE 5  Allocation of Labor

*The Gym Sock Company* manufactures cotton athletic socks. Production is partially automated through the use of robots. Daily operating costs amount to $50 per laborer and $30 per robot. The number of pairs of socks the company can manufacture in a day is given by a Cobb-Douglas* production formula

$$q = 50n^{0.6}r^{0.4},$$

where $q$ is the number of pairs of socks that can be manufactured by $n$ laborers and $r$ robots. Assuming that the company wishes to produce 1,000 pairs of socks per day at a minimum cost, how many laborers and how many robots should it use?

*Cobb-Douglas production formulas were discussed in Section 11.6.

**Solution**  The unknowns are the number of laborers $n$ and the number of robots $r$. The objective is to minimize the daily cost:

$$C = 50n + 30r.$$

The constraints are given by the daily quota

$$1,000 = 50n^{0.6}r^{0.4}$$

and the fact that $n$ and $r$ are nonnegative. We solve the constraint equation for one of the variables; let's solve for $n$:

$$n^{0.6} = \frac{1,000}{50r^{0.4}} = \frac{20}{r^{0.4}}.$$

Taking the $1/0.6$ power of both sides gives

$$n = \left(\frac{20}{r^{0.4}}\right)^{1/0.6} = \frac{20^{1/0.6}}{r^{0.4/0.6}} = \frac{20^{5/3}}{r^{2/3}} \approx \frac{147.36}{r^{2/3}}.$$

Substituting in the objective equation gives us the cost as a function of $r$:

$$C(r) \approx 50\left(\frac{147.36}{r^{2/3}}\right) + 30r$$
$$= 7,368r^{-2/3} + 30r.$$

The only remaining constraint on $r$ is that $r > 0$. To find the minimum value of $C(r)$, we first take the derivative:

$$C'(r) \approx -4,912r^{-5/3} + 30.$$

Setting this equal to zero, we solve for $r$:

$$r^{-5/3} \approx 0.006107$$
$$r \approx (0.006107)^{-3/5} \approx 21.3.$$

The corresponding cost is $C(21.3) \approx \$1,600$. There are no singular points or endpoints in the domain of $C$.

**Figure 25**

### using Technology

To see how $C$ behaves near its stationary point, let's draw its graph in a viewing window with an interval of $[0, 40]$ for $r$ and $[0, 2{,}000]$ for $C$. The result is Figure 25.

From the graph we can see that $C$ does have its minimum at the stationary point. The corresponding value of $n$ is

$$n \approx \frac{147.36}{r^{2/3}} \approx 19.2.$$

At this point, our solution appears to be this: Use (approximately) 19.2 laborers and (approximately) 21.3 robots to meet the manufacturing quota at a minimum cost. However, we are not interested in fractions of robots or people, so we need to find integer solutions for $n$ and $r$. If we round these numbers, we get the solution $(n, r) = (19, 21)$. However, a quick calculation shows that

$$q = 50(19)^{0.6}(21)^{0.4} \approx 989 \text{ pairs of socks,}$$

which fails to meet the quota of 1,000. Thus, we need to round at least one of the quantities $n$ and $r$ *upward* in order to meet the quota. The three possibilities, with corresponding values of $q$ and $C$, are as follows:

$$(n, r) = (20, 21), \text{ with } q \approx 1{,}020 \text{ and } C = \$1{,}630$$
$$(n, r) = (19, 22), \text{ with } q \approx 1{,}007 \text{ and } C = \$1{,}610$$
$$(n, r) = (20, 22), \text{ with } q \approx 1{,}039 \text{ and } C = \$1{,}660.$$

Of these, the solution that meets the quota at a minimum cost is $(n, r) = (19, 22)$. Thus, the Gym Sock Co. should use 19 laborers and 22 robots, at a cost of $50 \times 19 + 30 \times 22 = \$1{,}610$, to manufacture $50 \times 19^{0.6} \times 22^{0.4} \approx 1{,}007$ pairs of socks.

### FAQs

**Constraints and Objectives**

**Q** : *How do I know whether or not there are constraints in an applied optimization problem?*

**A** : There are usually at least *inequality* constraints; the variables usually represent real quantities, such as length or number of items, and so cannot be negative, leading to constraints like $x \geq 0$ (or $0 \leq x \leq 100$ in the event that there is an upper limit). *Equation* constraints usually arise when there is more than one unknown in the objective, and dictate how one unknown is related to others; as in, say "the length is twice the width," or "the demand is 8 divided by the price" (a demand equation).

**Q** : *How do I know what to use as the objective, and what to use as the constraint(s)?*

**A** : To identify the objective, look for a phrase such as "find the maximum (or minimum) value of." The amount you are trying to maximize or minimize is the objective. For example,

- ... at the least cost... The objective function is the equation for cost, $C = \dots$.
- ... the greatest area... The objective function is the equation for area, $A = \dots$.

To determine the constraint *inequalities*, ask yourself what limitations are placed on the unknown variables as above—are they nonnegative? are there upper limits? To identify the constraint *equations*, look for sentences that dictate restrictions in the form of relationships between the variables, as in the answer to the first question above.

## 12.2    EXERCISES

▼ more advanced        ◆ challenging
**T** indicates exercises that should be solved using technology

*Solve the optimization problems in Exercises 1–8.* HINT [See Example 2.]

**1.** Maximize $P = xy$ subject to $x + y = 10$.

**2.** Maximize $P = xy$ subject to $x + 2y = 40$.

**3.** Minimize $S = x + y$ subject to $xy = 9$ and both $x$ and $y > 0$.

**4.** Minimize $S = x + 2y$ subject to $xy = 2$ and both $x$ and $y > 0$.

**5.** Minimize $F = x^2 + y^2$ subject to $x + 2y = 10$.

**6.** Minimize $F = x^2 + y^2$ subject to $xy^2 = 16$.

**7.** Maximize $P = xyz$ subject to $x + y = 30$, $y + z = 30$, and $x, y, z \geq 0$.

**8.** Maximize $P = xyz$ subject to $x + z = 12$, $y + z = 12$, and $x, y, z \geq 0$.

**9.** For a rectangle with perimeter 20 to have the largest area, what dimensions should it have?

**10.** For a rectangle with area 100 to have the smallest perimeter, what dimensions should it have?

## APPLICATIONS

**11.** *Average Cost: iPods* Assume that it costs Apple approximately

$$C(x) = 22{,}500 + 100x + 0.01x^2$$

dollars to manufacture $x$ 32GB iPods in a day.[1] How many iPods should be manufactured in order to minimize average cost? What is the resulting average cost of an iPod? (Give your answer to the nearest dollar.) HINT [See Example 1.]

**12.** *Average Cost: Xboxes* Assume that it costs Microsoft approximately

$$C(x) = 14{,}400 + 550x + 0.01x^2$$

dollars to manufacture $x$ Xbox 360s in a day.[2] How many Xboxes should be manufactured in order to minimize average cost? What is the resulting average cost of an Xbox? (Give your answer to the nearest dollar.) HINT [See Example 1.]

**13.** *Pollution Control* The cost of controlling emissions at a firm rises rapidly as the amount of emissions reduced increases. Here is a possible model:

$$C(q) = 4{,}000 + 100q^2$$

where $q$ is the reduction in emissions (in pounds of pollutant per day) and $C$ is the daily cost to the firm (in dollars) of this reduction. What level of reduction corresponds to the lowest average cost per pound of pollutant, and what would be the resulting average cost to the nearest dollar?

**14.** *Pollution Control* Repeat the preceding exercise using the following cost function:

$$C(q) = 2{,}000 + 200q^2.$$

**15.** *Pollution Control* (Compare Exercise 13.) The cost of controlling emissions at a firm is given by

$$C(q) = 4{,}000 + 100q^2,$$

where $q$ is the reduction in emissions (in pounds of pollutant per day) and $C$ is the daily cost to the firm (in dollars) of this reduction. Government clean-air subsidies amount to $500 per pound of pollutant removed. How many pounds of pollutant should the firm remove each day in order to minimize *net* cost (cost minus subsidy)?

**16.** *Pollution Control* (Compare Exercise 14.) Repeat the preceding exercise, using the following cost function:

$$C(q) = 2{,}000 + 200q^2$$

with government subsidies amounting to $100 per pound of pollutant removed per day.

**17.** *Fences* I would like to create a rectangular vegetable patch. The fencing for the east and west sides costs $4 per foot, and the fencing for the north and south sides costs only $2 per foot. I have a budget of $80 for the project. What are the dimensions of the vegetable patch with the largest area I can enclose? HINT [See Example 2.]

**18.** *Fences* I would like to create a rectangular orchid garden that abuts my house so that the house itself forms the northern boundary. The fencing for the southern boundary costs $4 per foot, and the fencing for the east and west sides costs $2 per foot. If I have a budget of $80 for the project, what are the dimensions of the garden with the largest area I can enclose? HINT [See Example 2.]

**19.** *Fences* You are building a right-angled triangular flower garden along a stream as shown in the figure. (The borders can be in any directions as long as they are at right angles as shown.)

[1] Not the actual cost equation; the authors do not know Apple's actual cost equation. The marginal cost in the model given is in rough agreement with the actual marginal cost for reasonable values of $x$ for one of the 2007 models. Source for cost data: *Manufacturing & Technology News*, July 31, 2007 Volume 14, No. 14 (www.manufacturingnews.com).

[2] Not the actual cost equation; the authors do not know Microsoft's actual cost equation. The marginal cost in the model given is in rough agreement with the actual marginal cost for reasonable values of $x$. Source for estimate of marginal cost: iSuppli (www.isuppli.com).

The fencing of the left border costs $5 per foot, while the fencing of the lower border costs $1 per foot. (No fencing is required along the river.) You want to spend $100 and enclose as much area as possible. What are the dimensions of your garden, and what area does it enclose? HINT [The area of a right-triangle is given by $A = xy/2$.]

20. **Fences** Repeat Exercise 19, this time assuming that the fencing of the left border costs $8 per foot, while the fencing of the lower border costs $2 per foot, and that you can spend $400.

21. ▼ **Fences** (Compare Exercise 17.) For tax reasons, I need to create a rectangular vegetable patch with an area of exactly 242 sq. ft. The fencing for the east and west sides costs $4 per foot, and the fencing for the north and south sides costs only $2 per foot. What are the dimensions of the vegetable patch with the least expensive fence? HINT [Compare Exercise 3.]

22. ▼ **Fences** (Compare Exercise 18.) For reasons too complicated to explain, I need to create a rectangular orchid garden with an area of exactly 324 sq. ft. abutting my house so that the house itself forms the northern boundary. The fencing for the southern boundary costs $4 per foot, and the fencing for the east and west sides costs $2 per foot. What are the dimensions of the orchid garden with the least expensive fence? HINT [Compare Exercise 4.]

23. **Revenue** Hercules Films is deciding on the price of the video release of its film *Son of Frankenstein*. Its marketing people estimate that at a price of $p$ dollars, it can sell a total of $q = 200,000 - 10,000p$ copies. What price will bring in the greatest revenue? HINT [See Example 3.]

24. **Profit** Hercules Films is also deciding on the price of the video release of its film *Bride of the Son of Frankenstein*. Again, marketing estimates that at a price of $p$ dollars, it can sell $q = 200,000 - 10,000p$ copies, but each copy costs $4 to make. What price will give the greatest *profit*?

25. **Revenue: Cellphones** Worldwide quarterly sales of Nokia cellphones were approximately $q = -p + 156$ million phones when the wholesale price was $$p$. At what wholesale price should Nokia have sold its phones to maximize its quarterly revenue? What would have been the resulting revenue?[3]

26. **Revenue: Cellphones** Worldwide annual sales of all cellphones were approximately $-10p + 1,600$ million phones when the wholesale price was $$p$. At what wholesale price should cellphones have been sold to maximize annual revenue? What would have been the resulting revenue?[4]

27. **Revenue: Monorail Service** The demand, in rides per day, for monorail service in Las Vegas in 2005 can be approximated by $q = -4,500p + 41,500$ when the fare was $$p$. What price should have been charged to maximize total revenue?[5]

28. **Revenue: Mars Monorail** The demand, in rides per day, for monorail service in the three urbynes (or districts) of Utarek, Mars, can be approximated by $q = -2p + 24$ million riders when the fare is $\bar{\bar{Z}}p$. What price should be charged to maximize total revenue?[6]

29. **Revenue** Assume that the demand for tuna in a small coastal town is given by

$$p = \frac{500,000}{q^{1.5}},$$

where $q$ is the number of pounds of tuna that can be sold in a month at $p$ dollars per pound. Assume that the town's fishery wishes to sell at least 5,000 pounds of tuna per month.

a. How much should the town's fishery charge for tuna in order to maximize monthly revenue? HINT [See Example 3, and don't neglect endpoints.]

b. How much tuna will it sell per month at that price?

c. What will be its resulting revenue?

30. **Revenue** In the 1930s the economist Henry Schultz devised the following demand function for corn:

$$p = \frac{6,570,000}{q^{1.3}},$$

where $q$ is the number of bushels of corn that could be sold at $p$ dollars per bushel in one year.[7] Assume that at least 10,000 bushels of corn per year must be sold.

a. How much should farmers charge per bushel of corn to maximize annual revenue? HINT [See Example 3, and don't neglect endpoints.]

b. How much corn can farmers sell per year at that price?

c. What will be the farmers' resulting revenue?

31. **Revenue** During the 1950s the wholesale price for chicken in the United States fell from 25¢ per pound to 14¢ per pound, while per capita chicken consumption rose from 22 pounds per year to 27.5 pounds per year.[8] Assuming that the demand for chicken depended linearly on the price, what wholesale price for chicken would have maximized revenues for poultry farmers, and what would that revenue have amounted to?

32. **Revenue** Your underground used-book business is booming. Your policy is to sell all used versions of *Calculus and You* at the same price (regardless of condition). When you set the price at $10, sales amounted to 120 volumes during the first week of classes. The following semester, you set the price at $30 and sales dropped to zero. Assuming that the demand for

[3] Demand equation based on second- and fourth-quarter sales. Source: Embedded.com/Company reports December, 2004.

[4] Demand equation based on estimated 2004 sales and projected 2008 sales. Source: I-Stat/NDR, December 2004.

[5] Source for ridership data: *New York Times*, February 10, 2007, p. A9.

[6] The zonar ($\bar{\bar{Z}}$) is the official currency in the city-state of Utarek, Mars (formerly www.Marsnext.com, a now extinct virtual society).

[7] Based on data for the period 1915–1929. Source: Henry Schultz (1938), *The Theory and Measurement of Demand*, University of Chicago Press, Chicago.

[8] Data are provided for the years 1951–1958. Source: U.S. Department of Agriculture, *Agricultural Statistics*.

books depends linearly on the price, what price gives you the maximum revenue, and what does that revenue amount to?

**33. *Profit: Cellphones*** (Compare Exercise 25.) Worldwide quarterly sales of Nokia cellphones were approximately $q = -p + 156$ million phones when the wholesale price was $p$. Assuming that it cost Nokia \$40 to manufacture each cellphone, at what wholesale price should Nokia have sold its phones to maximize its quarterly profit? What would have been the resulting profit?[9] (The actual wholesale price was \$105 in the fourth quarter of 2004.) HINT [See Example 3, and recall that Profit = Revenue − Cost.]

**34. *Profit: Cellphones*** (Compare Exercise 26.) Worldwide annual sales of all cellphones were approximately $-10p + 1{,}600$ million phones when the wholesale price was $p$. Assuming that it costs \$30 to manufacture each cellphone, at what wholesale price should cellphones have been sold to maximize annual profit? What would have been the resulting profit?[10] HINT [See Example 3, and recall that Profit = Revenue − Cost.]

**35. ▼ *Profit*** The demand equation for your company's virtual reality video headsets is

$$p = \frac{1{,}000}{q^{0.3}},$$

where $q$ is the total number of headsets that your company can sell in a week at a price of $p$ dollars. The total manufacturing and shipping cost amounts to \$100 per headset.

**a.** What is the greatest profit your company can make in a week, and how many headsets will your company sell at this level of profit? (Give answers to the nearest whole number.)

**b.** How much, to the nearest \$1, should your company charge per headset for the maximum profit?

**36. ▼ *Profit*** Due to sales by a competing company, your company's sales of virtual reality video headsets have dropped, and your financial consultant revises the demand equation to

$$p = \frac{800}{q^{0.35}},$$

where $q$ is the total number of headsets that your company can sell in a week at a price of $p$ dollars. The total manufacturing and shipping cost still amounts to \$100 per headset.

**a.** What is the greatest profit your company can make in a week, and how many headsets will your company sell at this level of profit? (Give answers to the nearest whole number.)

**b.** How much, to the nearest \$1, should your company charge per headset for the maximum profit?

**37. *Paint Cans*** A company manufactures cylindrical paint cans with open tops with a volume of 27,000 cubic centimeters. What should be the dimensions of the cans in order to use the least amount of metal in their production? HINT [See Example 4.]

**38. *Metal Drums*** A company manufactures cylindrical metal drums with open tops with a volume of 1 cubic meter. What should be the dimensions of the drum in order to use the least amount of metal in their production? HINT [See Example 4.]

**39. *Tin Cans*** A company manufactures cylindrical tin cans with closed tops with a volume of 250 cubic centimeters. The metal used to manufacture the cans costs \$0.01 per square cm for the sides and \$0.02 per square cm for the (thicker) top and bottom. What should be the dimensions of the cans in order to minimize the cost of metal in their production? What is the ratio height/radius? HINT [See Example 4.]

**40. *Metal Drums*** A company manufactures cylindrical metal drums with open tops with a volume of 2 cubic meters. The metal used to manufacture the cans costs \$2 per square meter for the sides and \$3 per square meter for the (thicker) bottom. What should be the dimensions of the drums in order to minimize the cost of metal in their production? What is the ratio height/radius? HINT [See Example 4.]

**41. ▼ *Box Design*** *Chocolate Box Company* is going to make open-topped boxes out of $6 \times 16$-inch rectangles of cardboard by cutting squares out of the corners and folding up the sides. What is the largest volume box it can make this way?

**42. ▼ *Box Design*** *Vanilla Box Company* is going to make open-topped boxes out of $12 \times 12$-inch rectangles of cardboard by cutting squares out of the corners and folding up the sides. What is the largest volume box it can make this way?

**43. ▼ *Box Design*** A packaging company is going to make closed boxes, with square bases, that hold 125 cubic centimeters. What are the dimensions of the box that can be built with the least material?

**44. ▼ *Box Design*** A packaging company is going to make open-topped boxes, with square bases, that hold 108 cubic centimeters. What are the dimensions of the box that can be built with the least material?

**45. ▼ *Luggage Dimensions*** American Airlines requires that the total outside dimensions (length + width + height) of a checked bag not exceed 62 inches.[11] Suppose you want to check a bag whose height equals its width. What is the largest volume bag of this shape that you can check on an American flight?

**46. ▼ *Carry-on Dimensions*** American Airlines requires that the total outside dimensions (length + width + height) of a carry-on bag not exceed 45 inches.[12] Suppose you want to carry on a bag whose length is twice its height. What is the largest volume bag of this shape that you can carry on an American flight?

**47. ▼ *Luggage Dimensions*** Fly-by-Night Airlines has a peculiar rule about luggage: The length and width of a bag must add up to at most 45 inches, and the width and height must also add up to at most 45 inches. What are the dimensions of the bag with the largest volume that Fly-by-Night will accept?

---

[9] See Exercise 25.

[10] See Exercise 26.

[11] According to information on its Web site (www.aa.com/).

[12] *Ibid.*

48. ▼ *Luggage Dimensions* *Fair Weather Airlines* has a similar rule. It will accept only bags for which the sum of the length and width is at most 36 inches, while the sum of length, height, and twice the width is at most 72 inches. What are the dimensions of the bag with the largest volume that Fair Weather will accept?

49. ▼ *Package Dimensions* The U.S. Postal Service (USPS) will accept packages only if the length plus girth is no more than 108 inches.[13] (See the figure.)

Girth    Length

Assuming that the front face of the package (as shown in the figure) is square, what is the largest volume package that the USPS will accept?

50. ▼ *Package Dimensions* United Parcel Service (UPS) will accept only packages with a length of no more than 108 inches and length plus girth of no more than 165 inches.[14] (See figure for the preceding exercise.) Assuming that the front face of the package (as shown in the figure) is square, what is the largest volume package that UPS will accept?

51. ▼ *Cellphone Revenues* The number of cellphone subscribers in China in the years 2000–2005 was projected to follow the equation $N(t) = 39t + 68$ million subscribers in year $t$ ($t = 0$ represents January 2000). The average annual revenue per cellphone user was $350 in 2000.[15] If we assume that due to competition the revenue per cellphone user decreases continuously at an annual rate of 30%, we can model the annual revenue as

$$R(t) = 350(39t + 68)e^{-0.3t} \text{ million dollars.}$$

Determine **a.** when to the nearest 0.1 year the revenue was projected to peak and **b.** the revenue, to the nearest $1 million, at that time.

52. ▼ *Cellphone Revenues* (Refer to Exercise 51.) If we assume instead that the revenue per cellphone user decreases continuously at an annual rate of 20%, we obtain the revenue model

$$R(t) = 350(39t + 68)e^{-0.2t} \text{ million dollars.}$$

Determine **a.** when to the nearest 0.1 year the revenue was projected to peak and **b.** the revenue, to the nearest $1 million, at that time.

53. ▼ *Research and Development* Spending on research and development by drug companies in the United States $t$ years after 1970 can be modeled by

$$S(t) = 2.5e^{0.08t} \text{ billion dollars} \quad (0 \le t \le 31).$$

The number of new drugs approved by the Federal Drug Administration (FDA) over the same period can be modeled by

$$D(t) = 10 + t \text{ drugs per year}^{16} \quad (0 \le t \le 31).$$

When was the function $D(t)/S(t)$ at a maximum? What is the maximum value of $D(t)/S(t)$? What does the answer tell you about the cost of developing new drugs?

54. ▼ *Research and Development* (Refer to Exercise 53.) If the number of new drugs approved by the FDA had been $10 + 2t$ new drugs each year, when would the function $D(t)/S(t)$ have reached a maximum? What does the answer tell you about the cost of developing new drugs?

55. ▼ *Asset Appreciation* As the financial consultant to a classic auto dealership, you estimate that the total value (in dollars) of its collection of 1959 Chevrolets and Fords is given by the formula

$$v = 300,000 + 1,000t^2 \quad (t \ge 5),$$

where $t$ is the number of years from now. You anticipate a continuous inflation rate of 5% per year, so that the discounted (present) value of an item that will be worth $v$ in $t$ years' time is

$$p = ve^{-0.05t}.$$

When would you advise the dealership to sell the vehicles to maximize their discounted value?

56. ▼ *Plantation Management* The value of a fir tree in your plantation increases with the age of the tree according to the formula

$$v = \frac{20t}{1 + 0.05t},$$

where $t$ is the age of the tree in years. Given a continuous inflation rate of 5% per year, the discounted (present) value of a newly planted seedling is

$$p = ve^{-0.05t}.$$

At what age (to the nearest year) should you harvest your trees in order to ensure the greatest possible discounted value?

57. ▼ *Marketing Strategy* *FeatureRich Software Company* has a dilemma. Its new program, Doors-X 10.27, is almost ready to go on the market. However, the longer the company works on it, the better it can make the program and the more it can charge for it. The company's marketing analysts estimate that if it delays $t$ days, it can set the price at $100 + 2t$ dollars. On the other hand, the longer it delays, the more market share it will lose to its main competitor (see the next exercise) so that if it delays $t$ days it will be able to sell $400,000 - 2,500t$ copies of

---

[13] The requirement for packages sent other than Parcel Post, as of August 2011 (www.usps.com/).

[14] The requirement as of August, 2011 (www.ups.com/).

[15] Based on a regression of projected figures (coefficients are rounded). Source: Intrinsic Technology/*New York Times*, Nov. 24, 2000, p. C1.

[16] The exponential model for R&D is based on the 1970 and 2001 spending in constant 2001 dollars, while the linear model for new drugs approved is based on the 6-year moving average from data from 1970 to 2000. Source for data: Pharmaceutical Research and Manufacturers of America, FDA/*New York Times*, April 19, 2002, p. C1.

the program. How many days should FeatureRich delay the release in order to get the greatest revenue?

58. ▼ *Marketing Strategy* *FeatureRich Software's* main competitor (see previous exercise) is Moon Systems, and Moon is in a similar predicament. Its product, Walls-Y 11.4, could be sold now for $200, but for each day Moon delays, it could increase the price by $4. On the other hand, it could sell 300,000 copies now, but each day it waits will cut sales by 1,500. How many days should Moon delay the release in order to get the greatest revenue?

59. ▼ *Average Profit* The *FeatureRich Software Company* sells its graphing program, Dogwood, with a volume discount. If a customer buys $x$ copies, then he or she pays[17] $500\sqrt{x}$. It cost the company $10,000 to develop the program and $2 to manufacture each copy. If a single customer were to buy all the copies of Dogwood, how many copies would the customer have to buy for FeatureRich Software's average profit per copy to be maximized? How are average profit and marginal profit related at this number of copies?

60. ▼ *Average Profit* Repeat the preceding exercise with the charge to the customer $600\sqrt{x}$ and the cost to develop the program $9,000.

61. *Resource Allocation* Your company manufactures automobile alternators, and production is partially automated through the use of robots. Daily operating costs amount to $100 per laborer and $16 per robot. In order to meet production deadlines, the company calculates that the numbers of laborers and robots must satisfy the constraint

$$xy = 10,000,$$

where $x$ is the number of laborers and $y$ is the number of robots. Assuming that the company wishes to meet production deadlines at a minimum cost, how many laborers and how many robots should it use? HINT [See Example 5.]

62. *Resource Allocation* Your company is the largest sock manufacturer in the Solar System, and production is automated through the use of androids and robots. Daily operating costs amount to ₩200 per android and ₩8 per robot.[18] In order to meet production deadlines, the company calculates that the numbers of androids and robots must satisfy the constraint

$$xy = 1,000,000,$$

where $x$ is the number of androids and $y$ is the number of robots. Assuming that the company wishes to meet production deadlines at a minimum cost, how many androids and how many robots should it use? HINT [See Example 5.]

63. ▼ *Resource Allocation* Your automobile assembly plant has a Cobb-Douglas production function given by

$$q = x^{0.4}y^{0.6},$$

where $q$ is the number of automobiles it produces per year, $x$ is the number of employees, and $y$ is the daily operating budget (in dollars). Annual operating costs amount to an average of $20,000 per employee plus the operating budget of $365y$. Assume that you wish to produce 1,000 automobiles per year at a minimum cost. How many employees should you hire? HINT [See Example 5.]

64. ▼ *Resource Allocation* Repeat the preceding exercise using the production formula

$$q = x^{0.5}y^{0.5}.$$

HINT [See Example 5.]

65. ▼ *Incarceration Rate* The incarceration rate (the number of persons in prison per 100,000 residents) in the United States can be approximated by

$$N(t) = 0.04t^3 - 2t^2 + 40t + 460 \quad (0 \le t \le 18)$$

($t$ is the year since 1990).[19] When, to the nearest year, was the incarceration rate increasing most rapidly? When was it increasing least rapidly? HINT [You are being asked to find the extreme values of the rate of change of the incarceration rate.]

66. ▼ *Prison Population* The prison population in the United States can be approximated by

$$N(t) = 0.02t^3 - 2t^2 + 100t + 1,100 \text{ thousand people}$$
$$(0 \le t \le 18)$$

($t$ is the year since 1990).[20] When, to the nearest year, was the prison population increasing most rapidly? When was it increasing least rapidly? HINT [You are being asked to find the extreme values of the rate of change of the prison population.]

67. ▼ *Embryo Development* The oxygen consumption of a bird embryo increases from the time the egg is laid through the time the chick hatches. In a typical galliform bird, the oxygen consumption can be approximated by

$$c(t) = -0.065t^3 + 3.4t^2 - 22t + 3.6 \text{ milliliters per day}$$
$$(8 \le t \le 30),$$

where $t$ is the time (in days) since the egg was laid.[21] (An egg will typically hatch at around $t = 28$.) When, to the nearest day, is $c'(t)$ a maximum? What does the answer tell you?

68. ▼ *Embryo Development* The oxygen consumption of a turkey embryo increases from the time the egg is laid through the time the chick hatches. In a brush turkey, the oxygen consumption can be approximated by

$$c(t) = -0.028t^3 + 2.9t^2 - 44t + 95 \text{ milliliters per day}$$
$$(20 \le t \le 50),$$

---

[17] This is similar to the way site licenses have been structured for the program Maple®.

[18] ₩ are Neptunian Standard Solar Units of currency.

[19] Source for data: Sourcebook of Criminal Justice Statistics Online (www.albany.edu/sourcebook).

[20] *Ibid.*

[21] The model approximates graphical data published in the article "The Brush Turkey" by Roger S. Seymour, *Scientific American,* December, 1991, pp. 108–114.

where $t$ is the time (in days) since the egg was laid.[22] (An egg will typically hatch at around $t = 50$.) When, to the nearest day, is $c'(t)$ a maximum? What does the answer tell you?

69. ⬛ ▼ *Subprime Mortgages during the Housing Bubble* During the real estate run-up in 2000–2008, the percentage of mortgages issued in the United States that were subprime (normally classified as risky) could be approximated by

$$A(t) = \frac{15.0}{1 + 8.6(1.8)^{-t}} \text{ percent } \quad (0 \le t \le 8)$$

$t$ years after the start of 2000.[23] Graph the *derivative* of $A(t)$ and determine the year during which this derivative had an absolute maximum and also its value at that point. What does the answer tell you?

70. ⬛ ▼ *Subprime Mortgage Debt during the Housing Bubble* During the real estate run-up in 2000–2008, the value of subprime (normally classified as risky) mortgage debt outstanding in the United States was approximately

$$A(t) = \frac{1,350}{1 + 4.2(1.7)^{-t}} \text{ billion dollars } \quad (0 \le t \le 8)$$

$t$ years after the start of 2000.[24] Graph the *derivative* of $A(t)$ and determine the year during which this derivative had an absolute maximum and also its value at that point. What does the answer tell you?

71. ⬛ ▼ *Asset Appreciation* You manage a small antique company that owns a collection of Louis XVI jewelry boxes. Their value $v$ is increasing according to the formula

$$v = \frac{10,000}{1 + 500e^{-0.5t}},$$

where $t$ is the number of years from now. You anticipate an inflation rate of 5% per year, so that the present value of an item that will be worth $\$v$ in $t$ years' time is given by

$$p = v \cdot (1.05)^{-t}.$$

When (to the nearest year) should you sell the jewelry boxes to maximize their present value? How much (to the nearest constant dollar) will they be worth at that time?

72. ⬛ ▼ *Harvesting Forests* The following equation models the approximate volume in cubic feet of a typical Douglas fir tree of age $t$ years.[25]

$$V = \frac{22,514}{1 + 22,514t^{-2.55}}$$

The lumber will be sold at $10 per cubic foot, and you do not expect the price of lumber to appreciate in the foreseeable future. On the other hand, you anticipate a general inflation

rate of 5% per year, so that the present value of an item that will be worth $\$v$ in $t$ years' time is given by

$$p = v \cdot (1.05)^{-t}.$$

At what age (to the nearest year) should you harvest a Douglas fir tree in order to maximize its present value? How much (to the nearest constant dollar) will a Douglas fir tree be worth at that time?

73. ◆ *Agriculture* The fruit yield per tree in an orchard containing 50 trees is 100 pounds per tree each year. Due to crowding, the yield decreases by 1 pound per season for every additional tree planted. How many additional trees should be planted for a maximum total annual yield?

74. ◆ *Agriculture* Two years ago your orange orchard contained 50 trees and the yield per tree was 75 bags of oranges. Last year you removed 10 of the trees and noticed that the yield per tree increased to 80 bags. Assuming that the yield per tree depends linearly on the number of trees in the orchard, what should you do this year to maximize your total yield?

75. ◆ *Revenue* (based on a question on the GRE Economics Test[26]) If total revenue ($TR$) is specified by $TR = a + bQ - cQ^2$, where $Q$ is quantity of output and $a$, $b$, and $c$ are positive parameters, then $TR$ is maximized for this firm when it produces $Q$ equal to:

  **(A)** $b/2ac$. **(B)** $b/4c$. **(C)** $(a+b)/c$. **(D)** $b/2c$. **(E)** $c/2b$.

76. ◆ *Revenue* (based on a question on the GRE Economics Test) If total demand ($Q$) is specified by $Q = -aP + b$, where $P$ is unit price and $a$ and $b$ are positive parameters, then total revenue is maximized for this firm when it charges $P$ equal to:

  **(A)** $b/2a$. **(B)** $b/4a$. **(C)** $a/b$. **(D)** $a/2b$. **(E)** $-b/2a$.

## COMMUNICATION AND REASONING EXERCISES

77. You are interested in knowing the height of the tallest condominium complex that meets the city zoning requirements that the height $H$ should not exceed eight times the distance $D$ from the road and that it must provide parking for at least 50 cars. The objective function of the associated optimization problem is then:

  **(A)** $H$.  **(B)** $H - 8D$.  **(C)** $D$.  **(D)** $D - 8H$.

One of the constraints is:

  **(A)** $8H = D$.  **(B)** $8D = H$.
  **(C)** $H'(D) = 0$.  **(D)** $D'(H) = 0$.

78. You are interested in building a condominium complex with a height $H$ of at least 8 times the distance $D$ from the road and parking area of at least 1,000 sq. ft. at the cheapest cost $C$. The objective function of the associated optimization problem is then:

  **(A)** $H$.  **(B)** $D$.  **(C)** $C$.  **(D)** $H + D - C$.

[22] The model approximates graphical data published in the article "The Brush Turkey" by Roger S. Seymour, *Scientific American,* December, 1991, pp. 108–114.

[23] Sources: Mortgage Bankers Association, UBS.

[24] Source: www.data360.org/dataset.aspx?Data_Set_Id=9549.

[25] The model is the authors' and is based on data in *Environmental and Natural Resource Economics* by Tom Tietenberg, third edition (New York: HarperCollins, 1992), p. 282.

[26] Source: GRE Economics Test, by G. Gallagher, G. E. Pollock, W. J. Simeone, G. Yohe (Piscataway, NJ: Research and Education Association, 1989).

One of the constraints is:

(A) $H - 8D = 0$.          (B) $H + D - C = 0$.
(C) $C'(D) = 0$.          (D) $8H = D$.

**79.** Explain why the following problem is uninteresting: A packaging company wishes to make cardboard boxes with open tops by cutting square pieces from the corners of a square sheet of cardboard and folding up the sides. What is the box with the least surface area it can make this way?

**80.** Explain why finding the production level that minimizes a cost function is frequently uninteresting. What would a more interesting objective be?

**81.** Your friend Margo claims that all you have to do to find the absolute maxima and minima in applications is set the derivative equal to zero and solve. "All that other stuff about endpoints and so on is a waste of time just to make life hard for us," according to Margo. Explain why she is wrong, and find at least one exercise in this exercise set to illustrate your point.

**82.** You are having a hard time persuading your friend Marco that maximizing revenue is not the same as maximizing profit. "How on earth can you expect to obtain the largest profit if you are not taking in the largest revenue?" Explain why he is wrong, and find at least one exercise in this exercise set to illustrate your point.

**83.** ▼ If demand $q$ decreases as price $p$ increases, what does the minimum value of $dq/dp$ measure?

**84.** ▼ Explain how you would solve an optimization problem of the following form. Maximize $P = f(x, y, z)$ subject to $z = g(x, y)$ and $y = h(x)$.

## 12.3 Higher Order Derivatives: Acceleration and Concavity

The **second derivative** is simply the derivative of the derivative function. To explain why we would be interested in such a thing, we start by discussing one of its interpretations.

### Acceleration

Recall that if $s(t)$ represents the position of a car at time $t$, then its velocity is given by the derivative: $v(t) = s'(t)$. But one rarely drives a car at a constant speed; the velocity itself may be changing. The rate at which the velocity is changing is the **acceleration**. Because the derivative measures the rate of change, acceleration is the derivative of velocity: $a(t) = v'(t)$. Because $v$ is the derivative of $s$, we can express the acceleration in terms of $s$:

$$a(t) = v'(t) = (s')'(t) = s''(t).$$

That is, $a$ is the derivative of the derivative of $s$; in other words, the second derivative of $s$, which we write as $s''$. (In this context you will often hear the derivative $s'$ referred to as the **first derivative**.)

---

### Second Derivative, Acceleration

If a function $f$ has a derivative that is in turn differentiable, then its **second derivative** is the derivative of the derivative of $f$, written as $f''$. If $f''(a)$ exists, we say that $f$ is **twice differentiable at** $x = a$.

#### Quick Examples

1. If $f(x) = x^3 - x$, then $f'(x) = 3x^2 - 1$, so $f''(x) = 6x$ and $f''(-2) = -12$.
2. If $f(x) = 3x + 1$, then $f'(x) = 3$, so $f''(x) = 0$.
3. If $f(x) = e^x$, then $f'(x) = e^x$, so $f''(x) = e^x$ as well.

---

The **acceleration** of a moving object is the derivative of its velocity—that is, the second derivative of the position function.

If $t$ is time in hours and the position of a car at time $t$ is $s(t) = t^3 + 2t^2$ miles, then the car's velocity is $v(t) = s'(t) = 3t^2 + 4t$ miles per hour and its acceleration is $a(t) = s''(t) = v'(t) = 6t + 4$ miles per hour per hour.

## Differential Notation for the Second Derivative

We have written the second derivative of $f(x)$ as $f''(x)$. We could also use differential notation:

$$f''(x) = \frac{d^2 f}{dx^2}.$$

This notation comes from writing the second derivative as the derivative of the derivative in differential notation:

$$f''(x) = \frac{d}{dx}\left[\frac{df}{dx}\right] = \frac{d^2 f}{dx^2}.$$

Similarly, if $y = f(x)$, we write $f''(x)$ as $\frac{d}{dx}\left[\frac{dy}{dx}\right] = \frac{d^2 y}{dx^2}$. For example, if $y = x^3$, then $\frac{d^2 y}{dx^2} = 6x$.

An important example of acceleration is the acceleration due to gravity.

---

### EXAMPLE 1  Acceleration Due to Gravity

According to the laws of physics, the height of an object near the surface of the earth falling in a vacuum from an initial rest position 100 feet above the ground under the influence of gravity is approximately

$$s(t) = 100 - 16t^2 \text{ feet}$$

in $t$ seconds. Find its acceleration.

**Solution**  The velocity of the object is

$$v(t) = s'(t) = -32t \text{ ft/sec.} \qquad \text{Differential notation: } v = \frac{ds}{dt} = -32t \text{ ft/sec}$$

The reason for the negative sign is that the height of the object is decreasing with time, so its velocity is negative. Hence, the acceleration is

$$a(t) = s''(t) = -32 \text{ ft/sec}^2. \qquad \text{Differential notation: } a = \frac{d^2 s}{dt^2} = -32 \text{ ft/sec}^2$$

(We write ft/sec$^2$ as an abbreviation for feet/second/second—that is, feet per second per second. It is often read "feet per second squared.") Thus, the *downward* velocity is increasing by 32 ft/sec every second. We say that 32 ft/sec$^2$ is the **acceleration due to gravity**. In the absence of air resistance, all falling bodies near the surface of the earth, no matter what their weight, will fall with this acceleration.[*]

---

**Before we go on...**  In very careful experiments using balls rolling down inclined planes, Galileo made one of his most important discoveries—that the acceleration due to gravity is constant and does not depend on the weight or composition of the object falling.[†] A famous, though probably apocryphal, story has him dropping cannonballs of different weights off the Leaning Tower of Pisa to prove his point.[§] ∎

[*] On other planets the acceleration due to gravity is different. For example, on Jupiter, it is about three times as large as on Earth.

[†] An interesting aside: Galileo's experiments depended on getting extremely accurate timings. Because the timepieces of his day were very inaccurate, he used the most accurate time measurement he could: He sang and used the beat as his stopwatch.

[§] Here is a true story: The point was made again during the Apollo 15 mission to the moon (July 1971) when astronaut David R. Scott dropped a feather and a hammer from the same height. The moon has no atmosphere, so the two hit the surface of the moon simultaneously.

---

### EXAMPLE 2 **Acceleration of Sales**

For the first 15 months after the introduction of a new video game, the total sales can be modeled by the curve

$$S(t) = 20e^{0.4t} \text{ units sold,}$$

where $t$ is the time in months since the game was introduced. After about 25 months total sales follow more closely the curve

$$S(t) = 100,000 - 20e^{17-0.4t}.$$

How fast are total sales accelerating after 10 months? How fast are they accelerating after 30 months? What do these numbers mean?

**Solution** By acceleration we mean the rate of change of the rate of change, which is the second derivative. During the first 15 months, the first derivative of sales is

$$\frac{dS}{dt} = 8e^{0.4t}$$

and so the second derivative is

$$\frac{d^2S}{dt^2} = 3.2e^{0.4t}.$$

Thus, after 10 months the acceleration of sales is

$$\left.\frac{d^2S}{dt^2}\right|_{t=10} = 3.2e^4 \approx 175 \text{ units/month/month, or units/month}^2.$$

We can also compute total sales

$$S(10) = 20e^4 \approx 1,092 \text{ units}$$

and the rate of change of sales

$$\left.\frac{dS}{dt}\right|_{t=10} = 8e^4 \approx 437 \text{ units/month.}$$

What do these numbers mean? By the end of the tenth month, a total of 1,092 video games have been sold. At that time the game is selling at the rate of 437 units per month. This rate of sales is increasing by 175 units per month per month. More games will be sold each month than the month before.

Analysis of the sales after 30 months is done similarly, using the formula

$$S(t) = 100,000 - 20e^{17-0.4t}.$$

The derivative is

$$\frac{dS}{dt} = 8e^{17-0.4t}$$

and the second derivative is

$$\frac{d^2S}{dt^2} = -3.2e^{17-0.4t}.$$

After 30 months,

$$S(30) = 100,000 - 20e^{17-12} \approx 97,032 \text{ units}$$

$$\left. \frac{dS}{dt} \right|_{t=30} = 8e^{17-12} \approx 1,187 \text{ units/month}$$

$$\left. \frac{d^2 S}{dt^2} \right|_{t=30} = -3.2e^{17-12} \approx -475 \text{ units/month}^2.$$

By the end of the thirtieth month, 97,032 video games have been sold, the game is selling at a rate of 1,187 units per month, and the rate of sales is *decreasing* by 475 units per month. Fewer games are sold each month than the month before.

## Geometric Interpretation of Second Derivative: Concavity

The first derivative of $f$ tells us where the graph of $f$ is rising [where $f'(x) > 0$] and where it is falling [where $f'(x) < 0$]. The second derivative tells in what direction the graph of $f$ *curves* or *bends*. Consider the graphs in Figures 26 and 27.

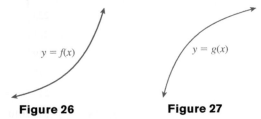

$y = f(x)$          $y = g(x)$

**Figure 26**          **Figure 27**

Think of a car driving from left to right along each of the roads shown in the two figures. A car driving along the graph of $f$ in Figure 26 will turn to the left (upward); a car driving along the graph of $g$ in Figure 27 will turn to the right (downward). We say that the graph of $f$ is **concave up** and the graph of $g$ is **concave down**. Now think about the derivatives of $f$ and $g$. The derivative $f'(x)$ starts small but *increases* as the graph gets steeper. Because $f'(x)$ is increasing, its derivative $f''(x)$ must be positive. On the other hand, $g'(x)$ *decreases* as we go to the right. Because $g'(x)$ is decreasing, its derivative $g''(x)$ must be negative. Summarizing, we have the following.

### Concavity and the Second Derivative

A curve is **concave up** if its slope is increasing, in which case the second derivative is positive. A curve is **concave down** if its slope is decreasing, in which case the second derivative is negative. A point in the domain of $f$ where the graph of $f$ changes concavity, from concave up to concave down or vice versa, is called a **point of inflection**. At a point of inflection, the second derivative is either zero or undefined.

### Locating Points of Inflection

To locate possible points of inflection, list points where $f''(x) = 0$ and also interior points where $f''(x)$ is not defined.

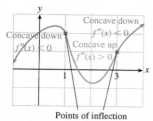

Points of inflection

**Figure 28**

Point of inflection

**Figure 29**

**Quick Examples**

**1.** The graph of the function $f$ shown in Figure 28 is concave up when $1 < x < 3$, so $f''(x) > 0$ for $1 < x < 3$. It is concave down when $x < 1$ and $x > 3$, so $f''(x) < 0$ when $x < 1$ and $x > 3$. It has points of inflection at $x = 1$ and $x = 3$.

**2.** Consider $f(x) = x^3 - 3x$, whose graph is shown in Figure 29. $f''(x) = 6x$ is negative when $x < 0$ and positive when $x > 0$. The graph of $f$ is concave down when $x < 0$ and concave up when $x > 0$. $f$ has a point of inflection at $x = 0$, where the second derivative is 0.

The following example shows one of the reasons it's useful to look at concavity.

### EXAMPLE 3  **Inflation**

Figure 30 shows the value of the U.S. Consumer Price Index (CPI) from January 2010 through April 2011.[27]

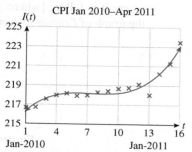

CPI Jan 2010–Apr 2011

**Figure 30**

The approximating curve shown on the figure is given by

$$I(t) = 0.0081t^3 - 0.18t^2 + 1.3t + 215 \qquad (1 \le t \le 16),$$

where $t$ is time in months ($t = 1$ represents January 2010). When the CPI is increasing, the U.S. economy is **experiencing inflation**. In terms of the model, this means that the derivative is positive: $I'(t) > 0$. Notice that $I'(t) > 0$ for most of the period shown (the graph is sloping upward), so the U.S. economy experienced inflation for most of $1 \le t \le 16$. We could measure **inflation** by the first derivative $I'(t)$ of the CPI, but we traditionally measure it as a ratio:

$$\text{Inflation rate} = \frac{I'(t)}{I(t)}, \qquad \text{Relative rate of change of the CPI}$$

expressed as a percentage per unit time (per month in this case).

**a.** Use the model to estimate the inflation rate in March 2010.

**b.** Was inflation slowing or speeding up in March 2010?

**c.** When was inflation slowing? When was inflation speeding up? When was inflation slowest?

[27]The CPI is compiled by the Bureau of Labor Statistics and is based upon a 1982 value of 100. For instance, a CPI of 200 means the CPI has doubled since 1982. Source: InflationData.com (www.inflationdata.com).

**Solution**

**a.** We need to compute $I'(t)$:

$$I'(t) = 0.0243t^2 - 0.36t + 1.3.$$

Thus, the inflation rate in March 2010 was given by

$$\text{Inflation rate} = \frac{I'(3)}{I(3)} = \frac{0.0243(3)^2 - 0.36(3) + 1.3}{0.0081(3)^3 - 0.18(3)^2 + 1.3(3) + 215}$$

$$= \frac{0.4387}{217.4987} \approx 0.0020,$$

* The 0.20% monthly inflation rate corresponds to a $12 \times 0.20 = 2.40\%$ annual inflation rate. This result could be obtained directly by changing the units of the $t$-axis from months to years and then redoing the calculation.

† When the CPI is falling, the inflation rate is negative and we experience *deflation*.

or 0.20% per month.*

**b.** We say that inflation is "slowing" when the CPI is decelerating ($I''(t) < 0$; the index rises at a slower rate or falls at a faster rate†). Similarly, inflation is "speeding up" when the CPI is accelerating ($I''(t) > 0$; the index rises at a faster rate or falls at a slower rate). From the formula for $I'(t)$, the second derivative is

$$I''(t) = 0.0486t - 0.36$$
$$I''(3) = 0.0486(3) - 0.36 = -0.2142.$$

Because this quantity is negative, we conclude that inflation was slowing down in March 2010.

**c.** When inflation is slowing, $I''(t)$ is negative, so the graph of the CPI is concave down. When inflation is speeding up, it is concave up. At the point at which it switches, there is a point of inflection (Figure 31).

**Figure 31**

The point of inflection occurs when $I''(t) = 0$; that is,

$$0.0486t - 0.36 = 0$$

$$t = \frac{0.36}{0.0486} \approx 7.4.$$

Thus, inflation was slowing when $t < 7.4$ (that is, until around the middle of July), and speeding up when $t > 7.4$ (after that time). Inflation was slowest at the point when it stopped slowing down and began to speed up, $t \approx 7.4$ (in fact there was slight deflation at that particular point as $I'(7.4)$ is negative); notice that the graph has the least slope at that point.

### EXAMPLE 4 **The Point of Diminishing Returns**

After the introduction of a new video game, the total worldwide sales are modeled by the curve

$$S(t) = \frac{1}{1 + 50e^{-0.2t}} \text{ million units sold,}$$

where $t$ is the time in months since the game was introduced (compare Example 2). The graphs of $S(t)$, $S'(t)$, and $S''(t)$ are shown in Figure 32. Where is the graph of $S$ concave up, and where is it concave down? Where are any points of inflection? What does this all mean?

(a) Graph of $S$    (b) Graph of $S'$    (c) Graph of $S''$

**Figure 32**

**Solution** Look at the graph of $S$. We see that the graph of $S$ is concave up in the early months and then becomes concave down later. The point of inflection, where the concavity changes, is somewhere between 15 and 25 months.

Now look at the graph of $S''$. This graph crosses the $t$-axis very close to $t = 20$, is positive before that point, and negative after that point. Because positive values of $S''$ indicate $S$ is concave up and negative values concave down, we conclude that the graph of $S$ is concave up for about the first 20 months; that is, for $0 < t < 20$ and concave down for $20 < t < 40$. The concavity switches at the point of inflection, which occurs at about $t = 20$ (when $S''(t) = 0$; a more accurate answer is $t \approx 19.56$).

What does this all mean? Look at the graph of $S'$, which shows sales per unit time, or monthly sales. From this graph we see that monthly sales are increasing for $t < 20$: more units are being sold each month than the month before. Monthly sales reach a peak of 0.05 million = 50,000 games per month at the point of inflection $t = 20$ and then begin to drop off. Thus, the point of inflection occurs at the time when monthly sales stop increasing and start to fall off; that is, the time when monthly sales peak. The point of inflection is sometimes called the **point of diminishing returns**. Although the total sales figure continues to rise (see the graph of $S$: game units continue to be sold), the *rate* at which units are sold starts to drop. (See Figure 33.)

**Figure 33**

## The Second Derivative Test for Relative Extrema

The second derivative often gives us a way of knowing whether or not a stationary point is a relative extremum. Figure 34 shows a graph with two stationary points: a relative maximum at $x = a$ and a relative minimum at $x = b$.

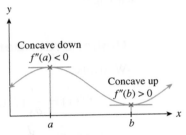

**Figure 34**

Notice that the curve is *concave down* at the relative maximum $(x = a)$, so that $f''(a) < 0$, and *concave up* at the relative minimum $(x = b)$, so that $f''(b) > 0$. This suggests the following (compare the First Derivative Test in Section 12.1).

---

### Second Derivative Test for Relative Extrema

Suppose that the function $f$ has a stationary point at $x = c$, and that $f''(c)$ exists. Determine the sign of $f''(c)$.

**1.** If $f''(c) > 0$, then $f$ has a relative minimum at $x = c$.

**2.** If $f''(c) < 0$, then $f$ has a relative maximum at $x = c$.

If $f''(c) = 0$, then the test is inconclusive and you need to use one of the methods of Section 12.1 (such as the first derivative test) to determine whether or not $f$ has a relative extremum at $x = c$.

---

**Quick Examples**

1. $f(x) = x^2 - 2x$ has $f'(x) = 2x - 2$ and hence a stationary point at $x = 1$. $f''(x) = 2$, and so $f''(1) = 2$, which is positive, so $f$ has a relative minimum at $x = 1$.

2. Let $f(x) = x^3 - 3x^2 - 9x$. Then

   $f'(x) = 3x^2 - 6x - 9 = 3(x + 1)(x - 3)$

   Stationary points at $x = -1, x = 3$

   $f''(x) = 6x - 6$

   $f''(-1) = -12$, so there is a relative maximum at $x = -1$

   $f''(3) = 12$, so there is a relative minimum at $x = 3$.

3. $f(x) = x^4$ has $f'(x) = 4x^3$ and hence a stationary point at $x = 0$. $f''(x) = 12x^2$ and so $f''(0) = 0$, telling us that the second derivative test is inconclusive. However, we can see from the graph of $f$ or the first derivative test that $f$ has a minimum at $x = 0$.

## Higher Order Derivatives

There is no reason to stop at the second derivative; we could once again take the *derivative* of the second derivative to obtain the **third derivative**, $f'''$, and we could take the derivative once again to obtain the **fourth derivative**, written $f^{(4)}$, and then continue to obtain $f^{(5)}$, $f^{(6)}$, and so on (assuming we get a differentiable function at each stage).

### Higher Order Derivatives

We define

$$f'''(x) = \frac{d}{dx}[f''(x)]$$

$$f^{(4)}(x) = \frac{d}{dx}[f'''(x)]$$

$$f^{(5)}(x) = \frac{d}{dx}[f^{(4)}(x)],$$

and so on, assuming all these derivatives exist.

### Different Notations

$$f'(x), f''(x), f'''(x), f^{(4)}(x), \ldots, f^{(n)}(x), \ldots$$

$$\frac{df}{dx}, \frac{d^2f}{dx^2}, \frac{d^3f}{dx^3}, \frac{d^4f}{dx^4}, \ldots, \frac{d^nf}{dx^n}, \ldots$$

$$\frac{dy}{dx}, \frac{d^2y}{dx^2}, \frac{d^3y}{dx^3}, \frac{d^4y}{dx^4}, \ldots, \frac{d^ny}{dx^n}, \ldots \qquad \text{When } y = f(x)$$

$$y, y', y'', y''', y^{(4)}, \ldots, y^{(n)}, \ldots \qquad \text{When } y = f(x)$$

### Quick Examples

1. If $f(x) = x^3 - x$, then $f'(x) = 3x^2 - 1$, $f''(x) = 6x$, $f'''(x) = 6$, $f^{(4)}(x) = f^{(5)}(x) = \cdots = 0$.

2. If $f(x) = e^x$, then $f'(x) = e^x$, $f''(x) = e^x$, $f'''(x) = e^x$, $f^{(4)}(x) = f^{(5)}(x) = \cdots = e^x$.

---

**Q**: *We know that the second derivative can be interpreted as acceleration. How do we interpret the third derivative; and the fourth, fifth, and so on?*

**A**: Think of a car traveling down the road (with position $s(t)$ at time $t$) in such a way that its acceleration $\frac{d^2s}{dt^2}$ is changing with time (for instance, the driver may be slowly increasing pressure on the accelerator, causing the car to accelerate at a greater and greater rate). Then $\frac{d^3s}{dt^3}$ is the rate of change of acceleration. $\frac{d^4s}{dt^4}$ would then be the *acceleration* of the acceleration, and so on.

**Q**: *How are these higher order derivatives reflected in the graph of a function f?*

**A**: Because the concavity is measured by $f''$, its derivative $f'''$ tells us the rate of change of concavity. Similarly, $f^{(4)}$ would tell us the *acceleration* of concavity, and so on. These properties are very subtle and hard to discern by simply looking at the curve; the higher the order, the more subtle the property. There is a remarkable theorem by Taylor* that tells us that, for a large class of functions (including polynomial, exponential, logarithmic, and trigonometric functions) the values of all orders of derivative $f(a)$, $f'(a)$, $f''(a)$, $f'''(a)$, and so on at the single point $x = a$ are enough to describe the entire graph (even at points very far from $x = a$)! In other words, the smallest piece of a graph near any point $a$ contains sufficient information to "clone" the entire graph!

\* Brook Taylor (1685–1731) was an English mathematician.

---

### FAQs

**Interpreting Points of Inflection and Using the Second Derivative Test**

**Q**: *It says in Example 4 that monthly sales reach a maximum at the point of inflection (second derivative is zero), but the second derivative test says that, for a maximum, the second derivative must be negative. What is going on here?*

**A**: What is a maximum in Example 4 is the *rate of change of* sales: which is measured in sales per unit time (monthly sales in the example). In other words, it is the *derivative* of the total sales function that is a maximum, so we located the maximum by setting its derivative (which is the *second* derivative of total sales) equal to zero. In general: To find relative (stationary) extrema of the *original* function, set $f'(x)$ equal to zero and solve for $x$ as usual. The second derivative test can then be used to test the stationary point obtained. To find relative (stationary) extrema of the *rate of change of f*, set $f''(x) = 0$ and solve for $x$.

**Q**: *I used the second derivative test and it was inconclusive. That means that there is neither a relative maximum nor a relative minimum at $x = a$, right?*

**A**: Wrong. If (as is often the case) the second derivative is zero at a stationary point, all it means is that the second derivative test itself cannot determine whether the given point is a relative maximum, minimum, or neither. For instance, $f(x) = x^4$ has a stationary minimum at $x = 0$, but the second derivative test is inconclusive. In such cases, one should use another test (such as the first derivative test) to decide if the point is a relative maximum, minimum, or neither.

---

## 12.3 EXERCISES

▼ more advanced   ◆ challenging
T indicates exercises that should be solved using technology

In Exercises 1–10, calculate $\dfrac{d^2y}{dx^2}$. HINT [See Quick Examples on page 892.]

**1.** $y = 3x^2 - 6$

**2.** $y = -x^2 + x$

**3.** $y = \dfrac{2}{x}$

**4.** $y = -\dfrac{2}{x^2}$

**5.** $y = 4x^{0.4} - x$

**6.** $y = 0.2x^{-0.1}$

**7.** $y = e^{-(x-1)} - x$

**8.** $y = e^{-x} + e^x$

**9.** $y = \dfrac{1}{x} - \ln x$

**10.** $y = x^{-2} + \ln x$

*In Exercises 11–16, the position s of a point (in feet) is given as a function of time t (in seconds). Find (a) its acceleration as a function of t and (b) its acceleration at the specified time.* HINT [See Example 1.]

**11.** $s = 12 + 3t - 16t^2$; $t = 2$

**12.** $s = -12 + t - 16t^2$; $t = 2$

**13.** $s = \dfrac{1}{t} + \dfrac{1}{t^2}$; $t = 1$     **14.** $s = \dfrac{1}{t} - \dfrac{1}{t^2}$; $t = 2$

**15.** $s = \sqrt{t} + t^2$; $t = 4$     **16.** $s = 2\sqrt{t} + t^3$; $t = 1$

*In Exercises 17–24, the graph of a function is given. Find the approximate coordinates of all points of inflection of each function (if any).* HINT [See Quick Examples on page 896.]

**17.**      **18.**

**19.**      **20.**

**21.**      **22.**

**23.**      **24.**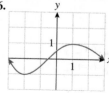

*In Exercises 25–28, the graph of the derivative, $f'(x)$, is given. Determine the x-coordinates of all points of inflection of $f(x)$, if any. (Assume that $f(x)$ is defined and continuous everywhere in $[-3, 3]$.)* HINT [See the **Before we go on** discussion in Example 4.]

**25.**      **26.**

**27.**      **28.**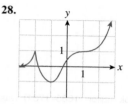

*In Exercises 29–32, the graph of the second derivative, $f''(x)$, is given. Determine the x-coordinates of all points of inflection of $f(x)$, if any. (Assume that $f(x)$ is defined and continuous everywhere in $[-3, 3]$.)* HINT [Remember that a point of inflection of $f$ corresponds to a point at which $f''$ changes sign, from positive to negative or vice versa. This could be a point where its graph crosses the x-axis, or a point where its graph is broken: positive on one side of the break and negative on the other.]

**29.**      **30.**

**31.**      **32.**

*In Exercises 33–44, find the x-coordinates of all critical points of the given function. Determine whether each critical point is a relative maximum, minimum, or neither by first applying the second derivative test, and, if the test fails, by some other method.* HINT [See Quick Examples on page 899.]

**33.** $f(x) = x^2 - 4x + 1$     **34.** $f(x) = 2x^2 - 2x + 3$

**35.** $g(x) = x^3 - 12x$     **36.** $g(x) = 2x^3 - 6x + 3$

**37.** $f(t) = t^3 - t$     **38.** $f(t) = -2t^3 + 3t$

**39.** $f(x) = x^4 - 4x^3$     **40.** $f(x) = 3x^4 - 2x^3$

**41.** $f(x) = e^{-x^2}$     **42.** $f(x) = e^{2-x^2}$

**43.** $f(x) = xe^{1-x^2}$     **44.** $f(x) = xe^{-x^2}$

*In Exercises 45–54, calculate the derivatives of all orders: $f'(x), f''(x), f'''(x), f^{(4)}(x), \ldots, f^{(n)}(x), \ldots$* HINT [See Quick Examples on page 900.]

**45.** $f(x) = 4x^2 - x + 1$     **46.** $f(x) = -3x^3 + 4x$

**47.** $f(x) = -x^4 + 3x^2$     **48.** $f(x) = x^4 + x^3$

**49.** $f(x) = (2x + 1)^4$     **50.** $f(x) = (-2x + 1)^3$

**51.** $f(x) = e^{-x}$     **52.** $f(x) = e^{2x}$

**53.** $f(x) = e^{3x-1}$     **54.** $f(x) = 2e^{-x+3}$

**APPLICATIONS**

**55. Acceleration on Mars** If a stone is dropped from a height of 40 meters above the Martian surface, its height in meters after $t$ seconds is given by $s = 40 - 1.9t^2$. What is its acceleration? HINT [See Example 1.]

**56. Acceleration on the Moon** If a stone is thrown up at 10 m per second from a height of 100 meters above the surface of the Moon, its height in meters after $t$ seconds is given by $s = 100 + 10t - 0.8t^2$. What is its acceleration? HINT [See Example 1.]

**57. Motion in a Straight Line** The position of a particle moving in a straight line is given by $s = t^3 - t^2$ ft after $t$ seconds. Find an expression for its acceleration after a time $t$. Is its velocity increasing or decreasing when $t = 1$?

**58. Motion in a Straight Line** The position of a particle moving in a straight line is given by $s = 3e^t - 8t^2$ ft after $t$ seconds. Find an expression for its acceleration after a time $t$. Is its velocity increasing or decreasing when $t = 1$?

**59. Bottled Water Sales** Annual sales of bottled water in the United States in the period 2000–2010 could be approximated by

$$R(t) = -45t^2 + 900t + 4{,}200 \text{ million gallons } (0 \le t \le 10),$$

where $t$ is time in years since 2000.[28] According to the model, were annual sales of bottled water accelerating or decelerating in 2009? How fast? HINT [See Example 2.]

**60. Bottled Water Sales** Annual U.S. per capita sales of bottled water in the period 2000–2010 could be approximated by

$$Q(t) = -0.18t^2 + 3t + 15 \text{ gallons } (0 \le t \le 10),$$

where $t$ is time in years since 2000.[29] According to the model, were annual U.S. per capita sales of bottled water accelerating or decelerating in 2009? How fast?

**61. Embryo Development** The daily oxygen consumption of a bird embryo increases from the time the egg is laid through the time the chick hatches. In a typical galliform bird, the oxygen consumption can be approximated by

$$c(t) = -0.065t^3 + 3.4t^2 - 22t + 3.6 \text{ ml } (8 \le t \le 30),$$

where $t$ is the time (in days) since the egg was laid.[30] (An egg will typically hatch at around $t = 28$.) Use the model to estimate the following (give the units of measurement for each answer and round all answers to two significant digits):

**a.** The daily oxygen consumption 20 days after the egg was laid

**b.** The rate at which the oxygen consumption is changing 20 days after the egg was laid

**c.** The rate at which the oxygen consumption is accelerating 20 days after the egg was laid

**62. Embryo Development** The daily oxygen consumption of a turkey embryo increases from the time the egg is laid through the time the chick hatches. In a brush turkey, the oxygen consumption can be approximated by

$$c(t) = -0.028t^3 + 2.9t^2 - 44t + 95 \text{ ml } (20 \le t \le 50),$$

where $t$ is the time (in days) since the egg was laid.[31] (An egg will typically hatch at around $t = 50$.) Use the model to estimate the following (give the units of measurement for each answer and round all answers to two significant digits):

**a.** The daily oxygen consumption 40 days after the egg was laid

**b.** The rate at which the oxygen consumption is changing 40 days after the egg was laid

**c.** The rate at which the oxygen consumption is accelerating 40 days after the egg was laid

**63. Inflation** The following graph shows the approximate value of the United States Consumer Price Index (CPI) from December 2006 through July 2007.[32]

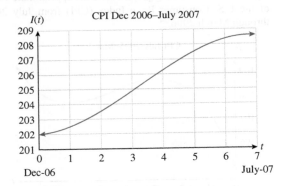

CPI Dec 2006–July 2007

The approximating curve shown on the figure is given by

$$I(t) = -0.04t^3 + 0.4t^2 + 0.1t + 202 \quad (0 \le t \le 7),$$

where $t$ is time in months ($t = 0$ represents December 2006).

**a.** Use the model to estimate the monthly inflation rate in February 2007 ($t = 2$). [Recall that the inflation *rate* is $I'(t)/I(t)$.]

**b.** Was inflation slowing or speeding up in February 2007?

**c.** When was inflation speeding up? When was inflation slowing? HINT [See Example 3.]

**64. Inflation** The following graph shows the approximate value of the U.S. Consumer Price Index (CPI) from September 2004 through November 2005.[33]

---

[28]Source for data: Beverage Marketing Corporation (www.bottledwater.org).

[29]*Ibid.*

[30]The model approximates graphical data published in the article "The Brush Turkey" by Roger S. Seymour, *Scientific American,* December, 1991, pp. 108–114.

[31]*Ibid.*

[32]The CPI is compiled by the Bureau of Labor Statistics and is based upon a 1982 value of 100. For instance, a CPI of 200 means the CPI has doubled since 1982. Source: InflationData.com (www.inflationdata.com).

[33]*Ibid.*

CPI Sep 2004–Nov 2005

The approximating curve shown on the figure is given by

$$I(t) = -0.005t^3 + 0.12t^2 - 0.01t + 190 \quad (0 \le t \le 14),$$

where $t$ is time in months ($t = 0$ represents September 2004).

a. Use the model to estimate the monthly inflation rate in July 2005 ($t = 10$). [Recall that the inflation *rate* is $I'(t)/I(t)$.]

b. Was inflation slowing or speeding up in July 2005?

c. When was inflation speeding up? When was inflation slowing? HINT [See Example 3.]

65. *Inflation* The following graph shows the approximate value of the U.S. Consumer Price Index (CPI) from July 2005 through March 2006.[34]

CPI July 2005–Mar 2006

The approximating curve shown on the figure is given by

$$I(t) = 0.06t^3 - 0.8t^2 + 3.1t + 195 \quad (0 \le t \le 8),$$

where $t$ is time in months ($t = 0$ represents July 2005).

a. Use the model to estimate the monthly inflation rates in December 2005 and February 2006 ($t = 5$ and $t = 7$).

b. Was inflation slowing or speeding up in February 2006?

c. When was inflation speeding up? When was inflation slowing?

66. *Inflation* The following graph shows the approximate value of the U.S. Consumer Price Index (CPI) from March 2006 through May 2007.[35]

---

CPI Mar 2006–May 2007

The approximating curve shown on the figure is given by

$$I(t) = 0.02t^3 - 0.38t^2 + 2t + 200 \quad (0 \le t \le 14),$$

where $t$ is time in months ($t = 0$ represents March, 2006).

a. Use the model to estimate the monthly inflation rates in September 2006 and January 2007 ($t = 6$ and $t = 10$).

b. Was inflation slowing or speeding up in January 2007?

c. When was inflation speeding up? When was inflation slowing?

67. *Scientific Research* The percentage of research articles in the prominent journal *Physical Review* that were written by researchers in the United States during the years 1983–2003 can be modeled by

$$P(t) = 25 + \frac{36}{1 + 0.06(0.7)^{-t}},$$

where $t$ is time in years since 1983.[36] The graphs of $P$, $P'$, and $P''$ are shown here:

Graph of $P$

Graph of $P'$

---

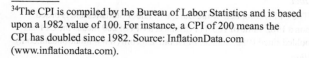

[34]The CPI is compiled by the Bureau of Labor Statistics and is based upon a 1982 value of 100. For instance, a CPI of 200 means the CPI has doubled since 1982. Source: InflationData.com (www.inflationdata.com).

[35]*Ibid.*

[36]Source: The American Physical Society/*New York Times*, May 3, 2003, p. A1.

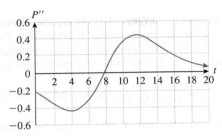

Graph of $P''$

Determine, to the nearest whole number, the values of $t$ for which the graph of $P$ is concave up and where it is concave down, and locate any points of inflection. What does the point of inflection tell you about science articles? HINT [See Example 4.]

**68. *Scientific Research*** The number of research articles in the prominent journal *Physical Review* that were written by researchers in Europe during the years 1983–2003 can be modeled by

$$P(t) = \frac{7.0}{1 + 5.4(1.2)^{-t}},$$

where $t$ is time in years since 1983.[37] The graphs of $P$, $P'$, and $P''$ are shown here:

Graph of $P$

Graph of $P'$

Graph of $P''$

[37] Source: The American Physical Society/*New York Times*, May 3, 2003, p. A1.

Determine, to the nearest whole number, the values of $t$ for which the graph of $P$ is concave up and where it is concave down, and locate any points of inflection. What does the point of inflection tell you about science articles?

**69. *Embryo Development*** Here are sketches of the graphs of $c$, $c'$, and $c''$ from Exercise 61:

Graph of $c$

Graph of $c'$

Graph of $c''$

Multiple choice:

**a.** The graph of $c'$ **(A)** has a point of inflection. **(B)** has no points of inflection.

**b.** At around 18 days after the egg is laid, daily oxygen consumption is: **(A)** at a maximum. **(B)** increasing at a maximum rate. **(C)** just beginning to decrease.

**c.** For $t > 18$ days, the oxygen consumption is **(A)** increasing at a decreasing rate. **(B)** decreasing at an increasing rate. **(C)** increasing at an increasing rate.

**70. *Embryo Development*** Here are sketches of the graphs of $c$, $c'$, and $c''$ from Exercise 62:

Graph of $c$

Graph of $c'$

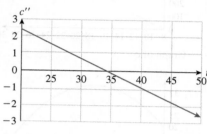

Graph of $c''$

Multiple choice:

a. The graph of $c$: **(A)** has points of inflection. **(B)** has no points of inflection. **(C)** may or may not have a point of inflection, but the graphs do not provide enough information.

b. At around 35 days after the egg is laid, the rate of change of daily oxygen consumption is: **(A)** at a maximum. **(B)** increasing at a maximum rate. **(C)** just becoming negative.

c. For $t < 35$ days, the oxygen consumption is: **(A)** increasing at an increasing rate. **(B)** increasing at a decreasing rate. **(C)** decreasing at an increasing rate.

71. **▥** **Subprime Mortgages during the Housing Bubble** During the real estate run-up in 2000–2008, the percentage of mortgages issued in the United States that were subprime (normally classified as risky) could be approximated by

$$A(t) = \frac{15.0}{1 + 8.6(1.8)^{-t}} \text{ percent} \quad (0 \le t \le 8)$$

$t$ years after the start of 2000.[38] Graph the function as well as its first and second derivatives. Determine, to the nearest whole number, the values of $t$ for which the graph of $A$ is concave up and concave down, and the $t$-coordinate of any points of inflection. What does the point of inflection tell you about subprime mortgages? HINT [To graph the second derivative, see the note in the margin after Example 4.]

72. **▥** **Subprime Mortgage Debt during the Housing Bubble** During the real estate run-up in 2000–2008, the value of subprime (normally classified as risky) mortgage debt outstanding in the United States was approximately

$$A(t) = \frac{1,350}{1 + 4.2(1.7)^{-t}} \text{ billion dollars} \quad (0 \le t \le 8)$$

$t$ years after the start of 2000.[39] Graph the function as well as its first and second derivatives. Determine, to the nearest whole number, the values of $t$ for which the graph of $A$ is concave up and concave down, and the $t$-coordinate of any points of inflection. What does the point of inflection tell you about subprime mortgages? HINT [To graph the second derivative, see the note in the margin after Example 4.]

73. **Epidemics** The following graph shows the total number $n$ of people (in millions) infected in an epidemic as a function of time $t$ (in years):

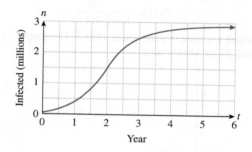

a. When to the nearest year was the rate of new infection largest?

b. When could the Centers for Disease Control and Prevention announce that the rate of new infection was beginning to drop? HINT [See Example 4.]

74. **Sales** The following graph shows the total number of *Pomegranate Q4* computers sold since their release ($t$ is in years):

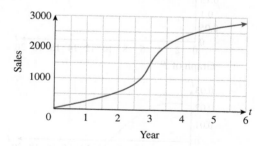

a. When were the computers selling fastest?

b. Explain why this graph might look as it does. HINT [See Example 4.]

---

[38] Sources: Mortgage Bankers Association, UBS.

[39] 2008 figure is an estimate.
Source: www.data360.org/dataset.aspx?Data_Set_Id=9549.

**75. Industrial Output** The following graph shows the yearly industrial output (measured in billions of zonars) of the city-state of Utarek, Mars over a seven-year period:

Industrial output (Republic of Mars)

Year since 2020

a. When to the nearest year did the rate of change of yearly industrial output reach a maximum?

b. When to the nearest year did the rate of change of yearly industrial output reach a minimum?

c. When to the nearest year does the graph first change from concave down to concave up? The result tells you that:

   **(A)** In that year the rate of change of industrial output reached a minimum compared with nearby years.

   **(B)** In that year the rate of change of industrial output reached a maximum compared with nearby years.

**76. Profits** The following graph shows the yearly profits of Gigantic Conglomerate, Inc. (GCI) from 2020 to 2035:

Yearly profit ($ billions)

Year since 2020

a. Approximately when were the profits rising most rapidly?

b. Approximately when were the profits falling most rapidly?

c. Approximately when could GCI's board of directors legitimately tell stockholders that they had "turned the company around"?

**77.** ▼ **Education and Crime** The following graph compares the total U.S. prison population and the average combined SAT score in the United States during the 1970s and 1980s:

Number of prisoners (thousands)

These data can be accurately modeled by

$$S(n) = 904 + \frac{1{,}326}{(n - 180)^{1.325}} \quad (192 \leq n \leq 563).$$

Here, $S(n)$ is the combined U.S. average SAT score at a time when the total U.S. prison population was $n$ thousand.[40]

a. Are there any points of inflection on the graph of $S$?

b. What does the concavity of the graph of $S$ tell you about prison populations and SAT scores?

**78.** ▼ **Education and Crime** Refer back to the model in the preceding exercise.

a. Are there any points of inflection on the graph of $S'$?

b. What does the concavity of the graph of $S'$ tell you about prison populations and SAT scores?

**79.** ▼ **Patents** In 1965, the economist F. M. Scherer modeled the number, $n$, of patents produced by a firm as a function of the size, $s$, of the firm (measured in annual sales in millions of dollars). He came up with the following equation based on a study of 448 large firms:[41]

$$n = -3.79 + 144.42s - 23.86s^2 + 1.457s^3.$$

a. Find $\dfrac{d^2 n}{ds^2}\bigg|_{s=3}$. Is the rate at which patents are produced as the size of a firm goes up increasing or decreasing with size when $s = 3$? Comment on Scherer's words, ". . . we find diminishing returns dominating."

b. Find $\dfrac{d^2 n}{ds^2}\bigg|_{s=7}$ and interpret the answer.

c. Find the $s$-coordinate of any points of inflection and interpret the result.

**80.** ▼ **Returns on Investments** A company finds that the number of new products it develops per year depends on the size of its annual R&D budget, $x$ (in thousands of dollars), according to the formula

$$n(x) = -1 + 8x + 2x^2 - 0.4x^3.$$

a. Find $n''(1)$ and $n''(3)$, and interpret the results.

b. Find the size of the budget that gives the largest rate of return as measured in new products per dollar (again, called the point of diminishing returns).

**81.** 🖥 ▼ **Oil Imports from Mexico** Daily oil production in Mexico and daily U.S. oil imports from Mexico during 2005–2009 can be approximated by

$$P(t) = 3.9 - 0.10t \text{ million barrels} \quad (5 \leq t \leq 9)$$
$$I(t) = 2.1 - 0.11t \text{ million barrels} \quad (5 \leq t \leq 9),$$

[40] Based on data for the years 1967–1989. Sources: *Sourcebook of Criminal Justice Statistics*, 1990, p. 604/Educational Testing Service.

[41] Source: F. M. Scherer, "Firm Size, Market Structure, Opportunity, and the Output of Patented Inventions," *American Economic Review* 55 (December 1965): pp. 1097–1125.

where $t$ is time in years since the start of 2000.[42]

Graph the function $I(t)/P(t)$ and its derivative. Is the graph of $I(t)/P(t)$ concave up or concave down? The concavity of $I(t)/P(t)$ tells you that

**(A)** the percentage of oil produced in Mexico that was exported to the United States was decreasing.

**(B)** the percentage of oil produced in Mexico that was not exported to the United States was increasing.

**(C)** the percentage of oil produced in Mexico that was exported to the United States was decreasing at a slower rate.

**(D)** the percentage of oil produced in Mexico that was exported to the United States was decreasing at a faster rate.

**82.** ▣ ▼ *Oil Imports from Mexico* Repeat Exercise 81 using instead the models for 2000–2004 shown below:

$$P(t) = 3.0 + 0.13t \text{ million barrels} \quad (0 \le t \le 4)$$
$$I(t) = 1.4 + 0.06t \text{ million barrels} \quad (0 \le t \le 4)$$

($t$ is time in years since the start of 2000).[43]

**83.** ◆ *Logistic Models* Let

$$f(x) = \frac{N}{1 + Ab^{-x}}$$

for constants $N$, $A$, and $b$ (*A* and *b* positive and $b \neq 1$). Show that $f$ has a single point of inflection at $x = \ln A / \ln b$.

**84.** ◆ *Logistic Models* Let

$$f(x) = \frac{N}{1 + Ae^{-kx}}$$

for constants $N$, $A$, and $k$ (*A* and *k* positive). Show that $f$ has a single point of inflection at $x = \ln A/k$.

**85.** ▣ *Population: Puerto Rico* The population of Puerto Rico in 1950–2025 can be approximated by

$$P(t) = \frac{4,500}{1 + 1.1466 \, (1.0357)^{-t}} \text{ thousand people} \ (0 \le t \le 75)$$

($t$ is the year since 1950).[44] Use the result of Exercise 83 to find the location of the point of inflection in the graph of $P$. What does the result tell you about the population of Puerto Rico?

**86.** ▣ *Population: Virgin Islands* The population of the Virgin Islands in 1950–2025 can be approximated by

$$P(t) = \frac{110}{1 + 2.3596 \, (1.0767)^{-t}} \text{ thousand people} \ (0 \le t \le 75)$$

($t$ is the year since 1950).[45] Use the result of Exercise 83 to find the location of the point of inflection in the graph of $P$. What does the result tell you about the population of the Virgin Islands?

**87.** ▣ ▼ *Asset Appreciation* You manage a small antique store that owns a collection of Louis XVI jewelry boxes. Their value $v$ is increasing according to the formula

$$v = \frac{10,000}{1 + 500e^{-0.5t}},$$

where $t$ is the number of years from now. You anticipate an inflation rate of 5% per year, so that the present value of an item that will be worth $\$v$ in $t$ years' time is given by

$$p = v \cdot (1.05)^{-t}.$$

What is the greatest rate of increase of the value of your antiques, and when is this rate attained?

**88.** ▣ ▼ *Harvesting Forests* The following equation models the approximate volume in cubic feet of a typical Douglas fir tree of age $t$ years[46]:

$$V = \frac{22,514}{1 + 22,514t^{-2.55}}.$$

The lumber will be sold at $\$10$ per cubic foot, and you do not expect the price of lumber to appreciate in the foreseeable

---

[44]Figures from 2010 on are U.S. census projections. Source for data: The 2008 Statistical Abstract (www.census.gov/).

[45]*Ibid.*

[46]The model is the authors', and is based on data in *Environmental and Natural Resource Economics* by Tom Tietenberg, third edition (New York: HarperCollins, 1992), p. 282.

---

[42] Source for data: Energy Information Administration/Pemex (www.eia.doe.gov).

[43] *Ibid.*

future. On the other hand, you anticipate a general inflation rate of 5% per year, so that the present value of an item that will be worth $v$ in $t$ years time is given by

$$p = v \cdot (1.05)^{-t}.$$

What is the largest rate of increase of the value of a fir tree, and when is this rate attained?

**89.** ☐ ▼ *Asset Appreciation* As the financial consultant to a classic auto dealership, you estimate that the total value of its collection of 1959 Chevrolets and Fords is given by the formula

$$v = 300,000 + 1,000t^2,$$

where $t$ is the number of years from now. You anticipate a continuous inflation rate of 5% per year, so that the discounted (present) value of an item that will be worth $v$ in $t$ years' time is given by

$$p = ve^{-0.05t}.$$

When is the value of the collection of classic cars increasing most rapidly? When is it decreasing most rapidly?

**90.** ☐ ▼ *Plantation Management* The value of a fir tree in your plantation increases with the age of the tree according to the formula

$$v = \frac{20t}{1 + 0.05t},$$

where $t$ is the age of the tree in years. Given a continuous inflation rate of 5% per year, the discounted (present) value of a newly planted seedling is

$$p = ve^{-0.05t}.$$

When is the discounted value of a tree increasing most rapidly? Decreasing most rapidly?

### COMMUNICATION AND REASONING EXERCISES

**91.** Complete the following: If the graph of a function is concave up on its entire domain, then its second derivative is _____ on the domain.

**92.** Complete the following: If the graph of a function is concave up on its entire domain, then its first derivative is _____ on the domain.

**93.** Daily sales of *Kent's Tents* reached a maximum in January 2002 and declined to a minimum in January 2003 before starting to climb again. The graph of daily sales shows a point of inflection at June 2002. What is the significance of the point of inflection?

**94.** The graph of daily sales of *Luddington's Wellington* boots is concave down, although sales continue to increase. What properties of the graph of daily sales versus time are reflected in the following behaviors?

**a.** a point of inflection next year
**b.** a horizontal asymptote

**95.** ▼ Company A's profits satisfy $P(0) = \$1$ million, $P'(0) = \$1$ million per year, and $P''(0) = -\$1$ million per year per year. Company B's profits satisfy $P(0) = \$1$ million, $P'(0) = -\$1$ million per year, and $P''(0) = \$1$ million per year per year. There are no points of inflection in either company's profit curve. Sketch two pairs of profit curves: one in which Company A ultimately outperforms Company B and another in which Company B ultimately outperforms Company A.

**96.** ▼ Company C's profits satisfy $P(0) = \$1$ million, $P'(0) = \$1$ million per year, and $P''(0) = -\$1$ million per year per year. Company D's profits satisfy $P(0) = \$0$ million, $P'(0) = \$0$ million per year, and $P''(0) = \$1$ million per year per year. There are no points of inflection in either company's profit curve. Sketch two pairs of profit curves: one in which Company C ultimately outperforms Company D and another in which Company D ultimately outperforms Company C.

**97.** ▼ Explain geometrically why the derivative of a function has a relative extremum at a point of inflection, if it is defined there. Which points of inflection give rise to relative maxima in the derivative?

**98.** ▼ If we regard position, $s$, as a function of time, $t$, what is the significance of the *third* derivative, $s'''(t)$? Describe an everyday scenario in which this arises.

# 12.4  Analyzing Graphs

Mathematical curves are beautiful—their subtle form can be imitated by only the best of artists—and calculus gives us the tools we need to probe their secrets. While it is easy to use graphing technology to draw a graph, we must use calculus to understand what we are seeing. Following is a list of some of the most interesting features of the graph of a function.

## Features of a Graph

1. **The x- and y-intercepts:** If $y = f(x)$, find the $x$-intercept(s) by setting $y = 0$ and solving for $x$; find the $y$-intercept by setting $x = 0$ and solving for $y$:

2. **Extrema:** Use the techniques of Section 12.1 to locate the maxima and minima:

3. **Points of inflection:** Use the techniques of Section 12.2 to locate the points of inflection:

4. **Behavior near points where the function is not defined:** If $f(x)$ is not defined at $x = a$, consider $\lim_{x \to a^-} f(x)$ and $\lim_{x \to a^+} f(x)$ to see how the graph of $f$ behaves as $x$ approaches $a$:

**5.** *Behavior at infinity:* Consider $\lim_{x \to -\infty} f(x)$ and $\lim_{x \to +\infty} f(x)$ if appropriate, to see how the graph of $f$ behaves far to the left and right:

**Note** It is sometimes difficult or impossible to solve all of the equations that come up in Steps 1, 2, and 3 of the previous analysis. As a consequence, we might not be able to say exactly where the $x$-intercept, extrema, or points of inflection are. When this happens, we will use graphing technology to assist us in determining accurate numerical approximations. ∎

---

**EXAMPLE 1 Analyzing a Graph**

Analyze the graph of $f(x) = \dfrac{1}{x} - \dfrac{1}{x^2}$.

$-50 \le x \le 50, -20 \le y \le 20$

$-10 \le x \le 10, -3 \le y \le 1$

**Figure 35**

**Solution** The graph, as drawn using graphing technology, is shown in Figure 35, using two different viewing windows. (Note that $x = 0$ is not in the domain of $f$.) The second window in Figure 35 seems to show the features of the graph better than the first. Does the second viewing window include *all* the interesting features of the graph? Or are there perhaps some interesting features to the right of $x = 10$ or to the left of $x = -10$? Also, where exactly do features like maxima, minima, and points of inflection occur? In our five-step process of analyzing the interesting features of the graph, we will be able to sketch the curve by hand, and also answer these questions.

**1. The x- and y-intercepts:** We consider $y = \dfrac{1}{x} - \dfrac{1}{x^2}$. To find the $x$-intercept(s), we set $y = 0$ and solve for $x$:

$$0 = \frac{1}{x} - \frac{1}{x^2}$$

$$\frac{1}{x} = \frac{1}{x^2}.$$

Multiplying both sides by $x^2$ (we know that $x$ cannot be zero, so we are not multiplying both sides by 0) gives

$$x = 1.$$

Thus, there is one $x$-intercept (which we can see in Figure 35) at $x = 1$.

For the $y$-intercept, we would substitute $x = 0$ and solve for $y$. However, we cannot substitute $x = 0$; because $f(0)$ is not defined, the graph does not meet the $y$-axis.

We add features to our freehand sketch as we go. Figure 36 shows what we have so far.

**Figure 36**

**2. *Relative extrema:*** We calculate $f'(x) = -\dfrac{1}{x^2} + \dfrac{2}{x^3}$. To find any stationary points, we set the derivative equal to 0 and solve for $x$:

$$-\frac{1}{x^2} + \frac{2}{x^3} = 0$$

$$\frac{1}{x^2} = \frac{2}{x^3}$$

$$x = 2.$$

Thus, there is one stationary point, at $x = 2$. We can use a test point to the right to determine that this stationary point is a relative maximum:

| $x$ | 1 (Intercept) | 2 | 3 (Test point) |
|---|---|---|---|
| $y = \dfrac{1}{x} - \dfrac{1}{x^2}$ | 0 | $\dfrac{1}{4}$ | $\dfrac{2}{9}$ |

The only possible singular point is at $x = 0$ because $f'(0)$ is not defined. However, $f(0)$ is not defined either, so there are no singular points. Figure 37 shows our graph so far.

**Figure 37**

**3. *Points of inflection:*** We calculate $f''(x) = \dfrac{2}{x^3} - \dfrac{6}{x^4}$. To find points of inflection, we set the second derivative equal to 0 and solve for $x$:

$$\frac{2}{x^3} - \frac{6}{x^4} = 0$$

$$\frac{2}{x^3} = \frac{6}{x^4}$$

$$2x = 6$$

$$x = 3.$$

Figure 35 confirms that the graph of $f$ changes from being concave down to being concave up at $x = 3$, so this is a point of inflection. $f''(x)$ is not defined at $x = 0$, but that is not in the domain, so there are no other points of inflection. In particular, the graph must be concave down in the whole region $(-\infty, 0)$, as we can see by calculating the second derivative at any one point in that interval: $f''(-1) = -8 < 0$.

**Figure 38**

Figure 38 shows our graph so far (we extended the curve near $x = 3$ to suggest a point of inflection at $x = 3$).

**4. *Behavior near points where $f$ is not defined:*** The only point where $f(x)$ is not defined is $x = 0$. From the graph, $f(x)$ appears to go to $-\infty$ as $x$ approaches 0 from either side. To calculate these limits, we rewrite $f(x)$:

$$f(x) = \frac{1}{x} - \frac{1}{x^2} = \frac{x-1}{x^2}.$$

Now, if $x$ is close to 0 (on either side), the numerator $x - 1$ is close to $-1$ and the denominator is a very small but positive number. The quotient is therefore a negative number of very large magnitude. Therefore,

$$\lim_{x \to 0^-} f(x) = -\infty$$

and

$$\lim_{x \to 0^+} f(x) = -\infty.$$

**Figure 39**

**Figure 40**

Technology:
2*x/3-((x-2)^2)^(1/3)

**Figure 41**

**Figure 42**

**Figure 43**

From these limits, we see the following:

**(1)** Immediately to the *left* of $x = 0$, the graph plunges down toward $-\infty$.

**(2)** Immediately to the *right* of $x = 0$, the graph also plunges down toward $-\infty$.

Figure 39 shows our graph with these features added. We say that $f$ has a **vertical asymptote** at $x = 0$, meaning that the points on the graph of $f$ get closer and closer to points on a vertical line (the $y$-axis in this case) further and further from the origin.

5. ***Behavior at infinity:*** Both $1/x$ and $1/x^2$ go to 0 as $x$ goes to $-\infty$ or $+\infty$; that is,

$$\lim_{x \to -\infty} f(x) = 0$$

and

$$\lim_{x \to +\infty} f(x) = 0.$$

Thus, on the extreme left and right of our picture, the height of the curve levels off toward zero. Figure 40 shows the completed freehand sketch of the graph.

We say that $f$ has a **horizontal asymptote** at $y = 0$. (Notice another thing: We haven't plotted a single point to the left of the $y$-axis, and yet we have a pretty good idea of what the curve looks like there! Compare the technology-drawn curve in Figure 35.)

In summary, there is one $x$-intercept at $x = 1$; there is one relative maximum (which, we can now see, is also an absolute maximum) at $x = 2$; there is one point of inflection at $x = 3$, where the graph changes from being concave down to concave up. There is a vertical asymptote at $x = 0$, on both sides of which the graph goes down toward $-\infty$, and a horizontal asymptote at $y = 0$.

## EXAMPLE 2 Analyzing a Graph

Analyze the graph of $f(x) = \dfrac{2x}{3} - (x - 2)^{2/3}$.

**Solution** Figure 41 shows a technology-generated version of the graph. Note that in the technology formulation $(x - 2)^{2/3}$ is written as $[(x - 2)^2]^{1/3}$ to avoid problems with some graphing calculators and Excel.

Let us now re-create this graph by hand, and in the process identify the features we see in Figure 41.

1. ***The x- and y-intercepts:*** We consider $y = \dfrac{2x}{3} - (x - 2)^{2/3}$. For the $y$-intercept, we set $x = 0$ and solve for $y$:

$$y = \frac{2(0)}{3} - (0 - 2)^{2/3} = -2^{2/3} \approx -1.59.$$

To find the $x$-intercept(s), we set $y = 0$ and solve for $x$. However, if we attempt this, we will find ourselves with a cubic equation that is hard to solve. (Try it!) Following the advice in the note on page 911, we use graphing technology to locate the $x$-intercept we see in Figure 41 by zooming in (Figure 42). From Figure 42, we find $x \approx 1.24$. We shall see in the discussion to follow that there can be no other $x$-intercepts.

Figure 43 shows our freehand sketch so far.

**2. *Relative extrema:*** We calculate

$$f'(x) = \frac{2}{3} - \frac{2}{3}(x-2)^{-1/3}$$

$$= \frac{2}{3} - \frac{2}{3(x-2)^{1/3}}.$$

To find any stationary points, we set the derivative equal to 0 and solve for $x$:

$$\frac{2}{3} - \frac{2}{3(x-2)^{1/3}} = 0$$

$$(x-2)^{1/3} = 1$$

$$x - 2 = 1^3 = 1$$

$$x = 3.$$

To check for singular points, look for points where $f(x)$ is defined and $f'(x)$ is not defined. The only such point is $x = 2$: $f'(x)$ is not defined at $x = 2$, whereas $f(x)$ is defined there, so we have a singular point at $x = 2$.

| $x$ | 2 (Singular point) | 3 (Stationary point) | 4 (Test point) |
|---|---|---|---|
| $y = \dfrac{2x}{3} - (x-2)^{2/3}$ | $\dfrac{4}{3}$ | 1 | 1.079 |

**Figure 44**

Figure 44 shows our graph so far.

We see that there is a singular relative maximum at $(2, 4/3)$ (we will confirm that the graph eventually gets higher on the right) and a stationary relative minimum at $x = 3$.

**3. *Points of inflection:*** We calculate

$$f''(x) = \frac{2}{9(x-2)^{4/3}}.$$

To find points of inflection, we set the second derivative equal to 0 and solve for $x$. But the equation

$$0 = \frac{2}{9(x-2)^{4/3}}$$

has no solution for $x$, so there are no points of inflection on the graph.

**4. *Behavior near points where f is not defined:*** Because $f(x)$ is defined everywhere, there are no such points to consider. In particular, there are no vertical asymptotes.

**5. *Behavior at infinity:*** We estimate the following limits numerically:

$$\lim_{x \to -\infty} \left[ \frac{2x}{3} - (x-2)^{2/3} \right] = -\infty$$

and

$$\lim_{x \to +\infty} \left[ \frac{2x}{3} - (x-2)^{2/3} \right] = +\infty.$$

**Figure 45**

Thus, on the extreme left the curve goes down toward $-\infty$, and on the extreme right the curve rises toward $+\infty$. In particular, there are no horizontal asymptotes. (There can also be no other $x$-intercepts.)

Figure 45 shows the completed graph.

## 12.4  EXERCISES

▼ more advanced    ◆ challenging
T indicates exercises that should be solved using technology

*In Exercises 1–26, sketch the graph of the given function, indicating (a) x- and y-intercepts, (b) extrema, (c) points of inflection, (d) behavior near points where the function is not defined, and (e) behavior at infinity. Where indicated, technology should be used to approximate the intercepts, coordinates of extrema, and/or points of inflection to one decimal place. Check your sketch using technology.* HINT [See Example 1.]

1. $f(x) = x^2 + 2x + 1$

2. $f(x) = -x^2 - 2x - 1$

3. $g(x) = x^3 - 12x$, domain $[-4, 4]$

4. $g(x) = 2x^3 - 6x$, domain $[-4, 4]$

5. $h(x) = 2x^3 - 3x^2 - 36x$ [Use technology for $x$-intercepts.]

6. $h(x) = -2x^3 - 3x^2 + 36x$ [Use technology for $x$-intercepts.]

7. $f(x) = 2x^3 + 3x^2 - 12x + 1$ [Use technology for $x$-intercepts.]

8. $f(x) = 4x^3 + 3x^2 + 2$ [Use technology for $x$-intercepts.]

9. $k(x) = -3x^4 + 4x^3 + 36x^2 + 10$ [Use technology for $x$-intercepts.]

10. $k(x) = 3x^4 + 4x^3 - 36x^2 - 10$ [Use technology for $x$-intercepts.]

11. $g(t) = \frac{1}{4}t^4 - \frac{2}{3}t^3 + \frac{1}{2}t^2$

12. $g(t) = 3t^4 - 16t^3 + 24t^2 + 1$

13. $f(x) = x + \frac{1}{x}$

14. $f(x) = x^2 + \frac{1}{x^2}$

15. $g(x) = x^3/(x^2 + 3)$

16. $g(x) = x^3/(x^2 - 3)$

17. $f(t) = \frac{t^2 + 1}{t^2 - 1}$, domain $[-2, 2]$, $t \neq \pm 1$

18. $f(t) = \frac{t^2 - 1}{t^2 + 1}$, domain $[-2, 2]$

19. $k(x) = \frac{2x}{3} + (x + 1)^{2/3}$ [Use technology for $x$-intercepts.
HINT [See Example 2.]

20. $k(x) = \frac{2x}{5} - (x - 1)^{2/5}$ [Use technology for $x$-intercepts.
HINT [See Example 2.]

21. $f(x) = x - \ln x$, domain $(0, +\infty)$

22. $f(x) = x - \ln x^2$, domain $(0, +\infty)$

23. $f(x) = x^2 + \ln x^2$ [Use technology for $x$-intercepts.]

24. $f(x) = 2x^2 + \ln x$ [Use technology for $x$-intercepts.]

25. $g(t) = e^t - t$, domain $[-1, 1]$

26. $g(t) = e^{-t^2}$

T *In Exercises 27–30, use technology to sketch the graph of the given function, labeling all relative and absolute extrema and points of inflection, and vertical and horizontal asymptotes. The coordinates of the extrema and points of inflection should be accurate to two decimal places.* HINT [To locate extrema accurately, plot the first derivative; to locate points of inflection accurately, plot the second derivative.]

27. ▼ $f(x) = x^4 - 2x^3 + x^2 - 2x + 1$

28. ▼ $f(x) = x^4 + x^3 + x^2 + x + 1$

29. ▼ $f(x) = e^x - x^3$

30. ▼ $f(x) = e^x - \frac{x^4}{4}$

## APPLICATIONS

**31. *Home Prices*** The following graph approximates historical and projected median home prices in the United States for the period 2000–2020:[47]

Here, $t$ is time in years since the start of 2000 and $C(t)$ is the median home price in thousands of dollars. The locations of extrema and points of inflection are indicated on the graph. Analyze the graph's important features and interpret each feature in terms of the median home price.

**32. *Housing Starts*** The following graph approximates historical and projected numbers of housing starts of single-family homes each year in the United States for the period 2000–2020:[48]

Here, $t$ is time in years since 2000 and $N(t)$ is the number, in thousands, of housing starts per year. The locations of extrema and points of inflection are indicated on the graph. Analyze the graph's important features and interpret each feature in terms of the number of housing starts.

**33. *Consumer Price Index*** The following graph shows the approximate value of the U.S. Consumer Price Index (CPI) from July 2005 through March 2006.[49]

The approximating curve shown on the figure is given by

$$I(t) = 0.06t^3 - 0.8t^2 + 3.1t + 195 \quad (0 \le t \le 8),$$

where $t$ is time in months ($t = 0$ represents July 2005).

**a.** Locate the intercepts, extrema, and points of inflection of the curve and interpret each feature in terms of the CPI. (Approximate all coordinates to one decimal place.) HINT [See Example 1.]

**b.** Recall from Section 12.2 that the inflation rate is defined to be $\dfrac{I'(t)}{I(t)}$. What do the stationary extrema of the curve shown above tell you about the inflation rate?

**34. *Consumer Price Index*** The following graph shows the approximate value of the U.S. Consumer Price Index (CPI) from March 2006 through May 2007.[50]

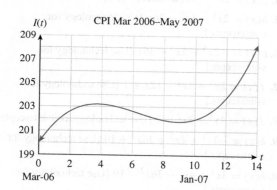

The approximating curve shown on the figure is given by

$$I(t) = 0.02t^3 - 0.38t^2 + 2t + 200 \quad (0 \le t \le 14),$$

where $t$ is time in months ($t = 0$ represents March, 2006).

**a.** Locate the intercepts, extrema, and points of inflection of the curve and interpret each feature in terms of the CPI. (Approximate all coordinates to one decimal place.) HINT [See Example 1.]

---

[47] Values from 2012 on are authors' projections. Source for data through 2011: Zillow (www.zillow.com).

[48] Values from 2011 on are authors' projections. Source for data through 2010: U.S. Census Bureau (www.census.gov).

[49] The CPI is compiled by the Bureau of Labor Statistics and is based upon a 1982 value of 100. For instance, a CPI of 200 means the CPI has doubled since 1982. Source: InflationData.com. (www.inflationdata.com).

[50] *Ibid.*

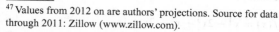

**b.** Recall from Section 12.2 that the inflation rate is defined to be $\dfrac{I'(t)}{I(t)}$. What do the stationary extrema of the curve shown above tell you about the inflation rate?

**35. *Motion in a Straight Line*** The distance of a UFO from an observer is given by $s = 2t^3 - 3t^2 + 100$ feet after $t$ seconds ($t \geq 0$). Obtain the extrema, points of inflection, and behavior at infinity. Sketch the curve and interpret these features in terms of the movement of the UFO.

**36. *Motion in a Straight Line*** The distance of the Mars orbiter from your location in Utarek, Mars is given by $s = 2(t - 1)^3 -3(t - 1)^2 + 100$ km after $t$ seconds ($t \geq 0$). Obtain the extrema, points of inflection, and behavior at infinity. Sketch the curve and interpret these features in terms of the movement of the Mars orbiter.

**37. *Average Cost: iPods*** Assume that it costs Apple approximately

$$C(x) = 22,500 + 100x + 0.01x^2$$

dollars to manufacture $x$ 32 GB iPods in a day.[51] Obtain the average cost function, sketch its graph, and analyze the graph's important features. Interpret each feature in terms of iPods. HINT [Recall that the average cost function is $\bar{C}(x) = C(x)/x$.]

**38. *Average Cost: Xboxes*** Assume that it costs Microsoft approximately

$$C(x) = 14,400 + 550x + 0.01x^2$$

dollars to manufacture $x$ Xbox 360s in a day.[52] Obtain the average cost function, sketch its graph, and analyze the graph's important features. Interpret each feature in terms of Xboxes. HINT [Recall that the average cost function is $\bar{C}(x) = C(x)/x$.]

**39.** ■▼ ***Subprime Mortgages during the Housing Bubble*** During the real estate run-up in 2000–2008, the percentage of mortgages issued in the United States that were subprime (normally classified as risky) could be approximated by

$$A(t) = \frac{15.0}{1 + 8.6(1.8)^{-t}} \text{ percent} \qquad (0 \leq t \leq 8)$$

$t$ years after the start of 2000.[53] Graph the *derivative* $A'(t)$ of $A(t)$ using an extended domain of $0 \leq t \leq 15$. Determine

the approximate coordinates of the maximum and determine the behavior of $A'(t)$ at infinity. What do the answers tell you?

**40.** ■▼ ***Subprime Mortgage Debt during the Housing Bubble*** During the real estate run-up in 2000–2008, the value of subprime (normally classified as risky) mortgage debt outstanding in the United States was approximately

$$A(t) = \frac{1,350}{1 + 4.2(1.7)^{-t}} \text{ billion dollars} \qquad (0 \leq t \leq 8)$$

$t$ years after the start of 2000.[54] Graph the *derivative* $A'(t)$ of $A(t)$ using an extended domain of $0 \leq t \leq 15$. Determine the approximate coordinates of the maximum and determine the behavior of $A'(t)$ at infinity. What do the answers tell you?

**COMMUNICATION AND REASONING EXERCISES**

**41.** A function is *bounded* if its entire graph lies between two horizontal lines. Can a bounded function have vertical asymptotes? Can a bounded function have horizontal asymptotes? Explain.

**42.** A function is *bounded above* if its entire graph lies below some horizontal line. Can a bounded above function have vertical asymptotes? Can a bounded above function have horizontal asymptotes? Explain.

**43.** If the graph of a function has a vertical asymptote at $x = a$ in such a way that $y$ increases to $+\infty$ as $x \to a$, what can you say about the graph of its derivative? Explain.

**44.** If the graph of a function has a horizontal asymptote at $y = a$ in such a way that $y$ decreases to $a$ as $x \to +\infty$, what can you say about the graph of its derivative? Explain.

**45.** Your friend tells you that he has found a continuous function defined on $(-\infty, +\infty)$ with exactly two critical points, each of which is a relative maximum. Can he be right?

**46.** Your other friend tells you that she has found a continuous function with two critical points, one a relative minimum and one a relative maximum, and no point of inflection between them. Can she be right?

**47.** ▼ By thinking about extrema, show that, if $f(x)$ is a polynomial, then between every pair of zeros ($x$-intercepts) of $f(x)$ there is a zero of $f'(x)$.

**48.** ▼ If $f(x)$ is a polynomial of degree 2 or higher, show that between every pair of relative extrema of $f(x)$ there is a point of inflection of $f(x)$.

---

[51] Not the actual cost equation; the authors do not know Apple's actual cost equation. The marginal cost in the model given is in rough agreement with the actual marginal cost for reasonable values of $x$ for one of the 2007 models. Source for cost data: *Manufacturing & Technology News*, July 31, 2007 Volume 14, No. 14 (www.manufacturingnews.com).

[52] Not the actual cost equation; the authors do not know Microsoft's actual cost equation. The marginal cost in the model given is in rough agreement with the actual marginal cost for reasonable values of $x$. Source for estimate of marginal cost: iSuppli: (www.isuppli.com/news/xbox/).

[53] 2009 figure is an estimate. Sources: Mortgage Bankers Association, UBS.

[54] 2008–2009 figure are estimates. Source: www.data360.org/dataset.aspx?Data_Set_Id=9549.

# 12.5   Related Rates

We start by recalling some basic facts about the rate of change of a quantity:

## Rate of Change of Q

If $Q$ is a quantity changing over time $t$, then the derivative $dQ/dt$ is the rate at which $Q$ changes over time.

### Quick Examples

1. If $A$ is the area of an expanding circle, then $dA/dt$ is the rate at which the area is increasing.
2. *Words:* The radius $r$ of a sphere is currently 3 cm and increasing at a rate of 2 cm/sec.

   *Symbols:* $r = 3$ cm and $dr/dt = 2$ cm/sec.

In this section we are concerned with what are called **related rates** problems. In such a problem we have two (sometimes more) related quantities, we know the rate at which one is changing, and we wish to find the rate at which another is changing. A typical example is the following.

### EXAMPLE 1   The Expanding Circle

The radius of a circle is increasing at a rate of 10 cm/sec. How fast is the area increasing at the instant when the radius has reached 5 cm?

**Solution**  We have two related quantities: the radius of the circle, $r$, and its area, $A$. The first sentence of the problem tells us that $r$ is increasing at a certain rate. When we see a sentence referring to speed or change, it is very helpful to rephrase the sentence using the phrase "the rate of change of." Here, we can say

*The rate of change of $r$ is 10 cm/sec.*

Because the rate of change is the derivative, we can rewrite this sentence as the equation

$$\frac{dr}{dt} = 10.$$

Similarly, the second sentence of the problem asks how $A$ is changing. We can rewrite that question:

*What is the rate of change of $A$ when the radius is 5 cm?*

Using mathematical notation, the question is:

*What is $\dfrac{dA}{dt}$ when $r = 5$?*

Thus, knowing one rate of change, $dr/dt$, we wish to find a related rate of change, $dA/dt$. To find exactly how these derivatives are related, we need the equation relating the variables, which is

$$A = \pi r^2.$$

To find the relationship between the derivatives, we take the derivative of both sides of this equation *with respect to t*. On the left we get $dA/dt$. On the right we need to remember that $r$ is a function of $t$ and use the chain rule. We get

$$\frac{dA}{dt} = 2\pi r \frac{dr}{dt}.$$

Now we substitute the given values $r = 5$ and $dr/dt = 10$. This gives

$$\left.\frac{dA}{dt}\right|_{r=5} = 2\pi(5)(10) = 100\pi \approx 314 \text{ cm}^2/\text{sec}.$$

Thus, the area is increasing at the rate of 314 cm²/sec when the radius is 5 cm.

We can organize our work as follows:

## Solving a Related Rates Problem

### A. The Problem

1. List the related, changing quantities.
2. Restate the problem in terms of rates of change. Rewrite the problem using mathematical notation for the changing quantities and their derivatives.

### B. The Relationship

1. Draw a diagram, if appropriate, showing the changing quantities.
2. Find an equation or equations relating the changing quantities.
3. Take the derivative with respect to time of the equation(s) relating the quantities to get the **derived equation(s)**, which relate the rates of change of the quantities.

### C. The Solution

1. Substitute into the derived equation(s) the given values of the quantities and their derivatives.
2. Solve for the derivative required.

We can illustrate the procedure with the "ladder problem" found in almost every calculus textbook.

### EXAMPLE 2 The Falling Ladder

Jane is at the top of a 5-foot ladder when it starts to slide down the wall at a rate of 3 feet per minute. Jack is standing on the ground behind her. How fast is the base of the ladder moving when it hits him if Jane is 4 feet from the ground at that instant?

**Solution** The first sentence talks about (the top of) the ladder sliding down the wall. Thus, one of the changing quantities is the height of the top of the ladder. The question asked refers to the motion of the base of the ladder, so another changing quantity is the distance of the base of the ladder from the wall. Let's record these variables and follow the outline above to obtain the solution.

## A. The Problem

**1.** The changing quantities are

$h$ = height of the top of the ladder

$b$ = distance of the base of the ladder from the wall

**2.** We rephrase the problem in words, using the phrase "rate of change":

*The rate of change of the height of the top of the ladder is* −3 *feet per minute. What is the rate of change of the distance of the base from the wall when the top of the ladder is* 4 *feet from the ground?*

We can now rewrite the problem mathematically:

$$\frac{dh}{dt} = -3. \text{ Find } \frac{db}{dt} \text{ when } h = 4.$$

## B. The Relationship

**Figure 46**

**1.** Figure 46 shows the ladder and the variables $h$ and $b$. Notice that we put in the figure the fixed length, 5, of the ladder, but any changing quantities, like $h$ and $b$, we leave as variables. We shall not use any specific values for $h$ or $b$ until the very end.

**2.** From the figure, we can see that $h$ and $b$ are related by the Pythagorean theorem:

$$h^2 + b^2 = 25.$$

**3.** Taking the derivative with respect to time of the equation above gives us the derived equation:

$$2h\frac{dh}{dt} + 2b\frac{db}{dt} = 0.$$

## C. The Solution

**1.** We substitute the known values $dh/dt = -3$ and $h = 4$ into the derived equation:

$$2(4)(-3) + 2b\frac{db}{dt} = 0.$$

We would like to solve for $db/dt$, but first we need the value of $b$, which we can determine from the equation $h^2 + b^2 = 25$, using the value $h = 4$:

$$16 + b^2 = 25$$
$$b^2 = 9$$
$$b = 3.$$

Substituting into the derived equation, we get

$$-24 + 2(3)\frac{db}{dt} = 0.$$

**2.** Solving for $db/dt$ gives

$$\frac{db}{dt} = \frac{24}{6} = 4.$$

Thus, the base of the ladder is sliding away from the wall at 4 ft/min when it hits Jack.

### EXAMPLE 3 Average Cost

The cost to manufacture $x$ cellphones in a day is

$$C(x) = 10,000 + 20x + \frac{x^2}{10,000} \text{ dollars.}$$

The daily production level is currently $x = 5,000$ cellphones and is increasing at a rate of 100 units per day. How fast is the average cost changing?

### Solution

#### A. The Problem

1. The changing quantities are the production level $x$ and the average cost, $\bar{C}$.

2. We rephrase the problem as follows:

> *The daily production level is $x = 5,000$ units and the rate of change of $x$ is 100 units/day. What is the rate of change of the average cost, $\bar{C}$?*

In mathematical notation,

$$x = 5,000 \text{ and } \frac{dx}{dt} = 100. \text{ Find } \frac{d\bar{C}}{dt}.$$

#### B. The Relationship

1. In this example the changing quantities cannot easily be depicted geometrically.

2. We are given a formula for the *total* cost. We get the *average* cost by dividing the total cost by $x$:

$$\bar{C} = \frac{C}{x}.$$

So,

$$\bar{C} = \frac{10,000}{x} + 20 + \frac{x}{10,000}.$$

3. Taking derivatives with respect to $t$ of both sides, we get the derived equation:

$$\frac{d\bar{C}}{dt} = \left( -\frac{10,000}{x^2} + \frac{1}{10,000} \right) \frac{dx}{dt}.$$

#### C. The Solution

Substituting the values from part A into the derived equation, we get

$$\frac{d\bar{C}}{dt} = \left( -\frac{10,000}{5,000^2} + \frac{1}{10,000} \right) 100$$

$$= -0.03 \text{ dollars/day.}$$

Thus, the average cost is decreasing by 3¢ per day.

The scenario in the following example is similar to Example 5 in Section 12.2.

### EXAMPLE 4 **Allocation of Labor**

*The Gym Sock Company* manufactures cotton athletic socks. Production is partially automated through the use of robots. The number of pairs of socks the company can manufacture in a day is given by a Cobb-Douglas production formula:

$$q = 50n^{0.6}r^{0.4},$$

where $q$ is the number of pairs of socks that can be manufactured by $n$ laborers and $r$ robots. The company currently produces 1,000 pairs of socks each day and employs 20 laborers. It is bringing one new robot on line every month. At what rate are laborers being laid off, assuming that the number of socks produced remains constant?

**Solution**

**A. The Problem**

**1.** The changing quantities are the number of laborers $n$ and the number of robots $r$.

**2.** $\dfrac{dr}{dt} = 1$. Find $\dfrac{dn}{dt}$ when $n = 20$.

**B. The Relationship**

**1.** No diagram is appropriate here.

**2.** The equation relating the changing quantities:

$$1,000 = 50n^{0.6}r^{0.4} \qquad \text{Productivity is constant at 1,000 pairs of socks each day.}$$

or

$$20 = n^{0.6}r^{0.4}.$$

**3.** The derived equation is

$$0 = 0.6n^{-0.4}\left(\frac{dn}{dt}\right)r^{0.4} + 0.4n^{0.6}r^{-0.6}\left(\frac{dr}{dt}\right)$$

$$= 0.6\left(\frac{r}{n}\right)^{0.4}\left(\frac{dn}{dt}\right) + 0.4\left(\frac{n}{r}\right)^{0.6}\left(\frac{dr}{dt}\right).$$

We solve this equation for $dn/dt$ because we shall want to find $dn/dt$ below and because the equation becomes simpler when we do this:

$$0.6\left(\frac{r}{n}\right)^{0.4}\left(\frac{dn}{dt}\right) = -0.4\left(\frac{n}{r}\right)^{0.6}\left(\frac{dr}{dt}\right)$$

$$\frac{dn}{dt} = -\frac{0.4}{0.6}\left(\frac{n}{r}\right)^{0.6}\left(\frac{n}{r}\right)^{0.4}\left(\frac{dr}{dt}\right)$$

$$= -\frac{2}{3}\left(\frac{n}{r}\right)\left(\frac{dr}{dt}\right).$$

**C. The Solution**

Substituting the numbers in A into the last equation in B, we get

$$\frac{dn}{dt} = -\frac{2}{3}\left(\frac{20}{r}\right) \quad (1).$$

We need to compute $r$ by substituting the known value of $n$ in the original formula:

$$20 = n^{0.6}r^{0.4}$$
$$20 = 20^{0.6}r^{0.4}$$
$$r^{0.4} = \frac{20}{20^{0.6}} = 20^{0.4}$$
$$r = 20.$$

Thus,

$$\frac{dn}{dt} = -\frac{2}{3}\left(\frac{20}{20}\right)(1) = -\frac{2}{3} \text{ laborers per month.}$$

The company is laying off laborers at a rate of $2/3$ per month, or two every three months.

We can interpret this result as saying that, at the current level of production and number of laborers, one robot is as productive as $2/3$ of a laborer, or 3 robots are as productive as 2 laborers.

## 12.5 EXERCISES

▼ more advanced    ◆ challenging
🚺 indicates exercises that should be solved using technology

*Rewrite the statements and questions in Exercises 1–8 in mathematical notation.* HINT [See Quick Examples on page 918.]

1. The population $P$ is currently 10,000 and growing at a rate of 1,000 per year.

2. There are presently 400 cases of Bangkok flu, and the number is growing by 30 new cases every month.

3. The annual revenue of your tie-dye T-shirt operation is currently $7,000 but is decreasing by $700 each year. How fast are annual sales changing?

4. A ladder is sliding down a wall so that the distance between the top of the ladder and the floor is decreasing at a rate of 3 feet per second. How fast is the base of the ladder receding from the wall?

5. The price of shoes is rising $5 per year. How fast is the demand changing?

6. Stock prices are rising $1,000 per year. How fast is the value of your portfolio increasing?

7. The average global temperature is 60°F and rising by 0.1°F per decade. How fast are annual sales of Bermuda shorts increasing?

8. The country's population is now 260,000,000 and is increasing by 1,000,000 people per year. How fast is the annual demand for diapers increasing?

### APPLICATIONS

9. *Sun Spots* The area of a circular sun spot is growing at a rate of 1,200 km²/sec.
   a. How fast is the radius growing at the instant when it equals 10,000 km? HINT [See Example 1.]
   b. How fast is the radius growing at the instant when the sun spot has an area of 640,000 km²? HINT [Use the area formula to determine the radius at that instant.]

10. *Puddles* The radius of a circular puddle is growing at a rate of 5 cm/sec.
   a. How fast is its area growing at the instant when the radius is 10 cm? HINT [See Example 1.]
   b. How fast is the area growing at the instant when it equals 36 cm²? HINT [Use the area formula to determine the radius at that instant.]

11. *Balloons* A spherical party balloon is being inflated with helium pumped in at a rate of 3 cubic feet per minute. How fast is the radius growing at the instant when the radius has reached 1 foot? (The volume of a sphere of radius $r$ is $V = \frac{4}{3}\pi r^3$.) HINT [See Example 1.]

12. *More Balloons* A rather flimsy spherical balloon is designed to pop at the instant its radius has reached 10 centimeters. Assuming the balloon is filled with helium at a rate of 10 cubic centimeters per second, calculate how fast the radius is growing at the instant it pops. (The volume of a sphere of radius $r$ is $V = \frac{4}{3}\pi r^3$.) HINT [See Example 1.]

13. **Sliding Ladders** The base of a 50-foot ladder is being pulled away from a wall at a rate of 10 feet per second. How fast is the top of the ladder sliding down the wall at the instant when the base of the ladder is 30 feet from the wall? HINT [See Example 2.]

14. **Sliding Ladders** The top of a 5-foot ladder is sliding down a wall at a rate of 10 feet per second. How fast is the base of the ladder sliding away from the wall at the instant when the top of the ladder is 3 feet from the ground? HINT [See Example 2.]

15. **Average Cost** The average cost function for the weekly manufacture of portable CD players is given by

$$\bar{C}(x) = 150{,}000x^{-1} + 20 + 0.0001x \text{ dollars per player,}$$

where $x$ is the number of CD players manufactured that week. Weekly production is currently 3,000 players and is increasing at a rate of 100 players per week. What is happening to the average cost? HINT [See Example 3.]

16. **Average Cost** Repeat the preceding exercise, using the revised average cost function

$$\bar{C}(x) = 150{,}000x^{-1} + 20 + 0.01x \text{ dollars per player.}$$

HINT [See Example 3.]

17. **Demand** Demand for your tie-dyed T-shirts is given by the formula

$$q = 500 - 100p^{0.5},$$

where $q$ is the number of T-shirts you can sell each month at a price of $p$ dollars. If you currently sell T-shirts for $15 each and you raise your price by $2 per month, how fast will the demand drop? (Round your answer to the nearest whole number.)

18. **Supply** The number of portable CD players you are prepared to supply to a retail outlet every week is given by the formula

$$q = 0.1p^2 + 3p,$$

where $p$ is the price it offers you. The retail outlet is currently offering you $40 per CD player. If the price it offers decreases at a rate of $2 per week, how will this affect the number you supply?

19. **Revenue** You can now sell 50 cups of lemonade per week at 30¢ per cup, but demand is dropping at a rate of 5 cups per week each week. Assuming that raising the price does not affect demand, how fast do you have to raise your price if you want to keep your weekly revenue constant? HINT [Revenue = Price × Quantity.]

20. **Revenue** You can now sell 40 cars per month at $20,000 per car, and demand is increasing at a rate of 3 cars per month each month. What is the fastest you could drop your price before your monthly revenue starts to drop? HINT [Revenue = Price × Quantity.]

21. ▼ **Oil Revenues** Daily oil production by **Pemex**, Mexico's national oil company, can be approximated by

$$q(t) = -0.022t^2 + 0.2t + 2.9 \text{ million barrels} \quad (1 \le t \le 9),$$

where $t$ is time in years since the start of 2000.[55]

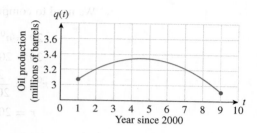

At the start of 2008 the price of oil was $90 per barrel and increasing at a rate of $80 per year.[56] How fast was Pemex's oil (daily) revenue changing at that time?

22. ▼ **Oil Expenditures** Daily oil imports to the United States from Mexico can be approximated by

$$q(t) = -0.015t^2 + 0.1t + 1.4 \text{ million barrels} \quad (0 \le t \le 8),$$

where $t$ is time in years since the start of 2000.[57]

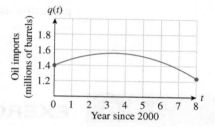

At the start of 2004 the price of oil was $30 per barrel and increasing at a rate of $40 per year.[58] How fast was (daily) oil expenditure for imports from Mexico changing at that time?

23. **Resource Allocation** Your company manufactures automobile alternators, and production is partially automated through the use of robots. In order to meet production deadlines, your company calculates that the numbers of laborers and robots must satisfy the constraint

$$xy = 10{,}000,$$

where $x$ is the number of laborers and $y$ is the number of robots. Your company currently uses 400 robots and is increasing robot deployment at a rate of 16 per month. How fast is it laying off laborers? HINT [See Example 4.]

24. **Resource Allocation** Your company is the largest sock manufacturer in the Solar System, and production is automated through the use of androids and robots. In order to meet production deadlines, your company calculates that the numbers of androids and robots must satisfy the constraint

$$xy = 1{,}000{,}000,$$

[55]Source for data: Energy Information Administration/Pemex (www.eia.doe.gov).

[56]Based on NYMEX crude oil futures; average rate of change during January–June, 2008.

[57]Source for data: Energy Information Administration/Pemex (www.eia.doe.gov).

[58]Based on NYMEX crude oil futures; average rate of change during 2004–2005.

where $x$ is the number of androids and $y$ is the number of robots. Your company currently uses 5000 androids and is increasing android deployment at a rate of 200 per month. How fast is it scrapping robots? HINT [See Example 4.]

25. **Production** The automobile assembly plant you manage has a Cobb-Douglas production function given by

$$P = 10x^{0.3}y^{0.7},$$

where $P$ is the number of automobiles it produces per year, $x$ is the number of employees, and $y$ is the daily operating budget (in dollars). You maintain a production level of 1,000 automobiles per year. If you currently employ 150 workers and are hiring new workers at a rate of 10 per year, how fast is your daily operating budget changing? HINT [See Example 4.]

26. **Production** Refer back to the Cobb-Douglas production formula in the preceding exercise. Assume that you maintain a constant workforce of 200 workers and wish to increase production in order to meet a demand that is increasing by 100 automobiles per year. The current demand is 1000 automobiles per year. How fast should your daily operating budget be increasing? HINT [See Example 4.]

27. **Demand** Assume that the demand equation for tuna in a small coastal town is

$$pq^{1.5} = 50,000,$$

where $q$ is the number of pounds of tuna that can be sold in one month at the price of $p$ dollars per pound. The town's fishery finds that the demand for tuna is currently 900 pounds per month and is increasing at a rate of 100 pounds per month each month. How fast is the price changing?

28. **Demand** The demand equation for rubies at *Royal Ruby Retailers* is

$$q + \frac{4}{3}p = 80,$$

where $q$ is the number of rubies RRR can sell per week at $p$ dollars per ruby. RRR finds that the demand for its rubies is currently 20 rubies per week and is dropping at a rate of one ruby per week. How fast is the price changing?

29. ▼ **Ships Sailing Apart** The H.M.S. *Dreadnaught* is 40 miles south of Montauk and steaming due south at 20 miles/hour, while the U.S.S. *Mona Lisa* is 50 miles east of Montauk and steaming due east at an even 30 miles/hour. How fast is their distance apart increasing?

30. ▼ **Near Miss** My aunt and I were approaching the same intersection, she from the south and I from the west. She was traveling at a steady speed of 10 miles/hour, while I was approaching the intersection at 60 miles/hour. At a certain instant in time, I was one tenth of a mile from the intersection, while she was one twentieth of a mile from it. How fast were we approaching each other at that instant?

31. ▼ **Baseball** A baseball diamond is a square with side 90 ft.

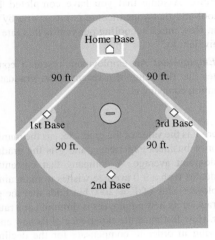

A batter at home base hits the ball and runs toward first base with a speed of 24 ft/sec. At what rate is his distance from third base increasing when he is halfway to first base?

32. ▼ **Baseball** Refer to Exercise 31. Another player is running from third base to home at 30 ft/sec. How fast is her distance from second base increasing when she is 60 feet from third base?

33. ▼ **Movement along a Graph** A point on the graph of $y = 1/x$ is moving along the curve in such a way that its $x$-coordinate is increasing at a rate of 4 units per second. What is happening to the $y$-coordinate at the instant the $y$-coordinate is equal to 2?

34. ▼ **Motion around a Circle** A point is moving along the circle $x^2 + (y - 1)^2 = 8$ in such a way that its $x$-coordinate is decreasing at a rate of 1 unit per second. What is happening to the $y$-coordinate at the instant when the point has reached $(-2, 3)$?

35. ▼ **Education** In 1991, the expected income of an individual depended on his or her educational level according to the following formula:

$$I(n) = 2.929n^3 - 115.9n^2 + 1,530n -$$
$$6,760 \text{ thousand dollars } (12 \le n \le 15).$$

Here, $n$ is the number of school years completed and $I(n)$ is the individual's expected income in thousands of dollars.[59] It is 1991, and you have completed 13 years of school and are currently a part-time student. Your schedule is such that you will complete the equivalent of one year of college every three years. Assuming that your salary is linked to the above model, how fast is your income going up? (Round your answer to the nearest $1.)

[59]The model is based on Table 358, U.S. Department of Education, *Digest of Education Statistics, 1991*, Washington, DC: Government Printing Office, 1991.

36. ▼ **Education** Refer back to the model in the preceding exercise. Assume that you have completed 14 years of school and that your income is increasing by $5,000 per year. How much schooling per year is this rate of increase equivalent to?

37. ▼ **Employment** An employment research company estimates that the value of a recent MBA graduate to an accounting company is

$$V = 3e^2 + 5g^3,$$

where $V$ is the value of the graduate, $e$ is the number of years of prior business experience, and $g$ is the graduate school grade-point average. A company that currently employs graduates with a 3.0 average wishes to maintain a constant employee value of $V = 200$, but finds that the grade-point average of its new employees is dropping at a rate of 0.2 per year. How fast must the experience of its new employees be growing in order to compensate for the decline in grade point average?

38. ▼ **Grades**[60] A production formula for a student's performance on a difficult English examination is given by

$$g = 4hx - 0.2h^2 - 10x^2,$$

where $g$ is the grade the student can expect to obtain, $h$ is the number of hours of study for the examination, and $x$ is the student's grade point average. The instructor finds that students' grade point averages have remained constant at 3.0 over the years, and that students currently spend an average of 15 hours studying for the examination. However, scores on the examination are dropping at a rate of 10 points per year. At what rate is the average study time decreasing?

39. ▼ **Cones** A right circular conical vessel is being filled with green industrial waste at a rate of 100 cubic meters per second. How fast is the level rising after $200\pi$ cubic meters have been poured in? The cone has a height of 50 m and a radius of 30 m at its brim. (The volume of a cone of height $h$ and cross-sectional radius $r$ at its brim is given by $V = \frac{1}{3}\pi r^2 h$.)

40. ▼ **More Cones** A circular conical vessel is being filled with ink at a rate of 10 cm³/sec. How fast is the level rising after 20 cm³ have been poured in? The cone has height 50 cm and radius 20 cm at its brim. (The volume of a cone of height $h$ and cross-sectional radius $r$ at its brim is given by $V = \frac{1}{3}\pi r^2 h$.)

41. ▼ **Cylinders** The volume of paint in a right cylindrical can is given by $V = 4t^2 - t$ where $t$ is time in seconds and $V$ is the volume in cm³. How fast is the level rising when the height is 2 cm? The can has a height of 4 cm and a radius of 2 cm. HINT [To get $h$ as a function of $t$, first solve the volume $V = \pi r^2 h$ for $h$.]

42. ▼ **Cylinders** A cylindrical bucket is being filled with paint at a rate of 6 cm³ per minute. How fast is the level rising when the bucket starts to overflow? The bucket has a radius of 30 cm and a height of 60 cm.

43. ▼ **Computers vs. Income** The demand for personal computers in the home goes up with household income. For a given community, we can approximate the average number of computers in a home as

$$q = 0.3454 \ln x - 3.047 \quad (10,000 \le x \le 125,000),$$

where $x$ is mean household income.[61] Your community has a mean income of $30,000, increasing at a rate of $2,000 per year. How many computers per household are there, and how fast is the number of computers in a home increasing? (Round your answer to four decimal places.)

44. ▼ **Computers vs. Income** Refer back to the model in the preceding exercise. The average number of computers per household in your town is 0.5 and is increasing at a rate of 0.02 computers per household per year. What is the average household income in your town, and how fast is it increasing? (Round your answers to the nearest $10).

*Education and Crime The following graph compares the total U.S. prison population and the average combined SAT score in the United States during the 1970s and 1980s:*

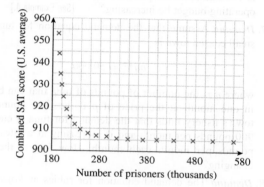

Combined SAT score (U.S. average) vs. Number of prisoners (thousands)

*Exercises 45 and 46 are based on the following model for these data:*

$$S(n) = 904 + \frac{1,326}{(n - 180)^{1.325}} \quad (192 \le n \le 563).$$

*Here, S(n) is the combined average SAT score at a time when the total prison population is n thousand.*[62]

45. ▼ In 1985, the U.S. prison population was 475,000 and increasing at a rate of 35,000 per year. What was the average SAT score, and how fast, and in what direction, was it changing? (Round your answers to two decimal places.)

---

[60] Based on an Exercise in *Introduction to Mathematical Economics* by A.L. Ostrosky Jr. and J.V. Koch (Waveland Press, Illinois, 1979).

[61] The model is a regression model. Source for data: Income distribution: Computer data: Forrester Research/*New York Times*, August 8, 1999, p. BU4.

[62] Based on data for the years 1967–1989. Sources: Sourcebook of Criminal Justice Statistics, 1990, p. 604/Educational Testing Service.

**46.** ▼ In 1970, the U.S. combined SAT average was 940 and dropping by 10 points per year. What was the U.S. prison population, and how fast, and in what direction, was it changing? (Round your answers to the nearest 100.)

*Divorce Rates A study found that the divorce rate d (given as a percentage) appears to depend on the ratio r of available men to available women.[63] This function can be approximated by*

$$d(r) = \begin{cases} -40r + 74 & \text{if } r \le 1.3 \\ \dfrac{130r}{3} - \dfrac{103}{3} & \text{if } r > 1.3 \end{cases}$$

*Exercises 47 and 48 are based on this model.*

**47.** ◆ There are currently 1.1 available men per available woman in Littleville, and this ratio is increasing by 0.05 per year. What is happening to the divorce rate?

**48.** ◆ There are currently 1.5 available men per available woman in Largeville, and this ratio is decreasing by 0.03 per year. What is happening to the divorce rate?

## COMMUNICATION AND REASONING EXERCISES

**49.** Why is this section titled "related rates"?

**50.** If you know how fast one quantity is changing and need to compute how fast a second quantity is changing, what kind of information do you need?

**51.** In a related rates problem, there is no limit to the number of changing quantities we can consider. Illustrate this by creating a related rates problem with four changing quantities.

---

[63]The cited study, by Scott J. South and associates, appeared in the *American Sociological Review* (February, 1995). Figures are rounded. Source: *New York Times*, February 19, 1995, p. 40.

**52.** If three quantities are related by a single equation, how would you go about computing how fast one of them is changing based on a knowledge of the other two?

**53.** ▼ The demand and unit price for your store's checkered T-shirts are changing with time. Show that the percentage rate of change of revenue equals the sum of the percentage rates of change of price and demand. (The percentage rate of change of a quantity $Q$ is $Q'(t)/Q(t)$.)

**54.** ▼ The number $N$ of employees and the total floor space $S$ of your company are both changing with time. Show that the percentage rate of change of square footage per employee equals the percentage rate of change of $S$ minus the percentage rate of change of $N$. (The percentage rate of change of a quantity $Q$ is $Q'(t)/Q(t)$.)

**55.** ▼ In solving a related rates problem, a key step is solving the derived equation for the unknown rate of change (once we have substituted the other values into the equation). Call the unknown rate of change $X$. The derived equation is what kind of equation in $X$?

**56.** ▼ On a recent exam, you were given a related rates problem based on an algebraic equation relating two variables $x$ and $y$. Your friend told you that the correct relationship between $dx/dt$ and $dy/dt$ was given by

$$\left(\frac{dx}{dt}\right) = \left(\frac{dy}{dt}\right)^2.$$

Could he be correct?

**57.** ▼ Transform the following into a mathematical statement about derivatives: If my grades are improving at twice the speed of yours, then your grades are improving at half the speed of mine.

**58.** ▼ If two quantities $x$ and $y$ are related by a linear equation, how are their rates of change related?

---

## 12.6  Elasticity

You manufacture an extremely popular brand of sneakers and want to know what will happen if you increase the selling price. Common sense tells you that demand will drop as you raise the price. But will the drop in demand be enough to cause your revenue to fall? Or will it be small enough that your revenue will rise because of the higher selling price? For example, if you raise the price by 1%, you might suffer only a 0.5% loss in sales. In this case, the loss in sales will be more than offset by the increase in price and your revenue will rise. In such a case, we say that the demand is **inelastic**, because it is not very sensitive to the increase in price. On the other hand, if your 1% price increase results in a 2% drop in demand, then raising the price will cause a drop in revenues. We then say that the demand is **elastic** because it reacts strongly to a price change.

✳ Coming up with a good demand equation is not always easy. We saw in Chapter 1 that it is possible to find a linear demand equation if we know the sales figures at two different prices. However, such an equation is only a first approximation. To come up with a more accurate demand equation, we might need to gather data corresponding to sales at several different prices and use curve-fitting techniques like regression. Another approach would be an analytic one, based on mathematical modeling techniques that an economist might use.

That said, we refer you again to *Camels and Rubber Duckies* by Joel Spolsky at www.joelon software.com/articles/Camelsand RubberDuckies.html just in case you think there is nothing more to demand curves.

We can use calculus to measure the response of demand to price changes if we have a demand equation for the item we are selling.✳ We need to know the *percentage drop in demand per percentage increase in price*. This ratio is called the **elasticity of demand**, or **price elasticity of demand**, and is usually denoted by $E$. Let's derive a formula for $E$ in terms of the demand equation.

Assume that we have a demand equation

$$q = f(p),$$

where $q$ stands for the number of items we would sell (per week, per month, or what have you) if we set the price per item at $p$. Now suppose we increase the price $p$ by a very small amount, $\Delta p$. Then our percentage increase in price is $(\Delta p / p) \times 100\%$. This increase in $p$ will presumably result in a decrease in the demand $q$. Let's denote this corresponding decrease in $q$ by $-\Delta q$ (we use the minus sign because, by convention, $\Delta q$ stands for the *increase* in demand). Thus, the percentage decrease in demand is $(-\Delta q / q) \times 100\%$.

Now $E$ is the ratio

$$E = \frac{\text{Percentage decrease in demand}}{\text{Percentage increase in price}}$$

so

$$E = \frac{-\dfrac{\Delta q}{q} \times 100\%}{\dfrac{\Delta p}{p} \times 100\%}.$$

Canceling the 100%s and reorganizing, we get

$$E = -\frac{\Delta q}{\Delta p} \cdot \frac{p}{q}.$$

**Q**: What small change in price will we use for $\Delta p$?

**A**: It should probably be pretty small. If, say, we increased the price of sneakers to $1 million per pair, the sales would likely drop to zero. But knowing this tells us nothing about how the market would respond to a modest increase in price. In fact, we'll do the usual thing we do in calculus and let $\Delta p$ approach 0.

In the expression for $E$, if we let $\Delta p$ go to 0, then the ratio $\Delta q / \Delta p$ goes to the derivative $dq/dp$. This gives us our final and most useful definition of the elasticity.

## Price Elasticity of Demand

The **price elasticity of demand** $E$ is the percentage rate of decrease of demand per percentage increase in price. $E$ is given by the formula

$$E = -\frac{dq}{dp} \cdot \frac{p}{q}.$$

We say that the demand is **elastic** if $E > 1$, is **inelastic** if $E < 1$, and has **unit elasticity** if $E = 1$.

> **Quick Example**
>
> Suppose that the demand equation is $q = 20,000 - 2p$, where $p$ is the price in dollars. Then
>
> $$E = -(-2)\frac{p}{20,000 - 2p} = \frac{p}{10,000 - p}.$$
>
> If $p = \$2,000$, then $E = 1/4$, and demand is inelastic at this price.
> If $p = \$8,000$, then $E = 4$, and demand is elastic at this price.
> If $p = \$5,000$, then $E = 1$, and the demand has unit elasticity at this price.

* For another—more rigorous—argument, see Exercise 29.

† See, for example, Exercise 53 in Section 12.5.

We are generally interested in the price that maximizes revenue and, in ordinary cases, the price that maximizes revenue must give unit elasticity. One way of seeing this is as follows:* If the demand is inelastic (which ordinarily occurs at a low unit price), then raising the price by a small percentage—1% say—results in a smaller percentage drop in demand. For example, in the Quick Example above, if $p = \$2,000$, then the demand would drop by only $\frac{1}{4}$% for every 1% increase in price. To see the effect on revenue, we use the fact† that, for small changes in price,

Percentage change in revenue $\approx$ Percentage change in price
+ Percentage change in demand

$$= 1 + \left(-\frac{1}{4}\right) = \frac{3}{4}\%.$$

Thus, the revenue will increase by about 3/4%. Put another way:

*If the demand is inelastic, raising the price increases revenue.*

On the other hand, if the price is elastic (which ordinarily occurs at a high unit price), then increasing the price slightly will lower the revenue, so:

*If the demand is elastic, lowering the price increases revenue.*

The price that results in the largest revenue must therefore be at unit elasticity.

### EXAMPLE 1  Price Elasticity of Demand: Dolls

Suppose that the demand equation for *Bobby Dolls* is given by $q = 216 - p^2$, where $p$ is the price per doll in dollars and $q$ is the number of dolls sold per week.

**a.** Compute the price elasticity of demand when $p = \$5$ and $p = \$10$, and interpret the results.

**b.** Find the range of prices for which the demand is elastic and the range for which the demand is inelastic.

**c.** Find the price at which the weekly revenue is maximized. What is the maximum weekly revenue?

### Solution

**a.** The price elasticity of demand is

$$E = -\frac{dq}{dp} \cdot \frac{p}{q}.$$

Taking the derivative and substituting for $q$ gives

$$E = 2p \cdot \frac{p}{216 - p^2} = \frac{2p^2}{216 - p^2}.$$

When $p = \$5$,

$$E = \frac{2(5)^2}{216 - 5^2} = \frac{50}{191} \approx 0.26.$$

Thus, when the price is set at $5, the demand is dropping at a rate of 0.26% per 1% increase in the price. Because $E < 1$, the demand is inelastic at this price, so raising the price will increase revenue.

When $p = \$10$,

$$E = \frac{2(10)^2}{216 - 10^2} = \frac{200}{116} \approx 1.72.$$

Thus, when the price is set at $10, the demand is dropping at a rate of 1.72% per 1% increase in the price. Because $E > 1$, demand is elastic at this price, so raising the price will decrease revenue; lowering the price will increase revenue.

**b.** and **c.** We answer part (c) first. Setting $E = 1$, we get

$$\frac{2p^2}{216 - p^2} = 1$$

$$p^2 = 72.$$

Thus, we conclude that the maximum revenue occurs when $p = \sqrt{72} \approx \$8.49$. We can now answer part (b): The demand is elastic when $p > \$8.49$ (the price is too high), and the demand is inelastic when $p < \$8.49$ (the price is too low). Finally, we calculate the maximum weekly revenue, which equals the revenue corresponding to the price of $8.49:

$$R = qp = (216 - p^2)p = (216 - 72)\sqrt{72} = 144\sqrt{72} \approx \$1,222.$$

---

The concept of elasticity can be applied in other situations. In the following example we consider *income* elasticity of demand—the percentage increase in demand for a particular item per percentage increase in personal income.

---

### EXAMPLE 2   Income Elasticity of Demand: Porsches

You are the sales director at *Suburban Porsche* and have noticed that demand for Porsches depends on income according to

$$q = 0.005e^{-0.05x^2 + x} \qquad (1 \le x \le 10).$$

Here, $x$ is the income of a potential customer in hundreds of thousands of dollars and $q$ is the probability that the person will actually purchase a Porsche.* The **income elasticity of demand** is

$$E = \frac{dq}{dx} \frac{x}{q}.$$

Compute and interpret $E$ for $x = 2$ and 9.

**using** Technology

See the Technology Guides at the end of the chapter to find out how to automate computations like those in part (a) of Example 1 using a graphing calculator or Excel. Here is an outline:

**TI-83/84 Plus**
Y₁=216-X²
Y₂=-nDeriv(Y₁,X,X)*X/Y₁
[2ND] [TABLE] Enter $x = 5$
[More details on page 943.]

**Spreadsheet**
Enter values of $p$: 4.9, 4.91, . . . , 5.0, 5.01, . . . , 5.1 in A5–A25.
In B5 enter 216-A5^2 and copy down to B25.
In C5 enter =(A6-A5)/A5 and paste the formula in C5–D24.
In E5 enter =-D5/C5 and copy down to E24. This column contains the values of $E$ for the values of $p$ in column A.
[More details on page 943.]

* In other words, $q$ is the fraction of visitors to your showroom having income $x$ who actually purchase a Porsche.

## Solution

Q : *Why is there no negative sign in the formula?*

A : Because we anticipate that the demand will increase as income increases, the ratio

$$\frac{\text{Percentage increase in demand}}{\text{Percentage increase in income}}$$

will be positive, so there is no need to introduce a negative sign.

Turning to the calculation, since $q = 0.005e^{-0.05x^2+x}$,

$$\frac{dq}{dx} = 0.005e^{-0.05x^2+x}(-0.1x + 1)$$

and so

$$E = \frac{dq}{dx}\frac{x}{q}$$

$$= 0.005e^{-0.05x^2+x}(-0.1x + 1)\frac{x}{0.005e^{-0.05x^2+x}}$$

$$= x(-0.1x + 1).$$

When $x = 2$, $E = 2[-0.1(2) + 1)] = 1.6$. Thus, at an income level of $200,000, the probability that a customer will purchase a Porsche increases at a rate of 1.6% per 1% increase in income.

When $x = 9$, $E = 9[-0.1(9) + 1)] = 0.9$. Thus, at an income level of $900,000, the probability that a customer will purchase a Porsche increases at a rate of 0.9% per 1% increase in income.

## 12.6 EXERCISES

▼ more advanced    ◆ challenging
🔢 indicates exercises that should be solved using technology

### APPLICATIONS

1. ***Demand for Oranges*** The weekly sales of *Honolulu Red Oranges* is given by $q = 1,000 - 20p$. Calculate the price elasticity of demand when the price is $30 per orange (yes, $30 per orange[64]). Interpret your answer. Also, calculate the price that gives a maximum weekly revenue, and find this maximum revenue. HINT [See Example 1.]

2. ***Demand for Oranges*** Repeat the preceding exercise for weekly sales of $1,000 - 10p$. HINT [See Example 1.]

3. ***Tissues*** The consumer demand equation for tissues is given by $q = (100 - p)^2$, where $p$ is the price per case of tissues and $q$ is the demand in weekly sales.

   a. Determine the price elasticity of demand $E$ when the price is set at $30, and interpret your answer.

   b. At what price should tissues be sold in order to maximize the revenue?

   c. Approximately how many cases of tissues would be demanded at that price?

4. ***Bodybuilding*** The consumer demand curve for *Professor Stefan Schwarzenegger* dumbbells is given by $q = (100 - 2p)^2$, where $p$ is the price per dumbbell, and $q$ is the demand in weekly sales. Find the price Professor Schwarzenegger should charge for his dumbbells in order to maximize revenue.

5. ***T-Shirts*** The Physics Club sells $E = mc^2$ T-shirts at the local flea market. Unfortunately, the club's previous administration has been losing money for years, so you decide to do an analysis of the sales. A quadratic regression based on old sales data reveals the following demand equation for the T-shirts:

   $$q = -2p^2 + 33p \quad (9 \le p \le 15).$$

   Here, $p$ is the price the club charges per T-shirt, and $q$ is the number it can sell each day at the flea market.

   a. Obtain a formula for the price elasticity of demand for $E = mc^2$ T-shirts.

[64]They are very hard to find, and their possession confers considerable social status.

**b.** Compute the elasticity of demand if the price is set at $10 per shirt. *Interpret the result.*

**c.** How much should the Physics Club charge for the T-shirts in order to obtain the maximum daily revenue? What will this revenue be?

**6. *Comics*** The demand curve for original *Iguanawoman* comics is given by

$$q = \frac{(400 - p)^2}{100} \quad (0 \le p \le 400),$$

where $q$ is the number of copies the publisher can sell per week if it sets the price at $p.

**a.** Find the price elasticity of demand when the price is set at $40 per copy.

**b.** Find the price at which the publisher should sell the books in order to maximize weekly revenue.

**c.** What, to the nearest $1, is the maximum weekly revenue the publisher can realize from sales of *Iguanawoman* comics?

**7. *College Tuition*** A study of about 1,800 U.S. colleges and universities resulted in the demand equation $q = 9,900 - 2.2p$, where $q$ is the enrollment at a college or university, and $p$ is the average annual tuition (plus fees) it charges.[65]

**a.** The study also found that the average tuition charged by universities and colleges was $2,900. What is the corresponding price elasticity of demand? Is the price elastic or inelastic? Should colleges charge more or less on average to maximize revenue?

**b.** Based on the study, what would you advise a college to charge its students in order to maximize total revenue, and what would the revenue be?

**8. *Demand for Fried Chicken*** A fried chicken franchise finds that the demand equation for its new roast chicken product, "Roasted Rooster," is given by

$$p = \frac{40}{q^{1.5}},$$

where $p$ is the price (in dollars) per quarter-chicken serving and $q$ is the number of quarter-chicken servings that can be sold per hour at this price. Express $q$ as a function of $p$ and find the price elasticity of demand when the price is set at $4 per serving. Interpret the result.

**9. *Paint-By-Number*** The estimated monthly sales of *Mona Lisa* paint-by-number sets is given by the formula $q = 100e^{-3p^2 + p}$, where $q$ is the demand in monthly sales and $p$ is the retail price in hundreds of yen.

**a.** Determine the price elasticity of demand $E$ when the retail price is set at ¥300 and interpret your answer.

**b.** At what price will revenue be a maximum?

**c.** Approximately how many paint-by-number sets will be sold per month at the price in part (b)?

**10. *Paint-By-Number*** Repeat the preceding exercise using the demand equation $q = 100e^{p - 3p^2/2}$.

**11.** ▼ *Linear Demand Functions* A general linear demand function has the form $q = mp + b$ ($m$ and $b$ constants, $m \ne 0$).

**a.** Obtain a formula for the price elasticity of demand at a unit price of $p$.

**b.** Obtain a formula for the price that maximizes revenue.

**12.** ▼ *Exponential Demand Functions* A general exponential demand function has the form $q = Ae^{-bp}$ ($A$ and $b$ nonzero constants).

**a.** Obtain a formula for the price elasticity of demand at a unit price of $p$.

**b.** Obtain a formula for the price that maximizes revenue.

**13.** ▼ *Hyperbolic Demand Functions* A general hyperbolic demand function has the form $q = \dfrac{k}{p^r}$ ($r$ and $k$ nonzero constants).

**a.** Obtain a formula for the price elasticity of demand at unit price $p$.

**b.** How does $E$ vary with $p$?

**c.** What does the answer to part (b) say about the model?

**14.** ▼ *Quadratic Demand Functions* A general quadratic demand function has the form $q = ap^2 + bp + c$ ($a$, $b$, and $c$ constants with $a \ne 0$).

**a.** Obtain a formula for the price elasticity of demand at a unit price $p$.

**b.** Obtain a formula for the price or prices that could maximize revenue.

**15.** ▼ *Modeling Linear Demand* You have been hired as a marketing consultant to *Johannesburg Burger Supply, Inc.*, and you wish to come up with a unit price for its hamburgers in order to maximize its weekly revenue. To make life as simple as possible, you assume that the demand equation for Johannesburg hamburgers has the linear form $q = mp + b$, where $p$ is the price per hamburger, $q$ is the demand in weekly sales, and $m$ and $b$ are certain constants you must determine.

**a.** Your market studies reveal the following sales figures: When the price is set at $2.00 per hamburger, the sales amount to 3,000 per week, but when the price is set at $4.00 per hamburger, the sales drop to zero. Use these data to calculate the demand equation.

**b.** Now estimate the unit price that maximizes weekly revenue and predict what the weekly revenue will be at that price.

**16.** ▼ *Modeling Linear Demand* You have been hired as a marketing consultant to *Big Book Publishing, Inc.*, and you have been approached to determine the best selling price for the hit calculus text by Whiner and Istanbul entitled *Fun with Derivatives*. You decide to make life easy and assume that the demand equation for *Fun with Derivatives* has the linear form $q = mp + b$, where $p$ is the price per book, $q$ is the demand in annual sales, and $m$ and $b$ are certain constants you must determine.

[65]Based on a study by A.L. Ostrosky Jr. and J.V. Koch, as cited in their book *Introduction to Mathematical Economics* (Waveland Press, Illinois, 1979) p. 133.

**a.** Your market studies reveal the following sales figures: When the price is set at $50.00 per book, the sales amount to 10,000 per year; when the price is set at $80.00 per book, the sales drop to 1000 per year. Use these data to calculate the demand equation.

**b.** Now estimate the unit price that maximizes annual revenue and predict what Big Book Publishing, Inc.'s annual revenue will be at that price.

**17. Income Elasticity of Demand: Live Drama** The likelihood that a child will attend a live theatrical performance can be modeled by

$$q = 0.01(-0.0078x^2 + 1.5x + 4.1) \qquad (15 \le x \le 100).$$

Here, $q$ is the fraction of children with annual household income $x$ thousand dollars who will attend a live dramatic performance at a theater during the year.[66] Compute the income elasticity of demand at an income level of $20,000 and interpret the result. (Round your answer to two significant digits.) HINT [See Example 2.]

**18. Income Elasticity of Demand: Live Concerts** The likelihood that a child will attend a live musical performance can be modeled by

$$q = 0.01(0.0006x^2 + 0.38x + 35) \qquad (15 \le x \le 100).$$

Here, $q$ is the fraction of children with annual household income $x$ who will attend a live musical performance during the year.[67] Compute the income elasticity of demand at an income level of $30,000 and interpret the result. (Round your answer to two significant digits.) HINT [See Example 2.]

**19. Income Elasticity of Demand: Broadband in 2010** The following graph shows the percentage $q$ of people in households with annual income $x$ thousand dollars using broadband Internet access in 2010, together with the exponential curve $q = -74e^{-0.021x} + 92$.[68]

**a.** Find an equation for the income elasticity of demand for broadband usage, and use it to compute the

elasticity for a household with annual income $100,000 to two decimal places. Interpret the result.

**b.** What does the model predict as the elasticity of demand for households with very large incomes?

**20. Income Elasticity of Demand: Broadband in 2007** The following graph shows the percentage $q$ of people in households with annual income $x$ thousand dollars using broadband Internet access in 2007, together with the exponential curve $q = -86e^{-0.013x} + 92$.[69]

**a.** Find an equation for the income elasticity of demand for broadband usage, and use it to compute the elasticity for a household with annual income $60,000 to two decimal places. Interpret the result.

**b.** What does the model predict as the elasticity of demand for households with very large incomes?

**21. Income Elasticity of Demand: Computer Usage in the 1990s** The following graph shows the probability $q$ that a household in the 1990s with annual income $x$ dollars had a computer, together with the logarithmic curve $q = 0.3454 \ln x - 3.047$.[70]

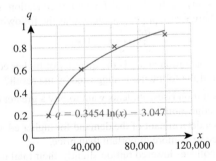

**a.** Compute the income elasticity of demand for computers, to two decimal places, for a household income of $60,000 and interpret the result.

**b.** As household income increases, how is income elasticity of demand affected?

**c.** How reliable is the given model of demand for incomes well above $120,000? Explain.

**d.** What can you say about $E$ for incomes much larger than those shown?

---

[66] Based on a quadratic regression of data from a 2001 survey. Source for data: New York Foundation of the Arts (www.nyfa.org/culturalblueprint).

[67] *Ibid.*

[68] Source for data: *Digital Nation: Expanding Internet Usage,* National Telecommunications and Information Administration (U.S. Department of Commerce) (http://search.ntia.doc.gov).

[69] *Ibid.*

[70] Source for data: Income distribution computer data: Forrester Research/*New York Times,* August 8, 1999, p. BU4.

**22. *Income Elasticity of Demand: Internet Usage in the 1990s*** The following graph shows the probability $q$ that a person in the 1990s with household annual income $x$ dollars used the Internet, together with the logarithmic curve $q = 0.2802 \ln x - 2.505$.[71]

**a.** Compute the income elasticity of demand for Internet usage, to two decimal places, for a household income of $60,000 and interpret the result.

**b.** As household income increases, how is income elasticity of demand affected?

**c.** The logarithmic model shown above is not appropriate for incomes well above $100,000. Suggest a model that might be more appropriate.

**d.** In the model you propose, how would $E$ behave for very large incomes?

**23.** ▼ *Income Elasticity of Demand (based on a question on the GRE Economics Test)* If $Q = a P^\alpha Y^\beta$ is the individual's demand function for a commodity, where $P$ is the (fixed) price of the commodity, $Y$ is the individual's income, and $a$, $\alpha$, and $\beta$ are parameters, explain why $\beta$ can be interpreted as the income elasticity of demand.

**24.** ▼ *College Tuition (from the GRE Economics Test)* A time-series study of the demand for higher education, using tuition charges as a price variable, yields the following result:
$$\frac{dq}{dp} \cdot \frac{p}{q} = -0.4,$$
where $p$ is tuition and $q$ is the quantity of higher education. Which of the following is suggested by the result?

**(A)** As tuition rises, students want to buy a greater quantity of education.

**(B)** As a determinant of the demand for higher education, income is more important than price.

**(C)** If colleges lowered tuition slightly, their total tuition receipts would increase.

**(D)** If colleges raised tuition slightly, their total tuition receipts would increase.

**(E)** Colleges cannot increase enrollments by offering larger scholarships.

**25.** ▼ *Modeling Exponential Demand* As the new owner of a supermarket, you have inherited a large inventory of unsold imported Limburger cheese, and you would like to set the price so that your revenue from selling it is as large as

possible. Previous sales figures of the cheese are shown in the following table:

| Price per Pound, $p$ | $3.00 | $4.00 | $5.00 |
|---|---|---|---|
| Monthly Sales in Pounds, $q$ | 407 | 287 | 223 |

**a.** Use the sales figures for the prices $3 and $5 per pound to construct a demand function of the form $q = Ae^{-bp}$, where $A$ and $b$ are constants you must determine. (Round $A$ and $b$ to two significant digits.)

**b.** Use your demand function to find the price elasticity of demand at each of the prices listed.

**c.** At what price should you sell the cheese in order to maximize monthly revenue?

**d.** If your total inventory of cheese amounts to only 200 pounds, and it will spoil one month from now, how should you price it in order to receive the greatest revenue? Is this the same answer you got in part (c)? If not, give a brief explanation.

**26.** ▼ *Modeling Exponential Demand* Repeat the preceding exercise, but this time use the sales figures for $4 and $5 per pound to construct the demand function.

## COMMUNICATION AND REASONING EXERCISES

**27.** Complete the following: When demand is inelastic, revenue will decrease if _____ .

**28.** Complete the following: When demand has unit elasticity, revenue will decrease if _____ .

**29.** ▼ Given that the demand $q$ is a differentiable function of the unit price $p$, show that the revenue $R = pq$ has a stationary point when
$$q + p\frac{dq}{dp} = 0.$$
Deduce that the stationary points of $R$ are the same as the points of unit price elasticity of demand. (Ordinarily, there is only one such stationary point, corresponding to the absolute maximum of $R$.) HINT [Differentiate $R$ with respect to $p$.]

**30.** ▼ Given that the demand $q$ is a differentiable function of income $x$, show that the quantity $R = q/x$ has a stationary point when
$$q - x\frac{dq}{dx} = 0.$$
Deduce that stationary points of $R$ are the same as the points of unit income elasticity of demand. HINT [Differentiate $R$ with respect to $x$.]

**31.** ◆ Your calculus study group is discussing price elasticity of demand, and a member of the group asks the following question: "Since elasticity of demand measures the response of demand to change in unit price, what is the difference between elasticity of demand and the quantity $-dq/dp$?" How would you respond?

**32.** ◆ Another member of your study group claims that unit price elasticity of demand need not always correspond to maximum revenue. Is he correct? Explain your answer.

---

[71] Sources: Luxembourg Income Study/*New York Times*, August 14, 1995, p. A9, Commerce Department, Deloitte & Touche Survey/*New York Times*, November 24, 1999, p. C1.

## KEY CONCEPTS

 Website www.WanerMath.com

Go to the Website at www.WanerMath
.com to find a comprehensive and
interactive Web-based summary
of Chapter 12.

### 12.1 Maxima and Minima

Relative maximum, relative minimum
*p. 865*

Absolute maximum, absolute
minimum *p. 865*

Stationary points, singular points,
endpoints *p. 867*

Finding and classifying maxima
and minima *p. 868*

First derivative test for relative
extrema *p. 869*

Extreme value theorem *p. 873*

Using technology to locate approximate
extrema *p. 874*

### 12.2 Applications of Maxima and Minima

Minimizing average cost *p. 878*

Maximizing area *p. 879*

Steps in solving optimization
problems *p. 881*

Maximizing revenue *p. 881*

Optimizing resources *p. 882*

Allocation of labor *p. 884*

### 12.3 Higher Order Derivatives: Acceleration and Concavity

The second derivative of a function $f$ is
the derivative of the derivative of $f$,
written as $f''$ *p. 892*

The acceleration of a moving object is
the second derivative of the position
function *p. 892*

Acceleration due to gravity *p. 893*

Acceleration of sales *p. 894*

Concave up, concave down, point of
inflection *p. 895*

Locating points of inflection *p. 895*

Application to inflation *p. 896*

Second derivative test for relative
extrema *p. 899*

Higher order derivatives *p. 900*

### 12.4 Analyzing Graphs

Features of a graph: $x$- and $y$-intercepts,
relative extrema, points of inflection;
behavior near points where the

function is not defined, behavior at
infinity *pp. 910–911*

Analyzing a graph *p. 911*

### 12.5 Related Rates

If $Q$ is a quantity changing over time $t$,
then the derivative $dQ/dt$ is the rate at
which $Q$ changes over time *p. 918*

The expanding circle *p. 918*

Steps in solving related rates
problems *p. 919*

The falling ladder *p. 919*

Average cost *p. 921*

Allocation of labor *p. 922*

### 12.6 Elasticity

Price elasticity of demand
$$E = -\frac{dq}{dp} \cdot \frac{p}{q}; \text{ demand is elastic}$$
if $E > 1$, inelastic if $E < 1$, has unit
elasticity if $E = 1$ *p. 928*

Computing and interpreting elasticity,
and maximizing revenue *p. 929*

Using technology to compute
elasticity *p. 930*

Income elasticity of demand *p. 930*

## REVIEW EXERCISES

*In Exercises 1–8, find all the relative and absolute extrema of
the given functions on the given domain (if supplied) or on the
largest possible domain (if no domain is supplied).*

1. $f(x) = 2x^3 - 6x + 1$ on $[-2, +\infty)$

2. $f(x) = x^3 - x^2 - x - 1$ on $(-\infty, \infty)$

3. $g(x) = x^4 - 4x$ on $[-1, 1]$

4. $f(x) = \dfrac{x+1}{(x-1)^2}$ for $-2 \leq x \leq 2, x \neq 1$

5. $g(x) = (x-1)^{2/3}$  6. $g(x) = x^2 + \ln x$ on $(0, +\infty)$

7. $h(x) = \dfrac{1}{x} + \dfrac{1}{x^2}$  8. $h(x) = e^{x^2} + 1$

*In Exercises 9–12, the graph of the function $f$ or its derivative is
given. Find the approximate $x$-coordinates of all relative extrema
and points of inflection of the original function $f$ (if any).*

9. Graph of $f$:

10. Graph of $f$:

11. Graph of $f'$:

12. Graph of $f'$:

*In Exercises 13 and 14, the graph of the second derivative of a
function $f$ is given. Find the approximate $x$-coordinates of all
points of inflection of the original function $f$ (if any).*

13. Graph of $f''$

14. Graph of $f''$

*In Exercises 15 and 16, the position s of a point (in meters) is given as a function of time t (in seconds). Find (a) its acceleration as a function of t and (b) its acceleration at the specified time.*

**15.** $s = \dfrac{2}{3t^2} - \dfrac{1}{t}; t = 1$     **16.** $s = \dfrac{4}{t^2} - \dfrac{3t}{4}; t = 2$

*In Exercises 17–22, sketch the graph of the given function, indicating all relative and absolute extrema and points of inflection. Find the coordinates of these points exactly, where possible. Also indicate any horizontal and vertical asymptotes.*

**17.** $f(x) = x^3 - 12x$ on $[-2, +\infty)$

**18.** $g(x) = x^4 - 4x$ on $[-1, 1]$

**19.** $f(x) = \dfrac{x^2 - 3}{x^3}$

**20.** $f(x) = (x - 1)^{2/3} + \dfrac{2x}{3}$

**21.** $g(x) = (x - 3)\sqrt{x}$

**22.** $g(x) = (x + 3)\sqrt{x}$

## APPLICATIONS: OHaganBooks.com

**23. *Revenue*** Demand for the latest best-seller at OHaganBooks.com, *A River Burns through It*, is given by

$$q = -p^2 + 33p + 9 \quad (18 \le p \le 28)$$

copies sold per week when the price is $p$ dollars. What price should the company charge to obtain the largest revenue?

**24. *Revenue*** Demand for *The Secret Loves of John O*, a romance novel by Margó Dufón that flopped after two weeks on the market, is given by

$$q = -2p^2 + 5p + 6 \quad (0 \le p \le 3.3)$$

copies sold per week when the price is $p$ dollars. What price should OHaganBooks charge to obtain the largest revenue?

**25. *Profit*** Taking into account storage and shipping, it costs OHaganBooks.com

$$C = 9q + 100$$

dollars to sell $q$ copies of *A River Burns through It* in a week (see Exercise 23).

**a.** If demand is as in Exercise 23, express the weekly profit earned by OHaganBooks.com from the sale of *A River Burns through It* as a function of unit price $p$.

**b.** What price should the company charge to get the largest weekly profit? What is the maximum possible weekly profit?

**c.** Compare your answer in part (b) with the price the company should charge to obtain the largest revenue (Exercise 23). Explain any difference.

**26. *Profit*** Taking into account storage and shipping, it costs OHaganBooks.com

$$C = 3q$$

dollars to sell $q$ copies of Margó Dufón's *The Secret Loves of John O* in a week (see Exercise 24).

**a.** If demand is as in Exercise 24, express the weekly profit earned by OHaganBooks.com from the sale of *The Secret Loves of John O* as a function of unit price $p$.

**b.** What price should the company charge to get the largest weekly profit? What is the maximum possible weekly profit?

**c.** Compare your answer in part (b) with the price the company should charge to obtain the largest revenue (Exercise 24). Explain any difference.

**27. *Box Design*** The sales department at OHaganBooks.com, which has decided to send chocolate lobsters to each of its customers, is trying to design a shipping box with a square base. It has a roll of cardboard 36 inches wide from which to make the boxes. Each box will be obtained by cutting out corners from a rectangle of cardboard as shown in the following diagram:

(Notice that the top and bottom of each box will be square, but the sides will not necessarily be square.) What are the dimensions of the boxes with the largest volume that can be made in this way? What is the maximum volume?

**28. *Box Redesign*** The sales department at OHaganBooks.com was not pleased with the result of the box design in the preceding exercise; the resulting box was too large for the chocolate lobsters, so, following a suggestion by a math major student intern, the department decided to redesign the boxes to meet the following specifications: As in Exercise 27, each box would be obtained by cutting out corners from a rectangle of cardboard as shown in the following diagram:

(Notice that the top and bottom of each box would be square, but not necessarily the sides.) The dimensions would be such that the total surface area of the sides plus the bottom of the box would be as large as possible. What are the dimensions of the boxes with the largest area that can be made in this way? How does this box compare with that obtained in Exercise 27?

**29. *Elasticity of Demand*** (Compare Exercise 23.) Demand for the latest best-seller at OHaganBooks.com, *A River Burns through It*, is given by

$$q = -p^2 + 33p + 9 \qquad (18 \le p \le 28)$$

copies sold per week when the price is $p$ dollars.

**a.** Find the price elasticity of demand as a function of $p$.
**b.** Find the elasticity of demand for this book at a price of $20 and at a price of $25. (Round your answers to two decimal places.) Interpret the answers.
**c.** What price should the company charge to obtain the largest revenue?

**30. *Elasticity of Demand*** (Compare Exercise 24.) Demand for *The Secret Loves of John O*, a romance novel by Margó Dufón that flopped after two weeks on the market, is given by

$$q = -2p^2 + 5p + 6 \qquad (0 \le p \le 3.3)$$

copies sold per week when the price is $p$ dollars.

**a.** Find the price elasticity of demand as a function of $p$.
**b.** Find the elasticity of demand for this book at a price of $2 and at a price of $3. (Round your answers to two decimal places.) Interpret the answers.
**c.** What price should the company charge to obtain the largest revenue?

**31. *Elasticity of Demand*** Last year OHaganBooks.com experimented with an online subscriber service, Red On Line (ROL), for its electronic book service. The consumer demand for ROL was modeled by the equation

$$q = 1{,}000e^{-p^2+p},$$

where $p$ was the monthly access charge and $q$ is the number of subscribers.

**a.** Obtain a formula for the price elasticity of demand, $E$, for ROL services.
**b.** Compute the elasticity of demand if the monthly access charge is set at $2 per month. Interpret the result.
**c.** How much should the company have charged in order to obtain the maximum monthly revenue? What would this revenue have been?

**32. *Elasticity of Demand*** JungleBooks.com (one of OHaganBooks' main competitors) responded with its own online subscriber service, Better On Line (BOL), for its electronic book service. The consumer demand for BOL was modeled by the equation

$$q = 2{,}000e^{-3p^2+2p},$$

where $p$ was the monthly access charge and $q$ is the number of subscribers.

**a.** Obtain a formula for the price elasticity of demand, $E$, for BOL services.
**b.** Compute the elasticity of demand if the monthly access charge is set at $2 per month. Interpret the result.
**c.** How much should the company have charged in order to obtain the maximum monthly revenue? What would this revenue have been?

**33. *Sales*** OHaganBooks.com modeled its weekly sales over a period of time with the function

$$s(t) = 6{,}053 + \frac{4{,}474}{1+e^{-0.55(t-4.8)}},$$

where $t$ is the time in weeks. Following are the graphs of $s$, $s'$, and $s''$:

Graph of $s$

Graph of $s'$

Graph of $s''$

**a.** Estimate when, to the nearest week, the weekly sales were growing fastest.
**b.** To what features on the graphs of $s$, $s'$, and $s''$ does your answer to part (a) correspond?
**c.** The graph of $s$ has a horizontal asymptote. What is the approximate value ($s$-coordinate) of this asymptote, and what is its significance in terms of weekly sales at OHaganBooks.com?

d. The graph of $s'$ has a horizontal asymptote. What is the value ($s'$-coordinate) of this asymptote, and what is its significance in terms of weekly sales at OHaganBooks.com?

34. *Sales* The quarterly sales of OHagan *oPods* (OHaganBooks' answer to the *iPod*; a portable audio book unit with an incidental music feature) from the fourth quarter of 2009 can be roughly approximated by the function

$$N(t) = \frac{1,100}{1 + 9(1.8)^{-t}} \, oPods \ (t \geq 0),$$

where $t$ is time in quarters since the fourth quarter of 2009. Following are the graphs of $N$, $N'$, and $N''$:

Graph of $N$

Graph of $N'$

Graph of $N''$

a. Estimate when, to the nearest quarter, the quarterly sales were growing fastest.
b. To what features on the graphs of $N$, $N'$, and $N''$ does your answer to part (a) correspond?
c. The graph of $N$ has a horizontal asymptote. What is the approximate value ($N$-coordinate) of this asymptote, and what is its significance in terms of quarterly sales of *oPods*?
d. The graph of $N'$ has a horizontal asymptote. What is the value ($N'$-coordinate) of this asymptote, and what is its significance in terms of quarterly sales of *oPods*?

35. *Chance Encounter* Marjory Duffin is walking north towards the corner entrance of OHaganBooks.com company headquarters at 5 ft/sec, while John O'Hagan is walking west toward the same entrance, also at 5 ft/sec. How fast is their distance apart decreasing when

a. each of them is 2 ft from the corner?
b. each of them is 1 ft from the corner?
c. each of them is $h$ ft from the corner?
d. they collide on the corner?

36. *Company Logos* OHaganBooks.com's Web site has an animated graphic with its name in a rectangle whose height and width change; on either side of the rectangle are semicircles, as in the figure, whose diameters are the same as the height of the rectangle.

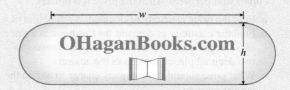

For reasons too complicated to explain, the designer wanted the combined area of the rectangle and semicircles to remain constant. At one point during the animation, the width of the rectangle is 1 inch, growing at a rate of 0.5 inches per second, while the height is 3 inches. How fast is the height changing?

## Case Study

### Production Lot Size Management

Your publishing company, *Knockem Dead Paperbacks, Inc.,* is about to release its next best-seller, *Henrietta's Heaving Heart* by Celestine A. Lafleur. The company expects to sell 100,000 books each month in the next year. You have been given the job of scheduling print runs to meet the anticipated demand and minimize total costs to the company. Each print run has a setup cost of $5,000, each book costs $1 to produce, and monthly storage costs for books awaiting shipment average 1¢ per book. What will you do?

If you decide to print all 1,200,000 books (the total demand for the year, 100,000 books per month for 12 months) in a single run at the start of the year and sales run as predicted, then the number of books in stock would begin at 1,200,000 and decrease to zero by the end of the year, as shown in Figure 47.

**Figure 47**

On average, you would be storing 600,000 books for 12 months at 1¢ per book, giving a total storage cost of $600,000 \times 12 \times .01 = \$72,000$. The setup cost for the single print run would be $5,000. When you add to these the total cost of producing 1,200,000 books at $1 per book, your total cost would be $1,277,000.

If, on the other hand, you decide to cut down on storage costs by printing the book in two runs of 600,000 each, you would get the picture shown in Figure 48.

**Figure 48**

Now, the storage cost would be cut in half because on average there would be only 300,000 books in stock. Thus, the total storage cost would be $36,000, and the setup cost would double to $10,000 (because there would now be two runs). The production costs would be the same: 1,200,000 books @ $1 per book. The total cost would therefore be reduced to $1,246,000, a savings of $31,000 compared to your first scenario.

"Aha!" you say to yourself, after doing these calculations. "Why not drastically cut costs by setting up a run every month?" You calculate that the setup costs alone would be $12 \times \$5,000 = \$60,000$, which is already more than the setup plus storage costs for two runs, so a run every month will cost too much. Perhaps, then, you should investigate three runs, four runs, and so on, until you find the lowest cost. This strikes you as too laborious a process, especially considering that you will have to do it all over again when planning for Lafleur's sequel, *Lorenzo's Lost Love,* due to be released next year. Realizing that this is an optimization problem, you decide to use some calculus to help you come up with a *formula* that you can use for all future plans. So you get to work.

Instead of working with the number 1,200,000, you use the letter $N$ so that you can be as flexible as possible. (What if *Lorenzo's Lost Love* sells more copies?) Thus, you have a total of $N$ books to be produced for the year. You now calculate the total cost of using $x$ print runs per year. Because you are to produce a total of $N$ books in $x$ print runs, you will have to produce $N/x$ books in each print run. $N/x$ is called the **lot size**. As you can see from the diagrams above, the average number of books in storage will be half that amount, $N/(2x)$.

Now you can calculate the total cost for a year. Write $P$ for the setup cost of a single print run ($P = \$5,000$ in your case) and $c$ for the *annual* cost of storing a book (to convert all of the time measurements to years; $c = \$0.12$ here). Finally, write $b$ for the cost of producing a single book ($b = \$1$ here). The costs break down as follows.

**Setup Costs:** $x$ print runs @ $P$ dollars per run:     $Px$

**Storage Costs:** $N/(2x)$ books stored @ $c$ dollars per year:    $cN/(2x)$

**Production Costs:** $N$ books @ $b$ dollars per book:    $\underline{\quad\quad Nb \quad\quad}$

**Total Cost:**    $Px + \dfrac{cN}{2x} + Nb$

Remember that $P$, $N$, $c$, and $b$ are all constants and $x$ is the only variable. Thus, your cost function is

$$C(x) = Px + \frac{cN}{2x} + Nb$$

and you need to find the value of $x$ that will minimize $C(x)$. But that's easy! All you need to do is find the relative extrema and select the absolute minimum (if any).

The domain of $C(x)$ is $(0, +\infty)$ because there is an $x$ in the denominator and $x$ can't be negative. To locate the extrema, you start by locating the critical points:

$$C'(x) = P - \frac{cN}{2x^2}.$$

The only singular point would be at $x = 0$, but $0$ is not in the domain. To find stationary points, you set $C'(x) = 0$ and solve for $x$:

$$P - \frac{cN}{2x^2} = 0$$

$$2x^2 = \frac{cN}{P}$$

so

$$x = \sqrt{\frac{cN}{2P}}.$$

There is only one stationary point, and there are no singular points or endpoints. To graph the function you will need to put in numbers for the various constants. Substituting $N = 1,200,000$, $P = 5,000$, $c = 0.12$, and $b = 1$, you get

$$C(x) = 5,000x + \frac{72,000}{x} + 1,200,000$$

with the stationary point at

$$x = \sqrt{\frac{(0.12)(1,200,000)}{2(5000)}} \approx 3.79.$$

The total cost at the stationary point is

$$C(3.79) \approx 1,237,900.$$

**Figure 49**

You now graph $C(x)$ in a window that includes the stationary point, say, $0 \le x \le 12$ and $1,100,000 \le C \le 1,500,000$, getting Figure 49.

From the graph, you can see that the stationary point is an absolute minimum. In the graph it appears that the graph is always concave up, which also tells you that your stationary point is a minimum. You can check the concavity by computing the second derivative:

$$C''(x) = \frac{cN}{x^3} > 0.$$

The second derivative is always positive because $c$, $N$, and $x$ are all positive numbers, so indeed the graph is always concave up. Now you also know that it works regardless of the particular values of the constants.

So now you are practically done! You know that the absolute minimum cost occurs when you have $x \approx 3.79$ print runs per year. Don't be disappointed that the answer is not a whole number; whole number solutions are rarely found in real scenarios. What the answer (and the graph) do indicate is that either three or four print runs per year will cost the least money. If you take $x = 3$, you get a total cost of

$$C(3) = \$1,239,000.$$

If you take $x = 4$, you get a total cost of

$$C(4) = \$1,238,000.$$

So, four print runs per year will allow you to minimize your total costs.

### EXERCISES

1. *Lorenzo's Lost Love* will sell 2,000,000 copies in a year. The remaining costs are the same. How many print runs should you use now?
2. In general, what happens to the number of runs that minimizes cost if both the setup cost and the total number of books are doubled?
3. In general, what happens to the number of runs that minimizes cost if the setup cost increases by a factor of 4?
4. Assuming that the total number of copies and storage costs are as originally stated, find the setup cost that would result in a single print run.
5. Assuming that the total number of copies and setup cost are as originally stated, find the storage cost that would result in a print run each month.

**6.** In Figure 48 we assumed that all the books in each run were manufactured in a very short time; otherwise the figure might have looked more like the following graph, which shows the inventory, assuming a slower rate of production.

How would this affect the answer?

**7.** Referring to the general situation discussed in the text, find the cost as a function of the total number of books produced, assuming that the number of runs is chosen to minimize total cost. Also find the average cost per book.

**8.** Let $\bar{C}$ be the average cost function found in the preceding exercise. Calculate $\lim_{N \to +\infty} \bar{C}(N)$ and interpret the result.

## TI-83/84 Plus    Technology Guide

### Section 12.6

**Example 1(a) (page 929)**  Suppose that the demand equation for *Bobby Dolls* is given by $q = 216 - p^2$, where $p$ is the price per doll in dollars and $q$ is the number of dolls sold per week. Compute the price elasticity of demand when $p = \$5$ and $p = \$10$, and interpret the results.

**Solution with Technology**

The TI-83/84 Plus function nDeriv can be used to compute approximations of the elasticity $E$ at various prices.

**1.** Set

$$Y_1 = 216-X^2 \qquad \text{Demand equation}$$
$$Y_2 = -\text{nDeriv}(Y_1,X,X) \ast \ X/Y_1 \quad \text{Formula for } E$$

**2.** Use the table feature to list the values of elasticity for a range of prices. For part (a) we chose values of X close to 5:

## SPREADSHEET    Technology Guide

### Section 12.6

**Example 1(a) (page 929)**  Suppose that the demand equation for *Bobby Dolls* is given by $q = 216 - p^2$, where $p$ is the price per doll in dollars and $q$ is the number of dolls sold per week. Compute the price elasticity of demand when $p = \$5$ and $p = \$10$, and interpret the results.

**Solution with Technology**

To approximate $E$ in a spreadsheet, we can use the following approximation of $E$.

$$E \approx \frac{\text{Percentage decrease in demand}}{\text{Percentage increase in price}} \approx -\frac{\left(\dfrac{\Delta q}{q}\right)}{\left(\dfrac{\Delta p}{p}\right)}$$

The smaller $\Delta p$ is, the better the approximation. Let's use $\Delta p = 1¢$, or 0.01 (which is small compared with the typical prices we consider—around \$5 to \$10).

**1.** We start by setting up our worksheet to list a range of prices, in increments of $\Delta p$, on either side of a price in which we are interested, such as $p_0 = \$5$:

We start in cell A5 with the formula for $p_0 - 10\Delta p$ and then successively add $\Delta p$ going down column A. You will find that the value $p_0 = 5$ appears midway down the list.

**2.** Next, we compute the corresponding values for the demand $q$ in column B.

**TECHNOLOGY GUIDE**

**3.** We add two new columns for the percentage changes in $p$ and $q$. The formula shown in cell C5 is copied down columns C and D, to Row 24. (Why not row 25?)

| | A | B | C | D |
|---|---|---|---|---|
| 1 | p0 | 5 | | |
| 2 | Delta p | 0.01 | | |
| 3 | | | | |
| 4 | p | q | Δp/p | Δq/q |
| 5 | 4.9 | 191.99 | =(A6-A5)/A5 | |
| 6 | 4.91 | 191.8919 | | |
| 7 | | | | |
| 24 | 5.09 | 190.0919 | | |
| 25 | 5.1 | 189.99 | | |

**4.** The elasticity can now be computed in column E as shown:

| | A | B | C | D | E |
|---|---|---|---|---|---|
| 1 | p0 | 5 | | | |
| 2 | Delta p | 0.01 | | | |
| 3 | | | | | |
| 4 | p | q | Δp/p | Δq/q | E |
| 5 | 4.9 | 191.99 | 0.00204082 | -0.00051096 | = -D5/C5 |
| 6 | 4.91 | 191.8919 | 0.00203666 | -0.00051227 | |
| 7 | | | | | |
| 24 | 5.09 | 190.0919 | 0.00196464 | -0.00053606 | |
| 25 | 5.1 | 189.99 | | | |

| | A | B | C | D | E |
|---|---|---|---|---|---|
| 1 | p0 | 5 | | | |
| 2 | Delta p | 0.01 | | | |
| 3 | | | | | |
| 4 | p | q | Δp/p | Δq/q | E |
| 5 | 4.9 | 191.99 | 0.00204082 | -0.00051096 | 0.25037242 |
| | | | | | |
| 14 | 4.99 | 191.0999 | 0.00200401 | -0.00052276 | 0.26083665 |
| 15 | 5 | 191 | 0.002 | -0.00052408 | 0.26204188 |
| 16 | 5.01 | 190.8999 | 0.00199601 | -0.00052541 | 0.26322853 |
| 17 | 5.02 | 190.7996 | 0.00199203 | -0.00052673 | 0.26441879 |

# 13

# The Integral

**W** Website

www.WanerMath.com

At the Website you will find:

- Section-by-section tutorials, including game tutorials with randomized quizzes

- A detailed chapter summary

- A true/false quiz

- Additional review exercises

- A numerical integration utility with Riemann sum grapher

- Graphing calculator programs for numerical integration

- Graphers, Excel tutorials, and other resources

- The following extra topic:
  Numerical Integration

## Case Study Spending on Housing Construction

It is March 2007, and *Time* magazine, in its latest edition, is asking "Will the Housing Bubble Burst in 2007?" You are a summer intern at *Ronald Ramp Real Estate Development, Inc.,* which is considering a major luxury condominium development in Tampa, Florida, nicknamed the "Ramp Towers Tampa." You have been asked to find formulas for monthly spending on housing construction in the United States and for the average spent per month starting one year ago. You have data about percentage spending changes. **How will you model the trend and estimate the total?**

Bill Varie/Flirt/Corbis

945

# Introduction

Roughly speaking, calculus is divided into two parts: **differential calculus** (the calculus of derivatives) and **integral calculus**, which is the subject of this chapter and the next. Integral calculus is concerned with problems that are in some sense the reverse of the problems seen in differential calculus. For example, where differential calculus shows how to compute the rate of change of a quantity, integral calculus shows how to find the quantity if we know its rate of change. This idea is made precise in the **Fundamental Theorem of Calculus**. Integral calculus and the Fundamental Theorem of Calculus allow us to solve many problems in economics, physics, and geometry, including one of the oldest problems in mathematics—computing areas of regions with curved boundaries.

## 13.1    The Indefinite Integral

Suppose that we knew the marginal cost to manufacture an item and we wanted to reconstruct the cost function. We would have to *reverse* the process of differentiation, to go from the derivative (the marginal cost function) back to the original function (the total cost). We'll first discuss how to do that and then look at some applications.

Here is an example: If the derivative of $F(x)$ is $4x^3$, what was $F(x)$? We recognize $4x^3$ as the derivative of $x^4$. So, we might have $F(x) = x^4$. However, $F(x) = x^4 + 7$ works just as well. In fact, $F(x) = x^4 + C$ works for any number $C$. Thus, there are *infinitely many* possible answers to this question.

In fact, we will see shortly that the formula $F(x) = x^4 + C$ covers *all* possible answers to the question. Let's give a name to what we are doing.

---

### Antiderivative

An **antiderivative** of a function $f$ is a function $F$ such that $F' = f$.

#### Quick Examples

1. An antiderivative of $4x^3$ is $x^4$.    Because the derivative of $x^4$ is $4x^3$
2. Another antiderivative of $4x^3$ is $x^4 + 7$.    Because the derivative of $x^4 + 7$ is $4x^3$
3. An antiderivative of $2x$ is $x^2 + 12$.    Because the derivative of $x^2 + 12$ is $2x$

---

Thus,

*If the derivative of A(x) is B(x), then an antiderivative of B(x) is A(x).*

We call the set of *all* antiderivatives of a function the **indefinite integral** of the function.

---

### Indefinite Integral

$$\int f(x)\, dx$$

is read "the **indefinite integral** of $f(x)$ with respect to $x$" and stands for the set of all antiderivatives of $f$. Thus, $\int f(x)\, dx$ is a *collection of functions*; it is not a single function or a number. The function *f* that is being **integrated** is called the **integrand**, and the variable *x* is called the **variable of integration**.

> **Quick Examples**
>
> 1. $\int 4x^3 \, dx = x^4 + C$    Every possible antiderivative of $4x^3$ has the form $x^4 + C$.
>
> 2. $\int 2x \, dx = x^2 + C$    Every possible antiderivative of $2x$ has the form $x^2 + C$.

The **constant of integration** *C* reminds us that we can add any constant and get a different antiderivative.

**Figure 1**

**Q**: If $F(x)$ is one antiderivative of $f(x)$, why must all other antiderivatives have the form $F(x) + C$?

**A**: Suppose $F(x)$ and $G(x)$ are both antiderivatives of $f(x)$, so that $F'(x) = G'(x)$. Consider what this means by looking at Figure 1. If $F'(x) = G'(x)$ for all $x$, then $F$ and $G$ have the *same slope* at each value of $x$. This means that their graphs must be *parallel* and hence remain exactly the same vertical distance apart. But that is the same as saying that the functions differ by a constant—that is, that $G(x) = F(x) + C$ for some constant $C$.*

\* This argument can be turned into a more rigorous proof—that is, a proof that does not rely on geometric concepts such as "parallel graphs." We should also say that the result (and our geometric argument as well!) requires that the domain of $F$ and $G$ be a single (possibly infinite) open interval.

## EXAMPLE 1  Indefinite Integral

Check that

**a.** $\int x \, dx = \dfrac{x^2}{2} + C$    **b.** $\int x^2 \, dx = \dfrac{x^3}{3} + C$    **c.** $\int x^{-1} \, dx = \ln|x| + C$

**Solution**  We check each equation by taking the derivative of its right-hand side and checking whether it equals the integrand on the left:

**a.** $\dfrac{d}{dx}\left(\dfrac{x^2}{2} + C\right) = \dfrac{2x}{2} + 0 = x$    ✔

**b.** $\dfrac{d}{dx}\left(\dfrac{x^3}{3} + C\right) = \dfrac{3x^2}{3} + 0 = x^2$    ✔

**c.** $\dfrac{d}{dx}(\ln|x| + C) = \dfrac{1}{x} + 0 = x^{-1}$.    ✔

† We are glossing over a subtlety in part (c): The constant of integration $C$ can be different for $x < 0$ and $x > 0$ because the graph breaks at $x = 0$. (See the comment at the end of the last marginal note.) In general, our understanding will be that the constant of integration may be different on disconnected intervals of the domain.

Because the derivative of the right-hand side is the integrand in each case, we can conclude that the given statements are all valid.†

➡ **Before we go on...** Example 1 gives us a very useful technique to check our answer every time we calculate an integral:

*Take the derivative of the answer and check that it equals the integrand.* ■

Now, we would like to make the process of finding indefinite integrals (anti-derivatives) more mechanical. For example, it would be nice to have a power rule for indefinite integrals similar to the one we already have for derivatives. Example 1 already suggests such a rule for us:

## Power Rule for the Indefinite Integral

$$\int x^n \, dx = \frac{x^{n+1}}{n+1} + C \qquad \text{This holds only if } n \neq -1.^*$$

$$\int x^{-1} \, dx = \ln |x| + C \qquad \text{For the special case } n = -1$$

**Equivalent Form of Second Formula:** $\int \frac{1}{x} \, dx = \ln |x| + C$   Because $x^{-1} = \frac{1}{x}$

**In Words**  For $n$ other than $-1$, to find the integral of $x^n$, add 1 to the exponent, and then divide by the new exponent. When $n = -1$, the answer is the natural logarithm of the absolute value of $x$.

### Quick Examples

**1.** $\int x^{55} \, dx = \dfrac{x^{56}}{56} + C$

**2.** $\int \dfrac{1}{x^{55}} \, dx = \int x^{-55} \, dx$   Exponent form

$\qquad = \dfrac{x^{-54}}{-54} + C$   When we add 1 to $-55$, we get $-54$, *not* $-56$.

$\qquad = -\dfrac{1}{54x^{54}} + C$

**3.** $\int 1 \, dx = x + C$   Because $1 = x^0$. This is an important special case.

**4.** $\int \sqrt{x} \, dx = \int x^{1/2} \, dx$   Exponent form

$\qquad = \dfrac{x^{3/2}}{3/2} + C$

$\qquad = \dfrac{2x^{3/2}}{3} + C$

## Notes

**1.** The integral $\int 1 \, dx$ is commonly written as $\int dx$. Similarly, the integral $\int \dfrac{1}{x^{55}} \, dx$ may be written as $\int \dfrac{dx}{x^{55}}$.

**2.** We can easily check the power rule formula by taking the derivative of the right-hand side:

$$\frac{d}{dx}\left( \frac{x^{n+1}}{n+1} + C \right) = \frac{(n+1)x^n}{n+1} = x^n \qquad ✔$$

**3.** Because the derivative of $\ln x$ is also $1/x$, you might be tempted to write $\int x^{-1} dx = \ln x + C$. But $\ln x$, being defined only for positive $x$, does not have the same domain as $1/x$, whereas $\ln |x|$ does. So, we must use $\ln |x| + C$ instead. ∎

Following are more indefinite integrals that come from formulas for differentiation we have encountered before:

---

### Indefinite Integral of $e^x$, $b^x$, and $|x|$

$$\int e^x \, dx = e^x + C \qquad \text{Because } \frac{d}{dx}(e^x) = e^x$$

If $b$ is any positive number other than 1, then

$$\int b^x \, dx = \frac{b^x}{\ln b} + C \qquad \text{Because } \frac{d}{dx}\left(\frac{b^x}{\ln b}\right) = \frac{b^x \ln b}{\ln b} = b^x$$

$$\int |x| \, dx = \frac{x|x|}{2} + C. \qquad \text{Because } \frac{d}{dx}\left(\frac{x|x|}{2}\right) = |x| \text{ (Check this yourself!)}$$

**Quick Example**

$$\int 2^x \, dx = \frac{2^x}{\ln 2} + C$$

---

For more complicated functions, like $2x^3 + 6x^5 - 1$, we need the following rules for integrating sums, differences, and constant multiples.

---

### Sums, Differences, and Constant Multiples

**Sum and Difference Rules**

$$\int [f(x) \pm g(x)] \, dx = \int f(x) \, dx \pm \int g(x) \, dx$$

**In Words:** The integral of a sum is the sum of the integrals, and the integral of a difference is the difference of the integrals.

**Constant Multiple Rule**

$$\int k f(x) \, dx = k \int f(x) \, dx \quad (k \text{ constant})$$

**In Words:** The integral of a constant times a function is the constant times the integral of the function. (In other words, the constant "goes along for the ride.")

Sum Rule: $\displaystyle\int (x^3 + 1)\,dx = \int x^3\,dx + \int 1\,dx = \frac{x^4}{4} + x + C$

$f(x) = x^3; g(x) = 1$

Constant Multiple Rule: $\displaystyle\int 5x^3\,dx = 5\int x^3\,dx = 5\frac{x^4}{4} + C$

$k = 5; f(x) = x^3$

Constant Multiple Rule: $\displaystyle\int 4\,dx = 4\int 1\,dx = 4x + C$

$k = 4; f(x) = 1$

Constant Multiple Rule: $\displaystyle\int 4e^x\,dx = 4\int e^x\,dx = 4e^x + C$

$k = 4; f(x) = e^x$

## Proof of the Sum Rule

We saw above that if two functions have the same derivative, they differ by a (possibly zero) constant. Look at the rule for sums:

$$\int [f(x) + g(x)]\,dx = \int f(x)\,dx + \int g(x)\,dx$$

If we take the derivative of the left-hand side with respect to $x$, we get the integrand, $f(x) + g(x)$. If we take the derivative of the right-hand side, we get

$$\frac{d}{dx}\left[\int f(x)\,dx + \int g(x)\,dx\right] = \frac{d}{dx}\left[\int f(x)\,dx\right] + \frac{d}{dx}\left[\int g(x)\,dx\right]$$

Derivative of a sum = Sum of derivatives.

$$= f(x) + g(x)$$

Because the left- and right-hand sides have the same derivative, they differ by a constant. But, because both expressions are indefinite integrals, adding a constant does not affect their value, so they are the same as indefinite integrals.

Notice that a key step in the proof was the fact that the derivative of a sum is the sum of the derivatives.

A similar proof works for the difference and constant multiple rules.

---

**EXAMPLE 2  Using the Sum and Difference Rules**

Find the integrals.

**a.** $\displaystyle\int (x^3 + x^5 - 1)\,dx$    **b.** $\displaystyle\int \left(x^{2.1} + \frac{1}{x^{1.1}} + \frac{1}{x} + e^x\right)dx$    **c.** $\displaystyle\int (e^x + 3^x - |x|)\,dx$

**Solution**

**a.** $\displaystyle\int (x^3 + x^5 - 1)\,dx = \int x^3\,dx + \int x^5\,dx - \int 1\,dx$    Sum/difference rule

$$= \frac{x^4}{4} + \frac{x^6}{6} - x + C$$    Power rule

**b.** $\int \left( x^{2.1} + \dfrac{1}{x^{1.1}} + \dfrac{1}{x} + e^x \right) dx$

$= \int (x^{2.1} + x^{-1.1} + x^{-1} + e^x)\,dx$ $\qquad\qquad$ Exponent form

$= \int x^{2.1}\,dx + \int x^{-1.1}\,dx + \int x^{-1}\,dx + \int e^x\,dx$ $\qquad$ Sum rule

$= \dfrac{x^{3.1}}{3.1} + \dfrac{x^{-0.1}}{-0.1} + \ln|x| + e^x + C$ $\qquad\qquad$ Power rule and exponential rule

$= \dfrac{x^{3.1}}{3.1} - \dfrac{10}{x^{0.1}} + \ln|x| + e^x + C$

**c.** $\int (e^x + 3^x - |x|)\,dx = \int e^x\,dx + \int 3^x\,dx - \int |x|\,dx$ $\quad$ Sum/difference rule

$= e^x + \dfrac{3^x}{\ln 3} - \dfrac{x|x|}{2} + C$ $\qquad$ Rules for powers, exponentials, and absolute value

---

➡ **Before we go on...** You should check each of the answers in Example 2 by differentiating.

---

**Q**: *Why is there only a single arbitrary constant C in each of the answers?*

**A**: We could have written the answer to part (a) as

$\dfrac{x^4}{4} + D + \dfrac{x^6}{6} + E - x + F$

where *D, E,* and *F* are all arbitrary constants. Now suppose, for example, we set $D = 1$, $E = -2$, and $F = 6$. Then the particular antiderivative we get is $x^4/4 + x^6/6 - x + 5$, which has the form $x^4/4 + x^6/6 - x + C$. Thus, we could have chosen the single constant *C* to be 5 and obtained the same answer. In other words, the answer $x^4/4 + x^6/6 - x + C$ is just as general as the answer $x^4/4 + D + x^6/6 + E - x + F$, but simpler.

---

In practice we do not explicitly write the integral of a sum as a sum of integrals but just "integrate term by term," much as we learned to differentiate term by term.

---

## EXAMPLE 3 Combining the Rules

Find the integrals.

**a.** $\int (10x^4 + 2x^2 - 3e^x)\,dx$ $\qquad$ **b.** $\int \left( \dfrac{2}{x^{0.1}} + \dfrac{x^{0.1}}{2} - \dfrac{3}{4x} \right) dx$

**c.** $\int (3e^x - 2(1.2^x) + 5|x|)\,dx$

### Solution

**a.** We need to integrate separately each of the terms $10x^4$, $2x^2$, and $3e^x$. To integrate $10x^4$ we use the rules for constant multiples and powers:

$$\int 10x^4 \, dx = 10 \int x^4 \, dx = 10\frac{x^5}{5} + C = 2x^5 + C.$$

The other two terms are similar. We get

$$\int (10x^4 + 2x^2 - 3e^x) \, dx = 10\frac{x^5}{5} + 2\frac{x^3}{3} - 3e^x + C = 2x^5 + \frac{2}{3}x^3 - 3e^x + C.$$

**b.** We first convert to exponent form and then integrate term by term:

$$\int \left( \frac{2}{x^{0.1}} + \frac{x^{0.1}}{2} - \frac{3}{4x} \right) dx = \int \left( 2x^{-0.1} + \frac{1}{2}x^{0.1} - \frac{3}{4}x^{-1} \right) dx \qquad \text{Exponent form}$$

$$= 2\frac{x^{0.9}}{0.9} + \frac{1}{2}\frac{x^{1.1}}{1.1} - \frac{3}{4}\ln|x| + C \qquad \begin{array}{l}\text{Integrate term by}\\\text{term.}\end{array}$$

$$= \frac{20x^{0.9}}{9} + \frac{x^{1.1}}{2.2} - \frac{3}{4}\ln|x| + C. \qquad \begin{array}{l}\text{Back to rational}\\\text{form}\end{array}$$

**c.** $\displaystyle\int (3e^x - 2(1.2^x) + 5|x|) \, dx = 3e^x - 2\frac{1.2^x}{\ln 1.2} + 5\frac{x|x|}{2} + C$

---

### EXAMPLE 4  **Different Variable Name**

Find $\displaystyle\int \left( \frac{1}{u} + \frac{1}{u^2} \right) du.$

**Solution**  This integral may look a little strange because we are using the letter $u$ instead of $x$, but there is really nothing special about $x$. Using $u$ as the variable of integration, we get

$$\int \left( \frac{1}{u} + \frac{1}{u^2} \right) du = \int (u^{-1} + u^{-2}) \, du \qquad \text{Exponent form.}$$

$$= \ln|u| + \frac{u^{-1}}{-1} + C \qquad \text{Integrate term by term.}$$

$$= \ln|u| - \frac{1}{u} + C. \qquad \text{Simplify the result.}$$

---

➡ **Before we go on...**  When we compute an indefinite integral, we want the independent variable in the answer to be the same as the variable of integration. Thus, if the integral in Example 4 had been written in terms of $x$ rather than $u$, we would have written

$$\int \left( \frac{1}{x} + \frac{1}{x^2} \right) dx = \ln|x| - \frac{1}{x} + C. \qquad ■$$

## APPLICATIONS

### EXAMPLE 5 Finding Cost from Marginal Cost

The marginal cost to produce baseball caps at a production level of $x$ caps is $4 - 0.001x$ dollars per cap, and the cost of producing 100 caps is $500. Find the cost function.

**Solution** We are asked to find the cost function $C(x)$, given that the *marginal* cost function is $4 - 0.001x$. Recalling that the marginal cost function is the derivative of the cost function, we can write

$$C'(x) = 4 - 0.001x$$

and must find $C(x)$. Now $C(x)$ must be an antiderivative of $C'(x)$, so

$$C(x) = \int (4 - 0.001x)\,dx$$
$$= 4x - 0.001\frac{x^2}{2} + K \qquad K \text{ is the constant of integration.}^*$$
$$= 4x - 0.0005x^2 + K.$$

✳ We used $K$ and not $C$ for the constant of integration because we are using $C$ for cost.

Now, unless we have a value for $K$, we don't really know what the cost function is. However, there is another piece of information we have ignored: The cost of producing 100 baseball caps is $500. In symbols

$$C(100) = 500.$$

Substituting in our formula for $C(x)$, we have

$$C(100) = 4(100) - 0.0005(100)^2 + K$$
$$500 = 395 + K$$
$$K = 105.$$

Now that we know what $K$ is, we can write down the cost function:

$$C(x) = 4x - 0.0005x^2 + 105.$$

➡ **Before we go on...** Let us consider the significance of the constant term 105 in Example 5. If we substitute $x = 0$ into the cost function, we get

$$C(0) = 4(0) - 0.0005(0)^2 + 105 = 105.$$

Thus, $105 is the cost of producing zero items; in other words, it is the **fixed cost**. ∎

### EXAMPLE 6 Total Sales from Annual Sales

By the start of 2008, Apple had sold a total of about 3.5 million iPhones. From the start of 2008 through the end of 2010, sales of iPhones were approximately

$$s(t) = 4.5t^2 + 5.8t + 6.5 \text{ million iPhones per year} \qquad (0 \le t \le 3),$$

where $t$ is time in years since the start of 2008.[1]

---

[1] Source for data: Apple quarterly press releases (www.apple.com/investor/).

**a.** Find an expression for the total sales of iPhones up to time $t$.

**b.** Use the answer to part (a) to estimate the total sales of iPhones by the end of 2010. (The actual figure was 89.7 million.)

**Solution**

**a.** Let $S(t)$ be the total sales of iPhones, in millions, up to time $t$, where $t$ is measured in years since the start of 2008, so we know that $S(0) = 3.5$. We are also given an expression for the number of iPhones sold per year. This function is the *derivative* of $S(t)$:

$$S'(t) = 4.5t^2 + 5.8t + 6.5.$$

Thus, the desired total sales function must be an antiderivative of $S'(t)$:

$$S(t) = \int (4.5t^2 + 5.8t + 6.5)\, dt$$

$$= \frac{4.5t^3}{3} + 5.8\frac{t^2}{2} + 6.5t + C$$

$$= 1.5t^3 + 2.9t^2 + 6.5t + C.$$

To calculate the value of the constant $C$, we can, as in the preceding example, use the known value of $S$: $S(0) = 3.5$.

$$S(0) = 1.5(0)^3 + 2.9(0)^2 + 6.5(0) + C = 3.5,$$

so

$$C = 3.5.$$

We can now write down the total sales function:

$$S(t) = 1.5t^3 + 2.9t^2 + 6.5t + 3.5 \text{ million iPhones.}$$

**b.** Because the end of 2010 (or the start of 2011) corresponds to $t = 3$, we calculate total sales as

$$S(3) = 1.5(3)^3 + 2.9(3)^2 + 6.5(3) + 3.5 = 89.6 \text{ million iPhones.}$$

## Motion in a Straight Line

An important application of the indefinite integral is to the study of motion. The application of calculus to problems about motion is an example of the intertwining of mathematics and physics. We begin by bringing together some facts, scattered through the last several chapters, that have to do with an object moving in a straight line, and then restating them in terms of antiderivatives.

### Position, Velocity, and Acceleration: Derivative Form

If $s = s(t)$ is the **position** of an object at time $t$, then its **velocity** is given by the derivative

$$v = \frac{ds}{dt}.$$

**In Words:** Velocity is the derivative of position.

The **acceleration** of an object is given by the derivative

$$a = \frac{dv}{dt}.$$

**In Words:** Acceleration is the derivative of velocity.

## Position, Velocity, and Acceleration: Integral Form

$$s(t) = \int v(t)\, dt \qquad \text{Because } v = \frac{ds}{dt}$$

$$v(t) = \int a(t)\, dt \qquad \text{Because } a = \frac{dv}{dt}$$

### Quick Examples

1. If the velocity of a particle moving in a straight line is given by $v(t) = 4t + 1$, then its position after $t$ seconds is given by $s(t) = \int v(t)\, dt = \int (4t + 1)\, dt = 2t^2 + t + C$.

2. If sales are accelerating at 2 golf balls/day$^2$, then the rate of change of sales ("velocity of sales") is $v(t) = \int a(t)\, dt = \int 2\, dt = 2t + C$ golf balls/day.

3. If the rate of change of sales is $v(t) = 2t + 5$ golf balls per day, then the total sales are $s(t) = \int v(t)\, dt = \int (2t + 5)\, dt = t^2 + 5t + C$ golf balls sold through time $t$.

---

### EXAMPLE 7  Motion in a Straight Line

a. The velocity of a particle moving along a straight line is given by $v(t) = 4t + 1$ m/sec. Given that the particle is at position $s = 2$ meters at time $t = 1$, find an expression for $s$ in terms of $t$.

b. For a freely falling body experiencing no air resistance and zero initial velocity, find an expression for the velocity $v$ in terms of $t$. [Note: On Earth, a freely falling body experiencing no air resistance accelerates downward at approximately 9.8 meters per second per second, or 9.8 m/sec$^2$ (or 32 ft/sec$^2$).]

### Solution

a. As we saw in the Quick Example above, the position of the particle after $t$ seconds is given by

$$s(t) = \int v(t)\, dt$$

$$= \int (4t + 1)\, dt = 2t^2 + t + C.$$

But what is the value of $C$? Now, we are told that the particle is at position $s = 2$ at time $t = 1$. In other words, $s(1) = 2$. Substituting this into the expression for $s(t)$ gives

$$2 = 2(1)^2 + 1 + C$$

so
$$C = -1.$$

Hence the position after $t$ seconds is given by

$$s(t) = 2t^2 + t - 1 \text{ meters.}$$

**b.** Let's measure heights above the ground as positive, so that a rising object has positive velocity and the acceleration due to gravity is negative. (It causes the upward velocity to decrease in value.) Thus, the acceleration of the object is given by

$$a(t) = -9.8 \text{ m/sec}^2.$$

We wish to know the velocity, which is an antiderivative of acceleration, so we compute

$$v(t) = \int a(t)\, dt = \int (-9.8)\, dt = -9.8t + C.$$

To find the value of $C$, we use the given information that at time $t = 0$ the velocity is 0: $v(0) = 0$. Substituting this into the expression for $v(t)$ gives

$$0 = -9.8(0) + C$$

so

$$C = 0.$$

Hence, the velocity after $t$ seconds is given by

$$v(t) = -9.8t \text{ m/sec.}$$

## EXAMPLE 8  Vertical Motion under Gravity

You are standing on the edge of a cliff and toss a stone upward at a speed of $v_0 = 30$ feet per second ($v_0$ is called the *initial velocity*).

**a.** Find the stone's velocity as a function of time. How fast and in what direction is it going after 5 seconds? (Neglect the effects of air resistance.)

**b.** Find the position of the stone as a function of time. Where will it be after 5 seconds?

**c.** When and where will the stone reach its zenith, its highest point?

### Solution

**a.** This is similar to Example 7(b): Measuring height above the ground as positive, the acceleration of the stone is given by $a(t) = -32$ ft/s², and so

$$v(t) = \int (-32)\, dt = -32t + C.$$

To obtain $C$, we use the fact that you tossed the stone upward at 30 ft/s; that is, when $t = 0$, $v = 30$, or $v(0) = 30$. Thus,

$$30 = v(0) = -32(0) + C.$$

So, $C = 30$ and the formula for velocity is

$$v(t) = -32t + 30 \text{ ft/sec.} \qquad v(t) = -32t + v_0$$

In particular, after 5 seconds the velocity will be

$$v(5) = -32(5) + 30 = -130 \text{ ft/sec.}$$

After 5 seconds the stone is *falling* with a speed of 130 ft/sec.

$s = 0$    $v_0 = 30$ ft/sec    $t = 0$

Zenith

Path of stone

$s = -250$    $t = 5$
$v = -130$ ft/sec

**Figure 2**

**b.** We wish to know the position, but position is an antiderivative of velocity. Thus,

$$s(t) = \int v(t)\, dt = \int (-32t + 30)\, dt = -16t^2 + 30t + C.$$

Now to find $C$, we need to know the initial position $s(0)$. We are not told this, so let's measure heights so that the initial position is zero. Then

$$0 = s(0) = C$$

and

$$s(t) = -16t^2 + 30t \text{ ft.}$$    $s(t) = -16t^2 + v_0 t + s_0$
$s_0 = $ initial position

In particular, after 5 seconds the stone has a height of

$$s(5) = -16(5)^2 + 30(5) = -250 \text{ ft.}$$

In other words, the stone is now 250 ft *below* where it was when you first threw it, as shown in Figure 2.

**c.** The stone reaches its zenith when its height $s(t)$ is at its maximum value, which occurs when $v(t) = s'(t)$ is zero. So we solve

$$v(t) = -32t + 30 = 0$$

getting $t = 30/32 = 15/16 = 0.9375$ sec. This is the time when the stone reaches its zenith. The height of the stone at that time is

$$s(15/16) = -16(15/16)^2 + 30(15/16) = 14.0625 \text{ ft.}$$

➡ **Before we go on...** Here again are the formulas we obtained in Example 8, together with their metric equivalents:

---

## Vertical Motion under Gravity: Velocity and Position

If we ignore air resistance, the vertical velocity and position of an object moving under gravity are given by

|  British Units  |  Metric Units  |
|---|---|
| Velocity: $v(t) = -32t + v_0$ ft/sec | $v(t) = -9.8t + v_0$ m/sec |
| Position: $s(t) = -16t^2 + v_0 t + s_0$ ft | $s(t) = -4.9t^2 + v_0 t + s_0$ m |

$v_0 = $ initial velocity $= $ velocity at time 0
$s_0 = $ initial position $= $ position at time 0

### Quick Example

If a ball is thrown down at 2 ft/sec from a height of 200 ft, then its velocity and position after $t$ seconds are $v(t) = -32t - 2$ ft/sec and $s(t) = -16t^2 - 2t + 200$ ft. ∎

## **13.1** EXERCISES

▼ more advanced   ◆ challenging
T indicates exercises that should be solved using technology

*Evaluate the integrals in Exercises 1–42.*
HINT [for 1–6: See Quick Examples on page 947.]

**1.** $\int x^5 \, dx$   **2.** $\int x^7 \, dx$   **3.** $\int 6 \, dx$

**4.** $\int (-5) \, dx$   **5.** $\int x \, dx$   **6.** $\int (-x) \, dx$

HINT [for 7–18: See Example 2.]

**7.** $\int (x^2 - x) \, dx$   **8.** $\int (x + x^3) \, dx$

**9.** $\int (1 + x) \, dx$   **10.** $\int (4 - x) \, dx$

**11.** $\int x^{-5} \, dx$   **12.** $\int x^{-7} \, dx$

**13.** $\int (x^{2.3} + x^{-1.3}) \, dx$   **14.** $\int (x^{-0.2} - x^{0.2}) \, dx$

**15.** $\int (u^2 - u^{-1}) \, du$ HINT [See Example 4.]

**16.** $\int (v^{-2} + 2v^{-1}) \, dv$ HINT [See Example 4.]

**17.** $\int \sqrt[4]{x} \, dx$   **18.** $\int \sqrt[3]{x} \, dx$

HINT [for 19–42: See Example 3.]

**19.** $\int (3x^4 - 2x^{-2} + x^{-5} + 4) \, dx$   **20.** $\int (4x^7 - x^{-3} + 1) \, dx$

**21.** $\int \left( \frac{2}{u} + \frac{u}{4} \right) du$   **22.** $\int \left( \frac{2}{u^2} + \frac{u^2}{4} \right) du$

**23.** $\int \left( \frac{1}{x} + \frac{2}{x^2} - \frac{1}{x^3} \right) dx$

**24.** $\int \left( \frac{3}{x} - \frac{1}{x^5} + \frac{1}{x^7} \right) dx$

**25.** $\int (3x^{0.1} - x^{4.3} - 4.1) \, dx$   **26.** $\int \left( \frac{x^{2.1}}{2} - 2.3 \right) dx$

**27.** $\int \left( \frac{3}{x^{0.1}} - \frac{4}{x^{1.1}} \right) dx$   **28.** $\int \left( \frac{1}{x^{1.1}} - \frac{1}{x} \right) dx$

**29.** $\int \left( 5.1t - \frac{1.2}{t} + \frac{3}{t^{1.2}} \right) dt$

**30.** $\int \left( 3.2 + \frac{1}{t^{0.9}} + \frac{t^{1.2}}{3} \right) dt$

**31.** $\int (2e^x + 5|x| + 1/4) \, dx$

**32.** $\int (-4e^x + |x|/3 - 1/8) \, dx$

**33.** $\int \left( \frac{6.1}{x^{0.5}} + \frac{x^{0.5}}{6} - e^x \right) dx$

**34.** $\int \left( \frac{4.2}{x^{0.4}} + \frac{x^{0.4}}{3} - 2e^x \right) dx$

**35.** $\int (2^x - 3^x) \, dx$

**36.** $\int (1.1^x + 2^x) \, dx$

**37.** $\int \left( 100(1.1^x) - \frac{2|x|}{3} \right) dx$

**38.** $\int \left( 1,000(0.9^x) + \frac{4|x|}{5} \right) dx$

**39.** ▼ $\int x^{-2} \left( x^4 - \frac{3}{2x^4} \right) dx$

**40.** ▼ $\int 3x^4 \left( \frac{2}{x^4} + \frac{3}{5x^6} \right) dx$

**41.** ▼ $\int \frac{x+2}{x^3} dx$   **42.** ▼ $\int \frac{x^2 - 2}{x} dx$

**43.** Find $f(x)$ if $f(0) = 1$ and the tangent line at $(x, f(x))$ has slope $x$. HINT [See Example 5.]

**44.** Find $f(x)$ if $f(1) = 1$ and the tangent line at $(x, f(x))$ has slope $\frac{1}{x}$. HINT [See Example 5.]

**45.** Find $f(x)$ if $f(0) = 0$ and the tangent line at $(x, f(x))$ has slope $e^x - 1$.

**46.** Find $f(x)$ if $f(1) = -1$ and the tangent line at $(x, f(x))$ has slope $2e^x + 1$.

## APPLICATIONS

**47.** *Marginal Cost* The marginal cost of producing the $x$th box of light bulbs is $5 - \frac{x}{10,000}$ and the fixed cost is \$20,000. Find the cost function $C(x)$. HINT [See Example 5.]

**48.** *Marginal Cost* The marginal cost of producing the $x$th box of DVDs is $10 + \frac{x^2}{100,000}$ and the fixed cost is \$100,000. Find the cost function $C(x)$. HINT [See Example 5.]

**49.** *Marginal Cost* The marginal cost of producing the $x$th roll of film is $5 + 2x + \frac{1}{x}$. The total cost to produce one roll is \$1,000. Find the cost function $C(x)$. HINT [See Example 5.]

**50.** *Marginal Cost* The marginal cost of producing the $x$th box of CDs is $10 + x + \frac{1}{x^2}$. The total cost to produce 100 boxes is \$10,000. Find the cost function $C(x)$. HINT [See Example 5.]

**51.** *Facebook Membership* At the start of 2005, Facebook had 1 million members. Since that time, new members joined at a rate of roughly

$$m(t) = 3.2t^3 - 12t + 10 \text{ million members per year}$$
$$(0 \le t \le 5),$$

where $t$ is time in years since the start of 2005.[2]

**a.** Find an expression for total Facebook membership $M(t)$ at time $t$. HINT [See Example 6.]

**b.** Use the answer to part (a) to estimate Facebook membership midway through 2008. (Round your answer to the nearest 1 million members. The actual figure was 80 million.)

**52.** *Uploads to YouTube* Since YouTube first became available to the public in mid-2005, the rate at which video has been uploaded to the site can be approximated by

$$v(t) = 525,600(0.42t^2 + 2.7t - 1) \text{ hours of video per year}$$
$$(0.5 \le t \le 7),$$

where $t$ is time in years since the start of 2005.[3]

**a.** Find an expression for the total number of hours $V(t)$ of video at time $t$ (starting from zero hours of video at $t = 0.5$). HINT [See Example 6.]

**b.** Use the answer to part (a) to estimate the total number of hours of video uploaded by the start of 2012. (Round your answer to the nearest million hours of video.)

**53.** *Median Household Income* From 2000 to 2007, median household income in the United States rose by an average of approximately \$1,200 per year.[4] Given that the median household income in 2000 was approximately \$42,000, use an indefinite integral to find a formula for the median household income $I$ as a function of the year $t$ ($t = 0$ represents 2000), and use your formula to estimate the median household income in 2005. (You can do this exercise without integration using the methods of Section 1.3, but here you should use an indefinite integral.)

**54.** *Mean Household Income* From 2000 to 2007, the mean household income in the United States rose by an average of approximately \$1,500 per year.[5] Given that the mean household income in 2000 was approximately \$57,000, use an indefinite integral to find a formula for the mean household income $I$ as a function of the year $t$ ($t = 0$ represents 2000), and use your formula to estimate the mean household income in 2006. (You can do this exercise without integration using the methods of Section 1.3, but here you should use an indefinite integral.)

**55.** *Bottled Water Sales* The rate of U.S. sales of bottled water for the period 2000–2010 can be approximated by

$$s(t) = -45t^2 + 900t + 4,200 \text{ million gallons per year}$$
$$(0 \le t \le 10),$$

where $t$ is time in years since the start of 2000.[6] Use an indefinite integral to approximate the total sales $S(t)$ of bottled water since the start of 2000. Approximately how much bottled water was sold from the start of 2000 to the start of 2008? HINT [At the start of 2000, sales since that time are zero.]

**56.** *Bottled Water Sales* The rate of U.S. per capita sales of bottled water for the period 2000–2010 can be approximated by

$$s(t) = -0.18t^2 + 3t + 15 \text{ gallons per year} \quad (0 \le t \le 10),$$

where $t$ is time in years since the start of 2000.[7] Use an indefinite integral to approximate the total per capita sales $S(t)$ of bottled water since the start of 2000. Approximately how much bottled water was sold, per capita, from the start of 2000 to the end of 2009? HINT [At the start of 2000, sales since that time are zero.]

**57.** ▽ *Health-Care Spending in the 1990s* Write $H(t)$ for the amount spent in the United States on health care in year $t$, where $t$ is measured in years since 1990. The rate of increase of $H(t)$ was approximately \$65 billion per year in 1990 and rose to \$100 billion per year in 2000.[8]

**a.** Find a linear model for the rate of change $H'(t)$.

**b.** Given that \$700 billion was spent on health care in the United States in 1990, find the function $H(t)$.

**58.** ▽ *Health-Care Spending in the 2000s* Write $H(t)$ for the amount spent in the United States on health care in year $t$, where $t$ is measured in years since 2000. The rate of increase of $H(t)$ was projected to rise from \$100 billion per year in 2000 to approximately \$190 billion per year in 2010.[9]

**a.** Find a linear model for the rate of change $H'(t)$.

**b.** Given that \$1,300 billion was spent on health care in the United States in 2000, find the function $H(t)$.

**59.** ▽ *Subprime Mortgage during the Housing Bubble* At the start of 2007, the percentage of U.S. mortgages that were subprime was about 13%, was increasing at a rate of 1 percentage point per year, but was decelerating at 0.4 percentage points per year per year.[10]

**a.** Find an expression for the rate of change (velocity) of this percentage at time $t$ in years since the start of 2007.

**b.** Use the result of part (a) to find an expression for the percentage of mortgages that were subprime at time $t$, and use it to estimate the rate at the start of 2008. HINT [See Quick Examples 2 and 3 on page 955.]

---

[2]Sources for data: www.facebook.com, www.insidefacebook.com.

[3]Source for data: www.YouTube.com.

[4]In current dollars, unadjusted for inflation. Source for data: U.S. Census Bureau (www.census.gov).

[5]*Ibid.*

[6]Source for data: Beverage Marketing Corporation (www.bottledwater.org).

[7]*Ibid.*

[8]Source: Centers for Medicare and Medicaid Services, "National Health Expenditures," 2002 version, released January 2004 (www.cms.hhs.gov/statistics/nhe/).

[9]Source: Centers for Medicare and Medicaid Services, "National Health Expenditures 1965–2013, History and Projections" (www.cms.hhs.gov/statistics/nhe/).

[10]Sources: Mortgage Bankers Association, UBS.

**60.** ▼ *Subprime Mortgage Debt during the Housing Bubble* At the start of 2008, the value of subprime mortgage debt outstanding in the U.S. was about $1,300 billion, was increasing at a rate of 40 billion dollars per year, but was decelerating at 20 billion dollars per year per year.[11]

  **a.** Find an expression for the rate of change (velocity) of the value of subprime mortgage debt at time $t$ in years since the start of 2008.

  **b.** Use the result of part (a) to find an expression for the value of subprime mortgage debt at time $t$, and use it to estimate the value at the start of 2009. HINT [See Quick Examples 2 and 3 on page 955.]

**61.** *Motion in a Straight Line* The velocity of a particle moving in a straight line is given by $v = t^2 + 1$.

  **a.** Find an expression for the position $s$ after a time $t$.

  **b.** Given that $s = 1$ at time $t = 0$, find the constant of integration $C$, and hence find an expression for $s$ in terms of $t$ without any unknown constants. HINT [See Example 7.]

**62.** *Motion in a Straight Line* The velocity of a particle moving in a straight line is given by $v = 3e^t + t$.

  **a.** Find an expression for the position $s$ after a time $t$.

  **b.** Given that $s = 3$ at time $t = 0$, find the constant of integration $C$, and hence find an expression for $s$ in terms of $t$ without any unknown constants. HINT [See Example 7.]

**63.** *Vertical Motion under Gravity* If a stone is dropped from a rest position above the ground, how fast (in feet per second) and in what direction will it be traveling after 10 seconds? (Neglect the effects of air resistance.) HINT [See Example 8.]

**64.** *Vertical Motion under Gravity* If a stone is thrown upward at 10 feet per second, how fast (in feet per second) and in what direction will it be traveling after 10 seconds? (Neglect the effects of air resistance.) HINT [See Example 8.]

**65.** *Vertical Motion under Gravity* Your name is Galileo Galilei and you toss a weight upward at 16 feet per second from the top of the Leaning Tower of Pisa (height 185 ft).

  **a.** Neglecting air resistance, find the weight's velocity as a function of time $t$ in seconds.

  **b.** Find the height of the weight above the ground as a function of time. Where and when will it reach its zenith? HINT [See Example 8 and the formulas that follow.]

**66.** *Vertical Motion under Gravity* Your name is Spaghettini Bologna (an assistant of Galileo Galilei) and, to impress your boss, you toss a weight upward at 24 feet per second from the top of the Leaning Tower of Pisa (height 185 ft).

  **a.** Neglecting air resistance, find the weight's velocity as a function of time $t$ in seconds.

  **b.** Find the height of the weight above the ground as a function of time. Where and when will it reach its zenith? HINT [See Example 8 and the formulas that follow.]

**67.** ▼ *Tail Winds* The ground speed of an airliner is obtained by adding its air speed and the tail-wind speed. On your recent trip from Mexico to the United States your plane was traveling at an air speed of 500 miles per hour and experienced tail winds of $25 + 50t$ miles per hour, where $t$ is the time in hours since takeoff.

  **a.** Obtain an expression for the distance traveled in terms of the time since takeoff. HINT [Ground speed = Air speed + Tail-wind speed.]

  **b.** Use the result of part (a) to estimate the time of your 1,800-mile trip.

  **c.** The equation solved in part (b) leads mathematically to two solutions. Explain the meaning of the solution you rejected.

**68.** ▼ *Head Winds* The ground speed of an airliner is obtained by subtracting its head-wind speed from its air speed. On your recent trip to Mexico from the United States your plane was traveling at an air speed of 500 miles per hour and experienced head winds of $25 + 50t$ miles per hour, where $t$ is the time in hours since takeoff. HINT [Ground speed = Air speed − Head-wind speed.]

  **a.** Obtain an expression for the distance traveled in terms of the time since takeoff.

  **b.** Use the result of part (a) to estimate the time of your 1,500-mile trip.

  **c.** The equation solved in part (b) leads mathematically to two solutions. Explain the meaning of the solution you rejected.

**69.** ▼ *Vertical Motion* Show that if a projectile is thrown upward with a velocity of $v_0$ ft/sec, then (neglecting air resistance) it will reach its highest point after $v_0/32$ seconds. HINT [See the formulas after Example 8.]

**70.** ▼ *Vertical Motion* Use the result of the preceding exercise to show that if a projectile is thrown upward with a velocity of $v_0$ ft/sec, its highest point will be $v_0^2/64$ feet above the starting point (if we neglect the effects of air resistance).

*Exercises 71–76 use the results in the preceding two exercises.*

**71.** ▼ I threw a ball up in the air to a height of 20 feet. How fast was the ball traveling when it left my hand?

**72.** ▼ I threw a ball up in the air to a height of 40 feet. How fast was the ball traveling when it left my hand?

**73.** ▼ A piece of chalk is tossed vertically upward by Prof. Schwarzenegger and hits the ceiling 100 feet above with a *BANG*.

  **a.** What is the minimum speed the piece of chalk must have been traveling to enable it to hit the ceiling?

  **b.** Assuming that Prof. Schwarzenegger in fact tossed the piece of chalk up at 100 ft/sec, how fast was it moving when it struck the ceiling?

  **c.** Assuming that Prof. Schwarzenegger tossed the chalk up at 100 ft/sec, and that it recoils from the ceiling with the same speed it had at the instant it hit, how long will it take the chalk to make the return journey and hit the ground?

---

[11]Source: www.data360.org/dataset.aspx?Data_Set_Id=9549.

**74.** ▼ A projectile is fired vertically upward from ground level at 16,000 feet per second.

   **a.** How high does the projectile go?
   **b.** How long does it take to reach its zenith (highest point)?
   **c.** How fast is it traveling when it hits the ground?

**75.** ▼ *Strength* Prof. Strong can throw a 10-pound dumbbell twice as high as Prof. Weak can. How much faster can Prof. Strong throw it?

**76.** ▼ *Weakness* Prof. Weak can throw a book three times as high as Prof. Strong can. How much faster can Prof. Weak throw it?

## COMMUNICATION AND REASONING EXERCISES

**77.** Why is this section called "The *Indefinite* Integral?"

**78.** If the derivative of Julius is Augustus, then Augustus is _____ of Julius.

**79.** Linear functions are antiderivatives of what kind of function? Explain.

**80.** Constant functions are antiderivatives of what kind of function? Explain.

**81.** If we know the *derivative* of a function, do we know the function? Explain. If not, what further information will suffice?

**82.** If we know an *antiderivative* of a function, do we know the function? Explain. If not, what further information will suffice?

**83.** If $F(x)$ and $G(x)$ are both antiderivatives of $f(x)$, how are $F(x)$ and $G(x)$ related?

**84.** Your friend Marco claims that once you have one antiderivative of $f(x)$, you have all of them. Explain what he means.

**85.** Complete the following: The total cost function is a(n) _____ of the _____ cost function.

**86.** Complete the following: The distance covered is an antiderivative of the _____ function, and the velocity is an antiderivative of the _____ function.

**87.** If $x$ represents the number of items manufactured and $f(x)$ represents dollars per item, what does $\int f(x)\,dx$ represent? In general, how are the units of $f(x)$ and the units of $\int f(x)\,dx$ related?

**88.** If $t$ represents time in seconds since liftoff and $g(t)$ represents the volume of rocket fuel burned per second, what does $\int g(t)\,dt$ represent?

**89.** Why was the following marked wrong? What is the correct answer?

$$\int (3x+1)\,dx = \frac{3x^2}{2} + 0 + C = \frac{3x^2}{2} + C \quad ✗ \; WRONG!$$

**90.** Why was the following marked wrong? What is the correct answer?

$$\int (3x^2 - 11x)\,dx = x^3 - 11 + C \quad ✗ \; WRONG!$$

**91.** Why was the following marked wrong? What is the correct answer?

$$\int (12x^5 - 4x)\,dx = \int 2x^6 - 2x^2 + C \quad ✗ \; WRONG!$$

**92.** Why was the following marked wrong? What is the correct answer?

$$\int 5\,dt = 5x + C \quad ✗ \; WRONG!$$

**93.** Why was the following marked wrong? What is the correct answer?

$$\int 4(e^x - 2x)\,dx = (4x)(e^x - x^2) + C \quad ✗ \; WRONG!$$

**94.** Why was the following marked wrong? What is the correct answer?

$$\int (2^x - 1)\,dx = \frac{2^{x+1}}{x+1} - x + C \quad ✗ \; WRONG!$$

**95.** Why was the following marked wrong? How should it be corrected?

$$\frac{1}{x} = \ln|x| + C \quad ✗ \; WRONG!$$

**96.** Why was the following marked wrong? What is the correct answer?

$$\int \frac{1}{x^3}\,dx = \ln|x^3| + C \quad ✗ \; WRONG!$$

**97.** ▼ Give an argument for the rule that the integral of a sum is the sum of the integrals.

**98.** ▼ Give an argument for the rule that the integral of a constant multiple is the constant multiple of the integrals.

**99.** ▼ Give an example to show that the integral of a product is not the product of the integrals.

**100.** ▼ Give an example to show that the integral of a quotient is not the quotient of the integrals.

**101.** ▼ Complete the following: If you take the _____ of the _____ of $f(x)$, you obtain $f(x)$ back. On the other hand, if you take the _____ of the _____ of $f(x)$, you obtain $f(x) + C$.

**102.** ▼ If a Martian told you that the *Institute of Alien Mathematics,* after a long and difficult search, has announced the discovery of a new antiderivative of $x - 1$ called $M(x)$ [the formula for $M(x)$ is classified information and cannot be revealed here], how would you respond?

## 13.2 Substitution

The chain rule for derivatives gives us an extremely useful technique for finding antiderivatives. This technique is called **change of variables** or **substitution**.

Recall that to differentiate a function like $(x^2 + 1)^6$, we first think of the function as $g(u)$, where $u = x^2 + 1$ and $g(u) = u^6$. We then compute the derivative, using the chain rule, as

$$\frac{d}{dx}g(u) = g'(u)\frac{du}{dx}.$$

Any rule for derivatives can be turned into a technique for finding antiderivatives by writing it in integral form. The integral form of the above formula is

$$\int g'(u)\frac{du}{dx}\,dx = g(u) + C.$$

But, if we write $g(u) + C = \int g'(u)\,du$, we get the following interesting equation:

$$\int g'(u)\frac{du}{dx}\,dx = \int g'(u)\,du.$$

This equation is the one usually called the *change of variables formula*. We can turn it into a more useful integration technique as follows. Let $f = g'(u)(du/dx)$. We can rewrite the above change of variables formula using $f$:

$$\int f\,dx = \int \left(\frac{f}{du/dx}\right)\,du.$$

In essence, we are making the formal substitution

$$dx = \frac{1}{du/dx}\,du.$$

Here's the technique:

### Substitution Rule

If $u$ is a function of $x$, then we can use the following formula to evaluate an integral:

$$\int f\,dx = \int \left(\frac{f}{du/dx}\right)\,du.$$

Rather than use the formula directly, we use the following step-by-step procedure:

1. Write $u$ as a function of $x$.
2. Take the derivative $du/dx$ and solve for the quantity $dx$ in terms of $du$.
3. Use the expression you obtain in step 2 to substitute for $dx$ in the given integral and substitute $u$ for its defining expression.

Now let's see how this procedure works in practice.

---

## EXAMPLE 1 Substitution

Find $\displaystyle\int 4x(x^2+1)^6\,dx$.

**Solution** To use substitution we need to choose an expression to be $u$. There is no hard and fast rule, but here is one hint that often works:

*Take $u$ to be an expression that is being raised to a power.*

In this case, let's set $u = x^2 + 1$. Continuing the procedure above, we place the calculations for Step 2 in a box.

| | |
|---|---|
| $u = x^2 + 1$ | Write $u$ as a function of $x$. |
| $\dfrac{du}{dx} = 2x$ | Take the derivative of $u$ with respect to $x$. |
| $dx = \dfrac{1}{2x}\,du$ | Solve for $dx$: $dx = \dfrac{1}{du/dx}\,du$. |

Now we *substitute $u$ for its defining expression and substitute for $dx$* in the original integral:

* This step is equivalent to using the formula stated in the Substitution Rule box. If it should bother you that the integral contains both $x$ and $u$, note that $x$ is now a function of $u$.

$$\int 4x(x^2+1)^6\,dx = \int 4xu^6\frac{1}{2x}\,du \qquad \text{Substitute* for } u \text{ and } dx.$$

$$= \int 2u^6\,du. \qquad \text{Cancel the } x\text{s and simplify.}$$

We have boiled the given integral down to the much simpler integral $\int 2u^6\,du$, and we can now write down the solution:

$$2\frac{u^7}{7} + C = \frac{2(x^2+1)^7}{7} + C. \qquad \text{Substitute } (x^2+1) \text{ for } u \text{ in the answer.}$$

---

➡ **Before we go on...** There are two points to notice in Example 1. First, before we can actually integrate with respect to $u$, *we must eliminate all $x$'s from the integrand.* If we cannot, we may have chosen the wrong expression for $u$. Second, after integrating, we must substitute back to obtain an expression involving $x$.

It is easy to check our answer. We differentiate:

$$\frac{d}{dx}\left[\frac{2(x^2+1)^7}{7}\right] = \frac{2(7)(x^2+1)^6(2x)}{7} = 4x(x^2+1)^6. \qquad ✔$$

Notice how we used the chain rule to check the result obtained by substitution. ■

When we use substitution, the first step is always to decide what to take as $u$. Again, there are no set rules, but we see some common cases in the examples.

---

### EXAMPLE 2  **More Substitution**

Evaluate the following:

**a.** $\int x^2(x^3+1)^2\,dx$  **b.** $\int 3xe^{x^2}\,dx$  **c.** $\int \dfrac{1}{2x+5}\,dx$  **d.** $\int \left(\dfrac{1}{2x+5}+4x^2+1\right)dx$

### Solution

**a.** As we said in Example 1, it often works to take $u$ to be an expression that is being raised to a power. We usually also want to see the derivative of $u$ as a factor in the integrand so that we can cancel terms involving $x$. In this case, $x^3+1$ is being raised to a power, so let's set $u = x^3+1$. Its derivative is $3x^2$; in the integrand we see $x^2$, which is missing the factor 3, but missing or incorrect constant factors are not a problem.

$$
\begin{array}{ll}
u = x^3+1 & \text{Write } u \text{ as a function of } x. \\[2mm]
\dfrac{du}{dx} = 3x^2 & \text{Take the derivative of } u \text{ with respect to } x. \\[2mm]
dx = \dfrac{1}{3x^2}\,du & \text{Solve for } dx: dx = \dfrac{1}{du/dx}\,du.
\end{array}
$$

$$
\begin{aligned}
\int x^2(x^3+1)^2\,dx &= \int x^2 u^2 \frac{1}{3x^2}\,du & \text{Substitute for } u \text{ and } dx.\\[2mm]
&= \int \frac{1}{3}u^2\,du & \text{Cancel the terms with } x.\\[2mm]
&= \frac{1}{9}u^3 + C & \text{Take the antiderivative.}\\[2mm]
&= \frac{1}{9}(x^3+1)^3 + C & \text{Substitute for } u \text{ in the answer.}
\end{aligned}
$$

**b.** When we have an exponential with an expression in the exponent, it often works to substitute $u$ for that expression. In this case, let's set $u = x^2$. (Notice again that we see a constant multiple of its derivative $2x$ as a factor in the integrand—a good sign.)

$$
\begin{array}{l}
u = x^2 \\[2mm]
\dfrac{du}{dx} = 2x \\[2mm]
dx = \dfrac{1}{2x}\,du
\end{array}
$$

Substituting into the integral, we have

$$
\int 3xe^{x^2}\,dx = \int 3xe^u \frac{1}{2x}\,du = \int \frac{3}{2}e^u\,du
$$

$$
= \frac{3}{2}e^u + C = \frac{3}{2}e^{x^2} + C.
$$

**c.** We begin by rewriting the integrand as a power:

$$
\int \frac{1}{2x+5}\,dx = \int (2x+5)^{-1}\,dx.
$$

Now we take our earlier advice and set $u$ equal to the expression that is being raised to a power:

$$u = 2x + 5$$

$$\frac{du}{dx} = 2$$

$$dx = \frac{1}{2} du$$

Substituting into the integral, we have

$$\int \frac{1}{2x+5} dx = \int \frac{1}{2} u^{-1} du = \frac{1}{2} \ln |u| + C$$

$$= \frac{1}{2} \ln |2x + 5| + C.$$

**d.** Here, the substitution $u = 2x + 5$ works for the first part of the integrand, $1/(2x + 5)$, but not for the rest of it, so we break up the integral:

$$\int \left( \frac{1}{2x+5} + 4x^2 + 1 \right) dx = \int \frac{1}{2x+5} dx + \int (4x^2 + 1)\, dx.$$

For the first integral we can use the substitution $u = 2x + 5$ [which we did in part (a)], and for the second, no substitution is necessary:

$$\int \frac{1}{2x+5} dx + \int (4x^2 + 1)\, dx = \frac{1}{2} \ln |2x + 5| + \frac{4x^3}{3} + x + C.$$

---

**EXAMPLE 3  Choosing $u$**

Evaluate $\int (x+3)\sqrt{x^2 + 6x}\, dx.$

**Solution**  There are two parenthetical expressions. Notice, however, that the derivative of the expression $(x^2 + 6x)$ is $2x + 6$, which is twice the term $(x + 3)$ in front of the radical. Recall that we would like the derivative of $u$ to appear as a factor. Thus, let's take $u = x^2 + 6x$.

$$u = x^2 + 6x$$

$$\frac{du}{dx} = 2x + 6 = 2(x+3)$$

$$dx = \frac{1}{2(x+3)} du$$

Substituting into the integral, we have

$$\int (x+3)\sqrt{x^2 + 6x}\, dx$$

$$= \int (x+3)\sqrt{u} \left( \frac{1}{2(x+3)} \right) du$$

$$= \int \frac{1}{2} \sqrt{u}\, du = \frac{1}{2} \int u^{1/2} du$$

$$= \frac{1}{2} \frac{2}{3} u^{3/2} + C = \frac{1}{3}(x^2 + 6x)^{3/2} + C.$$

---

Some cases require a little more work.

### EXAMPLE 4 **When the *x* Terms Do Not Cancel**

Evaluate $\int \dfrac{2x}{(x-5)^2}\, dx$.

**Solution**  We first rewrite

$$\int \frac{2x}{(x-5)^2}\, dx = \int 2x(x-5)^{-2}\, dx.$$

This suggests that we should set $u = x - 5$.

$$\boxed{\begin{aligned} u &= x - 5 \\ \frac{du}{dx} &= 1 \\ dx &= du \end{aligned}}$$

Substituting, we have

$$\int \frac{2x}{(x-5)^2}\, dx = \int 2xu^{-2}\, du.$$

Now, there is nothing in the integrand to cancel the *x* that appears. If, as here, there is still an *x* in the integrand after substituting, we go back to the expression for *u*, solve for *x*, and substitute the expression we obtain for *x* in the integrand. So, we take $u = x - 5$ and solve for $x = u + 5$. Substituting, we get

$$\int 2xu^{-2}\, du = \int 2(u+5)u^{-2}\, du$$

$$= 2 \int (u^{-1} + 5u^{-2})\, du$$

$$= 2 \ln|u| - \frac{10}{u} + C$$

$$= 2 \ln|x-5| - \frac{10}{x-5} + C.$$

### EXAMPLE 5 **Application: Bottled Water for Pets**

Annual sales of bottled spring water for pets can be modeled by the logistic function

$$s(t) = \frac{3{,}000e^{0.5t}}{3 + e^{0.5t}} \text{ million gallons per year} \qquad (0 \le t \le 12),$$

where *t* is time in years since the start of 2000.[12]

**a.** Find an expression for the total amount of bottled spring water for pets sold since the start of 2000.

**b.** How much bottled spring water for pets was sold from the start of 2005 to the start of 2008?

**Solution**

**a.** If we write the total amount of pet spring water sold since the start of 2000 as $S(t)$, then the information we are given says that

$$S'(t) = s(t) = \frac{3{,}000e^{0.5t}}{3 + e^{0.5t}}.$$

---

[12]Based on data through 2008 and the recovery in 2010 of the general bottled water market to pre-recession levels. Sources: "*Liquid Assets: America's Expensive Love Affair with Bottled Water*" Daniel Gross, April 26, 2011 (finance.yahoo.com), Beverage Marketing Corporation (www.beveragemarketing.com).

Thus,

$$S(t) = \int \frac{3{,}000e^{0.5t}}{3 + e^{0.5t}}\, dt$$

is the function we are after. To integrate the expression, take $u$ to be the denominator of the integrand:

$$u = 3 + e^{0.5t}$$

$$\frac{du}{dt} = 0.5e^{0.5t}$$

$$dt = \frac{1}{0.5e^{0.5t}}\, du$$

$$S(t) = \int \frac{3{,}000e^{0.5t}}{3 + e^{0.5t}}\, dt$$

$$= \int \frac{3{,}000e^{0.5t}}{u} \cdot \frac{1}{0.5e^{0.5t}}\, du$$

$$= \frac{3{,}000}{0.5} \int \frac{1}{u}\, du$$

$$= 6{,}000 \ln |u| + C = 6{,}000 \ln(3 + e^{0.5t}) + C.$$

(Why could we drop the absolute value in the last step?)

Now what is $C$? Because $S(t)$ represents the total amount of bottled spring water for pets sold *since time* $t = 0$, we have $S(0) = 0$ (because that is when we started counting). Thus,

$$0 = 6{,}000 \ln\left(3 + e^{0.5(0)}\right) + C$$

$$= 6{,}000 \ln 4 + C$$

$$C = -6{,}000 \ln 4 \approx -8{,}318.$$

Therefore, the total sales from the start of 2000 is approximately

$$S(t) = 6{,}000 \ln(3 + e^{0.5t}) - 8{,}318 \text{ million gallons.}$$

**b.** The period from the start of 2005 to the start of 2008 is represented by the interval [5, 8]. From part (a):

Sales through the start of 2005 = $S(5)$
$$= 6{,}000 \ln\left(3 + e^{0.5(5)}\right) - 8{,}318 \approx 8{,}003 \text{ million gallons.}$$

Sales through the start of 2008 = $S(8)$
$$= 6{,}000 \ln\left(3 + e^{0.5(8)}\right) - 8{,}318 \approx 16{,}003 \text{ million gallons.}$$

Therefore, sales over the period were about $16{,}003 - 8{,}003 = 8{,}000$ million gallons.

➡ **Before we go on...** You might wonder why we are writing a logistic function in the form we used in Example 5 rather than in one of the "standard" forms $\dfrac{N}{1 + Ab^{-t}}$ or $\dfrac{N}{1 + Ae^{-kt}}$. Our only reason for doing this is to make the substitution work. To convert from the second "standard" form to the form we used in the example, multiply top and bottom by $e^{kt}$. (See Exercises 81 and 82 in Section 13.4 for further discussion.) ∎

## Shortcuts

The following rule allows us to simply write down the antiderivative in cases where we would otherwise need the substitution $u = ax + b$, as in Example 2 ($a$ and $b$ are constants with $a \neq 0$):

> **Shortcut Rule: Integrals of Expressions Involving ($ax + b$)**
>
> If $\int f(x)\, dx = F(x) + C$ and $a$ and $b$ are constants, with $a \neq 0$, then[*]
>
> $$\int f(ax + b)\, dx = \frac{1}{a} F(ax + b) + C.$$
>
> **Quick Example**
>
> (Also see the examples in the table below.)
>
> Because $\int x^4\, dx = \dfrac{x^5}{5} + C$, it follows that
>
> $$\int (3x - 1)^4\, dx = \frac{1}{3} \frac{(3x - 1)^5}{5} + C = \frac{(3x - 1)^5}{15} + C.$$

[*] The shortcut rule can be justified by making the substitution $u = ax + b$ in the general case as stated.

Below are some instances of the shortcut rule with additional examples (their individual derivations using the substitution $u = ax + b$ will appear in the exercises):

| *Shortcut Rule* | *Example* |
|---|---|
| $\int (ax + b)^n\, dx = \dfrac{1}{a} \dfrac{(ax + b)^{n+1}}{n + 1} + C$ <br> (if $n \neq -1$) | $\int (3x - 1)^2\, dx = \dfrac{(3x - 1)^3}{3(3)} + C$ <br> $= \dfrac{(3x - 1)^3}{9} + C$ |
| $\int (ax + b)^{-1}\, dx = \dfrac{1}{a} \ln|ax + b| + C$ | $\int (3 - 2x)^{-1}\, dx = \dfrac{1}{(-2)} \ln|3 - 2x| + C$ <br> $= -\dfrac{1}{2} \ln|3 - 2x| + C$ |
| $\int e^{ax+b}\, dx = \dfrac{1}{a} e^{ax+b} + C$ | $\int e^{-x+4}\, dx = \dfrac{1}{(-1)} e^{-x+4} + C$ <br> $= -e^{-x+4} + C$ |
| $\int c^{ax+b}\, dx = \dfrac{1}{a \ln c} c^{ax+b} + C$ | $\int 2^{-3x+4}\, dx = \dfrac{1}{(-3 \ln 2)} 2^{-3x+4} + C$ <br> $= -\dfrac{1}{3 \ln 2} 2^{-3x+4} + C$ |
| $\int |ax + b|\, dx$ <br> $= \dfrac{1}{2a}(ax + b)|ax + b| + C$ | $\int |2x - 1|\, dx = \dfrac{1}{4}(2x - 1)|2x - 1| + C$ |

[†] Mike Fuschetto, who was a Hofstra business calculus student in Spring 2010.

The following more general version of the shortcut rule was suggested by a student.[†] We have marked it as optional so that you could skip it on a first reading, but we strongly suggest you come back to it afterward, as the rule will allow you to easily write down the answers in most of the exercises, as well as in almost all the examples of this section.

---

**(Optional) Mike's Shortcut Rule: Integrals of More General Expressions**

If $\int f(x)\,dx = F(x) + C$ and $g$ and $u$ are any differentiable functions of $x$, then

$$\int g \cdot f(u)\,dx = \frac{g}{u'} \cdot F(u) + C \text{ provided that } \frac{g}{u'} \text{ is constant.}$$

**Quick Example**

(Also see the examples in the table below.)

$$\int x^4\,dx = \frac{x^5}{5} + C, \text{ so}$$

$$\int 5x^2(x^3 - 1)^4\,dx = \frac{5x^2}{3x^2} \cdot \frac{(x^3 - 1)^5}{5} + C$$

$$= \frac{5}{3} \cdot \frac{(x^3 - 1)^5}{5} + C = \frac{(x^3 - 1)^5}{3} + C.$$

**Caution**

If $g/u'$ is not constant, then Mike's rule does not apply; for instance, the following calculation is *wrong*:

$$\int x(2x - 1)^2\,dx = \frac{x}{2} \cdot \frac{(2x - 1)^3}{3} + C \quad \text{✗ WRONG! because } \frac{x}{2} \text{ is not constant.}$$

---

Here are some instances of Mike's rule with additional examples.

| Shortcut Rule | Example |
|---|---|
| $\int g \cdot u^n\,dx = \frac{g}{u'}\frac{u^{n+1}}{n+1} + C$ <br><br> (if $n \neq -1$) | $\int 3x(x^2 - 1)^3\,dx = \frac{3x}{2x}\frac{(x^2 - 1)^4}{4} + C$ <br><br> $= \frac{3(x^2 - 1)^4}{8} + C$ |
| $\int g \cdot u^{-1}\,dx = \frac{g}{u'}\ln|u| + C$ | $\int e^x(3 - 2e^x)^{-1}\,dx = \frac{e^x}{-2e^x}\ln|3 - 2e^x| + C$ <br><br> $= -\frac{1}{2}\ln|3 - 2e^x| + C$ |
| $\int g \cdot e^u\,dx = \frac{g}{u'}e^u + C$ | $\int x^2 e^{-x^3+4}\,dx = \frac{x^2}{-3x^2}e^{-x^3+4} + C$ <br><br> $= -\frac{1}{3}e^{-x^3+4} + C$ |
| $\int g \cdot c^u\,dx = \frac{g}{u'\ln c}c^u + C$ | $\int x^3 2^{x^4-1}\,dx = \frac{x^3}{4x^3\ln 2}2^{x^4-1} + C$ <br><br> $= -\frac{1}{4\ln 2}2^{x^4-1} + C$ |
| $\int g \cdot |u|\,dx = \frac{g}{2u'}u|u| + C$ | $\int x|x^2 - 1|\,dx = \frac{x}{4x}(x^2 - 1)|x^2 - 1| + C$ <br><br> $= \frac{1}{4}(x^2 - 1)|x^2 - 1| + C$ |

**FAQs**

## When to Use Substitution and What to Use for *u*

**Q** : *If I am asked to calculate an antiderivative, how do I know when to use a substitution and when not to use one?*

**A** : Do *not* use substitution when integrating sums, differences, and/or constant multiples of powers of $x$ and exponential functions, such as $2x^3 - \dfrac{4}{x^2} + \dfrac{1}{2x} + 3^x + \dfrac{2^x}{3}$.

To recognize when you should try a substitution, pretend that you are *differentiating* the given expression instead of integrating it. If differentiating the expression would require use of the chain rule, then integrating that expression may well require a substitution, as in, say, $x(3x^2 - 4)^3$ or $(x + 1)e^{x^2+2x-1}$. (In the first we have a *quantity* cubed, and in the second we have $e$ raised to a *quantity*.)

**Q** : *If an integral seems to call for a substitution, what should I use for u?*

**A** : There are no set rules for deciding what to use for $u$, but the preceding examples show some common patterns:

- If you see a linear expression raised to a power, try setting $u$ equal to that linear expression. For example, in $(3x - 2)^{-3}$, set $u = 3x - 2$. (Alternatively, try using the shortcuts above.)
- If you see a constant raised to a linear expression, try setting $u$ equal to that linear expression. For example, in $3^{(2x+1)}$, set $u = 2x + 1$. (Alternatively, try a shortcut.)
- If you see an expression raised to a power multiplied by the derivative of that expression (or a constant multiple of the derivative), try setting $u$ equal to that expression. For example, in $x^2(3x^3 - 4)^{-1}$, set $u = 3x^3 - 4$.
- If you see a constant raised to an expression, multiplied by the derivative of that expression (or a constant multiple of its derivative), try setting $u$ equal to that expression. For example, in $5(x + 1)e^{x^2+2x-1}$, set $u = x^2 + 2x - 1$.
- If you see an expression in the denominator and its derivative (or a constant multiple of its derivative) in the numerator, try setting $u$ equal to that expression.
  For example, in $\dfrac{2^{3x}}{3 - 2^{3x}}$, set $u = 3 - 2^{3x}$.

Persistence often pays off: If a certain substitution does not work, try another approach or a different substitution.

## **13.2**    EXERCISES

▼ more advanced     ◆ challenging

Ⓣ indicates exercises that should be solved using technology

*In Exercises 1–10, evaluate the given integral using the substitution (or method) indicated.*

**1.** $\displaystyle\int (3x - 5)^3 \, dx; u = 3x - 5$

**2.** $\displaystyle\int (2x + 5)^{-2} \, dx; u = 2x + 5$

**3.** $\displaystyle\int (3x - 5)^3 \, dx;$ shortcut page 968

**4.** $\displaystyle\int (2x + 5)^{-2} \, dx;$ shortcut page 968

**5.** $\int e^{-x}\,dx; u = -x$

**6.** $\int e^{x/2}\,dx; u = x/2$

**7.** $\int e^{-x}\,dx;$ shortcut page 968

**8.** $\int e^{x/2}\,dx;$ shortcut page 968

**9.** $\int (x+1)e^{(x+1)^2}\,dx; u = (x+1)^2$

**10.** $\int (x-1)^2 e^{(x-1)^3}\,dx; u = (x-1)^3$

*In Exercises 11–52, decide on what substitution to use, and then evaluate the given integral using a substitution.* HINT [See the FAQ at the end of the section for advice on deciding on $u$, and the examples for the mechanics of doing the substitution.]

**11.** $\int (3x+1)^5\,dx$

**12.** $\int (-x-1)^7\,dx$

**13.** $\int 7.2\sqrt{3x-4}\,dx$

**14.** $\int 4.4e^{(-3x+4)}\,dx$

**15.** $\int 1.2e^{(0.6x+2)}\,dx$

**16.** $\int 8.1\sqrt{-3x+4}\,dx$

**17.** $\int x(3x^2+3)^3\,dx$

**18.** $\int x(-x^2-1)^3\,dx$

**19.** $\int 2x\sqrt{3x^2-1}\,dx$

**20.** $\int 3x\sqrt{-x^2+1}\,dx$

**21.** $\int \frac{x}{(x^2+1)^{1.3}}\,dx$

**22.** $\int \frac{x^2}{(1+x^3)^{1.4}}\,dx$

**23.** $\int x|4x^2-1|\,dx$

**24.** $\int x^2|4x^3+1|\,dx$

**25.** $\int (1+9.3e^{3.1x-2})\,dx$

**26.** $\int (3.2-4e^{1.2x-3})\,dx$

**27.** $\int xe^{-x^2+1}\,dx$

**28.** $\int xe^{2x^2-1}\,dx$

**29.** $\int (x+1)e^{-(x^2+2x)}\,dx$ HINT [See Example 2(b).]

**30.** $\int (2x-1)e^{2x^2-2x}\,dx$ HINT [See Example 2(b).]

**31.** $\int \frac{-2x-1}{(x^2+x+1)^3}\,dx$

**32.** $\int \frac{x^3-x^2}{3x^4-4x^3}\,dx$

**33.** $\int \frac{x^2+x^5}{\sqrt{2x^3+x^6-5}}\,dx$ HINT [See Example 3.]

**34.** $\int \frac{2(x^3-x^4)}{(5x^4-4x^5)^5}\,dx$ HINT [See Example 3.]

**35.** $\int x(x-2)^5\,dx$ HINT [See Example 4.]

**36.** $\int x(x-2)^{1/3}\,dx$ HINT [See Example 4.]

**37.** $\int 2x\sqrt{x+1}\,dx$

**38.** $\int \frac{x}{\sqrt{x+1}}\,dx$

**39.** $\int \frac{e^{-0.05x}}{1-e^{-0.05x}}\,dx$ HINT [See Example 5.]

**40.** $\int \frac{3e^{1.2x}}{2+e^{1.2x}}\,dx$ HINT [See Example 5.]

**41.** ▼ $\int \frac{3e^{-1/x}}{x^2}\,dx$

**42.** ▼ $\int \frac{2e^{2/x}}{x^2}\,dx$

**43.** ▼ $\int \frac{(4+1/x^2)^3}{x^3}\,dx$

**44.** ▼ $\int \frac{1}{x^2(2-1/x)}\,dx$

**45.** ▼ $\int \frac{e^x+e^{-x}}{2}\,dx$ HINT [See Example 2(d).]

**46.** ▼ $\int (e^{x/2}+e^{-x/2})\,dx$ HINT [See Example 2(d).]

**47.** ▼ $\int \frac{e^x-e^{-x}}{e^x+e^{-x}}\,dx$

**48.** ▼ $\int \frac{e^{x/2}+e^{-x/2}}{e^{x/2}-e^{-x/2}}\,dx$

**49.** ▼ $\int e^{3x-1}|1-e^{3x-1}|\,dx$

**50.** ▼ $\int |e^{-x-1}-1|(e^{-x-1})\,dx$

**51.** ▼ $\int \left((2x-1)e^{2x^2-2x}+xe^{x^2}\right)\,dx$

**52.** ▼ $\int \left(xe^{-x^2+1}+e^{2x}\right)\,dx$

*In Exercises 53–56, derive the given equation, where a and b are constants with $a \neq 0$.*

**53.** $\int (ax+b)^n\,dx = \dfrac{(ax+b)^{n+1}}{a(n+1)} + C$ (if $n \neq -1$)

**54.** $\int (ax+b)^{-1}\,dx = \dfrac{1}{a}\ln|ax+b| + C$

**55.** $\int |ax+b|\,dx = \dfrac{1}{2a}(ax+b)|ax+b| + C$

**56.** $\int e^{ax+b}\,dx = \dfrac{1}{a}e^{ax+b} + C$

*In Exercises 57–72, use the shortcut formulas (see page 968 and Exercises 53–56) to calculate the given integral.*

**57.** $\int e^{-x}\,dx$

**58.** $\int e^{x-1}\,dx$

**59.** $\int e^{2x-1}\,dx$

**60.** $\int e^{-3x}\,dx$

**61.** $\int (2x+4)^2\,dx$

**62.** $\int (3x-2)^4\,dx$

**63.** $\displaystyle\int \frac{1}{5x - 1}\, dx$    **64.** $\displaystyle\int (x - 1)^{-1}\, dx$

**65.** $\displaystyle\int (1.5x)^3\, dx$    **66.** $\displaystyle\int e^{2.1x}\, dx$

**67.** $\displaystyle\int 1.5^{3x}\, dx$    **68.** $\displaystyle\int 4^{-2x}\, dx$

**69.** $\displaystyle\int |2x + 4|\, dx$    **70.** $\displaystyle\int |3x - 2|\, dx$

**71.** $\displaystyle\int (2^{3x+4} + 2^{-3x+4})\, dx$    **72.** $\displaystyle\int (1.1^{-x+4} + 1.1^{x+4})\, dx$

**73.** Find $f(x)$ if $f(0) = 0$ and the tangent line at $(x, f(x))$ has slope $x(x^2 + 1)^3$.

**74.** Find $f(x)$ if $f(1) = 0$ and the tangent line at $(x, f(x))$ has slope $\dfrac{x}{x^2 + 1}$.

**75.** Find $f(x)$ if $f(1) = 1/2$ and the tangent line at $(x, f(x))$ has slope $xe^{x^2-1}$.

**76.** Find $f(x)$ if $f(2) = 1$ and the tangent line at $x$ has slope $(x - 1)e^{x^2-2x}$.

*In Exercises 77–84, use Mike's shortcut method (see p. 969 and the examples that follow) to calculate the given integral.*

**77.** $\displaystyle\int x(5x^2 - 3)^6\, dx$    **78.** $\displaystyle\int \frac{x}{(5x^2 - 3)^6}\, dx$

**79.** $\displaystyle\int \frac{e^x}{\sqrt{3e^x - 1}}\, dx$    **80.** $\displaystyle\int e^x\sqrt{1 + 2e^x}\, dx$

**81.** $\displaystyle\int x^3 e^{(x^4-8)}\, dx$    **82.** $\displaystyle\int \frac{x^3}{e^{(x^4-8)}}\, dx$

**83.** $\displaystyle\int \frac{e^{3x}}{1 + 2e^{3x}}\, dx$    **84.** $\displaystyle\int \frac{e^{-2x}}{e^{-2x} - 3}\, dx$

## APPLICATIONS

**85.** *Economic Growth* The value of sold goods in Mexico can be approximated by

$$v(t) = 210 - 62e^{-0.05t} \text{ trillion pesos per month} \quad (t \geq 0),$$

where $t$ is time in months since January 2005.[13] Find an expression for the total value $V(t)$ of sold goods in Mexico from January 2005 to time $t$. HINT [Use the shortcut on page 968.]

**86.** *Economic Contraction* The number of housing starts in the United States can be approximated by

$$n(t) = 1.1 + 1.2e^{-0.08t} \text{ million homes per year} \quad (t \geq 0),$$

where $t$ is time in years from the start of 2006.[14] Find an expression for the total number $N(t)$ of housing starts in the United States from January 2006 to time $t$. HINT [Use the shortcut on page 968.]

**87.** *Revenue* The annual revenue of Amazon.com over the period 2006–2010 can be approximated by

$$a(t) = 2e^{0.25t} \text{ billion dollars per year} \quad (6 \leq t \leq 10),$$

where $t$ is time in years since the start of 2000.[15] Find an expression for the total revenue $A(t)$ earned since the start of 2006, and hence estimate the total revenue earned from the start of 2006 to the start of 2010. (Round your answer to the nearest \$ billion.)

**88.** *Lobster Harvests* The size of the annual lobster harvest from Long Island Sound over the period 1997–2010 can be approximated by

$$h(t) = 9.1e^{-0.21t} \text{ million pounds of lobster per year} \quad (0 \leq t \leq 13),$$

where $t$ is time in years since mid-1997.[16] Find an expression for the total lobster harvest $H(t)$ since mid-1997, and hence estimate the total lobster harvest from mid-1997 to mid-2010. (Round your answer to the nearest million pounds.)

**89.** *Cost* The marginal cost of producing the $x$th roll of film is given by $5 + 1/(x + 1)^2$. The total cost to produce one roll is \$1,000. Find the total cost function $C(x)$.

**90.** *Cost* The marginal cost of producing the $x$th box of CDs is given by $10 - x/(x^2 + 1)^2$. The total cost to produce 2 boxes is \$1,000. Find the total cost function $C(x)$.

**91.** *Scientific Research* The number of research articles in the prominent journal *Physical Review* written by researchers in Europe can be approximated by

$$E(t) = \frac{7e^{0.2t}}{5 + e^{0.2t}} \text{ thousand articles per year} \quad (t \geq 0),$$

where $t$ is time in years ($t = 0$ represents 1983).[17]

**a.** Find an (approximate) expression for the total number of articles written by researchers in Europe since 1983 ($t = 0$). HINT [See Example 5.]

**b.** Roughly how many articles were written by researchers in Europe from 1983 to 2003? (Round your answer to the nearest 1,000 articles.)

---

[14]Source for data: *New York Times*, February 17, 2007, p. C3.

[15]Source for data: www.wikinvest.com.

[16]Authors' regression model. Source for data: Long Island Sound Study, data downloaded May 2011 from longislandsoundstudy.net/2010/07/lobster-landings.

[17]Based on data from 1983 to 2003. Source: The American Physical Society/*New York Times*, May 3, 2003, p. A1.

---

[13]Source: Instituto Nacional de Estadística y Geografía (INEGI) (www.inegi.gob.mx).

**92.** *Scientific Research* The number of research articles in the prominent journal *Physical Review* written by researchers in the United States can be approximated by

$$U(t) = \frac{4.6e^{0.6t}}{0.4 + e^{0.6t}} \text{ thousand articles per year} \quad (t \geq 0),$$

where *t* is time in years (*t* = 0 represents 1983).[18]

**a.** Find an (approximate) expression for the total number of articles written by researchers in the United States since 1983 (*t* = 0). HINT [See Example 5.]

**b.** Roughly how many articles were written by researchers in the United States from 1983 to 2003?

**93.** *Sales* The rate of sales of your company's *Jackson Pollock Advanced Paint-by-Number* sets can be modeled by

$$s(t) = \frac{900e^{0.25t}}{3 + e^{0.25t}} \text{ sets per month}$$

*t* months after their introduction. Find an expression for the total number of paint-by-number sets *S*(*t*) sold *t* months after their introduction, and use it to estimate the total sold in the first 12 months. HINT [See Example 5.]

**94.** *Sales* The rate of sales of your company's *Jackson Pollock Beginners Paint-by-Number* sets can be modeled by

$$s(t) = \frac{1{,}800e^{0.75t}}{10 + e^{0.75t}} \text{ sets per month}$$

*t* months after their introduction. Find an expression for the total number of paint-by-number sets *S*(*t*) sold *t* months after their introduction, and use it to estimate the total sold in the first 12 months. HINT [See Example 5.]

**95.** *Motion in a Straight Line* The velocity of a particle moving in a straight line is given by $v = t(t^2 + 1)^4 + t$.

**a.** Find an expression for the position *s* after a time *t*. HINT [See Example 2(d).]

**b.** Given that *s* = 1 at time *t* = 0, find the constant of integration *C* and hence an expression for *s* in terms of *t* without any unknown constants.

**96.** *Motion in a Straight Line* The velocity of a particle moving in a straight line is given by $v = 3te^{t^2} + t$.

**a.** Find an expression for the position *s* after a time *t*. HINT [See Example 2(d).]

**b.** Given that *s* = 3 at time *t* = 0, find the constant of integration *C* and hence an expression for *s* in terms of *t* without any unknown constants.

**97.** *Bottled Water Sales* The rate of U.S. sales of bottled water for the period 2000–2010 could be approximated by

$$s(t) = -45(t - 2{,}000)^2 + 900(t - 2{,}000) + 4{,}200 \text{ million}$$
$$\text{gallons per year} \quad (2{,}000 \leq t \leq 2{,}010),$$

where *t* is the year.[19] Use an indefinite integral to approximate the total sales *S*(*t*) of bottled water since 2003 (*t* = 2003). Approximately how much bottled water was sold from *t* = 2003 to *t* = 2008?

**98.** *Bottled Water Sales* The rate of U.S. per capita sales of bottled water for the period 2000–2010 could be approximated by

$$s(t) = -0.18(t - 2{,}000)^2 + 3(t - 2{,}000) + 15 \text{ gallons per year}$$
$$(2{,}000 \leq t \leq 2{,}010),$$

where *t* is the year.[20] Use an indefinite integral to approximate the total per capita sales *S*(*t*) of bottled water since the start of 2006. Approximately how much bottled water was sold, per capita, from *t* = 2006 to *t* = 2008?

## COMMUNICATION AND REASONING EXERCISES

**99.** Are there any circumstances in which you should use the substitution *u* = *x*? Illustrate your answer by giving an example that shows the effect of this substitution.

**100.** You are asked to calculate $\int \frac{u}{u^2 + 1}\, du$. What is wrong with the substitution $u = u^2 + 1$?

**101.** If the *x*'s do not cancel in a substitution, that means you chose the wrong expression for *u*, right?

**102.** At what stage of a calculation using a *u* substitution should you substitute back for *u* in terms of *x*: before or after taking the antiderivative?

**103.** Consider $\int \left( \frac{x}{x^2 - 1} + \frac{3x}{x^2 + 1} \right) dx$. To compute it, you should use which of the following?

**(A)** $u = x^2 - 1$   **(B)** $u = x^2 + 1$   **(C)** Neither   **(D)** Both

Explain your answer.

**104.** If the substitution $u = x^2 - 1$ works in $\int \frac{x}{x^2 - 1}\, dx$, why does it not work nearly so easily in $\int \frac{x^2 - 1}{x}\, dx$? How would you do the second integral most simply?

**105.** Why was the following calculation marked wrong? What is the correct answer?

$$\begin{array}{|c|}
\hline
u = x^2 - 1 \\
\\
\dfrac{du}{dx} = 2x \\
\\
dx = \dfrac{1}{2x}\, du \\
\hline
\end{array}
\qquad
\begin{aligned}
\int 3x(x^2 - 1) &= \int 3xu \\
&= 3x\frac{u^2}{2} + C = 3x\frac{(x^2 - 1)^2}{2} + C \\
&\qquad\qquad\qquad\qquad \textit{✗ WRONG!}
\end{aligned}$$

---

[18]Based on data from 1983 to 2003. Source: The American Physical Society/*New York Times*, May 3, 2003, p. A1.

[19]Source for data: Beverage Marketing Corporation (www.bottledwater .org).

[20]*Ibid.*

**106.** Why was the following calculation marked wrong? What is the correct answer?

$$u = x^3 - 1$$
$$\frac{du}{dx} = 3x^2$$
$$dx = \frac{1}{3x^2}\,du$$

$$\int x^2(x^3 - 1)^2\,dx = \int x^2 u^2 \frac{1}{3x^2}\,du$$
$$= \frac{1}{3}u^2 + C = \frac{1}{3}(x^3 - 1)^2 + C$$
$$✗ \text{ WRONG!}$$

**107.** Why was the following calculation marked wrong? What is the correct answer?

$$u = x^3 - 1$$
$$\frac{du}{dx} = 3x^2$$
$$dx = \frac{1}{3x^2}\,du$$

$$\int x^2(x^3 - 1)\,dx = \int x^2 u \frac{1}{3x^2}\,du$$
$$= \int \frac{1}{3}u\,du = \int \frac{1}{3}(x^3 - 1)\,dx$$
$$= \frac{1}{3}\left(\frac{x^4}{4} - x\right) + C \quad ✗ \text{ WRONG!}$$

**108.** Why was the following calculation marked wrong? What is the correct answer?

$$u = x^2 - 1$$
$$\frac{du}{dx} = 2x$$
$$dx = \frac{1}{2x}\,du$$

$$\int \frac{x}{x^2 - 1}\,dx = \int \frac{x}{u}\,du$$
$$= x \int \frac{1}{u}\,du = x \ln|u| + C$$
$$= x \ln|x^2 - 1| + C \quad ✗ \text{ WRONG!}$$

**109.** ▼ Show that *none* of the following substitutions work for $\int e^{-x^2}\,dx$: $u = -x$, $u = x^2$, $u = -x^2$. (The antiderivative of $e^{-x^2}$ involves the *error function* erf($x$).)

**110.** ▼ Show that *none* of the following substitutions work for $\int \sqrt{1 - x^2}\,dx$: $u = 1 - x^2$, $u = x^2$, and $u = -x^2$. (The antiderivative of $\sqrt{1 - x^2}$ involves inverse trigonometric functions, discussion of which is beyond the scope of this book.)

## 13.3  The Definite Integral: Numerical and Graphical Viewpoints

In Sections 13.1 and 13.2, we discussed the indefinite integral. There is an older, related concept called the **definite integral**. Let's introduce this new idea with an example. (We'll drop hints now and then about how the two types of integral are related. In Section 13.4 we discuss the exact relationship, which is one of the most important results in calculus.)

In Section 13.1, we used antiderivatives to answer questions of the form "Given the marginal cost, compute the total cost." (See Example 5 in Section 13.1.) In this section we approach such questions more directly, and we will forget about antiderivatives for now.

### EXAMPLE 1  Oil Spill

Your deep ocean oil rig has suffered a catastrophic failure, and oil is leaking from the ocean floor wellhead at a rate of

$$v(t) = 0.08t^2 - 4t + 60 \text{ thousand barrels per day } (0 \le t \le 20),$$

where $t$ is time in days since the failure.[21] Use a numerical calculation to estimate the total volume of oil released during the first 20 days.

---

[21]The model is consistent with the order of magnitude of the BP Deepwater Horizon oil spill of April 20–June 15, 2010, when the rate of flow of oil was estimated by the Federal Emergency Management Agency's Flow Rate Technical Group to be between 35,000 and 60,000 barrels per day (one barrel of oil is equivalent to about 0.16 cubic meters). Source: www.doi.gov/deepwaterhorizon.

**Solution** The graph of $v(t)$ is shown in Figure 3.

**Figure 3**

Let's start with a very crude estimate of the total volume of oil released, using the graph as a guide. The rate of change of this total volume at the beginning of the time period is $v(0) = 60$ thousand barrels per day. If this rate were to remain constant for the entire 20-day period, the total volume of oil released would be

$$\text{Total volume} = \text{Volume per day} \times \text{Number of days} = 60 \times 20$$
$$= 1{,}200 \text{ thousand barrels.}$$

Figure 4 shows how we can represent this calculation on the graph of $v(t)$.

**Figure 4**

The volume per day based on $v(0) = 60$ is represented by the $y$-coordinate of the graph at its left edge, while the number of days is represented by the width of the interval $[0, 20]$ on the $x$-axis. Therefore, computing the area of the shaded rectangle in the figure gives the same calculation:

$$\text{Area of rectangle} = \text{Volume per day} \times \text{Number of days}$$
$$= 60 \times 20 = 1{,}200 = \text{Total volume.}$$

But, as we see in the graph, the flow rate does not remain constant, but goes down quite significantly over the course of the 20-day interval. We can obtain a somewhat more accurate estimate of the total volume by re-estimating the volume using 10-day periods—that is, by dividing the interval $[0, 20]$ into two equal intervals, or subdivisions. We estimate the volume over each 10-day period using the flow rate at the beginning of that period.

$$\text{Volume in first period} = \text{Volume per day} \times \text{Number of days}$$
$$= v(0) \times 10 = 60 \times 10 = 600 \text{ thousand barrels}$$

$$\text{Volume in second period} = \text{Volume per day} \times \text{Number of days}$$
$$= v(10) \times 10 = 28 \times 10 = 280 \text{ thousand barrels}$$

Adding these volumes gives us the more accurate estimate

$$v(0) \times 10 + v(10) \times 10 = 880 \text{ thousand barrels.}$$   Calculation using 2 subdivisions

In Figure 5 we see that we are computing the combined area of two rectangles, each of whose heights is determined by the height of the graph at its left edge:

The areas of the rectangles are estimates of the volumes for successive 10-day periods.

**Figure 5**

Area of first rectangle = Volume per day × Number of days

$$= v(0) \times 10 = 60 \times 10 = 600 = \text{Volume for first 10 days}$$

Area of second rectangle = Volume per day × Number of days = $v(10) \times 10$

$$= 28 \times 10 = 280 = \text{Volume for second 10 days.}$$

We can get an even better estimate of the volume by using four divisions of [0, 20] instead of two:

$$v(0) \times 5 + v(5) \times 5 + v(10) \times 5 + v(15) \times 5$$   Calculation using 4 subdivisions
$$= 300 + 210 + 140 + 90 = 740 \text{ thousand barrels.}$$

As we see in Figure 6, we have now computed the combined area of *four* rectangles, each of whose heights is again determined by the height of the graph at its left edge.

Estimated Volume Using 4 Subdivisions
The areas of the rectangles are estimates of the volumes for successive 5-day periods.

**Figure 6**

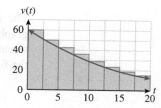

Estimated Volume Using 8 Subdivisions
The areas of the rectangles are estimates of the volumes for successive 2.5-day periods.

**Figure 7**

Notice how the volume seems to be decreasing as we use more subdivisions. More importantly, total volume seems to be getting closer to the area under the graph. Figure 7 illustrates the calculation for 8 equal subdivisions. The approximate total volume using 8 subdivisions is the total area of the shaded region in Figure 7:

$$v(0) \times 2.5 + v(2.5) \times 2.5 + v(5) \times 2.5 + \cdots + v(17.5) \times 2.5$$
$$= 675 \text{ thousand barrels.}$$   Calculation using 8 subdivisions

Looking at Figure 7, we still get the impression that we are overestimating the volume. If we want to be *really* accurate in our estimation of the volume, we should really be calculating the volume *continuously* every few hours or, better yet, minute by minute, as illustrated in Figure 8.

**using** Technology

**Website**
www.WanerMath.com
At the Website, select the Online Utilities tab, and choose the Numerical Integration Utility and Grapher. There, enter the formula

`0.08x^2-4x+60`

for *f*(*x*), enter 0 and 20 for the lower and upper limits, and press "Left Sum" to compute and see a Riemann sum. You can then adjust the number of subdivisions to see the effect on the graph and on the calculation, and thus replicate the information in Figure 8.

Every Six Hours (80 subdivisions)
Volume ≈ 619.35 thousand barrels

Every Minute (28,800 subdivisions)
Volume ≈ 613.35 thousand barrels

**Figure 8**

Figure 8 strongly suggests that the more accurately we estimate the total volume, the closer the answer gets to the exact area under the portion of the graph of $v(t)$ with $0 \le t \le 20$, and leads us to the conclusion that the *exact* total volume is the exact area under the rate of change of volume curve for $0 \le t \le 20$. In other words, we have made the following remarkable discovery:

*Total volume is the area under the rate of change of volume curve!*

➡ **Before we go on...** The 80-subdivision calculation in Example 1 is tedious to do by hand, and no one in his or her right mind would even *attempt* to do the minute-by-minute calculation by hand! Below we discuss ways of doing these calculations with the aid of technology. ∎

The type of calculation done in Example 1 is useful in many applications. Let's look at the general case and give the result a name.

In general, we have a function $f$ (such as the function $v$ in the example), and we consider an interval $[a, b]$ of possible values of the independent variable $x$. We subdivide the interval $[a, b]$ into some number of segments of equal length. Write $n$ for the number of segments, or **subdivisions**.

Next, we label the endpoints of these subdivisions $x_0$ for $a$, $x_1$ for the end of the first subdivision, $x_2$ for the end of the second subdivision, and so on until we get to $x_n$, the end of the $n$th subdivision, so that $x_n = b$. Thus,

$$a = x_0 < x_1 < \cdots < x_n = b.$$

The first subdivision is the interval $[x_0, x_1]$, the second subdivision is $[x_1, x_2]$, and so on until we get to the last subdivision, which is $[x_{n-1}, x_n]$. We are dividing the interval $[a, b]$ into $n$ subdivisions of equal length, so each segment has length $(b - a)/n$. We write $\Delta x$ for $(b - a)/n$ (Figure 9).

$$x_0 \quad x_1 \quad x_2 \quad x_3 \quad \cdots \quad x_{n-1} \quad x_n$$

$$a \qquad \vdash \Delta x \dashv \qquad\qquad\qquad b$$

**Figure 9**

Having established this notation, we can write the calculation that we want to do as follows: For each subdivision $[x_{k-1}, x_k]$, compute $f(x_{k-1})$, the value of the function

✱ After Georg Friedrich Bernhard
Riemann (1826–1866).

$f$ at the left endpoint. Multiply this value by the length of the interval, which is $\Delta x$. Then add together all $n$ of these products to get the number

$$f(x_0)\Delta x + f(x_1)\Delta x + \cdots + f(x_{n-1})\Delta x.$$

This sum is called a **(left) Riemann✱ sum** for $f$. In Example 1 we computed several different Riemann sums. Here is the computation for $n = 4$ we used in the oil spill example (see Figure 10):

**Figure 10**

Left Riemann sum $= f(x_0)\Delta x + f(x_1)\Delta x + \cdots + f(x_{n-1})\Delta x$
$$= f(0)(5) + f(5)(5) + f(10)(5) + f(15)(5)$$
$$= 60(5) + 42(5) + 28(5) + 18(5) = 740.$$

Because sums are often used in mathematics, mathematicians have developed a shorthand notation for them. We write

$$f(x_0)\Delta x + f(x_1)\Delta x + \cdots + f(x_{n-1})\Delta x \text{ as } \sum_{k=0}^{n-1} f(x_k)\Delta x.$$

The symbol $\sum$ is the Greek letter sigma and stands for **summation**. The letter $k$ here is called the index of summation, and we can think of it as counting off the segments. We read the notation as "the sum from $k = 0$ to $n - 1$ of the quantities $f(x_k)\Delta x$." Think of it as a set of instructions:

Set $k = 0$, and calculate $f(x_0)\Delta x$.          $f(0)(5)$ in the above calculation

Set $k = 1$, and calculate $f(x_1)\Delta x$.          $f(5)(5)$ in the above calculation

$\cdots$

Set $k = n - 1$, and calculate $f(x_{n-1})\Delta x$.          $f(15)(5)$ in the above calculation

Then sum all the quantities so calculated.

## Riemann Sum

If $f$ is a continuous function, the **left Riemann sum** with $n$ equal subdivisions for $f$ over the interval $[a, b]$ is defined to be

$$\text{Left Riemann sum} = \sum_{k=0}^{n-1} f(x_k)\Delta x$$
$$= f(x_0)\Delta x + f(x_1)\Delta x + \cdots + f(x_{n-1})\Delta x$$
$$= [f(x_0) + f(x_1) + \cdots + f(x_{n-1})]\Delta x,$$

where $a = x_0 < x_1 < \cdots < x_n = b$ are the endpoints of the subdivisions, and $\Delta x = (b - a)/n$.

### Interpretation of the Riemann Sum

If $f$ is the rate of change of a quantity $F$ (that is, $f = F'$), then the Riemann sum of $f$ approximates the total change of $F$ from $x = a$ to $x = b$. The approximation improves as the number of subdivisions increases toward infinity.

> **Quick Examples**
>
> 1. If $f(t)$ is the rate of change in the number of bats in a belfry and $[a, b] = [2, 3]$, then the Riemann sum approximates the total change in the number of bats in the belfry from time $t = 2$ to time $t = 3$.
> 2. If $c(x)$ is the marginal cost of producing the $x$th item and $[a, b] = [10, 20]$, then the Riemann sum approximates the cost of producing items 11 through 20.

### Visualizing a Left Riemann Sum (Non-negative Function)

Graphically, we can represent a left Riemann sum of a non-negative function as an approximation of the area under a curve:

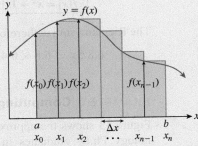

Riemann sum = Shaded area = Area of first rectangle + Area of second rectangle + $\cdots$ + Area of $n$th rectangle = $f(x_0)\Delta x + f(x_1)\Delta x + f(x_2)\Delta x + \cdots + f(x_{n-1})\Delta x$.

> **Quick Example**
>
> In Example 1 we computed several Riemann sums, including these:
>
> $n = 1$:  Riemann sum = $v(0)\Delta t = 60 \times 20 = 1{,}200$
>
> $n = 2$:  Riemann sum = $[v(t_0) + v(t_1)]\Delta t$
> $= [v(0) + v(10)](10) = 880$
>
> $n = 4$:  Riemann sum = $[v(t_0) + v(t_1) + v(t_2) + v(t_3)]\Delta t$
> $= [v(0) + v(5) + v(10) + v(15)](5) = 740$
>
> $n = 8$:  Riemann sum = $[v(t_0) + v(t_1) + \cdots + v(t_7)]\Delta t$
> $= [v(0) + v(2.5) + \cdots + v(17.5)](2.5) = 675$

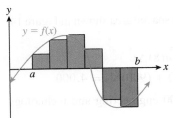

Riemann sum = Area above $x$-axis − Area below $x$-axis

**Figure 11**

**Note** To visualize the Riemann sum of a function that is negative, look again at the formula $f(x_0)\Delta x + f(x_1)\Delta x + f(x_2)\Delta x + \cdots + f(x_{n-1})\Delta x$ for the Riemann sum. Each term $f(x_k)\Delta x_k$ represents the area of one rectangle in the figure above. So, the areas of the rectangles with negative values of $f(x_k)$ are automatically counted as negative. They appear as red rectangles in Figure 11. ■

**Figure 12**

## EXAMPLE 2 Computing a Riemann Sum from a Formula

Compute the left Riemann sum for $f(x) = x^2 + 1$ over the interval $[-1, 1]$, using $n = 5$ subdivisions.

**Solution** Because the interval is $[a, b] = [-1, 1]$ and $n = 5$, we have

$$\Delta x = \frac{b - a}{n} = \frac{1 - (-1)}{5} = 0.4.$$   Width of subdivisions

Thus, the subdivisions of $[-1, 1]$ are determined by

$$-1 < -0.6 < -0.2 < 0.2 < 0.6 < 1.$$   Start with $-1$ and keep adding $\Delta x = 0.4$.

Figure 12 shows the graph with a representation of the Riemann sum.
The Riemann sum we want is

$$[f(x_0) + f(x_1) + \cdots + f(x_4)]\Delta x$$
$$= [f(-1) + f(-0.6) + f(-0.2) + f(0.2) + f(0.6)]0.4.$$

We can conveniently organize this calculation in a table as follows:

| $x$ | $-1$ | $-0.6$ | $-0.2$ | $0.2$ | $0.6$ | **Total** |
|---|---|---|---|---|---|---|
| $f(x) = x^2 + 1$ | 2 | 1.36 | 1.04 | 1.04 | 1.36 | 6.8 |

The Riemann sum is therefore

$$6.8\Delta x = 6.8 \times 0.4 = 2.72.$$

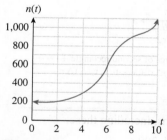

**Figure 13**

## EXAMPLE 3 Computing a Riemann Sum from a Graph

Figure 13 shows the approximate annual production $n(t)$ of engineering and technology Ph.D. graduates in Mexico during the period 2000–2010.[22] Use a left Riemann sum with four subdivisions to estimate the total number of Ph.D. graduates from 2002 to 2010.

**Solution** Let us represent the total number of (engineering and technology) Ph.D. graduates up to time $t$ (measured in years since 2000) by $N(t)$. The total number of Ph.D. graduates from 2002 to 2010 is then the total change in $N(t)$ over the interval $[2, 10]$. In view of the above discussion, we can approximate the total change in $N(t)$ using a Riemann sum of its rate of change $n(t)$. Because $n = 4$ subdivisions are specified, the width of each subdivision is

$$\Delta t = \frac{b - a}{n} = \frac{10 - 2}{4} = 2.$$

We can therefore represent the left Riemann sum by the shaded area shown in Figure 14.
From the graph,

$$\text{Left sum} = n(2)\Delta t + n(4)\Delta t + n(6)\Delta t + n(8)\Delta t$$
$$= (200)(2) + (300)(2) + (600)(2) + (900)(2) = 4,000.$$

So, we estimate that there was a total of about 4,000 engineering and technology Ph.D. graduates during the given period.

**Figure 14**

[22] Source for data: Instituto Nacional de Estadística y Geografía (www.inegi.org.mx).

Right riemann sum = (300)(2) + (600)(2) + (900)(2) + (1,100)(2) = 5,800

Height of each ractangle is determined by height of graph at right edge.

**Figure 15**

\* This applies to some other functions as well, including "piecewise continuous" functions discussed in the next section.

➡ **Before we go on...** A glance at Figure 14 tells us that we have significantly underestimated the actual number of graduates, as the actual area under the curve is considerably larger. Figure 15 shows a **right Riemann sum** and gives us a much larger estimate. For continuous functions, the difference between these two types of Riemann sums approaches zero as the number of subdivisions approaches infinity (see below) so we will focus primarily on only one type: left Riemann sums. ◼

As in Example 1, we're most interested in what happens to the Riemann sum when we let $n$ get very large. When $f$ is continuous,\* its Riemann sums will always approach a limit as $n$ goes to infinity. (This is not meant to be obvious. Proofs may be found in advanced calculus texts.) We give the limit a name.

## The Definite Integral

If $f$ is a continuous function, the **definite integral of $f$ from $a$ to $b$** is defined to be the limit of the Riemann sums as the number of subdivisions approaches infinity:

$$\int_a^b f(x)\,dx = \lim_{n \to \infty} \sum_{k=0}^{n-1} f(x_k)\,\Delta x.$$

**In Words:** The integral, from $a$ to $b$, of $f(x)\,dx$ equals the limit, as $n \to \infty$, of the Riemann Sum with a partition of $n$ subdivisions.

The function $f$ is called the **integrand**, the numbers $a$ and $b$ are the **limits of integration**, and the variable $x$ is the **variable of integration**. A Riemann sum with a large number of subdivisions may be used to approximate the definite integral.

### Interpretation of the Definite Integral
If $f$ is the rate of change of a quantity $F$ (that is, $f = F'$), then $\int_a^b f(x)\,dx$ is the (exact) total change of $F$ from $x = a$ to $x = b$.

> **Quick Examples**
>
> 1. If $f(t)$ is the rate of change in the number of bats in a belfry and $[a, b] = [2, 3]$, then $\int_2^3 f(t)\,dt$ is the total change in the number of bats in the belfry from time $t = 2$ to time $t = 3$.
>
> 2. If, at time $t$ hours, you are selling wall posters at a rate of $s(t)$ posters per hour, then
>
> $$\text{Total number of posters sold from hour 3 to hour 5} = \int_3^5 s(t)\,dt.$$

### Visualizing the Definite Integral
**Non-negative Functions:** If $f(x) \geq 0$ for all $x$ in $[a, b]$, then $\int_a^b f(x)\,dx$ is the area under the graph of $f$ over the interval $[a, b]$, as shaded in the figure.

**General Functions:** $\int_a^b f(x)\,dx$ is the area between $x = a$ and $x = b$ that is above the $x$-axis and below the graph of $f$, minus the area that is below the $x$-axis and above the graph of $f$:

$$\int_a^b f(x)\,dx = \text{Area above } x\text{-axis} - \text{Area below } x\text{-axis}$$

### Quick Examples

1.   $\displaystyle\int_1^4 2\,dx = \text{Area of rectangle} = 6$

2.   $\displaystyle\int_0^2 x\,dx = \text{Area of triangle} = \frac{1}{2}\text{ base} \times \text{height} = 2$

3.   $\displaystyle\int_{-1}^1 x\,dx = 0$   The areas above and below the $x$-axis are equal.

### Notes

1. Remember that $\int_a^b f(x)\,dx$ stands for a number that depends on $f$, $a$, and $b$. The variable of integration $x$ that appears is called a **dummy variable** because it has no effect on the answer. In other words,

$$\int_a^b f(x)\,dx = \int_a^b f(t)\,dt.$$   $x$ or $t$ is just a name we give the variable.

2. The notation for the definite integral (due to Leibniz) comes from the notation for the Riemann sum. The integral sign $\int$ is an elongated S, the Roman equivalent of the Greek $\sum$. The $d$ in $dx$ is the lowercase Roman equivalent of the Greek $\Delta$.

3. The definition above is adequate for continuous functions, but more complicated definitions are needed to handle other functions. For example, we broke the interval $[a, b]$ into $n$ subdivisions of equal length, but other definitions allow a **partition** of the interval into subdivisions of possibly unequal lengths. We have evaluated $f$ at the left endpoint of each subdivision, but we could equally well have used the right endpoint or any other point in the subdivision. All of these variations lead to the same answer when $f$ is continuous.

4. The similarity between the notations for the definite integral and the indefinite integral is no mistake. We will discuss the exact connection in the next section. ■

## Computing Definite Integrals

In some cases, we can compute the definite integral directly from the graph (see the quick examples above and the next example below). In general, the only method of computing definite integrals we have discussed so far is numerical estimation: Compute the Riemann sums for larger and larger values of $n$ and then estimate the number it seems to be approaching as we did in Example 1. (In the next section we will discuss an algebraic method for computing them.)

### EXAMPLE 4 Estimating a Definite Integral from a Graph

Figure 16 shows the graph of the (approximate) rate $f'(t)$, at which the United States consumed gasoline from 2000 through 2008. ($t$ is time in years since 2000.)[23]

$f'(t)$ = Gasoline consumption (100 million gals per year)

**Figure 16**

Use the graph to estimate the total U.S. consumption of gasoline over the period shown.

**Solution** The derivative $f'(t)$ represents the rate of change of the total U.S. consumption of gasoline, and so the total U.S. consumption of gasoline over the given period $[0, 8]$ is given by the definite integral:

$$\text{Total U.S. consumption of gasoline} = \text{Total change in } f(t) = \int_0^8 f'(t)\, dt$$

and is given by the area under the graph (Figure 17).

---

[23] Source: Energy Information Administration (Department of Energy). (www.eia.doe.gov).

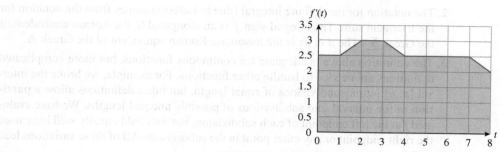

**Figure 17**

One way to determine the area is to count the number of filled rectangles as defined by the grid. Each rectangle has an area of $1 \times 0.5 = 0.5$ units (and the half-rectangles determined by diagonal portions of the graph have half that area). Counting rectangles, we find a total of 41.5 complete rectangles, so

$$\text{Total area} = 20.75.$$

Because $f'(t)$ is in 100 million gallons per year, we conclude that the total U.S. consumption of gasoline over the given period was about 2,075 million gallons, or 2.075 billion gallons.

While counting rectangles might seem easy, it becomes awkward in cases involving large numbers of rectangles or partial rectangles whose area is not easy to determine. In a case like this, in which the graph consists of straight lines, rather than counting rectangles, we can get the area by averaging the left and right Riemann sums whose subdivisions are determined by the grid:

$$\text{Left sum} = (2.5 + 2.5 + 3 + 3 + 2.5 + 2.5 + 2.5 + 2.5)(1) = 21$$
$$\text{Right sum} = (2.5 + 3 + 3 + 2.5 + 2.5 + 2.5 + 2.5 + 2)(1) = 20.5$$
$$\text{Average} = \frac{21 + 20.5}{2} = 20.75.$$

To see why this works, look at the single interval [1, 2]. The left sum contributes $2.5 \times 1 = 2.5$ and the right sum contributes $3 \times 1 = 3$. The exact area is their average, 2.75 (Figure 18).

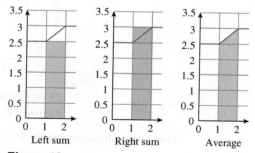

**Figure 18**

The average of the left and right Riemann sums is frequently a better estimate of the definite integral than either is alone.

⟹ **Before we go on...**   It is important to check that the units we are using in Example 4 match up correctly: $t$ is given in *years* and $f'(t)$ is given in 100 million gallons per *year*. The integral is then given in

$$\text{Years} \times \frac{100 \text{ million gallons}}{\text{Year}} = 100 \text{ million gallons}.$$

If we had specified $f'(t)$ in, say, 100 million gallons per *day* but $t$ in years, then we would have needed to convert either $t$ or $f'(t)$ so that the units of time match. ∎

The next example illustrates the use of technology in estimating definite integrals using Riemann sums.

**Figure 19**

 **using** Technology

See the Technology Guides at the end of the chapter to find out how to compute these sums on a TI-83/84 Plus and a spreadsheet.

**W** **Website**
www.WanerMath.com
At the Website, select the Online Utilities tab, and choose the Numerical Integration Utility and Grapher. This utility computes left and right Riemann sums, as well as a good numerical approximation of the integral. There is also a downloadable Excel spreadsheet that computes and graphs Riemann sums (Riemann Sum Grapher).

### EXAMPLE 5  ⓘ Using Technology to Approximate the Definite Integral

Use technology to estimate the area under the graph of $f(x) = 1 - x^2$ over the interval $[0, 1]$ using $n = 100$, $n = 200$, and $n = 500$ subdivisions.

**Solution**  We need to estimate the area under the parabola shown in Figure 19. From the discussion above,

$$\text{Area} = \int_0^1 (1 - x^2)\, dx.$$

The Riemann sum with $n = 100$ has $\Delta x = (b - a)/n = (1 - 0)/100 = 0.01$ and is given by

$$\sum_{k=0}^{99} f(x_k)\Delta x = [f(0) + f(0.01) + \cdots + f(0.99)](0.01).$$

Similarly, the Riemann sum with $n = 200$ has $\Delta x = (b - a)/n = (1 - 0)/200 = 0.005$ and is given by

$$\sum_{k=0}^{199} f(x_k)\Delta x = [f(0) + f(0.005) + \cdots + f(0.995)](0.005).$$

For $n = 500$, $x = (b - a)/n = (1 - 0)/500 = 0.002$ and the Riemann sum is

$$\sum_{k=0}^{499} f(x_k)\Delta x = [f(0) + f(0.002) + \cdots + f(0.998)](0.002).$$

Using technology to evaluate these Riemann sums, we find:

$$n = 100: \sum_{k=0}^{99} f(x_k)\Delta x = 0.67165$$

$$n = 200: \sum_{k=0}^{199} f(x_k)\Delta x = 0.6691625$$

$$n = 500: \sum_{k=0}^{499} f(x_k)\Delta x = 0.667666$$

so we estimate that the area under the curve is about 0.667. (The exact answer is 2/3, as we will be able to verify using the techniques in the next section.)

### EXAMPLE 6  Motion

A fast car has velocity $v(t) = 6t^2 + 10t$ ft/sec after $t$ seconds (as measured by a radar gun). Use several values of $n$ to find the distance covered by the car from time $t = 3$ seconds to time $t = 4$ seconds.

**Solution**  Because the velocity $v(t)$ is rate of change of position, the total change in position over the interval $[3, 4]$ is

$$\text{Distance covered} = \text{Total change in position} = \int_3^4 v(t)\,dt = \int_3^4 (6t^2 + 10t)\,dt.$$

As in Examples 1 and 5, we can subdivide the one-second interval $[3, 4]$ into smaller and smaller pieces to get more and more accurate approximations of the integral. By computing Riemann sums for various values of $n$, we get the following results.

**Website**
www.WanerMath.com
At the Website you will find the following optional online interactive section: Internet Topic: Numerical Integration.

$$n = 10: \sum_{k=0}^{9} v(t_k)\Delta t = 106.41 \qquad n = 100: \sum_{k=0}^{99} v(t_k)\Delta t \approx 108.740$$

$$n = 1{,}000: \sum_{k=0}^{999} v(t_k)\Delta t \approx 108.974 \quad n = 10{,}000: \sum_{k=0}^{9999} v(t_k)\Delta t \approx 108.997$$

These calculations suggest that the total distance covered by the car, the value of the definite integral, is approximately 109 feet.

➡ **Before we go on...**  Do Example 6 using antiderivatives instead of Riemann sums, as in Section 13.1. Do you notice a relationship between antiderivatives and definite integrals? This will be explored in the next section. ■

## 13.3   EXERCISES

▼ more advanced     ◆ challenging
🔳 indicates exercises that should be solved using technology

*Calculate the left Riemann sums for the given functions over the given intervals in Exercises 1–10, using the given values of n. (When rounding, round answers to four decimal places.)*
HINT [See Example 2.]

**1.** $f(x) = 4x - 1$ over $[0, 2]$, $n = 4$

**2.** $f(x) = 1 - 3x$ over $[-1, 1]$, $n = 4$

**3.** $f(x) = x^2$ over $[-2, 2]$, $n = 4$

**4.** $f(x) = x^2$ over $[1, 5]$, $n = 4$

**5.** $f(x) = \dfrac{1}{1 + x}$ over $[0, 1]$, $n = 5$

**6.** $f(x) = \dfrac{x}{1 + x^2}$ over $[0, 1]$, $n = 5$

**7.** $f(x) = e^{-x}$ over $[0, 10]$, $n = 5$

**8.** $f(x) = e^{-x}$ over $[-5, 5]$, $n = 5$

**9.** $f(x) = e^{-x^2}$ over $[0, 10]$, $n = 4$

**10.** $f(x) = e^{-x^2}$ over $[0, 100]$, $n = 4$

*In Exercises 11–18, use the given graph to estimate the left Riemann sum for the given interval with the stated number of subdivisions.* HINT [See Example 3.]

**11.** $[0, 5]$, $n = 5$

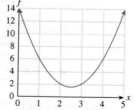

**12.** $[0, 8]$, $n = 4$

**13.** $[1, 9]$, $n = 4$

**14.** $[0.5, 2.5]$, $n = 4$

**15.** $[1, 3.5]$, $n = 5$

**16.** $[0.5, 3.5]$, $n = 3$

**31.** $[2, 6]$

**32.** $[0, 5]$

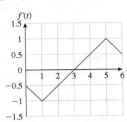

**17.** $[0, 3]$; $n = 3$

**18.** $[0.5, 3]$; $n = 5$

**33.** $[-1, 2]$

**34.** $[-1, 2]$

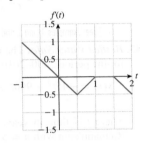

*Use geometry (not Riemann sums) to compute the integrals in Exercises 19–28.* HINT [See Quick Examples page 982.]

**19.** $\int_0^1 1\, dx$

**20.** $\int_0^2 5\, dx$

**21.** $\int_0^1 x\, dx$

**22.** $\int_1^2 x\, dx$

**23.** $\int_0^1 \frac{x}{2}\, dx$

**24.** $\int_1^2 \frac{x}{2}\, dx$

**25.** $\int_2^4 (x - 2)\, dx$

**26.** $\int_3^6 (x - 3)\, dx$

**27.** $\int_{-1}^1 x^3\, dx$

**28.** $\int_{-2}^2 \frac{x}{2}\, dx$

*In Exercises 29–34, the graph of the derivative $f'(t)$ of $f(t)$ is shown. Compute the total change of $f(t)$ over the given interval.* HINT [See Example 4.]

**29.** $[1, 5]$

**30.** $[2, 6]$

 *In Exercises 35–38, use technology to approximate the given integrals with Riemann sums, using (a) $n = 10$, (b) $n = 100$, and (c) $n = 1,000$. Round all answers to four decimal places.* HINT [See Example 5.]

**35.** $\int_0^1 4\sqrt{1 - x^2}\, dx$

**36.** $\int_0^1 \frac{4}{1 + x^2}\, dx$

**37.** $\int_2^3 \frac{2x^{1.2}}{1 + 3.5x^{4.7}}\, dx$

**38.** $\int_3^4 3xe^{1.3x}\, dx$

## APPLICATIONS

**39.** *Pumps* A pump is delivering water into a tank at a rate of

$$r(t) = 3t^2 + 5 \text{ liters/minute,}$$

where $t$ is time in minutes since the pump is turned on. Use a Riemann sum with $n = 5$ subdivisions to estimate the total volume of water pumped in during the first two minutes. HINT [See Examples 1 and 2. This exercise is also discussed in the tutorial on the Website.]

**40.** *Pumps* A pump is delivering water into a tank at a rate of

$$r(t) = 6t^2 + 40 \text{ liters/minute,}$$

where $t$ is time in minutes since the pump is turned on. Use a Riemann sum with $n = 6$ subdivisions to estimate the total volume of water pumped in during the first three minutes. HINT [See Examples 1 and 2. A similar exercise is also discussed in the tutorial on the Website.]

**41. Cost** The marginal cost function for the manufacture of portable MP3 players is given by

$$C'(x) = 20 - \frac{x}{200},$$

where $x$ is the number of MP3 players manufactured. Use a Riemann sum with $n = 5$ to estimate the cost of producing the first 5 MP3 players. HINT [See Examples 1 and 2 and Quick Example 2 on page 979.]

**42. Cost** Repeat the preceding exercise using the marginal cost function

$$C'(x) = 25 - \frac{x}{50}.$$

HINT [See Examples 1 and 2 and Quick Example 2 on page 979.]

**43. Bottled Water Sales** The rate of U.S. sales of bottled water for the period 2000–2010 could be approximated by

$$s(t) = -45t^2 + 900t + 4,200 \text{ million gallons per year}$$
$$(0 \le t \le 10),$$

where $t$ is time in years since the start of 2000.[24] Use a Riemann sum with $n = 5$ to estimate the total U.S. sales of bottled water from the start of 2005 to the start of 2010. (Round your answer to the nearest billion gallons.) HINT [See Example 2.]

**44. Bottled Water Sales** The rate of U.S. per capita sales of bottled water for the period 2000–2010 could be approximated by

$$s(t) = -0.18t^2 + 3t + 15 \text{ gallons per year} \quad (0 \le t \le 10),$$

where $t$ is time in years since the start of 2000.[25] Use a Riemann sum with $n = 5$ to estimate the total U.S. per capita sales of bottled water from the start of 2003 to the start of 2008. (Round your answer to the nearest gallon.) HINT [See Example 2.]

**45. Online Vehicle Sales** The following graph shows the rate of change $v(t)$ of the value of vehicle transactions on eBay, in billions of dollars per quarter from the first quarter of 2008 through the first quarter of 2011 ($t$ is time in quarters; $t = 1$ represents the first quarter of 2008).[26]

Use a left Riemann sum with four subdivisions to estimate the total value of vehicle transactions on eBay from the third quarter of 2008 to the third quarter of 2010 (the interval [3, 11]). HINT [See Example 3.]

**46. Online Auctions** The following graph shows the rate of change $n(t)$ of the number of active eBay users, in millions of users per quarter, from the first quarter of 2008 through the first quarter of 2011 ($t$ is time in quarters; $t = 1$ represents the first quarter of 2008).[27]

Use a left Riemann sum with four subdivisions to estimate the total change in the number of active users from the second quarter of 2008 to the second quarter of 2010 (the interval [2, 10]). HINT [See Example 3.]

**47. Scientific Research** The rate of change $r(t)$ of the total number of research articles in the prominent journal *Physical Review* written by researchers in Europe is shown in the following graph:

Here, $t$ is time in years ($t = 0$ represents the start of 1983).[28]

**a.** Use both left and right Riemann sums with eight subdivisions to estimate the total number of articles in *Physical Review* written by researchers in Europe during the 16-year period beginning at the start of 1983. (Estimate each value of $r(t)$ to the nearest 0.5.) HINT [See Example 3.]

**b.** Use the answers from part (a) to obtain an estimate of $\int_0^{16} r(t)\,dt$. HINT [See Example 4.] Interpret the result.

---

[24]Source for data: Beverage Marketing Corporation (www.bottledwater.org).

[25]*Ibid.*

[26]Source for data: eBay company reports (http://investor.ebay.com).

[27]*Ibid.*

[28]Based on data from 1983 to 2003. Source: The American Physical Society/*New York Times*, May 3, 2003, p. A1.

**48. *Scientific Research*** The rate of change $r(t)$ of the total number of research articles in the prominent journal *Physical Review* written by researchers in the United States is shown in the following graph:

Here, $t$ is time in years ($t = 0$ represents the start of 1983).[29]

**a.** Use both left and right Riemann sums with six subdivisions to estimate the total number of articles in *Physical Review* written by researchers in the United States during the 12-year period beginning at the start of 1993. (Estimate each value of $r(t)$ to the nearest 0.25.) HINT [See Example 3.]

**b.** Use the answers from part (a) to obtain an estimate of $\int_{10}^{22} r(t)\,dt$. HINT [See Example 4.] Interpret the result.

**49. *Graduate Degrees: Women*** The following graph shows the approximate number $f'(t)$ of doctoral degrees per year awarded to women in the United States during 2005–2014 ($t = 0$ represents 2005 and each unit of the $y$-axis represents 10,000 degrees).[30]

Use the graph to estimate the total number of doctoral degrees awarded to women from 2005 to 2010. HINT [See Example 4.]

**50. *Graduate Degrees: Men*** The following graph shows the approximate number $m'(t)$ of doctoral degrees per year awarded to men in the United States during 2005–2014 ($t = 0$ represents 2005 and each unit of the $y$-axis represents 10,000 degrees).[31]

Use the graph to estimate the total number of doctoral degrees awarded to men from 2009 to 2014. HINT [See Example 4.]

**51. *Visiting Students in the Early 2000s*** The aftermath of the September 11 attacks saw a decrease in the number of students visiting the United States. The following graph shows the approximate rate of change $c'(t)$ in the number of students from China who had taken the GRE exam required for admission to U.S. universities ($t$ is time in years since 2000):[32]

Use the graph to estimate, to the nearest 1,000, the total number of students from China who took the GRE exams during the period 2002–2004.

**52. *Visiting Students in the Early 2000s*** Repeat Exercise 51, using the following graph for students from India:[33]

[29]Based on data from 1983 to 2003. Source: The American Physical Society/*New York Times*, May 3, 2003, p. A1.

[30]Source for data and projections: National Center for Education Statistics (www.nces.ed.gov).

[31]*Ibid.*

[32]Source: Educational Testing Services/Shanghai and Jiao Tong University/*New York Times*, December 21, 2004, p. A25.

[33]*Ibid.*

**Net Income:** *Exercises 53 and 54 are based on the following graph, which shows* General Electric's *approximate net income in billions of dollars each year from 2005 to 2011.*[34]

53. Compute the left and right Riemann sum estimates of $\int_0^4 p'(t)\,dt$ using $\Delta t = 1$. Which of these two sums gives the actual total net income earned by GE during the years 2005 through 2008? Explain.

54. Compute the left and right Riemann sum estimates of $\int_2^6 p'(t)\,dt$ using $\Delta t = 1$. Which of these two sums gives the actual total net income earned by GE during the years 2008 through 2011? Explain.

55. **Petrochemical Sales: Mexico** The following table shows annual petrochemical sales in Mexico by Pemex, Mexico's national oil company, for 2005–2010 ($t = 0$ represents 2005).[35]

| Year $t$ (Year since 2005) | 0 | 1 | 2 | 3 | 4 | 5 |
|---|---|---|---|---|---|---|
| Petrochemical Sales $s$ (Billion metric tons) | 3.7 | 3.8 | 4.0 | 4.1 | 4.0 | 4.1 |

a. Use the table to compute the left and right Riemann sums for $s(t)$ over the interval [0, 4] using 4 subdivisions.
b. What does the *right* Riemann sum in part (a) tell you about petrochemical sales by Pemex?

56. **Petrochemical Production: Mexico** The following table shows annual petrochemical production in Mexico by Pemex, Mexico's national oil company, for 2005–2010 ($t = 0$ represents 2005):[36]

| Year $t$ (Year since 2005) | 0 | 1 | 2 | 3 | 4 | 5 |
|---|---|---|---|---|---|---|
| Petrochemical Production $p$ (Billion metric tons) | 10.8 | 11.0 | 11.8 | 12.0 | 12.0 | 13.3 |

a. Use the table to compute the left and right Riemann sums for $p(t)$ over the interval [1, 5] using 4 subdivisions.
b. What does the *left* Riemann sum in part (a) tell you about production of petrochemicals by Pemex?

57. **Motion under Gravity** The velocity of a stone moving under gravity $t$ seconds after being thrown up at 30 ft/s is given by $v(t) = -32t + 30$ ft/sec. Use a Riemann sum with 5 subdivisions to estimate $\int_0^4 v(t)\,dt$. What does the answer represent? HINT [See Example 6.]

58. **Motion under Gravity** The velocity of a stone moving under gravity $t$ seconds after being thrown up at 4 m/s is given by $v(t) = -9.8t + 4$ m/sec. Use a Riemann sum with 5 subdivisions to estimate $\int_0^1 v(t)\,dt$. What does the answer represent? HINT [See Example 6.]

59. ▣ **Motion** A model rocket has upward velocity $v(t) = 40t^2$ ft/sec, $t$ seconds after launch. Use a Riemann sum with $n = 10$ to estimate how high the rocket is 2 seconds after launch. (Use technology to compute the Riemann sum.)

60. ▣ **Motion** A race car has a velocity of $v(t) = 600(1 - e^{-0.5t})$ ft/sec, $t$ seconds after starting. Use a Riemann sum with $n = 10$ to estimate how far the car has traveled in the first 4 seconds. (Round your answer to the nearest whole number.) (Use technology to compute the Riemann sum.)

61. ▣ **Facebook Membership** Since the start of 2005, new members joined Facebook at a rate of roughly

$$m(t) = 3.2t^3 - 12t + 10 \text{ million members per year}$$
$$(0 \le t \le 5),$$

where $t$ is time in years since the start of 2005.[37] Estimate $\int_0^5 m(t)\,dt$ using a Riemann sum with $n = 150$. (Round your answer to the nearest whole number.) Interpret the answer. HINT [See Example 5.]

62. ▣ **Uploads to YouTube** Since YouTube first became available to the public in mid-2005, the rate at which video has been uploaded to the site can be approximated by

$$v(t) = 525{,}600(0.42t^2 + 2.7t - 1) \text{ hours of video per year}$$
$$(0.5 \le t \le 7),$$

where $t$ is time in years since the start of 2005.[38] Estimate $\int_1^7 v(t)\,dt$ using a Riemann sum with $n = 150$. (Round your answer to the nearest 1,000.) Interpret the answer. HINT [See Example 5.]

63. ▣ **Big Brother** The total number of wiretaps authorized each year by U.S. state and federal courts from 1990 to 2010 can be approximated by

$$w(t) = 820e^{0.051t} \quad (0 \le t \le 20)$$

($t$ is time in years since the start of 1990).[39] Estimate $\int_0^{15} w(t)\,dt$ using a (left) Riemann sum with $n = 100$. (Round your answer to the nearest 10.) Interpret the answer. HINT [See Example 5.]

---

[34]Source: Company reports (www.ge.com/investors).

[35]2010 figure is a projection based on data through May 2010. Source: www.pemex.com (July 2010).

[36]*Ibid.*

[37]Sources for data: www.facebook.com, www.insidefacebook.com.

[38]Source for data: www.YouTube.com.

[39]Source for data: Wiretap Reports, Administrative Office of the United States Courts (www.uscourts.gov/Statistics/WiretapReports.aspx).

**64.** 🔲 *Big Brother* The number of wiretaps authorized each year by U.S. state courts from 1990 to 2010 can be approximated by

$$w(t) = 430e^{0.065t} \quad (0 \le t \le 20)$$

($t$ is time in years since the start of 1990).[40] Estimate $\int_{10}^{15} w(t)\,dt$ using a (left) Riemann sum with $n = 100$. (Round your answer to the nearest 10.) Interpret the answer. HINT [See Example 5.]

**65.** ▼ *Surveying* My uncle intends to build a kidney-shaped swimming pool in his small yard, and the town zoning board will approve the project only if the total area of the pool does not exceed 500 square feet. The accompanying figure shows a diagram of the planned swimming pool, with measurements of its width at the indicated points. Will my uncle's plans be approved? Use a (left) Riemann sum to approximate the area.

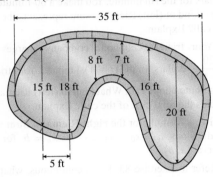

**66.** ▼ *Pollution* An aerial photograph of an ocean oil spill shows the pattern in the accompanying diagram. Assuming that the oil slick has a uniform depth of 0.01 m, how many cubic meters of oil would you estimate to be in the spill? (Volume = Area × Thickness. Use a (left) Riemann sum to approximate the area.)

**67.** 🔲 ▼ *Oil Consumption: United States* During the period 1980–2008 the United States was consuming oil at a rate of about

$$q(t) = 76t + 5{,}540 \text{ million barrels per year} \quad (0 \le t \le 28),$$

where $t$ is time in years since the start of 1980.[41] During the same period, the price per barrel of crude oil in constant 2008 dollars was about

$$p(t) = 0.45t^2 - 12t + 105 \text{ dollars}[42] \quad (0 \le t \le 28).$$

**a.** Graph the function $r(t) = p(t)q(t)$ for $0 \le t \le 28$, indicating the area that represents $\int_{10}^{20} r(t)\,dt$. What does this area signify?

**b.** Estimate the area in part (a) using a Riemann sum with $n = 200$. (Round the answer to 3 significant digits.) Interpret the answer.

**68.** 🔲 ▼ *Oil Consumption: China* Repeat Exercise 67 using instead the rate of consumption of oil in China:

$$q(t) = 82t + 221 \text{ million barrels per year}[43] \quad (0 \le t \le 28).$$

*The Normal Curve The normal distribution curve, which models the distributions of data in a wide range of applications, is given by the function*

$$p(x) = \frac{1}{\sqrt{2\pi}\,\sigma}\, e^{-(x-\mu)^2/2\sigma^2},$$

*where $\pi = 3.14159265\ldots$ and $\sigma$ and $\mu$ are constants called the **standard deviation** and the **mean**, respectively. The graph of the normal distribution (when $\sigma = 1$ and $\mu = 2$) is shown in the figure. Exercises 69 and 70 illustrate its use.*

**69.** 🔲 ▼ *Test Scores* Enormous State University's Calculus I test scores are modeled by a normal distribution with $\mu = 72.6$ and $\sigma = 5.2$. The percentage of students who obtained scores between $a$ and $b$ on the test is given by

$$\int_a^b p(x)\,dx.$$

**a.** Use a Riemann sum with $n = 40$ to estimate the percentage of students who obtained between 60 and 100 on the test.

**b.** What percentage of students scored less than 30?

[40] Source for data: Wiretap Reports, Administrative Office of the United States Courts (www.uscourts.gov/Statistics/WiretapReports.aspx).

[41] Source for data: BP Statistical Review of World Energy (www.bp.com/statisticalreview).

[42] Source for data: www.inflationdata.com.

[43] Source for data: BP Statistical Review of World Energy (www.bp.com/statisticalreview).

**70.** ⚏ ▼ *Consumer Satisfaction* In a survey, consumers were asked to rate a new toothpaste on a scale of 1–10. The resulting data are modeled by a normal distribution with $\mu = 4.5$ and $\sigma = 1.0$. The percentage of consumers who rated the toothpaste with a score between $a$ and $b$ on the test is given by

$$\int_a^b p(x)\,dx.$$

**a.** Use a Riemann sum with $n = 10$ to estimate the percentage of customers who rated the toothpaste 5 or above. (Use the range 4.5 to 10.5.)

**b.** What percentage of customers rated the toothpaste 0 or 1? (Use the range −0.5 to 1.5.)

## COMMUNICATION AND REASONING EXERCISES

**71.** If $f(x) = 6$, then the left Riemann sum _____ (increases/decreases/stays the same) as $n$ increases.

**72.** If $f(x) = -1$, then the left Riemann sum _____ (increases/decreases/stays the same) as $n$ increases.

**73.** If $f$ is an increasing function of $x$, then the left Riemann sum _____ (increases/decreases/stays the same) as $n$ increases.

**74.** If $f$ is a decreasing function of $x$, then the left Riemann sum _____ (increases/decreases/stays the same) as $n$ increases.

**75.** If $\int_a^b f(x)\,dx = 0$, what can you say about the graph of $f$?

**76.** Sketch the graphs of two (different) functions $f$ and $g$ such that $\int_a^b f(x)\,dx = \int_a^b g(x)\,dx$.

**77.** ▼ The definite integral counts the area under the $x$-axis as negative. Give an example that shows how this can be useful in applications.

**78.** ▼ Sketch the graph of a nonconstant function whose Riemann sum with $n = 1$ gives the exact value of the definite integral.

**79.** ▼ Sketch the graph of a nonconstant function whose Riemann sums with $n = 1, 5$, and 10 are all zero.

**80.** ▼ Besides left and right Riemann sums, another approximation of the integral is the **midpoint** approximation, in which we compute the sum

$$\sum_{k=1}^n f(\bar{x}_k)\,\Delta x$$

where $\bar{x}_k = (x_{k-1} + x_k)/2$ is the point midway between the left and right endpoints of the interval $[x_{k-1}, x_k]$. Why is it true that the midpoint approximation is exact if $f$ is linear? (Draw a picture.)

**81.** ▼ Your cellphone company charges you $c(t) = \dfrac{20}{t + 100}$ dollars for the $t$th minute. You make a 60-minute phone call. What kind of (left) Riemann sum represents the total cost of the call? Explain.

**82.** ▼ Your friend's cellphone company charges her $c(t) = \dfrac{20}{t + 100}$ dollars for the $(t + 1)$st minute. Your friend makes a 60-minute phone call. What kind of (left) Riemann sum represents the total cost of the call? Explain.

**83.** ▼ Give a formula for the **right Riemann Sum** with $n$ equal subdivisions $a = x_0 < x_1 < \cdots < x_n = b$ for $f$ over the interval $[a, b]$.

**84.** ▼ Refer to Exercise 83. If $f$ is continuous, what happens to the difference between the left and right Riemann sums as $n \to \infty$? Explain.

**85.** ▼ Sketch the graph of a nonzero function whose left Riemann sum with $n$ subdivisions is zero for every *even* number $n$.

**86.** ▼ When approximating a definite integral by computing Riemann sums, how might you judge whether you have chosen $n$ large enough to get your answer accurate to, say, three decimal places?

## 13.4 The Definite Integral: Algebraic Viewpoint and the Fundamental Theorem of Calculus

In Section 13.3 we saw that the definite integral of the marginal cost function gives the total cost. However, in Section 13.1 we used antiderivatives to recover the cost function from the marginal cost function, so we *could* use antiderivatives to compute total cost. The following example, based on Example 5 in Section 13.1, compares these two approaches.

## EXAMPLE 1 Finding Cost from Marginal Cost

The marginal cost of producing baseball caps at a production level of $x$ caps is $4 - 0.001x$ dollars per cap. Find the total change of cost if production is increased from 100 to 200 caps.

### Solution

**Method 1: Using an Antiderivative** (based on Example 5 in Section 13.1): Let $C(x)$ be the cost function. Because the marginal cost function is the derivative of the cost function, we have $C'(x) = 4 - 0.001x$ and so

$$C(x) = \int (4 - 0.001x)\,dx$$

$$= 4x - 0.001\frac{x^2}{2} + K \qquad \text{$K$ is the constant of integration.}$$

$$= 4x - 0.0005x^2 + K.$$

Although we do not know what to use for the value of the constant $K$, we can say:

Cost at production level of 100 caps $= C(100)$

$$= 4(100) - 0.0005(100)^2 + K$$

$$= \$395 + K$$

Cost at production level of 200 caps $= C(200)$

$$= 4(200) - 0.0005(200)^2 + K$$

$$= \$780 + K.$$

Therefore,

Total change in cost $= C(200) - C(100)$

$$= (\$780 + K) - (\$395 + K) = \$385.$$

Notice how the constant of integration simply canceled out! So, we could choose any value for $K$ that we wanted (such as $K = 0$) and still come out with the correct total change. Put another way, we could use *any antiderivative* of $C'(x)$, such as

$$F(x) = 4x - 0.0005x^2 \qquad \text{$F(x)$ is any antiderivative of $C'(x)$}$$
$$\text{whereas $C(x)$ is the actual cost function.}$$

or

$$F(x) = 4x - 0.0005x^2 + 4,$$

compute $F(200) - F(100)$, and obtain the total change, \$385.

Summarizing this method: To compute the total change of $C(x)$ over the interval $[100, 200]$, use any antiderivative $F(x)$ of $C'(x)$, and compute $F(200) - F(100)$.

**Method 2: Using a Definite Integral** (based on the interpretation of the definite integral as total change discussed in Section 13.3): Because the marginal cost $C'(x)$ is the rate of change of the total cost function $C(x)$, the total change in $C(x)$ over the interval $[100, 200]$ is given by

Total change in cost $=$ Area under the marginal cost function curve

$$= \int_{100}^{200} C'(x)\,dx$$

$$= \int_{100}^{200} (4 - 0.001x)\,dx \qquad \text{See Figure 20.}$$

$$= \$385. \qquad \text{Using geometry or Riemann sums}$$

**Figure 20**

Putting these two methods together gives us the following surprising result:

$$\int_{100}^{200} C'(x)\, dx = F(200) - F(100),$$

where $F(x)$ is any antiderivative of $C'(x)$.

Now, there is nothing special in Example 1 about the specific function $C'(x)$ or the choice of endpoints of integration. So if we replace $C'(x)$ by a general continuous function $f(x)$, we can write

$$\int_a^b f(x)\, dx = F(b) - F(a),$$

where $F(x)$ is any antiderivative of $f(x)$. This result is known as the **Fundamental Theorem of Calculus**.

## The Fundamental Theorem of Calculus (FTC)

Let $f$ be a continuous function defined on the interval $[a, b]$ and let $F$ be *any* antiderivative of $f$ defined on $[a, b]$. Then

$$\int_a^b f(x)\, dx = F(b) - F(a).$$

Moreover, an antiderivative of $f$ is guaranteed to exist.

**In Words:** Every continuous function has an antiderivative. To compute the definite integral of $f(x)$ over $[a, b]$, first find an antiderivative $F(x)$, then evaluate it at $x = b$, evaluate it at $x = a$, and subtract the two answers.

### Quick Example

Because $F(x) = x^2$ is an antiderivative of $f(x) = 2x$,

$$\int_0^1 2x\, dx = F(1) - F(0) = 1^2 - 0^2 = 1.$$

**Note** The Fundamental Theorem of Calculus actually applies to some other functions besides the continuous ones. The function $f$ is **piecewise continuous** on $[a, b]$ if it is defined and continuous at all but finitely many points in the interval, and at each point where the function is not defined or is discontinuous, the left and right limits of $f$ exist and are finite. (See Figure 21.)

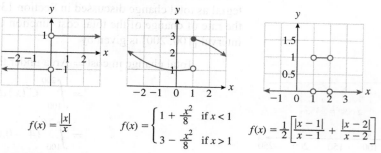

$$f(x) = \frac{|x|}{x} \qquad f(x) = \begin{cases} 1 + \dfrac{x^2}{8} & \text{if } x < 1 \\[2mm] 3 - \dfrac{x^2}{8} & \text{if } x > 1 \end{cases} \qquad f(x) = \frac{1}{2}\left[\frac{|x-1|}{x-1} + \frac{|x-2|}{x-2}\right]$$

**Figure 21**

The FTC also applies to any piecewise continuous function $f$, as long as we specify that the antiderivative $F$ that we choose be continuous. To be precise, to say that $F$ is an antiderivative of $f$ means here that $F'(x) = f(x)$ except at the points at which $f$ is discontinuous or not defined, where $F'(x)$ may not exist. For example, if $f$ is the step function $f(x) = |x|/x$, shown on the left in Figure 21, then we can use $F(x) = |x|$. (Note that $F$ is continuous, and $F'(x) = |x|/x$ except when $x = 0$.) ∎

---

### EXAMPLE 2  Using the FTC to Calculate a Definite Integral

Calculate $\displaystyle\int_0^1 (1 - x^2)\, dx$.

**Solution**  To use the FTC, we need to find an antiderivative of $1 - x^2$. But we know that

$$\int (1 - x^2)\, dx = x - \frac{x^3}{3} + C.$$

We need only one antiderivative, so let's take $F(x) = x - x^3/3$. The FTC tells us that

$$\int_0^1 (1 - x^2)\, dx = F(1) - F(0) = \left(1 - \frac{1}{3}\right) - (0) = \frac{2}{3}$$

which is the value we estimated in Section 13.3.

---

* There seem to be several notations in use, actually. Another common notation is $\left.F(x)\right|_a^b$.

➡ **Before we go on...**  A useful piece of notation is often used here. We write*

$$\left[F(x)\right]_a^b = F(b) - F(a).$$

Thus, we can rewrite the computation in Example 2 more simply as follows:

$$\int_0^1 (1 - x^2)\, dx = \left[x - \frac{x^3}{3}\right]_0^1$$

Substitute $x = 1$.     Substitute $x = 0$.

$$= \left(1 - \frac{1}{3}\right) - \left(0 - \frac{0}{3}\right)$$

$$= \left(1 - \frac{1}{3}\right) - (0) = \frac{2}{3}.$$     ∎

---

### EXAMPLE 3  More Use of the FTC

Compute the following definite integrals.

**a.** $\displaystyle\int_0^1 (2x^3 + 10x + 1)\, dx$          **b.** $\displaystyle\int_1^5 \left(\frac{1}{x^2} + \frac{1}{x}\right) dx$

**Solution**

**a.** $\displaystyle\int_0^1 (2x^3 + 10x + 1)\, dx = \left[\frac{1}{2}x^4 + 5x^2 + x\right]_0^1$

Substitute $x = 1$.     Substitute $x = 0$.

$$= \left(\frac{1}{2} + 5 + 1\right) - \left(\frac{1}{2}(0) + 5(0) + 0\right)$$

$$= \left(\frac{1}{2} + 5 + 1\right) - (0) = \frac{13}{2}$$

**b.** $\int_1^5 \left(\frac{1}{x^2} + \frac{1}{x}\right) dx = \int_1^5 (x^{-2} + x^{-1}) dx$

$\qquad\qquad\qquad\qquad\; = \left[-x^{-1} + \ln|x|\right]_1^5$

$\qquad\qquad\qquad\qquad\qquad$ Substitute $x = 5$.    Substitute $x = 1$.

$\qquad\qquad\qquad\qquad\; = \left(-\frac{1}{5} + \ln 5\right) - (-1 + \ln 1)$

$\qquad\qquad\qquad\qquad\; = \frac{4}{5} + \ln 5$

When calculating a definite integral, we may have to use substitution to find the necessary antiderivative. We could substitute, evaluate the indefinite integral with respect to $u$, express the answer in terms of $x$, and then evaluate at the limits of integration. However, there is a shortcut, as we shall see in the next example.

**EXAMPLE 4    Using the FTC with Substitution**

Evaluate $\int_1^2 (2x - 1)e^{2x^2 - 2x} \, dx$.

**Solution**    The shortcut we promised is to put *everything* in terms of $u$, including the limits of integration.

$u = 2x^2 - 2x$

$\dfrac{du}{dx} = 4x - 2$

$dx = \dfrac{1}{4x - 2} \, du$

When $x = 1, u = 0$.        Substitute $x = 1$ in the formula for $u$.
When $x = 2, u = 4$.        Substitute $x = 2$ in the formula for $u$.

We get the value $u = 0$, for example, by substituting $x = 1$ in the equation $u = 2x^2 - 2x$. We can now rewrite the integral.

$$\int_1^2 (2x - 1)e^{2x^2 - 2x} \, dx = \int_0^4 (2x - 1)e^u \frac{1}{4x - 2} \, du$$

$$= \int_0^4 \frac{1}{2} e^u \, du$$

$$= \left[\frac{1}{2} e^u\right]_0^4 = \frac{1}{2} e^4 - \frac{1}{2}$$

➡ **Before we go on...**    The alternative, longer calculation in Example 4 is first to calculate the indefinite integral:

$$\int (2x - 1)e^{2x^2 - 2x} \, dx = \int \frac{1}{2} e^u \, du$$

$$= \frac{1}{2} e^u + C = \frac{1}{2} e^{2x^2 - 2x} + C.$$

Then we can say that

$$\int_1^2 (2x - 1)e^{2x^2 - 2x}\, dx = \left[\frac{1}{2}e^{2x^2 - 2x}\right]_1^2 = \frac{1}{2}e^4 - \frac{1}{2}.$$    ■

# APPLICATIONS

### EXAMPLE 5  Oil Spill

**using** Technology

**TI-83/84 Plus**
Home Screen:
`fnInt(0.08x^2-4x+60,x,0,20)`
(`fnInt` is `MATH` → `9`)

**Website**
www.WanerMath.com
At the Website, select the Online Utilities tab, and choose the Numerical Integration Utility and Grapher. There, enter the formula

`0.08x^2-4x+60`

for $f(x)$, enter 0 and 20 for the lower and upper limits, and press "Integral".

In Section 13.3 we considered the following example: Your deep ocean oil rig has suffered a catastrophic failure, and oil is leaking from the ocean floor wellhead at a rate of

$$v(t) = 0.08t^2 - 4t + 60 \text{ thousand barrels per day } (0 \le t \le 20),$$

where $t$ is time in days since the failure. Compute the total volume of oil released during the first 20 days.

**Solution**  We calculate

$$\text{Total volume} = \int_0^{20} (0.08t^2 - 4t + 60)\, dt = \left[0.08\frac{t^3}{3} - 2t^2 + 60t\right]_0^{20}$$

$$= \left[0.08\frac{20^3}{3} - 2(20)^2 + 60(20)\right] - [0.08(0) - 2(0) + 60(0)]$$

$$= \frac{640}{3} - 800 + 1{,}200 \approx 613.3 \text{ thousand barrels.}$$

### EXAMPLE 6  Computing Area

Find the total area of the region enclosed by the graph of $y = xe^{x^2}$, the $x$-axis, and the vertical lines $x = -1$ and $x = 1$.

**Solution**  The region whose area we want is shown in Figure 22. Notice the symmetry of the graph. Also, half the region we are interested in is above the $x$-axis, while the other half is below. If we calculated the integral $\int_{-1}^1 xe^{x^2}\, dx$, the result would be

$$\text{Area above } x\text{-axis} - \text{Area below } x\text{-axis} = 0,$$

which does not give us the total area. To prevent the area below the $x$-axis from being combined with the area above the axis, we do the calculation in two parts, as illustrated in Figure 23.

(In Figure 23 we broke the integral at $x = 0$ because that is where the graph crosses the $x$-axis.) These integrals can be calculated using the substitution $u = x^2$:

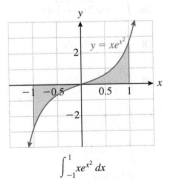

$$\int_{-1}^1 xe^{x^2}\, dx$$

**Figure 22**

$$\int_{-1}^0 xe^{x^2}\, dx = \frac{1}{2}\left[e^{x^2}\right]_{-1}^0 = \frac{1}{2}(1 - e) \approx -0.85914 \qquad \text{Why is it negative?}$$

$$\int_0^1 xe^{x^2}\, dx = \frac{1}{2}\left[e^{x^2}\right]_0^1 = \frac{1}{2}(e - 1) \approx 0.85914$$

$$\int_{-1}^{0} xe^{x^2}\, dx \qquad\qquad \int_{0}^{1} xe^{x^2}\, dx$$

**Figure 23**

To obtain the total area, we should add the *absolute* values of these answers because we don't wish to count any area as negative. Thus,

$$\text{Total area} \approx 0.85914 + 0.85914 = 1.71828.$$

## 13.4  EXERCISES

▼ more advanced    ◆ challenging
T indicates exercises that should be solved using technology

*Evaluate the integrals in Exercises 1–44.* HINT [See Example 2.]

**1.** $\displaystyle\int_{-1}^{1} (x^2 + 2)\, dx$    **2.** $\displaystyle\int_{-2}^{1} (x - 2)\, dx$

**3.** $\displaystyle\int_{0}^{1} (12x^5 + 5x^4 - 6x^2 + 4)\, dx$

**4.** $\displaystyle\int_{0}^{1} (4x^3 - 3x^2 + 4x - 1)\, dx$

**5.** $\displaystyle\int_{-2}^{2} (x^3 - 2x)\, dx$    **6.** $\displaystyle\int_{-1}^{1} (2x^3 + x)\, dx$

**7.** $\displaystyle\int_{1}^{3} \left(\frac{2}{x^2} + 3x\right) dx$    **8.** $\displaystyle\int_{2}^{3} \left(x + \frac{1}{x}\right) dx$

**9.** $\displaystyle\int_{0}^{1} (2.1x - 4.3x^{1.2})\, dx$    **10.** $\displaystyle\int_{-1}^{0} (4.3x^2 - 1)\, dx$

**11.** $\displaystyle\int_{0}^{1} 2e^x\, dx$    **12.** $\displaystyle\int_{-1}^{0} 3e^x\, dx$

**13.** $\displaystyle\int_{0}^{1} \sqrt{x}\, dx$    **14.** $\displaystyle\int_{-1}^{1} \sqrt[3]{x}\, dx$

**15.** $\displaystyle\int_{0}^{1} 2^x\, dx$    **16.** $\displaystyle\int_{0}^{1} 3^x\, dx$

HINT [In Exercises 17–44 use a shortcut or see Example 4.]

**17.** $\displaystyle\int_{0}^{1} 18(3x + 1)^5\, dx$    **18.** $\displaystyle\int_{0}^{1} 8(-x + 1)^7\, dx$

**19.** $\displaystyle\int_{-1}^{1} e^{2x-1}\, dx$    **20.** $\displaystyle\int_{0}^{2} e^{-x+1}\, dx$

**21.** $\displaystyle\int_{0}^{2} 2^{-x+1}\, dx$    **22.** $\displaystyle\int_{-1}^{1} 3^{2x-1}\, dx$

**23.** $\displaystyle\int_{0}^{4} |-3x + 4|\, dx$    **24.** $\displaystyle\int_{-4}^{4} |-x - 2|\, dx$

**25.** $\displaystyle\int_{0}^{1} 5x(8x^2 + 1)^{-1/2}\, dx$    **26.** $\displaystyle\int_{0}^{\sqrt{2}} x\sqrt{2x^2 + 1}\, dx$

**27.** $\displaystyle\int_{-\sqrt{2}}^{\sqrt{2}} 3x\sqrt{2x^2 + 1}\, dx$    **28.** $\displaystyle\int_{-2}^{2} xe^{-x^2+1}\, dx$

**29.** $\displaystyle\int_{0}^{1} 5xe^{x^2+2}\, dx$    **30.** $\displaystyle\int_{0}^{2} \frac{3x}{x^2 + 2}\, dx$

**31.** $\displaystyle\int_{2}^{3} \frac{x^2}{x^3 - 1}\, dx$    **32.** $\displaystyle\int_{2}^{3} \frac{x}{2x^2 - 5}\, dx$

**33.** $\displaystyle\int_{0}^{1} x(1.1)^{-x^2}\, dx$    **34.** $\displaystyle\int_{0}^{1} x^2(2.1)^{x^3}\, dx$

**35.** $\displaystyle\int_{1}^{2} \frac{e^{1/x}}{x^2}\, dx$    **36.** $\displaystyle\int_{1}^{2} \frac{\sqrt{\ln x}}{x}\, dx$

**37.** $\displaystyle\int_{0}^{2} \frac{e^{-2x}}{1 + 3e^{-2x}}\, dx$ (Round the answer to four decimal places.)

**38.** $\displaystyle\int_{0}^{1} \frac{e^{2x}}{1 - 3e^{2x}}\, dx$ (Round the answer to four decimal places.)

**39.** ▼ $\displaystyle\int_0^2 \frac{x}{x+1}\, dx$    **40.** ▼ $\displaystyle\int_{-1}^1 \frac{2x}{x+2}\, dx$

**41.** ▼ $\displaystyle\int_1^2 x(x-2)^5\, dx$    **42.** ▼ $\displaystyle\int_1^2 x(x-2)^{1/3}\, dx$

**43.** ▼ $\displaystyle\int_0^1 x\sqrt{2x+1}\, dx$    **44.** ▼ $\displaystyle\int_{-1}^0 2x\sqrt{x+1}\, dx$

*Calculate the total area of the regions described in Exercises 45–54. Do not count area beneath the x-axis as negative.* HINT [See Example 6.]

**45.** Bounded by the line $y = x$, the $x$-axis, and the lines $x = 0$ and $x = 1$

**46.** Bounded by the line $y = 2x$, the $x$-axis, and the lines $x = 1$ and $x = 2$

**47.** Bounded by the curve $y = \sqrt{x}$, the $x$-axis, and the lines $x = 0$ and $x = 4$

**48.** Bounded by the curve $y = 2\sqrt{x}$, the $x$-axis, and the lines $x = 0$ and $x = 16$

**49.** Bounded by the graph of $y = |2x - 3|$, the $x$-axis, and the lines $x = 0$ and $x = 3$

**50.** Bounded by the graph of $y = |3x - 2|$, the $x$-axis, and the lines $x = 0$ and $x = 3$

**51.** ▼ Bounded by the curve $y = x^2 - 1$, the $x$-axis, and the lines $x = 0$ and $x = 4$

**52.** ▼ Bounded by the curve $y = 1 - x^2$, the $x$-axis, and the lines $x = -1$ and $x = 2$

**53.** ▼ Bounded by the $x$-axis, the curve $y = xe^{x^2}$, and the lines $x = 0$ and $x = (\ln 2)^{1/2}$

**54.** ▼ Bounded by the $x$-axis, the curve $y = xe^{x^2-1}$, and the lines $x = 0$ and $x = 1$

## APPLICATIONS

**55.** *Cost* The marginal cost of producing the $x$th box of light bulbs is $5 + x^2/1{,}000$ dollars. Determine how much is added to the total cost by a change in production from $x = 10$ to $x = 100$ boxes. HINT [See Example 5.]

**56.** *Revenue* The marginal revenue of the $x$th box of flash cards sold is $100e^{-0.001x}$ dollars. Find the revenue generated by selling items 101 through 1,000. HINT [See Example 5.]

**57.** *Motion* A car traveling down a road has a velocity of $v(t) = 60 - e^{-t/10}$ mph at time $t$ hours. Find the distance it has traveled from time $t = 1$ hour to time $t = 6$ hours. (Round your answer to the nearest mile.)

**58.** *Motion* A ball thrown in the air has a velocity of $v(t) = 100 - 32t$ ft/sec at time $t$ seconds. Find the total displacement of the ball between times $t = 1$ second and $t = 7$ seconds, and interpret your answer.

**59.** *Motion* A car slows to a stop at a stop sign, then starts up again, in such a way that its speed at time $t$ seconds after it starts to slow is $v(t) = |-10t + 40|$ ft/sec. How far does the car travel from time $t = 0$ to time $t = 10$ sec?

**60.** *Motion* A truck slows, doesn't quite stop at a stop sign, and then speeds up again in such a way that its speed at time $t$ seconds is $v(t) = 10 + |-5t + 30|$ ft/sec. How far does the truck travel from time $t = 0$ to time $t = 10$?

**61.** *Bottled Water Sales* (Compare Exercise 55 in Section 13.1.) The rate of U.S. sales of bottled water for the period 2000–2010 could be approximated by

$$s(t) = -45t^2 + 900t + 4{,}200 \text{ million gallons per year} \\ (0 \le t \le 10),$$

where $t$ is time in years since the start of 2000.[44] Use a definite integral to estimate the total sales $S(t)$ of bottled water from the start of 2000 to the start of 2008.

**62.** *Bottled Water Sales* (Compare Exercise 56 in Section 13.1.) The rate of U.S. per capita sales of bottled water for the period 2000–2010 could be approximated by

$$s(t) = -0.18t^2 + 3t + 15 \text{ gallons per year} \quad (0 \le t \le 10),$$

where $t$ is time in years since the start of 2000.[45] Use a definite integral to estimate the total U.S. per capita sales of bottled water from the start of 2000 to the end of 2009.

**63.** *Facebook Membership* At the start of 2005, Facebook had 1 million members. Since that time, new members joined at a rate of roughly

$$m(t) = 3.2t^3 - 12t + 10 \text{ million members per year} \\ (0 \le t \le 5),$$

where $t$ is time in years since the start of 2005.[46] Use a definite integral to estimate the total number of new Facebook members from the start of 2007 to the start of 2010.

**64.** *Uploads to YouTube* Since YouTube first became available to the public in mid-2005, the rate at which video has been uploaded to the site can be approximated by

$$v(t) = 525{,}600(0.42t^2 + 2.7t - 1) \text{ hours of video per year} \\ (0.5 \le t \le 7),$$

where $t$ is time in years since the start of 2005.[47] Use a definite integral to estimate the total number of hours of video uploaded from the start of 2006 to the start of 2012. (Round your answer to the nearest million hours of video.)

**65.** *Big Brother* (Compare Exercise 69 in Section 13.3.) The total number of wiretaps authorized each year by U.S. state and federal courts from 1990 to 2010 can be approximated by

$$w(t) = 820e^{0.051t} \quad (0 \le t \le 20)$$

($t$ is time in years since the start of 1990).[48] Compute $\int_0^{15} w(t)\, dt$. (Round your answer to the nearest 10.) Interpret the answer.

———————
[44] Source for data: Beverage Marketing Corporation (www.bottledwater.org).

[45] *Ibid.*

[46] Sources for data: www.facebook.com, www.insidefacebook.com.

[47] Source for data: www.YouTube.com.

[48] Source for data: 2007 Wiretap Report, Administrative Office of the United States Courts (www.uscourts.gov/wiretap07/2007WTText.pdf).

**66. *Big Brother*** (Compare Exercise 64 in Section 13.3.) The number of wiretaps authorized each year by U.S. state courts from 1990 to 2010 can be approximated by

$$w(t) = 430e^{0.065t} \quad (0 \le t \le 20)$$

($t$ is time in years since the start of 1990).[49] Compute $\int_{10}^{15} w(t)\,dt$. (Round your answer to the nearest 10.) Interpret the answer.

**67. *Economic Growth*** (Compare Exercise 85 in Section 13.2.) The value of sold goods in Mexico can be approximated by

$$v(x) = 210 - 62e^{-0.05x} \text{ trillion pesos per month } (x \ge 0),$$

where $x$ is time in months since January 2005.[50] Find an expression for the total value $V(t)$ of sold goods in Mexico from January 2005 to time $t$.

**68. *Economic Contraction*** (Compare Exercise 86 in Section 13.2.) Housing starts in the United States can be approximated by

$$n(x) = 1.1 + 1.2e^{-0.08x} \text{ million homes per year}$$
$$(x \ge 0),$$

where $x$ is time in years since January 2006.[51] Find an expression for the total number $N(t)$ of housing starts in the United States from January 2006 to time $t$.

**69. *Fuel Consumption*** The way Professor Waner drives, he burns gas at the rate of $1 - e^{-t}$ gallons each hour, $t$ hours after a fill-up. Find the number of gallons of gas he burns in the first 10 hours after a fill-up.

**70. *Fuel Consumption*** The way Professor Costenoble drives, he burns gas at the rate of $1/(t+1)$ gallons each hour, $t$ hours after a fill-up. Find the number of gallons of gas he burns in the first 10 hours after a fill-up.

**71. ▼ *Sales*** Weekly sales of your *Lord of the Rings* T-shirts have been falling by 5% per week. Assuming you are now selling 50 T-shirts per week, how many shirts will you sell during the coming year? (Round your answer to the nearest shirt.)

**72. ▼ *Sales*** Annual sales of fountain pens in Littleville are 4,000 per year and are increasing by 10% per year. How many fountain pens will be sold over the next five years?

**73. ▣ *Embryo Development*** The oxygen consumption of a bird embryo increases from the time the egg is laid through the time the chick hatches. In a typical galliform bird, the oxygen consumption can be approximated by

$$c(t) = -0.065t^3 + 3.4t^2 - 22t + 3.6 \text{ milliliters per day}$$
$$(8 \le t \le 30),$$

where $t$ is the time (in days) since the egg was laid.[52] (An egg will typically hatch at around $t = 28$.) Use technology to estimate the total amount of oxygen consumed during the ninth and tenth days ($t = 8$ to $t = 10$). Round your answer to the nearest milliliter. HINT [See the technology note in the margin on page 997.]

**74. ▣ *Embryo Development*** The oxygen consumption of a turkey embryo increases from the time the egg is laid through the time the chick hatches. In a brush turkey, the oxygen consumption can be approximated by

$$c(t) = -0.028t^3 + 2.9t^2 - 44t + 95 \text{ milliliters per day}$$
$$(20 \le t \le 50),$$

where $t$ is the time (in days) since the egg was laid.[53] (An egg will typically hatch at around $t = 50$.) Use technology to estimate the total amount of oxygen consumed during the 21st and 22nd days ($t = 20$ to $t = 22$). Round your answer to the nearest 10 milliliters. HINT [See the technology note in the margin on page 997.]

**75. ▣ *Online Vehicle Sales*** The rate of change $v(t)$ of the value of vehicle transactions on eBay could be approximated by

$$v(t) = -0.002t^3 + 0.06t^2 - 0.55t + 3.9 \text{ billion dollars}$$
$$\text{per quarter } (1 \le t \le 13)$$

($t$ is time in quarters; $t = 1$ represents the first quarter of 2008).[54] Use technology to estimate $\int_1^9 v(t)\,dt$. Interpret your answer.

**76. ▣ *Online Auctions*** The rate of change $n(t)$ of the number of active eBay users could be approximated by

$$n(t) = -0.002t^4 + 0.06t^3 - 0.55t^2 + 1.9t - 1 \text{ million}$$
$$\text{users per quarter } (1 \le t \le 13)$$

($t$ is time in quarters; $t = 1$ represents the first quarter of 2008).[55] Use technology to compute $\int_5^{13} n(t)\,dt$ correct to the nearest whole number. Interpret your answer.

**77. ▼ *Cost*** Use the Fundamental Theorem of Calculus to show that if $m(x)$ is the marginal cost at a production level of $x$ items, then the cost function $C(x)$ is given by

$$C(x) = C(0) + \int_0^x m(t)\,dt.$$

What do we call $C(0)$?

**78. ▼ *Cost*** The total cost of producing $x$ items is given by

$$C(x) = 246.76 + \int_0^x 5t\,dt.$$

Find the fixed cost and the marginal cost of producing the 10th item.

---

[49]Source for data: Wiretap Reports, Administrative Office of the United States Courts (www.uscourts.gov/Statistics/WiretapReports.aspx).

[50]Source: Instituto Nacional de Estadística y Geografía (INEGI) (www.inegi.org.mx).

[51]Source for data: *New York Times*, February 17, 2007, p. C3.

[52]The model approximates graphical data published in the article "The Brush Turkey" by Roger S. Seymour, *Scientific American*, December 1991, pp. 108–114.

[53]*Ibid.*

[54]Source for data: eBay company reports (http://investor.ebay.com).

[55]*Ibid.*

79. *Scientific Research* (Compare Exercise 91 in Section 13.2.) The number of research articles in the prominent journal *Physical Review* written by researchers in Europe can be approximated by

$$E(t) = \frac{7e^{0.2t}}{5 + e^{0.2t}} \text{ thousand articles per year} \quad (t \geq 0),$$

where $t$ is time in years ($t = 0$ represents 1983).[56] Use a definite integral to estimate the number of articles written by researchers in Europe from 1983 to 2003. (Round your answer to the nearest 1,000 articles.) HINT [See Example 5 in Section 13.2.]

80. *Scientific Research* (Compare Exercise 92 in Section 13.2.) The number of research articles in the prominent journal *Physical Review* written by researchers in the United States can be approximated by

$$U(t) = \frac{4.6e^{0.6t}}{0.4 + e^{0.6t}} \text{ thousand articles per year} \quad (t \geq 0),$$

where $t$ is time in years ($t = 0$ represents 1983).[57] Use a definite integral to estimate the total number of articles written by researchers in the United States from 1983 to 2003. (Round your answer to the nearest 1,000 articles.) HINT [See Example 5 in Section 13.2.]

81. ▼ *The Logistic Function and High School Graduates*
   a. Show that the logistic function $f(x) = \dfrac{N}{1 + Ab^{-x}}$ can be written in the form

   $$f(x) = \frac{Nb^x}{A + b^x}.$$

   HINT [See the note after Example 5 in Section 13.2.]
   b. Use the result of part (a) and a suitable substitution to show that

   $$\int \frac{N}{1 + Ab^{-x}} \, dx = \frac{N \ln(A + b^x)}{\ln b} + C.$$

   c. The rate of graduation of private high school students in the United States for the period 1994–2008 was approximately

   $$r(t) = 220 + \frac{110}{1 + 3.8(1.27)^{-t}} \text{ thousand students per year}$$
   $$(0 \leq t \leq 14)$$

   $t$ years since 1994.[58] Use the result of part (b) to estimate the total number of private high school graduates over the period 2000–2008.

82. ▼ *The Logistic Function and Grant Spending*
   a. Show that the logistic function $f(x) = \dfrac{N}{1 + Ae^{-kx}}$ can be written in the form

   $$f(x) = \frac{Ne^{kx}}{A + e^{kx}}.$$

   HINT [See the note after Example 5 in Section 13.2.]

   b. Use the result of part (a) and a suitable substitution to show that

   $$\int \frac{N}{1 + Ae^{-kx}} \, dx = \frac{N \ln(A + e^{kx})}{k} + C.$$

   c. The rate of spending on grants by U.S. foundations in the period 1993–2003 was approximately

   $$s(t) = 11 + \frac{20}{1 + 1{,}800e^{-0.9t}} \text{ billion dollars per year}$$
   $$(3 \leq t \leq 13),$$

   where $t$ is the number of years since 1990.[59] Use the result of part (b) to estimate, to the nearest $10 billion, the total spending on grants from 1998 to 2003.

83. ◆ *Kinetic Energy* The work done in accelerating an object from velocity $v_0$ to velocity $v_1$ is given by

   $$W = \int_{v_0}^{v_1} v \frac{dp}{dv} \, dv,$$

   where $p$ is its momentum, given by $p = mv$ ($m = $ mass). Assuming that $m$ is a constant, show that

   $$W = \frac{1}{2}mv_1^2 - \frac{1}{2}mv_0^2.$$

   The quantity $\frac{1}{2}mv^2$ is referred to as the **kinetic energy** of the object, so the work required to accelerate an object is given by its change in kinetic energy.

84. ◆ *Einstein's Energy Equation* According to the special theory of relativity, the apparent mass of an object depends on its velocity according to the formula

   $$m = \frac{m_0}{\left(1 - \dfrac{v^2}{c^2}\right)^{1/2}},$$

   where $v$ is its velocity, $m_0$ is the "rest mass" of the object (that is, its mass when $v = 0$), and $c$ is the velocity of light: approximately $3 \times 10^8$ meters per second.

   a. Show that, if $p = mv$ is the momentum,

   $$\frac{dp}{dv} = \frac{m_0}{\left(1 - \dfrac{v^2}{c^2}\right)^{3/2}}.$$

   b. Use the integral formula for $W$ in the preceding exercise, together with the result in part (a) to show that the work required to accelerate an object from a velocity of $v_0$ to $v_1$ is given by

   $$W = \frac{m_0 c^2}{\sqrt{1 - \dfrac{v_1^2}{c^2}}} - \frac{m_0 c^2}{\sqrt{1 - \dfrac{v_0^2}{c^2}}}.$$

---

[56]Based on data from 1983 to 2003. Source: The American Physical Society/*New York Times*, May 3, 2003, p. A1.

[57]*Ibid.*

[58]Based on a logistic regression. Source for data: National Center for Educational Statistics (www.nces.ed.gov).

[59]Based on a logistic regression. Source for data: The Foundation Center, *Foundation Growth and Giving Estimates*, 2004, downloaded from the Center's Web site (www.fdncenter.org).

We call the quantity $\dfrac{m_0 c^2}{\sqrt{1 - \frac{v^2}{c^2}}}$ the **total relativistic energy** of an object moving at velocity $v$. Thus, the work to accelerate an object from one velocity to another is given by the change in its total relativistic energy.

**c.** Deduce (as Albert Einstein did) that the total relativistic energy $E$ of a body at rest with rest mass $m$ is given by the famous equation

$$E = mc^2.$$

## COMMUNICATION AND REASONING EXERCISES

**85.** Explain how the indefinite integral and the definite integral are related.

**86.** What is "definite" about the definite integral?

**87.** The total change of a quantity from time $a$ to time $b$ can be obtained from its rate of change by doing what?

**88.** Complete the following: The total sales from time $a$ to time $b$ are obtained from the marginal sales by taking its _____ _____ from _____ to _____ .

**89.** What does the Fundamental Theorem of Calculus permit one to do?

**90.** If Felice and Philipe have different antiderivatives of $f$ and each uses his or her own antiderivative to compute $\int_a^b f(x)\,dx$, they might get different answers, right?

**91.** ▼ Give an example of a nonzero velocity function that will produce a displacement of 0 from time $t = 0$ to time $t = 10$.

**92.** ▼ Give an example of a nonzero function whose definite integral over the interval [4, 6] is zero.

**93.** ▼ Give an example of a decreasing function $f(x)$ with the property that $\int_a^b f(x)\,dx$ is positive for every choice of $a$ and $b > a$.

**94.** ▼ Explain why, in computing the total change of a quantity from its rate of change, it is useful to have the definite integral subtract area below the $x$-axis.

**95.** ◆ If $f(x)$ is a continuous function defined for $x \geq a$, define a new function $F(x)$ by the formula

$$F(x) = \int_a^x f(t)\,dt.$$

Use the Fundamental Theorem of Calculus to deduce that $F'(x) = f(x)$. What, if anything, is interesting about this result?

**96.** ▊ ◆ Use the result of Exercise 95 and technology to compute a table of values for $x = 1, 2, 3$ for an antiderivative $A(x)$ of $e^{-x^2}$ with the property that $A(0) = 0$. (Round answers to two decimal places.)

# CHAPTER 13 REVIEW

## KEY CONCEPTS

 Website www.WanerMath.com
Go to the Website at www.WanerMath
.com to find a comprehensive and
interactive Web-based summary
of Chapter 13.

### 13.1 The Indefinite Integral
An antiderivative of a function $f$ is a
function $F$ such that $F' = f$. *p. 946*
Indefinite integral $\int f(x)\,dx$  *p. 946*
Power rule for the indefinite integral:

$$\int x^n\,dx = \frac{x^{n+1}}{n+1} + C$$

$$(\text{if } n \neq -1)\ \ p.\ 948$$

$$\int x^{-1}\,dx = \ln|x| + C\ \ p.\ 948$$

Indefinite integral of $e^x$ and $b^x$:

$$\int e^x\,dx = e^x + C$$

$$\int b^x\,dx = \frac{b^x}{\ln b} + C\ \ p.\ 949$$

Indefinite integral of $|x|$:

$$\int |x|\,dx = \frac{x|x|}{2} + C\ \ p.\ 949$$

Sums, differences, and constant
multiples:

$$\int [f(x) \pm g(x)]\,dx$$

$$= \int f(x)\,dx \pm \int g(x)\,dx$$

$$\int k f(x)\,dx = k \int f(x)\,dx$$

$$(k\ \text{constant})\ \ p.\ 949$$

Combining the rules  *p. 951*
Position, velocity, and acceleration:

$$v = \frac{ds}{dt} \qquad s(t) = \int v(t)\,dt$$

$$a = \frac{dv}{dt} \qquad v(t) = \int a(t)\,dt$$

*pp. 954–955*

Motion in a straight line  *p. 955*
Vertical motion under gravity  *p. 957*

### 13.2 Substitution
Substitution rule:

$$\int f\,dx = \int \left(\frac{f}{du/dx}\right) du\ \ p.\ 962$$

Using the substitution rule  *p. 963*
Shortcuts: integrals of expressions
involving $(ax + b)$:

$$\int (ax+b)^n\,dx = \frac{(ax+b)^{n+1}}{a(n+1)} + C$$

$$(\text{if } n \neq -1)$$

$$\int (ax+b)^{-1}\,dx = \frac{1}{a}\ln|ax+b| + C$$

$$\int e^{ax+b}\,dx = \frac{1}{a}e^{ax+b} + C$$

$$\int c^{ax+b}\,dx = \frac{1}{a\ln c}c^{ax+b} + C$$

$$\int |ax+b|\,dx$$

$$= \frac{1}{2a}(ax+b)|ax+b| + C\ \ p.\ 968$$

Mike's shortcut rule:

If $\int f(x)\,dx = F(x) + C$ and $g$ and $u$

are differentiable functions of $x$, then

$$\int g \cdot f(u)\,dx = \frac{g}{u'} \cdot F(u) + C$$

provided that $\dfrac{g}{u'}$ is constant.  *p. 969*

### 13.3 The Definite Integral: Numerical and Graphical Viewpoints
Left Riemann sum:

$$\sum_{k=0}^{n-1} f(x_k)\Delta x$$

$$= [f(x_0) + f(x_1) + \cdots + f(x_{n-1})]\Delta x$$

*p. 978*

Computing the Riemann sum from a
formula  *p. 980*
Computing the Riemann sum from a
graph  *p. 980*
Definite integral of $f$ from $a$ to $b$:

$$\int_a^b f(x)\,dx = \lim_{n\to\infty} \sum_{k=0}^{n-1} f(x_k)\Delta x.$$

*p. 981*

Estimating the definite integral from a
graph  *p. 983*
Estimating the definite integral using
technology  *p. 985*
Application to motion in a straight line
*p. 985*

### 13.4 The Definite Integral: Algebraic Viewpoint and the Fundamental Theorem of Calculus
Computing total cost from marginal
cost  *p. 993*
The Fundamental Theorem of Calculus
(FTC)  *p. 994*
Using the FTC to compute definite
integrals  *p. 995*
Using the FTC with substitution  *p. 996*
Computing area  *p. 997*

## REVIEW EXERCISES

*Evaluate the indefinite integrals in Exercises 1–18.*

**1.** $\int (x^2 - 10x + 2)\,dx$

**2.** $\int (e^x + \sqrt{x})\,dx$

**3.** $\int \left(\frac{4x^2}{5} - \frac{4}{5x^2}\right) dx$

**4.** $\int \left(\frac{3x}{5} - \frac{3}{5x}\right) dx$

**5.** $\int (2x)^{-1}\,dx$

**6.** $\int (-2x + 2)^{-2}\,dx$

**7.** $\int e^{-2x+11}\,dx$

**8.** $\int \frac{dx}{(4x-3)^2}$

**9.** $\int x(x^2 + 1)^{1.3}\,dx$

**10.** $\int x(x^2 + 4)^{10}\,dx$

**11.** $\int \frac{4x}{(x^2-7)}\,dx$

**12.** $\int \frac{x}{(3x^2-1)^{0.4}}\,dx$

**13.** $\int (x^3 - 1)\sqrt{x^4 - 4x + 1}\,dx$

1003

**14.** $\int \dfrac{x^2+1}{(x^3+3x+2)^2}\,dx$

**15.** $\int (-xe^{x^2/2})\,dx$

**16.** $\int xe^{-x^2/2}\,dx$

**17.** $\int \dfrac{x+1}{x+2}\,dx$

**18.** $\int x\sqrt{x-1}\,dx$

*In Exercises 19 and 20, use the given graph to estimate the left Riemann sum for the given interval with the stated number of subdivisions.*

**19.** $[0,3]$, $n=6$

**20.** $[1,3]$, $n=4$

*Calculate the left Riemann sums for the given functions over the given interval in Exercises 21–24, using the given values of n. (When rounding, round answers to four decimal places.)*

**21.** $f(x)=x^2+1$ over $[-1,1]$, $n=4$

**22.** $f(x)=(x-1)(x-2)-2$ over $[0,4]$, $n=4$

**23.** $f(x)=x(x^2-1)$ over $[0,1]$, $n=5$

**24.** $f(x)=\dfrac{x-1}{x-2}$ over $[0,1.5]$, $n=3$

⊤ *In Exercises 25 and 26, use technology to approximate the given definite integrals using left Riemann sums with $n=10$, 100, and 1,000. (Round answers to four decimal places.)*

**25.** $\int_0^1 e^{-x^2}\,dx$

**26.** $\int_1^3 x^{-x}\,dx$

*In Exercises 27 and 28, the graph of the derivative $f'(x)$ of $f(x)$ is shown. Compute the total change of $f(x)$ over the given interval.*

**27.** $[-1,2]$

**28.** $[0,2]$

*Evaluate the definite integrals in Exercises 29–38, using the Fundamental Theorem of Calculus.*

**29.** $\int_{-1}^1 (x-x^3+|x|)\,dx$

**30.** $\int_0^9 (x+\sqrt{x})\,dx$

**31.** $\int_{-1}^1 \dfrac{3}{(2x-5)^2}\,dx$

**32.** $\int_0^9 \dfrac{1}{x+1}\,dx$

**33.** $\int_0^{50} e^{-0.02x-1}\,dx$

**34.** $\int_{-20}^0 3e^{2.2x}\,dx$

**35.** $\int_0^2 x^2\sqrt{x^3+1}\,dx$

**36.** $\int_0^2 \dfrac{x^2}{\sqrt{x^3+1}}\,dx$

**37.** $\int_0^{\ln 2} \dfrac{e^{-2x}}{1+4e^{-2x}}\,dx$

**38.** $\int_0^{\ln 3} e^{2x}(1-3e^{2x})^2\,dx$

*In Exercises 39–42, find the areas of the specified regions. (Do not count area below the x-axis as negative.)*

**39.** The area bounded by $y=4-x^2$, the $x$-axis, and the lines $x=-2$ and $x=2$

**40.** The area bounded by $y=4-x^2$, the $x$-axis, and the lines $x=0$ and $x=5$

**41.** The area bounded by $y=xe^{-x^2}$, the $x$-axis, and the lines $x=0$ and $x=5$

**42.** The area bounded by $y=|2x|$, the $x$-axis, and the lines $x=-1$ and $x=1$

## APPLICATIONS: OHaganBooks.com

**43.** *Sales* The rate of net sales (sales minus returns) of *The Secret Loves of John O*, a romance novel by Margó Dufón, can be approximated by

$$n(t)=196+t^2-0.16t^5 \text{ copies per week}$$

$t$ weeks since its release.

**a.** Find the total net sales $N$ as a function of time $t$.

**b.** How many books are still held by customers after 6 weeks? (Round your answer to the nearest book.)

**44.** *Demand* If OHaganBooks.com were to give away its latest bestseller, *A River Burns through It*, the demand $q$ would be 100,000 books. The marginal demand $(dq/dp)$ for the book is $-20p$ at a price of $p$ dollars.

**a.** What is the demand function for this book?

**b.** At what price does demand drop to zero?

**45.** *Motion under Gravity* Billy-Sean O'Hagan's friend Juan (Billy-Sean is John O'Hagan's son, currently a senior in college) says he can throw a baseball vertically upward at 100 feet per second. Assuming Juan's claim is true,

**a.** Where would the baseball be at time $t$ seconds?

**b.** How high would the ball go?

**c.** When would it return to Juan's hand?

**46.** *Motion under Gravity* An overworked employee at OHaganBooks.com goes to the top of the company's 100-foot-tall headquarters building and flings a book up into the air at a speed of 60 feet per second.

**a.** When will the book hit the ground 100 feet below? (Neglect air resistance.)

**b.** How fast will it be traveling when it hits the ground?

**c.** How high will the book go?

**47. Sales** Sales at the OHaganBooks.com Web site of *Larry Potter and the Riemann Sum* fluctuated rather wildly in the first 5 months of last year as the following graph shows:

Puzzled by the graph, CEO John O'Hagan asks Jimmy Duffin[60] to estimate the total sales over the entire 5-month period shown. Jimmy decides to use a left Riemann sum with 10 partitions to estimate the total sales. What does he find?

**48. Sales** The following graph shows the approximate rate of change $s(t)$ of the total value, in thousands of dollars, of Spanish books sold online at OHaganBooks.com ($t$ is the number of months since January 1):

Use the graph to estimate the total value of Spanish books sold from March 1 through June 1. (Use a left Riemann sum with three subdivisions.)

**49. Promotions** Unlike sales of *Larry Potter and the Riemann Sum*, sales at OHaganBooks.com of the special leather-bound gift editions of *Calculus for Vampires* have been suffering lately, as shown in the following graph (negative sales indicate returns by dissatisfied customers; $t$ is time in months since January 1 of this year):

Use the graph to compute the total (net) sales over the period shown.

**50. Sales** Even worse than with the leather-bound *Calculus for Vampires*, sales of *Real Estate for Werewolves* have been dismal, as shown in the following graph (negative sales

indicate returns by dissatisfied customers; $t$ is time in months since January 1 of this year):

Use the graph to compute the total (net) sales over the period shown.

**51. Web Site Activity** The number of "hits" on the OHaganBooks.com Web site has been steadily increasing over the past month in response to recent publicity over a software glitch that caused the company to pay customers for buying books online. The activity can be modeled by

$$n(t) = 1,000t - 10t^2 + t^3 \text{ hits per day,}$$

where $t$ is time in days since news about the software glitch was first publicized on GrungeReport.com. Use a left Riemann sum with five partitions to estimate the total number of hits during the first 10 days of the period.

**52. Web Site Crashes** The latest DoorsXL servers OHaganBooks.com has been using for its Web site have been crashing with increasing frequency lately. One of the student summer interns has estimated the number of crashes to be

$$q(t) = 0.05t^2 + 0.4t + 9 \text{ crashes per week} \quad (0 \le t \le 10),$$

where $t$ is the number of weeks since the DoorsXL system was first installed. Use a Riemann sum with five partitions to estimate the total number of crashes from the start of week 5 to the start of week 10. (Round your answer to the nearest crash.)

**53. Student Intern Costs** The marginal monthly cost of maintaining a group of summer student interns at OHaganBooks.com is calculated to be

$$c(x) = \frac{1,000(x+3)^2}{(8 + (x+3)^3)^{3/2}} \text{ thousand dollars per additional student.}$$

Compute, to the nearest $100, the total monthly cost if O'HaganBooks increases the size of the student intern program from five students to seven students.

**54. Legal Costs** The legal team maintained by OHaganBooks.com to handle the numerous lawsuits brought against the company by disgruntled clients may have to be expanded. The marginal monthly cost to maintain a team of $x$ lawyers is estimated (by a method too complicated to explain) as

$$c(x) = (x-2)^2[8 - (x-2)^3]^{3/2} \text{ thousand dollars per additional lawyer.}$$

Compute, to the nearest $1,000, the total monthly cost if O'HaganBooks goes ahead with a proposal to increase the size of the legal team from two to four.

[60]Marjory Duffin's nephew, currently at OHaganBooks.com on a summer internship.

55. *Projected Sales* When OHaganBooks.com was about to go online, it estimated that its weekly sales would begin at about 6,400 books per week, with sales increasing at such a rate that weekly sales would double about every 2 weeks. If these estimates had been correct, how many books would the company have sold in the first 5 weeks? (Round your answer to the nearest 1,000 books.)

56. *Projected Sales* Once OHaganBooks.com actually went online, its weekly sales began at about 7,500 books per week, with weekly sales doubling every 3 weeks. How many books did the company actually sell in the first 5 weeks? (Round your answer to the nearest 1,000 books.)

57. *Actual Sales* OHaganBooks.com modeled its revised weekly sales over a period of time after it went online with the function

$$s(t) = 6,053 + \frac{4,474e^{0.55t}}{e^{0.55t} + 14.01},$$

where $t$ is the time in weeks after it went online. According to this model, how many books did it actually sell in the first 5 weeks?

58. *Computer Usage* A consultant recently hired by OHaganBooks.com estimates total weekly computer usage as

$$w(t) = 620 + \frac{900e^{0.25t}}{3 + e^{0.25t}} \text{ hours} \quad (0 \le t \le 20),$$

where $t$ is time in weeks since January 1 of this year. Use the model to estimate the total computer usage during the first 14 weeks of the year.

Bill Varie/Flirt/Corbis

**Case Study**

## Spending on Housing Construction

It is March 2007, and *Time* magazine, in its latest edition, is asking "Will the Housing Bubble Burst in 2007?"[61] You are a summer intern at *Ronald Ramp Real Estate Development, Inc.,* which is considering a major luxury condominium development in Tampa, Florida nicknamed the "Ramp Towers Tampa."

Yesterday, you received the following memo from your supervisor:

**DATE:** March 15, 2007
**TO:** SW@EnormousStateU.edu
**FROM:** SC@RonaldRampDev.com (Junior VP Development)
**CC:** SGLombardoVP@RonaldRampDev.com (S. G. Lombardo, Senior VP Development)
**SUBJECT:** Residential Construction Trends. Urgent!

Help! There is a management meeting in two hours and Michelle Homestead, who, as you know, is spearheading the latest Ramp Towers Tampa feasibility study, must report to "the Ronald" by tomorrow and has asked me to immediately produce some mathematical formulas to (1) model the trend in residential construction spending since January 2006, when it was $618.7 billion, and (2) estimate the average spent per month on residential construction over a specified period of time. All I have on hand so far is data giving the month-over-month percentage changes (attached). Do you have any ideas?

[61]*Time,* February, 2007 (www.time.com/time/business/article/0,8599,1592751,00.html).

**ATTACHMENT\***

| Month | % Change | Month | % Change |
|-------|----------|-------|----------|
| 1 | 1.16 | 16 | −1.59 |
| 2 | 1.17 | 17 | −1.62 |
| 3 | 1.33 | 18 | −1.58 |
| 4 | 0.67 | 19 | −1.67 |
| 5 | 0.42 | 20 | −1.77 |
| 6 | −0.04 | 21 | −1.92 |
| 7 | −0.24 | 22 | −2.06 |
| 8 | −0.44 | 23 | −2.19 |
| 9 | −0.70 | 24 | −2.45 |
| 10 | −0.94 | 25 | −2.50 |
| 11 | −1.31 | 26 | −2.65 |
| 12 | −1.34 | 27 | −2.85 |
| 13 | −1.49 | 28 | −2.65 |
| 14 | −1.75 | 29 | −2.66 |
| 15 | −1.74 | 30 | −2.66 |

\* Based on 12-month moving average; Source for data: U.S. Census Bureau: Manufacturing, Mining and Construction Statistics, Data 360 (www.data360.org/dataset.aspx? Data_Set_Id=3627).

Getting to work, you decide that the first thing to do is fit these data to a mathematical curve that you can use to project future changes in construction spending. You graph the data to get a sense of what mathematical models might be appropriate (Figure 24).

**Figure 24**

The graph suggests a decreasing trend, possibly concave up. You recall that there are a variety of curves that can behave this way, one of the simplest being

$$y = at^c + b \quad (t > 0),$$

where $a$, $b$, and $c$ are constants.

You convert all the percentages to decimals, giving the following table of data:

| t | y | t | y |
|---|---|---|---|
| 1 | 0.0116 | 16 | −0.0159 |
| 2 | 0.0117 | 17 | −0.0162 |
| 3 | 0.0133 | 18 | −0.0158 |
| 4 | 0.0067 | 19 | −0.0167 |
| 5 | 0.0042 | 20 | −0.0177 |
| 6 | −0.0004 | 21 | −0.0192 |
| 7 | −0.0024 | 22 | −0.0206 |
| 8 | −0.0044 | 23 | −0.0219 |
| 9 | −0.0070 | 24 | −0.0245 |
| 10 | −0.0094 | 25 | −0.0250 |
| 11 | −0.0131 | 26 | −0.0265 |
| 12 | −0.0134 | 27 | −0.0285 |
| 13 | −0.0149 | 28 | −0.0265 |
| 14 | −0.0175 | 29 | −0.0266 |
| 15 | −0.0174 | 30 | −0.0266 |

✱ To do this, you can use Excel's Solver or the Website function evaluator and grapher with model `$1*x^$2+$3`.

You then find the values of $a$, $b$, and $c$ that best fit the given data:✱

$$a = -0.0200974, \ b = 0.0365316, \ c = 0.345051.$$

These values give you the following model for construction spending (with figures rounded to five significant digits):

$$y = -0.020097t^{0.34505} + 0.036532.$$

Figure 25 shows the graph of $y$ superimposed on the data.

**Figure 25**

Now that you have a model for the month-over-month change in construction spending, you must use it to find the actual spending on construction. First, you realize that the model gives the *fractional rate of increase* of construction spending (because it is specified as a percentage, or fraction, of the total spending). In other words, if $p(t)$ represents the construction cost in month $t$, then

$$y = \frac{dp/dt}{p} = \frac{d}{dt}(\ln p). \qquad \text{By the chain rule for derivatives}$$

You find an equation for actual monthly construction cost at time $t$ by solving for $p$:

$$\ln p = \int y\,dt$$
$$= \int (at^c + b)\,dt$$
$$= \frac{at^{c+1}}{c+1} + bt + K$$
$$= dt^{c+1} + bt + K,$$

where

$$d = \frac{a}{c+1} = \frac{-0.020097}{0.34505 + 1} \approx -0.014941,$$

$b$ and $c$ are as above, and $K$ is the constant of integration. So

$$p(t) = e^{dt^{c+1}+bt+K}.$$

To compute $K$, you substitute the initial data from the memo: $p(1) = 618.7$. Thus,

$$618.7 = e^{d+b+K} = e^{-0.014941+0.036532+K} = e^{0.021591+K}.$$

Thus,

$$\ln(618.7) = 0.021591 + K,$$

which gives

$$K = \ln(618.7) - 0.021591 \approx 6.4060 \text{ (to five significant digits)}.$$

Now you can write down the following formula for the monthly spending on residential construction as a function of $t$, the number of months since the beginning of 2006:

$$p(t) = e^{dt^{c+1}+bt+K} = e^{-0.014941t^{0.34505+1}+0.036532t+6.4060}.$$

What remains is the calculation of the average spent per month over a specified period $[r, s]$. Since $p$ is the rate of change of the total spent, the total spent on housing construction over this period is

$$P = \int_r^s p(t)\,dt$$

and so the average spent per month is

$$\bar{P} = \frac{1}{s-r} \int_r^s p(t)\,dt. \qquad \frac{1}{\text{Number of months}} \times \text{Total spent}$$

Substituting the formula for $p(t)$ gives

$$\bar{P} = \frac{1}{s-r} \int_r^s e^{-0.014941t^{0.34505+1}+0.036532t+6.4060}\,dt.$$

You cannot find an explicit antiderivative for the integrand, so you decide that the only way to compute it is numerically. You send the following memo to SC.

**DATE:** March 15, 2007

**TO:** SC@RonaldRampDev.com (Junior VP Development)

**FROM:** SW@EnormousStateU.edu

**CC:** SGLombardoVP@RonaldRampDev.com (S. G. Lombardo, Senior VP Development)

**SUBJECT:** The formula you wanted

Spending in the U.S. on housing construction in the $t$th month of 2006 can be modeled by

$$p(t) = e^{-0.014941t^{0.34505+1}+0.036532t+6.4060} \text{ million dollars.}$$

Further, the average spent per month from month $r$ to month $s$ (since the start of January 2006) can be computed as

$$\bar{P} = \frac{1}{s-r}\int_r^s e^{-0.014941t^{0.34505+1}+0.036532t+6.4060}\,dt$$

To calculate it easily (and impress the subcommittee members), I suggest you have a graphing calculator on hand and enter the following on your graphing calculator (watch the parentheses!):

`Y₁=1/(S-R)*fnInt(e^(-0.014941T^(0.34505+1)+0.036532T+6.4060),T,R,S)`

Then suppose, for example, you need to estimate the average for the period March 1, 2006 ($t = 3$) to February 1, 2007 ($t = 14$). All you do is enter

`3→R`
`14→S`
`Y₁`

and your calculator will give you the result: The average spending was $628 million per month.

Good luck with the meeting!

## EXERCISES

**1.** Use the actual January 2006 spending figure of $618.7 million and the percentage changes in the table to compute the actual spending in February, March, and April of that year. Also use the model of monthly spending to estimate those figures, and compare the predicted values with the actual figures. Is it unacceptable that the April figures agree to only one significant digit? Explain.

**2.** Use the model developed above to estimate the average monthly spending on residential construction over the 12-month period beginning June 1, 2006. (Round your answer to the nearest $1 million.)

**3.** What (if any) advantages are there to using a model for residential construction spending when the actual residential construction spending figures are available?

**4.** The formulas for $p(t)$ and $\bar{P}$ were based on the January 2006 spending figure of $618.7 million. Change the models to allow for a possibly revised January 2006 spending figure of $p_0$ million.

**5.** If we had used quadratic regression to model the construction spending data, we would have obtained

$$y = 0.00005t^2 - 0.0028t + 0.0158.$$

(See the graph.)

Use this formula and the given January 2006 spending figure to obtain corresponding models for $p(t)$ and $\bar{P}$.

6. Compare the model in the text with the quadratic model in Exercise 5 in terms of both short- and long-term predictions; in particular, when does the quadratic model predict construction spending will have reached its biggest monthly decrease? Are either of these models realistic in the near term? in the long term?

TECHNOLOGY GUIDE

## TI-83/84 Plus    **Technology Guide**

### Section 13.3

**Example 5 (page 985)**    Estimate the area under the graph of $f(x) = 1 - x^2$ over the interval $[0, 1]$ using $n = 100$, $n = 200$, and $n = 500$ partitions.

#### Solution with Technology

There are several ways to compute Riemann sums with a graphing calculator. We illustrate one method. For $n = 100$, we need to compute the sum

$$\sum_{k=0}^{99} f(x_k)\Delta x = [f(0) + f(0.01) + \cdots + f(0.99)](0.01).$$

See discussion in Example 5.

Thus, we first need to calculate the numbers $f(0)$, $f(0.01)$, and so on, and add them up. The TI-83/84 Plus has a built-in sum function (available in the LIST MATH menu), which, like the SUM function in a spreadsheet, sums the entries in a list.

1. To generate a list that contains the numbers we want to add together, use the seq function (available in the LIST OPS menu). If we enter

    seq(1-X^2,X,0,0.99,0.01)

    seq:  2ND   LIST   OPS   5

    the calculator will calculate a list by evaluating 1-X^2 for values of X from 0 to 0.99 in steps of 0.01.

2. To take the sum of all these numbers, we wrap the seq function in a call to sum:

    sum(seq(1-X^2,X,0,0.99,0.01))

    sum:  2ND   LIST   MATH   5

    This gives the sum

    $$f(0) + f(0.01) + \cdots + f(0.99) = 67.165.$$

3. To obtain the Riemann sum, we need to multiply this sum by $\Delta x = 0.01$, and we obtain the estimate of $67.165 \times 0.01 = 0.67165$ for the Riemann sum:

We obtain the other Riemann sums similarly, as shown here:

$n = 200$                $n = 500$

One disadvantage of this method is that the TI-83/84 Plus can generate and sum a list of at most 999 entries. The LEFTSUM program below calculates left Riemann sums for any $n$. The TI-83/84 Plus also has a built-in function fnInt, which finds a very accurate approximation of a definite integral, using a more sophisticated technique than the one we are discussing here.

### The LEFTSUM program for the TI-83/84 Plus

The following program calculates (left) Riemann sums for any $n$. The latest version of this program (and others) is available at the Website.

```
PROGRAM: LEFTSUM
:Input "LEFT ENDPOINT? ",A
```
Prompts for the left endpoint $a$
```
:Input "RIGHT ENDPOINT? ",B
```
Prompts for the right endpoint $b$
```
:Input "N? ",N
```
Prompts for the number of rectangles
```
:(B-A)/N→D
```
D is $\Delta x = (b-a)/n$.
```
:∅→L
```
L will eventually be the left sum.
```
:A→X
```
X is the current $x$-coordinate.
```
:For(I,1,N)
```
Start of a loop—recall the sigma notation.
```
:L+Y₁→L
```
Add $f(x_{i-1})$ to L.
```
:A+I*D→X
```
Uses formula $x_i = a + i\Delta x$
```
:End
```
End of loop
```
:L*D→L
```
Multiply by $\Delta x$.
```
:Disp "LEFT SUM IS ",L
:Stop
```

# SPREADSHEET Technology Guide

## Section 13.3

**Example 5 (page 985)** Estimate the area under the graph of $f(x) = 1 - x^2$ over the interval $[0, 1]$ using $n = 100$, $n = 200$, and $n = 500$ partitions.

### Solution with Technology

We need to compute various sums.

$$\sum_{k=0}^{99} f(x_k)\Delta x = [f(0) + f(0.01) + \cdots + f(0.99)](0.01)$$

See discussion in Example 5.

$$\sum_{k=0}^{199} f(x_k)\Delta x = [f(0) + f(0.005) + \cdots + f(0.995)](0.005)$$

$$\sum_{k=0}^{499} f(x_k)\Delta x = [f(0) + f(0.002) + \cdots + f(0.998)](0.002).$$

Here is how you can compute them all on the same spreadsheet.

1. Enter the values for the endpoints $a$ and $b$, the number of subdivisions $n$, and the formula $\Delta x = (b - a)/n$:

| | A | B | C | D |
|---|---|---|---|---|
| 1 | x | f(x) | a | 0 |
| 2 | | | b | 1 |
| 3 | | | n | 100 |
| 4 | | | Delta x | =(D2-D1)/D3 |

2. Next, we compute all the $x$-values we might need in column A. Because the largest value of $n$ that we will be using is 500, we will need a total of 501 values of $x$. Note that the value in each cell below A3 is obtained from the one above by adding $\Delta x$.

| | A | B | C | D |
|---|---|---|---|---|
| 1 | x | f(x) | a | 0 |
| 2 | =D1 | | b | 1 |
| 3 | =A2+$D$4 | | n | 100 |
| 4 | | | Delta x | 0.01 |
| 5 | | | | |
| 501 | | | | |
| 502 | | | | |

| | A | B | C | D |
|---|---|---|---|---|
| 1 | x | f(x) | a | 0 |
| 2 | 0 | | b | 1 |
| 3 | 0.01 | | n | 100 |
| 4 | 0.02 | | Delta x | 0.01 |
| 5 | 0.03 | | | |
| 501 | 4.99 | | | |
| 502 | 5 | | | |

(The fact that the values of $x$ currently go too far will be corrected in the next step.)

3. We need to calculate the numbers $f(0)$, $f(0.01)$, and so on, but only those for which the corresponding $x$-value is less than $b$. To do this, we use a logical formula like we did with piecewise-defined functions in Chapter 1:

| | A | B | C | D |
|---|---|---|---|---|
| 1 | x | f(x) | a | 0 |
| 2 | 0 | =(1-A2^2)*(A2<$D$2) | b | 1 |
| 3 | 0.01 | | n | 100 |
| 4 | 0.02 | | Delta x | 0.01 |
| 5 | 0.03 | | | |
| 501 | 4.99 | | | |
| 502 | 5 | | | |

When the value of $x$ is $b$ or above, the function will evaluate to zero, because we do not want to count it.

4. Finally, we compute the Riemann sum by adding up everything in column B and multiplying by $\Delta x$:

| | A | B | C | D |
|---|---|---|---|---|
| 1 | x | f(x) | a | 0 |
| 2 | 0 | 1 | b | 1 |
| 3 | 0.01 | 0.9999 | n | 100 |
| 4 | 0.02 | 0.9996 | Delta x | 0.01 |
| 5 | 0.03 | 0.9991 | Left Sum | =SUM(B:B)*D4 |
| 6 | 0.04 | 0.9984 | | |

| | A | B | C | D |
|---|---|---|---|---|
| 1 | x | f(x) | a | 0 |
| 2 | 0 | 1 | b | 1 |
| 3 | 0.01 | 0.9999 | n | 100 |
| 4 | 0.02 | 0.9996 | Delta x | 0.01 |
| 5 | 0.03 | 0.9991 | Left Sum | 0.67165 |
| 6 | 0.04 | 0.9984 | | |

Now it is easy to obtain the sums for $n = 200$ and $n = 500$: Simply change the value of $n$ in cell D3:

| | A | B | C | D |
|---|---|---|---|---|
| 1 | x | f(x) | a | 0 |
| 2 | 0 | 1 | b | 1 |
| 3 | 0.005 | 0.999975 | n | 200 |
| 4 | 0.01 | 0.9999 | Delta x | 0.005 |
| 5 | 0.015 | 0.999775 | Left Sum | 0.6691625 |
| 6 | 0.02 | 0.9996 | | |

| | A | B | C | D |
|---|---|---|---|---|
| 1 | x | f(x) | a | 0 |
| 2 | 0 | 1 | b | 1 |
| 3 | 0.002 | 0.999996 | n | 500 |
| 4 | 0.004 | 0.999984 | Delta x | 0.002 |
| 5 | 0.006 | 0.999964 | Left Sum | 0.667666 |
| 6 | 0.008 | 0.999936 | | |

# 14

# Further Integration Techniques and Applications of the Integral

 **Website**
www.WanerMath.com
At the Website you will find:

- A detailed chapter summary

- A true/false quiz

- Additional review exercises

- A numerical integration utility

- Graphing calculator programs for numerical integration

- Graphers, Excel tutorials, and other resources

## Case Study  Estimating Tax Revenues

You have just been hired by the incoming administration to coordinate national tax policy, and the so-called experts on your staff can't seem to agree on which of two tax proposals will result in more revenue for the government. The data you have are the two income tax proposals (graphs of tax vs. income) and the distribution of incomes in the country. **How do you use this information to decide which tax policy will result in more revenue?**

Dmitriy Shironosov/Shutterstock.com

1015

# Introduction

In the preceding chapter, we learned how to compute many integrals and saw some of the applications of the integral. In this chapter, we look at some further techniques for computing integrals and then at more applications of the integral. We also see how to extend the definition of the definite integral to include integrals over infinite intervals, and show how such integrals can be used for long-term forecasting. Finally, we introduce the beautiful theory of differential equations and some of its numerous applications.

## 14.1  Integration by Parts

**Integration by parts** is an integration technique that comes from the product rule for derivatives. The tabular method we present here has been around for some time and makes integration by parts quite simple, particularly in problems where it has to be used several times.*

* The version of the tabular method we use was developed and taught to us by Dan Rosen at Hofstra University.

We start with a little notation to simplify things while we introduce integration by parts. (We use this notation only in the next few pages.) If $u$ is a function, denote its derivative by $D(u)$ and an antiderivative by $I(u)$. Thus, for example, if $u = 2x^2$, then

$$D(u) = 4x$$

and

$$I(u) = \frac{2x^3}{3}.$$

[If we wished, we could instead take $I(u) = \frac{2x^3}{3} + 46$, but we usually opt to take the simplest antiderivative.]

### Integration by Parts

If $u$ and $v$ are continuous functions of $x$, and $u$ has a continuous derivative, then

$$\int u \cdot v \, dx = u \cdot I(v) - \int D(u)I(v) \, dx.$$

#### Quick Example

(Discussed more fully in Example 1 below)

$$\int x \cdot e^x \, dx = xI(e^x) - \int D(x)I(e^x) \, dx$$

$$= xe^x - \int 1 \cdot e^x \, dx \qquad I(e^x) = e^x; \; D(x) = 1$$

$$= xe^x - e^x + C. \qquad \int e^x \, dx = e^x + C$$

As the Quick Example shows, although we could not immediately integrate $u \cdot v = x \cdot e^x$, we could easily integrate $D(u)I(v) = 1 \cdot e^x = e^x$.

# Derivation of Integration-by-Parts Formula

As we mentioned, the integration-by-parts formula comes from the product rule for derivatives. We apply the product rule to the function $uI(v)$

$$D[u \cdot I(v)] = D(u)I(v) + uD(I(v))$$
$$= D(u)I(v) + uv$$

because $D(I(v))$ is the derivative of an antiderivative of $v$, which is $v$. Integrating both sides gives

$$u \cdot I(v) = \int D(u)I(v)\, dx + \int uv\, dx.$$

A simple rearrangement of the terms now gives us the integration-by-parts formula.

The integration-by-parts formula is easiest to use via the tabular method illustrated in the following example, where we repeat the calculation we did in the Quick Example above.

---

## EXAMPLE 1 Integration by Parts: Tabular Method

Calculate $\int xe^x\, dx$.

**Solution** First, the reason we *need* to use integration by parts to evaluate this integral is that none of the other techniques of integration that we've talked about up to now will help us. Furthermore, we cannot simply find antiderivatives of $x$ and $e^x$ and multiply them together. [You should check that $(x^2/2)e^x$ is *not* an antiderivative of $xe^x$.] However, as we saw above, this integral can be found by integration by parts. We want to find the integral of the *product* of $x$ and $e^x$. We must make a decision: Which function will play the role of $u$ and which will play the role of $v$ in the integration-by-parts formula? Because the derivative of $x$ is just 1, differentiating makes it simpler, so we try letting $x$ be $u$ and letting $e^x$ be $v$. We need to calculate $D(u)$ and $I(v)$, which we record in the following table.

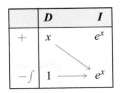

The table is read as
$+x \cdot e^x - \int 1 \cdot e^x\, dx.$

Below $x$ in the $D$ column, we put $D(x) = 1$; below $e^x$ in the $I$ column, we put $I(e^x) = e^x$. The arrow at an angle connecting $x$ and $I(e^x)$ reminds us that the product $xI(e^x)$ will appear in the answer; the plus sign on the left of the table reminds us that it is $+xI(e^x)$ that appears. The integral sign and the horizontal arrow connecting $D(x)$ and $I(e^x)$ remind us that the *integral* of the product $D(x)I(e^x)$ also appears in the answer; the minus sign on the left reminds us that we need to subtract this integral. Combining these two contributions, we get

$$\int xe^x\, dx = xe^x - \int e^x\, dx.$$

The integral that appears on the right is much easier than the one we began with, so we can complete the problem:

$$\int xe^x\, dx = xe^x - \int e^x\, dx = xe^x - e^x + C.$$

➡ **Before we go on...** In Example 1, what if we had made the opposite decision and put $e^x$ in the $D$ column and $x$ in the $I$ column? Then we would have had the following table:

| | $D$ | $I$ |
|---|---|---|
| $+$ | $e^x$ | $x$ |
| $-\int$ | $e^x$ | $\longrightarrow x^2/2$ |

This gives

$$\int xe^x \, dx = \frac{x^2}{2}e^x - \int \frac{x^2}{2}e^x \, dx.$$

The integral on the right is harder than the one we started with, not easier! How do we know beforehand which way to go? We don't. We have to be willing to do a little trial and error: We try it one way, and if it doesn't make things simpler, we try it another way. *Remember, though, that the function we put in the I column must be one that we can integrate.* ■

---

### EXAMPLE 2   Repeated Integration by Parts

Calculate $\int x^2 e^{-x} \, dx.$

**Solution**  Again, we have a product—the integrand is the product of $x^2$ and $e^{-x}$. Because differentiating $x^2$ makes it simpler, we put it in the $D$ column and get the following table:

| | $D$ | $I$ |
|---|---|---|
| $+$ | $x^2$ | $e^{-x}$ |
| $-\int$ | $2x$ | $\longrightarrow -e^{-x}$ |

This table gives us

$$\int x^2 e^{-x} \, dx = x^2(-e^{-x}) - \int 2x(-e^{-x}) \, dx.$$

The last integral is simpler than the one we started with, but it still involves a product. It's a good candidate for another integration by parts. The table we would use would start with $2x$ in the $D$ column and $-e^{-x}$ in the $I$ column, which is exactly what we see in the last row of the table we've already made. Therefore, we *continue the process,* elongating the table above:

| | $D$ | $I$ |
|---|---|---|
| $+$ | $x^2$ | $e^{-x}$ |
| $-$ | $2x$ | $-e^{-x}$ |
| $+\int$ | $2$ | $\longrightarrow e^{-x}$ |

(Notice how the signs on the left alternate. Here's why: To compute $-\int 2x(-e^{-x})\,dx$ we use the negative of the following table:

|   | D | I |
|---|---|---|
| + | $2x$ | $-e^{-x}$ |
| $-\int$ | $2$ | $e^{-x}$ |

so we reverse all the signs.)

Now, we still have to compute an integral (the integral of the product of the functions in the bottom row) to complete the computation. But why stop here? Let's continue the process one more step:

|   | D | I |
|---|---|---|
| + | $x^2$ | $e^{-x}$ |
| − | $2x$ | $-e^{-x}$ |
| + | $2$ | $e^{-x}$ |
| $-\int$ | $0$ | $-e^{-x}$ |

In the bottom line we see that all that is left to integrate is $0(-e^{-x}) = 0$. Because the indefinite integral of $0$ is $C$, we can read the answer from the table as

$$\int x^2 e^{-x}\,dx = x^2(-e^{-x}) - 2x(e^{-x}) + 2(-e^{-x}) + C$$
$$= -x^2 e^{-x} - 2x e^{-x} - 2e^{-x} + C$$
$$= -e^{-x}(x^2 + 2x + 2) + C.$$

In Example 2 we saw a technique that we can summarize as follows:

## Integrating a Polynomial Times a Function

If one of the factors in the integrand is a polynomial and the other factor is a function that can be integrated repeatedly, put the polynomial in the $D$ column and keep differentiating until you get zero. Then complete the $I$ column to the same depth, and read off the answer.

For practice, redo Example 1 using this technique.

It is not always the case that the integrand is a polynomial times something easy to integrate, so we can't always expect to end up with a zero in the $D$ column. In that case we hope that at some point we will be able to integrate the product of the functions in the last row. Here are some examples.

**EXAMPLE 3  Polynomial Times a Logarithm**

Calculate:   **a.** $\int x \ln x \, dx$    **b.** $\int (x^2 - x) \ln x \, dx$    **c.** $\int \ln x \, dx$

**Solution**

**a.** This is a product and therefore a good candidate for integration by parts. Our first impulse is to differentiate $x$, but that would mean integrating $\ln x$, and we do not (yet) know how to do that. So we try it the other way around and hope for the best.

| | $D$ | $I$ |
|---|---|---|
| $+$ | $\ln x$ | $x$ |
| $-\int$ | $\dfrac{1}{x}$ | $\dfrac{x^2}{2}$ |

Why did we stop? If we continued the table, both columns would get more complicated. However, if we stop here we get

$$\int x \ln x \, dx = (\ln x)\left(\frac{x^2}{2}\right) - \int \left(\frac{1}{x}\right)\left(\frac{x^2}{2}\right) dx$$

$$= \frac{x^2}{2} \ln x - \frac{1}{2} \int x \, dx$$

$$= \frac{x^2}{2} \ln x - \frac{x^2}{4} + C.$$

**b.** We can use the same technique we used in part (a) to integrate any polynomial times the logarithm of $x$:

| | $D$ | $I$ |
|---|---|---|
| $+$ | $\ln x$ | $x^2 - x$ |
| $-\int$ | $\dfrac{1}{x}$ | $\dfrac{x^3}{3} - \dfrac{x^2}{2}$ |

$$\int (x^2 - x) \ln x \, dx = (\ln x)\left(\frac{x^3}{3} - \frac{x^2}{2}\right) - \int \left(\frac{1}{x}\right)\left(\frac{x^3}{3} - \frac{x^2}{2}\right) dx$$

$$= \left(\frac{x^3}{3} - \frac{x^2}{2}\right) \ln x - \int \left(\frac{x^2}{3} - \frac{x}{2}\right) dx$$

$$= \left(\frac{x^3}{3} - \frac{x^2}{2}\right) \ln x - \frac{x^3}{9} + \frac{x^2}{4} + C.$$

**c.** The integrand $\ln x$ is not a product. We can, however, *make* it into a product by thinking of it as $1 \cdot \ln x$. Because this is a polynomial times $\ln x$, we proceed as in parts (a) and (b):

| | $D$ | $I$ |
|---|---|---|
| $+$ | $\ln x$ | $1$ |
| $-\int$ | $1/x$ | $x$ |

We notice that the product of $1/x$ and $x$ is just 1, which we know how to integrate, so we can stop here:

$$\int \ln x \, dx = x \ln x - \int \left(\frac{1}{x}\right) x \, dx$$

$$= x \ln x - \int 1 \, dx$$

$$= x \ln x - x + C.$$

---

### FAQs

### Whether to Use Integration by Parts, and What Goes in the *D* and *I* Columns

**Q**: *Will integration by parts always work to integrate a product?*

**A**: No. Although integration by parts often works for products in which one factor is a polynomial, it will almost *never* work in the examples of products we saw when discussing substitution in Section 13.2. For example, although integration by parts can be used to compute $\int (x^2 - x)e^{2x-1} \, dx$ (put $x^2 - x$ in the $D$ column and $e^{2x-1}$ in the $I$ column), it *cannot* be used to compute $\int (2x - 1)e^{x^2-x} \, dx$ (put $u = x^2 - x$). Recognizing when to use integration by parts is best learned by experience.

**Q**: *When using integration by parts, which expression goes in the D column, and which in the I column?*

**A**: Although there is no general rule, the following guidelines are useful:

- To integrate a product in which one factor is a polynomial and the other can be integrated several times, put the polynomial in the $D$ column and the other factor in the $I$ column. Then differentiate the polynomial until you get zero.
- If one of the factors is a polynomial but the other factor cannot be integrated easily, put the polynomial in the $I$ column and the other factor in the $D$ column. Stop when the product of the functions in the bottom row can be integrated.
- If neither factor is a polynomial, put the factor that seems easier to integrate in the $I$ column and the other factor in the $D$ column. Again, stop the table as soon as the product of the functions in the bottom row can be integrated.
- If your method doesn't work, try switching the functions in the $D$ and $I$ columns or try breaking the integrand into a product in a different way. If none of this works, maybe integration by parts isn't the technique to use on this problem.

---

## 14.1     EXERCISES

▼ more advanced     ◆ challenging
🔲 indicates exercises that should be solved using technology

*Evaluate the integrals in Exercises 1–40 using integration by parts where possible.* HINT [See the Examples in the text.]

**1.** $\int 2xe^x \, dx$

**2.** $\int 3xe^{-x} \, dx$

**3.** $\int (3x - 1)e^{-x} \, dx$

**4.** $\int (1 - x)e^x \, dx$

**5.** $\int (x^2 - 1)e^{2x} \, dx$

**6.** $\int (x^2 + 1)e^{-2x} \, dx$

**7.** $\int (x^2 + 1)e^{-2x+4} \, dx$

**8.** $\int (x^2 + 1)e^{3x+1} \, dx$

9. $\int (2-x)2^x \, dx$

10. $\int (3x-2)4^x \, dx$

11. $\int (x^2-1)3^{-x} \, dx$

12. $\int (1-x^2)2^{-x} \, dx$

13. ▼ $\int \dfrac{x^2-x}{e^x} \, dx$

14. ▼ $\int \dfrac{2x+1}{e^{3x}} \, dx$

15. $\int x(x+2)^6 \, dx$ (See note.[1])  16. $\int x^2(x-1)^6 \, dx$ (See note.[1])

17. ▼ $\int \dfrac{x}{(x-2)^3} \, dx$

18. ▼ $\int \dfrac{x}{(x-1)^2} \, dx$

19. $\int x^3 \ln x \, dx$

20. $\int x^2 \ln x \, dx$

21. $\int (t^2+1)\ln(2t) \, dt$

22. $\int (t^2-t)\ln(-t) \, dt$

23. $\int t^{1/3} \ln t \, dt$

24. $\int t^{-1/2} \ln t \, dt$

25. $\int \log_3 x \, dx$

26. $\int x \log_2 x \, dx$

27. ▼ $\int (xe^{2x}-4e^{3x}) \, dx$

28. ▼ $\int (x^2e^{-x}+2e^{-x+1}) \, dx$

29. ▼ $\int \left(x^2e^x - xe^{x^2}\right) dx$

30. ▼ $\int \left[ (2x+1)e^{x^2+x} - x^2e^{2x+1} \right] dx$

31. ▼ $\int (3x-4)\sqrt{2x-1} \, dx$ (See note.[1])

32. ▼ $\int \dfrac{2x+1}{\sqrt{3x-2}} \, dx$ (See note.[1])

33. $\int_0^1 (x+1)e^x \, dx$

34. $\int_{-1}^1 (x^2+x)e^{-x} \, dx$

35. $\int_0^1 x^2(x+1)^{10} \, dx$

36. $\int_0^1 x^3(x+1)^{10} \, dx$

37. $\int_1^2 x \ln(2x) \, dx$

38. $\int_1^2 x^2 \ln(3x) \, dx$

39. $\int_0^1 x \ln(x+1) \, dx$

40. $\int_0^1 x^2 \ln(x+1) \, dx$

41. Find the area bounded by the curve $y = xe^{-x}$, the $x$-axis, and the lines $x=0$ and $x=10$.

42. Find the area bounded by the curve $y = x \ln x$, the $x$-axis, and the lines $x=1$ and $x=e$.

43. Find the area bounded by the curve $y = (x+1)\ln x$, the $x$-axis, and the lines $x=1$ and $x=2$.

44. Find the area bounded by the curve $y = (x-1)e^x$, the $x$-axis, and the lines $x=0$ and $x=2$.

---

[1] Several exercises, including these, can also be done using substitution, although integration by parts is easier and should be used instead.

***Integrals of Functions Involving Absolute Values*** In Exercises 45–52, use integration by parts to evaluate the given integral using the following integral formulas where necessary. *(You have seen some of these before; all can be checked by differentiating.)*

| Integral Formula | Shortcut Version |
|---|---|
| $\int \dfrac{\lvert x \rvert}{x} \, dx = \lvert x \rvert + C$ | $\int \dfrac{\lvert ax+b \rvert}{ax+b} \, dx = \dfrac{1}{a}\lvert ax+b \rvert + C$ |
| | Because $\dfrac{d}{dx}\lvert x \rvert = \dfrac{\lvert x \rvert}{x}$. |
| $\int \lvert x \rvert \, dx = \dfrac{1}{2}x\lvert x \rvert + C$ | $\int \lvert ax+b \rvert \, dx$ $= \dfrac{1}{2a}(ax+b)\lvert ax+b \rvert + C$ |
| $\int x\lvert x \rvert \, dx = \dfrac{1}{3}x^2\lvert x \rvert + C$ | $\int (ax+b)\lvert ax+b \rvert \, dx$ $= \dfrac{1}{3a}(ax+b)^2\lvert ax+b \rvert + C$ |
| $\int x^2\lvert x \rvert \, dx = \dfrac{1}{4}x^3\lvert x \rvert + C$ | $\int (ax+b)^2\lvert ax+b \rvert \, dx$ $= \dfrac{1}{4a}(ax+b)^3\lvert ax+b \rvert + C$ |

45. $\int x\lvert x-3 \rvert \, dx$

46. $\int x\lvert x+4 \rvert \, dx$

47. $\int 2x\dfrac{\lvert x-3 \rvert}{x-3} \, dx$

48. $\int 3x\dfrac{\lvert x+4 \rvert}{x+4} \, dx$

49. ▼ $\int 2x^2\lvert -x+4 \rvert \, dx$

50. ▼ $\int 3x^2\lvert 2x-3 \rvert \, dx$

51. ▼ $\int (x^2-2x+3)\lvert x-4 \rvert \, dx$

52. ▼ $\int (x^2-x+1)\lvert 2x-4 \rvert \, dx$

## APPLICATIONS

53. ***Displacement*** A rocket rising from the ground has a velocity of $2{,}000te^{-t/120}$ ft/sec, after $t$ seconds. How far does it rise in the first two minutes?

54. ***Sales*** Weekly sales of graphing calculators can be modeled by the equation
$$s(t) = 10 - te^{-t/20},$$
where $s$ is the number of calculators sold per week after $t$ weeks. How many graphing calculators (to the nearest unit) will be sold in the first 20 weeks?

55. ***Total Cost*** The marginal cost of the $x$th box of light bulbs is $10 + [\ln(x+1)]/(x+1)^2$, and the fixed cost is \$5,000. Find the total cost to make $x$ boxes of bulbs.

56. ***Total Revenue*** The marginal revenue for selling the $x$th box of light bulbs is $10 + 0.001x^2e^{-x/100}$. Find the total revenue generated by selling 200 boxes of bulbs.

57. ***Spending on Gasoline*** During 2000–2008, the United States consumed gasoline at a rate of about
$$q(t) = -3.5t + 280 \text{ billion gallons per year.} \quad (0 \le t \le 8)$$

($t$ is the number of years since 2000.)[2] During the same period, the price of gasoline was approximately

$$p(t) = 1.2e^{0.12t} \text{ dollars per gallon.}$$

Use an integral to estimate, to the nearest billion dollars, the total spent on gasoline during the given period. HINT [Rate of spending = $p(t)q(t)$.]

**58. Spending on Gasoline** During 1992–2000, the United States consumed gasoline at a rate of about

$$q(t) = 3.2t + 240 \text{ billion gallons per year.} \quad (0 \le t \le 8)$$

($t$ is the number of years since 1992.)[3] During the same period, the price of gasoline was approximately

$$p(t) = 1.0e^{0.02t} \text{ dollars per gallon.}$$

Use an integral to estimate, to the nearest billion dollars, the total spent on gasoline during the given period. HINT [Rate of spending = $p(t)q(t)$.]

**59. Housing** The following graph shows the annual number of housing starts in the United States during 2000–2008 together with a quadratic approximating model:

$$s(t) = -30t^2 + 240t + 800 \text{ thousand homes per year}$$
$$(0 \le t \le 8).$$

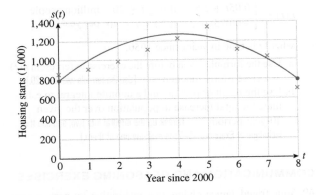

Year since 2000

($t$ is the time in years since 2000.)[4] At the same time, the homes being built were getting larger: The average area per home was approximately

$$a(t) = 40t + 2,000 \text{ square feet.}$$

Use the given models to estimate the total housing area under construction over the given period. (Use integration by parts to evaluate the integral and round your answer to the nearest billion square feet.) HINT [Rate of change of area under construction = $s(t)a(t)$.]

**60. Housing for Sale** The following graph shows the number of housing starts for sale purposes in the United States

during 2000–2008 together with a quadratic approximating model:

$$s(t) = -33t^2 + 240t + 700 \text{ thousand homes per year}$$
$$(0 \le t \le 8).$$

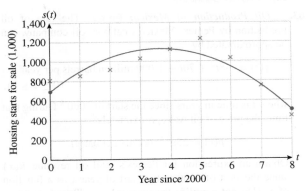

Year since 2000

($t$ is the time in years since 2000.)[5] At the same time, the homes being built were getting larger: The average area per home was approximately

$$a(t) = 40t + 2,000 \text{ square feet.}$$

Use the given models to estimate the total housing area under construction for sale purposes over the given period. (Use integration by parts to evaluate the integral and round your answer to the nearest billion square feet.) HINT [Rate of change of area under construction = $s(t)a(t)$.]

**61.** ▼ **Bottled Water Sales** The rate of U.S. sales of bottled water for the period 2000–2010 can be approximated by

$$s(t) = -45t^2 + 900t + 4,200 \text{ million gallons per year}$$
$$(0 \le t \le 10),$$

where $t$ is time in years since the start of 2000.[6] After conducting a survey of sales in your town, you estimate that consumption in gyms accounts for a fraction

$$f(t) = \sqrt{0.1 + 0.02t}$$

of all bottled water sold in year $t$. Assuming your model is correct, estimate, to the nearest hundred million gallons, the total amount of bottled water consumed in gyms from the start of 2005 to the start of 2010. HINT [Rate of consumption = $s(t)f(t)$. Also see Exercise 31.]

**62. Bottled Water Sales** The rate of U.S. per capita sales of bottled water for the period 2000–2010 can be approximated by

$$s(t) = -0.18t^2 + 3t + 15 \text{ gallons per year} \quad (0 \le t \le 10),$$

where $t$ is the time in years since the start of 2000.[7] After conducting a survey of sales in your state, you estimate that consumption in gyms accounts for a fraction

$$f(t) = \sqrt{0.2 + 0.04t}$$

[2] Source for data: Energy Information Administration (Department of Energy) (www.eia.gov).

[3] Ibid.

[4] Source for data: U.S. Census Bureau (www.census.gov).

[5] Ibid.

[6] Source for data: Beverage Marketing Corporation/ www.bottledwater.org.

[7] Ibid.

of all bottled water consumed in year $t$. Assuming your model is correct, estimate, to the nearest gallon, the total amount of bottled water consumed per capita in gyms from the start of 2005 to the start of 2010. HINT [Rate of consumption = $s(t)f(t)$. Also see Exercise 31.]

63. ▼ *Oil Production in Mexico:* Pemex The rate of oil production by Pemex, Mexico's national oil company, can be approximated by

$$q(t) = -8t^2 + 70t + 1,000 \text{ million barrels per year}$$
$$(0 \le t \le 9),$$

where $t$ is time in years since the start of 2000.[8] During that time, the price of oil was approximately[9]

$$p(t) = 25e^{0.1t} \text{ dollars per barrel.}$$

Obtain an expression for Pemex's total oil revenue $R(x)$ since the start of 2000 to the start of year $x$ as a function of $x$. (Do not simplify the answer.) HINT [Rate of revenue = $p(t)q(t)$.]

64. ▼ *Oil Imports from Mexico* The rate of oil imports to the United States from Mexico can be approximated by

$$r(t) = -5t^2 + 40t + 500 \text{ million barrels per year}$$
$$(0 \le t \le 8),$$

where $t$ is time in years since the start of 2000.[10] During that time, the price of oil was approximately[11]

$$p(t) = 25e^{0.1t} \text{ dollars per barrel.}$$

Obtain an expression for the total oil revenue $R(x)$ Mexico earned from the United States since the start of 2000 to the start of year $x$ as a function of $x$. (Do not simplify the answer.) HINT [Rate of revenue = $p(t)r(t)$.]

65. ▼ *Revenue* You have been raising the price of your *Lord of the Fields* T-shirts by 50¢ per week, and sales have been falling continuously at a rate of 2% per week. Assuming you are now selling 50 T-shirts per week and charging $10 per T-shirt, how much revenue will you generate during the coming year? (Round your answer to the nearest dollar.) HINT [Weekly revenue = weekly sales × price per T-shirt.]

66. ▼ *Revenue* Luckily, sales of your *Star Wars and Piece* T-shirts are now 50 T-shirts per week and increasing continuously at a rate of 5% per week. You are now charging $10 per T-shirt and are decreasing the price by 50¢ per week. How much revenue will you generate during the next six weeks?

---

[8] Source for data: Energy Information Administration/Pemex (www.eia.gov).

[9] Source for data: BP Statistical Review of World Energy (www.bp.com/statisticalreview).

[10] Source for data: Energy Information Administration/Pemex (www.eia.gov).

[11] Source for data: BP Statistical Review of World Energy (www.bp.com/statisticalreview).

*Integrals of Piecewise-Linear Functions* Exercises 67 and 68 are based on the following formula that can be used to represent a piecewise linear function as a closed-form function:

$$\begin{cases} p(x) & \text{if } x < a \\ q(x) & \text{if } x > a \end{cases} = p(x) + \frac{1}{2}[q(x) - p(x)]\left[1 + \frac{|x - a|}{x - a}\right]$$

*Such functions can then be integrated using the technique of Exercises 45–52.*

67. ◆ *Population: Mexico* The rate of change of population in Mexico over 1990–2010 was approximately

$$r(t) = \begin{cases} -0.1t + 3 & \text{if } 0 \le t \le 10 \\ -0.05t + 2.5 & \text{if } 10 \le t \le 20 \end{cases} \begin{array}{l} \text{million people} \\ \text{per year,} \end{array}$$

where $t$ is time in years since 1990.

a. Use the formula given before the exercise to represent $r(t)$ as a closed-form function. HINT [Use the formula with $a = 10$.]

b. Use the result of part (a) and a definite integral to estimate the total increase in population over the given 20-year period. HINT [Break up the integral into two, and use the technique of Exercises 45–52 to evaluate one of them.]

68. ◆ *Population: Mexico* The rate of change of population in Mexico over 1950–1990 was approximately

$$r(t) = \begin{cases} 0.05t + 2.5 & \text{if } 0 \le t \le 20 \\ -0.075t + 5 & \text{if } 20 \le t \le 40 \end{cases} \begin{array}{l} \text{million people} \\ \text{per year,} \end{array}$$

where $t$ is time in years since 1950.

a. Use the formula given before the exercise to represent $r(t)$ as a closed-form function. HINT [Use the formula with $a = 20$.]

b. Use the result of part (a) and a definite integral to estimate the total increase in population over the given 40-year period. HINT [Break up the integral into two, and use the technique of Exercises 45–52 to evaluate one of them.]

## COMMUNICATION AND REASONING EXERCISES

69. Your friend Janice claims that integration by parts allows one to integrate any product of two functions. Prove her wrong by giving an example of a product of two functions that cannot be integrated using integration by parts.

70. Complete the following sentence in words: The integral of $u$ $v$ is the first times the integral of the second minus the integral of _____.

71. Give an example of an integral that can be computed in two ways: by substitution or integration by parts.

72. Give an example of an integral that can be computed by substitution but not by integration by parts. (You need not compute the integral.)

In Exercises 73–80, indicate whether the given integral calls for integration by parts or substitution.

73. $\int (6x - 1)e^{3x^2 - x} dx$    74. $\int \frac{x^2 - 3x + 1}{e^{2x - 3}} dx$

**75.** $\int (3x^2 - x)e^{6x-1}\, dx$

**76.** $\int \dfrac{2x - 3}{e^{x^2-3x+1}}\, dx$

**77.** $\int \dfrac{1}{(x+1)\ln(x+1)}\, dx$

**78.** $\int \dfrac{\ln(x+1)}{x+1}\, dx$

**79.** $\int \ln(x^2)\, dx$

**80.** $\int (x+1)\,\ln(x+1)\, dx$

**81.** ▼ If $p(x)$ is a polynomial of degree $n$ and $f(x)$ is some function of $x$, how many times do we generally have to integrate $f(x)$ to compute $\int p(x)f(x)\, dx$?

**82.** ▼ Use integration by parts to show that $\int (\ln x)^2\, dx = x(\ln x)^2 - 2x \ln x + 2x + C$.

**83.** ◆ *Hermite's Identity* If $f(x)$ is a polynomial of degree $n$, show that

$$\int_0^b f(x)e^{-x}\, dx = F(0) - F(b)e^{-b},$$

where $F(x) = f(x) + f'(x) + f''(x) + \cdots + f^{(n)}(x)$. (This is the sum of $f$ and all of its derivatives.)

**84.** ◆ Write down a formula similar to Hermite's identity for $\int_0^b f(x)e^x\, dx$ when $f(x)$ is a polynomial of degree $n$.

# 14.2 Area between Two Curves and Applications

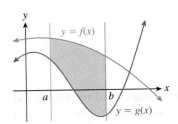

**Figure 1**

As we saw in the preceding chapter, we can use the definite integral to calculate the area between the graph of a function and the $x$-axis. With only a little more work, we can use it to calculate the area between two graphs. Figure 1 shows the graphs of two functions, $f(x)$ and $g(x)$, with $f(x) \geq g(x)$ for every $x$ in the interval $[a, b]$.

To find the shaded area between the graphs of the two functions, we use the following formula.

**Area between Two Graphs**

If $f(x) \geq g(x)$ for all $x$ in $[a, b]$ (so that the graph of $f$ does not move below that of $g$), then the area of the region between the graphs of $f$ and $g$ and between $x = a$ and $x = b$ is given by

$$A = \int_a^b [f(x) - g(x)]\, dx. \quad \text{Integral of (Top − Bottom)}$$

**Caution** If the graphs of $f$ and $g$ cross in the interval, the above formula does not hold; for instance, if $f(x) = x$ and $g(x) = -x$, then the total area shown in the figure is 2 square units, whereas $\int_{-1}^{1}[f(x) - g(x)]\, dx = 0$.

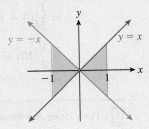

Let's look at an example and then discuss why the formula works.

**Figure 2**

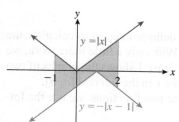

**Figure 3**

### EXAMPLE 1  The Area between Two Curves

Find the areas of the following regions:

**a.** Between $f(x) = -x^2 - 3x + 4$ and $g(x) = x^2 - 3x - 4$ and between $x = -1$ and $x = 1$

**b.** Between $f(x) = |x|$ and $g(x) = -|x - 1|$ over $[-1, 2]$

### Solution

**a.** The area in question is shown in Figure 2. Because the graph of $f$ lies above the graph of $g$ in the interval $[-1, 1]$, we have $f(x) \geq g(x)$ for all $x$ in $[-1, 1]$. Therefore, we can use the formula given above and calculate the area as follows:

$$A = \int_{-1}^{1} [f(x) - g(x)] \, dx$$

$$= \int_{-1}^{1} [(-x^2 - 3x + 4) - (x^2 - 3x - 4)] \, dx$$

$$= \int_{-1}^{1} (8 - 2x^2) \, dx$$

$$= \left[ 8x - \frac{2}{3}x^3 \right]_{-1}^{1}$$

$$= \frac{44}{3}.$$

**b.** The given area (see Figure 3) can be broken up into triangles and rectangles, but we already know a formula for the antiderivative of $|ax + b|$ for constants $a$ and $b$, so we can use calculus instead:

$$A = \int_{-1}^{2} [f(x) - g(x)] \, dx$$

$$= \int_{-1}^{2} [|x| - (-|x - 1|)] \, dx$$

$$= \int_{-1}^{2} [|x| + |x - 1|] \, dx$$

$$= \frac{1}{2}[x|x| + (x - 1)|x - 1|]_{-1}^{2} \qquad \int |ax + b| \, dx = \frac{1}{2a}(ax + b)|ax + b| + C$$

$$= \frac{1}{2}[(4 + 1) - (-1 - 4)]$$

$$= \frac{1}{2}(10) = 5.$$

**Q:** *Why does the formula for the area between two curves work?*

**A:** Let's go back once again to the general case illustrated in Figure 1, where we were given two functions $f$ and $g$ with $f(x) \geq g(x)$ for every $x$ in the interval $[a, b]$. To avoid complicating the argument by the fact that the graph of $g$, or $f$, or both may dip below the $x$-axis in the interval $[a, b]$ (as occurs in Figure 1 and also in Example 1), we shift both graphs vertically upward by adding a big enough constant $M$ to lift them both above the $x$-axis in the interval $[a, b]$, as shown in Figure 4.

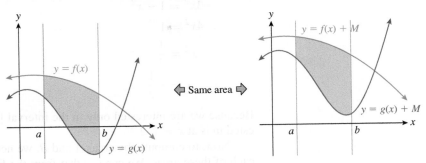

**Figure 4**

As the figure illustrates, the area of the region between the graphs is not affected, so we will calculate the area of the region shown on the right of Figure 4. That calculation is shown in Figure 5.

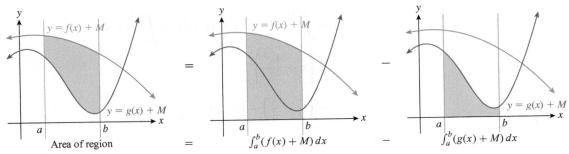

**Figure 5**

From the figure, the area we want is

$$\int_a^b (f(x) + M)\, dx - \int_a^b (g(x) + M)\, dx = \int_a^b [(f(x) + M) - (g(x) + M)]\, dx$$

$$= \int_a^b [f(x) - g(x)]\, dx,$$

which is the formula we gave originally.

So far, we've been assuming that $f(x) \geq g(x)$, so that the graph of $f$ never dips below the graph of $g$ and so the graphs cannot cross (although they can touch). Example 2 shows how we compute the area between graphs that *do* cross.

## EXAMPLE 2  Regions Enclosed by Crossing Curves

Find the area of the region between $y = 3x^2$ and $y = 1 - x^2$ and between $x = 0$ and $x = 1$.

**Solution**  The area we wish to calculate is shown in Figure 6. From the figure, we can see that neither graph lies above the other over the whole interval. To get around this, we break the area into the two pieces on either side of the point at which the graphs cross and then compute each area separately. To do this, we need to know

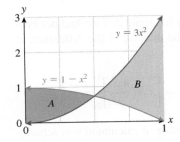

**Figure 6**

exactly where that crossing point is. The crossing point is where $3x^2 = 1 - x^2$, so we solve for $x$:

$$3x^2 = 1 - x^2$$

$$4x^2 = 1$$

$$x^2 = \frac{1}{4}$$

$$x = \pm\frac{1}{2}.$$

Because we are interested only in the interval $[0, 1]$, the crossing point we're interested in is at $x = 1/2$.

Now, to compute the areas $A$ and $B$, we need to know which graph is on top in each of these areas. We can see that from the figure, but what if the functions were more complicated and we could not easily draw the graphs? To be sure, we can test the values of the two functions at some point in each region. But we really need not worry. If we make the wrong choice for the top function, the integral will yield the negative of the area (why?), so we can simply take the absolute value of the integral to get the area of the region in question. For this example, we have

$$A = \int_0^{1/2} [(1 - x^2) - 3x^2]\, dx = \int_0^{1/2} (1 - 4x^2)\, dx$$

$$= \left[ x - \frac{4x^3}{3} \right]_0^{1/2}$$

$$= \left( \frac{1}{2} - \frac{1}{6} \right) - (0 - 0) = \frac{1}{3}$$

and

$$B = \int_{1/2}^1 [3x^2 - (1 - x^2)]\, dx = \int_{1/2}^1 (4x^2 - 1)\, dx$$

$$= \left[ \frac{4x^3}{3} - x \right]_{1/2}^1$$

$$= \left( \frac{4}{3} - 1 \right) - \left( \frac{1}{6} - \frac{1}{2} \right) = \frac{2}{3}.$$

This gives a total area of $A + B = \dfrac{1}{3} + \dfrac{2}{3} = 1$.

---

➡ **Before we go on…** What would have happened in Example 2 if we had not broken the area into two pieces but had just calculated the integral of the difference of the two functions? We would have calculated

$$\int_0^1 [(1 - x^2) - 3x^2]\, dx = \int_0^1 [1 - 4x^2]\, dx = \left[ x - \frac{4x^3}{3} \right]_0^1 = -\frac{1}{3},$$

which is not even close to the right answer. What this integral calculated was actually $A - B$ rather than $A + B$. Why? ∎

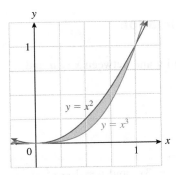

**Figure 7**

## EXAMPLE 3 The Area Enclosed by Two Curves

Find the area enclosed by $y = x^2$ and $y = x^3$.

**Solution** This example has a new wrinkle: We are not told what interval to use for $x$. However, if we look at the graph in Figure 7, we see that the question can have only one meaning.

We are being asked to find the area of the shaded sliver, which is the only region that is actually *enclosed* by the two graphs. This sliver is bounded on either side by the two points where the graphs cross, so our first task is to find those points. They are the points where $x^2 = x^3$, so we solve for $x$:

$$x^2 = x^3$$
$$x^3 - x^2 = 0$$
$$x^2(x - 1) = 0$$
$$x = 0 \quad \text{or} \quad x = 1$$

Thus, we must integrate over the interval [0, 1]. Although we see from the diagram (or by substituting $x = 1/2$) that the graph of $y = x^2$ is above that of $y = x^3$, if we didn't notice that we might calculate

$$\int_0^1 (x^3 - x^2)\, dx = \left[\frac{x^4}{4} - \frac{x^3}{3}\right]_0^1 = -\frac{1}{12}.$$

This tells us that the required area is 1/12 square units and also that we had our integral reversed. Had we calculated $\int_0^1 (x^2 - x^3)\, dx$ instead, we would have found the correct answer, 1/12.

We can summarize the procedure we used in the preceding two examples.

---

### Finding the Area between the Graphs of $f(x)$ and $g(x)$

1. Find all points of intersection by solving $f(x) = g(x)$ for $x$. This either determines the interval over which you will integrate or breaks up a given interval into regions between the intersection points.
2. Determine the area of each region you found by integrating the difference of the larger and the smaller function. (If you accidentally take the smaller minus the larger, the integral will give the negative of the area, so just take the absolute value.)
3. Add together the areas you found in step 2 to get the total area.

---

**Q**: *Is there any quick and easy method to find the area between two graphs without having to find all points of intersection? What if it is hard or impossible to find out where the curves intersect?*

**A**: We can use technology to give the approximate area between two graphs. First recall that, if $f(x) \geq g(x)$ for all $x$ in [a, b], then the area between their graphs over [a, b] is given by $\int_a^b [f(x) - g(x)]\, dx$, whereas if $g(x) \geq f(x)$, the area is given by $\int_a^b [g(x) - f(x)]\, dx$. Notice that both expressions are equal to

$$\int_a^b |f(x) - g(x)|\, dx$$

telling us that we can use this same formula in both cases.

**Figure 8**

## ⊤ Area between Two Graphs: Approximation Using Technology

The area of the region between the graphs of $f$ and $g$ and between $x = a$ and $x = b$ is given by

$$A = \int_a^b |f(x) - g(x)|\, dx.$$

### Quick Example

To approximate the area of the region between $y = 3x^2$ and $y = 1 - x^2$ and between $x = 0$ and $x = 1$ we calculated in Example 2, use technology to compute

$$\int_0^1 |3x^2 - (1 - x^2)|\, dx = 1.$$

**TI-83/84 Plus:**   Enter `fnInt(abs(3x^2-(1-x^2)),X,0,1)`

**Website:**   Online Utilities → Numerical Integration Utility
Enter `abs(3x^2-(1-x^2))` for $f(x)$ and 0 and 1 for
the lower and upper limits, and press "Integral".
(See Figure 8.)

## 14.2   EXERCISES

▼ more advanced    ◆ challenging
⊤ indicates exercises that should be solved using technology

*Find the area of the shaded region in Exercises 1–8. (We suggest you use technology to check your answers.)*

**1.**

**2.**

**3.**

**4.**

**5.**

**6.**

**7.** ▼

**8.** ▼

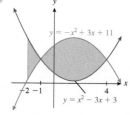

*Find the area of the indicated region in Exercises 9–42. We suggest you graph the curves to check whether one is above the other or whether they cross, and that you use technology to check your answers.*

**9.** Between $y = x^2$ and $y = -1$ for $x$ in $[-1, 1]$ HINT [See Example 1.]

**10.** Between $y = x^3$ and $y = -1$ for $x$ in $[-1, 1]$ HINT [See Example 1.]

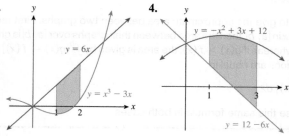

**11.** Between $y = -x$ and $y = x$ for $x$ in $[0, 2]$

**12.** Between $y = -x$ and $y = x/2$ for $x$ in $[0, 2]$

**13.** Between $y = |x|$ and $y = x^2$ in $[-1, 1]$

**14.** Between $y = -|x|$ and $y = x^2 - 2$ in $[-1, 1]$

**15.** Between $y = x$ and $y = x^2$ for $x$ in $[-1, 1]$ HINT [See Example 2.]

**16.** Between $y = x$ and $y = x^3$ for $x$ in $[-1, 1]$ HINT [See Example 2.]

**17.** Between $y = x^2 - 2x$ and $y = -x^2 + 4x - 4$ for $x$ in $[0, 2]$

**18.** Between $y = x^2 - 4x + 2$ and $y = -x^2 + 4x - 4$ for $x$ in $[0, 3]$

**19.** Between $y = 2x^2 + 10x - 5$ and $y = -x^2 + 4x + 4$ for $x$ in $[-3, 2]$

**20.** Between $y = 2x^2 + 7x - 2$ and $y = -x^2 + 4x + 4$ for $x$ in $[-2, 2]$

**21.** Between $y = e^x$ and $y = x$ for $x$ in $[0, 1]$

**22.** Between $y = e^{-x}$ and $y = -x$ for $x$ in $[0, 1]$

**23.** Between $y = (x - 1)^2$ and $y = -(x - 1)^2$ for $x$ in $[0, 1]$

**24.** Between $y = x^2(x^3 + 1)^{10}$ and $y = -x(x^2 + 1)^{10}$ for $x$ in $[0, 1]$

**25.** Enclosed by $y = x$ and $y = x^4$ HINT [See Example 3.]

**26.** Enclosed by $y = x$ and $y = -x^4$ HINT [See Example 3.]

**27.** Enclosed by $y = x^3$ and $y = x^4$

**28.** Enclosed by $y = x$ and $y = x^3$

**29.** Enclosed by $y = x^2$ and $y = x^4$

**30.** Enclosed by $y = x^4 - x^2$ and $y = x^2 - x^4$

**31.** Enclosed by $y = x^2 - 2x$ and $y = -x^2 + 4x - 4$

**32.** Enclosed by $y = x^2 - 4x + 2$ and $y = -x^2 + 4x - 4$

**33.** Enclosed by $y = 2x^2 + 10x - 5$ and $y = -x^2 + 4x + 4$

**34.** Enclosed by $y = 2x^2 + 7x - 2$ and $y = -x^2 + 4x + 4$

**35.** Enclosed by $y = e^x$, $y = 2$, and the $y$-axis

**36.** Enclosed by $y = e^{-x}$, $y = 3$, and the $y$-axis

**37.** Enclosed by $y = \ln x$, $y = 2 - \ln x$, and $x = 4$

**38.** Enclosed by $y = \ln x$, $y = 1 - \ln x$, and $x = 4$

**39.** ▣ Enclosed by $y = e^x$, $y = 2x + 1$, $x = -1$, and $x = 1$ (Round answer to four significant digits.) HINT [See Quick Example page 1030.]

**40.** ▣ Enclosed by $y = 2^x$, $y = x + 2$, $x = -2$, and $x = 2$ (Round answer to four significant digits.) HINT [See Quick Example page 1030.]

**41.** ▣ Enclosed by $y = \ln x$ and $y = \dfrac{x}{2} - \dfrac{1}{2}$ (Round answer to four significant digits.) [First use technology to determine approximately where the graphs cross.]

**42.** ▣ Enclosed by $y = \ln x$ and $y = x - 2$ (Round answer to four significant digits.) [First use technology to determine approximately where the graphs cross.]

### APPLICATIONS

**43.** *Revenue and Cost* Suppose your daily revenue from selling used DVDs is

$$R(t) = 100 + 10t \quad (0 \le t \le 5)$$

dollars per day, where $t$ represents days from the beginning of the week, while your daily costs are

$$C(t) = 90 + 5t \quad (0 \le t \le 5)$$

dollars per day. Find the area between the graphs of $R(t)$ and $C(t)$ for $0 \le t \le 5$. What does your answer represent?

**44.** *Income and Expenses* Suppose your annual income is

$$I(t) = 50,000 + 2,000t \quad (0 \le t \le 3)$$

dollars per year, where $t$ represents the number of years since you began your job, while your annual expenses are

$$E(t) = 45,000 + 1,500t \quad (0 \le t \le 3)$$

dollars per year. Find the area between the graphs of $I(t)$ and $E(t)$ for $0 \le t \le 3$. What does your answer represent?

**45.** *Housing* The total number of housing starts in the United States during 2000–2008 was approximately

$$h(t) = -30t^2 + 240t + 800 \text{ thousand homes per year}$$
$$(0 \le t \le 8),$$

where $t$ is time in years since the start of 2000.[12] During that time, the number of housing starts for sale purposes in the United States was approximately

$$s(t) = -33t^2 + 240t + 700 \text{ thousand homes per year}$$
$$(0 \le t \le 8).$$

Compute the area shown in the graph, and interpret the answer.

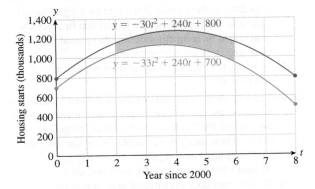

**46.** *Oil Production in Mexico:* Pemex The rate of oil production by Pemex, Mexico's national oil company, can be approximated by

$$q(t) = -8t^2 + 70t + 1,000 \text{ million barrels per year}$$
$$(0 \le t \le 9),$$

---

[12] Source for data: U.S. Census Bureau (www.census.gov).

where $t$ is time in years since the start of 2000.[13] During that time, Mexico exported oil to the United States at a rate of

$$r(t) = -5t^2 + 40t + 500 \text{ million barrels per year}$$
$$(0 \le t \le 8).$$

Compute the area shown in the graph, and interpret the answer.

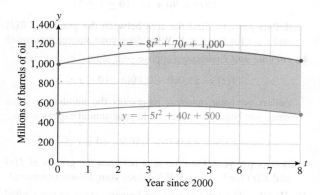

**MySpace _and_ Facebook** _Exercises 47 and 48 are based on the following models, which show the rate at which new members joined Facebook and MySpace in the period from 2005 to the middle of 2008:_

Facebook:[14]  $f(t) = 12t^2 - 20t + 10$ _million members per year_  $(0 \le t \le 3.5)$

MySpace:[15]  $m(t) = 10.5t^2 + 25t + 18.5$ _million members per year_  $(0 \le t \le 3.5)$

(_t is time in years since the start of 2005._)

**47. a.** Use a graph to determine which of the two Internet sites was experiencing a larger influx of new members from June 2005 to the start of 2007. Use an integral to estimate how many more people joined that Internet site than joined its competitor during that period. (Round your answer to the nearest million.)
   **b.** To what area does the integral used in part (a) correspond?

**48. a.** Use a graph to determine which of the two Internet sites was experiencing a larger influx of new members from the start of 2007 to June 2008. Use an integral to estimate how many more people joined that Internet site than joined its competitor during that period. (Round your answer to the nearest million.)
   **b.** To what area does the integral used in part (a) correspond?

---

**Big Brother** _Exercises 49 and 50 are based on the following models, which show the total number of wiretaps authorized per year by all state and federal courts in the United States and the number authorized by state courts._

State and federal courts: $a(t) = 820e^{0.051t}$  $(0 \le t \le 20)$
State courts: $s(t) = 440e^{0.06t}$  $(0 \le t \le 20)$

(_t is time in years since the start of_ 1990.)[16]

**49.** ▼ Estimate the area between the graphs of the two functions over the interval $[0, t]$. Interpret your answer.

**50.** ▼ Estimate the area between the graphs of the two functions over the interval $[t, 20]$. Interpret your answer.

## COMMUNICATION AND REASONING EXERCISES

**51.** If $f$ and $g$ are continuous functions with $\int_a^b [f(x) - g(x)]\, dx = 0$, it follows that the area between the graphs of $f$ and $g$ is zero, right?

**52.** You know that $f$ and $g$ are continuous and their graphs do not cross on the interval $[a, b]$, so you calculate $\int_a^b [f(x) - g(x)]\, dx$ and find that the answer is $-40$. Why is it negative, and what is the area between the curves?

**53.** The following graph shows annual U.S. exports and imports for the period 1960–2007.[17]

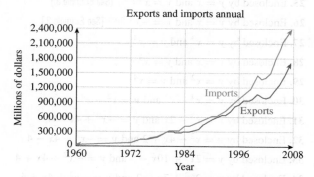

What does the area between the export and import curves represent?

**54.** The following graph shows a fictitious country's monthly exports and imports for the period 1997–2001.

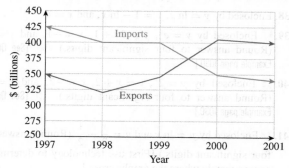

---

[13] Source for data: Energy Information Administration/Pemex (www.eia.gov).

[14] Sources for data: Some data are interpolated (www.facebook.com/, www.insidehighered.com).

[15] Source for data: www.swivel.com/data_sets.

[16] Source for data: Wiretap Reports, Administrative Office of the United States Courts (www.uscourts.gov/Statistics/WiretapReports.aspx).

[17] Source: Data 360 (www.data360.org).

What does the total area enclosed by the export and import curves represent, and what does the definite integral of the difference, Exports − Imports, represent?

**55.** ▼ The following graph shows the daily revenue and cost in your *Adopt-a-Chia* operation *t* days from its inception.

Multiple choice: The area between the cost and revenue curves represents:

**(A)** the accumulated loss through day 4 plus the accumulated profit for days 5 through 7.

**(B)** the accumulated profit for the week.

**(C)** the accumulated loss for the week.

**(D)** the accumulated cost through day 4 plus the accumulated revenue for days 5 through 7.

**56.** ▼ The following graph shows daily orders and inventory (stock on hand) for your *Jackson Pollock Paint by Number* sets *t* days into last week.

a. Multiple choice: Which is greatest?

**(A)** $\int_0^7 (\text{Orders} - \text{Inventory})\, dt$

**(B)** $\int_0^7 (\text{Inventory} - \text{Orders})\, dt$

**(C)** The area between the Orders and Inventory curves

**b.** Multiple choice: The answer to part (a) measures

**(A)** the accumulated gap between orders and inventory.

**(B)** the accumulated surplus through day 3 minus the accumulated shortage for days 3 through 5 plus the accumulated surplus through days 6 through 7.

**(C)** the total net surplus.

**(D)** the total net loss.

**57.** ▼ What is wrong with the following claim? "I purchased Novartis AG shares for $50 at the beginning of March 2008. My total profit per share from this investment from March through August is represented by the area between the stock price curve and the purchase price curve as shown on the following graph, where *t* is time in months since March 1, 2008."[18]

**58.** ▼ Your pharmaceutical company monitors the amount of medication in successive batches of 100 mg tetracycline capsules, and obtains the following graph.

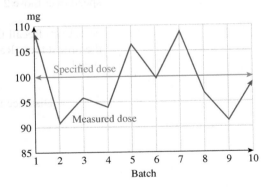

Your production manager claims that the batches of tetracycline conform to the exact dosage requirement because half of the area between the graphs is above the Specified dose line and half is below it. Comment on this reasoning.

---

[18] Source: www.finance.google.com.

# 14.3  Averages and Moving Averages

## Averages

To find the average of, say, 20 numbers, we simply add them up and divide by 20. More generally, if we want to find the **average**, or **mean**, of the $n$ numbers $y_1, y_2, y_3, \ldots y_n$, we add them up and divide by $n$. We write this average as $\bar{y}$ ("$y$-bar").

### Average, or Mean, of a Collection of Values

$$\bar{y} = \frac{y_1 + y_2 + \cdots + y_n}{n}$$

### Quick Example

The average of $\{0, 2, -1, 5\}$ is $\bar{y} = \dfrac{0 + 2 - 1 + 5}{4} = \dfrac{6}{4} = 1.5.$

But, we also use the word *average* in other senses. For example, we speak of the average speed of a car during a trip.

### EXAMPLE 1  Average Speed

Over the course of 2 hours, my speed varied from 50 miles per hour to 60 miles per hour, following the function $v(t) = 50 + 2.5t^2$, $0 \le t \le 2$. What was my average speed over those 2 hours?

**Solution**  Recall that average speed is simply the total distance traveled divided by the time it took. Recall, also, that we can find the distance traveled by integrating the speed:

$$\text{Distance traveled} = \int_0^2 v(t)\, dt$$

$$= \int_0^2 (50 + 2.5t^2)\, dt$$

$$= \left[ 50t + \frac{2.5}{3}t^3 \right]_0^2$$

$$= 100 + \frac{20}{3}$$

$$\approx 106.67 \text{ miles.}$$

It took 2 hours to travel this distance, so the average speed was

$$\text{Average speed} \approx \frac{106.67}{2} \approx 53.3 \text{ mph.}$$

In general, if we travel with velocity $v(t)$ from time $t = a$ to time $t = b$, we will travel a distance of $\int_a^b v(t)\,dt$ in time $b - a$, which gives an average velocity of

$$\text{Average velocity} = \frac{1}{b-a} \int_a^b v(t)\,dt.$$

Thinking of this calculation as finding the average value of the velocity function, we generalize and make the following definition.

---

### Average Value of a Function

The **average**, or **mean**, of a function $f(x)$ on an interval $[a, b]$ is

$$\bar{f} = \frac{1}{b-a} \int_a^b f(x)\,dx.$$

**Quick Example**

The average of $f(x) = x$ on $[1, 5]$ is

$$\bar{f} = \frac{1}{b-a} \int_a^b f(x)\,dx$$

$$= \frac{1}{5-1} \int_1^5 x\,dx$$

$$= \frac{1}{4}\left[\frac{x^2}{2}\right]_1^5 = \frac{1}{4}\left(\frac{25}{2} - \frac{1}{2}\right) = 3.$$

---

**Figure 9**

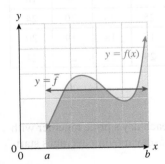

**Figure 10**

## Interpreting the Average of a Function Geometrically

The average of a function has a geometric interpretation. Referring to the Quick Example above, we can compare the graph of $y = f(x)$ with the graph of $y = 3$, both over the interval $[1, 5]$ (Figure 9). We can find the area under the graph of $f(x) = x$ by geometry or by calculus; it is 12. The area in the rectangle under $y = 3$ is also 12.

In general, the average $\bar{f}$ of a positive function over the interval $[a, b]$ gives the height of the rectangle over the interval $[a, b]$ that has the same area as the area under the graph of $f(x)$ as illustrated in Figure 10. The equality of these areas follows from the equation

$$(b-a)\bar{f} = \int_a^b f(x)\,dx.$$

### EXAMPLE 2 **Average Balance**

A savings account at the People's Credit Union pays 3% interest, compounded continuously, and at the end of the year you get a bonus of 1% of the average balance in the account during the year. If you deposit $10,000 at the beginning of the year, how much interest and how large a bonus will you get?

**Solution** We can use the continuous compound interest formula to calculate the amount of money you have in the account at time $t$:

$$A(t) = 10{,}000e^{0.03t},$$

where $t$ is measured in years. At the end of 1 year, the account will have

$$A(1) = \$10{,}304.55$$

so you will have earned $304.55 interest. To compute the bonus, we need to find the average amount in the account, which is the average of $A(t)$ over the interval $[0, 1]$. Thus,

$$\bar{A} = \frac{1}{b-a} \int_a^b A(t)\, dt$$

$$= \frac{1}{1-0} \int_0^1 10{,}000e^{0.03t}\, dt = \frac{10{,}000}{0.03}\left[e^{0.03t}\right]_0^1$$

$$\approx \$10{,}151.51.$$

The bonus is 1% of this, or $101.52.

➡ **Before we go on...** The 1% bonus in Example 2 was one third of the total interest. Why did this happen? What fraction of the total interest would the bonus be if the interest rate was 4%, 5%, or 10%? ■

## Moving Averages

Suppose you follow the performance of a company's stock by recording the daily closing prices. The graph of these prices may seem jagged or "jittery" due to random day-to-day fluctuations. To see any trends, you would like a way to "smooth out" these data. The **moving average** is one common way to do that.

### EXAMPLE 3  Stock Prices

The following table shows Colossal Conglomerate's closing stock prices for 20 consecutive trading days.

| Day | 1 | 2 | 3 | 4 | 5 | 6 | 7 | 8 | 9 | 10 |
|---|---|---|---|---|---|---|---|---|---|---|
| Price | 20 | 22 | 21 | 24 | 24 | 23 | 25 | 26 | 20 | 24 |
| Day | 11 | 12 | 13 | 14 | 15 | 16 | 17 | 18 | 19 | 20 |
| Price | 26 | 26 | 25 | 27 | 28 | 27 | 29 | 27 | 25 | 24 |

Plot these prices and the 5-day moving average.

**Solution** The 5-day moving average is the average of each day's price together with the prices of the preceding 4 days. We can compute the 5-day moving averages starting on the fifth day. We get these numbers:

| Day | 1 | 2 | 3 | 4 | 5 | 6 | 7 | 8 | 9 | 10 |
|---|---|---|---|---|---|---|---|---|---|---|
| Moving Average | | | | | 22.2 | 22.8 | 23.4 | 24.4 | 23.6 | 23.6 |
| Day | 11 | 12 | 13 | 14 | 15 | 16 | 17 | 18 | 19 | 20 |
| Moving Average | 24.2 | 24.4 | 24.2 | 25.6 | 26.4 | 26.6 | 27.2 | 27.6 | 27.2 | 26.4 |

See the Technology Guides at the end of the chapter to find out how to tabulate and graph moving averages using a graphing calculator or a spreadsheet. Outline:

**TI-83/84 Plus**
STAT _EDIT; days in $L_1$, prices in $L_2$.
Home screen:
`seq((L₂(X)+L₂(X-1)`
`+L₂(X-2)+L₂(X-3)`
`+L₂(X-4))/5,X,5,20)`
`→L₃`
[More details on page 1079.]

**Spreadsheet**
Day data in A2–A21
Price data in B2–B21
Enter `=AVERAGE(B2:B6)` in C6, copy down to C21.
Graph the data in columns A–C.
[More details on page 1080.]

The closing stock prices and moving averages are plotted in Figure 11.

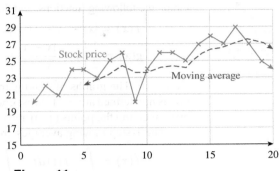

**Figure 11**

As you can see, the moving average is less volatile than the closing price. Because the moving average incorporates the stock's performance over 5 days at a time, a single day's fluctuation is smoothed out. Look at day 9 in particular. The moving average also tends to lag behind the actual performance because it takes past performance into account. Look at the downturns at days 6 and 18 in particular.

The period of 5 days for a moving average, as used in Example 3, is arbitrary. Using a longer period of time would smooth the data more but increase the lag. For data used as economic indicators, such as housing prices or retail sales, it is common to compute the four-quarter moving average to smooth out seasonal variations.

It is also sometimes useful to compute moving averages of continuous functions. We may want to do this if we use a mathematical model of a large collection of data. Also, some physical systems have the effect of converting an input function (an electrical signal, for example) into its moving average. By an **$n$-unit moving average** of a function $f(x)$ we mean the function $\bar{f}$ for which $\bar{f}(x)$ is the average of the value of $f(x)$ on $[x - n, x]$. Using the formula for the average of a function, we get the following formula.

---

**$n$-Unit Moving Average of a Function**

The $n$-unit moving average of a function $f$ is

$$\bar{f}(x) = \frac{1}{n} \int_{x-n}^{x} f(t) \, dt.$$

**Quick Example**

The 2-unit moving average of $f(x) = x^2$ is

$$\bar{f}(x) = \frac{1}{2} \int_{x-2}^{x} t^2 \, dt = \frac{1}{6} \left[ t^3 \right]_{x-2}^{x} = x^2 - 2x + \frac{4}{3}.$$

The graphs of $f(x)$ and $\bar{f}(x)$ are shown in Figure 12.

**Figure 12**

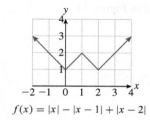

$$f(x) = |x| - |x-1| + |x-2|$$

$$g(x) = \frac{1}{2}\left[1 + \frac{|x-1|}{x-1}\right]$$

**Figure 13**

**TI-83/84 Plus**

We can graph the moving average of $f$ in Example 4 on a TI-83/84 Plus as follows:

```
Y₁=abs(X)-abs(X-1)
+abs(X-2)
Y₂=fnInt(Y1(T),T,
X-1,X)
ZOOM 0
```

For $g$, change $Y_1$ to

```
Y₁=.5*(1+abs(X-1)/
(X-1))
```

## EXAMPLE 4  Moving Averages: Sawtooth and Step Functions

Graph the following functions, and then compute and graph their 1-unit moving averages.

$$f(x) = |x| - |x-1| + |x-2| \qquad \text{Sawtooth}$$

$$g(x) = \frac{1}{2}\left[1 + \frac{|x-1|}{x-1}\right] \qquad \text{Unit step at } x = 1$$

**Solution**  The graphs of $f$ and $g$ are shown in Figure 13. (Notice that the step function is not defined at $x = 1$. Most graphers will show the step function as an actual step by connecting the points $(1, 0)$ and $(1, 1)$ with a vertical line.)

The 1-step moving averages are:

$$\bar{f}(x) = \int_{x-1}^{x} f(t)\,dt = \int_{x-1}^{x} [|t| - |t-1| + |t-2|]\,dt$$

$$= \frac{1}{2}[t|t| - (t-1)|t-1| + (t-2)|t-2|]_{x-1}^{x}$$

$$= \frac{1}{2}([x|x| - (x-1)|x-1| + (x-2)|x-2|]$$
$$- [(x-1)|x-1| - (x-2)|x-2| + (x-3)|x-3|])$$

$$= \frac{1}{2}[x|x| - 2(x-1)|x-1| + 2(x-2)|x-2| - (x-3)|x-3|].$$

$$\bar{g}(x) = \int_{x-1}^{x} g(t)\,dt = \int_{x-1}^{x} \frac{1}{2}\left[1 + \frac{|t-1|}{t-1}\right]dt$$

$$= \frac{1}{2}[t + |t-1|]_{x-1}^{x}$$

$$= \frac{1}{2}[(x + |x-1|) - (x-1+|x-2|)]$$

$$= \frac{1}{2}[1 + |x-1| - |x-2|]$$

The graphs of $\bar{f}$ and $\bar{g}$ are shown in Figure 14.

**Figure 14**

Notice how the graph of $\bar{f}$ smooths out the zig-zags of the sawtooth function.

**Figure 15**

➡ **Before we go on...**  Figure 15 shows the 2-unit moving average of $f$ in Example 4,

$$\bar{f}(x) = \frac{1}{2}\int_{x-2}^{x} f(t)\,dt$$

$$= \frac{1}{4}(x|x| - (x-1)|x-1| + (x-3)|x-3| - (x-4)|x-4|).$$

Notice how the 2-point moving average has completely eliminated the zig-zags, illustrating how moving averages can be used to remove seasonal fluctuations in real-life situations. ∎

## 14.3 EXERCISES

▼ more advanced   ◆ challenging

🔲 indicates exercises that should be solved using technology

*Find the averages of the functions in Exercises 1–8 over the given intervals. Plot each function and its average on the same graph (as in Figure 8).* HINT [See Quick Example page 1035.]

**1.** $f(x) = x^3$ over $[0, 2]$   **2.** $f(x) = x^3$ over $[-1, 1]$

**3.** $f(x) = x^3 - x$ over $[0, 2]$   **4.** $f(x) = x^3 - x$ over $[0, 1]$

**5.** $f(x) = e^{-x}$ over $[0, 2]$   **6.** $f(x) = e^x$ over $[-1, 1]$

**7.** $f(x) = |2x - 5|$ over $[0, 4]$

**8.** $f(x) = |-x + 2|$ over $[-1, 3]$

*In Exercises 9 and 10, complete the given table with the values of the 3-unit moving average of the given function.* HINT [See Example 3.]

**9.**

| $x$ | 0 | 1 | 2 | 3 | 4 | 5 | 6 | 7 |
|------|---|---|---|---|---|---|---|---|
| $r(x)$ | 3 | 5 | 10 | 3 | 2 | 5 | 6 | 7 |
| $\bar{r}(x)$ | | | | | | | | |

**10.**

| $x$ | 0 | 1 | 2 | 3 | 4 | 5 | 6 | 7 |
|------|---|---|---|---|---|---|---|---|
| $s(x)$ | 2 | 9 | 7 | 3 | 2 | 5 | 7 | 1 |
| $\bar{s}(x)$ | | | | | | | | |

*In Exercises 11 and 12, some values of a function and its 3-unit moving average are given. Supply the missing information.*

**11.**

| $x$ | 0 | 1 | 2 | 3 | 4 | 5 | 6 | 7 |
|------|---|---|---|---|---|---|---|---|
| $r(x)$ | 1 | 2 | | | 11 | | 10 | 2 |
| $\bar{r}(x)$ | | | 3 | 5 | | 11 | | |

**12.**

| $x$ | 0 | 1 | 2 | 3 | 4 | 5 | 6 | 7 |
|------|---|---|---|---|---|---|---|---|
| $s(x)$ | 1 | 5 | | 1 | | | | |
| $\bar{s}(x)$ | | | 5 | | 5 | 2 | 3 | 2 |

*Calculate the 5-unit moving average of each function in Exercises 13–24. Plot each function and its moving average on the same graph, as in Example 4. (You may use graphing technology for these plots, but you should compute the moving averages analytically.)* HINT [See Quick Example page 1037, and Example 4.]

**13.** $f(x) = x^3$   **14.** $f(x) = x^3 - x$

**15.** $f(x) = x^{2/3}$   **16.** $f(x) = x^{2/3} + x$

**17.** $f(x) = e^{0.5x}$   **18.** $f(x) = e^{-0.02x}$

**19.** $f(x) = \sqrt{x}$   **20.** $f(x) = x^{1/3}$

**21.** $f(x) = 1 - \dfrac{|2x - 1|}{2x - 1}$   **22.** $f(x) = 2 + \dfrac{|3x + 1|}{3x + 1}$

**23.** ▼ $f(x) = 2 - |x + 1| + |x|$   [Do not simplify the answer.]

**24.** ▼ $f(x) = |2x + 1| - |2x| - 2$   [Do not simplify the answer.]

🔲 *In Exercises 25–34, use graphing technology to plot the given functions together with their 3-unit moving averages.* HINT [See Technology Note for Example 4.]

**25.** $f(x) = \dfrac{10x}{1 + 5|x|}$   **26.** $f(x) = \dfrac{1}{1 + e^x}$

**27.** $f(x) = \ln(1 + x^2)$   **28.** $f(x) = e^{1-x^2}$

**29.** $f(x) = |x| - |x - 1| + |x - 2| - |x - 3| + |x - 4|$

**30.** $f(x) = |x| - 2|x - 1| + 2|x - 2| - 2|x - 3| + |x - 4|$

**31.** $f(x) = \dfrac{|x|}{x} - \dfrac{|x - 1|}{x - 1} + \dfrac{|x - 2|}{x - 2} - \dfrac{|x - 3|}{x - 3}$

**32.** $f(x) = \dfrac{|x|}{x} - 2\dfrac{|x - 1|}{x - 1} + 2\dfrac{|x - 2|}{x - 2} - \dfrac{|x - 3|}{x - 3}$

**33.** $f(x) = \dfrac{|x|}{x} + \dfrac{|x - 1|}{x - 1} + \dfrac{|x - 2|}{x - 2} + \dfrac{|x - 3|}{x - 3}$

**34.** $f(x) = 4 - \dfrac{|x|}{x} - \dfrac{|x - 1|}{x - 1} - \dfrac{|x - 2|}{x - 2} - \dfrac{|x - 3|}{x - 3}$

## APPLICATIONS

**35. Television Advertising** The cost, in millions of dollars, of a 30-second television ad during the Super Bowl in the years 2000 to 2010 can be approximated by

$$C(t) = 0.14t + 1.1 \text{ million dollars} \quad (0 \le t \le 10)$$

($t = 0$ represents 2000).[19] What was the average cost of a Super Bowl ad during the given period? HINT [See Example 1.]

**36. Television Advertising** The cost, in millions of dollars, of a 30-second television ad during the Super Bowl in the years 1980 to 2000 can be approximated by

$$C(t) = 0.044t + 0.222 \text{ million dollars} \quad (0 \le t \le 20)$$

($t = 0$ represents 1980).[20] What was the average cost of a Super Bowl ad during the given period? HINT [See Example 1.]

**37. Membership: Facebook** The number of new members joining Facebook each year in the period from 2005 to the middle of 2008 can be modeled by

$$m(t) = 12t^2 - 20t + 10 \text{ million members per year} \quad (0 \le t \le 3.5),$$

---

[19] Source for data: en.wikipedia.org/wiki/Super_Bowl_advertising.
[20] *Ibid.*

where $t$ is time in years since the start of 2005.[21] What was the average number of new members joining Facebook each year from the start of 2005 to the start of 2008?

**38. *Membership:* MySpace** The number of new members joining Myspace each year in the period from 2004 to the middle of 2007 can be modeled by

$$m(t) = 10.5t^2 + 14t - 6 \text{ million members per year}$$
$$(0 \le t \le 3.5),$$

where $t$ is time in years since the start of 2004.[22] What was the average number of new members joining MySpace each year from the start of 2004 to the start of 2007?

**39. *Freon Production*** Annual production of ozone-layer-damaging Freon 22 (chlorodifluoromethane) in developing countries from 2000 to 2010 can be modeled by

$$F(t) = 97.2(1.20)^t \text{ million tons} \quad (0 \le t \le 10)$$

($t$ is the number of years since 2000).[23] What was the average annual production over the period shown? (Round your answer to the nearest million tons.) HINT [See Example 2.]

**40. *Health Expenditures*** Annual expenditures on health in the United States from 1980 to 2010 could be modeled by

$$F(t) = 296(1.08)^t \text{ billion dollars} \quad (0 \le t \le 30)$$

($t = 0$ represents 1980).[24] What was the average annual expenditure over the period shown? (Round your answer to the nearest billion dollars.) HINT [See Example 2.]

**41. *Investments*** If you invest $10,000 at 8% interest compounded continuously, what is the average amount in your account over one year?

**42. *Investments*** If you invest $10,000 at 12% interest compounded continuously, what is the average amount in your account over one year?

**43. ▼ *Average Balance*** Suppose you have an account (paying no interest) into which you deposit $3,000 at the beginning of each month. You withdraw money continuously so that the amount in the account decreases linearly to 0 by the end of the month. Find the average amount in the account over a period of several months. (Assume that the account starts at $0 at $t = 0$ months.)

**44. ▼ *Average Balance*** Suppose you have an account (paying no interest) into which you deposit $4,000 at the beginning of each month. You withdraw $3,000 during the course of each month, in such a way that the amount decreases linearly. Find the average amount in the account in the first two months. (Assume that the account starts at $0 at $t = 0$ months.)

[21] Sources for data: www.facebook.com, insidehighered.com. (Some data are interpolated.)

[22] Source for data: www.swivel.com/data_sets.

[23] Figures are approximate. Source: Lampert Kuijpers (Panel of the Montreal Protocol), National Bureau of Statistics in China, via CEIC Data/*New York Times*, February 23, 2007, p. C1.

[24] Source for data: U.S. Department of Health & Human Services/Centers for Medicare & Medicaid Services, National Health Expenditure Data, downloaded April 2011 from www.cms.gov.

**45. ▯ *Online Vehicle Sales*** The value of vehicle transactions on eBay each quarter from the first quarter of 2008 through the last quarter of 2010 could be approximated by

$$v(t) = -0.002x^3 + 0.06x^2 - 0.55x + 3.9 \text{ billion dollars}$$
$$\text{per quarter} \quad (1 \le t \le 13)$$

($t$ is time in quarters; $t = 1$ represents the first quarter of 2008).[25] Use technology to estimate the average quarterly value of vehicle transactions on eBay during the given period. (Round your answer to the nearest hundred million dollars.)

**46. ▯ *Online Auctions*** The net number of new active eBay users each quarter from the first quarter of 2008 through the last quarter of 2010 could be approximated by

$$n(t) = -0.002x^4 + 0.06x^3 - 0.55x^2 + 1.9x - 1$$
$$\text{million users per quarter} \quad (1 \le t \le 13)$$

($t$ is time in quarters; $t = 1$ represents the first quarter of 2008).[26] Use technology to estimate the average number of new active eBay users each quarter during the given period. (Round your answer to the nearest hundred thousand users.)

**47. *Stock Prices:* Exxon Mobil** The following table shows the approximate price of Exxon Mobil stock in December of each year from 2001 through 2010.[27] Complete the table by computing the 4-year moving averages. (Note the peak in 2007 and the drop in subsequent years.) Round each average to the nearest dollar.

| Year $t$ | 2001 | 2002 | 2003 | 2004 | 2005 | 2006 | 2007 | 2008 | 2009 | 2010 |
|---|---|---|---|---|---|---|---|---|---|---|
| Stock Price | 39 | 35 | 41 | 51 | 56 | 77 | 94 | 80 | 68 | 73 |
| Moving Average (rounded) | | | | | | | | | | |

The stock price spiked in 2007 and then dropped steeply in 2008. What happened to the corresponding moving average? HINT [See Example 3.]

**48. *Stock Prices:* Nokia** The following table shows the approximate price of Nokia stock in December of each year from 2001 through 2010.[28] Complete the table by computing the 4-year moving averages. (Note the peak in 2007 and the drop in subsequent years.) Round each average to the nearest dollar.

| Year $t$ | 2001 | 2002 | 2003 | 2004 | 2005 | 2006 | 2007 | 2008 | 2009 | 2010 |
|---|---|---|---|---|---|---|---|---|---|---|
| Stock Price | 25 | 16 | 17 | 16 | 18 | 20 | 38 | 16 | 13 | 10 |
| Moving Average (rounded) | | | | | | | | | | |

[25] Source for data: eBay company reports http://investor.ebay.com/.

[26] *Ibid.*

[27] Source: finance.yahoo.com.

[28] *Ibid.*

How does the average year-by-year change in the moving average compare with the average year-by-year change in the stock price? HINT [See Example 3.]

49. ▼ *Cancun* The *Playa Loca Hotel* in Cancun has an advertising brochure with the following chart, showing the year-round temperature.[29]

a. Estimate and plot the year-round 6-month moving average. (Use graphing technology, if available, to check your graph.)

b. What can you say about the 12-month moving average?

50. ▼ *Reykjavik* Repeat the preceding exercise, using the following data from the brochure of the *Tough Traveler Lodge* in Reykjavik.[30]

51. ⬛ ▼ *Sales: Apple* The following table shows approximate quarterly sales of iPods in millions of units, starting in the first quarter of 2006.[31]

| Quarter | 2006 Q1 | 2006 Q2 | 2006 Q3 | 2006 Q4 | 2007 Q1 | 2007 Q2 | 2007 Q3 | 2007 Q4 | 2008 Q1 | 2008 Q2 |
|---|---|---|---|---|---|---|---|---|---|---|
| Sales (millions) | 8.5 | 8.1 | 8.7 | 21.1 | 10.5 | 9.8 | 10.2 | 22.1 | 10.6 | 11.0 |

| Quarter | 2008 Q3 | 2008 Q4 | 2009 Q1 | 2009 Q2 | 2009 Q3 | 2009 Q4 | 2010 Q1 | 2010 Q2 | 2010 Q3 | 2010 Q4 |
|---|---|---|---|---|---|---|---|---|---|---|
| Sales (millions) | 11.1 | 22.7 | 11.0 | 10.2 | 10.2 | 21.0 | 10.9 | 9.4 | 9.1 | 19.5 |

a. Use technology to compute and plot the four-quarter moving average of these data.

b. The graph of the moving average for the last eight quarters will appear almost linear during 2009 and 2010. Use the 2009 Q1 and 2010 Q4 figures of the moving average to give an estimate (to the nearest 0.1 million units) of the rate of change of iPod sales during 2009–2010.

52. ⬛ ▼ *Housing Starts* The following table shows the number of housing starts for one-family units, in thousands of units, starting in the first quarter of 2006.[32]

| Quarter | 2006 Q1 | 2006 Q2 | 2006 Q3 | 2006 Q4 | 2007 Q1 | 2007 Q2 | 2007 Q3 | 2007 Q4 | 2008 Q1 | 2008 Q2 |
|---|---|---|---|---|---|---|---|---|---|---|
| Housing Starts (1,000) | 382 | 433 | 372 | 278 | 260 | 333 | 265 | 188 | 162 | 194 |

| Quarter | 2008 Q3 | 2008 Q4 | 2009 Q1 | 2009 Q2 | 2009 Q3 | 2009 Q4 | 2010 Q1 | 2010 Q2 | 2010 Q3 | 2010 Q4 |
|---|---|---|---|---|---|---|---|---|---|---|
| Housing Starts (1,000) | 163 | 103 | 78 | 124 | 138 | 105 | 114 | 142 | 119 | 96 |

a. Use technology to compute and plot the four-quarter moving average of these data.

b. The graph of the moving average for the eight quarters in 2007 and 2008 will appear almost linear. Use the 2007 Q1 and 2008 Q4 figures of the moving average to give an estimate (to the nearest thousand units) of the rate of change of housing starts during 2007–2008.

53. *Bottled Water Sales* The rate of U.S. sales of bottled water for the period 2000–2010 could be approximated by

$$s(t) = -45t^2 + 900t + 4{,}200 \text{ million gallons per year}$$
$$(0 \le t \le 10),$$

where $t$ is time in years since the start of 2000.[33]

a. Estimate the average annual sales of bottled water over the period 2000–2010, to the nearest 100 million gallons per year. HINT [See Quick Example page 1035.]

b. Compute the two-year moving average of $s$. (You need not simplify the answer.) HINT [See Quick Example page 1037.]

c. Without simplifying the answer in part (b), say what kind of function the moving average is.

54. *Bottled Water Sales* The rate of U.S. per capita sales of bottled water for the period 2000–2010 coud be approximated by

$$s(t) = -0.18t^2 + 3t + 15 \text{ gallons per year} \quad (0 \le t \le 10),$$

where $t$ is the time in years since the start of 2000.[34] Repeat the preceding exercise as applied to per capita sales. (Give your answer to (a) to the nearest gallon per year.)

[29] Source: www.holiday-weather.com. (Temperatures are rounded.)

[30] *Ibid.*

[31] Source: Apple quarterly press releases, www.apple.com/investor.

[32] Source: U.S. Census Bureau, www.census.gov/const/www/newresconstindex.html.

[33] Source for data: Beverage Marketing Corporation/www.bottledwater.org.

[34] *Ibid.*

**55. *Medicare Spending*** Annual spending on Medicare was projected to increase from $526 billion in 2010 to around $977 billion in 2021.[35]

   **a.** Use this information to express $s$, the annual spending on Medicare (in billions of dollars), as a linear function of $t$, the number of years since 2010.

   **b.** Find the 4-year moving average of your model.

   **c.** What can you say about the slope of the moving average?

**56. *Pasta Imports in the 1990s*** In 1990, the United States imported 290 million pounds of pasta. From 1990 to 2000 imports increased by an average of 40 million pounds per year.[36]

   **a.** Use these data to express $q$, the annual U.S. imports of pasta (in millions of pounds), as a linear function of $t$, the number of years since 1990.

   **b.** Find the 4-year moving average of your model.

   **c.** What can you say about the slope of the moving average?

**57.** ▼ ***Moving Average of a Linear Function*** Find a formula for the $a$-unit moving average of a general linear function $f(x) = mx + b$.

**58.** ▼ ***Moving Average of an Exponential Function*** Find a formula for the $a$-unit moving average of a general exponential function $f(x) = Ae^{kx}$.

---

[35]Source: Congressional Budget Office, *March 2011 Medicare Baseline* (www.cbo.gov).

[36]Data are rounded. Sources: Department of Commerce/*New York Times*, September 5, 1995, p. D4, International Trade Administration (www.ita.doc.gov) March 31, 2002.

## COMMUNICATION AND REASONING EXERCISES

**59.** Explain why it is sometimes more useful to consider the moving average of a stock price rather than the stock price itself.

**60.** Sales this month were sharply lower than they were last month, but the 12-unit moving average this month was higher than it was last month. How can that be?

**61.** Your company's 6-month moving average of sales is constant. What does that say about the sales figures?

**62.** Your monthly salary has been increasing steadily for the past year, and your average monthly salary over the past year was $x$ dollars. Would you have earned more money if you had been paid $x$ dollars per month? Explain your answer.

**63.** ▼ What property does the graph of a (nonconstant) function have if its average value over an interval is zero? Give an example of such a function.

**64.** ▼ Can the average value of a function $f$ on an interval be greater than its value at every point in that interval? Explain.

**65.** ▼ Criticize the following claim: The average value of a function on an interval is midway between its highest and lowest value.

**66.** ▼ Your manager tells you that 12-month moving averages give at least as much information as shorter-term moving averages and very often more. How would you argue that he is wrong?

**67.** ▼ Which of the following most closely approximates the original function: (A) its 10-unit moving average, (B) its 1-unit moving average, or (C) its 0.8-unit moving average? Explain your answer.

**68.** ▼ Is an increasing function larger or smaller than its 1-unit moving average? Explain.

---

| **14.4** | **Applications to Business and Economics: Consumers' and Producers' Surplus and Continuous Income Streams** |

### Consumers' Surplus

**Figure 16**

Consider a general demand curve presented, as is traditional in economics, as $p = D(q)$, where $p$ is unit price and $q$ is demand measured, say, in annual sales (Figure 16). Thus, $D(q)$ is the price at which the demand will be $q$ units per year. The price $p_0$ shown on the graph is the highest price that customers are willing to pay.

Suppose, for example, that the graph in Figure 16 is the demand curve for a particular new model of computer. When the computer first comes out and supplies are low ($q$ is small), "early adopters" will be willing to pay a high price. This is the part of the graph on the left, near the $p$-axis. As supplies increase and the price drops, more consumers will be willing to pay and more computers will be sold. We can ask the following question: How much are consumers willing to spend for the first $\bar{q}$ units?

## Consumers' Willingness to Spend

We can approximate consumers' willingness to spend on the first $\bar{q}$ units as follows. We partition the interval $[0, \bar{q}]$ into $n$ subintervals of equal length, as we did when discussing Riemann sums. Figure 17 shows a typical subinterval, $[q_{k-1}, q_k]$.

The price consumers are willing to pay for each of units $q_{k-1}$ through $q_k$ is approximately $D(q_{k-1})$, so the total that consumers are willing to spend for these units is approximately $D(q_{k-1})(q_k - q_{k-1}) = D(q_{k-1})\Delta q$, the area of the shaded region in Figure 17. Thus, the total amount that consumers are willing to spend for items $0$ through $\bar{q}$ is

**Figure 17**

$$W \approx D(q_0)\Delta q + D(q_1)\Delta q + \cdots + D(q_{n-1})\Delta q = \sum_{k=0}^{n-1} D(q_k)\Delta q,$$

which is a Riemann sum. The approximation becomes better the larger $n$ becomes, and in the limit the Riemann sums converge to the integral

$$W = \int_0^{\bar{q}} D(q)\,dq.$$

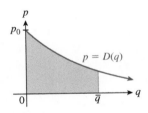

**Figure 18**

This quantity, the area shaded in Figure 18, is the total consumers' willingness to spend to buy the first $\bar{q}$ units.

## Consumers' Expenditure

Now suppose that the manufacturer simply sets the price at $\bar{p}$, with a corresponding demand of $\bar{q}$, so $D(\bar{q}) = \bar{p}$. Then the amount that consumers will actually spend to buy these $\bar{q}$ is $\bar{p}\,\bar{q}$, the product of the unit price and the quantity sold. This is the area of the rectangle shown in Figure 19. Notice that we can write $\bar{p}\,\bar{q} = \int_0^{\bar{q}} \bar{p}\,dq$, as suggested by the figure.

The difference between what consumers are willing to pay and what they actually pay is money in their pockets and is called the **consumers' surplus**.

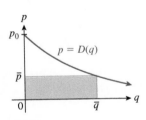

**Figure 19**

### Consumers' Surplus

If demand for an item is given by $p = D(q)$, the selling price is $\bar{p}$, and $\bar{q}$ is the corresponding demand [so that $D(\bar{q}) = \bar{p}$], then the **consumers' surplus** is the difference between willingness to spend and actual expenditure:✱

$$CS = \int_0^{\bar{q}} D(q)\,dq - \bar{p}\,\bar{q} = \int_0^{\bar{q}} (D(q) - \bar{p})\,dq.$$

Graphically, it is the area between the graphs of $p = D(q)$ and $p = \bar{p}$, as shown in the figure.

## EXAMPLE 1  Consumers' Surplus

Your used-CD store has an exponential demand equation of the form

$$p = 15e^{-0.01q}$$

where $q$ represents daily sales of used CDs and $p$ is the price you charge per CD. Calculate the daily consumers' surplus if you sell your used CDs at \$5 each.

**Solution**  We are given $D(q) = 15e^{-0.01q}$ and $\bar{p} = 5$. We also need $\bar{q}$. By definition,

$$D(\bar{q}) = \bar{p}$$

or   $15e^{-0.01\bar{q}} = 5,$

which we must solve for $\bar{q}$:

$$e^{-0.01\bar{q}} = \frac{1}{3}$$

$$-0.01\bar{q} = \ln\left(\frac{1}{3}\right) = -\ln 3$$

$$\bar{q} = \frac{\ln 3}{0.01} \approx 109.8612.$$

We now have

$$CS = \int_0^{\bar{q}} (D(q) - \bar{p})\, dq$$

$$= \int_0^{109.8612} (15e^{-0.01q} - 5)\, dq$$

$$= \left[\frac{15}{-0.01}e^{-0.01q} - 5q\right]_0^{109.8612}$$

$$\approx (-500 - 549.31) - (-1,500 - 0)$$

$$= \$450.69 \text{ per day.}$$

## Producers' Surplus

We can also calculate extra income earned by producers. Consider a supply equation of the form $p = S(q)$, where $S(q)$ is the price at which a supplier is willing to supply $q$ items (per time period). Because a producer is generally willing to supply more units at a higher price per unit, a supply curve usually has a positive slope, as shown in Figure 20. The price $p_0$ is the lowest price that a producer is willing to charge.

Arguing as before, we see that the minimum amount of money producers are willing to receive in exchange for $\bar{q}$ items is $\int_0^{\bar{q}} S(q)\, dq$. On the other hand, if the producers charge $\bar{p}$ per item for $\bar{q}$ items, their actual revenue is $\bar{p}\bar{q} = \int_0^{\bar{q}} \bar{p}\, dq$.

The difference between the producers' actual revenue and the minimum they would have been willing to receive is the **producers' surplus**.

**Figure 20**

### Producers' Surplus

The **producers' surplus** is the extra amount earned by producers who were willing to charge less than the selling price of $\bar{p}$ per unit and is given by

$$PS = \int_0^{\bar{q}} [\bar{p} - S(q)]\,dq,$$

where $S(\bar{q}) = \bar{p}$. Graphically, it is the area of the region between the graphs of $p = \bar{p}$ and $p = S(q)$ for $0 \le q \le \bar{q}$, as in the figure.

### EXAMPLE 2  Producers' Surplus

My tie-dye T-shirt enterprise has grown to the extent that I am now able to produce T-shirts in bulk, and several campus groups have begun placing orders. I have informed one group that I am prepared to supply $20\sqrt{p-4}$ T-shirts at a price of $p$ dollars per shirt. What is my total surplus if I sell T-shirts to the group at $8 each?

**Solution** We need to calculate the producers' surplus when $\bar{p} = 8$. The supply equation is

$$q = 20\sqrt{p-4}$$

but in order to use the formula for producers' surplus, we need to express $p$ as a function of $q$. First, we square both sides to remove the radical sign:

$$q^2 = 400(p-4)$$

so

$$p - 4 = \frac{q^2}{400}$$

giving

$$p = S(q) = \frac{q^2}{400} + 4.$$

We now need the value of $\bar{q}$ corresponding to $\bar{p} = 8$. Substituting $p = 8$ in the original equation gives

$$\bar{q} = 20\sqrt{8-4} = 20\sqrt{4} = 40.$$

Thus,

$$PS = \int_0^{\bar{q}} (\bar{p} - S(q))\,dq$$

$$= \int_0^{40} \left[ 8 - \left( \frac{q^2}{400} + 4 \right) \right] dq$$

$$= \int_0^{40} \left( 4 - \frac{q^2}{400} \right) dq$$

$$= \left[ 4q - \frac{q^3}{1{,}200} \right]_0^{40} \approx \$106.67.$$

Thus, I earn a surplus of $106.67 if I sell T-shirts to the group at $8 each.

**EXAMPLE 3 Equilibrium**

To continue the preceding example: A representative informs me that the campus group is prepared to order only $\sqrt{200(16 - p)}$ T-shirts at $p$ dollars each. I would like to produce as many T-shirts for them as possible but avoid being left with unsold T-shirts. Given the supply curve from the preceding example, what price should I charge per T-shirt, and what are the consumers' and producers' surpluses at that price?

**Solution**   The price that guarantees neither a shortage nor a surplus of T-shirts is the **equilibrium price**, the price where supply equals demand. We have

$$\text{Supply:} \quad q = 20\sqrt{p - 4}.$$

$$\text{Demand:} \quad q = \sqrt{200(16 - p)}.$$

Equating these gives

$$20\sqrt{p - 4} = \sqrt{200(16 - p)}$$

$$400(p - 4) = 200(16 - p),$$

$$400p - 1{,}600 = 3{,}200 - 200p$$

$$600p = 4{,}800$$

$$p = \$8 \text{ per T-shirt.}$$

We therefore take $\bar{p} = 8$ (which happens to be the price we used in the preceding example). We get the corresponding value for $q$ by substituting $p = 8$ into either the demand or supply equation:

$$\bar{q} = 20\sqrt{8 - 4} = 40.$$

Thus, $\bar{p} = 8$ and $\bar{q} = 40$.

We must now calculate the consumers' surplus and the producers' surplus. We calculated the producers' surplus for $\bar{p} = 8$ in the preceding example:

$$PS = \$106.67.$$

For the consumers' surplus, we must first express $p$ as a function of $q$ for the demand equation. Thus, we solve the demand equation for $p$ as we did for the supply equation and we obtain

$$\text{Demand:} \quad D(q) = 16 - \frac{q^2}{200}.$$

Therefore,

$$CS = \int_0^{\bar{q}} (D(q) - \bar{p}) \, dq$$

$$= \int_0^{40} \left[ \left( 16 - \frac{q^2}{200} \right) - 8 \right] dq$$

$$= \int_0^{40} \left( 8 - \frac{q^2}{200} \right) dq$$

$$= \left[ 8q - \frac{q^3}{600} \right]_0^{40} \approx \$213.33.$$

➡ **Before we go on...** Figure 21 shows both the consumers' surplus (top portion) and the producers' surplus (bottom portion) from Example 3. Because extra money in people's pockets is a good thing, the total of the consumers' and the producers' surpluses is called the **total social gain**. In this case it is

$$\text{Social gain} = CS + PS = 213.33 + 106.67 = \$320.00.$$

As you can see from the figure, the total social gain is also the area between two curves and equals

$$\int_0^{40} (D(q) - S(q))\, dq.$$

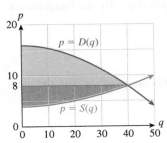

**Figure 21**

## Continuous Income Streams

For purposes of calculation, it is often convenient to assume that a company with a high sales volume receives money continuously. In such a case, we have a function $R(t)$ that represents the rate at which money is being received by the company at time $t$.

### EXAMPLE 4 **Continuous Income**

An ice cream store's business peaks in late summer; the store's summer revenue is approximated by

$$R(t) = 300 + 4.5t - 0.05t^2 \text{ dollars per day} \quad (0 \le t \le 92),$$

where $t$ is measured in days after June 1. What is its total revenue for the months of June, July, and August?

**Solution** Let's approximate the total revenue by breaking up the interval $[0, 92]$ representing the three months into $n$ subintervals $[t_{k-1}, t_k]$, each with length $\Delta t$. In the interval $[t_{k-1}, t_k]$ the store receives money at a rate of approximately $R(t_{k-1})$ dollars per day for $\Delta t$ days, so it will receive a total of $R(t_{k-1})\Delta t$ dollars. Over the whole summer, then, the store will receive approximately

$$R(t_0)\Delta t + R(t_1)\Delta t + \cdots + R(t_{n-1})\Delta t \text{ dollars.}$$

As we let $n$ become large to better approximate the total revenue, this Riemann sum approaches the integral

$$\text{Total revenue} = \int_0^{92} R(t)\, dt.$$

Substituting the function we were given, we get

$$\text{Total revenue} = \int_0^{92} (300 + 4.5t - 0.05t^2)\, dt$$

$$= \left[ 300t + 2.25t^2 - \frac{0.05}{3}t^3 \right]_0^{92}$$

$$\approx \$33{,}666.$$

➡ **Before we go on...** We could approach the calculation in Example 4 another way: $R(t) = S'(t)$, where $S(t)$ is the total revenue earned up to day $t$. By the Fundamental Theorem of Calculus,

$$\text{Total revenue} = S(92) - S(0) = \int_0^{92} R(t)\, dt.$$

We did the calculation using Riemann sums mainly as practice for the next example. ∎

Generalizing Example 4, we can say the following:

---

## Total Value of a Continuous Income Stream

If the rate of receipt of income is $R(t)$ dollars per unit of time, then the total income received from time $t = a$ to $t = b$ is

$$\text{Total value} = TV = \int_a^b R(t)\, dt.$$

---

### EXAMPLE 5  Future Value

Suppose the ice cream store in the preceding example deposits its receipts in an account paying 5% interest per year compounded continuously. How much money will it have in its account at the end of August?

**Solution**  Now we have to take into account not only the revenue but also the interest it earns in the account. Again, we break the interval $[0, 92]$ into $n$ subintervals. During the interval $[t_{k-1}, t_k]$, approximately $R(t_{k-1})\Delta t$ dollars are deposited in the account. That money will earn interest until the end of August, a period of $92 - t_{k-1}$ days, or $(92 - t_{k-1})/365$ years. The formula for continuous compounding tells us that by the end of August, those $R(t_{k-1})\Delta t$ dollars will have turned into

$$R(t_{k-1})\Delta t\, e^{0.05(92-t_{k-1})/365} = R(t_{k-1})e^{0.05(92-t_{k-1})/365}\Delta t \text{ dollars.}$$

(Recall that 5% is the *annual* interest rate.) Adding up the contributions from each subinterval, we see that the total in the account at the end of August will be approximately

$$R(t_0)e^{0.05(92-t_0)/365}\Delta t + R(t_1)e^{0.05(92-t_1)/365}\Delta t + \cdots + R(t_{n-1})e^{0.05(92-t_{n-1})/365}\Delta t.$$

This is a Riemann sum; as $n$ gets large the sum approaches the integral

$$\text{Future value} = FV = \int_0^{92} R(t)e^{0.05(92-t)/365}\, dt.$$

Substituting $R(t) = 300 + 4.5t - 0.05t^2$, we obtain

$$FV = \int_0^{92} (300 + 4.5t - 0.05t^2)e^{0.05(92-t)/365}\, dt$$

$$\approx \$33,\!880. \qquad\qquad\qquad \text{Using technology or integration by parts}$$

➡ **Before we go on...** The interest earned in the account in Example 5 was fairly small. (Compare this answer to that in Example 4.) Not only was the money in the account for only three months, but much of it was put in the account toward the end of that period, so had very little time to earn interest. ◼

Generalizing again, we have the following:

### Future Value of a Continuous Income Stream

If the rate of receipt of income from time $t = a$ to $t = b$ is $R(t)$ dollars per unit of time and the income is deposited as it is received in an account paying interest at rate $r$ per unit of time, compounded continuously, then the amount of money in the account at time $t = b$ is

$$\text{Future value} = FV = \int_a^b R(t)e^{r(b-t)}dt.$$

---

### EXAMPLE 6  Present Value

You are thinking of buying the ice cream store discussed in the preceding two examples. What is its income stream worth to you on June 1? Assume that you have access to the same account paying 5% per year compounded continuously.

**Solution**  The value of the income stream on June 1 is the amount of money that, if deposited June 1, would give you the same future value as the income stream will. If we let $PV$ denote this "present value," its value after 92 days will be

$$PVe^{0.05\times92/365}.$$

We equate this with the future value of the income stream to get

$$PVe^{0.05\times92/365} = \int_0^{92} R(t)e^{0.05(92-t)/365}\,dt$$

so

$$PV = \int_0^{92} R(t)e^{-0.05t/365}\,dt.$$

Substituting the formula for $R(t)$ and integrating using technology or integration by parts, we get

$$PV \approx \$33{,}455.$$

---

The general formula is the following:

### Present Value of a Continuous Income Stream

If the rate of receipt of income from time $t = a$ to $t = b$ is $R(t)$ dollars per unit of time and the income is deposited as it is received in an account paying interest at rate $r$ per unit of time, compounded continuously, then the value of the income stream at time $t = a$ is

$$\text{Present value} = PV = \int_a^b R(t)e^{r(a-t)}dt.$$

We can derive this formula from the relation

$$FV = PVe^{r(b-a)}$$

because the present value is the amount that would have to be deposited at time $t = a$ to give a future value of $FV$ at time $t = b$.

**Note** These formulas are more general than we've said. They still work when $R(t) < 0$ if we interpret negative values as money flowing *out* rather than in. That is, we can use these formulas for income we receive or for payments that we make, or for situations where we sometimes receive money and sometimes pay it out. These formulas can also be used for flows of quantities other than money. For example, if we use an exponential model for population growth and we let $R(t)$ represent the rate of immigration [$R(t) > 0$] or emigration [$R(t) < 0$], then the future value formula gives the future population. ∎

## 14.4  EXERCISES

▼ more advanced   ◆ challenging
**T** indicates exercises that should be solved using technology

*Calculate the consumers' surplus at the indicated unit price $\bar{p}$ for each of the demand equations in Exercises 1–12.* HINT [See Example 1.]

**1.** $p = 10 - 2q;\ \bar{p} = 5$

**2.** $p = 100 - q;\ \bar{p} = 20$

**3.** $p = 100 - 3\sqrt{q};\ \bar{p} = 76$

**4.** $p = 10 - 2q^{1/3};\ \bar{p} = 6$

**5.** $p = 500e^{-2q};\ \bar{p} = 100$

**6.** $p = 100 - e^{0.1q};\ \bar{p} = 50$

**7.** $q = 100 - 2p;\ \bar{p} = 20$

**8.** $q = 50 - 3p;\ \bar{p} = 10$

**9.** $q = 100 - 0.25p^2;\ \bar{p} = 10$

**10.** $q = 20 - 0.05p^2;\ \bar{p} = 5$

**11.** $q = 500e^{-0.5p} - 50;\ \bar{p} = 1$

**12.** $q = 100 - e^{0.1p};\ \bar{p} = 20$

*Calculate the producers' surplus for each of the supply equations in Exercises 13–24 at the indicated unit price $\bar{p}$.* HINT [See Example 2.]

**13.** $p = 10 + 2q;\ \bar{p} = 20$

**14.** $p = 100 + q;\ \bar{p} = 200$

**15.** $p = 10 + 2q^{1/3};\ \bar{p} = 12$

**16.** $p = 100 + 3\sqrt{q};\ \bar{p} = 124$

**17.** $p = 500e^{0.5q};\ \bar{p} = 1,000$

**18.** $p = 100 + e^{0.01q};\ \bar{p} = 120$

**19.** $q = 2p - 50;\ \bar{p} = 40$

**20.** $q = 4p - 1,000;\ \bar{p} = 1,000$

**21.** $q = 0.25p^2 - 10;\ \bar{p} = 10$

**22.** $q = 0.05p^2 - 20;\ \bar{p} = 50$

**23.** $q = 500e^{0.05p} - 50;\ \bar{p} = 10$

**24.** $q = 10(e^{0.1p} - 1);\ \bar{p} = 5$

*In Exercises 25–30, find the total value of the given income stream and also find its future value (at the end of the given interval) using the given interest rate.* HINT [See Examples 4 and 5.]

**25.** $R(t) = 30,000,\ 0 \le t \le 10$, at 7%

**26.** $R(t) = 40,000,\ 0 \le t \le 5$, at 10%

**27.** $R(t) = 30,000 + 1,000t,\ 0 \le t \le 10$, at 7%

**28.** $R(t) = 40,000 + 2,000t,\ 0 \le t \le 5$, at 10%

**29.** $R(t) = 30,000e^{0.05t},\ 0 \le t \le 10$, at 7%

**30.** $R(t) = 40,000e^{0.04t},\ 0 \le t \le 5$, at 10%

*In Exercises 31–36, find the total value of the given income stream and also find its present value (at the beginning of the given interval) using the given interest rate.* HINT [See Examples 4 and 6.]

**31.** $R(t) = 20,000,\ 0 \le t \le 5$, at 8%

**32.** $R(t) = 50,000,\ 0 \le t \le 10$, at 5%

**33.** $R(t) = 20,000 + 1,000t$, $0 \le t \le 5$, at 8%

**34.** $R(t) = 50,000 + 2,000t$, $0 \le t \le 10$, at 5%

**35.** $R(t) = 20,000e^{0.03t}$, $0 \le t \le 5$, at 8%

**36.** $R(t) = 50,000e^{0.06t}$, $0 \le t \le 10$, at 5%

## APPLICATIONS

**37.** *College Tuition* A study of U.S. colleges and universities resulted in the demand equation $q = 20,000 - 2p$, where $q$ is the enrollment at a public college or university and $p$ is the average annual tuition (plus fees) it charges.[37] Officials at Enormous State University have developed a policy whereby the number of students it will accept per year at a tuition level of $p$ dollars is given by $q = 7,500 + 0.5p$. Find the equilibrium tuition price $\bar{p}$ and the consumers' and producers' surpluses at this tuition level. What is the total social gain at the equilibrium price? HINT [See Example 3.]

**38.** *Fast Food* A fast-food outlet finds that the demand equation for its new side dish, "Sweetdough Tidbit," is given by

$$p = \frac{128}{(q+1)^2},$$

where $p$ is the price (in cents) per serving and $q$ is the number of servings that can be sold per hour at this price. At the same time, the franchise is prepared to sell $q = 0.5p - 1$ servings per hour at a price of $p$ cents. Find the equilibrium price $\bar{p}$ and the consumers' and producers' surpluses at this price level. What is the total social gain at the equilibrium price? HINT [See Example 3.]

**39.** *Revenue:* Nokia The annual net sales (revenue) earned by Nokia in the years January 2004 to January 2010 can be approximated by

$$R(t) = -1.75t^2 + 12.5t + 30 \text{ billion euros per year}$$
$$(0 \le t \le 6),$$

where $t$ is time in years ($t = 0$ represents January 2004).[38] Estimate, to the nearest €10 billion, Nokia's total revenue from January 2006 to January 2010. HINT [See Example 4.]

**40.** *Revenue:* Nintendo The annual net sales (revenue) earned by Nintendo Co., Ltd., in the fiscal years from April 1, 2004, to April 1, 2010, can be approximated by

$$R(t) = -14.5t^3 + 38t^2 + 380t + 510 \text{ billion yen per year}$$
$$(0 \le t \le 6),$$

where $t$ is time in years ($t = 0$ represents April 1, 2004).[39] Estimate, to the nearest ¥100 billion, Nintendo's total revenue from April 1, 2006, to April 1, 2010. HINT [See Example 4.]

**41.** *Revenue:* Walmart The annual revenue earned by Walmart in the years from January 2004 to January 2010 can be approximated by

$$R(t) = 242e^{0.098t} \text{ billion dollars per year} \quad (0 \le t \le 6),$$

where $t$ is time in years ($t = 0$ represents January 2004).[40] Estimate, to the nearest $10 billion, Wal-Mart's total revenue from January 2004 to January 2008.

**42.** *Revenue:* Target The annual revenue earned by Target for fiscal years 2004 through 2010 can be approximated by

$$R(t) = 41e^{0.094t} \text{ billion dollars per year} \quad (0 \le t \le 6),$$

where $t$ is time in years ($t = 0$ represents the beginning of fiscal year 2004).[41] Estimate, to the nearest $10 billion, Target's total revenue from the beginning of fiscal year 2006 to the beginning of fiscal year 2010.

**43.** ▼ *Revenue* Refer back to Exercise 39. Suppose that, from January 2004 on, Nokia invested its revenue in an investment yielding 4% compounded continuously. What, to the nearest €10 billion, would the total value of Nokia's revenue from January 2006 to January 2010 have been in January 2010? HINT [See Example 5.]

**44.** ▼ *Revenue* Refer back to Exercise 40. Suppose that, from April 2004 on, Nintendo invested its revenue in an investment yielding 5% compounded continuously. What, to the nearest ¥100 billion, would the total value of Nintendo's revenue from April 1, 2006, to April 1, 2010, have been in April 2010? HINT [See Example 5.]

**45.** ▼ *Revenue* Refer back to Exercise 41. Suppose that, from January 2004 on, Walmart invested its revenue in an investment that depreciated continuously at a rate of 5% per year. What, to the nearest $10 billion, would the total value of Walmart's revenues from January 2004 to January 2008 have been in January 2008?

**46.** ▼ *Revenue* Refer back to Exercise 42. Suppose that, from the start of fiscal year 2004 on, Target invested its revenue in an investment that depreciated continuously at a rate of 3% per year. What, to the nearest $10 billion, would the total value of Target's revenue from the beginning of fiscal year 2006 to the beginning of fiscal year 2010 have been at the beginning of fiscal year 2010?

---

[37] Idea based on a study by A. L. Ostrosky, Jr. and J. V. Koch, as cited in their book *Introduction to Mathematical Economics* (Waveland Press, Illinois, 1979, p. 133). The data used here are fictitious, however.

[38] Source for data: Nokia financial statements (www.investors.nokia.com).

[39] Source for data: Nintendo annual reports (www.nintendo.com/corp).

[40] Source for data: Wal-Mart annual reports (www.Walmartstores.com/Investors).

[41] Source for data: Target annual reports (investors.target.com).

**47.** ▼ *Saving for Retirement* You are saving for your retirement by investing $700 per month in an annuity with a guaranteed interest rate of 6% per year. With a continuous stream of investment and continuous compounding, how much will you have accumulated in the annuity by the time you retire in 45 years?

**48.** ▼ *Saving for College* When your first child is born, you begin to save for college by depositing $400 per month in an account paying 12% interest per year. With a continuous stream of investment and continuous compounding, how much will you have accumulated in the account by the time your child enters college 18 years later?

**49.** ▼ *Saving for Retirement* You begin saving for your retirement by investing $700 per month in an annuity with a guaranteed interest rate of 6% per year. You increase the amount you invest at the rate of 3% per year. With continuous investment and compounding, how much will you have accumulated in the annuity by the time you retire in 45 years?

**50.** ▼ *Saving for College* When your first child is born, you begin to save for college by depositing $400 per month in an account paying 12% interest per year. You increase the amount you save by 2% per year. With continuous investment and compounding, how much will have accumulated in the account by the time your child enters college 18 years later?

**51.** ▼ *Bonds* The U.S. Treasury issued a 30-year bond on February 10, 2011, paying 4.750% interest.[42] Thus, if you bought $100,000 worth of these bonds you would receive $4,750 per year in interest for 30 years. An investor wishes to buy the rights to receive the interest on $100,000 worth of these bonds. The amount the investor is willing to pay is the present value of the interest payments, assuming a 5% rate of return. Assuming (incorrectly, but approximately) that the interest payments are made continuously, what will the investor pay? HINT [See Example 6.]

**52.** ▼ *Bonds* The Megabucks Corporation is issuing a 20-year bond paying 7% interest. (See the preceding exercise.) An investor wishes to buy the rights to receive the interest on $50,000 worth of these bonds, and seeks a 6% rate of return. Assuming that the interest payments are made continuously, what will the investor pay? HINT [See Example 6.]

**53.** ▼ *Valuing Future Income* Inga was injured and can no longer work. As a result of a lawsuit, she is to be awarded the present value of the income she would have received over the next 20 years. Her income at the time she was injured was $100,000 per year, increasing by $5,000 per year. What will be the amount of her award, assuming continuous income and a 5% interest rate?

**54.** ▼ *Valuing Future Income* Max was injured and can no longer work. As a result of a lawsuit, he is to be awarded the present value of the income he would have received over the next 30 years. His income at the time he was injured was $30,000 per year, increasing by $1,500 per year. What will be the amount of his award, assuming continuous income and a 6% interest rate?

## COMMUNICATION AND REASONING EXERCISES

**55.** Complete the following: The future value of a continuous income stream earning 0% interest is the same as the _____ value.

**56.** Complete the following: The present value of a continuous income stream earning 0% interest is the same as the _____ value.

**57.** ▼ *Linear Demand* Given a linear demand equation $q = -mp + b$ $(m > 0)$, find a formula for the consumers' surplus at a price level of $\bar{p}$ per unit.

**58.** ▼ *Linear Supply* Given a linear supply equation of the form $q = mp + b$ $(m > 0)$, find a formula for the producers' surplus at a price level of $\bar{p}$ per unit.

**59.** ▼ Your study group friend says that the future value of a continuous stream of income is always greater than the total value, assuming a positive rate of return. Is she correct? Why?

**60.** ▼ Your other study group friend says that the present value of a continuous stream of income can sometimes be greater than the total value, depending on the (positive) interest rate. Is he correct? Explain.

**61.** ▼ Arrange from smallest to largest: Total Value, Future Value, Present Value of a continuous stream of income (assuming a positive income and positive rate of return).

**62.** **a.** Arrange the following functions from smallest to largest $R(t)$, $R(t)e^{r(b-t)}$, $R(t)e^{r(a-t)}$, where $a \le t \le b$, and $r$ and $R(t)$ are positive.

   **b.** Use the result from part (a) to justify your answers in Exercises 59–61.

---

[42]Source: The Bureau of the Public Debt (www.publicdebt.treas.gov).

# 14.5 Improper Integrals and Applications

All the definite integrals we have seen so far have had the form $\int_a^b f(x)\,dx$, with $a$ and $b$ finite and $f(x)$ piecewise continuous on the closed interval $[a, b]$. If we relax one or both of these requirements somewhat, we obtain what are called **improper integrals**. There are various types of improper integrals.

## Integrals in Which a Limit of Integration Is Infinite

Integrals in which one or more limits of integration are infinite can be written as

$$\int_a^{+\infty} f(x)\,dx, \quad \int_{-\infty}^b f(x)\,dx, \quad \text{or} \quad \int_{-\infty}^{+\infty} f(x)\,dx.$$

Let's concentrate for a moment on the first form, $\int_a^{+\infty} f(x)\,dx$. What does the $+\infty$ mean here? As it often does, it means that we are to take a limit as something gets large. Specifically, it means the limit as the upper bound of integration gets large.

### Improper Integral with an Infinite Limit of Integration

We define

$$\int_a^{+\infty} f(x)\,dx = \lim_{M \to +\infty} \int_a^M f(x)\,dx,$$

provided the limit exists. If the limit exists, we say that $\int_a^{+\infty} f(x)\,dx$ **converges**. Otherwise, we say that $\int_a^{+\infty} f(x)\,dx$ **diverges**. Similarly, we define

$$\int_{-\infty}^b f(x)\,dx = \lim_{M \to -\infty} \int_M^b f(x)\,dx,$$

provided the limit exists. Finally, we define

$$\int_{-\infty}^{+\infty} f(x)\,dx = \int_{-\infty}^a f(x)\,dx + \int_a^{+\infty} f(x)\,dx$$

for some convenient $a$, provided *both* integrals on the right converge.

#### Quick Examples

1. $\displaystyle \int_1^{+\infty} \frac{dx}{x^2} = \lim_{M \to +\infty} \int_1^M \frac{dx}{x^2} = \lim_{M \to +\infty} \left[ -\frac{1}{x} \right]_1^M = \lim_{M \to +\infty} \left( -\frac{1}{M} + 1 \right) = 1$

   Converges

2. $\displaystyle \int_1^{+\infty} \frac{dx}{x} = \lim_{M \to +\infty} \int_1^M \frac{dx}{x} = \lim_{M \to +\infty} [\ln |x|]_1^M = \lim_{M \to +\infty} (\ln M - \ln 1) = +\infty$

   Diverges

3. $\displaystyle \int_{-\infty}^{-1} \frac{dx}{x^2} = \lim_{M \to -\infty} \int_M^{-1} \frac{dx}{x^2} = \lim_{M \to -\infty} \left[ -\frac{1}{x} \right]_M^{-1} = \lim_{M \to -\infty} \left( 1 + \frac{1}{M} \right) = 1$

   Converges

**4.** $\displaystyle\int_{-\infty}^{+\infty} e^{-x}\,dx = \int_{-\infty}^{0} e^{-x}\,dx + \int_{0}^{+\infty} e^{-x}\,dx$

$\displaystyle\qquad = \lim_{M\to-\infty}\int_{M}^{0} e^{-x}\,dx + \lim_{M\to+\infty}\int_{0}^{M} e^{-x}\,dx$

$\displaystyle\qquad = \lim_{M\to-\infty} -[e^{-x}]_{M}^{0} + \lim_{M\to+\infty} -[e^{-x}]_{0}^{M}$

$\displaystyle\qquad = \lim_{M\to-\infty} (e^{-M} - 1) + \lim_{M\to+\infty} (1 - e^{-M})$

$\displaystyle\qquad = +\infty + 1$           Diverges

**5.** $\displaystyle\int_{-\infty}^{+\infty} xe^{-x^2}\,dx = \int_{-\infty}^{0} xe^{-x^2}\,dx + \int_{0}^{+\infty} xe^{-x^2}\,dx$

$\displaystyle\qquad = \lim_{M\to-\infty}\int_{M}^{0} xe^{-x^2}\,dx + \lim_{M\to+\infty}\int_{0}^{M} xe^{-x^2}\,dx$

$\displaystyle\qquad = \lim_{M\to-\infty}\left[-\frac{1}{2}e^{-x^2}\right]_{M}^{0} + \lim_{M\to+\infty}\left[-\frac{1}{2}e^{-x^2}\right]_{0}^{M}$

$\displaystyle\qquad = \lim_{M\to-\infty}\left(-\frac{1}{2} + \frac{1}{2}e^{-M^2}\right) + \lim_{M\to+\infty}\left(-\frac{1}{2}e^{-M^2} + \frac{1}{2}\right)$

$\displaystyle\qquad = -\frac{1}{2} + \frac{1}{2} = 0$           Converges

**Q**: *We learned that the integral can be interpreted as the area under the curve. Is this still true for improper integrals?*

**A**: Yes. Figure 22 illustrates how we can represent an improper integral as the area of an infinite region.

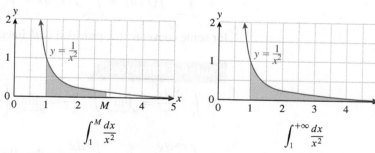

$$\int_{1}^{M} \frac{dx}{x^2} \qquad\qquad\qquad \int_{1}^{+\infty} \frac{dx}{x^2}$$

**Figure 22**

On the left we see the area represented by $\int_{1}^{M} dx/x^2$. As $M$ gets larger, the integral approaches $\int_{1}^{+\infty} dx/x^2$. In the picture, think of $M$ being moved farther and farther along the $x$-axis in the direction of increasing $x$, resulting in the region shown on the right.

**Q** : *Wait! We calculated $\int_1^{+\infty} dx/x^2 = 1$. Does this mean that the infinitely long area in Figure 22 has an area of only 1 square unit?*

**A** : That is exactly what it means. If you had enough paint to cover 1 square unit, you would never run out of paint while painting the region in Figure 22. This is one of the places where mathematics seems to contradict common sense. But common sense is notoriously unreliable when dealing with infinities.

Andy Crawford/Dorling Kindersley/
Getty Images

### EXAMPLE 1 Future Sales of CDs

By 2009, music downloads were making serious inroads into the sales of physical CDs. Approximately 290 million CD albums were sold in 2009 and sales declined by about 23% per year the following year.[43] Suppose that this rate of decrease were to continue indefinitely and continuously. How many CD albums, total, would be sold from 2009 on?

**Solution** Recall that the total sales between two dates can be computed as the definite integral of the rate of sales. So, if we wanted the sales between 2009 and a time far in the future, we would compute $\int_0^M s(t)\,dt$ with a large $M$, where $s(t)$ is the annual sales $t$ years after 2009. Because we want to know the *total* number of CD albums sold from 2009 on, we let $M \to +\infty$; that is, we compute $\int_0^{+\infty} s(t)\,dt$.

Because sales of CD albums are decreasing by 23% per year, we can model $s(t)$ by

$$s(t) = 290(0.77)^t \text{ million CD albums per year,}$$

where $t$ is the number of years since 2009.

 **using Technology**

You can estimate the integral in Example 1 with technology by computing $\int_0^M 290(0.77)^t\,dt$ for $M = 10, 100, 1000, \ldots$. You will find that the resulting values appear to converge to about 1,110. (Stop when the effect of further increases of $M$ has no effect at this level of accuracy.)

**TI-83/84 Plus**
$Y_1=290*0.77^X$
Home screen:
fnInt(Y$_1$,X,0,10)
fnInt(Y$_1$,X,0,100)
fnInt(Y$_1$,X,0,1000)

**Website**
www.WanerMath.org:

On-Line Utilities

→ Numerical Integration
   Utility and Grapher
Enter
290*0.77^X
for $f(x)$. Enter 0 and 10 for the lower and upper limits and press "Integral" for the most accurate estimate of the integral.
Repeat with the upper limit set to 100, 1000, and higher.

$$
\begin{aligned}
\text{Total sales from 2009 on} &= \int_0^{+\infty} 290(0.77)^t\,dt \\
&= \lim_{M \to +\infty} \int_0^M 290(0.77)^t\,dt \\
&= \frac{290}{\ln 0.77} \lim_{M \to +\infty} [(0.77)^t]_0^M \\
&= \frac{290}{\ln 0.77} \lim_{M \to +\infty} (0.77^M - 0.77^0) \\
&= \frac{290}{\ln 0.77}(-1) \\
&\approx 1{,}110 \text{ million CD albums}
\end{aligned}
$$

## Integrals in Which the Integrand Becomes Infinite

We can sometimes compute integrals $\int_a^b f(x)\,dx$ in which $f(x)$ becomes infinite. As we'll see in Example 4, the Fundamental Theorem of Calculus does not work for such integrals. The first case to consider is when $f(x)$ approaches $\pm\infty$ at either $a$ or $b$.

───────

[43]Source: *2010 Year-End Shipment Statistics*, Recording Industry Association of America (www.riaa.com).

**Figure 23**

**Figure 24**

### EXAMPLE 2 Integrand Infinite at One Endpoint

Calculate $\int_0^1 \dfrac{1}{\sqrt{x}}\, dx$.

**Solution** Notice that the integrand approaches $+\infty$ as $x$ approaches 0 from the right and is not defined at 0. This makes the integral an improper integral. Figure 23 shows the region whose area we are trying to calculate; it extends infinitely vertically rather than horizontally.

Now, if $0 < r < 1$, the integral $\int_r^1 (1/\sqrt{x})\, dx$ is a proper integral because we avoid the bad behavior at 0. This integral gives the area shown in Figure 24. If we let $r$ approach 0 from the right, the area in Figure 24 will approach the area in Figure 23. So, we calculate

$$\int_0^1 \frac{1}{\sqrt{x}}\, dx = \lim_{r \to 0^+} \int_r^1 \frac{1}{\sqrt{x}}\, dx$$

$$= \lim_{r \to 0^+} \left[ 2\sqrt{x} \right]_r^1$$

$$= \lim_{r \to 0^+} (2 - 2\sqrt{r})$$

$$= 2.$$

Thus, we again have an infinitely long region with finite area.

Generalizing, we make the following definition.

---

### Improper Integral in Which the Integrand Becomes Infinite

If $f(x)$ is defined for all $x$ with $a < x \le b$ but approaches $\pm\infty$ as $x$ approaches $a$, we define

$$\int_a^b f(x)\, dx = \lim_{r \to a^+} \int_r^b f(x)\, dx$$

provided the limit exists. Similarly, if $f(x)$ is defined for all $x$ with $a \le x < b$ but approaches $\pm\infty$ as $x$ approaches $b$, we define

$$\int_a^b f(x)\, dx = \lim_{r \to b^-} \int_a^r f(x)\, dx$$

provided the limit exists. In either case, if the limit exists, we say that $\int_a^b f(x)\, dx$ **converges**. Otherwise, we say that $\int_a^b f(x)\, dx$ **diverges**.

---

**Note** We saw in Chapter 13 that the Fundamental Theorem of Calculus applies to piecewise continuous functions as well as continuous ones. Examples are $f(x) = |x|/x$ and $(x^2 - 1)/(x - 1)$. The integrals of such functions are not improper, and we can use the Fundamental Theorem of Calculus to evaluate such integrals in the usual way. ■

## EXAMPLE 3  Testing for Convergence

Does $\int_{-1}^{3} \dfrac{x}{x^2 - 9}\, dx$ converge? If so, to what?

**Solution**  We first check to see where, if anywhere, the integrand approaches $\pm\infty$. That will happen where the denominator becomes 0, so we solve $x^2 - 9 = 0$.

$$x^2 - 9 = 0$$
$$x^2 = 9$$
$$x = \pm 3$$

The solution $x = -3$ is outside of the range of integration, so we ignore it. The solution $x = 3$ is, however, the right endpoint of the range of integration, so the integral is improper. We need to investigate the following limit:

$$\int_{-1}^{3} \frac{x}{x^2 - 9}\, dx = \lim_{r \to 3^-} \int_{-1}^{r} \frac{x}{x^2 - 9}\, dx.$$

Now, to calculate the integral we use a substitution:

$$u = x^2 - 9$$
$$\frac{du}{dx} = 2x$$
$$dx = \frac{1}{2x}\, du$$
when $x = r$, $u = r^2 - 9$
when $x = -1$, $u = (-1)^2 - 9 = -8$

Thus,

$$\int_{-1}^{r} \frac{x}{x^2 - 9}\, dx = \int_{-8}^{r^2 - 9} \frac{1}{2u}\, du$$

$$= \frac{1}{2}[\ln |u|]_{-8}^{r^2 - 9}$$

$$= \frac{1}{2}(\ln |r^2 - 9| - \ln 8).$$

Now we take the limit:

$$\int_{-1}^{3} \frac{x}{x^2 - 9}\, dx = \lim_{r \to 3^-} \int_{-1}^{r} \frac{x}{x^2 - 9}\, dx$$

$$= \lim_{r \to 3^-} \frac{1}{2}(\ln |r^2 - 9| - \ln 8)$$

$$= -\infty$$

because, as $r \to 3$, $r^2 - 9 \to 0$, and so $\ln |r^2 - 9| \to -\infty$. Thus, this integral diverges.

---

### EXAMPLE 4 **Integrand Infinite between the Endpoints**

Does $\displaystyle\int_{-2}^{3} \frac{1}{x^2}\, dx$ converge? If so, to what?

**Solution** Again we check to see if there are any points at which the integrand approaches $\pm\infty$. There is such a point, at $x = 0$. This is between the endpoints of the range of integration. To deal with this we break the integral into two integrals:

$$\int_{-2}^{3} \frac{1}{x^2}\, dx = \int_{-2}^{0} \frac{1}{x^2}\, dx + \int_{0}^{3} \frac{1}{x^2}\, dx.$$

Each integral on the right is an improper integral with the integrand approaching $\pm\infty$ at an endpoint. If both of the integrals on the right converge, we take the sum as the value of the integral on the left. So now we compute

$$\int_{-2}^{0} \frac{1}{x^2}\, dx = \lim_{r \to 0^-} \int_{-2}^{r} \frac{1}{x^2}\, dx$$

$$= \lim_{r \to 0^-} \left[ -\frac{1}{x} \right]_{-2}^{r}$$

$$= \lim_{r \to 0^-} \left( -\frac{1}{r} - \frac{1}{2} \right),$$

which diverges to $+\infty$. There is no need now to check $\int_{0}^{3} (1/x^2)\, dx$; because one of the two pieces of the integral diverges, we simply say that $\int_{-2}^{3} (1/x^2)\, dx$ diverges.

---

➡ **Before we go on...** What if we had been sloppy in Example 4 and had not checked first whether the integrand approached $\pm\infty$ somewhere? Then we probably would have applied the Fundamental Theorem of Calculus and done the following:

$$\int_{-2}^{3} \frac{1}{x^2}\, dx = \left( -\frac{1}{x} \right)_{-2}^{3} = \left( -\frac{1}{3} - \frac{1}{2} \right) = -\frac{5}{6}. \qquad ✗ \quad \textit{WRONG!}$$

Notice that the answer this "calculation" gives is patently ridiculous. Because $1/x^2 > 0$ for all $x$ for which it is defined, any definite integral of $1/x^2$ must give a positive answer. *Moral:* Always check to see whether the integrand blows up anywhere in the range of integration. If it does, the FTC does not apply, and we must use the methods of this example. ∎

We end with an example of what to do if an integral is improper for more than one reason.

---

### EXAMPLE 5 **An Integral Improper in Two Ways**

Does $\displaystyle\int_{0}^{+\infty} \frac{1}{\sqrt{x}}\, dx$ converge? If so, to what?

**Solution** This integral is improper for two reasons. First, the range of integration is infinite. Second, the integrand blows up at the endpoint 0. In order to separate these two problems, we break up the integral at some convenient point:

$$\int_0^{+\infty} \frac{1}{\sqrt{x}}\, dx = \int_0^1 \frac{1}{\sqrt{x}}\, dx + \int_1^{+\infty} \frac{1}{\sqrt{x}}\, dx$$

We chose to break the integral at 1. Any positive number would have sufficed, but 1 is generally easier to use in calculations.

The first piece, $\int_0^1 (1/\sqrt{x})\, dx$, we discussed in Example 2; it converges to 2. For the second piece we have

$$\int_1^{+\infty} \frac{1}{\sqrt{x}}\, dx = \lim_{M \to +\infty} \int_1^M \frac{1}{\sqrt{x}}\, dx$$

$$= \lim_{M \to +\infty} [2\sqrt{x}]_1^M$$

$$= \lim_{M \to +\infty} (2\sqrt{M} - 2),$$

which diverges to $+\infty$. Because the second piece of the integral diverges, we conclude that $\int_0^{+\infty} (1/\sqrt{x})\, dx$ diverges.

---

# 14.5 EXERCISES

▼ more advanced    ◆ challenging
T indicates exercises that should be solved using technology

*Note: For some of the exercises in this section you need to assume the fact that* $\lim_{M \to +\infty} M^n e^{-M} = 0$ *for all* $n \geq 0$. *(See Exercises 59 and 60 in Section 10.1 and Exercise 103 in Section 10.3.)*

*In Exercises 1–26, decide whether or not the given integral converges. If the integral converges, compute its value.*
HINT [See Quick Examples page 1053.]

**1.** $\int_1^{+\infty} x\, dx$

**2.** $\int_0^{+\infty} e^{-x}\, dx$

**3.** $\int_{-2}^{+\infty} e^{-0.5x}\, dx$

**4.** $\int_1^{+\infty} \frac{1}{x^{1.5}}\, dx$

**5.** $\int_{-\infty}^2 e^x\, dx$

**6.** $\int_{-\infty}^{-1} \frac{1}{x^{1/3}}\, dx$

**7.** $\int_{-\infty}^{-2} \frac{1}{x^2}\, dx$

**8.** $\int_{-\infty}^0 e^{-x}\, dx$

**9.** $\int_0^{+\infty} x^2 e^{-6x}\, dx$

**10.** $\int_0^{+\infty} (2x - 4)e^{-x}\, dx$

**11.** $\int_0^5 \frac{2}{x^{1/3}}\, dx$    HINT [See Example 2.]

**12.** $\int_0^2 \frac{1}{x^2}\, dx$

**13.** $\int_{-1}^2 \frac{3}{(x+1)^2}\, dx$    HINT [See Example 3.]

**14.** $\int_{-1}^2 \frac{3}{(x+1)^{1/2}}\, dx$

**15.** $\int_{-1}^2 \frac{3x}{x^2 - 1}\, dx$    HINT [See Example 4.]

**16.** $\int_{-1}^2 \frac{3}{x^{1/3}}\, dx$

**17.** $\int_{-2}^2 \frac{1}{(x+1)^{1/5}}\, dx$

**18.** $\int_{-2}^2 \frac{2x}{\sqrt{4 - x^2}}\, dx$

**19.** $\int_{-1}^1 \frac{2x}{x^2 - 1}\, dx$

**20.** $\int_{-1}^2 \frac{2x}{x^2 - 1}\, dx$

**21.** $\int_{-\infty}^{+\infty} xe^{-x^2}\, dx$

**22.** $\int_{-\infty}^{\infty} xe^{1-x^2}\, dx$

**23.** $\int_0^{+\infty} \frac{1}{x \ln x}\, dx$    HINT [See Example 5.]

**24.** $\int_0^{+\infty} \ln x\, dx$

**25.** ▼ $\int_0^{+\infty} \frac{2x}{x^2 - 1}\, dx$

**26.** ▼ $\int_{-\infty}^0 \frac{2x}{x^2 - 1}\, dx$

**T** *In Exercises 27–34, use technology to approximate the given integrals with* $M = 10, 100, 1,000, \ldots$ *and hence decide whether the associated improper integral converges and estimate its value to four significant digits if it does.* HINT [See the Technology Note for Example 1.]

**27.** $\int_1^M \dfrac{1}{x^2}\,dx$

**28.** $\int_0^M e^{-x^2}\,dx$

**29.** $\int_0^M \dfrac{x}{1+x}\,dx$

**30.** $\int_{1/M}^1 \dfrac{1}{\sqrt{x}}\,dx$

**31.** $\int_{1+1/M}^2 \dfrac{1}{\sqrt{x-1}}\,dx$

**32.** $\int_1^M \dfrac{1}{x}\,dx$

**33.** $\int_0^{1-1/M} \dfrac{1}{(1-x)^2}\,dx$

**34.** $\int_0^{2-1/M} \dfrac{1}{(2-x)^3}\,dx$

## APPLICATIONS

**35. New Home Sales** Sales of new homes in the United States decreased dramatically from 2005 to 2010 as shown in the model

$$n(t) = 1.33e^{-0.299t} \text{ million homes per year} \quad (0 \le t \le 5),$$

where $t$ is the year since 2005.[44] If this trend were to have continued into the indefinite future, estimate the total number of new homes that would have been sold in the United States from 2005 on. HINT [See Example 1.]

**36. Revenue from New Home Sales** Revenue from the sale of new homes in the United States decreased dramatically from 2005 to 2010 as shown in the model

$$r(t) = 412e^{-0.323t} \text{ billion dollars per year} \quad (0 \le t \le 5),$$

where $t$ is the year since 2005.[45] If this trend were to have continued into the indefinite future, estimate the total revenue from the sale of new homes in the United States from 2005 on. HINT [See Example 1.]

**37. Cigarette Sales** According to data published by the Federal Trade Commission, the number of cigarettes sold domestically has been decreasing by about 3% per year from the 2000 total of about 415 billion.[46] Use an exponential model to forecast the total number of cigarettes sold from 2000 on. (Round your answer to the nearest 100 billion cigarettes.) HINT [Use a model of the form $Ab^t$.]

**38. Sales** Sales of the text *Calculus and You* have been declining continuously at a rate of 5% per year. Assuming that

*Calculus and You* currently sells 5,000 copies per year and that sales will continue this pattern of decline, calculate total future sales of the text. HINT [Use a model of the form $Ae^{rt}$.]

**39. ▼ Sales** My financial adviser has predicted that annual sales of Frodo T-shirts will continue to decline by 10% each year. At the moment, I have 3,200 of the shirts in stock and am selling them at a rate of 200 per year. Will I ever sell them all?

**40. ▼ Revenue** Alarmed about the sales prospects for my Frodo T-shirts (see the preceding exercise), I will try to make up lost revenues by increasing the price by $1 each year. I now charge $10 per shirt. What is the total amount of revenue I can expect to earn from sales of my T-shirts, assuming the sales levels described in the previous exercise? (Give your answer to the nearest $1,000.)

**41. ▼ Education** Let $N(t)$ be the number of high school students graduated in the United States in year $t$. This number has been changing at a rate of about

$$N'(t) = 0.22t^{-0.91} \text{ million graduates per year} \quad (2 \le t \le 9),$$

where $t$ is time in years since 2000.[47] In 2002, there were about 2.7 million high school students graduated. By extrapolating the model, what can you say about the number of high school students graduated in a year far in the future?

**42. ▼ Education, Martian** Let $M(t)$ be the number of high school students graduated in the Republic of Mars in year $t$. This number is projected to change at a rate of about

$$M'(t) = 0.321t^{-1.10} \text{ thousand graduates per year}$$
$$(1 \le t \le 50),$$

where $t$ is time in years since 2020. In 2021, there were about 1,300 high school students graduated. By extrapolating the model, what can you say about the number of high school students graduated in a year far in the future?

**43. ▼ Cellphone Revenues** The number of cellphone subscribers in China in the early 2000s was projected to follow the equation,[48]

$$N(t) = 39t + 68 \text{ million subscribers}$$

in year $t$ ($t = 0$ represents 2000). The average annual revenue per cellphone user was $350 in 2000.

**a.** Assuming that, due to competition, the revenue per cellphone user decreases continuously at an annual rate of 10%, give a formula for the annual revenue in year $t$.

**b.** Using the model you obtained in part (a) as an estimate of the rate of change of total revenue, estimate the total revenue from 2000 into the indefinite future.

---

[44]Based on new home sales data from the U.S. Census Bureau (www.census.gov/const/www/newressalesindex.html).

[45]*Ibid.*

[46]Source for data: Federal Trade Commission Cigarette Report for 2007 and 2008, issued July 2011 (www.ftc.gov).

[47]Based on a regression model. Source for Data: U.S. Department of Education, National Center for Education Statistics, *Digest of Education Statistics: 2010*, April 2011 (nces.ed.gov).

[48]Based on a regression of projected figures (coefficients are rounded). Source: Intrinsic Technology/*New York Times*, Nov. 24, 2000, p. C1.

**44.** ▼ *Vidphone Revenues* The number of vidphone subscribers in the Republic of Mars for the period 2200–2300 was projected to follow the equation

$$N(t) = 18t - 10 \text{ thousand subscribers}$$

in year $t$ ($t = 0$ represents 2200). The average annual revenue per vidphone user was $\overline{\overline{Z}}40$ in 2200.[49]

**a.** Assuming that, due to competition, the revenue per vidphone user decreases continuously at an annual rate of 20%, give a formula for the annual revenue in year $t$.

**b.** Using the model you obtained in part (a) as an estimate of the rate of change of total revenue, estimate the total revenue from 2200 into the indefinite future.

**45.** 🔢 *Development Assistance* According to data published by the World Bank, development assistance to low-income countries from 2000 through 2008 was approximately

$$q(t) = 0.2t^2 + 3.5t + 60 \text{ billion dollars per year}$$

where $t$ is time in years since 2000.[50] Assuming a worldwide inflation rate of 3% per year, and that the above model remains accurate into the indefinite future, find the value of all development assistance to low-income countries from 2000 on in constant dollars. (The constant dollar value of $q(t)$ dollars $t$ years from now is given by $q(t)e^{-rt}$, where $r$ is the fractional rate of inflation. Give your answer to the nearest \$100 billion.) HINT [See Technology Note for Example 1.]

**46.** 🔢 *Humanitarian Aid* Repeat the preceding exercise, using the following model for humanitarian aid.[51]

$$q(t) = 0.07t^2 + 0.3t + 5 \text{ billion dollars per year}$$

**47.** ▼ *Hair Mousse Sales* The amount of extremely popular hair mousse sold online at your Web site can be approximated by

$$N(t) = \frac{80(7)^t}{20 + 7^t} \text{ million gallons per year.}$$

($t = 0$ represents the current year.) Investigate the integrals $\int_0^{+\infty} N(t)\, dt$ and $\int_{-\infty}^0 N(t)\, dt$ and interpret your answers.

**48.** ▼ *Chocolate Mousse Sales* The weekly demand for your company's *Lo-Cal Chocolate Mousse* is modeled by the equation

$$q(t) = \frac{50e^{2t-1}}{1 + e^{2t-1}} \text{ gallons per week,}$$

where $t$ is time from now in weeks. Investigate the integrals $\int_0^{+\infty} q(t)\, dt$ and $\int_{-\infty}^0 q(t)\, dt$ and interpret your answers.

🔢 *The Normal Curve* Exercises 49–52 require the use of a graphing calculator or computer programmed to do numerical integration. The normal distribution curve, which models the distributions of data in a wide range of applications, is given by the function

$$p(x) = \frac{1}{\sqrt{2\pi}\,\sigma}e^{-(x-\mu)^2/2\sigma^2},$$

where $\pi = 3.14159265\ldots$ and $\sigma$ and $\mu$ are constants called the standard deviation and the mean, respectively. Its graph (for $\sigma = 1$ and $\mu = 2$) is shown in the figure.

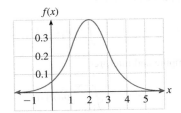

**49.** ▼ With $\sigma = 4$ and $\mu = 1$, approximate $\int_{-\infty}^{+\infty} p(x)\, dx$. HINT [See Example 5 and Technology Note for Example 1.]

**50.** ▼ With $\sigma = 1$ and $\mu = 0$, approximate $\int_0^{+\infty} p(x)\, dx$.

**51.** ▼ With $\sigma = 1$ and $\mu = 0$, approximate $\int_1^{+\infty} p(x)\, dx$.

**52.** ▼ With $\sigma = 1$ and $\mu = 0$, approximate $\int_{-\infty}^1 p(x)\, dx$.

**53.** ◆ *Variable Sales* The value of your Chateau Petit Mont Blanc 1963 vintage burgundy is increasing continuously at an annual rate of 40%, and you have a supply of 1,000 bottles worth \$85 each at today's prices. In order to ensure a steady income, you have decided to sell your wine at a diminishing rate—starting at 500 bottles per year, and then decreasing this figure continuously at a fractional rate of 100% per year. How much income (to the nearest dollar) can you expect to generate by this scheme? HINT [Use the formula for continuously compounded interest.]

**54.** ◆ *Panic Sales* Unfortunately, your large supply of Chateau Petit Mont Blanc is continuously turning to vinegar at a fractional rate of 60% per year! You have thus decided to sell off your Petit Mont Blanc at \$50 per bottle, but the market is a little thin, and you can only sell 400 bottles per year. Because you have no way of knowing which bottles now contain vinegar until they are opened, you shall have to give refunds for all the bottles of vinegar. What will your net income be before all the wine turns to vinegar?

**55.** ◆ *Meteor Impacts* The frequency of meteor impacts on Earth can be modeled by

$$n(k) = \frac{1}{5.6997k^{1.081}}$$

where $n(k) = N'(k)$, and $N(k)$ is the average number of meteors of energy less than or equal to $k$ megatons that will

---
[49]The zonar ($\overline{\overline{Z}}$) is the official currency in the city-state of Utarek, Mars. Source: www.Marsnext.com, a now extinct virtual society.

[50]The authors' approximation, based on data from the World Bank, obtained from www.worldbank.org.

[51]*Ibid.*

hit the Earth in one year.[52] (A small nuclear bomb releases on the order of one megaton of energy.)

**a.** How many meteors of energy at least $k = 0.2$ hit the Earth each year?

**b.** Investigate and interpret the integral $\int_0^1 n(k)\,dk$.

**56.** ◆ *Meteor Impacts* (continuing the previous exercise)

**a.** Explain why the integral

$$\int_a^b kn(k)\,dk$$

computes the total energy released each year by meteors with energies between $a$ and $b$ megatons.

**b.** Compute and interpret

$$\int_0^1 kn(k)\,dk.$$

**c.** Compute and interpret

$$\int_1^{+\infty} kn(k)\,dk.$$

**57.** ◆ *The Gamma Function* The gamma function is defined by the formula

$$\Gamma(x) = \int_0^{+\infty} t^{x-1}e^{-t}\,dt.$$

**a.** Find $\Gamma(1)$ and $\Gamma(2)$.

**b.** Use integration by parts to show that for every positive integer $n$, $\Gamma(n+1) = n\Gamma(n)$.

**c.** Deduce that $\Gamma(n) = (n-1)!\,[= (n-1)(n-2)\cdots 2\cdot 1]$ for every positive integer $n$.

**58.** ◆ *Laplace Transforms* The Laplace transform $F(x)$ of a function $f(t)$ is given by the formula

$$F(x) = \int_0^{+\infty} f(t)e^{-xt}\,dt \quad (x > 0).$$

**a.** Find $F(x)$ for $f(t) = 1$ and for $f(t) = t$.

**b.** Find a formula for $F(x)$ if $f(t) = t^n$ $(n = 1, 2, 3, \ldots)$.

**c.** Find a formula for $F(x)$ if $f(t) = e^{at}$ ($a$ constant).

---

[52]The authors' model, based on data published by NASA International Near-Earth-Object Detection Workshop (*The New York Times*, Jan. 25, 1994, p. C1).

## COMMUNICATION AND REASONING EXERCISES

**59.** Why can't the Fundamental Theorem of Calculus be used to evaluate $\int_{-1}^{1} \dfrac{1}{x}\,dx$?

**60.** Why can't the Fundamental Theorem of Calculus be used to evaluate $\int_{1}^{+\infty} \dfrac{1}{x^2}\,dx$?

**61.** It sometimes happens that the Fundamental Theorem of Calculus gives the correct answer for an improper integral. Does the FTC give the correct answer for improper integrals of the form

$$\int_{-a}^{a} \frac{1}{x^{1/r}}\,dx$$

if $r = 3, 5, 7, \ldots$?

**62.** Does the FTC give the correct answer for improper integrals of the form

$$\int_{-a}^{a} \frac{1}{x^r}\,dx$$

if $r = 3, 5, 7, \ldots$?

**63.** Which of the following integrals are improper, and why? (Do not evaluate any of them.)

**a.** $\displaystyle\int_{-1}^{1} \frac{|x|}{x}\,dx$    **b.** $\displaystyle\int_{-1}^{1} x^{-1/3}\,dx$

**c.** $\displaystyle\int_{0}^{2} \frac{x-2}{x^2-4x+4}\,dx$

**64.** Which of the following integrals are improper, and why? (Do not evaluate any of them.)

**a.** $\displaystyle\int_{-1}^{1} \frac{|x-1|}{x-1}\,dx$    **b.** $\displaystyle\int_{0}^{1} \frac{1}{x^{2/3}}\,dx$

**c.** $\displaystyle\int_{0}^{2} \frac{x^2-4x+4}{x-2}\,dx$

**65.** ▯ ▼ How could you use technology to approximate improper integrals? (Your discussion should refer to each type of improper integral.)

**66.** ▯ ▼ Use technology to approximate the integrals $\int_0^M e^{-(x-10)^2}\,dx$ for larger and larger values of $M$, using Riemann sums with 500 subdivisions. What do you find? Comment on the answer.

**67.** ▼ Make up an interesting application whose solution is $\int_{10}^{+\infty} 100te^{-0.2t}\,dt = \$1,015.01$.

**68.** ▼ Make up an interesting application whose solution is $\int_{100}^{+\infty} \frac{1}{r^2}\,dr = 0.01$.

# 14.6 Differential Equations and Applications

A **differential equation** is an equation that involves a derivative of an unknown function. A **first-order differential equation** involves only the first derivative of the unknown function. A **second-order differential equation** involves the second derivative of the unknown function (and possibly the first derivative). Higher order differential equations are defined similarly. In this book, we will deal only with first-order differential equations.

To **solve** a differential equation means to find the unknown function. Many of the laws of science and other fields describe how things change. When expressed mathematically, these laws take the form of equations involving derivatives—that is, differential equations. The field of differential equations is a large and very active area of study in mathematics, and we shall see only a small part of it in this section.

---

### EXAMPLE 1 Motion

A dragster accelerates from a stop so that its speed $t$ seconds after starting is $40t$ ft/sec. How far will the car go in 8 seconds?

**Solution** We wish to find the car's position function $s(t)$. We are told about its speed, which is $ds/dt$. Precisely, we are told that

$$\frac{ds}{dt} = 40t.$$

This is the differential equation we have to solve to find $s(t)$. But we already know how to solve this kind of differential equation; we integrate:

$$s(t) = \int 40t \, dt = 20t^2 + C.$$

We now have the **general solution** to the differential equation. By letting $C$ take on different values, we get all the possible solutions. We can specify the one **particular solution** that gives the answer to our problem by imposing the **initial condition** that $s(0) = 0$. Substituting into $s(t) = 20t^2 + C$, we get

$$0 = s(0) = 20(0)^2 + C = C$$

so $C = 0$ and $s(t) = 20t^2$. To answer the question, the car travels $20(8)^2 = 1,280$ feet in 8 seconds.

---

We did not have to work hard to solve the differential equation in Example 1. In fact, any differential equation of the form $dy/dx = f(x)$ can (in theory) be solved by integrating. (Whether we can actually carry out the integration is another matter!)

### Simple Differential Equations

A **simple** differential equation has the form

$$\frac{dy}{dx} = f(x).$$

Its general solution is

$$y = \int f(x)\,dx.$$

**Quick Example**

The differential equation

$$\frac{dy}{dx} = 2x^2 - 4x^3$$

is simple and has general solution

$$y = \int f(x)\,dx = \frac{2x^3}{3} - x^4 + C.$$

Not all differential equations are simple, as the next example shows.

**EXAMPLE 2   Separable Differential Equation**

Consider the differential equation $\dfrac{dy}{dx} = \dfrac{x}{y^2}$.

**a.** Find the general solution.

**b.** Find the particular solution that satisfies the initial condition $y(0) = 2$.

**Solution**

**a.** This is not a simple differential equation because the right-hand side is a function of both $x$ and $y$. We cannot solve this equation by just integrating; the solution to this problem is to "separate" the variables.

**Step 1:** *Separate the variables algebraically.* We rewrite the equation as

$$y^2\,dy = x\,dx.$$

**Step 2:** *Integrate both sides.*

$$\int y^2\,dy = \int x\,dx$$

giving

$$\frac{y^3}{3} = \frac{x^2}{2} + C.$$

**Step 3:** *Solve for the dependent variable.* We solve for $y$:

$$y^3 = \frac{3}{2}x^2 + 3C = \frac{3}{2}x^2 + D$$

(Rewriting $3C$ as $D$, an equally arbitrary constant), so

$$y = \left(\frac{3}{2}x^2 + D\right)^{1/3}.$$

This is the general solution of the differential equation.

**b.** We now need to find the value for $D$ that will give us the solution satisfying the condition $y(0) = 2$. Substituting 0 for $x$ and 2 for $y$ in the general solution, we get

$$2 = \left(\frac{3}{2}(0)^2 + D\right)^{1/3} = D^{1/3}$$

so

$$D = 2^3 = 8.$$

Thus, the particular solution we are looking for is

$$y = \left(\frac{3}{2}x^2 + 8\right)^{1/3}.$$

➡ **Before we go on...** We can check the general solution in Example 2 by calculating both sides of the differential equation and comparing.

$$\frac{dy}{dx} = \frac{d}{dx}\left(\frac{3}{2}x^2 + D\right)^{1/3} = x\left(\frac{3}{2}x^2 + D\right)^{-2/3}$$

$$\frac{x}{y^2} = \frac{x}{\left(\frac{3}{2}x^2 + 8\right)^{2/3}} = x\left(\frac{3}{2}x^2 + D\right)^{-2/3} \qquad ✔ \qquad ■$$

**Q** : In Example 2, we wrote $y^2\, dy$ and $x\, dx$. What do they mean?

**A** : Although it is possible to give meaning to these symbols, for us they are just a notational convenience. We could have done the following instead:

$$y^2\frac{dy}{dx} = x.$$

Now we integrate both sides with respect to $x$.

$$\int y^2\frac{dy}{dx}\, dx = \int x\, dx$$

We can use substitution to rewrite the left-hand side:

$$\int y^2\frac{dy}{dx}\, dx = \int y^2\, dy,$$

which brings us back to the equation

$$\int y^2\, dy = \int x\, dx.$$

We were able to separate the variables in the preceding example because the right-hand side, $x/y^2$, was a *product* of a function of $x$ and a function of $y$—namely,

$$\frac{x}{y^2} = x\left(\frac{1}{y^2}\right).$$

In general, we can say the following:

---

### Separable Differential Equation

A **separable** differential equation has the form

$$\frac{dy}{dx} = f(x)g(y).$$

We solve a separable differential equation by separating the $x$s and the $y$s algebraically, writing

$$\frac{1}{g(y)}\,dy = f(x)\,dx$$

and then integrating:

$$\int \frac{1}{g(y)}\,dy = \int f(x)\,dx.$$

---

### EXAMPLE 3  Rising Medical Costs

Spending on Medicare from 2010 to 2021 was projected to rise continuously at an instantaneous rate of 5.6% per year.[53] Find a formula for Medicare spending $y$ as a function of time $t$ in years since 2010.

**Solution**  When we say that Medicare spending $y$ was going up continuously at an instantaneous rate of 5.6% per year, we mean that

*the instantaneous rate of increase of y was 5.6% of its value*

or    $\dfrac{dy}{dt} = 0.056y.$

This is a separable differential equation. Separating the variables gives

$$\frac{1}{y}dy = 0.056\,dt.$$

Integrating both sides, we get

$$\int \frac{1}{y}\,dy = \int 0.056\,dt$$

so    $\ln y = 0.056t + C.$

(We should write $\ln |y|$, but we know that the medical costs are positive.) We now solve for $y$.

$$y = e^{0.056t+C} = e^{C}e^{0.056t} = Ae^{0.056t},$$

where $A$ is a positive constant. This is the formula we used before for continuous percentage growth.

---

[53]Spending is in constant 2010 dollars. Source for projected data: Congressional Budget Office, *March 2011 Medicare Baseline* (www.cbo.gov).

➡ **Before we go on...** To determine $A$ in Example 3 we need to know, for example, Medicare spending at time $t = 0$ (the initial condition). The source cited gives Medicare spending as \$525.6 billion in 2010. Substituting $t = 0$ in the equation above gives

$$525.6 = Ae^0 = A.$$

Thus, projected Medicare spending is

$$y = 525.6e^{0.056t} \text{ billion dollars}$$

$t$ years after 2010. ∎

---

## EXAMPLE 4  Newton's Law of Cooling

Newton's Law of Cooling states that a hot object cools at a rate proportional to the difference between its temperature and the temperature of the surrounding environment (the **ambient temperature**). If a hot cup of coffee, at 170°F, is left to sit in a room at 70°F, how will the temperature of the coffee change over time?

**Solution**  We let $H(t)$ denote the temperature of the coffee at time $t$. Newton's Law of Cooling tells us that $H(t)$ *decreases* at a rate proportional to the difference between $H(t)$ and 70°F, the ambient temperature. In other words,

$$\frac{dH}{dt} = -k(H - 70),$$

* When we say that a quantity $Q$ is *proportional* to a quantity $R$, we mean that $Q = kR$ for some constant $k$. The constant $k$ is referred to as the **constant of proportionality**.

where $k$ is some positive constant.* Note that $H \geq 70$: The coffee will never cool to less than the ambient temperature.

The variables here are $H$ and $t$, which we can separate as follows:

$$\frac{dH}{H - 70} = -k\,dt.$$

Integrating, we get

$$\int \frac{dH}{H - 70} = \int (-k)\,dt$$

so        $\ln(H - 70) = -kt + C.$

(Note that $H - 70$ is positive, so we don't need absolute values.) We now solve for $H$:

$$H - 70 = e^{-kt+C}$$
$$= e^C e^{-kt}$$
$$= Ae^{-kt}$$

so

$$H(t) = 70 + Ae^{-kt},$$

where $A$ is some positive constant. We can determine the constant $A$ using the initial condition $H(0) = 170$:

$$170 = 70 + Ae^0 = 70 + A$$

so        $A = 100.$

Therefore,

$$H(t) = 70 + 100e^{-kt}.$$

**Figure 25**

**Q:** *But what is k?*

**A:** The constant $k$ determines the rate of cooling. Its value depends on the units of time we are using, on the substance cooling—in this case the coffee—and on its container. Because $k$ depends so heavily on the particular circumstances, it's usually easiest to determine it experimentally. Figure 25 shows two possible graphs, one with $k = 0.1$ and the other with $k = 0.03$ ($k \approx 0.03$ would be reasonable for a cup of coffee in a polystyrene container with $t$ measured in minutes).

In any case, we can see from the graph or the formula for $H(t)$ that the temperature of the coffee will approach the ambient temperature exponentially.

➡ **Before we go on...** Notice that the calculation in Example 4 shows that the temperature of an object cooling according to Newton's Law is given in general by

$$H(t) = T_a + (T_0 - T_a)e^{-kt},$$

where $T_a$ is the ambient temperature (70° in the example) and $T_0$ is the initial temperature (170° in the example). The formula also holds if the ambient temperature is higher than the initial temperature ("Newton's Law of Heating"). ■

# 14.6  EXERCISES

▼ more advanced   ◆ challenging
**T** indicates exercises that should be solved using technology

*Find the general solution of each differential equation in Exercises 1–10. Where possible, solve for y as a function of x.*

**1.** $\dfrac{dy}{dx} = x^2 + \sqrt{x}$

HINT [See Quick Example page 1064.]

**2.** $\dfrac{dy}{dx} = \dfrac{1}{x} + 3$

HINT [See Quick Example page 1064.]

**3.** $\dfrac{dy}{dx} = \dfrac{x}{y}$

HINT [See Example 2(a).]

**4.** $\dfrac{dy}{dx} = \dfrac{y}{x}$

HINT [See Example 2(a).]

**5.** $\dfrac{dy}{dx} = xy$

**6.** $\dfrac{dy}{dx} = x^2 y$

**7.** $\dfrac{dy}{dx} = (x + 1)y^2$

**8.** $\dfrac{dy}{dx} = \dfrac{1}{(x + 1)y^2}$

**9.** $x\dfrac{dy}{dx} = \dfrac{1}{y}\ln x$

**10.** $\dfrac{1}{x}\dfrac{dy}{dx} = \dfrac{1}{y}\ln x$

*For each differential equation in Exercises 11–20, find the particular solution indicated.* HINT [See Example 2(b).]

**11.** $\dfrac{dy}{dx} = x^3 - 2x$; $y = 1$ when $x = 0$

**12.** $\dfrac{dy}{dx} = 2 - e^{-x}$; $y = 0$ when $x = 0$

**13.** $\dfrac{dy}{dx} = \dfrac{x^2}{y^2}$; $y = 2$ when $x = 0$

**14.** $\dfrac{dy}{dx} = \dfrac{y^2}{x^2}$; $y = \dfrac{1}{2}$ when $x = 1$

**15.** $x\dfrac{dy}{dx} = y$; $y(1) = 2$

**16.** $x^2\dfrac{dy}{dx} = y$; $y(1) = 1$

**17.** $\dfrac{dy}{dx} = x(y + 1)$; $y(0) = 0$

**18.** $\dfrac{dy}{dx} = \dfrac{y + 1}{x}$; $y(1) = 2$

**19.** $\dfrac{dy}{dx} = \dfrac{xy^2}{x^2 + 1}$; $y(0) = -1$

**20.** $\dfrac{dy}{dx} = \dfrac{xy}{(x^2 + 1)^2}$; $y(0) = 1$

## APPLICATIONS

**21.** *Sales* Your monthly sales of Green Tea Ice Cream are falling at an instantaneous rate of 5% per month. If you currently sell 1,000 quarts per month, find the differential equation describing your change in sales, and then solve it to predict your monthly sales. HINT [See Example 3.]

**22.** *Profit* Your monthly profit on sales of Avocado Ice Cream is rising at an instantaneous rate of 10% per month. If you currently make a profit of $15,000 per month, find the differential equation describing your change in profit, and solve it to predict your monthly profits. HINT [See Example 3.]

**23.** *Newton's Law of Cooling* For coffee in a ceramic cup, suppose $k \approx 0.05$ with time measured in minutes. **(a)** Use Newton's Law of Cooling to predict the temperature of the coffee, initially at a temperature of 200°F, that is left to sit in a room at 75°F. **(b)** When will the coffee have cooled to 80°F? HINT [See Example 4.]

24. *Newton's Law of Cooling* For coffee in a paper cup, suppose $k \approx 0.08$ with time measured in minutes. (a) Use Newton's Law of Cooling to predict the temperature of the coffee, initially at a temperature of 210°F, that is left to sit in a room at 60°F. (b) When will the coffee have cooled to 70°F? HINT [See Example 4.]

25. *Cooling* A bowl of clam chowder at 190°F is placed in a room whose air temperature is 75°F. After 10 minutes, the soup has cooled to 150°F. Find the value of $k$ in Newton's Law of Cooling, and hence find the temperature of the chowder as a function of time.

26. *Heating* Suppose that a pie, at 20°F, is put in an oven at 350°F. After 15 minutes, its temperature has risen to 80°F. Find the value of $k$ in Newton's Law of Heating (see the note after Example 4), and hence find the temperature of the pie as a function of time.

27. *Market Saturation* You have just introduced a new 3D monitor to the market. You predict that you will eventually sell 100,000 monitors and that your monthly rate of sales will be 10% of the difference between the saturation value of 100,000 and the total number you have sold up to that point. Find a differential equation for your total sales (as a function of the month) and solve. (What are your total sales at the moment when you first introduce the monitor?)

28. *Market Saturation* Repeat the preceding exercise, assuming that monthly sales will be 5% of the difference between the saturation value (of 100,000 monitors) and the total sales to that point, and assuming that you sell 5,000 monitors to corporate customers before placing the monitor on the open market.

29. *Determining Demand* Nancy's Chocolates estimates that the elasticity of demand for its dark chocolate truffles is $E = 0.05p - 1.5$ where $p$ is the price per pound. Nancy's sells 20 pounds of truffles per week when the price is $20 per pound. Find the formula expressing the demand $q$ as a function of $p$. Recall that the elasticity of demand is given by

$$E = -\frac{dq}{dp} \times \frac{p}{q}.$$

30. *Determining Demand* Nancy's Chocolates estimates that the elasticity of demand for its chocolate strawberries is $E = 0.02p - 0.5$ where $p$ is the price per pound. It sells 30 pounds of chocolate strawberries per week when the price is $30 per pound. Find the formula expressing the demand $q$ as a function of $p$. Recall that the elasticity of demand is given by

$$E = -\frac{dq}{dp} \times \frac{p}{q}.$$

*Linear Differential Equations* Exercises 31–36 are based on first-order linear differential equations with constant coefficients. These have the form

$$\frac{dy}{dt} + py = f(t) \quad (p \text{ constant})$$

and the general solution is

$$y = e^{-pt} \int f(t) e^{pt} \, dt. \quad (\text{Check this by substituting!})$$

31. Solve the linear differential equation

$$\frac{dy}{dt} + y = e^{-t}; \, y = 1 \text{ when } t = 0.$$

32. Solve the linear differential equation

$$\frac{dy}{dt} - y = e^{2t}; \, y = 2 \text{ when } t = 0.$$

33. Solve the linear differential equation

$$2\frac{dy}{dt} - y = 2t; \, y = 1 \text{ when } t = 0.$$

HINT [First rewrite the differential equation in the form $\frac{dy}{dt} + py = f(t)$.]

34. Solve the linear differential equation

$$2\frac{dy}{dt} + y = -t. \quad y = 1 \text{ when } t = 0$$

HINT [First rewrite the differential equation in the form $\frac{dy}{dt} + py = f(t)$.]

35. ▼ *Electric Circuits* The flow of current $i(t)$ in an electric circuit without capacitance satisfies the linear differential equation

$$L\frac{di}{dt} + Ri = V(t),$$

where $L$ and $R$ are constants (the *inductance* and *resistance,* respectively) and $V(t)$ is the applied voltage. (See figure.)

If the voltage is supplied by a 10-volt battery and the switch is turned on at time $t = 1$, then the voltage $V$ is a step function that jumps from 0 to 10 at $t = 1$: $V(t) = 5\left[1 + \frac{|t-1|}{t-1}\right]$. Find the current as a function of time for $L = R = 1$. Use a grapher to plot the resulting current as a function of time. (Assume there is no current flowing at time $t = 0$.) [Use the following integral formula: $\int \left[1 + \frac{|t-1|}{t-1}\right]e^t \, dt = \left[1 + \frac{|t-1|}{t-1}\right](e^t - e) + C.$]

36. ▼ *Electric Circuits* Repeat Exercise 35 for $L = 1$, $R = 5$, and $V(t) = 5\left[1 + \frac{|t-2|}{t-2}\right]$. (The switch flipped on at time $t = 2$.) [Use the following integral formula: $\int \left[1 + \frac{|t-2|}{t-2}\right]e^{5t} \, dt = \left(1 + \frac{|t-2|}{t-2}\right)\left(\frac{e^{5t} - e^{10}}{5}\right) + C.$]

**37.** ▼ *Approach to Equilibrium* The *Extrasoft Toy Co.* has just released its latest creation, a plush platypus named "Eggbert." The demand function for Eggbert dolls is $D(p) = 50,000 - 500p$ dolls per month when the price is $p$ dollars. The supply function is $S(p) = 30,000 + 500p$ dolls per month when the price is $p$ dollars. This makes the equilibrium price $20. The **Evans price adjustment model** assumes that if the price is set at a value other than the equilibrium price, it will change over time in such a way that its rate of change is proportional to the shortage $D(p) - S(p)$.

a. Write the differential equation given by the Evans price adjustment model for the price $p$ as a function of time.
b. Find the general solution of the differential equation you wrote in (a). (You will have two unknown constants, one being the constant of proportionality.)
c. Find the particular solution in which Eggbert dolls are initially priced at $10 and the price rises to $12 after one month.

**38.** ▼ *Approach to Equilibrium* Spacely Sprockets has just released its latest model, the Dominator. The demand function is $D(p) = 10,000 - 1,000p$ sprockets per year when the price is $p$ dollars. The supply function is $S(p) = 8,000 + 1,000p$ sprockets per year when the price is $p$ dollars.

a. Using the Evans price adjustment model described in the preceding exercise, write the differential equation for the price $p(t)$ as a function of time.
b. Find the general solution of the differential equation you wrote in (a).
c. Find the particular solution in which Dominator sprockets are initially priced at $5 each but fall to $3 each after one year.

**39.** ▼ *Logistic Equation* There are many examples of growth in which the rate of growth is slow at first, becomes faster, and then slows again as a limit is reached. This pattern can be described by the differential equation

$$\frac{dy}{dt} = ay(L - y),$$

where $a$ is a constant and $L$ is the limit of $y$. Show by substitution that

$$y = \frac{CL}{e^{-aLt} + C}$$

is a solution of this equation, where $C$ is an arbitrary constant.

**40.** ▼ *Logistic Equation* Using separation of variables and integration with a table of integrals or a symbolic algebra program, solve the differential equation in the preceding exercise to derive the solution given there.

▣ *Exercises 41–44 require the use of technology.*

**41.** ▼ *Market Saturation* You have just introduced a new model of Blu-ray disc player. You predict that the market will saturate at 2,000,000 Blu-ray disc players and that your total sales will be governed by the equation

$$\frac{dS}{dt} = \frac{1}{4}S(2 - S),$$

where $S$ is the total sales in millions of Blu-ray disc players and $t$ is measured in months. If you give away 1,000 Blu-ray disc players when you first introduce them, what will $S$ be? Sketch the graph of $S$ as a function of $t$. About how long will it take to saturate the market? (See Exercise 39.)

**42.** ▼ *Epidemics* A certain epidemic of influenza is predicted to follow the function defined by

$$\frac{dA}{dt} = \frac{1}{10}A(20 - A),$$

where $A$ is the number of people infected in millions and $t$ is the number of months after the epidemic starts. If 20,000 cases are reported initially, find $A(t)$ and sketch its graph. When is $A$ growing fastest? How many people will eventually be affected? (See Exercise 39.)

**43.** ▼ *Growth of Tumors* The growth of tumors in animals can be modeled by the Gompertz equation:

$$\frac{dy}{dt} = -ay \ln\left(\frac{y}{b}\right),$$

where $y$ is the size of a tumor, $t$ is time, and $a$ and $b$ are constants that depend on the type of tumor and the units of measurement.

a. Solve for $y$ as a function of $t$.
b. If $a = 1$, $b = 10$, and $y(0) = 5$ cm$^3$ (with $t$ measured in days), find the specific solution and graph it.

**44.** ▼ *Growth of Tumors* Refer back to the preceding exercise. Suppose that $a = 1$, $b = 10$, and $y(0) = 15$ cm$^3$. Find the specific solution and graph it. Comparing its graph to the one obtained in the preceding exercise, what can you say about tumor growth in these instances?

## COMMUNICATION AND REASONING EXERCISES

**45.** What is the difference between a particular solution and the general solution of a differential equation? How do we get a particular solution from the general solution?

**46.** Why is there always an arbitrary constant in the general solution of a differential equation? Why are there not two or more arbitrary constants in a first-order differential equation?

**47.** ▼ Show by example that a **second-order** differential equation, one involving the second derivative $y''$, usually has two arbitrary constants in its general solution.

**48.** ▼ Find a differential equation that is not separable.

**49.** ▼ Find a differential equation whose general solution is $y = 4e^{-x} + 3x + C$.

**50.** ▼ Explain how, knowing the elasticity of demand as a function of either price or demand, you may find the demand equation. (See Exercise 29.)

# CHAPTER 14 REVIEW

## KEY CONCEPTS

Website www.WanerMath.com
Go to the Website at www.WanerMath
.com to find a comprehensive and
interactive Web-based summary
of Chapter 14.

### 14.1 Integration by Parts

Integration-by-parts formula:

$$\int u \cdot v \, dx = u \cdot I(v) - \int D(u)I(v) \, dx$$

p. 1016

Tabular method for integration by parts
p. 1017

Integrating a polynomial times a
logarithm p. 1020

### 14.2 Area between Two Curves and Applications

If $f(x) \geq g(x)$ for all $x$ in $[a, b]$, then
the area of the region between the
graphs of $f$ and $g$ and between $x = a$
and $x = b$ is given by

$$A = \int_a^b [f(x) - g(x)] \, dx \quad p. 1025$$

Regions enclosed by crossing curves
p. 1027

Area enclosed by two curves p. 1029

General instructions for finding the area
between the graphs of $f(x)$ and $g(x)$
p. 1029

Approximating the area between two
curves using technology:

$$A = \int_a^b |f(x) - g(x)| \, dx \quad p. 1030$$

### 14.3 Averages and Moving Averages

Average, or mean, of a collection of
values

$$\bar{y} = \frac{y_1 + y_2 + \cdots + y_n}{n} \quad p. 1034$$

The *average*, or *mean*, of a function
$f(x)$ on an interval $[a, b]$ is

$$\bar{f} = \frac{1}{b - a} \int_a^b f(x) \, dx. \quad p. 1035$$

Average balance p. 1035

Computing the moving average of a set
of data p. 1036

$n$-Unit moving average of a function:

$$\bar{f}(x) = \frac{1}{n} \int_{x-n}^x f(t) \, dt \quad p. 1037$$

Computing moving averages of
sawtooth and step functions p. 1038

### 14.4 Applications to Business and Economics: Consumers' and Producers' Surplus and Continuous Income Streams

Consumers' surplus:

$$CS = \int_0^{\bar{q}} [D(q) - \bar{p}] \, dq \quad p. 1043$$

Producers' surplus:

$$PS = \int_0^{\bar{q}} [\bar{p} - S(q)] \, dq \quad p. 1045$$

Equilibrium price p. 1046

Social gain $= CS + PS$ p. 1047

Total value of a continuous income

stream: $TV = \int_a^b R(t) \, dt$ p. 1048

Future value of a continuous income

stream: $FV = \int_a^b R(t)e^{r(b-t)} \, dt$
p. 1049

Present value of a continuous income

stream: $PV = \int_a^b R(t)e^{r(a-t)} \, dt$
p. 1049

### 14.5 Improper Integrals and Applications

Improper integral with an infinite limit
of integration:

$$\int_a^{+\infty} f(x) \, dx, \int_{-\infty}^b f(x) \, dx,$$

$$\int_{-\infty}^{+\infty} f(x) \, dx \quad p. 1053$$

Improper integral in which the
integrand becomes infinite p. 1056

Testing for convergence p. 1057

Integrand infinite between the endpoints
p. 1058

Integral improper in two ways p. 1058

### 14.6 Differential Equations and Applications

Simple differential equations:

$$\frac{dy}{dx} = f(x) \quad p. 1063$$

Separable differential equations:

$$\frac{dy}{dx} = f(x)g(y) \quad p. 1066$$

Newton's Law of Cooling p. 1067

## REVIEW EXERCISES

Evaluate the integrals in Exercises 1–10.

**1.** $\int (x^2 + 2)e^x \, dx$

**2.** $\int (x^2 - x)e^{-3x+1} \, dx$

**3.** $\int x^2 \ln(2x) \, dx$

**4.** $\int \log_5 x \, dx$

**5.** $\int 2x|2x + 1| \, dx$

**6.** $\int 3x|-x + 5| \, dx$

**7.** $\int 5x \frac{|-x + 3|}{-x + 3} \, dx$

**8.** $\int 2x \frac{|3x + 1|}{3x + 1} \, dx$

**9.** $\int_{-2}^2 (x^3 + 1)e^{-x} \, dx$

**10.** $\int_1^e x^2 \ln x \, dx$

In Exercises 11–14, find the areas of the given regions.

**11.** Between $y = x^3$ and $y = 1 - x^3$ for $x$ in $[0, 1]$

**12.** Between $y = e^x$ and $y = e^{-x}$ for $x$ in $[0, 2]$

**13.** Enclosed by $y = 1 - x^2$ and $y = x^2$

**14.** Between $y = x$ and $y = xe^{-x}$ for $x$ in $[0, 2]$

In Exercises 15–18, find the average value of the given function over the indicated interval.

**15.** $f(x) = x^3 - 1$ over $[-2, 2]$

**16.** $f(x) = \frac{x}{x^2 + 1}$ over $[0, 1]$

**17.** $f(x) = x^2 e^x$ over $[0, 1]$

**18.** $f(x) = (x + 1) \ln x$ over $[1, 2e]$

*In Exercises 19–22, find the 2-unit moving averages of the given function.*

**19.** $f(x) = 3x + 1$

**20.** $f(x) = 6x^2 + 12$

**21.** $f(x) = x^{4/3}$

**22.** $f(x) = \ln x$

*In Exercises 23 and 24, calculate the consumers' surplus at the indicated unit price $\bar{p}$ for the given demand equation.*

**23.** $p = 50 - \dfrac{1}{2}q; \ \bar{p} = 10$

**24.** $p = 10 - q^{1/2}; \ \bar{p} = 4$

*In Exercises 25 and 26, calculate the producers' surplus at the indicated unit price $\bar{p}$ for the given supply equation.*

**25.** $p = 50 + \dfrac{1}{2}q; \ \bar{p} = 100$

**26.** $p = 10 + q^{1/2}; \ \bar{p} = 40$

*In Exercises 27–32, decide whether the given integral converges. If the integral converges, compute its value.*

**27.** $\displaystyle\int_1^{\infty} \dfrac{1}{x^5}\, dx$

**28.** $\displaystyle\int_0^1 \dfrac{1}{x^5}\, dx$

**29.** $\displaystyle\int_{-1}^{1} \dfrac{x}{(x^2 - 1)^{5/3}}\, dx$

**30.** $\displaystyle\int_0^2 \dfrac{x}{(x^2 - 1)^{1/3}}\, dx$

**31.** $\displaystyle\int_0^{+\infty} 2xe^{-x^2}\, dx$

**32.** $\displaystyle\int_0^{+\infty} x^2 e^{-6x^3}\, dx$

*Solve the differential equations in Exercises 33–36.*

**33.** $\dfrac{dy}{dx} = x^2 y^2$

**34.** $\dfrac{dy}{dx} = xy + 2x$

**35.** $xy\dfrac{dy}{dx} = 1; \ y(1) = 1$

**36.** $y(x^2 + 1)\dfrac{dy}{dx} = xy^2; \ y(0) = 2$

## APPLICATIONS: OHaganBooks.com

**37. *Spending on Stationery*** Alarmed by the volume of pointless memos and reports being copied and circulated by management at OHaganBooks.com, John O'Hagan ordered a 5-month audit of paper usage at the company. He found that management consumed paper at a rate of

$$q(t) = 45t + 200 \text{ thousand sheets per month} \quad (0 \le t \le 5)$$

($t$ is the time in months since the audit began). During the same period, the price of paper was escalating; the company was charged approximately

$$p(t) = 9e^{0.09t} \text{ dollars per thousand sheets.}$$

Use an integral to estimate, to the nearest hundred dollars, the total spent on paper for management during the given period.

**38. *Spending on Shipping*** During the past 10 months, OHaganBooks.com shipped orders at a rate of about

$$q(t) = 25t + 3{,}200 \text{ packages per month} \quad (0 \le t \le 10)$$

($t$ is the time in months since the beginning of the year). During the same period, the cost of shipping a package averaged approximately

$$p(t) = 4e^{0.04t} \text{ dollars per package.}$$

Use an integral to estimate, to the nearest thousand dollars, the total spent on shipping orders during the given period.

**39. *Education Costs*** Billy-Sean O'Hagan, having graduated *summa cum laude* from college, has been accepted by the doctoral program in biophysics at Oxford. John O'Hagan estimates that the total cost (minus scholarships) he will need to pay is $2,000 per month, but that this cost will escalate at a continuous compounding rate of 1% per month.

**a.** What, to the nearest dollar, is the average monthly cost over the course of two years?

**b.** Find the four-month moving average of the monthly cost.

**40. *Investments*** OHaganBooks.com keeps its cash reserves in a hedge fund paying 6% compounded continuously. It starts a year with $1 million in reserves and does not withdraw or deposit any money.

**a.** What is the average amount it will have in the fund over the course of two years?

**b.** Find the one-month moving average of the amount it has in the fund.

**41. *Consumers' and Producers' Surplus*** Currently, the hottest selling item at OHaganBooks.com is *Mensa for Dummies*[54] with a demand curve of $q = 20{,}000(28 - p)^{1/3}$ books per week, and a supply curve of $q = 40{,}000(p - 19)^{1/3}$ books per week.

**a.** Find the equilibrium price and demand.

**b.** Find the consumers' and producers' surpluses at the equilibrium price.

**42. *Consumers' and Producers' Surplus*** OHaganBooks.com is about to start selling a new coffee table book, *Computer Designs of the Late Twentieth Century*. It estimates the demand curve to be $q = 1{,}000\sqrt{200 - 2p}$, and its willingness to order books from the publisher is given by the supply curve $q = 1{,}000\sqrt{10p - 400}$.

**a.** Find the equilibrium price and demand.

**b.** Find the consumers' and producers' surpluses at the equilibrium price.

**43. *Revenue*** Sales of the bestseller *A River Burns through It* are dropping at OHaganBooks.com. To try to bolster sales, the company is dropping the price of the book, now $40, at a rate of $2 per week. As a result, this week OHaganBooks.com will sell 5,000 copies, and it estimates that sales will fall continuously at a rate of 10% per week. How much revenue will it earn on sales of this book over the next 8 weeks?

**44. *Foreign Investments*** Panicked by the performance of the U.S. stock market, Marjory Duffin is investing her 401(k) money in a Russian hedge fund at a rate of approximately

$$q(t) = 1.7t^2 - 0.5t + 8 \text{ thousand shares per month,}$$

---

[54]The actual title is: *Let Us Just Have A Ball! Mensa for Dummies*, by Wendu Mekbib, Silhouette Publishing Corporation.

where $t$ is time in months since the stock market began to plummet. At the time she started making the investments, the hedge fund was selling for $1 per share, but subsequently declined in value at a continuous rate of 5% per month. What was the total amount of money Marjory Duffin invested after one year? (Answer to the nearest $1,000.)

**45. *Investments*** OHaganBooks.com CEO John O'Hagan has started a gift account for the Marjory Duffin Foundation. The account pays 6% compounded continuously and is initially empty. OHaganBooks.com deposits money continuously into it, starting at the rate of $100,000 per month and increasing continuously by $10,000 per month.

  **a.** How much money will the company have in the account at the end of two years?

  **b.** How much of the amount you found in part (a) was principal deposited and how much was interest earned? (Round answers to the nearest $1,000.)

**46. *Savings*** John O'Hagan had been saving money for Billy-Sean's education since he had been a wee lad. O'Hagan began depositing money at the rate of $1,000 per month and increased his deposits continuously by $50 per month. If the account earned 5% compounded continuously and O'Hagan continued these deposits for 15 years,

  **a.** How much money did he accumulate?

  **b.** How much was money deposited and how much was interest?

**47. *Acquisitions*** The Megabucks Corporation is considering buying OHaganBooks.com. It estimates OHaganBooks.com's revenue stream at $50 million per year, growing continuously at a 10% rate. Assuming interest rates of 6%, how much is OHaganBooks.com's revenue for the next year worth now?

**48. *More Acquisitions*** OHaganBooks.com is thinking of buying JungleBooks and would like to recoup its investment after three years. The estimated net profit for JungleBooks is $40 million per year, growing linearly by $5 million per year. Assuming interest rates of 4%, how much should OHaganBooks.com pay for JungleBooks?

**49. *Incompetence*** OHaganBooks.com is shopping around for a new bank. A junior executive at one bank offers it the following interesting deal: The bank will pay OHaganBooks.com interest continuously at a rate numerically equal to 0.01% of the square of the amount of money it has in the account at any time. By considering what would happen if $10,000 was deposited in such an account, explain why the junior executive was fired shortly after this offer was made.

**50. *Shrewd Bankers*** The new junior officer at the bank (who replaced the one fired in the preceding exercise) offers OHaganBook.com the following deal for the $800,000 they plan to deposit: While the amount in the account is less than $1 million, the bank will pay interest continuously at a rate equal to 10% of the difference between $1 million and the amount of money in the account. When it rises over $1 million, the bank will pay interest of 20%. Why should OHaganBooks.com not take this offer?

---

**Case Study**

## Estimating Tax Revenues

You have just been hired by the incoming administration of your country as chief consultant for national tax policy, and you have been getting conflicting advice from the finance experts on your staff. Several of them have come up with plausible suggestions for new tax structures, and your job is to choose the plan that results in more revenue for the government.

Before you can evaluate their plans, you realize that it is essential to know your country's income distribution—that is, how many people earn how much money.[*] You might think that the most useful way of specifying income distribution would be to use a function that gives the exact number $f(x)$ of people who earn a given salary $x$. This would necessarily be a discrete function—it makes sense only if $x$ happens to be a whole number of cents. There is, after all, no one earning a salary of exactly $22,000.142567! Furthermore, this function would behave rather erratically, because there are, for example, probably many more people making a salary of exactly $30,000 than exactly $30,000.01. Given these problems, it is far more convenient to start with the function defined by

$N(x) =$ *the total number of people earning between* 0 *and* $x$ *dollars.*

[*] To simplify our discussion, we are assuming that (1) all tax revenues are based on earned income and that (2) everyone in the population we consider earns some income.

Actually, you would want a "smoothed" version of this function. The graph of $N(x)$ might look like the one shown in Figure 26.

**Figure 26**

If we take the *derivative* of $N(x)$, we get an income distribution function. Its graph might look like the one shown in Figure 27.

**Figure 27**

* A very similar idea is used in probability. See the optional chapter "Calculus Applied to Probability and Statistics" on the Website.

† Gamma distributions are often good models for income distributions. The one used in the text is the authors' approximation of the income distribution in the United States in 2009. Source for data: U.S. Census Bureau, Current Population Survey, 2010 Annual Social and Economic Supplement. (www.census.gov).

Because the derivative measures the rate of change, its value at $x$ is the additional number of taxpayers per \$1 increase in salary. Thus, the fact that $N'(25,000) \approx 3,700$ tells us that approximately 3,700 people are earning a salary of between \$25,000 and \$25,001. In other words, $N'$ shows the distribution of incomes among the population—hence, the name "distribution function."*

You thus send a memo to your experts requesting the income distribution function for the nation. After much collection of data, they tell you that the income distribution function is

$$N'(x) = 14x^{0.672}e^{-x/20,400}.$$

This is in fact the function whose graph is shown in Figure 27 and is an example of a **gamma distribution**.† (You might find it odd that you weren't given the original function $N$, but it will turn out that you don't need it. How would you compute it?)

Given this income distribution, your financial experts have come up with the two possible tax policies illustrated in Figures 28 and 29.

**Figure 28**

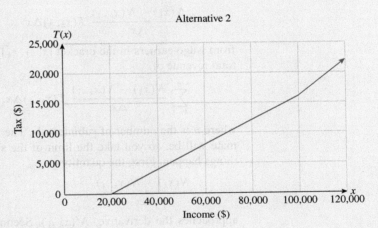

**Figure 29**

In the first alternative, all taxpayers pay 20% of their income in taxes, except that no one pays more than $20,000 in taxes. In the second alternative, there are three tax brackets, described by the following table:

| Income | Marginal tax rate |
| --- | --- |
| $0–20,000 | 0% |
| $20,000–100,000 | 20% |
| Above $100,000 | 30% |

Now you must determine which alternative will generate more annual tax revenue.

Each of Figures 28 and 29 is the graph of a function, $T$. Rather than using the formulas for these particular functions, you begin by working with the general situation. You have an income distribution function $N'$ and a tax function $T$, both functions of annual income. You need to find a formula for total tax revenues. First you decide to use a cutoff so that you need to work only with incomes in some finite interval $[0, M]$; you might use, for example, $M = \$10$ million. (Later you will let $M$ approach $+\infty$.) Next, you subdivide the interval $[0, M]$ into a large number of intervals of small width, $\Delta x$. If $[x_{k-1}, x_k]$ is a typical such interval, you wish to calculate the approximate tax revenue from people whose total incomes lie between $x_{k-1}$ and $x_k$. You will then sum over $k$ to get the total revenue.

You need to know how many people are making incomes between $x_{k-1}$ and $x_k$. Because $N(x_k)$ people are making incomes *up to* $x_k$ and $N(x_{k-1})$ people are making incomes up to $x_{k-1}$, the number of people making incomes between $x_{k-1}$ and $x_k$ is $N(x_k) - N(x_{k-1})$. Because $x_k$ is very close to $x_{k-1}$, the incomes of these people are all approximately equal to $x_{k-1}$ dollars, so each of these taxpayers is paying an annual tax of about $T(x_{k-1})$. This gives a tax revenue of

$$[N(x_k) - N(x_{k-1})]T(x_{k-1}).$$

Now you do a clever thing. You write $x_k - x_{k-1} = \Delta x$ and replace $N(x_k) - N(x_{k-1})$ by

$$\frac{N(x_k) - N(x_{k-1})}{\Delta x}\Delta x.$$

This gives you a tax revenue of about

$$\frac{N(x_k) - N(x_{k-1})}{\Delta x}T(x_{k-1})\Delta x$$

from wage-earners in the bracket $[x_{k-1}, x_k]$. Summing over $k$ gives an approximate total revenue of

$$\sum_{k=1}^{n}\frac{N(x_k) - N(x_{k-1})}{\Delta x}T(x_{k-1})\Delta x,$$

where $n$ is the number of subintervals. The larger $n$ is, the more accurate your estimate will be, so you take the limit of the sum as $n \to \infty$. When you do this, two things happen. First, the quantity

$$\frac{N(x_k) - N(x_{k-1})}{\Delta x}$$

approaches the derivative, $N'(x_{k-1})$. Second, the sum, which you recognize as a Riemann sum, approaches the integral

$$\int_0^M N'(x)T(x)\,dx.$$

You now take the limit as $M \to +\infty$ to get

$$\text{Total tax revenue} = \int_0^{+\infty} N'(x)T(x)\,dx.$$

This improper integral is fine in theory, but the actual calculation will have to be done numerically, so you stick with the upper limit of \$10 million for now. You will have to check that it is reasonable at the end. (Notice that, by the graph of $N'$, it appears that extremely few, if any, people earn that much.) Now you already have a formula for $N'(x)$, but you still need to write formulas for the tax functions $T(x)$ for both alternatives.

**Alternative 1** The graph in Figure 28 rises linearly from 0 to 20,000 as $x$ ranges from 0 to 100,000, and then stays constant at 20,000. The slope of the first part is $20,000/100,000 = 0.2$. The taxation function is therefore

$$T_1(x) = \begin{cases} 0.2x & \text{if } 0 \leq x < 100,000 \\ 20,000 & \text{if } x \geq 100,000 \end{cases}.$$

✱ To see how to obtain the formula, consult the introduction to
Exercises 67–68 in Section 14.1.

For use of technology, it's convenient to express this in closed form using absolute values:✱

$$T_1(x) = 0.2x + \frac{1}{2}\left(1 + \frac{|x - 100,000|}{x - 100,000}\right)(20,000 - 0.2x).$$

The total revenue generated by this tax scheme is, therefore,

$$R_1 = \int_0^{10,000,000} (14x^{0.672}e^{-x/20,400})$$
$$\times \left[0.2x + \frac{1}{2}\left(1 + \frac{|x - 100,000|}{x - 100,000}\right)(20,000 - 0.2x)\right] dx.$$

You decide not to attempt this by hand! You use numerical integration software to obtain a grand total of $R_1 = \$1,360,990,000,000$, or $1.36099 trillion (rounded to six significant digits).

(If you use the Numerical Integration utility on the Website, enter

```
14x^(0.672)*exp(-x/20400)*(0.2x+0.5*(20000-0.2x)*
              (1+abs(x-100000)/(x-100000)))
```

for $f(x)$, and 0 and 10000000 for $a$ and $b$ respectively, and press "Integral".)

**Alternative 2** The graph in Figure 29 rises with a slope of 0.2 from 0 to 16,000 as $x$ ranges from 20,000 to 100,000, then rises from that point on with a slope of 0.3. (This is why we say that the *marginal* tax rates are 20% and 30%, respectively.) The taxation function is therefore

$$T_2(x) = \begin{cases} 0 & \text{if } 0 \le x < 20,000 \\ 0.2(x - 20,000) & \text{if } 20,000 \le x < 100,000. \\ 16,000 + 0.3(x - 100,000) & \text{if } x \ge 100,000 \end{cases}$$

Again, you express this in closed form using absolute values:

$$T_2(x) = [0.2(x - 20,000)]\frac{1}{2}\left(\frac{|x - 20,000|}{x - 20,000} - \frac{|x - 100,000|}{x - 100,000}\right)$$
$$+ [16,000 + 0.3(x - 100,000)]\frac{1}{2}\left(1 + \frac{|x - 100,000|}{x - 100,000}\right)$$
$$= 0.1(x - 20,000)\left(\frac{|x - 20,000|}{x - 20,000} - \frac{|x - 100,000|}{x - 100,000}\right)$$
$$+ [8,000 + 0.15(x - 100,000)]\left(1 + \frac{|x - 100,000|}{x - 100,000}\right).$$

Values of $x$ between 0 and 20,000 do not contribute to the integral, so

$$R_2 = \int_{20,000}^{10,000,000} 14x^{0.672}e^{-x/20,400}T_2(x)\, dx$$

with $T_2(x)$ as above. Numerical integration software gives $R_2 = \$0.713465$ trillion—considerably less than Alternative 1. Thus, even though Alternative 2 taxes the wealthy more heavily, it yields less total revenue.

Now what about the cutoff at $10 million annual income? If you try either integral again with an upper limit of $100 million, you will see no change in either result to six significant digits. There simply are not enough taxpayers earning an income above $10,000,000 to make a difference. You conclude that your answers are sufficiently accurate and that the first alternative provides more tax revenue.

## EXERCISES

In Exercises 1–4, calculate the total tax revenue for a country with the given income distribution and tax policies (all currency in dollars).

1. ⊤ $N'(x) = 100x^{0.466}e^{-x/23,000}$; 25% tax on all income

2. ⊤ $N'(x) = 100x^{0.4}e^{-x/30,000}$; 45% tax on all income

3. ⊤ $N'(x) = 100x^{0.466}e^{-x/23,000}$; tax brackets as in the following tax table:

| Income | Marginal Tax Rate |
|---|---|
| $0–30,000 | 0% |
| $30,000–250,000 | 10% |
| Above $250,000 | 80% |

4. ⊤ $N'(x) = 100x^{0.4}e^{-x/30,000}$; no tax on any income below $250,000, 100% marginal tax rate on any income above $250,000

5. Let $N'(x)$ be an income distribution function.
   a. If $0 \leq a < b$, what does $\int_a^b N'(x)\,dx$ represent? HINT [Use the Fundamental Theorem of Calculus.]
   b. What does $\int_0^{+\infty} N'(x)\,dx$ represent?

6. Let $N'(x)$ be an income distribution function. What does $\int_0^{+\infty} xN'(x)\,dx$ represent? HINT [Argue as in the text.]

7. Let $P(x)$ be the number of people earning more than $x$ dollars.
   a. What is $N(x) + P(x)$?
   b. Show that $P'(x) = -N'(x)$.
   c. Use integration by parts to show that, if $T(0) = 0$, then the total tax revenue is

$$\int_0^{+\infty} P(x)T'(x)\,dx.$$

[Note: You may assume that $T'(x)$ is continuous, but the result is still true if we assume only that $T(x)$ is continuous and piecewise continuously differentiable.]

8. Income tax functions $T$ are most often described, as in the text, by tax brackets and marginal tax rates.
   a. If one tax bracket is $a < x \leq b$, show that $\int_a^b P(x)\,dx$ is the total income earned in the country that falls into that bracket ($P$ as in the preceding exercise).
   b. Use (a) to explain directly why $\int_0^{+\infty} P(x)T'(x)\,dx$ gives the total tax revenue in the case where $T$ is described by tax brackets and constant marginal tax rates in each bracket.

## TI-83/84 Plus   Technology Guide

### Section 14.3

**Example 3 (page 1036)**   The following table shows Colossal Conglomerate's closing stock prices for 20 consecutive trading days:

| Day | 1 | 2 | 3 | 4 | 5 | 6 | 7 | 8 | 9 | 10 |
|-----|---|---|---|---|---|---|---|---|---|----|
| Price | 20 | 22 | 21 | 24 | 24 | 23 | 25 | 26 | 20 | 24 |
| Day | 11 | 12 | 13 | 14 | 15 | 16 | 17 | 18 | 19 | 20 |
| Price | 26 | 26 | 25 | 27 | 28 | 27 | 29 | 27 | 25 | 24 |

Plot these prices and the 5-day moving average.

### Solution with Technology

Here is how to automate this calculation on a TI-83/84 Plus.

1. Use

   $\text{seq}(X, X, 1, 20) \rightarrow L_1$   2ND STAT → OPS → 5
   STO 2ND STAT → L₁

   to enter the sequence of numbers 1 through 20 into the list $L_1$, representing the trading days.

2. Using the list editor accessible through the STAT menu, enter the daily stock prices in list $L_2$.

3. Calculate the list of 5-day moving averages by using the following command:

   $\text{seq}((L_2(X) + L_2(X-1) + L_2(X-2) + L_2(X-3) + L_2(X-4))/5, X, 5, 20) \rightarrow L_3$

   This has the effect of putting the moving averages into elements 1 through 15 of list $L_3$.

4. If you wish to plot the moving average on the same graph as the daily prices, you will want the averages in $L_3$ to match up with the prices in $L_2$. One way to do this is to put four more entries at the beginning of $L_3$—say, copies of the first four entries of $L_2$. The following command accomplishes this:

   $\text{augment}(\text{seq}(L_2(X), X, 1, 4), L_3) \rightarrow L_3$
   2ND STAT → OPS → 9

5. You can now graph the prices and moving averages by creating an xyLine scatter plot through the STAT PLOT menu, with $L_1$ being the Xlist and $L_2$ being the Ylist for Plot1, and $L_1$ being the Xlist and $L_3$ the Ylist for Plot2:

## SPREADSHEET   Technology Guide

### Section 14.3

**Example 3 (page 1036)**   The following table shows Colossal Conglomerate's closing stock prices for 20 consecutive trading days:

| Day | 1 | 2 | 3 | 4 | 5 | 6 | 7 | 8 | 9 | 10 |
|-----|---|---|---|---|---|---|---|---|---|----|
| Price | 20 | 22 | 21 | 24 | 24 | 23 | 25 | 26 | 20 | 24 |
| Day | 11 | 12 | 13 | 14 | 15 | 16 | 17 | 18 | 19 | 20 |
| Price | 26 | 26 | 25 | 27 | 28 | 27 | 29 | 27 | 25 | 24 |

Plot these prices and the 5-day moving average.

### Solution with Technology

**1.** Compute the moving averages in a column next to the daily prices, as shown here:

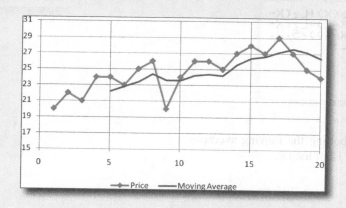

**2.** You can then graph the price and moving average using a scatter plot.

# Appendix A  Logic

## Introduction

Logic is the underpinning of all reasoned argument. The ancient Greeks recognized its role in mathematics and philosophy, and studied it extensively. Aristotle, in his *Organon*, wrote the first systematic treatise on logic. His work had a heavy influence on philosophy, science and religion through the Middle Ages.

But Aristotle's logic was expressed in ordinary language, so was subject to the ambiguities of ordinary language. Philosophers came to want to express logic more formally and symbolically, more like the way that mathematics is written (Leibniz, in the 17th century, was probably the first to envision and call for such a formalism). It was with the publication in 1847 of G. Boole's *The Mathematical Analysis of Logic* and A. DeMorgan's *Formal Logic* that **symbolic logic** came into being, and logic became recognized as part of mathematics. Since Boole and DeMorgan, logic and mathematics have been inextricably intertwined. Logic is part of mathematics, but at the same time it is the language of mathematics.

The study of symbolic logic is usually broken into several parts. The first and most fundamental is the **propositional logic**. Built on top of this is the **predicate logic**, which is the language of mathematics. In this appendix we give an introduction to propositional logic.

## A.1  Statements and Logical Operators

**Propositional logic** is the study of *propositions*. A **statement**, or **proposition**, is any declarative sentence which is either true (T) or false (F). We refer to T or F as the **truth value** of the statement.

### EXAMPLE 1  Statements

**a.** "$2 + 2 = 4$" is a statement because it can be either true or false.[1] Because it happens to be a true statement, its truth value is T.

**b.** "$1 = 0$" is also a statement, but its truth value is F.

**c.** "It will rain tomorrow" is a statement. To determine its truth value, we shall have to wait for tomorrow.

**d.** "Solve the following equation for $x$" is not a statement, because it cannot be assigned any truth value whatsoever. It is an imperative, or command, rather than a declarative sentence.

**e.** "The number 5" is not a statement, because it is not even a complete sentence.

For a much more extensive interactive treatment of logic, including discussion of proofs, rules of inference, and an introduction to the predicate calculus, go online and follow:

Online Text
→ On Line Topics in Finite Mathematics
→ Introduction to Logic

---

[1] Is "$2 + 2 = 4$" a sentence? Read it aloud: "Two plus two equals four," is a perfectly respectable English sentence.

A1

**f.** "This statement is false" gets us into a bind: If it were true, then, because it is declaring itself to be false, it must be false. On the other hand, if it were false, then its declaring itself false is a lie, so it is true! In other words, if it is true, then it is false, and if it is false, then it is true, and we go around in circles. We get out of this bind by saying that because the sentence cannot be either true or false, we refuse to call it a statement. An equivalent pseudo-statement is: "I am lying," so this sentence is known as **the liar's paradox.**

---

**Note** Sentences that refer to themselves, or *self-referential sentences*, as illustrated in Example 1(f), are not permitted to be statements. This eliminates the liar's paradox and several similar problems. ■

We shall use letters like $p, q, r$, and so on to stand for statements. Thus, for example, we might decide that $p$ should stand for the statement "the moon is round." We write

$p$: "the moon is round"     $p$ is the statement that the moon is round

to express this.

We can form new statements from old ones in several different ways. For example, starting with $p$: "I am an Anchovian," we can form the **negation** of $p$: "It is not the case that I am an Anchovian" or simply "I am not an Anchovian."

### Negation of a Statement

If $p$ is a statement, then its **negation** is the statement "not $p$" and is denoted by $\sim p$. We mean by this that, if $p$ is true, then $\sim p$ is false, and vice versa.

#### Quick Examples

1. If $p$: "$2 + 2 = 4$," then $\sim p$: "It is not the case that $2 + 2 = 4$," or, more simply, $\sim p$: "$2 + 2 \neq 4$."
2. If $q$: "$1 = 0$," then $\sim q$: "$1 \neq 0$."
3. If $r$: "Diamonds are a pearl's best friend," then $\sim r$: "Diamonds are not a pearl's best friend."
4. If $s$: "All politicians are crooks," then $\sim s$: "Not all politicians are crooks."
5. **Double Negation:** If $p$ is any statement, then the negation of $\sim p$ is $\sim(\sim p)$: "not (not $p$)," or, in other words, $p$. Thus $\sim(\sim p)$ has the same meaning as $p$.

### Notes

1. Notice in Quick Example 1 above that $\sim p$ is false, because $p$ is true. However, in Quick Example 2, $\sim q$ is true, because $q$ is false. A statement of the form $\sim q$ can very well be true; it is a common mistake to think it must be false.
2. Saying that not all politicians are crooks is not the same as saying that no politicians are crooks, but is the same as saying that some (meaning one or more) politicians are not crooks.
3. The symbol $\sim$ is our first example of a **logical operator**.
4. When we say in Quick Example 5 above that $\sim(\sim p)$ has the same meaning as $p$, we mean that they are *logically equivalent*—a notion we will make precise below. ■

Here is another way we can form a new statement from old ones. Starting with $p$: "I am wise," and $q$: "I am strong," we can form the statement "I am wise and I am strong." We denote this new statement by $p \wedge q$, read "$p$ and $q$." In order for $p \wedge q$ to be true, *both p and q* must be true. Thus, for example, if I am wise but not strong, then $p \wedge q$ is false. The symbol $\wedge$ is another logical operator. The statement $p \wedge q$ is called the **conjunction** of $p$ and $q$.

### Conjunction

The **conjunction** of $p$ and $q$ is the statement $p \wedge q$, which we read "$p$ and $q$." It can also be said in a number of different ways, such as "$p$ even though $q$." The statement $p \wedge q$ is true when both $p$ and $q$ are true and false otherwise.

#### Quick Examples

1. If $p$: "This galaxy will ultimately disappear into a black hole" and $q$: "$2 + 2 = 4$," then $p \wedge q$ is the statement "Not only will this galaxy ultimately disappear into a black hole, but $2 + 2 = 4$!"
2. If $p$: "$2 + 2 = 4$" and $q$: "$1 = 0$," then $p \wedge q$: "$2 + 2 = 4$ and $1 = 0$." Its truth value is F because $q$ is F.
3. With $p$ and $q$ as in Quick Example 1, the statement $p \wedge (\sim q)$ says: "This galaxy will ultimately disappear into a black hole and $2 + 2 \neq 4$," or, more colorfully, as "Contrary to your hopes, this galaxy is doomed to disappear into a black hole; moreover, two plus two is decidedly *not* equal to four!"

**Notes**

1. We sometimes use the word "but" as an emphatic form of "and." For instance, if $p$: "It is hot," and $q$: "It is not humid," then we can read $p \wedge q$ as "It is hot but not humid." There are always many ways of saying essentially the same thing in a natural language; one of the purposes of symbolic logic is to strip away the verbiage and record the underlying logical structure of a statement.
2. A **compound statement** is a statement formed from simpler statements via the use of logical operators. Examples are $\sim p$, $(\sim p) \wedge (q \wedge r)$ and $p \wedge (\sim p)$. A statement that cannot be expressed as a compound statement is called an **atomic statement**.[2] For example, "I am clever" is an atomic statement. In a compound statement such as $(\sim p) \wedge (q \wedge r)$, we refer to $p$, $q$, and $r$ as the **variables** of the statement. Thus, for example, $\sim p$ is a compound statement in the single variable $p$. ∎

Before discussing other logical operators, we pause for a moment to talk about **truth tables**, which give a convenient way to analyze compound statements.

### Truth Table

The **truth table** for a compound statement shows, for each combination of possible truth values of its variables, the corresponding truth value of the statement.

---

[2]"Atomic" comes from the Greek for "not divisible." Atoms were originally thought to be the indivisible components of matter, but the march of science proved that wrong. The name stuck, though.

**Quick Examples**

**1.** The truth table for negation, that is, for $\sim p$, is:

| $p$ | $\sim p$ |
|---|---|
| T | F |
| F | T |

Each row shows a possible truth value for $p$ and the corresponding value of $\sim p$.

**2.** The truth table for conjunction, that is, for $p \wedge q$, is:

| $p$ | $q$ | $p \wedge q$ |
|---|---|---|
| T | T | T |
| T | F | F |
| F | T | F |
| F | F | F |

Each row shows a possible combination of truth values of $p$ and $q$ and the corresponding value of $p \wedge q$.

---

**EXAMPLE 2 Construction of Truth Tables**

Construct truth tables for the following compound statements.

**a.** $\sim(p \wedge q)$       **b.** $(\sim p) \wedge q$

**Solution**

**a.** Whenever we encounter a complex statement, we work from the inside out, just as we might do if we had to evaluate an algebraic expression like $-(a + b)$. Thus, we start with the $p$ and $q$ columns, then construct the $p \wedge q$ column, and finally, the $\sim(p \wedge q)$ column.

| $p$ | $q$ | $p \wedge q$ | $\sim(p \wedge q)$ |
|---|---|---|---|
| T | T | T | F |
| T | F | F | T |
| F | T | F | T |
| F | F | F | T |

Notice how we get the $\sim(p \wedge q)$ column from the $p \wedge q$ column: we reverse all the truth values.

**b.** Because there are two variables, $p$ and $q$, we again start with the $p$ and $q$ columns. We then evaluate $\sim p$, and finally take the conjunction of the result with $q$.

| $p$ | $q$ | $\sim p$ | $(\sim p) \wedge q$ |
|---|---|---|---|
| T | T | F | F |
| T | F | F | F |
| F | T | T | T |
| F | F | T | F |

Because we are "and-ing" $\sim p$ with $q$, we look at the values in the $\sim p$ and $q$ columns and combine these according to the instructions for "and." Thus, for example, in the first row we have $F \wedge T = F$ and in the third row we have $T \wedge T = T$.

Here is a third logical operator. Starting with $p$: "You are over 18" and $q$: "You are accompanied by an adult," we can form the statement "You are over 18 or are accompanied by an adult," which we write symbolically as $p \vee q$, read "$p$ or $q$." Now in English the word "or" has several possible meanings, so we have to agree on which one we want here. Mathematicians have settled on the **inclusive or:** $p \vee q$ means $p$ is true or $q$ is true *or both are true.*[3] With $p$ and $q$ as above, $p \vee q$ stands for "You are over 18 or are accompanied by an adult, or both." We shall sometimes include the phrase "or both" for emphasis, but even if we leave it off we still interpret "or" as inclusive.

## Disjunction

The **disjunction** of $p$ and $q$ is the statement $p \vee q$, which we read "$p$ or $q$." Its truth value is defined by the following truth table.

| $p$ | $q$ | $p \vee q$ |
|---|---|---|
| T | T | T |
| T | F | T |
| F | T | T |
| F | F | F |

This is the **inclusive** or, so $p \vee q$ is true when $p$ is true or $q$ is true *or both* are true.

> **Quick Examples**
>
> 1. Let $p$: "The butler did it" and let $q$: "The cook did it." Then $p \vee q$: "Either the butler or the cook did it."
> 2. Let $p$: "The butler did it," and let $q$: "The cook did it," and let $r$: "The lawyer did it." Then $(p \vee q) \wedge (\sim r)$: "Either the butler or the cook did it, but not the lawyer."

**Note** The only way for $p \vee q$ to be false is for *both* $p$ and $q$ to be false. For this reason, we can say that $p \vee q$ also means "$p$ and $q$ are not both false." ∎

To introduce our next logical operator, we ask you to consider the following statement: "If you earn an A in logic, then I'll buy you a new car." It seems to be made up out of two simpler statements,

    $p$: "You earn an A in logic," and
    $q$: "I will buy you a new car."

---

[3]There is also the **exclusive or:** "$p$ or $q$ but not both." This can be expressed as $(p \vee q) \wedge \sim(p \wedge q)$. Do you see why?

The original statement says: *if p is true, then q is true*, or, more simply, **if** *p*, **then** *q*. We can also phrase this as *p* **implies** *q*, and we write the statement symbolically as $p \to q$.

Now let us suppose for the sake of argument that the original statement: "If you earn an A in logic, then I'll buy you a new car," is true. This does *not* mean that you *will* earn an A in logic. All it says is that *if* you do so, then I will buy you that car. Thinking of this as a promise, the only way that it can be broken is if you *do* earn an A and I do *not* buy you a new car. With this in mind, we define the logical statement $p \to q$ as follows.

## Conditional

The **conditional** $p \to q$, which we read "if *p*, then *q*" or "*p* implies *q*," is defined by the following truth table:

| $p$ | $q$ | $p \to q$ |
|-----|-----|-----------|
| T | T | T |
| T | F | F |
| F | T | T |
| F | F | T |

The arrow "$\to$" is the **conditional** operator, and in $p \to q$ the statement *p* is called the **antecedent**, or **hypothesis**, and *q* is called the **consequent**, or **conclusion.** A statement of the form $p \to q$ is also called an **implication.**

### Quick Examples

1. "If $1 + 1 = 2$ then the sun rises in the east" has the form $p \to q$ where *p*: "$1 + 1 = 2$" is true and *q*: "the sun rises in the east." is also true. Therefore the statement is true.

2. "If the moon is made of green cheese, then I am Arnold Schwartzenegger" has the form $p \to q$ where *p* is false. From the truth table, we see that $p \to q$ is therefore true, regardless of whether or not I am Arnold Schwartzenegger.

3. "If $1 + 1 = 2$ then $0 = 1$" has the form $p \to q$ where this time *p* is true but *q* is false. Therefore, by the truth table, the given statement is false.

### Notes

1. The only way that $p \to q$ can be false is if *p* is true and *q* is false—this is the case of the "broken promise" in the car example above.

2. If you look at the last two rows of the truth table, you see that we say that "$p \to q$" is true when *p* is false, *no matter what the truth value of q*. Think again about the promise—if you don't get that A, then whether or not I buy you a new car, I have not broken my promise. It may seem strange at first to say that $F \to T$ is T and $F \to F$ is also T, but, as they did in choosing to say that "or" is always inclusive, mathematicians agreed that the truth table above gives the most useful definition of the conditional. ■

It is usually misleading to think of "if *p* then *q*" as meaning that *p causes q*. For instance, tropical weather conditions cause hurricanes, but one cannot claim that if there are tropical weather conditions, then there are (always) hurricanes. Here is a list of some English phrases that *do* have the same meaning as $p \to q$.

Appendix A

## Some Phrasings of the Conditional

We interpret each of the following as equivalent to the conditional $p \to q$.

| | |
|---|---|
| If $p$ then $q$. | $p$ implies $q$. |
| $q$ follows from $p$. | Not $p$ unless $q$. |
| $q$ if $p$. | $p$ only if $q$. |
| Whenever $p$, $q$. | $q$ whenever $p$. |
| $p$ is sufficient for $q$. | $q$ is necessary for $p$. |
| $p$ is a sufficient condition for $q$. | $q$ is a necessary condition for $p$. |

### Quick Example

"If it's Tuesday, this must be Belgium" can be rephrased in several ways as follows:

"Its being Tuesday implies that this is Belgium."
"This is Belgium if it's Tuesday."
"It's Tuesday only if this is Belgium."
"It can't be Tuesday unless this is Belgium."
"Its being Tuesday is sufficient for this to be Belgium."
"That this is Belgium is a necessary condition for its being Tuesday."

Notice the difference between "if" and "only if." We say that "$p$ only if $q$" means $p \to q$ because, assuming that $p \to q$ is true, $p$ can be true only if $q$ is also. In other words, the only line of the truth table that has $p \to q$ true and $p$ true also has $q$ true. The phrasing "$p$ is a sufficient condition for $q$" says that it suffices to know that $p$ is true to be able to conclude that $q$ is true. For example, it is sufficient that you get an A in logic for me to buy you a new car. Other things might induce me to buy you the car, but an A in logic would suffice. The phrasing "$q$ is necessary for $p$" says that for $p$ to be true, $q$ must be true (just as we said for "$p$ only if $q$").

---

**Q**: *Does the commutative law hold for the conditional? In other words, is $p \to q$ the same as $q \to p$?*

**A**: No, as we can see in the following truth table:

| $p$ | $q$ | $p \to q$ | $q \to p$ |
|---|---|---|---|
| T | T | T | T |
| T | F | F | T |
| F | T | T | F |
| F | F | T | T |

not the same

---

## Converse and Contrapositive

The statement $q \to p$ is called the **converse** of the statement $p \to q$. A conditional and its converse are *not* the same.

The statement $\sim q \to \sim p$ is the **contrapositive** of the statement $p \to q$. A conditional and its contrapositive are logically equivalent in the sense we define below: they have the same truth value for all possible values of $p$ and $q$.

Appendix A

### EXAMPLE 3 **Converse and Contrapositive**

Give the converse and contrapositive of the statement "If you earn an A in logic, then I'll buy you a new car."

**Solution** This statement has the form $p \to q$ where $p$: "you earn an A" and $q$: "I'll buy you a new car." The converse is $q \to p$. In words, this is "If I buy you a new car then you earned an A in logic."

The contrapositive is $(\sim q) \to (\sim p)$. In words, this is "If I don't buy you a new car, then you didn't earn an A in logic."

Assuming that the original statement is true, notice that the converse is not necessarily true. There is nothing in the original promise that prevents me from buying you a new car if you do not earn the A. On the other hand, the contrapositive is true. If I don't buy you a new car, it must be that you didn't earn an A; otherwise I would be breaking my promise.

It sometimes happens that we do want both a conditional and its converse to be true. The conjunction of a conditional and its converse is called a **biconditional**.

## Biconditional

The **biconditional**, written $p \leftrightarrow q$, is defined to be the statement $(p \to q) \wedge (q \to p)$. Its truth table is the following:

| $p$ | $q$ | $p \leftrightarrow q$ |
|---|---|---|
| T | T | T |
| T | F | F |
| F | T | F |
| F | F | T |

### Phrasings of the Biconditional
We interpret each of the following as equivalent to $p \leftrightarrow q$.

$p$ if and only if $q$.
$p$ is necessary and sufficient for $q$.
$p$ is equivalent to $q$.

### Quick Example

"I teach math if and only if I am paid a large sum of money" can be rephrased in several ways as follows:

"I am paid a large sum of money if and only if I teach math."
"My teaching math is necessary and sufficient for me to be paid a large sum of money."
"For me to teach math, it is necessary and sufficient that I be paid a large sum of money."

## A.2 | Logical Equivalence

We mentioned above that we say that two statements are **logically equivalent** if for all possible truth values of the variables involved the two statements always have the same truth values. If $s$ and $t$ are equivalent, we write $s \equiv t$. This is *not* another logical statement. It is simply the claim that the two statements $s$ and $t$ are logically equivalent. Here are some examples.

**EXAMPLE 4  Logical Equivalence**

Use truth tables to show the following:

**a.** $p \equiv \sim(\sim p)$. This is called **double negation**.

**b.** $\sim(p \wedge q) \equiv (\sim p) \vee (\sim q)$. This is one of **DeMorgan's Laws**.

**Solution**

**a.** To demonstrate the logical equivalence of these two statements, we construct a truth table with columns for both $p$ and $\sim(\sim p)$.

same

| $p$ | $\sim p$ | $\sim(\sim p)$ |
|---|---|---|
| T | F | T |
| F | T | F |

Because the $p$ and $\sim(\sim p)$ columns contain the same truth values in all rows, the two statements are logically equivalent.

**b.** We construct a truth table showing both $\sim(p \wedge q)$ and $(\sim p) \vee (\sim q)$.

same

| $p$ | $q$ | $p \wedge q$ | $\sim(p \wedge q)$ | $\sim p$ | $\sim q$ | $(\sim p) \vee (\sim q)$ |
|---|---|---|---|---|---|---|
| T | T | T | F | F | F | F |
| T | F | F | T | F | T | T |
| F | T | F | T | T | F | T |
| F | F | F | T | T | T | T |

Because the $\sim(p \wedge q)$ column and $(\sim p) \vee (\sim q)$ column agree, the two statements are equivalent.

➡ **Before we go on...** The statement $\sim(p \wedge q)$ can be read as "It is not the case that both $p$ and $q$ are true" or "$p$ and $q$ are not both true." We have just shown that this is equivalent to "Either $p$ is false or $q$ is false." ∎

Here are the two equivalences known as DeMorgan's Laws.

### DeMorgan's Laws

If $p$ and $q$ are statements, then

$$\sim(p \wedge q) \equiv (\sim p) \vee (\sim q)$$
$$\sim(p \vee q) \equiv (\sim p) \wedge (\sim q)$$

---

**Quick Example**

Let $p$: "the President is a Democrat," and $q$: "the President is a Republican." Then the following two statements say the same thing:

$\sim(p \wedge q)$: "the President is not both a Democrat and a Republican."

$(\sim p) \vee (\sim q)$: "either the President is not a Democrat, or he is not a Republican (or he is neither)."

---

Here is a list of some important logical equivalences, some of which we have already encountered. All of them can be verified using truth tables as in Example 4. (The verifications of some of these are in the exercise set.)

---

## Important Logical Equivalences

| | |
|---|---|
| $\sim(\sim p) \equiv p$ | the Double Negative Law |
| $p \wedge q \equiv q \wedge p$ | the Commutative Law for Conjunction |
| $p \vee q \equiv q \vee p$ | the Commutative Law for Disjunction |
| $(p \wedge q) \wedge r \equiv p \wedge (q \wedge r)$ | the Associative Law for Conjunction |
| $(p \vee q) \vee r \equiv p \vee (q \vee r)$ | the Associative Law for Disjunction |
| $\sim(p \vee q) \equiv (\sim p) \wedge (\sim q)$ | DeMorgan's Laws |
| $\sim(p \wedge q) \equiv (\sim p) \vee (\sim q)$ | |
| $p \wedge (q \vee r) \equiv (p \wedge q) \vee (p \wedge r)$ | the Distributive Laws |
| $p \vee (q \wedge r) \equiv (p \vee q) \wedge (p \vee r)$ | |
| $p \wedge p \equiv p$ | Absorption Laws |
| $p \vee p \equiv p$ | |
| $p \to q \equiv (\sim q) \to (\sim p)$ | Contrapositive Law |

---

Note that these logical equivalences apply to *any* statement. The $p$s, $q$s, and $r$s can stand for atomic statements or compound statements, as we see in the next example.

---

### EXAMPLE 5  Applying Logical Equivalences

**a.** Apply DeMorgan's law (once) to the statement $\sim([p \wedge (\sim q)] \wedge r)$.

**b.** Apply the distributive law to the statement $(\sim p) \wedge [q \vee (\sim r)]$.

**c.** Consider: "You will get an A if either you are clever and the sun shines, or you are clever and it rains." Rephrase the condition more simply using the distributive law.

### Solution

**a.** We can analyze the given statement from the outside in. It is first of all a negation, but further, it is the negation $\sim(A \wedge B)$, where $A$ is the compound statement $[p \wedge (\sim q)]$ and $B$ is $r$:

$$\sim(\overbrace{\quad A \quad}^{} \wedge \; B)$$
$$\sim(\; [p \wedge (\sim q)] \wedge \quad r)$$

Now one of DeMorgan's laws is

$$\sim(A \wedge B) \equiv (\sim A) \vee (\sim B)$$

Applying this equivalence gives

$$\sim([p \wedge (\sim q)] \wedge r) \equiv (\sim[p \wedge (\sim q)]) \vee (\sim r)$$

**b.** The given statement has the form $A \wedge [B \vee C]$, where $A = (\sim p)$, $B = q$, and $C = (\sim r)$. So, we apply the distributive law $A \wedge [B \vee C], \equiv [A \wedge B] \vee [A \wedge C]$:

$$(\sim p) \wedge [q \vee (\sim r)] \equiv [(\sim p) \wedge q] \vee [(\sim p) \wedge (\sim r)]$$

(We need not stop here: The second expression on the right is just begging for an application of DeMorgan's law. . .)

**c.** The condition is "either you are clever and the sun shines, or you are clever and it rains." Let's analyze this symbolically: Let $p$: "You are clever," $q$: "The sun shines," and $r$: "It rains." The condition is then $(p \wedge q) \vee (p \wedge r)$. We can "factor out" the $p$ using one of the distributive laws in reverse, getting

$$(p \wedge q) \vee (p \wedge r) \equiv p \wedge (q \vee r)$$

We are taking advantage of the fact that the logical equivalences we listed can be read from right to left as well as from left to right. Putting $p \wedge (q \vee r)$ back into English, we can rephrase the sentence as "You will get an A if you are clever and either the sun shines or it rains."

➡ **Before we go on...** In part (a) of Example 5 we could, if we wanted, apply DeMorgan's law again, this time to the statement $\sim[p \wedge (\sim q)]$ that is part of the answer. Doing so gives

$$\sim[p \wedge (\sim q)] \equiv (\sim p) \vee \sim(\sim q) \equiv (\sim p) \vee q$$

Notice that we've also used the double negative law. Therefore, the original expression can be simplified as follows:

$$\sim([p \wedge (\sim q)] \wedge r) \equiv (\sim[p \wedge (\sim q)]) \vee (\sim r) \equiv ((\sim p) \vee q) \vee (\sim r)$$

which we can write as

$$(\sim p) \vee q \vee (\sim r)$$

because the associative law tells us that it does not matter which two expressions we "or" first. ∎

# A.3 Tautologies, Contradictions, and Arguments

## Tautologies and Contradictions

A compound statement is a **tautology** if its truth value is always T, regardless of the truth values of its variables. It is a **contradiction** if its truth value is always F, regardless of the truth values of its variables.

## Quick Examples

**1.** $p \vee (\sim p)$ has truth table

| $p$ | $\sim p$ | $p \vee (\sim p)$ |
|---|---|---|
| T | F | T |
| F | T | T |

— all T's

and is therefore a tautology.

**2.** $p \wedge (\sim p)$ has truth table

| $p$ | $\sim p$ | $p \wedge (\sim p)$ |
|---|---|---|
| T | F | F |
| F | T | F |

and is therefore a contradiction.

When a statement is a tautology, we also say that the statement is **tautological**. In common usage this sometimes means simply that the statement is self-evident. In logic it means something stronger: that the statement is always true, under all circumstances. In contrast, a contradiction, or **contradictory** statement, is *never* true, under any circumstances.

Some of the most important tautologies are the **tautological implications**, tautologies that have the form of implications. We look at two of them: Direct Reasoning, and Indirect Reasoning:

## *Modus Ponens* or Direct Reasoning

The following tautology is called *modus ponens* or **direct reasoning**:

$$[(p \rightarrow q) \wedge p] \rightarrow q$$

### In Words
If an implication and its antecedent ($p$) are both true, then so is its consequent ($q$).

## Quick Example

*If my loving math implies that I will pass this course, and if I do love math, then I will pass this course.*

**Note** You can check that the statement $[(p \rightarrow q) \wedge p] \rightarrow q$ is a tautology by drawing its truth table. ∎

Tautological implications are useful mainly because they allow us to check the validity of **arguments**.

## Argument

An **argument** is a list of statements called **premises** followed by a statement called the **conclusion**. If the premises are $P_1, P_2, \ldots, P_n$ and the conclusion is $C$, then we say that the argument is **valid** if the statement $(P_1 \wedge P_2 \wedge \ldots \wedge P_n) \rightarrow C$ is a tautology. In other words, an argument is valid if the truth of all its premises logically implies the truth of its conclusion.

### Quick Examples

1. The following is a valid argument:

$$p \rightarrow q$$
$$\underline{p}$$
$$\therefore \quad q$$

(This is the traditional way of writing an argument: We list the premises above a line and then put the conclusion below; the symbol "$\therefore$" stands for the word "therefore.") This argument is valid because the statement $[(p \rightarrow q) \wedge p] \rightarrow q$ is a tautology, namely *modus ponens*.

2. The following is an invalid argument:

$$p \rightarrow q$$
$$\underline{q}$$
$$\therefore \quad p$$

The argument is invalid because the statement $[(p \rightarrow q) \wedge q] \rightarrow p$ is not a tautology. In fact, if $p$ is F and $q$ is T, then the whole statement is F.

The argument in Quick Example 2 above is known as the *fallacy of affirming the consequent*. It is a common invalid argument and not always obviously flawed at first sight, so is often exploited by advertisers. For example, consider the following claim: All Olympic athletes drink Boors, so you should too. The suggestion is that, if you drink Boors, you will be an Olympic athlete:

| | |
|---|---|
| If you are an Olympic Athlete you drink Boors. | Premise (Let's pretend this is True) |
| You drink Boors. | Premise (True) |
| $\therefore$  You are an Olympic Athlete. | Conclusion (May be false!) |

This is an error that Boors hopes you will make!

There is, however, a correct argument in which we *deny* the consequent:

## *Modus Tollens* or Indirect Reasoning

The following tautology is called *modus tollens* or **indirect reasoning**:

$$[(p \rightarrow q) \wedge (\sim q)] \rightarrow (\sim p)$$

**In Words**
If an implication is true but its consequent ($q$) is false, then its antecedent ($p$) is false.

**In Argument Form**

$$p \to q$$
$$\sim q$$
$$\therefore \quad \sim p$$

**Quick Example**

*If my loving math implies that I will pass this course, and if I do not pass
the course, then it must be the case that I do not love math.*
In argument form:

If I love math, then I will pass this course.

I will not pass the course.

Therefore, I do not love math.

**Note** This argument is not as direct as *modus ponens;* it contains a little twist: "If
I loved math I would pass this course. However, I will not pass this course. Therefore, it must be that I don't love math (else I *would* pass this course)." Hence the
name "indirect reasoning."

Note that, again, there is a similar, but fallacious argument to avoid, for instance:
"If I were an Olympic athlete then I would drink Boors ($p \to q$). However, I am not
an Olympic athlete ($\sim p$). Therefore, I won't drink Boors. ($\sim q$)." This is a mistake
Boors certainly hopes you do *not* make! ■

There are other interesting tautologies that we can use to justify arguments. We
mention one more and refer the interested reader to the website for more examples
and further study.

For an extensive list of tautologies
go online and follow:

Chapter L Logic

→ List of Tautologies and
   Tautological Implications

## Disjunctive Syllogism or "One or the Other"

The following tautologies are both known as the **disjunctive syllogism** or
**one-or-the-other:**

$$[(p \vee q) \wedge (\sim p)] \to q \qquad [(p \vee q) \wedge (\sim q)] \to p$$

**In Words**
If one or the other of two statements is true, but one is known to be false, then
the other must be true.

**In Argument Form**

$$p \vee q \qquad\qquad p \vee q$$
$$\sim p \qquad\qquad\qquad \sim q$$
$$\therefore \quad q \qquad\qquad \therefore \quad p$$

**Quick Example**

*The butler or the cook did it. The butler didn't do it. Therefore, the cook
did it.*
In argument form:

The butler or the cook did it.

The butler did not do it.

Therefore, the cook did it.

# A     EXERCISES

*Which of Exercises 1–10 are statements? Comment on the truth values of all the statements you encounter. If a sentence fails to be a statement, explain why.* HINT [See Example 1.]

**1.** All swans are white.          **2.** The fat cat sat on the mat.

**3.** Look in thy glass and tell whose face thou viewest.[4]

**4.** My glass shall not persuade me I am old.[5]

**5.** There is no largest number.

**6.** 1,000,000,000 is the largest number.

**7.** Intelligent life abounds in the universe.

**8.** There may or may not be a largest number.

**9.** This is exercise number 9.     **10.** This sentence no verb.[6]

*Let p: "Our mayor is trustworthy," q: "Our mayor is a good speller," and r = "Our mayor is a patriot." Express each of the statements in Exercises 11–16 in logical form:* HINT [See Quick Examples on pages A2, A3, A5.]

**11.** Although our mayor is not trustworthy, he is a good speller.

**12.** Either our mayor is trustworthy, or he is a good speller.

**13.** Our mayor is a trustworthy patriot who spells well.

**14.** While our mayor is both trustworthy and patriotic, he is not a good speller.

**15.** It may or may not be the case that our mayor is trustworthy.

**16.** Our mayor is either not trustworthy or not a patriot, yet he is an excellent speller.

*Let p: "Willis is a good teacher," q: "Carla is a good teacher," r: "Willis' students hate math," s: "Carla's students hate math." Express the statements in Exercises 17–24 in words.*

**17.** $p \wedge (\sim r)$          **18.** $(\sim p) \wedge (\sim q)$

**19.** $q \vee (\sim q)$          **20.** $((\sim p) \wedge (\sim s)) \vee q$

**21.** $r \wedge (\sim r)$          **22.** $(\sim s) \vee (\sim r)$

**23.** $\sim(q \vee s)$          **24.** $\sim(p \wedge r)$

*Assume that it is true that "Polly sings well," it is false that "Quentin writes well," and it is true that "Rita is good at math." Determine the truth of each of the statements in Exercises 25–32.*

**25.** Polly sings well and Quentin writes well.

**26.** Polly sings well or Quentin writes well.

**27.** Polly sings poorly and Quentin writes well.

**28.** Polly sings poorly or Quentin writes poorly.

**29.** Either Polly sings well and Quentin writes poorly, or Rita is good at math.

**30.** Either Polly sings well and Quentin writes poorly, or Rita is not good at math.

**31.** Either Polly sings well or Quentin writes well, or Rita is good at math.

**32.** Either Polly sings well and Quentin writes well, or Rita is bad at math.

*Find the truth value of each of the statements in Exercises 33–48.* HINT [See Quick Examples on page A6.]

**33.** "If $1 = 1$, then $2 = 2$."          **34.** "If $1 = 1$, then $2 = 3$."

**35.** "If $1 \neq 0$, then $2 \neq 2$."          **36.** "If $1 = 0$, then $1 = 1$."

**37.** "A sufficient condition for 1 to equal 2 is $1 = 3$."

**38.** "$1 = 1$ is a sufficient condition for 1 to equal 0."

**39.** "$1 = 0$ is a necessary condition for 1 to equal 1."

**40.** "$1 = 1$ is a necessary condition for 1 to equal 2."

**41.** "If I pay homage to the great Den, then the sun will rise in the east."

**42.** "If I fail to pay homage to the great Den, then the sun will still rise in the east."

**43.** "In order for the sun to rise in the east, it is necessary that it sets in the west."

**44.** "In order for the sun to rise in the east, it is sufficient that it sets in the west."

**45.** "The sun rises in the west only if it sets in the west."

**46.** "The sun rises in the east only if it sets in the east."

**47.** "In order for the sun to rise in the east, it is necessary and sufficient that it sets in the west."

**48.** "In order for the sun to rise in the west, it is necessary and sufficient that it sets in the east."

*Construct the truth tables for the statements in Exercises 49–62.* HINT [See Example 2.]

**49.** $p \wedge (\sim q)$          **50.** $p \vee (\sim q)$

**51.** $\sim(\sim p) \vee p$          **52.** $p \wedge (\sim p)$

**53.** $(\sim p) \wedge (\sim q)$          **54.** $(\sim p) \vee (\sim q)$

**55.** $(p \wedge q) \wedge r$          **56.** $p \wedge (q \wedge r)$

**57.** $p \wedge (q \vee r)$          **58.** $(p \wedge q) \vee (p \wedge r)$

**59.** $p \rightarrow (q \vee p)$          **60.** $(p \vee q) \rightarrow \sim p$

**61.** $p \leftrightarrow (p \vee q)$          **62.** $(p \wedge q) \leftrightarrow \sim p$

*Use truth tables to verify the logical equivalences given in Exercises 63–72.*

**63.** $p \wedge p \equiv p$          **64.** $p \vee p \equiv p$

**65.** $p \vee q \equiv q \vee p$
(Commutative law for disjunction)          **66.** $p \wedge q \equiv q \wedge p$
(Commutative law for conjunction)

---

[4] William Shakespeare Sonnet 3.

[5] *Ibid.*, Sonnet 22.

[6] From *Metamagical Themas: Questing for the Essence of Mind and Pattern* by Douglas R. Hofstadter (Bantam Books, New York 1986).

**67.** $\sim(p \vee q) \equiv (\sim p) \wedge (\sim q)$

**68.** $\sim(p \wedge (\sim q)) \equiv (\sim p) \vee q$

**69.** $(p \wedge q) \wedge r \equiv p \wedge (q \wedge r)$
(Associative law for conjunction)

**70.** $(p \vee q) \vee r \equiv p \vee (q \vee r)$
(Associative law for disjunction)

**71.** $p \to q \equiv (\sim q) \to (\sim p)$    **72.** $\sim(p \to q) \equiv p \wedge (\sim q)$

*In Exercises 73–78, use truth tables to check whether the given statement is a tautology, a contradiction, or neither.* **HINT** [See Quick Examples on page A11.]

**73.** $p \wedge (\sim p)$           **74.** $p \wedge p$

**75.** $p \wedge \sim(p \vee q)$      **76.** $p \vee \sim(p \vee q)$

**77.** $p \vee \sim(p \wedge q)$      **78.** $q \vee \sim(p \wedge (\sim p))$

*Apply the stated logical equivalence to the given statement in Exercises 79–84.* **HINT** [See Example 5a, b.]

**79.** $p \vee (\sim p)$; the commutative law

**80.** $p \wedge (\sim q)$; the commutative law

**81.** $\sim(p \wedge (\sim q))$; DeMorgan's law

**82.** $\sim(q \vee (\sim q))$; DeMorgan's law

**83.** $p \vee ((\sim p) \wedge q)$; the distributive law

**84.** $(\sim q) \wedge ((\sim p) \vee q)$; the distributive law

*In Exercises 85–88, use the given logical equivalence to rewrite the given sentence.* **HINT** [See Example 5c.]

**85.** It is not true that both I am Julius Caesar and you are a fool. DeMorgan's law.

**86.** It is not true that either I am Julius Caesar or you are a fool. DeMorgan's law.

**87.** Either it is raining and I have forgotten my umbrella, or it is raining and I have forgotten my hat. The distributive law.

**88.** I forgot my hat or my umbrella, and I forgot my hat or my glasses. The distributive law.

*Give the contrapositive and converse of each of the statements in Exercises 89 and 90, phrasing your answers in words.*

**89.** "If I think, then I am."

**90.** "If these birds are of a feather, then they flock together."

*Exercises 91 and 92 are multiple choice. Indicate which statement is equivalent to the given statement, and say why that statement is equivalent to the given one.*

**91.** "In order for you to worship Den, it is necessary for you to sacrifice beasts of burden."
  **(A)** "If you are not sacrificing beasts of burden, then you are not worshiping Den."
  **(B)** "If you are sacrificing beasts of burden, then you are worshiping Den."
  **(C)** "If you are not worshiping Den, then you are not sacrificing beasts of burden."

**92.** "In order to read the Tarot, it is necessary for you to consult the Oracle."
  **(A)** "In order to consult the Oracle, it is necessary to read the Tarot."
  **(B)** "In order not to consult the Oracle, it is necessary not to read the Tarot."
  **(C)** "In order not to read the Tarot, it is necessary not to read the Oracle."

*In Exercises 93–102, write the given argument in symbolic form (use the underlined letters to represent the statements containing them), then decide whether it is valid or not, If it is valid, name the validating tautology.* **HINT** [See Quick Examples on pages A12, A13, A14.]

**93.** If I am <u>h</u>ungry I am also <u>t</u>hirsty. I am hungry. Therefore, I am thirsty.

**94.** If I am not <u>h</u>ungry, then I certainly am not <u>t</u>hirsty either. I am not thirsty, and so I cannot be hungry.

**95.** For me to bring my <u>u</u>mbrella, it's sufficient that it <u>r</u>ain. It is not raining. Therefore, I will not bring my umbrella.

**96.** For me to bring my <u>u</u>mbrella, it's necessary that it <u>r</u>ain. But it is not raining. Therefore, I will not bring my umbrella.

**97.** For me to pass <u>m</u>ath, it is sufficient that I have a <u>g</u>ood teacher. I will not pass math. Therefore, I have a bad teacher.

**98.** For me to pass <u>m</u>ath, it is necessary that I have a <u>g</u>ood teacher. I will pass math. Therefore, I have a good teacher.

**99.** I will either pass <u>m</u>ath or I have a <u>b</u>ad teacher. I have a good teacher. Therefore, I will pass math.

**100.** Either <u>r</u>oses are not red or <u>v</u>iolets are not blue. But roses are red. Therefore, violets are not blue.

**101.** I am either <u>s</u>mart or <u>a</u>thletic, and I am athletic. So I must not be smart.

**102.** The president is either <u>w</u>ise or <u>s</u>trong. She is strong. Therefore, she is not wise.

*In Exercises 103–108, use the stated tautology to complete the argument.*

**103.** If John is a swan, it is necessary that he is green. John is indeed a swan. Therefore, ____. (*Modus ponens.*)

**104.** If Jill had been born in Texas, then she would be able to ride horses. But Jill cannot ride horses. Therefore, ___. (*Modus tollens.*)

**105.** If John is a swan, it is necessary that he is green. But John is not green. Therefore, ____. (*Modus tollens.*)

**106.** If Jill had been born in Texas, then she would be able to ride horses. Jill was born in Texas. Therefore, ___ (*Modus ponens.*)

**107.** Peter is either a scholar or a gentleman. He is not, however, a scholar. Therefore, ___. (Disjunctive syllogism.)

**108.** Pam is either a plumber or an electrician. She is not, however, an electrician. Therefore, ___ (Disjunctive syllogism.)

**COMMUNICATION AND REASONING EXERCISES**

**109.** If two statements are logically equivalent, what can be said about their truth tables?

**110.** If a proposition is neither a tautology nor a contradiction, what can be said about its truth table?

**111.** If $A$ and $B$ are two compound statements such that $A \vee B$ is a contradiction, what can you say about $A$ and $B$?

**112.** If $A$ and $B$ are two compound statements such that $A \wedge B$ is a tautology, what can you say about $A$ and $B$?

**113.** Give an example of an instance where $p \to q$ means that $q$ causes $p$.

**114.** Complete the following. If $p \to q$, then its converse, ____ , is the statement that ____ and (is/is not) logically equivalent to $p \to q$.

**115.** Give an instance of a biconditional $p \leftrightarrow q$ where neither one of $p$ or $q$ causes the other.

# Appendix B

# Area Under a Normal Curve

$P(0 \leq Z \leq b)$

The table below gives the probabilities $P(0 \leq Z \leq b)$ where $Z$ is a standard normal variable. For example, to find $P(0 \leq Z \leq 2.43)$, write 2.43 as $2.4 + 0.03$, and read the entry in the row labeled 2.4 and the column labeled 0.03. From the portion of the table shown at left, you will see that $P(0 \leq Z \leq 2.43) = .4925$.

| Z | 0.00 | 0.01 | 0.02 | 0.03 |
|---|------|------|------|------|
| 2.3 | .4893 | .4896 | .4898 | .4901 |
| 2.4 | .4918 | .4920 | .4922 | .4925 |
| 2.5 | .4938 | .4940 | .4941 | .4943 |

| Z | 0.00 | 0.01 | 0.02 | 0.03 | 0.04 | 0.05 | 0.06 | 0.07 | 0.08 | 0.09 |
|-----|-------|-------|-------|-------|-------|-------|-------|-------|-------|-------|
| 0.0 | .0000 | .0040 | .0080 | .0120 | .0160 | .0199 | .0239 | .0279 | .0319 | .0359 |
| 0.1 | .0398 | .0438 | .0478 | .0517 | .0557 | .0596 | .0636 | .0675 | .0714 | .0753 |
| 0.2 | .0793 | .0832 | .0871 | .0910 | .0948 | .0987 | .1026 | .1064 | .1103 | .1141 |
| 0.3 | .1179 | .1217 | .1255 | .1293 | .1331 | .1368 | .1406 | .1443 | .1480 | .1517 |
| 0.4 | .1554 | .1591 | .1628 | .1664 | .1700 | .1736 | .1772 | .1808 | .1844 | .1879 |
| 0.5 | .1915 | .1950 | .1985 | .2019 | .2054 | .2088 | .2123 | .2157 | .2190 | .2224 |
| 0.6 | .2257 | .2291 | .2324 | .2357 | .2389 | .2422 | .2454 | .2486 | .2517 | .2549 |
| 0.7 | .2580 | .2611 | .2642 | .2673 | .2704 | .2734 | .2764 | .2794 | .2823 | .2852 |
| 0.8 | .2881 | .2910 | .2939 | .2967 | .2995 | .3023 | .3051 | .3078 | .3106 | .3133 |
| 0.9 | .3159 | .3186 | .3212 | .3238 | .3264 | .3289 | .3315 | .3340 | .3365 | .3389 |
| 1.0 | .3413 | .3438 | .3461 | .3485 | .3508 | .3531 | .3554 | .3577 | .3599 | .3621 |
| 1.1 | .3643 | .3665 | .3686 | .3708 | .3729 | .3749 | .3770 | .3790 | .3810 | .3830 |
| 1.2 | .3849 | .3869 | .3888 | .3907 | .3925 | .3944 | .3962 | .3980 | .3997 | .4015 |
| 1.3 | .4032 | .4049 | .4066 | .4082 | .4099 | .4115 | .4131 | .4147 | .4162 | .4177 |
| 1.4 | .4192 | .4207 | .4222 | .4236 | .4251 | .4265 | .4279 | .4292 | .4306 | .4319 |
| 1.5 | .4332 | .4345 | .4357 | .4370 | .4382 | .4394 | .4406 | .4418 | .4429 | .4441 |
| 1.6 | .4452 | .4463 | .4474 | .4484 | .4495 | .4505 | .4515 | .4525 | .4535 | .4545 |
| 1.7 | .4554 | .4564 | .4573 | .4582 | .4591 | .4599 | .4608 | .4616 | .4625 | .4633 |
| 1.8 | .4641 | .4649 | .4656 | .4664 | .4671 | .4678 | .4686 | .4693 | .4699 | .4706 |
| 1.9 | .4713 | .4719 | .4726 | .4732 | .4738 | .4744 | .4750 | .4756 | .4761 | .4767 |
| 2.0 | .4772 | .4778 | .4783 | .4788 | .4793 | .4798 | .4803 | .4808 | .4812 | .4817 |
| 2.1 | .4821 | .4826 | .4830 | .4834 | .4838 | .4842 | .4846 | .4850 | .4854 | .4857 |
| 2.2 | .4861 | .4864 | .4868 | .4871 | .4875 | .4878 | .4881 | .4884 | .4887 | .4890 |
| 2.3 | .4893 | .4896 | .4898 | .4901 | .4904 | .4906 | .4909 | .4911 | .4913 | .4916 |
| 2.4 | .4918 | .4920 | .4922 | .4925 | .4927 | .4929 | .4931 | .4932 | .4934 | .4936 |
| 2.5 | .4938 | .4940 | .4941 | .4943 | .4945 | .4946 | .4948 | .4949 | .4951 | .4952 |
| 2.6 | .4953 | .4955 | .4956 | .4957 | .4959 | .4960 | .4961 | .4962 | .4963 | .4964 |
| 2.7 | .4965 | .4966 | .4967 | .4968 | .4969 | .4970 | .4971 | .4972 | .4973 | .4974 |
| 2.8 | .4974 | .4975 | .4976 | .4977 | .4977 | .4978 | .4979 | .4979 | .4980 | .4981 |
| 2.9 | .4981 | .4982 | .4982 | .4983 | .4984 | .4984 | .4985 | .4985 | .4986 | .4986 |
| 3.0 | .4987 | .4987 | .4987 | .4988 | .4988 | .4989 | .4989 | .4989 | .4990 | .4990 |

# Answers to Selected Exercises

## Chapter 0

### Section 0.1

**1.** $-48$ **3.** $2/3$ **5.** $-1$ **7.** $9$ **9.** $1$ **11.** $33$ **13.** $14$
**15.** $5/18$ **17.** $13.31$ **19.** $6$ **21.** $43/16$ **23.** $0$
**25.** `3*(2-5)` **27.** `3/(2-5)` **29.** `(3-1)/(8+6)`
**31.** `3-(4+7)/8` **33.** `2/(3+x)-x*y^2`
**35.** `3.1x^3-4x^(-2)-60/(x^2-1)`
**37.** `(2/3)/5` **39.** `3^(4-5)*6`
**41.** `3*(1+4/100)^(-3)` **43.** `3^(2*x-1)+4^x-1`
**45.** `2^(2x^2-x+1)` **47.** `4*e^(-2*x)/`
`(2-3e^(-2*x))` or `4(*e^(-2*x))/`
`(2-3e^(-2*x))` **49.** `3(1-(-1/2)^2)^2+1`

### Section 0.2

**1.** $27$ **3.** $-36$ **5.** $4/9$ **7.** $-1/8$ **9.** $16$ **11.** $2$
**13.** $32$ **15.** $2$ **17.** $x^5$ **19.** $-\dfrac{y}{x}$ **21.** $\dfrac{1}{x}$ **23.** $x^3 y$
**25.** $\dfrac{z^4}{y^3}$ **27.** $\dfrac{x^6}{y^6}$ **29.** $\dfrac{x^4 y^6}{z^4}$ **31.** $\dfrac{3}{x^4}$ **33.** $\dfrac{3}{4x^{2/3}}$
**35.** $1 - 0.3x^2 - \dfrac{6}{5x}$ **37.** $2$ **39.** $1/2$ **41.** $4/3$
**43.** $2/5$ **45.** $7$ **47.** $5$ **49.** $-2.668$ **51.** $3/2$ **53.** $2$
**55.** $2$ **57.** $ab$ **59.** $x + 9$ **61.** $x\sqrt[3]{a^3 + b^3}$ **63.** $\dfrac{2y}{\sqrt{x}}$
**65.** $3^{1/2}$ **67.** $x^{3/2}$ **69.** $(xy^2)^{1/3}$ **71.** $x^{3/2}$
**73.** $\dfrac{3}{5}x^{-2}$ **75.** $\dfrac{3}{2}x^{-1.2} - \dfrac{1}{3}x^{-2.1}$
**77.** $\dfrac{2}{3}x - \dfrac{1}{2}x^{0.1} + \dfrac{4}{3}x^{-1.1}$
**79.** $\dfrac{3}{4}x^{1/2} - \dfrac{5}{3}x^{-1/2} + \dfrac{4}{3}x^{-3/2}$ **81.** $\dfrac{3}{4}x^{2/5} - \dfrac{7}{2}x^{-3/2}$
**83.** $(x^2 + 1)^{-3} - \dfrac{3}{4}(x^2 + 1)^{-1/3}$ **85.** $\sqrt[3]{2^2}$ **87.** $\sqrt[3]{x^4}$
**89.** $\sqrt[5]{\sqrt{x}\sqrt[3]{y}}$ **91.** $-\dfrac{3}{2\sqrt[4]{x}}$ **93.** $\dfrac{0.2}{\sqrt[3]{x^2}} + \dfrac{3\sqrt{x}}{7}$
**95.** $\dfrac{3}{4\sqrt{(1-x)^5}}$ **97.** $64$ **99.** $\sqrt{3}$ **101.** $1/x$
**103.** $xy$ **105.** $\left(\dfrac{y}{x}\right)^{1/3}$ **107.** $\pm 4$ **109.** $\pm 2/3$
**111.** $-1, -1/3$ **113.** $-2$ **115.** $16$ **117.** $\pm 1$
**119.** $33/8$

### Section 0.3

**1.** $4x^2 + 6x$ **3.** $2xy - y^2$ **5.** $x^2 - 2x - 3$
**7.** $2y^2 + 13y + 15$ **9.** $4x^2 - 12x + 9$

**11.** $x^2 + 2 + 1/x^2$ **13.** $4x^2 - 9$ **15.** $y^2 - 1/y^2$
**17.** $2x^3 + 6x^2 + 2x - 4$
**19.** $x^4 - 4x^3 + 6x^2 - 4x + 1$
**21.** $y^5 + 4y^4 + 4y^3 - y$
**23.** $(x + 1)(2x + 5)$
**25.** $(x^2 + 1)^5(x + 3)^3(x^2 + x + 4)$
**27.** $-x^3(x^3 + 1)\sqrt{x + 1}$ **29.** $(x + 2)\sqrt{(x + 1)^3}$
**31. a.** $x(2 + 3x)$ **b.** $x = 0, -2/3$ **33. a.** $2x^2(3x - 1)$
**b.** $x = 0, 1/3$ **35. a.** $(x - 1)(x - 7)$ **b.** $x = 1, 7$
**37. a.** $(x - 3)(x + 4)$ **b.** $x = 3, -4$
**39. a.** $(2x + 1)(x - 2)$ **b.** $x = -1/2, 2$
**41. a.** $(2x + 3)(3x + 2)$ **b.** $x = -3/2, -2/3$
**43. a.** $(3x - 2)(4x + 3)$ **b.** $x = 2/3, -3/4$
**45. a.** $(x + 2y)^2$ **b.** $x = -2y$
**47. a.** $(x^2 - 1)(x^2 - 4)$ **b.** $x = \pm 1, \pm 2$

### Section 0.4

**1.** $\dfrac{2x^2 - 7x - 4}{x^2 - 1}$ **3.** $\dfrac{3x^2 - 2x + 5}{x^2 - 1}$ **5.** $\dfrac{x^2 - x + 1}{x + 1}$
**7.** $\dfrac{x^2 - 1}{x}$ **9.** $\dfrac{2x - 3}{x^2 y}$ **11.** $\dfrac{(x + 1)^2}{(x + 2)^4}$ **13.** $\dfrac{-1}{\sqrt{(x^2 + 1)^3}}$
**15.** $\dfrac{-(2x + y)}{x^2(x + y)^2}$

### Section 0.5

**1.** $-1$ **3.** $5$ **5.** $13/4$ **7.** $43/7$ **9.** $-1$
**11.** $(c - b)/a$ **13.** $x = -4, 1/2$ **15.** No solutions
**17.** $\pm\sqrt{\dfrac{5}{2}}$ **19.** $-1$ **21.** $-1, 3$ **23.** $\dfrac{1 \pm \sqrt{5}}{2}$ **25.** $1$
**27.** $\pm 1, \pm 3$ **29.** $\pm\sqrt{\dfrac{-1 \pm \sqrt{5}}{2}}$ **31.** $-1, -2, -3$
**33.** $-3$ **35.** $1$ **37.** $-2$ **39.** $1, \pm\sqrt{5}$ **41.** $\pm 1, \pm\dfrac{1}{\sqrt{2}}$
**43.** $-2, -1, 2, 3$

### Section 0.6

**1.** $0, 3$ **3.** $\pm\sqrt{2}$ **5.** $-1, -5/2$ **7.** $-3$ **9.** $0, -1, 1$
**11.** $x = -1$ ($x = -2$ is not a solution.)
**13.** $-2, -3/2, -1$ **15.** $-1$ **17.** $\pm\sqrt[4]{2}$ **19.** $\pm 1$
**21.** $\pm 3$ **23.** $2/3$ **25.** $-4, -1/4$

### Section 0.7

**1.** $P(0, 2)$, $Q(4, -2)$, $R(-2, 3)$, $S(-3.5, -1.5)$,
$T(-2.5, 0)$, $U(2, 2.5)$

**3.**

**5.**

**7.**

**9.**

**11.**

**13.** $\sqrt{2}$

**15.** $\sqrt{a^2 + b^2}$

**17.** 1/2

**19.** Circle with center (0, 0) and radius 3

# Chapter 1

## Section 1.1

**1. a.** 2  **b.** 0.5   **3. a.** −1.5  **b.** 8  **c.** −8    **5. a.** 20  **b.** 30
**c.** 30  **d.** 20  **e.** 0    **7. a.** −1  **b.** 1.25  **c.** 0  **d.** 1  **e.** 1
**9. a.** Yes; $f(4) = 63/16$  **b.** Not defined  **c.** Not defined
**11. a.** Not defined  **b.** Not defined  **c.** Yes, $f(-10) = 0$
**13. a.** −7  **b.** −3  **c.** 1  **d.** $4y − 3$  **e.** $4(a + b) − 3$
**15. a.** 3  **b.** 6  **c.** 2  **d.** 6  **e.** $a^2 + 2a + 3$
**f.** $(x + h)^2 + 2(x + h) + 3$   **17. a.** 2  **b.** 0  **c.** 65/4
**d.** $x^2 + 1/x$  **e.** $(s + h)^2 + 1/(s + h)$
**f.** $(s + h)^2 + 1/(s + h) − (s^2 + 1/s)$
**19.**

`-(x^3)`

**21.**

`x^4`

**23.**

`1/x^2`

**25. a.** (I)  **b.** (IV)  **c.** (V)  **d.** (VI)  **e.** (III)  **f.** (II)
**27.** `0.1*x^2-4*x+5`

| $x$ | 0 | 1 | 2 | 3 |
|---|---|---|---|---|
| $f(x)$ | 5 | 1.1 | −2.6 | −6.1 |
| $x$ | 4 | 5 | 6 | 7 |
| $f(x)$ | −9.4 | −12.5 | −15.4 | −18.1 |
| $x$ | 8 | 9 | 10 | |
| $f(x)$ | −20.6 | −22.9 | −25 | |

**29.** `(x^2-1)/(x^2+1)`

| $x$ | 0.5 | 1.5 | 2.5 | 3.5 |
|---|---|---|---|---|
| $h(x)$ | −0.6000 | 0.3846 | 0.7241 | 0.8491 |
| $x$ | 4.5 | 5.5 | 6.5 | 7.5 |
| $h(x)$ | 0.9059 | 0.9360 | 0.9538 | 0.9651 |
| $x$ | 8.5 | 9.5 | 10.5 | |
| $h(x)$ | 0.9727 | 0.9781 | 0.9820 | |

**31. a.** −1  **b.** 2  **c.** 2     **33. a.** 1  **b.** 0  **c.** 1

`x* (x<0) +2* (x>=0)`

`(x^2) * (x<=0) + (1/x) * (0<x)`

**35. a.** 0  **b.** 2  **c.** 3  **d.** 3

`x* (x<=0) + (x+1) * (0<x) * (x<=2) + x* (2<x)`

**37. a.** $h(2x + h)$  **b.** $2x + h$    **39. a.** $-h(2x + h)$
**b.** $-(2x + h)$    **41. a.** $s(1) = 3.8$. In 2006, Pemex sold
3.8 billion metric tons of petrochemicals domestically
(i.e., in Mexico). $s(4) = 4.0$. In 2009, Pemex sold

4.0 billion metric tons of petrochemicals domestically. $s(5) = 4.1$. In 2010, Pemex sold 4.1 billion metric tons of petrochemicals domestically. **b.** [0, 5] **c.** Graph:

s(t)

$s(1.5) \approx 3.9$. Pemex sold about 3.9 billion metric tons of petrochemicals domestically in the year ending June 30, 2007.   **43.** $f(4) \approx 1,600$, $f(5) \approx 1,700$, $f(7.5) \approx 800$. There were 1.6 million housing starts in 2004, 1.7 million housing starts in 2005, and 800,000 housing starts in the year beginning July 2007.
**45.** $f(8) - f(0)$. The change in the number of housing starts from 2000 to 2008 was larger in magnitude than the change from 2000 to 2005.
**47. a.** $P(0) \approx -100$; $P(4) \approx -400$; $P(1.5) \approx 400$. Continental Airlines lost about $100 million in 2005, lost about $400 million in 2009, and made a profit of about $400 million in the year starting July 1, 2006.
**b.** 0; Continental's net income was increasing most rapidly in 2005.  **c.** 2.5; Continental's net income was decreasing most rapidly midway through 2007.
**49. a.** [0, 10]. $t \geq 0$ is not an appropriate domain because it would predict U.S. trade with China into the indefinite future with no basis. **b.** $280 billion; U.S. trade with China in 2004 was valued at approximately $280 billion.   **51. a.** `100*(1-12200/t^4.48)`
**b.** Graph:

**c.** Table:

| t | 9 | 10 | 11 | 12 | 13 | 14 |
|---|---|----|----|----|----|----|
| p(t) | 35.2 | 59.6 | 73.6 | 82.2 | 87.5 | 91.1 |
| t | 15 | 16 | 17 | 18 | 19 | 20 |
| p(t) | 93.4 | 95.1 | 96.3 | 97.1 | 97.7 | 98.2 |

**d.** 82.2% **e.** 14 months
**53. a.** $v(10) \approx 58$, $v(16) = 200$, $v(28) = 3,800$. Processor speeds were about 58 MHz in 1990, 200 MHz in 1996, and 3,800 MHz in 2008.
**b.** `(8*(1.22)^x)*(x<16)+(400*x-6200)*`
`(x>=16)*(x<25)+3800*(x>=25)`

**c.** Graph:

v(t)

Table:

| t | 0 | 2 | 4 | 6 | 8 | 10 |
|---|---|---|---|---|---|----|
| v(t) | 8.0 | 12 | 18 | 26 | 39 | 58 |
| t | 12 | 14 | 16 | 18 | 20 | 22 |
| v(t) | 87 | 130 | 200 | 1,000 | 1,800 | 2,600 |
| t | 24 | 26 | 28 | 30 | | |
| v(t) | 3,400 | 3,800 | 3,800 | 3,800 | | |

**d.** 2003
**55. a.**

$$T(x) = \begin{cases} 0.10x & \text{if } 0 < x \leq 8,375 \\ 837.50 + 0.15(x - 8,375) & \text{if } 8,375 < x \leq 34,000 \\ 4,663.25 + 0.25(x - 34,000) & \text{if } 34,000 < x \leq 82,400 \\ 16,763.25 + 0.28(x - 82,400) & \text{if } 82,400 < x \leq 171,850 \\ 41,809.25 + 0.33(x - 171,850) & \text{if } 171,850 < x \leq 373,650 \\ 108,403.25 + 0.35(x - 373,650) & \text{if } 373,650 < x \end{cases}$$

**b.** $5,163.25   **57.** $t$; $m$   **59.** $y(x) = 4x^2 - 2$
(or $f(x) = 4x^2 - 2$)   **61.** False. A graph usually gives infinitely many values of the function while a numerical table will give only a finite number of values.
**63.** False. In a numerically specified function, only certain values of the function are specified so we cannot know its value on every real number in [0, 10], whereas an algebraically specified function would give values for every real number in [0, 10].   **65.** False: Functions with infinitely many points in their domain (such as $f(x) = x^2$) cannot be specified numerically.   **67.** As the text reminds us: to evaluate $f$ of a quantity (such as $x + h$) replace $x$ everywhere by the *whole quantity* $x + h$, getting $f(x + h) = (x + h)^2 - 1$.
**69.** They are different portions of the graph of the associated equation $y = f(x)$.   **71.** The graph of $g$ is the same as the graph of $f$, but shifted 5 units to the right.

## Section 1.2

**1. a.** $s(x) = x^2 + x$  **b.** Domain: All real numbers  **c.** 6
**3. a.** $p(x) = (x - 1)\sqrt{x + 10}$  **b.** Domain: $[-10, 0)$
**c.** $-14$  **5. a.** $q(x) = \frac{\sqrt{10 - x}}{x - 1}$
**b.** Domain: $0 \leq x \leq 10; x \neq 1$
**c.** Undefined  **7. a.** $m(x) = 5(x^2 + 1)$  **b.** Domain: All
real numbers  **c.** 10  **9.** $N(t) = 200 + 10t$
($N$ = number of music files, $t$ = time in days)
**11.** $A(x) = x^2/2$  **13.** $C(x) = 12x$

**15.** $h(n) = \begin{cases} 4 & \text{if } 1 \leq n \leq 5 \\ 0 & \text{if } n > 5 \end{cases}$

**17.** $C(x) = 1,500x + 1,200$ per day  **a.** $5,700
**b.** $1,500  **c.** $1,500  **d.** Variable cost = $1,500x$; Fixed
cost = $1,200; Marginal cost = $1,500 per piano
**e.** Graph:

**19. a.**
$C(x) = 0.4x + 70$, $R(x) = 0.5x$, $P(x) = 0.1x - 70$
**b.** $P(500) = -20$; a loss of $20  **c.** 700 copies
**21.** $R(x) = 100x$, $P(x) = -2,000 + 90x - 0.2x^2$; at
least 24 jerseys.  **23.** $P(x) = -1.7 - 0.02x + 0.0001x^2$;
approximately 264 thousand square feet
**25.** $P(x) = 100x - 5,132$, with domain $[0, 405]$.
For profit, $x \geq 52$  **27.** 5,000 units
**29.** $FC/(SP - VC)$  **31.** $P(x) = 579.7x - 20,000$,
with domain $x \geq 0$; $x = 34.50$ g per day for breakeven
**33. a.** Graph:

**b.** Ridership drops by about 3,070 rides per day.
**35. a.** Graph:

**b.** 210 million units  **c.** 233 million units
**d.** $R(p) = 0.0009p^3 - 0.63p^2 + 245p$ million dollars/
year; $13 billion/year  **37.** $240 per skateboard.
**39. a.** $110 per phone.  **b.** Shortage of 25 million phones

**41. a.** $3.50 per ride.
Graph:

**b.** A surplus of around 9,170 rides.  **43. a.** $12,000
**b.** $N(q) = 2,000 + 100q^2 - 500q$; This is the cost of
removing $q$ pounds of PCPs per day after the subsidy is
taken into account.  **c.** $N(20) = $32,000$; The net cost
of removing 20 pounds of PCPs per day is $32,000.
**45. a. (B)  b.** $36.8 billion  **47. a. (C)  b.** $20.80 per
shirt if the team buys 70 shirts
Graph:

**49. a.** Graph:

(A), (D)  **b.** Model (A); Approximately $0.0021
**51.** A linear model (A) is the best choice; a plot of the
given points gives a straight line.  **53.** Model (D) is
the best choice; the exponential models predict either
perpetually increasing or perpetually decreasing sales,
and Model (C) would give a concave-up parabola.
**55.** $A(t) = 5,000(1 + 0.003/12)^{12t}$; $5,106  **57.** 2024
**59.** 31.0 grams, 9.25 grams, 2.76 grams  **61.** 20,000
years  **63. a.** 1,000 years: 65%, 2,000 years: 42%,
3,000 years: 27%  **b.** 1,600 years  **65.** 30
**67.** Curve fitting. The model is based on fitting a
curve to a given set of observed data.

**69.** The cost of downloading a movie was $4 in January and is decreasing by 20¢ per month.    **71.** Variable; marginal.    **73.** Yes, as long as the supply is going up at a faster rate, as illustrated by the following graph:

**75.** Extrapolate both models and choose the one that gives the most reasonable predictions.
**77.** They are $\geq 0$.    **79.** Books per person

## Section 1.3

**1.** 11; $m = 3$    **3.** $-4$; $m = -1$    **5.** 7; $m = 3/2$
**7.** $f(x) = -x/2 - 2$    **9.** $f(0) = -5$, $f(x) = -x - 5$
**11.** $f$ is linear: $f(x) = 4x + 6$    **13.** $g$ is linear:
$g(x) = 2x - 1$    **15.** $-3/2$    **17.** 1/6    **19.** Undefined
**21.** 0    **23.** $-4/3$

**25.**

**27.**

**29.**

**31.**

**33.**

**35.**

**37.**

**39.** 2    **41.** 2    **43.** $-2$    **45.** Undefined    **47.** 1.5
**49.** $-0.09$    **51.** 1/2    **53.** $(d - b)/(c - a)$
**55.** Undefined    **57.** $-b/a$    **59. a.** 1  **b.** 1/2  **c.** 0
**d.** 3  **e.** $-1/3$  **f.** $-1$  **g.** Undefined  **h.** $-1/4$  **i.** $-2$
**61.** $y = 3x$    **63.** $y = \dfrac{1}{4}x - 1$
**65.** $y = 10x - 203.5$    **67.** $y = -5x + 6$
**69.** $y = -3x + 2.25$    **71.** $y = -x + 12$
**73.** $y = 2x + 4$    **75.** $y = \dfrac{q}{p}x$    **77.** $y = q$
**79.** Fixed cost = $8,000, marginal cost = $25 per bicycle    **81.** $C = 188x + 20$; $188 per iPhone; $7,540
**83.** $q = -40p + 2,000$    **85. a.** $q = -2p + 273$; 123 million phones  **b.** $1; 2 million
**87. a.** $q = -4,500p + 41,500$  **b.** Rides/day per $1 increase in the fare; ridership decreases by 4,500 rides per day for every $1 increase in the fare.
**c.** 14,500 rides/day    **89. a.** $y = 40t + 290$ million pounds of pasta  **b.** 890 million pounds
**91. a.** $N = 5t - 20$  **b.** Billions of dollars per year; Amazon's net income grew at a rate of $5 billion per year.  **c.** $20 billion    **93. a.** 2.5 ft/sec  **b.** 20 feet along the track  **c.** after 6 seconds    **95. a.** 130 miles per hour  **b.** $s = 130t - 1,300$    **97. a.** $L = 55n + 475$
**b.** Pages per edition; *Applied Calculus* is growing at a rate of 55 pages per edition.  **c.** 19th edition
**99.** $F = 1.8C + 32$; 86°F; 72°F; 14°F; 7°F
**101. a.** $A = 2.4C - 730$  **b.** $100 million lower than the actual $450 million net income American earned in 2007  **c.** Millions of dollars of American's net income per million dollars of Continental's net income; American earned an additional net income of $2.40 per $1 additional net income earned by Continental.
**103.** $I(N) = 0.05N + 50,000$; $N = $1,000,000$; marginal income is $m = 5$¢ per dollar of net profit
**105.** Increasing at 400 MHz per year
**107. a.** $y = 31.1t + 78$  **b.** $y = 90t - 1,100$

**c.** $y = \begin{cases} 31.1t + 78 & \text{if } 0 \leq t < 20 \\ 90t - 1,100 & \text{if } 20 \leq t \leq 40 \end{cases}$ or

$y = \begin{cases} 31.1t + 78 & \text{if } 0 \leq t \leq 20 \\ 90t - 1,100 & \text{if } 20 < t \leq 40 \end{cases}$

**d.** $1,960,000, considerably lower than the actual shown on the graph. The actual cost is not linear in the range 1990–2010.

**109.** $N = \begin{cases} 0.22t + 3 & \text{if } 0 \leq t \leq 5 \\ -0.15t + 4.85 & \text{if } 5 < t \leq 9 \end{cases}$

3.8 million jobs

**111.** Compute the corresponding successive changes $\Delta x$ in $x$ and $\Delta y$ in $y$, and compute the ratios $\Delta y/\Delta x$. If the answer is always the same number, then the values in the table come from a linear function.

**113.** $f(x) = -\dfrac{a}{b}x + \dfrac{c}{b}$. If $b = 0$, then $\dfrac{a}{b}$ is undefined, and $y$ cannot be specified as a function of $x$. (The graph of the resulting equation would be a vertical line.)
**115.** slope, 3.    **117.** If $m$ is positive, then $y$ will increase as $x$ increases; if $m$ is negative, then $y$ will decrease as $x$ increases; if $m$ is zero, then $y$ will not change as $x$ changes.    **119.** The slope increases, because an increase in the $y$-coordinate of the second point increases $\Delta y$ while leaving $\Delta x$ fixed.
**121.** Bootlags per zonar; bootlags    **123.** It must increase by 10 units each day, including the third.
**125.** (B)    **127.** It is linear with slope $m + n$.
**129.** Answers may vary. For example, $f(x) = x^{1/3}$, $g(x) = x^{2/3}$    **131.** Increasing the number of items from the break-even number results in a profit: Because the slope of the revenue graph is larger than the slope of the cost graph, it is higher than the cost graph to the right of the point of intersection, and hence corresponds to a profit.

## Section 1.4

**1.** 6    **3.** 86    **5. a.** 0.5 (better fit)  **b.** 0.75
**7. a.** 27.42  **b.** 27.16 (better fit)
**9.** $y = 1.5x - 0.6667$
Graph:

**11.** $y = 0.7x + 0.85$
Graph:

**13. a.** $r = 0.9959$ (best, not perfect)  **b.** $r = 0.9538$
**c.** $r = 0.3273$ (worst)
**15.**

| $x$ | $y$ | $xy$ | $x^2$ |
|-----|-----|------|-------|
| 2 | 0 | 0 | 4 |
| 6 | 70 | 420 | 36 |
| 10 | 900 | 9,000 | 100 |
| 18 | 970 | 9,420 | 140 |

$y = 112.5x - 351.7$; 998.3 million

**17.** $y = 3.4t + 5$; \$25.4 billion    **19.** $y = 0.135x + 0.15$; 6.9 million jobs    **21. a.** $I = 44S - 220$  **b.** Amazon.com earned \$44 million in net income per billion dollars in net sales.  **c.** \$28 billion
**d.** Graph:

The graph shows a close fit, so the linear model seems reasonable.    **23. a.** $L = 52.70n + 486.30$
Graph:

**b.** *Applied Calculus* is growing at a rate of 52.7 pages per edition.    **25. a.** $y = 1.62x - 23.87$.
Graph:

**b.** Each acre of cultivated land produces about 1.62 tons of soybeans.
**27. a.** $y = -11.85x + 797.71$; $r \approx -0.414$
**b.** Continental's net income is not correlated to the price of oil.  **c.** The points are nowhere near the regression line, confirming the conclusion in (b).
Graph:

**29. a.** Regression line: $y = 0.582x + 35.4$ Graph:

**b.** There are about 582 additional doctorates in engineering per additional 1,000 doctorates in the natural sciences. **c.** $r \approx 0.949$; a strong correlation. **d.** No, the data points suggest a concave down curve rather than a straight line. **31. a.** $y = 44.0t - 67.8$; $r \approx 0.912$ Graph:

**b.** The number of natural science doctorates has been increasing at a rate of about 44 per year. **c.** Faster and faster rate; the slopes of successive pairs of points increase as we go from left to right. **d.** No; If $r$ had been equal to 1, then the points would lie exactly on the regression line, which would indicate that the number of doctorates is growing at a constant rate.
**33. a.** $p = 0.13t + 0.22$; $r \approx 0.97$
Graph:

**b.** Yes; the first and last points lie above the regression line, while the central points lie below it, suggesting a curve.
**c.**

|   | A | B | C | D |
|---|---|---|---|---|
| 1 | t | p (Observed) | p (predicted) | Residual |
| 2 | 0 | 0.38 | 0.22 | 0.16 |
| 3 | 2 | 0.4 | 0.48 | -0.08 |
| 4 | 4 | 0.6 | 0.74 | -0.14 |
| 5 | 6 | 0.95 | 1 | -0.05 |
| 6 | 8 | 1.2 | 1.26 | -0.06 |
| 7 | 10 | 1.6 | 1.52 | 0.08 |
| 8 |   |   |   |   |

Notice that the residuals are positive at first, then become negative, and then become positive, confirming the impression from the graph.

**35.** The line that passes through $(a, b)$ and $(c, d)$ gives a sum-of-squares error SSE $= 0$, which is the smallest value possible. **37.** The regression line is the line passing through the given points. **39.** 0 **41.** No. The regression line through $(-1, 1)$, $(0, 0)$, and $(1, 1)$ passes through none of these points. **43.** (Answers may vary.) The data in Exercise 33 give $r \approx 0.97$, yet the plotted points suggest a curve, not a straight line.

## Chapter 1 Review

**1. a.** 1 **b.** $-2$ **c.** 0 **d.** $-1$ **3. a.** 1 **b.** 0 **c.** 0 **d.** $-1$
**5.**

**7.**

**9.** Absolute value **11.** Linear **13.** Quadratic
**15.** $y = -3x + 11$ **17.** $y = 1.25x - 4.25$
**19.** $y = (1/2)x + 3/2$ **21.** $y = 4x - 12$
**23.** $y = -x/4 + 1$ **25.** $y = -0.214x + 1.14$,
$r \approx -0.33$
**27. a.** Exponential. Graph:

**b.** The ratios (rounded to 1 decimal place) are:

| $V(1)/V(0)$ | $V(2)/V(1)$ | $V(3)/V(2)$ | $V(4)/V(3)$ | $V(5)/V(4)$ | $V(6)/V(5)$ |
|---|---|---|---|---|---|
| 3 | 3.3 | 3.3 | 3.2 | 3.2 | 3.2 |

They are close to 3.2. **c.** About 343,700 visits/day
**29. a.** 2.3; 3.5; 6 **b.** For Web site traffic of up to 50,000 visits per day, the number of crashes is increasing by 0.03 per additional thousand visits.
**c.** 140,000 **31. a.** (A) **b.** (A) Leveling off
(B) Rising (C) Rising; begins to fall after 7 months
(D) Rising **33. a.** The number of visits would increase by 30 per day. **b.** No; it would increase at a slower and slower rate and then begin to decrease.
**c.** Probably not. This model predicts that Web site popularity will start to decrease as advertising increases beyond \$8,500 per month, and then drop toward zero. **35. a.** $v = 0.05c + 1,800$
**b.** 2,150 new visits per day **c.** \$14,000 per month
**37.** $d = 0.95w + 8$; 86 kg **39. a.** Cost:
$C = 5.5x + 500$; Revenue: $R = 9.5x$; Profit
$P = 4x - 500$ **b.** More than 125 albums per week
**c.** More than 200 albums per week

**41. a.** $q = -80p + 1,060$ **b.** 100 albums per week
**c.** \$9.50, for a weekly profit of \$700
**43. a.** $q = -74p + 1,015.5$ **b.** 239 albums per week

# Chapter 2

## Section 2.1

**1.** $INT = \$120, FV = \$2,120$
**3.** $INT = \$505, FV = \$20,705$
**5.** $INT = \$250, FV = \$10,250$ **7.** $PV = \$9,090.91$
**9.** $PV = \$966.18$ **11.** $PV = \$14,457.83$ **13.** \$5,200
**15.** \$787.40 **17.** 5% **19.** About 0.2503%
**21.** In 2 years **23.** 3.775% **25.** 65% **27.** 10%
**29.** 86.957% **31.** 48.04% **33.** 58.96% if you had sold
in November 2010 **35.** No. Simple interest increase is
linear. The graph is visibly not linear in that time period.
Further, the slopes of the lines through the successive
pairs of marked points are quite different. **37.** 9.2%
**39.** 3,260,000 **41.** $P = 500 + 46t$ thousand
($t$ = time in years since 1950)

**43.** Graph (A) is the only possible choice, because the
equation $FV = PV(1 + rt) = PV + PVrt$ gives the
future value as a linear function of time. **45.** Wrong. In
simple interest growth, the change each year is a fixed percentage of the *starting* value, and not the preceding
year's value. (Also see Exercise 46.)
**47.** Simple interest is always calculated on a constant
amount, *PV*. If interest is paid into your account, then
the amount on which interest is calculated does not
remain constant.

## Section 2.2

**1.** \$13,439.16 **3.** \$11,327.08 **5.** \$19,154.30
**7.** \$12,709.44 **9.** \$613.91 **11.** \$810.65
**13.** \$1,227.74 **15.** 5.09% **17.** 10.47% **19.** 10.52%
**21.** \$268.99 **23.** \$728.91 **25.** \$2,927.15
**27.** \$21,161.79 **29.** \$163,414.56
**31.** \$55,526.45 per year **33.** \$174,110 **35.** \$750.00
**37.** \$27,171.92 **39.** \$111,678.96 **41.** \$1,039.21

**43.** The one earning 11.9% compounded monthly
**45.** Yes. The investment will have grown to about
\$394,020 million. **47.** 136 reals **49.** 245 bolivianos
**51.** 223 bolivars **53.** The Nicaragua investment is
better: It is worth about 1.006 units of currency (in
constant units) per unit invested as opposed to about
1.003 units for Mexico. **55.** 53.81%
**57.** 65.99% if you had sold in November 2010
**59.** No. Compound interest increase is exponential,
and exponential curves either increase continually (in
the case of appreciation) or decrease continually (in
the case of depreciation). The graph of the stock price
has both increases and decreases during the given
period, so the curve cannot model compound interest
change. **61.** 31 years; about \$26,100 **63.** 2.3 years
**65. a.** \$1,510.31 **b.** \$54,701.29 **c.** 23.51%
**67.** The function $y = P(1 + r/m)^{mx}$ is not a linear
function of $x$, but an exponential function. Thus, its
graph is not a straight line. **69.** Wrong. Its growth is
exponential and can be modeled by $0.01(1.10)^t$.
**71.** The graphs are the same because the formulas
give the same function of $x$; a compound interest
investment behaves as though it was being
compounded once a year at the effective rate.
**73.** The effective rate exceeds the nominal rate when
the interest is compounded more than once a year
because then interest is being paid on interest accumulated during each year, resulting in a larger effective
rate. Conversely, if the interest is compounded less
often than once a year, the effective rate is less than
the nominal rate. **75.** Compare their future
values in constant dollars. The investment with the
larger future value is the better investment. **77.** The
graphs are approaching a particular curve as $m$ gets
larger, approximately the curve given by the largest
two values of $m$.

## Section 2.3

**1.** \$15,528.23 **3.** \$171,793.82 **5.** \$23,763.28
**7.** \$147.05 **9.** \$491.12 **11.** \$105.38 **13.** \$90,155.46
**15.** \$69,610.99 **17.** \$95,647.68 **19.** \$554.60
**21.** \$1,366.41 **23.** \$524.14 **25.** \$248.85
**27.** \$1,984.65 **29.** \$999.61 **31.** \$998.47
**33.** \$917.45 **35.** 3.617% **37.** 3.059% **39.** 6.038%
**41.** You should take the loan from Solid Savings & Loan:
It will have payments of \$248.85 per month. The
payments on the other loan would be more than
\$300 per month. **43.** \$973.54 **45.** \$7,451.49

**47.** Answers using correctly rounded intermediate results:

| Year | Interest | Payment on Principal |
|------|----------|----------------------|
| 1 | $3,934.98 | $1,798.98 |
| 2 | $3,785.69 | $1,948.27 |
| 3 | $3,623.97 | $2,109.99 |
| 4 | $3,448.84 | $2,285.12 |
| 5 | $3,259.19 | $2,474.77 |
| 6 | $3,053.77 | $2,680.19 |
| 7 | $2,831.32 | $2,902.64 |
| 8 | $2,590.39 | $3,143.57 |
| 9 | $2,329.48 | $3,404.48 |
| 10 | $2,046.91 | $3,687.05 |
| 11 | $1,740.88 | $3,993.08 |
| 12 | $1,409.47 | $4,324.49 |
| 13 | $1,050.54 | $4,683.42 |
| 14 | $661.81 | $5,072.15 |
| 15 | $240.84 | $5,491.80 |

**49.** 1st 5 years: $402.62/month; last 25 years: $601.73
**51.** Original monthly payments were $824.79. The new monthly payments will be $613.46. You will save $36,481.77 in interest.   **53.** 10.81%   **55.** 13 years
**57.** 4.5 years   **59.** 24 years   **61.** He is wrong because his estimate ignores the interest that will be earned by your annuity—both while it is increasing and while it is decreasing. Your payments will be considerably smaller (depending on the interest earned).   **63.** He is not correct. For instance, the payments on a $100,000 10-year mortgage at 12% are $1,434.71, while for a 20-year mortgage at the same rate, they are $1,101.09, which is a lot more than half the 10-year mortgage payment.   **65.** $PV = FV(1 + i)^{-n} =$

$$PMT\frac{(1 + i)^n - 1}{i}(1 + i)^{-n} = PMT\frac{1 - (1 + i)^{-n}}{i}$$

## Chapter 2 Review

**1.** $7,425.00   **3.** $7,604.88   **5.** $6,757.41
**7.** $4,848.48   **9.** $4,733.80   **11.** $5,331.37
**13.** $177.58   **15.** $112.54   **17.** $187.57
**19.** $9,584.17   **21.** 5.346%   **23.** 14.0 years
**25.** 10.8 years   **27.** 7.0 years   **29.** 168.85%
**31.** 85.28% if she sold in February 2010.   **33.** No. Simple interest increase is linear. We can compare slopes between successive points to see if the slope remained roughly constant: From December 2002 to August 2004 the slope was $(16.31 - 3.28)/(20/12) = 7.818$ while from August 2004 to March 2005 the slope was $(33.95 - 16.31)/(7/12) = 30.24$. These slopes are quite different.   **35.** 2003

| Year | 2000 | 2001 | 2002 | 2003 | 2004 |
|------|------|------|------|------|------|
| Revenue | $180,000 | $216,000 | $259,200 | $311,040 | $373,248 |

**37.** At least 52,515 shares   **39.** $3,234.94
**41.** $231,844   **43.** 7.75%   **45.** $420,275
**47.** $140,778   **49.** $1,453.06   **51.** $2,239.90 per month
**53.** $53,055.66   **55.** 5.99%

# Chapter 3

## Section 3.1

**1.** (2, 2)   **3.** (3, 1)   **5.** (6, 6)   **7.** (5/3, −4/3)
**9.** (0, −2)   **11.** $(x, (1 − 2x)/3)$ or $((1 − 3y)/2, y)$
**13.** No solution   **15.** (5, 0)   **17.** (0.3, −1.1)
**19.** (116.6, −69.7)   **21.** (3.3, 1.8)   **23.** (3.4, 1.9)
**25.** 200 quarts of vanilla and 100 quarts of mocha
**27.** 2 servings of Mixed Cereal and 1 serving of Mango Tropical Fruit   **29. a.** 4 servings of beans and 5 slices of bread   **b.** No. One of the variables in the solution of the system has a negative value.
**31.** Mix 12 servings of Designer Whey and 2 servings of Muscle Milk for a cost of $9.20.   **33.** 65 g
**35.** 20 AAPL, 30 GOOG   **37.** 200 BOH, 400 JPM
**39.** 242 in favor and 193 against   **41.** 5 soccer games and 7 football games   **43.** 7   **45.** $1.50 each
**47.** 55 widgets   **49.** Demand: $q = −4p + 47$; supply: $q = 4p − 29$; equilibrium price: $9.50
**51.** 33 pairs of dirty socks and 11 T-shirts   **53.** $1,200
**55.** The three lines in a plane must intersect in a single point for there to be a unique solution. This can happen in two ways: (1) The three lines intersect in a single point, or (2) two of the lines are the same, and the third line intersects it in a single point.
**57.** Yes. Even if two lines have negative slope, they will still intersect if the slopes differ.   **59.** You cannot round both of them up, since there will not be sufficient eggs and cream. Rounding both answers down will ensure that you will not run out of ingredients. It may be possible to round one answer down and the other up, and this should be tried.   **61.** (B)   **63.** (B)
**65.** Answers will vary.   **67.** It is very likely. Two randomly chosen straight lines are unlikely to be parallel.

## Section 3.2

**1.** (3, 1)   **3.** (6, 6)   **5.** $\left(\frac{1}{2}(1 − 3y), y\right)$; $y$ arbitrary
**7.** No solution   **9.** (1/4, 3/4)   **11.** No solution
**13.** (10/3, 1/3)   **15.** (4, 4, 4)   **17.** $\left(−1, −3, \frac{1}{2}\right)$
**19.** $(z, z, z)$; $z$ arbitrary   **21.** No solution
**23.** (−1, 1, 1)   **25.** $(1, z − 2, z)$; $z$ arbitrary
**27.** $(4 + y, y, −1)$; $y$ arbitrary

**29.** $(4 - y/3 + z/3, y, z)$; $y$ arbitrary, $z$ arbitrary
**31.** $(-17, 20, -2)$　**33.** $\left(-\frac{3}{2}, 0, 1/2, 0\right)$
**35.** $(-3z, 1 - 2z, z, 0)$; $z$ arbitrary　**37.** $(7/5 - 17z/5 + 8w/5, 1/5 - 6z/5 - 6w/5, z, w)$; $z$, $w$ arbitrary
**39.** $(1, 2, 3, 4, 5)$　**41.** $(-2, -2 + z - u, z, u, 0)$; $z$, $u$ arbitrary　**43.** $(16, 12/7, -162/7, -88/7)$
**45.** $(-8/15, 7/15, 7/15, 7/15, 7/15)$　**47.** $(1.0, 1.4, 0.2)$
**49.** $(-5.5, -0.9, -7.4, -6.6)$　**51.** A pivot is an entry in a matrix that is selected to "clear a column"; that is, use the row operations of a certain type to obtain zeros everywhere above and below it. "Pivoting" is the procedure of clearing a column using a designated pivot.　**53.** $2R_1 + 5R_4$, or $6R_1 + 15R_4$ (which is less desirable)　**55.** It will include a row of zeros.
**57.** The claim is wrong. If there are more equations than unknowns, there can be a unique solution as well as row(s) of zeros in the reduced matrix, as in Example 6.　**59.** Two　**61.** The number of pivots must equal the number of variables, since no variable will be used as a parameter.　**63.** A simple example is: $x = 1$; $y - z = 1$; $x + y - z = 2$.
**65.** It has to be the zero solution (each unknown is equal to zero): Putting each unknown equal to zero causes each equation to be satisfied because the right-hand sides are zero. Thus, the zero solution is in fact a solution. Because the solution is unique, this solution is the *only* solution.
**67.** No: As pointed out in Exercise 65, every homogeneous system has at least one solution (namely, the zero solution), and hence cannot be inconsistent.

## Section 3.3

**1.** 100 batches of vanilla, 50 batches of mocha, and 100 batches of strawberry　**3.** 3 sections of Finite Math, 2 sections of Applied Calculus, and 1 section of Computer Methods　**5.** $25 million for regional music, $16 million for pop/rock music, and $5 million for tropical music　**7.** 4 Airbus A330-300s, 4 Boeing 767-200ERs, and 8 Dreamliners　**9.** 22 tons from Cheesy Cream, 56 tons from Super Smooth & Sons, and 22 tons from Bagel's Best Friend　**11.** 10 evil sorcerers, 50 trolls, and 500 orcs　**13.** $600 to each of the MPBF and the SCN, and $1,200 to the Jets
**15.** Continental: 120; American: 40; Southwest: 50
**17.** $5,000 in SHPIX, $2,000 in RYURX, $2,000 in RYIHX　**19.** 100 shares of KT, 50 shares of IFF, 50 shares of QSII　**21.** Microsoft: 88 million, Time Warner: 79 million, Yahoo: 75 million, Google: 42 million
**23.** The third equation is $x + y + z + w = 100$.
General Solution: $x = 50.5 - 0.5w$, $y = 33.5 - 0.3w$,

$z = 16 - 0.2w$, $w$ arbitrary. State Farm is most impacted by other companies.
**25. a.** Brooklyn to Long Island: 500 books; Brooklyn to Manhattan: 500 books; Queens to Long Island: 1,000 books; Queens to Manhattan: 1,000 books.
**b.** Brooklyn to Long Island: none; Brooklyn to Manhattan: 1,000 books; Queens to Long Island: 1,500 books; Queens to Manhattan: 500 books, giving a total cost of $8,000　**27. a.** The associated system of equations has infinitely many solutions.　**b.** No; the associated system of equations still has infinitely many solutions.　**c.** Yes; North America to Australia: 440,000, North America to South Africa: 190,000, Europe to Australia: 950,000, Europe to South Africa: 950,000.　**29. a.** $x + y = 14,000$; $z + w = 95,000$; $x + z = 63,550$, $y + w = 45,450$. The system does not have a unique solution, indicating that the given data are insufficient to obtain the missing data.
**b.** $(x, y, z, w) = (5,600, 8,400, 57,950, 37,050)$
**31. a.** No; The general solution is: Eastward Blvd.: $S + 200$; Northwest La.: $S + 50$; Southwest La.: $S$, where $S$ is arbitrary. Thus it would suffice to know the traffic along Southwest La.　**b.** Yes, as it leads to the solution Eastward Blvd.: 260; Northwest La.: 110; Southwest La.: 60.　**c.** 50 vehicles per day
**33. a.** With $x = $ traffic on middle section of Bree, $y = $ traffic on middle section of Jeppe, $z = $ traffic on middle section of Simmons, $w = $ traffic on middle section of Harrison, the general solution is $x = w - 100$, $y = w$, $z = w$, $w$ arbitrary.
**b.** No; there are infinitely many possible values for $y$.
**c.** 500 cars/minute　**d.** 100 cars/minute　**e.** No; a large number of cars can circulate around the middle block without affecting the numbers shown.　**35. a.** No; the corresponding system of equations is underdetermined. The net flow of traffic along any of the three stretches of Broadway would suffice.　**b.** West
**37.** $10 billion　**39.** $x = $ water, $y = $ gray matter, $z = $ tumor　**41.** $x = $ water, $y = $ bone, $z = $ tumor, $u = $ air　**43.** tumor　**45.** 200 Democrats, 20 Republicans, 13 of other parties　**47.** Yes; $20m in Company X, $5m in Company Y, $10m in Company Z, and $30m in Company W　**49.** It is not realistic to expect to use exactly all of the ingredients. Solutions of the associated system may involve negative numbers or not exist. Only solutions with nonnegative values for all the unknowns correspond to being able to use up all of the ingredients.　**51.** Yes; $x = 100$
**53.** Yes; $0.3x - 0.7y + 0.3z = 0$ is one form of the equation.　**55.** No; represented by an inequality rather than an equation.　**57.** Answers will vary.

## Chapter 3 Review

**1.** One solution

**3.** Infinitely many solutions

**5.** One solution

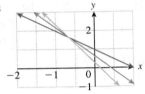

**7.** $(6/5, 7/5)$  **9.** $(3y/2, y)$; $y$ arbitrary
**11.** $(-0.7, 1.7)$  **13.** $(-1, -1, -1)$
**15.** $(z - 2, 4(z - 1), z)$; $z$ arbitrary  **17.** No solution
**19.** $-40°$  **21.** It is impossible; setting $F = 1.8C$
leads to an inconsistent system of equations.
**23.** $x + y + z + w = 10$; Linear  **25.** $w = 0$; Linear
**27.** $-1.3y + z = 0$ or $1.3y - z = 0$; Linear
**29.** 550 packages from Duffin House, 350 from
Higgins Press  **31.** 1,200 packages from Duffin
House, 200 from Higgins Press  **33.** $40
**35.** 7 of each  **37.** 5,000 hits per day at
OHaganBooks.com, 1,250 at JungleBooks.com,
3,750 at FarmerBooks.com  **39.** 100 shares of HAL,
20 shares of POM, and 80 shares of WELL
**41.** Billy-Sean is forced to take exactly the following
combination: Liberal Arts: 52 credits, Sciences:
12 credits, Fine Arts: 12 credits, Mathematics: 48
credits.  **43. a.** $x = 100$, $y = 100 + w$, $z = 300 - w$,
$w$ arbitrary **b.** 100 book orders per day **c.** 300 book
orders per day **d.** $x = 100$, $y = 400$, $z = 0$, $w = 300$
**e.** 100 book orders per day
**45.** Yes; New York to OHaganBooks.com: 450 packages, New York to FantasyBooks.com: 50 packages,
Illinois to OHaganBooks.com: 150 packages, Illinois
to FantasyBooks.com: 150 packages

## Chapter 4

### Section 4.1

**1.** $1 \times 4$; 0  **3.** $4 \times 1$; 5/2  **5.** $p \times q$; $e_{22}$  **7.** $2 \times 2$; 3
**9.** $1 \times n$; $d_r$  **11.** $x = 1$, $y = 2$, $z = 3$, $w = 4$

**13.** $\begin{bmatrix} 0.25 & -2 \\ 1 & 0.5 \\ -2 & 5 \end{bmatrix}$  **15.** $\begin{bmatrix} -0.75 & -1 \\ 0 & -0.5 \\ -1 & 6 \end{bmatrix}$

**17.** $\begin{bmatrix} -1 & -1 \\ 1 & -1 \\ -1 & 5 \end{bmatrix}$  **19.** $\begin{bmatrix} 0 & 2 & -2 \\ -2 & 0 & 4 \end{bmatrix}$

**21.** $\begin{bmatrix} 4 & -1 & -1 \\ 5 & 1 & 0 \end{bmatrix}$  **23.** $\begin{bmatrix} -2+x & 0 & 1+w \\ -5+z & 3+r & 2 \end{bmatrix}$

**25.** $\begin{bmatrix} -1 & -2 & 1 \\ -5 & 5 & -3 \end{bmatrix}$  **27.** $\begin{bmatrix} 9 & 15 \\ 0 & -3 \\ -3 & 3 \end{bmatrix}$

**29.** $\begin{bmatrix} -8.5 & -22.35 & -24.4 \\ 54.2 & 20 & 42.2 \end{bmatrix}$

**31.** $\begin{bmatrix} 1.54 & 8.58 \\ 5.94 & 0 \\ 6.16 & 7.26 \end{bmatrix}$  **33.** $\begin{bmatrix} 7.38 & 76.96 \\ 20.33 & 0 \\ 29.12 & 39.92 \end{bmatrix}$

**35.** $\begin{bmatrix} -19.85 & 115.82 \\ -50.935 & 46 \\ -57.24 & 94.62 \end{bmatrix}$

**37.** Q1 2010: [3.3  21  8.7];  Q1 2011:
[4.1  19.4  16.2]

**39.** Sales $= \begin{bmatrix} 700 & 1,300 & 2,000 \\ 400 & 300 & 500 \end{bmatrix}$;

Inventory $-$ Sales $= \begin{bmatrix} 300 & 700 & 3,000 \\ 600 & 4,700 & 1,500 \end{bmatrix}$

**41.** Profit $=$ Revenue $-$ Cost $= \begin{bmatrix} 8,000 & 7,200 & 8,800 \\ 5,600 & 5,760 & 7,040 \\ 2,800 & 3,500 & 4,000 \end{bmatrix}$

**43.** 2000 distribution $= A = [53.6 \ 64.4 \ 100.2 \ 63.2]$;
2010 distribution $= B = [55.3 \ 66.9 \ 114.6 \ 71.9]$;
Net change 2000 to 2010 $= D = B - A =$
$[1.7 \ 2.5 \ 14.4 \ 8.7]$; 2020 Distribution $= B + D =$
$[57.0 \ 69.4 \ 129.0 \ 80.6]$
**45.** Total Foreclosures $=$ Foreclosures in California $+$
Foreclosures in Florida $+$ Foreclosures in Texas $=$
[55,900  51,900  54,100  56,200  59,400] $+$
[19,600  19,200  23,800  22,400  23,600] $+$
[8,800  9,100  9,300  10,600  10,100] $=$
[84,300  80,200  87,200  89,200  93,100]
**47.** Difference $=$ Foreclosures in California $-$
Foreclosures in Florida $=$ [55,900  51,900  54,100
56,200  59,400] $-$ [19,600  19,200  23,800  22,400
23,600] $=$ [36,300  32,700  30,300  33,800  35,800]
The difference was greatest in April.

**49. a.** Use = 
|  | Proc | Mem | Tubes |
Pom II $\begin{bmatrix} 2 & 16 & 20 \\ 1 & 4 & 40 \end{bmatrix}$ ; Pom Classic

Inventory = $\begin{bmatrix} 500 & 5{,}000 & 10{,}000 \\ 200 & 2{,}000 & 20{,}000 \end{bmatrix}$ ;

Inventory $- 100 \times$ Use $= \begin{bmatrix} 300 & 3{,}400 & 8{,}000 \\ 100 & 1{,}600 & 16{,}000 \end{bmatrix}$

**b.** After 4 months

**51. a.** $A = \begin{bmatrix} 440 & 190 \\ 950 & 950 \\ 1{,}790 & 200 \end{bmatrix}$ $D = \begin{bmatrix} -20 & 40 \\ 50 & 50 \\ 0 & 100 \end{bmatrix}$

2008 Tourism $= A + D = \begin{bmatrix} 420 & 230 \\ 1{,}000 & 1{,}000 \\ 1{,}790 & 300 \end{bmatrix}$

**b.** $\frac{1}{2}(A + B)$; $\begin{bmatrix} 430 & 210 \\ 975 & 975 \\ 1{,}790 & 250 \end{bmatrix}$

**53.** No; for two matrices to be equal, they must have the same dimension.    **55.** The $ij$th entry of the sum $A + B$ is obtained by adding the $ij$th entries of $A$ and $B$.
**57.** It would have zeros down the main diagonal:

$A = \begin{bmatrix} 0 & \# & \# & \# & \# \\ \# & 0 & \# & \# & \# \\ \# & \# & 0 & \# & \# \\ \# & \# & \# & 0 & \# \\ \# & \# & \# & \# & 0 \end{bmatrix}$

The symbols # indicate arbitrary numbers.
**59.** $(A^T)_{ij} = A_{ji}$    **61.** Answers will vary.

**a.** $\begin{bmatrix} 0 & -4 \\ 4 & 0 \end{bmatrix}$  **b.** $\begin{bmatrix} 0 & -4 & 5 \\ 4 & 0 & 1 \\ -5 & -1 & 0 \end{bmatrix}$

**63.** The associativity of matrix addition is a consequence of the associativity of addition of numbers, since we add matrices by adding the corresponding entries (which are real numbers).
**65.** Answers will vary.

## Section 4.2

**1.** [13]  **3.** [5/6]  **5.** $[-2y + z]$  **7.** Undefined
**9.** $[3 \quad 0 \quad -6 \quad -2]$  **11.** $[-6 \quad 37 \quad 7]$

**13.** $\begin{bmatrix} -4 & -7 & -1 \\ 9 & 17 & 0 \end{bmatrix}$  **15.** $\begin{bmatrix} 0 & 1 \\ 0 & 0 \end{bmatrix}$  **17.** $\begin{bmatrix} 1 & -1 \\ 1 & -1 \end{bmatrix}$

**19.** $\begin{bmatrix} 0 & 0 \\ 0 & 0 \end{bmatrix}$  **21.** Undefined  **23.** $\begin{bmatrix} 1 & -5 & 3 \\ 0 & 0 & 9 \\ 0 & 4 & 1 \end{bmatrix}$

**25.** $\begin{bmatrix} 3 \\ -4 \\ 0 \\ 3 \end{bmatrix}$  **27.** $\begin{bmatrix} 0.23 & 5.36 & -21.65 \\ -13.18 & -5.82 & -16.62 \\ -11.21 & -9.9 & 0.99 \\ -2.1 & 2.34 & 2.46 \end{bmatrix}$

**29.** $A^2 = \begin{bmatrix} 0 & 0 & 1 & 2 \\ 0 & 0 & 0 & 1 \\ 0 & 0 & 0 & 0 \\ 0 & 0 & 0 & 0 \end{bmatrix}$ ; $A^3 = \begin{bmatrix} 0 & 0 & 0 & 1 \\ 0 & 0 & 0 & 0 \\ 0 & 0 & 0 & 0 \\ 0 & 0 & 0 & 0 \end{bmatrix}$ ;

$A^4 = \begin{bmatrix} 0 & 0 & 0 & 0 \\ 0 & 0 & 0 & 0 \\ 0 & 0 & 0 & 0 \\ 0 & 0 & 0 & 0 \end{bmatrix}$ ; $\ldots$ ; $A^{100} = \begin{bmatrix} 0 & 0 & 0 & 0 \\ 0 & 0 & 0 & 0 \\ 0 & 0 & 0 & 0 \\ 0 & 0 & 0 & 0 \end{bmatrix}$ ;

**31.** $\begin{bmatrix} 4 & -1 \\ -1 & -7 \end{bmatrix}$  **33.** $\begin{bmatrix} 4 & -1 \\ -12 & 2 \end{bmatrix}$

**35.** $\begin{bmatrix} -2 & 1 & -2 \\ 10 & -2 & 2 \\ -10 & 2 & -2 \end{bmatrix}$

**37.** $\begin{bmatrix} -2+x-z & 2-r & -6+w \\ 10+2z & -2+2r & 10 \\ -10-2z & 2-2r & -10 \end{bmatrix}$

**39. a.–d.** $P^2 = P^4 = P^8 = P^{1,000} = \begin{bmatrix} 0.2 & 0.8 \\ 0.2 & 0.8 \end{bmatrix}$

**41. a.** $P^2 = \begin{bmatrix} 0.01 & 0.99 \\ 0 & 1 \end{bmatrix}$

**b.** $P^4 = \begin{bmatrix} 0.0001 & 0.9999 \\ 0 & 1 \end{bmatrix}$

**c and d.** $P^8 \approx P^{1,000} \approx \begin{bmatrix} 0 & 1 \\ 0 & 1 \end{bmatrix}$

**43. a.–d.** $P^2 = P^4 = P^8 = P^{1,000} = \begin{bmatrix} 0.3 & 0.3 & 0.4 \\ 0.3 & 0.3 & 0.4 \\ 0.3 & 0.3 & 0.4 \end{bmatrix}$

The rows of $P$ are the same, and the entries in each row add up to 1. If $P$ is any square matrix with identical rows such that the entries in each row add up to 1, then $P \cdot P = P$.

**45.** $2x - y + 4z = 3$; $-4x + \frac{3}{4}y + \frac{1}{3}z = -1$; $-3x = 0$
**47.** $x - y + w = -1$; $x + y + 2z + 4w = 2$

**49.** $\begin{bmatrix} 1 & -1 \\ 2 & -1 \end{bmatrix}\begin{bmatrix} x \\ y \end{bmatrix} = \begin{bmatrix} 4 \\ 0 \end{bmatrix}$

**51.** $\begin{bmatrix} 1 & 1 & -1 \\ 2 & 1 & 1 \\ \frac{3}{4} & 0 & \frac{1}{2} \end{bmatrix}\begin{bmatrix} x \\ y \\ z \end{bmatrix} = \begin{bmatrix} 8 \\ 4 \\ 1 \end{bmatrix}$

**53.** Revenue = Price $\times$ Quantity =
$[15 \quad 10 \quad 12]\begin{bmatrix} 50 \\ 40 \\ 30 \end{bmatrix} = [1{,}510]$

**55.** $32.5 million
**57.** Revenue = Quantity $\times$ Price =
$\begin{bmatrix} 700 & 1{,}300 & 2{,}000 \\ 400 & 300 & 500 \end{bmatrix}\begin{bmatrix} 30 \\ 10 \\ 15 \end{bmatrix} = \begin{bmatrix} 64{,}000 \\ 22{,}500 \end{bmatrix}$

**59.** \$5,100 billion (or \$5.1 trillion)
**61.** $D = N(F - M)$, where $N$ is the income per person, and $F$ and $M$ are, respectively, the female and male populations in 2020; \$400 billion
**63.** $[1.2 \quad 1.0]$, which represents the amount, in billions of pounds, by which cheese production in north central states exceeded that in western states.
**65.** Number of foreclosures filings handled by firm = Percentage handled by firm × Total number = $[8,460 \quad 8,860 \quad 9,140]$
**67.** The number of foreclosures in California and Florida combined in each of the months shown.

**69.** $[1 \quad -1 \quad 1] \begin{bmatrix} 54,100 & 56,200 & 59,400 \\ 23,800 & 22,400 & 23,600 \\ 9,300 & 10,600 & 10,100 \end{bmatrix} \begin{bmatrix} 1 \\ 1 \\ 1 \end{bmatrix}$

$= [129,900]$

**71.** $\begin{bmatrix} 2 & 16 & 20 \\ 1 & 4 & 40 \end{bmatrix} \begin{bmatrix} 100 & 150 \\ 50 & 40 \\ 10 & 15 \end{bmatrix} = \begin{bmatrix} 1,200 & 1,240 \\ 700 & 910 \end{bmatrix}$

**73.** $AB = \begin{bmatrix} 29.6 \\ 85.5 \\ 97.5 \end{bmatrix}$   $AC = \begin{bmatrix} 22 & 7.6 \\ 47.5 & 38 \\ 89.5 & 8 \end{bmatrix}$

The entries of $AB$ give the number of people from each of the three regions who settle in Australia or South Africa, while the entries in $AC$ break those figures down further into settlers in South Africa and settlers in Australia.   **75.** $[54.6 \quad 66.0 \quad 111.8 \quad 70.6]$
**77.** Answers will vary. One example:

$A = [1 \quad 2]$, $B = \begin{bmatrix} 1 & 2 & 3 \\ 4 & 5 & 6 \end{bmatrix}$.

Another example: $A = [1]$, $B = [1 \quad 2]$
**79.** We finnd that the addition and multiplication of $1 \times 1$ matrices is identical to the addition and multiplication of numbers.   **81.** The claim is correct. Every matrix equation represents the equality of two matrices. When two matrices are equal, each of their corresponding entries must be equal. Equating the corresponding entries gives a system of equations.
**83.** Here is a possible scenario: costs of items A, B, and C in 2013 = $[10 \quad 20 \quad 30]$, percentage increases in these costs in 2014 = $[0.5 \quad 0.1 \quad 0.20]$, actual increases in costs = $[10 \times 0.5 \quad 20 \times 0.1 \quad 30 \times 0.20]$
**85.** It produces a matrix whose $ij$ entry is the product of the $ij$ entries of the two matrices.

**Section 4.3**
**1.** Yes   **3.** Yes   **5.** No   **7.** $\begin{bmatrix} -1 & 1 \\ 2 & -1 \end{bmatrix}$   **9.** $\begin{bmatrix} 0 & 1 \\ 1 & 0 \end{bmatrix}$

**11.** $\begin{bmatrix} 1 & -1 \\ -1 & 2 \end{bmatrix}$   **13.** Singular   **15.** $\begin{bmatrix} 1 & -1 & 0 \\ 0 & 1 & -1 \\ 0 & 0 & 1 \end{bmatrix}$

**17.** $\begin{bmatrix} 1 & -1 & 1 \\ \frac{1}{2} & 0 & -\frac{1}{2} \\ -\frac{1}{2} & 1 & -\frac{1}{2} \end{bmatrix}$

**19.** $\begin{bmatrix} 1 & \frac{1}{3} & -\frac{1}{3} \\ 1 & -\frac{2}{3} & -\frac{1}{3} \\ -1 & \frac{1}{3} & \frac{2}{3} \end{bmatrix}$

**21.** Singular   **23.** $\begin{bmatrix} 0 & 1 & -2 & 1 \\ 0 & 1 & -1 & 0 \\ 1 & -1 & 2 & -1 \\ 0 & 1 & -1 & 1 \end{bmatrix}$

**25.** $\begin{bmatrix} 1 & -2 & 1 & 0 \\ 0 & 1 & -2 & 1 \\ 0 & 0 & 1 & -2 \\ 0 & 0 & 0 & 1 \end{bmatrix}$   **27.** $-2$; $\begin{bmatrix} \frac{1}{2} & \frac{1}{2} \\ \frac{1}{2} & -\frac{1}{2} \end{bmatrix}$

**29.** $-2$; $\begin{bmatrix} -2 & 1 \\ \frac{3}{2} & -\frac{1}{2} \end{bmatrix}$   **31.** $1/36$; $\begin{bmatrix} 6 & 6 \\ 0 & 6 \end{bmatrix}$

**33.** 0; Singular

**35.** $\begin{bmatrix} 0.38 & 0.45 \\ 0.49 & -0.41 \end{bmatrix}$   **37.** $\begin{bmatrix} 0.00 & -0.99 \\ 0.81 & 2.87 \end{bmatrix}$

**39.** Singular

**41.** $\begin{bmatrix} 91.35 & -8.65 & 0 & -71.30 \\ -0.07 & -0.07 & 0 & 2.49 \\ 2.60 & 2.60 & -4.35 & 1.37 \\ 2.69 & 2.69 & 0 & -2.10 \end{bmatrix}$

**43.** $(5/2, 3/2)$
**45.** $(6, -4)$   **47.** $(6, 6, 6)$   **49. a.** $(10, -5, -3)$
**b.** $(6, 1, 5)$ **c.** $(0, 0, 0)$   **51. a.** $10/3$ servings of beans, and $5/6$ slices of bread

**b.** $\begin{bmatrix} -1/2 & 1/6 \\ 7/8 & -5/24 \end{bmatrix} \begin{bmatrix} A \\ B \end{bmatrix} = \begin{bmatrix} -A/2 + B/6 \\ 7A/8 - 5B/24 \end{bmatrix}$; that is,

$-A/2 + B/6$ servings of beans and $7A/8 - 5B/24$ slices of bread   **53. a.** 100 batches of vanilla, 50 batches of mocha, 100 batches of strawberry   **b.** 100 batches of vanilla, no mocha, 200 batches of strawberry

**c.** $\begin{bmatrix} 1 & -1/3 & -1/3 \\ -1 & 0 & 1 \\ 0 & 2/3 & -1/3 \end{bmatrix} \begin{bmatrix} A \\ B \\ C \end{bmatrix}$, or

$A - B/3 - C/3$ batches of vanilla, $-A + C$ batches of mocha, and $2B/3 - C/3$ batches of strawberry
**55.** \$5,000 in SHPIX, \$2,000 in RYURX, \$2,000 in RYIHX   **57.** 100 shares of KT, 50 shares of IFF, 50 shares of QSII   **59.** $[54.1 \quad 65.7 \quad 112.9 \quad 70.3]$
**61. a.** $(-0.7071, 3.5355)$ **b.** $R^2$, $R^3$ **c.** $R^{-1}$
**63.** $[37 \quad 81 \quad 40 \quad 80 \quad 15 \quad 45 \quad 40 \quad 96 \quad 29 \quad 59 \quad 4 \quad 8]$
**65.** CORRECT ANSWER   **67.** (A)   **69.** The inverse does not exist; the matrix is singular. (If two rows of a matrix are the same, then row reducing it will lead to a row of zeros, and so it cannot be reduced to the identity.)

**71.** Calculation (See the Student Solutions Manual.)
**73.** When one or more of the $d_i$ are zero. If that is the case, then the matrix $[D \mid I]$ easily reduces to a matrix that has a row of zeros on the left-hand portion, so that $D$ is singular. Conversely, if none of the $d_i$ are zero, then $[D \mid I]$ easily reduces to a matrix of the form $[I \mid E]$, showing that $D$ is invertible.
**75.** $(AB)(B^{-1}A^{-1}) = A(BB^{-1})A^{-1} = AIA^{-1} = AA^{-1} = I$    **77.** If $A$ has an inverse, then every system of equations $AX = B$ has a unique solution, namely $X = A^{-1}B$. But if $A$ reduces to a matrix with a row of zeros, then such a system has either infinitely many solutions or no solution at all.

## Section 4.4

**1.** $-1$    **3.** $-0.25$    **5.** $[0 \quad 0 \quad 1 \quad 0]$; $e = 2.25$
**7.** $[1 \quad 0 \quad 0]^T$ or $[0 \quad 1 \quad 0]^T$; $e = 1/4$

**9.** $\begin{array}{c} \\ a \\ b \end{array} \begin{array}{c} p \quad\quad r \\ \left[ \begin{array}{cc} 1 & 10 \\ 2 & -4 \end{array} \right] \end{array}$    **11.** $\begin{array}{c} c \\ 3[-1] \end{array}$    **13.** $\begin{array}{c} b \\ q\,[0] \end{array}$

**15.** Strictly determined. The row player's optimal strategy is $a$; the column player's optimal strategy is $q$; Value: 1    **17.** Not strictly determined Circle the row minima and box the row maxima:

$$\begin{array}{c} a \\ b \end{array} \begin{array}{c} p \quad\;\; q \quad\;\; r \\ \left[ \begin{array}{ccc} 2 & 0 & \boxed{-2} \\ \boxed{-1} & 3 & 0 \end{array} \right] \end{array} \qquad \begin{array}{c} a \\ b \end{array} \begin{array}{c} p \quad\;\; q \quad\;\; r \\ \left[ \begin{array}{ccc} \boxed{2} & 0 & -2 \\ -1 & \boxed{3} & \boxed{0} \end{array} \right] \end{array}.$$

Since no entry is both circled and boxed, there are no saddle points, and so the game is not strictly determined.

**19.** Not strictly determined
Circle the row minima and box the row maxima:

$$\begin{array}{c} P \\ Q \\ R \\ S \end{array} \begin{array}{c} a \quad\;\; b \quad\;\; c \\ \left[ \begin{array}{ccc} 1 & -1 & \boxed{-5} \\ 4 & \boxed{-4} & 2 \\ 3 & -3 & \boxed{-10} \\ 5 & \boxed{-5} & -4 \end{array} \right] \end{array} \quad \begin{array}{c} P \\ Q \\ R \\ S \end{array} \begin{array}{c} a \quad\;\; b \quad\;\; c \\ \left[ \begin{array}{ccc} 1 & \boxed{-1} & -5 \\ 4 & -4 & \boxed{2} \\ 3 & -3 & -10 \\ \boxed{5} & -5 & -4 \end{array} \right] \end{array}.$$

Since no entry is both circled and boxed, there are no saddle points, and so the game is not strictly determined.

**21.** $R = [1/4 \quad 3/4]$, $C = [3/4 \quad 1/4]^T$, $e = -1/4$
**23.** $R = [3/4 \quad 1/4]$, $C = [3/4 \quad 1/4]^T$, $e = -5/4$
**25.** Row player: You; Column player: Your friend;

$$\begin{array}{c} H \\ T \end{array} \begin{array}{c} H \quad\;\; T \\ \left[ \begin{array}{cc} -1 & 1 \\ 1 & -1 \end{array} \right] \end{array}$$

**27.** $F = $ France; $S = $ Sweden; $N = $ Norway

**Your Opponent Defends**

$$\text{You Invade} \quad \begin{array}{c} F \\ S \\ N \end{array} \begin{array}{c} F \quad\;\; S \quad\;\; N \\ \left[ \begin{array}{ccc} -1 & 1 & 1 \\ 1 & -1 & 1 \\ 1 & 1 & -1 \end{array} \right] \end{array}$$

**29.** Row player: You, Column player: Your opponent; B = Brakpan, N = Nigel, S = Springs;

**Your Opponent**

$$\text{You} \quad \begin{array}{c} B \\ N \\ S \end{array} \begin{array}{c} B \quad\quad N \quad\quad S \\ \left[ \begin{array}{ccc} 0 & 0 & 1{,}000 \\ 0 & 0 & 1{,}000 \\ -1{,}000 & -1{,}000 & 0 \end{array} \right] \end{array}$$

**31.** P = PleasantTap; T = Thunder Rumble; S = Strike the Gold, N = None;

**Winner**

$$\text{You Bet} \quad \begin{array}{c} P \\ T \\ S \end{array} \begin{array}{c} P \quad\;\; T \quad\;\; S \quad\;\; N \\ \left[ \begin{array}{cccc} 25 & -10 & -10 & -10 \\ -10 & 35 & -10 & -10 \\ -10 & -10 & 40 & -10 \end{array} \right] \end{array}$$

**33.** You can expect to lose 39 customers.
**35.** Option 2: Move to the suburbs.    **37. a.** About 66%
**b.** Yes; spend the whole night studying game theory; 75%
**c.** Game theory; 57.5%    **39. a.** Lay off 10 workers; cost: $40,000  **b.** 60 inches of snow, costing $350,000
**c.** Lay off 15 workers.    **41. a.** CE should charge $1,000, and GCS should charge $900; 15% gain in market share for CE.  **b.** CE should charge $1,200 (the more CE can charge for the same market, the better!).
**43.** Pablo vs. Noto; evenly matched    **45.** Both commanders should use the northern route; 2 days.
**47.** Confess

**49. a.** $\begin{array}{c} F \\ O \end{array} \begin{array}{c} F \quad\;\; O \\ \left[ \begin{array}{cc} 24 & 21 \\ 25 & 24 \end{array} \right] \end{array}$

**b.** Both candidates should visit Ohio, leaving Romney with a 24% chance of winning the election.
**51.** Allocate 1/7 of the budget to WISH and the rest (6/7) to WASH. Softex will lose approximately $2,860.
**53.** Like a saddle point in a payoff matrix, the center of a saddle is a low point (minimum height) in one direction and a high point (maximum) in a perpendicular direction.    **55.** Although there is a saddle point in the (2, 4) position, you would be wrong to use saddle points (based on the minimax criterion) to reach the conclusion that row strategy 2 is best. One reason is that the entries in the matrix do not represent payoffs, since high numbers of employees in an area do not necessarily represent benefit to the row player. Another reason for this is that there is no opponent deciding what your job will be in such a way as to force you into the least populated job.

**57.** If you strictly alternate the two strategies, the column player will know which pure strategy you will play on each move and can choose a pure strategy accordingly. For example, consider the game

$$\begin{array}{cc} & \begin{array}{cc} a & b \end{array} \\ \begin{array}{c} A \\ B \end{array} & \begin{bmatrix} 1 & 0 \\ 0 & 1 \end{bmatrix}. \end{array}$$

By the analysis of Example 3 (or the symmetry of the game), the best strategy for the row player is [0.5  0.5] and the best strategy for the column player is $[0.5 \ 0.5]^T$. This gives an expected value of 0.5 for the game. However, suppose that the row player alternates $A$ and $B$ strictly and that the column player catches on to this. Then, whenever the row player plays $A$ the column player will play $b$, and whenever the row player plays $B$ the column player will play $a$. This gives a payoff of 0 each time, worse for the row player than the expected value of 0.5.

## Section 4.5

**1. a.** 0.8 **b.** 0.2 **c.** 0.05   **3.** $\begin{bmatrix} 0.2 & 0.1 \\ 0.5 & 0 \end{bmatrix}$

**5.** $[\,52{,}000 \quad 40{,}000\,]^T$   **7.** $[\,50{,}000 \quad 50{,}000\,]^T$
**9.** $[\,2{,}560 \quad 2{,}800 \quad 4{,}000\,]^T$
**11.** $[\,27{,}000 \quad 28{,}000 \quad 17{,}000\,]^T$
**13.** Increase of 100 units in each sector.
**15.** Increase of $[\,1.5 \quad 0.2 \quad 0.1\,]^T$; the $i$th column of $(I - A)^{-1}$ gives the change in production necessary to meet an increase in external demand of one unit for the product of Sector $i$.

**17.** $A = \begin{bmatrix} 0.2 & 0.4 & 0.5 \\ 0 & 0.8 & 0 \\ 0 & 0.2 & 0.5 \end{bmatrix}$

**19.** Main DR: $80,000, Bits & Bytes: $38,000
**21.** Equipment Sector production approximately $86,000 million, Components Sector production approximately $140,000 million   **23. a.** 0.006
**b.** textiles; clothing and footwear
**25.** Columns of

$$\begin{bmatrix} 1{,}140.99 & 2.05 & 13.17 & 20.87 \\ 332.10 & 1{,}047.34 & 26.05 & 111.18 \\ 0.12 & 0.13 & 1{,}031.19 & 1.35 \\ 93.88 & 95.69 & 215.50 & 1{,}016.15 \end{bmatrix}$$

(in millions of dollars)
**27. a.** $0.78 **b.** Other food products   **29.** It would mean that all of the sectors require neither their own product nor the product of any other sector.   **31.** It would mean that all of the output of that sector was used internally in the economy; none of the output was available for export and no importing was necessary.

**33.** If an entry in the matrix $(I - A)^{-1}$ is zero, then an increase in demand for one sector (the column sector) has no effect on the production of another sector (the row sector).   **35.** Usually, to produce one unit of one sector requires less than one unit of input from another. We would expect then that an increase in demand of one unit for one sector would require a smaller increase in production in another sector.

## Chapter 4 Review

**1.** Undefined **3.** $\begin{bmatrix} 1 & 8 \\ 5 & 11 \\ 6 & 13 \end{bmatrix}$ **5.** $\begin{bmatrix} 1 & 3 \\ 2 & 3 \\ 3 & 3 \end{bmatrix}$ **7.** $\begin{bmatrix} 1 & -2 \\ 0 & 1 \end{bmatrix}$

**9.** $\begin{bmatrix} 2 & 4 \\ 1 & 12 \end{bmatrix}$ **11.** $\begin{bmatrix} 1 & 1 \\ 0 & 1 \end{bmatrix}$ **13.** $\begin{bmatrix} 1 & -1/2 & -5/2 \\ 0 & 1/4 & -1/4 \\ 0 & 0 & 1 \end{bmatrix}$

**15.** Singular

**17.** $\begin{bmatrix} 1 & 2 \\ 3 & 4 \end{bmatrix}\begin{bmatrix} x \\ y \end{bmatrix} = \begin{bmatrix} 0 \\ 2 \end{bmatrix}; \begin{bmatrix} x \\ y \end{bmatrix} = \begin{bmatrix} 2 \\ -1 \end{bmatrix}$

**19.** $\begin{bmatrix} 1 & 1 & 1 \\ 1 & 2 & 1 \\ 1 & 1 & 2 \end{bmatrix}\begin{bmatrix} x \\ y \\ z \end{bmatrix} = \begin{bmatrix} 2 \\ 3 \\ 1 \end{bmatrix}; \begin{bmatrix} x \\ y \\ z \end{bmatrix} = \begin{bmatrix} 2 \\ 1 \\ -1 \end{bmatrix}$

**21.** $R = [1 \quad 0 \quad 0], C = [0 \quad 1 \quad 0 \quad 0]^T, e = 1$
**23.** $R = [0 \quad 0.8 \quad 0.2], C = [0.2 \quad 0 \quad 0.8],$
$e = -0.2$

**25.** $\begin{bmatrix} 1{,}100 \\ 700 \end{bmatrix}$ **27.** $\begin{bmatrix} 48{,}125 \\ 22{,}500 \\ 10{,}000 \end{bmatrix}$

**29.** Inventory − Sales + Purchases =

$$\begin{bmatrix} 2{,}500 & 4{,}000 & 3{,}000 \\ 1{,}500 & 3{,}000 & 1{,}000 \end{bmatrix} - \begin{bmatrix} 300 & 500 & 100 \\ 100 & 600 & 200 \end{bmatrix} +$$
$$\begin{bmatrix} 400 & 400 & 300 \\ 200 & 400 & 300 \end{bmatrix} = \begin{bmatrix} 2{,}600 & 3{,}900 & 3{,}200 \\ 1{,}600 & 2{,}800 & 1{,}100 \end{bmatrix}$$

**31.** $N =$

$$\begin{bmatrix} 2{,}600 & 3{,}900 & 3{,}200 \\ 1{,}600 & 2{,}800 & 1{,}100 \end{bmatrix} - x\begin{bmatrix} 280 & 550 & 100 \\ 50 & 500 & 120 \end{bmatrix} +$$
$$x\begin{bmatrix} 400 & 400 & 300 \\ 200 & 400 & 300 \end{bmatrix} = \begin{bmatrix} 2{,}600 & 3{,}900 & 3{,}200 \\ 1{,}600 & 2{,}800 & 1{,}100 \end{bmatrix} +$$
$$x\begin{bmatrix} 120 & -150 & 200 \\ 150 & -100 & 180 \end{bmatrix}; \text{28 months from now (July 1)}$$

**33.** Revenue = Quantity × Price =

$$\begin{bmatrix} 280 & 550 & 100 \\ 50 & 500 & 120 \end{bmatrix}\begin{bmatrix} 5 \\ 6 \\ 5.5 \end{bmatrix} = \begin{bmatrix} 5{,}250 \\ 3{,}910 \end{bmatrix} \begin{array}{l} \text{Texas} \\ \text{Nevada} \end{array}$$

**35.** July 1: 1,000 shares, August 1: 2,000 shares, September 1: 2,000 shares

**37.** Loss = Number of shares × (Purchase price −
Dividends − Selling price) = [1,000   2,000   2,000]

$$\left( \begin{bmatrix} 20 \\ 10 \\ 5 \end{bmatrix} - \begin{bmatrix} 0.10 \\ 0.10 \\ 0 \end{bmatrix} - \begin{bmatrix} 3 \\ 1 \\ 1 \end{bmatrix} \right) = [42,700]$$

**39.** $[2,000 \quad 4,000 \quad 4,000] \begin{bmatrix} 0.8 & 0.1 & 0.1 \\ 0.4 & 0.6 & 0 \\ 0.2 & 0 & 0.8 \end{bmatrix} =$

[4,000   2,600   3,400]   **41.** The matrix shows that
no JungleBooks customers switched directly to
FarmerBooks, so the only way to get to FarmerBooks
is via OHaganBooks.   **43.** Go with the "3 for 1"
promotion and gain 20,000 customers from JungleBooks.
**45.** JungleBooks will go with "3 for 2" and
OHaganBooks will go with Finite Math, resulting
in a gain of 15,000 customers to OHaganBooks.
**47.** Choose between the "3 for 1" and Finite Math
promotions with probabilities 60% and 40%,
respectively. You would expect to gain 12,000 customers from JungleBooks (if you played this game
many times).

**49.** $A = \begin{bmatrix} 0.1 & 0.5 \\ 0.01 & 0.05 \end{bmatrix}$

**51.** $1,190 worth of paper, $1,802 worth of books

# Chapter 5

## Section 5.1

**1.**
Unbounded

**3.**
Unbounded

**5.**
Unbounded

**7.**
Unbounded

**9.**
Unbounded

**11.**
Unbounded

**13.**
Unbounded;
corner point: (2, 0)

**15.**
Unbounded;
corner points:
(2, 0), (0, 3)

**17.**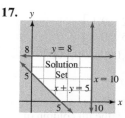
Bounded; corner points:
(5, 0), (10, 0), (10, 8),
(0, 8), (0, 5)

**19.**
Bounded; corner points:
(0, 0), (5, 0), (0, 5),
(2, 4), (4, 2)

**21.**
Unbounded; corner points:
(0, 10), (10, 0), (2, 6),
(6, 2)

**23.**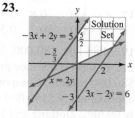
Unbounded;
corner points:(0, 0),
(0, 5/2), (3, 3/2)

**25.**
Unbounded;
corner point: (0, 0)

**27.**

**29.**

Corner point:
$(-7.74, 2.50)$

**31.**

Corner points:
$(0.36, -0.68)$, $(1.12, 0.61)$

**33.** $x = $ # quarts of Creamy Vanilla,
$y = $ # quarts of Continental Mocha

Corner points: $(0, 0)$, $(250, 0)$, $(0, 300)$, $(200, 100)$

**35.** $x = $ # ounces of chicken, $y = $ # ounces of grain

Corner points: $(30, 0)$, $(10, 50)$, $(0, 100)$

**37.** $x = $ # servings of Mixed Cereal for Baby,
$y = $ # servings of Mango Tropical Fruit Dessert

Corner points: $(0, 7/4)$, $(1, 1)$, $(32/11, 0)$

**39.** $x = $ # dollars in PNF, $y = $ # dollars in FDMMX.

Corner points: $(70{,}000, 0)$, $(80{,}000, 0)$, $(20{,}000, 60{,}000)$

**41.** Let $x = $ number of shares of BOH; $y = $ number of shares of JPM.

Corner points: $(200, 400)$, $(556, 0)$, $(422, 0)$

**43.** $x = $ # full-page ads in Sports Illustrated,
$y = $ # full-page ads in GQ

Corner points: $(3, 7)$, $(4, 3)$ (Rounded)

**45.** An example is $x \geq 0$, $y \geq 0$, $x + y \geq 1$.
**47.** The given triangle can be described as the solution set of the system $x \geq 0$, $y \geq 0$, $x + 2y \leq 2$.
**49.** Answers may vary. One limitation is that the method is only suitable for situations with two unknown quantities. Accuracy is also limited when graphing.  **51. (C)**  **53. (B)**  **55.** There are no feasible solutions; that is, it is impossible to satisfy all the constraints.  **57.** Answers will vary.

### Section 5.2

**1.** $p = 6$, $x = 3$, $y = 3$  **3.** $c = 4$, $x = 2$, $y = 2$
**5.** $p = 24$, $x = 7$, $y = 3$  **7.** $p = 16$, $x = 4$, $y = 2$
**9.** $c = 1.8$, $x = 6$, $y = 2$  **11.** Max: $p = 16$, $x = 4$, $y = 6$. Min: $p = 2$, $x = 2$, $y = 0$  **13.** No optimal solution; objective function unbounded  **15.** $c = 28$; $(x, y) = (14, 0)$ and $(6, 4)$ and the line connecting them  **17.** $c = 3$, $x = 3$, $y = 2$  **19.** No solution; feasible region empty  **21.** You should make 200 quarts of vanilla and 100 quarts of mocha.  **23.** Ruff, Inc., should use 100 oz of grain and no chicken.
**25.** Feed your child 1 serving of cereal and 1 serving of dessert.  **27.** Purchase 60 compact fluorescent light bulbs and 960 square feet of insulation for a saving of $312 per year in energy costs.  **29.** Mix 5 servings of Cell-Tech and 6 servings of RiboForce HP for a cost of $20.60.  **31.** Make 200 Dracula Salamis and 400 Frankenstein Sausages, for a profit of $1,400.
**33.** 30 spots on *WWE Raw* and 30 spots on *Family Guy*  **35.** 88.89 shares of FDX and 100 shares of WU

**37.** 422.2 shares of BOH and no shares of JPM
**39.** He should instruct in diplomacy for 10 hours per week and in battle for 40 hours per week, giving a weekly profit of 2,400 ducats.    **41.** Gillian could expend a minimum of 360,000 pico-shirleys of energy by using 480 sleep spells and 160 shock spells. (There is actually a whole line of solutions joining the one above with $x = 2,880/7$, $y = 1,440/7$.)    **43.** 100 hours per week for new customers and 60 hours per week for old customers.    **45. (A)**    **47.** Every point along the line connecting them is also an optimal solution.
**49.** Answers may vary. Maximize $p = x + y$ subject to $x + y \leq 10$; $x + y \geq 11$, $x \geq 0$, $y \geq 0$.
**51.** Answers may vary.
**53.** A simple example is the following: Maximize profit $p = 2x + y$ subject to $x \geq 0$, $y \geq 0$. Then $p$ can be made as large as we like by choosing large values of $x$ and/or $y$. Thus there is no optimal solution to the problem.    **55.** Mathematically, this means that there are infinitely many possible solutions: one for each point along the line joining the two corner points in question. In practice, select those points with integer solutions (since $x$ and $y$ must be whole numbers in this problem) that are in the feasible region and close to this line, and choose the one that gives the largest profit.    **57.** Proof

## Section 5.3

**1.** $p = 8$; $x = 4$, $y = 0$    **3.** $p = 4$; $x = 4$, $y = 0$
**5.** $p = 80$; $x = 10$, $y = 0$, $z = 10$
**7.** $p = 53$; $x = 5$, $y = 0$, $z = 3$
**9.** $z = 14,500$; $x_1 = 0$, $x_2 = 500/3$, $x_3 = 5,000/3$
**11.** $p = 6$; $x = 2$, $y = 1$, $z = 0$, $w = 3$
**13.** $p = 7$; $x = 1$, $y = 0$, $z = 2$, $w = 0$, $v = 4$
(or: $x = 1$, $y = 0$, $z = 2$, $w = 1$, $v = 3$.)
**15.** $p = 21$; $x = 0$, $y = 2.27$, $z = 5.73$
**17.** $p = 4.52$; $x = 1$, $y = 0$, $z = 0.67$, $w = 1.52$; or: $x = 1.67$, $y = 0.67$, $z = 0$, $w = 1.52$    **19.** $p = 7.7$; $x = 1.1$, $y = 0$, $z = 2.2$, $w = 0$, $v = 4$    **21.** You should purchase 500 calculus texts, no history texts, and no marketing texts. The maximum profit is $5,000 per semester.    **23.** The company can make a maximum profit of $650 by making 100 gallons of PineOrange, 200 gallons of PineKiwi, and 150 gallons of OrangeKiwi.    **25.** The department should offer no Ancient History, 30 sections of Medieval History, and 15 sections of Modern History, for a profit of $1,050,000. There will be 500 students without classes, but all sections and professors are used.    **27.** Plant 80 acres of tomatoes and leave the other 20 acres unplanted. This will give you a profit of $160,000.    **29.** It can make a profit of $10,000 by selling 1,000 servings

of granola, 500 servings of nutty granola, and no nuttiest granola. It is left with 2,000 oz almonds.
**31.** Allocate 5 million gals to process A and 45 million gals to process C. Another solution: Allocate 10 million gals to process B and 40 million gals to process C.
**33.** Use 15 servings of RiboForce HP and none of the others for a maximum of 75 g creatine.    **35.** She is right; you should buy 142.9 shares of DO and no others.    **37.** Allocate $2,250,000 to automobile loans, $500,000 to signature loans, and $2,250,000 to any combination of furniture loans and other secured loans.
**39.** Invest $75,000 in Universal, none in the rest. Another optimal solution is: Invest $18,750 in Universal, and $75,000 in EMI.    **41.** Tucson to Honolulu: 290 boards; Tucson to Venice Beach: 330 boards; Toronto to Honolulu: 0 boards; Toronto to Venice Beach: 200 boards, giving 820 boards shipped.
**43.** Fly 10 people from Chicago to Los Angeles, 5 people from Chicago to New York, and 10 people from Denver to New York.    **45.** Yes; the given problem can be stated as: Maximize $p = 3x - 2y$ subject to $-x + y - z \leq 0$, $x - y - z \leq 6$, $x \geq 0$, $y \geq 0$, $z \geq 0$.    **47.** The graphical method applies only to LP problems in two unknowns, whereas the simplex method can be used to solve LP problems with any number of unknowns.    **49.** She is correct. There are only two constraints, so there can be only two active variables, giving two or fewer nonzero values for the unknowns at each stage.    **51.** A basic solution to a system of linear equations is a solution in which all the nonpivotal variables are taken to be zero; that is, all variables whose values are arbitrary are assigned the value zero. To obtain a basic solution for a given system of linear equations, one can row reduce the associated augmented matrix, write down the general solution, and then set all the parameters (variables with "arbitrary" values) equal to zero.    **53.** No. Let us assume for the sake of simplicity that all the pivots are 1s. (They may certainly be changed to 1s without affecting the value of any of the variables.) Because the entry at the bottom of the pivot column is negative, the bottom row gets replaced by itself plus a positive multiple of the pivot row. The value of the objective function (bottom-right entry) is thus replaced by itself plus a positive multiple of the nonnegative rightmost entry of the pivot row. Therefore, it cannot decrease.

## Section 5.4

**1.** $p = 20/3$; $x = 4/3$, $y = 16/3$    **3.** $p = 850/3$; $x = 50/3$, $y = 25/3$    **5.** $p = 750$; $x = 0$, $y = 150$, $z = 0$    **7.** $p = 450$; $x = 10$, $y = 10$, $z = 10$

**9.** $p = 260/3$; $x = 50/3$, $y = 0$, $z = 70/3$, $w = 0$
**11.** $c = 80$; $x = 20/3$, $y = 20/3$
**13.** $c = 100$; $x = 0$, $y = 100$, $z = 0$
**15.** $c = 111$; $x = 1$, $y = 1$, $z = 1$
**17.** $c = 200$; $x = 200$, $y = 0$, $z = 0$, $w = 0$
**19.** $p = 136.75$; $x = 0$, $y = 25.25$, $z = 0$, $w = 15.25$
**21.** $c = 66.67$; $x = 0$, $y = 66.67$, $z = 0$
**23.** $c = -250$; $x = 0$, $y = 500$, $z = 500$,
$w = 1,500$   **25.** Plant 100 acres of tomatoes and no
other crops. This will give you a profit of $200,000.
(You will be using all 100 acres of your farm.)
**27.** 10 mailings to the East Coast, none to the
Midwest, 10 to the West Coast. Cost: $900. Another
solution resulting in the same cost is no mailings to
the East Coast, 15 to the Midwest, none to the West
Coast.   **29.** 10,000 quarts of orange juice and 2,000
quarts of orange concentrate   **31.** Sell 25,000 regional
music albums, 10,000 pop/rock music albums, and
5,000 tropical music albums per day for a maximum
revenue of $195,000.   **33.** One serving of cereal, one
serving of juice, and no dessert!   **35.** 15 bundles from
Nadir, 5 from Sonny, and none from Blunt. Cost:
$70,000. Another solution resulting in the same cost
is 10 bundles from Nadir, none from Sonny, and
10 from Blunt.   **37.** Mix 6 servings of Riboforce HP
and 10 servings of Creatine Transport for a cost of
$15.60.   **39. a.** Build 1 convention-style hotel,
4 vacation-style hotels, and 2 small motels. The total
cost will amount to $188 million.   **b.** Because 20% of
this is $37.6 million, you will still be covered by the
subsidy.   **41.** Tucson to Honolulu: 500 boards/week;
Tucson to Venice Beach: 120 boards/week; Toronto to
Honolulu: 0 boards/week; Toronto to Venice Beach:
410 boards/week. Minimum weekly cost is $9,700.
**43.** $2,500 from Congressional Integrity Bank,
$0 from Citizens' Trust, $7,500 from Checks R Us.
**45.** Fly 5 people from Chicago to LA, 15 from
Chicago to New York, 5 from Denver to LA, none
from Denver to New York at a total cost of $4,000.
**47.** Hire no more cardiologists, 12 rehabilitation
specialists, and 5 infectious disease specialists.
**49.** The solution $x = 0$, $y = 0$, ... , represented
by the initial tableau may not be feasible. In Phase I
we use pivoting to arrive at a basic solution that is
feasible.   **51.** The basic solution corresponding to
the initial tableau has all the unknowns equal to zero,
and this is not a feasible solution because it does not
satisfy the given inequality.   **53. (C)**   **55.** Answers
may vary. Examples are Exercises 1 and 2.
**57.** Answers may vary. A simple example is:
Maximize $p = x + y$ subject to $x + y \le 10$,
$x + y \ge 20$, $x \ge 0$, $y \ge 0$.

## Section 5.5

**1.** Minimize $c = 6s + 2t$ subject to $s - t \ge 2$,
$2s + t \ge 1$, $s \ge 0$, $t \ge 0$
**3.** Maximize $p = 100x + 50y$ subject to $x + 2y \le 2$,
$x + y \le 1$, $x \le 3$, $x \ge 0$, $y \ge 0$.   **5.** Minimize
$c = 3s + 4t + 5u + 6v$ subject to $s + u + v \ge 1$,
$s + t + v \ge 1$, $s + t + u \ge 1$, $t + u + v \ge 1$,
$s \ge 0$, $t \ge 0$, $u \ge 0$, $v \ge 0$.   **7.** Maximize
$p = 1,000x + 2,000y + 500z$ subject to $5x + z \le 1$,
$-x + z \le 3$, $y \le 1$, $x - y \le 0$, $x \ge 0$, $y \ge 0$, $z \ge 0$.
**9.** $c = 4$; $s = 2$, $t = 2$   **11.** $c = 80$; $s = 20/3$,
$t = 20/3$   **13.** $c = 1.8$; $s = 6$, $t = 2$
**15.** $c = 25$; $s = 5$, $t = 15$   **17.** $c = 30$; $s = 30$, $t = 0$,
$u = 0$   **19.** $c = 100$; $s = 0$, $t = 100$, $u = 0$
**21.** $c = 30$; $s = 10$, $t = 10$, $u = 10$
**23.** $R = [3/5 \quad 2/5]$, $C = [2/5 \quad 3/5 \quad 0]^T$, $e = 1/5$
**25.** $R = [1/4 \quad 0 \quad 3/4]$, $C = [1/2 \quad 0 \quad 1/2]^T$,
$e = 1/2$   **27.** $R = [0 \quad 3/11 \quad 3/11 \quad 5/11]$,
$C = [8/11 \quad 0 \quad 2/11 \quad 1/11]^T$, $e = 9/11$
**29.** 4 ounces each of fish and cornmeal, for a total cost
of 40¢ per can; 5/12¢ per gram of protein, 5/12¢ per
gram of fat.   **31.** 100 oz of grain and no chicken, for
a total cost of $1; 1/2¢ per gram of protein, 0¢ per
gram of fat.   **33.** One serving of cereal, one serving
of juice, and no dessert! for a total cost of 37¢; 1/6¢
per calorie and 17/120¢ per % U.S. RDA of vitamin C.
**35.** 10 mailings to the East coast, none to the Midwest,
10 to the West Coast. Cost: $900; 20¢ per Democrat
and 40¢ per Republican. OR 15 mailings to the
Midwest and no mailing to the coasts. Cost: $900;
20¢ per Democrat and 40¢ per Republican.
**37.** Gillian should use 480 sleep spells and 160 shock
spells, costing 360,000 pico-shirleys of energy OR
2,880/7 sleep spells and 1,440/7 shock spells.
**39.** T. N. Spend should spend about 73% of the days
in Littleville, 27% in Metropolis, and skip Urbantown.
T. L. Down should spend about 91% of the days in
Littleville, 9% in Metropolis, and skip Urbantown. The
expected outcome is that T. L. Down will lose about
227 votes per day of campaigning.   **41.** Each player
should show one finger with probability 1/2, two fingers with probability 1/3, and three fingers with probability 1/6. The expected outcome is that player A will
win 2/3 point per round, on average.   **43.** Write
moves as $(x, y)$, where $x$ represents the number of regi-
ments sent to the first location and $y$ represents the
number sent to the second location. Colonel Blotto
should play $(0, 4)$ with probability 4/9, $(2, 2)$ with
probability 1/9, and $(4, 0)$ with probability 4/9.
Captain Kije has several optimal strategies, one of
which is to play $(0, 3)$ with probability 1/30, $(1, 2)$
with probability 8/15, $(2, 1)$ with probability 16/45,

and (3, 0) with probability 7/90. The expected outcome is that Colonel Blotto will win 14/9 points on average. **45.** The dual of a standard minimization problem satisfying the nonnegative objective condition is a standard maximization problem, which can be solved using the standard simplex algorithm, thus avoiding the need to do Phase I. **47.** Answers will vary. An example is: Minimize $c = x - y$ subject to $x - y \geq 100$, $x + y \geq 200$, $x \geq 0$, $y \geq 0$. This problem can be solved using the techniques in Section 5.4. **49.** If the given problem is a standard minimization problem satisfying the nonnegative objective condition, its dual is a standard maximization problem, and so can be solved using a single-phase simplex method. Otherwise, dualizing may not save any labor, since the dual will not be a standard maximization problem.

## Chapter 5 Review

**1.**

Solution Set

$2x - 3y \leq 12$

Unbounded

**3.**

$3x + 2y = 30$

$x + 2y = 20$

Solution Set

Bounded; corner points: (0, 0), (0, 10), (5, 15/2), (10, 0)

**5.** $p = 21$; $x = 9$, $y = 3$   **7.** $c = 22$; $x = 8$, $y = 6$
**9.** $p = 45$;  $x = 0$, $y = 15$, $z = 15$
**11.** $p = 220$; $x = 20$, $y = 20$, $z = 60$
**13.** $c = 30$; $x = 30$, $y = 0$, $z = 0$
**15.** No solution; feasible region unbounded
**17.** $c = 50$; $x = 20$, $y = 10$, $z = 0$, $w = 20$, OR
$x = 30$, $y = 0$, $z = 0$, $w = 20$
**19.** $c = 60$; $x = 24$, $y = 12$ OR $x = 0$, $y = 60$
**21.** $c = 20$; $x = 0$, $y = 20$
**23.** $R = [1/2 \quad 1/2 \quad 0]$, $C = [0 \quad 1/3 \quad 2/3]^T$, $e = 0$
**25.** $R = [1/27 \quad 7/9 \quad 5/27]$,
$C = [8/27 \quad 5/27 \quad 14/27]^T$, $e = 8/27$
**27. (A)**   **29.** 35   **31.** 400 packages from each for a minimum cost of $52,000   **33. a.** (B), (D) **b.** 450 packages from Duffin House, 375 from Higgins Press for a minimum cost of $52,500   **35.** 220 shares of EEE and 20 shares of RRR. The minimum total risk index is 500.   **37.** 240 Sprinkles, 120 Storms, and no Hurricanes   **39.** Order 600 packages from Higgins and none from the others, for a total cost of $90,000.
**41. a.** Let $x = $ # science credits, $y = $ # fine arts credits, $z = $ # liberal arts credits, and $w = $ # math credits.

Minimize $C = 300x + 300y + 200z + 200w$ subject to: $x + y + z + w \geq 120$; $x - y \geq 0$; $-2x + w \leq 0$; $-y + 3z - 3w \leq 0$; $x \geq 0$, $y \geq 0$, $z \geq 0$, $w \geq 0$.
**b.** Billy-Sean should take the following combination: Sciences—24 credits, Fine Arts—no credits, Liberal Arts—48 credits, Mathematics—48 credits, for a total cost of $26,400.   **43.** Smallest cost is $20,000; New York to OHaganBooks.com: 600 packages, New York to FantasyBooks.com: 0 packages, Illinois to OHaganbooks.com: 0 packages, Illinois to FantasyBooks.com: 200 packages.
**45.** FantasyBooks.com should choose between "2 for 1" and "3 for 2" with probabilities 20% and 80%, respectively. OHaganBooks.com should choose between "3 for 1" and "Finite Math" with probabilities 60% and 40%, respectively. OHaganBooks.com expects to gain 12,000 customers from FantasyBooks.com.

# Chapter 6

## Section 6.1

**1.** $F = \{$spring, summer, fall, winter$\}$
**3.** $I = \{1, 2, 3, 4, 5, 6\}$   **5.** $A = \{1, 2, 3\}$
**7.** $B = \{2, 4, 6, 8\}$   **9. a.** $S = \{$(H, H), (H, T), (T, H), (T, T)$\}$ **b.** $S = \{$(H, H), (H, T), (T, T)$\}$
**11.** $S = \{$(1, 5), (2, 4), (3, 3), (4, 2), (5, 1)$\}$
**13.** $S = \{$(1, 5), (2, 4), (3, 3)$\}$   **15.** $S = \emptyset$
**17.**

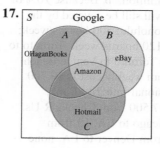

S

Google

A   B

OHaganBooks   eBay

Amazon

Hotmail

C

**19.**

S   OHaganBooks

A   B

eBay
Google

Hotmail
Amazon

C

**21.** $A$   **23.** $A$   **25.** {June, Janet, Jill, Justin, Jeffrey, Jello, Sally, Solly, Molly, Jolly}   **27.** {Jello}   **29.** Ø
**31.** {Jello}   **33.** {Janet, Justin, Jello, Sally, Solly,

Molly, Jolly}    **35.** {(small, triangle), (small, square), (medium, triangle), (medium, square), (large, triangle), (large, square)}    **37.** {(small, blue), (small, green), (medium, blue), (medium, green), (large, blue), (large, green)}

**39.**

|   | A | B | C |
|---|---|---|---|
| 1 |   | Triangle | Square |
| 2 | Blue | Blue Triangle | Blue Square |
| 3 | Green | Green Triangle | Green Square |

**41.**

|   | A | B | C |
|---|---|---|---|
| 1 |   | Blue | Green |
| 2 | Small | Small Blue | Small Green |
| 3 | Medium | Medium Blue | Medium Green |
| 4 | Large | Large Blue | Large Green |

**43.** $B \times A = \{1H, 1T, 2H, 2T, 3H, 3T, 4H, 4T, 5H, 5T, 6H, 6T\}$
**45.** $A \times A \times A = \{HHH, HHT, HTH, HTT, THH, THT, TTH, TTT\}$
**47.** {(1,1), (1,3), (1,5), (3,1), (3,3), (3,5), (5,1), (5,3), (5,5)}    **49.** Ø    **51.** {(1,1), (1,3), (1,5), (3,1), (3,3), (3,5), (5,1), (5,3), (5,5), (2,2), (2,4), (2,6), (4,2), (4,4), (4,6), (6,2), (6,4), (6,6)}    **61.** $A \cap B = \{$Acme, Crafts$\}$
**63.** $B \cup C = \{$Acme, Brothers, Crafts, Dion, Effigy, Global, Hilbert$\}$    **65.** $A' \cap C = \{$Dion, Hilbert$\}$
**67.** $A \cap B' \cap C' = \emptyset$
**69.** {2003, 2004, 2005, 2006} × {Sail Boats, Motor Boats, Yachts}

|   | A | B | C | D |
|---|---|---|---|---|
| 1 |   | **Sail Boats** | **Motor Boats** | **Yachts** |
| 2 | **2003** | (2003 Sail Boats) | (2003 Motor Boats) | (2003 Yachts) |
| 3 | **2004** | (2004 Sail Boats) | (2004 Motor Boats) | (2004 Yachts) |
| 4 | **2005** | (2005 Sail Boats) | (2005 Motor Boats) | (2005 Yachts) |
| 5 | **2006** | (2006 Sail Boats) | (2006 Motor Boats) | (2006 Yachts) |

**71.** $I \cup J$    **73.** (B)    **75.** Answers may vary. Let $A = \{1\}$, $B = \{2\}$, and $C = \{1, 2\}$. Then $(A \cap B) \cup C = \{1, 2\}$ but $A \cap (B \cup C) = \{1\}$. In general, $A \cap (B \cup C)$ must be a subset of $A$, but $(A \cap B) \cup C$ need not be; also, $(A \cap B) \cup C$ must contain $C$ as a subset, but $A \cap (B \cup C)$ need not.
**77.** A universal set is a set containing all "things" currently under consideration. When discussing sets of positive integers, the universe might be the set of all positive integers, or the set of all integers (positive, negative, and 0), or any other set containing the set of all positive integers.    **79.** $A$ is the set of suppliers who deliver components on time, $B$ is the set of suppliers

whose components are known to be of high quality, and $C$ is the set of suppliers who do not promptly replace defective components.    **81.** Let $A = \{$movies that are violent$\}$, $B = \{$movies that are shorter than two hours$\}$, $C = \{$movies that have a tragic ending$\}$, and $D = \{$movies that have an unexpected ending$\}$. The given sentence can be rewritten as "She prefers movies in $A' \cap B \cap (C \cup D)'$." It can also be rewritten as "She prefers movies in $A' \cap B \cap C' \cap D'$."
**83.** Removing the comma would cause the statement to be ambiguous, as it could then correspond to either WWII $\cup$ (Comix $\cap$ Aliens$'$) or to (WWII $\cup$ Comix) $\cap$ Aliens$'$. (See Exercise 76.)

## Section 6.2

**1.** 9    **3.** 7    **5.** 4    **7.** $n(A \cup B) = 7$, $n(A) + n(B) - n(A \cap B) = 4 + 5 - 2 = 7$
**9.** 4    **11.** 18    **13.** 72    **15.** 60    **17.** 20    **19.** 6    **21.** 9
**23.** 4    **25.** $n((A \cap B)') = 9$
$n(A') + n(B') - n((A \cup B)') = 6 + 7 - 4 = 9$

**27.**

**29.**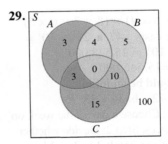

**31.** 115.4 million    **33.** 2    **35.** $C \cap N$ is the set of authors who are both successful and new. $C \cup N$ is the set of authors who are either successful or new (or both). $n(C) = 30$; $n(N) = 20$; $n(C \cup N) = 5$; $n(C \cup N) = 45$; $45 = 30 + 20 - 5$    **37.** $C \cap N'$ is the set of authors who are successful but not new. $n(C \cap N') = 25$    **39.** 31.25%; 83.33%
**41.** $M \cap C$; $n(M \cap C) = 1.1$ million
**43.** $B \cap T'$; $n(B \cap T') = 3.1$ million
**45.** $B \cap (W \cup M)$; $n(B \cap (W \cup M)) = 2.3$ million
**47.** $V \cap I'$; $n(V \cap I') = 15$    **49.** 80; The number of stocks that were either not pharmaceutical stocks, or were unchanged in value after a year (or both).

**51.** 3/8; the fraction of Internet stocks that increased in value   **53. a.** 931  **b.** 382
**55. a.**

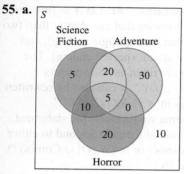

**b.** 37.5%   **57.** 17   **59.** $n(A) < n(B)$   **61.** The number of elements in the Cartesian product of two finite sets is the product of the number of elements in the two sets. **65.** When $A \cap B \neq \emptyset$   **67.** When $B \subseteq A$
**69.** $n(A \cup B \cup C) = n(A) + n(B) + n(C) - n(A \cap B) - n(B \cap C) - n(A \cap C) + n(A \cap B \cap C)$

## Section 6.3

**1.** 10   **3.** 30   **5.** 6 outcomes   **7.** 15 outcomes
**9.** 13 outcomes   **11.** 25 outcomes   **13.** 4   **15.** 93
**17.** 16   **19.** 30   **21.** 13   **23.** 18   **25.** 25,600
**27.** 3,381   **29. a.** 288  **b.** 288   **31.** 256   **33.** 10
**35.** 286   **37.** 4   **39. a.** 8,000,000  **b.** 30,000
**c.** 4,251,528   **41. a.** $4^3 = 64$  **b.** $4^n$  **c.** $4^{2.1 \times 10^{10}}$
**43. a.** $16^6 = 16,777,216$  **b.** $16^3 = 4,096$
**c.** $16^2 = 256$  **d.** 766
**45.** $(10 \times 9 \times 8 \times 7 \times 6 \times 5 \times 4) \times (8 \times 7 \times 6 \times 5)$
$= 1,016,064,000$ possible casts   **47. a.** $26^3 \times 10^3 = 17,576,000$ **b.** $26^2 \times 23 \times 10^3 = 15,548,000$
**c.** $15,548,000 - 3 \times 10^3 = 15,545,000$
**49. a.** 4   **b.** 4   **c.** There would be an infinite number of routes.   **51. a.** 72   **b.** 36   **53.** 96
**55. a.** 36 **b.** 37   **57.** Step 1: Choose a day of the week on which Jan. 1 will fall: 7 choices. Step 2: Decide whether or not it is a leap year; 2 choices. Total: $7 \times 2 = 14$ possible calendars.   **59.** 1,900   **61.** Step 1: Choose a position in the left-right direction; $m$ choices. Step 2: Choose a position in the front-back direction; $n$ choices. Step 3: Choose a position in the up-down direction; $r$ choices. Hence, there are $m \cdot n \cdot r$ possible outcomes.
**63.** 4   **65.** Cartesian product   **67.** The decision algorithm produces every pair of shirts twice, first in one order and then in the other.   **69.** Think of placing the five squares in a row of five empty slots. Step 1: Choose a slot for the blue square; five choices. Step 2: Choose a slot for the green square; four choices. Step 3: Choose the remaining three slots for the yellow squares; one choice. Hence, there are 20 possible five-square sequences.

## Section 6.4

**1.** 720   **3.** 56   **5.** 360   **7.** 15   **9.** 3   **11.** 45   **13.** 20
**15.** 4,950   **17.** 360   **19.** 35   **21.** 120   **23.** 120
**25.** 20   **27.** 60   **29.** 210   **31.** 7   **33.** 35   **35.** 24
**37.** 126   **39.** 196   **41.** 105

**43.** $\dfrac{C(30, 5) \times 5^{25}}{6^{30}} \approx 0.192$

**45.** $\dfrac{C(30, 15) \times 3^{15} \times 3^{15}}{6^{30}} \approx 0.144$

**47.** $C(11, 1)C(10, 4)C(6, 4)C(2, 2)$
**49.** $C(11, 2)C(9, 1)C(8, 1)C(7, 3)C(4, 1)$
$C(3, 1)C(2, 1)C(1, 1)$
**51.** $C(10, 2)C(8, 4)C(4, 1)C(3, 1)C(2, 1)C(1, 1)$
**53.** 24   **55.** $C(13, 2)C(4, 2)C(4, 2) \times 44 = 123,552$
**57.** $13 \times C(4, 2)C(12, 3) \times 4 \times 4 \times 4 = 1,098,240$
**59.** 10,200   **61. a.** 252  **b.** 20  **c.** 26   **63. a.** 300 **b.** 3
**c.** 1 in 100 or .01   **65. a.** 210 **b.** 91 **c.** No   **67. a.** 23!
**b.** 18! **c.** $19 \times 18!$   **69.** (A)   **71.** (D)   **73. a.** 9,880
**b.** 1,560 **c.** 11,480
**75. a.** $C(20, 2) = 190$  **b.** $C(n, 2)$
**77.** The multiplication principle; it can be used to solve all problems that use the formulas for permutations.
**79.** (A), (D)   **81.** A permutation. Changing the order in a list of driving instructions can result in a different outcome; for instance, "1. Turn left. 2. Drive one mile." and "1. Drive one mile. 2. Turn left." will take you to different locations.   **83.** Urge your friend not to focus on formulas, but instead learn to formulate decision algorithms and use the principles of counting.
**85.** It is ambiguous on the following point: Are the three students to play different characters, or are they to play a group of three, such as "three guards"? This should be made clear in the exercise.

## Chapter 6 Review

**1.** $N = \{-3, -2, -1\}$   **3.** $S = \{(1, 2), (1, 3), (1, 4),$
$(1, 5), (1, 6), (2, 1), (2, 3), (2, 4), (2, 5), (2, 6), (3, 1),$
$(3, 2), (3, 4), (3, 5), (3, 6), (4, 1), (4, 2), (4, 3), (4, 5),$
$(4, 6), (5, 1), (5, 2), (5, 3), (5, 4), (5, 6), (6, 1), (6, 2),$
$(6, 3), (6, 4), (6, 5)\}$   **5.** $A \cup B' = \{a, b, d\},$
$A \times B' = \{(a, a), (a, d), (b, a), (b, d)\}$   **7.** $A \times B$
**9.** $(P \cap E' \cap Q)'$ or $P' \cup E \cup Q$   **11.** $n(A \cup B) =$
$n(A) + n(B) - n(A \cap B), n(C') = n(S) - n(C); 100$
**13.** $n(A \times B) = n(A)n(B), n(A \cup B) = n(A) +$
$n(B) - n(A \cap B), n(A') = n(S) - n(A); 21$
**15.** $C(12, 1)C(4, 2)C(11, 3)C(4, 1)C(4, 1)C(4, 1)$
**17.** $C(4, 1)C(10, 1)$   **19.** 6   **21.** $C(4, 4)C(8, 1) = 8$
**23.** $C(3, 2)C(9, 3) + C(3, 3)C(9, 2) = 288$
**25.** The set of books that are either sci-fi or stored

in Texas (or both); $n(S \cup T) = 112,000$   **27.** The set of books that are either stored in California or not sci-fi; $n(C \cup S') = 175,000$   **29.** The romance books that are also horror books or stored in Texas; $n(R \cap (T \cup H)) = 20,000$   **31.** 1,000
**33.** FarmerBooks.com; 1,800   **35.** FarmerBooks.com; 5,400   **37.** $26 \times 26 \times 26 = 17,576$
**39.** $26 \times 25 \times 9 \times 10 = 58,500$   **41.** 60,000
**43.** 19,600

# Chapter 7

## Section 7.1

**1.** $S = \{\text{HH, HT, TH, TT}\}$; $E = \{\text{HH, HT, TH}\}$
**3.** $S = \{\text{HHH, HHT, HTH, HTT, THH, THT, TTH, TTT}\}$; $E = \{\text{HTT, THT, TTH, TTT}\}$

**5.** $S = \left\{ \begin{array}{l} (1, 1), (1, 2), (1, 3), (1, 4), (1, 5), (1, 6), \\ (2, 1), (2, 2), (2, 3), (2, 4), (2, 5), (2, 6), \\ (3, 1), (3, 2), (3, 3), (3, 4), (3, 5), (3, 6), \\ (4, 1), (4, 2), (4, 3), (4, 4), (4, 5), (4, 6), \\ (5, 1), (5, 2), (5, 3), (5, 4), (5, 5), (5, 6), \\ (6, 1), (6, 2), (6, 3), (6, 4), (6, 5), (6, 6) \end{array} \right\}$;

$E = \{(1, 4), (2, 3), (3, 2), (4, 1)\}$

**7.** $S = \left\{ \begin{array}{l} (1, 1), (1, 2), (1, 3), (1, 4), (1, 5), (1, 6), \\ (2, 2), (2, 3), (2, 4), (2, 5), (2, 6), \\ (3, 3), (3, 4), (3, 5), (3, 6), \\ (4, 4), (4, 5), (4, 6), \\ (5, 5), (5, 6), \\ (6, 6) \end{array} \right\}$;

$E = \{(1, 3), (2, 2)\}$
**9.** $S$ as in Exercise 7; $E = \{(2, 2), (2, 3), (2, 5), (3, 3),$ $(3, 5), (5, 5)\}$   **11.** $S = \{\text{m, o, z, a, r, t}\}$: $E = \{\text{o, a}\}$
**13.** $S = \{(\text{s, o}), (\text{s, r}), (\text{s, e}), (\text{o, s}), (\text{o, r}), (\text{o, e}), (\text{r, s}),$ $(\text{r, o}), (\text{r, e}), (\text{e, s}), (\text{e, o}), (\text{e, r})\}$; $E = \{(\text{o, s}), (\text{o, r}),$ $(\text{o, e}), (\text{e, s}), (\text{e, o}), (\text{e, r})\}$   **15.** $S = \{01, 02, 03, 04,$ $10, 12, 13, 14, 20, 21, 23, 24, 30, 31, 32, 34, 40, 41,$ $42, 43\}$; $E = \{10, 20, 21, 30, 31, 32, 40, 41, 42, 43\}$
**17.** $S = \{\text{domestic car, imported car, van, antique car,}$ $\text{antique truck}\}$; $E = \{\text{van, antique truck}\}$
**19. a.** All sets of 4 gummy candies chosen from the packet of 12   **b.** All sets of four gummy candies in which two are strawberry and two are black currant
**21. a.** All lists of 15 people chosen from 20
**b.** All lists of 15 people chosen from 20, in which Hillary Clinton occupies the eleventh position
**23.** $A \cap B$; $n(A \cap B) = 1$   **25.** $B'$; $n(B') = 33$
**27.** $B' \cap D'$; $n(B' \cap D') = 2$
**29.** $C \cup B$; $n(C \cup B) = 12$
**31.** $W \cap I$   **33.** $E \cup I'$   **35.** $I \cup (W \cap E')$
**37.** $(I \cup W) \cap E'$   **39.** 56; 4
**41.** $C(4, 1) C(2, 1) C(2, 1) = 16$

**43.** $E = \{\text{Pacific, Mountain, South Atlantic}\}$
**45.** $E \cup F$ is the event that you choose a region that saw a decrease in housing prices of 7% or more or is on the east coast. $E \cup F = \{\text{Pacific, Mountain, New England,}$ $\text{Middle Atlantic, South Atlantic}\}$. $E \cap F$ is the event that you choose a region that saw a decrease in housing prices of 7% or more and is on the east coast. $E \cap F = \{\text{South Atlantic}\}$.   **47. a.** Not mutually exclusive   **b.** Mutually exclusive   **49.** $S \cap N$ is the event that an author is successful and new. $S \cup N$ is the event that an author is either successful or new; $n(S \cap N) = 5$; $n(S \cup N) = 45$
**51.** $N$ and $E$   **53.** $S \cap N'$ is the event that an author is successful but not a new author. $n(S \cap N') = 25$
**55.** 31.25%; 83.33%   **57.** $V \cap I'$; $n(V \cap I') = 15$
**59.** 80; the number of stocks that were either not pharmaceutical stocks, or were unchanged in value after a year (or both)   **61.** $P$ and $E$, $P$ and $I$, $E$ and $I$, $N$ and $E$, $V$ and $N$, $V$ and $D$, $N$ and $D$   **63.** 3/8; the fraction of Internet stocks that increased in value
**65. a.** $E' \cap H$   **b.** $E \cup H$   **c.** $(E \cup G)' = E' \cap G'$
**67. a.** $\{9\}$   **b.** $\{6\}$   **69. a.** The dog's fight drive is weakest.   **b.** The dog's fight and flight drives are either both strongest or both weakest.   **c.** Either the dog's fight drive is strongest, or its flight drive is strongest.   **71.** $C(6, 4) = 15$; $C(1, 1)C(5, 3) = 10$
**73. a.** $n(S) = P(7, 3) = 210$   **b.** $E \cap F$ is the event that Celera wins and Electoral College is in second or third place. In other words, it is the set of all lists of three horses in which Celera is first and Electoral College is second or third. $n(E \cap F) = 10$   **75.** Subset of the sample space   **77.** $E$ and $F$ do not both occur
**79.** (B)   **81.** True; consider the following experiment: Select an element of the set $S$ at random.
**83.** Answers may vary. Cast a die and record the remainder when the number facing up is divided by 2.
**85.** Yes. For instance, $E = \{(2, 5), (5, 1)\}$ and $F = \{(4, 3)\}$ are two such events.

## Section 7.2

**1.** .4   **3.** .8   **5.** .6
**7.**

| Outcome | HH | HT | TH | TT |
|---|---|---|---|---|
| Rel. Frequency | .275 | .2375 | .3 | .1875 |

**9.** .575   **11.** The second coin *seems* slightly biased in favor of heads, as heads comes up approximately 58% of the time. On the other hand, it is conceivable that the coin is fair and that heads came up 58% of the time purely by chance. Deciding which conclusion is more reasonable requires some knowledge of inferential statistics.   **13.** Yes   **15.** No; relative frequencies cannot be negative.   **17.** Yes   **19.** Missing value: .3 **a.** .6
**b.** .4   **21.** Answers will vary.   **23.** Answers will vary.
**25. a.** .5   **b.** .17   **c.** .94

**27. a.**

| Mortgage Status | Current | Past Due | In Foreclosure | Repossessed |
|---|---|---|---|---|
| Rel. Frequency | .67 | .26 | .045 | .025 |

**b.** .33
**29. a.**

| Age | 0–14 | 15–29 | 30–64 | >64 |
|---|---|---|---|---|
| Rel. Frequency | .30 | .27 | .37 | .06 |

**b.** .36
**31. a.**

| Test Rating | 3 | 2 | 1 | 0 |
|---|---|---|---|---|
| Rel. Frequency | .1 | .4 | .4 | .1 |

**b.** .5    **33.** Dial-up: 1256, Cable Modem: 412,
DSL: 302, Other: 30
**35.**

| Outcome | Surge | Plunge | Steady |
|---|---|---|---|
| Rel. Frequency | .2 | .3 | .5 |

**37.** .25    **39.** .2    **41.** .7    **43.** 5/6    **45.** 5/16
**47.**

| Outcome | U | C | R |
|---|---|---|---|
| Rel. Frequency | .2 | .64 | .16 |

**49.**

| Conventional | No pesticide | Single pesticide | Multiple pesticide |
|---|---|---|---|
| Probability | .27 | .13 | .60 |
| Organic | No pesticide | Single pesticide | Multiple pesticide |
| Probability | .77 | .13 | .10 |

**51.** $P(\text{false negative}) = 10/400 = .025$,
$P(\text{false positive}) = 10/200 = .05$    **53.** Answers will
vary.    **55.** The fraction of times $E$ occurs    **57.** 101;
$fr(E)$ can be any number between 0 and 100 inclusive,
so the possible answers are $0/100 = 0$, $1/100 = .01$,
$2/100 = .02, \ldots, 99/100 = .99$, $100/100 = 1$.
**59.** Wrong. For a pair of fair dice, the probability of a
pair of matching numbers is 1/6, as Ruth says.
However, it is quite possible, although not very likely,
that if you cast a pair of fair dice 20 times, you will
never obtain a matching pair (in fact, there is approximately a 2.6% chance that this will happen). In
general, a nontrivial claim about probability can never
be absolutely validated or refuted experimentally. All
we can say is that the evidence suggests that the dice
are not fair.    **61.** For a (large) number of days, record
the temperature prediction for the next day and then
check the actual high temperature the next day.
Record whether the prediction was accurate (within,
say, 2°F of the actual temperature). The fraction of
times the prediction was accurate is the relative
frequency.

## Section 7.3

**1.** $P(e) = .2$ **a.** .9 **b.** .95 **c.** .1 **d.** .8
**3.** $P(E) = 1/4$  **5.** $P(E) = 1$  **7.** $P(E) = 3/4$
**9.** $P(E) = 3/4$  **11.** $P(E) = 1/2$  **13.** $P(E) = 1/9$
**15.** $P(E) = 0$  **17.** $P(E) = 1/4$
**19.** $1/12$; $\{(4, 4), (2, 3), (3, 2)\}$
**21.**

| Outcome | 1 | 2 | 3 | 4 | 5 | 6 |
|---|---|---|---|---|---|---|
| Probability | 1/9 | 2/9 | 1/9 | 2/9 | 1/9 | 2/9 |

$P(\{1, 2, 3\}) = 4/9$
**23.**

| Outcome | 1 | 2 | 3 | 4 |
|---|---|---|---|---|
| Probability | 8/15 | 4/15 | 2/15 | 1/15 |

**25.** .65  **27.** .1  **29.** .7  **31.** .4  **33.** .25  **35.** 1.0
**37.** .3  **39.** 1.0  **41.** No; $P(A \cup B)$ should be $\leq$
$P(A) + P(B)$.  **43.** Yes  **45.** No; $P(A \cup B)$
should be $\geq P(A)$.  **47. a.** .93 **b.** .33
**49.**

| Outcome | Hispanic or Latino | White (not Hispanic) | African American | Asian | Other |
|---|---|---|---|---|---|
| Probability | 0.48 | 0.29 | 0.08 | 0.08 | 0.07 |

$P(\text{Neither White nor Asian}) = .63$
**51. a.** $S = \{$Stock market success, Sold to other concern, Fail$\}$
**b.**

| Outcome | Stock Market Success | Sold to Other Concern | Fail |
|---|---|---|---|
| Probability | .2 | .3 | .5 |

**c.** .5
**53.**

| Outcome | SUV | Pickup | Passenger Car | Minivan |
|---|---|---|---|---|
| Probability | .25 | .15 | .50 | .10 |

**55.** $P(1) = 0$, $P(6) = 0$; $P(2) = P(3) = P(4) =$
$P(5) = 1/4 = .25$; $P(\text{odd}) = .5$    **57.** $P(1) = P(6) =$
$1/10$; $P(2) = P(3) = P(4) = P(5) = 1/5$, $P(\text{odd}) =$
$1/2$    **59.** $P(1, 1) = P(2, 2) = \ldots = P(6, 6) = 1/66$;
$P(1, 2) = \ldots = P(6, 5) = 1/33$, $P(\text{odd sum}) = 6/11$
**61.** $P(2) = 15/38$; $P(4) = 3/38$, $P(1) = P(3) =$
$P(5) = P(6) = 5/38$, $P(\text{odd}) = 15/38$    **63.** 5/6
**65.** .39  **67.** .25  **69.** .25  **71.** .00  **73.** .75  **75.** .45
**77.** .55  **79.** .01  **81.** .53  **83.** .44  **85.** 22%; 43%
**87.** All of them  **89.** 88.4%  **91.** Here is one possible
experiment: Roll a die and observe which of the
following outcomes occurs. Outcome $A$: 1 or 2 facing
up; $P(A) = 1/3$, Outcome $B$: 3 or 4 facing up;
$P(B) = 1/3$, Outcome $C$: 5 or 6 facing up;
$P(C) = 1/3$.    **93.** He is wrong. It is possible to have
a run of losses of any length. Tony may have grounds
to *suspect* that the game is rigged, but no proof.
**95.** They are mutually exclusive.    **97.** Wrong. For
example, the modeled probability of winning a state

lotto is small but nonzero. However, the vast majority of people who play lotto every day of their lives never win no matter how frequently they play, so the relative frequency is zero for these people.   **99.** When $A \cap B = \emptyset$ we have $P(A \cap B) = P(\emptyset) = 0$, so $P(A \cup B) = P(A) + P(B) - P(A \cap B) = P(A) + P(B) - 0 = P(A) + P(B)$.   **101.** Zero. According to the assumption, no matter how many thunderstorms occur, lightning cannot strike a given spot more than once, and so, after $n$ trials the relative frequency will never exceed $1/n$, and so will approach zero as the number of trials gets large. Since the modeled probability models the limit of relative frequencies as the number of trials gets large, it must therefore be zero.   **103.** $P(A \cup B \cup C) = P(A) + P(B) + P(C) - P(A \cap B) - P(A \cap C) - P(B \cap C) + P(A \cap B \cap C)$

## Section 7.4

**1.** $1/42$   **3.** $7/9$   **5.** $1/7$   **7.** $1/2$   **9.** $41/42$
**11.** $1/15$   **13.** $4/15$   **15.** $1/5$   **17.** $1/(2^8 \times 5^5 \times 5!)$
**19.** $.4226$   **21.** $.0475$   **23.** $.0020$   **25.** $1/27^{39}$
**27.** $1/7$   **29.** Probability of being a Big Winner $= 1/2,118,760 \approx .000000472$. Probability of being a Small-Fry Winner $= 225/2,118,760 \approx .000106194$. Probability of being either a Big Winner or a Small-Fry Winner $= 226/2,118,760 \approx .000106666$.
**31. a.** $C(600,300)/C(700,400)$   **b.** $C(699,399)/C(700,400)$ or $400/700$   **33.** $P(10, 3)/10^3 = 18/25 = .72$   **35.** $8!/8^8$   **37.** $1/8$   **39.** $1/8$
**41.** $37/10,000$   **43. a.** $90,720$   **b.** $25,200$
**c.** $25,200/90,720 = 25/90 \approx .28$   **45.** The four outcomes listed are not equally likely; for example, (red, blue) can occur in four ways. The methods of this section yield a probability for (red, blue) of $C(2, 2)/C(4, 2) = 1/6$   **47.** No. If we do not pay attention to order, the probability is $C(5, 2)/C(9, 2) = 10/36 = 5/18$. If we do pay attention to order, the probability is $P(5, 2)/P(9, 2) = 20/72 = 5/18$ again. The difference between permutations and combinations cancels when we compute the probability.
**49.** Answers will vary.

## Section 7.5

**1.** $.4$   **3.** $.08$   **5.** $.75$   **7.** $.2$   **9.** $.5$
**11.**

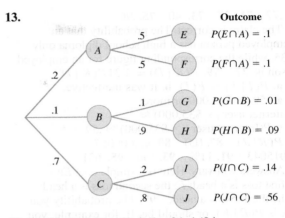

| | Outcome |
| --- | --- |
| | $P(A \cap C) = .15$ |
| | $P(A \cap D) = .15$ |
| | $P(B \cap C) = .14$ |
| | $P(B \cap D) = .56$ |

**13.**

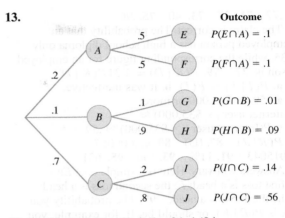

| | Outcome |
| --- | --- |
| | $P(E \cap A) = .1$ |
| | $P(F \cap A) = .1$ |
| | $P(G \cap B) = .01$ |
| | $P(H \cap B) = .09$ |
| | $P(I \cap C) = .14$ |
| | $P(J \cap C) = .56$ |

**15.** $1/10$   **17.** $1/5$   **19.** $2/9$   **21.** $1/84$
**23.** $5/21$   **25.** $24/175$   **27.** Mutually exclusive
**29.** Neither   **31.** $\dfrac{1}{2} \cdot \dfrac{1}{2} = \dfrac{1}{4}$ Independent
**33.** $\dfrac{5}{18} \cdot \dfrac{1}{2} \neq \dfrac{1}{9}$ Dependent   **35.** $\dfrac{25}{36} \cdot \dfrac{5}{18} \neq \dfrac{2}{9}$
Dependent   **37.** $(1/2)^{11} = 1/2,048$
**39.** $P(D) = .10$; $P(D|M) = .30$   **41.** $P(A|L) = .30$; $P(L|A) = .10$   **43.** $P(E|M) = .55$; $P(E|M') = .05$
**45.** $.8$   **47. a.** $.24$   **b.** $.059$   **49.** $.34$   **51.** Not independent; $P(\text{giving up} \mid \text{used Brand X}) = .1$ is larger than $P(\text{giving up}) = .05$.   **53.** $.00015$
**55.** $5/6$   **57.** $3/4$   **59.** $11/16$   **61.** $11/14$
**63.**

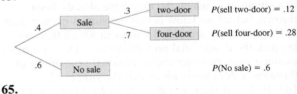

| | |
| --- | --- |
| | $P(\text{sell two-door}) = .12$ |
| | $P(\text{sell four-door}) = .28$ |
| | $P(\text{No sale}) = .6$ |

**65.**

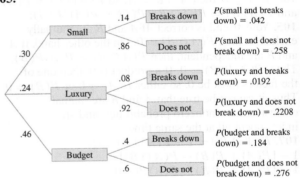

| | |
| --- | --- |
| | $P(\text{small and breaks down}) = .042$ |
| | $P(\text{small and does not break down}) = .258$ |
| | $P(\text{luxury and breaks down}) = .0192$ |
| | $P(\text{luxury and does not break down}) = .2208$ |
| | $P(\text{budget and breaks down}) = .184$ |
| | $P(\text{budget and does not break down}) = .276$ |

**67.**

| | Probability |
| --- | --- |
| Rain tomorrow | .25 |
| No rain tomorrow | .25 |
| Rain tomorrow | .25 |
| No rain tomorrow | .25 |

**69.** .77 **71.** .37 **73.** .40 **75.** .96
**77.** The claim is correct. The probability that an unemployed person has a high school diploma only is .35, while the corresponding figure for an employed person is .27. **79.** $P(K \mid D) = 1.31 P(K \mid D')$
**81. a.** $P(I \mid T) > P(I)$ **b.** It was ineffective.
**83. a.** .59 **b.** \$35,000 or more:
$P(\text{Internet user} \mid < \$35,000) \approx$
$.27 < P(\text{Internet user} \mid \geq \$35,000) \approx .59.$
**85.** $P(R \mid J)$ **87.** (D) **89. a.** .000057
**b.** .015043 **91.** 11% **93.** 106 **95.** .631
**97.** Answers will vary. Here is a simple one: $E$: the first toss is a head; $F$: the second toss is a head; $G$: the third toss is a head. **99.** The probability you seek is $P(E \mid F)$, or should be. If, for example, you were going to place a wager on whether $E$ occurs or not, it is crucial to know that the sample space has been reduced to $F$ (you know that $F$ did occur). If you base your wager on $P(E)$ rather than $P(E \mid F)$ you will misjudge your likelihood of winning.
**101.** You might explain that the conditional probability of $E$ is not the *a priori* probability of $E$, but it is the probability of $E$ in a hypothetical world in which the outcomes are restricted to be what is given. In the example she is citing, yes, the probability of throwing a double-six is 1/36, in the absence of any other knowledge. However by the "conditional probability" of throwing a double-six given that the sum is larger than 7, we might mean the probability of a double-six in a situation in which the dice have already been thrown, but all we know is that the sum is greater than 7. Because there are only 15 ways in which that can happen, the conditional probability is 1/15. For a more extreme case, consider the conditional probability of throwing a double-six given that the sum is 12.
**103.** If $A \subseteq B$ then $A \cap B = A$, so $P(A \cap B) = P(A)$ and $P(A \mid B) = P(A \cap B)/P(B) = P(A)/P(B)$.
**105.** Your friend is correct. If $A$ and $B$ are mutually exclusive, then $P(A \cap B) = 0$. On the other hand, if $A$ and $B$ are independent, then $P(A \cap B) = P(A)P(B)$. Thus, $P(A)P(B) = 0$. If a product is 0, then one of the factors must be 0, so either $P(A) = 0$ or $P(B) = 0$. Thus, it cannot be true that $A$ and $B$ are mutually exclusive, have nonzero probabilities, and are independent all at the same time.
**107.** $P(A' \cap B') = 1 - P(A \cup B) =$
$1 - [P(A) + P(B) - P(A \cap B)] =$
$1 - [P(A) + P(B) - P(A)P(B)] =$
$(1 - P(A))(1 - P(B)) = P(A')P(B')$

## Section 7.6

**1.** .4 **3.** .7887 **5.** .7442 **7.** .1163 **9.** 26.8%
**11.** .1724 **13.** 85% **15.** .73 **17.** .71 **19.** .165
**21.** 82% **23.** 12% **25. a.** 14.43%; **b.** 19.81% of single homeowners have pools. Thus they should go after the single homeowners. **27.** 9 **29.** .9310

**31.** 1.76% **33.** .20 **35.** $K$: child killed; $D$: Airbag deployed; $P(K \mid D) = 1.31 P(K \mid D')$; $P(D \mid K) = 1.31(.25)/[1.31(.25) + .75] = .30$
**37.** Show him an example like Example 1 of this section, where $P(T \mid A) = .95$ but $P(A \mid T) \approx .64$.
**39.** Suppose that the steroid test gives 10% false negatives and that only 0.1% of the tested population uses steroids. Then the probability that an athlete uses steroids, given that he or she has tested positive, is
$$\frac{(.9)(.001)}{(.9)(.001) + (.01)(.999)} \approx .083. \quad \textbf{41.} \text{ Draw a tree}$$
in which the first branching shows which of $R_1$, $R_2$, or $R_3$ occurred, and the second branching shows which of $T$ or $T'$ then occurred. There are three final outcomes in which $T$ occurs: $P(R_1 \cap T) = P(T \mid R_1)P(R_1)$, $P(R_2 \cap T) = P(T \mid R_2)P(R_2)$, and $P(R_3 \cap T) = P(T \mid R_3)P(R_3)$. In only one of these, the first, does $R_1$ occur. Thus,
$$P(R_1 \mid T) = \frac{P(R_1 \cap T)}{P(T)}$$
$$= \frac{P(T \mid R_1)P(R_1)}{P(T \mid R_1)P(R_1) + P(T \mid R_2)P(R_2) + P(T \mid R_3)P(R_3)}$$
**43.** The reasoning is flawed. Let $A$ be the event that a Democrat agrees with Safire's column, and let $F$ and $M$ be the events that a Democrat reader is female and male, respectively. Then A. D. makes the following argument:
$P(M \mid A) = 0.9$, $P(F \mid A') = 0.9$. Therefore, $P(A \mid M) = 0.9$.

According to Bayes' theorem, we cannot conclude anything about $P(A \mid M)$ unless we know $P(A)$, the percentage of all Democrats who agreed with Safire's column. This was not given.

## Section 7.7

**1.** $\begin{bmatrix} 1/4 & 3/4 \\ 1/2 & 1/2 \end{bmatrix}$ **3.** $\begin{bmatrix} 0 & 1 \\ 1/6 & 5/6 \end{bmatrix}$

**5.** $\begin{bmatrix} 0 & .8 & .2 \\ .9 & 0 & .1 \\ 0 & 0 & 1 \end{bmatrix}$ **7.** $\begin{bmatrix} 1 & 0 & 0 \\ 0 & 1 & 0 \\ 0 & 0 & 1 \end{bmatrix}$

**9.** $\begin{bmatrix} 1 & 0 & 0 & 0 & 0 & 0 \\ 2/3 & 0 & 1/3 & 0 & 0 & 0 \\ 0 & 2/3 & 0 & 1/3 & 0 & 0 \\ 0 & 0 & 2/3 & 0 & 1/3 & 0 \\ 0 & 0 & 0 & 2/3 & 0 & 1/3 \\ 0 & 0 & 0 & 0 & 1 & 0 \end{bmatrix}$

**11. a.** $\begin{bmatrix} .25 & .75 \\ 0 & 1 \end{bmatrix}$

**b.** distribution after one step: [.5 .5]; after two steps: [.25 .75]; after three steps: [.125 .875]

**13. a.** $\begin{bmatrix} .36 & .64 \\ .32 & .68 \end{bmatrix}$ **b.** distribution after one step:
[.3   .7]; after two steps: [.34   .66]; after three steps:
[.332   .668]

**15. a.** $\begin{bmatrix} 3/4 & 1/4 \\ 1/2 & 1/2 \end{bmatrix}$ **b.** distribution after one step:
[2/3   1/3]; after two steps: [2/3   1/3]; after three
steps: [2/3   1/3]

**17. a.** $\begin{bmatrix} 3/4 & 1/4 \\ 3/4 & 1/4 \end{bmatrix}$ **b.** distribution after one step:
[3/4   1/4]; after two steps: [3/4   1/4]; after three
steps: [3/4   1/4]

**19. a.** $\begin{bmatrix} .25 & .75 & 0 \\ 0 & 1 & 0 \\ 0 & .75 & .25 \end{bmatrix}$ **b.** distribution after one step:
[.5   .5   0]; after two steps: [.25   .75   0]; after three
steps: [.125   .875   0]

**21. a.** $\begin{bmatrix} 1/3 & 1/3 & 1/3 \\ 4/9 & 4/9 & 1/9 \\ 0 & 1 & 0 \end{bmatrix}$ **b.** distribution after one
step: [1/2   1/2   0]; after two steps: [1/6   2/3   1/6];
after three steps: [7/18   7/18   2/9]

**23. a.** $\begin{bmatrix} .01 & .99 & 0 \\ 0 & 1 & 0 \\ 0 & .36 & .64 \end{bmatrix}$ **b.** distribution after one step:
[.05   .55   .4]; after two steps: [.005   .675   .32]; after
three steps: [.0005   .7435   .256]
**25.** [2/3   1/3]   **27.** [3/7   4/7]   **29.** [2/5   3/5]
**31.** [2/5   1/5   2/5]   **33.** [1/3   1/2   1/6]
**35.** [0   1   0]   **37.** 1 = Sorey state, 2 = C&T;

$P = \begin{bmatrix} \frac{1}{2} & \frac{1}{2} \\ \frac{1}{4} & \frac{3}{4} \end{bmatrix}$; 3/8 = .375   **39. a.** 1 =

not checked in; 2 = checked in

$P = \begin{bmatrix} .4 & .6 \\ 0 & 1 \end{bmatrix}$, $P^2 = \begin{bmatrix} .16 & .84 \\ 0 & 1 \end{bmatrix}$,

$P^3 = \begin{bmatrix} .064 & .936 \\ 0 & 1 \end{bmatrix}$ **b.** 1 hour: .6; 2 hours: .84;

3 hours: .936 **c.** Eventually, all the roaches will have
checked in. **41.** 16.67% fall into the high-risk
category and 83.33% into the low-risk category.
**43. a.** 47/300 ≈ .156667 **b.** 3/13 **45.** 41.67% of the
customers will be in the Paid up category, 41.67% in
the 0–90 days category, and 16.67% in the bad debt
category.

**47. a.** $P = \begin{bmatrix} .729 & .271 & 0 \\ .075 & .84 & .085 \\ 0 & .304 & .696 \end{bmatrix}$ **b.** 2.3%

**c.** Affluent: 17.8%; middle class: 64.3%; poor: 18.0%
**49.** Long-term income distribution (top to bottom):
[8.43%, 41.57%, 41.57%, 8.43%]

**51. a.** $P = \begin{bmatrix} .981 & .005 & .005 & .009 \\ .01 & .972 & .006 & .012 \\ .01 & .006 & .973 & .011 \\ .008 & .006 & .005 & .981 \end{bmatrix}$

**b.** Verizon: 29.6%, Cingular: 19.4%, AT&T: 18.2%,
Other: 32.8% **c.** Verizon: 30.3%, Cingular: 18.6%,
AT&T: 17.6%, Other: 33.5%. The biggest gainers
are Verizon and Other, each gaining 0.6%.
**53.** [1/5   1/5   1/5   1/5   1/5]   **55.** Answers
will vary.   **57.** There are two assumptions made by
Markov systems that may not be true about the stock
market: the assumption that the transition probabilities
do not change over time, and the assumption that the
transition probability depends only on the current
state.   **59.** If $q$ is a row of $Q$, then by assumption
$qP = q$. Thus, when we multiply the rows of $Q$ by $P$,
nothing changes, and $QP = Q$.   **61.** At each step,
only 0.4 of the population in state 1 remains there,
and nothing enters from any other state. Thus, when
the first entry in the steady-state distribution vector is
multiplied by 0.4 it must remain unchanged. The only
number for which this true is 0.
**63.** An example is

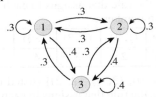

**65.** If $vP = v$ and $wP = w$, then $\frac{1}{2}(v + w)P =$
$\frac{1}{2}vP + \frac{1}{2}wP = \frac{1}{2}v + \frac{1}{2}w = \frac{1}{2}(v + w)$. Further, if the
entries of $v$ and $w$ add up to 1, then so do the entries
of $(v + w)/2$.

## Chapter 7 Review

**1.** $n(S) = 8$, $E = $ {HHT, HTH, HTT, THH, THT,
TTH, TTT}, $P(E) = 7/8$   **3.** $n(S) = 36$; $E =$
{(1, 6), (2, 5), (3, 4), (4, 3), (5, 2), (6, 1)}; $P(E) = 1/6$
**5.** $n(S) = 6$; $E = $ {2}; $P(E) = 1/8$   **7.** .76   **9.** .25
**11.** .5   **13.** 7/15   **15.** 8/792   **17.** 48/792
**19.** 288/792   **21.** $C(8, 5)/C(52, 5)$
**23.** $C(4, 3)C(1, 1)C(3, 1)/C(52, 5)$
**25.** $C(9, 1)C(8, 1)C(4, 3)C(4, 2)/C(52, 5)$   **27.** 1/5;
dependent   **29.** 1/6; independent   **31.** 1; dependent

**33.** $P = \begin{bmatrix} 1/2 & 1/2 \\ 1/4 & 3/4 \end{bmatrix}$   **35.** Brand A: 65/192 ≈ .339,

Brand B: 127/192 ≈ .661   **37.** 14/25   **39.** 15/94
**41.** 79/167   **43.** 98%   **45.** 6.9%   **47.** .931
**49.** $P(H \cap C)$, since $P(H|C) > P(H)$ gives
$P(H \cap C) > P(H)P(C)$   **51.** .246   **53.** 91%

**55.** $[.2 \quad .4 \quad .4] \begin{bmatrix} 0.8 & 0.1 & 0.1 \\ 0.4 & 0.6 & 0 \\ 0.2 & 0 & 0.8 \end{bmatrix} = [.4 \quad .26 \quad .34]$,

so 40% for OHaganBooks.com, 26% for JungleBooks.com, and 34% for FarmerBooks.com
**57.** Here are three: (1) it is possible for someone to be a customer at two different enterprises; (2) some customers may stop using all three of the companies; and (3) new customers can enter the field.

# Chapter 8

## Section 8.1

**1.** Finite; $\{2, 3, \ldots, 12\}$    **3.** Discrete infinite; $\{0, 1, -1, 2, -2, \ldots\}$ (negative profits indicate loss)
**5.** Continuous; $X$ can assume any value between 0 and 60.    **7.** Finite; $\{0, 1, 2, \ldots, 10\}$    **9.** Discrete infinite $\{k/1, k/4, k/9, k/16, \ldots\}$
**11. a.** $S = \{HH, HT, TH, TT\}$
**b.** $X$ is the rule that assigns to each outcome the number of tails.
**c.**

| Outcome | HH | HT | TH | TT |
|---------|----|----|----|----|
| Value of $X$ | 0 | 1 | 1 | 2 |

**13. a.** $S = \{(1, 1), (1, 2), \ldots, (1, 6), (2, 1), (2, 2), \ldots, (6, 6)\}$ **b.** $X$ is the rule that assigns to each outcome the sum of the two numbers.
**c.**

| Outcome | (1, 1) | (1, 2) | (1, 3) | ... | (6, 6) |
|---------|--------|--------|--------|-----|--------|
| Value of $X$ | 2 | 3 | 4 | ... | 12 |

**15. a.** $S = \{(4, 0), (3, 1), (2, 2)\}$ (listed in order (red, green)) **b.** $X$ is the rule that assigns to each outcome the number of red marbles.
**c.**

| Outcome | (4, 0) | (3, 1) | (2, 2) |
|---------|--------|--------|--------|
| Value of $X$ | 4 | 3 | 2 |

**17. a.** $S =$ the set of students in the study group.
**b.** $X$ is the rule that assigns to each student his or her final exam score. **c.** The values of $X$, in the order given, are 89%, 85%, 95%, 63%, 92%, 80%.
**19. a.** $P(X = 8) = P(X = 6) = .3$ **b.** .7; .5
**21.**

| $x$ | 1 | 2 | 3 | 4 | 5 | 6 |
|-----|---|---|---|---|---|---|
| $P(X = x)$ | $\frac{1}{6}$ | $\frac{1}{6}$ | $\frac{1}{6}$ | $\frac{1}{6}$ | $\frac{1}{6}$ | $\frac{1}{6}$ |

$P(X < 5) = 2/3$

**23.**

| $x$ | 0 | 1 | 4 | 9 |
|-----|---|---|---|---|
| $P(X = x)$ | $\frac{1}{8}$ | $\frac{3}{8}$ | $\frac{3}{8}$ | $\frac{1}{8}$ |

$P(1 \le X \le 9) = 7/8$

**25.**

| $x$ | 2 | 3 | 4 | 5 | 6 | 7 | 8 | 9 | 10 | 11 | 12 |
|-----|---|---|---|---|---|---|---|---|----|----|----|
| $P(X = x)$ | $\frac{1}{36}$ | $\frac{2}{36}$ | $\frac{3}{36}$ | $\frac{4}{36}$ | $\frac{5}{36}$ | $\frac{6}{36}$ | $\frac{5}{36}$ | $\frac{4}{36}$ | $\frac{3}{36}$ | $\frac{2}{36}$ | $\frac{1}{36}$ |

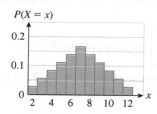

$P(X \ne 7) = 5/6$

**27.**

| $x$ | 1 | 2 | 3 | 4 | 5 | 6 |
|-----|---|---|---|---|---|---|
| $P(X = x)$ | $\frac{1}{36}$ | $\frac{3}{36}$ | $\frac{5}{36}$ | $\frac{7}{36}$ | $\frac{9}{36}$ | $\frac{11}{36}$ |

$P(X \le 3) = 1/4$

**29. a.**

| $x$ | 10,000 | 30,000 | 50,000 | 70,000 | 90,000 |
|-----|--------|--------|--------|--------|--------|
| $P(X = x)$ | .24 | .29 | .18 | .17 | .12 |

**b.** .29 Histogram:

**31.** The random variable is $X =$ age of a resident in Mexico.

| $x$ | 7 | 22 | 47 | 70 |
|---|---|---|---|---|
| $P(X = x)$ | .300 | .272 | .368 | .060 |

**33. a.** 2,000, 3,000, 4,000, 5,000, 6,000, 7,000, 8,000 (7,000 is optional)
**b.**

| $x$ | 2,000 | 3,000 | 4,000 | 5,000 | 6,000 | 7,000 | 8,000 |
|---|---|---|---|---|---|---|---|
| Freq. | 2 | 1 | 1 | 1 | 2 | 0 | 3 |
| $P(X = x)$ | .2 | .1 | .1 | .1 | .2 | 0 | .3 |

**c.** $P(X \le 5,000) = .5$  **35. a.** $-700, -600, -500,$ $-400, -300, -200, -100, 0, 100, 200, 300, 400, 500,$ $600, 700, 800, 900$ $(-600, 0, 100, 300, 500, 600, 700,$ and 800 are optional.)
**b.**

| $x$ | $-700$ | $-600$ | $-500$ | $-400$ | $-300$ | $-200$ | $-100$ | 0 |
|---|---|---|---|---|---|---|---|---|
| Freq. | 2 | 0 | 2 | 1 | 1 | 4 | 4 | 0 |
| $P(X = x)$ | .1 | 0 | .1 | .05 | .05 | .2 | .2 | 0 |

| $x$ | 100 | 200 | 300 | 400 | 500 | 600 | 700 | 800 | 900 |
|---|---|---|---|---|---|---|---|---|---|
| Freq. | 0 | 2 | 0 | 2 | 0 | 0 | 0 | 0 | 2 |
| $P(X = x)$ | 0 | .1 | 0 | .1 | 0 | 0 | 0 | 0 | .1 |

**c.** .3
**37.**

| Class | $1.1 - 2.0$ | $2.1 - 3.0$ | $3.1 - 4.0$ |
|---|---|---|---|
| Freq. | 4 | 7 | 9 |

| $x$ | 1.5 | 2.5 | 3.5 |
|---|---|---|---|
| $P(X = x)$ | .20 | .35 | .45 |

**39.** 95.5%
**41.**

| $x$ | 3 | 2 | 1 | 0 |
|---|---|---|---|---|
| $P(X = x)$ | .0625 | .6875 | .125 | .125 |

**43.** .75 The relative frequency that a randomly selected small car is rated Good or Acceptable is .75.
**45.** $P(Y \ge 2) = .50$, $P(Z \ge 2) \approx .53$, suggesting that medium SUVs are safer than small SUVs in frontal crashes  **47.** Small cars  **49.** .375
**51.**

| $x$ | 1 | 2 | 3 | 4 |
|---|---|---|---|---|
| $P(X = x)$ | $\dfrac{4}{35}$ | $\dfrac{18}{35}$ | $\dfrac{12}{35}$ | $\dfrac{1}{35}$ |

$P(X \ge 2) = 31/35 \approx .886$
**53.** Answers will vary.  **55.** No; for instance, if $X$ is the number of times you must toss a coin until heads comes up, then $X$ is infinite but not continuous.
**57.** By measuring the values of $X$ for a large number of outcomes, and then using the estimated probability

(relative frequency)  **59.** Here is an example: Let $X$ be the number of days a diligent student waits before beginning to study for an exam scheduled in 10 days' time.  **61.** The bars should be 1 unit wide, so that their height is numerically equal to their area.  **63.** Answers may vary. If we are interested in exact page counts, then the number of possible values is very large and the values are (relatively speaking) close together, so using a continuous random variable might be advantageous. In general, the finer and more numerous the measurement classes, the more likely it becomes that a continuous random variable could be advantageous.

## Section 8.2

**1.** .0729  **3.** .59049  **5.** .00001  **7.** .99144  **9.** .00856
**11.** .27648  **13.** .54432  **15.** .04096  **17.** .77414
**19.** $P(X = x)$

**21.** $P(X = x)$

$P(X \le 2) = .8889$
**23.** .2637  **25.** .8926  **27.** .875  **29. a.** .0081
**b.** .08146  **31.** .41  **33. a.** .0509  **b.** Probability distribution (entries rounded to four decimal places):

| $x$ | 0 | 1 | 2 | 3 | 4 |
|---|---|---|---|---|---|
| $P(X = x)$ | .0643 | .2030 | .2885 | .2429 | .1343 |
| $x$ | 5 | 6 | 7 | 8 | 9 | 10 |
| $P(X = x)$ | .0509 | .0134 | .0024 | .0003 | .0000 | .0000 |

**c.** 2  **35.** .000298  **37.** .8321  **39. a.** 21  **b.** 20
**c.** The graph for $n = 50$ trials is more widely distributed than the graph for $n = 20$.  **41.** 69 trials
**43.** $.562 \times 10^{-5}$  **45.** .0266; because there is only a 2.66% chance of detecting the disease in a given year, the government's claim seems dubious.  **47.** No; in a sequence of Bernoulli trials, the occurrence of one success does not affect the probability of success on the next attempt.  **49.** No; if life is a sequence of Bernoulli trials, then the occurrence of one misfortune ("success") does not affect the probability of a misfortune on the next trial. Hence, misfortunes may very

well not "occur in threes."  **51.** Think of performing the experiment as a Bernoulli trial with "success" being the occurrence of E. Performing the experiment n times independently in succession would then be a sequence of n Bernoulli trials.  **53.** The probability of selecting a red marble changes after each selection, as the number of marbles left in the bag decreases. This violates the requirement that, in a sequence of Bernoulli trials, the probability of "success" does not change.

## Section 8.3

**1.** $\bar{x} = 6$, median $= 5$, mode $= 5$
**3.** $\bar{x} = 3$, median $= 3.5$, mode $= -1$
**5.** $\bar{x} = -0.1875$, median $= 0.875$, every value is a mode  **7.** $\bar{x} = 0.2$, median $= -0.1$, mode $= -0.1$
**9.** Answers may vary. Two examples are: 0, 0, 0, 0, 0, 6 and 0, 0, 0, 1, 2, 3  **11.** 0.9  **13.** 21  **15.** $-0.1$
**17.** 3.5  **19.** 1  **21.** 4.472  **23.** 2.667  **25.** 2
**27.** 0.385  **29.** $\bar{x} = -150$, $m = -150$; $-150$
**31.** $\bar{x} = 1{,}708.4$, median $= 1{,}704$; mode $= 1{,}704$;
Over the 10-business-day period sampled, the price of gold averaged $1,708.4 per ounce. It was above $1,704 as many times as it was below that price, and stood at $1,704 per ounce more often than at any other price.
**33. a.** 6.5; there were, on average, 6.5 checkout lanes in each supermarket that was surveyed.
**b.** $P(X < \mu) = .42$; $P(X > \mu) = .58$, and is thus larger. Most supermarkets have more than the average number of checkout lanes.
**35.**

| X | 5 | 10 | 15 | 20 | 25 | 35 |
|---|---|----|----|----|----|----|
| P(X) | .17 | .33 | .21 | .19 | .03 | .07 |

$E(X) = 14.3$;

the average age of a school goer in 1998 was 14.3.
**37.** 29.6  **39.** $43,000
**41.**

| x | 3 | 2 | 1 | 0 |
|---|---|---|---|---|
| P(X = x) | .0625 | .6875 | .125 | .125 |

$E(X) = 1.6875$

| y | 3 | 2 | 1 | 0 |
|---|---|---|---|---|
| P(Y = y) | .1 | .4 | .4 | .1 |

$E(Y) = 1.5$;

small cars
**43.** Large cars  **45.** Expect to lose 5.3¢.  **47.** 25.2 students  **49. a.** Two defective airbags **b.** 120 airbags
**51.**

| x | 1 | 2 | 3 | 4 |
|---|---|---|---|---|
| P(X = x) | $\frac{4}{35}$ | $\frac{18}{35}$ | $\frac{12}{35}$ | $\frac{1}{35}$ |

$E(X) = 16/7 \approx 2.2857$ tents

**53.** FastForward: 3.97%; SolidState: 5.51%; SolidState gives the higher expected return.  **55.** A loss of $29,390
**57.** (A)  **59.** He is wrong; for example, the collection 0, 0, 300 has mean 100 and median 0.  **61.** No. The expected number of times you will hit the dart-board is the average number of times you will hit the bull's eye per 50 shots; the average of a set of whole numbers need not be a whole number.  **63.** Wrong. It might be the case that only a small fraction of people in the class scored better than you but received exceptionally high scores that raised the class average. Suppose, for instance, that there are 10 people in the class. Four received 100%, you received 80%, and the rest received 70%. Then the class average is 83%, 5 people have lower scores than you, but only 4 have higher scores.  **65.** No; the mean of a very large sample is only an *estimate* of the population mean. The means of larger and larger samples *approach* the population mean as the sample size increases.
**67.** Wrong. The statement attributed to President Bush asserts that the mean tax refund would be $1,000, whereas the statements referred to as "The Truth" suggest that the *median* tax refund would be close to $100 [and that the 31st percentile would be zero].
**69.** Select a U.S. household at random, and let X be the income of that household. The expected value of X is then the population mean of all U.S. household incomes.

## Section 8.4

**1.** $s^2 = 29$; $s = 5.39$  **3.** $s^2 = 12.4$; $s = 3.52$
**5.** $s^2 = 6.64$; $s = 2.58$  **7.** $s^2 = 13.01$; $s = 3.61$
**9.** 1.04  **11.** 9.43  **13.** 3.27  **15.** Expected value $= 3.5$, variance $= 2.918$, standard deviation $= 1.71$
**17.** Expected value $= 1$, variance $= 0.5$, standard deviation $= 0.71$  **19.** Expected value $= 4.47$, variance $= 1.97$, standard deviation $= 1.40$
**21.** Expected value $= 2.67$, variance $= 0.36$, standard deviation $= 0.60$  **23.** Expected value $= 2$, variance $= 1.8$, standard deviation $= 1.34$
**25. a.** $\bar{x} = 3$, $s = 3.54$ **b.** [0, 6.54] We must assume that the population distribution is bell shaped and symmetric.  **27. a.** $\bar{x} = 5.0$, $s = 0.6$ **b.** 3.8, 6.2
**29. a.** $\bar{x} = -150$, $s \approx 495$ **b.** 645, 20%  **31. a.** 2.18
**b.** [11.22, 24.28] **c.** 100%; Empirical rule
**33.** $\mu = 1.5$, $\sigma = 1.43$; 100%  **35.** $\mu = 43$, $\sigma = 27$; $54,000 **37. a.** $\mu \approx 30.2$ yrs. old, $\sigma = 11.78$ years
**b.** 18–42  **39.** At most 6.25%  **41.** At most; 12.5%
**43. a.** $\mu = 25.2$, $\sigma = 3.05$ **b.** 31  **45. a.** $\mu = 780$, $\sigma \approx 13.1$ **b.** 754, 806  **47. a.** $\mu = 6.5$, $\sigma^2 = 4.0$, $\sigma = 2.0$ **b.** [2.5, 10.5]; three checkout lanes
**49.** $10,700 or less  **51.** $65,300 or more  **53.** U.S.
**55.** U.S.  **57.** 16%  **59.** 0–$76,000  **61.** 2000 data: $\mu = 12.56\%$, $\sigma \approx 1.8885\%$; 2010 data: $\mu = 13.30\%$,

$\sigma \approx 1.6643\%$   **63.** (B)   **65.** 72%;. The empirical rule predicts 68% The associated probability distribution is roughly bell shaped but not symmetric.   **67.** 94%. Chebyshev's rule is valid, since it predicts that *at least* 75% of the scores are in this range.   **69.** (B), (D)   **71.** (B), (E)   **73.** The sample standard deviation is bigger; the formula for sample standard deviation involves division by the smaller term $n-1$ instead of $n$, which makes the resulting number larger.   **75.** The grades in the first class were clustered fairly close to 75. By Chebyshev's inequality, at least 88% of the class had grades in the range 60–90. On the other hand, the grades in the second class were widely dispersed. The second class had a much wider spread of ability than did the first class.   **77.** The variable must take on only the value 10, with probability 1.   **79.** $(y-x)/2$

### Section 8.5

**1.** .1915   **3.** .5222   **5.** .6710   **7.** .2417   **9.** .8664
**11.** .8621   **13.** .2286   **15.** .3830   **17.** .5028   **19.** 108.4
**21.** 117.5   **23.** .35   **25.** .05   **27.** .3830   **29.** .8185
**31.** 26%   **33.** 33,100,000   **35.** 665   **37.** 0
**39.** About 6,680   **41.** 28%   **43.** 5%   **45.** The U.S.
**47.** Wechsler. As this test has a smaller standard deviation, a greater percentage of scores fall within 20 points of the mean.   **49.** This is surprising, because the time between failures was more than five standard deviations away from the mean, which happens with an extremely small probability.   **51.** .6103   **53.** .6103 × .5832 ≈ .3559
**55.** .6255   **57.** .7881   **59.** .9049   **61.** .2912
**63.** Probability that a person will say Goode = .54. Probability that Goode polls more than 52% ≈ .8925.
**65.** 23.4   **67.** When the distribution is normal
**69.** Neither. They are equal.   **71.** $1/(b-a)$
**73.** A normal distribution with standard deviation 0.5, because it is narrower near the mean but must enclose the same amount of area as the standard curve, and so it must be higher.

### Chapter 8 Review

**1.**

| $x$ | 0 | 1 | 2 |
|---|---|---|---|
| $P(X = x)$ | 1/4 | 1/2 | 1/4 |

**3.**

| $x$ | 2 | 3 | 4 | 5 | 6 | 7 | 8 |
|---|---|---|---|---|---|---|---|
| $P(X = x)$ | 1/16 | 2/16 | 3/16 | 4/16 | 3/16 | 2/16 | 1/16 |

**5.**

| $x$ | 0 | 1 | 2 | 3 |
|---|---|---|---|---|
| $P(X = x)$ | .2071 | .4439 | .2908 | .0582 |

**7.** $\bar{x} = 2, m = 2, s \approx 2.7386$   **9.** Two examples are: 0, 0, 0, 4 and $-1, -1, 1, 5$   **11.** An example is $-1, -1, -1, 1, 1, 1, 1$   **13.** .4165   **15.** .9267   **17.** .0067
**19.** .7330   **21.** $\mu = 1.5, \sigma = 0.8660; 2$
**23.**

| $x$ | $-3$ | $-2$ | $-1$ | 0 | 1 | 2 | 3 |
|---|---|---|---|---|---|---|---|
| $P(X = x)$ | 1/16 | 2/16 | 3/16 | 4/16 | 3/16 | 2/16 | 1/16 |

$\mu = 0, \sigma = 1.5811$; within 1.3 standard deviations of the mean   **25.** [49.4, 150.6]   **27.** 260   **29.** .4332
**31.** .0358   **33.** .3085   **35.** $12.15   **37.** $27,210
**39.** False; let $X$ = price and $Y$ = weekly sales. Then weekly Revenue = $XY$. However, $27,210 \neq 12.15 \times 2,620$. In other words, $E(XY) \neq E(X)E(Y)$.
**41. a.**

| $x$ | 2 | 4 | 6 | 8 | 10 |
|---|---|---|---|---|---|
| $P(X = x)$ | .25 | .35 | .15 | .15 | .10 |

$\mu = 5, \sigma = 2.5690$  **b.** Between 2.431 and 7.569 orders per million residents; the empirical rule does not apply because the distribution is not symmetric.
**c.** (A)   **43.** .190   **45.** .060   **47.** 0.5   **49.** .284
**51.** .108   **53.** Using normal distribution table: 407,000 people; more accurate answer: 423,000 people
**55.** 148

# Chapter 9

## Section 9.1

**1.** Vertex: $(-3/2, -1/4)$,
y-intercept: 2,
x-intercepts: $-2, -1$

**3.** Vertex: $(2,0)$,
y-intercept: $-4$,
x-intercept: 2

**5.** Vertex: $(-20, 900)$,
y-intercept: 500,
x-intercepts: $-50, 10$

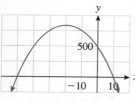

**7.** Vertex: $(-1/2, -5/4)$,
y-intercept: $-1$,
x-intercepts: $-1/2 \pm \sqrt{5}/2$

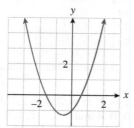

**9.** Vertex: $(0, 1)$,
y-intercept: 1,
no x-intercepts

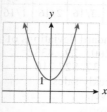

**11.** $R = -4p^2 + 100p$;
Maximum revenue
when $p = \$12.50$

**13.** $R = -2p^2 + 400p$; Maximum revenue when $p = \$100$

**15.** $y = -0.7955x^2 + 4.4591x - 1.6000$
**17.** $y = -1.1667x^2 - 6.1667x - 3.0000$
**19. a.** Positive because the data suggest a curve that is concave up. **b.** (C) **c.** 1998. Extrapolating in the positive direction leads one to predict more and more steeply rising military expenditure, which may or may not

occur; extrapolating in the negative direction predicts continually more an more steeply rising military expenditure as we go back in time, contradicting history. **21.** 2003; about 1.6 million barrels/day **23.** 2006; About \$21 billion; No, as the model predicts net income dropping without bound. **25.** Maximum revenue when $p = \$140$, $R = \$9,800$
**27.** Maximum revenue with 70 houses, $R = \$9,800,000$
**29. a.** $q = -4,500p + 41,500$ **b.** \$4.61 for a daily revenue of \$95,680.55 **c.** No
**31. a.** $q = -560x + 1,400$; $R = -560x^2 + 1,400x$
**b.** $P = -560x^2 + 1,400x - 30$; $x = \$1.25$;
$P = \$845$ per month **33.** $C = -200x + 620$;
$P = -400x^2 + 1,400x - 620$; $x = \$1.75$ per log-on;
$P = \$605$ per month **35. a.** $q = -10p + 400$
**b.** $R = -10p^2 + 400p$ **c.** $C = -30p + 4,200$
**d.** $P = -10p^2 + 430p - 4,200$; $p = \$21.50$
**37.** $f(t) = 6.25t^2 - 100t + 1,200$; \$1,425 billion, which is \$25 billion lower than the actual value.
**39. a.** $S(t) = 3.0t^2 - 37t + 120$;
Graph:

**b.** 30 million units. If we look at more historical data, we see that iPod sales were highly seasonal, with sales in the last quarter of each year being significantly higher than in other quarters. The model's figure of 30 million units for the first quarter of 2011 is therefore probably much too high. **41.** The graph is a straight line.
**43.** (C) **45.** Positive; the x coordinate of the vertex is negative, so $-b/(2a)$ must be negative. Because $a$ is positive (the parabola is concave up), this means that $b$ must also be positive to make $-b/(2a)$ negative. **47.** The x-coordinate of the vertex represents the unit price that leads to the maximum revenue, the y-coordinate of the vertex gives the maximum possible revenue, the x-intercepts give the unit prices that result in zero revenue, and the y-intercept gives the revenue resulting from zero unit price (which is obviously zero).
**49.** Graph the data to see whether the points suggest a curve rather than a straight line. It the curve suggested by the graph is concave up or concave down, then a quadratic model would be a likely candidate.
**51.** No; the graph of a quadratic function is a parabola. In the case of a concave-up parabola, the curve would unrealistically predict sales increasing without bound in

the future. In the case of a concave-down parabola, the curve would predict "negative" sales from some point on.  **53.** If $q = mp + b$ (with $m < 0$), then the revenue is given by $R = pq = mp^2 + bp$. This is the equation of a parabola with $a = m < 0$, and so is concave down. Thus, the vertex is the highest point on the parabola, showing that there is a single highest value for $R$, namely, the $y$-coordinate of the vertex.  **55.** Since $R = pq$, the demand must be given by

$$q = \frac{R}{p} = \frac{-50p^2 + 60p}{p} = -50p + 60.$$

## Section 9.2

**1.** `4^x`

| $x$ | −3 | −2 | −1 | 0 | 1 | 2 | 3 |
|---|---|---|---|---|---|---|---|
| $f(x)$ | $\frac{1}{64}$ | $\frac{1}{16}$ | $\frac{1}{4}$ | 1 | 4 | 16 | 64 |

**3.** `3^(-x)`

| $x$ | −3 | −2 | −1 | 0 | 1 | 2 | 3 |
|---|---|---|---|---|---|---|---|
| $f(x)$ | 27 | 9 | 3 | 1 | $\frac{1}{3}$ | $\frac{1}{9}$ | $\frac{1}{27}$ |

**5.** `2*2^x` or `2*(2^x)`

| $x$ | −3 | −2 | −1 | 0 | 1 | 2 | 3 |
|---|---|---|---|---|---|---|---|
| $f(x)$ | $\frac{1}{4}$ | $\frac{1}{2}$ | 1 | 2 | 4 | 8 | 16 |

**7.** `-3*2^(-x)`

| $x$ | −3 | −2 | −1 | 0 | 1 | 2 | 3 |
|---|---|---|---|---|---|---|---|
| $f(x)$ | −24 | −12 | −6 | −3 | $-\frac{3}{2}$ | $-\frac{3}{4}$ | $-\frac{3}{8}$ |

**9.** `2^x-1`

| $x$ | −3 | −2 | −1 | 0 | 1 | 2 | 3 |
|---|---|---|---|---|---|---|---|
| $f(x)$ | $-\frac{7}{8}$ | $-\frac{3}{4}$ | $-\frac{1}{2}$ | 0 | 1 | 3 | 7 |

**11.** `2^(x-1)`

| $x$ | −3 | −2 | −1 | 0 | 1 | 2 | 3 |
|---|---|---|---|---|---|---|---|
| $f(x)$ | $\frac{1}{16}$ | $\frac{1}{8}$ | $\frac{1}{4}$ | $\frac{1}{2}$ | 1 | 2 | 4 |

**13.**   **15.**   **17.**

$y = 3^{-x}$

$y = 2(2^x)$   $y = -3(2^{-x})$

**19.** Both; $f(x) = 4.5(3^x)$. $g(x) = 2(1/2)^x$, or $2(2^{-x})$
**21.** Neither   **23.** $g$; $g(x) = 4(0.2)^x$

**25.** `e^(-2*x)` or `EXP(-2*x)`

| $x$ | −3 | −2 | −1 | 0 | 1 | 2 | 3 |
|---|---|---|---|---|---|---|---|
| $f(x)$ | 403.4 | 54.60 | 7.389 | 1 | 0.1353 | 0.01832 | 0.002479 |

**27.** `1.01*2.02^(-4*x)`

| $x$ | −3 | −2 | −1 | 0 | 1 | 2 | 3 |
|---|---|---|---|---|---|---|---|
| $f(x)$ | 4662 | 280.0 | 16.82 | 1.01 | 0.06066 | 0.003643 | 0.0002188 |

**29.** `50*(1+1/3.2)^(2*x)`

| $x$ | −3 | −2 | −1 | 0 | 1 | 2 | 3 |
|---|---|---|---|---|---|---|---|
| $f(x)$ | 9.781 | 16.85 | 29.02 | 50 | 86.13 | 148.4 | 255.6 |

**31.** `2^(x-1)` *not* `2^x-1`   **33.** `2/(1-2^(-4*x))` *not* `2/1-2^-4*x and not 2/1-2^(-4*x)`
**35.** `(3+x)^(3*x)/(x+1)` or `((3+x)^(3*x))/(x+1)` *not* `(3+x)^(3*x)/x+1 and not (3+x^(3*x))/(x+1)`
**37.** `2*e^((1+x)/x)` or `2*EXP((1+x)/x)` *not* `2*e^1+x/x and not 2*e^(1+x)/x and not 2*EXP(1+x)/x`

**39.**   **41.**

$y1 = 1.6$^x   $y2 = 1.8$^x

$y1 = 300*1.1$^x
$y2 = 300*1.1$^(2*x)

**43.**   **45.**

$y1 = 2.5$^(1.02*x)
$y2 = e$^(1.02*x)
or `exp(1.02*x)`

$y1 = 1000*1.045$^(−3*x)
$y2 = 1000*1.045$^(3*x)

**47.** $f(x) = 500(0.5)^x$   **49.** $f(x) = 10(3)^x$
**51.** $f(x) = 500(0.45)^x$   **53.** $f(x) = -100(1.1)^x$
**55.** $y = 4(3^x)$   **57.** $y = -1(0.2^x)$
**59.** $y = 2.1213(1.4142^x)$   **61.** $y = 3.6742(0.9036^x)$
**63.** $f(t) = 5,000e^{0.10t}$   **65.** $f(t) = 1,000e^{-0.063t}$
**67.** $y = 1.0442(1.7564)^x$   **69.** $y = 15.1735(1.4822)^x$
**71.** $f(t) = 300(0.5)^t$; 9.375 mg   **73. a.** Linear model: $F = 65t - 60$. Exponential model: $F = 97.2(1.20)^t$. The exponential model is more appropriate. **b.** 418 tons, not too far off the projected figure

**75. a.** $P = 180(1.01087)^t$ million **b.** 6 significant digits **c.** 344 million **77. a.** $y = 50,000(1.5^{t/2})$, $t =$ time in years since two years ago **b.** 91,856 tags **79.** $y = 1,000(2^{t/3})$; 65,536,000 bacteria after 2 days **81.** $A(t) = 167(1.18)^t$; 1,695 cases **83.** $A(t) = 5,000(1 + 0.003/12)^{12t}$; $5,106 **85.** 2024 **87.** 491.82 **89.** $A(t) = 4.9e^{0.032t}$ million homes; 2010: 5.2 million homes; 2012: 5.6 million homes **91. a.**

| Year | 1950 | 2000 | 2050 | 2100 |
|---|---|---|---|---|
| $C(t)$ parts per million | 355 | 377 | 400 | 425 |

**b.** 2030
**93. a.** $P(t) = 0.339(1.169)^t$.
Graph:

**b.** $1.9 million
**95. a.** $n = 1.127(3.544)^t$
Graph:

**b.** 3.544 **c.** 178 million **97.** (B) **99.** Exponential functions of the form $f(x) = Ab^x$ ($b > 1$) increase rapidly for large values of $x$. In real-life situations, such as population growth, this model is reliable only for relatively short periods of growth. Eventually, population growth tapers off because of pressures such as limited resources and overcrowding. **101.** The article was published about a year before the "housing bubble" burst in 2006, whereupon house prices started to fall, contrary to the prediction of the graph—they continued to drop for several years, as documented in Exercise 90. This shows the danger of using any mathematical model to extrapolate. The blogger was, however, cautious in the choice of words, claiming only to be estimating what the future U.S. median house price "might be." **103.** Linear functions better: cost models where there is a fixed cost and a variable cost; simple interest, where interest is paid on the original amount invested. Exponential models better: compound interest, population growth. (In both of these, the rate of growth depends on the

present number of items, rather than on some fixed quantity.) **105.** Take the ratios $y_2/y_1$ and $y_3/y_2$. If they are the same, the points fit on an exponential curve. **107.** This reasoning is suspect—the bank need not use its computer resources to update all the accounts every minute, but can instead use the continuous compounding formula to calculate the balance in any account at any time.

## Section 9.3

**1.**

| Logarithmic Form | $\log_{10} 10,000 = 4$ | $\log_4 16 = 2$ | $\log_3 27 = 3$ | $\log_5 5 = 1$ | $\log_7 1 = 0$ | $\log_4 \frac{1}{16} = -2$ |
|---|---|---|---|---|---|---|

**3.**

| Exponential Form | $(0.5)^2 = 0.25$ | $5^0 = 1$ | $10^{-1} = 0.1$ | $4^3 = 64$ | $2^8 = 256$ | $2^{-2} = \frac{1}{4}$ |
|---|---|---|---|---|---|---|

**5.** 1.4650 **7.** −1.1460 **9.** −0.7324 **11.** 6.2657
**13.**                              **15.**

**17.**

**19.** $Q = 1,000e^{-t \ln 2}$ **21.** $Q = 1,000e^{t(\ln 2)/2}$
**23.** Doubling time $= 2 \ln 2$ **25.** Half-life $= (\ln 2)/4$
**27.** $f(x) = 4(7.389)^x$ **29.** $f(t) = 2.1e^{0.000\ 9995t}$
**31.** $f(t) = 10e^{-0.01309t}$. **33.** 3.36 years **35.** 11 years
**37.** 23.1% **39.** 63,000 years old **41.** 12 years
**43.** 238 months **45.** 13 years **47.** 20.39 years
**49.** 1,600 years **51. a.** $b = 3^{1/6} \approx 1.20$
**b.** 3.8 months **53. a.** $Q(t) = Q_0 e^{-0.139t}$ **b.** 3 years
**55.** 2,360 million years **57.** 3.2 hours **59.** 3.89 days
**61. a.** $P(t) = 6.591 \ln(t) - 17.69$ **b.** 1 digit **c.** (A)
**63.** $S(t) = 55.49 \ln t + 39.67$
Graph:

Positive direction; extrapolating in the negative direction eventually leads to negative values, which do not model reality.
**65. a.** About $4.467 \times 10^{23}$ ergs **b.** About 2.24% **c.** $E = 10^{1.5R+11.8}$ **d.** Proof **e.** 1,000 **67. a.** 75 dB, 69 dB, 61 dB **b.** $D = 95 - 20\log r$ **c.** 57,000 feet **69.** Graph:

The green curve is $y = \ln x$. The blue curve is $y = 2\ln x$, and the red curve is $y = 2\ln x + 0.5$. Multiplying by $A$ stretches the graph in the $y$-direction by a factor of $A$. Adding $C$ moves the graph $C$ units vertically up. **71.** The logarithm of a negative number, were it defined, would be the power to which a base must be raised to give that negative number. But raising a base to a power never results in a negative number, so there can be no such number as the logarithm of a negative number. **73.** Any logarithmic curve $y = \log_b t + C$ will eventually surpass 100%, and hence not be suitable as a long-term predictor of market share. **75.** $\log_4 y$ **77.** 8 **79.** $x$ **81.** Time is increasing logarithmically with population; solving $P = Ab^t$ for $t$ gives $t = \log_b(P/A) = \log_b P - \log_b A$, which is of the form $t = \log_b P + C$. **83.** (Proof)

## Section 9.4

**1.** $N = 7, A = 6, b = 2$;  **3.** $N = 10, A = 4, b = 0.3$;

7/(1+6*2^-x)  10/(1+4*0.3^-x)

**5.** $N = 4, A = 7, b = 1.5$;

4/(1+7*1.5^-x)

**7.** $f(x) = \dfrac{200}{1 + 19(2^{-x})}$  **9.** $f(x) = \dfrac{6}{1 + 2^{-x}}$

**11.** (B)  **13.** (B)  **15.** (C)

**17.** $y = \dfrac{7.2}{1 + 2.4(1.04)^{-x}}$  **19.** $y = \dfrac{97}{1 + 2.2(0.942)^{-x}}$

**21. a.** (A) **b.** 2003  **23. a.** (A) **b.** 20% per year
**25. a.** 95.0% **b.** $P(x) \approx 25.13(1.064)^x$ **c.** \$18,000

**27.** $N(t) = \dfrac{10,000}{1 + 9(1.25)^{-t}}$; $N(7) \approx 3,463$ cases

**29.** $N(t) = \dfrac{3,000}{1 + 29(2^{1/5})^{-t}}$; $t = 16$ days

**31. a.** $A(t) = \dfrac{6.3}{1 + 4.8(1.2)^{-t}}$; 6,300 articles

**b.** 5,200 articles  **33. a.** $B(t) = \dfrac{1,090}{1 + 0.410(1.09)^{-t}}$;

1,090 teams **b.** $t \approx -10.3$. According to the model, the number of teams was rising fastest about 10.3 years *prior* to 1990; that is, sometime during 1979. **c.** The number of men's basketball teams was growing by about 9% per year in the past, well before 1979.

**35.** $y = \dfrac{4,500}{1 + 1.1466\,(1.0357)^{-t}}$; 2013  **37.** Just as diseases are communicated via the spread of a pathogen (such as a virus), new technology is communicated via the spread of information (such as advertising and publicity). Further, just as the spread of a disease is ultimately limited by the number of susceptible individuals, so the spread of a new technology is ultimately limited by the size of the potential market. **39.** It can be used to predict where the sales of a new commodity might level off. **41.** The curve is still a logistic curve, but decreases when $b > 1$ and increases when $b < 1$. **43.** (Proof)

## Chapter 9 Review

**1.**

**3.** $f: f(x) = 5(1/2)^x$, or $5(2^{-x})$

**5.**                                    **7.**

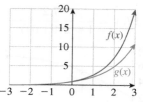

**9.** $3,484.85    **11.** $3,705.48    **13.** $3,485.50

**15.** $f(x) = 4.5(9^x)$    **17.** $f(x) = \dfrac{2}{3}3^x$    **19.** $-\dfrac{1}{2}\log_3 4$

**21.** $\dfrac{1}{3}\log 1.05$

**23.**

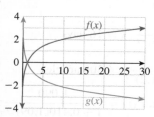

**25.** $Q = 5e^{-0.00693t}$    **27.** $Q = 2.5e^{0.347t}$
**29.** 10.2 years    **31.** 10.8 years
**33.** $f(x) = \dfrac{900}{1 + 8(1.5)^{-x}}$
**35.** $f(x) = \dfrac{20}{1 + 3(0.8)^{-x}}$    **37. a.** $8,500 per month;

an average of approximately 2,100 hits per day
**b.** $29,049 per month  **c.** The fact that $-0.000005$,
the coefficient of $c^2$, is negative.
**39. a.** $R = -60p^2 + 950p$; $p = $7.92$ per novel,
Monthly revenue = $3,760.42
**b.** $P = -60p^2 + 1,190p - 4,700$; $p = $9.92$ per
novel, Monthly profit = $1,200.42    **41. a.** 9.1, 19
**b.** About 310,000 pounds    **43.** 2016
**45.** 1.12 million pounds    **47.** $n(t) = 9.6(0.80^t)$ million
pounds of lobster    **49.** (C)

# Chapter 10

## Section 10.1

**1.** 0    **3.** 4    **5.** Does not exist    **7.** 1.5    **9.** 0.5
**11.** Diverges to $+\infty$    **13.** 0    **15.** 1    **17.** 0    **19.** 0
**21. a.** $-2$ **b.** $-1$    **23. a.** 2 **b.** 1 **c.** 0 **d.** $+\infty$
**25. a.** 0 **b.** 2 **c.** $-1$ **d.** Does not exist **e.** 2 **f.** $+\infty$
**27. a.** 1 **b.** 1 **c.** 2 **d.** Does not exist **e.** 1 **f.** 2
**29. a.** 1 **b.** Does not exist **c.** Does not exist **d.** 1
**31. a.** 1 **b.** $+\infty$ **c.** $+\infty$ **d.** $+\infty$ **e.** not defined **f.** $-1$
**33. a.** $-1$ **b.** $+\infty$ **c.** $-\infty$ **d.** Does not exist **e.** 2 **f.** 1
**35.** 210 trillion pesos per month. In the long term, the

model predicts that the value of sold goods in Mexico
will approach 210 trillion pesos per month.    **37.** $+\infty$;
In the long term, Amazon's revenue will grow
without bound. **39.** 7.0; In the long term, the number
of research articles in *Physical Review* written by
researchers in Europe approaches 7,000 per year.
**41.** 573. This suggests that students with an exceptionally
large household income earn an average of 573 on
the math SAT test.    **43. a.** $\lim_{t \to 14.75^-} r(t) = 21$,
$\lim_{t \to 14.75^+} r(t) = 21$, $\lim_{t \to 14.75} r(t) = 21$,
$r(14.75) = 0.01$ **b.** Just prior to 2:45 pm, the stock
was approaching $21, but then fell suddenly to a penny
($0.01) at 2:45 exactly, after which time it jumped back
to values close to $21.    **45.** 100; In the long term, the
home price index will level off at 100 points.
**47.** $\lim_{t \to 1^-} C(t) = 0.06$, $\lim_{t \to 1^+} C(t) = 0.08$, so
$\lim_{t \to 1} C(t)$ does not exist.    **49.** $\lim_{t \to +\infty} I(t) = +\infty$,
$\lim_{t \to +\infty}(I(t)/E(t)) \approx 2.5$. In the long term, U.S.
imports from China will rise without bound and be 2.5
times U.S. exports to China. In the real world, imports
and exports cannot rise without bound. Thus, the given
models should not be extrapolated far into the future.
**51.** To approximate $\lim_{x \to a} f(x)$ numerically, choose
values of $x$ closer and closer to, and on either side of
$x = a$, and evaluate $f(x)$ for each of them. The limit
(if it exists) is then the number that these values of $f(x)$
approach. A disadvantage of this method is that it may
never give the exact value of the limit, but only an
approximation. (However, we can make this as accurate
as we like.)    **53.** Any situation in which there is a
sudden change can be modeled by a function in which
$\lim_{t \to a^+} f(t)$ is not the same as $\lim_{t \to a^-} f(t)$. One
example is the value of a stock market index before and
after a crash: $\lim_{t \to a^-} f(t)$ is the value immediately
before the crash at time $t = a$, while $\lim_{t \to a^+} f(t)$ is the
value immediately after the crash. Another example
might be the price of a commodity that is suddenly
increased from one level to another.    **55.** It is possible
for $\lim_{x \to a} f(x)$ to exist even though $f(a)$ is not

defined. An example is $\displaystyle \lim_{x \to 1} \frac{x^2 - 3x + 2}{x - 1}$.

**57.** An example is $f(x) = (x - 1)(x - 2)$.
**59.** These limits are all 0.

## Section 10.2

**1.** Continuous on its domain    **3.** Continuous
on its domain    **5.** Discontinuous at $x = 0$
**7.** Discontinuous at $x = -1$    **9.** Continuous
on its domain    **11.** Continuous on its domain
**13.** Discontinuous at $x = -1$ and 0    **15.** (A), (B),
(D), (E)    **17.** 0    **19.** $-1$    **21.** No value possible
**23.** $-1$    **25.** Continuous on its domain

**27.** Continuous on its domain   **29.** Discontinuity at $x = 0$   **31.** Discontinuity at $x = 0$   **33.** Continuous on its domain   **35.** Not unless the domain of the function consists of all real numbers. (It is impossible for a function to be continuous at points not in its domain.) For example, $f(x) = 1/x$ is continuous on its domain—the set of nonzero real numbers—but not at $x = 0$.   **37.** True. If the graph of a function has a break in its graph at any point $a$, then it cannot be continuous at the point $a$.   **39.** Answers may vary. $f(x) = 1/[(x - 1)(x - 2)(x - 3)]$ is such a function; it is undefined at $x = 1, 2, 3$ and so its graph consists of three distinct curves.   **41.** Answers may vary.

**43.** Answers may vary. The price of OHaganBooks.com stocks suddenly drops by \$10 as news spreads of a government investigation. Let $f(x)$ = Price of OHaganBooks.com stocks.

## Section 10.3

**1.** $x = 1$   **3.** 2   **5.** Determinate; diverges to $+\infty$
**7.** Determinate; does not exist   **9.** Determinate; diverges to $-\infty$   **11.** Determinate; 0
**13.** Indeterminate; $-1/3$   **15.** Indeterminate; 0
**17.** Determinate; 0   **19.** Determinate; $-60$   **21.** 1
**23.** 2   **25.** 0   **27.** 6   **29.** 4   **31.** 2   **33.** 0   **35.** 0
**37.** 12   **39.** $+\infty$   **41.** Does not exist; left and right (infinite) limits differ.   **43.** $-\infty$   **45.** Does not exist
**47.** $+\infty$   **49.** 0   **51.** 1/6   **53.** 3/2   **55.** 1/2
**57.** $+\infty$   **59.** 3/2   **61.** 1/2   **63.** $-\infty$   **65.** 0   **67.** 12
**69.** 0   **71.** $+\infty$   **73.** 0   **75.** Discontinuity at $x = 0$
**77.** Continuous everywhere   **79.** Discontinuity at $x = 0$   **81.** Discontinuity at $x = 0$
**83. a.** $\lim_{t \to 15^-} v(t) = 3{,}800$, $\lim_{t \to 15^+} v(t) = 3{,}800$; Shortly before 2005 the speed of Intel processors was approaching 3,800 MHz. Shortly after 2005 the speed of Intel processors was close to 3,800 MHz
**b.** Continuous at $t = 15$; No abrupt changes.
**85. a.** 0.49, 1.16. Shortly before 1999 annual advertising expenditures were close to \$0.49 billion. Shortly after 1999 annual advertising expenditures were close to \$1.16 billion.   **b.** Not continuous; movie advertising expenditures jumped suddenly in 1999.
**87.** 1.59; if the trend continued indefinitely, the annual spending on police would be 1.59 times the annual spending on courts in the long run.   **89.** 573. This suggests that students with an exceptionally large

household income earn an average of 573 on the math SAT test.   **91.** $\lim_{t \to +\infty} I(t) = +\infty$, $\lim_{t \to +\infty}(I(t)/E(t)) = 2.5$.   In the long term, U.S. imports from China will rise without bound and be 2.5 times U.S. exports to China. In the real world, imports and exports cannot rise without bound. Thus, the given models should not be extrapolated far into the future.   **93.** $\lim_{t \to +\infty} p(t) = 100$. The percentage of children who learn to speak approaches 100% as their age increases.   **95.** To evaluate $\lim_{x \to a} f(x)$ algebraically, first check whether $f(x)$ is a closed-form function. Then check whether $x = a$ is in its domain. If so, the limit is just $f(a)$; that is, it is obtained by substituting $x = a$. If not, then try to first simplify $f(x)$ in such a way as to transform it into a new function such that $x = a$ is in its domain, and then substitute. A disadvantage of this method is that it is sometimes extremely difficult to evaluate limits algebraically, and rather sophisticated methods are often needed.   **97.** $x = 3$ is not in the domain of the given function $f$, so, yes, the *function* is undefined at $x = 3$. However, the *limit* may well be defined. In this case, it leads to the indeterminate form 0/0, telling us that we need to try to simplify, and that leads us to the correct limit of 27.   **99.** She is wrong. Closed-form functions are continuous only at points in their domains, and $x = 2$ is not in the domain of the closed-form function $f(x) = 1/(x - 2)^2$.
**101.** Answers may vary. (1) See Example 1(b): $\lim_{x \to 2} \dfrac{x^3 - 8}{x - 2}$, which leads to the indeterminate form 0/0 but the limit is 12. (2) $\lim_{x \to +\infty} \dfrac{60x}{2x}$, which leads to the indeterminate form $\infty/\infty$, but where the limit exists and equals 30.   **103.** $\pm\infty/\infty$; The limits are zero.
This suggests that limits resulting in $\dfrac{p(\infty)}{e^{\infty}}$ are zero.

**105.** The statement may not be true, for instance, if $f(x) = \begin{cases} x + 2 & \text{if } x < 0 \\ 2x - 1 & \text{if } x \geq 0 \end{cases}$, then $f(0)$ is defined and equals $-1$, and yet $\lim_{x \to 0} f(x)$ does not exist. The statement can be corrected by requiring that $f$ be a closed-form function: "If $f$ is a closed-form function, and $f(a)$ is defined, then $\lim_{x \to a} f(x)$ exists and equals $f(a)$."   **107.** Answers may vary, for example
$f(x) = \begin{cases} 0 & \text{if } x \text{ is any number other than 1 or 2} \\ 1 & \text{if } x = 1 \text{ or } 2 \end{cases}$
**109.** Answers may vary.
(1) $\lim_{x \to +\infty} [(x + 5) - x] = \lim_{x \to +\infty} 5 = 5$
(2) $\lim_{x \to +\infty} [x^2 - x] = \lim_{x \to +\infty} x(x - 1) = +\infty$
(3) $\lim_{x \to +\infty} [(x - 5) - x] = \lim_{x \to +\infty} -5 = -5$

## Section 10.4

**1.** −3    **3.** 0.3    **5.** −$25,000 per month    **7.** −200 items per dollar    **9.** $1.33 per month    **11.** 0.75 percentage point increase in unemployment per 1 percentage point increase in the deficit    **13.** 4    **15.** 2    **17.** 7/3

**19.**

| h | Ave. Rate of Change |
|---|---|
| 1 | 2 |
| 0.1 | 0.2 |
| 0.01 | 0.02 |
| 0.001 | 0.002 |
| 0.0001 | 0.0002 |

**21.**

| h | Ave. Rate of Change |
|---|---|
| 1 | −0.1667 |
| 0.1 | −0.2381 |
| 0.01 | −0.2488 |
| 0.001 | −0.2499 |
| 0.0001 | −0.24999 |

**23.**

| h | Ave. Rate of Change |
|---|---|
| 1 | 9 |
| 0.1 | 8.1 |
| 0.01 | 8.01 |
| 0.001 | 8.001 |
| 0.0001 | 8.0001 |

**25. a.** $60 billion per year; World military expenditure increased at an average rate of $60 billion per year during 2005–2010. **b.** $50 billion per year; World military expenditure increased at an average rate of $50 billion per year during 2000–2010.    **27. a.** −20,000 barrels/year; during 2002–2007, daily oil production by Pemex was decreasing at an average rate of 20,000 barrels of oil per year. **b.** (C)    **29. a.** 1.7; the percentage of mortgages classified as subprime was increasing at an average rate of around 1.7 percentage points per year between 2000 and 2006. **b.** 2004–2006
**31. a.** 2007–2009; During 2007–2009 immigration to Ireland was decreasing at an average rate of 22,500 people per year. **b.** 2008–2010; During 2008–2010 immigration to Ireland was decreasing at an average rate of 7,500 people per year.    **33. a.** [3, 5]; −0.25 thousand articles per year. During the period 1993–1995, the number of articles authored by U.S. researchers decreased at an average rate of 250 articles per year. **b.** Percentage rate ≈ −0.1765; Average rate = −0.09 thousand articles/year. Over the period 1993–2003, the number of articles authored by U.S. researchers decreased at an average rate of 90 per year, representing a 17.65% decrease over that period.
**35. a.** 12 teams per year **b.** Decreased    **37. a.** (A) **b.** (C) **c.** (B) **d.** Approximately −$0.088 per year (if we round to two significant digits). This is less than the slope of the regression line, about −$0.063 per year.

**39.** Answers may vary. Graph:

**41.** The index was increasing at an average rate of 300 points per day.    **43. a.** $0.15 per year **b.** No; according to the model, during that 25-year period the price of oil went down from around $93 to a low of around $25 in 1993 before climbing back up.
**45. a.** 47.3 new cases per day; the number of SARS cases was growing at an average rate of 47.3 new cases per day over the period March 17 to March 23.
**b.** (A)    **47. a.** 8.85 manatee deaths per 100,000 boats; 23.05 manatee deaths per 100,000 boats **b.** More boats result in more manatee deaths per additional boat.    **49. a.** The average rates of change are shown in the following table

| Interval | [0, 40] | [40, 80] | [80, 120] | [120, 160] | [160, 200] |
|---|---|---|---|---|---|
| Average Rate of Change of S | 1.35 | 0.80 | 0.48 | 0.28 | 0.17 |

**b.** For household incomes between $40,000 and $80,000, a student's math SAT increases at an average rate of 0.80 points per $1,000 of additional income.
**c.** (A) **d.** (B)    **51.** The average rate of change of $f$ over an interval $[a, b]$ can be determined numerically, using a table of values, graphically, by measuring the slope of the corresponding line segment through two points on the graph, or algebraically, using an algebraic formula for the function. Of these, the least precise is the graphical method, because it relies on reading coordinates of points on a graph.    **53.** No, the formula for the average rate of a function $f$ over $[a, b]$ depends only on $f(a)$ and $f(b)$, and not on any values of $f$ between $a$ and $b$.    **55.** Answers will vary. Graph:

**57.** 6 units of quantity A per unit of quantity C
**59.** (A)

**61.** Yes. Here is an example:

| Year | 2000 | 2001 | 2002 | 2003 |
|---|---|---|---|---|
| Revenue ($ billion) | $10 | $20 | $30 | $5 |

**63.** (A)

## Section 10.5

**1.** 6   **3.** −5.5

**5.**

| h | 1 | 0.1 | 0.01 |
|---|---|---|---|
| Ave. rate | 39 | 39.9 | 39.99 |

Instantaneous rate = $40 per day

**7.**

| h | 1 | 0.1 | 0.01 |
|---|---|---|---|
| Ave. Rate | 140 | 66.2 | 60.602 |

Instantaneous rate = $60 per day

**9.**

| h | 10 | 1 |
|---|---|---|
| $C_{ave}$ | 4.799 | 4.7999 |

**11.**

| h | 10 | 1 |
|---|---|---|
| $C_{ave}$ | 99.91 | 99.90 |

$C'(1,000) = \$4.80$ per item   $C'(100) = \$99.90$ per item
**13.** 1/2   **15.** 0   **17. a.** $R$ **b.** $P$   **19. a.** $P$ **b.** $R$
**21. a.** $Q$ **b.** $P$   **23. a.** $Q$ **b.** $R$ **c.** $P$   **25. a.** $R$
**b.** $Q$ **c.** $P$   **27. a.** $(1, 0)$ **b.** None **c.** $(-2, 1)$
**29. a.** $(-2, 0.3)$, $(0, 0)$, $(2, -0.3)$ **b.** None **c.** None
**31.** $(a, f(a))$; $f'(a)$   **33.** (B)   **35. a.** (A) **b.** (C)
**c.** (B) **d.** (B) **e.** (C)   **37.** −2   **39.** −1.5   **41.** −5
**43.** 16   **45.** 0   **47.** −0.0025

**49. a.** 3 **b.** $y = 3x + 2$     **51. a.** $\dfrac{3}{4}$ **b.** $y = \dfrac{3}{4}x + 1$

**53. a.** $\dfrac{1}{4}$ **b.** $y = \dfrac{1}{4}x + 1$

**55.** 1.000   **57.** 1.000   **59.** (C)   **61.** (A)   **63.** (F)
**65.** Increasing for $x < 0$; decreasing for $x > 0$.
**67.** Increasing for $x < -1$ and $x > 1$; decreasing for
$-1 < x < 1$   **69.** Increasing for $x > 1$; decreasing
for $x < 1$.   **71.** Increasing for $x < 0$; decreasing
for $x > 0$.

**73.** $x = -1.5, x = 0$
Graph:

**75.** Note: Answers depend on the form of technology
used. Excel ($h = 0.1$):

| | A | B | C | D | E | F |
|---|---|---|---|---|---|---|
| 1 | x | f(x) | f'(x) | xmin | | 4 |
| 2 | 4 | 6 | -4.545454545 | h | | 0.1 |
| 3 | 4.1 | 5.545454545 | -3.787878788 | | | |
| 4 | 4.2 | 5.166666667 | -3.205128205 | | | |
| 5 | 4.3 | 4.846153846 | -2.747252747 | | | |
| 6 | 4.4 | 4.571428571 | -2.380952381 | | | |
| 7 | 4.5 | 4.333333333 | -2.083333333 | | | |
| 8 | 4.6 | 4.125 | -1.838235294 | | | |
| 9 | 4.7 | 3.941176471 | -1.633986928 | | | |
| 10 | 4.8 | 3.777777778 | -1.461988304 | | | |
| 11 | 4.9 | 3.631578947 | -1.315789474 | | | |
| 12 | 5 | 3.5 | | | | |
| 13 | | | | | | |
| 14 | | | | | | |

Graphs:

**77.** $q(100) = 50,000$, $q'(100) = -500$. A total of
50,000 pairs of sneakers can be sold at a price of $100,
but the demand is decreasing at a rate of 500 pairs per
$1 increase in the price.   **79. a.** −0.05; daily oil imports
from Mexico in 2005 were 1.6 million barrels and
declining at a rate of 0.05 million barrels (or 50,000
barrels) per year.   **b.** Decreasing   **81. a.** (B) **b.** (B)
**c.** (A) **d.** 2004 **e.** 0.033; in 2004 the U.S. prison
population was increasing at a rate of 0.033 million
prisoners (33,000 prisoners) per year.   **83. a.** −96 ft/sec
**b.** −128 ft/sec   **85. a.** $0.60 per year; the price per
barrel of crude oil in constant 2008 dollars was
growing at an average rate of about 60¢ per year over
the 28-year period beginning at the start of 1980.
**b.** −$12 per year; the price per barrel of crude oil in
constant 2008 dollars was dropping at an instantaneous
rate of about $12 per year at the start of 1980.
**c.** The price of oil was decreasing in January 1980, but
eventually began to increase (making the average rate
of change in part (a) positive).   **87. a.** 144.7 new
cases per day; the number of SARS cases was growing
at a rate of about 144.7 new cases per day on March 27.
**b.** (A)

**89.** $S(5) \approx 109$, $\left.\dfrac{dS}{dt}\right|_{t=5} \approx 9.1$. After 5 weeks, sales are 109 pairs of sneakers per week, and sales are increasing at a rate of 9.1 pairs per week each week. **91.** $A(0) = 4.5$ million; $A'(0) = 60,000$ subscribers/week **93. a.** 60% of children can speak at the age of 10 months. At the age of 10 months, this percentage is increasing by 18.2 percentage points per month. **b.** As $t$ increases, $p$ approaches 100 percentage points (all children eventually learn to speak), and $dp/dt$ approaches zero because the percentage stops increasing. **95. a.** $A(6) \approx 12.0$; $A'(6) \approx 1.4$; at the start of 2006, about 12% of U.S. mortgages were subprime, and this percentage was increasing at a rate of about 1.4 percentage points per year **b.** Graphs:

Graph of $A$:

Graph of $A'$:

From the graphs, $A(t)$ approaches 15 as $t$ becomes large (in terms of limits, $\lim_{x\to+\infty} A(t) = 15$) and $A'(t)$ approaches 0 as $t$ becomes large (in terms of limits, $\lim_{x\to+\infty} A'(t) = 0$). Interpretation: If the trend modeled by the function $A$ had continued indefinitely, in the long term 15% of U.S. mortgages would have been subprime, and this percentage would not be changing. **97. a.** (D) **b.** 33 days after the egg was laid **c.** 50 days after the egg was laid. Graph:

**99.** $L(.95) \approx 31.2$ meters and $L'(.95) \approx -304.2$ meters/warp. Thus, at a speed of warp 0.95, the spaceship has an observed length of 31.2 meters and its length is decreasing at a rate of 304.2 meters per unit warp, or 3.042 meters per increase in speed of 0.01 warp. **101.** The difference quotient is not defined when $h = 0$ because there is no such number as $0/0$. **103.** (D) **105.** The derivative is positive and decreasing toward zero. **107.** Company B. Although the company is currently losing money, the derivative is

positive, showing that the profit is increasing. Company A, on the other hand, has profits that are declining. **109.** (C) is the only graph in which the instantaneous rate of change on January 1 is greater than the one-month average rate of change. **111.** The tangent to the graph is horizontal at that point, and so the graph is almost horizontal near that point. **113.** Answers may vary.

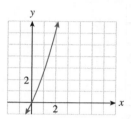

**115.** If $f(x) = mx + b$, then its average rate of change over any interval $[x, x + h]$ is $\dfrac{m(x + h) + b - (mx + b)}{h} = m$. Because this does not depend on $h$, the instantaneous rate is also equal to $m$. **117.** Increasing because the average rate of change appears to be rising as we get closer to 5 from the left. (See the bottom row.)

**119.** Answers may vary.

**121.** Answers may vary.

**123.** (B)    **125.** Answers will vary.

## Section 10.6

**1.** 4 **3.** 3 **5.** 7 **7.** 4 **9.** 14 **11.** 1 **13.** $m$ **15.** $2x$
**17.** 3 **19.** $6x + 1$ **21.** $2 - 2x$ **23.** $3x^2 + 2$
**25.** $1/x^2$ **27.** $m$ **29.** $-1.2$ **31.** 30.6 **33.** $-7.1$
**35.** 4.25 **37.** $-0.6$ **39.** $y = 4x - 7$
**41.** $y = -2x - 4$ **43.** $y = -3x - 1$
**45.** $s'(t) = -32t$; $s'(4) = -128$ ft/sec
**47.** $dI/dt = -0.030t + 0.1$; daily oil imports were decreasing at a rate of 0.11 million barrels per year.
**49.** $R'(t) = -90t + 900$; Increasing at a rate of 450 million gallons per year **51.** $f'(8) = 26.6$ manatee deaths per 100,000 boats. At a level of 800,000 boats, the number of manatee deaths is increasing at a rate of 26.6 manatees per 100,000 additional boats. **53.** Yes; $\lim_{t \to 20^-} C(t) = \lim_{t \to 20^+} C(t) = 700 = C(20)$. **b.** No; $\lim_{t \to 20^-} C'(t) = 31.1$ while $\lim_{t \to 20^+} C'(t) = 90$. Until 1990, the cost of a Super Bowl ad was increasing at a rate of \$31,100 per year. Immediately thereafter, it was increasing at a rate of \$90,000 per year. **55.** The algebraic method because it gives the exact value of the derivative. The other two approaches give only approximate values (except in some special cases).
**57.** The error is in the second line: $f(x + h)$ is *not* equal to $f(x) + h$. For instance, if $f(x) = x^2$, then $f(x + h) = (x + h)^2$, whereas $f(x) + h = x^2 + h$.
**59.** The error is in the second line: One could only cancel the $h$ if it were a *factor* of both the numerator and denominator; it is not a factor of the numerator.
**61.** Because the algebraic computation of $f'(a)$ is exact and not an approximation, it makes no difference whether one uses the balanced difference quotient or the ordinary difference quotient in the algebraic computation. **63.** The computation results in a limit that cannot be evaluated.

## Chapter 10 Review

**1.** 5 **3.** Does not exist **5. a.** $-1$ **b.** 3 **c.** Does not exist **7.** $-4/5$ **9.** $-1$ **11.** $-1$ **13.** Does not exist
**15.** 10/7 **17.** Does not exist **19.** $+\infty$ **21.** 0
**23.** Diverges to $-\infty$ **25.** 0 **27.** 2/5 **29.** 1
**31.**

| $h$ | 1 | 0.01 | 0.001 |
|---|---|---|---|
| Ave. Rate of Change | $-0.5$ | $-0.9901$ | $-0.9990$ |

Slope $\approx -1$
**33.**

| $h$ | 1 | 0.01 | 0.001 |
|---|---|---|---|
| Avg. Rate of Change | 6.3891 | 2.0201 | 2.0020 |

Slope $\approx 2$
**35. a.** $P$ **b.** $Q$ **c.** $R$ **d.** $S$ **37. a.** $Q$ **b.** None **c.** None **d.** None **39. a.** (B) **b.** (B) **c.** (B) **d.** (A) **e.** (C) **41.** $2x + 1$ **43.** $2/x^2$

**45.**

**47.**

**49. a.** $P(3) = 25$ : O'Hagan purchased the stock at \$25. $\lim_{t \to 3^-} P(t) = 25$: The value of the stock had been approaching \$25 up to the time he bought it. $\lim_{t \to 3^+} P(t) = 10$: The value of the stock dropped to \$10 immediately after he bought it. **b.** Continuous but not differentiable. Interpretation: the stock price changed continuously but suddenly reversed direction (and started to go up) the instant O'Hagan sold it.
**51. a.** $\lim_{t \to 3} p(t) \approx 40$; $\lim_{t \to +\infty} p(t) = +\infty$. Close to 2007 ($t = 3$), the home price index was about 40. In the long term, the home price index will rise without bound. **b.** 10 (The slope of the linear portion of the curve is 10.) In the long term, the home price index will rise about 10 points per year.
**53. a.** 500 books per week **b.** [3, 4], [4, 5] **c.** [3, 5]; 650 books per week **55. a.** 3 percentage points per year **b.** 0 percentage points per year **c.** (D)
**57. a.** $72t + 250$ **b.** 322 books per week **c.** 754 books per week.

## Chapter 11

## Section 11.1

**1.** $5x^4$ **3.** $-4x^{-3}$ **5.** $-0.25x^{-0.75}$ **7.** $8x^3 + 9x^2$
**9.** $-1 - 1/x^2$ **11.** $\dfrac{dy}{dx} = 10(0) = 0$ (constant multiple and power rule) **13.** $\dfrac{dy}{dx} = \dfrac{d}{dx}(x^2) + \dfrac{d}{dx}(x)$ (sum rule) $= 2x + 1$ (power rule)
**15.** $\dfrac{dy}{dx} = \dfrac{d}{dx}(4x^3) + \dfrac{d}{dx}(2x) - \dfrac{d}{dx}(1)$ (sum and difference) $= 4\dfrac{d}{dx}(x^3) + 2\dfrac{d}{dx}(x) - \dfrac{d}{dx}(1)$ (constant multiples) $= 12x^2 + 2$ (power rule)
**17.** $f'(x) = 2x - 3$ **19.** $f'(x) = 1 + 0.5x^{-0.5}$
**21.** $g'(x) = -2x^{-3} + 3x^{-2}$ **23.** $g'(x) = -\dfrac{1}{x^2} + \dfrac{2}{x^3}$
**25.** $h'(x) = -\dfrac{0.8}{x^{1.4}}$ **27.** $h'(x) = -\dfrac{2}{x^3} - \dfrac{6}{x^4}$

**29.** $r'(x) = -\dfrac{2}{3x^2} + \dfrac{0.1}{2x^{1.1}}$

**31.** $r'(x) = \dfrac{2}{3} - \dfrac{0.1}{2x^{0.9}} - \dfrac{4.4}{3x^{2.1}}$

**33.** $t'(x) = |x|/x - 1/x^2$   **35.** $s'(x) = \dfrac{1}{2\sqrt{x}} - \dfrac{1}{2x\sqrt{x}}$

**37.** $s'(x) = 3x^2$   **39.** $t'(x) = 1 - 4x$

**41.** $2.6x^{0.3} + 1.2x^{-2.2}$   **43.** $1.2(1 - |x|/x)$

**45.** $3at^2 - 4a$   **47.** $5.15x^{9.3} - 99x^{-2}$

**49.** $-\dfrac{2.31}{t^{2.1}} - \dfrac{0.3}{t^{0.4}}$   **51.** $4\pi r^2$   **53.** 3   **55.** $-2$   **57.** $-5$

**59.** $y = 3x + 2$                **61.** $y = \dfrac{3}{4}x + 1$

**63.** $y = \dfrac{1}{4}x + 1$                **65.** $x = -3/4$
**67.** No such values
**69.** $x = 1, -1$
**71.** See Solutions Manual.
**73. a.** 2/3  **b.** Not differentiable at 0
**75. a.** 5/2  **b.** Not differentiable at 0

**77.** Yes; 0   **79.** Yes; 12   **81.** No; 3   **83.** Yes; 3/2
**85.** Yes; diverges to $-\infty$   **87.** Yes; diverges to $-\infty$
**89.** $P'(t) = 0.9t - 12$;  $P'(20) = 6$; the price of a
barrel of crude oil was increasing at a rate of $6 per
year in 2000.   **91.** 0.55   **93. a.** $s'(t) = -32t$;
0, $-32$, $-64$, $-96$, $-128$ ft/s  **b.** 5 seconds;
downward at 160 ft/s   **95. a.** $P'(t) = -4.0t + 6.6$;
Decreasing at a rate of $5.4 billion per year  **b.** (B)
**97. a.** $f'(x) = 7.1x - 30.2$; $f'(8) = 26.6$ manatees
per 100,000 boats; At a level of 800,000 boats, manatee deaths are increasing at a rate of 26.6 deaths each
year per 100,000 additional boats.  **b.** Increasing; the
number of manatees killed per additional 100,000
boats increases as the number of boats increases.
**99. a.** $I - A$ measures the amount by which the
iPhone market share exceeds the Android market
share. $(I - A)'$ measures the rate at which this difference is changing.  **b.** (B)  **c.** $-0.2t + 0.5$ percentage
points per month; (C); The vertical distance between
the graphs at first increases, and then decreases; the
iPhone increases its advantage over the Android share
at first, but then the Android begins to catch up.
**d.** $-0.3$ percentage points per month; In May 2009,

the iPhone's advantage over Android was decreasing at
a rate of 0.3 percentage points per month.
**101.** After graphing the curve $y = 3x^2$, draw the line
passing through $(-1, 3)$ with slope $-6$.   **103.** The
slope of the tangent line of $g$ is twice the slope of the
tangent line of $f$.   **105.** $g'(x) = -f'(x)$   **107.** The
left-hand side is not equal to the right-hand side. The
*derivative* of the left-hand side is equal to the right-
hand side, so your friend should have written
$\dfrac{d}{dx}(3x^4 + 11x^5) = 12x^3 + 55x^4$.   **109.** $\dfrac{1}{2x}$ is not
equal to $2x^{-1}$. Your friend should have written
$y = \dfrac{1}{2x} = \dfrac{1}{2}x^{-1}$, so $\dfrac{dy}{dx} = -\dfrac{1}{2}x^{-2}$.   **111.** The
derivative of a constant times a function is the
constant times the derivative of the function, so that
$f'(x) = (2)(2x) = 4x$. Your enemy mistakenly computed the *derivative* of the constant times the derivative of the function. (The derivative of a product of
two functions is not the product of the derivative of the
two functions. The rule for taking the derivative of a
product is discussed later in the chapter.).
**113.** For a general function $f$, the derivative of $f$ is
defined to be $f'(x) = \lim\limits_{h \to 0} \dfrac{f(x + h) - f(x)}{h}$. One
then finds by calculation that the derivative of the
specific function $x^n$ is $nx^{n-1}$. In short, $nx^{n-1}$ is the
derivative of a specific function: $f(x) = x^n$, it is not
the *definition* of the derivative of a general function or
even the definition of the derivative of the function
$f(x) = x^n$.   **115.** Answers may vary.

## Section 11.2

**1.** $C'(1,000) = \$4.80$ per item   **3.** $C'(100) = \$99.90$
per item   **5.** $C'(x) = 4$; $R'(x) = 8 - 0.002$;
$P'(x) = 4 - 0.002x$; $P'(x) = 0$ when $x = 2,000$.
Thus, at a production level of 2,000, the profit is
stationary (neither increasing nor decreasing) with
respect to the production level. This may indicate a
maximum profit at a production level of 2,000.
**7. a.** (B)  **b.** (C)  **c.** (C)   **9. a.** $C'(x) = 2,500 - 0.04x$;
The cost is going up at a rate of $2,499,840 per televi-
sion commercial. The exact cost of airing the fifth

television commercial is \$2,499,820.  **b.** $\overline{C}(x) = 150/x + 2,500 - 0.02x$; $\overline{C}(4) = 2,537.42$ thousand dollars. The average cost of airing the first four television commercials is \$2,537,420.   **11. a.** $R'(x) = 0.90$, $P'(x) = 0.80 - 0.002x$  **b.** Revenue: \$450, Profit: \$80, Marginal revenue: \$0.90, Marginal profit: $-\$0.20$. The total revenue from the sale of 500 copies is \$450. The profit from the production and sale of 500 copies is \$80. Approximate revenue from the sale of the 501st copy is 90¢. Approximate loss from the sale of the 501st copy is 20¢.  **c.** $x = 400$. The profit is a maximum when you produce and sell 400 copies. **13.** The profit on the sale of 1,000 DVDs is \$3,000, and is decreasing at a rate of \$3 per additional DVD sold.   **15.** Profit $\approx$ \$257.07; Marginal profit $\approx$ 5.07. Your current profit is \$257.07 per month, and this would increase at a rate of \$5.07 per additional magazine sold.   **17. a.** \$2.50 per pound **b.** $R(q) = 20,000/q^{0.5}$  **c.** $R(400) = \$1,000$. This is the monthly revenue that will result from setting the price at \$2.50 per pound. $R'(400) = -\$1.25$ per pound of tuna. Thus, at a demand level of 400 pounds per month, the revenue is decreasing at a rate of \$1.25 per pound.  **d.** The fishery should raise the price (to reduce the demand).   **19.** $P'(50) = \$350$. This means that, at an employment level of 50 workers, the firm's daily profit will increase at a rate of \$350 per additional worker it hires.   **21. a.** (B) **b.** (B)  **c.** (C)

**23. a.** $C(x) = 500,000 + 1,600,000x - 100,000\sqrt{x}$;

$C'(x) = 1,600,000 - \dfrac{50,000}{\sqrt{x}}$;

$\overline{C}(x) = \dfrac{500,000}{x} + 1,600,000 - \dfrac{100,000}{\sqrt{x}}$

**b.** $C'(3) \approx \$1,570,000$ per spot, $\overline{C}(3) \approx \$1,710,000$ per spot. The average cost will decrease as $x$ increases.   **25. a.** \$2,000 per one-pound reduction in emissions.  **b.** 2.5 pounds per day reduction. **c.** $N(q) = 100q^2 - 500q + 4,000$; 2.5 pounds per day reduction. The value of $q$ is the same as that for part (b). The net cost to the firm is minimized at the reduction level for which the cost of controlling emissions begins to increase faster than the subsidy. This is why we get the answer by setting these two rates of increase equal to each other. **27.** $M'(10) \approx 0.0002557$ mpg/mph. This means that, at a speed of 10 mph, the fuel economy is increasing at a rate of 0.0002557 miles per gallon per 1-mph increase in speed. $M'(60) = 0$ mpg/mph. This means that, at a speed of 60 mph, the fuel economy is neither increasing nor decreasing with increasing speed.

$M'(70) \approx -0.00001799$. This means that, at 70 mph, the fuel economy is decreasing at a rate of 0.00001799 miles per gallon per 1-mph increase in speed. Thus 60 mph is the most fuel-efficient speed for the car.   **29.** (C)   **31.** (D)   **33.** (B)   **35.** Cost is often measured as a function of the number of items $x$. Thus, $C(x)$ is the cost of producing (or purchasing, as the case may be) $x$ items.  **a.** The average cost function $\overline{C}(x)$ is given by $\overline{C}(x) = C(x)/x$. The marginal cost function is the derivative, $C'(x)$, of the cost function. **b.** The average cost $\overline{C}(r)$ is the slope of the line through the origin and the point on the graph where $x = r$. The marginal cost of the $r$th unit is the slope of the tangent to the graph of the cost function at the point where $x = r$.  **c.** The average cost function $\overline{C}(x)$ gives the average cost of producing the first $x$ items. The marginal cost function $C'(x)$ is the rate at which cost is changing with respect to the number of items $x$, or the incremental cost per item, and approximates the cost of producing the $(x + 1)$st item.   **37.** Answers may vary. An example is $C(x) = 300x$.   **39.** The marginal cost   **41.** Not necessarily. For example, it may be the case that the marginal cost of the 101st item is larger than the average cost of the first 100 items (even though the marginal cost is decreasing). Thus, adding this additional item will *raise* the average cost.   **43.** The circumstances described suggest that the average cost function is at a relatively low point at the current production level, and so it would be appropriate to advise the company to maintain current production levels; raising or lowering the production level will result in increasing average costs.

## Section 11.3

**1.** 3   **3.** $3x^2$   **5.** $2x + 3$   **7.** $210x^{1.1}$
**9.** $-2/x^2$   **11.** $2x/3$   **13.** $36x^2 - 3$   **15.** $3x^2 - 5x^4$
**17.** $8x + 12$   **19.** $-8/(5x - 2)^2$   **21.** $-14/(3x - 1)^2$
**23.** 0   **25.** $-|x|/x^3$   **27.** $3\sqrt{x}/2$
**29.** $(x^2 - 1) + 2x(x + 1) = (x + 1)(3x - 1)$
**31.** $(x^{-0.5} + 4)(x - x^{-1}) + (2x^{0.5} + 4x - 5)(1 + x^{-2})$
**33.** $8(2x^2 - 4x + 1)(x - 1)$
**35.** $(1/3.2 - 3.2/x^2)(x^2 + 1) + 2x(x/3.2 + 3.2/x)$
**37.** $2x(2x + 3)(7x + 2) + 2x^2(7x + 2) + 7x^2(2x + 3)$
**39.** $5.3(1 - x^{2.1})(x^{-2.3} - 3.4) - 2.1x^{1.1}(5.3x - 1)(x^{-2.3} - 3.4) - 2.3x - 3.3(5.3x - 1)(1 - x^{2.1})$

**41.** $\dfrac{1}{2\sqrt{x}}\left(\sqrt{x} + \dfrac{1}{x^2}\right) + (\sqrt{x} + 1)\left(\dfrac{1}{2\sqrt{x}} - \dfrac{2}{x^3}\right)$

**43.** $\dfrac{(4x+4)(3x-1) - 3(2x^2+4x+1)}{(3x-1)^2} = (6x^2 - 4x - 7)/(3x-1)^2$

**45.** $\dfrac{(2x-4)(x^2+x+1) - (x^2-4x+1)(2x+1)}{(x^2+x+1)^2} = (5x^2 - 5)/(x^2+x+1)^2$

**47.** $\dfrac{(0.23x^{-0.77} - 5.7)(1 - x^{-2.9}) - 2.9x^{-3.9}(x^{0.23} - 5.7x)}{(1 - x^{-2.9})^2}$

**49.** $\dfrac{\frac{1}{2}x^{-1/2}(x^{1/2} - 1) - \frac{1}{2}x^{-1/2}(x^{1/2} + 1)}{(x^{1/2} - 1)^2} = \dfrac{-1}{\sqrt{x}\left(\sqrt{x} - 1\right)^2}$    **51.** $-3/x^4$

**53.** $\dfrac{[(x+1) + (x+3)](3x-1) - 3(x+3)(x+1)}{(3x-1)^2} = (3x^2 - 2x - 13)/(3x-1)^2$

**55.** $\dfrac{[(x+1)(x+2) + (x+3)(x+2) + (x+3)(x+1)](3x-1) - 3(x+3)(x+1)(x+2)}{(3x-1)^2}$

**57.** $4x^3 - 2x$   **59.** $64$   **61.** $3$   **63.** Difference; $4x^3 - 12x^2 + 2x - 480$   **65.** Sum; $1 + 2/(x+1)^2$

**67.** Product; $\left[\dfrac{x}{x+1}\right] + (x+2)\dfrac{1}{(x+1)^2}$

**69.** Difference; $2x - 1 - 2/(x+1)^2$   **71.** $y = 12x - 8$

**73.** $y = x/4 + 1/2$   **75.** $y = -2$   **77.** $q'(5) = 1{,}000$ units/month (sales are increasing at a rate of 1,000 units per month); $p'(5) = -\$10$/month (the price of a sound system is dropping at a rate of \$10 per month); $R'(5) = 900{,}000$ (revenue is increasing at a rate of \$900,000 per month).   **79.** \$703 million; increasing at a rate of \$67 million per year   **81.** Decreasing at a rate of \$1 per day   **83.** Decreasing at a rate of approximately \$0.10 per month   **85.** $M'(x) = \dfrac{3{,}000(3{,}600x^{-2} - 1)}{\left(x + 3{,}600x^{-1}\right)^2}$;

$M'(10) \approx 0.7670$ mpg/mph. This means that, at a speed of 10 mph, the fuel economy is increasing at a rate of 0.7670 miles per gallon per one mph increase in speed. $M'(60) = 0$ mpg/mph. This means that, at a speed of 60 mph, the fuel economy is neither increasing nor decreasing with increasing speed. $M'(70) \approx -0.0540$. This means that, at 70 mph, the fuel economy is decreasing at a rate of 0.0540 miles per gallon per one mph increase in speed. 60 mph is the most fuel-efficient speed for the car. (In the next chapter we shall discuss how to locate largest values in general.)   **87. a.** $P(t) - I(t)$ represents the daily production of oil in Mexico that was not exported to the United States. $I(t)/P(t)$ represents U.S. imports of oil from Mexico as a fraction of the total produced there.   **b.** $-0.023$ per year; at the start of 2008, the fraction of oil produced in

Mexico that was imported by the United States was decreasing at a rate of 0.023 (or 2.3 percentage points) per year.   **89.** Increasing at a rate of about \$3,420 million per year.   **91.** $R'(p) = -\dfrac{5.625}{(1 + 0.125p)^2}$;

$R'(4) = -2.5$ thousand organisms per hour, per 1,000 organisms. This means that the reproduction rate of organisms in a culture containing 4,000 organisms is declining at a rate of 2,500 organisms per hour, per 1,000 additional organisms.   **93.** Oxygen consumption is decreasing at a rate of 1,600 milliliters per day. This is due to the fact that the number of eggs is decreasing, because $C'(25)$ is positive.   **95.** 20; 33   **97.** $5/4$; $-17/16$   **99.** The analysis is suspect, as it seems to be asserting that the annual increase in revenue, which we can think of as $dR/dt$, is the product of the annual increases, $dp/dt$ in price, and $dq/dt$ in sales. However, because $R = pq$, the product rule implies that $dR/dt$ is not the product of $dp/dt$ and $dq/dt$, but is instead $\dfrac{dR}{dt} = \dfrac{dp}{dt} \cdot q + p \cdot \dfrac{dq}{dt}$.   **101.** Answers will vary $q = -p + 1{,}000$ is one example.   **103.** Mine; it is increasing twice as fast as yours. The rate of change of revenue is given by $R'(t) = p'(t)q(t)$ because $q'(t) = 0$. Thus, $R'(t)$ does not depend on the selling price $p(t)$.   **105.** (A)

### Section 11.4

**1.** $4(2x+1)$   **3.** $-(x-1)^{-2}$   **5.** $2(2-x)^{-3}$
**7.** $(2x+1)^{-0.5}$   **9.** $-3/(3x-1)^2$

**11.** $4(x^2 + 2x)^3(2x + 2)$  **13.** $-4x(2x^2 - 2)^{-2}$

**15.** $-5(2x - 3)(x^2 - 3x - 1)^{-6}$  **17.** $-6x/(x^2 + 1)^4$

**19.** $1.5(0.2x - 4.2)(0.1x^2 - 4.2x + 9.5)^{0.5}$

**21.** $4(2s - 0.5s^{-0.5})(s^2 - s^{0.5})^3$  **23.** $-x/\sqrt{1 - x^2}$

**25.** $\dfrac{3|3x - 6|}{3x - 6}$  **27.** $\dfrac{(-3x^2 + 5)|-x^3 + 5x|}{-x^3 + 5x}$

**29.** $-[(x + 1)(x^2 - 1)]^{-3/2}(3x - 1)(x + 1)$

**31.** $6.2(3.1x - 2) + 6.2/(3.1x - 2)^3$

**33.** $2[(6.4x - 1)^2 + (5.4x - 2)^3] \times$
$[12.8(6.4x - 1) + 16.2(5.4x - 2)^2]$

**35.** $-2(x^2 - 3x)^{-3}(2x - 3)(1 - x^2)^{0.5}$
$-x(x^2 - 3x)^{-2}(1 - x^2)^{-0.5}$

**37.** $-56(x + 2)/(3x - 1)^3$  **39.** $3z^2(1 - z^2)/(1 + z^2)^4$

**41.** $3[(1 + 2x)^4 - (1 - x)^2]^2[8(1 + 2x)^3 + 2(1 - x)]$

**43.** $6|3x - 1|$  **45.** $\dfrac{|x - (2x - 3)^{1/2}|}{x - (2x - 3)^{1/2}}[1 - (2x - 3)^{-1/2}]$

**47.** $-\dfrac{\left(\dfrac{1}{\sqrt{2x + 1}} - 2x\right)}{\left(\sqrt{2x + 1} - x^2\right)^2}$

**49.** $54(1 + 2x)^2(1 + (1 + 2x)^3)^2(1 + (1 + (1 + 2x)^3)^3)^2$

**51.** $(100x^{99} - 99x^{-2})\,dx/dt$

**53.** $(-3r^{-4} + 0.5r^{-0.5})\,dr/dt$  **55.** $4\pi r^2 dr/dt$

**57.** $-47/4$  **59.** $1/3$  **61.** $-5/3$  **63.** $1/4$

**65.** $\dfrac{dP}{dn}\bigg|_{n=10} = 146{,}454.9$. At an employment

level of 10 engineers, Paramount will increase its profit at a rate of $146,454.90 per additional engineer hired.
**67.** $y = 35(7 + 0.2t)^{-0.25}$; $-0.11$ percentage points per month.  **69.** $-\$30$ per additional ruby sold. The revenue is decreasing at a rate of $30 per additional ruby sold.
**71.** $\dfrac{dy}{dt} = \dfrac{dy}{dx}\dfrac{dx}{dt} = (1.5)(-2) = -3$ murders per
100,000 residents/yr each year.  **73.** $5/6 \approx 0.833$;
relative to the 2003 levels, home sales were changing at a rate of about 0.833 percentage points per percentage point change in price. (Equivalently, home sales in 2008 were dropping at a rate of about 0.833 percentage points per percentage point drop in price.)  **75.** $12\pi$ mi$^2$/h  **77.** $\$200{,}000\pi$/week $\approx$
$628,000/week  **79. a.** $q'(4) \approx 333$ units per month
**b.** $dR/dq = \$800$/unit  **c.** $dR/dt \approx \$267{,}000$ per month
**81.** 3% per year  **83.** 8% per year  **85.** The glob squared, times the derivative of the glob.  **87.** The derivative of a quantity cubed is three times the *original quantity* squared, times the derivative of the quantity, not three times the derivative of the quantity squared. Thus, the correct answer is $3(3x^3 - x)^2(9x^2 - 1)$.

**89.** First, the derivative of a quantity cubed is three times the *original quantity* squared times the derivative of the quantity, not three times the derivative of the quantity squared. Second, the derivative of a quotient is not the quotient of the derivatives; the quotient rule needs to be used in calculating the derivative of $\dfrac{3x^2 - 1}{2x - 2}$. Thus, the correct result (before simplifying) is

$$3\left(\frac{3x^2 - 1}{2x - 2}\right)^2 \left(\frac{6x(2x - 2) - (3x^2 - 1)(2)}{(2x - 2)^2}\right).$$

**91.** Following the calculation thought experiment, pretend that you were evaluating the function at a specific value of $x$. If the last operation you would perform is addition or subtraction, look at each summand separately. If the last operation is multiplication, use the product rule first; if it is division, use the quotient rule first; if it is any other operation (such as raising a quantity to a power or taking a radical of a quantity) then use the chain rule first.  **93.** An

example is $f(x) = \sqrt{x + \sqrt{x + \sqrt{x + \sqrt{x + \sqrt{x + 1}}}}}$.

### Section 11.5

**1.** $1/(x - 1)$  **3.** $1/(x \ln 2)$  **5.** $2x/(x^2 + 3)$
**7.** $e^{x+3}$  **9.** $-e^{-x}$  **11.** $4^x \ln 4$  **13.** $2^{x^2-1}2x \ln 2$
**15.** $1 + \ln x$  **17.** $2x\ln x + (x^2 + 1)/x$
**19.** $10x(x^2 + 1)^4\ln x + (x^2 + 1)^5/x$
**21.** $3/(3x - 1)$  **23.** $4x/(2x^2 + 1)$
**25.** $(2x - 0.63x^{-0.7})/(x^2 - 2.1x^{0.3})$
**27.** $-2/(-2x + 1) + 1/(x + 1)$
**29.** $3/(3x + 1) - 4/(4x - 2)$
**31.** $1/(x + 1) + 1/(x - 3) - 2/(2x + 9)$
**33.** $5.2/(4x - 2)$
**35.** $2/(x + 1) - 9/(3x - 4) - 1/(x - 9)$

**37.** $\dfrac{1}{(x + 1)\ln 2}$  **39.** $\dfrac{1 - 1/t^2}{(t + 1/t)\ln 3}$  **41.** $\dfrac{2 \ln |x|}{x}$

**43.** $\dfrac{2}{x} - \dfrac{2\ln(x - 1)}{x - 1}$  **45.** $e^x(1 + x)$

**47.** $1/(x + 1) + 3e^x(x^3 + 3x^2)$  **49.** $e^x(\ln |x| + 1/x)$
**51.** $2e^{2x+1}$  **53.** $(2x - 1)e^{x^2-x+1}$  **55.** $2xe^{2x-1}(1 + x)$
**57.** $4(e^{2x-1})^2$  **59.** $2 \cdot 3^{2x-4} \ln 3$
**61.** $2 \cdot 3^{2x+1} \ln 3 + 3e^{3x+1}$

**63.** $\dfrac{2x3^{x^2}[(x^2 + 1)\ln 3 - 1]}{(x^2 + 1)^2}$

**65.** $-4/(e^x - e^{-x})^2$

**67.** $5e^{5x-3}$    **69.** $-\dfrac{\ln x + 1}{(x \ln x)^2}$    **71.** $2(x-1)$    **73.** $\dfrac{1}{x \ln x}$

**75.** $\dfrac{1}{2x \ln x}$    **77.** $y = (e/\ln 2)(x-1) \approx 3.92(x-1)$

**79.** $y = x$    **81.** $y = -[1/(2e)](x-1)+e$    **83.** $163 billion and increasing at a rate of $5.75 billion per year    **85.** $163 billion and increasing at a rate of $5.75 billion per year.    **87.** $-1,653$ years per gram; the age of the specimen is decreasing at a rate of about 1,653 years per additional one gram of carbon 14 present in the sample. (Equivalently, the age of the specimen is increasing at a rate of about 1,653 years per additional one gram less of carbon 14 in the sample.)    **89.** Average price: $1.4 million; increasing at a rate of about $220,000 per year.
**91. a.** $N(15) \approx 1,762 \approx 1,800$ (rounded to 2 significant digits) wiretap orders; $N'(15) \approx 89.87 \approx 90$ wiretap orders per year (rounded to 2 significant digits). The constants in the model are specified to 2 significant digits, so we cannot expect the answer to be accurate to more than 2 digits. **b.** In 2005, the number of people whose communications were intercepted was about 180,000 and increasing at a rate of about 9,000 people per year. **c.** (C)    **93.** $451.00 per year
**95.** $446.02 per year    **97.** $A(t) = 167(1.18)^t$; 280 new cases per day    **99. a.** (A) **b.** The math SAT increases by approximately 0.97 points. **c.** $S'(x)$ decreases with increasing $x$, so that as parental income increases, the effect on math SAT scores decreases.
**101. a.** $-6.25$ years/child; when the fertility rate is 2 children per woman, the average age of a population is dropping at a rate of 6.25 years per one-child increase in the fertility rate. **b.** 0.160
**103.** 3,300,000 cases/week; 11,000,000 cases/week; 640,000 cases/week    **105.** 2.1 percentage points per year; the rate of change is the slope of the tangent at $t = 3$. This is also approximately the average rate of change over [2, 4], which is about $4/2 = 2$, in approximate agreement with the answer.
**107. a.** 2.1 percentage points per year
**b.** $\lim_{t \to +\infty} A(t) = 15$; Had the trend continued indefinitely, the percentage of mortgages that were subprime would have approached 15% in the long term. $\lim_{t \to +\infty} A'(t) = 0$; Had the trend continued indefinitely, the rate of change of the percentage of mortgages that were subprime would have approached 0 percentage points per year in the long term.
**109.** 277,000 people per year    **111.** 0.000283 g/yr
**113.** $R(t) = 350e^{-0.1t}(39t + 68)$ million dollars; $R(2) \approx 42$ billion; $R'(2) \approx 7$ billion per year

**115.** $e$ raised to the glob, times the derivative of the glob.    **117.** 2 raised to the glob, times the derivative of the glob, times the natural logarithm of 2.
**119.** The derivative of $\ln |u|$ is not $\dfrac{1}{|u|}\dfrac{du}{dx}$; it is $\dfrac{1}{u}\dfrac{du}{dx}$. Thus, the correct derivative is $\dfrac{3}{3x+1}$.
**121.** The power rule does not apply when the exponent is not constant. The derivative of 3 raised to a quantity is 3 raised to the quantity, times the derivative of the quantity, times $\ln 3$. Thus, the correct answer is $3^{2x}\, 2 \ln 3$.    **123.** No. If $N(t)$ is exponential, so is its derivative.    **125.** If $f(x) = e^{kx}$, then the fractional rate of change is $\dfrac{f'(x)}{f(x)} = \dfrac{ke^{kx}}{e^{kx}} = k$, the fractional growth rate.    **127.** If $A(t)$ is growing exponentially, then $A(t) = A_0 e^{kt}$ for constants $A_0$ and $k$. Its percentage rate of change is then
$$\frac{A'(t)}{A(t)} = \frac{kA_0 e^{kt}}{A_0 e^{kt}} = k,\text{ a constant.}$$

## Section 11.6

**1.** $-2/3$    **3.** $x$    **5.** $(y-2)/(3-x)$    **7.** $-y$
**9.** $-\dfrac{y}{x(1+\ln x)}$    **11.** $-x/y$    **13.** $-2xy/(x^2 - 2y)$
**15.** $-(6+9x^2 y)/(9x^3 - x^2)$    **17.** $3y/x$
**19.** $(p + 10p^2 q)/(2p - q - 10pq^2)$
**21.** $(ye^x - e^y)/(xe^y - e^x)$    **23.** $se^{st}/(2s - te^{st})$
**25.** $ye^x/(2e^x + y^3 e^y)$    **27.** $(y - y^2)/(-1 + 3y - y^2)$
**29.** $-y/(x + 2y - xye^y - y^2 e^y)$    **31. a.** 1
**b.** $y = x - 3$    **33. a.** $-2$ **b.** $y = -2x$
**35. a.** $-1$ **b.** $y = -x + 1$    **37. a.** $-2,000$
**b.** $y = -2,000x + 6,000$    **39. a.** 0 **b.** $y = 1$
**41. a.** $-0.1898$ **b.** $y = -0.1898x + 1.4721$

**43.** $\dfrac{2x+1}{4x-2}\left[\dfrac{2}{2x+1} - \dfrac{4}{4x-2}\right]$

**45.** $\dfrac{(3x+1)^2}{4x(2x-1)^3}\left[\dfrac{6}{3x+1} - \dfrac{1}{x} - \dfrac{6}{2x-1}\right]$

**47.** $(8x-1)^{1/3}(x-1)\left[\dfrac{8}{3(8x-1)} + \dfrac{1}{x-1}\right]$

**49.** $(x^3+x)\sqrt{x^3+2}\left[\dfrac{3x^2+1}{x^3+x} + \dfrac{1}{2}\dfrac{3x^2}{x^3+2}\right]$

**51.** $x^x(1 + \ln x)$    **53.** $-$$3,000 per worker. The monthly budget to maintain production at the fixed level $P$ is decreasing by approximately $3,000 per additional worker at an employment level of 100 workers and a monthly operating budget of $200,000.

**55.** $-125$ T-shirts per dollar; when the price is set at $5, the demand is dropping by 125 T-shirts per $1 increase in price.    **57.** $\left.\dfrac{dk}{de}\right|_{e=15} = -0.307$ carpenters per electrician. This means that, for a $200,000 house whose construction employs 15 electricians, adding one more electrician would cost as much as approximately 0.307 additional carpenters. In other words, one electrician is worth approximately 0.307 carpenters. **59. a.** 22.93 hours. (The other root is rejected because it is larger than 30.)    **b.** $\dfrac{dt}{dx} = \dfrac{4t - 20x}{0.4t - 4x}; \left.\dfrac{dt}{dx}\right|_{x=3.0} \approx$ $-11.2$ hours per grade point. This means that, for a 3.0 student who scores 80 on the examination, 1 grade point is worth approximately 11.2 hours. **61.** $\dfrac{dr}{dy} = 2\dfrac{r}{y}$, so $\dfrac{dr}{dt} = 2\dfrac{r}{y}\dfrac{dy}{dt}$ by the chain rule. **63.** $x, y, y, x$    **65.** Let $y = f(x)g(x)$. Then $\ln y = \ln f(x) + \ln g(x)$, and $\dfrac{1}{y}\dfrac{dy}{dx} = \dfrac{f'(x)}{f(x)} + \dfrac{g'(x)}{g(x)}$,

so $\dfrac{dy}{dx} = y\left(\dfrac{f'(x)}{f(x)} + \dfrac{g'(x)}{g(x)}\right) =$

$f(x)g(x)\left(\dfrac{f'(x)}{f(x)} + \dfrac{g'(x)}{g(x)}\right) = f'(x)g(x) + f(x)g'(x)$.

**67.** Writing $y = f(x)$ specifies $y$ as an explicit function of $x$. This can be regarded as an equation giving $y$ as an *implicit* function of $x$. The procedure of finding $dy/dx$ by implicit differentiation is then the same as finding the derivative of $y$ as an explicit function of $x$: We take $d/dx$ of both sides.    **69.** Differentiate both sides of the equation $y = f(x)$ with respect to $y$ to get $1 = f'(x) \cdot \dfrac{dx}{dy}$, giving $\dfrac{dx}{dy} = \dfrac{1}{f'(x)} = \dfrac{1}{dy/dx}$.

### Chapter 11 Review

**1.** $50x^4 + 2x^3 - 1$    **3.** $9x^2 + x^{-2/3}$    **5.** $1 - 2/x^3$

**7.** $-\dfrac{4}{3x^2} + \dfrac{0.2}{x^{1.1}} + \dfrac{1.1x^{0.1}}{3.2}$    **9.** $e^x(x^2 + 2x - 1)$

**11.** $(-3x|x| + |x|/x - 6x)/(3x^2 + 1)^2$

**13.** $-4(4x - 1)^{-2}$    **15.** $20x(x^2 - 1)^9$

**17.** $-0.43(x + 1)^{-1.1}[2 + (x + 1)^{-0.1}]^{3.3}$

**19.** $e^x(x^2 + 1)^9(x^2 + 20x + 1)$

**21.** $3^x[(x - 1)\ln 3 - 1]/(x - 1)^2$    **23.** $2xe^{x^2-1}$

**25.** $2x/(x^2 - 1)$    **27.** $x = 7/6$    **29.** $x = \pm 2$

**31.** $x = (1 - \ln 2)/2$    **33.** None    **35.** $\dfrac{2x - 1}{2y}$

**37.** $-y/x$    **39.** $\dfrac{(2x - 1)^4(3x + 4)}{(x + 1)(3x - 1)^3} \times$

$\left[\dfrac{8}{2x - 1} + \dfrac{3}{3x + 4} - \dfrac{1}{x + 1} - \dfrac{9}{3x - 1}\right]$

**41.** $y = -x/4 + 1/2$    **43.** $y = -3ex - 2e$

**45.** $y = x + 2$    **47. a.** 274 books per week  **b.** 636 books per week  **c.** The function $w$ begins to decrease more and more rapidly after $t = 14$ Graph:

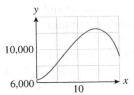

**d.** Because the data suggest an upward curving parabola, the long-term prediction of sales for a quadratic model would be that sales will increase without bound, in sharp contrast to (c)    **49. a.** $2.88 per book **b.** $3.715 per book  **c.** Approximately $-$0.000104 per book, per additional book sold.  **d.** At a sales level of 8,000 books per week, the cost is increasing at a rate of $2.88 per book (so that the 8,001st book costs approximately $2.88 to sell), and it costs an average of $3.715 per book to sell the first 8,000 books. Moreover, the average cost is decreasing at a rate of $0.000104 per book, per additional book sold. **51. a.** $3,000 per week (rising)  **b.** 300 books per week **53.** $R = pq$ gives $R' = p'q + pq'$. Thus, $R'/R = R'/(pq) = (p'q + pq')/pq = p'/p + q'/q$

**55.** $110 per year    **57. a.** $s'(t) = \dfrac{2{,}475e^{-0.55(t-4.8)}}{(1 + e^{-0.55(t-4.8)})^2}$; 556 books per week  **b.** 0; In the long term, the rate of increase of weekly sales slows to zero.    **59.** 616.8 hits per day per week.    **61. a.** $-17.24$ copies per $1. The demand for the gift edition of *The Complete Larry Potter* is dropping at a rate of about 17.24 copies per $1 increase in the price.  **b.** $138 per dollar is positive, so the price should be raised.

## Chapter 12

### Section 12.1

**1.** Absolute min: $(-3, -1)$, relative max: $(-1, 1)$, relative min: $(1, 0)$, absolute max: $(3, 2)$    **3.** Absolute min: $(3, -1)$ and $(-3, -1)$, absolute max: $(1, 2)$ **5.** Absolute min: $(-3, 0)$ and $(1, 0)$, absolute max: $(-1, 2)$ and $(3, 2)$    **7.** Relative min: $(-1, 1)$ **9.** Absolute min: $(-3, -1)$, relative max: $(-2, 2)$, relative min: $(1, 0)$, absolute max: $(3, 3)$

**11.** Relative max: $(-3, 0)$, absolute min: $(-2, -1)$, stationary nonextreme point: $(1, 1)$ **13.** Absolute max: $(0, 1)$, absolute min: $(2, -3)$, relative max: $(3, -2)$ **15.** Absolute min: $(-4, -16)$, absolute max: $(-2, 16)$, absolute min: $(2, -16)$, absolute max: $(4, 16)$ **17.** Absolute min: $(-2, -10)$, absolute max: $(2, 10)$ **19.** Absolute min: $(-2, -4)$, relative max: $(-1, 1)$, relative min: $(0, 0)$ **21.** Relative max: $(-1, 5)$, absolute min: $(3, -27)$ **23.** Absolute min: $(0, 0)$ **25.** Absolute maxima at $(0, 1)$ and $(2, 1)$, absolute min at $(1, 0)$ **27.** Relative maximum at $(-2, -1/3)$, relative minimum at $(-1, -2/3)$, absolute maximum at $(0, 1)$ **29.** Relative min: $(-2, 5/3)$, relative max: $(0, -1)$, relative min: $(2, 5/3)$ **31.** Relative max: $(0, 0)$; absolute min: $(1/3, -2\sqrt{3}/9)$ **33.** Relative max: $(0, 0)$, absolute min: $(1, -3)$ **35.** No relative extrema **37.** Absolute min: $(1, 1)$ **39.** Relative max: $(-1, 1 + 1/e)$, absolute min: $(0, 1)$, absolute max: $(1, e - 1)$ **41.** Relative max: $(-6, -24)$, relative min: $(-2, -8)$ **43.** Absolute max $(1/\sqrt{2}, \sqrt{e/2})$, absolute min: $(-1/\sqrt{2}, -\sqrt{e/2})$ **45.** Relative min at $(0.15, -0.52)$ and $(2.45, 8.22)$, relative max at $(1.40, 0.29)$ **47.** Absolute max at $(-5, 700)$, relative max at $(3.10, 28.19)$ and $(6, 40)$, absolute min at $(-2.10, -392.69)$ and relative min at $(5, 0)$. **49.** Stationary minimum at $x = -1$ **51.** Stationary minima at $x = -2$ and $x = 2$, stationary maximum at $x = 0$ **53.** Singular minimum at $x = 0$, stationary nonextreme point at $x = 1$ **55.** Stationary minimum at $x = -2$, singular nonextreme points at $x = -1$ and $x = 1$, stationary maximum at $x = 2$ **57.** Answers will vary.    **59.** Answers will vary.

**61.** Not necessarily; it could be neither a relative maximum nor a relative minimum, as in the graph of $y = x^3$ at the origin.
**63.** Answers will vary.

**65.** The graph oscillates faster and faster above and below zero as it approaches the end-point at 0, so 0 cannot be either a relative minimum or maximum.

## Section 12.2

**1.** $x = y = 5$; $P = 25$   **3.** $x = y = 3$; $S = 6$
**5.** $x = 2$, $y = 4$; $F = 20$   **7.** $x = 20$, $y = 10$, $z = 20$; $P = 4,000$   **9.** $5 \times 5$   **11.** 1,500 per day for an average cost of \$130 per iPod   **13.** $\sqrt{40} \approx 6.32$ pounds of pollutant per day, for an average cost of about \$1,265 per pound   **15.** 2.5 lb   **17.** $5 \times 10$
**19.** $50 \times 10$ for an area of 250 sq. ft.   **21.** $11 \times 22$
**23.** \$10   **25.** \$78 for a quarterly revenue of \$6,084 million, or \$6.084 billion   **27.** \$4.61 for a daily revenue of \$95,680.55   **29. a.** \$1.41 per pound **b.** 5,000 pounds **c.** \$7,071.07 per month
**31.** 34.5¢ per pound, for an annual (per capita) revenue of \$5.95   **33.** \$98 for an annual profit of \$3,364 million, or \$3.364 billion   **35. a.** 656 headsets, for a profit of \$28,120 **b.** \$143 per headset
**37.** Height = Radius of base $\approx 20.48$ cm.
**39.** Height $\approx 10.84$ cm; Radius $\approx 2.71$ cm; Height/Radius = 4   **41.** $13\frac{1}{3}$ in $\times 3\frac{1}{3}$ in $\times 1\frac{1}{3}$ in for a volume of $1,600/27 \approx 59$ cubic inches
**43.** $5 \times 5 \times 5$ cm   **45.** $l = w = h \approx 20.67$ in, volume $\approx 8,827$ in$^3$   **47.** $l = 30$ in, $w = 15$ in, $h = 30$ in   **49.** $l = 36$ in, $w = h = 18$ in, $V = 11,664$ in$^3$   **51. a.** 1.6 years, or year 2001.6; **b.** $R_{max} = \$28,241$ million   **53.** $t = 2.5$ or midway through 1972; $D(2.5)/S(2.5) \approx 4.09$. The number of new (approved) drugs per \$1 billion of spending on research and development reached a high of around four approved drugs per \$1 billion midway through 1972.   **55.** 30 years from now   **57.** 55 days
**59.** 1,600 copies. At this value of $x$, average profit equals marginal profit; beyond this the marginal profit is smaller than the average.   **61.** 40 laborers and 250 robots   **63.** 71 employees   **65.** Increasing most rapidly in 1990; increasing least rapidly in 2007
**67.** Maximum when $t = 17$ days. This means that the embryo's oxygen consumption is increasing most rapidly 17 days after the egg is laid.
**69.** Graph of derivative:

The absolute maximum occurs at approximately (3.7, 2.2) during the year 2003. The percentage of mortgages that were subprime was increasing most rapidly during 2003, when it increased at a rate of around 2.2 percentage points per year.   **71.** You should sell them in 17 years' time, when they will be worth approximately \$3,960.   **73.** 25 additional trees   **75.** (D)   **77.** (A); (B)   **79.** The problem is uninteresting because the company can accomplish the objective by cutting away the entire sheet of cardboard, resulting in a box with surface area zero.   **81.** Not all absolute extrema occur at stationary points; some may occur at an endpoint or singular point of the domain, as in Exercises 29, 30, 65, and 66.   **83.** The minimum of $dq/dp$ is the fastest that the demand is dropping in response to increasing price.

### Section 12.3

**1.** 6   **3.** $4/x^3$   **5.** $-0.96x^{-1.6}$   **7.** $e^{-(x-1)}$
**9.** $2/x^3 + 1/x^2$   **11. a.** $a = -32$ ft/sec$^2$
**b.** $a = -32$ ft/sec$^2$   **13. a.** $a = 2/t^3 + 6/t^4$ ft/sec$^2$
**b.** $a = 8$ ft/sec$^2$   **15. a.** $a = -1/(4t^{3/2}) + 2$ ft/sec$^2$
**b.** $a = 63/32$ ft/sec$^2$   **17.** (1, 0)   **19.** (1, 0)   **21.** None
**23.** $(-1, 0), (1, 1)$   **25.** Points of inflection at $x = -1$ and $x = 1$   **27.** One point of inflection, at $x = -2$
**29.** Points of inflection at $x = -2$, $x = 0$, $x = 2$
**31.** Points of inflection at $x = -2$ and $x = 2$
**33.** $x = 2$; minimum   **35.** Maximum at $x = -2$, minimum at $x = 2$   **37.** Maximum at $t = -1/\sqrt{3}$, minimum at $t = 1/\sqrt{3}$   **39.** Nonextreme stationary point at $x = 0$ minimum at $x = 3$   **41.** Maximum at $x = 0$   **43.** Minimum at $x = -1/\sqrt{2}$; maximum at $x = 1/\sqrt{2}$   **45.** $f'(x) = 8x - 1$; $f''(x) = 8$;
$f'''(x) = f^{(4)}(x) = \ldots = f^{(n)}(x) = 0$
**47.** $f'(x) = -4x^3 + 6x$; $f''(x) = -12x^2 + 6$;
$f'''(x) = -24x$; $f^{(4)}(x) = -24$;
$f^{(5)}(x) = f^{(6)}(x) = \ldots = f^{(n)}(x) = 0$
**49.** $f'(x) = 8(2x + 1)^3$; $f''(x) = 48(2x + 1)^2$;
$f'''(x) = 192(2x + 1)$; $f^{(4)}(x) = 384$;
$f^{(5)}(x) = f^{(6)}(x) = \ldots = f^{(n)}(x) = 0$
**51.** $f'(x) = -e^{-x}$; $f''(x) = e^{-x}$; $f'''(x) = -e^{-x}$;
$f^{(4)}(x) = e^{-x}$; $f^{(n)}(x) = (-1)^n e^{-x}$
**53.** $f'(x) = 3e^{3x-1}$; $f''(x) = 9e^{3x-1}$;
$f'''(x) = 27e^{3x-1}$; $f^{(4)}(x) = 81e^{3x-1}$;
$f^{(n)}(x) = 3^n e^{3x-1}$   **55.** $-3.8$ m/s$^2$
**57.** $6t - 2$ ft/s$^2$; increasing   **59.** Decelerating by 90 million gals/yr$^2$   **61. a.** 400 ml   **b.** 36 ml/day
**c.** $-1$ ml/day$^2$   **63. a.** 0.6%   **b.** Speeding up
**c.** Speeding up for $t < 3.33$ (prior to 1/3 of the way through March) and slowing for $t > 3.33$ (after that time)   **65. a.** December 2005: $-0.202\%$ (deflation rate of 0.202%) February 2006: 0.363%   **b.** Speeding up

**c.** Speeding up for $t > 4.44$ (after mid-November) and decreasing for $t < 4.44$ (prior to that time).
**67.** Concave up for $8 < t < 20$, concave down for $0 < t < 8$, point of inflection around $t = 8$. The percentage of articles written by researchers in the United States was decreasing most rapidly at around $t = 8$ (1991).   **69. a.** (B)  **b.** (B)  **c.** (A)
**71.** Graphs:
$A(t)$:

$A'(t)$:

$A''(t)$:

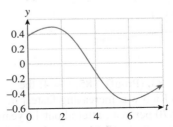

Concave up when $t < 4$; concave down when $t > 4$; point of inflection when $t \approx 4$. The percentage of U.S. mortgages that were subprime was increasing fastest at the beginning of 2004.   **73. a.** 2 years into the epidemic  **b.** 2 years into the epidemic   **75. a.** 2024
**b.** 2026  **c.** 2022; (A)   **77. a.** There are no points of inflection in the graph of $S$.  **b.** Because the graph is concave up, the derivative of $S$ is increasing, and so the rate of *decrease* of SAT scores with increasing numbers of prisoners was diminishing. In other words, the apparent effect of more prisoners on SAT scores was diminishing.   **79. a.** $\left.\dfrac{d^2n}{ds^2}\right|_{s=3} = -21.494$. Thus, for a firm with annual sales of \$3 million, the rate at which new patents are produced decreases with

increasing firm size. This means that the returns (as measured in the number of new patents per increase of $1 million in sales) are diminishing as the firm size increases. **b.** $\left.\dfrac{d^2n}{ds^2}\right|_{s=7} = 13.474$. Thus, for a firm with annual sales of $7 million, the rate at which new patents are produced increases with increasing firm size by 13.474 new patents per $1 million increase in annual sales. **c.** There is a point of inflection when $s \approx 5.4587$, so that in a firm with sales of $5,458,700 per year, the number of new patents produced per additional $1 million in sales is a minimum.
**81.** Graphs:

$I(t)/P(t)$:

$[I(t)/P(t)]'$:

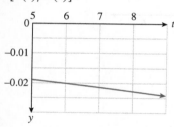

Concave down; (D)   **83.** (Proof)   **85.** $t \approx 4$; the population of Puerto Rico was increasing fastest in 1954.   **87.** About $570 per year, after about 12 years   **89.** Increasing most rapidly in 17.64 years, decreasing most rapidly now (at $t = 0$)   **91.** Non-negative   **93.** Daily sales were decreasing most rapidly in June 2002.
**95.**

 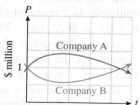

**97.** At a point of inflection, the graph of a function changes either from concave up to concave down, or

vice versa. If it changes from concave up to concave down, then the derivative changes from increasing to decreasing, and hence has a relative maximum. Similarly, if it changes from concave down to concave up, the derivative has a relative minimum.

## Section 12.4

**1. a.** $x$-intercept: $-1$; $y$-intercept: 1  **b.** Absolute min at $(-1, 0)$  **c.** None  **d.** None  **e.** $y \to +\infty$ as $x \to \pm\infty$
Graph:

**3. a.** $x$-intercepts: $-\sqrt{12}, 0, \sqrt{12}$; $y$-intercept: 0
**b.** Absolute min at $(-4, -16)$ and $(2, -16)$, absolute max at $(-2, 16)$ and $(4, 16)$  **c.** $(0, 0)$  **d.** None
**e.** None
Graph:

**5. a.** $x$-intercepts: $-3.6, 0, 5.1$; $y$-intercept: 0
**b.** Relative max at $(-2, 44)$, relative min at $(3, -81)$
**c.** $(0.5, -18.5)$  **d.** None  **e.** $y \to -\infty$ as $x \to -\infty$; $y \to +\infty$ as $x \to +\infty$
Graph:

**7. a.** $x$-intercepts: $-3.3, 0.1, 1.8$; $y$-intercept: 1
**b.** Relative max at $(-2, 21)$, relative min at $(1, -6)$
**c.** $(-1/2, 15/2)$  **d.** None  **e.** $y \to -\infty$ as $x \to -\infty$; $y \to +\infty$ as $x \to +\infty$

Graph:

**9. a.** $x$-intercepts: $-2.9$, $4.2$; $y$-intercept: $10$
**b.** Relative max at $(-2, 74)$, relative min at $(0, 10)$, absolute max at $(3, 199)$ **c.** $(-1.12, 44.8)$, $(1.79, 117.3)$
**d.** None **e.** $y \to -\infty$ as $x \to -\infty$; $y \to -\infty$ as $x \to +\infty$
Graph:

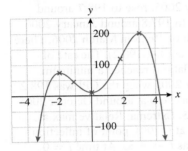

**11. a.** $t$-intercepts: $t = 0$; $y$-intercept: $0$ **b.** Absolute min at $(0, 0)$ **c.** $(1/3, 11/324)$ and $(1, 1/12)$ **d.** None
**e.** $y \to +\infty$ as $t \to -\infty$; $y \to +\infty$ as $t \to +\infty$
Graph:

**13. a.** $x$-intercepts: None; $y$-intercept: None
**b.** Relative min at $(1, 2)$, relative max at $(-1, -2)$
**c.** None **d.** $y \to -\infty$ as $x \to 0^-$; $y \to +\infty$ as $x \to 0^+$, so there is a vertical asymptote at $x = 0$. **e.** $y \to -\infty$ as $x \to -\infty$; $y \to +\infty$ as $x \to +\infty$
Graph:

**15. a.** $x$-intercept: $0$; $y$-intercept: $0$ **b.** None **c.** $(0, 0)$, $(-3, -9/4)$, and $(3, 9/4)$ **d.** None **e.** $y \to -\infty$ as $x \to -\infty$; $y \to +\infty$ as $x \to +\infty$

Graph:

**17. a.** $t$-intercepts: None; $y$-intercept: $-1$ **b.** Relative min at $(-2, 5/3)$ and $(2, 5/3)$, relative max at $(0, -1)$ **c.** None **d.** $y \to +\infty$ as $t \to -1^-$; $y \to -\infty$ as $t \to -1^+$; $y \to -\infty$ as $t \to 1^-$; $y \to +\infty$ as $t \to 1^+$; so there are vertical asymptotes at $t = \pm 1$. **e.** None
Graph:

**19. a.** $x$-intercepts: $-0.6$; $y$-intercept: $1$ **b.** Relative maximum at $(-2, -1/3)$, relative minimum at $(-1, -2/3)$ **c.** None **d.** None. **e.** $y \to -\infty$ as $x \to -\infty$; $y \to +\infty$ as $x \to +\infty$
Graph:

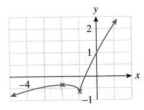

**21. a.** $x$-intercepts: None; $y$-intercept: None
**b.** Absolute min at $(1, 1)$ **c.** None **d.** Vertical asymptote at $x = 0$ **e.** $y \to +\infty$ as $x \to +\infty$
Graph:

**23. a.** $x$-intercepts: $\pm 0.8$; $x$-intercept: None **b.** None **c.** $(1,1)$ and $(-1, 1)$ **d.** $y \to -\infty$ as $x \to 0$; vertical asymptote at $x = 0$ **e.** $y \to +\infty$ as $x \to \pm \infty$

Graph:

**25. a.** *t*-intercepts: None; *y*-intercept: 1 **b.** Absolute min at $(0, 1)$. Absolute max at $(1, e - 1)$, relative max at $(-1, e^{-1} + 1)$. **c.** None **d.** None **e.** None
Graph:

**27.** Absolute min at $(1.40, -1.49)$; points of inflection: $(0.21, 0.61)$, $(0.79, -0.55)$
Graph:

**29.** $f(x) = e^x - x^3$. Relative min at $(-0.46, 0.73)$, relative max at $(0.91, 1.73)$, absolute min at $(3.73, -10.22)$; points of inflection at $(0.20, 1.22)$ and $(2.83, -5.74)$
Graph:

**31.** *y*-intercept: 125; *t*-intercepts: None. The median home price was about $125,000 at the start of 2000 $(t = 0)$. Extrema: Absolute minimum at $(0, 125)$; absolute maximum at $(6, 225)$. The median home price was lowest in 2000 $(t = 0)$ when it stood at $125,000; the median home price was at its highest at the start of

2006 $(t = 6)$ at $225,000. Points of inflection at $(3, 175)$ and $(10, 200)$. The median home price was increasing most rapidly at the start of 2003 $(t = 3)$ when it was $175,000, and decreasing most rapidly at the start of 2010 when it was $200,000. Points where the function is not defined: None. Behavior at infinity: As $t \to +\infty$, $y \to 175$. Assuming the trend shown in the graph continues indefinitely, the median home price will approach a value of $175,000 in the long term.    **33. a.** Intercepts: No *t*-intercept; *y*-intercept at $I(0) = 195$. The CPI was never zero during the given period; in July 2005 the CPI was 195. Absolute min at $(0, 195)$, absolute max at $(8, 199.3)$, relative max at $(2.9, 198.7)$, relative min at $(6.0, 197.8)$. The CPI was at a low of 195 in July 2005, rose to 198.7 around October 2005, dipped to 197.8 around January 2006, and then rose to a high of 199.3 in March 2006. There is a point of inflection at $(4.4, 198.2)$. The rate of change of the CPI (inflation) reached a minimum around mid-November 2005 when the CPI was 198.2. **b.** The inflation rate was zero at around October 2005 and January 2006.    **35.** Extrema: Relative max at $(0, 100)$, absolute min at $(1, 99)$; point of inflection $(0.5, 99.5)$; $s \to +\infty$ as $t \to +\infty$. At time $t = 0$ seconds, the UFO is 100 ft away from the observer, and begins to move closer. At time $t = 0.5$ seconds, when the UFO is 99.5 feet away, its distance is decreasing most rapidly (it is moving toward the observer most rapidly). It then slows down to a stop at $t = 1$ sec when it is at its closest point (99 ft away) and then begins to move further and further away.
Graph:

**37.** Intercepts: None; Absolute minimum at $(1,500, 130)$; No points of inflection; vertical asymptote at $x = 0$. As $x \to +\infty$, $y \to +\infty$. The average cost is never zero, nor is it defined for zero iPods. The average cost is a minimum ($130) when 1,500 iPods are manufactured per day. The average cost becomes extremely large for very small or very large numbers of iPods.

Graph:

**39.** Graph of derivative:

The absolute maximum occurs at approximately (3.7, 2.2); during the year 2003. The percentage of mortgages that were subprime was increasing most rapidly during 2003, when it increased at a rate of around 2.2 percentage points per year. As $t \to +\infty$, $A'(t) \to 0$; In the long term, assuming the trend shown in the model continues, the rate of change of the percentage of mortgages that were subprime approaches zero; that is, the percentage of mortgages that were subprime approaches a constant value. **41.** No; yes. Near a vertical asymptote the value of $y$ increases without bound, and so the graph could not be included between two horizontal lines; hence, no vertical asymptotes are possible. Horizontal asymptotes are possible, as for instance in the graph in Exercise 31.    **43.** It too has a vertical asymptote at $x = a$; the magnitude of the derivative increases without bound as $x \to a$.    **45.** No. If the leftmost critical point is a relative maximum, the function will decrease from there until it reaches the rightmost critical point, so can't have a relative maximum there. **47.** Between every pair of zeros of $f(x)$ there must be a local extremum, which must be a stationary point of $f(x)$, hence a zero of $f'(x)$.

### Section 12.5

**1.** $P = 10,000$; $\frac{dP}{dt} = 1,000$    **3.** Let $R$ be the annual revenue of my company, and let $q$ be annual sales. $R = 7,000$ and $\frac{dR}{dt} = -700$. Find $\frac{dq}{dt}$.    **5.** Let $p$ be the price of a pair of shoes, and let $q$ be the demand

for shoes. $\frac{dp}{dt} = 5$. Find $\frac{dq}{dt}$.    **7.** Let $T$ be the average global temperature, and let $q$ be the number of Bermuda shorts sold per year. $T = 60$ and $\frac{dT}{dt} = 0.1$. Find $\frac{dq}{dt}$.    **9. a.** $6/(100\pi) \approx 0.019$ km/sec **b.** $6/(8\sqrt{\pi}) \approx 0.4231$ km/sec    **11.** $3/(4\pi) \approx$ 0.24 ft/min    **13.** 7.5 ft/sec    **15.** Decreasing at a rate of \$1.66 per player per week    **17.** Monthly sales will drop at a rate of 26 T-shirts per month.    **19.** Raise the price by 3¢ per week.    **21.** Increasing at a rate of \$233.68 million per year    **23.** 1 laborer per month **25.** Dropping at a rate of \$2.40 per year.    **27.** The price is decreasing at a rate of approximately 31¢ per pound per month.    **29.** $2,300/\sqrt{4,100} \approx 36$ miles/hour. **31.** About 10.7 ft/sec    **33.** The $y$ coordinate is decreasing at a rate of 16 units per second. **35.** \$534 per year    **37.** Their prior experience must increase at a rate of approximately 0.97 years every year.    **39.** $\frac{2,500}{9\pi}\left(\frac{3}{5,000}\right)^{2/3} \approx 0.63$ m/sec **41.** $\frac{\sqrt{1+128\pi}}{4\pi} \approx 1.6$ cm/sec    **43.** 0.5137 computers per household, and increasing at a rate of 0.0230 computers per household per year.    **45.** The average SAT score was 904.71 and decreasing at a rate of 0.11 per year.    **47.** Decreasing by 2 percentage points per year **49.** The section is called "related rates" because the goal is to compute the rate of change of a quantity based on a knowledge of the rate of change of a related quantity.    **51.** Answers may vary: A rectangular solid has dimensions 2 cm × 5 cm × 10 cm, and each side is expanding at a rate of 3 cm/second. How fast is the volume increasing?    **53.** (Proof)    **55.** Linear **57.** Let $x =$ my grades and $y =$ your grades. If $dx/dt = 2\,dy/dt$, then $dy/dt = (1/2)\,dx/dt$.

### Section 12.6

**1.** $E = 1.5$; the demand is going down 1.5% per 1% increase in price at that price level; revenue is maximized when $p = \$25$; weekly revenue at that price is \$12,500.    **3. a.** $E = 6/7$; the demand is going down 6% per 7% increase in price at that price level; thus, a price increase is in order.    **b.** Revenue is maximized when $p = 100/3 \approx \$33.33$    **c.** 4,444 cases per week **5. a.** $E = (4p - 33)/(-2p + 33)$    **b.** 0.54; the demand for $E = mc^2$ T-shirts is going down by about 0.54% per 1% increase in the price.    **c.** \$11 per shirt for a daily revenue of \$1,331    **7. a.** $E = 1.81$. Thus, the demand is elastic at the given tuition level, showing that a decrease in tuition will result in an increase in revenue. **b.** They should charge an average of $2,250 per student, and this will result in an enrollment of about 4,950 students, giving a revenue of about $11,137,500.    **9. a.** $E = 51$; the demand is going down 51% per 1% increase in price at that price level; thus, a large price decrease is advised. **b.** ¥50  **c.** About 78 paint-by-number sets per month **11. a.** $E = -\dfrac{mp}{mp + b}$  **b.** $p = -\dfrac{b}{2m}$  **13. a.** $E = r$ **b.** $E$ is independent of $p$. **c.** If $r = 1$, then the revenue is not affected by the price. If $r > 1$, then the revenue is always elastic, while if $r < 1$, the revenue is always inelastic. This is an unrealistic model because there should always be a price at which the revenue is a maximum.    **15. a.** $q = -1,500p + 6,000$. **b.** $2 per hamburger, giving a total weekly revenue of $6,000 **17.** $E \approx 0.77$. At a family income level of $20,000, the fraction of children attending a live theatrical performance is increasing by 0.77% per 1% increase in household income.    **19. a.** $E = \dfrac{1.554xe^{-0.021x}}{-74e^{-0.021x} + 92}$; $E(100) \approx 0.23$: At a household income level of $100,000, the percentage of people using broadband in 2010 was increasing by 0.23% per 1% increase in household income. **b.** The model predicts elasticity approaching zero for households with large incomes.    **21. a.** $E \approx 0.46$. The demand for computers is increasing by 0.46% per 1% increase in household income. **b.** $E$ decreases as income increases. **c.** Unreliable; it predicts a likelihood greater than 1 at incomes of $123,000 and above. In a more appropriate model, we would expect the curve to level off at or below 1. **d.** $E \approx 0$ **23.** The income elasticity of demand is

$$\frac{dQ}{dY} \cdot \frac{Y}{Q} = a\beta P^\alpha Y^{\beta-1} \frac{Y}{aP^\alpha Y^\beta} = \beta.$$

**25. a.** $q = 1,000e^{-0.30p}$  **b.** At $p = $3$, $E = 0.9$; at $p = $4$, $E = 1.2$; at $p = $5$, $E = 1.5$  **c.** $p = $3.33$ **d.** $p = $5.36$. Selling at a lower price would increase demand, but you cannot sell more than 200 pounds anyway. You should charge as much as you can and still be able to sell all 200 pounds.    **27.** The price is lowered.    **29.** Start with $R = pq$, and differentiate with respect to $p$ to obtain $\dfrac{dR}{dp} = q + p\dfrac{dq}{dp}$. For a stationary point, $dR/dp = 0$, and so $q + p\dfrac{dq}{dp} = 0$. Rearranging this result gives $p\dfrac{dq}{dp} = -q$, and hence $-\dfrac{dq}{dp} \cdot \dfrac{p}{q} = 1$, or $E = 1$, showing that stationary

points of $R$ correspond to points of unit elasticity. **31.** The distinction is best illustrated by an example. Suppose that $q$ is measured in weekly sales and $p$ is the unit price in dollars. Then the quantity $-dq/dp$ measures the drop in weekly sales per $1 increase in price. The elasticity of demand $E$, on the other hand, measures the *percentage* drop in sales per 1% increase in price. Thus, $-dq/dp$ measures absolute change, while $E$ measures fractional, or percentage, change.

## Chapter 12 Review

**1.** Relative max: $(-1, 5)$, absolute min: $(-2, -3)$ and $(1, -3)$    **3.** Absolute max: $(-1, 5)$, absolute min: $(1, -3)$    **5.** Absolute min: $(1, 0)$    **7.** Absolute min: $(-2, -1/4)$    **9.** Relative max at $x = 1$, point of inflection at $x = -1$    **11.** Relative max at $x = -2$, relative min at $x = 1$, point of inflection at $x = -1$ **13.** One point of inflection, at $x = 0$ **15. a.** $a = 4/t^4 - 2/t^3$ m/sec$^2$  **b.** 2 m/sec$^2$ **17.** Relative max: $(-2, 16)$; absolute min: $(2, -16)$; point of inflection: $(0, 0)$; no horizontal or vertical asymptotes

**19.** Relative min: $(-3, -2/9)$; relative max: $(3, 2/9)$; inflection: $(-3\sqrt{2}, -5\sqrt{2}/36)$, $(3\sqrt{2}, 5\sqrt{2}/36)$; vertical asymptote: $x = 0$; horizontal asymptote: $y = 0$

**21.** Relative max at $(0, 0)$, absolute min at $(1, -2)$, no asymptotes

**23.** $22.14 per book    **25. a.** Profit $= -p^3 + 42p^2 - 288p - 181$  **b.** $24 per copy; $3,275  **c.** For

maximum revenue, the company should charge $22.14 per copy. At this price, the cost per book is decreasing with increasing price, while the revenue is not decreasing (its derivative is zero). Thus, the profit is increasing with increasing price, suggesting that the maximum profit will occur at a higher price.
**27.** 12 in $\times$ 12 in $\times$ 6 in, for a volume of 864 in$^3$
**29. a.** $E = \dfrac{2p^2 - 33p}{-p^2 + 33p + 9}$ **b.** 0.52, 2.03; when the
price is $20, demand is dropping at a rate of 0.52% per 1% increase in the price; when the price is $25, demand is dropping at a rate of 2.03% per 1% increase in the price. **c.** $22.14 per book **31. a.** $E = 2p^2 - p$
**b.** 6; the demand is dropping at a rate of 6% per 1% increase in the price. **c.** $1.00, for a monthly revenue of $1,000 **33. a.** Week 5 **b.** Point of inflection on the graph of $s$; maximum on the graph of $s'$, $t$-intercept in the graph of $s''$. **c.** 10,500; if weekly sales continue as predicted by the model, they will level off at around 10,500 books per week in the long term. **d.** 0; if weekly sales continue as predicted by the model, the rate of change of sales approaches zero in the long term. **35. a.–d.** $10/\sqrt{2}$ ft/sec

## Chapter 13

### Section 13.1

**1.** $x^6/6 + C$ **3.** $6x + C$ **5.** $x^2/2 + C$
**7.** $x^3/3 - x^2/2 + C$ **9.** $x + x^2/2 + C$
**11.** $-x^{-4}/4 + C$
**13.** $x^{3.3}/3.3 - x^{-0.3}/0.3 + C$ **15.** $u^3/3 - \ln|u| + C$
**17.** $\dfrac{4x^{5/4}}{5} + C$ **19.** $3x^5/5 + 2x^{-1} - x^{-4}/4 + 4x + C$
**21.** $2\ln|u| + u^2/8 + C$ **23.** $\ln|x| - \dfrac{2}{x} + \dfrac{1}{2x^2} + C$
**25.** $3x^{1.1}/1.1 - x^{5.3}/5.3 - 4.1x + C$
**27.** $\dfrac{x^{0.9}}{0.3} + \dfrac{40}{x^{0.1}} + C$
**29.** $2.55t^2 - 1.2\ln|t| - \dfrac{15}{t^{0.2}} + C$
**31.** $2e^x + 5x|x|/2 + x/4 + C$
**33.** $12.2x^{0.5} + x^{1.5}/9 - e^x + C$ **35.** $\dfrac{2^x}{\ln 2} - \dfrac{3^x}{\ln 3} + C$
**37.** $\dfrac{100(1.1^x)}{\ln(1.1)} - \dfrac{x|x|}{3} + C$ **39.** $x^3/3 + 3x^{-5}/10 + C$
**41.** $1/x - 1/x^2 + C$ **43.** $f(x) = x^2/2 + 1$
**45.** $f(x) = e^x - x - 1$
**47.** $C(x) = 5x - x^2/20,000 + 20,000$
**49.** $C(x) = 5x + x^2 + \ln x + 994$

**51. a.** $M(t) = 0.8t^4 - 6t^2 + 10t + 1$
**b.** 83 million members
**53.** $I(t) = 42,000 + 1200t$; $48,000
**55.** $S(t) = -15t^3 + 450t^2 + 4,200t$; 54,720 million gallons **57. a.** $H'(t) = 3.5t + 65$ billion dollars per year **b.** $H(t) = 1.75t^2 + 65t + 700$ billion dollars **59. a.** $-0.4t + 1$ percentage points per year
**b.** $-0.2t^2 + t + 13$; 13.8 **61. a.** $s = t^3/3 + t + C$
**b.** $C = 1$; $s = t^3/3 + t + 1$ **63.** 320 ft/sec downward **65. a.** $v(t) = -32t + 16$ **b.** $s(t) = -16t^2 + 16t + 185$; zenith at $t = 0.5$ sec $s = 189$ feet, 4 feet above the top of the tower.
**67. a.** $s = 525t + 25t^2$ **b.** 3 hours **c.** The negative solution indicates that, at time $t = -24$, the tail wind would have been large and negative, causing the plane to be moving backward through that position 24 hours prior to departure and arrive at the starting point of the flight at time 0! **69.** (Proof) **71.** $(1,280)^{1/2} \approx$ 35.78 ft/sec **73. a.** 80 ft/sec **b.** 60 ft/sec
**c.** 1.25 seconds **75.** $\sqrt{2} \approx 1.414$ times as fast
**77.** The term *indefinite* refers to the arbitrary constant term in the indefinite integral; we do not obtain a definite value for $C$ and hence the integral is "not definite." **79.** Constant; because the derivative of a linear function is constant, linear functions are antiderivatives of constant functions. **81.** No; there are infinitely many antiderivatives of a given function, each pair of them differing by a constant. Knowing the value of the function at a specific point suffices.
**83.** They differ by a constant, $G(x) - F(x) =$ Constant **85.** Antiderivative, marginal **87.** Up to a constant, $\int f(x)\,dx$ represents the total cost of manufacturing $x$ items. The units of $\int f(x)\,dx$ are the product of the units of $f(x)$ and the units of $x$.
**89.** The indefinite integral of the constant 1 is not zero; it is $x$ (+ constant). Correct answer: $3x^2/2 + x + C$
**91.** There should be no integral sign ($\int$) in the answer. Correct answer: $2x^6 - 2x^2 + C$ **93.** The integral of a constant times a function is the constant times the integral of the function; not the *integral* of the constant times the integral of the function. (In general, the integral of a product is *not* the product of the integrals.) Correct answer: $4(e^x - x^2) + C$ **95.** It is the *integral* of $1/x$ that equals $\ln|x| + C$, and not $1/x$ itself. Correct answer: $\int \frac{1}{x}dx = \ln|x| + C$
**97.** $\int (f(x) + g(x))\,dx$ is, by definition, an antiderivative of $f(x) + g(x)$. Let $F(x)$ be an antiderivative of $f(x)$ and let $G(x)$ be an antiderivative of $g(x)$. Then, because the derivative of $F(x) + G(x)$ is $f(x) + g(x)$ (by the rule for sums of derivatives), this means that $F(x) + G(x)$ is an antiderivative of $f(x) + g(x)$. In

symbols, $\int (f(x) + g(x))\, dx = F(x) + G(x) + C = \int f(x)\, dx + \int g(x)\, dx$, the sum of the indefinite integrals.   **99.** Answers will vary. $\int x \cdot 1\, dx = \int x\, dx = x^2/2 + C$, whereas $\int x\, dx \cdot \int 1\, dx = (x^2/2 + D) \cdot (x + E)$, which is not the same as $x^2/2 + C$, no matter what values we choose for the constants $C$, $D$, and $E$.   **101.** Derivative; indefinite integral; indefinite integral; derivative

## Section 13.2

**1.** $(3x - 5)^4/12 + C$   **3.** $(3x - 5)^4/12 + C$
**5.** $-e^{-x} + C$   **7.** $-e^{-x} + C$   **9.** $\dfrac{1}{2}e^{(x+1)^2} + C$
**11.** $(3x + 1)^6/18 + C$   **13.** $1.6(3x - 4)^{3/2} + C$
**15.** $2e^{(0.6x+2)} + C$   **17.** $(3x^2 + 3)^4/24 + C$
**19.** $2(3x^2 - 1)^{3/2}/9 + C$   **21.** $-(x^2 + 1)^{-0.3}/0.6 + C$
**23.** $(4x^2 - 1)|4x^2 - 1|/16 + C$   **25.** $x + 3e^{3.1x-2} + C$
**27.** $-(1/2)e^{-x^2+1} + C$   **29.** $-(1/2)e^{-(x^2+2x)} + C$
**31.** $(x^2 + x + 1)^{-2}/2 + C$
**33.** $(2x^3 + x^6 - 5)^{1/2}/3 + C$
**35.** $(x - 2)^7/7 + (x - 2)^6/3 + C$
**37.** $4[(x + 1)^{5/2}/5 - (x + 1)^{3/2}/3] + C$
**39.** $20 \ln |1 - e^{-0.05x}| + C$   **41.** $3e^{-1/x} + C$
**43.** $-\dfrac{(4 + 1/x^2)^4}{8} + C$   **45.** $(e^x - e^{-x})/2 + C$
**47.** $\ln(e^x + e^{-x}) + C$
**49.** $(1 - e^{3x-1})|1 - e^{3x-1}|/6 + C$
**51.** $(e^{2x^2-2x} + e^{x^2})/2 + C$   **53.** Derivation
**55.** Derivation   **57.** $-e^{-x} + C$   **59.** $(1/2)e^{2x-1} + C$
**61.** $(2x + 4)^3/6 + C$   **63.** $(1/5) \ln |5x - 1| + C$
**65.** $(1.5x)^4/6 + C$   **67.** $\dfrac{1.5^{3x}}{3 \ln(1.5)} + C$
**69.** $\dfrac{1}{4}(2x + 4)|2x + 4| + C$   **71.** $\dfrac{2^{3x+4} - 2^{-3x+4}}{3 \ln 2} + C$
**73.** $f(x) = (x^2 + 1)^4/8 - 1/8$
**75.** $f(x) = (1/2)e^{x^2-1}$   **77.** $(5x^2 - 3)^7/70 + C$
**79.** $2(3e^x - 1)^{1/2}/3 + C$   **81.** $e^{(x^4-8)}/4 + C$
**83.** $\dfrac{1}{6} \ln |1 + 2e^{3x}| + C$
**85.** $210t + 1{,}240e^{-0.05t} - 1{,}240$
**87.** $8e^{0.25t} - 35.854$; \$62 billion
**89.** $C(x) = 5x - 1/(x + 1) + 995.5$
**91. a.** $N(t) = 35 \ln(5 + e^{0.2t}) - 63$   **b.** 80,000 articles
**93.** $S(t) = 3{,}600[\ln(3 + e^{0.25t}) - \ln 4]$; 6,310 sets
**95. a.** $s = (t^2 + 1)^5/10 + t^2/2 + C$   **b.** $C = 9/10$; $s = (t^2 + 1)^5/10 + t^2/2 + 9/10$
**97. a.** $S(t) = -15(t - 2{,}000)^3 + 450(t - 2{,}000)^2 + 4{,}200(t - 2{,}000) - 16{,}245$ million gallons   **b.** 38,475 million gallons   **99.** None; the substitution $u = x$ simply replaces the letter $x$ throughout by the letter $u$,

and thus does not change the integral at all. For instance, the integral $\int x(3x^2 + 1)\, dx$ becomes $\int u(3u^2 + 1)\, du$ if we substitute $u = x$.   **101.** It may mean that, but it may not; see Example 4.   **103.** (D); to compute the integral, first break it up into a sum of two integrals: $\displaystyle\int \frac{x}{x^2 - 1}\, dx + \int \frac{3x}{x^2 + 1}\, dx$, and then compute the first using $u = x^2 - 1$ and the second using $u = x^2 + 1$.   **105.** There are several errors: First, the term "$dx$" is missing in the integral, and this affects all the subsequent steps (since we must substitute for $dx$ when changing to the variable $u$). Second, when there is a noncanceling $x$ in the integrand, we cannot treat it as a constant. Correct answer: $3(x^2 - 1)^2/4 + C$   **107.** In the fourth step, $u$ was substituted back for $x$ before the integral was taken and $du$ was just changed to $dx$; they're not equal. Correct answer: $(x^3 - 1)^2/6 + C$   **109.** Proof

## Section 13.3

**1.** 4   **3.** 6   **5.** 0.7456   **7.** 2.3129   **9.** 2.5048   **11.** 30
**13.** 22   **15.** $-2$   **17.** 0   **19.** 1   **21.** 1/2   **23.** 1/4
**25.** 2   **27.** 0   **29.** 6   **31.** 0   **33.** 0.5   **35.** 3.3045, 3.1604, 3.1436   **37.** 0.0275, 0.0258, 0.0256
**39.** 15.76 liters   **41.** \$99.95   **43.** 41 billion gallons
**45.** \$19 billion   **47. a.** Left sum: about 46,000 articles, right sum: about 55,000 articles   **b.** 50.5; a total of about 50,500 articles in *Physical Review* were written by researchers in Europe in the 16-year period beginning 1983.   **49.** 118,750 degrees   **51.** 54,000 students   **53.** Left sum: 76; right sum: 71; Left sum, as it is the sum of the annual net incomes for the given years.   **55. a.** Left sum $= 15.6$; right sum $= 15.9$   **b.** Pemex sold a total of 15.9 billion tons of petrochemicals in the period 2006 through 2009.
**57.** $-84.8$; after 4 seconds, the stone is about 84.8 ft below where it started.   **59.** 91.2 ft   **61.** 394; A total of about 394 million members joined Facebook from the start of 2005 to the start of 2010.   **63.** 18,400; a total of 18,400 wiretaps were authorized by U.S. state and federal courts during the 15-year period starting January 1990.   **65.** Yes. The Riemann sum gives an estimated area of 420 square feet.
**67. a.** Graph:

Total expenditure on oil in the United States from 1990 to 2000. **b.** 2,010,000; a total of $2,010,000 million, or $2.01 trillion, was spent on oil in the U.S. from 1990 to 2000. **69. a.** 99.4% **b.** 0 (to at least 15 decimal places) **71.** Stays the same **73.** Increases **75.** The area under the curve and above the $x$ axis equals the area above the curve and below the $x$ axis. **77.** Answers will vary. One example: Let $r(t)$ be the rate of change of net income at time $t$. If $r(t)$ is negative, then the net income is decreasing, so the change in net income, represented by the definite integral of $r(t)$, is negative. **79.** Answers may vary.

**81.** The total cost is $c(1) + c(2) + \cdots + c(60)$, which is represented by the Riemann sum approximation of $\int_1^{61} c(t)\, dt$ with $n = 60$. **83.** $[f(x_1) + f(x_2) + \cdots + f(x_n)]\Delta x = \sum_{k=1}^{n} f(x_k)\Delta x$

**85.** Answers may vary:

## Section 13.4

**1.** 14/3 **3.** 5 **5.** 0 **7.** 40/3 **9.** −0.9045
**11.** $2(e − 1)$ **13.** 2/3 **15.** $1/\ln 2$
**17.** $4^6 − 1 = 4{,}095$ **19.** $(e^1 − e^{-3})/2$ **21.** $3/(2\ln 2)$
**23.** 40/3 **25.** 5/4 **27.** 0 **29.** $(5/2)(e^3 − e^2)$
**31.** $(1/3)[\ln 26 − \ln 7]$ **33.** $\dfrac{0.1}{2.2\ln(1.1)}$
**35.** $e − e^{1/2}$ **37.** 0.2221 **39.** $2 − \ln 3$ **41.** −4/21
**43.** $3^{5/2}/10 − 3^{3/2}/6 + 1/15$ **45.** 1/2 **47.** 16/3
**49.** 9/2 **51.** 56/3 **53.** 1/2 **55.** $783
**57.** 296 miles **59.** 260 ft **61.** 54,720 million gallons
**63.** 391.2 million members **65.** 18,470; a total of about 18,470 wiretaps were authorized by U.S. state and federal courts during the 15-year period starting January 1990. **67.** $210t + 1{,}240e^{-0.05t} − 1{,}240$
**69.** 9 gallons **71.** 907 T-shirts **73.** 68 milliliters

**75.** 20.48; The total value of vehicle transactions on eBay during 2008 and 2009 was about $20.48 billion. **77.** Change in cost $= \int_0^x m(t)\, dt = C(x) − C(0)$ by the FTC, so $C(x) = C(0) + \int_0^x m(t)\, dt$. $C(0)$ is the *fixed cost.* **79.** 80,000 articles **81. a., b.** (Proofs) **c.** 2,401 thousand students **85.** They are related by the Fundamental Theorem of Calculus, which (briefly) states that the definite integral of a suitable function can be calculated by evaluating the indefinite integral at the two endpoints and subtracting. **87.** Computing its definite integral from $a$ to $b$. **89.** Calculate definite integrals by using antiderivatives. **91.** An example is $v(t) = t − 5$. **93.** An example is $f(x) = e^{-x}$. **95.** By the FTC, $\int_a^x f(t)\, dt = G(x) − G(a)$, where $G$ is an antiderivative of $f$. Hence, $F(x) = G(x) − G(a)$. Taking derivatives of both sides, $F'(x) = G'(x) + 0 = f(x)$, as required. The result gives us a formula, in terms of area, for an antiderivative of any continuous function.

### Chapter 13 Review

**1.** $x^3/3 − 5x^2 + 2x + C$ **3.** $4x^3/15 + 4/(5x) + C$
**5.** $(1/2)\ln|2x| + C$ or $(1/2)\ln|x| + C$
**7.** $-e^{-2x+11}/2 + C$ **9.** $(x^2 + 1)^{2.3}/4.6 + C$
**11.** $2\ln|x^2 − 7| + C$ **13.** $\dfrac{1}{6}(x^4 − 4x + 1)^{3/2} + C$
**15.** $-e^{x^2/2} + C$ **17.** $(x + 2) − \ln|x + 2| + C$ or $x − \ln|x + 2| + C$ **19.** 1 **21.** 2.75
**23.** −0.24 **25.** 0.7778, 0.7500, 0.7471
**27.** 0 **29.** 1 **31.** 2/7 **33.** $50(e^{-1} − e^{-2})$
**35.** 52/9 **37.** $[\ln 5 − \ln 2]/8 = \ln(2.5)/8$
**39.** 32/3 **41.** $(1 − e^{-25})/2$
**43. a.** $N(t) = 196t + t^2/2 − 0.16t^6/6$ **b.** 4 books
**45. a.** At a height of $-16t^2 + 100t$ ft **b.** 156.25 ft
**c.** After 6.25 seconds **47.** 25,000 copies. **49.** 0 books
**51.** 39,200 hits **53.** $8,200 **55.** About 86,000 books
**57.** About 35,800 books

## Chapter 14

### Section 14.1

**1.** $2e^x(x − 1) + C$ **3.** $-e^{-x}(2 + 3x) + C$
**5.** $e^{2x}(2x^2 − 2x − 1)/4 + C$
**7.** $-e^{-2x+4}(2x^2 + 2x + 3)/4 + C$
**9.** $2^x[(2 − x)/\ln 2 + 1/(\ln 2)^2] + C$
**11.** $-3^{-x}[(x^2 − 1)/\ln 3 + 2x/(\ln 3)^2 + 2/(\ln 3)^3] + C$
**13.** $-e^{-x}(x^2 + x + 1) + C$
**15.** $\dfrac{1}{7}x(x + 2)^7 − \dfrac{1}{56}(x + 2)^8 + C$

**17.** $-\dfrac{x}{2(x-2)^2} - \dfrac{1}{2(x-2)} + C$

**19.** $(x^4 \ln x)/4 - x^4/16 + C$

**21.** $(t^3/3 + t)\ln(2t) - t^3/9 - t + C$

**23.** $(3/4)t^{4/3}(\ln t - 3/4) + C$

**25.** $x \log_3 x - x/\ln 3 + C$

**27.** $e^{2x}(x/2 - 1/4) - 4e^{3x}/3 + C$

**29.** $e^x(x^2 - 2x + 2) - e^{x^2}/2 + C$

**31.** $\dfrac{1}{3}(3x - 4)(2x - 1)^{3/2} - \dfrac{1}{5}(2x - 1)^{5/2} + C$

**33.** $e$    **35.** $38,229/286$

**37.** $(7/2)\ln 2 - 3/4$    **39.** $1/4$

**41.** $1 - 11e^{-10}$    **43.** $4\ln 2 - 7/4$

**45.** $\dfrac{1}{2}x(x-3)|x-3| - \dfrac{1}{6}(x-3)^2|x-3| + C$

**47.** $2x|x-3| - (x-3)|x-3| + C$

**49.** $-x^2(-x+4)|-x+4| - \dfrac{2}{3}x(-x+4)^2|-x+4| - \dfrac{1}{6}(-x+4)^3|-x+4| + C$

**51.** $\dfrac{1}{2}(x^2 - 2x + 3)(x-4)|x-4| - \dfrac{1}{3}(x-1)(x-4)^2|x-4| + \dfrac{1}{12}(x-4)^3|x-4| + C$

**53.** $28,800,000(1 - 2e^{-1})$ ft

**55.** $5,001 + 10x - 1/(x+1) - [\ln(x+1)]/(x+1)$

**57.** $4,252$ billion    **59.** $19$ billion square feet

**61.** $20,800$ million gallons

**63.** $e^{0.1x}(250[-8x^2 + 70x + 1,000] - 2,500[-16x + 70] - 400,000) + 325,000$    **65.** $33,598$

**67. a.** $r(t) = -0.075t + 2.75 + [0.025t - 0.25]\dfrac{|t - 10|}{t - 10}$

**b.** $42.5$ million people    **69.** Answers will vary. Examples are $xe^{x^2}$ and $e^{x^2} = 1 \cdot e^{x^2}$.    **71.** Answers will vary. Examples are Exercises 31 and 32, or, more simply, integrals like $\int x(x+1)^5\, dx$.

**73.** Substitution    **75.** Parts    **77.** Substitution

**79.** Parts    **81.** $n + 1$ times    **83.** Proof.

## Section 14.2

**1.** $16/3$    **3.** $9.75$    **5.** $2$    **7.** $31/3$    **9.** $8/3$    **11.** $4$
**13.** $1/3$    **15.** $1$    **17.** $2$    **19.** $39$    **21.** $e - 3/2$    **23.** $2/3$
**25.** $3/10$    **27.** $1/20$    **29.** $4/15$    **31.** $1/3$    **33.** $32$
**35.** $2\ln 2 - 1$    **37.** $8\ln 4 + 2e - 16$    **39.** $0.9138$
**41.** $0.3222$    **43.** $112.5$. This represents your total profit for the week, \$112.50.    **45.** $608$; there were approximately $608,000$ housing starts from the start of 2002 to the start of 2006 not for sale purposes.

**47. a.** Graph: (The upper curve is Myspace.)

Myspace; about 93 million members    **b.** The area between the curves $y = f(t)$ and $y = m(t)$ over [0.5, 2]
**49.** $16,078(e^{0.051t} - 1) - 7,333.3(e^{0.06t} - 1)$; the total number of wiretaps authorized by federal courts from the start of 1990 up to time $t$ was about $16,078(e^{0.051t} - 1) - 7,333.3(e^{0.06t} - 1)$.
**51.** Wrong: It could mean that the graphs of $f$ and $g$ cross, as shown in the caution at the start of this topic in the textbook.    **53.** The area between the export and import curves represents the United States's accumulated trade deficit (that is, the total excess of imports over exports) from 1960 to 2007.    **55.** (A)    **57.** The claim is wrong because the area under a curve can only represent income if the curve is a graph of income *per unit time*. The value of a stock price is not income per unit time—the income can be realized only when the stock is sold, and it amounts to the current market price. The total net income (per share) from the given investment would be the stock price on the date of sale minus the purchase price of \$50.

## Section 14.3

**1.** Average $= 2$    **3.** Average $= 1$

**5.** Average $= (1 - e^{-2})/2$    **7.** Average $= 17/8$

**9.**

| $x$ | 0 | 1 | 2 | 3 | 4 | 5 | 6 | 7 |
|-----|---|---|----|---|---|------|------|---|
| $r(x)$ | 3 | 5 | 10 | 3 | 2 | 5 | 6 | 7 |
| $\bar{r}(x)$ |  |  | 6 | 6 | 5 | 10/3 | 13/3 | 6 |

**11.**

| $x$ | 0 | 1 | 2 | 3 | 4 | 5 | 6 | 7 |
|---|---|---|---|---|---|---|---|---|
| $r(x)$ | 1 | 2 | 6 | 7 | 11 | 15 | 10 | 2 |
| $\bar{r}(x)$ | | | 3 | 5 | 8 | 11 | 12 | 9 |

**13.** Moving average:
$$\bar{f}(x) = x^3 - (15/2)x^2 + 25x - 125/4$$

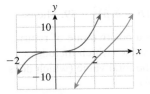

**15.** Moving average:
$$\bar{f}(x) = (3/25)[x^{5/3} - (x-5)^{5/3}]$$

**17.** $\bar{f}(x) = \dfrac{2}{5}(e^{0.5x} - e^{0.5(x-5)})$

**19.** $\bar{f}(x) = \dfrac{2}{15}(x^{3/2} - (x-5)^{3/2})$

**21.** $\bar{f}(x) = \dfrac{1}{5}\left[5 - \dfrac{1}{2}|2x-1| + \dfrac{1}{2}|2x-11|\right]$

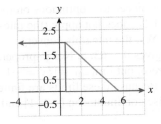

**23.** $\bar{f}(x) = \dfrac{1}{5}\left[2x - \dfrac{1}{2}(x+1)|x+1| + \dfrac{1}{2}x|x| - \right.$
$$\left. 2(x-5) + \dfrac{1}{2}(x-4)|x-4| - \dfrac{1}{2}(x-5)|x-5|\right]$$

**25.**

**27.**

**29.**

**31.**

**33.**
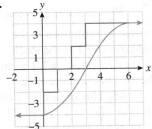

**35.** \$1.8 million   **37.** 16 million members per year
**39.** 277 million tons   **41.** \$10,410.88   **43.** \$1,500
**45.** \$2.5 billion per quarter
**47.**

| Year $t$ | 2001 | 2002 | 2003 | 2004 | 2005 | 2006 | 2007 | 2008 | 2009 | 2010 |
|---|---|---|---|---|---|---|---|---|---|---|
| Stock Price | 39 | 35 | 41 | 51 | 56 | 77 | 94 | 80 | 68 | 73 |
| Moving Average (rounded) | | | | 42 | 46 | 56 | 70 | 77 | 80 | 79 |

The moving average continued to rise, at a lower rate, until it began to fall in 2010.
**49. a.** To obtain the moving averages from January to June, use the fact that the data repeats every 12 months.

Graph:

**b.** The 12-month moving average is constant and equal to the year-long average of approximately 79°.

**51. a.** Graph:

— Sales — Moving average

**b.** Approximately $-0.2$ million iPods per quarter
**53. a.** 7,200 million gallons per year
**b.** $\frac{1}{2}[-15[t^3 - (t-2)^3] + 450[t^2 - (t-2)^2] + 8,400]$
**c.** The function is quadratic because the $t^3$ terms cancel.
**55. a.** $s = 41t + 526$  **b.** $\bar{s}(t) = 41t + 444$
**c.** The slope of the moving average is the same as the slope of the original function.
**57.** $\bar{f}(x) = mx + b - \frac{ma}{2}$   **59.** The moving average "blurs" the effects of short-term oscillations in the price and shows the longer-term trend of the stock price.   **61.** They repeat every 6 months.   **63.** The area above the $x$-axis equals the area below the $x$-axis. Example: $y = x$ on $[-1,1]$   **65.** This need not be the case; for instance, the function $f(x) = x^2$ on $[0, 1]$ has average value $1/3$, whereas the value midway between the maximum and minimum is $1/2$.
**67.** (C) A shorter-term moving average most closely approximates the original function because it averages the function over a shorter period, and continuous functions change by only a small amount over a small period.

## Section 14.4

**1.** $6.25   **3.** $512   **5.** $119.53   **7.** $900   **9.** $416.67
**11.** $326.27   **13.** $25   **15.** $0.50   **17.** $386.29
**19.** $225   **21.** $25.50   **23.** $12,684.63
**25.** $TV = \$300,000, FV = \$434,465.45$
**27.** $TV = \$350,000, FV = \$498,496.61$
**29.** $TV = \$389,232.76, FV = \$547,547.16$

**31.** $TV = \$100,000, PV = \$82,419.99$
**33.** $TV = \$112,500, PV = \$92,037.48$
**35.** $TV = \$107,889.50, PV = \$88,479.69$
**37.** $\bar{p} = \$5,000, \bar{q} = 10,000, CS = \$25$ million, $PS = \$100$ million. The total social gain is $125 million.   **39.** €200 billion   **41.** $1,190 billion
**43.** €220 billion **45.** $1,080 billion   **47.** $1,943,162.44
**49.** $3,086,245.73   **51.** $73,802.63   **53.** $1,792,723.35
**55.** total   **57.** $CS = \frac{1}{2m}(b - m\bar{p})^2$   **59.** She is
correct, provided there is a positive rate of return, in which case the future value (which includes interest) is greater than the total value (which does not).   **61.** $PV < TV < FV$

## Section 14.5

**1.** Diverges   **3.** Converges to $2e$   **5.** Converges to $e^2$
**7.** Converges to ½   **9.** Converges to $1/108$
**11.** Converges to $3 \times 5^{2/3}$   **13.** Diverges
**15.** Diverges   **17.** Converges to $\frac{5}{4}(3^{4/5} - 1)$
**19.** Diverges   **21.** Converges to 0   **23.** Diverges
**25.** Diverges   **27.** 0.9, 0.99, 0.999, . . . ; converges to 1.   **29.** 7.602, 95.38, 993.1, . . . ; diverges.
**31.** 1.368, 1.800, 1.937, 1.980, 1.994, 1.998, 1.999, 2.000, . . . ; converges to 2.   **33.** 9.000, 99.00, 999.0, . . . ; diverges to $+\infty$.   **35.** 4.45 million homes
**37.** 13,600 billion cigarettes   **39.** No; you will not sell more than 2,000 of them.   **41.** The number of graduates each year will rise without bound.
**43. a.** $R(t) = 350e^{-0.1t}(39t + 68)$ million dollars/yr;
**b.** $1,603,000 million   **45.** $20,700 billion
**47.** $\int_0^{+\infty} N(t)\, dt$ diverges, indicating that there is no bound to the expected future total online sales of mousse. $\int_{-\infty}^0 N(t)\, dt$ converges to approximately 2.006, indicating that total online sales of mousse prior to the current year amounted to approximately 2 million gallons.   **49.** 1   **51.** 0.1587   **53.** $70,833
**55. a.** 2.468 meteors on average **b.** The integral diverges. We can interpret this as saying that the number of impacts by meteors smaller than 1 megaton is very large. (This makes sense because, for example, this number includes meteors no larger than a grain of dust.)   **57. a.** $\Gamma(1) = 1; \Gamma(2) = 1$   **59.** The integrand is neither continuous nor piecewise continuous on the interval $[-1, 1]$, so the FTC does not apply (the integral is improper).   **61.** Yes; the integrals converge to 0, and the FTC also gives 0.   **63. a.** Not improper. $|x|/x$ is not defined at zero, but $\lim_{x \to 0^-} |x|/x = -1$ and $\lim_{x \to 0^+} |x|/x = 1$. Because these limits are finite, the integrand is piecewise continuous on $[-1, 1]$ and

so the integral is not improper. **b.** Improper, because $x^{-1/3}$ has infinite left and right limits at 0. **c.** Improper, since $(x - 2)/(x^2 - 4x + 4) = 1/(x - 2)$, which has an infinite left limit at 2. **65.** In all cases, you need to rewrite the improper integral as a limit and use technology to evaluate the integral of which you are taking the limit. Evaluate for several values of the endpoint approaching the limit. In the case of an integral in which one of the limits of integration is infinite, you may have to instruct the calculator or computer to use more subdivisions as you approach $+\infty$. **67.** Answers will vary.

## Section 14.6

**1.** $y = \dfrac{x^3}{3} + \dfrac{2x^{3/2}}{3} + C$   **3.** $\dfrac{y^2}{2} = \dfrac{x^2}{2} + C$

**5.** $y = Ae^{x^2/2}$   **7.** $y = -\dfrac{2}{(x + 1)^2 + C}$

**9.** $y = \pm\sqrt{(\ln x)^2 + C}$   **11.** $y = \dfrac{x^4}{4} - x^2 + 1$

**13.** $y = (x^3 + 8)^{1/3}$   **15.** $y = 2x$   **17.** $y = e^{x^2/2} - 1$

**19.** $y = -\dfrac{2}{\ln(x^2 + 1) + 2}$   **21.** With $s(t) =$ monthly

sales after $t$ months, $\dfrac{ds}{dt} = -0.05s$; $s = 1,000$ when $t = 0$. Solution: $s = 1,000e^{-0.05t}$ quarts per month

**23. a.** $75 + 125e^{-0.05t}$   **b.** 64.4 minutes

**25.** $k \approx 0.04274$; $H(t) = 75 + 115e^{-0.04274t}$ degrees Fahrenheit after $t$ minutes   **27.** With $S(t) =$ total

sales after $t$ months, $\dfrac{ds}{dt} = 0.1(100,000 - S)$;

$S(0) = 0$. Solution: $S = 100,000(1 - e^{-0.1t})$ monitors after $t$ months.   **29.** $q = 0.6078e^{-0.05p}p^{1.5}$

**31.** $y = e^{-t}(t + 1)$

**33.** $y = e^{t/2}\left[-2te^{-t/2} - 4e^{-t/2} + 5\right]$

**35.** $i = 5e^{-t}(e^t - e)\left[1 + \dfrac{|t - 1|}{t - 1}\right]$

Graph:

**37. a.** $\dfrac{dp}{dt} = k(D(p) - S(p)) = k(20,000 - 1,000p)$

**b.** $p = 20 - Ae^{-kt}$ **c.** $p = 20 - 10e^{-0.2231t}$ dollars after $t$ months   **39.** Verification

**41.** $S = \dfrac{2/1,999}{e^{-0.5t} + 1/1,999}$

Graph:

It will take about 27 months to saturate the market.

**43. a.** $y = be^{Ae^{-at}}$, $A =$ constant   **b.** $y = 10e^{-0.69315e^{-t}}$

Graph:

**45.** A general solution gives all possible solutions to the equation, using at least one arbitrary constant. A particular solution is one specific function that satisfies the equation. We obtain a particular solution by substituting specific values for any arbitrary constants in the general solution.   **47.** Example: $y'' = x$ has general solution $y = \frac{1}{6}x^3 + Cx + D$ (integrate twice).   **49.** $y' = -4e^{-x} + 3$

## Chapter 14 Review

**1.** $(x^2 - 2x + 4)e^x + C$   **3.** $(1/3)x^3 \ln 2x - x^3/9 + C$

**5.** $\dfrac{1}{2}x(2x + 1)|2x + 1| - \dfrac{1}{12}(2x + 1)^2|2x + 1| + C$

**7.** $-5x|-x + 3| - \dfrac{5}{2}(-x + 3)|-x + 3| + C$

**9.** $-e^2 - 39/e^2$   **11.** $\dfrac{3}{2 \cdot 2^{1/3}} - \dfrac{1}{2}$   **13.** $\dfrac{2\sqrt{2}}{3}$   **15.** $-1$

**17.** $e - 2$   **19.** $3x - 2$   **21.** $\dfrac{3}{14}[x^{7/3} - (x - 2)^{7/3}]$

**23.** \$1,600   **25.** \$2,500   **27.** 1/4   **29.** Diverges   **31.** 1

**33.** $y = -\dfrac{3}{x^3 + C}$   **35.** $y = \sqrt{2\ln|x| + 1}$

**37.** \$18,200   **39. a.** \$2,260
**b.** $50,000e^{0.01t}(1 - e^{-0.04}) \approx 1,960.53e^{0.01t}$
**41. a.** $\bar{p} = 20$, $\bar{q} = 40,000$   **b.** $CS = \$240,000$, $PS = \$30,000$   **43.** Approximately \$910,000
**45. a.** \$5,549,000   **b.** Principal: \$5,280,000, interest: \$269,000   **47.** \$51 million **49.** The amount in the account would be given by $y = 10,000/(1 - t)$, where $t$ is time in years, so would approach infinity 1 year after the deposit.

# Chapter 15

## Section 15.1

**1. a.** 1  **b.** 1  **c.** 2  **d.** $a^2 - a + 5$  **e.** $y^2 + x^2 - y + 1$
**f.** $(x+h)^2 + (y+k)^2 - (x+h) + 1$    **3. a.** 0  **b.** 0.2
**c.** $-0.1$  **d.** $0.18a + 0.2$  **e.** $0.1x + 0.2y - 0.01xy$
**f.** $0.2(x+h) + 0.1(y+k) - 0.01(x+h)(y+k)$
**5. a.** 1  **b.** $e$  **c.** $e$  **d.** $e^{x+y+z}$  **e.** $e^{x+h+y+k+z+l}$
**7. a.** Does not exist  **b.** 0  **c.** 0  **d.** $xyz/(x^2 + y^2 + z^2)$
**e.** $(x+h)(y+k)(z+l)/[(x+h)^2 + (y+k)^2 + (z+l)^2]$    **9. a.** Increases; 2.3  **b.** Decreases; 1.4
**c.** Decreases; 1 unit increase in $z$    **11.** Neither
**13.** Linear  **15.** Linear  **17.** Interaction
**19. a.** 107  **b.** $-14$  **c.** $-113$

**21.**

|  | $x \rightarrow$ | | | |
|---|---|---|---|---|
|  | **10** | **20** | **30** | **40** |
| $y$ ↓  **10** | 52 | 107 | 162 | 217 |
| **20** | 94 | 194 | 294 | 394 |
| **30** | 136 | 281 | 426 | 571 |
| **40** | 178 | 368 | 558 | 748 |

**25.** 18, 4, 0.0965, 47,040    **27.** 6.9078, 1.5193, 5.4366, 0
**29.** Let $z$ = annual sales of Z (in millions of dollars),
$x$ = annual sales of X, and $y$ = annual sales of Y.
The model is $z = -2.1x + 0.4y + 16.2$.

**31.**

**33.**

**35.**

**37.**

**39.**

**41.** (H)    **43.** (B)    **45.** (F)    **47.** (C)
**49.**

**51.**

**53.**

**55. a.** 4  **b.** 5  **c.** $-1$
**57.** (2, 2) and $(-2, -2)$
**59.**

$z = 1 - x - y$

**61.**

$z = 2x + y - 2$

**63.**

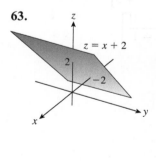

$z = x + 2$

**65.**

$z = 2x^2 + 2y^2$
$z = 2$
$x^2 + y^2 = 1$
$z = 1$
$x^2 + y^2 = \frac{1}{2}$

**67.**   **69.**

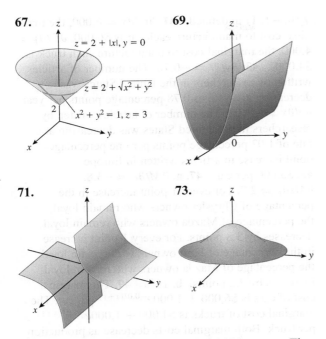

$z = 2 + |x|, y = 0$

$z = 2 + \sqrt{x^2 + y^2}$

$x^2 + y^2 = 1, z = 3$

**71.**   **73.**

**75. a.** The marginal cost of cars is $6,000 per car. The marginal cost of trucks is $4,000 per truck.   **b.** The graph is a plane with $x$-intercept $-40$, $y$-intercept $-60$, and $z$-intercept 240,000.   **c.** The slice $x = 10$ is the straight line with equation $z = 300,000 + 4,000y$. It describes the cost function for the manufacture of trucks if car production is held fixed at 10 cars per week.   **d.** The level curve $z = 480,000$ is the straight line $6,000x + 4,000y = 240,000$. It describes the number of cars and trucks you can manufacture to maintain weekly costs at $480,000.
**77.** $C(x, y) = 10 + 0.03x + 0.04y$, where $C$ is the cost in dollars, $x =$ # video clips sold per month, $y =$ # audio clips sold per month   **79. a.** 28%   **b.** 21%   **c.** Percentage points per year   **81.** The graph is a plane with $x_1$-intercept 0.3, $x_2$-intercept 33, and $x_3$-intercept 0.66. The slices by $x_1 =$ constant are straight lines that are parallel to each other. Thus, the rate of change of General Motors' share as a function of Ford's share does not depend on Chrysler's share. Specifically, GM's share decreases by 0.02 percentage points per one percentage-point increase in Ford's market share, regardless of Chrysler's share.   **83. a.** 189 thousand prisoners   **b.** $y = 300$: $N = -0.02x + 51$; $y = 500$: $N = 0.02x + 67$; when there are 300,000 prisoners in local jails, the number in federal prisons decreases by 20 per 1,000 additional prisoners in state prisons. When there are 500,000 prisoners in local

jails, the number in federal prisons increases by 20 per 1,000 additional prisoners in state prisons.
**85. a.** The slices $x =$ constant and $y =$ constant are straight lines.   **b.** No. Even though the slices $x =$ constant and $y =$ constant are straight lines, the level curves are not, and so the surface is not a plane.   **c.** The slice $x = 10$ has a slope of 3,800. The slice $x = 20$ has a slope of 3,600. Manufacturing more cars lowers the marginal cost of manufacturing trucks.   **87. a.** $9,980   **b.** $R(z) = 9,850 + 0.04z$
**89.** $f(x, t) = 2.5x + 17.5t + 145$; $311 billion
**91.** $U(11, 10) - U(10, 10) \approx 5.75$. This means that, if your company now has 10 copies of Macro Publish and 10 copies of Turbo Publish, then the purchase of one additional copy of Macro Publish will result in a productivity increase of approximately 5.75 pages per day.   **93. a.** Answers will vary. $(a, b, c) = (3, 1/4, 1/\pi)$; $(a, b, c) = (1/\pi, 3, 1/4)$.
**b.** $a = \left(\frac{3}{4\pi}\right)^{1/3}$. The resulting ellipsoid is a sphere with radius $a$.   **95.** 7,000,000
**97. a.** $100 = K(1,000)^a(1,000,000)^{1-a}$; $10 = K(1,000)^a(10,000)^{1-a}$   **b.** $\log K - 3a = -4$; $\log K - a = -3$   **c.** $a = 0.5$, $K \approx 0.003162$   **d.** $P = 71$ pianos (to the nearest piano)   **99. a.** $4 \times 10^{-3}$ gram per square meter   **b.** The total weight of sulfates in the Earth's atmosphere   **101. a.** The value of $N$ would be doubled.   **b.** $N(R, f_p, n_e, f_l, f_i, L) = R f_p n_e f_l f_i L$, where $L$ is the average lifetime of an intelligent civilization   **c.** Take the logarithm of both sides, since this would yield the linear function $\ln(N) = \ln(R) + \ln(f_p) + \ln(n_e) + \ln(f_l) + \ln(f_i) + \ln(f_c) + \ln(L)$.
**103.** They are reciprocals of each other.   **105.** For example, $f(x, y) = x^2 + y^2$.   **107.** For example, $f(x, y, z) = xyz$.   **109.** For example, take $f(x, y) = x + y$. Then setting $y = 3$ gives $f(x, 3) = x + 3$. This can be viewed as a function of the single variable $x$. Choosing other values for $y$ gives other functions of $x$.   **111.** The slope is independent of the choice of $y = k$.   **113.** That CDs cost more than cassettes   **115.** plane
**117. (B)** Traveling in the direction B results in the shortest trip to nearby isotherms, and hence the fastest rate of increase in temperature.   **119.** Agree: Any slice through a plane is a straight line.   **121.** The graph of a function of three or more variables lives in four-dimensional (or higher) space, which makes it difficult to draw and visualize.   **123.** We need one dimension for each of the variables plus one dimension for the value of the function.

## Section 15.2

**1.** $f_x(x, y) = -40$; $f_y(x, y) = 20$; $f_x(1, -1) = -40$; $f_y(1, -1) = 20$   **3.** $f_x(x, y) = 6x + 1$; $f_y(x, y) = -3y^2$; $f_x(1, -1) = 7$; $f_y(1, -1) = -3$
**5.** $f_x(x, y) = -40 + 10y$; $f_y(x, y) = 20 + 10x$; $f_x(1, -1) = -50$; $f_y(1, -1) = 30$
**7.** $f_x(x, y) = 6xy$; $f_y(x, y) = 3x^2$; $f_x(1, -1) = -6$; $f_y(1, -1) = 3$   **9.** $f_x(x, y) = 2xy^3 - 3x^2y^2 - y$; $f_y(x, y) = 3x^2y^2 - 2x^3y - x$; $f_x(1, -1) = -4$; $f_y(1, -1) = 4$
**11.** $f_x(x, y) = 6y(2xy + 1)^2$; $f_y(x, y) = 6x(2xy + 1)^2$; $f_x(1, -1) = -6$; $f_y(1, -1) = 6$   **13.** $f_x(x, y) = e^{x+y}$; $f_y(x, y) = e^{x+y}$; $f_x(1, -1) = 1$; $f_y(1, -1) = 1$
**15.** $f_x(x, y) = 3x^{-0.4}y^{0.4}$; $f_y(x, y) = 2x^{0.6}y^{-0.6}$; $f_x(1, -1)$ undefined; $f_y(1, -1)$ undefined
**17.** $f_x(x, y) = 0.2ye^{0.2xy}$; $f_y(x, y) = 0.2xe^{0.2xy}$; $f_x(1, -1) = -0.2e^{-0.2}$; $f_y(1, -1) = 0.2e^{-0.2}$
**19.** $f_{xx}(x, y) = 0$; $f_{yy}(x, y) = 0$; $f_{xy}(x, y) = f_{yx}(x, y) = 0$; $f_{xx}(1, -1) = 0$; $f_{yy}(1, -1) = 0$; $f_{xy}(1, -1) = f_{yx}(1, -1) = 0$
**21.** $f_{xx}(x, y) = 0$; $f_{yy}(x, y) = 0$; $f_{xy}(x, y) = f_{yx}(x, y) = 10$; $f_{xx}(1, -1) = 0$; $f_{yy}(1, -1) = 0$; $f_{xy}(1, -1) = f_{yx}(1, -1) = 10$   **23.** $f_{xx}(x, y) = 6y$; $f_{yy}(x, y) = 0$; $f_{xy}(x, y) = f_{yx}(x, y) = 6x$; $f_{xx}(1, -1) = -6$; $f_{yy}(1, -1) = 0$; $f_{xy}(1, -1) = f_{yx}(1, -1) = 6$   **25.** $f_{xx}(x, y) = e^{x+y}$; $f_{yy}(x, y) = e^{x+y}$; $f_{xy}(x, y) = f_{yx}(x, y) = e^{x+y}$; $f_{xx}(1, -1) = 1$; $f_{yy}(1, -1) = 1$; $f_{xy}(1, -1) = f_{yx}(1, -1) = 1$   **27.** $f_{xx}(x, y) = -1.2x^{-1.4}y^{0.4}$; $f_{yy}(x, y) = -1.2x^{0.6}y^{-1.6}$; $f_{xy}(x, y) = f_{yx}(x, y) = 1.2x^{-0.4}y^{-0.6}$; $f_{xx}(1, -1)$ undefined; $f_{yy}(1, -1)$ undefined; $f_{xy}(1, -1)$ & $f_{yx}(1, -1)$ undefined
**29.** $f_x(x, y, z) = yz$; $f_y(x, y, z) = xz$; $f_z(x, y, z) = xy$; $f_x(0, -1, 1) = -1$; $f_y(0, -1, 1) = 0$; $f_z(0, -1, 1) = 0$
**31.** $f_x(x, y, z) = 4/(x + y + z^2)^2$; $f_y(x, y, z) = 4/(x + y + z^2)^2$; $f_z(x, y, z) = 8z/(x + y + z^2)^2$; $f_x(0, -1, 1)$ undefined; $f_y(0, -1, 1)$ undefined; $f_z(0, -1, 1)$ undefined   **33.** $f_x(x, y, z) = e^{yz} + yze^{xz}$; $f_y(x, y, z) = xze^{yz} + e^{xz}$; $f_z(x, y, z) = xy(e^{yz} + e^{xz})$; $f_x(0, -1, 1) = e^{-1} - 1$; $f_y(0, -1, 1) = 1$; $f_z(0, -1, 1) = 0$   **35.** $f_x(x, y, z) = 0.1x^{-0.9}y^{0.4}z^{0.5}$; $f_y(x, y, z) = 0.4x^{0.1}y^{-0.6}z^{0.5}$; $f_z(x, y, z) = 0.5x^{0.1}y^{0.4}z^{-0.5}$; $f_x(0, -1, 1)$ undefined; $f_y(0, -1, 1)$ undefined, $f_z(0, -1, 1)$ undefined
**37.** $f_x(x, y, z) = yze^{xyz}$, $f_y(x, y, z) = xze^{xyz}$, $f_z(x, y, z) = xye^{xyz}$; $f_x(0, -1, 1) = -1$; $f_y(0, -1, 1) = f_z(0, -1, 1) = 0$   **39.** $f_x(x, y, z) = 0$; $f_y(x, y, z) = -\dfrac{600z}{y^{0.7}(1 + y^{0.3})^2}$; $f_z(x, y, z) = \dfrac{2{,}000}{1 + y^{0.3}}$; $f_x(0, -1, 1)$ undefined; $f_y(0, -1, 1)$ undefined;

$f_z(0, -1, 1)$ undefined   **41.** $\partial C/\partial x = 6{,}000$, the marginal cost to manufacture each car is \$6,000. $\partial C/\partial y = 4{,}000$, the marginal cost to manufacture each truck is \$4,000.   **43.** $\partial y/\partial t = -0.78$. The number of articles written by researchers in the United States was decreasing at a rate of 0.78 percentage points per year. $\partial y/\partial x = -1.02$. The number of articles written by researchers in the United States was decreasing at a rate of 1.02 percentage points per one percentage-point increase in articles written in Europe.
**45.** \$5,600 per car   **47. a.** $\partial M/\partial c = -3.8$, $\partial M/\partial f = 2.2$. For every 1 point increase in the percentage of Chrysler owners who remain loyal, the percentage of Mazda owners who remain loyal decreases by 3.8 points. For every 1 point increase in the percentage of Ford owners who remain loyal, the percentage of Mazda owners who remain loyal increases by 2.2 points. **b.** 16%   **49.** The marginal cost of cars is $\$6{,}000 + 1{,}000e^{-0.01(x+y)}$ per car. The marginal cost of trucks is $\$4{,}000 + 1{,}000e^{-0.01(x+y)}$ per truck. Both marginal costs decrease as production rises.   **51. a.** \$36,600; \$52,900 **b.** \$270 per year; \$410 per year **c.** Widening **d.** The rate at which the income gap is widening
**53.** $\bar{C}(x, y) = \dfrac{200{,}000 + 6{,}000x + 4{,}000y - 100{,}000e^{-0.01(x+y)}}{x + y}$;
$\bar{C}_x(50, 50) = -\$2.64$ per car. This means that at a production level of 50 cars and 50 trucks per week, the average cost per vehicle is decreasing by \$2.64 for each additional car manufactured.
$\bar{C}_y(50, 50) = -\$22.64$ per truck. This means that at a production level of 50 cars and 50 trucks per week, the average cost per vehicle is decreasing by \$22.64 for each additional truck manufactured.
**55.** No; your marginal revenue from the sale of cars is $\$15{,}000 - \dfrac{2{,}500}{\sqrt{x + y}}$ per car and
$\$10{,}000 - \dfrac{2{,}500}{\sqrt{x + y}}$ per truck from the sale of trucks. These increase with increasing $x$ and $y$. In other words, you will earn more revenue per vehicle with increasing sales, and so the rental company will pay more for each additional vehicle it buys.
**57.** $P_z(10, 100{,}000, 1{,}000{,}000) \approx 0.0001010$ papers/\$
**59. a.** $U_x(10, 5) = 5.18$, $U_y(10, 5) = 2.09$. This means that if 10 copies of Macro Publish and 5 copies of Turbo Publish are purchased, the company's daily productivity is increasing at a rate of 5.18 pages per day for each additional copy of Macro purchased and by 2.09 pages per day for each additional copy of Turbo purchased. **b.** $\dfrac{U_x(10, 5)}{U_y(10, 5)} \approx 2.48$ is the ratio

of the usefulness of one additional copy of Macro to one of Turbo. Thus, with 10 copies of Macro and 5 copies of Turbo, the company can expect approximately 2.48 times the productivity per additional copy of Macro compared to Turbo.   **61.** $6 \times 10^9$ N/sec
**63. a.** $A_P(100, 0.1, 10) = 2.59$; $A_r(100, 0.1, 10) = 2,357.95$; $A_t(100, 0.1, 10) = 24.72$. Thus, for a $100 investment at 10% interest, after 10 years the accumulated amount is increasing at a rate of $2.59 per $1 of principal, at a rate of $2,357.95 per increase of 1 in $r$ (note that this would correspond to an increase in the interest rate of 100%), and at a rate of $24.72 per year. **b.** $A_P(100, 0.1, t)$ tells you the rate at which the accumulated amount in an account bearing 10% interest with a principal of $100 is growing per $1 increase in the principal, $t$ years after the investment.
**65. a.** $P_x = Ka\left(\dfrac{y}{x}\right)^b$ and $P_y = Kb\left(\dfrac{x}{y}\right)^a$. They are

equal precisely when $\dfrac{a}{b} = \left(\dfrac{x}{y}\right)^b \left(\dfrac{x}{y}\right)^a$. Substituting

$b = 1 - a$ now gives $\dfrac{a}{b} = \dfrac{x}{y}$. **b.** The given information

implies that $P_x(100, 200) = P_y(100, 200)$. By part (a), this occurs precisely when $a/b = x/y = 100/200 = 1/2$. But $b = 1 - a$, so $a/(1 - a) = 1/2$, giving $a = 1/3$ and $b = 2/3$.   **67.** Decreasing at 0.0075 parts of nutrient per part of water/sec   **69.** $f$ is increasing at a rate of __s__ units per unit of $x$, $f$ is increasing at a rate of __t__ units per unit of $y$, and the value of $f$ is __r__ when $x = $ __a__ and $y = $ __b__   **71.** the marginal cost of building an additional orbicus; zonars per unit.
**73.** Answers will vary. One example is $f(x, y) = -2x + 3y$. Others are $f(x, y) = -2x + 3y + 9$ and $f(x, y) = xy - 3x + 2y + 10$.   **75. a.** $b$ is the $z$-intercept of the plane. $m$ is the slope of the intersection of the plane with the $xz$-plane. $n$ is the slope of the intersection of the plane with the $yz$-plane.
**b.** Write $z = b + rx + sy$. We are told that $\partial z/\partial x = m$, so $r = m$. Similarly, $s = n$. Thus, $z = b + mx + ny$. We are also told that the plane passes through $(h, k, l)$. Substituting gives $l = b + mh + nk$. This gives $b$ as $l - mh - nk$. Substituting in the equation for $z$ therefore gives $z = l - mh - nk + mx + ny = l + m(x - h) + n(y - k)$, as required.

## Section 15.3

**1.** $P$: relative minimum; $Q$: none of the above; $R$: relative maximum   **3.** $P$: saddle point; $Q$: relative maximum; $R$: none of the above

**5.** Relative minimum    **7.** Neither
**9.** Saddle point    **11.** Relative minimum at $(0, 0, 1)$
**13.** Relative maximum at $(-1/2, 1/2, 3/2)$
**15.** Saddle point at $(0, 0, 0)$   **17.** Minimum at $(4, -3, -10)$   **19.** Maximum at $(-7/4, 1/4, 19/8)$
**21.** Relative maximum at $(0, 0, 0)$, saddle points at $(\pm 4, 2, -16)$   **23.** Relative minimum at $(0, 0, 0)$, saddle points at $(-1, \pm 1, 1)$   **25.** Relative minimum at $(0, 0, 1)$   **27.** Relative minimum at $(-2, \pm 2, -16)$, $(0, 0)$ a critical point that is not a relative extremum
**29.** Saddle point at $(0, 0, -1)$   **31.** Relative maximum at $(-1, 0, e)$   **33.** Relative minimum at $(2^{1/3}, 2^{1/3}, 3(2^{2/3}))$   **35.** Relative minimum at $(1, 1, 4)$ and $(-1, -1, 4)$   **37.** Absolute minimum at $(0, 0, 1)$
**39.** None; the relative maximum at $(0, 0, 0)$ is not absolute.   **41.** Minimum of $1/3$ at $(c, f) = (2/3, 2/3)$. Thus, at least $1/3$ of all Mazda owners would choose another new Mazda, and this lowest loyalty occurs when $2/3$ of Chrysler and Ford owners remain loyal to their brands.   **43.** It should remove 2.5 pounds of sulfur and 1 pound of lead per day.
**45.** You should charge $580.81 for the Ultra Mini and $808.08 for the Big Stack.   **47.** $l = w = h \approx 20.67$ in, volume $\approx 8,827$ cubic inches   **49.** 18 in $\times$ 18 in $\times$ 36 in, volume $= 11,664$ cubic inches

**51.**

**53.**   Continues up indefinitely

Continues down indefinitely
Function not defined on circle

**55.** $H$ must be positive.    **57.** No. In order for there to be a relative maximum at $(a, b)$, *all* vertical planes through $(a, b)$ should yield a curve with a relative maximum at $(a, b)$. It could happen that a slice by another vertical plane through $(a, b)$ (such as $x - a = y - b$) does not yield a curve with a relative maximum at $(a, b)$. [An example is $f(x, y) = x^2 + y^2 - \sqrt{xy}$, at the point $(0, 0)$. Look at the slices through $x = 0$, $y = 0$ and $y = x$.]

**59.** $\bar{C}_x = \dfrac{\partial}{\partial x}\left(\dfrac{C}{x+y}\right) = \dfrac{(x+y)C_x - C}{(x+y)^2}$. If this is

zero, then $(x+y)C_x = C$, or $C_x = \dfrac{C}{x+y} = \bar{C}$.

Similarly, if $\bar{C}_y = 0$, then $C_y = \bar{C}$. This is reasonable because if the average cost is decreasing with increasing $x$, then the average cost is greater than the marginal cost $C_x$. Similarly, if the average cost is increasing with increasing $x$, then the average cost is less than the marginal cost $C_x$. Thus, if the average cost is stationary with increasing $x$, then the average cost equals the marginal cost $C_x$. (The situation is similar for the case of increasing $y$.)    **61.** The equation of the tangent plane at the point $(a, b)$ is $z = f(a, b) + f_x(a, b)(x - a) + f_y(a, b)(y - b)$. If $f$ has a relative extremum at $(a, b)$, then $f_x(a, b) = 0 = f_y(a, b)$. Substituting these into the equation of the tangent plane gives $z = f(a, b)$, a constant. But the graph of $z = constant$ is a plane parallel to the $xy$-plane.

## Section 15.4

**1.** 1; $(0, 0, 0)$   **3.** 1.35; $(1/10, 3/10, 1/2)$   **5.** Minimum value $= 6$ at $(1, 2, 1/2)$   **7.** 200; $(20, 10)$   **9.** 16; $(2, 2)$ and $(-2, -2)$   **11.** 20; $(2, 4)$   **13.** 1; $(0, 0, 0)$
**15.** 1.35; $(1/10, 3/10, 1/2)$   **17.** Minimum value $= 6$ at $(1, 2, 1/2)$   **19. a.** $f(x, y, z)$ is the square of the distance from the point $(x, y, z)$ to $(3, 0, 0)$, and the constraint tells us that $(x, y, z)$ must lie on the paraboloid $z = x^2 + y^2$. Because there must be such a point (or points) on the paraboloid closest to $(3, 0, 0)$, the given problem must have at least one solution. Solution: $(x, y, z) = (1, 0, 1)$ for a minimum value of 5. **b.** Same solution as part (a) **c.** There are no critical points using this method. **d.** The constraint equation $y^2 = z - x^2$ tells us that $z - x^2$ cannot be negative, and thus restricts the domain of $f$ to the set of points $(x, y, z)$ with $z - x^2 \geq 0$. However, this information is lost when $z - x^2$ is substituted in the expression for $f$, and so the substitution in part (c) results in a different optimization problem; one in which there is no requirement that $z - x^2$ be $\geq 0$. If we pay attention to this constraint we can see that the minimum will occur when $z = x^2$, which will then lead us to the correct solution.   **21.** $5 \times 10 = 50$ sq. ft.   **23.** $10
**25.** $(1/\sqrt{3}, 1/\sqrt{3}, 1/\sqrt{3})$, $(-1/\sqrt{3}, -1/\sqrt{3}, 1/\sqrt{3})$, $(1/\sqrt{3}, -1/\sqrt{3}, -1/\sqrt{3})$, $(-1/\sqrt{3}, 1/\sqrt{3}, -1/\sqrt{3})$
**27.** $(0, 1/2, -1/2)$   **29.** $(-5/9, 5/9, 25/9)$
**31.** $l \times w \times h = 1 \times 1 \times 2$   **33.** 18 in $\times$ 18 in $\times$ 36 in, volume $= 11{,}664$ cubic inches   **35.** $(2l/h)^{1/3} \times (2l/h)^{1/3} \times 2^{1/3}(h/l)^{2/3}$, where $l = $ cost of

lightweight cardboard and $h = $ cost of heavy-duty cardboard per square foot   **37.** $1 \times 1 \times 1/2$
**39.** 6 laborers, 10 robots for a productivity of 368 pairs of socks per day   **41.** Method 1: Solve $g(x, y, z) = 0$ for one of the variables and substitute in $f(x, y, z)$. Then find the maximum value of the resulting function of two variables. Advantage (answers may vary): We can use the second derivative test to check whether the resulting critical points are maxima, minima, saddle points, or none of these. Disadvantage (answers may vary): We may not be able to solve $g(x, y, z) = 0$ for one of the variables. Method 2: Use the method of Lagrange multipliers. Advantage (answers may vary): We do not need to solve the constraint equation for one of the variables. Disadvantage (answers may vary): The method does not tell us whether the critical points obtained are maxima, minima, points of inflection, or none of these.   **43.** If the only constraint is an equality constraint, and if it is impossible to eliminate one of the variables in the objective function by substitution (solving the constraint equation for a variable or some other method).   **45.** Answers may vary: Maximize $f(x, y) = 1 - x^2 - y^2$ subject to $x = y$.   **47.** Yes. There may be relative extrema at points on the boundary of the domain. The partial derivatives of the function need not be 0 at such points.   **49.** In a linear programming problem, the objective function is linear, and so the partial derivatives can never all be zero. (We are ignoring the simple case in which the objective function is constant.) It follows that the extrema cannot occur in the interior of the domain (since the partial derivatives must be zero at such points).

## Section 15.5

**1.** $-1/2$   **3.** $e^2/2 - 7/2$   **5.** $(e^3 - 1)(e^2 - 1)$
**7.** $7/6$   **9.** $[e^3 - e - e^{-1} + e^{-3}]/2$   **11.** $1/2$
**13.** $(e - 1)/2$   **15.** $45/2$   **17.** $8/3$   **19.** $4/3$   **21.** 0
**23.** $2/3$   **25.** $2/3$   **27.** $2(e - 2)$   **29.** $1/3$

**31.** $\displaystyle\int_0^1 \int_0^{1-x} f(x, y)\,dy\,dx$

**33.** $\int_0^1 \int_0^{1-x} f(x, y)\, dy\, dx$

**35.** $\int_1^4 \int_1^{2/\sqrt{y}} f(x, y)\, dx\, dy$

**37.** 4/3   **39.** 1/6   **41.** 162,000 gadgets
**43.** \$312,750   **45.** \$17,500   **47.** 8,216   **49.** 1 degree
**51.** The area between the curves $y = r(x)$ and
$y = s(x)$ and the vertical lines $x = a$ and $x = b$ is
given by $\int_a^b \int_{r(x)}^{s(x)} dy\, dx$ assuming that $r(x) \leq s(x)$ for
$a \leq x \leq b$.   **53.** The first step in calculating an
integral of the form $\int_a^b \int_{r(x)}^{s(x)} f(x, y)\, dy\, dx$ is to evaluate
the integral $\int_{r(x)}^{s(x)} f(x, y)\, dy$, obtained by holding $x$
constant and integrating with respect to $y$.
**55.** Paintings per picasso per dali   **57.** Left-hand side
is $\int_a^b \int_c^d f(x)\, g(y)\, dx\, dy = \int_a^b \left( g(y) \int_c^d f(x)\, dx \right) dy$
(because $g(y)$ is treated as a constant in the inner
integral) $= \left( \int_c^d f(x)\, dx \right) \int_a^b g(y)\, dy$ (because
$\int_c^d f(x)\, dx$ is a constant and can therefore be taken
outside the integral). $\int_0^1 \int_1^2 ye^x\, dx\, dy = \frac{1}{2}(e^2 - e)$
no matter how we compute it.

## Chapter 15 Review

**1.** 0; 14/3; 1/2; $\dfrac{1}{1+z} + z^3$; $\dfrac{x+h}{y+k+(x+h)(z+l)} +$
$(x+h)^2(y+k)$   **3.** Decreases by 0.32 units; increases
by 12.5 units   **5.** Reading left to right, starting
at the top: 4, 0, 0, 3, 0, 1, 2, 0, 2   **7.** Answers may
vary; two examples are $f(x, y) = 3(x - y)/2$ and
$f(x, y) = 3(x - y)^3/8$.

**9.**

**11.**

**13.**

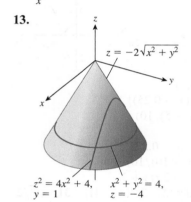

**15.** $f_x = 2x + y$, $f_y = x$, $f_{yy} = 0$   **17.** 0
**19.** $\dfrac{\partial f}{\partial x} = \dfrac{-x^2 + y^2 + z^2}{(x^2 + y^2 + z^2)^2}$, $\dfrac{\partial f}{\partial y} = \dfrac{2xy}{(x^2 + y^2 + z^2)^2}$,
$\dfrac{\partial f}{\partial z} = -\dfrac{2xz}{(x^2 + y^2 + z^2)^2}$, $\dfrac{\partial f}{\partial x}\bigg|_{(0,1,0)} = 1$
**21.** Absolute minimum at $(1, 3/2)$   **23.** Maximum at
$(0, 0)$, saddle points at $(\pm\sqrt{2}, 1)$   **25.** Saddle point at
$(0, 0)$   **27.** 1/27 at $(1/3, 1/3, 1/3)$   **29.** $(0, 2, \sqrt{2})$
**31.** 4; $(\sqrt{2}, \sqrt{2})$ and $(-\sqrt{2}, -\sqrt{2})$   **33.** $(0, 2, \sqrt{2})$
**35.** 2   **37.** ln 5   **39.** 1   **41. a.** $h(x, y) = 5,000 -$
$0.8x - 0.6y$ hits per day ($x =$ number of new
customers at JungleBooks.com, $y =$ number of
new customers at FarmerBooks.com)   **b.** 250
**c.** $h(x, y, z) = 5,000 - 0.8x - 0.6y + 0.0001z$
($z =$ number of new Internet shoppers)   **d.** 1.4 million
**43. a.** 2,320 hits per day   **b.** $0.08 + 0.00003x$ hits
(daily) per dollar spent on television advertising per
month; increases with increasing $x$   **c.** \$4,000 per
month   **45.** (A)   **47. a.** About 15,800 additional
orders per day   **b.** 11   **49.** \$23,050

# Chapter 16

## Section 16.1

**1.**

**3.**

**5.**

**7.**

**9.**

**11.**

**13.** $f(x) = \sin(2\pi x) + 1$
**15.** $f(x) = 1.5 \sin(4\pi(x - 0.25))$
**17.** $f(x) = 50 \sin(\pi(x - 5)/10) - 50$
**19.** $f(x) = \cos(2\pi x)$
**21.** $f(x) = 1.5 \cos(4\pi(x - 0.375))$
**23.** $f(x) = 40 \cos(\pi(x - 10)/10) + 40$
**25.** $f(t) = 4.2 \sin(\pi/2 - 2\pi t) + 3$
**27.** $g(x) = 4 - 1.3 \sin[\pi/2 - 2.3(x - 4)]$   **31.** $\sqrt{3}/2$
**37.** $\tan(x + \pi) = \tan(x)$   **39. a.** $2\pi/0.602 \approx 10.4$
years.  **b.** Maximum: $58.8 + 57.7 = 116.5 \approx 117$;
minimum: $58.8 - 57.7 = 1.1 \approx 1$
**c.** $1.43 + P/4 + 2P = 1.43 + 23.48 \approx 25$ years, or
about the beginning of 2022   **41. a.** Sales were highest when $t \approx 0$, 4, 8, and 12, which correspond to the
end of the last quarter or beginning of the first quarter
of each year. Sales were lowest when $t \approx 2$, 6, and 10,
which correspond to the beginning of the third quarter
of each year. **b.** The maximum quarterly sales were
approximately 21 million iPods per quarter; minimum
quarterly sales were approximately 9 million iPods
per quarter. **c.** Maximum: $15 + 6 = 21$; minimum:
$15 - 6 = 9$   **43.** Amplitude $= 6.00$, vertical
offset $= 15$, phase shift $= -1.85/1.51 \approx -1.23$,
angular frequency $= 1.51$, period $\approx 4.16$. From 2008
through 2010, Apple's sales of iPods fluctuated in
cycles of 4.16 quarters about a baseline of 15 million
iPods per quarter. Every cycle, sales peaked at
$15 + 6 = 21$ million iPods per quarter and dipped to a
low of $15 - 6 = 9$ million iPods per quarter. Sales first

peaked at $t = -1.23 + (5/4) \times 4.16 = 3.97$, the end
of 2008.   **45.** $P(t) = 7.5 \sin[\pi(t - 13)/26] + 12.5$
**47.** $T(t) = 3.5 \cos[\pi(t - 7)/6] + 78.5$
**49.** $n(t) = 1.25 \sin[\pi(t - 1)/2] + 5.75$
**51.** $n(t) = 1.25 \cos[\pi(t - 2)/2] + 5.75$
**53.** $d(t) = 5 \sin(2\pi(t - 1.625)/13.5) + 10$
**55. a.** $u(t) = 2.5 \sin(2\pi(t - 0.75)) + 7.5$
**b.** $c(t) = 1.04^t[2.5 \sin(2\pi(t - 0.75)) + 7.5]$
**57. a.** $C \approx 50$, $A \approx 8$, $P \approx 12$, $\beta \approx 6$

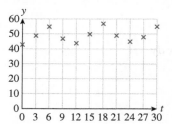

**b.** $f(t) = 5.882 \cos[2\pi(t - 5.696)/12.263] + 49.238$

**c.** 12; 43; 55
**59. a.**

**b.** $y_{11} = \dfrac{2}{\pi} \cos x + \dfrac{2}{3\pi} \cos 3x + \dfrac{2}{5\pi} \cos 5x$

$\qquad + \dfrac{2}{7\pi} \cos 7x + \dfrac{2}{9\pi} \cos 9x + \dfrac{2}{11\pi} \cos 11x$

**c.** Multiply the amplitudes by 3 and change $\omega$ to $1/2$:

$y_{11} = \dfrac{6}{\pi} \cos \dfrac{x}{2} + \dfrac{6}{3\pi} \cos \dfrac{3x}{2} + \dfrac{6}{5\pi} \cos \dfrac{5x}{2}$

$\qquad + \dfrac{6}{7\pi} \cos \dfrac{7x}{2} + \dfrac{6}{9\pi} \cos \dfrac{9x}{2} + \dfrac{6}{11\pi} \cos \dfrac{11x}{2}.$

**61.** The period is approximately 12.6 units.

**63.** Lows: $B - A$; highs: $B + A$.  **65.** He is correct. The other trig functions can be obtained from the sine function by first using the formula $\cos x = \sin(x + \pi/2)$ to obtain cosine, and then using the formulas

$$\tan x = \frac{\sin x}{\cos x}, \quad \cot x = \frac{\cos x}{\sin x}, \quad \sec x = \frac{1}{\cos x},$$

$\csc x = \dfrac{1}{\sin x}$ to obtain the rest  **67.** The largest $B$ can be is $A$. Otherwise, if $B$ is larger than $A$, the low figure for sales would have the negative value of $A - B$.

## Section 16.2

**1.** $\cos x + \sin x$  **3.** $(\cos x)(\tan x) + (\sin x)(\sec^2 x)$
**5.** $-2 \csc x \cot x - \sec x \tan x + 3$
**7.** $\cos x - x \sin x + 2x$
**9.** $(2x - 1)\tan x + (x^2 - x + 1)\sec^2 x$
**11.** $-[\csc^2 x(1 + \sec x) + \cot x \sec x \tan x]/(1 + \sec x)^2$
**13.** $-2\cos x \sin x$  **15.** $2\sec^2 x \tan x$
**17.** $3\cos(3x - 5)$  **19.** $2\sin(-2x + 5)$

**21.** $\pi \cos\left[\dfrac{\pi}{5}(x - 4)\right]$  **23.** $-(2x - 1)\sin(x^2 - x)$

**25.** $(2.2x^{1.2} + 1.2)\sec(x^{2.2} + 1.2x - 1)$
$\tan(x^{2.2} + 1.2x - 1)$  **27.** $\sec x \tan x \tan(x^2 - 1) + 2x \sec x \sec^2(x^2 - 1)$  **29.** $e^x[-\sin(e^x) + \cos x - \sin x]$
**31.** $\sec x$  **37.** $e^{-2x}[-2\sin(3\pi x) + 3\pi \cos(3\pi x)]$
**39.** $1.5[\sin(3x)]^{-0.5}\cos(3x)$

**41.** $\dfrac{x^4 - 3x^2}{(x^2 - 1)^2}\sec\left(\dfrac{x^3}{x^2 - 1}\right)\tan\left(\dfrac{x^3}{x^2 - 1}\right)$

**43.** $\dfrac{\cot(2x - 1)}{x} - 2\ln|x|\csc^2(2x - 1)$

**45. a.** Not differentiable at 0  **b.** $f'(1) \approx 0.5403$
**47.** 0  **49.** 2  **51.** Does not exist  **53.** $1/\sec^2 y$
**55.** $-[1 + y\cos(xy)]/[1 + x\cos(xy)]$
**57.** $c'(t) = 7\pi \cos[2\pi(t - 0.75)]$; $c'(0.75) \approx \$21.99$ per *year* $\approx \$0.42$ per week  **59.** $N'(6) \approx -32.12$ On January 1, 2003, the number of sunspots was decreasing at a rate of 32.12 sunspots per year.
**61.** $c'(t) = 1.035^t \times [\ln(1.035)(0.8\sin(2\pi t) + 10.2) +$

$1.6\pi \cos(2\pi t)]$; $c'(1) = 1.035[10.2\ln|1.035| + 1.6\pi] \approx$ $\$5.57$ per year, or $\$0.11$ per week.
**63. a.** $d(t) = 5\cos(2\pi t/13.5) + 10$  **b.** $d'(t) = -(10\pi/13.5)\sin(2\pi t/13.5)$; $d'(7) \approx 0.270$. At noon, the tide was rising at a rate of 0.270 feet per hour.

**65.** $\dfrac{dV}{dt} = 11{,}000\pi \dfrac{|\sin(100\pi t)|}{\sin(100\pi t)}\cos(100\pi t)$
Graph:

The sudden jumps in the graph are due to the nondifferentiability of $V$ at the times 0, $\pm 0.01$, $\pm 0.02$, .... The derivative is negative immediately to the left and positive immediately to the right of these points.  **67. a.** 1.2 cm above the rest position  **b.** 0 cm/sec; not moving; moving downward at 18.85 cm/sec  **c.** 2.5 cycles per second  **69. a.** Moving downward at 0.12 cm/sec; moving downward at 18.66 cm/sec  **b.** 0.1 sec
Graphs: of $p$:

Graphs of $p'$:

**71. a. (III) b.** Increasing at a rate of 0.157 degrees per thousand years   **73.** −6; 6   **75.** Answers will vary. Examples: $f(x) = \sin x$; $f(x) = \cos x$   **77.** Answers will vary. Examples: $f(x) = e^{-x}$; $f(x) = -2e^{-x}$
**79.** The graph of $\cos x$ slopes down over the interval $(0, \pi)$, so that its derivative is negative over that interval. The function $-\sin x$, and not $\sin x$, has this property.   **81.** The velocity is $p'(t) = A\omega \cos(\omega t + d)$, which is a maximum when its derivative, $p''(t) = -A\omega^2 \sin(\omega t + d)$, is zero. But this occurs when $\sin(\omega t + d) = 0$, so that $p(t)$ is zero as well, meaning that the stock is at yesterday's close.
**83.** The derivative of $\sin x$ is $\cos x$. When $x = 0$, this is $\cos(0) = 1$. Thus, the tangent to the graph of $\sin x$ at the point (0, 0) has slope 1, which means it slopes upward at 45°.

## Section 16.3

**1.** $-\cos x - 2\sin x + C$
**3.** $2\sin x + 4.3\cos x - 9.33x + C$
**5.** $3.4\tan x + (\sin x)/1.3 - 3.2e^x + C$
**7.** $(7.6/3)\sin(3x - 4) + C$
**9.** $-(1/6)\cos(3x^2 - 4) + C$
**11.** $-2\cos(x^2 + x) + C$
**13.** $(1/6)\tan(3x^2 + 2x^3) + C$
**15.** $-(1/6)\ln|\cos(2x^3)| + C$
**17.** $3\ln|\sec(2x - 4) + \tan(2x - 4)| + C$
**19.** $(1/2)\sin(e^{2x} + 1) + C$   **21.** −2   **23.** $\ln(2)$
**25.** 0   **27.** 1   **33.** $-\dfrac{1}{4}\cos(4x) + C$
**35.** $-\sin(-x + 1) + C$
**37.** $[\cos(-1.1x - 1)]/1.1 + C$
**39.** $-\dfrac{1}{4}\ln|\sin(-4x)| + C$
**41.** 0   **43.** $2\pi$   **45.** $-x\cos x + \sin x + C$
**47.** $\left[\dfrac{x^2}{2} - \dfrac{1}{4}\right]\sin(2x) + \dfrac{x}{2}\cos(2x) + C$
**49.** $-\dfrac{1}{2}e^{-x}\cos x - \dfrac{1}{2}e^{-x}\sin x + C$   **51.** $\pi^2 - 4$

**53.** Average = $2/\pi$

**55.** Diverges   **57.** Converges to 1/2
**59.** $C(t) = 0.04t + \dfrac{2.6}{\pi}\cos\left[\dfrac{\pi}{26}(t - 25)\right] + 1.02$
**61.** 12 feet   **63.** 79 sunspots
**65.** $P(t) = 7.5\sin[(\pi/26(t - 13)] + 12.5$; 7.7%
**67. a.** Average voltage over [0, 1/6] is zero; 60 cycles per second.
**b.**

**c.** 116.673 volts   **69.** \$50,000   **71.** It is always zero.
**73.** 1   **75.** $s = -\dfrac{K}{\omega^2}\sin(\omega t - \alpha) + Lt + M$ for
constants $L$ and $M$

## Chapter 16 Review

**1.** $f(x) = 1 + 2\sin x$
**3.** $f(x) = 2 + 2\sin[\pi(x - 1)] = 2 + 2\sin[\pi(x + 1)]$
**5.** $f(x) = 1 + 2\cos(x - \pi/2)$
**7.** $f(x) = 2 + 2\cos[\pi(x + 1/2)] =$
$2 + 2\cos[\pi(x - 3/2)]$   **9.** $-2x\sin(x^2 - 1)$
**11.** $2e^x \sec^2(2e^x - 1)$   **13.** $4x\sin(x^2)\cos(x^2)$
**15.** $2\sin(2x - 1) + C$   **17.** $\tan(2x^2 - 1) + C$
**19.** $-\dfrac{1}{2}\ln|(\cos(x^2 + 1)| + C$   **21.** 1
**23.** $-x^2\cos x + 2x\sin x + 2\cos x + C$
**25.** $s(t) = 10{,}500 + 1{,}500\sin[(2\pi/52)t - \pi] =$
$10{,}500 + 1{,}500\sin(0.12083t - 3.14159)$
**27.** Decreasing at a rate of \$3,852 per month
**29.** \$2,029,700
**31.** $150t - \dfrac{100}{\pi}\cos\left[\dfrac{\pi}{2}(t - 1)\right]$ grams